Communications and Control Engineering

Communications and Control Engineering is a high-level academic monograph series publishing research in control and systems theory, control engineering and communications. It has worldwide distribution to engineers, researchers, educators (several of the titles in this series find use as advanced textbooks although that is not their primary purpose), and libraries.

The series reflects the major technological and mathematical advances that have a great impact in the fields of communication and control. The range of areas to which control and systems theory is applied is broadening rapidly with particular growth being noticeable in the fields of finance and biologically inspired control. Books in this series generally pull together many related research threads in more mature areas of the subject than the highly specialised volumes of *Lecture Notes in Control and Information Sciences*. This series's mathematical and control-theoretic emphasis is complemented by *Advances in Industrial Control* which provides a much more applied, engineering-oriented outlook.

Indexed by SCOPUS and Engineering Index.

Publishing Ethics: Researchers should conduct their research from research proposal to publication in line with best practices and codes of conduct of relevant professional bodies and/or national and international regulatory bodies. For more details on individual ethics matters please see: https://www.springer.com/gp/authors-editors/journal-author/journal-author-helpdesk/publishing-ethics/14214

Andrey Polyakov

Generalized Homogeneity in Systems and Control Volume I

Finite-Dimensional Systems

Second Edition

Springer

Andrey Polyakov
Inria Centre
University of Lille
Lille, France

ISSN 0178-5354 ISSN 2197-7119 (electronic)
Communications and Control Engineering
ISBN 978-3-031-76245-1 ISBN 978-3-031-76246-8 (eBook)
https://doi.org/10.1007/978-3-031-76246-8

Mathematics Subject Classification: 93-02, 93C10, 93C15, 93C20, 93C25, 93C35, 93C55, 93C73

This Springer imprint is published by the registered company Springer Nature Switzerland AG
The registered company address is: Gewerbestrasse 11, 6330 Cham, Switzerland

To my family

Preface

Homogeneity is an invariance of a mathematical object (e.g., a function or a set) with respect to a class of transformations called dilations. All linear and a lot of essentially nonlinear models of mathematical physics are homogeneous (symmetric) in some sense. Homogeneous models can be utilized as local approximations of dynamics if, for example, linearization is too conservative, non-informative or simply impossible. The dilation symmetry simplifies stability and robustness analysis, controllers and observers design, discretization and digital implementation of nonlinear systems.

Homogeneous control systems appear as solutions of various nonlinear problems such as finite/fixed-time stabilization, high-order sliding-mode control design, the minimum time regulation, etc. Homogeneous controllers studied in this monograph are parameterized by the so-called homogeneity degree. Conventional linear and certain sliding-mode controllers are particular cases of the homogeneous feedback law with the degrees 0 and -1, respectively. Being essentially nonlinear, the developed homogeneous controllers admit simple tuning rules similar to linear algorithms. In many cases, homogeneous feedback laws may demonstrate faster convergence, better robustness and smaller overshoots than linear controllers. These features of homogeneous systems are important for optimal and safety-critical control design.

The monograph presents an introduction to the generalized homogeneous control systems theory for both finite and infinite-dimensional spaces. It contains 25 chapters collected in two volumes.

The first volume is devoted to homogeneous systems in Euclidean (finite-dimensional) spaces. Ideas of homogeneity-based analysis and design as well as some historical remarks about an evolution of the homogeneity theory are presented. Finite-dimensional models of control systems are considered. In particular, ordinary differential equations with continuous and discontinuous right-hand sides as well as differential inclusions are studied. Stability notions and some tools of stability analysis (e.g., the Lyapunov function method) are surveyed. The concepts of *finite-time and fixed-time stability* (typical for homogeneous systems) are studied with details. A class of linear dilations in the Euclidean space is introduced. Various mathematical objects (spheres, balls, cones, norms, etc.) induced by generalized dilations are considered. Homogeneous mappings in finite-dimensional spaces are studied. A special attention is paid to homogeneous vector fields in Euclidean spaces. Stability, robustness, completeness and positive invariance of homogeneous systems are investigated. Issues of consistent discretization of continuous-time homogeneous models are studied. Robust finite/fixed/prescribed-time stabilizers and the generalized homogeneous sliding-mode algorithms are designed. Digitization (sampled-time implementation and quantization) of homogeneous controllers is elaborated. Algorithms for upgrading linear regulators to homogeneous controllers are proposed. An essential improvement of the control quality after the upgrade is confirmed on real experiments with electro-mechanical systems. Homogeneous safety-critical systems are introduced. Some homogeneous safety filters are designed. Various homogeneous optimal control problems are investigated. Homogeneous observers for linear and nonlinear plants are designed. Algorithms for regulation of nonlinear plants are developed. An introduction to homogeneous artificial and differential neural networks is presented. Homogeneous formation and consensus control protocols for multi-agent systems are considered.

Infinite-dimensional homogeneous systems are studied in the second volume of the monograph. Elements of the theory of evolution equations in Banach and Hilbert spaces are surveyed. Features of stability analysis of infinite-dimensional systems are discussed. Linear dilations in abstract spaces are introduced. A special attention is paid to linear dilations in Lebesgue and Sobolev spaces. The theory of homogeneous operators in Banach and Hilbert spaces is elaborated. Homogeneous evolution equations are studied. Consistent discretization of homogeneous systems is developed based on the so-called homogeneous Galerkin method. Various homogeneous controllers for infinite-dimensional systems are designed.

The second edition of the monograph summarizes both well-known and recent results about homogeneous control systems design, analysis and applications. Most chapters are substantially revised in order to present a more complete picture of the state of the art. The four new chapters survey recent contributions of the homogeneity theory to control digitization, safety, machine learning and multi-agent systems. The key feature of the second edition is splitting finite-dimensional and infinite-dimensional homogeneity theory in two volumes of the monograph. This allows more systematic presentation of the results and makes the monograph friendly to the reader.

I acknowledge my colleagues for their strong support, for fruitful ideas, suggestions, criticism and always interesting discussions. I am also very grateful to my family for their patience and personal sacrifices which they have given to my work.

Lille, France Andrey Polyakov
June 2024

Contents

Notation

- $\mathbb{N}$ (resp., $\mathbb{Z}$, $\mathbb{Q}$) is the set of natural (resp., integer, rational) numbers.
- $\mathbb{R}$ is the field of reals; $\overline{\mathbb{R}} = \mathbb{R} \cup \{-\infty\} \cup \{+\infty\}$ and $\mathbb{R}_+ = \{x \in \mathbb{R} : x \geq 0\}$.
- $[a, b] = \{x \in \mathbb{R} : a \leq x \leq b\}$, $(a, b) = \{x \in \mathbb{R} : a < x < b\}$, $[a, b) = \{x \in \mathbb{R} : a \leq x < b\}$, $(a, b] = \{x \in \mathbb{R} : a < x \leq b\}$ are intervals of $\mathbb{R}$.
- $\mathbb{C}$ is the field of complex numbers with the imaginary unit $\mathbf{i} = \sqrt{-1}$; $\mathfrak{R}(\lambda) = \alpha$ and $\mathfrak{I}(\lambda) = \beta$ denote the real and imaginary parts of the complex number $\lambda = \alpha + \mathbf{i}\beta \in \mathbb{C}$, where $\alpha, \beta \in \mathbb{R}$.
- $\mathbb{R}^n$ is the Euclidean space of real column vectors: $z = (z_1, ..., z_n)^\top \in \mathbb{R}^n$, where $z^\top$ denotes the transposition of the vector $z \in \mathbb{R}^n$.
- $\mathbb{R}^{n\times m}$ is the Euclidean space of real $n \times m$ matrices, $\mathbb{R}^n = \mathbb{R}^{n\times 1}$.
- $\begin{pmatrix} a_{11} & .... & a_{1m} \\ ... & ... & ... \\ a_{n1} & ... & a_{nm} \end{pmatrix} \in \mathbb{R}^{n\times m}$ is a matrix with $a_{i,j} \in \mathbb{R}$, $i = 1, ..., m, j = 1, ..., n$.
- $B^\top$ denotes the transposition of $B \in \mathbb{R}^{n\times m}$ and $A^{-\top} = (A^{-1})^\top$ for an invertible matrix $A \in \mathbb{R}^{n\times n}$.
- $\lambda_{\min}(P)$ and $\lambda_{\max}(P)$ denotes minimum and maximum eigenvalues of a symmetric matrix $P = P^\top \in \mathbb{R}^n$.
- If $P = P^\top \in \mathbb{R}^{n\times n}$ then the order relation $P \succ 0$ means that the matrix P is positive definite, i.e., $\lambda_{\min}(P) > 0$. The order relation $P \prec 0$ means that the matrix P is negative definite, i.e., $\lambda_{\max}(P) < 0$. The order relations $\succeq 0$ and $\preceq 0$ are defined similarly. By definition, the order relation $P \succ Q$ (resp., $\prec$, $\succeq$, $\preceq$) means that $P - Q \succ 0$ (resp., $\prec$, $\succeq$, $\preceq$).
- rank(A) denotes the rank of the matrix $A \in \mathbb{R}^{m\times n}$, which is defined as a number of linearly dependent column (if $n \geq m$) or rows (if $n \leq m$) of A.
- tr(A) denotes the trace (sum of diagonal elements) of $A \in \mathbb{R}^{n\times n}$.
- $A = \text{diag}\{a_{11}, ..., a_{nn}\} \in \mathbb{R}^{n\times n}$ is the so-called *diagonal matrix* with $a_{ij} = 0$ for $i \neq j$.
- $I_n = \text{diag}\{1, ..., 1\} \in \mathbb{R}^{n\times n}$ is the identity matrix.

- $\mathrm{vec}(q_1, q_2, ...) = \begin{pmatrix} q_1 \\ q_2 \\ \dots \end{pmatrix}$ and $A \otimes B = \begin{pmatrix} a_{11}B \; a_{12}B \; ... \\ a_{21}B \; a_{22}B \; ... \\ \dots \quad \dots \quad \dots \end{pmatrix}$.
- $|x| = \sqrt{x^\top x}$ is the *canonical Euclidean norm* for $x \in \mathbb{R}^n$.
- $\|\cdot\|$ is a norm in $\mathbb{R}^n$. By default, $\|x\| = \sqrt{x^\top P x}$ with $0 \prec P = P^\top \in \mathbb{R}^{n\times n}$ is the *weighted Euclidean norm.*
- $S = \{x \in \mathbb{R}^n : \|x\| = 1\}$ is the unit sphere.
- $\mathbf{0}$ is the zero element of a vector space, in particular, $\mathbf{0} \in \mathbb{R}^n$ is a zero vector, $\mathbf{0} \in \mathbb{R}^{n\times m}$ is the zero matrix, $\mathbf{0} \in C(\mathbb{R}^n)$ is the zero function, etc.
- The inequalities $y < \mathbf{0}$, $y \le \mathbf{0}$, $y > \mathbf{0}$, $y \ge \mathbf{0}$ for $y \in \mathbb{R}^n$ (or $y \in \mathbb{R}^{n\times m}$) are understood in the component-wise sense.
- $\|A\| = \sup_{x\neq 0} \frac{\|Ax\|}{\|x\|}$ is the matrix norm. The matrix norm induced by the canonical Euclidean norm $|\cdot|$ is denoted by $|A| = \sup_{x\neq 0} \frac{|Ax|}{|x|}$, respectively.
- The symbol $\mapsto$ is utilized for notation of mappings/functions. We write $f : X \mapsto Y$ or $x \in X \mapsto f(x) \in Y$ if a function f maps a set X into a set Y.
- The symbol $\to$ denotes a limit. We write $q(p) \to q_0$ as $p \to p_0$ if a distance between $q(p)$ and q tends to zero as a distance between p and p_0 tends to zero, where $q(p)$, q_0 are elements of a metric space $\mathcal{Q}$ and p, p_0 are elements of a metric space $\mathcal{P}$ (see Appendix A). For normed vector spaces $(\mathcal{Q}, \|\cdot\|_{\mathcal{Q}})$ and $(\mathcal{P}, \|\cdot\|_{\mathcal{Q}})$, the above limit means $\|q(p) - q_0\|_{\mathcal{Q}} \to 0$ as $\|p - p_0\|_{\mathcal{P}} \to 0$. An alternative notation for the limit is $\lim\limits_{p\to p_0} q(p) = q_0$.
- $f_1(f_2)$ and $f_1 \circ f_2$ denote the composition of nonlinear operators (functions) f_1 and f_2.
- $X \times Y$ denotes the Cartesian product of sets X and Y.
- $\overline{X}$ denotes the closure of the set $X \subset \mathbb{R}^n$.
- $C(X; Y)$ is the space of continuous functions $X \subset \mathbb{R}^n \mapsto Y \subset \mathbb{R}^m$. We write shortly $C(X)$ if $Y = \mathbb{R}$. We also write shortly $f \in C$ if a domain and a co-domain are clear from the context or not important for considerations. $C_c(X; Y) \subset C(X; Y)$ is the subset of compactly supported in X continuous functions.
- $L^p(\Omega; \mathbb{R}^m)$ is the Lebesgue space of measurable functions $\Omega \mapsto \mathbb{R}^m$:

$$L^p(\Omega; \mathbb{R}^m) := \{u : \|u\|_p < +\infty\},$$

$$\|u\|_{L^p} := \left(\int_\Omega |u(x)|^p\, dx\right)^{1/p}, \; 1 \le p < \infty$$

$$\|u\|_{L^\infty} := ess \sup_{x\in\Omega} \|u(x)\|, \; p = \infty,$$

where *ess* sup means essential supremum (see Appendix B). We write shortly $L^p(\Omega)$ if $m = 1$. We also write shortly $u \in L^p$ if a domain and a co-domain are clear from the context or not important for considerations. For $q \in L^\infty(\mathbb{R}; \mathbb{R}^k)$ we also use the notation.

$$\|q\|_{L^{\infty}(t_1,t_1)} = ess \sup_{t \in (t_1,t_2)} |q(t)|.$$

- The set consisting of elements $x_1, x_2, ..., x_n$ is denoted $\{x_1, x_2, ..., x_n\}$.
- 2^M is the power set (i.e., the set of all subsets) of a set M.
- If $\Omega_1 \subset \mathbb{R}^n$ and $\Omega_2 \subset \mathbb{R}^n$ then, by definition, the identity $\Omega_1 = \Omega_2$ means $\Omega_1 \subset \Omega_2$ and $\Omega_2 \subset \Omega_1$.
- The geometric sum of two sets is denoted "$\dot{+}$" and given by

$$\mathbf{M}_1 \dot{+} \mathbf{M}_2 = \bigcup_{x_1 \in \mathbf{M}_1, x_2 \in \mathbf{M}_2} x_1 + x_2, \tag{1}$$

  where $\mathrm{M}_1 \subset \mathbb{R}^n$, $\mathrm{M}_2 \subset \mathbb{R}^n$.
- The product of a scalar $\alpha \in \mathbb{R}$ and a set $\mathbf{M} \subset \mathbb{R}^n$ is defined as follows:

$$\alpha \mathbf{M} = \mathbf{M}\alpha = \bigcup_{x \in \mathbf{M}} \alpha x. \tag{2}$$

- The product of a set $\mathbf{N} \subset \mathbb{R}$ and a set $\mathrm{M} \subset \mathbb{R}^n$ is defined as follows:

$$\mathbf{N} \cdot \mathbf{M} = \bigcup_{\alpha \in \mathbf{N}} \alpha \mathbf{M}. \tag{3}$$

- An application of an operator $f : \mathbb{R}^n \mapsto \mathbb{R}^m$ to a set $\mathrm{M} \subset \mathbb{R}^n$ is given by

$$f(\mathbf{M}) := \bigcup_{x \in \mathbf{M}} f(x). \tag{4}$$

- $B(r) := \{x \in \mathbb{R}^n : \|x\| < r\}$ is the open ball in $\mathbb{R}^n$ of the radius $r \in \mathbb{R}_+$ with the center at the origin. Under introduced notation, $B(y, r) = y \dot{+} B(\varepsilon)$ is an open ball of the radius $\varepsilon > 0$ centered at $y \in \mathbb{R}^n$.
- $\partial\Omega$ is the boundary of a set $\Omega \subset \mathbb{R}^n$.
- $\mathrm{int}(\Omega)$ denotes the interior of a set $\Omega \subset \mathbb{R}^n$, i.e., $x \in \mathrm{int}(\Omega)$ if and only if $\exists r \in \mathbb{R}_+ : x \dot{+} B(r) \subset \Omega$.
- $\mathrm{span}\{e_1, e_2, ...\}$ denotes a linear hull of vectors $e_1, e_2, ...$, i.e.,

$$\mathrm{span}\{e_1, e_2, ...\} = \{y : \ y = \alpha_1 e_1 + \alpha_2 e_2 + ..., \alpha_1, \alpha_2, ... \in \mathbb{R}\}.$$

- The convex hull of $A \subset \mathbb{R}^n$, denoted by $co(A)$, is the smallest convex set containing A. The closed convex hull of A, denoted $\overline{co}(A)$, is the smallest closed convex set containing A.
- The sign function is defined by

$$\mathrm{sign}_\sigma(\rho) := \begin{cases} 1 & \text{if } \rho > 0, \\ -1 & \text{if } \rho < 0, \\ \sigma & \text{if } \rho = 0, \end{cases} \tag{5}$$

where $\sigma \in \mathbb{R} : -1 \leq \sigma \leq 1$. By default, $\mathrm{sign}(\rho) = \mathrm{sign}_0(\rho)$.

- The upper semi-continuous regularization of the sign function is given by

$$\overline{\mathrm{sign}}(\rho) := \begin{cases} \{1\} & \text{if } \rho > 0, \\ \{-1\} & \text{if } \rho < 0, \\ [-1, 1] & \text{if } \rho = 0. \end{cases} \tag{6}$$

- The notation $O(h)$ with $h \in \mathbb{R}$ is utilized for functions of the order k at zero. This means that $O(h)/h$ is bounded in a small enough neighborhood of zero. We use this notation for shortness and without specification of co-domain of the function.
- $\lfloor x \rceil^\alpha := |x|^\alpha \mathrm{sign}(x)$ with $\alpha \in \mathbb{R}$ is a power operation, which preserves the sign of the number $x \in \mathbb{R}$. For example, $\lfloor -2 \rceil^2 = -\lfloor 2 \rceil^2 = -4$, but $2^2 = (-2)^2 = 4$.
- $SO(3) \subset \mathbb{R}^{3\times 3}$ is a special orthogonal group of rotation matrices ($RR^\top = R^\top R = I_3$) and $\mathfrak{so}(3)$ is the corresponding Lie algebra consisting of the skew-symmetric 3×3 matrices ($W^\top = -W$) (see Chap. 15);
- the mapping $(\cdot)^\wedge : \mathbb{R}^3 \mapsto \mathfrak{so}(3)$ is given by $x^\wedge = \begin{pmatrix} 0 & -x_3 & x_2 \\ x_3 & 0 & -x_1 \\ -x_2 & x_1 & 0 \end{pmatrix}$, where $x = (x_1, x_2, x_3)^\top \in \mathbb{R}^3$; $(\cdot)^\vee : \mathfrak{so}(3) \mapsto \mathbb{R}^3$ is the inverse mapping to $(\cdot)^\wedge$
- A continuous function $\sigma : \mathbb{R}_+ \mapsto \mathbb{R}_+$ belongs to the class $\mathcal{K}$ if $\sigma(0) = 0$ and σ is strictly increasing (i.e., $t_1 < t_2 \Rightarrow \sigma(t_1) < \sigma(t_2)$ for $t_1, t_2 \in \mathbb{R}_+$).
- A function $\sigma \in \mathcal{K}$ belongs to the class $\mathcal{K}^\infty$ if $\sigma(t) \to +\infty$ as $t \to +\infty$.
- A continuous function $\xi : \mathbb{R}_+ \times \mathbb{R}_+ \mapsto \mathbb{R}_+$ belongs to the class $\mathcal{KL}$ if $\xi(\cdot, t) \in \mathcal{K}$ for any fixed $t \geq 0$ and $\xi(s, \cdot) : \mathbb{R}_+ \mapsto \mathbb{R}_+$ is strictly decreasing to zero for any fixed $s \geq 0$.
- $\mathbf{d}$ denotes a linear continuous dilation in $\mathbb{R}^n$ (see Chap. 5).
- $\|\cdot\|_{\mathbf{d}} : \mathbb{R}^n \mapsto \mathbb{R}_+$ is the canonical homogeneous norm induced by the linear continuous dilation $\mathbf{d}$ being monotone with respect to a norm $\|\cdot\|$ in $\mathbb{R}^n$.
- $\pi_{\mathbf{d}} : \mathbb{R}^n \backslash \{0\} \mapsto S$ is the $\mathbf{d}$-homogeneous projector on the unit sphere $S = \{x \in \mathbb{R}^n : \|x\| = 1\}$.
- $\mathcal{H}_{\mathbf{d}}(\mathbb{R}^n)$ is a set of $\mathbf{d}$-homogeneous functions $\mathbb{R}^n \mapsto \mathbb{R}$ and $\deg_{\mathbf{d}}(h) \in \mathbb{R}$ is a homogeneity degree of $h \in \mathcal{H}_{\mathbf{d}}(\mathbb{R}^n)$ (see Chap. 6).
- $\mathcal{F}_{\mathbf{d}}(\mathbb{R}^n)$ is a set of $\mathbf{d}$-homogeneous vector fields $\mathbb{R}^n \mapsto \mathbb{R}^n$ and $\deg_{\mathbf{d}}(f) \in \mathbb{R}$ is a homogeneity degree of $f \in \mathcal{F}_{\mathbf{d}}(\mathbb{R}^n)$.

Chapter 1
Introduction

This chapter introduces some basics of the theory of homogeneous dynamical systems. A symmetry of solutions and convergence rates of homogeneous differential equations are studied in simple examples. An evolution of the homogeneity theory is briefly surveyed. Possible advantages of the homogeneity-based approach to control systems design are discussed.

1.1 Homogeneity versus Linearity

1.1.1 Dilation Symmetry

In mathematics, an invariance of some characteristics of an object with respect to a certain group of transformations is known as symmetry. The simplest example of the symmetry is the invariance of geometric figures with respect to rotation, translation, or dilation. For example, the size of the figure is invariant with respect to rotations, while the dilation (scaling) does not change its shape (see, Fig. 1.1).

The symmetry of shapes with respect to the dilation can be discovered for level sets of the so-called homogeneous functions studied by Leonhard Euler in the eighteenth century. *A function $f : \mathbb{R}^n \mapsto \mathbb{R}^m$ is said to be homogeneous if it is symmetric with respect to scaling of its argument:*

$$f(\lambda x) = \lambda^{\nu} f(x), \quad \forall \lambda > 0, \quad \forall x \in \mathbb{R}^n,$$

A. Polyakov, *Generalized Homogeneity in Systems and Control Volume I*,
Communications and Control Engineering,
https://doi.org/10.1007/978-3-031-76246-8_1

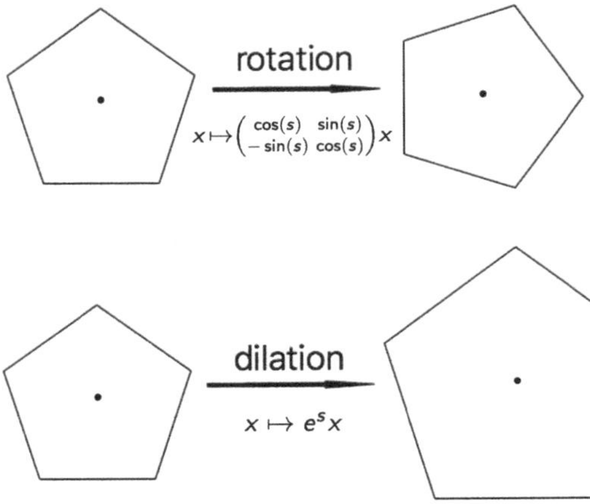

Fig. 1.1 Rotation on the angle $s \in \mathbb{R}$ and standard dilation with the scaling factor $\lambda = e^s$

where $\nu \in \mathbb{R}$ is the so-called homogeneity degree and $\lambda > 0$ is the scaling factor. Such a symmetry is known today as the standard (or Euler) homogeneity and the scaling of the argument $x \mapsto \lambda x$ by a positive scalar is referred as to the standard (or Euler) dilation. By definition, *the homogeneity is a dilation symmetry.*

The dilation symmetry is an important property for various mathematical objects. For example, a norm in a vector space is a standard homogeneous function of degree 1, but any positive cone is a standard homogeneous set (i.e., a set being invariant with respect to the standard dilation) of a vector space. The standard homogeneity is well known in algebra, for example, in the context of the so-called *homogeneous polynomials* such as $p_1(x) = x_1 + x_2$ or $p_2(x) = x_1^2 + x_1x_2 + x_2^2$. In this case, the homogeneity degree corresponds to the degree of each monomial and the polynomial itself. Indeed, the monomials x_1, x_2 and the polynomial p_1 are standard homogeneous of degree 1, but the monomials x_1^2, x_1x_2, x_2^2 and the polynomial p_2 are standard homogeneous of degree 2. Obviously, any linear function is standard homogeneous of degree 1. In the view of Noether's theorem [1], all conservation laws are symmetric. A dilation symmetry (possibly in a generalized sense) is expected to be discovered in many physical models. Various examples considered in the monograph support this conjecture.

Euler's Homogeneous Function Theorem given below is one of the most famous results of homogeneity theory.

A continuously differentiable function $f : \mathbb{R}^n \mapsto \mathbb{R}^m$ is homogeneous of degree k if and only if $\frac{\partial f}{\partial x} x = kf(x), x \in \mathbb{R}^n$.

As a consequence of this theorem, the standard homogeneity of a function is inherited by its derivatives. In fact, *a symmetry of a function is inherited somehow by any other mathematical objects induced by this function.* For example, solutions of homogeneous differential equations are also symmetric (homogeneous) in a certain sense.

The aim of this chapter is to show that homogeneous nonlinear systems are rather similar to linear ones. However, they have their own specific features which could be useful for an advanced control design and systems analysis.

1.1.2 Homogeneous Differential Equations

Let us consider the scalar ordinary differential equation (ODE) with the standard homogeneous right-hand side

$$\dot{x} = -x^{1+\mu}, \quad t > 0, \quad \mu = p/q,$$

where t is a time variable, $x(t) \in \mathbb{R}$ is a state variable, $\dot{x}$ is time derivative of the state, $p \neq 0$ is an even natural number and q is an odd natural number, $|p| \leq q$. The unique solution of this equation with the initial condition $x(0) = x_0 \in \mathbb{R}$ is given by

$$x(t, x_0) = \frac{x_0}{(1+\mu t|x_0|^{\mu})^{1/\mu}}.$$

Hence, we derive the symmetry represented by the following identity

$$x\left(\lambda^{-\mu}t, \lambda x_0\right) = \lambda x\left(t, x_0\right), \quad \lambda > 0,$$

which means that all solutions of the considered homogeneous ODE are homogeneous (symmetric) with respect to the simultaneous scaling of the initial condition and the time variable t. Notice that the time scaling in the above formula depends on the homogeneity degree. In fact, it is easy to check that the mentioned symmetry of solutions can be established for any standard homogeneous differential equation [2].

Let $f : \mathbb{R}^n \mapsto \mathbb{R}^n$ be a standard homogeneous function of degree $\mu + 1$

$$f(\lambda x) = \lambda^{1+\mu} f(x), \quad \forall\lambda > 0, \quad \forall x \in \mathbb{R}^n,$$

where $\mu \in \mathbb{R}$. If the Cauchy problem

$$\dot{x} = f(x), \quad x(0) = x_0 \in \mathbb{R}^n \tag{1.1}$$

has a solution $x(t, x_0)$ defined for $t > 0$ *then for any $\lambda > 0$ it holds*

$$x(\lambda^{-\mu}t, \lambda x_0) = \lambda x(t, x_0), \tag{1.2}$$

where $x(\cdot, \lambda x_0)$ is a solution to the same ODE with the scaled initial condition $x(0) = \lambda x_0$.

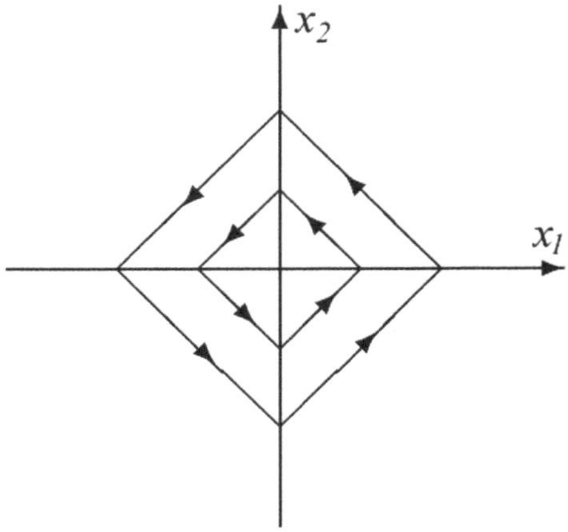

Fig. 1.2 A phase portrait of the relay oscillator

The identity shows that the standard homogeneity of an ODE implies a standard homogeneity of its solutions. The homogeneity degree plays in important role in this case. It characterizes the scaling factor of the time variable. Geometrically, the discovered dilation symmetry of solutions of ODEs is illustrated in Fig. 1.2, where the trajectories of the relay (standard homogeneous) oscillator

$$\begin{cases} \dot{x}_1 = -\operatorname{sign}(x_2), \\ \dot{x}_2 = \operatorname{sign}(x_1), \end{cases} \quad t > 0, \quad x_1(t), x_2(t) \in \mathbb{R}$$

are depicted. The parameter μ for this homogeneous system equals to -1. Solutions of the considered *discontinuous* ODE are understood in the sense of Filippov (see [3] or Chap. 2). They are symmetric and characterize the cyclic dynamics of the relay oscillator. In the view of (1.2), the period depends on the initial state as follows: the larger initial state, the larger period of oscillations is. In Chap. 7, a symmetry of solutions is proven for dynamical systems, which are homogeneous in a generalized sense.

Homogeneous systems have many properties similar to linear systems. For example, the origin is always an equilibrium $f(\mathbf{0}) = \mathbf{0}$ of any continuous homogeneous system (1.1). If there exists another equilibrium $x^* \neq \mathbf{0} : f(x^*) = \mathbf{0}$ then, due to homogeneity, the set $M = \{x \in \mathbb{R}^n : x = \lambda x^*, \lambda > 0\}$ is, at least, weakly invariant (since $f(x) = \mathbf{0}$ for all $x \in M$). Recall that, in the case of the linear system $f(x) = Ax$ with $A \in \mathbb{R}^{n\times n}$, the set M is an invariant subspace of the linear vector field too.

Due to the dilation symmetry of solutions, any local property (e.g., stability or existence of solutions for small initial data) can always be expanded globally in the state space. Similarly to linear systems, the robustness (Input-to-State Stability [4]) of a homogeneous system can be granted by its stability in the disturbance-free case (see Chap. 7 or [5]). Table 1.1 compares some basic properties of linear and nonlinear homogeneous control systems. All mentioned (any many other) properties

Table 1.1 Some properties of linear and homogeneous systems

	Linear system $\dot{x} = Ax$ $A \in \mathbb{R}^{n\times n}$	Homogeneous system $\dot{x} = f(x)$ $f(\lambda x) = \lambda^{1+\mu} f(x)$
Trajectory scaling	$x(t, \lambda x_0) = \lambda x(t, x_0)$	$x(t, \lambda x_0) = \lambda x(\lambda^{\mu} t, x_0)$
Stability Local ⇔ Global	✓	✓
Lyapunov function	Quadratic $V = x^{\top} P x$, $P \succ 0$	Homogeneous (quadratic-like) $V = x^{\top} \Xi^{\top}(x) P\, \Xi(x) x$, $P \succ 0,\ \Xi(\lambda x) = \Xi(x) \in \mathbb{R}^{n\times n}$
Stability ⇒ Robustness (Input-to-state stability)	$\dot{x} = Ax + Dw$ $w \in L^{\infty}$	$\dot{x} = f(x, w),\quad w \in L^{\infty}$, $\tilde{f} = \begin{pmatrix} f \\ 0 \end{pmatrix}$ is homogeneous
Convergence rate	Exponential	Finite-time (if $\mu < 0$) Exponential (if $\mu = 0$) Nearly fixed-time (if $\mu > 0$)
Consistent discretization preserves the convergence rate	✓	✓
Optimal control design with quadratic functional	Algebraic Riccati Equation for infinite horizon	Algebraic Riccati Equation for finite horizon

of homogeneous systems are studied for finite-dimensional setting in Volume I and for infinite-dimensional setting in Volume II of this monograph.

The homogeneity degree allows some additional qualitative analysis of ODEs to be done easily. For instance, the so-called *finite-time stability* and the so-called nearly *fixed-time stability* of homogeneous systems are characterized by the homogeneity degree. Indeed, if $\mu < 0$ then the solution of the scalar homogeneous system

$$\dot{x} = -x^{\mu+1}, \quad t > 0, \quad x(0) = x_0$$

is given by

$$x(t, x_0) = \begin{cases} \left(|x_0|^{-\mu} + \mu t\right)^{\frac{1}{-\mu}} \operatorname{sign}(x_0) & \text{if } t \in \left[0, \frac{|x_0|^{-\mu}}{-\mu}\right), \\ 0 & \text{if } t \geq \frac{|x_0|^{-\mu}}{-\mu}. \end{cases}$$

It converges to zero in a finite time: $T(x_0) = \frac{|x_0|^{-\mu}}{1-\mu}$. However, if $\mu > 0$ then each trajectory of the system reaches any neighborhood[1] of the origin in a fixed-time independent of the initial condition, namely,

[1] A set $\mathbb{U} \subset \mathbb{R}^n$ is a neighborhood of a set $\Omega \subset \mathcal{U}$ if any point $x \in \Omega$ belongs to $\mathcal{U}$ together with an open ball centered at x, i.e., there exists $r > 0$ such that $x \dot{+} B(r) \subset \mathcal{U}$, where the radius of the ball may depend of x.

$$|x(t, x_0)| \leq r, \quad \forall t \geq \frac{1}{r^{\nu-1}(\nu-1)}, \quad \forall x_0 \in \mathbb{R}^n, \quad \forall r > 0.$$

It can be shown that if the origin of a standard homogeneous system (1.1) is asymptotically stable then each trajectory of the system

(a) vanishes in a finite time provided that $\mu < 0$ (*finite-time stability* [6]);
(b) converges to zero exponentially if $\mu = 0$ (*exponential stability*, [7]);
(c) converges to any neighborhood of the origin in a fixed-time independent of the initial condition if $\mu > 0$ (*nearly fixed-time stability*, [8]).

The finite-time stability of homogeneous systems was first discovered by Vladimir Zubov in 1957 (see, the book [2, page 109]).

The above example highlights the important feature of homogeneous dynamical systems being an acceleration of the convergence rate dependently on the homogeneity degree. The convergence rate is accelerated close to zero (i.e., for small initial data) if $\mu < 0$ or close to infinity (i.e., for large initial data) if $\mu > 0$. By combining homogeneous systems of different degrees, the convergence rate can be accelerated globally.

1.1.3 Homogeneous Approximations

An approximation by a homogeneous function (*homogeneous approximation*) can be utilized for a local analysis of dynamical systems if linearization[2] is non-informative or simply impossible [9, 10]. For example, the scalar ODE $\dot{x} = -x^3 + x^5$ is locally asymptotically stable such that for any solution $x(t)$ with $|x(0)| < 1$ we have $x(t) \to 0$ as $t \to +\infty$. This system does not admit a local asymptotic stability analysis by the first-order approximation, since the linearization at the origin gives $\dot{x} = 0$. However, the cubic term is dominating close to zero, so the homogeneous approximation of the system is $\dot{x} = -x^3$. This approximation allows us to make the correct conclusion about the local asymptotic stability of the original system. These intuitive considerations are mathematically formalized in Chaps. 6 and 7.

As an example of a locally homogeneous dynamical model, which does not admit linearization, let us consider a mechanical system consisting of a rigid body moving laterally on a contact surface in a viscous environment (fluid). The simplest example of such a mechanical system is a car moving on a flat road. The modeling is based on classical (Newtonian) physics, which is appropriate for the illustration of local homogeneity.

Let $z(t)$ be the position of the center of mass of the body at the time instant $t \in \mathbb{R}$ in an inertial frame. Assume that the body has the unit mass. The dynamics of this system is governed by the following system of ordinary differential equations

[2] The first-order approximation (linearization) of an ODE $\dot{x} = f(x)$ with $f(\mathbf{0}) = \mathbf{0}$ is the linear ODE $\dot{x} = Ax$ with $A = \left.\frac{\partial f(x)}{\partial x}\right|_{x=\mathbf{0}}$.

$$\dot{z}(t) = v(t), \quad \dot{v}(t) = F(t), \quad t > 0, \quad z(t) \in \mathbb{R},$$

where $v(t)$ is the velocity and F is the sum of external forces.

Let us study only the deceleration phase of the motion assuming that, at the initial instant of time $t = 0$, this mechanical system has some nonzero velocity $\dot{z}(0) = v(0) \neq 0$. Dissipation of the energy in the system is caused, basically, by two external forces:

- the *drag force* (fluid resistance) is proportional to the velocity squared [11]

$$F_{drag}(t) = -k_{drag}\, v^2(t)\, \operatorname{sign}(v(t)),$$

  where $k_{drag} > 0$ is the coefficient of fluid (air) resistance and the sign function is given by

$$\operatorname{sign}(\rho) = \begin{cases} 1 & \text{if } \rho > 0, \\ 0 & \text{if } \rho = 0, \\ -1 & \text{if } \rho < 0; \end{cases}$$

- the *dry friction force* is nearly independent of the magnitude velocity and can be modeled as follows (see, e.g., [12])

$$F_{dry}(t) = -k_{dry}\, \operatorname{sign}(v(t)),$$

  where $k_{dry} > 0$ is the coefficient of dry friction.

The sum of external forces $F(t)$ can be represented as follows

$$F(t) = F_{drag}(t) + F_{dry}(t) = -\left(k_{dry} + k_{drag}\, v^2(t)\right) \operatorname{sign}(v(t))$$

and the differential equation describing an evolution of the velocity of the body has the form:

$$\dot{v}(t) = -\left(k_{dry} + k_{drag}\, v^2(t)\right) \operatorname{sign}(v(t)).$$

It is not difficult to see that $v = 0$ is the equilibrium of the above equation, which is globally asymptotically stable, $v(t) \to 0$ as $t \to +\infty$. The solution of this ODE can be found explicitly :

$$v(t) = \sqrt{\frac{k_{dry}}{k_{drag}}} \tan\left(\arctan\left(\sqrt{\frac{k_{drag}}{k_{dry}}}|v(0)|\right) - \sqrt{k_{dry}\, k_{drag}} t\right) \operatorname{sign}(v(0)).$$

This immediately implies $v(t) = 0$ for $t \geq \frac{\arctan(|v(0)|\sqrt{k_{drag}/k_{dry}})}{\sqrt{k_{dry}\, k_{drag}}}$. The function arctan is globally uniformly bounded. We conclude that *independently of the initial velocity, the motion of the body terminates no later than the following instant of time*

$$T_{\max} = \frac{\pi}{2\sqrt{k_{dry}\,k_{drag}}}.$$

The model of the friction force F is not a homogeneous function of v, but locally (close to the origin or close to infinity) it is. Namely,

$$F \approx -k_{dry}\,\mathrm{sign}(v) \text{ as } v \to 0 \quad \text{and} \quad F \approx -k_{drag}v^2\,\mathrm{sign}(v) \text{ as } v \to \infty.$$

In other words, the approximations of F at zero and at infinity are a standard homogeneous functions of different degrees. The first term allows us to conclude a finite-time convergence of $v(t)$ to zero for small initial conditions, while the second one guarantees a nearly fixed-time convergence independently of the initial velocity. Such a combination yields a fixed-time deceleration of the mechanical system independently of the initial velocity. In [13] such a property was called the *fixed-time stability*. In the theory of evolution semigroups, the similar property is also known as *super stability* [14]. It can be discovered for nilpotent semigroups generated even by linear PDEs (Partial Differential Equations). For example, all solutions of the transport equation with homogeneous Dirichlet boundary condition (see, e.g., [15, page 537]) and all solutions of the wave equation with the so-called "transparent" boundary conditions (see, e.g., [16, 17]) vanish to zero after a fixed instant of time independently of the initial state of the system. In this monograph, we call the property discovered above by the *fixed-time stability*, since the super stability has completely another meaning (see, e.g., [18]) in the theory of control systems.

1.2 Generalized Homogeneity

1.2.1 Weighted Dilations

The standard homogeneity considered above is introduced by means of the isotropic scaling $x \mapsto \lambda x$, $\lambda > 0$ of the vector x. To introduce a generalized homogeneity, it is sufficient to replace the standard dilation with an anisotropic one. The *weighted dilation* (studied by V.I. Zubov in 1958, [9]) is the most popular example of the generalized dilation:

$$(x_1, x_2, \ldots, x_n) \quad \mapsto \quad (\lambda^{r_1}x_1, \lambda^{r_2}x_2, \ldots, \lambda^{r_n}x_n),$$

where $\lambda > 0$, as before, is the scaling factor, and the positive numbers $r_1, r_2, \ldots, r_n$ are the weights, which specify the dilation rates along different coordinates. The introduced transformation of coordinates

$$x \mapsto \Lambda x, \quad x = (x_1, x_2, \ldots, x_n)^\top \in \mathbb{R}^n$$

is a linear mapping of $\mathbb{R}^n$ onto $\mathbb{R}^n$ given by the dilation matrix

$$\Lambda = \begin{pmatrix} \lambda^{r_1} & 0 & \dots & 0 \\ 0 & \lambda^{r_2} & \dots & 0 \\ \dots & \dots & \dots & \dots \\ 0 & 0 & \dots & \lambda^{r_n} \end{pmatrix}.$$

If $r_1 = \dots = r_n = 1$ then $\Lambda = \lambda I_n$ and the weighted dilation becomes standard.

A symmetry of a scalar-valued function with respect to the weighted dilation can be defined analogously to the standard homogeneity.

> *A scalar-valued function* $h : \mathbb{R}^n \mapsto \mathbb{R}$ is Λ*-homogeneous (i.e., homogeneous with respect to the weighted dilation* Λ *) if*
>
> $$h(\Lambda x) = \lambda^\nu h(x), \quad \forall x \in \mathbb{R}^n, \quad \forall \lambda > 0,$$
>
> *where* $\nu \in \mathbb{R}$ *is the homogeneity degree of the function* h.

The weighted dilation extends the class of homogeneous functions under consideration. For example, the polynomial function $h : \mathbb{R}^2 \mapsto \mathbb{R}$ given by

$$h(x_1, x_2) = x_1^2 + x_1 x_2^2 + x_2^4$$

is homogeneous with respect to the weighted dilation $(x_1, x_2) \mapsto (\lambda^2 x_1, \lambda x_2)$, but it is not symmetric with respect to the standard dilation $(x_1, x_2) \mapsto (\lambda x_1, \lambda x_2)$.

In this monograph, we use the notion of a *vector field* $f : \mathbb{R}^n \mapsto \mathbb{R}^n$ in order to indicate that f defines a dynamical system governed by the ODE:

$$\dot{x} = f(x), \quad t > 0.$$

A symmetry of the ordinary differential equation with respect to the weighted dilation is understood as a commutativity of f and Λ up to a scalar factor λ^μ [9, 19–21].

> *A vector field* $f : \mathbb{R}^n \mapsto \mathbb{R}^n$ *is said to be* Λ*-homogeneous if*
>
> $$f(\Lambda x) = \lambda^\mu \Lambda f(x), \quad \forall x \in \mathbb{R}^n, \quad \forall \lambda > 0,$$
>
> *where* $\mu \in \mathbb{R}$ *is a homogeneity degree of the vector field* f.

Due to the non-uniformity of scaling, the definition of a weighted homogeneous vector field $f : \mathbb{R}^n \mapsto \mathbb{R}^n$ differs from the definition of a weighted homogeneous function $h : \mathbb{R}^n \mapsto \mathbb{R}^m$ as follows:

> *each component* f_i *of a weighted homogeneous vector field* $f = (f_1, \dots, f_n)^\top$ *of degree* $\mu \in \mathbb{R}$ *is a weighted homogeneous function of degree* $\mu + r_i$.

Being the particular case of the weighted homogeneity, the standard homogeneity ($r_1 = \dots = r_n = 1$) has to be introduced differently for functions and vector fields. Namely, *if* $\nu \in \mathbb{R}$ *is a homogeneity degree of the standard homogeneous function*

$f(\lambda x) = \lambda^{\nu} f(x)$ then, by definition, $\mu = \nu - 1$ is the homogeneity of the standard homogeneous ODE: $\dot{x} = f(x)$. This perfectly fits the analysis presented in the previous section.

In Chap. 7 it is shown that *a generalized homogeneous system is topologically equivalent (homeomorphic) to a standard homogeneous one*. In other words, there exists a continuous (on $\mathbb{R}^n$) transformation of coordinates $y = \Phi(x)$ such that the solution set of a generalized homogeneous system $\dot{x} = f(x)$ of degree μ coincides (up to the coordinate transformation) with the solution set of a standard homogeneous system $\dot{y} = g(y)$ of the same degree μ. This means that generalized homogeneous systems have all the properties discovered above for standard homogeneous systems such as a dilation symmetry of solutions, the equivalence of local and global properties, a finite-time convergence for $\mu<0$, an exponential convergence for $\mu=0$, a nearly fixed-time convergence for $\mu>0$, etc.

Weighted homogeneous models frequently appear in control theory and applications [19, 20, 22, 23].

Example 1.1 The classical minimum-time control problem

$$T \to \min_{u}$$

subject to

$$\begin{cases} \dot{x}_1 = x_2, & u \in L^{\infty}((0,T);\mathbb{R}) \text{ such that } |u(t)| \leq 1 \\ \dot{x}_2 = u, & x_1(T) = x_2(T) = 0, \end{cases}$$

has the solution (see, [23] or Chap. 13) in the form of the weighted homogeneous feedback

$$u = -\operatorname{sign}(|x_2|x_2 + 2x_1).$$

Indeed, $u(\lambda^2 x_1, \lambda x_2) = u(x_1, x_2)$, i.e., u is the weighted homogeneous function of the zero degree. The closed-loop system is the weighted homogeneous function of the degree -1. The negative homogeneity degree of an asymptotically stable system implies the finite-time stability (see Chap. 7).

In control systems theory the weighted homogeneity is utilized for

- stability analysis and Lyapunov function design [9, 21, 24, 25];
- stabilizability/controllability/observability analysis [19, 26–28];
- controllers/observers design [29–31], [10, 32–35];
- robustness analysis of finite-dimensional [5, 10, 36] and infinite-dimensional systems [37–39];
- sliding-mode control and estimation [40–44];
- optimal and model predictive control [45–47];
- digitization of the control systems [48–53].

All mentioned and many other problems of homogeneous control theory are studied in this monograph using the so-called linear dilations.

1.2.2 Linear Dilations in Euclidean Spaces

Let the scaling factor $\lambda > 0$ be defined by $\lambda = e^s$, where $s \in \mathbb{R}$ and $e = 2.71828...$ is the Euler number. In this case, the weighted dilation $x \mapsto \Lambda x$ becomes

$$x \mapsto \mathbf{d}(s)x, \quad s \in \mathbb{R},$$

where

$$\mathbf{d}(s) = \begin{pmatrix} e^{r_1 s} & 0 & \cdots & 0 \\ 0 & e^{r_2 s} & \cdots & 0 \\ \cdots & \cdots & \cdots & \cdots \\ 0 & 0 & \cdots & e^{r_n s} \end{pmatrix}, \quad r_i > 0.$$

With such a redefinition, the one-parameter family of linear transformations $\mathbf{d}(s) \in \mathbb{R}^{n\times n}$, $s \in \mathbb{R}$ satisfies the following properties

- **Group property:** $\mathbf{d}(0) = I$ and $\mathbf{d}(s+t) = \mathbf{d}(s)\mathbf{d}(t)$, $\forall s, t \in \mathbb{R}$;
- **Limit property:** $\lim_{s\to\pm\infty} \|\mathbf{d}(s)x\| = e^{\pm\infty}$ for all $x \neq \mathbf{0}$.

Inspired by [20], a dilation in a normed vector space can be introduced as any one-parameter family of transformations satisfying the group and limit properties given above. In the general case [54], a dilation group may be a multi-parametric group. Such dilations are not considered in this monograph.

Once a homogeneity of the right-hand side of ODE with respect to dilation is established, many important properties of the system can be characterized easily. The crucial point of the homogeneity-based analysis is the selection of a dilation group $\mathbf{d}(s)$, $s \in \mathbb{R}$. The limit property characterizes one-parametric groups of transformations being dilations in normed vector spaces. Generalized dilations in Euclidean spaces are studied in [20, 55, 56], where the authors suggest to generate a group $\mathbf{d}$ as a flow of the so-called Euler vector field (see Chap. 5 for more details). The corresponding dilation symmetry is known as *coordinate-free or geometric homogeneity*. Being a particular case of the geometric dilation, the *linear dilation* (see, e.g., [35]) is given by the matrix exponential:

$$\mathbf{d}(s) = e^{sG_{\mathbf{d}}} = \sum_{i=0}^{+\infty} \frac{s^i G_{\mathbf{d}}^i}{i!}, \tag{1.3}$$

where $s \in \mathbb{R}$ is the group parameter and $G_{\mathbf{d}} \in \mathbb{R}^{n\times n}$ is an *anti-Hurwitz matrix*,[3] which is called the *generator of the dilation* $\mathbf{d}$. Usually (see, e.g., [9, 37, 57]), the generator $G_{\mathbf{d}}$ of a linear dilation $\mathbf{d}$ in $\mathbb{R}^n$ is assumed to be diagonalizable. In this monograph, we do not make this assumption allowing the linear dilation to be merged even with rotations. Such "mixed" dilation may appear, for example, in homogeneous optimal control (see Chap. 13) or in finite-dimensional homogeneous

[3] A matrix $G_{\mathbf{d}}$ is said to be anti-Hurwitz if all its eigenvalues are placed in the open right half-plane of $\mathbb{C}$.

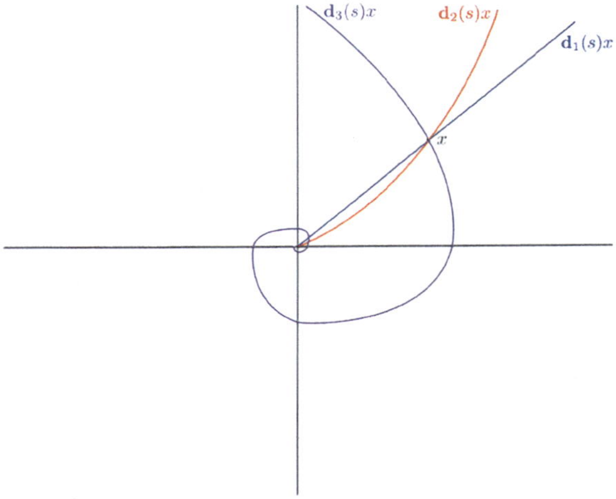

Fig. 1.3 $\mathbf{d}_1(s)$—standard dilation, $\mathbf{d}_2(s)$—weighted dilation, $\mathbf{d}_3$—linear dilation

approximations of infinite-dimensional homogeneous systems (see Volume II of this monograph). However, in many other cases (see, e.g., Chaps. 9, 10, 14) the restriction of our considerations to the class of diagonalizable linear dilations does not introduce a conservatism to homogeneous control systems design.

The standard and weighted dilations considered above are linear dilations with the generators $G_{\mathbf{d}} = I_n$ and $G_{\mathbf{d}} = \mathrm{diag}\{r_1, \ldots, r_n\}$, $r_i > 0$, respectively. In the planar case, the difference between standard, weighted and linear dilations is illustrated in Fig. 1.3, where the so-called homogeneous curves $\{\mathbf{d}_i(s)x : s \in \mathbb{R}\}$, $i = 1, 2, 3$ for

$$\mathbf{d}_1(s) = e^s I_2, \qquad \mathbf{d}_2(s) = \begin{pmatrix} e^{2s} & 0 \\ 0 & e^s \end{pmatrix}, \qquad \mathbf{d}_3(s) = e^{sG_{\mathbf{d}}}$$

are schematically depicted.

The linear homogeneity (dilation symmetry) of a function $h : \mathbb{R}^n \mapsto \mathbb{R}^m$ and a vector field $f : \mathbb{R}^n \mapsto \mathbb{R}^n$ can be introduced similarly to the weighted homogeneity:

$$h(\mathbf{d}(s)x) = e^{\nu s} h(x), \qquad f(\mathbf{d}(s)x) = e^{\mu s}\mathbf{d}(s) f(x), \qquad \forall s \in \mathbb{R},\ \forall x \in \mathbb{R}^n,$$

where $\mathbf{d}$ is a linear dilation in $\mathbb{R}^n$, but $\nu \in \mathbb{R}$ and $\mu \in \mathbb{R}$ are homogeneity degrees of h and f, respectively.

Example 1.2 The vector field $f : \mathbb{R}^3 \mapsto \mathbb{R}^3$ given by

$$f(x) = \begin{pmatrix} -|x_1|x_1 + x_2^2 + x_3^2 \\ (x_2^2+x_3^2)\cos(\ln|x_1|) + x_1^2\sin(\ln|x_1|) \\ -(x_2^2+x_3^2)\sin(\ln|x_1|) + x_1^2\cos(\ln|x_1|) \end{pmatrix}, \quad x = (x_1, x_2, x_3)^\top \in \mathbb{R}^3$$

is homogeneous (symmetric) with respect to the linear dilation

$$\mathbf{d}(s) = e^s \begin{pmatrix} 1 & 0 & 0 \\ 0 & \cos(s) & \sin(s) \\ 0 & -\sin(s) & \cos(s) \end{pmatrix}.$$

Indeed, using the identities

$$\begin{pmatrix} \cos(\alpha) & \sin(\alpha) \\ -\sin(\alpha) & \cos(\alpha) \end{pmatrix} \begin{pmatrix} \cos(\beta) & \sin(\beta) \\ -\sin(\beta) & \cos(\beta) \end{pmatrix} = \begin{pmatrix} \cos(\alpha+\beta) & \sin(\alpha+\beta) \\ -\sin(\alpha+\beta) & \cos(\alpha+\beta) \end{pmatrix}, \quad \forall \alpha, \beta \in \mathbb{R}$$

and

$$(\cos(\alpha)x_2+\sin(\alpha)x_3)^2+(-\sin(\alpha)x_2+\cos(\alpha)x_3)^2=x_2^2+x_3^2, \quad \forall \alpha \in \mathbb{R}, \forall x \in \mathbb{R}^3$$

we derive $f(\mathbf{d}(s)x) = e^s\mathbf{d}(s)f(x)$ for all $s \in \mathbb{R}$, $x \in \mathbb{R}^3$.

This monograph studies various problems of systems analysis and control design using linear dilations in both finite-dimensional Euclidean spaces (Volume I) and infinite-dimensional Banach/Hilbert spaces (Volume II).

1.2.3 Homogeneous Stabilization

As an example illustrating a possible application of the generalized homogeneity in control systems design, let us consider a plant modeled by the following ODE:

$$\dot{x} = Ax + Bu, \quad t > 0, \quad x(0) = x_0 \in \mathbb{R}^n, \tag{1.4}$$

where $x(t) \in \mathbb{R}^n$ is the system state, $u(t) \in \mathbb{R}^m$ is the control input, $A \in \mathbb{R}^{n\times n}$, $B \in \mathbb{R}^{n\times m}$ are constant matrices of the system.

An asymptotic stabilization of the system (1.4) is the classical control problem [58–61]. The traditional (exponential[4]) stabilizer for the system (1.4) is the linear control

$$u(t) = Kx(t), \tag{1.5}$$

where the feedback gain $K \in \mathbb{R}^{m\times n}$ is selected such that the matrix $A + BK$ is Hurwitz.[5] The desired gain K always exists if the pair $\{A, B\}$ is controllable [58, 62]:

$$\text{rank}(B, AB, ..., A^{n-1}B) = n.$$

Nonlinear stabilizers for linear models like (1.4) are employed, for example, if some additional requirements about the robustness of the control system [59] and/or some restrictions to a response time [13] have to be fulfilled.

---

4 The system is exponentially stable if $\exists C \geq 1, \exists \rho > 0 : \|x(t)\| \leq Ce^{-\rho t}\|x(0)\|, \forall t \geq 0$.

5 A matrix is Hurwitz if its eigenvalues have negative real parts.

Let us study a possibility of a nonlinear feedback design

$$u(t) = u_{\text{nonlin}}(x(t)), \quad u_{\text{nonlin}} : \mathbb{R}^n \mapsto \mathbb{R}^m, \tag{1.6}$$

which, in addition, to asymptotic stabilization ensures a homogeneity of the closed-loop system

$$\dot{x} = f(x) := Ax + Bu_{\text{nonlin}}(x) \tag{1.7}$$

with a desired homogeneity degree $\mu \in \mathbb{R}$. In the view of Table 1.1, the homogeneity provides some useful properties to the control system (see Sect. 1.3 for more inspirations). By definition, *homogeneous stabilization is a homogenization of the control system together with its asymptotic stabilization* (see [63] or Chap. 9 for formal definitions).

Standard Homogeneous Stabilization of Linear Plant

Let us study, first, a possibility of the *standard homogeneous stabilization* of the linear plant. The vector field f defined above has the linear component $f_{\text{lin}}(x) = Ax$ and the (possibly) nonlinear component $f_{\text{nonlin}}(x) = Bu_{\text{nonlin}}(x)$. The linear vector field f_{lin} is always standard homogeneous of the zero degree :

$$f_{\text{lin}}(\mathbf{d}_1(s)x) = e^{\mu s}\mathbf{d}_1(s) f_{\text{lin}}(x), \quad \forall s \in \mathbb{R}, \quad \forall x \in \mathbb{R}^n, \quad \mu = 0,$$

where $\mathbf{d}_1(s) = e^s I_n$ is the standard dilation and $\mu = 0$ is the standard homogeneity degree. Hence, we conclude that the linear feedback (1.5) is a *zero-degree* homogeneous stabilizer of the linear system (1.4). To ensure a faster (e.g., finite-time) stabilization (see Table 1.1), the homogeneity degree of the closed-loop system should be nonzero.

The system (1.7) may be standard homogeneous of a *nonzero degree* $\mu \neq 0$ only if either $f_{\text{lin}}(z) \equiv \mathbf{0}$ or the nonlinear component f_{nonlin} completely rejects the linear term f_{lin}. In the general case, the latter is possible only if

$$n = m \quad \text{and} \quad \det(B) \neq 0.$$

So, taking a Hurwitz matrix $K \in \mathbb{R}^{n\times n}$, we design the stabilizer

$$u_{\text{nonlin}}(x) = \tilde{K}(\|x\|)x, \tag{1.8}$$

$$\tilde{K}(\|x\|) := B^{-1}\left(-A + \|x\|^{\mu} K\right),$$

which results in the standard homogeneous closed-loop system

$$\dot{x} = \|x\|^{\mu} Kx, \tag{1.9}$$

of a given degree $\mu \in \mathbb{R}$. The right-hand side of the ODE (1.9) is continuous for $\mu > -1$, so the system is well-posed in this case (see Chap. 2). Since K is a Hurwitz matrix then, by Lyapunov theorem (see [7] or Example 4.14), there exists a positive definite symmetric matrix $P = P^\top \in \mathbb{R}^{n\times n}$ satisfying the *linear matrix inequality (LMI)*:

$$PK + K^\top P + 2\rho P \preceq 0, \quad P \succ 0$$

with some $\rho > 0$. Then, for the weighted Euclidean norm $\|x\| = \sqrt{x^\top Px}$ in (1.8), we have

$$\frac{d\|x\|}{dt} = \frac{x^\top P\dot{x}}{\|x\|} = \|x\|^{\mu}\frac{x^\top(PK + K^\top P)x}{2\|x\|} \leq -\rho\|x\|^{1+\mu}, \quad x \neq \mathbf{0}.$$

This means that the closed-loop system (1.9) is finite-time stable for $\mu \in (-1, 0)$

$$\|x(t)\| = 0, \quad \forall t \geq \frac{\|x_0\|^{-\mu}}{-\mu\rho}$$

and nearly fixed-time stable for $\mu > 0$:

$$\|x(t)\| \leq r, \quad \forall t \geq \frac{1}{\rho\mu r}, \quad \forall x_0 \in \mathbb{R}^n.$$

Therefore, any controllable linear plant (1.4) admits a zero-degree standard homogeneous stabilizer in the form of the linear feedback, but the standard homogeneous stabilization with a nonzero degree requires the additional restriction $n = m$, $\det(B) \neq 0$. *This very conservative restriction can be eliminated considering the generalized homogeneity.*

Generalized Homogeneous Stabilization of Linear Plant

The linear vector field f_{lin} may be standard homogeneous of a nonzero degree if and only if $f_{\text{lin}}(x) \equiv \mathbf{0}$. However, Example 1.1 shows that the generalized homogeneity relaxes this restriction. Indeed, considering the single-input control system (1.4) in the Brunovsky (canonical) form:

$$A = \begin{pmatrix} 0 & 1 & 0 & \ldots & 0 & 0 & 0 \\ 0 & 0 & 1 & \ldots & 0 & 0 & 0 \\ \ldots & \ldots & \ldots & \ldots & \ldots & \ldots & \ldots \\ 0 & 0 & 0 & \ldots & 0 & 1 & 0 \\ 0 & 0 & 0 & \ldots & 0 & 0 & 1 \\ a_1 & a_2 & a_3 & \ldots & a_{n-2} & a_{n-1} & a_n \end{pmatrix}, \quad B = \begin{pmatrix} 0 \\ 0 \\ \vdots \\ 0 \\ 0 \\ 1 \end{pmatrix}, \quad a_i \in \mathbb{R}, i = 1, 2, \ldots, n,$$

let us take

$$u_{\text{nonlin}}(x) = -\begin{pmatrix} a_1 & a_2 & \ldots & a_{n-1} & a_n \end{pmatrix} x + u_{\text{hom}}(x), \quad u_{\text{hom}} : \mathbb{R}^n \mapsto \mathbb{R}. \tag{1.10}$$

In this case, the closed-loop system (1.7) has the form

$$\dot{x} = f(x) := A_0 x + B u_{\text{hom}}(x), \tag{1.11}$$

where $A_0 \in \mathbb{R}^{n\times n}$ is the following *nilpotent matrix*:

$$A_0 = \begin{pmatrix} 0 & 1 & 0 & \ldots & 0 & 0 & 0 \\ 0 & 0 & 1 & \ldots & 0 & 0 & 0 \\ \ldots & \ldots & \ldots & \ldots & \ldots & \ldots & \ldots \\ 0 & 0 & 0 & \ldots & 0 & 1 & 0 \\ 0 & 0 & 0 & \ldots & 0 & 0 & 1 \\ 0 & 0 & 0 & \ldots & 0 & 0 & 0 \end{pmatrix}.$$

For $\mu < 1/(n-1)$ let us introduce the *weighted dilation* in $\mathbb{R}^n$ as follows

$$\mathbf{d}_2(s) = \begin{pmatrix} e^{(1+\mu(n-1))s} & 0 & \ldots & 0 & 0 \\ 0 & e^{(1+\mu(n-2))s} & \ldots & 0 & 0 \\ \ldots & \ldots & \ldots & \ldots & \ldots \\ 0 & 0 & \ldots & e^{(1+\mu)s} & 0 \\ 0 & 0 & \ldots & 0 & e^{s} \end{pmatrix}, \quad s \in \mathbb{R}.$$

The linear component $A_0 x$ of the vector field f is weighted homogeneous of degree μ. Indeed, simple calculations show that

$$A_0 \mathbf{d}_2(s) = e^{\mu s} \mathbf{d}_2(s) A_0, \quad \forall s \in \mathbb{R}.$$

Therefore, taking some weighted homogeneous feedback law of degree $\mu + 1$:

$$u_{\text{hom}}(\mathbf{d}_2(s)x) = e^{(\mu+1)s} u_{\text{hom}}(x), \quad \forall s \in \mathbb{R}, \quad \forall x \in \mathbb{R}^2, \tag{1.12}$$

we derive the weighted homogeneous closed-loop system

$$f(\mathbf{d}_2(s)x) = e^{\mu s} \mathbf{d}_2(s) f(x), \quad \forall s \in \mathbb{R} \quad \forall x \in \mathbb{R}^n.$$

If the system (1.11), (1.12) is asymptotically stable then the feedback (1.10) is a (weighted) homogeneous stabilizer with degree $\mu < 1/(n-1)$ for the linear control system in the Brunovsky form. The existence of such a stabilizer is proven, for example, in [10, 24, 64].

Therefore, a nonzero-degree homogeneous stabilization of a *controllable* single-input linear plant in the Brunovsky form can always realized using the weighted homogeneity. The corresponding stabilizer (1.10) contains two components. The linear component homogenizes the system. Namely, it makes the matrix A_0 nilpotent and weighted homogeneous of a desired degree $\mu \neq 0$. The nonlinear component u_{hom} is a weighted homogeneous feedback of the degree $\mu + 1$, which stabilizes the homogenized system. In Chap. 9 we show that the linear homogeneity allows a similar design for any controllable *multi-input multi-output (MIMO)* linear plant (1.4). The corresponding generalized homogeneous stabilizer has the form

$$u = K_0 x + \|x\|_{\mathbf{d}}^{1+\mu} K \mathbf{d}(-\ln \|x\|_{\mathbf{d}}) x, \quad K_0, K \in \mathbb{R}^{m\times n}, \tag{1.13}$$

where $\mathbf{d}$ is a linear continuous dilation in $\mathbb{R}^n$ and $\|\cdot\|_{\mathbf{d}} \in C(\mathbb{R}^n) \cap C^\infty(\mathbb{R}^n \backslash \{\mathbf{0}\})$ is a function being a norm in a special (generalized homogeneous) Euclidean space $\mathbb{R}^n_{\mathbf{d}}$ homeomorphic to $\mathbb{R}^n$.

1.3 From Linearity to Homogeneity in Control Systems

A quality of any control system is estimated by some *quantitative* indexes (see, e.g., [59, 65, 66]), which reflect control precision, optimality of motion, energetic effectiveness, robustness of a closed-loop system with respect to disturbances, etc. From mathematical point of view, the design of a "good" control law is a multi-objective optimization problem. The mentioned objectives may contradict to each other, e.g., a time optimal feedback control could not be energetically optimal but it may be efficient for a disturbance rejection [23]. Moreover, some criteria (such as a simplicity of practical implementation and tuning) cannot be even clearly mathematically formulated. The control practice frequently relaxes mathematical criteria for simplicity of the design and tuning. For example, an adjustment of a guaranteed (small enough) convergence time can be considered instead of the minimum-time control design, but an exact convergence of systems states to a set-point could be relaxed to a convergence into a sufficiently small neighborhood of this point, etc.

A well-tuned linear controller, such as PID (Proportional-Integral-Differential) algorithm, guarantees a good enough control quality in many practical cases [65]. However, the theory of linear control systems reaches its peak of maintenance and further improvements of quality using the same linear strategy seem impossible. Being a relaxation of the linearity, the homogeneity could provide some additional tools for control improvement. In this context, it is important to know which features of homogeneous systems may be useful for advanced analysis and design of control systems.

1.3.1 Convergence Rates of Homogeneous Algorithms

Finite-Time and Fixed-Time Stabilization

Finite-time and fixed-time stabilization (see above) may simplify an adjustment of a response time of the control system. The typical example, where this time is critically important, is a missile control problem [67]. The control plant disappears after a certain time instant. A control algorithm, which guarantees just an asymptotic convergence without any adjustment of the response time, is not appropriate in this case.

The stability properties of homogeneous and locally homogeneous systems discussed above allow us to propose a simple way to fulfill time constraints in control systems. For example, to stabilize a state in a fixed-time independently of the initial condition, a stable closed-loop system should be homogeneous of a positive degree close to infinity and homogeneous of a negative degree close to the origin. This idea can be illustrated on the simplest scalar example

$$\dot{x}(t) = u(t), \quad t > 0, \quad x(0) = x_0,$$

where $x(t) \in \mathbb{R}$ is the state variable and $u(t) \in \mathbb{R}$ is the control signal. The control aim is to stabilize this system at the origin such that the condition $|u(x)| \leq 1$ must be fulfilled for $|x| \leq 1$.

- The classical approach gives the standard *linear* proportional feedback algorithm

$$u_{\text{lin}}(x) = -x,$$

  which guarantees an asymptotic (in fact, the *exponential*) convergence to the origin of any solution $x(t, x_0)$ of the Cauchy problem with $u = u_{\text{lin}}(x)$:

$$x(t, x_0) = e^{-t}x_0 \to 0 \quad \text{as} \quad t \to +\infty.$$

- The globally homogeneous feedback of the form

$$u_{\text{ft}}(x) = -\sqrt{|x|}\text{sign}[x]$$

  stabilizes the system to the origin in a *finite time*:

$$x(t, x_0) = 0 \quad \text{for} \quad t \geq T(x_0).$$

  The convergence time T depends on the initial condition $x(0) = x_0$, in particular, $T(x_0) = 2\sqrt{|x_0|}$ for the considered control law. Obviously, $T(x_0)$ tends to infinity as $|x_0|$ tends to infinity.
- The *fixed-time* stabilizing controller can be selected locally homogeneous in the form:

$$u_{\text{fxt}}(x) = \begin{cases} -|x|^{1/2}\text{sign}[x] & \text{if } |x| \leq 1, \\ -|x|^{3/2}\text{sign}[x] & \text{if } |x| > 1. \end{cases}$$

  It stabilizes the system in a fixed time, namely,

$$x(t) = 0 \quad \text{for} \quad t \geq 4$$

  independently of $x_0 \in \mathbb{R}$.

The results of numerical simulations for the considered controllers are presented in Figs. 1.4, 1.5 and 1.6. They show that the fixed-time controller is always faster than the linear one, while the finite-time controller shows a faster convergence only close to the origin (in the zone $|x| < 1$). A globally homogeneous fixed-time stabilization

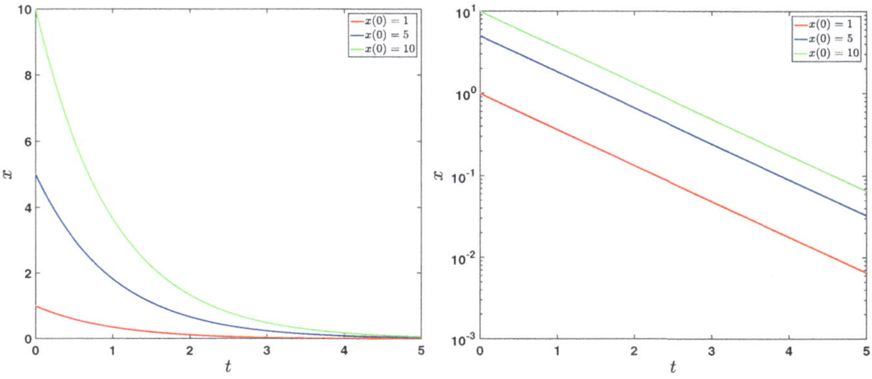

Fig. 1.4 Trajectories of the *exponentially* stable system with $u = u_{\mathrm{lin}}(x)$ in linear and logarithmic scales

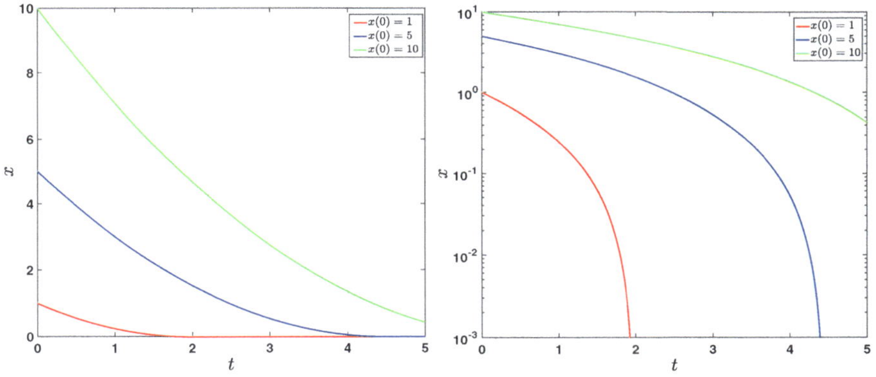

Fig. 1.5 Trajectories of the *finite-time* stable system with $u = u_{\mathrm{ft}}(x)$ in linear and logarithmic scales

can be realized by means of a feedback, which gain is dependent on the initial state of the system (see Chap. 9).

Finite-Time and Fixed-Time State Estimation

The finite-time stability of homogeneous systems can also be utilized for observers design. Indeed, let us consider the simplest observation problem

$$\begin{cases} \dot{x}_1 = x_2, \\ \dot{x}_2 = u, \qquad t > 0, \\ y = x_1 \end{cases}$$

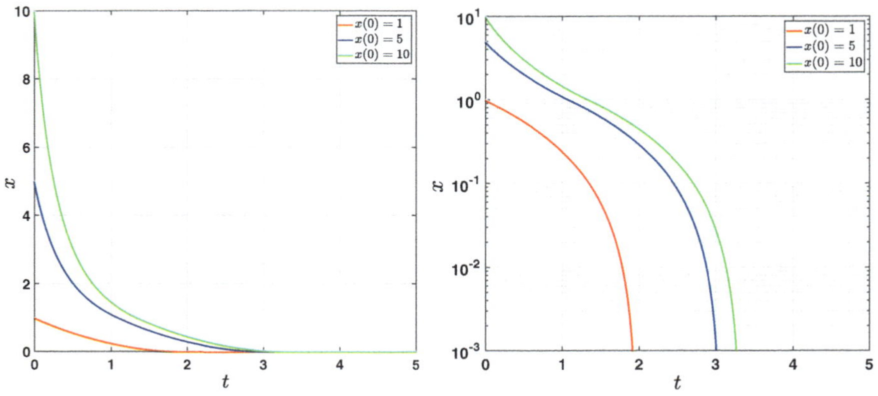

Fig. 1.6 Trajectories of the *fixed-time* stable system with $u{=}u_{\text{fxt}}(x)$ in linear and logarithmic scales

where $x_1(t), x_2(t) \in \mathbb{R}$ are system states, $u(t) \in \mathbb{R}$ is a known input, $y(t) \in \mathbb{R}$ is a measured output. The aim is to estimate the unknown state variable x_2 in a finite time. Recall that, the conventional *Luenberger observer* [68]

$$\begin{cases} \frac{d}{dt}\hat{x}_1 = k_1(\hat{x}_1 - y) + \hat{x}_2, & \hat{x}_1, \hat{x}_2 \in \mathbb{R}, \\ \frac{d}{dt}\hat{x}_2 = k_2(\hat{x}_1 - y) + u, & k_1, k_2 \text{ are constant gains} \end{cases}$$

estimates the desired variable x_2 asymptotically, i.e., $\hat{x}_2(t) \to x_2(t)$ as $t \to +\infty$. However, if we allow the gains k_1 and k_2 to be depended on $\hat{x}_1 - y$ as follows

$$k_1 = \tilde{k}_1|\hat{x}_1 - y|^{0.5(\alpha-1)}, \quad k_2 = \tilde{k}_2|\hat{x}_1 - y|^{\alpha-1},$$

$$\alpha > 0, \quad \tilde{k}_1, \tilde{k}_2 \text{ are constants}$$

then the error system with $e_1 = \hat{x}_1 - x_1$ and $e_2 = \hat{x}_2 - x_2$ becomes

$$\begin{cases} \dot{e}_1 = \tilde{k}_1|e_1|^{\frac{1+\alpha}{2}} \operatorname{sign}(e_1) + e_2, \\ \dot{e}_2 = \tilde{k}_2|e_1|^{\alpha} \operatorname{sign}(e_1). \end{cases}$$

The vector field $f : \mathbb{R}^2 \mapsto \mathbb{R}^2$ given by

$$f(e) = \begin{pmatrix} \tilde{k}_1|e_1|^{\frac{1+\alpha}{2}} \operatorname{sign}(e_1) + e_2 \\ \tilde{k}_2|e_1|^{\alpha} \operatorname{sign}(e_1) \end{pmatrix}, \quad e = (e_1, \ e_2)^\top$$

is $\mathbf{d}$-homogeneous of the degree $\mu = \alpha - 1$ with respect to the weighted dilation

$$\mathbf{d}(s) = \begin{pmatrix} e^{2s} & 0 \\ 0 & e^{(1+\alpha)s} \end{pmatrix}, \quad s \in \mathbb{R}^n.$$

Indeed, for any $s \in \mathbb{R}$ and any $e \in \mathbb{R}^2$ we derive

$$f(\mathbf{d}(s)e) = e^{(\alpha-1)s}\mathbf{d}(s)f(e).$$

The estimation error converges to zero provided that the gains $\tilde{k}_1, \tilde{k}_2$ and the homogeneity degree $\mu = \alpha - 1 \geq -1$ are properly selected. Indeed, the obtained homogeneous observer coincides with the Luenberger observer if $\mu = 0$. By continuity, the gains of a well-tuned Luenberger observer are expected to be appropriate [10, 24, 32] for the homogeneous observer, at least, if μ is close to 0. As discovered above, the convergence rate of an asymptotically stable homogeneous system depends on the sign of the homogeneity degree μ. Therefore, the finite-time (nearly fixed-time) convergence of the homogeneous observer can always be guaranteed, at least, for $\mu \in [-1, 0)$ (resp., $\mu > 0$) being sufficiently close to 0. The case $\mu = -1$ corresponds to the so-called exact (Levant's) differentiator [64]. Combining observers with both positive and negative homogeneity degrees, a globally fixed-time convergent observer can also be designed [34, 69]. More rigorously, these issues are studied in Chap. 14, where homogeneous finite-time and fixed-time observers are designed for finite-dimensional linear and nonlinear plants.

1.3.2 *Robustness Issues*

Homogeneous Domination

Let us consider the simplest stabilization problem

$$\dot{x} = \lambda x + u,$$

where $x(t) \in \mathbb{R}$ is the system state, $\lambda > 0$ is an unknown constant parameter and $u(t) \in \mathbb{R}$ is a control law to be designed. Since λ is unknown then a static linear feedback

$$u = -kx, \quad k > 0$$

cannot guarantee an a priori boundedness of system trajectories. However, the homogeneous feedback

$$u = -kx^2 \operatorname{sign}(x), \qquad k > 0$$

always ensures a practical stabilization of the system independently of the parameter λ. Indeed, estimating the time derivative of the "energy" $V = x^2/2$ along trajectories of the system we derive

$$\frac{dx^2}{dt} = 2x\dot{x} \leq \lambda x^2 - k|x|^3 < 0 \quad \text{for} \quad |x| > \lambda/k.$$

This guarantees a boundedness of system trajectories and their convergence to a zone:

$$\limsup_{t\to+\infty} |x(t)| \leq \frac{\lambda}{k}.$$

In this example, the homogeneous control dominates the uncertain term at infinity, so the local approximation of the closed-loop system at ∞

$$\dot{x} = -kx^2 \operatorname{sign}(x)$$

is asymptotically stable. This always implies a practical fixed-time stability of the original system (see Chap. 7). Sliding-mode controllers are homogeneous of negative degree. They allow small enough bounded perturbations to be rejected [59, 70]. Therefore, a homogeneous control may be more robust with respect to system uncertainties than a linear control law.

Notice that the homogeneity degree allows us to identify a dominating term at ∞ or at 0, respectively. Indeed, if $f_1 : \mathbb{R} \mapsto \mathbb{R}$ and $f_2 : \mathbb{R} \mapsto \mathbb{R}$ (such that $f_i(x) \neq 0$ for $x \neq 0$ and $i = 1, 2$) are continuous homogeneous functions of degree $\nu_1 < \nu_2$ then f_2 *dominates* f_1 at $+\infty$:

$$\frac{|f_2(x)|}{|f_1(x)|} = \frac{|x|^{\nu_2}\left|f_2\left(\frac{x}{|x|}\right)\right|}{|x|^{\nu_1}\left|f_1\left(\frac{x}{|x|}\right)\right|} \to +\infty \quad \text{as} \quad x \to \infty$$

and, inversely, f_1 *dominates* f_2 at 0, i.e.,

$$\frac{|f_1(x)|}{|f_2(x)|} \to +\infty \quad \text{as} \quad x \to 0.$$

This *homogeneous domination* principle (known since 1950s [9]) underlies the robustness analysis of homogeneous control systems (see Chap. 7).

Input-to-State Stability versus Asymptotic Stability

Input-to-state stability [4] can be interpreted as a robustness of a dynamical system with respect to disturbances. Let us consider the model

$$\dot{x}(t) = f(x(t), q(t)), \quad t > 0, \tag{1.14}$$

where $x(t)$ is the system state and $q \in L^\infty$ is an exogenous input, which can be treated, in particular, as a perturbation or a measurement noise of a closed-loop control system. Recall that the classical (Lyapunov) stability [7]

$$\exists \varepsilon \in \mathcal{K}_\infty \quad : \quad |x(t, x_0)| \leq \varepsilon(|x_0|), \quad \forall t \geq 0, \quad \forall x_0 \in \mathbb{R}^n$$

means a continuous dependence (for the infinite horizon[6]) of the zero solution on the initial condition. The input-to-state stability (ISS)

$$\exists \beta \in \mathcal{KL}, \exists \gamma \in \mathcal{K}_\infty \ : \ |x(t,x_0)| \leq \beta(|x_0|,t) + \gamma(|q|_{L^\infty(0,t)}), \quad \forall t \geq 0, \ \forall x_0$$

can be interpreted as a continuous dependence of the stable zero solution (of unperturbed system) on a magnitude of the exogenous input. Homogeneity of f simplifies the ISS analysis [5, 10, 36, 39, 63]. Indeed, it can be shown that the homogeneous (on both variables x and q) system is ISS if and only if it is asymptotically stable for $q = \mathbf{0}$. In other words, asymptotically stable homogeneous system is always robust (in ISS sense) with respect to homogeneously involved perturbations. This important result holds for both finite-dimensional (see Chap. 7) and infinite-dimensional systems (see Volume II of this monograph).

1.3.3 Elimination of an Unbounded Peaking Effect

The finite/fixed-time stability is an interesting theoretical feature of homogeneous systems. However, any controllable linear system can be stabilized to an arbitrary small neighborhood of the origin even by means of a static linear feedback. A time of convergence of trajectories from a unit ball into this small neighborhood can be prescribed in advance by means of an appropriate tuning of the linear feedback gain. Such a stabilization is sufficient for many practical problems. The reasonable question in this case:

> *Is there any advantage of a homogeneous controller in comparison with the classical linear feedback in the case of practical stabilization?*

Let us consider the control system

$$\dot{x} = Ax + bu(x), \quad t > 0, \quad x(0) = x_0,$$

$$A = \begin{pmatrix} 0 & 1 & 0 & \ldots & 0 \\ 0 & 0 & 1 & \cdots & 0 \\ \cdots & \cdots & \cdots & \cdots & \cdots \\ 0 & 0 & 0 & \cdots & 1 \\ 0 & 0 & 0 & \cdots & 0 \end{pmatrix}, \quad B = \begin{pmatrix} 0 \\ 0 \\ \cdots \\ 0 \\ 1 \end{pmatrix}$$

where $x = (x_1, x_2, \ldots, x_n)^\top$ is the state vector and $u : \mathbb{R}^n \mapsto \mathbb{R}$ is the feedback control. Initial conditions of the system are assumed to be in the unit ball.

[6] Continuous dependence of solutions on initial conditions for a finite horizon follows from the well-posedness of the system (see Chap. 2).

The control aim is to stabilize the state vector $x(t)$ of the system into the ball of a small radius $\varepsilon > 0$ in a prescribed time $T > 0$, i.e.,

$$|x_0| \leq 1 \quad \Rightarrow \quad |x(t, x_0)| \leq \varepsilon, \quad \forall t \geq T.$$

Let us consider the static linear feedback

$$u_\ell(x) = kx, \quad k = (k_1, k_2, ..., k_n).$$

The eigenvalues $\{\lambda_1, ..., \lambda_n\}$ of the closed-loop linear system

$$\dot{x} = (A + bk)x$$

can be placed in any given set of the complex plane $\mathbb{C}$ by choosing the vector k (see, e.g., [58]). Therefore, it is possible to design a closed-loop system with an arbitrary fast damping speed such that

$$\forall \varepsilon > 0, \quad \exists k \in \mathbb{R}^{1\times n} : \sup_{|x_0|\leq 1} |x(t, x_0)| < \varepsilon, \quad \forall t > T.$$

Trajectories of this system converge to the origin exponentially fast

$$|x(t, x_0)| \leq Ce^{-\sigma t}|x_0|, \quad t > 0,$$

where the constant $C \geq 1$ depends on λ_i, $i = 1, 2, .., n$ and $\Re(\lambda_i) < -\sigma$. Hence, the smaller $\varepsilon > 0$, the larger $\sigma > 0$ has to be assigned in order to solve the control problem provided that T is fixed. Therefore, we conclude that the linear state feedback is, indeed, a possible solution of the considered stabilization problem for any fixed $\varepsilon > 0$ and any fixed $T > 0$.

However, the trajectories of the closed-loop linear system with fast decays have large deviations from the origin during the initial phase of the stabilization. This phenomenon is called the peaking effect and the large deviation is referred to as an overshoot (see, e.g., [71] for more details). In particular, it was shown [72] that there exists $\gamma > 0$ independent of λ_i such that

$$\sup_{0\leq t\leq \sigma^{-1}} \sup_{|x_0|\leq 1} |x(t, x_0)| \geq \gamma\sigma^{n-1}.$$

For $n > 1$ *the linear closed-loop system has an infinite overshoot as* $\varepsilon \to 0$:

$$\sup_{0\leq t\leq T} \sup_{|x_0|\leq 1} |x(t, x_0)| \to +\infty \quad \text{as} \quad \varepsilon \to 0.$$

This means that for a sufficiently small $\varepsilon > 0$ the linear system may have so huge overshoot that practical (physical) restrictions to the system states or to a control input would not allow it. The static linear control needs to be somehow modified to overcome this difficulty. The simplest way is to use some input saturation, which, in fact, anyway must be taken into account in practice. However, in this case, it is not clear if the saturated feedback could solve the considered stabilization problem with the prescribed time $T > 0$ provided that the saturation would not destroy the stability of the system initiated in the unit ball.

Another possible way to eliminate the infinite peaking effect is to transform ("upgrade") a linear controller to a homogeneous one. Indeed, let us consider the following feedback law

$$u_{\mathrm{h}}(x) = \tilde{k}\mathbf{d}(-\ln \|x\|_{\mathbf{d}})x, \quad x \neq 0 \quad \text{and} \quad u_{\mathrm{h}}(\mathbf{0}) = \mathbf{0},$$

where $\mathbf{d}$ is the weighted dilation

$$\mathbf{d}(s) = \begin{pmatrix} e^{ns} & 0 & \cdots & 0 \\ 0 & e^{(n-1)s} & \cdots & 0 \\ \cdots & \cdots & \cdots & \cdots \\ 0 & 0 & \cdots & e^{s} \end{pmatrix}, \quad s \in \mathbb{R}$$

and $\|\cdot\|_{\mathbf{d}} : \mathbb{R}^n \mapsto \mathbb{R}_+$ is the so-called canonical homogeneous norm studied in Chap. 5 and satisfying the identity $\|\mathbf{d}(s)x\|_{\mathbf{d}} = e^s\|x\|_{\mathbf{d}}$. The vector field f given by

$$f(x) := Ax + bu_{\mathrm{h}}(x)$$

is weighted homogeneous of degree -1, i.e., $f(\mathbf{d}(s)x) = e^{-s}\mathbf{d}(s)f(x)$. The asymptotic stability of homogeneous system with negative degree implies its finite-time stability.

In Chap. 9 it is shown that the vector $\tilde{k} = (\tilde{k}_1, \tilde{k}_2, ..., \tilde{k}_n)^\top$ can be easily selected to guarantee

$$\sup_{|x(0)|\leq 1} |x(t)| = 0, \quad t \geq T$$

for any fixed $T > 0$. In addition, one can be shown that the feedback law u_h is globally bounded:

$$\sup_{x\in\mathbb{R}^n} |u_{\mathrm{h}}(x)| \leq M < +\infty,$$

where M depends on T as follows: the smaller T, the larger M is. The homogeneous control stabilizes the considered system globally and in a finite time. It solves the stabilization problem considered above independently of $\varepsilon > 0$. Due to the global boundedness of the controller it does not have the unbounded peaking effect discovered for the linear system as $\varepsilon \to 0$. Moreover, given M the controller gains can be optimized (using Semi-Definite Programming) to guarantee minimal settling time T (see Chap. 13).

The simulation results for the linear controller $u_\ell(x) = kx, k = (-100 - 20)$ and the homogeneous controller $u_{\mathrm{h}}(x) = \tilde{k}\,\mathrm{diag}(\|x\|_{\mathbf{d}}^{-2}, \|x\|_{\mathbf{d}}^{-1})x, \tilde{k} = (-4.1721 - 2.8718)$ are depicted in Fig. 1.7. Initial conditions $x(0)$ for the numerical simulations have been taken from the unit sphere. In both cases, trajectories of the

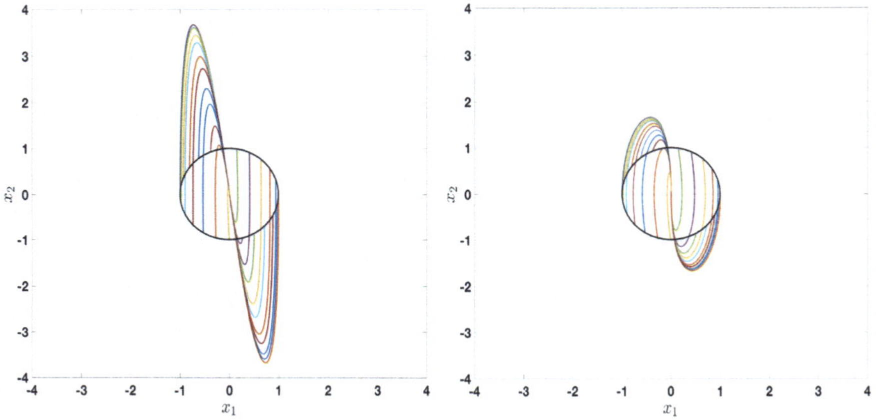

Fig. 1.7 Comparison of "overshoots" for linear (left) and homogeneous (right) controllers

closed-loop system converge to the origin. The homogeneous controller guarantees the (theoretically) exact stabilization of any solution of the closed-loop system with $|x(0)| \leq 1$ in the time $T = 1$, i.e., $x(t) = 0$ for all $t \geq 1$ and for all $|x(0)| \leq 1$. The linear controller allows only the practical stabilization $|x(t)| \leq \varepsilon = 0.005$ for $t \geq 1$. Even in this case the overshoot of the homogeneous controller is twice smaller (see Fig. 1.7). The overshoot of the linear controller increases drastically for smaller ε.

For x belonging to the unit sphere $\|x\|_{\mathbf{d}} = 1$, we have $u_{\rm h}(x) = \tilde{k}x$. This means that the homogeneous controller $u_{\rm h}$ is designed by a certain scaling of a linear stabilizer $u(x) = \tilde{k}x$. In Chap. 11 we show that any linear stabilizer can be upgraded to a nonlinear (generalized homogeneous) one with an improved control quality. An efficiency of this scheme is confirmed by real experiments with robotic systems (see Chap. 11 for more details). The discovered improvements perfectly fit the above theoretical conclusions:

> *Homogeneous controllers may provide faster convergence, better robustness and smaller overshoots than liner algorithms.*

Concluding Remarks

Homogeneity (dilation symmetry) is a common property of many physical models. This monograph presents an introduction to the theory of generalized homogeneous control systems, studies homogeneous differential equations in Euclidean $\mathbb{R}^n$ (Volume I) and in Banach/Hilbert (Volume II) spaces and solves several control and estimation problems using the method of the linear (geometric) homogeneity.

Generalized homogeneous controllers/observers could demonstrate a better regulation/estimation quality (namely, *faster convergence, better robustness and smaller overshoots*) than linear algorithms. A systematic study of this issue is one of the main aims of this monograph.

Homogeneity is just one of the symmetries, which may be useful for control systems design. We refer the reader to [73] for more inspirations about symmetry-based analysis of dynamical systems.

References

1. Noether E (1918) Invariante variationsprobleme. Kgl. Ges. d. Wiss. Nachrichten, Math.-phys. Klasse, pp 235–257 (in German)
2. Zubov VI (1964) Methods of A.M. Lyapunov and their applications. Noordhoff, Leiden. (Translated from V.I. Zubov, Metody A.M. Lyapunova i ih primenenie, Izdatelstvo Leningradskogo universiteta, 1957) edition
3. Filippov AF (1988) Differential equations with discontinuous right-hand sides. Kluwer Academic Publishers
4. Sontag ED (1989) Smooth stabilization implies coprime factorization. IEEE Trans Autom Control 34:435–443
5. Ryan EP (1995) Universal stabilization of a class of nonlinear systems with homogeneous vector fields. Syst & Control Lett 26:177–184
6. Bhat SP, Bernstein DS (2000) Finite time stability of continuous autonomous systems. SIAM J Control Optim 38(3):751–766
7. Lyapunov AM (1992) The general problem of the stability of motion. Taylor & Francis
8. Efimov D, Polyakov A (2021) Finite-time stability tools for control and estimation. Found Trends Syst Control. Now Publishers Inc, Netherlands
9. Zubov VI (1958) On systems of ordinary differential equations with generalized homogeneous right-hand sides. Izvestia vuzov. Mathematica (available on-line: http://mi.mathnet.ru/ivm2874) 1:80–88
10. Andrieu V, Praly L, Astolfi A (2008) Homogeneous approximation, recursive observer design, and output feedback. SIAM J Control Optim 47(4):1814–1850
11. Falkovich G (2011) Fluid mechanics (a short course for physicists). Cambridge University Press
12. Armstrong-Helouvry B (1991) Control of machines with friction. Springer
13. Polyakov A (2012) Nonlinear feedback design for fixed-time stabilization of linear control systems. IEEE Trans Autom Control 57(8):2106–2110
14. Balakrishnan AV (2005) Superstability of systems. Appl Math Comput 164:321–326
15. Hille E, Phillips R (1957) Functional analysis and semi-groups. AMS
16. Majda A (1975) Disappearing solutions for the dissipative wave equation. Indiana Univ Math J 24(12):1119–1133
17. Perrollaz V, Rosier L (2014) Finite-time stabilization of 2x2 hyperbolic systems on tree-shaped networks. SIAM J Control Optim 52(1):143–163
18. Polyak BT, Scherbakov PS (2002) Superstable linear control systems. Autom Remote Control 63(8):1239–1254
19. Hermes H (1986) Nilpotent approximations of control systems and distributions. SIAM J Control Optim 24(4):731
20. Kawski M (1991) Families of dilations and asymptotic stability. In: Analysis of controlled dynamical systems, pp 285–294
21. Rosier L (1992) Homogeneous Lyapunov function for homogeneous continuous vector field. Syst & Control Lett 19:467–473

22. Fuller A (1960) Relay control systems optimized for various performance criteria. In: Proceedings of the 1st IFAC world congress, pp 510–519
23. Chernous'ko FL, Ananievski IM, Reshmin SA (2008) Control of nonlinear dynamical systems. Springer
24. Bhat SP, Bernstein DS (2005) Geometric homogeneity with applications to finite-time stability. Math Control Signals Syst 17:101–127
25. Aleksandrov AY, Kosov AA, Platonov AV (2012) On the asymptotic stability of switched homogeneous systems. Syst & Control Lett 61(1):127–133
26. Kawski M (1990) Homogeneous stabilizing feedback laws. Control Theory Adv Technol 6(4):497–516
27. Sepulchre R, Aeyels D (1996) Homogeneous Lyapunov functions and necessary conditions for stabilization. Math Control Signals Syst 9:34–58
28. Sepulchre R, Aeyels D (1996) Stabilizability does not imply homogeneous stabilizability for controllable homogeneous systems. SIAM J Control Optim 34(5):1798–1813
29. Andreini A, Bacciotti A, Stefani G (1988) Global stabilizability of homogeneous vector fields of odd degree. Syst & Control Lett 10:251–256
30. Coron J-M, Praly L (1991) Adding an integrator for the stabilization problem. Syst & Control Lett 17(2):89–104
31. Grüne L (2000) Homogeneous state feedback stabilization of homogeneous systems. SIAM J Control Optim 38(4):1288–1308
32. Perruquetti W, Floquet T, Moulay E (2008) Finite-time observers: application to secure communication. IEEE Trans Autom Control 53(1):356–360
33. Polyakov A, Efimov D, Perruquetti W (2015) Finite-time and fixed-time stabilization: implicit Lyapunov function approach. Automatica 51(1):332–340
34. Lopez-Ramirez F, Polyakov A, Efimov D, Perruquetti W (2018) Finite-time and fixed-time observer design: implicit Lyapunov function approach. Automatica 87(1):52–60
35. Polyakov A (2019) Sliding mode control design using canonical homogeneous norm. Int J Robust Nonlinear Control 29(3):682–701
36. Bernuau E, Polyakov A, Efimov D, Perruquetti W (2013) Verification of ISS, iISS and IOSS properties applying weighted homogeneity. Syst & Control Lett 62(12):1159–1167
37. Folland G (1975) Subelliptic estimates and function spaces on nilpotent Lie groups. Arkiv for Matematik 13(1–2):161–207
38. Zimenko K, Efimov D, Polyakov A, Perruquetti W (2017) A note on delay robustness for homogeneous systems with negative degree. Automatica 79(5):178–184
39. Polyakov A (2021) Input-to-state stability of homogeneous infinite dimensional systems with locally Lipschitz nonlinearities. Automatica 129:109615
40. Levant A (2005) Homogeneity approach to high-order sliding mode design. Automatica 41(5):823–830
41. Orlov Y (2005) Finite time stability and robust control synthesis of uncertain switched systems. SIAM J Control Optim 43(4):1253–1271
42. Moreno J, Osorio M (2012) Strict Lyapunov functions for the super-twisting algorithm. IEEE Trans Autom Control 57:1035–1040
43. Bernuau E, Efimov D, Perruquetti W, Polyakov A (2014) On homogeneity and its application in sliding mode control. J Frankl Inst 351(4):1866–1901
44. Cruz-Zavala E, Moreno JA (2017) Homogeneous high order sliding mode design: a Lyapunov approach. Automatica 80(6):232–238
45. Zelikin MI, Borisov VF (1994) Theory of chattering control. Springer
46. Coron J-M, Grüne L, Worthmann K (2020) Model predictive control, cost controllability, and homogeneity. SIAM J Control Optim 58(5):2979–2996
47. Polyakov A, Efimov D, Ping X (2021) On energetically optimal finite-time stabilization. In: Conference on decision and control
48. Livne M, Levant A (2014) Proper discretization of homogeneous differentiators. Automatica 50(8):2007–2014

49. Efimov D, Polyakov A, Levant A, Perruquetti W (2017) Realization and discretization of asymptotically stable homogeneous systems. IEEE Trans Autom Control 62(11):5962–5969
50. Polyakov A, Efimov D, Brogliato B (2019) Consistent discretization of finite-time and fixed-time stable systems. SIAM J Control Optim 57(1):78–103
51. Koch S, Reichhartinger M, Horn M, Fridman L (2019) Discrete-time implementation of homogeneous differentiators. IEEE Trans Autom Control 65(2):757–762
52. Brogliato B, Polyakov A, Efimov D (2020) The implicit discretization of the super-twisting sliding-mode control algorithm. IEEE Trans Autom Control 65(8):3707–3713
53. Polyakov A, Efimov D, Ping X (2023) Consistent discretization of homogeneous finite/fixed-time controllers for LTI systems. Automatica 155:111118
54. Husch LS (1970) Topological characterization of the dilation and the translation in frechet spaces. Math Ann 190:1–5
55. Khomenuk VV (1961) On systems of ordinary differential equations with generalized homogenous right-hand sides. Izvestia vuzov. Mathematica 3(22):157–164 (in Russian)
56. Rosier L (1993) Etude de quelques problèmes de stabilization. PhD Thesis, Ecole Normale Superieure de Cachan (France)
57. Fischer V, Ruzhansky M (2016) Quantization on nilpotent Lie groups. Springer
58. Wonham WM (1985) Linear multivariable control: a geometric approach. Springer
59. Utkin VI (1992) Sliding modes in control optimization. Springer-Verlag, Berlin
60. Boyd S, Ghaoui E, Feron E, Balakrishnan V (1994) Linear matrix inequalities in system and control theory. SIAM, Philadelphia
61. Bacciotti A, Rosier L (2001) Lyapunov functions and stability in control theory. Springer
62. Kalman R (1960) On the general theory of control systems. In: Proceedings of the first international congress of the IFAC, Moscow
63. Hong Y (2001) H_∞ control, stabilization, and input-output stability of nonlinear systems with homogeneous properties. Automatica 37(7):819–829
64. Levant A (2003) Higher-order sliding modes, differentiation and output-feedback control. Int J Control 76(9–10):924–941
65. Boiko I (2013) Non-parametric tuning of PID controllers. Springer-Verlag, London
66. Sontag ED (2007) Input to state stability: basic concepts and results. In: Nistri PP, Stefani G (eds) Nonlinear and optimal control theory. Springer-Verlag, Berlin, pp 163–220
67. Shinar J, Glizer VY, Turetsky V (2014) Capture zone of linear strategies in interception problems with variable structure dynamics. J Frankl Inst 351(4):2378–2395
68. Luenberger D (1964) Observing the state of a linear system. IEEE Trans Mil Electron 8(2):74–80
69. Cruz-Zavala E, Moreno JA, Fridman LM (2011) Uniform robust exact differentiator. IEEE Trans Autom Control 56(11):2727–2733
70. Shtessel Y, Edwards C, Fridman L, Levant A (2014) Sliding mode control and observation. Birkhauser
71. Polyak BT, Smirnov G (2016) Large deviations for non-zero initial conditions in linear systems. Automatica 74:297–307
72. Izmailov R (1987) The peak effect in stationary linear systems with scalar inputs and outputs. Autom Remote Control 48:1018–1024
73. Olver P (1993) Applications of Lie groups to differential equations. Springer

Chapter 2
Finite-Dimensional Dynamical Systems

This chapter considers ordinary differential equations and inclusions, which are common mathematical models of control systems. Elements of the theory of differential equations with continuous and discontinuous right-hand sides are presented. The existence and uniqueness of solutions as well as their continuous dependence on parameters are briefly studied. Some issues of modeling control systems with uncertainties and disturbances are discussed.

2.1 Ordinary Differential Equation

Many real physical, biological, economical and even social systems and processes can be described by means of *mathematical models*. Various issues of mathematical modeling and model validation are studied in the literature (see, for example, [1, 2] and references therein). Since the end of the seventeenth century, many modeling activities were conducted within the framework of infinitesimal calculus (known today as *differential calculus*, see Appendix D). The corresponding models describe relations between a vector of system variables ξ, being a function of a special scalar variable t named "time", and its derivatives with respect to t:

$$Q\left(t, \xi, \dot{\xi}, ..., \xi^{(q)}\right) = \mathbf{0}, \quad t \in \mathbb{R}, \quad q \in \mathbb{N}.$$

This relation is called the *ordinary differential equation (ODE)* of the order q provided that $\xi(t) \in \mathbb{R}^p$. The simple notation

A. Polyakov, *Generalized Homogeneity in Systems and Control Volume I*,
Communications and Control Engineering,
https://doi.org/10.1007/978-3-031-76246-8_2

$$\tilde{Q}(t, x, \dot{x}) = \mathbf{0}, \qquad x = \begin{pmatrix} \xi \\ \dot{\xi} \\ \dots \\ \xi^{(q-1)} \end{pmatrix}$$

allows us to reduce (virtually) the number of derivatives in the ODE representation. The equation $\tilde{Q} = 0$ includes both the original equation $Q = 0$ and the differential relation $\left(\begin{smallmatrix} I & 0 \\ 0 & 0 \end{smallmatrix}\right)\dot{x} - \left(\begin{smallmatrix} 0 & I \\ 0 & 0 \end{smallmatrix}\right)x = 0$ between components of x, where I is an identity matrix of an appropriate dimension. For example, the second-order differential equation $(\ddot{\xi} + \xi\dot{\xi})^{|t|+1} + \xi = 0$ can be rewritten as the following system of the first-order ODEs: $(\dot{x}_2 + x_1 x_2)^{|t|+1} + x_1 = 0, \dot{x}_1 = x_2$, where $x_1 = \xi, x_2 = \dot{\xi}$.

The vector $x(t) \in \mathbb{R}^n, n = pq$ is known as the *state vector*. In control theory, it is also assumed that the ODE model may depend on parameters $u(t) \in \mathbb{R}^m$ and $d(t) \in \mathbb{R}^r$:

$$\tilde{Q}(t, x, \dot{x}, d, u) = \mathbf{0},$$

where the vector u is the *control input* to be modified (dynamically if needed) in order to guarantee a desired behavior of the system, while the vector d describes *system uncertainties* and *disturbances*, which may be time-varying.

If the above equation is uniquely solvable with respect to the *vector of time derivatives* $\dot{x}$ then we come up with the following canonical model of the control system

$$\dot{x} = f(t, x, u, d), \qquad f : \mathbb{R} \times \mathbb{R}^n \times \mathbb{R}^m \times \mathbb{R}^r \mapsto \mathbb{R}^n. \tag{2.1}$$

Frequently, the whole state vector x cannot be measured directly, while its part (or a function of x) is available for control design purposes. To complete the mathematical description of the control system, a model of measurements must be taken into account:

$$y = h(t, x, d), \qquad h : \mathbb{R} \times \mathbb{R}^n \times \mathbb{R}^r \mapsto \mathbb{R}^k, \tag{2.2}$$

where $y(t) \in \mathbb{R}^k$ is the *system output* which contains available measurements of the system state. The output measurements are noisy if y depends on d. This monograph deals only with *deterministic models* assuming that d is an unknown Lebesgue measurable function of time. Possible stochastic characteristics of the disturbance vector d are not considered.

2.1.1 Continuous Differential Equations

The mathematical model like

$$\dot{x} = f(t, x), \quad f \in C(\mathbb{R} \times \mathbb{R}^n; \mathbb{R}^n) \tag{2.3}$$

defines the rule of an evolution of the state x in time.

A continuous vector-valued function $x : \mathbb{R} \mapsto \mathbb{R}^n$ satisfying this rule (in a certain sense) is called a *solution of ODE* (2.3). Let $\mathcal{I}$ be one of the following intervals: $[a, b]$, (a, b), $[a, b)$ or $(a, b]$, where $a, b \in [-\infty, +\infty]$, $a < b$.

Definition 2.1 *(Classical solution of ODE)* A differentiable function $x : \mathcal{I} \subset \mathbb{R} \mapsto \mathbb{R}^n$ is said to be a *classical solution of ODE* (2.3) if it satisfies Eq. (2.3) on a time interval $\mathcal{I}$.

ODE models serve for various practical needs. For example, a solution $x(t)$ of ODE in the forward time $t \geq t_0$ (*forward solution*) can be utilized for prediction a future behavior of the system based on its current (initial) state $x(t_0) = x_0$, while a solution in the backward time $t \leq t_0$(*backward solution*) may characterize prehistory of the system. To achieve these tasks, the so-called Cauchy problem (or, equivalently, the *initial value problem*) has to be solved. It consists in finding a solution of ODE (2.3) which has a given initial state $x_0 \in \mathbb{R}^n$ at a given initial time $t_0 \in \mathbb{R}$:

$$x(t_0) = x_0. \tag{2.4}$$

The *Cauchy problem* for (2.3) can be formulated as follows:

Find a solution of (2.3) *satisfying the identity* (2.4) *for given* $t_0 \in \mathbb{R}$ *and given* $x_0 \in \mathbb{R}^n$.

Below we use *the notation* $x(t, t_0, x_0)$ *for a solution of the Cauchy problem* (2.3), (2.4). The solution may be defined for $t \leq t_0$ and/or $t \geq t_0$ such that $x(t_0, t_0, x_0) = x_0$. In *autonomous*[1] case, a solution of the Cauchy problem is independent of the initial time $t_0 \in \mathbb{R}$ in the sense that $x(t, t_0, x_0) = x(t - t_0, 0, x_0)$. For an autonomous system with $t_0 = 0$, we denote a solution by $x(t, x_0)$. For shortness, we also use the notation $x(t)$ if the dependence of t_0 and x_0 is not important for considerations.

The function f in the right-hand side of ODE (2.3) is called *vector field*. Geometrically, in each point of the state space (also known as phase space), it defines the speed and direction of a moving point $x(t) = x(t, t_0, x_0) \in \mathbb{R}^n$. The curve $\{y = x(t) \in \mathbb{R}^m, t \geq t_0\}$ generated by the moving point in $\mathbb{R}^n$ is called *phase curve* (see Fig. 2.1). The vector field f may be time-varying $f = f(t, x)$ or time-invariant $f = f(x)$.

The Cauchy problem (2.3)–(2.4) can be equivalently represented in the form of the integral equation

$$x(t) = x_0 + \int_{t_0}^{t} f(s, x(s))ds. \tag{2.5}$$

Since $\int_{t_0}^{t_0} f(s, x(s))ds = \mathbf{0}$ and $\frac{d}{dt} \int_{t_0}^{t} f(s, x(s))ds = f(t, x(t))$ for any continuous x and any continuous f, then any $x \in C^1(\mathcal{I}; \mathbb{R}^n)$ satisfying the integral equation (2.5) is a classical solution to the Cauchy problem and vise versa.

---

[1] The system (2.3) is said to be *autonomous* (or *time-invariant*) if f is independent of t.

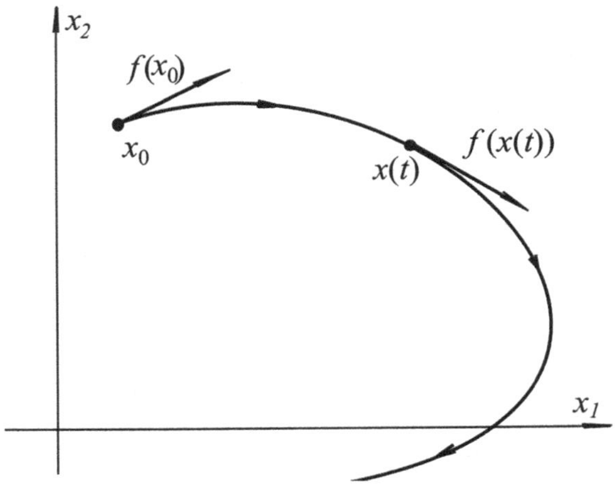

Fig. 2.1 Phase curve of a planar autonomous system $\dot{x} = f(x), t > 0, x(t_0) = x_0$

In the general case, the Cauchy problem (2.3), (2.4) may not have classical solutions or it may have many solutions. To guarantee the existence and uniqueness of solutions, some restrictions on the right-hand side of the system (2.3) must be imposed.

Theorem 2.1 (Peano existence theorem) *Let a vector field* $f : \mathbb{R} \times \mathbb{R}^n \mapsto \mathbb{R}^n$ *be continuous and* $|f(t,x)| \le M$ *for* $\forall (t,x) \in \mathcal{G}(t_0, x_0)$,

$$\mathcal{G}(t_0, x_0) := \left\{(t,x) : |t - t_0| \le \alpha, |x - x_0| \le \beta, t \in \mathbb{R}, x \in \mathbb{R}^n\right\}.$$

Then the Cauchy problem (2.3), (2.4) *has, at least, one classical solution defined on the segment* $[t_0 - \min\{\alpha, \beta/M\}, t_0 + \min\{\alpha, \beta/M\}]$.

The proof is based on *Euler approximates* given by

$$\begin{cases} \phi_n(t_0) = x_0, \\ \phi_n(t) = \phi_n(t_{i-1}) + f(t_{i-1}, \phi_n(t_{i-1}))(t - t_{i-1}), \quad t_{i-1} < t \le t_i, \\ t_i = t_0 + \frac{i}{n}\alpha, i = \{0, \dots, n\}. \end{cases}$$

The sequence $\{\phi_n\}$ satisfies all conditions of Arzelà–Ascoli theorem (see Appendix D) on the time interval $[t_0 - \alpha, t_0 + \alpha]$. So, there exists a subsequence ϕ_{n_k} that converges uniformly to a continuous function ϕ satisfying:

$$\begin{aligned} \phi(t) &= \lim_{n_k \to +\infty} \phi_{n_k}(t) = x_0 + \int_{t_0}^{t} \lim_{n_k \to +\infty} f(s, \phi_{n_k}(s))ds \\ &= x_0 + \int_{t_0}^{t} f(s, \phi(s))ds. \end{aligned}$$

Since f is continuous on both variables then ϕ is continuously differentiable.

Uniqueness of solutions can always be guaranteed under a more strong restriction to the right-hand side of (2.3). The following result is also known in the literature as Picard existence theorem, Picard–Lindelof existence theorem, or Cauchy–Lipschitz–Picard existence theorem.

Theorem 2.2 (Cauchy–Lipschitz theorem) *Let a vector field* $f : \mathcal{G}(t_0, x_0) \mapsto \mathbb{R}^n$ *be continuous and there exists* $L > 0$ *such that*

$$|f(t,x) - f(t,y)| \leq L|x - y|, \quad \forall (t,x), (t,y) \in \mathcal{G}(t_0, x_0). \tag{2.6}$$

Then the Cauchy problem (2.3), (2.4) *has a unique classical solution defined, at least, on the segment* $[t_0 - \delta, t_0 + \delta]$ *for* $\delta < \min\{\alpha, L^{-1}, \beta/M\}$, *where* $M = \sup_{(t,x)\in\mathcal{G}(t_0,x_0}|f(t,x)|$.

The condition (2.6) is known as *Lipschitz condition*. The proof can be based on Banach Fixed-Point theorem (see Appendix C). Indeed, the operator $F : C([t_0 - \delta, t_0 + \delta]; \mathbb{R}^n) \mapsto C([t_0 - \delta, t_0 + \delta]; \mathbb{R}^n)$ defined as

$$(F(x))(t) = x_0 + \int_{t_0}^{t} f(s, x(s))ds$$

satisfies the Lipschitz condition

$$\|F(x) - F(y)\|_C \leq \delta L \, \|x - y\|_C, \quad \forall x, y \in \mathbf{X},$$

where $\|\cdot\|_C$ denotes the supremum norm in $C([t_0 - \delta, t_0 + \delta]; \mathbb{R}^n)$ and the set

$$\mathbf{X} = \{x \in C([t_0 - \delta, t_0 + \delta]; \mathbb{R}^n) : \|x - x_0\|_C \leq \beta\}$$

is a complete metric space (see Appendix A) with the metric

$$\rho(x, y) = \max_{t\in[t_0-\delta, t_0+\delta]} |x(t) - y(t)|.$$

Since $\delta \leq \beta/M$ then $\|F(x) - x_0\|_C \leq \beta$, $\forall x \in \mathbf{X}$, i.e., $F : \mathbf{X} \mapsto \mathbf{X}$. For $\delta L < 1$, Banach Fixed-Point theorem guarantees the existence of a unique solution in $\mathbf{X}$ for the functional equation $x = F(x)$. This implies the existence of a unique solution of the integral equation (2.5) on the compact interval of time $[t_0 - \delta, t_0 + \delta]$.

Piano's theorem ensures the existence of a solution in both the forward $t > t_0$ and the backward times $t < t_0$. Any solution $x : [t_0 - \delta, t_0 + \delta] \mapsto \mathbb{R}^n$ in the backward time $t < t_0 + \delta$ defines a solution in the forward time $t > t_0 - \delta$ and vise versa. The Cauchy–Lipschitz theorem guarantees the uniqueness of this solution. However, an ODE may have unique solutions in the forward (or in the backward) time even if it does not satisfy the Lipschitz condition (2.6). In this case, the uniqueness in the forward time does not imply the uniqueness in the backward time and vice versa.

Moreover, the Cauchy problem may have a unique solution in the forward time and infinitely many solutions in the backward time and vice versa.

Example 2.1 Any solution of the Cauchy problem

$$\dot{x} = -x^{\frac{1}{3}}, \qquad x(0) = x_0 \in \mathbb{R}$$

is unique in the forward time and given by

$$x(t, x_0) = \begin{cases} \left(x_0^{\frac{2}{3}} - \frac{2}{3}t\right)^{\frac{3}{2}} \text{sign}(x_0) & \text{if } t \in \left[0, \frac{3}{2}x_0^{\frac{2}{3}}\right], \\ 0 & \text{if } \quad t > \frac{3}{2}x_0^{\frac{2}{3}}. \end{cases}$$

However, for any scalar parameter $\alpha \in \left(0, \frac{3}{2}\right)$, the function $\tilde{x}_\alpha : \left[-\frac{3}{2}, 0\right] \mapsto \mathbb{R}$ defined as follows

$$\tilde{x}(t) = \begin{cases} 0 & \text{if } \ t \in [-\alpha, 0], \\ x\left(\frac{3}{2} + t + \alpha, 1\right) & \text{if } \ -\frac{3}{2} < t < -\alpha \end{cases}$$

is a solution in the backward time (i.e., for $t < 0$) for the considered Cauchy problem with $x_0 = 0$. Therefore, the zero solution is not unique in the backward time. Moreover, there exist infinitely many solutions in the backward time such that $x(0) = 0$.

Notice that the zero solution of the slightly modified system

$$\dot{x} = -|x|^{1/3}$$

is neither forward nor backward unique.

The above theorem proves the local-in-time existence of solutions. Solutions of the ODE (2.3) may exist just locally and tend to infinity as time goes to some finite instant of time. This situation is called a *finite-time blow up*. An ODE system, which does not have a finite-time bow up in the *forward* (in the backward time) is called *forward (resp., backward) complete*. The system is *complete* it is both forward and backward complete.

Well-Posedness of ODEs

A mathematical problem is said to be well-posed in Hadamard sense if the following properties holds: *(1) the problem has a solution; (2) the solution is unique; (3) the solution changes continuously with the initial condition.* The following theorem is useful for the well-posedness analysis of ODEs.

Theorem 2.3 (Continuity of classical solutions on parameters) *Let a vector field* $f : \mathcal{G}(t_0, x_0) \times \mathcal{I}_\mu \mapsto \mathbb{R}^n$ *be continuous, where*

$$\mathcal{I}_\mu := \left\{\mu \in \mathbb{R}^k : \|\mu - \mu_0\| \le c\right\}, \qquad \mu_0 \in \mathbb{R}^k, \quad c > 0.$$

Let for $\mu = \mu_0$ *the Cauchy problem*

$$\dot{x} = f(t, x, \mu), \quad x(t_0) = \xi \tag{2.7}$$

have a unique solution x_{μ_0} *on an interval* $[t_0 - \alpha, t_0 + \alpha]$. *Then there exists* $\delta > 0$ *such that*

- *for any* $\mu : \|\mu - \mu_0\| < \delta$, *a classical solution* x_μ *of the above Cauchy problem exists, at least, on* $[t_0 - \alpha, t_0 + \alpha]$;
- $x_\mu \to x_{\mu_0}$ *as* $\mu \to \mu_0$ *uniformly on* $[t_0 - \alpha, t_0 + \alpha]$.

Moreover, if $f \in C^1$ *then the dependence of* $x(t, x_0, \mu)$ *on* μ *is smooth (i.e., the function* $\mu \mapsto \frac{\partial x_\mu}{\partial \mu}$ *is continuous too).*

For $\mu \neq \mu_0$ the system (2.7) may have non-unique solutions but, anyway, these solutions depend continuously on the parameter μ at the point $\mu = \mu_0$.

The proof of the above theorem (see, e.g., [3]) is based on the fact that the set of solutions

$$x_\mu(t) = \xi + \int_{t_0}^{t} f(s, x_\mu(s), \mu)\, ds, \quad \mu \in \mathcal{I}_\mu$$

satisfy Arzelà–Ascoli theorem on some time interval $|t - t_0| < \alpha_0 \le \alpha$. Hence, the solution x_{μ_0} is the only limit of x_μ as $\mu \to \mu_0$ due to uniqueness of x_{μ_0}. For $f \in C^1$, the differential equation for $\frac{\partial x}{\partial \mu}$ is given by [3]

$$\frac{d}{dt}\frac{\partial x_\mu}{\partial \mu} = \frac{\partial f(t, x_\mu, \mu)}{\partial \mu}. \tag{2.8}$$

Applying the proven continuous dependence of solutions to this equation we derive a smooth dependence of x_μ on μ.

A continuous dependence of solutions on initial conditions can be studied by means of the above theorem and the change of coordinates $x_{\text{new}} = x - x_0$. Therefore, *any ODE with a locally Lipschitz continuous right-hand side is well-posed in Hadamard sense*. ODE models of uncertain control systems/processes may violate the property of the uniqueness of solutions. However, the existence of solutions and their continuous dependence on initial data are two typical properties of most dynamical systems studied in this monograph.

Linear ODEs with Time-Invariant Coefficients

Linear ODEs form an important class of dynamical systems. In addition to plenty of linear physical models, they frequently describe the local behavior of nonlinear systems of various natures. Many schemes of control design are developed for systems modeled by linear ODEs. Linear ODE admits an explicit representation (see, e.g., [4]) of solution in terms of the matrix exponent, which is utilized also for a definition of the linear dilation in $\mathbb{R}^n$ (see Chap. 5).

Proposition 2.1 (Representation of solutions of linear ODEs) *The linear ODE*

$$\dot{x}(t) = Ax(t), \quad t \in \mathbb{R}, \quad A \in \mathbb{R}^{n\times n} \tag{2.9}$$

has a unique solution for any initial condition $x(0) = x_0$, which admits the representation

$$x(t, x_0) = e^{At}x_0 = S\begin{pmatrix} e^{\alpha_1 t}X_{\omega_1}(t) & \mathbf{0} & \dots & \mathbf{0} \\ \mathbf{0} & e^{\alpha_2 t}X_{\omega_2}(t) & \dots & \mathbf{0} \\ \dots & \dots & \dots & \dots \\ \mathbf{0} & \mathbf{0} & \dots & e^{\alpha_k t}X_{\omega_k}(t) \end{pmatrix} S^{-1}x_0, \tag{2.10}$$

where

- $\lambda_j = \alpha_j + \mathbf{i}\omega_j \in \mathbb{C}$ *are eigenvalues of the matrix A with $\omega_j \geq 0$;*
- $k \leq n$ *(with $k = n$ if all eigenvalues are distinct and real);*
- $S \in \mathbb{R}^{n\times n}$ *is the real Jordan transformation:*

$$S^{-1}AS = \begin{pmatrix} J_{\lambda_1} & \mathbf{0} & \dots & \mathbf{0} \\ \mathbf{0} & J_{\lambda_2} & \dots & \mathbf{0} \\ \dots & \dots & \dots & \dots \\ \mathbf{0} & \mathbf{0} & \dots & J_{\lambda_k} \end{pmatrix},$$

where $J_{\lambda_j} \in \mathbb{R}^{n_j\times n_j}$ is the real Jordan block:

$$J_{\lambda_j} = \begin{cases} \alpha_j I_{n_j} + N_{n_j} & \text{if } \omega_j = 0, \\ \alpha_j I_{n_j} + \omega_j E_{n_j} + M_{n_j} & \text{if } \omega_j > 0, \end{cases}$$

with $n_j \in \mathbb{N}$ being even for $\omega_j > 0$, $n_1 + \dots + n_k = n$ and the matrices $E_{n_j}, N_{n_j}, M_{n_j} \in \mathbb{R}^{n_j\times n_j}$ defined as follows

$$E_{n_j} = \begin{pmatrix} 0 & 1 & 0 & 0 & 0 & 0 & \dots \\ -1 & 0 & 0 & 0 & 0 & 0 & \dots \\ 0 & 0 & 0 & 1 & 0 & 0 & \dots \\ 0 & 0 & -1 & 0 & 0 & 0 & \dots \\ 0 & 0 & 0 & 0 & 0 & 1 & \dots \\ 0 & 0 & 0 & 0 & -1 & 0 & \dots \\ \dots & \dots & \dots & \dots & \dots & & \dots \end{pmatrix}, \quad N_{n_j} = \begin{pmatrix} \mathbf{0} & I_{n_j-1} \\ 0 & \mathbf{0} \end{pmatrix}, \quad M_{n_j} = \begin{pmatrix} \mathbf{0} & I_{n_j-2} \\ \mathbf{0} & \mathbf{0} \end{pmatrix};$$

- *the matrix-valued function X_{ω_j} is given by*

$$X_{\omega_j}(t) = \begin{cases} \sum_{i=0}^{n_j-1} \frac{t^i N_{n_j}^i}{i!} & \text{if } \omega_j = 0, \\ R_{\omega_j}(t) \sum_{i=0}^{\frac{n_j}{2}-1} \frac{t^i M_{n_j}^i}{i!} & \text{if } \omega_j \neq 0, \end{cases}$$

$$R_{\omega_j}(t) = \begin{pmatrix} \cos(\omega_j t) & \sin(\omega_j t) & \dots & \dots & 0 & 0 \\ -\sin(\omega_j t) & \cos(\omega_j t) & \dots & \dots & 0 & 0 \\ \dots & \dots & \dots & \dots & \dots & \dots \\ 0 & 0 & \dots & \dots & \cos(\omega_j t) & \sin(\omega_j t) \\ 0 & 0 & \dots & \dots & -\sin(\omega_j t) & \cos(\omega_j t) \end{pmatrix} \in \mathbb{R}^{n_i \times n_i}.$$

Moreover, for any $\epsilon > 0$ there exist $C_1 > 0$ and $C_2 \geq 0$ such that

$$C_1 e^{t(-\epsilon + \min \alpha_i)} \|x_0\| \leq \|x(t, x_0)\| \leq C_2 e^{t(\epsilon + \max \alpha_i)} \|x_0\|, \quad \forall t \geq 0. \tag{2.11}$$

Proof On the one hand, since $\left|\frac{e^{Ah}-I}{h} - A\right| \leq \left|\frac{A^2 h}{2!} + \frac{A^3 h^2}{3!} + \dots\right| \leq |A|(e^{|Ah|} - 1)|x_0| \to 0$ as $h \to 0$ then using the identity $e^{A(t+s)} = e^{At}e^{As}$, $\forall s, t \in \mathbb{R}$ we derive

$$\lim_{h\to 0} \left|\frac{e^{A(t+h)} - e^{At}}{h} - Ae^{At}\right| = \lim_{h \to 0} \left|\left(\frac{e^{Ah}-I}{h} - A\right) e^{At}\right| = 0$$

or, equivalently, $\frac{d}{dt}e^{At} = \lim_{h\to 0} \frac{e^{A(t+h)} - e^{At}}{h} = Ae^{At}$. This means that $x(t, x_0) = e^{tA}x_0$ is the unique solution of the linear ODE.

On the other hand, applying real Jordan transformation (see, Appendix E) we derive

$$Se^{At}S^{-1} = \sum_{i=0}^{\infty} \frac{t^i S A^i S^{-1}}{i!} = e^{tSAS^{-1}} = e^{t\begin{pmatrix} J_{\lambda_1} & \dots & 0 \\ \dots & \dots & \dots \\ 0 & \dots & J_{\lambda_k} \end{pmatrix}} = \begin{pmatrix} e^{tJ_{\lambda_1}} & \dots & 0 \\ \dots & \dots & \dots \\ 0 & \dots & e^{tJ_{\lambda_k}} \end{pmatrix}.$$

Since $e^{A_1 + A_2} = e^{A_1}e^{A_2}$ for any commuting pair of matrices $A_1 A_2 = A_2 A_1$ then for $J_{\lambda_j} = \alpha_j I_{n_j} + N_{n_j} \in \mathbb{R}^{n_j \times n_j}$ we have

$$e^{tJ_{\lambda_j}} = e^{\alpha_j t I_{n_j} + tN_{n_j}} = e^{\alpha_j t I_{n_j}} e^{tN_{n_j}} = e^{\alpha_j t} \sum_{i=0}^{\infty} \frac{t^i N_{n_j}^i}{i!},$$

where the matrix N_{n_j} is nilpotent. So, $N_{n_j}^{n_j} = \mathbf{0}$ and

$$X_{\omega_j}(t) = e^{tN_{n_j}} = \sum_{j=0}^{n_i - 1} \frac{t^i N_{n_i}^i}{i!}.$$

If $J_{\lambda_j} = \alpha_j I_{n_j} + \omega_j E_{n_j} + M_j \in \mathbb{R}^{n_j \times n_j}$ then

$$e^{J_{\lambda_j}\tau} = e^{\alpha_j I_{n_j}} e^{\omega_j t E_{n_j} + tM_{n_j}} = e^{\alpha_j t} e^{\omega_j t E_{n_j}} e^{tM_{n_j}},$$

where the identity $E_{n_j} M_{n_j} = M_{n_j} E_{n_j}$ is utilized on the last step. Since M_{n_j} is nilpotent with $M_{n_j}^{\frac{n_j}{2}} = \mathbf{0}$ and $e^{\rho\left(\begin{smallmatrix} 0 & 1 \\ -1 & 0 \end{smallmatrix}\right)} = \left(\begin{smallmatrix} \cos(\rho) & \sin(\rho) \\ -\sin(\rho) & \cos(\rho) \end{smallmatrix}\right)$, $\forall \rho \in \mathbb{R}$ then

$$X_{\omega_j}(t) = e^{\omega_j t E_{n_j}} e^{t M_{n_j}} = R_{\omega_j}(t) \sum_{i=0}^{\frac{n_j}{2}-1} \frac{t^i M_{n_i}^i}{i!}.$$

Finally, since $\alpha_i - \epsilon - \max \alpha_i < 0$ and $\alpha_i + \epsilon - \min \alpha_i > 0$ then the estimate (2.11) follows from

$$\inf_{t \geq 0} \|e^{t(\alpha_i + \epsilon - \min \alpha_i)} X_{\omega_i}(t)\| > 0 \quad \text{and} \quad \sup_{t \geq 0} \|e^{t(\alpha_i - \epsilon - \max \alpha_i)} X_{\omega_i}(t)\| < +\infty.$$

The proof is complete. □

As a trivial corollary of the above proposition, we derive that any classical solution of the system

$$\dot{x}(t) = Ax(t) + f(t, x(t)), \quad t \in \mathbb{R}, \quad x(t_0) = x_0, \quad A \in \mathbb{R}^{n \times n}, \quad f \in C \tag{2.12}$$

admits the representation

$$x(t) = e^{A(t-t_0)} x_0 + \int_{t_0}^{t} e^{A(t-s)} f(s, x(s)) ds, \tag{2.13}$$

which is known as the *Cauchy formula*. This can be proven simply by the substitution of (2.13) into (2.12).

Linear ODEs with Time-Varying Coefficients

Let $A : \mathbb{R} \mapsto \mathbb{R}^{n \times n}$ be a continuous matrix-valued function. The linear system with time-varying coefficients

$$\dot{x} = A(t)x, \quad t > t_0, \quad x(t_0) = x_0 \tag{2.14}$$

admit the following representation [5, Vol. 2, Chap. XIV, Sect. 5] of solution

$$x(t, t_0, x_0) = \Phi(t, t_0) x_0, \tag{2.15}$$

where[2]

[2] Formula (2.16) is inspired by the recursive procedure for finding solutions of (2.14):

$$x_{k+1}(t) = x_0 + \int_{t_0}^{t} A(\tau) x_k(\tau) d\tau, \quad x_1(t) = x_0, \quad t > t_0, \quad k = 1, 2, \ldots.$$

$$\Phi(t,t_0) = I_n + \int_{t_0}^{t} A(\tau_1)d\tau_1 + \int_{t_0}^{t} A(\tau_1)\left(\int_{t_0}^{\tau_1} A(\tau_2)d\tau_2\right)d\tau_1 + \ldots \tag{2.16}$$

is the so-called fundamental matrix [3, 5] satisfying

$$\Phi(t,t) = I_n, \quad \frac{d\Phi(t,s)}{dt} = A(t)\Phi(t,s), \quad \frac{d\Phi(t,s)}{ds} = -A(t)\Phi(t,s), \quad t,s \in \mathbb{R}. \tag{2.17}$$

For the time-invariant matrix $A(t) = A$, the above formula gives $\Phi(t,t_0) = e^{(t-t_0)A}$.

The quasi-linear system

$$\dot{x} = A(t)x + g(t,x), \quad t > t_0, \quad x(t_0) = x_0, \quad A \in C(\mathbb{R}; \mathbb{R}^{n\times n}), \quad g \in C(\mathbb{R}^{n+1}; \mathbb{R}^n) \tag{2.18}$$

admits the equivalent representation

$$x(t) = \Phi(t,t_0)x_0 + \int_{t_0}^{t} \Phi(t,s)g(s,x(s))ds, \tag{2.19}$$

which can be proven by substitution of (2.19) into (2.18).

Finding an analytical representation of the matrix-valued function Φ is a nontrivial problem in the general case. However, for the particular matrix

$$A(t) = p(t)A, \quad p \in C(\mathbb{R}) \tag{2.20}$$

the explicit formula can be easily obtained

$$\Phi(t,s) = e^{A\int_s^t p(\tau)d\tau}, \quad \forall t,s, \in \mathbb{R} \tag{2.21}$$

using the matrix exponential.

Positive Systems

The so-called *positive system* is an ODE (2.3), which solutions satisfy

$$x(t_0) \geq \mathbf{0} \quad \Rightarrow \quad x(t) \geq \mathbf{0}, \quad \forall t \geq t_0, \tag{2.22}$$

where the order relation "$\geq$" for the vectors $x(t_0), x(t), \mathbf{0} \in \mathbb{R}^n$ is understood in the component-wise manner. Such ODEs appear, for example, as models of physical or biological systems having positive quantities as state variables.

Positive linear systems are well studied in the literature (see, e.g., [6, 7]). *The linear system* (2.9) *is positive if the matrix* A *is Metzler.*[3] Indeed, in this case, the matrix $\lambda I_n + A$ has non-negative entries provided that $\lambda > 0$ is sufficiently large, so the matrix

[3] A matrix A is Metzler if all its off-diagonal elements are non-negative.

$$e^{At} = e^{-\lambda t} e^{(A+\lambda I_n)t} = e^{-\lambda t} \sum_{k=0}^{+\infty} \frac{t^k (A+\lambda I)^k}{k!}$$

has non-negative entries for all $t \geq 0$. This results in (2.22) due to (2.10).

A similar condition can be obtained for a nonlinear system of the form

$$\dot{x} = A(t,x)x + g(t,x), \quad t > t_0, \tag{2.23}$$

where $A(t,x) \in \mathbb{R}^{n \times n}$ and $g(t,x) \in \mathbb{R}^n$.

Lemma 2.1 *Let $g : \mathbb{R} \times \mathbb{R}^n \mapsto \mathbb{R}^n$ be a continuous non-negative function (i.e., $g(t,x) \geq 0$ for all $t \in \mathbb{R}$ and $x \in \mathbb{R}^n$). Let a matrix-valued function $(t,x) \mapsto A(t,x)$ be continuous on $\mathbb{R} \times (\mathbb{R}^n \backslash \{\mathbf{0}\})$ and $\|A(t,x)x\| \to 0$ as $x \to \mathbf{0}$ for any fixed $t \in \mathbb{R}$. If $A(t,x) \in \mathbb{R}^{n\times n}$ is a Metzler matrix for any $(t,x) \in \mathbb{R} \times (\mathbb{R}^n \backslash \{\mathbf{0}\})$ then the implication (2.22) holds as long as $0 < |x(t)| < +\infty, t \geq t_0$. Moreover, if the system (2.23) is forward complete and one of conditions*

(a) $g(t,\mathbf{0}) > \mathbf{0}$

or

(b) $g(t,\mathbf{0}) = \mathbf{0}$ for all $t \in \mathbb{R}$ and the zero solution is unique

holds then the system is positive.

Proof Let $f(t,x) = A(t,x)x$. Let us assign $f(t,\mathbf{0}) = \mathbf{0}$. The vector field f is continuous on $\mathbb{R} \times \mathbb{R}^n$ and the system (2.23) has classical solutions for all initial conditions. Since, by assumption, the system is forward complete, then any classical solution of (2.23) satisfies the equation

$$\dot{x}(t) = \tilde{A}(t)x(t) + \tilde{g}(t), \quad \tilde{A}(t) = A(t, x(t)), \quad \tilde{g}(t) = g(t, x(t)), \quad t > t_0.$$

Let $t_1 > t_0$ be an arbitrary instants of time such that $0 < |x(t)| < +\infty$. Then the matrix-valued function $t \mapsto \tilde{A}(t)$ is continuous for all $t \in [t_0, t_1]$.

Let $a_{ij}(t)$ be the elements of the matrix $\tilde{A}(t)$ and

$$\lambda = -\min_{t\in[t_0,t_1], i=1,2,\ldots,n} a_{ii}(t) \in \mathbb{R}. \tag{2.24}$$

Since, by assumption, $\tilde{A}(t)$ is a Metzler matrix then the matrix $\tilde{A}(t) + \lambda I_n$ has non-negative entries for all $t \in [t_0, t_1]$. Let us consider

$$\tilde{x}(t) = e^{\lambda(t-t_0)} x(t), \quad t \in [t_0, t_1].$$

Obviously, $\tilde{x}(t_0) = x(t_0)$ and the function $\tilde{x}$ satisfies the following ODE

$$\frac{d}{dt}\tilde{x}(t) = (\tilde{A}(t) + \lambda I_n)\tilde{x}(t) + e^{\lambda(t-t_0)}\tilde{g}(t), \quad t \in (t_0, t_1].$$

In this case, by formula (2.19), we derive

$$\tilde{x}(t) = \tilde{\Phi}(t, t_0)x(t_0) + \int_{t_0}^{t} \tilde{\Phi}(t, s)e^{\lambda(s-t_0)}\tilde{g}(s)ds,$$

where the fundamental matrix $\tilde{\Phi}$ is given by

$$\tilde{\Phi}(t, t_0) = I_n + \int_{t_0}^{t} \tilde{A}(\tau_1) + \lambda I_n d\tau_1 + \int_{t_0}^{t} (\tilde{A}(\tau_1) + \lambda I_n)\left(\int_{t_0}^{\tau_1} \tilde{A}(\tau_2) + \lambda I_n d\tau_2\right) d\tau_1 + \ldots \quad (2.25)$$

Since $\tilde{A}(t) + \lambda I_n$ has non-negative entries for all $t \in [t_0, t_1]$ then $\tilde{\Phi}(t, t_0)$ also has non-negative entries for all $t \in [t_0, t_1]$. Therefore, $\tilde{x}(t) \geq \mathbf{0}$ for all $t \in [t_0, t_1]$ provided that $\tilde{x}(t_0) \geq \mathbf{0}$. Since $t_1 > t_0$ is an arbitrary instant of time then the implication (2.22) holds as long as $0 < |x(t)| < +\infty$. Moreover, if the system is forward complete then the implication holds as long as $|x(t)| > 0$.

(a) If $x(t) = \mathbf{0}$ and $g(t, \mathbf{0}) > \mathbf{0}$ then $\dot{x}(t) = g(t, \mathbf{0}) > \mathbf{0}$ and the system enters into the interior of the positive cone (i.e., into $\operatorname{int} \mathbb{R}^n_+$) right after the time instant $t : x(t) = \mathbf{0}$. So, the system is positive in this case.
(b) If $g(t, \mathbf{0}) = \mathbf{0}$ then the system has the zero solution. If the zero solution of the unperturbed ($g = \mathbf{0}$) system is unique, then any solution, which reaches zero (or initiated at zero) remains in the origin while $g(t, \mathbf{0}) = \mathbf{0}$. In this case, the system is positive too.

The proof is complete. □

Notice that, the restriction about uniqueness of the zero solution for the system with $g(t, \mathbf{0}) = \mathbf{0}$ cannot be omitted in the general case.

Example 2.2 All restrictions of Lemma 2.1 on A and g are fulfilled for

$$A(x) = \frac{1}{\sqrt{x_1^2+x_2^2}} \begin{pmatrix} -\sqrt{|x_1|}\,\mathrm{sign}(x_1) & 0.5\sqrt{|x_2|} \\ 0.5\sqrt{|x_1|} & -\sqrt{|x_2|}\,\mathrm{sign}(x_2) \end{pmatrix}, \quad g = \frac{1}{\sqrt{x_1^2+x_2^2}} \begin{pmatrix} |x_2|^{\frac{3}{2}} \\ |x_1|^{\frac{3}{2}} \end{pmatrix}.$$

Indeed, the matrix $A(x)$ is Metzler for any $x \in \mathbb{R}^n$. The function g is non-negative on $\mathbb{R}^n$. The functions g and

$$x \mapsto A(x)x = \begin{pmatrix} \frac{-\sqrt{|x_1|}|x_1| + 0.5\sqrt{|x_2|}x_2}{\sqrt{x_1^2+x_2^2}} \\ \frac{0.5\sqrt{|x_1|}x_1 - \sqrt{|x_2|}|x_2|}{\sqrt{x_1^2+x_2^2}} \end{pmatrix}$$

are continuous on $\mathbb{R}^n$, so the system (2.23) has classical solutions. For any $x(t_0) \geq \mathbf{0}$ the solution remains positive $x(t) \geq 0$ as long as $0 < |x(t)| < +\infty, t \geq t_0$. However, for $x(t_0) = \mathbf{0}$ the zero solution is not unique. In particular,

$$x(t) = -\frac{1}{32}(t - t_0)^2 \begin{pmatrix} 1 \\ 1 \end{pmatrix}, \quad t \geq t_0$$

is a solution of the system $\dot{x} = A(x)x + g(x)$ with $x(t_0) = \mathbf{0}$ as well. Therefore, the implication (2.22) does not hold for $x(t_0) = \mathbf{0}$ and the system is not positive. Notice that this system is standard homogeneous.

A less conservative condition of positiveness of a nonlinear system (2.23) is given by the following remark.

Remark 2.1 Lemma 2.1 remains valid even if the matrix $A(t, x)$ is Metzler in a neighborhood $U(z) \subset \mathbb{R}^n$ of any point $z \in \partial\mathbb{R}^n_+$. Indeed, considering solutions initiated in $U(z) \cap \mathbb{R}^n_+$ we conclude (repeating the proof of Lemma 2.1) that the solutions remain positive as long as they belong to a neighborhood of $\partial\mathbb{R}^n_+$. Since any solution initiated in int $\mathbb{R}^n_+$ may leave $\mathbb{R}^n_+$ only by crossing $\partial\mathbb{R}^n_+$ then all solutions initiated in $\mathbb{R}^n_+$ are non-negative, i.e., the system is positive.

For example, the system (2.23) with

$$A(x) = \begin{pmatrix} -\sqrt{|x_1|}\,\mathrm{sign}(x_1) & 1-|x_1x_2| \\ 1-|x_1x_2| & -\sqrt{|x_2|}\,\mathrm{sign}(x_2) \end{pmatrix}, \quad g(t, x) \geq 0$$

is positive, but the matrix $A(x)$ is not Metzler for $|x_1x_2| > 1$. Positive homogeneous systems are studied in Chaps. 7 and 12.

2.1.2 *Discontinuous Differential Equations*

Strong Solution

Equation (2.5) with discontinuous right-hand side f may have solutions provided that the integral in (2.5) is understood in the sense of Lebesgue (see Appendix B). Indeed, in this case, the identity (2.5) may hold if $x : \mathcal{I} \mapsto \mathbb{R}^n$ is an *absolutely continuous function*[4] and the function $s \mapsto f(s, x(s)), s \in \mathcal{I}$ is Lebesgue integrable on $\mathcal{I}$. Since any absolutely continuous function on $\mathcal{I}$ is differentiable almost everywhere on $\mathcal{I}$, the function x should satisfy (2.3) almost everywhere on $\mathcal{I}$.

Definition 2.2 *(Strong solution of ODE)* An absolutely continuous function $x : \mathcal{I} \subset \mathbb{R}^n \mapsto \mathbb{R}^n$ is said to be a strong (or Carathéodory) solution of ODE (2.3) if it satisfies Eq. (2.3) almost everywhere on the time interval $\mathcal{I}$.

The existence of strong solutions can be proven under some relaxed restrictions on f. In particular, the existence theorem given below allows the right-hand side of ODE to be discontinuous on the time variable t.

[4] A continuous function $x : \mathcal{I} \mapsto \mathbb{R}^n$ is said to be absolutely continuous if for any $\varepsilon > 0$ there exists $\delta > 0$ such that whenever a finite sequence of pairwise disjoint intervals (α_k, β_k) of I with $\alpha_k, \beta_k \in \mathcal{I}$ satisfies $\sum_k (\beta_k - \alpha_k) < \delta$ then $\sum_k |x(\beta_k) - x(\alpha_k)| < \varepsilon$.

Theorem 2.4 (Carathéodory existence theorem, 1918) *Let a vector field* $f : \mathcal{G}(t_0, x_0) \mapsto \mathbb{R}^n$ *be defined on*

$$\mathcal{G}(t_0, x_0) := \left\{(t, x) : |t - t_0| \leq \alpha, |x - x_0| \leq \beta, t \in \mathbb{R}, x \in \mathbb{R}^n\right\},$$

measurable on t *for any fixed* x, *continuous on* x *for any fixed* t *and* $|f(t, x)| \leq m(t)$ *for all* $(t, x) \in \mathcal{G}(t_0, x_0)$, *where* $m : [t_0 - \alpha, t_0 + \alpha] \mapsto \mathbb{R}_+$ *is a Lebesgue integrable function. Then the Cauchy problem* (2.3), (2.4) *has, at least, one strong solution* $x(t, x_0)$ *defined on the segment*

$$t \in \left[t_0 - \min\{\alpha, \beta/\bar{M}\}, t_0 + \min\{\alpha, \beta/\bar{M}\}\right],$$

where $\bar{M} = \int_{t_0-\alpha}^{t_0+\alpha} m(s)ds$. *Moreover, if* f *satisfies Lipschitz condition* (2.6) *for almost* t *then the strong solution is unique.*

The proof uses the Picard–Lindelöf approximates:

$$x_j(t) = x_0 \quad \text{for} \quad t \in [t_0, t_0 + \alpha/j],$$

$$x_j(t) = \xi + \int_{t_0}^{t-\alpha/j} f(s, x_j(s))ds, \quad \text{for} \quad t \in (t_0 + \alpha/j, t_0 + \alpha),$$

where $j \geq 1$. Obviously, $|x_j(t) - x_0| \leq M(t - \alpha/j)$, where $M(t) = \int_{t_0}^{t} m(s)ds$. Moreover, for $t_1, t_2 \in [t_0, t_0 + \alpha]$ one has $|x_j(t_1) - x_j(t_2)| \leq |M(t_1 - \alpha/j) - M(t_2 - \alpha/j)|$. Since M is a uniformly continuous function then the sequence $\{x_j\}$ is equicontinuous and uniformly bounded. Hence, by Arzelà–Ascoli theorem, there exists a subsequence $\{x_{j_k}\}$, which converges uniformly to a continuous function ϕ as $k \to +\infty$. Since f is continuous on x for any fixed t then $f(t, x_{j_k}(t)) \to f(t, \phi(t))$ as $k \to +\infty$ and using Lebesgue's Dominated Convergence theorem (see Appendix B) we conclude $\int_{t_0}^{t} f(s, x_{j_k}(s))ds \to \int_{t_0}^{t} f(s, \phi(s))ds$ as $k \to +\infty$ for all $t \in [t_0, t_0 + \alpha]$.

To prove uniqueness we assume that there exists two solutions $x_1(t, t_0, x_0)$ and $x_2(t, t_0, x_0)$ with the same initial condition $x(t_0) = x_0$. In this case, the Lipschitz condition gives

$$\begin{aligned} \|x_1(t, t_0, x_0) - x_2(t, t_0, x_0)\| &\leq \int_{t_0}^{t} \|f(\tau, x_1(\tau, t_0, x_0)) - f(\tau, x_2(\tau, t_0, x_0))\|d\tau \\ &\leq \int_{t_0}^{t} L\|x_1(\tau, t_0, x_0) - x_2(\tau, t_0, x_0)\|d\tau. \end{aligned}$$

Using Grönwall–Bellman inequality (see Appendix F) we derive $x_1(\tau, t_0, x_0) = x_2(\tau, t_0, x_0)$.

Remark 2.2 If the function m in Theorem 2.4 is locally Lipschtiz continuous then the strong (Carathéodory) solution is locally Lipschitz continuous too. The simplest sufficient condition of this is the local boundedness of the vector field f on both arguments.

The notion of strong (Carathéodory) solutions allows some deterministic noise and perturbations to be modeled using ODEs. For example, the system $\dot{x} = -x + g(t)$ does not have a classical solution if $g \in L^{\infty}$ is a discontinuous signal, but the strong solution of the system exists and satisfies the differential equation almost everywhere.

The strong solutions depend continuously on initial conditions and on parameters in the right-hand side of ODE.

Theorem 2.5 (Continuity of strong solutions on parameters) *Let a vector field $f : \mathcal{G}(t_0, x_0) \times \mathcal{I}_{\mu} \mapsto \mathbb{R}^n$ be measurable on t for any fixed x and μ, where*

$$\mathcal{I}_{\mu} := \left\{\mu \in \mathbb{R}^k : \|\mu - \mu_0\| \leq c\right\}, \qquad \mu_0 \in \mathbb{R}^k, \quad c > 0.$$

Let f be uniformly continuous in (x, μ) almost everywhere in t, i.e., there exists $\omega \in \mathcal{K}$ such that inequality

$$|f(t, x_1, \mu_1) - f(t, x_2, \mu_2)| \leq \omega(|x_1 - x_2| + \|\mu_1 - \mu_2\|), \tag{2.26}$$

holds for all most t and all x_i, μ_i satisfying

$$(t, x_i, \mu_i) \in \mathcal{G}(t_0, x_0) \times \mathcal{I}_{\mu}, \quad i = 1, 2.$$

Let there exist a function $m : \mathbb{R} \mapsto \mathbb{R}$ being Lebesgue integrable on the time interval $[t_0 - \alpha, t_0 + \alpha]$ such that

$$|f(t, x, \mu)| \leq m(t), \quad \forall (t, x, \mu) \in \mathcal{G}(t_0, x_0) \times \mathcal{I}_{\mu}. \tag{2.27}$$

If, for $\mu = \mu_0$, the Cauchy problem

$$\dot{x} = f(t, x, \mu), \quad x(\tau) = \xi \tag{2.28}$$

has a unique strong solution x_{μ_0} on an interval $[t_0 - \alpha, t_0 + \alpha]$, then there exists $\delta > 0$ such that

- *for any $\mu : \|\mu - \mu_0\| < \delta$, a strong solution x_{μ} of the above Cauchy problem exists, at least, on $[t_0 - \alpha, t_0 + \alpha]$;*
- *$x_{\mu} \to x_{\mu_0}$ as $\mu \to \mu_0$ uniformly on $[t_0 - \alpha, t_0 + \alpha]$.*

This theorem can be proven similarly to Theorem 2.3. Indeed, the condition (2.27) guarantees that the family of strong solutions

$$x_\mu(t) = \xi + \int_{t_0}^{t} f(t, x_\mu(\tau), \mu(\tau))d\tau, \quad \mu \in \mathcal{I}_\mu$$

is uniformly bounded, but the uniform continuity of f implies the equicontinuity of the family in the view of Bihari–LaSalle inequality (see Appendix F).

Remark 2.3 Theorem 2.5 remains valid even if $\mathbb{R}^k$ is replaced with $L^\infty(\mathbb{R}; \mathbb{R}^k)$ and $\|\cdot\|$ is replaced with $\|\cdot\|_{L^\infty}$. Indeed, the uniform continuity of f guarantees that such a replacement does not destroy equicontinuity of the family $\{x_\mu\}$, where $\mu \in \mathcal{I}_\mu = \left\{\mu \in L^\infty(\mathbb{R}; \mathbb{R}^k) : \|\mu - \mu_0\|_{L^\infty} \leq c\right\}$ and $c > 0$ is small enough. So, Arzelà–Ascoli theorem remains applicable in this case.

Making the change of variable $y = x - x_0$ and applying the above theorem a continuous dependence of strong (Carathéodory) solutions on initial conditions can be proven.

Remark 2.4 If a solution x_{μ_0} exists globally in time then continuous dependence of solutions discovered in Theorem 2.5 (resp., Remark 2.3) takes the place on any compact interval of time.

Filippov Solution

The control theory frequently deals with dynamical systems, which are modeled by ODEs with discontinuous right-hand sides [8–10]. For example, sliding-mode control methodology studies models like

$$\dot{x} = u + 0.5\sin(3t), \quad t > 0, \quad u = -\operatorname{sign}(x), \quad x(0) = x_0, \tag{2.29}$$

where x is the state variable, u is the control variable and $\operatorname{sign}(\cdot)$ is the sign function. The considered ODE admits a classical solution on $\{x \in \mathbb{R} : x > 0\}$ and on $\{x \in \mathbb{R} : x < 0\}$ separately. Each classical solution with $x_0 \neq 0$ reaches zero in a finite time $t = T(t_0, x_0)$ (see Fig. 2.2), since $\frac{d|x(t)|}{dt} \leq -0.5$ for $x(t) \neq 0$. Hence, the natural continuation of a solution for $t > T(t_0, x_0)$ is $x(t) = 0$, but neither Piano nor Carathéodory's existence theorem is applicable in this case. In addition, formally, if $x(t) = 0$ for all $t > T(t_0, x_0)$ then $\dot{x}(t) = 0 = u(t) + 0.5\sin(3t)$ for $t > T(t_0, x_0)$, but $u(t) = -\operatorname{sign}(x(t)) = 0$ for $t > T(t_0, x_0)$. Therefore, ODE (2.3) with piece-wise continuous (on state variable) right-hand side is ill-posed in the general case. It needs some regularization on the discontinuity set $x = 0$, which can be based, for example, on an extension of ODE to a *differential inclusion* (DI) [11].

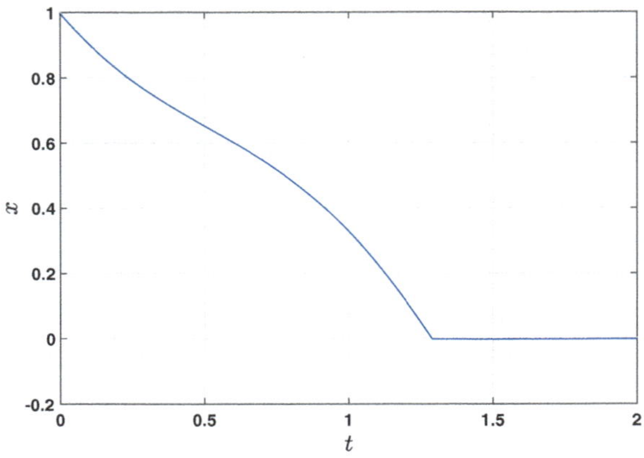

Fig. 2.2 Trajectory of the sliding-mode control system (2.29) with $x_0 = 1$

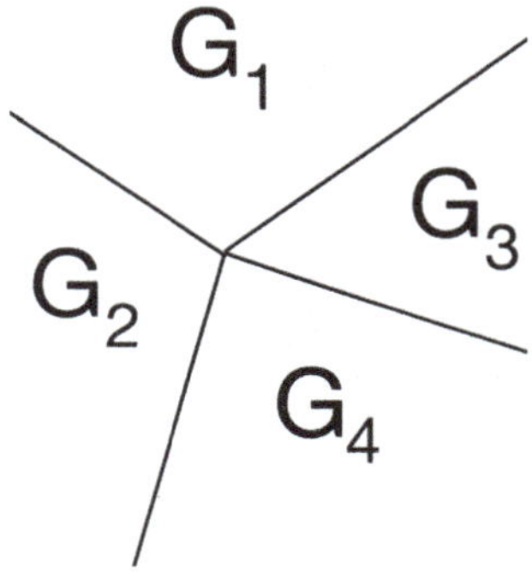

Fig. 2.3 Partition of $\mathbb{R}^2$ on domains of continuity

Recall that a vector field $f : \mathbb{R} \times \mathbb{R}^n \mapsto \mathbb{R}^n$ is *piece-wise continuous* if $\mathbb{R}^{n+1}$ can be split into a finite number of open connected sets (see, Fig. 2.3)

$$G_j \subset \mathbb{R}^{n+1},\ j = 1, 2, ..., N, \quad G_i \bigcap G_j = \emptyset \text{ for } i \neq j \tag{2.30}$$

with the boundary set

$$\mathcal{S} = \bigcup_{i=1}^{N} \partial G_j$$

of measure zero such that

- f is continuous in G_j;
- for each $(t^*, x^*) \in \partial G_j$ there exists a vector $f^j(t^*, x^*) \in \mathbb{R}^n$ possibly depended on $j = 1, 2, ..., N$, such that for any sequence $(t^k, x^k) \in G_j : (t^k, x^k) \to (t^*, x^*)$ we have $f(t^k, x^k) \to f^j(t^*, x^*)$.

Let the vector field $f^j : \overline{G}_j \subset \mathbb{R} \times \mathbb{R}^n \mapsto \mathbb{R}^n$ coincide with f on G_j and be prolonged on the boundary ∂G_j according to the limiting process:

$$f^j(t,x) = \lim_{(t^j,x^j)\to(t,x)} f(t^j,x^j), \qquad (t^j,x^j)\in G_j, \qquad (t,x)\in\partial G_j,$$

where the above limit for the piece-wise continuous function is assumed to be unique at any point $(t,x)\in\partial G_j$. If $\partial G_j\cap\partial G_i\neq\emptyset$ then for $(t^*,x^*)\in\partial G_j\cap\partial G_i$ we may have $f^j(t^*,x^*)\neq f^i(t^*,x^*)$. In this case, $(t^*,x^*)\in\mathcal{S}$ is the discontinuity point of f. Due to the continuity of f on $\mathbb{R}^n\backslash\mathcal{S}$, solutions of the system (2.3) are well defined away from $\mathcal{S}$. To define solutions on the discontinuity set, a special regularization procedure known as the *Filippov method* [11] can be utilized.

Let us consider the following differential inclusion

$$\dot{x}\in K[f](t,x), \quad t\in\mathbb{R}, \tag{2.31}$$

$$K[f](t,x) = \begin{cases} \{f(t,x)\} & \text{if } (t,x)\in\mathbb{R}^{n+1}\backslash\mathcal{S}, \\ \overline{\mathrm{co}}\left(\bigcup_{j\in\mathcal{N}(t,x)}\{f^j(t,x)\}\right) & \text{if } (t,x)\in\mathcal{S}, \end{cases} \tag{2.32}$$

where $\overline{\mathrm{co}}(M)$ is the closed convex hull of M (see Appendix A) and the set-valued function $\mathcal{N}:\mathcal{S}\mapsto 2^{\{1,2,\ldots,N\}}$ indicates the domains, which have a common boundary point $(t,x)\in\mathcal{S}$, i.e.,

$$\mathcal{N}(t,x) = \left\{j\in\{1,2,\ldots,N\} : (t,x)\in\partial G_j\right\}.$$

The set $K[f](t,x)$ is *convex and compact* for any $(t,x)\in\mathcal{S}$. The set-valued vector field $K[f]$ is called a *Filippov regularization* of the discontinuous vector field f. The existence of solution of the *differential inclusion* (2.31) is studied in Sect. 2.2.

Definition 2.3 *(Filippov solution)* An absolutely continuous function $x:\mathcal{I}\mapsto\mathbb{R}^n$ defined on a time interval (or a segment) $\mathcal{I}$ is said to be a *Filippov solution* of (2.3) if it satisfies the differential inclusion (2.31) almost everywhere on $\mathcal{I}$.

This definition allows any solution of the discontinuous differential equation (2.29) to be prolonged for $t > T(x_0)$, since the corresponding differential inclusion

$$\dot{x}\in\begin{cases} -1+0.5\sin(3t) & \text{if } x>0, \\ [-1+0.5\sin(3t), 1+0.5\sin(3t)] & \text{if } x=0, \\ 1+0.5\sin(3t) & \text{if } x<0, \end{cases}$$

obviously, admits the zero solution. Moreover, this solution is unique in the forward time.

The prolongation of solutions to the discontinuity set $\mathcal{S}$ by means of Filippov's method has the following simple geometric interpretation. Let us consider a system (2.3) with a function f having discontinuities only on a smooth surface $\mathcal{S}=\{x\in\mathbb{R}^n : s(x)=0\}$, $s\in C^1$, which splits $\mathbb{R}^n$ into domains $G^+=\{x\in\mathbb{R}^n : s(x)>0\}$ and $G^-=\{x\in\mathbb{R}^n : s(x)<0\}$.

Let $P(x)$ be a tangential plane to the surface $\mathcal{S}$ at a point $x \in \mathcal{S}$ and

$$f^+(t,x) = \lim_{x_i \to x, x_i \in G^+} f(t,x_i) \quad \text{and} \quad f^-(t,x) = \lim_{x_i \to x, x_i \in G^-} f(t,x_i),$$

where the above limits are assumed to be unique for any $x \in \mathcal{S}$. For $x \in \mathcal{S}$, the set $K[f](t,x)$ defines a segment connecting the vectors $f^+(t,x)$ and $f^-(t,x)$ (see Fig. 2.4). If this segment crosses the plane $P(x)$ then the cross point is a velocity vector which defines the system motion on the surface $\mathcal{S}$ (see Fig. 2.4b). In this case the system (2.31) has trajectories, which "slide" on the surface $\mathcal{S}$. The *sliding motion* equation is given by

$$\dot{x} = f_0(t,x), \tag{2.33}$$

where the function

$$f_0(t,x) = \frac{\langle \nabla s(x), f^-(t,x)\rangle f^+(t,x) + \langle \nabla s(x), f^+(t,x)\rangle f^-(t,x)}{\langle \nabla s(x), f^+(t,x) - f^-(t,x)\rangle} \tag{2.34}$$

is the velocity vector defined by an intersection point of the segment and the plane $P(x)$, i.e., $f_0(t,x) = \mu f^+(t,x) + (1-\mu) f^-(t,x)$ with $\mu \in [0,1]$ such that

$$\langle \nabla s(x), \mu f^+(t,x) + (1-\mu) f^-(t,x)\rangle = 0.$$

In the control literature (see, e.g., [12, 13]) such a motion on the surface $\mathcal{S}$ is known as *sliding mode*.

If $\nabla s(x) \not\perp \big(\mu f^-(t,x) + (1-\mu) f^+(t,x)\big)$ for every $\mu \in [0,1]$ then a trajectory of (2.31) crosses the surface (see Fig. 2.4a, (a)) implying an isolated in time "switching" of the right-hand side of the system (2.3).

Seemingly, Filippov's method is the most simple regularization of ODEs with discontinuous right-hand sides. However, this approach was severely criticized by many authors [12, 14–16] since its appearance in the 1960s. In fact, Filippov's regularization does not describe correctly some real control systems with discontinuous models

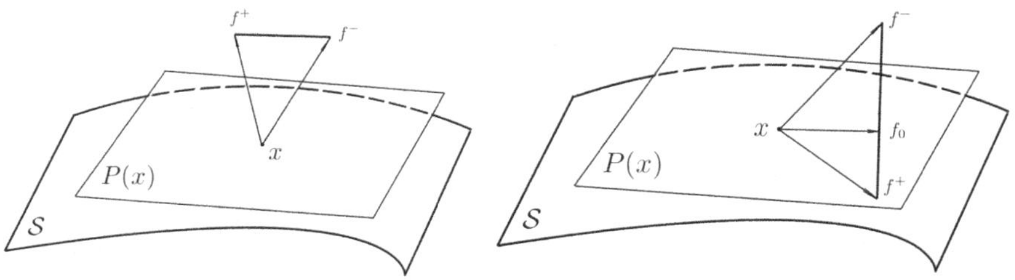

(a) Switching case. (b) Sliding-mode case.

Fig. 2.4 Geometrical illustration of Filippov's Definition

of actuators. All contradictions to the reality usually appear due to an inadequate modeling. However, some problems can be avoided by means of a proper modification of Filippov's definition. A simple example of an inadequate (from practical point of view) description of the system behavior is given below.

Example 2.3 Let us consider the discontinuous single-input planar control system

$$\begin{cases} \dot{x}_1 = u, \\ \dot{x}_2 = (\varepsilon u^2 + \varepsilon^2 |u| - \varepsilon) x_2, \end{cases} \qquad u = -\operatorname{sign}(x_1), \tag{2.35}$$

where $x_1, x_2 \in \mathbb{R}$ are components of the system state, $\varepsilon \in \mathbb{R}_+$ is a small parameter $0 < \varepsilon \ll 1$, $u \in \mathbb{R}$ is the relay ("on"-"off") control with the sign function defined by formula (0.5).

If we apply Filippov's method only to the first equation of the system (2.35), we obtain

$$\dot{x}_1 \in \begin{cases} 1 & \text{if } x_1 < 0, \\ [-1, 1] & \text{if } x_1 = 0, \\ -1 & \text{if } x_1 > 0. \end{cases}$$

So, the system has a sliding mode (Filippov's solution) on the surface $x_1 = 0$ with the following sliding motion equation $\dot{x}_1 = 0$, which implicitly implies $u = 0$ for $x_1 = 0$. In this case, the expectable sliding motion equation for the whole system (2.35) is

$$\begin{cases} \dot{x}_1 = 0, \\ \dot{x}_2 = -\varepsilon x_2, \end{cases} \qquad \text{for} \quad x_1 = 0, \quad t \geq T, \tag{2.36}$$

where $T \in \mathbb{R}$ is such that $x_1(T) = 0$. Therefore, $x_2(t) = e^{-\varepsilon(t-T)} x_2(T) \to 0$ as $t \to +\infty$ and the considered system is *stable*.

However, according to Fillipov method, the regularization has to be applied to the whole system (2.35). This yields

$$f^+(x_1, x_2) = \begin{pmatrix} -1 \\ \varepsilon^2 x_2 \end{pmatrix} \text{ for } x_1 \to +0$$

$$f^-(x_1, x_2) = \begin{pmatrix} 1 \\ \varepsilon^2 x_2 \end{pmatrix} \text{ for } x_1 \to -0$$

and formula (2.34) for $s(x) = x_1$ gives another sliding motion equation:

$$\begin{pmatrix} \dot{x}_1 \\ \dot{x}_2 \end{pmatrix} = \frac{\langle \nabla s(x), f^-(t, x) \rangle f^+(t, x) + \langle \nabla s(x), f^+(t, x) \rangle f^-(t, x)}{\langle \nabla s(x), f^+(t, x) - f^-(t, x) \rangle} = \begin{pmatrix} 0 \\ \varepsilon^2 x_2 \end{pmatrix},$$

which is *unstable*: $x_2(t) = e^{\varepsilon^2 (t-T)} x_2(T)$. From the practical point of view the sliding motion equation (2.36) looks more realistic from the control point of view. Indeed, in practice we usually do not have ideal relays, so the model of switchings like (0.5) is just a comfortable approximation of a real relay element (e.g., a transistor). A "real" sliding mode is, in fact, a switching regime of a high (but always bounded) frequency. In this case, an average value of the control input in the "real" sliding mode can be estimated as follows

$$|u|_{average} = \frac{1}{t - t_0} \int_{t_0}^{t} |u(\tau)| d\tau, \quad t > t_0 : x_1(t_0) = 0.$$

So, in the "real" sliding mode, we have $|u(\tau)| < 1$ almost everywhere. This means $|u|_{average} < 1$ and, particularly, $|u|_{average} \leq 1 - \varepsilon$ for a small enough $\varepsilon \in (0, 1)$. Hence, $\varepsilon |u|^2_{average} + \varepsilon^2 |u|_{average} - \varepsilon \leq -\varepsilon^2$ and $|x_2(t)| \leq e^{-\varepsilon^2 (t-T)} \to 0$ as $t \to +\infty$, but Filippov solutions are unstable. Such inconsistency in Filippov solutions with reality may appear if the control input u is incorporated into the system model in a nonlinear way [17].

The above example shows that *stability analysis of a system with discontinuous right-hand side may depend on a regularization algorithm.*

Another criticism of Filippov's and all other methods based on an extension of ODE to DI is related to an artificial extension of the solution set. For example, if, on Fig. 2.4a, b, the vector field f is oriented toward the discontinuity set $\mathcal{S}$, then, indeed, the natural continuation of solutions to the discontinuity set is a sliding mode on $\mathcal{S}$. However, if, considering the same figure, we assume that the vector field is oriented away from S, then any solution initiated on $\mathcal{S}$ should immediately leave this repelling manifold, but Filippov's method guarantees the existence of solution sliding on the discontinuity set $\mathcal{S}$. This solution is an "artifact" of Filippov's method, which may not have any relation with the real physical behavior of the system modeled by the discontinuous ODE. Moreover, in some particular cases, this "artifact" complicates a strong stability analysis of dynamical systems with discontinuities.

Remark 2.5 *(On Filippov's regularization in the general case)* The regularization of the ODE system with discontinuous right-hand side is possible even if the vector field f in (2.3) is not piece-wise continuous, but it is measurable and locally bounded. In this case, the differential inclusion (2.31) is defined by means of the set-valued vector field given by

$$K[f](t, x) = \bigcap_{\delta > 0} \bigcap_{\mu(N)=0} \overline{\text{co}} f(t, x \dot{+} B(\delta) \setminus N), \tag{2.37}$$

where $\overline{\text{co}}(M)$ is the closed convex hull of M, $\dot{+}$ denotes the geometric sum of sets (see (0.1)), $B(\delta)$ is an open ball of the radius δ centered at the origin, and the intersections are taken over all sets $N \subset \mathbb{R}^n$ of measure zero ($\mu(N) = 0$) and all $\delta > 0$. More intuitive form of the Filippov regulation can be defined as follows [18]

$$K[f](t, x) = \lim_{\delta \to 0^+} \overline{\text{co}} f(t, x \dot{+} B(\delta) \setminus N),$$

where N is a set of discontinuity points of f and the limit is understood in the Hausdorff topology.

We refer the reader to [11, 18] for more details about Filippov's regularization in the general case.

Utkin's Solution ***(Equivalent Control Method)***

Let us consider the control system

$$\dot{x} = f(t, x, u(t, x)), \quad t \in \mathbb{R}, \tag{2.38}$$

where $f : \mathbb{R} \times \mathbb{R}^n \times \mathbb{R}^m \mapsto \mathbb{R}^n$ is a continuous vector field and a piece-wise continuous function

$$u : \mathbb{R} \times \mathbb{R}^n \mapsto \mathbb{R}^m, \quad u(t, x) = (u_1(t, x), u_2(t, x), ..., u_m(t, x))^\top$$

can be treated as a discontinuous feedback control.

Assumption 2.1 Each component $u_i(t, x)$ is discontinuous only on a surface $\mathcal{S}_i = \{(t, x) \in \mathbb{R}^n : s_i(t, x) = 0\}$, where functions $s_i : \mathbb{R}^{n+1} \mapsto \mathbb{R}$ are smooth, i.e., $s_i \in C^1(\mathbb{R}^{n+1}; \mathbb{R})$.

Let us introduce the following differential inclusion

$$\dot{x} \in f(t, x, K[u](t, x)), \quad t \in \mathbb{R}, \tag{2.39}$$

where

$$K[u](t, x) = (K[u_1](t, x), ..., K[u_m](t, x))^\top,$$

$$K[u_i](t,x) = \begin{cases} \{u_i(t, x)\}, & s_i(t, x) \neq 0, \\ \overline{\text{co}} \left\{ \lim\limits_{\substack{(t_j,x_j)\to(t,x) \\ s_i(t_j,x_j)>0}} u_i(t_j,x_j), \lim\limits_{\substack{(t_j,x_j)\to(t,x) \\ s_i(t_j,x_j)<0}} u_i(t_j,x_j) \right\}, & s_i(t, x) = 0, \end{cases} \tag{2.40}$$

where the limits in the above formulas are assumed to be unique for any $(t, x) \in \mathcal{S}_i$ (otherwise, the general Filippov regulations (2.37) has to be applied in order to obtain $K[u]$). Notice that the set $f(t, x, K[u](t, x)$ may be *non-convex* [17] in the general case.

Definition 2.4 *(Utkin's solution)* An absolutely continuous function $x : \mathcal{I} \mapsto \mathbb{R}^n$ defined on a time interval $\mathcal{I}$ is said to be an *Utkin's solution* of (2.38) if there exists a measurable function $u_{eq} : \mathcal{I} \mapsto \mathbb{R}^m$ such that $u_{eq}(t) \in K[u](t, x(t))$ and $\dot{x}(t) = f(t, x(t), u_{eq}(t))$ almost everywhere on $\mathcal{I}$.

The above definition introduces a solution of the differential equation (2.38), which we call *Utkin's solution*, since it follows the basic idea of the *equivalent control method* introduced by V.I. Utkin (see [12], page 14 or [11], page 54).

Obviously, for $(t, x(t)) \notin \mathcal{S}$ we have $u_{eq}(t) = u(t, x(t))$. So, the only issue is to define $u_{eq}(t)$ on a switching surface. The scheme proposed in [12] is based on solving of the algebraic equation

$$\frac{\partial s}{\partial t} + \frac{\partial s}{\partial x} f(t, x, u_{eq}) = \mathbf{0} \quad \text{(or, equivalently, } \dot{s}(t, x) = \mathbf{0})$$

with respect to the variable u_{eq}. The solution $u_{eq}(t, x)$ of the above algebraic equation (if it exists) is called the *equivalent control* [12].

In order to show a difference between Utkin's and Filippov's definitions, we consider the system (2.38) with a scalar ($m = 1$) control u and a *time-invariant switching surface* $\mathcal{S} = \{x \in \mathbb{R}^n : s(x) = 0\}$, $s \in C^1$. For $x \in \mathcal{S}$, let us denote

$$u^+(t, x) = \lim_{x_j \to x, s(x_j) > 0} u(t, x_j) \quad \text{and} \quad u^-(t, x) = \lim_{x_j \to x, s(x_j) < 0} u(t, x_j),$$

$$f^+(t, x) = f(t, x, u^+(t, x)) \quad \text{and} \quad f^-(t, x) = f(t, x, u^-(t, x)),$$

where the above limits are assumed to be unique for any $x \in \mathcal{S}$. The sliding-mode existence condition

$$\exists \mu \in [0, 1] : \nabla s(x) \perp \left(\mu f^-(t, x) + (1 - \mu) f^+(t, x)\right)$$

is the same for both definitions.

The sliding motion equation obtained by Filippov's definition has the form (2.33) recalled here as follows

$$\dot{x} = f_0(t, x),$$

$$f_0(t, x) = \frac{\langle \nabla s(x), f^-(t, x) \rangle f^+(t, x) + \langle \nabla s(x), f^+(t, x) \rangle f^-(t, x)}{\langle \nabla s(x), f^+(t, x) - f^-(t, x) \rangle}.$$

The corresponding vector $f_0(t, x)$ is defined by an intersection point of the tangential plane at the point $x \in \mathcal{S}$ and a *segment* connecting the ends of the vectors $f^+(t, x)$ and $f^-(t, x)$ (see Fig. 2.5a).

Utkin's definition considers the set $K[u](t, x)$, which is the convex closure of the set of limit values of the discontinuous control function u. For different $u_1, u_2, u_3, \ldots \in K[u](t, x)$ the vectors $f(t, x, u_1)$, $f(t, x, u_2)$, $f(t, x, u_3)$, ... end on a curve connecting the ends of the vectors $f^+(t, x)$ and $f^-(t, x)$ (see Fig. 2.5b). In this case the vector $f(t, x, u_{eq})$, which defines the right-hand side of the sliding

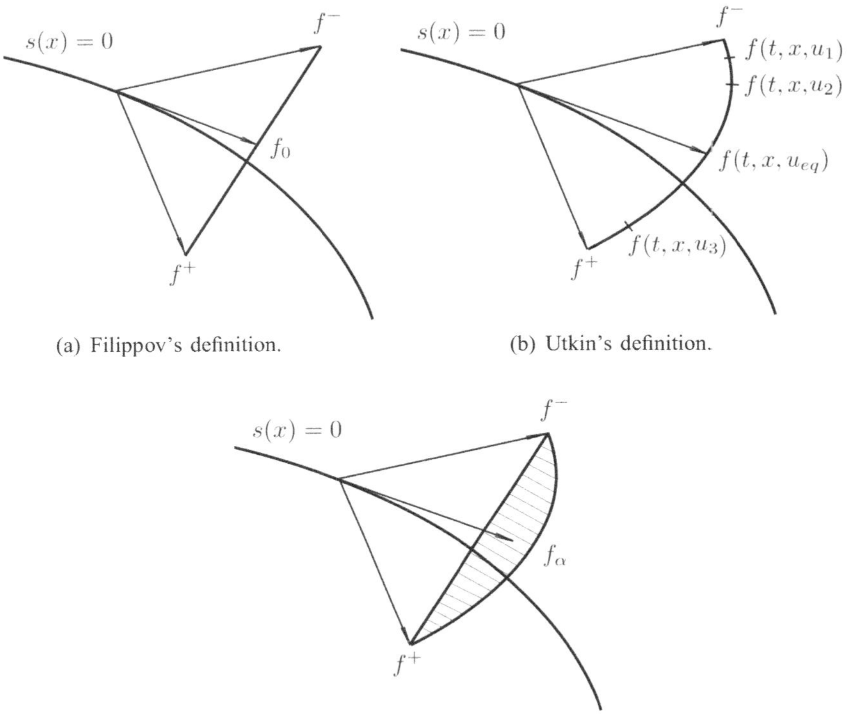

Fig. 2.5 The sliding vector for different definitions

motion equation, is given by an intersection point of this curve and the tangential plane at the point $x \in \mathcal{S}$ (see Fig. 2.5b), i.e.,

$$\dot{x} = f(t, x, u_{eq}(t, x)), \quad x \in \mathcal{S}, \tag{2.41}$$

where $u_{eq}(t, x) \in K[u](t, x) : \nabla s(x) \perp f(t, x, u_{eq}(t, x))$.

Sometimes the equivalent control method gives quite strange, from a mathematical point of view, results, but they are consistent with real applications. The following example is inspired by [11, page 53].

Example 2.4 Let us consider the system

$$\dot{x} = Ax + bu_1 + cu_2, \quad u_1 = \text{sign}(x_1), \quad u_2 = \text{sign}(x_1), \tag{2.42}$$

where $x = (x_1, x_2, ..., x_n)^\top \in \mathbb{R}^n$, $A \in \mathbb{R}^{n\times n}$, $c, b \in \mathbb{R}^n$, $c \neq b$. Filippov's method gives the inclusion

$$\dot{x} \in \{Ax\} \dot{+} (b + c)\overline{\text{sign}}(x_1), \tag{2.43}$$

where $\dot{+}$ is the geometric sum of sets (see formula (0.1)), $\overline{\text{sign}}$ is the set-valued regularization ("convex closure") of the sign function (see formula (0.6)), the product of a vector and a set is defined by formula (0.4).

If the functions u_1 and u_2 are independent control inputs, then Utkin's method gives

$$\dot{x} \in \{Ax\}\dot{+}b\overline{\text{sign}}[x_1]\dot{+}c\overline{\text{sign}}[x_1]. \tag{2.44}$$

The right-hand sides of (2.43) and (2.44) coincide if the vectors c and b are collinear, otherwise the proposed methods generate different set-valued mappings. For example, if $x = (x_1, x_2)^\top \in \mathbb{R}^2$, $A = 0$, $b = (-1, 0)^\top$ and $c = (0, -1)^\top$, then

(a) Filippov's method gives $K[f](x) = [-1, 1]\begin{pmatrix}1\\1\end{pmatrix}$ for $x_1 = 0$, i.e., $K[f](x)$ is a segment connecting the points $(-1, -1)$ and $(1, 1)$, and the corresponding sliding motion equation is

$$\dot{x} = \mathbf{0} \quad \text{for} \quad x_1 = 0;$$

(b) Utkin's method generates the square box $f(x, K[u](x)) = [-1, 1] \times [-1, 1]$ for $x_1 = 0$, and the sliding motion equation has the form

$$\dot{x} = \begin{pmatrix}0\\u_{eq}(t)\end{pmatrix} \quad \text{for} \quad x_1 = 0,$$

where $u_{eq} : \mathbb{R} \mapsto \mathbb{R}$ is an arbitrary locally measurable function such that $|u_{eq}(t)| \leq 1$ for every $t \in \mathbb{R}$.

In this example, Filippov's definition corresponds to the case when the relays (modeled by sign functions) switch absolutely synchronously. This is impossible in practice if u_1 and u_2 correspond to physically different relay elements. In this particular case, Utkin's definition is more realistic than Filippov's one.

In fact, an adequate mathematical description of a relay control system depends on the properties of relay elements. In practice, there is no ideal switching device. A more precise model of the relay has a hysteresis and/or some small zone of linearity implying continuity at the switching point, e.g.,

$$u_\delta(s) = \begin{cases} 1 & \text{if} \quad s > \delta, \\ \frac{s}{\delta} & \text{if} \quad s \in [-\delta, \delta], \\ -1 & \text{if} \quad s < -\delta, \end{cases}$$

where $\delta > 0$ is a small parameter. A mathematical model obtained by Filippov's method may be "more" or "less" adequate compared with Utkin's approach dependently on "nonideality" (delay, hysteresis, etc.) of the relay element [12]. In the general case, the "nonideality" of switching elements may vary in time (e.g., a change of temperature changes the characteristics of transistors) implying a change of an adequate modeling scheme.

Aizerman–Pyatnitskii's Solution

Let us consider the following differential inclusion

$$\dot{x} \in \overline{\text{co}} f(t, x, K[u](t, x)), \quad t \in \mathbb{R} \tag{2.45}$$

for the system (2.38), which is obtained by means of a convexification of Utkin's regularization.

Definition 2.5 *(Aizerman–Pyatnitskii's definition)* An absolutely continuous function $x : \mathcal{I} \mapsto \mathbb{R}^n$ defined on a time interval (or a segment) $\mathcal{I}$ is called an *Aizerman–Pyatnitskii's solution* of (2.3) if it satisfies the differential inclusion (2.45) almost everywhere on $\mathcal{I}$.

Aizerman–Pyatnitskii's definition (see [11], page 55 and [15, 16]) covers both Filippov's and Utkin's methods. Indeed, for the control system studied above

$$\dot{x} = f(t, x, u), \quad t \in \mathbb{R}, \quad x(t) \in \mathbb{R}^n, \quad u(t) \in \mathbb{R},$$

the Aizerman–Pyatnitskii's definition results in the following differential inclusion

$$\dot{x} \in F_{SM}(t, x) = \overline{\text{co}}\{f_0(t, x), f(t, x, u_{eq}(t, x))\},$$

which describes the motion of the discontinuous differential equation (2.38) in a sliding mode on the surface $s(x) = 0$ (see Fig. 2.5c with $f_\alpha \in F_{SM}(t, x)$ as an illustration).

Obviously, any Filippov's and Utkin's solution is an Aizerman–Pyatnitskii's solution too. The main criticism of Definition 2.5 is related to the nonuniqueness of solutions even in the simplest nonlinear cases. However, the Aizerman–Pyatnitskii's approach is more "robust" for system analysis in the following sense.

If some (e.g., stability) property of the system is proven using the Aizerman–Pyatnitskii's solutions, then the same property holds for both Filippov's and Utkin's solutions.

Utkin's and Aizerman–Pyatnitskii's solutions are inspired by discontinuous control theory, such as Sliding-Mode Control [12, 13, 18]. So, it is important to characterize a class of nonlinear discontinuous control systems, which admits the equivalence of all considered regularization algorithms.

The following theorem (see also [12] or [19, Theorem 14, page 44]) proves that all definitions may be equivalent in the case of affine-in-control systems.

Theorem 2.6 *Let a right-hand side of the system* (2.3) *be affine with respect to the control input:*

$$f(t,x) = a(t,x) + b(t,x)u(t,x),$$

where $a : \mathbb{R}^{n+1} \mapsto \mathbb{R}^n$ *is a continuous vector-valued function,* $b : \mathbb{R}^{n+1} \mapsto \mathbb{R}^{n\times m}$ *is a continuous matrix-valued function and the control law* $u : \mathbb{R}^{n+1} \mapsto \mathbb{R}^m$ *is a piece-wise continuous function*

$$u(t,x) = (u_1(t,x), ..., u_m(t,x))^\top,$$

where u_i *has a unique time-invariant switching surface* $s_i(x) = 0, s_i \in C(\mathbb{R}^n, \mathbb{R})$*, such that* $\mathbb{R}^n$ *can be split in* 2^m *disjoint nonempty open connected sets*

$$G_j = \Pi_1^{\beta_1^j} \cap \Pi_2^{\beta_2^j} ... \cap \Pi_m^{\beta_m^j}, \quad j = 1 + \sum_{i=1}^m 2^{i-1}\beta_i^j, \quad \beta_i^j \in \{0,1\},$$

where $\Pi_i^1 = \{x \in \mathbb{R}^n : s_i(x) > 0\}$, $\Pi_i^0 = \{x \in \mathbb{R}^n : s_i(x) < 0\}$ *with*

$$\bigcup_{j=1}^{2^m} \overline{G_j} = \mathbb{R}^n, \qquad G_j \bigcap G_i = \emptyset, \quad j \neq i,$$

where $\overline{G_j}$ *denotes a closure of the set* G_j*. Then Filippov's, Utkin's and Aizerman–Pyatnitskii's definitions are equivalent.*

Proof Let us denote as before $\mathcal{S} = \bigcup_{j=1}^{2^m} \partial G_j$. According to Filippov's definition (see (2.31)) we have

$$K[f](t,x) = \text{co}\left(\bigcup_{j\in\mathcal{N}(t,x)} \{f^j(t,x)\} \right)$$

for any $x \in \mathcal{S}$. On the one hand, Carathéodory's lemma (see Appendix A) yields

$$K[f](t,x) = \left\{ z \in \mathbb{R}^n : z = \sum_{j\in\mathcal{N}(t,x)} \alpha_j f^j(t,x), \quad \alpha_j > 0, \quad \sum_{j\in\mathcal{N}(t,x)} \alpha_j = 1 \right\}.$$

Due to continuity of a and b we obtain $f^j(t,x) = a(t,x) + b(t,x)u^j(t,x)$ for any $x \in \overline{G_j}$, where

$$u^j(t,x) = \lim_{x_k \to x \,:\, x_k \in G_j} u(t,x_k).$$

Hence, we derive

$$\sum_j \alpha_j f^j(t,x) = a(t,x) + b(t,x) \sum_{j\in\mathcal{N}(t,x)} \alpha_j u^j(t,x).$$

On the other hand, Utkin's definition gives

$$f(t,x,K[u](t,x)) = a(t,x) + b(t,x)K[u](t,x)$$

where $K[u]$ is defined by formula (2.40). Let us show that the set

$$U(t,x) = \left\{ v \in \mathbb{R}^m : \sum_{j\in\mathcal{N}(t,x)} \alpha_j u^j(t,x), \quad \alpha_j > 0, \quad \sum_{j\in\mathcal{N}(t,x)} \alpha_j = 1 \right\}$$

coincides with $K[u](t,x)$. If we denote

$$u_i^1(t,x) := \lim_{\substack{x_j\to x \\ s_i(x_j)>0}} u_i(t,x_j), \qquad u_i^0(t,x) = \lim_{\substack{x_j\to x \\ s_i(x_j)<0}} u_i(t,x_j)$$

then, by definition of G_j, we derive

$$u^j(t,x) = (u_1^{\beta_1^j}(t,x), \ldots, u_m^{\beta_m^j}(t,x))$$

for any $x \in \overline{G_j}$. This immediately implies that the vertexes of the parallelotope $K[u](t,x)$ coincide with the vertexes of the polytope $U(t,x)$. By construction, the set $U(t,x)$ is the minimal convex set containing the limit points of the discontinuous vector field $u : \mathbb{R} \times \mathbb{R}^n \mapsto \mathbb{R}^m$, so $K[u](t,x) = U(t,x)$ and

$$K[f](t,x) = f(t,x,K[u](t,x)).$$

Finally, since $K[f](t,x)$ is the convex set then $K[f](t,x) = \mathrm{co}(K[f](t,x)) = \mathrm{co}(f(t,x,K[u](t,x)))$. The proof is complete. □

Filippov's and Aizerman–Pyatnitskii's methods extend the discontinuous (ill-posed) ODEs to differential inclusions, which are well-posed, namely, *compact-valued, convex-valued and upper semi-continuous* (see below). The existence of solutions of such ordinary differential inclusions is discussed in the next section.

2.2 Ordinary Differential Inclusion

2.2.1 Models with Disturbances

Some modifications of a definition of the solution may be required again if the ODE model of a control system includes disturbances. However, if a maximum magnitude (L^∞ norm) of disturbances is known a priori then a problem of analysis and design of the control system with disturbances can be reduced to a similar problem with a dynamical model given by a *differential inclusion* (see, e.g., [20, 21]).

Example 2.5 Let us consider the following sliding-mode system with disturbances:

$$\dot{x} = -d_1(t, x)\,\mathrm{sign}(x) + d_2(t, x), \tag{2.46}$$

where $x \in \mathbb{R}$, unknown functions $d_i : \mathbb{R} \mapsto \mathbb{R}$ are bounded by

$$d_i^{\min} \leq d_i(t, x) \leq d_i^{\max}, \qquad i = 1, 2, \tag{2.47}$$

and the sign function is defined by formula (0.5).

If x is a classical (or a strong) solution of the system (2.46) exists then, obviously, it satisfies (almost everywhere) the following *extended differential inclusion*

$$\dot{x} \in -[d_1^{\min}, d_1^{\max}] \cdot \overline{\mathrm{sign}}[x] \dot{+} [d_2^{\min}, d_2^{\max}], \tag{2.48}$$

i.e., x is a classical (resp., strong) solution of the differential inclusion (see the definition below). In other words, the set of solutions of Eq. (2.46) is embedded into a set of solutions of the differential inclusion (2.48). In this case, a stability of the system (2.48) implies the same property for the system (2.46). In particular, for $d_1^{\min} > \max\{|d_2^{\min}|, |d_2^{\max}|\}$ both these systems are asymptotically stable. This example shows that a stability discovered for the extended differential inclusion guarantees a *robust stability* of the original differential equation.

Let a model of control systems have the form

$$\dot{x} = f(t, x, u(t, x), d(t, x)), \quad t \in \mathbb{R}, \tag{2.49}$$

where $x(t) \in \mathbb{R}^n$ is the vector of system states, $u(t) \in \mathbb{R}^m$ is the vector of control inputs, $d(t) \in \mathbb{R}^k$ is the vector of disturbances, the function $f : \mathbb{R}^{n+m+k+1} \mapsto \mathbb{R}^n$ is assumed to be continuous, the control function $u : \mathbb{R}^{n+1} \mapsto \mathbb{R}^m$ is piece-wise continuous, the unknown vector-valued function $d : \mathbb{R}^{n+1} \mapsto \mathbb{R}^k$ is assumed to be bounded as follows:

$$d_i^{\min} \leq d_i(t, x) \leq d_i^{\max}, \tag{2.50}$$

where $d = (d_1, d_2, ..., d_k)^T$, $t \in \mathbb{R}$, $x \in \mathbb{R}^n$.

The finite-dimensional model of the closed-loop control system with disturbances can be defined as follows

$$\dot{x} \in F(t, x), \quad t \in \mathbb{R}, \tag{2.51}$$

where $F(t, x) = \text{co}(f(t, x, K[u](t, x), D))$, the set-valued function $K[u](t, x)$ is given by (2.40) and

$$D = \left\{(d_1, d_2, ..., d_k)^T \in \mathbb{R}^k : d_i \in [d_i^{\min}, d_i^{\max}], i = 1, 2, ..., k\right\}. \tag{2.52}$$

If u is continuous then $K[u](t, x) = u(t, x)$. The same extended differential inclusion can be utilized if the vector d (or its part) has a sense of *parametric uncertainties*. Notice that solution existence theorems for the system (2.49) would require d to be at least measurable. Formally, the model (2.51) eliminates this restriction, but it implicitly appears again in the definition of solutions of DI.

For further analysis of set-valued mappings, the Hausdorff metric needs to be introduced. Let the distance from a point to a set be defined as follows

$$\rho_0(x, \mathbf{M}) = \inf_{\forall y \in \mathbf{M}} |x - y|, \quad x \in \mathbb{R}^n, \quad \mathbf{M} \subset \mathbb{R}^n,$$

and the Hausdorff distance between sets is given by

$$\overline{\rho}(\mathbf{M}_1, \mathbf{M}_2) = \max\{\rho(\mathbf{M}_1, \mathbf{M}_2), \rho(\mathbf{M}_2, \mathbf{M}_1)\},$$

where

$$\rho(\mathbf{M}_1, \mathbf{M}_2) = \sup_{x \in \mathbf{M}_1} \rho_0(x, \mathbf{M}_2), \quad \mathbf{M}_1 \subset \mathbb{R}^n, \quad \mathbf{M}_2 \subset \mathbb{R}^n.$$

Definition 2.6 A set-valued mapping $F : \mathbb{R}^k \mapsto 2^{\mathbb{R}^m}$ is said to be continuous (upper semi-continuous) at a point $v^* \in \mathbb{R}^k$ if

$$\overline{\rho}(F(v), F(v^*)) \to 0 \quad \text{as} \quad v \to v^*$$

$$\left(\text{resp.,} \quad \rho(F(v), F(v^*)) \to 0 \quad \text{as} \quad v \to v^*\right).$$

A mapping is continuous (upper semi-continuous) on a domain $\mathcal{D} \subset \mathbb{R}^k$ if it is continuous (upper semi-continuous) at each point of this domain $\mathcal{D}$.

Obviously, a composition $F_1(F_2)$ of continuous compact-valued mappings F_1 and F_2 with properly defined domains and co-domains gives a continuous compact-valued mapping, but this composition is upper semi-continuous if F_2 is upper semi-continuous compact-valued mapping.

A set-valued mapping $F : \mathcal{D} \subset \mathbb{R}^k \mapsto 2^{\mathbb{R}^m}$ is locally Lipschitz continuous on an open nonempty set $\mathcal{D}$ if for any $x_0 \in \mathcal{D}$ there exist $r > 0$ and $L > 0$:

$$F(x_1) \subset F(x_2)\dot{+}\overline{B}(L|x_1 - x_2|), \forall x_1, x_2 \in \overline{B}(x_0, r) \subset \mathcal{D},$$

where $\overline{B}(x_0, r)$ is a closed ball of the radius $r > 0$ centered at x_0.

The procedure of set-valued extension of ODE with disturbances given above results in a compact-valued and continuous set-valued mapping F if the original system is continuous.

Lemma 2.2 *If $f \in C(\mathbb{R}^n \times \mathbb{R}^k, \mathbb{R}^n)$ and $Q \subset \mathbb{R}^n$ is a compact set then*

- *the mapping $x \mapsto F(x) := f(x, Q)$ is compact-valued and continuous;*
- *the mapping $x \mapsto \overline{\text{co}} f(x, Q)$ is compact-valued, convex-valued and continuous.*

Moreover, if f is locally Lipschitz continuous on the first argument then $F(\cdot)$ and co $F(\cdot)$ *are locally Lipschitz continuous too.*

Proof Since f is continuous then, by Heine–Cantor theorem, it is uniformly continuous on any compact, and for any $r > 0$ there exists $\omega \in \mathcal{K}$ such that

$$|f(x_1, q) - f(x_2, q)| \leq \omega(|x_1 - x_2), \quad \forall x_1, x_2 \in \overline{B}(r), \forall q \in Q.$$

Hence, we derive

$$f(x_1, Q) \subset f(x_2\dot{+}\overline{B}(|x_1 - x_2|), Q) \subset f(x_2, Q)\dot{+}B(\omega(|\delta|)), \quad \forall x_1, x_2 \in \overline{B}(r),$$

i.e., $F(\cdot) = f(\cdot, Q)$ is continuous. Continuity of f on both arguments implies compactness of $f(x, Q)$. Taking into account (see Appendix A)

$$\overline{\text{co}} f(x_1, Q) \subset \overline{\text{co}}\left(f(x_2, Q)\dot{+}\overline{B}(\omega(|\delta|))\right) = \overline{\text{co}} f(x_2, Q)\dot{+}\overline{B}(\omega(|\delta|)),$$

we conclude that $x \mapsto \text{co}\, F(x)$ is compact-valued, convex-valued and continuous too. Taking into account that $\omega(|\delta|) = L|\delta|$ in the case of locally Lipschitz continuous f (on the first argument), we complete the proof. □

For compact-valued mappings, the (local Lipschitz) continuity always implies upper semi-continuity.

The procedure of set-valued extension given above results in a convex-valued, compact-valued and upper semi-continuous set-valued mapping F if the control input is a piece-wise continuous feedback and the plant model is continuous.

Theorem 2.7 *If* $f \in C(\mathbb{R}^{n+m+k+1}, \mathbb{R}^n)$*,* $\mathcal{D}$ *is a compact in* $\mathbb{R}^k$ *and* $u : \mathbb{R}^{n+1} \mapsto \mathbb{R}^m$ *is piece-wise continuous function then* $F : \mathbb{R}^{n+1} \mapsto 2^{\mathbb{R}^n}$ *given by* $F = \mathrm{co}(f(t, x, K[u](t, x), D))$ *is compact-valued, convex-valued and upper semi-continuous.*

Proof The set-valued mapping $(t, x) \mapsto K[u](t, x)$ is convex-valued, compact-valued and upper semi-continuous by construction (see, formula (2.31) and the definition of the upper semi-continuity). By Lemma 2.2, the set-valued mapping $(t, x, u) \mapsto \mathrm{co}\, f(t, x, u, D)$ is compact-valued, convex-valued and continuous on (t, x, u). Hence, F is upper semi-continuous as a composition of continuous compact-valued mapping and upper semi-continuous compact-valued mapping. The proof is complete. □

The Filippov regularization (2.37) in the general case results in the same properties of F but with respect to the variable x only [11]. Upper semi-continuity on the time variable t can be relaxed to a Lebesgue integrability condition.

2.2.2 *Existence of Solutions*

As it is shown above, a regularization of the differential equation with discontinuous right-hand side as well as a robustness analysis of uncertain control system can be based on representations of systems in the form of a differential inclusion

$$\dot{x} \in F(t, x), \quad t \in \mathbb{R}, \quad x(t) \in \mathbb{R}^n, \tag{2.53}$$

where $F : \mathbb{R}^{n+1} \mapsto 2^{\mathbb{R}^{n+1}}$ is a set-valued vector field.

Definition 2.7 A differentiable (absolutely continuous) function $x : \mathcal{I} \mapsto \mathbb{R}^n$ is said to be classical (strong) solution to the differential inclusion (2.53) if it satisfies (2.53) for all (almost all) $t \in \mathcal{I}$.

Similarly to ODE, the set-valued vector field on the right-hand side of the differential inclusion defines a type of solution. The system (2.53) has classical solutions if F is continuous, but the upper semi-continuity guarantees the existence of strong solutions [11].

Theorem 2.8 (Filippov 1960, [11], page 77) *Let a set-valued vector field* $F : \mathcal{G}(t_0, x_0) \mapsto 2^{\mathbb{R}^n}$ *be upper semi-continuous at each point of the set*

$$\mathcal{G}(t_0, x_0) := \left\{(t, x) \in \mathbb{R}^{n+1} : |t - t_0| < a \text{ and } |x - x_0| < b\right\},$$

where $a, b \in \mathbb{R}_+$, $t_0 \in \mathbb{R}$, $x_0 \in \mathbb{R}^n$, *and the set* $F(t, x)$ *be nonempty, compact and convex for any* $(t, x) \in \mathcal{G}$. *If there exists* $M > 0 : \rho(0, F(t, x)) \leq M$ *for all* $(t, x) \in \mathcal{G}$ *then the differential inclusion (2.53) has, at least, one strong solution* $x(t) : x(t_0) = x_0$ *defined, at least, on the segment* $[t_0 - \alpha, t_0 + \alpha]$, *where* $\alpha = \min\{a, b/M\}$.

Proof For $k = 1, 2, \dots$ let us take

$$h^k = \alpha/k, \quad t_i^k = t_0 + ih^k, \quad i = 0, 1, \dots, k.$$

Let us define Euler approximates $x^k(t)$ with $t \in [t_0, t_0 + \alpha]$ as follows $x^k(t_0) = x_0$ and

$$x^k(t) = x(t_i^k) + (t - t_i^k)v_i^k \quad \text{for} \quad t_i^k < t \leq t_{i+1}^k,$$

where $i = 0, 1, \dots, k - 1$ and v_i^k is an arbitrary vector from the set $F\left(t_i^k, x^k(t_i^k)\right)$. Since $\rho(0, F(t, x)) \leq M$ then $|v_i^k| \leq M$ and

$$|x^k(t) - x_0| \leq M|t - t_0| \quad t \in [t_0, t_0 + \alpha].$$

Since $\alpha = \min\{a, b/M\}$ then the curve $(t, x^k(t))$, $t \in [t_0.t_0 + \alpha]$ belongs to the set $\mathcal{G}(t_0, x_0)$. Moreover, x^k is continuous and $|\dot{x}^k(t)| \leq M$ for $t \neq t_i^k$. The sequence of functions $\{x^k\}$ is uniformly bounded and equicontinuous. Thus from Arzelá–Ascoli theorem (see Appendix D), we conclude that there exists a subsequence $\{x^{k_j}\}$ which converges uniformly to a continuous function ϕ as $j \to +\infty$. Since for any $t', t'' \in [t_0, t_0 + \alpha]$ we have

$$\left|x^k(t') - x^k(t'')\right| = \left|\int_{t''}^{t'} \dot{x}^k(s)ds\right| \leq M|t' - t''|$$

then taking the limit $k \to +\infty$ in the left-hand side of the above inequality we conclude that the same inequality holds for ϕ, i.e., ϕ is an absolutely continuous function that is differentiable almost everywhere. Let $\tilde{t} \in (t_0, t_0 + \alpha)$ be an arbitrary point where the derivative $\dot{\phi}(\tilde{t})$ exists. Let us consider the set

$$\tilde{F}_\varepsilon = F(\tilde{t}, \phi(\tilde{t})) \dot{+} B(\varepsilon),$$

where $\varepsilon > 0$ is an arbitrary small real number, $B(\varepsilon)$ is the unit ball of the radius ε centered at the origin and $\dot{+}$ is the geometric sum of sets (see (0.1)). Being the geometric sum of convex and compact sets, the set $\tilde{F}_\varepsilon$ is also convex and compact.

Due to upper semi-continuity of F and continuity of ϕ there exists $\gamma = \gamma(\varepsilon) \in (0, \varepsilon)$ such that

$$F(t, \phi(t)) \subset \tilde{F}_{\varepsilon}, \qquad \forall t \in [\tilde{t}, \tilde{t} + 2\gamma].$$

Hence, the uniform convergence of the subsequence $\{x^{k_j}\}$ to ϕ implies $F(t, x^{k_j}(t)) \subset \tilde{F}_{\varepsilon}, \forall t \in [\tilde{t}, \tilde{t} + \gamma]$ and $\forall j > N$, where $N = N(\varepsilon)$ is a sufficiently large natural number. Moreover, if $[t_i^{k_j}, t_{i+1}^{k_j}] \in [\tilde{t}, \tilde{t} + \gamma]$ then by construction we have $\dot{x}^{k_j}(t) = v_i^{k_j} \in \tilde{F}_{\varepsilon}$ for all $t \in (t_i^{k_j}, t_{i+1}^{k_j})$. For any fixed j let use define

$$\mu_i = \int_{\tilde{t}}^{\tilde{t}+\gamma} I\left(\tau, [t_i^{k_j}, t_{i+1}^{k_j}] \cap [\tilde{t}, \tilde{t} + \gamma\right) d\tau,$$

where $I(\tau, S)$ is the indicator function of the set S, i.e., $I(\tau, S) = 1$ if $\tau \in S$ and $I(\tau, S)$ if $i \notin S$. Since $[t_i^{k_j}, t_{i+1}^{k_j}]$ and $[\tilde{t}, \tilde{t} + \gamma]$ are intervals then $I\left(\tau, [t_i^{k_j}, t_{i+1}^{k_j}] \cap [\tilde{t}, \tilde{t} + \gamma\right)$ is piece-wise constant and, consequently, Riemann integrable. Hence, μ_i is well defined for all i and

$$\frac{1}{\gamma} \int_{\tilde{t}}^{\tilde{t}+\gamma} \dot{x}^{k_j}(s)ds = \sum_i \mu_i v_i^{k_j}, \qquad \sum_i \mu_i = 1.$$

Thus taking into account convexity of $\tilde{F}_{\varepsilon}$ we conclude

$$\frac{x^{k_j}(\tilde{t} + \gamma) - x^{k_j}(\tilde{t})}{\gamma} = \frac{1}{\gamma} \int_{\tilde{t}}^{\tilde{t}+\gamma} \dot{x}^{k_j}(s)ds \in \tilde{F}_{\varepsilon}$$

and for $j \to +\infty$ we derive $\frac{\phi(\tilde{t}+\gamma)-\phi(\tilde{t})}{\gamma} \in \tilde{F}_{\varepsilon}$ Finally, since $\frac{\phi(\tilde{t}+\gamma)-\phi(\tilde{t})}{\gamma} \to \dot{\phi}(\tilde{t})$ as $\varepsilon \to 0$ and $\rho(\tilde{F}_{\varepsilon}, F(\tilde{t}, \phi(\tilde{t}))) \to 0$ as $\varepsilon \to 0$ then $\dot{\phi}(\tilde{t}) \in F(\tilde{t}, \phi(\tilde{t}))$ due to compactness of the set $F(\tilde{t}, \phi(\tilde{t})) \subset \mathbb{R}^n$. Therefore, ϕ satisfies the differential inclusion almost everywhere on $[t_0, t_0 + \alpha]$. The existence of a solution on the segment $[t_0 - \alpha, t_0]$ can be proven similarly considering the differential inclusion in the backward time. The proof is complete. □

By construction, Filippov's and Aizerman–Pyatnitskii's set-valued extensions (see (2.31) and (2.45)) of a piece-wise continuous ODE satisfy the conditions of Theorem 2.8. The Filippov regularization (2.37) of a locally measurable function f also satisfies Theorem 2.8 in the *autonomous* case (i.e., if f is independent of time). The non-autonomous case is treated by the following remark [11].

Remark 2.6 The conditions of Theorem 2.8 can be relaxed as follows

- F is compact-valued, convex-valued and upper semi-continuous on x for almost all t;
- $\exists f : \mathbb{R} \times \mathbb{R}^n \mapsto \mathbb{R}^n$ such that $f(t,x) \in F(t,x)$ for almost all t and all x satisfying $(t,x) \in \mathcal{G}(t_0, x_0)$;
- f is measurable by t for any fixed x;
- $\exists m : \mathbb{R} \mapsto \mathbb{R}$ such that m is Lebesgue integrable function and $|f(t,x)| \leq m(t)$ for almost all t and all x.

Such a relaxation is useful for an analysis of discontinuous ODEs with time-varying locally bounded measurable right-hand sides. We refer the reader to [11, page 83] for more details about Filippov regularization and the existence of Filippov solutions in the general case.

Theorem 2.8 proves the existence of local-in-time solutions of differential inclusion. A sufficient condition of the global-in-time existence of solutions is given, for example, by the so-called Wintner's theorem being a consequence of the Bihari–LaSalle inequality (known also as a nonlinear Grönwall–Bellman inequality, see Appendix F). For differential inclusions, the theorem is formulated as follows [22, page 169].

Theorem 2.9 (Wintner's theorem for differential inclusions) *Let F satisfy conditions of Theorem 2.8 on $\mathcal{G}(t_0, x_0) \subset \mathbb{R}^{n+1}$ with any $a > 0$ and any $b > 0$. If there exists a continuous function $L : [0, +\infty) \mapsto (0, +\infty)$ such that*

$$\rho_0(0, F(t,x)) \leq L(|x|), \quad \forall t \in \mathbb{R}, \quad \forall x \in \mathbb{R}^n \quad \text{and} \quad \int_0^{+\infty} \frac{dr}{L(r)} = +\infty,$$

then any solution of the differential inclusion (2.53) *exists for all $t \in \mathbb{R}$.*

Similarly to ordinary differential equations, a continuous dependence of solutions on initial conditions and parameters on the right-hand side can be discovered for differential inclusions.

Theorem 2.10 (Continuity of Filippov solutions on parameters) *Let F satisfy all conditions of Theorem 2.8. Let all solutions of the differential inclusion* (2.53) *with $x(t_0) = x_0$ exist on $[t - \alpha, t + \alpha]$ and their graphs belong to $\mathcal{G}(t_0, x_0)$. Then for any $\varepsilon > 0$ there exists $\delta > 0$ such that for any $t_0^* \in \mathbb{R}$, any $x_0^* \in \mathbb{R}^n$ and $F^* : \mathcal{G}(t_0, x_0) \mapsto 2^{\mathbb{R}^n}$ satisfying the inequalities*

$$|t_0^* - t_0| < \delta, \quad |x_0^* - x_0| \leq \delta, \quad \bar{\rho}(F, F^*) < \delta,$$

and satisfying all conditions of Theorem 2.8, each solution of the system

$$\dot{x}^* \in F^*(t, x^*), \qquad x^*(t_0^*) = x_0^*$$

exists on $[t_0 - \alpha, t_0 + \alpha]$ *and*

$$\max_{t \in [t_0 - \alpha, t_0 + \alpha]} |x^*(t) - x(t)| \leq \varepsilon,$$

where x *is some solution of* (2.53) *with* $x(t_0) = x_0$.

The proof of this theorem considers the differential inclusion

$$\dot{x}^\delta \in F^\delta(t, x)$$

defined as follows

$$F^\delta(t, x) = \overline{\text{co}} \bigcup_{|s| \leq \delta, |u| \leq \delta} F(t + s, x + u) \dot{+} \delta B.$$

Repeating considerations from the proof of Theorem 2.8, it can be shown that for any sequence $\delta_i \to 0$ a sequence of solutions x^{δ_i} contains a subsequence which converges to a solution of the original differential inclusion with $\delta = 0$. We refer the reader to [11, page 87] for more details.

The set-valued extension (2.39) of a system $\dot{x} = f(t, x, u(x))$ with a discontinuous control u may be non-convex in the general case, so it does not satisfy the conditions of Theorem 2.8. The exception is the system being affine in control u, for which Utkin's solutions coincide with Filippov's solutions (see Theorem 2.6). Most homogeneous discontinuous systems studied in this monograph are affine in control, so they have Filippov's solution which coincide with Utkin's and Aizerman–Pyatnitskii's solution. In order to complete the introduction to the regularization problem of discontinuous ODEs, some remarks about non-convex and non-compact differential inclusions are given below.

Theorem 2.8 is not applicable if the set-valued map is non-convex or non-compact. This may happen if a non-affine-in-control system has discontinuous feedback control [17]. However, non-convexity or non-compactness of a set-valued vector field does not mean that the corresponding differential equation has no solution. For example, if $d : \mathbb{R} \mapsto \mathbb{R}$ is an unknown continuous function such that $d(x) \in M = (1, 2) \cup (3, 4)$, $\forall x \in \mathbb{R}$ then due to Theorem 2.1 the uncertain differential equation

$$\dot{x} = -d(x)\sqrt[3]{x}, \quad x \in \mathbb{R}, \quad t > 0$$

has a solution for any initial condition $x(0) = x_0$. This solution satisfies the differential inclusion

$$\dot{x} \in -M\sqrt[3]{x},$$

which has no uncertainties anymore. Therefore, a differential inclusion may have strong and even classical solutions if its right-hand side has non-convex, non-compact and, in fact, even unbounded values.

For stability analysis and control design, we may extend artificially a continuous ODE to a differential inclusion, which has some desirable properties (e.g., homogeneity). The mentioned differential inclusion may be non-convex and/or non-compact but it, obviously, would have, at least, one solution which corresponds to a solution of the original ODE.

Lemma 2.3 *If there exists a vector field* $f : \mathcal{G}(t_0, x_0) \mapsto \mathbb{R}^n$ *satisfying Theorem 2.1 (resp.,Theorem 2.4) such that* $f(t, x) \in F(t, x)$ *for all* $(t, x) \in \mathcal{G}(t_0, x_0)$, *where* $F : \mathcal{G}(t_0, x_0) \mapsto 2^{\mathbb{R}^n}$ *is a set-valued vector field, then the differential inclusion* (2.53) *has at least one classical (resp., strong) solution defined, at least, on* $[t_0 - \min\{\alpha, \beta/M\}, t_0 + \min\{\alpha, \beta/M\}]$.

The vector field f mentioned above is called *selector*. The proof of the existence of such a selector is a nontrivial problem in the general case [11, 17, 23].

Lemma 2.4 (Filippov 1962, [24]) *Let a set-valued mapping* $U : \mathbb{R}^{n+1} \mapsto 2^{\mathbb{R}^m}$ *be nonempty-valued, compact-valued, convex-valued and upper semi-continuous, a function* $f : \mathbb{R}^{n+m+1} \mapsto \mathbb{R}^n$ *be continuous, a function* $x : \mathcal{I} \mapsto \mathbb{R}^n$ *be absolutely continuous and*

$$\dot{x}(t) \in f(t, x(t), U(t, x(t)))$$

almost everywhere on $\mathcal{I}$. *Then there exists a measurable function* $u_{eq} : \mathbb{R} \mapsto \mathbb{R}^m$ *such that* $\dot{x}(t) = f(t, x(t), u_{eq}(t))$ *almost everywhere on* $\mathcal{I}$.

Proof First of all, notice that the mapping $U(\cdot, x(\cdot)) : \mathcal{I} \to \mathbb{R}^m$ is nonempty-valued, compact-valued, convex-valued and upper semi-continuous. Let us denote $\tilde{f}(t, u) = f(t, x(t), u)$ and let $y : \mathcal{I} \to \mathbb{R}^n$ be an arbitrary measurable function satisfying $y(t) \in \tilde{f}(t, U(t, x(t)))$ for all $t \in \mathcal{I}$. Since the function $\tilde{f}$ is continuous then for any $t \in \mathcal{I}$ the set $\tilde{U}(t) \subset \mathbb{R}^m$ of solutions of the algebraic equation $\tilde{f}(t, u) = y(t)$ with respect to $u \in U(t, x(t))$ is compact. Let us select $\tilde{u}(t) = (u_1(t), ..., u_m(t))^\top \in \tilde{U}(t)$ such $u_1(t)$ has the smallest value. If there is more than one such point, we take that one for which the coordinate $u_2(t)$ has the smallest value, etc.

Let us show *by induction* that the obtained function u is measurable. Suppose that $u_1, ..., u_{i-1}$ are measurable and prove that u_i is measurable as well. Indeed, according to Luzin's theorem (see Appendix B) for any $\varepsilon > 0$ there exists a compact set $E \subset \mathcal{I}$ of the measure $\mu(E) \geq \mu(\mathcal{I}) - \varepsilon$ such that the functions $y, u_1, ..., u_{i-1}$ are continuous almost everywhere on E.

Let us show now that the set $\{t \in E : u_i(t) \leq a\}$ is closed, where a is an arbitrary positive number. Let us suppose that there exists a sequence $\{t_k\}$ such that $t_k \to \bar{t}$ and $u_s(t_k) \leq u_i(\bar{t}) - \epsilon$ for some $\epsilon > 0$. Since $|u_s(t)| \leq$const for $s = 1, 2, ..., m$ then there exists a subsequence t_{k_j} such that $t_{k_j} \to \bar{t}$ and $u_s(t_{j_k}) \to \tilde{u}_s$ for $s \in 1, ..., m$. Since $U(\cdot, x(\cdot))$ is upper semi-continuous and $u(t_{j_k}) \in U(t_{j_k}, x(t_{j_k}))$ then $\tilde{u} \in U(\tilde{t}, x(\tilde{t}))$ and $u(\bar{t}) = u_s(\bar{t})$ for $s = 1, 2, .., i-1$ (due to continuity on E) and $\tilde{u}_i \leq u(\bar{t}) - \epsilon$. Therefore, we derive $f(\bar{t}, \tilde{u}_1(\bar{t}), ..., \tilde{u}_{i-1}(\bar{t}), \tilde{u}_i, ...\tilde{u}_m) = y(\bar{t})$ that contradicts with our selection $u_i(\bar{t})$ as the smallest value fulfilling identity. The lemma is proven. □

Filippov's lemma given above justifies the *equivalent control method* [12] and the consistency of Definition 2.4 in the general case. For more details about the existence of Utkin's solutions in a non-affine-in-control case, we refer the reader to [17]. This monograph deals mainly with affine-in-control systems.

2.3 Models of Discontinuous Dynamics

ODE is one of the most popular models of dynamics. However, it cannot model well any kind of dynamic system. Systems with distributed parameters (infinite-dimensional systems) are discussed in the second volume of this monograph. Their typical models are *Partial Differential Equations*. However, even some finite-dimensional systems cannot be described well by ODEs. Indeed, classical, strong and Filippov solutions of ODEs are, by definition, continuous functions of time. This means that they cannot be utilized for the description of discontinuous (e.g., jump) dynamics.

2.3.1 Impulsive Systems

An impulsive system is a system, in which the state has jumps (resets) at certain instances of time, for example, when the system state reaches a certain set $\mathcal{D} \subset \mathbb{R}$. Such jumps in the state can be observed in dynamics of mechanical systems with impacts. For example, a bouncing ball instantly inverses its velocity at the instant of impact. Since the velocity is a part of the state of the mechanical system, the corresponding dynamical system must model correctly a jump of the state vector at a certain instant of time.

Inspired by [9, 25–27] and others, let us consider the following model of the impulsive system

$$\begin{cases} \dot{x} = f(t,x) & \text{if } x^- \in \mathcal{C}, \\ x = \Pi(t,x^-) & \text{if } x^- \in \mathcal{D}, \end{cases} \quad t > t_0, \quad x(t_0) = x_0, \tag{2.54}$$

where $x(t) \in \mathbb{R}^n$ is the system state at the time t, $\mathcal{C} \subset \mathbb{R}^n$ is the set of admissible system states, $f \in C(\mathbb{R}^{n+1}; \mathbb{R}^n)$ is a vector field governing the continuous dynamics of the system, the function $\Pi : \mathbb{R} \times \mathbb{R}^n \mapsto \mathbb{R}^n$ is a reset map, which defines the jump dynamics of the systems and

$$x^- = \lim_{h \to 0^-} x(t+h). \tag{2.55}$$

The reset (a jump of the state) appears when the system trajectory reaches the set $\mathcal{D}$. In most practical cases, it is reasonable to assume that the state after the jump should not belong to $\mathcal{D}$, i.e., $\Pi(t, \mathcal{D}) \cap \mathcal{D} = \emptyset, \forall t \geq t_0$.

Inspired by [27], a solution of (2.54) is a right-continuous function x satisfying (2.54) for all $t > t_0$ and having left limits at each $t > t_0$.

The well-posedness in Hadamard sense is not always appropriate for impulsive systems, since a small variation of initial condition near $\partial\mathcal{D}$ may imply a large variation of solutions in the general case. Therefore, among the three conditions of well-posedness by Hadamard we may have just the first one (the existence of solutions). Therefore, we say that the impulsive system (2.54) is well-posed if for any $x_0 \in \mathcal{C} \cup \mathcal{D}$ it has a solution defined, at least, locally in time.

In the general case, the impulsive systems may have the so-called *Zeno behavior* (see, e.g., [9] and references therein), when the sequence of time instances $t_i \geq 0$ with state jumps $x(t_i^-) \in \mathcal{D}$ has an accumulation point: $\exists t^* \in (t_0, +\infty)$ such that $t_i \to t^*$ as $i \to +\infty$. We say that the system is *Zeno-free* if such a behavior is excluded. Existence of solutions for zeno-free impulsive systems with $\Pi(t, \mathcal{D}) \cap \mathcal{D} = \emptyset, \forall t \geq t_0$ straightforwardly follows from classical existence theorems for ODEs.

Impulsive systems frequently appear as models of some mechanical systems. In Chap. 15 such a model is utilized for homogeneous rigid body attitude control design.

2.3.2 Difference Equations

A system, which can change its state only at discrete instances of time, is called *discrete-time system*. The digital computers are typical examples of such systems. Modern control engineering deals with computer-based controllers. The control signals generated by such controllers are sampled in time. This inspires a wide use of discrete-time modeling in control theoretic framework (see, e.g., [28, 29] and references therein). The *difference equation*

$$x_{k+1} = g(k, x_k), \quad k \geq k_0, \quad x_k \in \mathbb{R}^n, \quad f : \mathbb{R} \times \mathbb{R}^n \mapsto \mathbb{R}^n \tag{2.56}$$

is one of the most popular models of discrete-time systems. The solution existence and uniqueness problem for the difference equation is trivial, since any solution $x(k, k_0, x_0)$ is given explicitly by the formula

$$x(k, k_0, x_0) = g\left(k, g(k-1, g(..., g(k_0, x_0)))\right), \quad k > k_0. \tag{2.57}$$

Difference equations are frequently utilized as discrete-time approximations of ODEs [30]. Indeed, since

$$\dot{x}(t) \approx \frac{x(t+h) - x(t)}{h}$$

for a small enough $h > 0$, then the ODE model (2.49) may be approximated by the difference equation

$$x_{k+1} = x_k + hf(t_k, x_k, u_k, d_k), \quad k \geq k_0, \tag{2.58}$$

where $t_k = t_0 + kh$, $x_k \approx x(t_k)$, $u_k = u(t_k, x_k)$, $d_k = d(t_k, x_k)$. Such discretization of a continuous-time system is known as *Euler method*. Application of this method to generalized homogeneous systems is studied in Chap. 8.

Concluding Remarks

Ordinary differential equations (ODEs) are popular models of control systems. Elements of the theory of continuous and discontinuous ODEs are considered in this chapter. The existence and uniqueness of solutions and their dependence on initial conditions and parameters are studied. Classical, strong (Carathéodory) and Filippov solutions of continuous and discontinuous ODEs exist under rather mild assumptions on the right-hand sides of ODEs.

We complete this chapter with the following simple remark.

Remark 2.7 Let $x(t, t_0, x_0)$ with $t \geq t_0$ denote a classical or strong forward (backward) solution of the ODE (2.3) with the initial condition $x(t_0) = x_0$. If the solution is forward (resp., backward) unique then the identity

$$x(t, t_0, x_0) = x(t, t_1, x(t_1, t_0, x_0)), \quad t \geq t_1 \geq t_0 \tag{2.59}$$

(resp., $t \leq t_1 \leq t_0$) holds as long as the solution exists.

The identity (2.59) is known as *semigroup property of solutions*. The same semigroup property holds for differential inclusions (2.53) and ODEs with non-unique solutions, but it is understood in the set-theoretic sense: $x(t, t_0, x_0)$ denotes the set of all solutions initiated in the set x_0. Notice that the set of solutions of differential inclusions satisfying standard conditions is compact in C (as long as all solutions exist) provided that the set of initial conditions is compact in $\mathbb{R}^n$. We refer reader to [11, 23] for more details about differential inclusions and ODEs with discontinuous right-hand sides, and to [9, 12, 13, 31] for their applications in systems and control theory.

References

1. Lin CC, Segel LA (1988) Mathematics applied to deterministic problems in the natural sciences. SIAM, Philadelphia
2. Bender EA (2000) An introduction to mathematical modeling. Dover, New York

3. Coddington EA, Levinson N (1955) Theory of ordinary differential equations. McGraw-Hill, New York
4. Poznyak AS (2008) Advanced mathematical tools for automatic control engineers. Volume 1: deterministic technique. Elsevier
5. Gantmacher FR (1959) The theory of matrices. AMS Chelsea Publishing
6. Krasnoselskii MA, Lifshits JA, Sobolev AV (1989) Positive linear systems: the method of positive operators. Heldermann Verlag
7. Farina L, Rinaldi S (2000) Positive linear systems: theory and applications. John Wiley
8. Utkin V, Guldner J, Shi J (2009) Sliding mode control in electro-mechanical systems. CRC Press
9. Liberzon D (2003) Switching in systems and control. Birkhauser, Boston
10. Gelig AK, Leonov GA, Yakubovich VA (2004) Stability of stationary sets in control systems with discontinuous nonlinearities. World Scientific
11. Filippov AF (1988) Differential equations with discontinuous right-hand sides. Kluwer Academic Publishers
12. Utkin VI (1992) Sliding modes in control optimization. Springer-Verlag, Berlin
13. Shtessel Y, Edwards C, Fridman L, Levant A (2014) Sliding mode control and observation. Birkhauser
14. Neimark YI (1961) Note on A. Filippov paper. In: Proceedings of 1st IFAC congress II. Butterworths, London
15. Aizerman MA, Pyatnitskii ES (1974) Fundamentals of the theory of discontinuous systems i. Autom. Remote. Control. 35(7):33–47 (in Russian)
16. Aizerman MA, Pyatnitskii ES (1974) Fundamentals of the theory of discontinuous systems ii. Autom. Remote. Control. 35(8):38–61 (in Russian)
17. Bartolini G, Zolezzi T (1985) Variable structure systems nonlinear in the control law. IEEE Trans. Autom. Control. 30(7):681–684
18. Utkin V, Poznyak A, Orlov Y, Polyakov A (2020) Road map for sliding mode control design. Springer
19. Zolezzi T (2002) Differential inclusions and sliding mode control. In: Perruquetti W, Barbot JP (eds) Sliding mode control in engineering. Marcel Dekker, pp 29–52
20. Orlov Y (2005) Finite time stability and robust control synthesis of uncertain switched systems. SIAM J. Control. Optim. 43(4):1253–1271
21. Levant A (2005) Quasi-continuous high-order sliding-mode controllers. IEEE Trans. Autom. Control. 50(11):1812–1816
22. Gelig AK, Leonov GA, Yakubovich VA (1978) Stability of nonlinear systems with nonunique equilibrium position. Nauka (in Russian)
23. Aubin J-P, Cellina A (1984) Differential inclusions. Springer
24. Filippov AF (1962) On certain questions in the theory of optimal control. J. SIAM Control 1(1):76–84
25. Haddad WM, Chellaboina VS, Nersesov SG (2006) Impulsive and hybrid dynamical systems: stability dissipativity and control. Princeton University Press, NJ, USA
26. Tuna S, Teel A (2006) Homogeneous hybrid systems and a converse Lyapunov theorem. In: Conference on decision and control, pp 6235–6240
27. Hespanha J, Liberzon D, Teel A (2008) Lyapunov conditions for input-to-state stability of impulsive systems. Automatica 44:2735–2744
28. van Schuppen JH (2021) Control and system theory of discrete-time stochastic systems. Springer
29. Palani S (2021) Discrete time systems and signal processing. Springer
30. Butcher JC (2016) Numerical methods for ordinary differential equations. Wiley
31. Edwards C, Spurgeon S (1998) Sliding mode control: theory and applications. Taylor and Francis

Chapter 3
Stability and Robustness

The concept of *stability* introduced in the famous thesis of Aleksandr Lyapunov [1] is one of the central notions of the modern systems and control theory. Many problems of state estimation and control can be reduced to a stability analysis or to a stabilization problem. This chapter surveys some elements of the Lyapunov theory and its generalizations such as finite-time stability [2], fixed-time stability [3] and input-to-state stability [4].

3.1 Basic Definitions

In the previous chapter, we have considered a dynamical system modeled by a differential law like

$$\dot{x} \in F(t, x), \quad t > t_0 \tag{3.1}$$

with the initial condition

$$x(t_0) = x_0, \tag{3.2}$$

where $x(t) \in \mathbb{R}^n$ is a system state and F is a (possibly set-valued) vector field. The ordinary differential inclusion (3.1) becomes an ordinary differential equation (ODE) if F is single valued:

$$F(t, x) = f(t, x), \quad f : \mathbb{R} \times \mathbb{R}^n \mapsto \mathbb{R}^n.$$

The system (3.1) has solutions under some restrictions on F. We refer the reader to Chap. 2 of this monograph or to the books [5, 6] for more details about the existence/uniqueness of solutions and other well-posedness issues of ordinary differential equations and inclusions.

A. Polyakov, *Generalized Homogeneity in Systems and Control Volume I*,
Communications and Control Engineering,
https://doi.org/10.1007/978-3-031-76246-8_3

The stability of the system (3.1) is a special property of its solutions (to be discussed below). Since solutions of the system (3.1), (3.2) may be non-unique, then two types of stability is possible: *weak stability* (a property holds for *a solution*) and *strong stability* (a property holds for *all solutions*). The weak stability usually is not enough for control systems design. In this book, we deal only with the strong stability analysis. *All conditions given in definitions below are assumed to be fulfilled for all solutions* $x(t, t_0, x_0)$ *of the system* (3.1), where t_0, x_0 indicates the initial condition of the system and means that $x(t_0, t_0, x_0) = x_0$. We use the notation $x(t, x_0)$ for a solution of an autonomous (time-invariant) system with $t_0 = 0$. We also write shortly $x(t)$ if the dependence of t_0 and x_0 is not important or clear from the context.

3.1.1 Lyapunov and Asymptotic Stability

The stability introduced by Aleksandr Lyapunov [1] considers some *nominal motion* $x^*(t) = x(t, t_0, x_0)$ of the system (3.1) and studies a dependence of the system trajectories on a perturbation Δx_0 of the initial condition x_0. If small perturbations Δx_0 imply small deviations of the perturbed motions $x(t, t_0, x_0 + \Delta x_0)$ from $x^*(t)$ for all $t > t_0$ then the nominal motion is called *stable*. In other words, the conventional Lyapunov stability is a continuous dependence of solutions on the initial data for the *infinite horizon* of time. Recall that continuous dependence of solutions on initial condition for a *finite horizon* of time is one of the properties required for the well-posedness of the system (see Chap. 2).

A constant vector $\tilde{x} \in \mathbb{R}^n$ of the state space is said to be an *equilibrium* of the system (3.1) if $\mathbf{0} \in F(t, \tilde{x})$ for (almost) all $t \geq t_0$. In this case, taking into account $\frac{d}{dt}\tilde{x} = \mathbf{0}$ we conclude that $x^*(t) = \tilde{x}, \forall t \geq t_0$ is a classical (resp., strong) solution of the system, which can be treated as a nominal motion whose stability has to be analyzed. Many practical control problems can be reduced to stability analysis of an equilibrium. Below we deal only with the stability analysis of the zero equilibrium (or, equivalently, of *the origin*) of the system, since, making the change of variables $y = x - x^*$, we can always transform a problem of stability analysis of a nontrivial solution $x^*(t)$ to the same problem for the trivial (zero) solution. Below we say "the system (3.1) is stable" if we mean that its zero solution is stable.

Definition 3.1 *(Lyapunov stability)* The system (3.1) is said to be locally (globally) *Lyapunov stable* if for any $t_0 \in \mathbb{R}$ there exists $\varepsilon_{t_0} \in \mathcal{K}_\infty$:

$$|x(t, t_0, x_0)| \leq \varepsilon_{t_0}(|x_0|), \qquad \forall t > t_0$$

for any solution $x(t, t_0, x_0)$ of (3.1) and any $x_0 \in U(t_0)$, where $U(t_0)$ is an open connected neighborhood of $\mathbf{0}$ (resp., $U(t_0) = \mathbb{R}^n$, $\forall t_0 \in \mathbb{R}$).

If the function ε_{t_0} and the set $U(t_0)$ do not depend on t_0 then the system is locally/globally *uniformly Lyapunov stable*.

Example 3.1 The system

$$\dot{x}(t) = (1 - 2t)x(t), \quad x(t_0) = x_0$$

has the unique solution

$$x(t, t_0, x_0) = e^{\int_{t_0}^{t} 1-2s\, ds} x_0 = e^{t-t_0-t^2+t_0^2} x_0.$$

The zero solution of the system is globally Lyapunov stable. Indeed, for

$$\varepsilon_{t_0}(|x_0|) = e^{t_0^2-t_0+1} |x_0|$$

we have

$$|x(t, t_0, x_0)| = e^{-1+t-t^2} e^{1-t_0+t_0^2} |x_0| = e^{-(1-0.5t)^2-0.75t^2} \varepsilon_{t_0}(|x_0|) \leq \varepsilon_{t_0}(|x_0|)$$

for all $t \geq t_0$.

Obviously, if $F(t, x)$ is independent of t and the system (3.1) is Lyapunov stable, then it is uniformly Lyapunov stable. The system is *unstable* if it does not satisfy Definition 3.1.

Proposition 3.1 (On uniqueness of the zero solution of stable system) *If the system* (3.1) *is Lyapunov stable then* $x(t, t_0, \mathbf{0}) \equiv \mathbf{0}$ *is the unique solution of the system* (3.1) *with the initial condition* $x(t_0) = \mathbf{0}$.

Proof Suppose that there exists a solution $x(t, t_0, \mathbf{0})$ and a time instant $t' > t_0$ such that $x(t', t_0, \mathbf{0}) \neq \mathbf{0}$. Denote $\varepsilon := |x(t', t_0, \mathbf{0})|/2 > 0$. By definition of Lyapunov stability, for any $\varepsilon > 0$ there exits $\delta > 0$ such that $|x_0| < \delta$ always implies $|x(t, t_0, x_0)| < \varepsilon$ for all $t > t_0$. Taking into account that, in our case, $|x_0| = |\,\mathbf{0}\,| = 0 < \delta$ we obtain the contradiction. The proof is complete. □

The Lyapunov stability is a desirable property for control systems design, since it guarantees certain "robustness" of the system behavior with respect to perturbation of initial condition. However, usually the control problem consists in a regulation of a system to a set-point, at least, asymptotically (as $t \to +\infty$). The following definition introduces the related stability property of the dynamical system.

Definition 3.2 *(Asymptotic convergence)* We say that the system (3.1) locally (globally) asymptotically converges to the origin if for any $t_0 \in \mathbb{R}$ there exists an open connected neighborhood $U(t_0)$ of the origin $x = \mathbf{0}$ (resp., $U(t_0) = \mathbb{R}^n$) such that for any $x_0 \in U(t_0)$

- each solution $x(t, t_0, x_0)$ exists for all $t > t_0$;
- $\lim_{t \to +\infty} |x(t, t_0, x_0)| = 0$.

The open connected neighborhood $U(t_0)$ of the origin introduced in the above definition is called an *attraction domain* (a *domain* of attraction). In the general case, the attraction domain $U(t_0)$ may not be forward invariant.[1] However, if $U(t_0)$ is an attraction domain then any open connected neighborhood of the origin belonging to $U(t_0)$ is an attraction domain as well. An estimate of a maximal attraction domain is important for many practical control applications. Analytical methods for this estimation can be based on Lyapunov functions considered in Chap. 4.

Definition 3.3 *(Asymptotic stability)* The locally (globally) Lyapunov stable system (3.1) is said to be locally (globally) *asymptotically stable* if it locally (globally) asymptotically converges to the origin.

In the general case, the Lyapunov stability cannot be omitted in Definition 3.3, since even a *global asymptotic convergence to the origin may not imply the Lyapunov stability of the system.*

Example 3.2 *(see [7], page 433 or [8], page 191)* The ODE system

$$\dot{x}_1 = \frac{x_1^2(x_2-x_1)+x_2^5}{(x_1^2+x_2^2)\left(1+(x_1^2+x_2^2)^2\right)} \quad \text{and} \quad \dot{x}_2 = \frac{x_2^2(x_2-2x_1)}{(x_1^2+x_2^2)\left(1+(x_1^2+x_2^2)^2\right)}$$

globally asymptotically converges to the origin. However, it is not Lyapunov stable, since this system has the so-called homoclinic trajectories, which start in an arbitrary small neighborhood of the origin and always leave the ball $B(\varepsilon_0)$ of a fixed radius $\varepsilon_0 > 0$, i.e., the function ε_{t_0} from Definition 3.1 does not belong to the class $\mathcal{K}$.

In Chap. 7, we show that, for homogeneous systems with positive degrees, the asymptotic convergence is equivalent to the asymptotic stability. This means that convergent homogeneous systems cannot have homoclinic trajectories mentioned above.

Uniform asymptotic stability requests a more strong convergence property.

Definition 3.4 *(Uniform asymptotic convergence)* We say that the system (3.1) locally (globally) uniformly asymptotically converges to the origin if there exist $U \subset \mathbb{R}^n$ (resp., $U = \mathbb{R}^n$) and a *locally bounded function*[a] $T : U \times (0, +\infty) \mapsto [0, +\infty)$ such that

$$|x(t, t_0, x_0)| \leq \varepsilon, \quad \forall t > t_0 + T(x_0, \varepsilon), \quad \forall \varepsilon > 0, \quad \forall x_0 \in U.$$

[a] A function is locally bounded on a domain with nonempty interior if it is bounded on any compact from the interior of the domain.

[1] A set U is said to be forward invariant for a dynamical system if for any its trajectory one holds $x(t_0) \in U \Rightarrow x(t) \in U, \forall t > t_0$.

> **Definition 3.5** *(Uniform asymptotic stability)* The locally (globally) uniformly Lyapunov stable system (3.1) is said to be locally (globally) *uniformly asymptotically stable* if it locally (globally) uniformly converges to the origin.

The uniform asymptotic stability always implies asymptotic stability. The converse proposition holds for time-invariant ordinary differential equations and inclusions in $\mathbb{R}^n$ (see, [9]).

Example 3.3 Let us consider the ordinary differential equation for the dynamics of a pendulum:

$$\ddot{\theta}(t) + \frac{g}{r}\sin(\theta(t)) = 0, \tag{3.3}$$

where θ is an inclination angle, r is a length of pendulum, g is the gravitation constant. The model of the system can be rewritten as follows

$$\dot{x} = \begin{pmatrix} x_2 \\ -\frac{g}{r}\sin(x_1) \end{pmatrix}, \quad t > 0, \quad x = (x_1, x_2)^\top,$$

where $x_1(t) = \theta(t)$, $x_2(t) = \dot{\theta}(t)$. The pendulum system has the equilibriums

$$(\pi i, 0)^\top, \quad i \in \mathbb{Z}.$$

The upper positions of the pendulum are unstable equilibriums $(\pi + 2\pi i, 0)$. The lower positions are Lyapunov stable equilibriums $(2\pi i, 0)$. Adding the friction term $k\dot{\theta}$ with $k > 0$ into the left-hand side of the pendulum model (3.3), the Lyapunov stable equilibriums become asymptotically stable.

3.1.2 Input-to-State Stability

Lyapunov stability can be interpreted as a continuous dependence of solutions of ODE on the initial condition for the *infinite horizon* of time. For control systems, it is also important to analyze the continuous dependence of solutions on time-varying parameters on the right-hand side. These time-varying parameters may model, for example, some perturbations of the system. Recall that, for a finite horizon of time, the continuity of solutions with respect to parameters on the right-hand side is usually guaranteed by well-posedness (see Chap. 2). For the case of the infinite horizon of time, the corresponding property was introduced by Eduardo Sontag in [4]. Today it is known as *Input-to-State Stability* (ISS). An ISS analysis is developed for various classes of dynamical systems [10–14]. In the first volume of this monograph, we deal mainly with ISS analysis of ODE models.

Let the vector field F in (3.1) have the special structure:

$$\dot{x}(t) \in F(x(t), q(t)), \quad t > t_0, \quad x(t_0) = x_0, \tag{3.4}$$

where $x(t)$ is as before, and $q \in L^\infty(\mathbb{R}; \mathbb{R}^k)$ is a uniformly essentially bounded measurable function of time. The (possibly) time-varying parameter q is called *input* but, in practice, it can be interpreted as an exogenous perturbation, a measurement noise, or parametric uncertainties of the system. Let $x_q(t, t_0, x_0)$ denote a solution (3.4) with $q \in L^\infty(\mathbb{R}; \mathbb{R}^k)$.

Definition 3.6 *(Input-to-state stability (ISS))* The system (3.4) is said to be *input-to-state stable* if there exist functions $\xi \in \mathcal{KL}$ and $\theta \in \mathcal{K}$:

$$|x_q(t, t_0, x_0)| \leq \xi(|x_0|, t - t_0) + \theta(|q|_{L^\infty(t_0,t)}), \quad \forall t \geq 0 \tag{3.5}$$

for any solution $x_q(t, t_0, x_0)$ of (3.4), any $q \in L^\infty(\mathbb{R}; \mathbb{R}^k)$, any $t_0 \in \mathbb{R}$ and any initial state $x_0 \in \mathbb{R}^n$.

Obviously, the ISS implies global uniform asymptotic stability of the unperturbed system ($q = \mathbf{0}$). In Chap. 7 we show that the converse proposition holds for homogeneous systems.

Example 3.4 The system

$$\dot{x}(t) = Ax(t) + q(t), \quad t > t_0, \quad A \in \mathbb{R}^{n\times n}$$

is ISS if and only if A is the Hurwitz matrix. Indeed, using the representation (2.13) we derive

$$|x_q(t, t_0, x_0)| \leq |e^{A(t-t_0}| \cdot |x(t_0)| + \int_{t_0}^{t} \left|e^{A(t-s)}\right| \cdot |q(s)|ds.$$

If the matrix A is Hurwitz[2] then Proposition 2.1 implies that $\exists M \geq 1, \lambda > 0$ such that $\left|e^{A\tau}\right| \leq Me^{-\lambda\tau}$ for all $\tau \geq 0$. This gives (3.5) with $\xi(\sigma, t) = M\sigma e^{-\lambda t}$ and $\theta(\sigma) = \lambda^{-1}M\sigma$ for $\sigma \geq 0$. If A is not Hurwitz then a required function $\xi \in \mathcal{KL}$ does not exist even if $q = \mathbf{0}$.

3.2 Convergence Rates

In control theory, a settling (or transient) time is a time required for reaching of a steady state by a control system. For the mathematical model of the system, this may correspond to a time of convergence of solution to (a small enough) neighborhood of a stable equilibrium. The settling time is one of the performance indexes of a control system, which defines the so-called time response. The classical concept of asymptotic stability does not characterize this time. The "rated" stability studied in this section proposes some direct or indirect mathematical characterizations of the time response of the system.

[2] A matrix is Hurwitz if its eigenvalues have negative real parts.

3.2.1 Exponential Stability

An exponential stability [1] is the classical example of the "rated" stability.

Definition 3.7 *(Exponential stability)* A system (3.1) is said to be locally (globally) exponentially stable if it is locally (globally) asymptotically stable with an attraction domain $U(t_0)$ and for any $t_0 \in \mathbb{R}$ there exist $C_{t_0} \geq 1$ and $r_{t_0} > 0$ such that

$$|x(t, t_0, x_0)| \leq C_{t_0}|x_0|e^{-r_{t_0}(t-t_0)}, \quad \forall t > t_0, \quad \forall x_0 \in U(t_0). \qquad (3.6)$$

The exponential stability is *uniform* if U_{t_0}, C_{t_0}, r_{t_0} are independent of t_0. The parameter r_{t_0} defines the so-called *exponential decay rate*. The exponential stability is commonly exploited by linear control theory. If the parameters C_{t_0} and r_{t_0} are known then a transient (convergence) time $\tilde{T}_\varepsilon(x_0)$ of any trajectory to a ball of the radius $\varepsilon > 0$ can be estimated as follows

$$\tilde{T}_\varepsilon(x_0) \leq t_0 + \frac{1}{r_{t_0}} \ln \frac{\max\{\varepsilon, C_{t_0}|x_0|\}}{\varepsilon}.$$

Example 3.5 The linear ODE

$$\dot{x} = Ax, \quad t > 0$$

with a Hurwitz matrix[3] $A \in \mathbb{R}^{n\times n}$ is the simplest example of exponentially stable system. Indeed, in the view of Proposition 2.1 for any $\varepsilon \in (0, \sigma)$ there exists $C_\varepsilon \geq 1$ such that

$$|x(t, x_0)| = |e^{At}x_0| \leq C_\varepsilon e^{(-\sigma+\varepsilon)t}|x_0|,$$

where the number $\sigma = -\max_i \Re(\lambda_i)$ corresponds to the real part of an eigenvalue λ_i of the matrix A, which is closest to the imaginary axes.

3.2.2 Finite-Time Stability

In control theory, a regulation of a state (or an output) of a control system to a desired set-point in a finite time is desirable in some applications. The corresponding property is known as finite-time reaching.

Definition 3.8 *(Finite-time reaching)* We say that the system (3.1) locally (globally) reaches the origin in a finite time if for any $t_0 \in \mathbb{R}$ there exists an open connected neighborhood $U(t_0)$ of the origin (resp., $U(t_0) = \mathbb{R}^n$) such

[3] The matrix is Hurwitz if all its eigenvalues have negative real parts.

that, for any $x_0 \in U(t_0)$ and for any solution $x(t, t_0, x_0)$ of the system, there exists a finite instant of time $t_{x_0} \in [t_0, +\infty)$ such that $x(t_{x_0}, t_0, x_0) = \mathbf{0}$.

The first instant of time $t_{x_0} \geq t_0$ satisfying the above definition is known as a *reaching time*. In the general case, the finite-time reaching of the origin does not imply asymptotic convergence to the origin. However, if a trajectory $x(t, t_0, x_0)$ of a Lyapunov stable system reaches the origin in a finite time $t_{x_0} \geq 0$ then, by Proposition 3.1, we have $x(t, t_0, x_0) = \mathbf{0}$ for all $t \geq t_{x_0}$ and all $x_0 \in U(t_0)$. In Chap. 7, it is shown that the finite-time reaching of the origin by homogeneous system with negative degree implies the Lyapunov stability.

Definition 3.9 *(Finite-time convergence)* We say that the system (3.1) locally (globally) converges to the origin in a finite time if for any $t_0 \in \mathbb{R}$ there exists an open connected neighborhood $U(t_0)$ of the origin such that for any $x_0 \in U(t_0)$

- each solution $x(t, t_0, x_0)$ exists for all $t > t_0$;
- $\exists T : \mathbb{R} \times \mathbb{R}^n \mapsto \mathbb{R}_+$ such that the implication

$$\forall x_0 \in U(t_0) \quad \Rightarrow \quad |x(t, t_0, x_0)| = 0, \;\; \forall t \geq t_0 + T(t_0, x_0)$$

holds for any solution $x(t, t_0, x_0)$ of the system.

Finite-time reaching does not imply to finite-time convergence in the general case, but the finite-time convergence always implies asymptotic convergence.

If a function T satisfies the above definition then any function $T + T^+$ with $T^+ \geq 0$ also does. So, to characterize a time response (convergence time) of the system, it is reasonable to consider a "minimal" function T satisfying the above definition. Such a function is known as a *settling-time function* [15].

Definition 3.10 *(Settling-time function)* A function $(t_0, x_0) \mapsto T(t_0, x_0)$ satisfying Definition 3.11 with $x_0 \in U(t_0)$ is said to be a *settling-time function* of the system (3.1) if, for any nonzero function $(t_0, x_0) \mapsto \delta(t_0, x_0)$, the function $(t_0, x_0) \mapsto T^{new}(t_0, x_0) := T(t_0, x_0) - |\delta(t_0, x_0)|$ does not satisfy Definition 3.11 with the same attraction domain $U(t_0)$.

If solutions of the system (3.1) are unique and the system converges to the origin in a finite time then the settling time corresponds to the first instant of time when a trajectory reaches the origin. For the case of non-unique solutions, a maximum over settling times of particular solutions having the same initial condition has to be taken in order to compute a value of the settling-time function for the given initial state.

Asymptotically stable systems, which may reach the stable equilibrium in a finite time, are considered in Sects. 1.1.3 and 2.1.2. In the literature, such systems are known as *finite-time stable* (FTS) [2], [15–17].

Definition 3.11 *(Finite-time stability)* A locally (globally) Lyapunov stable system (3.1) is said to be locally (globally) finite-time stable if it locally (globally) converges to the origin in a finite time.

Example 3.6 Let us consider the so-called sliding-mode system

$$\dot{x} = -\frac{2}{\sqrt{\pi}}\,\mathrm{sign}(x) + 2tx, \quad t > t_0, \quad x_{(t)} \in \mathbb{R},$$

which, according to Filippov's method, is regularized by the differential inclusion

$$\dot{x} \in -\frac{2}{\sqrt{\pi}} \cdot \overline{\mathrm{sign}}(x) \dot{+} \{2tx\}, \quad t > t_0, \quad x_{(t)} \in \mathbb{R}, \tag{3.7}$$

where $t_0 \in \mathbb{R}$. Since

$$|x(t)| = e^{t^2 - t_0^2}|x(0)| - \frac{2}{\sqrt{\pi}} \int_{t_0}^{t} e^{t^2 - s^2} ds$$

as long as $|x(t)| \neq 0$ then the system locally converges to zero in a finite time with the attraction domain $U_{t_0} = B\left(e^{t_0^2}(1 - \mathrm{erf}(|t_0|))\right)$, where

$$\mathrm{erf}(z) = \frac{2}{\sqrt{\pi}} \int_0^z e^{-\tau^2} d\tau, \quad z \in \mathbb{R}$$

is the so-called *Gauss error function*: $\mathrm{erf}(s) \to 1$ as $s \to +\infty$. Moreover, the origin of the considered system is Lyapunov stable: $\forall \varepsilon > 0$ and $\forall t_0 \in \mathbb{R}$ we can select

$$\delta = \delta(t_0) = \min\left\{\varepsilon, e^{t_0^2}(1 - \mathrm{erf}(t_0))\right\}.$$

Therefore, the system is finite-time stable. In particular, for $t_0 > 0$ the settling-time function has the form

$$T(t_0, x_0) = \mathrm{erf}^{-1}\left(|x_0|e^{-t_0^2} + \mathrm{erf}(t_0)\right) - t_0,$$

where $\mathrm{erf}^{-1}(\cdot)$ denotes the inverse function to $\mathrm{erf}(\cdot)$. The results of the numerical simulation of the considered system with $t_0 = 0$ and $x_0 = 0.4$ are depicted in Fig. 3.1. The plot is presented in the logarithmic scale for $|x|$ in order to highlight the typical behavior of the finite-time stable system. Indeed, in the logarithmic scale, the trajectory of the finite-time stable system has the vertical asymptote at the settling instant of time. This asymptote can always be observed on numerical simulation provided that the finite-time stable system is correctly discretized (see Chap. 8). This can be considered as a numerical test of finite-time stability.

The settling-time function T of a time-invariant finite-time stable system is independent of t_0. However, T may be discontinuous even if the right-hand side of an autonomous finite-time stable ODE is continuous and its solutions are unique. The settling-time function may be unbounded near the origin.

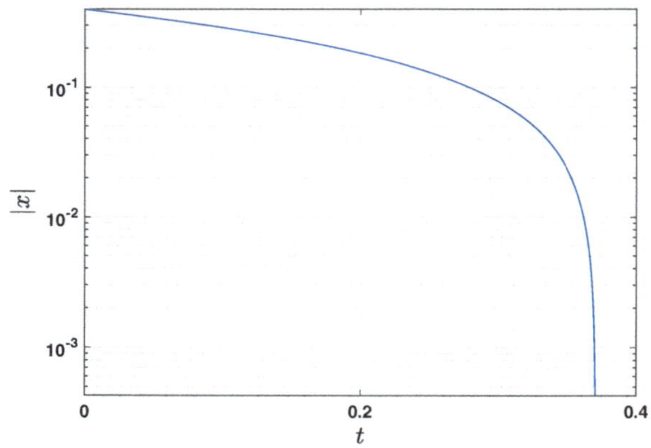

Fig. 3.1 Evolution of a finite-time stable system (in the logarithmic scale)

Example 3.7 *([15], page 756)* Let a vector field $f : \mathbb{R}^2 \mapsto \mathbb{R}^2$ of a time-invariant system (see Fig. 3.2) have the following representation in the polar coordinates $x = (x_1, x_2)^\top = (r\cos(\theta), r\sin(\theta))^\top$, $r > 0$, $\theta \in [0, 2\pi)$:

$$\begin{cases} \dot{\theta} = -\sqrt{\theta} \\ \dot{r} = -\sqrt{r}\cos\theta \end{cases} \quad \text{if } x_1 \geq 0,\ x_2 \geq 0,$$
$$\begin{cases} \dot{\theta} = r\cos\theta - \sqrt{\frac{\pi}{2}}\sin\theta \\ \dot{r} = 0 \end{cases} \quad \text{if } x_1 < 0,\ x_2 \geq 0,$$
$$\begin{cases} \dot{\theta} = -r \\ \dot{r} = 0 \end{cases} \quad \text{if } x_1 < 0,\ x_2 < 0,$$
$$\begin{cases} \dot{x}_1 = -\sqrt{x_1} - x_2^2 \\ \dot{x}_2 = 0 \end{cases} \quad \text{if } x_1 \geq 0,\ x_2 < 0.$$

The vector field f is continuous and locally Lipschitz on $\mathbb{R}^2 \backslash \{x_1 = 0, x_2 \geq 0\}$, $f(0) = 0$. This system has unique solutions in forward time for all initial conditions. Moreover, the system is globally *uniformly asymptotically stable and globally FTS*. Any solution reaches the forward invariant quadrant $\{x_1 \geq 0, x_2 \geq 0\}$ (see, Fig. 3.2) and, next, converges to the origin in a finite time due to the equation $\dot{r} = -\sqrt{r}\cos(\theta)$. However, for the sequence of the initial conditions $x_0^i = (0, -1/i)^\top$, $i = 1, 2, ...$ we have $x_0^i \to 0$ and $T(x_0^i) \to +\infty$ as $i \to +\infty$. This follows from the equations $\dot{\theta} = -r$, $\dot{r} = 0$, which imply that the closer x_0^i to the origin more time the system will spent crossing the quadrant $\{x_1 < 0, x_2 < 0\}$. So, considering an open ball $B(r)$ of the radius r with the center at the origin, we have that

$$\sup_{x_0 \in B(r)} T(x_0) = +\infty, \quad \forall r > 0,$$

i.e., all trajectories of the considered system converge to zero in a finite time, but non-uniformly with respect to initial conditions.

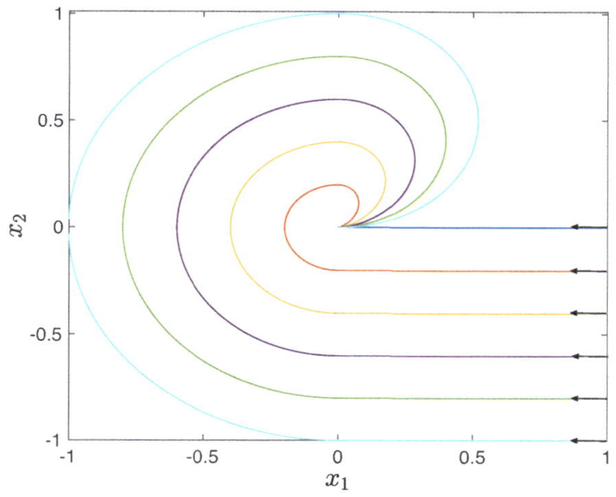

Fig. 3.2 The example of the finite-time stable system with a settling-time function unbounded close to the origin

The above example also shows that, contrary to asymptotic and Lyapunov stability, finite-time stability of time-invariant system, in general, does not imply a "uniformity" of the finite-time stability with respect to initial states. Various versions of uniform finite-time stability [2, 15, 16] are introduced by the following definition.

Definition 3.12 *(Uniform FTS, 0-FTS and continuous FTS)* The system (3.1) is said to be locally (globally)

- uniformly finite-time stable if it is FTS with a time-invariant attraction domain U and the settling-time function $T : \mathbb{R} \times U \mapsto \mathbb{R}$ is *locally bounded uniformly on the first argument*:

$$\forall y \in U, \quad \exists \varepsilon > 0 \quad \text{such that} \quad \sup_{t_0 \in \mathbb{R},\ |x_0 - y| < \varepsilon} T(t_0, x_0) < +\infty;$$

- 0-FTS if it uniformly FTS and the settling-time function T is continuous at $\mathbf{0}$;
- continuously FTS if it is 0-FTS and the settling-time function T is continuous on U.

Example 3.8 The uniformly FTS system can be obtained from Example 3.7 by means of the following modification

$$
\begin{cases}
\dot{\theta} = -\sqrt{\theta} \\
\dot{r} = -\sqrt{r}\cos\theta
\end{cases} \quad \text{if } x_1 \geq 0,\ x_2 \geq 0,
$$
$$
\begin{cases}
\dot{\theta} = \cos\theta - \sqrt{\frac{\pi}{2}}\sin(\theta) \\
\dot{r} = 0
\end{cases} \quad \text{if } x_1 < 0,\ x_2 \geq 0,
$$
$$
\begin{cases}
\dot{\theta} = -1 \\
\dot{r} = 0
\end{cases} \quad \text{if } x_1 < 0,\ x_2 < 0,
$$
$$
\begin{cases}
\dot{x}_1 = -\sqrt{x_1} + x_2 \\
\dot{x}_2 = 0
\end{cases} \quad \text{if } x_1 \geq 0,\ x_2 < 0.
$$

The right-hand side of the system remains continuous on $\mathbb{R}^2$ after such a modification, but the settling-time function becomes locally bounded. Indeed, the phase portrait of the modified system repeats Fig. 3.2. However, independently of the initial condition in the quadrant $\{x_1 < 0, x_2 < 0\}$ the system trajectory will leave it in a time uniformly bounded by $\frac{\pi}{2}$ (due to $\dot{\theta} = -1$). The presented modification of Example 3.7 eliminates the unboundedness of the settling-time function T but remains discontinuous at the origin. Indeed, on the one hand, for the sequence of the initial conditions $x_0^i = (0, 1/i)^\top, i = 1, 2, \ldots$ we have $T(x_0^i) \to 0$ as $i \to +\infty$. On the other hand, it is easy to show that $\pi/2 \leq T(-x_0^i) \leq \pi + T(x_0^i)$. Hence, $\limsup_{i\to+\infty} T(-x_0^i) = T^* \in [\pi/2, \pi]$, i.e., T is discontinuous (but bounded) at zero.

Therefore, in the general case, the uniformity of the finite-time convergence with respect to initial conditions does not imply a continuity of the settling-time function T at zero. In practice, a finite-time stable system with a discontinuous settling-time function can be designed for a very specific control problem, when the settling time has to be prescribed by some constant for all nonzero initial conditions, i.e., $T(x_0) = \text{const}$ for all $x_0 \neq \mathbf{0}$ (see Chap. 9). In all other cases, we usually expect that the settling time should tend to zero as the initial state tends to the origin. In this monograph, the corresponding property is denoted by 0-FTS. If $T \in C(U; \mathbb{R})$ then the FTS system is called continuously FTS. The continuity at the origin is necessary (but not sufficient) for the settling-time function to be continuous on U. The sufficient condition of the continuity of T on $\mathbb{R}^n$ was first discovered in [15]. It requires the well-posedness of the system (3.1).

Proposition 3.2 *Let the system* (3.1) *be autonomous (i.e., F is independent of t) and 0-FTS with an attraction domain U being a positively invariant set*[4] *of the system. If the system* (3.1) *has unique solutions in the forward time, which depend continuously on the initial state in U, then the settling-time function is continuous on U.*

Proof Since solutions are assumed to be unique then for any solution $x(t, z)$ of the autonomous system (3.1) with the initial condition $x(0) = z \in U$ we have

$$x(a+b, x(0)) = x(b, x(a)), \forall a, b \geq 0.$$

Hence, $0 = x(T(z), z) = x(t + T(z) - t, z) = x(T(z) - t, x(t, z)), \forall t \in [0, T(z)]$ and

$$T(x(t, z)) = \max\{T(z) - t, 0\}, \quad \forall t \geq 0. \tag{3.8}$$

Let us consider a sequence z_i such that $z_i \to z$ as $i \to \infty$. Since the system is assumed to be 0-FTS (consequently, uniformly FTS) then the sequence $T(z_i)$ is uniformly bounded.

Let $\tau^+ = \limsup_{i\to+\infty} T(z_i)$ and $\tau^- = \liminf_{i\to+\infty} T(z_i)$ and z_i^+, z_i^- be subsequences of z_i such that $T(z_i^+) \to \tau^+$ as $i \to +\infty$ and $T(z_i^-) \to \tau^-$ as $i \to +\infty$. Since it is assumed that the solutions of the system depend continuously on initial conditions then $x(t, z_i^+) \to x(t, z)$ as $i \to +\infty$ uniformly on the time interval $[0, T(z)]$, and, in particular, $x_i := x(T(z), z_i^+) \to x(T(z), z) = \mathbf{0}$ as $i \to +\infty$. Since T is assumed to be continuous at the origin then $T(x_i) \to T(\mathbf{0}) = 0$ as $i \to +\infty$. Using (3.8) with $z = z_i^+$ and $t = T(z)$ we derive

$$T(x_i) = \max\{T(z_i^+) - T(z), 0\} \to 0 \quad \text{as} \quad i \to +\infty.$$

This means that $\tau^+ \leq T(z)$.

On the other hand, a continuous dependence of solutions on time argument and on the initial data implies that $x(T(z_i^-), z_i^-) \to x(\tau^-, z)$ as $i \to +\infty$. Since, by definition of the settling-time function, we have $x(T(z_i^-), z_i^-) = \mathbf{0}$ then $x(\tau^-, z) = \mathbf{0}$ and $T(z) \leq \tau^-$. Taking into account that the inequality $T(z) \leq \tau^- \leq \tau^+ \leq T(z)$ may hold only as identity, we complete the proof. □

In Chap. 7 we show that the settling-time function of any finite-time stable homogeneous system is continuous at zero. However, in the general case, the homogeneity does not guarantee the uniqueness of solutions even if the right-hand side of the system is continuous. As shown in the example below, the settling-time function of a finite-time stable homogeneous system may be discontinuous away from the origin if solutions are not unique.

Example 3.9 The system

$$\begin{aligned} \dot{x}_1 &= |x_1 x_2|^{\frac{1}{4}}(2\,\mathrm{sign}(x_2) + 3\,\mathrm{sign}(x_1 x_2)), \\ \dot{x}_2 &= -|x_2|^{\frac{1}{2}}\,\mathrm{sign}(x_2) - |x_1|^{\frac{1}{2}}(3 + 2\,\mathrm{sign}(x_1)) \end{aligned} \tag{3.9}$$

has a continuous right-hand side. Since

$$\begin{aligned} \frac{d}{dt}\left(\frac{4}{5}|x_1|^{\frac{5}{4}} + \frac{4}{5}x_2^{\frac{5}{4}}\right) &= |x_1|^{\frac{1}{4}} \cdot |x_1 x_2|^{\frac{1}{4}}(2\,\mathrm{sign}(x_1 x_2) + 3\,\mathrm{sign}(x_2)) - |x_2|^{\frac{3}{4}} \\ &\quad - |x_2|^{\frac{1}{4}}|x_1|^{\frac{1}{2}}(3\,\mathrm{sign}(x_2) + 2\,\mathrm{sign}(x_1 x_2)) \\ &= -\,|x_2|^{\frac{3}{4}} \end{aligned}$$

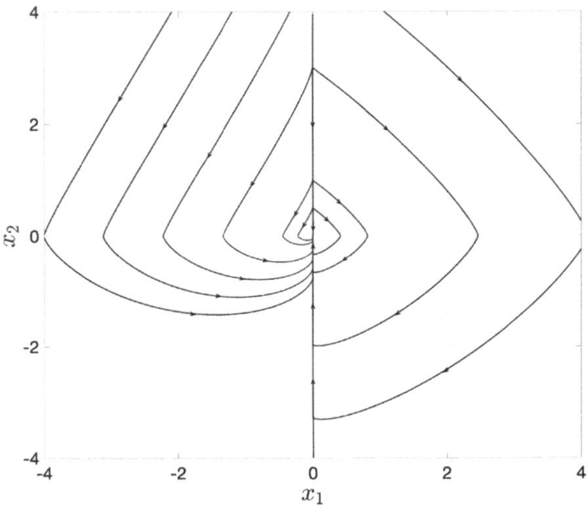

Fig. 3.3 Phase portrait of the system (3.9)

then using the LaSalle invariance principle (see, [18] or Chap. 4) we conclude that this system is asymptotically stable. The finite-time stability of the system follows from the negative homogeneity degree of its right-hand side (see, Chaps. 1 and 7 for more details). However, a trajectory of the system initiated on the ray $x_1 = 0, x_2 > 0$ is not unique. It may slide on the line $x_1 = 0$ (while $x_2 > 0$) or enter to either the positive ($x_1 > 0$) or the negative half-plane ($x_1 < 0$). The ray $x_1 = 0, x_2 < 0$ is the forward invariant set of the system since

$$\begin{aligned} \dot{x}_1 &= 5|x_1 x_2|^{\frac{1}{4}} \operatorname{sign}(x_2), \\ \dot{x}_2 &= -|x_2|^{\frac{1}{2}} \operatorname{sign}(x_2) - 5|x_1|^{\frac{1}{2}} \end{aligned} \quad \text{for} \quad x_1 > 0,$$

$$\begin{aligned} \dot{x}_1 &= -|x_1 x_2|^{\frac{1}{4}} \operatorname{sign}(x_2), \\ \dot{x}_2 &= -|x_2|^{\frac{1}{2}} \operatorname{sign}(x_2) - |x_1|^{\frac{1}{2}} \end{aligned} \quad \text{for} \quad x_1 < 0.$$

The discontinuity of the settling-time function follows from a difference in convergence times for solutions initiated in left and right half-planes (see Fig. 3.3) close to the line $x_1 = 0, x_2 > 0$.

3.2.3 Fixed-Time Stability

Let us consider the ODE model of dynamics

$$\dot{v}(t) = -(k_{dry} + k_{drag} v^2(t)) \operatorname{sign}(v(t)), \quad t > 0, \quad v(t) \in \mathbb{R}, \quad k_{dry}, k_{drag} > 0$$

of a mechanical system studied in Sect. 1.1.3. The system has Filippov solutions (see Sect. 2.1.2) and its origin is uniformly finite-time stable even if the drag force is assumed to be equal zero ($k_{drag} = 0$). The quadratic term inspired by the fluid resistance provides a fast decay "at ∞" characterized below.

Definition 3.13 *(Nearly fixed-time convergence)* We say that a system (3.1) locally (globally) converges to the origin nearly in a fixed time if for any $t_0 \in \mathbb{R}$ there exists an attraction domain $U(t_0) \subset \mathbb{R}^n$ being a neighborhood of the origin (resp., $U(t_0) = \mathbb{R}^n$) such that

$$\forall r > 0, \ \exists T_{t_0,r} > 0 \ : \ |x(t, t_0, x_0)| \leq r, \quad \forall t \geq t_0 + T_{t_0,r}, \forall x_0 \in U(t_0). \quad (3.10)$$

In other words, the system converges to the origin nearly in a fixed time if for any neighborhood of the origin there exists a fixed time T such that all trajectories of the system converge into this neighborhood no later than the time instant $t_0 + T$ *independently of the initial condition* $x(t_0) = x_0 \in U(t_0)$. Such a property can be observed, for example, in systems with polynomial nonlinearities and homogeneous systems [19].

Definition 3.14 *(Nearly fixed-time stability)* A locally (globally) Lyapunov stable system (3.1) is said to be locally (globally) nearly fixed-time stable (nFxTS) if it converges to the origin nearly in a fixed time.

In the first edition of this monograph, the above stability property was called "practical fixed-time stability". Such a name is not appropriate for the characterization of the corresponding behavior of the system, since the practical fixed-time stability [3] corresponds to the case $\exists r > 0$ (instead of $\forall r > 0$) in formula (3.10). The name "nearly fixed-time stability" has appeared for the first time in the survey [20].

Example 3.10 The system

$$\dot{x} = -x^3, \quad t > t_0 \quad (3.11)$$

is nearly fixed-time stable (see, Fig. 3.4). Its solution for $x(t_0) = x_0 \in \mathbb{R}$ is given by

$$x(t, t_0, x_0) = \frac{x_0}{\sqrt{2(t-t_0)x_0^2+1}}, \quad t \geq t_0,$$

so $|x(t, t_0, x_0)| < r$ for $t \geq t_0 + \frac{1}{2r^2}$ independently of x_0.

All other considerations in this chapter are valid only for the system (3.1) *in* $\mathbb{R}^n$ *satisfying the conditions of Theorem* 2.8 *(or Remark* 2.6*), which guarantees existence of solutions in the forward time and in the backward time, simultaneously.* In this case, the property of nFxTS is "adjoint" to FTS in the sense that a globally nearly fixed-time stable system has a finite-time escape [21] to the boundary of U (or to infinity if $U = \mathbb{R}^n$) in the backward time (see, Fig. 3.4). Indeed, otherwise a fixed-time reaching of a neighborhood of zero in the forward time is impossible.

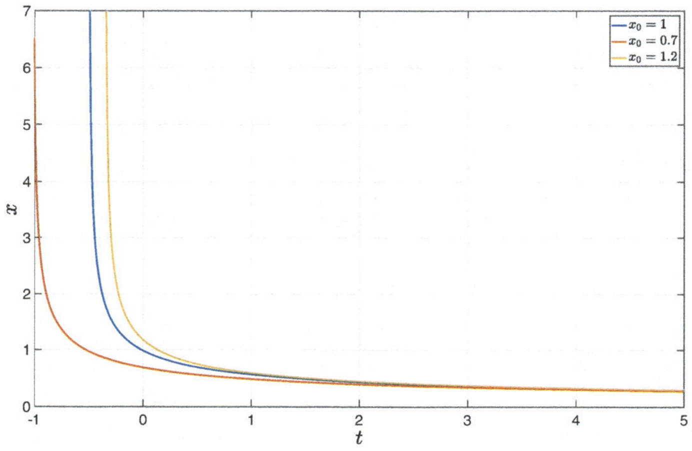

Fig. 3.4 A trajectories of the system (3.11)

Recall that a boundary of a set U is a set ∂U such that any neighborhood of a point $x \in \partial U$ contains some points belonging to $U \backslash \{x\}$ and some points belonging to $\mathbb{R}^n \backslash U$. Let, as before (see Chap. 2),

$$\rho_0(x, U) = \inf_{y \in U} |x - y|$$

denote the distance between the point x and the set U.

Proposition 3.3 *The system* (3.1) *converges to the origin nearly in a fixed time if and only if there exists a function*

$$\tau : \mathbb{R} \times \mathbb{R}^n \backslash \{\mathbf{0}\} \mapsto [0, +\infty)$$

such that

1. *for any* $t_0 \in \mathbb{R}$*, for any* $x_0 \in U(t_0)$ *and for any solution* $x(t, t_0, x_0)$ *of the system* (3.1) *in the backward time there exists* $\tilde{\tau} \in (0, \tau(t_0, x_0)]$*:*

$$\lim_{t \to t_0 - \tilde{\tau}} \rho_0(x(t, t_0, x_0), \partial U(t_0)) = 0, \quad or \quad \lim_{t \to t_0 - \tilde{\tau}} |x(t, t_0, x_0)| = +\infty; \tag{3.12}$$

2. $\sup_{x_0 \in U(t_0) \backslash B(r)} \tau(x_0) < +\infty$ *for any* $t_0 \in \mathbb{R}$ *and any* $r > 0$*,*

where $B(r) \subset \mathbb{R}^n$ *is a ball of the radius* $r > 0$ *centered at the origin.*

Proof *Sufficiency*. Let $x(t, t_0, x_0), t \in (t_0 - \tilde{\tau}, t)$ be a solution of the system (3.1) in the backward time, where $\tilde{\tau} \in (0, \tau(t_0, x_0)]$. Then $\tilde{x}(t, t_0, \tilde{x}_0) = x(t, t_0, x_0)$ with $\tilde{x}_0 = x(t_0 - \tilde{\tau})$, is a solution of the system in the forward time. Similarly, any solution in the forward time, that is defined on $(t_0, t_0 + T)$ with $T > 0$, is a solution in the backward time initiated at the final time $t_0 + T$. Therefore, despite that the function τ is defined based on solutions in the backward time, it characterizes the behavior of the system in the forward time as well. Hence, the boundedness of the function τ (see the property 2) implies the nearly fixed-time stability of the system. Indeed, the property 2 guarantees that for any $r > 0$ all backward time solutions initiated outside the ball $B(r)$ blow up or reach the boundary of $U(t_0)$ in a finite time $T(t_0, r) = \sup_{x_0 \in U(t_0) \backslash B(r)} \tau(x_0)$. This means that the corresponding forward time solutions initiated in $U(t_0) \backslash B(r)$ will reach the ball $B(t)$ no later that $t_0 + T(t_0, r)$.

Necessity. By definition, if the system converges to the origin nearly in a fixed time then for any $r > 0$ there exists $T = T(t_0, r) > 0$ such that for all solutions $x(t, t_0, x_0)$ with $x_0 \in U(t_0)$ we have $\|x(t, t_0, x_0)\| < r$ for all $t \geq t_0 + T_{t_0,r}$.

Let us show that, in this case, for each solution $x(t, t_0, x_0)$ in the backward time there exists $\tilde{\tau} \in (0, +\infty)$ such that the condition (3.12) is fulfilled. Suppose the contrary, i.e., the system (3.1) converges to the origin of nearly in fixed-time, but there exists a function $x^-(\cdot, t_0, x_0) : (-\infty, t_0] \mapsto \mathbb{R}^n$ being a solution of the system (3.1) such that $\rho(x^-(t, t_0, x_0), \partial U(t_0)) > 0$ for all $t \in (-\infty, t_0]$ and $|x^-(t, t_0, x_0)| < +\infty$ for all $t \in (-\infty, t_0]$. This means that $x^-(t, t_0, x_0) \in \operatorname{int} U(t_0)$ for all $t \in (-\infty, t_0]$. Indeed, since the function $t \to x^-(t, t_0, x_0)$ with $x^-(t_0, t_0, x_0) \in U(t_0)$ is continuous, then the trajectory $x^-(t, t_0, x_0)$ may leave the open connected set $U(t_0)$ only by crossing its boundary $\partial U(t_0)$ (implying the zero distance to the boundary $\partial U(t_0)$ at the crossing point).

Let $r := |x_0|$ and $T_{t_0,r} > 0$ be the settling time to the ball of the radius $r > 0$ (see, Definition 3.14). Let $x^+(\cdot, t_0, x_0) : (t_0, +\infty) \mapsto \mathbb{R}^n$ be a solution of the system (3.1) in the forward time with the same initial condition $x(t_0) = x_0$ and let the function $x(\cdot, t_0, x_0) : (-\infty, +\infty) \mapsto \mathbb{R}^n$ merge $x^-(t, t_0, x_0)$ and $x^+(t, t_0, x_0)$. Obviously, any restriction of $x(t, t_0, x_0)$ to a time interval $[t_1, +\infty)$ defines a solution of the system (3.1). For $t_1 < t_0$ and $x_1 = x(t_1, t_0, x_0)$ we have that, at least, one solution of the system (3.1) is defined as $x(t, t_1, x_1) = x(t, t_0, x_0)$ for all $t \geq t_1$. Moreover, $|x(t, t_1, x_1)| \leq r$ for all $t \geq T_{r,t_0} + t_0 - t_1$ and $|x(t_0, t_1, x_1)| = |x_0| = r$. Since, by our supposition, $x_1 = x^-(t_1, t_0, x_0) \in U(t_0)$ for any $t_1 < t_0$ then we derive the settling-time $T_{r,t_0} + t_0 - t_1$ tends to $+\infty$ as $t_1 \to -\infty$. This contradicts to the nearly fixed-time convergence. Therefore, for any solution $x(t, t_0, x_0)$ in the backward time with $x_0 \in U(t_0)$ there exists $\tilde{\tau} \in (0, +\infty)$ such that the condition (3.12) is fulfilled.

Let us show that even when solutions of the system in the backward time are not unique, their escape times $\tilde{\tau}$ are uniformly bounded by some number $\tau(t_0, x_0) < +\infty$. Suppose the contrary, i.e., for some $t_0 \in \mathbb{R}$ and $x_0 \in \mathbb{R}^n \backslash \{\mathbf{0}\}$, there exists a sequence of solutions $x^i(t, t_0, x_0)$ initiated as $x(t_0) = x_0$ such that the corresponding escape times $\tilde{\tau}_i \to +\infty$ as $i \to +\infty$. In this case, repeating the above considerations we conclude that there exists a sequence of time instances $t_1^i > t_0 - \tilde{\tau}_i$

such that $x^i(t_1^i, t_0, x_0) \in U(t_0)$ and $t_1^i \to -\infty$ as $i \to +\infty$, so the settling times $T_{r,t_0} + t_0 - t_1^i$ tend to $+\infty$ as $i \to +\infty$. This contradicts to the nearly fixed-time convergence to the origin. Therefore, $\tau(t_0, x_0) < +\infty$ for any $t_0 \in \mathbb{R}$ and any $x_0 \in U(t_0)$. The uniform boundedness $\tau(t_0, x_0)$ on $x_0 \in U(t_0)\backslash B(r)$ can be proven analogously. Indeed, assuming that there exists a sequence of initial states $x_i \in U(t_0)\backslash B(r)$ such that $\tau(t_0, x_i) \to +\infty$, we derive a contradiction to nearly fixed-time convergence to the origin. The proof is complete. □

By analogy with the settling-time function (see Definition 3.10), a "minimal" function τ satisfying the properties proven in Proposition 3.3 is called *escape-time function*.

For finite-time stable systems, we distinguish the notions of uniform and continuous finite-time stability dependently on the properties of the settling-time function. The same classification can be proposed for nearly fixed-time stable systems using the escape-time function [21].

Definition 3.15 The system (3.1) is said to be

- uniformly nFxTS if it is nFxTS with a time-invariant attraction domain U and the escape-time function is locally bounded on $U\backslash\{\mathbf{0}\}$ uniformly in $t_0 \in \mathbb{R}$;
- ∞-nFxTS if it is uniformly nFxTS and the escape-time function is "continuous at infinity", i.e., it holds

$$\tau(x_0) \to 0 \quad \text{as} \quad \rho(x_0, \partial U) \to 0$$

  and, for unbounded U, it additionally holds

$$\tau(x_0) \to 0 \quad \text{as} \quad |x_0| \to +\infty, \quad x_0 \in U;$$

- continuously nFxTS if it ∞-nFxTS and the escape-time function is continuous on $U\backslash\{\mathbf{0}\}$.

In Chap. 7, we show that all stable homogeneous systems with positive degree are ∞-nFxTS. In the general case, the continuity at infinity of the escape-time function cannot be always guaranteed even if the system is autonomous.

Example 3.11 Let $f : \mathbb{R}^2 \mapsto \mathbb{R}^2$ be a vector field of a time-invariant system, which has the following representation in the polar coordinates $x = (x_1, x_2)^\top = (r\cos(\theta), r\sin(\theta))^\top$, $r > 0$, $\theta \in [0, 2\pi)$:

$$\begin{cases} \dot{\theta} = -\theta \\ \dot{r} = -(\sqrt{r} + r^{\frac{3}{2}}) \cos\theta \end{cases} \quad \text{if } x_1 \geq 0,\ x_2 > 0,$$
$$\begin{cases} \dot{\theta} = (1 - e^{-r}) \cos\theta - \frac{\pi}{2} \sin(\theta) \\ \dot{r} = 0 \end{cases} \quad \text{if } x_1 < 0,\ x_2 \geq 0,$$
$$\begin{cases} \dot{\theta} = -1 + e^{-r} \\ \dot{r} = 0 \end{cases} \quad \text{if } x_1 < 0,\ x_2 < 0,$$
$$\begin{cases} \dot{x}_1 = -\sqrt{x_1} - x_1^{\frac{3}{2}} + (1 - e^{x_2})x_2 \\ \dot{x}_2 = 0 \end{cases} \quad \text{if } x_1 \geq 0,\ x_2 \leq 0.$$

It can be shown that f is continuous on $\mathbb{R}^n$. Moreover, the considered system has unique solutions in the forward time for all initial conditions in $\mathbb{R}^n$. In the backward time solutions are not unique for $x_0 \in \{x : x_1 \geq 0, x_2 = \mathbf{0}\}$. The phase portrait of the system is similar to the phase portrait of the system studied in Example 3.7 and presented in Fig. 3.2.

Let us show that for any $r_0 > 0$ there exists $T_{r_0} > 0$ such that $|x_{x_0}(t)| \leq r_0$ for all $t \geq T_{r_0}$ independently of the initial state $x(0) = x_0 \in \mathbb{R}^2$. To prove this we study the behavior of the system in each of four quadrants:

$$\mathcal{Q}_1 = \{x : x_1 \geq 0,\ x_2 \geq 0\}, \qquad \mathcal{Q}_2 = \{x : x_1 < 0,\ x_2 \geq 0\},$$

$$\mathcal{Q}_3 = \{x : x_1 < 0,\ x_2 < 0\}, \qquad \mathcal{Q}_2 = \{x : x_1 \leq 0,\ x_2 < 0\}.$$

First of all notice, that $\dot{r} \leq 0$ in each quadrant. This means that the ball $\{x : |x| \leq r_0\}$ is invariant and the system is, at least, Lyapunov stable.

If $x_0 \in \mathcal{Q}_4$ then the trajectory of the system remains in $\mathcal{Q}_4$ till it reaches the quadrant $\mathcal{Q}_3$ or the origin. Since for $x \in \mathcal{Q}_4$ we have $\dot{x}_1 \leq -\sqrt{x_1} - x_1^{\frac{3}{2}}$ then the reaching time is bounded by the finite number

$$T_4 = \pi$$

independently of $x_0 \in \mathcal{Q}_4$. If $x_0 \in \mathcal{Q}_3$ then due to $\dot{r} = 0$ we have $|x(t, x_0)| = |x_0|$ till the trajectory remains in the third quadrant. In this case, $\dot{\theta} = -1 + e^{-|x_0|}$ for any $x_0 \in \mathcal{Q}_3$. This means that the maximum time which the system may spend in the quadrant $\mathcal{Q}_3$ is

$$T_3 = \frac{\pi}{2(1 - e^{-|x_0|})}.$$

If $x_0 \in \mathcal{Q}_2$ then, by the same arguments, $|x(t, x_0)| = |x_0|$ as long as solution belongs to the quadrant $\mathcal{Q}_2$. In this case, we have $\dot{\theta} = (1 - e^{-|x_0|}) \cos\theta - \frac{\pi}{2} \sin(\theta)$. So, we have $\dot{\theta} \leq -\frac{1-e^{-|x_0|}}{\sqrt{2}}$ for $\theta \in (3\pi/4, \pi]$ and we have $\dot{\theta} \leq -\frac{\frac{\pi}{2}}{\sqrt{2}}$ for $\theta \in (\pi/2, 3\pi/4]$. Therefore, the maximum time, which the system may stay in the second quadrant, is estimated by

$$T_2 = \left(\frac{2\sqrt{2}}{\pi} + \frac{\sqrt{2}}{1 - e^{-|x_0|}}\right)\frac{\pi}{4}.$$

Finally, if $x_0 \in \mathcal{Q}_1$ then the trajectory will never leave the first quadrant, since $\dot{\theta} = -\theta$ and $\theta(t) = e^{-t}\theta_0 \neq 0$ for all $t \geq 0$. Moreover, we have $\dot{\theta} \leq -\frac{\pi}{4}$ as long as $\theta \geq \pi/4$. This means that in one second we would have $\theta \leq \pi/4$ independently of $x_0 \in \mathcal{Q}_1$ and $\dot{r} \leq -(\sqrt{r} + r^{3/2})/\sqrt{2}$ for all $t \geq 1$. This means any trajectory of the system initiated at $x_0 \in \mathcal{Q}_1$ will reach zero no later than the time

$$T_1 = 1 + \frac{\pi}{\sqrt{2}}.$$

Combining the results for all four quadrants we derive that for any initial condition in $\mathbb{R}^2$ the system will reach the origin no later than the time instant

$$\tilde{T}(|x_0|) = T_1 + T_2 + T_3 + T_4 = 1 + \tfrac{\pi}{\sqrt{2}} + \left(\tfrac{2\sqrt{2}}{\pi} + \tfrac{\sqrt{2}}{1-e^{-|x_0|}}\right)\tfrac{\pi}{4} + \tfrac{\pi}{2(1-e^{-|x_0|})} + \pi.$$

Since the function $r \mapsto \tilde{T}(r)$ is monotone decreasing then, taking into account the positive invariance[5] of the ball $\{x : |x| \leq r_0\}$, we conclude that independently of $x_0 \in \mathbb{R}^2$ the system will converge to the ball of the radius r_0 no later than the time instant $T_{r_0} = \tilde{T}(r_0)$. Moreover, it holds

$$\tilde{T}(r) \to 1 + \tfrac{\sqrt{2}}{2} + \tfrac{3\pi}{2\sqrt{2}} + \tfrac{3\pi}{2} \text{ as } r \to +\infty.$$

So, the system is globally uniformly nearly fixed-time stable. Let τ be its escape-time function, which exists due to Proposition 3.3.

For the sequence of initial conditions $x_i = (-i, 0)^\top \in \mathbb{R}^2$, $i \geq 1$, the system has unique solutions in the backward time, which remains the quadrant $\mathcal{Q}_3$ in exactly $\frac{\pi}{2(1-e^{-i})}$ seconds and next the backward solution "escapes" to infinity in the finite-time tending to π as $i \to +\infty$. This means that

$$\limsup_{i \to +\infty} \tau(x_i) = \pi + \lim_{i \to +\infty} \frac{\pi}{2(1 - e^{-i})} = \pi + \frac{\pi}{2}.$$

However, for initial conditions $\tilde{x}_i = (0, -i)$ the escape time to infinity is

$$\limsup_{i \to +\infty} \tau(\tilde{x}_i) = \pi.$$

Therefore, the system is globally uniformly nFxTS but it is not globally ∞-nFxTS since the escape-time function is discontinuous at infinity.

---

[5] A set Ω is positively invariant for a system $\dot{x} \in F(t, x)$ if $x(t_0) \in \Omega \Rightarrow x(t) \in \Omega, \forall t \geq t_0$.

Similarly to the case of the finite-time stability, a continuity of the escape-time function at infinity and a well-posedness of the system in the backward time imply continuity of the escape-time function on $U\backslash\{\mathbf{0}\}$.

Proposition 3.4 *Let an autonomous system* (3.1) *be* ∞*-nFxTS with an attraction domain U being an invariant set of the system in the backward time. If solutions of the system are unique in the backward time, and, on compact intervals of time, they depend continuously on initial states in* $U\backslash\{\mathbf{0}\}$*, then the escape-time function* τ *is continuous on* $U\backslash\{\mathbf{0}\}$.

Proof Let us prove the claim for $U = \mathbb{R}^n$. Since solutions of (3.1) are assumed to be unique in the backward time then for any $x_0 \in \mathbb{R}^n\backslash\{\mathbf{0}\}$ the solution $x(\cdot, x_0)$ is defined on $(-\tau(x_0), 0)$ and $|x(-t, x_0)| \to +\infty$ as $t \to \tau(x_0)$.

Let us "swap" the zero and the infinity points as follows. Let $y_0 = |x_0|^{-2}x_0$. Since $|y_0| = |x_0|^{-1}$ then $x_0 = |y_0|^{-2}y_0$. The function

$$t \mapsto y(t, y_0) := \frac{x(-t, x_0)}{|x(-t, x_0)|^2}$$

tends to zero in a finite time (in the forward time). Let us define the continuous prolongation of $y(t, y_0)$ as follows:

$$y(t, y_0) = \mathbf{0} \quad \text{for all} \quad t \geq \tau(x_0).$$

By definition, let us assign $y(t, \mathbf{0}) = \mathbf{0}$ for all $t \geq 0$. Notice that the functions y constructed in such a way have the semigroup property

$$y(t + \tau, y_0) = y(t, y(\tau, y_0)), \quad \forall t, \tau \in \mathbb{R}_+, \quad \forall y_0 \in \mathbb{R}^n.$$

So, they can be considered as trajectories of some dynamical system (referred below as y-system).

Let $T(y_0) = \tau(|y_0|^{-2}y_0)$ for $y_0 \neq \mathbf{0}$ and $T(\mathbf{0}) = 0$. If τ is continuous at infinity (i.e., $\sup_{|x_0|\geq r} \tau(x_0) \to 0$ as $r \to \infty$) then, obviously, T is continuous at $\mathbf{0}$. By construction, the function $y_0 \to T(y_0)$ is the settling-time function for $y(t, y_0)$.

Since the system (3.1) is assumed to be stable then for any $x_0 \neq \mathbf{0}$ we have

$$|x_0| \leq \varepsilon(|x(-t, x_0)|), \quad \forall t \in [0, T(y_0)),$$

where $\varepsilon \in \mathcal{K}_\infty$ is introduced in the definition of Lyapunov stability. Hence, taking into account $y(t, y_0) = \mathbf{0}$ for $t \geq T(y_0)$, we derive

$$|y(t, y_0)| \leq \tilde{\varepsilon}(|y_0|) := \frac{1}{\varepsilon^{-1}(1/|y_0|)}, \quad \forall t \geq 0,$$

where $\varepsilon^{-1} \in \mathcal{K}_\infty$ is the inverse function to $\varepsilon \in \mathcal{K}_\infty$. Therefore, the y-system is globally Lyapunov stable.

Since, but assumption, the backward solutions $x(t, x_0)$ of the system (3.1) depend continuously of initial state $x_0 \in \mathbb{R}^n\backslash\{\mathbf{0}\}$ on compact intervals of time (while solutions exist), then for any $\epsilon > 0$ solutions of the y-system depend continuously on $y_0 \in \mathbb{R}^n\backslash\{\mathbf{0}\}$ on compact intervals of time such that $y(t, y_0) \neq \mathbf{0}$. Continuous dependence at $\mathbf{0}$ follows from Lyapunov stability of y-solutions proven above. Therefore, $y(t, y_0)$ depend continuously on $y_0 \in \tilde{\mathbb{R}}^n$ on any compact interval of time. Repeating the proof of Proposition 3.2 we conclude that the continuity of T at zero yields the continuity of T on $\mathbb{R}^n$. This means that a continuity of τ at infinity guarantees the continuity of τ on $\mathbb{R}^n\backslash\{\mathbf{0}\}$. The proof for $U \neq \mathbb{R}^n$ can be done similarly. The proof is complete. □

Merging finite-time and nearly fixed-time stability we introduce the notion of fixed-time stability (FxTS) [3].

Definition 3.16 *(Fixed-time stability)* The system (3.1) is said to be

- fixed-time stable (FxTS) if it FTS and nFxTS simultaneously;
- uniformly FxTS if it is uniformly FTS and uniformly nFxTS;
- 0-FxTS if it is 0-FTS and uniformly nFxTS;
- ∞-FxTS if it is uniformly FTS and ∞-nFxTS;
- $\{0, \infty\}$-FxTS if it is 0-FTS and ∞-nFxTS;
- continuously FxTs if it is continuously FTS and continuously nFxTS;
- constant-time stable if it is Lyapunov stable and $\exists C \in \mathbb{R}_+ : T(x_0) = C$ for all $x_0 \neq \mathbf{0}$.

Example 3.12 To highlight some differences between constant-time and fixed-time stability, let us consider the following two controllers

$$u_{\text{ct}}(t) = \begin{cases} -\frac{x(t)}{T-t+t_0} & \text{if } t \in [t_0, t_0+T), \\ 0 & \text{if } \quad t \geq t_0 + T, \end{cases} \quad \text{and } u_{\text{fx}}(t) = -\frac{\pi}{2T}(1 + x^2(t))\,\text{sign}(x(t))$$

for the controlled single integrator

$$\dot{x} = u, \quad t > t_0, \quad x(t_0) = x_0,$$

where $T > 0$ is any positive constant (the prescribed settling time). The solution of the system $\dot{x} = u_{\text{ct}}(t), t > t_0, x(t_0) = x_0$ is given by

$$x(t, t_0, x_0) = e^{-\int_{t_0}^{t} \frac{1}{T-\tau+t_0} d\tau} x_0 = \frac{(T - t + t_0)x_0}{T}, \quad t \in [t_0, t_0 + T]$$

and $x(t, t_0, x_0) = 0$ for $t \geq t_0 + T$. Hence, we have $x(t_0 + T, t_0, x_0) = 0$ independently of initial value x_0 and initial time t_0. The regulator $u_{\text{ct}}(t)$ steers the state of the system to zero in the constant-time $T > 0$ (see [22]). The settling-time function of such a system is equal to T for any $x_0 \neq 0$.

The solution of the system $\dot{x} = u_{\text{fx}}(t), t > t_0, x(t_0) = x_0$ has the form

$$x(t, t_0, x_0) = \tan\left(\arctan|x_0| - \frac{\pi}{2T}(t - t_0)\right) \operatorname{sign}(x_0)$$

for $t \in \left[t_0, t_0 + \arctan(|x_0|)\frac{2T}{\pi}\right]$ and $x(t, t_0, x_0) = 0$ for all $t \geq t_0 + \arctan(|x_0|)\frac{2T}{\pi}$. Therefore, the closed-loop system with the feedback u_{fx} is the fixed-time stable, and its settling-time function is uniformly bounded as follows

$$\tilde{T}(x_0) = \arctan(|x_0|)\frac{2T}{\pi} \leq T, \quad \forall x_0 \in \mathbb{R}.$$

Therefore, both the prescribed-time (time-varying) regulator and the fixed-time (static) stabilizer guarantee

$$x(t, t_0, x_0) = 0, \quad \forall t \geq t_0 + T, \quad \forall x_0 \in \mathbb{R}, \quad \forall t_0 \in \mathbb{R}.$$

However, these controllers have different settling-time functions and totally different robustness properties (see in Chap. 9 or [23]).

Corollary 3.1 • *A uniformly FTS system* (3.1) *is uniformly FxTS if and only if* $\exists T_{\max} \in (0, +\infty)$ *such that*

$$\sup_{t_0 \in \mathbb{R}, x_0 \in U} T(t_0, x_0) \leq T_{\max},$$

where T *is the settling-time function of the system.*

• *A uniformly nFxTS system* (3.1) *is uniformly FxTS if and only if* $\exists T_{\max} \in (0, +\infty)$ *such that*

$$\sup_{t_0 \in \mathbb{R}, x_0 \in U \setminus \{\mathbf{0}\}} \tau(t_0, x_0) \leq T_{\max},$$

where τ *is the escape-time function of the system.*

Proof *Necessity*. Let $x(t, t_0, x_0)$ be a solution of the system (3.1) with $x_0 \in U$ defined in both the forward and the backward times for $t \in (t_0 - \tilde{\tau}, t_0 + \tilde{T})$ such that $x(t, t_0, x_0) \to \infty$ as $t \to t_0 - \tilde{\tau}$ and $x(t, t_0, x_0) \to 0$ as $t \to t_0 + \tilde{T}$. Obviously, FxTS guarantees that $\tilde{\tau} + \tilde{T} \leq T_{\max}$ for any $t_0 \in \mathbb{R}$ and any $x_0 \in U$. This means that the functions T and τ are uniformly bounded by $T_{\max}$.

Sufficiency. Suppose the contrary, i.e., there exists a sequence of initial conditions $x(t_0) = x_i$ and numbers $\tau_i \geq 0$ (resp., $T_i \geq 0$) such that $x_i \in U$, $\tau_i \to +\infty$ (resp., $T_i \to +\infty$) as $i \to +\infty$ and $(-\tau_i + t_0, t_0, x_i) \in U$ (resp., $x(t_0 + T_i, t_0, x_i) \neq 0$), where $x(t, t_0, x_i)$ is an arbitrary solution of the system (3.1) with the initial condition $x(t_i) = x_i$ in the backward (resp., forward) time. Obviously, this situation is impossible if the escape-time function (resp., the settling-time function T) is uniformly bounded on U. The proof is complete. □

In Chap. 7 it is shown that any globally asymptotically stable locally homogeneous system with negative degree at 0-limit and positive degree at ∞-limit is $\{0, \infty\}$-FxTS. Moreover, in Chap. 9 we design a globally homogeneous control system with a constant settling-time function. Therefore, finite-time and fixed-time stability are natural properties of stable locally/globally homogeneous systems studied in this monograph.

3.2.4 Rated ISS

Input-to-State Stability (see Definition 3.6) is introduced as a generalization of asymptotic stability to ODEs with exogenous inputs. A more detailed structure and quantitative properties of the class $\mathcal{K}$ and $\mathcal{KL}$ functions are not important for the conventional ISS analysis. However, the control engineering practice requires quantitative estimates and a more detailed characterization of ISS gains.

To take into account convergence rates in robustness analysis, let us introduce some generalizations of $\mathcal{K}$ and $\mathcal{KL}$ functions [24, 25] inspired by finite-time and fixed-time stability properties.

Definition 3.17 *(Generalized class $\mathcal{K}$ function)* An increasing function $\gamma : [0, +\infty) \mapsto [0, +\infty)$ is said to be of the class $\mathcal{GK}$ if $\exists s_0 \geq 0$ such that the function $s \mapsto \gamma(s + s_0)$ is of the class $\mathcal{K}$.

Obviously, if $s_0 = 0$ then a class-$\mathcal{GK}$ function becomes $\mathcal{K}$ function. The use of a class-$\mathcal{GK}$ functions allows a uniformity ("non-sensitivity") of the system with respect to perturbations q to be included in the ISS analysis. Indeed, if the function γ in Definition 3.6 is of the class $\mathcal{GK}$ then such an ISS yields that the asymptotic stability of the system to zero is preserved up to some magnitude of the perturbation (namely, for $\|q\| \leq s_0$ with $s_0 \geq 0$ from the above definition).

Definition 3.18 *(Generalized $\mathcal{KL}$ functions)* A function $\beta : [0, +\infty) \times [0, +\infty) \mapsto [0, +\infty)$ is of the class

- $\mathfrak{e}\mathcal{KL}$ (exponential $\mathcal{KL}$ functions) if $\beta \in \mathcal{KL}$ and

$$\exists C \geq 1, \exists \lambda > 0 \; : \; \beta(s, t) \leq C e^{-\lambda t} s, \;\; \forall s \geq 0, \; \forall t \geq 0;$$

- $\mathfrak{f}\mathcal{KL}$ (finite-time $\mathcal{KL}$ functions) if $\beta \in \mathcal{KL}$ and

$$\forall s \geq 0, \quad \exists T_s \in [0, +\infty) \; : \; \beta(s, t) = 0, \;\; \forall t \geq T_s;$$

- $\mathfrak{n}\mathcal{KL}$ (nearly fixed-time $\mathcal{KL}$ functions) if $\beta \in \mathcal{KL}$ and

$$\forall r > 0, \quad \exists T_r > 0 \; : \; \beta(s, t) \leq r, \; \forall s \geq 0, \; \forall t \geq T_r;$$

- $\mathbf{x}\mathcal{KL}$ (fixed-time $\mathcal{KL}$ functions) if $\beta \in \mathcal{KL}$ and

$$\exists T_{\max} > 0 \; : \; \beta(s,t) = 0, \;\; \forall s \geq 0, \; \forall t \geq T_{\max};$$

- $\mathfrak{c}\mathcal{KL}$ (constant-time $\mathcal{KL}$ functions) if $\beta \in \mathcal{KL}$ and

$$\exists T > 0 \; : \; \begin{matrix} \beta(s,t) \neq 0, \; \forall t \in [0,T], \\ (s,t) = 0, \; \forall t \geq T \end{matrix} \quad \forall s \geq 0.$$

Obviously, $\mathfrak{c}\mathcal{KL} \subset \mathbf{x}\mathcal{KL} \subset (\mathfrak{f}\mathcal{KL} \cap \mathfrak{n}\mathcal{KL})$. Inspired by [24–28] we introduce a rated ISS as follows.

Definition 3.19 *(Rated ISS)* A system (3.4) is said to be exponentially (resp., finite time, nearly fixed time, fixed time, constant time) ISS if there exist $\gamma \in \mathcal{GK}$ and $\beta \in \mathfrak{e}\mathcal{KL}$ (resp., $\beta \in \mathfrak{f}\mathcal{KL}$, $\beta \in \mathfrak{n}\mathcal{KL}$,$\beta \in \mathbf{x}\mathcal{KL}$, $\beta \in \mathfrak{c}\mathcal{KL}$) such that the inequality (3.5) holds for any solution $x_q(t, t_0, x_0)$ of (3.4) with $q \in L^\infty(\mathbb{R}; \mathbb{R}^k)$.

Notice that any function from $\mathbf{x}\mathcal{KL}$ can be upper bounded by a function from $\mathfrak{c}\mathcal{KL}$, so the fixed-time and constant-time ISS properties are almost indistinguishable. The notion of constant-time ISS is introduced for the completeness of presentation.

Example 3.13 The simplest example illustrating the introduced definition is

$$\dot{x} = -2|x|^{1+\alpha}\operatorname{sign}(x) + q(t), \quad t > t_0, \quad x(t_0) = x_0$$

where $\alpha > -1$, $x(t) \in \mathbb{R}$ and $q \in L^\infty(\mathbb{R}; \mathbb{R})$. For $\alpha = 0$, the solution of the considered ODE has the form

$$x_q(t, t_0, x_0) = e^{-2(t-t_0)}x_0 + \int_{t_0}^{t} e^{2s-2t}q(s)ds,$$

and

$$|x_q(t, t_0, x_0)| \leq e^{-2(t-t_0)}|x_0| + \|q\|_{L^\infty(t_0,t)}.$$

This means that the system is *exponentially ISS* for $\alpha = 1$. If $\alpha \neq 1$ then

$$\frac{d}{dt}|x(t)| \overset{a.e}{\leq} -|x(t)|^{1+\alpha} \quad \text{for} \quad |x(t)| \geq \left(\|q\|_{L^\infty(t_0,t)}\right)^{\frac{1}{1+\alpha}},$$

and

$$|x_q(t, t_0, x_0)| \leq \max\left\{\frac{|x_0|}{(1+\alpha(t-t_0)|x_0|^\alpha)^{1/\alpha}}, \left(\|q\|_{L^\infty(t_0,t)}\right)^{\frac{1}{\alpha}}\right\}.$$

Taking into account, that $\max\{a, b\} \leq a + b$ for $a \geq 0$ and $b \geq 0$, we conclude that the considered system finite-time ISS for $-1 < \alpha < 0$ and nearly fixed-time ISS for $\alpha > 0$.

Repeating the above considerations for

$$\dot{x} = -2(1 + |x|^2)\,\mathrm{sign}(x) + q(t), \quad 0 < \alpha < 1, \quad \beta > 1 \tag{3.13}$$

we derive (see, Example 3.12)

$$|x_q(t, t_0, x_0)| \leq \max\{0, \tan(\arctan(|x_0|) - t + t_0)\} + \sqrt{\|q\|_{L^\infty(t_0,t)}}, \tag{3.14}$$

i.e., the system is fixed-time ISS. However, the system (3.13) is globally fixed-time stable if $\|q\|_{L^\infty} \leq 1$, so the estimate (3.14) is rather conservative characterization of the robustness. In fact, the system (3.13) satisfies Definition 3.6 with $\beta \in \mathbf{x}\mathcal{KL}$ and $\gamma \in \mathcal{GK}$:

$$\beta(s,t) = \max\{0, \tan(\arctan(s) - t)\}, \quad \gamma(s) = \sqrt{\max\{0, s-1\}}, \quad s \geq 0, t \geq 0.$$

Therefore, the system (3.13) is fixed-time ISS with a generalized class $\mathcal{K}$ function γ. The estimate (3.5) with β and γ given above is less conservative in comparison with (3.14).

A design of functions β and γ, which give a sharp ISS estimate (3.5), is an important practical problem, which can be studied based on a Lyapunov function method (see, Chap. 4).

Concluding Remarks

Reaching versus Convergence

Trajectories of finite-dimensional dynamical systems modeled by ODEs and DIs may have various qualitative properties when time grows. In particular, they may converge asymptotically or in a finite time to an equilibrium, to a periodic orbit, to a set, etc. Making a change of the variables the analysis of convergence to a trajectory or point can be reduced to the analysis of convergence to the origin. Convergence rates to the origin are summarized in the Table 3.1. In Chap. 7, it is shown that only these convergence properties are possible for asymptotically stable locally or globally homogeneous systems.

Notice that the convergence is a property introduced on *infinite* horizon of time. We distinguish convergence and reaching properties, since the system may reach a desired point in a finite time, but it may never return back to this point as time grows. Similarly to convergence to the origin, the reaching of the origin may also be asymptotic or finite-time. However, the asymptotic reaching without asymptotic convergence is a rather "exotic" property. It was not studied in this chapter, but Table 3.2 presents the corresponding definitions for completeness.

Table 3.1 Convergence types of autonomous systems

Type of convergence	Property of solutions
Asymptotic convergence:	$\limsup_{t\to+\infty} \lvert x(t, x_0)\rvert = 0$
Uniform asymptotic convergence	$\forall r>0, \exists T_{r,R}>0$: $\lvert x_0\rvert \le R \Rightarrow \lvert x(t, x_0)\rvert \le r, \forall t \ge T_{r,R}$
Exponential convergence	$\exists \alpha, C > 0 : \lvert x(t, x_0)\rvert \le C\lvert x_0\rvert e^{-\alpha t}, \ \forall t \ge 0$
Nearly fixed-time convergence	$\forall r>0, \ \exists T_r>0 \ : \ \lvert x(t, x_0)\rvert \le r, \ \forall t \ge T_r, \ \forall x_0$
Finite-time convergence:	$\exists T_{x_0} > 0 : \lvert x(t, x_0)\rvert = 0, \ \forall t \ge T_{x_0}$
Uniform finite-time convergence	$\exists T_R>0 : \lvert x(t, x_0)\rvert \le R \ \Rightarrow \ x(t, x_0) = \mathbf{0}, \ \forall t \ge T_R$
Fixed-time convergence	$\exists T_{\max} > 0 : \lvert x(t, x_0)\rvert = 0, \ \forall t \ge T_{\max}, \forall x_0$
Constant-time convergence[a]	$\exists T > 0 : \begin{array}{l} \lvert x(t, x_0)\rvert \ne 0, \forall t \ge [0, T), \\ \lvert x(t, x_0)\rvert = 0, \forall t \ge T, \end{array} \ \forall x_0 \ne \mathbf{0}$

[a]The constant-time convergence of an autonomous system is possible if the right-hand side of ODE depends of x_0 (see Chap. 7)

Table 3.2 Types of reaching of the origin by autonomous system

Type of reaching property	Property of solutions
Asymptotic reaching:	$\liminf_{t\to+\infty} \lvert x(t, x_0)\rvert = 0$
Uniform asymptotic reaching	$\forall r>0, \exists T_{r,R}>0$:$\lvert x_0\rvert \le R \Rightarrow \inf_{\tau\in[0,T_{r,R}]} \lvert x(\tau, x_0)\rvert \le r$
Exponential reaching	$\inf_{\tau\in[0,t]} \lvert x(\tau, x_0)\rvert \le C\lvert x_0\rvert e^{-\alpha t}, \ \forall t \ge 0$
Nearly fixed-time reaching	$\forall r>0, \ \exists T_r>0 : \inf_{\tau\in[0,T_r]} \lvert x(\tau, x_0)\rvert \le r, \forall x_0$
Finite-time reaching:	$\exists T_{x_0} > 0 : \inf_{\tau\in[0,T_{x_0}]} \lvert x(\tau, x_0)\rvert = 0$
Uniform finite-time reaching	$\exists T_R>0 : \lvert x(t, x_0)\rvert \le R \ \Rightarrow \ \inf_{\tau\in[0,T]} \lvert x(\tau, x_0)\rvert = 0$
Fixed-time reaching	$\exists T_{\max} > 0 : \inf_{\tau\in[0,T_{\max}]} \lvert x(\tau, x_0)\rvert = 0, \forall x_0$
Constant time reaching	$\exists T > 0 : \begin{array}{l} \lvert x(t, x_0)\rvert \ne 0, \forall t \ge [0, T), \\ \lvert x(T, x_0)\rvert = 0 \end{array} \ \forall x_0 \ne \mathbf{0}$

Stability Definitions The following definitions of stability of dynamical systems are surveyed:

LS=Lyapunov stability,	AS=Asymtotic Stability,
ES=Exponential Stability,	FTS=Finite-time Stability,
FxTS=Fixed-Time Stability,	CTS=Constant-Time Stability,

$$\text{nFxTS} = \text{nearly Fixed-Time Stability.}$$

The brief summary of the stability definitions is presented in Table 3.3.

The following scheme

$$\text{CTS} \subset \text{FxTS} \subset \begin{array}{c}\text{FTS}\\ \text{ES}\\ \text{nFxTS}\end{array} \subset \text{AS} \subset \text{LS}$$

Table 3.3 Summary of stability definitions

Type of stability	Property
Laypunov stability	$\exists \varepsilon \in \mathcal{K} : \|x(t, x_0)\| \leq \varepsilon(\|x_0\|), \forall t \geq 0, \forall x_0$
Asymptotic stability	Lyapunov stability + Asymptotic convergence
Exponential stability	Lyapunov stability + Exponential convergence
Nearly fixed-time stability	Lyapunov stability + Nearly fixed-time convergence
Finite-time stability	Lyapunov stability + Finite-time convergence
Uniform finite-time stability	Lyapunov stability + Uniform finite-time convergence
Fixed-time stability	Lyapunov stability + Fixed-time convergence
Constant-time stability	Lyapunov stability + Constant-time convergence

summarizes some relations between presented definitions for both local and global stability properties. Further embeddings

$$\text{local FTS} \subset \text{local ES} \subset \text{local nFxTS}$$

take a place in the local case. The inverse inclusions take a place for the case of the so-called practical stability (see Remark 3.1):

$$\text{global practical nFxTS} \subset \text{global practical ES} \subset \text{global practical FTS}$$

Remark 3.1 Each definition of stability (resp., convergence or reaching) given above can be generalized to a stability of a set M (resp., convergence to a set M or reaching of a set M) provided that the Euclidean norm $|x(t, t_0, x_0)|$ (being a distance to the origin) is replaced with $\rho_0(x(t, t_0, x_0), M)$ (being a distance to the set M). If there exists a compact set $M \subset \mathbb{R}^n : \mathbf{0} \in \text{int}(M)$ satisfying such a definition (locally or globally) then the system (3.1) is called (locally or globally) **practically stable**.

Robustness

The robustness analysis of the homogeneous systems studied in this monograph is mainly based on the concept of Input-to-State Stability (ISS):

$$|x(t, t_0, x_)| \leq \beta(|x_0|, t - t_0) + \gamma(\|q\|_{L^\infty(t_0,t)}).$$

The rated ISS characterizes the robustness of homogeneous control systems taking into account their convergence rates (see Table 3.4).

Table 3.4 Summary of ISS definitions

Type of stability	Property
(Asymptotic) ISS:	$\beta \in \mathcal{KL}, \gamma \in \mathcal{K}$
Exponential ISS	$\beta \in \mathfrak{e}\mathcal{KL}, \gamma \in \mathcal{GK}$
Nearly fixed-time ISS	$\beta \in \mathfrak{n}\mathcal{KL}, \gamma \in \mathcal{GK}$
Finite-time ISS:	$\beta \in \mathfrak{f}\mathcal{KL}, \gamma \in \mathcal{GK}$
Fixed-time ISS	$\beta \in \mathfrak{x}\mathcal{KL}, \gamma \in \mathcal{GK}$
Constant-time ISS	$\beta \in \mathfrak{c}\mathcal{KL}, \gamma \in \mathcal{GK}$

Regulation versus Stabilization

The stability (convergence) analysis is the problem of the system theory, which exists out of context of control theory. However, they may be inspired by well-known control problems formalized below.

Remark 3.2 Let us consider a control system

$$\dot{x} \in F(t, x, u), \quad t > 0, \quad x(t_0) = x_0, \tag{3.15}$$

where x is a system state and u is a control input.

- The **regulation** problem consists in finding a control function u (possibly dependent of t and x) such that the controlled system (3.15) **reaches zero** asymptotically or in a finite time.
- The **stabilization** problem consists in finding a control function u (possibly dependent of t and x) such that the controlled system (3.15) is asymptotically or finite-time **stable**.

Following this remark, a regulation (= regulator design) problem deals with an analysis of asymptotic or finite-time reaching of a set-point (e.g., the origin), but a stabilization (= stabilizer design) problem requires an analysis of asymptotic or finite-time stability, respectively. The regulation/stabilization may be local or global. Table 3.5 summarizes the control problems and their counterparts in the systems theory.

By definition, the finite-time regulation means that a control law u ensures a finite-time reaching of the origin by the closed-loop system (3.15). However, the control engineering practice needs some tools for tuning of this reaching time. Notice that, in the general case, fixed-time stabilization does not mean that the settling time estimate $T_{\max} > 0$ can be tuned arbitrary small, since the system may have a fixed-time stable subspace of the state space with a nonadjustable settling time. Therefore, a stabilization/regulation with prescribed convergence time is a special control problem, which is known as *prescribed-time stabilization/regulation* (see Table 3.6). Prescribed-time stabilizers are designed, for example, in [22, 27, 29, 30].

Table 3.5 Convergence/Stability versus Regulation/Stabilization

Property of system	Control problem
Asymptotic reaching:	*Asymptotic regulation*:
Uniform asymptotic reaching	Uniform asymptotic regulation
Exponential reaching	Exponential regulation
Nearly fixed-time reaching	Nearly fixed-time regulation
Asymptotic stability:	*Asymptotic stabilization*:
Uniform asymptotic stability	Uniform asymptotic stabilization
Exponential stability	Exponential stabilization
Nearly fixed-time stability	Nearly fixed-time stabilization
Finite-time reaching:	*Finite-time regulation*:
Uniform finite-time reaching	Uniform finite-time regulation
Fixed-time reaching	Fixed-time regulation
Constant-time reaching	Constant-time regulation
Finite-time stability:	*Finite-time stabilization*:
Uniform finite-time stability	Uniform finite-time stabilization
Fixed-time stability	Fixed-time stabilization
Constant-time stability	Constant-time stabilization

Table 3.6 Prescribed-time regulation/stabilization

Notion	Property
Prescribed-time regulation:	
Prescribed nearly fixed-time regulation	Nearly fixed-time regulation with prescribed T_r
Prescribed finite-time regulation	Finite-time regulation with prescribed T_{x_0}
Prescribed uniform finite-time regulation	Uniform finite-time regulation with prescribed T_R
Prescribed fixed-time regulation	Fixed-time regulation with prescribed $T_{\max}$
Prescribed constant-time regulation	Constant-time regulation with prescribed T
Prescribed-time stabilization:	
Prescribed nearly fixed-time stabilization	Nearly fixed-time stabilization with prescribed T_r
Prescribed finite-time stabilization	Finite-time stabilization with prescribed T_{x_0}
Prescribed uniform finite-time stabilization	Uniform finite-time stabilization with prescribed T_R
Prescribed fixed-time stabilization	Fixed-time stabilization with prescribed $T_{\max}$
Prescribed constant-time stabilization	Constant-time stabilization with prescribed T

The notions like prescribed-time stability [20] or predefined-time stability are meaningless, since any problem of adjustment of a settling time is a control problem, but not a stability analysis problem. Notice that $u \in \mathbb{R}^m$ in (3.15) may be a vector of constant adjustable parameters of the system.

In Chap. 9, we solve all prescribed-time stabilization problems mentioned in Table 3.6 for controllable linear MIMO systems using the homogeneity theory. In Chap. 15, some finite/fixed-time stabilization algorithms are developed for nonlinear systems.

References

1. Lyapunov AM (1992) The general problem of the stability of motion. Taylor & Francis
2. Roxin E (1966) On finite stability in control systems. Rendiconti del Circolo Matematico di Palermo 15:273–283
3. Polyakov A (2012) Nonlinear feedback design for fixed-time stabilization of linear control systems. IEEE Trans Autom Control 57(8):2106–2110
4. Sontag ED (1989) Smooth stabilization implies coprime factorization. IEEE Trans Autom Control 34:435–443
5. Coddington EA, Levinson N (1955) Theory of ordinary differential equations. McGraw-Hill, New York
6. Filippov AF (1988) differential equations with discontinuous right-hand sides. Kluwer Academic Publishers
7. Vinograd RE (1957) The inadequacy of the method of characteristic exponents for the study of nonlinear differential equations. Math. Sbornik 41(83):431–438
8. Hahn W (1967) Stability of motion. Springer, Berlin Heidelberg, New York
9. Clarke FH, Ledyaev YuS, Stern RJ (1998) Asymptotic stability and smooth Lyapunov functions. J Differ Equ 149:69–114
10. Jiang Z-P, Wang Y (2001) Input-to-state stability for discrete-time nonlinear systems. Automatica 37(6):856–869
11. Sontag ED (2007) Input to state stability: basic concepts and results. In: Nistri PP, Stefani G (eds) Nonlinear and optimal control theory. Springer, Berlin, pp 163–220
12. Jayawardhana B, Logemann H, Ryan EP, the circle criterion and input-to-state stability (2008) Infinite-dimensional feedback systems. Commun Inf Syst 8:403–404
13. Karafyllis I, Krstic M (2018) Input-to-state stability for PDEs. Springer
14. Mironchenko A (2023) Input-to-state stability: theory and applications. Springer
15. Bhat SP, Bernstein DS (2000) Finite time stability of continuous autonomous systems. SIAM J Control Optim 38(3):751–766
16. Orlov Y (2005) Finite time stability and robust control synthesis of uncertain switched systems. SIAM J Control Optim 43(4):1253–1271
17. Polyakov A, Fridman L (2014) Stability notions and Lyapunov functions for sliding mode control systems. J Franklin Inst 351(4):1831–1865
18. La Salle J, Lefschetz S (1961) Stability by Liapunov's direct method with applications. Elsevier Science
19. Nakamura H, Yamashita Y, Nishitani H (2002) Smooth Lyapunov functions for homogeneous differential inclusions. In: Proceedings of the 41st SICE annual conference, pp 1974–1979
20. Efimov D, Polyakov A (2021) Finite-time stability tools for control and estimation. Foundations and trends in systems and control. Netherlands: now Publishers Inc
21. Lopez-Ramirez F, Efimov D, Polyakov A, Perruquetti W (2019) Conditions for fixed-time stability and stabilization of continuous autonomous systems. Syst Control Lett 129:26–35

22. Holloway JC (2018) Prescribed time stabilization and estimation for linear systems with applications in tactical missile guidance. PhD thesis, University of California, San Diego
23. Aldana-Lopez R, Seeber R, Haimovich H, Gomez-Gutierrez D (2023) On inherent limitations in robustness and performance for a class of prescribed-time algorithms. Automatica 158:111284
24. Hong Y, Jiang Z, Feng G (2010) Finite-time input-to-state stability and applications to finite-time control design. SIAM J Control Optim 48(7):4395–4418
25. Lopez-Ramirez F, Efimov D, Polyakov A, Perruquetti W (2020) Finite-time and fixed-time input-to-state stability: explicit and implicit approaches. Syst Control Lett 144:104775
26. Grüne L, Sontag ED, Wirth FR (1999) Asymptotic stability equals exponential stability, and is equals finite energy gain - if you twist your eyes. Syst Control Lett 38:127–134
27. Song Y, Wang Y, Holloway J, Krstic M (2017) Time-varying feedback for regulation of normal-form nonlinear systems in prescribed finite time. Automatica 83:243–251
28. Efimov D, Polyakov A (2024) On converse lyapunov theorem for fixed-time input-to-state stability. SIAM J Control Optim 62(1):118–134
29. Orlov Y (2022) Time space deformation approach to prescribed-time stabilization: Synergy of time-varying and non-lipschitz feedback designs. Automatica 144(10):110485
30. Polyakov A, Krstic M (2024) Fixed-time stabilization with a prescribed constant settling time by static feedback for delay-free and input delay systems. Int J Robust Nonlinear Control 34(10):6393–6415

Chapter 4
Method of Lyapunov Function

The celebrated Lyapunov function method is the main tool for stability analysis of nonlinear differential equations. The advantage of this method lies in the fact that a decision on the stability or instability of the system can be made by an investigation of the right-hand side of the differential equation without finding a solution. Initially, the Lyapunov function method was limited by a regular class of ODEs with continuous right-hand sides. The later evolution of the theory of differential equations and its applications had required extensions of this method to differential equations with discontinuous right-hand sides, functional differential equations, PDEs and evolution equations in Banach spaces. This chapter surveys some elements of the Lyapunov functions theory for finite-dimensional systems.

4.1 Energetic Approach to Stability Analysis

The Lyapunov function method is founded on the so-called energetic approach to the stability analysis. It considers a *positive definite function* as a possible energetic characteristic (energy) of a dynamical system and studies an evolution of this function in time. If this energetic function is decreasing (strongly decreasing or bounded) along any system trajectory then the system has some stability property and the corresponding energetic function is called *Lyapunov function*. The modern stability theory [1] uses this name for any function satisfying the above properties, despite that the classical Lyapunov theory [3] was developed only for smooth Lyapunov functions.

To prove the uniform asymptotic stability of the origin of the system

$$\dot{x} = f(x), \quad t > 0, \quad f \in C(\mathbb{R}^n; \mathbb{R}^n), \tag{4.1}$$

A. Polyakov, *Generalized Homogeneity in Systems and Control Volume I*,
Communications and Control Engineering,
https://doi.org/10.1007/978-3-031-76246-8_4

it is sufficient to find a positive definite function $V : \mathbb{R}^n \mapsto \mathbb{R}$ such that V is continuous at zero and the function $t \mapsto V(x(t))$ monotonically decreases to zero as $t \to +\infty$ for each solution x of the system (4.1). The existence of such a *Lyapunov function* is the necessary and sufficient condition [4, 5] for the uniform asymptotic stability of the system (4.1). If the function V is continuously differentiable then the required monotonicity property can be rewritten in the classical form [3]:

$$\dot{V}(x) = \frac{\partial V}{\partial x} f(x) < 0. \tag{4.2}$$

The inequality (4.2) is more useful in practice, since, in this case, the stability analysis does not require a knowledge of solutions of (4.1).

The Lyapunov function method has also a great practical value. The control theory uses it for qualitative analysis of system dynamics, for estimation of attraction and/or attracting domains, for robust control design and for solving many other control problems (see [1]). For instance, the previous chapter studies the so-called finite-time (fixed-time) stability of dynamical systems, which means that solutions of a stable system reach the origin in a finite (independent of an initial condition) instant of time called *the settling time*. The Lyapunov function method is the main tool for an estimation of the settling time (see [1, 2, 6] and references within).

Due to Kurzweil's theorem [4], the existence of a C^∞ Lyapunov function is necessary for asymptotic stability of ODE with a continuous right-hand side. Therefore, a smooth Lyapunov function always exists for any asymptotically stable system (4.1). However, the key difficulty of the method is finding an appropriate Lyapunov function for a concrete system. There is no universal and efficient way to design a practically useful smooth Lyapunov function despite that its existence is theoretically proven. For non-smooth control systems, sometimes, it is much simpler to construct a non-smooth or even a discontinuous Lyapunov function (see, e.g., [2, 7, 29]). The problem of a reformulation of the condition (4.2) for non-differentiable functions arises in this case.

Inspired by [5, 8, 9], the next section presents some tools for both smooth and non-smooth Lyapunov analysis.

4.2 Generalized Derivatives and Monotonicity

Since the fundamental works of Isaac Newton and Gottfried Leibniz published at the end of the seventeenth century, differential calculus became an important field of mathematics. The notion of derivative has undergone numerous changes. Inspired by the ideas of Augustin Louis Cauchy, Benard Riemann, Karl Weierstrass, Ulisse Dini, Gaston Darboux, Stanislaw Zaremba and many others, various generalized derivatives have been introduced and studied. This section surveys a few results about generalized derivatives [8–10].

4.2.1 Contingent Derivatives

Let $\mathcal{I}$ be one of the following sets: $[a, b]$, (a, b), $[a, b)$ or $(a, b]$, where $a, b \in [-\infty, +\infty]$, $a < b$. Let $\mathbb{K}$ be the set of sequences of real nonzero numbers convergent to zero:

$$\{h_n\} \in \mathbb{K} \quad \Leftrightarrow \quad h_n \to 0,\ h_n \neq 0 \text{ as } n \to +\infty.$$

Inspired by [8], we introduce derivative numbers and a contingent derivative of a real-valued function as follows.

Definition 4.1 • A number

$$D_{\{h_n\}}\varphi(t) = \lim_{n\to+\infty} \frac{\varphi(t+h_n) - \varphi(t)}{h_n}, \qquad \{h_n\} \in \mathbb{K} : t + h_n \in \mathcal{I}$$

is called a *derivative number* of a function $\varphi : \mathcal{I} \mapsto \mathbb{R}$ at $t \in \mathcal{I}$, if the above limit exists (both finite and infinite limits are admissible).

- A set of all derivative numbers of the function φ at the point $t \in \mathcal{I}$ is called a *contingent derivative*:

$$D_{\mathbb{K}}\varphi(t) = \bigcup_{\{h_n\}\in\mathbb{K}:\exists D_{\{h_n\}}\varphi(t)} \{D_{\{h_n\}}\varphi(t)\} \subset [-\infty, +\infty].$$

- The function φ is said to be strongly differentiable at $t \in \mathcal{I}$ if a finite limit

$$\lim_{h\to 0} \frac{\phi(t+h) - \phi(t)}{h} = \dot{\phi}(t) \in \mathbb{R}$$

exists.

For the *strong derivative of the function* ϕ at $t \in \mathcal{I}$ we use the classical notation $\dot{\varphi}(t)$ or $\varphi'(t)$. In mathematical analysis, the word "strong" is usually omitted in the definition of the strong derivative. We would follow this tradition if the context is clear. Obviously, if ϕ is differentiable at $t \in \mathcal{I}$ then $D_{\mathbb{K}}\varphi(t) = \{\dot{\phi}(t)\}$.

Example 4.1 If $\varphi(t) = (2 - \text{sign}(t))t\cos(1/t)$ then $D_{\mathbb{K}}\varphi(0) = [-3, 1]$. Indeed, for $h_n > 0$ we have

$$\frac{\varphi(h_n) - \varphi(0)}{h_n} = \lim_{n\to+\infty} \cos(1/h_n)$$

but for $h_n < 0$ we derive

$$\frac{\varphi(h_n) - \varphi(0)}{h_n} = \lim_{n\to+\infty} 3\cos(1/h_n).$$

On the one hand, due to the boundedness of the cosine function, the derivative number $D_{\{h_n\}}\varphi(0) = D_{\{h_n\}}\varphi(0) = \lim_{n\to+\infty} \frac{\varphi(h_n)-\varphi(0)}{h_n}$ may belong only to the interval

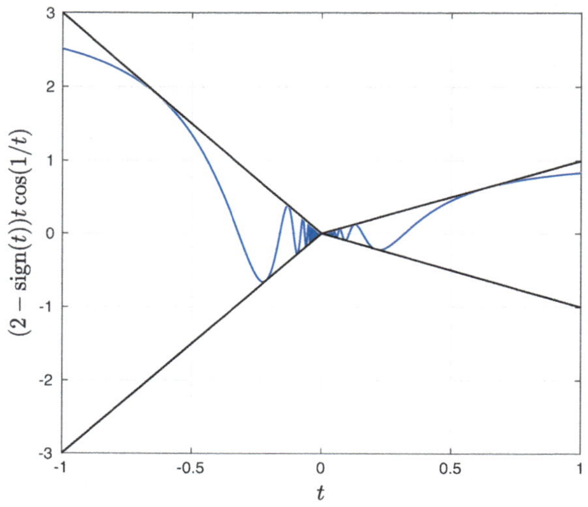

Fig. 4.1 A function with a compact set of derivative numbers at $t = 0$

$[-3, 3]$. On the other hand, for any $\alpha \in [-1, 1]$ we select $h_n = \frac{1}{\arccos(\alpha)-2\pi n}$ with $n \in \mathbb{N}$ and derive $D_{\{h_n\}}\varphi(0) = 3\alpha \in \mathcal{D}_{\mathbb{K}}\varphi(0)$ if $h_n < 0$, but $D_{\{h_n\}}\varphi(0) = \alpha \in \mathcal{D}_{\mathbb{K}}\varphi(0)$ if $h_n > 0$ (Fig. 4.1).

Partial derivatives numbers and a partial contingent derivative of a function $f : \mathbb{R}^n \mapsto \mathbb{R}$ can be introduced similarly considering the derivatives of $t \mapsto \phi(t) = f(x_1, \ldots, x_{i-1}, t, x_{i+1}, \ldots, x_n)$ at $t = x_i$, where $x = (x_1, \ldots, x_n)^\top \in \mathbb{R}^n$. If all strong partial derivatives exist then f is called differentiable, and the derivative is denoted $\frac{\partial f(x)}{\partial x} = \left(\frac{\partial f(x)}{\partial x_1}, \ldots, \frac{\partial f(x)}{\partial x_n}\right)$, where $\frac{\partial f(x)}{\partial x_i}$ is the corresponding partial derivative, $i = 1, 2, \ldots, n$.

As shown in [8], the contingent derivative exists for any function φ.

Lemma 4.1 *The set $D_{\mathbb{K}}\varphi(t) \subseteq \overline{\mathbb{R}}$ is nonempty for any $\varphi : \mathcal{I} \subset \mathbb{R} \mapsto \mathbb{R}$ and any $t \in \mathcal{I}$.*

Proof Let us consider an arbitrary sequence $\{h_n\} \in \mathbb{K}$ and $\phi_n = \frac{\varphi(t+h_n)-\varphi(t)}{h_n}$, $n \in \mathbb{N}$. If the sequence $\{\phi_n\}$ is bounded, then, by Bolzano–Weierstrass theorem, it contains a convergent subsequence. This means that there exists a subsequence $\{h_{n_k}\}$ of the sequence $\{h_n\}$ such that $D_{\{h_{n_k}\}}\varphi(t)$ is finite. On the other hand, if the sequence $\{\phi_n\}$ is unbounded then it contains a subsequence that tends to $+\infty$ or $-\infty$, so the corresponding derivative number is $+\infty$ or $-\infty$, respectively. □

If $\varphi : \mathcal{I} \mapsto \mathbb{R}$ then the notation $D_{\mathbb{K}}\varphi(t) < 0$ means $y < 0, \forall y \in D_{\mathbb{K}}\varphi(t)$. Similarly, we denote $|D_{\mathbb{K}}\varphi(t)| < L$ with $L \in \mathbb{R}_+$ if $|y| < L$ for $\forall y \in D_{\mathbb{K}}\varphi(t)$. Other ordering relations $\leq, >, \geq$ for contingent (set-valued) derivatives are interpreted analogously. The inequality $|D_{\mathbb{K}}\varphi(t)| < +\infty$ means that there exists $L \in \mathbb{R}_+$ such that $|D_{\mathbb{K}}\varphi(t)| \leq L$.

Contingent derivatives characterize *monotonicity* of non-differentiable functions. Recall that a function $\varphi : \mathcal{I} \mapsto \mathbb{R}$ is called *decreasing* on $\mathcal{I}$ if

$$\forall t_1, t_2 \in \mathcal{I} : t_1 \leq t_2 \ \Rightarrow \ \varphi(t_1) \geq \varphi(t_2).$$

Lemma 4.2 *A function $\varphi : \mathcal{I} \subset \mathbb{R} \mapsto \mathbb{R}$ is decreasing on $\mathcal{I}$ if and only if $D_{\mathbb{K}}\varphi(t) \leq 0$ holds for all $t \in \mathcal{I}$.*

Proof *Necessity* immediately follows from the definition of the decreasing function. Let us prove the *Sufficiency*. Suppose the contrary: $\exists t_0, t_1 \in \mathcal{I} : t_0 < t_1$ and $\varphi(t_0) < \varphi(t_1)$. Let us denote

$$\Delta(a, b) = \left\{t \in [a, b] : \varphi(t) > \varphi(b)\tfrac{t-a}{b-a} + \varphi(a)\tfrac{b-t}{b-a}\right\}.$$

If $\Delta(t_0, t_1)$ is empty then

$$\varphi(t) - \varphi(t_1) \leq \varphi(t_1)\tfrac{t-t_1}{t_1-t_0} + \varphi(t_0)\tfrac{t_1-t}{t_1-t_0} - \varphi(t_1) < 0, \quad \forall t \in [t_0, t_1)$$

and taking any sequence $\{h_k\} \in \mathbb{K}$, $h_k < 0$ we derive

$$\begin{aligned} D_{\{h_n\}}\varphi(t_1) &= \lim_{n \to +\infty} \tfrac{\varphi(t_1+h_n)-\varphi(t_1)}{h_n} \\ &\geq \lim_{n \to +\infty} \frac{\varphi(t_1)\frac{t_1+h_k-t_0}{t_1-t_0} + \varphi(t_2)\frac{t_1-t_1-h_k}{t_1-t_0} - \varphi(t_1)}{h_n} \\ &= \frac{\varphi(t_1) - \varphi(t_0)}{t_1 - t_0} > 0. \end{aligned}$$

This contradicts to the condition $D_{\mathbb{K}}\varphi(t) \leq 0, \forall t \in \mathcal{I}$. For $\Delta(t_0, t_1) \neq \emptyset$ we take

$$t_2 = \inf_{t \in \Delta(t_0, t_1)} t.$$

Since $\Delta(t_0, t_1) \neq \emptyset$ then $t_2 \in [t_0, t_1)$. If $t_2 \in (t_0, t_1)$ then, by construction, $t_0 < t_2$, $\varphi(t_0) < \varphi(t_2)$ and $\Delta(t_0, t_2)$ is empty. In this case, repeating the above considerations, we derive $D_{\{h_n\}}\phi(t_2) = \frac{\varphi(t_2)-\varphi(t_0)}{t_2-t_0} > 0$ for some $\{h_n\} \in \mathbb{K}$ that contradicts to the condition $D_{\mathbb{K}}\varphi(t) \leq 0, \forall t \in \mathcal{I}$. If $t_2 = t_0$ then by definition of inf there exists $\{h_n\} \in \mathbb{K}$, $h_n > 0$ such that $t_2 + h_n \in \Delta(t_0, t_1)$. This means that

$$\varphi(t_2 + h_n) > \varphi(t_1)\tfrac{t_2+h_n-t_1}{t_1-t_0} + \varphi(t_0)\tfrac{t_1-t_2-h_n}{t_1-t_0}, \quad \forall n \in \mathbb{N}.$$

In this case, we have

$$
\begin{aligned}
D_{h_n}\varphi(t_2) &= \lim_{n\to+\infty} \tfrac{\varphi(t_2+h_n)-\varphi(t_2)}{h_n} \\
&\geq \lim_{n\to+\infty} \tfrac{\varphi(t_1)-\varphi(t_0)}{t_1-t_0} + \tfrac{\varphi(t_1)\frac{t_2-t_1}{t_1-t_0}+\varphi(t_0)\frac{t_1-t_2}{t_1-t_0}-\varphi(t_2)}{h_n} \\
&\geq \tfrac{\varphi(t_1)-\varphi(t_0)}{t_1-t_0} > 0.
\end{aligned}
$$

This also contradicts to the condition $D_{\mathbb{K}}\varphi(t) \leq 0, \forall t \in \mathcal{I}$. The proof is complete. □

Remark 4.1 Lemma 4.2 *requires neither the continuity of the function* $\varphi(t)$ *nor the boundedness of its derivative numbers.* Notice that if $+\infty \notin D_{\mathbb{K}}\varphi(t)$ for all $t \in \mathcal{I}$ then the monotonicity condition can be relaxed asking $D_{\mathbb{K}}\varphi(t) \leq 0$ for almost all $t \in \mathcal{I}$ (see [8, p. 266]).

Example 4.2 The function $\phi(t) = -t - \text{sign}_\sigma(t)$ (see Fig. 4.2) has a negative contingent derivative for all $t \in \mathbb{R}$ and for any $\sigma \in [-1, 1]$, where the function sign_σ is defined by formula (0.5). Indeed, $D_{\mathbb{K}}\phi(t) = \{-1\}$ for $t \neq 0$, $D_{\mathbb{K}}\phi(0) = \{-\infty\}$ if $\sigma \in (-1, 1)$ and $D_{\mathbb{K}}\phi(0) = \{-\infty, -1\}$ if $\sigma \in \{-1, 1\}$. If $\sigma \notin [-1, 1]$ then the function ϕ is non-monotone and $D_{\mathbb{K}}\phi(0) = \{-\infty, +\infty\}$.

Let us study some simple properties of contingent derivatives.

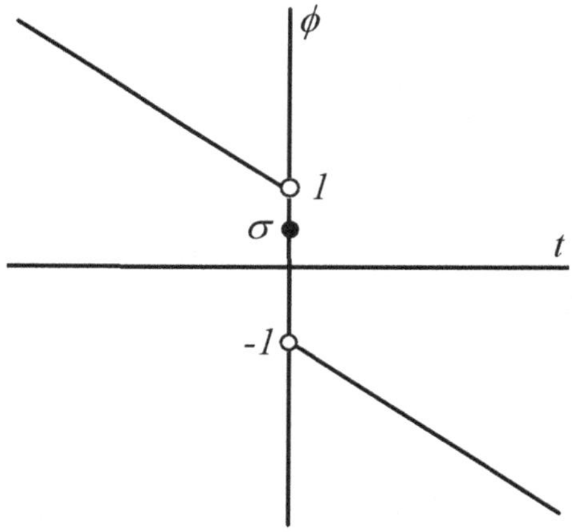

Fig. 4.2 Example of a discontinuous monotone function

Lemma 4.3 (On arithmetics of contingent derivatives)

- *If* $\alpha \in \mathbb{R}$, $\varphi : \mathcal{I} \mapsto \mathbb{R}$ *then* $D_{\mathbb{K}}(\alpha\varphi)(t) = \alpha D_K\varphi(t)$, $t \in \mathcal{I}$.
- *If* $\varphi_1, \varphi_2 : \mathcal{I} \mapsto \mathbb{R}$ *and* φ_1 *is differentiable at* $t \in \mathcal{I}$ *then*

$$D_{\mathbb{K}}(\varphi_1 + \varphi_2)(t) = \dot{\varphi}_1(t) \dot{+} D_{\mathbb{K}}\varphi_2(t),$$

where $\dot{+}$ *denotes the geometric sum of sets (see, formula (0.1)).*
- *If* $\varphi_1 : \mathcal{I} \mapsto [0, \bar{\varphi}_1]$ *with* $\bar{\varphi}_1 < +\infty$ *and* $\varphi_2 : \mathcal{I} \mapsto [0, \bar{\varphi}_2]$ *with* $\bar{\varphi}_2 < +\infty$ *are such that* $D_{\mathbb{K}}\varphi_1(t) \le -c_1 \le 0$ *and* $D_{\mathbb{K}}\varphi_2(t) \le -c_2 \le 0$ *then*

$$D_{\mathbb{K}}(\varphi_1\varphi_2)(t) \le -c_1 \liminf_{\tau \to t} \varphi_2(\tau) - c_2 \liminf_{\tau \to t} \varphi_1(t), \quad t \in \mathcal{I}, \tag{4.3}$$

$$D_{\mathbb{K}}(\varphi_1\varphi_2)(t) \le \tfrac{-c_1\varphi_2(\tau) - c_2\varphi_1(t)}{2}, \quad t \in \mathcal{I}. \tag{4.4}$$

Proof The first two claims are obvious. Let us prove the third one. Since

$$\begin{aligned}
\tfrac{\varphi_1(t+h_n)\varphi_2(t+h_n) - \varphi_1(t)\varphi_2(t)}{h_n} &= \tfrac{\varphi_1(t+h_n)\varphi_2(t+h_n) - \varphi_2(t)\varphi_1(t+h_n)}{h_n} \\
&\quad + \tfrac{\varphi_2(t)\varphi_1(t+h_n) - \varphi_1(t)\varphi_2(t)}{h_n} \\
&= \tfrac{\varphi_1(t+h_n)(\varphi_2(t+h_n) - \varphi_2(t))}{h_n} + \varphi_2(t)\tfrac{\varphi_1(t+h_n) - \varphi_1(t)}{h_n}
\end{aligned}$$

then under our assumptions we derive

$$D_{\mathbb{K}}(\varphi_1\varphi_2)(t) \le -c_2 \liminf_{\tau \to t} \varphi_1(\tau) - c_1\varphi_2(t) \le -c_2 \liminf_{\tau \to t} \varphi_1(\tau) - c_1 \liminf_{\tau \to t} \varphi_2(\tau).$$

Similarly we obtain

$$D_{\mathbb{K}}(\varphi_1\varphi_2)(t) \le -c_1 \liminf_{\tau \to t} \varphi_2(\tau) - c_2\varphi_1(t).$$

and

$$\begin{aligned}
2D_{\mathbb{K}}(\varphi_1\varphi_2)(t) &\le -c_1 \liminf_{\tau \to t} \varphi_2(\tau) - c_2 \liminf_{\tau \to t} \varphi_1(\tau) - c_1\varphi_2(t) - c_2\varphi_1(t) \\
&\le -c_1\varphi_2(t) - c_2\varphi_1(t).
\end{aligned}$$

The proof is complete. □

A local Lipschitz continuity of a function φ can be characterized by contingent derivatives (see, e.g., [11]).

Lemma 4.4 *A function* $\varphi : (a, b) \mapsto \mathbb{R}$ *is locally Lipschitz continuous*[a] *on* (a, b) *if and only if* $|D_{\mathbb{K}}\varphi(t)| < +\infty, \forall t \in (a, b)$.

[a] A function is said to be locally Lipschitz continuous on an open domain if for any point of the domain there exists its neighborhood where the function satisfies a Lipschitz condition.

Proof *Necessity*. Let the function φ satisfy the Lipschitz condition with a constant $L \in \mathbb{R}_+$ in a neighborhood $U(t)$ of a point $t \in (a, b)$. Since $U(t)$ is a neighborhood then for any sequence $\{h_n\} \in \mathbb{K} : t + h_n \in (a, b)$ there exists $N^* \in \mathbb{N}$ such that $h_n \in U(t)$ for all $n \geq N^*$. In this case, using the Lipschitz condition

$$\left|\frac{\varphi(t+h_n)-\varphi(t)}{h_n}\right| \leq L, \forall n \geq N^*.$$

This means that $|D_{\{h_n\}}\varphi(t)| \leq L$.

Sufficiency. Let $|D_{\{h_n\}}\varphi(t)| \leq L$ us show that there exists a neighborhood of t, where φ satisfies the Lipschitz condition with the constant $2L + 2\delta$ for any $\delta > 0$. Suppose the contrary, i.e., for any $\varepsilon_n > 0$ there exists $t^1_{\varepsilon_n}, t^2_{\varepsilon_n} \in [t - \varepsilon_n, t + \varepsilon_n]$ such that $t^1_{\varepsilon_n} \neq t^2_{\varepsilon_n}$ and $|\varphi(t^1_{\varepsilon_n}) - \varphi(t^2_{\varepsilon_n})| > 2L + 2\delta > 0$. In this case, we have

$$0 < 2L + 2\delta < |\varphi(t^1_{\varepsilon_n}) - \varphi(t^2_{\varepsilon_n})| \leq |\varphi(t^1_{\varepsilon_n}) - \varphi(t)| + |\varphi(t^2_{\varepsilon_n}) - \phi(t)|.$$

If

$$h_n = \begin{cases} t^1_{\varepsilon_n} - t & \text{for } |\varphi(t^1_{\varepsilon_n}) - \varphi(t)| \geq |\varphi(t^2_{\varepsilon_n}) - \varphi(t)|, \\ t^2_{\varepsilon_n} - t & \text{for } |\varphi(t^1_{\varepsilon_n}) - \varphi(t)| < |\varphi(t^2_{\varepsilon_n}) - \varphi(t)|, \end{cases}$$

then $h_n \neq 0$ and for $\varepsilon_n \to 0$ as $n \to +\infty$ we have $\{h_n\} \in \mathbb{K}$ such that

$$\left|\frac{\varphi(t+h_n)-\varphi(t)}{h_n}\right| \geq \frac{L+\delta}{|h_n|} \to +\infty \text{ as } n \to +\infty.$$

This contradicts the condition $|D_{\{h_n\}}\varphi(t)| \leq L < +\infty$. □

The contingent derivatives can be introduced for vector-valued functions $\mathcal{I} \mapsto \mathbb{R}^n$ in a component-wise manner. The next result is inspired by [11]. It relates a strong solution of a finite-dimensional differential inclusion and its contingent derivative.

Lemma 4.5 (Wazewski's lemma) *Let a set-valued function* $F : \mathbb{R}^{n+1} \mapsto 2^{\mathbb{R}^n}$ *be defined, upper semi-continuous on a closed nonempty set* $\Omega \subset \mathbb{R}^{n+1}$ *and the set* $F(t, x)$ *be nonempty, compact and convex for all* $(t, x) \in \Omega$. *Let an absolutely continuous function* $x : \mathbb{R} \mapsto \mathbb{R}^n$ *be defined on* $\mathcal{I}$ *such that* $(t, x(t)) \in \Omega$ *for* $t \in \mathcal{I}$. *Then*

$$\left.\begin{matrix} \dot{x}(t) \in F(t, x(t)) \\ \textit{almost everywhere on } \mathcal{I} \end{matrix}\right\} \quad \Leftrightarrow \quad \begin{matrix} D_{\mathbb{K}}x(t) \subseteq F(t, x(t)) \\ \textit{everywhere on } \mathcal{I}. \end{matrix}$$

Proof Let $t_0 \in \operatorname{int} \mathcal{I}$. Since the set-valued function F is upper semi-continuous and x is continuous then for any $\varepsilon > 0$ there exists $\delta \in (0, \varepsilon)$ such that

$$F(t, x(t)) \subset F^{\varepsilon}_{t_0, x(t_0)} \quad \text{for} \quad |t - t_0| < \delta,$$

where $F^{\varepsilon}_{t_0,x(t_0)} = \operatorname{co} F(t_0, x(t_0) \dot{+} B(\varepsilon))$, $B(\varepsilon)$ is (as before) the open ball in $\mathbb{R}^n$ of the radius ε centered at the origin and $\dot{+}$ is the geometric sum of the sets. Since the function x is absolutely continuous then for nonzero $h \in (-\delta, \delta)$ we have

$$\frac{x(t_0 + h) - x(t_0)}{h} = \frac{1}{h} \int_{t_0}^{t_0+h} \dot{x}(s)\, ds \subseteq F^{\varepsilon}(t_0, x(t_0)).$$

This implies that $D_{\mathbb{K}} x(t_0) \subset F^{\varepsilon}(t_0, x(t_0))$ (see, the definition of a contingent derivative). We complete the proof taking the limit $\varepsilon \to 0$. □

Notice, if $\mathcal{I} = [a, b]$ then one-sided derivatives are assumed to exist at $t = a$ and $t = b$ in the statement of the above lemma. One of the conclusions of Wazewski's lemma is a local Lipschitz continuity of solutions of the ordinary differential inclusions studied in Chap. 2. For more details about Wazewski's lemma and its corollaries, we refer the reader to [11].

4.2.2 Dini Derivatives

A monotonicity can be also studied using the so-called Dini derivatives [9]:

Definition 4.2 • *Right-hand Upper Dini derivative:*

$$D^{+}\varphi(t) = \limsup_{h \to 0^{+}} \frac{\varphi(t + h) - \varphi(t)}{h}.$$

• *Right-hand Lower Dini derivative:*

$$D_{+}\varphi(t) = \liminf_{h \to 0^{+}} \frac{\varphi(t + h) - \varphi(t)}{h}.$$

• *Left-hand Upper Dini derivative:*

$$D^{-}\varphi(t) = \limsup_{h \to 0^{-}} \frac{\varphi(t + h) - \varphi(t)}{h}.$$

• *Left-hand Lower Dini derivative:*

$$D_{-}\varphi(t) = \liminf_{h \to 0^{-}} \frac{\varphi(t + h) - \varphi(t)}{h}.$$

Even if the function φ is continuous at $t \in \mathcal{I}$ all four Dini derivatives may have different values at t.

Example 4.3 If $\varphi(t) = (2 - \text{sign}(t))t\cos(1/t)$ then

$$D_{-}\phi(0) = -3, \quad D^{-}\phi(0) = 3, \quad D^{+}\phi(0) = 1, \quad D_{+}\phi(0) = -1.$$

Obviously, the Dini derivatives satisfy the relations $D_{+}\varphi(t) \leq D^{+}\varphi(t)$ and $D_{-}\varphi(t) \leq D^{-}\varphi(t)$. Moreover, they always belong to $D_{\mathbb{K}}\varphi(t)$.

Corollary 4.1 *For any function* $\varphi : \mathcal{I} \mapsto \mathbb{R}$ *one holds*

$$\inf D_{\pm}\varphi(t) \leq D_{\mathbb{K}}\varphi(t) \leq \sup D^{\pm}\varphi(t), \quad \forall t \in \text{int}\,\mathcal{I}. \tag{4.5}$$

Proof If $\{h_n\} \in \mathbb{K}$ then both sequences $h_n^+ = \max_{n\geq m}\{h_m, 0\}$ and $h_n^- = \min_{n\geq m}\{0, h_m\}$ are defined for all $n \in \mathbb{N}$. Since $h_n \neq 0$ for all $n \in \mathbb{N}$, then either $\{h_n^+\} \in \mathbb{K}$ or $\{h_n^-\} \in \mathbb{K}$ or both. Moreover, if a derivative number $D_{\{h_n\}}\varphi(t)$ exists then either $D_{\{h_n\}}\varphi(t) = D_{\{h_n^+\}}\varphi(t)$ or $D_{\{h_n\}}\varphi(t) = D_{\{h_n^-\}}\varphi(t)$ or both identities take the place. In the first case, we have

$$\begin{aligned} D^{+}\varphi(t) &= \limsup_{h\to 0^+} \tfrac{\varphi(t+h)-\varphi(t)}{h} = \lim_{h\to 0^+} \sup_{0<\delta<h} \tfrac{\varphi(t+\delta)-\varphi(t)}{\delta} \\ &\geq \lim_{h\to 0^+} \sup_{h_n^+\leq h} \tfrac{\varphi(t+h_n^+)-\varphi(t)}{h^+} = D_{\{h_n\}}\varphi(t). \end{aligned}$$

For the second case, we derive

$$\begin{aligned} D^{-\varphi}(t) &= \limsup_{h\to 0^-} \tfrac{\varphi(t+h)-\varphi(t)}{h} = \lim_{h\to 0^-} \sup_{h<\delta<0} \tfrac{\varphi(t+\delta)-\varphi(t)}{\delta} \\ &\geq \lim_{h\to 0^-} \sup_{h<h_n^-<0} \tfrac{\varphi(t+h_n^-)-\varphi(t)}{h^-} = D_{\{h_n\}}\varphi(t). \end{aligned}$$

Since the above considerations hold for any $\{h_n\} \in \mathbb{K}$ then $D_{\mathbb{K}}\varphi(t) \leq \sup D^{\pm}\varphi(t) = \sup\{D^{-}\varphi(t), D^{+}\varphi(t)\}$. The case of lower Dini derivatives can be treated similarly. The proof is complete. □

On boundaries of $\mathcal{I}$, one of inequalities (4.5) is fulfilled. Therefore, the condition of Lemma 4.2 can be rewritten using the two Dini derivatives: $\sup D^{\pm}\varphi(t) \leq 0$. Moreover, for a continuous function ϕ it is enough to consider just one of them [9].

Lemma 4.6 *Let $\varphi : \mathcal{I} \subset \mathbb{R} \mapsto \mathbb{R}$ be a continuous function. The following claims are equivalent*

- φ *is decreasing on* $\mathcal{I}$;
- $D_+\varphi(t) \leq 0$ *for every* $t \in \operatorname{int} \mathcal{I}$;
- $D_-\varphi(t) \leq 0$ *for every* $t \in \operatorname{int} \mathcal{I}$;
- $D^-\varphi(t) \leq 0$ *for every* $t \in \operatorname{int} \mathcal{I}$;
- $D^+\varphi(t) \leq 0$ *for every* $t \in \operatorname{int} \mathcal{I}$.

Proof The definition of the decreasing function immediately yields required properties for the Dini derivatives. Let us prove the converse result. Let us consider the function φ_ε defined as follows

$$\varphi_\varepsilon(t) = \varphi(t) - \varepsilon t, \quad t \in \mathcal{I},$$

where $\varepsilon > 0$ is an arbitrary small number. In this case, we have $D_+\phi_\varepsilon(t) \leq -\varepsilon$ for all $t \in \mathcal{I}$. Let us show that the function φ_ε is decreasing on $\mathcal{I}$. Suppose the contrary: $\exists t_0, t_1 \in \mathcal{I} : t_0 < t_1$ and $\varphi_\varepsilon(t_0) < \varphi_\varepsilon(t_1)$. Let $y \in (\varphi_\varepsilon(t_0), \varphi_\varepsilon(t_1))$, $\Delta = \{t \in (t_0, t_1) : y = \varphi_\varepsilon(t)\}$ and $t_2 = \sup_{t \in \Delta} t$. Then, due to the continuity of φ_ε, we have $t_2 \in (t_0, t_1)$, $\phi(t_2) = y$ and $\varphi_\varepsilon(t) > y$, $\forall t \in (t_2, t_1)$. This means $D_+\phi_\varepsilon(t_2) \geq 0$ and we obtain the contradiction. Finally, tending ε to zero using continuity of ϕ we complete the proof. The other Dini derivative can be treated similarly. The proof is complete. □

If a function φ is differentiable at t then $\dot{\varphi}(t) = D^+\varphi(t) = D^-\varphi(t) = D_+\varphi(t) = D_-\varphi(t)$.

Example 4.4 If $\phi(t) = -\sqrt{|t|}\,\mathrm{sign}(t)$ then $\dot{\phi}(t) = -\frac{1}{\sqrt{|t|}} < 0$ for $t \neq \mathbf{0}$ and

$$D_+\phi(0) = \liminf_{h \to 0^+} \frac{-1}{\sqrt{|h|}} = -\infty, \quad D^+\phi(0) = \limsup_{h \to 0^+} \frac{-1}{\sqrt{|h|}} = -\infty,$$

$$D^-\phi(0) = \limsup_{h \to 0^-} \frac{-1}{\sqrt{|-h|}} = -\infty, \quad D_-\phi(0) = \liminf_{h \to 0^-} \frac{-1}{\sqrt{|-h|}} = -\infty.$$

The example shows that all four Dini derivatives may coincide even if a function φ is not differentiable at t. The following theorem proves that a set of such points is of measure zero for any function φ.

Theorem 4.1 (Denjoy–Young–Saks theorem) *If $\varphi : \mathcal{I} \mapsto \mathbb{R}$ then the Dini derivatives of $\varphi(t)$ almost everywhere on $\mathcal{I}$ satisfy one of the following four conditions:*

- $\varphi(t)$ *has a finite derivative;*
- $D^+\varphi(t) = D_-\varphi(t)$ *is finite and* $D^-\varphi(t) = +\infty$, $D_+\varphi(t) = -\infty$;
- $D^-\varphi(t) = D_+\varphi(t)$ *is finite and* $D^+\varphi(t) = +\infty$, $D_-\varphi(t) = -\infty$;
- $D^-\varphi(t) = D^+\varphi(t) = +\infty$, $D_-\varphi(t) = D_+\varphi(t) = -\infty$.

The proof of this theorem requires rather cumbersome constructions. We refer the reader to [10, p. 65] for more details.

Denjoy–Young–Saks theorem has the following simple corollary, which is useful for Lyapunov analysis of non-smooth systems [12].

Corollary 4.2 *If a function* $\varphi : \mathbb{R} \mapsto \mathbb{R}$ *is defined on* $\mathcal{I}$, *then the equality* $D_{\mathbb{K}}\varphi(t) = \{-\infty\}$ *(resp.,* $D_{\mathbb{K}}\varphi(t) = \{+\infty\}$*) may hold only on a set* $\Delta \subseteq \mathcal{I}$ *of measure zero.*

4.2.3 Monotone Functions

As discussed in Chap. 2, the differential equations are traditional models of dynamical systems. They characterize an evolution (growth/decay) of system states. The differentiability is an expectable property of a monotone function. The following corollary [8] follows immediately from Theorem 4.1 and Lemma 4.6.

Corollary 4.3 *Any monotone (i.e., decreasing or increasing) function is differentiable almost everywhere.*

A composition of two monotone functions is monotone, so it satisfies the above corollary. If one of the functions is differentiable then the derivative of the composition admits the chain rule.

Lemma 4.7 *Let a function* $f : [0, T) \mapsto \mathbb{R}$ *be continuously differentiable and increasing, where* $T \leq +\infty$. *If* $\varphi : \mathcal{I} \mapsto [0, T)$ *is a decreasing function then for any* $t \in \operatorname{int}\mathcal{I}$ *we have*

$$D_{\mathbb{K}} f(\varphi(t)) \leq D_{\mathbb{K}}\varphi(t) \cdot \left(\liminf_{\varepsilon \to 0} f'(\varphi(t+\varepsilon)) \right) \tag{4.6}$$

and

$$\frac{d}{dt} f(\varphi(t)) \leq f'(\varphi(t))\dot{\varphi}(t) \text{ almost everywhere on } \mathcal{I}.$$

Proof Since φ is decreasing and f is differentiable with the derivative $f' \geq 0$ then for $h_n > 0$ we have

$$\frac{f(\varphi(t+h_n)) - f(\varphi(t))}{h_n} = \frac{-\int_{\varphi(t+h_n)}^{\varphi(t)} f'(x)dx}{h_n} \leq \frac{\varphi(t+h_n) - \varphi(t)}{h_n} \min_{x \in [\varphi(t+h_n), \varphi(t)]} f'(x) \leq 0,$$

but for $h_n < 0$ we derive

$$\frac{f(\varphi(t+h_n)) - f(\varphi(t))}{h_n} = \frac{\int_{\varphi(t)}^{\varphi(t+h_n)} f'(x)dx}{h_n} \leq \frac{\varphi(t+h_n) - \varphi(t)}{h_n} \min_{x \in [\varphi(t), \varphi(t+h_n)]} f'(x) \leq 0.$$

This proves formula (4.6). According to Corollary 4.3 the function φ is differentiable almost everywhere on $\mathcal{I}$. Hence, it is continuous almost everywhere on $\mathcal{I}$ and

$$D_{\mathbb{K}}\phi(t) \liminf_{\varepsilon \to \mathbf{0}} f'(\varphi(t+\varepsilon)) = f'(\varphi(t))\dot{\varphi}(t)$$

almost everywhere on $\mathcal{I}$. The proof is complete. □

According to Newton–Leibniz Formula (see Appendix B), the integration is an inverse operation to the differentiation. For any monotone function, this formula holds, at least, as an inequality.

Lemma 4.8 ([8], Chapt. VII, Theorem 5) *If a function* $\varphi : [a, b] \mapsto \mathbb{R}$ *is decreasing then*

$$\varphi(b) - \varphi(a) \leq \int_a^b \dot{\varphi}(\tau) d\tau, \tag{4.7}$$

where the integral is understood in the sense of Lebesgue.

The above lemma can be proven using Lebesgue's theorem about the differentiability of monotone functions and Fatou's lemma for non-negative Lebesgue measurable functions (see Appendix B).

The inequality (4.7) holds even for discontinuous monotone functions. However, the continuity of φ is not sufficient to fulfill (4.7) as the identity.

Example 4.5 Let $\mathcal{C} \subset [0, 1]$ be the so-called Cantor set (see Fig. 4.3):

$$\mathcal{C} = \bigcap_{n \in \mathbb{N}} C_n, \qquad C_{n+1} = \left(\tfrac{1}{3} C_n\right) \cup \left(\tfrac{2}{3} \dot{+} \tfrac{1}{3} C_n\right),$$

where $C_0 = [0, 1]$ and $\dot{+}$ denotes the geometric sum of sets. Let us consider the so-called Cantor function $\phi : [0, 1] \mapsto [0, 1]$ defined as follows:

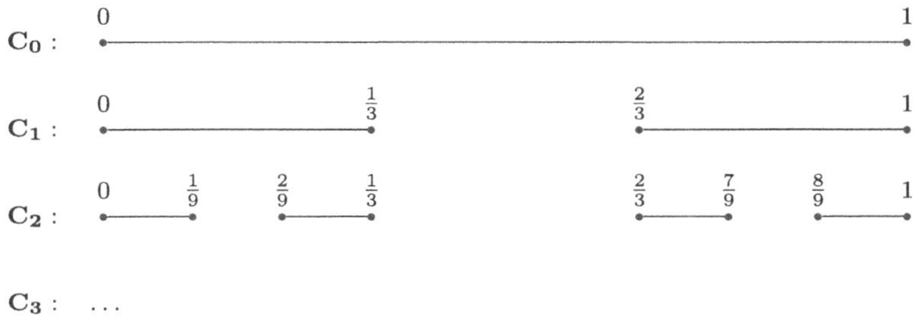

Fig. 4.3 The Cantor set (red points)

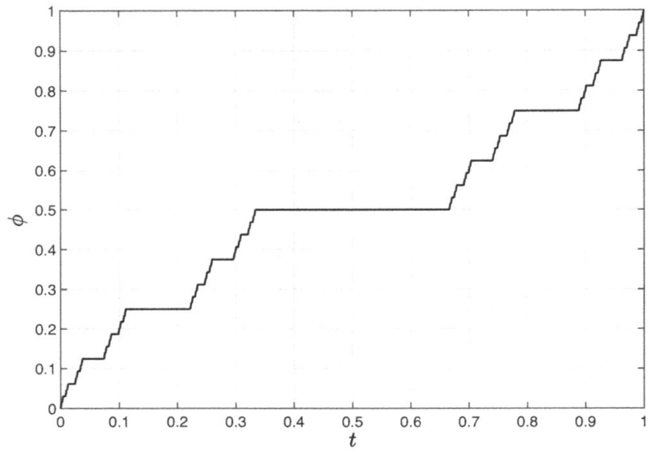

Fig. 4.4 The Cantor function

- if $t = \sum_{n=1}^{\infty} \frac{2a_n}{3^n} \in \mathcal{C}$ with $a_n \in \{0, 1\}$ then $\phi(t) = \sum_{n=1}^{\infty} \frac{a_n}{2^n}$;
- if $t \notin \mathcal{C}$ then $\phi(t) = \inf_{\delta \in [0,t] \cap \mathcal{C}} \phi(\delta)$.

The cantor function (see, Fig. 4.4) is continuous, increasing, $\phi(0) = 0$ and $\phi(1) = 1$ and its derivative equals to zero on $[0, 1] \backslash \mathcal{C}$. The function $\varphi = -\phi$ is continuous, decreasing, $\varphi(0) = 0$ and $\varphi(1) = -1$ and its derivative equals to zero on $[0, 1] \backslash \mathcal{C}$ as well. Since the cantor set is of measure zero then the integral in the right-hand side of (4.7) equals zero and the inequality is strict despite that the function φ is continuous.

Absolutely continuous functions[1] belongs to the class of functions, which excludes the singular functions such as the Cantor function considered above. For an absolutely continuous function, the sign-definiteness of its derivative almost everywhere (i.e., except a set of points of Lebesgue measure zero) is necessary and sufficient for monotonicity (see, e.g., [8, 9]).

Lemma 4.9 *An absolutely continuous function* $\varphi : \mathcal{I} \mapsto \mathbb{R}$ *is decreasing if and only if* $\dot{\varphi}(t) \leq 0$ *almost everywhere on* $\mathcal{I}$.

Proof The *Necessity* follows from Corollary 4.3 and Lemma 4.2. Let us prove the *Sufficiency*. Let the function φ_ε be defined as in the proof of Lemma 4.6. Let us show that φ_ε is decreasing for any $\varepsilon > 0$. Suppose the contrary: $\exists t_0, t_0 \in \mathcal{I} : t_0 < t_1$, $\varphi_\varepsilon(t_0) < \varphi_\varepsilon(t_1)$. Since $\dot{\varphi}_\varepsilon(t) = \dot{\varphi}(t) - \varepsilon$ then the set $\mathcal{I} \backslash Z$, where $Z = \{t \in \mathcal{I} : \dot{\varphi}_\varepsilon < 0\}$, is of measure zero. Since the function φ is absolutely continuous then the function φ_ε is absolutely continuous too and the set $\varphi(\mathcal{I} \backslash Z)$ is of measure zero. Therefore,

[1] A continuous function $x : \mathcal{I} \mapsto \mathbb{R}^n$ is said to be absolutely continuous if for any $\varepsilon > 0$ there exists $\delta > 0$ such that whenever a finite sequence of pairwise disjoint intervals (α_k, β_k) of I with $\alpha_k, \beta_k \in \mathcal{I}$ satisfies $\sum_k (\beta_k - \alpha_k) < \delta$ then $\sum_k \|x(\beta_k) - x(\alpha_k)\| < \varepsilon$.

there exists $y_0 \in (\varphi_\varepsilon(t_0), \varphi_\varepsilon(t_1))$ such that $y_0 \notin \varphi(\mathcal{I}\backslash\Delta)$. This means that $t \in (t_0, t_1)$: $\varphi_\varepsilon(t) = y_0 \Rightarrow t \in Z$. Let the set $\Delta(t_0, t_1)$ and the point $t_2 \in \Delta(t_0, t_1)$ be defined as in the proof of Lemma 4.6. Since $\varphi_\varepsilon(t_2) = y_0$ then $t_2 \in Z$, i.e., $\dot{\varphi}_\varepsilon(t_2) < 0$. On the other hand, by construction, $\varphi_\varepsilon(t) > \varphi_\varepsilon(t_2), \forall t \in (t_2, t_1)$, i.e., $D_+(t_2) \geq 0$. So, we derive a contradiction. Tending ε to zero and using the continuity of φ we complete the proof. □

For an absolutely continuous function $\phi : [a, b] \mapsto \mathbb{R}$ the inequality (4.7) *is always fulfilled as the identity known as the Newton–Leibniz formula (see Appendix B or* [8, Chap. IX, Sect. 4, Theorem 2]).

4.2.4 Directional Derivatives

Let $\mathbb{M}$ be the set of sequences converging to $\mathbf{0} \in \mathbb{R}^n$, i.e.,

$$\{v_n\} \in \mathbb{M} \quad \Leftrightarrow \quad |v_n| \to 0 \text{ as } n \to +\infty, \; v_n \in \mathbb{R}^n.$$

As a straightforward extension of Definitions (4.1) and 4.2 generalized directional derivatives can be introduced as follows.

Definition 4.3 Let Ω be an open nonempty set in $\mathbb{R}^n$.

- The number

$$D_{\{h_n\},\{v_n\}} V(x; d) = \lim_{n \to +\infty} \frac{V(x + h_n(d + v_n)) - V(x)}{h_n},$$

where $\{h_n\} \in \mathbb{K}$, $\{v_n\} \in \mathbb{M} : x + h_n(d + v_n) \in \Omega$, is called a *directional derivative number* of a function $V : \Omega \subset \mathbb{R}^n \mapsto \mathbb{R}$ at the point $x \in \Omega$ in the direction $d \in \mathbb{R}^n$, if a finite or an infinite limit exists.
- The set of all directional derivative numbers of the function V at the point $x \in \Omega$ in the direction $d \in \mathbb{B}$ is called a **directional contingent derivative:**

$$D_{\mathbb{K},\mathbb{M}} V(x; d) = \bigcup_{\{h_n\} \in \mathbb{K}, \{v_n\} \in \mathbb{M}} \{D_{\{h_n\},\{v_n\}} V(x; d) \subset \overline{\mathbb{R}}, \qquad d \in \mathbb{R}^n.$$

- The number

$$DV(x; d) := \lim_{h \to 0} \frac{V(x + hd) - V(x)}{h}$$

is called a **directional derivative** of V at a point $x \in \Omega$ in the direction $d \in \mathbb{R}^n$, if a finite or infinite limit exists.
- The finite or infinite number

$$DV^{+}(x;d) = \limsup_{h \to 0^{+}} \frac{V(x+hd)-V(x)}{h}$$

is called a **Right-hand Upper Dini directional derivative** of V at a point $x \in \Omega$ in the direction $d \in \mathbb{R}^n$.

- The finite or infinite number

$$DV_{+}(x;d) = \liminf_{h \to 0^{+}} \frac{V(x+hd)-V(x)}{h}$$

is called a **Right-hand Lower Dini directional derivative** of V at a point $x \in \Omega$ in the direction $d \in \mathbb{R}^n$.

- The finite or infinite number

$$DV^{-}(x;d) = \limsup_{h \to 0^{-}} \frac{V(x+hd)-V(x)}{h}$$

is called a **Left-hand Upper Dini directional derivative** of V at a point $x \in \Omega$ in the direction $d \in \mathbb{R}^n$.

- The finite or infinite number

$$DV_{-}(x;d) = \liminf_{h \to 0^{-}} \frac{V(x+hd)-V(x)}{h}$$

is called a **Left-hand Lower Dini directional derivative** of V at a point $x \in \Omega$ in the direction $d \in \mathbb{R}^n$.

A Lyapunov function-based stability analysis requires a calculation of derivatives of positive definite functions along trajectories of a system. If the Lyapunov function is non-differentiable, the above generalized directional derivatives can be used in this case (see, for example, [13–15]).

The directional contingent derivative always exists, i.e., the set $D_{\mathbb{K},\mathbb{M}}V(x;d)$ is nonempty for any function $V : \Omega \subset \mathbb{R}^n \mapsto \mathbb{R}$, any $x \in \Omega$, any $d \in \mathbb{R}^n$. The proof of the following lemma follows the proof of Lemma 4.1.

Lemma 4.10 *Let a functional $V : \Omega \subset \mathbb{R}^n \mapsto \mathbb{R}$ be defined on a neighborhood Ω of a point $x \in \mathbb{R}^n$. Then $D_{\mathbb{K},\mathbb{M}}V(x)d \neq \emptyset$ for any $d \in \mathbb{R}^n$.*

Proof Let us consider a sequence $\{h_n\}$ from $\mathbb{K}$ and a sequence $\{v_n\}$ from $M(d)$. Since $d \in \mathbb{R}^n$ then $\|d\| < +\infty$ and $\|h_n v_h\| \to 0$ as $n \to +\infty$. The set Ω is a neighborhood of x, so for a sufficiently large $N^* \in \mathbb{N}$ we have $x + h_n v_n \in \Omega$, $\forall n \geq N^*$. Therefore, the sequence of real numbers $\phi_n = \frac{V(x+h_n v_n)-V(x)}{h_n}$ is well defined. If the sequence of real numbers $\{\phi_n\}$ is bounded then, according to Bolzano–Weierstrass theorem, it contains a convergent subsequence. This means that there exist subsequence $\{n_k\}$ such that $\{\phi_{n_k}\}$ converges to a finite number. On the other hand, if the sequence $\{\phi_n\}$ is unbounded then it contains a subsequence that tends to $+\infty$ or $-\infty$, so the corresponding directional derivative number is $+\infty$ or $-\infty$, respectively. The proof is complete. □

Repeating the proof of Corollary 4.1 one can be shown

$$\inf DV_{\pm}(x; d) \le D_{\mathbb{K},\mathbb{M}}V(x; d) \le \sup DV^{\pm}(x; d).$$

Chain rule is an important result of calculus allowing to compute derivatives of compositions of functions. It exists even for generalized derivatives (see, e.g., [8, 16]). Inspired by [12], we present the following chain rule for the contingent derivatives of discontinuous functions.

Lemma 4.11 (Chain rule for discontinuous functions) *Let $t \in \operatorname{int}\mathcal{I}$, $x : \mathcal{I} \mapsto \mathbb{R}^n$ and the set $D_{\mathbb{K}}x(t) \subset \mathbb{R}^n$ be bounded. If $V : \Omega \subset \mathbb{R}^n \mapsto \mathbb{R}$ and $x(t) \in \operatorname{int}\Omega$ then*

$$D_{\mathbb{K}}V(x(t)) \subset \bigcup_{d \in D_{\mathbb{K}}x(t)} D_{\mathbb{K},\mathbb{M}}V(x(t); d). \tag{4.8}$$

Proof Since $D_{\mathbb{K}}x(t) \subset \mathbb{R}^n$ be bounded then x is continuous at t. Since $x(t) \in \operatorname{int}\Omega$ and $t \in \operatorname{int}\mathcal{I}$ then there exists a neighborhood $\tilde{\mathcal{I}}(t)$ of the point t such that $V(x(\cdot)) : \tilde{\mathcal{I}}(t) \mapsto \mathbb{R}$. Lemma 4.1 implies that $D_{\mathbb{K}}V(x(t))$ is nonempty for any $t \in \tilde{\mathcal{I}}(t)$. Let $D_{\{h_n\}}V(x(t)) \in D_{\mathbb{K}}V(x(t))$ be an arbitrary derivative number, i.e., by Definition 4.1 a finite or infinite limit

$$\lim_{n\to+\infty} \tfrac{V(x(t+h_n))-V(x(t))}{h_n} = D_{\{h_n\}}V(x(t)), \quad \{h_n\} \in \mathbb{K} : t + h_n \in \mathcal{I}$$

exists. Since $D_{\mathbb{K}}x(t) \subset \mathbb{R}^n$ is bounded then the sequence

$$\phi_n = \tfrac{x(t+h_n)-x(t)}{h_n}$$

is bounded as well and, by Bolzano–Weierstrass theorem, it contains a convergent subsequence $\phi_{n_k} \to d = D_{\{h_{n_k}\}}x(t) \in D_{\mathbb{K}}x(t) \subset \mathbb{R}^n$ as $k \to +\infty$. Then, denoting $v_{n_k} = \phi_{n_k} - d$, we derive

$$D_{\{h_n\}}V(x(t)) = \lim_{n\to+\infty} \tfrac{V(x(t+h_n))-V(x(t))}{h_n} = \lim_{k\to+\infty} \tfrac{V(x(t+h_{n_k}))-V(x(t))}{h_{n_k}} =$$

$$\lim_{k\to+\infty} \tfrac{V(x(t)+h_{n_k}(d+v_{n_k}))-V(x(t))}{h_{n_k}} = D_{\{h_{n_k}\},\{v_{n_k}\}}V(x(t), d) \in D_{\mathbb{K},\mathbb{M}(d)}V(x(t); d).$$

The proof is complete. □

A similar result can be proven for directional Dini derivatives.

Lemma 4.12 (Chain rule for continuous functions) *If under conditions of Lemma 4.11 the function $V : \mathbb{R}^n \mapsto \mathbb{R}$ is continuous then*

$$\exists d^{\pm} \in D_{\mathbb{K}}x(t) \quad : \quad D_{\pm}V(x(t)) \le D^{\pm}V(x(t); d^{\pm}), \tag{4.9}$$

$$\exists d_{\pm} \in D_{\mathbb{K}}x(t) \quad : \quad D^{\pm}V(x(t)) \ge D_{\pm}V(x(t); d_{\pm}). \tag{4.10}$$

Proof Let $\{h_n\} \in \mathbb{K}$ such that $h_n > 0$. On the one hand, since $D_{\mathbb{K}}x(t)$ is bounded then there exists $N \in \mathbb{N}$ such that $\frac{x(t+h_n)-x(t)}{h_n}$ is bounded for all $n \geq N^*$, and, by Bolzano–Weierstrass theorem, there exists a subsequence $\{h_{n_k}\}$ such that $\lim_{k\to+\infty} \frac{x(t+h_{n_k})-x(t)}{h_{n_k}} = d^+ \in D_{\mathbb{K}}x(t) \in \mathbb{R}^n$. The continuity of V implies that for any fixed $\delta_1 \neq 0$ it holds

$$\inf_{0<\delta_2<h} \frac{V\left(x(t)+\delta_1 \frac{x(t+\delta_2)-x(t)}{\delta_2}\right)-V(x(t)+\delta_1 d^+)}{\delta_1} \leq 0.$$

On the other hand, we have

$$\begin{aligned}
D_+V(x(t)) &= \liminf_{h\to 0^+} \frac{V(x(t+h))-V(x(t))}{h} \\
&= \lim_{h\to 0^+} \inf_{0<\delta<h} \frac{V(x(t+\delta))-V(x(t))}{\delta} \\
&= \lim_{h\to 0^+} \inf_{0<\delta<h} \frac{V\left(x(t)+\delta \frac{x(t+\delta)-x(t)}{\delta}\right)-V(x(t))}{\delta} \\
&\leq \lim_{h\to 0^+} \sup_{0<\delta_1<h} \inf_{0<\delta_2<h} \frac{V\left(x(t)+\delta_1 \frac{x(t+\delta_2)-x(t)}{\delta_2}\right)-V(x(t))}{\delta_1} \\
&\leq \lim_{h\to 0^+} \sup_{0<\delta_1<h} \frac{V\left(x(t)+\delta_1 d^+\right)-V(x(t))}{\delta_1} \\
&\quad + \inf_{0<\delta_2<h} \frac{V\left(x(t)+\delta_1 \frac{x(t+\delta_2)-x(t)}{\delta_2}\right)-V(x(t)+\delta_1 d^+)}{\delta_1} \\
&\leq \lim_{h\to 0^+} \sup_{0<\delta_1<h} \frac{V\left(x(t)+\delta_1 d^+\right)-V(x(t))}{\delta_1} = D^+V(x(t); d^+).
\end{aligned}$$

The relations for other Dini derivatives can be derived similarly. □

Taking into account Lemma 4.6 we conclude that to study the monotonicity of a composition of continuous functions it is enough to consider just one of the directional Dini derivatives.

Example 4.6 Let $V(x) = \sqrt{|x_1|}\,\text{sign}(x_1) + x_2$ with $x = (x_1, x_2) \in \mathbb{R}^2$ and $x(t) = t\left[\begin{smallmatrix}-1\\1\end{smallmatrix}\right]$. Let us check if the continuous function $t \mapsto V(x(t))$ is decreasing on the time interval $[-1, 1]$. The function V is not differentiable on the line $x_1 = 0$ and not locally Lipschitz. However, according to Lemma 4.6 it is necessary and sufficient to show that one of the Dini derivatives of the function is non-positive for all $t \in (-1, 1)$. Since the function $t \mapsto x(t)$ is differentiable and $D_{\mathbb{K}}x(t) = \left[\begin{smallmatrix}1\\-1\end{smallmatrix}\right]$ then in the view of Lemma 4.12 it is sufficient to show that $D^+V\left(x(t), \left[\begin{smallmatrix}1\\-1\end{smallmatrix}\right]\right) \leq 0$ for $t \in (-1, 1)$. Simple calculations show

$$D^+V\left(x, \left[\begin{smallmatrix}1\\-1\end{smallmatrix}\right]\right) = \limsup_{h\to 0^+} \frac{\sqrt{|x_1-h|}\,\text{sign}(x_1-h)+x_2+h-\sqrt{|x_1|}\,\text{sign}(x_1)-x_2}{h} =$$

$$1 + \limsup_{h\to 0^+} \frac{|x_1-h|-|x_1|}{h(\sqrt{|x_1-h|}\,\text{sign}(x_1-h)+\sqrt{|x_1|}\,\text{sign}(x_1))} = \begin{cases} -\frac{1}{\sqrt{|x_1|}}+1 & \text{if } x_1 \neq 0, \\ -\infty & \text{if } x_1 = 0. \end{cases}$$

Hence, if $|x_1(t)| \le 1$ for $t \in [-1, 1]$ then $D^+V\left(x(t), \left[\begin{smallmatrix} 1 \\ -1 \end{smallmatrix}\right]\right) \le 0$ for $t \in (-1, 1)$ and the continuous function $t \mapsto V(x(t))$ is decreasing on $[-1, 1]$.

If a function x is differentiable at t, then the inclusion (4.8) becomes

$$D_{\mathbb{K}}V(x(t)) \subset D_{\mathbb{K},\mathbb{M}}V(x(t); \dot{x}(t)).$$

As a corollary of the Rademacher theorem (see Appendix D) one can be shown that this inclusion is fulfilled as the identity if V is locally Lipschitz continuous.

Lemma 4.13 (Chain rule for locally Lipschitz continuous functions) *If under conditions of Lemma 4.11 the function V satisfies a Lipschitz condition in a neighborhood of $x(t)$ and x is differentiable at t then*

$$D_{\mathbb{K}}V(x(t)) = D_{\mathbb{K},\mathbb{M}}V(x(t); \dot{x}(t)) \subset \mathbb{R}, \tag{4.11}$$

$$\inf D_{\pm}V(x(t); \dot{x}(t)) \le D_{\mathbb{K}}V(x(t)) \le \sup D^{\pm}V(x(t); \dot{x}(t)) \in \mathbb{R}. \tag{4.12}$$

Proof If $\{h_n\} \in \mathbb{K}$ and $\{v_n\} \in \mathbb{M}$ then

$$\begin{aligned} D_{\{h_n\},\{v_n\}}V(x(t), \dot{x}(t)) &= \lim_{n\to+\infty} \tfrac{V(x(t)+h_n(\dot{x}(t)+v_n))-V(x(t))}{h_n} \\ &= \lim_{n\to+\infty} \tfrac{V(x(t)+h_n(\dot{x}(t)+v_n))-V(x(t+h_n))}{h_n} \\ &+ \tfrac{V(x(t+h_n))-V(x(t))}{h_n}. \end{aligned}$$

Since V satisfies the Lipschitz condition then

$$\left|\tfrac{V(x(t)+h_n(\dot{x}(t)+v_n))-V(x(t+h_n))}{h_n}\right| \le L|v_n| + L\left|\tfrac{x(t+h_n)-x(t)}{h_n} - \dot{x}(t)\right| \to 0$$

as $n \to +\infty$, where $L > 0$ is a Lipschitz constant. This means that

$$D_{\{h_n\},\{v_n\}}V(x(t); \dot{x}(t)) = \lim_{n\to+\infty} \tfrac{V(x(t)+h_n\dot{x}(t)))-V(x(t))}{h_n} = D_{\{h_n\}}V(x(t))$$

independently of $\{v_n\} \in \mathbb{M}$. Hence, taking into account the inclusion (4.8), we conclude that the set-valued identity (4.11) holds. Similarly, we have

$$\begin{aligned} D_{\{h_n\},\{v_n\}}V(x(t), \dot{x}(t)) &= \lim_{n\to+\infty} \tfrac{V(x(t)+h_n(\dot{x}(t)+v_n))-V(x(t))}{h_n} = \\ &= \lim_{n\to+\infty} \tfrac{V(x(t)+h_n(\dot{x}(t)+v_n))-V(x(t)+h_n\dot{x}(t))}{h_n} \\ &+ \tfrac{V(x(t)+h_n\dot{x}(t))-V(x(t)}{h_n} \end{aligned}$$

and using the Lipschitz condition we derive

$$\left|\frac{V(x(t)+h_n(\dot{x}(t)+v_n))-V(x(t)+h_n\dot{x}(t))}{h_n}\right| \leq L\|v_n\| \to 0 \text{ as } n \to +\infty.$$

Therefore, we obtain

$$\begin{aligned}\inf D_{\pm}V(x(t);\dot{x}(t)) \leq & D_{\{h_n\},\{v_n\}}V(x(t),\dot{x}(t)) = \lim_{n\to+\infty} \tfrac{V(x(t)+h_n\dot{x}(t))-V(x(t)}{h_n} \\ \leq & \sup D^{\pm}V(x(t);\dot{x}(t)).\end{aligned}$$

The proof is complete. □

The above lemma provides an alternative formula for the chain rule of Lipschitz continuous functions [17] written in terms of Clarke's Gradient (see [15] or Appendix D). If the functions $V:\mathbb{R}^n \mapsto \mathbb{R}$ and $x:\mathbb{R}\mapsto\mathbb{R}^n$ are differentiable then the composition function $t\mapsto V(x(t))$ is differentiable and formula (4.11) becomes the classical chain rule of calculus:

$$\dot{V}(x(t)) = \left.\frac{\partial V(z)}{\partial z}\right|_{z=x(t)} \dot{x}(t).$$

The generalized derivatives and chain rules considered above are utilized in this chapter for the non-smooth Lyapunov analysis.

4.3 Lyapunov Function Theorems

4.3.1 Positive Definiteness and Generalized Properness

Let $\Omega \subset \mathbb{R}^n$ be a neighborhood of zero. A function $W:\Omega\subset\mathbb{R}^n\mapsto\mathbb{R}$ is said to be *positive definite* if

$$W(0)=0 \quad \text{and} \quad W(x)>0 \quad \text{for} \quad x\in\Omega\backslash\{\mathbf{0}\}.$$

Lyapunov function candidates in $\mathbb{R}^n$ are positive definite and proper (see, e.g., [3, 18]). A mapping $V:\mathbb{R}^n\mapsto\mathbb{R}$ is *proper* if an inverse image of any compact set is a compact set. For "generalized" proper functions introduced in Definition 4.4, an inverse image of any compact set belongs to a bounded set (which may not be closed in the general case). Below the word "generalized" is omitted for shortness.

Definition 4.4 A positive definite function $V:\mathbb{R}\times\mathbb{R}^n\mapsto\mathbb{R}$ is said to be

- proper at **0** (**locally proper**) if there exist $\underline{V},\overline{V}\in\mathcal{K}_\infty$ such that

$$\underline{V}(|x|)\leq V(t,x)\leq\overline{V}(|x|), \qquad \forall t\in\mathbb{R}, \quad \forall x\in\Omega,$$

  where Ω is a neighborhood of **0**.
- **globally proper** if the above inequalities hold for $\Omega=\mathbb{R}^n$.

The locally and globally proper positive definite functions are conventional Lyapunov function candidates for stability analysis of dynamical systems (see [1, 19]). To analyze a decay of a non-smooth function V along trajectories of a system, the generalized derivatives considered above can be utilized.

Example 4.7 The function $V : \mathbb{R}^2 \mapsto \mathbb{R}$ defined as

$$V(x, y) = (2 + \text{sign}(xy))\sqrt{x^2 + y^2}$$

is globally proper (in generalized sense).

If $V : \mathbb{R}^n \mapsto \mathbb{R}$ is continuous on $\mathbb{R}^n$, then the generalized properness is equivalent to the classical one. The property introduced by Definition 4.4 is related with the notion of *coercivity* of Lyapunov functions discussed, for example, in [20].

4.3.2 Lyapunov Stability

Let us consider the nonlinear system

$$\dot{x} \in F(t, x), \quad t > t_0, \quad x(t_0) = x_0, \tag{4.13}$$

where $t_0 \in \mathbb{R}$ is an initial instant of time, $x(t) \in \mathbb{R}^n$ is the system state, and the (possibly time-varying) vector field $F : \mathbb{R} \times \mathbb{R}^n \mapsto 2^{\mathbb{R}^n}$ satisfy one of the following assumptions

A1 (*locally Lipschitz continuous solutions*): F is compact-valued, convex-valued and upper semi-continuous;
A2 (*absolutely continuous solutions*):
$F(t, x) = K[f](t, x)$, where $K[f]$ is a Filippov regularization of the vector field $f : \mathbb{R} \times \mathbb{R}^n \mapsto \mathbb{R}^n$ being measurable in t, piece-wise continuous in x such that on any compact in $\mathbb{R} \times \mathbb{R}^n$ there exists $m \in L^1$ satisfying $|f(t, x)| \leq m(t)$.

Notice that under each of the above assumptions, the system (4.13) may correspond to an ODE

$$\dot{x} = f(t, x), \quad t > t_0$$

with continuous or discontinuous right-hand side. Indeed, if f is continuous then Assumption **A1** (as well as Assumption **A2**) is fulfilled for $F(t, x) = \{f(t, x)\}$. If f is piece-wise continuous then the Filippov regularization $K[f]$ of f is compact-valued, convex-valued and upper semi-continuous on both variables t and x, so Assumption **A1** is fulfilled for $F(t, x) = K[f](t, x)$. However, if f is measurable on t and continuous on x then Assumption **A2** is fulfilled for $F(t, x) = \{f(t, x)\}$.

The system (4.13) under Assumption **A1** or **A2** has, at least, strong solution[2] defined, at least, locally in time. Due to Lemma 4.5, Assumption **A1** guarantees that any solution of the system (4.13) is locally Lipschitz continuous. For more details about solution existence and uniqueness problems, we refer the reader to Chap. 2 or to the monographs [11, 21]. Inspired by [22] we prove the following theorem.

Theorem 4.2 *The system* (4.13) *is locally (globally) uniformly Lyapunov stable if and only if there exists a locally (globally) proper function* $V : \mathbb{R} \times \mathbb{R}^n \mapsto \mathbb{R}_+$ *such that the inequality*

$$D_{\mathbb{K}} V(t, x(t, t_0, x_0)) \leq 0, \qquad \forall t > t_0 \tag{4.14}$$

holds for any $t_0 \in \mathbb{R}$ *and any solution* $x(\cdot, t_0, x_0)$ *of* (4.13) *as long as*

$$x(t, t_0, x_0) \in \Omega \backslash \{\mathbf{0}\},$$

where Ω *is a neighborhood of the origin (resp.,* $\Omega = \mathbb{R}^n$*).*

Proof *Sufficiency*. Given $x_0 \in \Omega$ let us consider an arbitrary solution $x_{t_0,x_0}(t)$ of (4.13). The inequality (4.14) and Lemma 4.2 imply that the function $t \mapsto V(x(t, t_0, x_0))$ is decreasing for $t > t_0$ as long as $x(t, t_0, x_0) \in \Omega\backslash\{\mathbf{0}\}$. The properness of V implies

$$\underline{V}(|x(t, t_0, x_0)|) \leq V(t, x(t, t_0, x_0)) \leq V(t_0, x_0) \leq \overline{V}(|x_0|). \tag{4.15}$$

Since $\underline{V} \in \mathcal{K}^\infty$ then there exists $\underline{V}^{-1} \in \mathcal{K}^\infty$ such that $\underline{V}^{-1}(\underline{V}(\rho))=\rho, \forall \rho \in [0, \rho^{\max})$ for some $\rho^{\max} > 0$ (resp., $\forall \rho \in \mathbb{R}_+$ in the case of global stability analysis). Hence, for $\varepsilon := \underline{V}^{-1}(\overline{V}) \in \mathcal{K}^\infty$ we have

$$|x(t, t_0, x_0)| \leq \varepsilon(|x_0|)$$

as long as $x(t, t_0, x_0) \in \Omega\backslash\{\mathbf{0}\}$ (resp., $x(t, t_0, x_0) \neq \mathbf{0}$). Therefore, the function ε (independent of t_0) satisfies Definition 3.1 for any $x_0 \in B(R)$ (resp., any $x_0 \in \mathbb{R}^n$), where $R > 0$ is such that $\overline{V}(R) < \rho^{\max}$ and $B(\varepsilon(R)) \subset \Omega$. If $\exists t' \geq t_0 : x(t', t_0, x_0) = \mathbf{0}$ then repeating the proof of Proposition 3.1 we derive $x(t, t_0, x_0) = \mathbf{0}, \forall t \geq t'$.

Necessity. Let S_{t_0,x_0} denote a set of solutions of the system (4.13) with the initial condition $x(t_0) = x_0$. Let the system (4.13) be uniformly Lyapunov stable. In this case, the function $V : \mathbb{R} \times \Omega \mapsto \mathbb{R}_+$ given by

$$V(t_0, x_0) := \sup_{x(\cdot,t_0,x_0)\in S_{t_0,x_0}} \sup_{t\geq t_0} |x(t, t_0, x_0)| \tag{4.16}$$

[2] A strong solution of a differential equation/inclusion is an absolutely continuous function satisfying a differential equation/inclusion almost everywhere in time.

is locally (globally) proper. Indeed, by construction, V is non-negative and $\underline{V}(|x_0|) := |x_0| \leq V(t_0, x_0)$. The definition of local (global) uniform Lyapunov stability (see Definition 3.1 with ε independent of t_0) implies that $V(t_0, x_0) \leq \varepsilon(|x_0|)$ for all $x_0 \in \Omega$ and all $t_0 \in \mathbb{R}$, where $\varepsilon \in \mathcal{K}$ (resp., $\varepsilon \in \mathcal{K}^{\infty}$) and Ω is a neighborhood of $\mathbf{0}$ (resp., $\Omega = \mathbb{R}^n$). Finally, due to the uniform Lyapunov stability of the origin, the function $t \mapsto V(t, x(t, t_0, x_0))$ is decreasing and, by Lemma 4.2, $D_{\mathbb{K}} V(t, x(t, t_0, x_0)) \leq 0$ as long as $x(t, t_0, x_0) \in \Omega \backslash \{\mathbf{0}\}$. □

The above theorem is universal in the sense that it does not require Assumption **A1** or **A2**. It uses just continuity and the semigroup property of solutions to prove the Lyapunov stability. So, it remains valid even for abstract evolution systems studied in Volume II of this monograph. However, formally, this theorem needs a knowledge of the solutions of the system to analyze the stability. Under Assumptions **A1** or **A2** the condition (4.14) may be checked without finding a solution of (4.13) even if the function V is not differentiable (see Example 4.10). Indeed, to prove the Lyapunov stability it is sufficient to show that the function $t \mapsto V(x_{t_0,x_0}(t))$ is decreasing. Various chain rules obtained in the previous sections can be utilized for this purpose.

Let us introduce *a time derivative of the function V along trajectories of the system* (4.13) dependently on assumption on its right-hand side.

Definition 4.5 Let $V : \mathbb{R}^n \mapsto \mathbb{R}$ be a locally or globally proper function.

- Under assumption **A1** let us define

$$\dot{V}(t, x) := \begin{cases} \sup\limits_{d \in F(t,x)} \frac{\partial V(x)}{\partial x} d & \text{if } V \in C^1(\mathbb{R}^n \backslash \{\mathbf{0}\}; \mathbb{R}), \\ \sup\limits_{d \in F(t,x)} D^+ V(x; d) & \text{if } V \in C(\mathbb{R}^n; \mathbb{R}), \\ \bigcup\limits_{d \in F(t,x)} D_{\mathbb{K},\mathbb{M}} V(x; d) & \text{otherwise.} \end{cases} \tag{4.17}$$

- Under assumption **A2** let us define

$$\dot{V}(t, x) := \begin{cases} \frac{\partial V(x)}{\partial x} f(t,x) & \text{if } V \in C^1(\mathbb{R}^n \backslash \{\mathbf{0}\}; \mathbb{R}), \\ \sup\limits_{d \in K[f](t,x)} D^+ V(x; d) & \text{if } V \text{ is locally Lipschitz on } \mathbb{R}^n \backslash \{\mathbf{0}\}. \end{cases} \tag{4.18}$$

Obviously, if $F(t, x) = \{f(t, x)\}$ and f is continuous then

$$\dot{V}(t, x) := \begin{cases} \frac{\partial V(x)}{\partial x} f(t,x) & \text{if } V \in C^1(\mathbb{R}^n \backslash \{\mathbf{0}\}; \mathbb{R}), \\ D^+ V(x; f(t,x)) & \text{if } V \in C(\mathbb{R}^n; \mathbb{R}) \\ D_{\mathbb{K},\mathbb{M}} V(x; f(t,x)) & \text{otherwise.} \end{cases} \tag{4.19}$$

It is worth stressing that under Assumption **A2** an admissible function V has to be at least locally Lipschitz continuous. If the system (4.13) is autonomous (i.e., if F is independent of time t) then $\dot{V}(t, x) = \dot{V}(x)$ is time-invariant. Using the function $\dot{V}$ introduced in Definition 4.5, the stability of various models of dynamical systems can be analyzed without finding solutions for the system. The following corollary merges Lyapunov theorems for various models and various classes of Lyapunov functions (see, e.g., [12, 22–24]).

Corollary 4.4 *If there exists a locally (globally) proper function* $V : \mathbb{R}^n \mapsto \mathbb{R}$ *such that*

$$\dot{V}(t, x) \leq 0, \quad \forall x \in \Omega \backslash \{\mathbf{0}\}, \quad \forall t \in \mathbb{R}, \tag{4.20}$$

then the system (4.13) *is locally (globally) uniformly Lyapunov stable, where* Ω *is a neighborhood of the origin (resp.,* $\Omega = \mathbb{R}^n$*). Moreover, if the condition* (4.20) *is fulfilled for almost all* $t \in \mathbb{R}$ *then the claim remains valid provided that* V *is locally Lipschitz continuous on* $\Omega \backslash \{\mathbf{0}\}$.

Proof If x_{t_0,x_0} is a solution of (4.13) under Assumption **A1** then by Lemma 4.5 the set $D_{\mathbb{K}}x(t) \in F(t, x(t))$ is bounded for any $t \geq t_0$ and the solution $x(\cdot, t_0, x_0)$ is locally Lipschitz continuous (see Lemma 4.4).

The case of a discontinuous function V**.** Using the chain rule (4.8) we derive $D_{\mathbb{K}}V(x(t, t_0, x_0)) \leq \dot{V}(t, x(t, t_0, x_0)) \leq 0$ as long as $x(t, t_0, x_0) \in \Omega \backslash \{\mathbf{0}\}$.

The case of a continuous function V**.** Using the chain rule (4.9) we derive $D_{+}V(x(t, t_0, x_0)) \leq \dot{V}(t, x(t, t_0, x_0))) \leq 0$ as long as $x(t, t_0, x_0) \in \Omega \backslash \{\mathbf{0}\}$. By Lemma 4.6 we conclude that the function $t \mapsto V(x(t, t_0, x_0))$ is decreasing and, by Lemma 4.2, $D_{\mathbb{K}}V(x(t, t_0, x_0)) \leq 0$ as long as $x(t, t_0, x_0) \in \Omega \backslash \{\mathbf{0}\}$.

The case of a locally Lipschitz continuous function V**.** Under Assumption **A2** any solution of (4.13) is, at least, absolutely continuous (see, Chap. 2). A composition of an absolutely continuous function $x(\cdot, t_0, x_0)$ and a locally Lipschitz continuous function V is an absolutely continuous function $t \mapsto V(x(t, t_0, x_0))$. Since $x(t, t_0, x_0)$ satisfies (4.13) almost everywhere then using the chain rule (4.11) we derive $D_{\mathbb{K}}V(x(t, t_0, x_0)) = D_{\mathbb{K},\mathbb{M}}V(x(t, t_0, x_0); \dot{x}(t, t_0, x_0)) \leq \dot{V}(t, x(t, t_0, x_0)) \leq 0$ for almost all $t > t_0$. By Lemma 4.9 we conclude that the function $t \mapsto V(x(t, t_0, x_0))$ is decreasing. Hence, using Lemma 4.2 we have $D_{\mathbb{K}}V(x(t, t_0, x_0)) \leq 0$ as long as $x(t, t_0, x_0) \in \Omega \backslash \{\mathbf{0}\}$.

The case of $V \in C^1(\Omega \backslash \{\mathbf{0}\}; \mathbb{R})$**.** An analysis under Assumptions **A1** literally repeats the case of a locally Lipschitz continuous function V. The analysis of the system (4.13) satisfying Assumption **A2** is inspired by [11]. By definition of the Filippov regularization for almost all t we have

$$K[f](t, x) = \left\{ \sum_{i=1}^{N} \alpha_i f^i(t, x) : \quad \forall \alpha_i \geq 0 : \sum_{i=1}^{N} \alpha_i = 1 \right\},$$

where $f^i(t, x) = \lim_{x_* \to x} f(t, x_*)$ are limit vectors of the piece-wise continuous vector field f at the discontinuity point $x \in \mathbb{R}^n$ (see, Chap. 2). On the one hand, we have

$$\begin{aligned} \frac{\partial V(x)}{\partial x} \sum_{i=1}^{N} \alpha_i f^i(t, x) &= \sum_{i=1}^{N} \alpha_i \lim_{x_* \to x} \frac{\partial V(x)}{\partial x} f(t, x_*) \\ &= \sum_{i=1}^{N} \alpha_i \lim_{x_* \to x} \left(\tfrac{\partial V(x_*)}{\partial x_*} + w\,(x - x_*) \right) f(t, x_*), \end{aligned}$$

where

$$w(x, x_*) = \frac{\partial V(x)}{\partial x} - \frac{\partial V(x_*)}{\partial x_*}$$

and due to continuity of $\frac{\partial V}{\partial x}$ there exists $\exists \omega \in \mathcal{K}$ such that $\|w(x, x_*)\| \leq \omega(\|x - x_*\|)$ Since we have $\dot{V}(t, x) = \frac{\partial V(x)}{\partial x} f(t, x) \leq 0$ for all $x \in \Omega\backslash\{\mathbf{0}\}$ and almost all $t > 0$, then

$$\sup_{d \in K[f](t,x)} \frac{\partial V(x)}{\partial x} d = \sup_{\forall \alpha_i \geq 0 : \sum_{i=1}^{N} \alpha_i = 1} \frac{\partial V(x)}{\partial x} \sum_{i=1}^{N} \alpha_i f^i(t, x) \leq 0$$

for all $x \in \Omega\backslash\{\mathbf{0}\}$ and almost all $t > 0$. On the other hand, since $\dot{x}(t, t_0, x_0) \in K[f](t, x(t, t_0, x_0))$ for almost all $t > t_0 \geq 0$ then

$$\dot{V}(x(t, t_0, x_0)) = \left.\frac{\partial V(z)}{\partial z} \dot{z}\right|_{z = x(t, t_0, x_0)} \leq \sup_{d \in K[f](t, x(t, t_0, x_0))} \left.\frac{\partial V(z)}{\partial z}\right|_{z = x(t, t_0, x_0)} d \leq 0$$

for almost all $t \geq t_0$. By Lemma 4.9 we conclude that the function $t \mapsto V(x(t, t_0, x_0))$ is decreasing. Hence, by Lemma 4.2 we have $D_{\mathbb{K}} V(x(t, t_0, x_0)) \leq 0$ for $t > t_0$ as long as $x(t, t_0, x_0) \in \Omega\backslash\{\mathbf{0}\}$. Applying Theorem 4.2 we complete the proof. □

Example 4.8 Let us consider the following differential inclusion

$$\dot{x} \in F(x) := [-1, 1] \begin{pmatrix} 0 & 1 \\ -1 & 0 \end{pmatrix} x, \quad x \in \mathbb{R}^2.$$

For $V(x) = x^\top x$ simple calculations show

$$\dot{V}(x) = 2 \sup_{y \in F(x)} x^\top y = 0, \quad \forall x \in \mathbb{R}^2.$$

According to Corollary 4.4 this guarantees the uniform Lyapunov stability of the origin of the considered differential inclusion.

For a continuously differentiable function $V \in C^1$, the time derivative $\dot{V}$ (see Definition 4.5) can be computed using (possibly discontinuous) vector field f only. However, for a locally Lipschitz continuous function $V \notin C^1$, the use of a Filippov regularization $K[f]$ of f in the definition of time derivative $\dot{V}$ is crucial (despite of almost everywhere differentiability of V). The computation of $\dot{V}$ using f (instead of $K[f]$) may lead to a wrong conclusion in this case.

Example 4.9 ([25]) Let us consider the system

$$\begin{cases} \dot{x}_1 = -2\,\mathrm{sign}(x_1) - \mathrm{sign}(x_2), \\ \dot{x}_2 = -2\,\mathrm{sign}(x_1) + \mathrm{sign}(x_2) \end{cases} \tag{4.21}$$

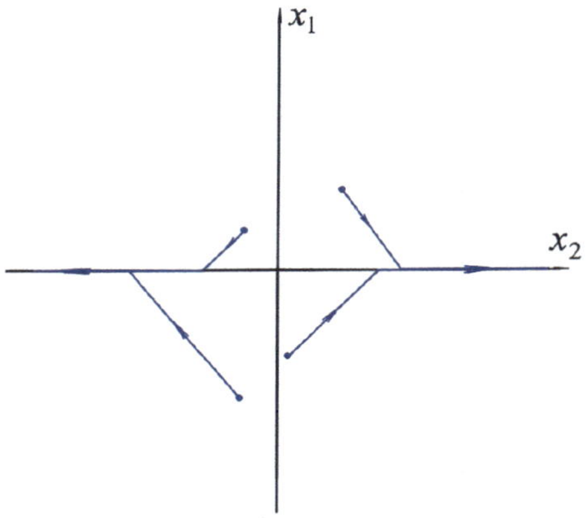

Fig. 4.5 Trajectories of the system (4.21)

and a globally Lipschitz continuous function $V : \mathbb{R}^2 \mapsto \mathbb{R}$ defined as

$$V = 4|x_1| + |x_2|.$$

For $x_1 x_2 \neq 0$ simple computations show

$$\begin{aligned}\dot{V} &= \tfrac{\partial V}{\partial x_1}\dot{x}_1 + \tfrac{\partial V}{\partial x_2}\dot{x}_2 \\ &= 4\,\mathrm{sign}(x_1)(-2\,\mathrm{sign}(x_1) - \mathrm{sign}(x_2)) + \mathrm{sign}(x_2)(-2\,\mathrm{sign}(x_1) + \mathrm{sign}(x_2)) \\ &= -7 - 6\,\mathrm{sign}(x_1 x_2) \le -1.\end{aligned}$$

Therefore, the globally proper positive definition function V has a negative derivative $\dot{V}$ almost everywhere (except a set of measure zero) in $\mathbb{R}^2$. For $V \in C^1$ the obtained estimate would be enough to guarantee both Lyapunov and asymptotic stability of the system (see the proof of Corollary 4.4 or [11, p. 155]). However, for the Lipschitz continuous function V, the negative definiteness of $\dot{V}$ almost everywhere in state space is not sufficient to make a right conclusion about the stability of discontinuous ODE. For instance, the considered system is unstable and it has solutions tending to infinity along the line $x_2 = 0$ (see Fig. 4.5).

4.3.3 *Finite-Time Stability*

The finite-time stability (FTS), by definition, is a combination of two properties: Lyapunov stability and finite-time convergence. The theorem below uses two non-negative functions in order to characterize this type of stability.

Theorem 4.3 *The system* (4.13) *is locally (globally) uniformly FTS if and only if there exists a locally (globally) proper function* $V : \mathbb{R} \times \mathbb{R}^n \mapsto \mathbb{R}$ *and a locally bounded function* $T : \mathbb{R}^n \mapsto [0, +\infty)$ *such that the inequalities*

$$D_{\mathbb{K}} V(t, x(t, t_0, x_0)) \leq 0, \qquad \forall t > t_0,$$

$$D_{\mathbb{K}} T(x(t, t_0, x_0)) \leq -1, \qquad \forall t > t_0 \tag{4.22}$$

hold for any $t_0 \in \mathbb{R}$ *and any solution* $x(\cdot, t_0, x_0)$ *of* (4.13) *as long as*

$$x(t, t_0, x_0) \in \Omega \backslash \{\mathbf{0}\},$$

where Ω *is a neighborhood of the origin (resp.,* $\Omega = \mathbb{R}^n$*). Moreover, the set* $U = \{x \in \mathbb{R}^n : \overline{V}(|x|) < r\}$*, where* $r > 0$ *is a maximal number such that* $U_r = \{x \in \mathbb{R}^n : \underline{V}(|x|) \leq r\} \subset \Omega$*, belongs to the finite-time attraction domain of the system* (4.13)*, where the functions* $\underline{V}, \overline{V} \in \mathcal{K}_\infty$ *are introduced in Definition 4.4.*

Proof *Sufficiency*. Theorem 4.2 implies the locally (globally) Lyapunov stability of the system (4.13). By construction, the set $U \subset U_r \subset \Omega$ is positively invariant for the system (4.13) due to the estimates (4.15).

Let us consider the interval $[t_0, t_1 + \varepsilon]$, $\varepsilon > 0$, $t_1 = t_0 + T(x_0)$ and show that there exists an instant of time $t^* \in [t_0, t_1 + \varepsilon]$ such that $x(t^*, t_0, x_0) = \mathbf{0}$. Suppose that $x(t, t_0, x_0) \neq 0$ for $\forall t \in [t_0, t_1 + \varepsilon]$.

The inequality (4.22) and Lemma 4.2 imply that the function $t \mapsto T(x(t, t_0, x_0))$ is decreasing on $[t_0, t_1 + \varepsilon]$ and, consequently, differentiable almost everywhere on $[t_0, t_1 + \varepsilon]$. Then using Lemma 4.8 we derive

$$\begin{aligned} T(x(t_1 + \varepsilon, t_0, x_0)) - T(x(t_0, t_0, x_0)) &\leq \int_{t_0}^{t_1+\varepsilon} \dot{T}(x(\tau, t_0, x_0)) d\tau \\ &\leq -(t_1 + \varepsilon - t_0) = -T(x_0) - \varepsilon, \end{aligned}$$

i.e., $T(x(t_1 + \varepsilon, t_0, x_0)) \leq T(x(t_0, t_0, x_0)) - T(x_0) - \varepsilon = -\varepsilon$. This contradicts $T(x) \geq 0$ for $x \in U \subset \Omega$. Therefore, there exists $t^* \in [t_0, t_1 + \varepsilon]$ such that $x(t^*, t_0, x_0) = \mathbf{0}$. Since $\varepsilon > 0$ can be selected arbitrary small then $t^* \in [t_0, t_1]$, i.e., $t^* \leq t_0 + T(x_0)$. Proposition 3.1 guarantees the uniqueness of the zero solution of any Lyapunov stable system. So, the system (4.13) is uniformly finite-time stable and the set U belongs to the finite-time attraction domain of the system.

Necessity. Let the origin of the system (4.13) be locally (globally) uniformly finite-time stable. Notice that Proposition 3.1 implies that $x(t, t_0, \mathbf{0}) \equiv \mathbf{0}$ is the unique solution of our system with the zero initial condition. The existence of a function V follows from Theorem 4.2.

Let S_{t_0, x_0} denote a set of solutions of the system (4.13) in the forward time with the initial condition $x(t_0) = x_0$ and introduce the *settling-time function* T as follows

$$T(x_0) = \sup_{t_0 \in \mathbb{R}} \sup_{x \in S_{t_0,x_0}} \inf_{\tau > 0 : x(t_0+\tau, t_0, x_0) = \mathbf{0}} \tau.$$

Since the origin is uniformly finite-time stable then there exists a ball $\Omega := B(\delta)$, $\delta > 0$ such that the function $T : \Omega \mapsto \mathbb{R}_+$ given above is locally bounded on Ω. Notice that for any $h \in \mathbb{R}$ one holds

$$\sup_{t_0 \in \mathbb{R}} \sup_{x \in S_{t_0,x_0}} \inf_{\tau > 0 : x(t_0+\tau, t_0, x_0) = \mathbf{0}} \tau = \sup_{t_0 \in \mathbb{R}} \sup_{x \in S_{t_0+h,x_0}} \inf_{\tau > 0 : x(t_0+h+\tau, t_0+h, x_0) = \mathbf{0}} \tau.$$

Let $x_0 \in U \backslash \{\mathbf{0}\}$, let $h > 0$ be a real number and let x be an arbitrary solution of the system (4.13) such that $x(p, t_0, x_0) \in \Omega \backslash \{\mathbf{0}\}$ for all $p \in [t_0, t_0 + h]$. In this case, there exists $\tilde{x} \in S_{t_0, x(t_0)}$ such that $\inf\limits_{\tau > 0 : \tilde{x}(t_0+\tau) = \mathbf{0}} \tau \geq h$ and

$$T(x_0) - h = \sup_{t_0 \in \mathbb{R}, x \in S_{t_0,x_0}} \inf_{\tau > 0 : x(t_0+\tau, t_0, x_0) = \mathbf{0}} \tau - h$$

$$= \sup_{t_0 \in \mathbb{R}, x \in S_{t_0,x_0}} \inf_{\tilde{\tau} > -h : x(t_0+h+\tilde{\tau}, t_0, x_0) = \mathbf{0}} \tilde{\tau}$$

$$= \sup_{t_0 \in \mathbb{R}, x \in S_{t_0,x_0}} \inf_{\tilde{\tau} > 0 : x(t_0+h+\tilde{\tau}, t_0, x_0) = \mathbf{0}} \tilde{\tau}.$$

Due to the local (global) Lyapunov stability of the system with $x_0 \in U$ (resp, $x_0 \in \mathbb{R}^n$), any solution of the system $x(t, t_0, x_0)$ is forward complete. Let us denote

$$S^h_{t_0,x_0} = \left\{ x(\cdot, t_0 + h, x_0) \in C([t_0 + h, +\infty); \mathbb{R}^n) : x \in S_{t_0,x_0} \right\},$$

where $h > 0$ and $x(\cdot, t_0 + h, +\infty)$ is the restriction of a solution x to the time interval $[t_0 + h, +\infty)$. Since $S_{t_0+h, x(t_0+h, t_0, x_0)} \subset S^h_{t_0,x_0}$ then

$$\begin{aligned} T(x(t_0+h, t_0, x_0)) &= \sup_{t_0 \in \mathbb{R}} \sup_{x \in S_{t_0+h, x(t_0+h, t_0, x_0)}} \inf_{\tilde{\tau} > 0 : x(t_0+h+\tilde{\tau}, t_0+h, x(t_0+h, t_0, x_0)) = \mathbf{0}} \tilde{\tau} \\ &\leq \sup_{t_0 \in \mathbb{R}} \sup_{x \in S^h_{t_0,x_0}} \inf_{\tilde{\tau} > 0 : x(t_0+h+\tilde{\tau}, t_0, x_0) = \mathbf{0}} \tilde{\tau} \\ &= \sup_{t_0 \in \mathbb{R}} \sup_{x \in S_{t_0,x_0}} \inf_{\tilde{\tau} > 0 : x(t_0+h+\tilde{\tau}) = \mathbf{0}} \tilde{\tau} = T(x_0) - h. \end{aligned}$$

Hence, we obtain

$$\frac{T(x(t_0+h, t_0, x_0)) - T(x(t_0, t_0, x_0))}{h} \leq -1$$

for small $h > 0$. Similarly, one can be shown

$$\frac{T(x(t_0, t_0, x_0)) - T(x(t_0-h, t_0, x_0))}{h} \leq -1$$

for small enough $h > 0$. Therefore, $D_{\mathbb{K}} T(x(t, t_0, x_0)) \leq -1$ as long as $x(t, t_0, x_0) \in \Omega\backslash\{\mathbf{0}\}$. The proof is complete. □

The proven theorem characterizes the finite-time stability in terms of two functions V and T. If the settling-time function T is continuous at the origin (i.e., the system is 0-FTS) and the system (4.13) is autonomous[3] then the finite-time stability can be characterized using just one strict Lyapunov function [6, 26].

Corollary 4.5 *Let all solutions of the autonomous system* (4.13) *be unique in the forward time. The autonomous system* (4.13) *is locally (globally) 0-FTS if and only if there exists a locally (globally) proper continuous function* $V : \mathbb{R}^n \mapsto \mathbb{R}$ *such that the inequality*

$$D_{\mathbb{K}} V(x(t, t_0, x_0)) \leq -1, \qquad \forall t > t_0 = 0 \tag{4.23}$$

holds for any solution $x(\cdot, t_0, x_0)$ *of* (4.13) *as long as*

$$x(t, t_0, x_0) \in \Omega\backslash\{\mathbf{0}\},$$

where $\Omega \subset \mathbb{R}^n$ *is a neighborhood of the origin (resp.,* $\Omega = \mathbb{R}^n$*). The settling time of the finite-time stable system* (4.13) *admits the estimate*

$$T(x_0) \leq V(x_0), \quad \forall x_0 \in U,$$

where $U = \{x \in \mathbb{R}^n : V(x) < r\}$ *with* $r > 0$ *being a maximal number such that* $U \subset \Omega$ *(resp.,* $U = \mathbb{R}^n$*) is a positively invariant attraction domain of the system.*

Proof *Sufficiency* can be proven similarly to Theorem 4.3 using V instead of T in all considerations, so the Lyapunov function V is an upper estimate of the settling-time function. Since V is continuous and locally (globally) proper then the set U is an open neighborhood of the origin (resp., $U = \mathbb{R}^n$). Moreover, U is a positively invariant connected set. Indeed, the positive invariance follows from $V(x(t, t_0, x_0)) \leq V(x_0)$, $\forall t \geq t_0$, $\forall x_0 \in U$, but finite-time convergence of all trajectories to zero guarantees that the set U is connected. Therefore, U is a positively invariant attraction domain of the system.

Necessity. Let the system (4.13) be locally (globally) 0-FTS with an attraction domain $\tilde{U} \subset \mathbb{R}^n$. On the one hand, since the system is assumed to be autonomous then repeating the proof of Theorem 4.3 we derive the Lyapunov function

$$V(x_0) = \sup_{\tau \geq t_0} |x(t, t_0, x_0)|$$

satisfying $D_{\mathbb{K}} V(x(t, t_0, x_0)) \leq 0$ and a settling-time function T such that $D_{\mathbb{K}} T(x(t, t_0, x_0)) \leq -1$, where V is locally (globally) proper, T is non-negative definite locally

[3] The system (4.13) is autonomous if $F(t, x)$ is independent of t.

bounded. Moreover, V is continuous. Indeed, the uniform FTS of the system guarantees that for any $x_0 \in \tilde{U}$ there exists a compact neighborhood $U_{x_0} \subset \tilde{U}$ of x_0 and a number $T_{U_{x_0}}$ such that the supremum in the definition of V is reached on a finite-time interval $[0, T_{U_{x_0}}]$ for any initial value from U_{x_0}. Hence, the uniqueness of solutions and their continuous dependence on initial data on compact intervals of time (see, Theorem 2.10) yield the continuity of V at any $x_0 \in \tilde{U}$. On the other hand, by the definition of 0-FTS, the locally bounded positive definite function T is continuous at $\mathbf{0}$. Hence, the function $V^{new} := V + T$ is globally proper. Due to Lemma 4.3 and $D_{\mathbb{K}}T(x(t, t_0, x_0)) \leq -1$ we have

$$D_{\mathbb{K}}V^{new}(t, x(t, t_0, x_0)) \leq -1$$

as long as $x(t, t_0, x_0) \neq 0$. Finally, if solutions of the system are unique then T (and, consequently, V_{new}) is continuous on $\tilde{U}$ due to Proposition 3.2. The proof is complete. □

The above theorem highlights that the settling time and an attraction domain of the system can be estimated using a continuous Lyapunov function.

Example 4.10 *(Twisting algorithm)* Let us consider the so-called "twisting" second-order sliding-mode system [27]

$$\dot{x} = f(x_1, x_2) = \begin{pmatrix} x_2 \\ -2\,\mathrm{sign}(x_1) - \mathrm{sign}(x_2) \end{pmatrix}, \quad x = (x_1, x_2)^\top \in \mathbb{R}^2. \tag{4.24}$$

The Filippov regularization $K[f]$ can be easily obtained from f by replacing the sign function with the set-valued function $\overline{\mathrm{sign}}$ given by (0.6). Since the vector field $F = K[f]$ is nonempty-valued, compact-valued, convex-valued and upper semi-continuous then Assumption **A1** for the system (4.13) is fulfilled. Let us show that the function $V : \mathbb{R}^n \mapsto \mathbb{R}$ given by [7]:

$$V(x) = p\sqrt{|x_1| + \frac{x_2^2}{2(2 + \mathrm{sign}_1(x_1x_2))}} + \frac{|x_2|\,\mathrm{sign}_1(x_1x_2)}{2 + \mathrm{sign}_1(x_1x_2)}, \quad p = \frac{4\sqrt{2}}{3 - \sqrt{3}}$$

is a Lyapunov function for the considered system. Indeed, it is positively defined and continuous (but not differentiable on the lines $x_1 = 0$ and $x_2 = 0$). For any solution x and for all instants of time $t > 0$ such that $x_1(t)x_2(t) \neq 0$ we have

$$\dot{V}(x(t)) = \frac{\partial V}{\partial x_1}x_2(t) + \frac{\partial V}{\partial x_2}(-2\,\mathrm{sign}(x_1(t)) - \mathrm{sign}(x_2(t))) = -1.$$

For $t > 0$ such that $x_1(t) \neq 0$ or $x_2(t) \neq 0$ one can be shown

$$D_{\mathbb{K}}V(x(t)) = \left\{\lim_{n\to+\infty} \frac{V(x(t + h_n)) - V(x(t))}{h_n} : \{h_n\} \in \mathbb{K}\right\} \leq -1.$$

Therefore, the origin of the considered system is globally uniformly finite-time stable and V is, in fact, the settling-time function of the "twisting" algorithm.

A sufficient condition of finite-time stability in terms of the time derivative along trajectories of the system (4.13) (see Definition 4.5) can be obtained using the chain rules given in the previous section. The following corollary is inspired by [26, 28].

Corollary 4.6 *If there exists a locally (globally) proper positive definite function* $V : \mathbb{R}^n \mapsto \mathbb{R}$ *such that*

$$\dot{V}(t, x) \leq -\rho V^{\alpha}(x), \quad \forall x \in \Omega\backslash\{\mathbf{0}\}, \quad \forall t \in \mathbb{R}, \tag{4.25}$$

where $\rho > 0$, $\alpha \in [0, 1)$ *and* $\dot{V}(t, x)$ *is introduced by Definition 4.5, then the system* (4.13) *is locally (globally)* 0*-FTS with a finite-time attraction domain* $U \subset \Omega$*, and the settling time* $T(x_0)$ *admits the estimate*

$$T(x_0) \leq V(x_0), \quad \forall x_0 \in U, \tag{4.26}$$

where Ω *and* U *are neighborhoods of the origin (resp.,* $U = \Omega = \mathbb{R}^n$*).*

Proof Taking $V_1 = \frac{1}{\rho(1-\alpha)} V^{1-\alpha}$ we derive $D_{\mathbb{K}} V_1(x(t, t_0, x_0)) \leq -1$, so, using Theorem 4.3, we complete the proof. □

Example 4.11 *(Unit sliding-mode control)* Let us consider the system

$$\dot{x} = g(t, x) + u(x), \quad t > 0, \quad x(0) = x_0 \in \mathbb{R}^n, \tag{4.27}$$

where

$$u(x) = -\rho \frac{x}{|x|}, \quad x \in \mathbb{R}^n, \quad |x| = \sqrt{x^\top x}, \quad \rho > 1$$

and g is continuous on x, locally integrable on t such that

$$x^\top g(t, x) \leq |x|, \quad \text{for all } x \in \mathbb{R}^n \quad \text{and} \quad \text{for almost all} \quad t > 0.$$

The considered system has Filippov solutions and Assumption **A2** is fulfilled for $f = g + u$. Moreover, for $x \neq \mathbf{0}$ we have $K[f](t, x) = \{g(t, x) + u(x)\}$, where $K[f]$ is the Filippov regularization of the right-hand side of the system. The regularization is set-valued only for $x = \mathbf{0}$. The function $V : \mathbb{R}^n \mapsto [0, +\infty)$ given by

$$V(x) = \sqrt{x^\top x}$$

is differentiable on $\mathbb{R}^n\backslash\{\mathbf{0}\}$ and for any $x \neq \mathbf{0}$ we have

$$\dot{V}(x) = \frac{x^\top g(t, x) - \rho|x|}{\sqrt{x^\top x}} \leq -(\rho - 1) < 0.$$

The origin of the considered system is globally finite-time stable and the settling-time T admits the estimate $T(x_0) \leq -(\rho-1)|x_0|$ for any $x_0 \in \mathbb{R}^n$.

Let us complete this section with one more example in $\mathbb{R}^n$ that demonstrates a finite-time stability analysis for a differential inclusion.

Example 4.12 *(Super-twisting algorithm)* Let us consider the so-called "super-twisting" system [27]:

$$\begin{pmatrix} \dot{x} \\ \dot{y} \end{pmatrix} \in F(x,y) = \begin{pmatrix} -\alpha \lfloor x \rceil^{\frac{1}{2}} + y \\ -\beta \cdot \mathrm{sign}(x) \end{pmatrix}, \tag{4.28}$$

where $x \in \mathbb{R}$, $y \in \mathbb{R}$, $\alpha > 0$, $\beta > 0$ and $\lfloor x \rceil^{\mu} = |x|^{\mu} \,\mathrm{sign}(x)$, $\mu \in \mathbb{R}_+$.

Let us show that the function [29]

$$V(x,y) = \frac{2\sqrt{V_0(x,y)}}{\gamma}, \quad V_0(x,y) = (2\beta + \alpha^2/2)|x| + y^2 - \alpha y \lfloor x \rceil^{\frac{1}{2}}$$

is a Lyapunov function for the system (4.28), where γ is a positive number defined as follows

$$\gamma = \alpha \inf_{z \in \mathbb{R}} \tfrac{\beta + \alpha^2/2 - \alpha z + z^2/2}{\sqrt{2\beta + \alpha^2/2 - \alpha z + z^2}}.$$

Obviously, such a selection of γ guarantees that

$$0 < \gamma \leq \alpha \tfrac{(\beta+\alpha^2/2)|x| - \alpha \lfloor x \rceil^{\frac{1}{2}} y + y^2/2}{|x|^{\frac{1}{2}} \sqrt{V_0(x,y)}}, \quad \forall x, y \in \mathbb{R}.$$

For $x \neq 0$ this function is differentiable and

$$\dot{V} = \frac{\partial V}{\partial x} \begin{pmatrix} \dot{x} \\ \dot{y} \end{pmatrix} = \tfrac{(2\beta+\alpha^2/2)\,\mathrm{sign}(x)\dot{x} + 2y\dot{y} - \alpha \dot{y} \lfloor x \rceil^{\frac{1}{2}} - \alpha y |x|^{-\frac{1}{2}} \dot{x}/2}{\gamma \sqrt{V_0(x,y)}} =$$

$$\tfrac{-\alpha(\beta+\alpha^2/2)|x|^{\frac{1}{2}} + \alpha^2 \,\mathrm{sign}(x) y - \alpha y^2 |x|^{-\frac{1}{2}}/2}{\gamma \sqrt{V_0(x,y)}} \leq -1.$$

To calculate $D_{\mathbb{K}} V(x(t), y(t))$ for $x(t) = 0$ and $y(t) \neq 0$ we use Lemma 4.11. According to Lemma 4.5 we have $D_{\mathbb{K}} V(x(t), y(t)) \subset F(x(t), y(t))$. Let us study the limit

$$D_{\{h_n\},\{u_n\}} V(0,y) = \lim_{n \to \infty} \frac{V(h_n(d_x + u_n^x), y + h_n(d_y + u_n^y)) - V(0,y)}{h_n},$$

where $\{h_n\} \in \mathbb{K}$, $u_n = (u_n^x, u_n^y)^\top$, $\{u_n\} \in \mathbb{M}$, $d = (d_x, d_y)^\top \in F(0,y)$. Using continuity of V we conclude

$$D_{\{h_n\},\{u_n\}} V(0,y) = 2 \lim_{n \to \infty} \tfrac{\sqrt{(2\beta+\alpha^2/2)|h_n y| + (y+h_n q)^2 - \alpha \lfloor h_n y \rceil^{\frac{1}{2}} (y + h_n q)} - |y|}{\gamma h_n} = -\infty,$$

where $q \in [-\beta, \beta]$. Therefore, we have

$$\sup_{f \in F(x,y)} D_{\mathbb{K},\mathbb{M}} V(x,y) f \leq -1, \quad (x,y) \neq \mathbf{0},$$

and we conclude that all conditions Corollary 4.5 are fulfilled, the origin of the super-twisting system is globally uniformly finite-time stable and the settling time admits the estimate $T(x_0, y_0) \leq V(x_0, y_0)$, $x_0, y_0 \in \mathbb{R}$.

4.3.4 Fixed-Time Stability

Being "adjoint" to the finite-time stability (see Chap. 3), the nearly fixed-time stability can be characterized in the similar way using the escape-time function.

Theorem 4.4 *The system* (4.13) *is locally (globally) uniformly nFxTS if and only if there exist a locally (globally) proper positive definite function* $V : \mathbb{R} \times \mathbb{R}^n \mapsto \mathbb{R}$ *and a function* $\tau : \mathbb{R}^n \mapsto [0, +\infty)$ *such that* τ *is bounded on* $\mathbb{R}^n \backslash B(r)$ *for any* $r > 0$ *and the inequalities*

$$D_{\mathbb{K}} V(t, x(t, t_0, x_0)) \leq 0, \qquad \forall t > t_0,$$

$$D_{\mathbb{K}} \tau(x(t, t_0, x_0)) \geq 1, \qquad \forall t > t_0 \tag{4.29}$$

hold for any $t_0 \in \mathbb{R}$ *and any solution* $x(\cdot, t_0, x_0)$ *of* (4.13) *as long as*

$$x(t, t_0, x_0) \in \Omega \backslash \{\mathbf{0}\},$$

where Ω *is a neighborhood of the origin (resp.,* $\Omega = \mathbb{R}^n$*). Moreover, the set* $U = \{x \in \mathbb{R}^n : \overline{V}(|x|) < r\}$*, where* $r > 0$ *is a maximal number such that* $U_r = \{x \in \mathbb{R}^n : \underline{V}(|x|) \leq r\} \subset \Omega$*, is an attraction domain of the system* (4.13)*, where the functions* $\underline{V}, \overline{V} \in \mathcal{K}_\infty$ *are introduced in Definition 4.4.*

Proof *Sufficiency.* Theorem 4.2 implies that the origin of the system (4.13) is locally (globally) Lyapunov stable and $V(t, x(t, t_0, x_0)) \leq V(t_0, x_0)$ as long as $x(t, t_0, x_0) \in \Omega$. For $x_0 \in U \subset \Omega$ we have $V(t_0, x_0) \leq \overline{V}(|x_0|) \leq r$. In this case, due to the properness of V, the inequality $V(t, x(t, t_0, x_0)) \leq V(t_0, x_0)$ implies that $\underline{V}(|x(t, t_0, x_0)|) \leq r$. Therefore, for any $x_0 \in U$ we have $x(t, t_0, x_0) \in U_r \subset \Omega$ for all $t \geq 0$. Let us prove nearly fixed-time convergence.

Let $x_0 \in U$ and $t_0 \in \mathbb{R}$ be arbitrary. Let us consider the interval $[t_1 - \varepsilon, t_0]$, $\varepsilon > 0$, $t_1 = t_0 - \tau(x_0)$ and show that there exists an instant of time $t^* \in [t_1 - \varepsilon, t_0]$ such that $x(t^*, t_0, x_0) \notin U$ (or $x(t, t_0, x_0)$ as $t \to t^*$), where $x(\cdot, t_0, x_0)$ is an arbitrary solution of the system in the backward time for the initial condition $x(t_0) = x_0$. Suppose the contrary, i.e., $x(t, t_0, x_0) \in U$ and $|x(t, t_0, x_0)| < +\infty$ for all $t \in [t_1 -$

$\varepsilon, t_0]$. Denote $x^* = x(t_1 - \varepsilon, t_0, x_0)$. Notice that $x(t, t_0, x_0)$ defines a solution in the forward time with the initial condition $x(t_1 - \varepsilon) = x^*$.

The inequality (4.29) and Lemma 4.2 imply that the function $t \to \tau(x(t, t_0, x_0))$ is increasing on $[t_1 - \varepsilon, t_0]$ and, consequently, differentiable almost everywhere on $[t_1 - \varepsilon, t_0]$. Then using Lemma 4.8 we derive

$$\tau(x_0) - \tau(x(t_1 - \varepsilon, t_0, x_0)) \geq \int_{t_1-\varepsilon}^{t_0} \dot{\tau}(x(\tau, t_0, x_0))d\tau \geq t_0 - t_1 + \varepsilon = \tau(x_0) + \varepsilon,$$

i.e., $\tau(x(t_1 - \varepsilon, t_0, x_0)) \leq -\varepsilon$. This contradicts to the assumption $\tau(x) \geq 0$ for $x \in U \subset \Omega$. Since $\varepsilon > 0$ can be selected arbitrary small then $t^* \in [t_0 - \tau(x_0), \tau]$. This means that τ is an upper estimate of the escape-time function of the system. Using Proposition (3.3) we complete the proof of the nearly fixed-time convergence to the origin.

Necessity. Let the system (4.13) be locally (globally) uniformly nFxTS. The existence of a functional V follows from Theorem 4.2.

Let S_{t_0,x_0} denote a set of solutions of the system (4.13) in the backward time with the initial condition $x(t_0) = x_0$ and introduce the *escape-time function* τ as follows

$$\tau(x_0) = \sup_{t_0\in\mathbb{R}} \sup_{x\in S_{t_0,x_0}} \inf_{\tilde{\tau}>0\,:\,x(t_0-\tilde{\tau},t_0,x_0)\notin U} \tilde{\tau},$$

where U is nearly the fixed-time attraction domain. In the case of global nFxTS, we have $U = \mathbb{R}^n$ and the condition $x(t_0 - \tilde{\tau}) \notin U = \mathbb{R}^n$ means the finite-time blow up (in the backward time). By construction (due to nFxTS), τ is uniformly bounded on $U \backslash B(r)$ for any $r > 0$.

Let $h > 0$ be a real number and let x be an arbitrary solution of the system (4.13) such that $x(p) \in \Omega\backslash\{\mathbf{0}\}$ for all $p \in [t_0, t_0 - h]$. In this case, there exists $\tilde{x} \in S_{t_0,x_0}$ such that $\inf\limits_{\tau>0\,:\,\tilde{x}(t_0+\tau)\notin U} \tau \geq h$ and

$$\begin{aligned}
\tau(x(t_0)) - h &= \sup_{t_0\in\mathbb{R}} \sup_{x\in S_{t_0,x_0}} \inf_{\tilde{\tau}>0\,:\,x(t_0-\tilde{\tau})\notin U} \tilde{\tau} - h \\
&= \sup_{t_0\in\mathbb{R}} \sup_{x\in S_{t_0,x_0}} \inf_{\tilde{\tau}>-h\,:\,x(t_0-h-\tilde{\tau})\notin U} \tilde{\tau} \\
&= \sup_{t_0\in\mathbb{R}} \sup_{x\in S_{t_0,x_0}} \inf_{\tilde{\tau}>0\,:\,x(t_0-h-\tilde{\tau})\notin U} \tilde{\tau}.
\end{aligned}$$

Let us denote $S^h_{t_0,x_0}$ be restriction (see above) of the set of solutions S_{t_0,x_0} to the time interval $(-\infty, t_0 - h]$. Since $S_{t_0-h,x(t_0-h,t_0,x_0)} \subset S^h_{t_0,x(t_0)}$ then

$$\begin{aligned}\tau(x(t_0-h,t_0,x_0)) &= \sup_{t_0\in\mathbb{R}}\ \sup_{x\in S_{t_0-h,x(t_0-h,t_0,x_0)}}\ \inf_{\tilde\tau>0:x(t_0-h-\tilde\tau,t_0-h,x(t_0-h,t_0,x_0))\notin U}\tilde\tau\\ &\le \sup_{t_0\in\mathbb{R}}\ \sup_{x\in S^h_{t_0,x_0}}\ \inf_{\tilde\tau>0:x(t_0-h-\tilde\tau,t_0,x_0)\notin U}\tilde\tau\\ &= \sup_{t_0\in\mathbb{R},x\in S_{t_0,x_0}}\ \inf_{\tilde\tau>0:x(t_0-h-\tilde\tau,t_0,x_0)\notin U}\tilde\tau\\ &=\tau(x(t_0))-h.\end{aligned}$$

Hence, we obtain $h^{-1}(\tau(x(t_0,t_0,x_0))-\tau(x(t_0-h,t_0,x_0)))\ge 1$. Similarly, one can be shown $h^{-1}(\tau(x(t_0+h,t_0,x_0))-\tau(x(t_0,t_0,x_0)))\ge 1$ for $h>0$. Therefore, $D_{\mathbb{K}}\tau(x(t,t_0,x_0))\ge 1$ as long as $x(t,t_0,x_0)\in U\backslash\{\mathbf{0}\}$. The proof is complete. □

Similarly to the finite-time stability, if the escape-time function is continuous at infinity then the nFxTS can be characterized using just one strict Lyapunov function.

Corollary 4.7 *Let solutions of the autonomous system* (4.13) *be unique in the forward time and in the backward time for all* $x_0\neq\mathbf{0}$. *The autonomous system* (4.13) *is locally (globally)* ∞*-nFxTS if and only if there exist a locally (globally) proper continuous function* $V:\mathbb{R}^n\mapsto\mathbb{R}$ *such that*

$$D_K V(x(t,t_0,x_0))\le -V^2(x(t,t_0,x_0))),\qquad \forall t>t_0=0 \tag{4.30}$$

hold for any solution $x(\cdot,t_0,x_0)$ *of* (4.13) *as long as*

$$x(t,t_0,x_0)\in\Omega\backslash\{\mathbf{0}\},$$

where $\Omega\subset\mathbb{R}^n$ *is a neighborhood of the origin (resp.,* $\Omega=\mathbb{R}^n$*). Moreover,* $U=\{x\in\mathbb{R}^n : V(x)<r\}$ *with* $r>0$ *being a maximal number such that* $U\subset\Omega$ *(resp.,* $r=+\infty$ *and* $U=\mathbb{R}^n$*) is a positively invariant attraction domain of the system:*

$$|x(t,t_0,x_0)|\le \underline{V}^{-1}\left(\tfrac{1}{t-t_0+1/r}\right),\quad \forall t\ge t_0,\ \forall x_0\in U,\ (resp.,\ \forall x_0\in\mathbb{R}^n,\ r=+\infty),$$

where $\overline{V}^{-1}$ *is the inverse function of the class-*K_∞ *function* $\underline{V}$ *introduced in Definition 4.4.*

Proof *Sufficiency*. Let us consider the function $\tilde\tau=\frac{1}{V}-\frac{1}{r}$, where $r\in(0,+\infty)$: $U\subset\Omega$ (resp., $r=+\infty$ and $\tilde\tau=\frac{1}{V}$ in the case of global stability analysis). By Lemma 4.7 we derive

$$D_{\mathbb{K}}\tilde\tau(x(t,t_0,x_0))\ge\frac{-D_{\mathbb{K}}V(x(t,t_0,x_0))}{V^2(x(t,t_0,x_0))}\ge 1. \tag{4.31}$$

Using Theorem 4.2 we conclude that all conditions of Theorem 4.4 are fulfilled, so the system is locally (globally) uniformly nFxTS. Moreover, U is a positively invariant attraction domain of the system. Indeed, since V is continuous and locally proper then the set U is an open neighborhood of the origin and $\partial U = \{x \in \mathbb{R}^n : V(x) = r\}$. The set U is positively invariant since $V(x(t, t_0, x_0)) \leq V(x_0), \forall t \geq t_0, \forall x_0 \in U$. Moreover, since U is a neighborhood of the origin then there exists a closed ball $\bar{B} \subset U$ centered at the origin, such that, by definition of uniform nFxTS, the system convergence into this ball in a fixed-time dependent of $x_0 \in U \backslash \bar{B}$. This means that U is a connected set. Therefore, U is, indeed, an attraction domain of the system.

If τ is the escape-time function of the uniformly nFxTS system then $\tau(x) \leq \tilde{\tau}(x)$ for all $x \in \mathbb{R}^n$ (see the proof of Theorem 4.4). Since the function V is globally proper then

$$0 \leq \tau(x) \leq \tilde{\tau}(x) = \frac{1}{V(x)} - \frac{1}{r} \to 0 \quad \text{as} \quad \rho(x, \partial U) \to 0$$

(resp., $\tilde{\tau}(x) = \frac{1}{V(x)} \to 0$ as $|x| \to +\infty$), i.e., the system is locally (globally) ∞-nFxTS.

Finally, using the estimate (4.31) and Lemma (4.8) we conclude that $\tilde{\tau}(x(t, t_0, x_0)) - \tilde{\tau}(x_0) \geq t - t_0$ and for all $t \geq t_0$ we have

$$\underline{V}(|x(t, t_0, x_0)|) \leq V(x(t, t_0, x_0)) \leq \frac{1}{1/V(x_0) + t - t_0} \leq \frac{1}{r^{-1} + t - t_0}.$$

Necessity. Let the system (4.13) be locally (globally) ∞-nFxTS with an attraction domain $U \subset \mathbb{R}^n$ (resp., $U = \mathbb{R}^n$). By Theorem 4.2 the globally proper function

$$V_0(x_0) = \sup_{\tau \geq t_0} |x(t, t_0, x_0)|$$

is a Lyapunov function of the system satisfying $D_{\mathbb{K}} V_0(x(t, t_0, x_0)) \leq 0$ as long as $x(t, t_0, x_0) \neq \mathbf{0}$. Moreover, solutions are unique in the forward time. Indeed, the uniform nFxTS of the system grantees that for any $x_0 \in \mathbb{R}^n$ there exists a compact neighborhood U_{x_0} of x_0 and a number $T_{U_{x_0}}$ such that the supremum in the definition of V_0 is reached on a finite-time interval $[0, T_{U_{x_0}}]$ for any initial value x_0 from U_{x_0}. Hence, the uniqueness of solutions and their continuous dependence on initial data on compact intervals of time (see, Theorem 2.10) yield the continuity of V_0 at x_0.

Let τ be the escape-time function. Inspired by [30] let us define the function V_0 as follows

$$V(x) = \frac{\min\{V_0(x), 1\}}{\tau(x)}, \quad x \in U.$$

The global proneness V follows from the continuity of τ at infinity (see Definition 3.15). For any solution $x(\cdot, t_0, x_0)$ of the nearly fixed-time stable system we have that the function the function $t \mapsto \tau(x(t, t_0, x_0))$ is increasing such that

$$D_{\mathbb{K}} \tau(x(t, t_0, x_0)) \geq 1, \quad \forall t \geq t_0$$

as long as $x(t, t_0, x_0) \neq \mathbf{0}$ (see Theorem 4.4). Hence, using Lemma 4.3 we derive

$$D_{\mathbb{K}} V_0(x(t, t_0, x_0)) \leq -\frac{\min\{V(x), 1\}}{\tau^2(x(t, t_0, x_0))} \leq -\frac{V_0^2(x(t, t_0, x_0))}{\min\{V(x), 1\}} \leq -V_0^2(x(t, t_0, x_0)).$$

By Proposition (3.4) the escape-function τ is continuous on $\mathbb{R}^n \backslash \{\mathbf{0}\}$ if solutions of the system are unique in the backward time. Therefore, the function V_0 is continuous by construction. The proof is complete. □

A sufficient condition of ∞-nFxTS in terms of $\dot{V}(t, x)$ (see Definition 4.5) is given by the following corollary inspired by [31].

Corollary 4.8 *If there exists a locally (globally) proper function* $V : \mathbb{R}^n \mapsto \mathbb{R}$ *such that*

$$\dot{V}(t, x) \leq -\rho V^{\beta}(x)), \quad \forall x \in \Omega \backslash \{\mathbf{0}\}, \quad \forall t \in \mathbb{R}, \tag{4.32}$$

where $\rho > 0$, $\beta > 1$ *and* $\dot{V}(t, x)$ *is introduced by Definition 4.5, then the system* (4.13) *is locally (globally)* ∞*-nFnTS and*

$$|x(t, t_0, x_0)| \leq \underline{V}^{-1}\left(\frac{1}{(V^{1-\beta}(x_0) + \rho(\beta - 1)(t - t_0))^{1/(\beta-1)}}\right), \quad \forall t \geq t_0. \tag{4.33}$$

Proof From (4.32) we derive

$$\frac{1}{1-\beta}(V^{1-\beta}(x(t, t_0, x_0)) - V^{1-\beta}(x_0)) \leq -\rho(t - t_0),$$

or equivalently,

$$V(x(t, t_0, x_0)) \leq \frac{1}{(V^{1-\beta}(x_0) + \rho(\beta - 1)(t - t_0))^{1/(\beta-1)}}, \quad \forall t \geq t_0.$$

This implies the nearly fixed-time stability, but the estimate (4.33) follows from $\underline{V}(|x|) \leq V(x)$ (see the definition of the locally(globally) proper function). The proof is complete. □

Being a mixture of both finite-time and nearly fixed-time stability, the *fixed-time stability* can be characterized in the similar way.

Theorem 4.5 *The system* (4.13) *is locally (globally) uniformly FxTS if and only if there exist a locally (globally) proper positive definite function* $V : \mathbb{R} \times \mathbb{R}^n \mapsto \mathbb{R}$ *and a function* $T : \mathbb{R}^n \mapsto [0, +\infty)$ *such that* $T(x) \leq T_{\max} < +\infty$ *for all* $x \in \Omega$ *and the inequalities*

$$D_{\mathbb{K}} V(t, x(t, t_0, x_0)) \leq 0, \qquad \forall t > t_0,$$

$$D_{\mathbb{K}} T(x(t, t_0, x_0)) \leq -1, \qquad \forall t > t_0$$

hold for any $t_0 \in \mathbb{R}$ *and any solution* $x(\cdot, t_0, x_0)$ *of* (4.13) *as long as*

$$x(t, t_0, x_0) \in \Omega \backslash \{\mathbf{0}\},$$

where Ω *is a neighborhood of the origin (resp.,* $\Omega = \mathbb{R}^n$*).*

In the view of Corollary 3.1, the proof of this (resp., the following) theorem just literally repeats the proof of Theorem 4.3 (resp., Theorem 4.4) using the property of the uniform boundedness of the settling-time function T (resp., the escape-time function).

Theorem 4.6 *The system* (4.13) *is locally (globally) uniformly FxTS if and only if there exist a locally (globally) proper positive definite function* $V : \mathbb{R} \times \mathbb{R}^n \mapsto \mathbb{R}$ *and a function* $\tau : \mathbb{R}^n \mapsto [0, +\infty)$ *such that* $\tau(x) \leq T_{\max} < +\infty$ *for all* $x \in \Omega \backslash \{\mathbf{0}\}$ *and the inequalities*

$$D_{\mathbb{K}} V(t, x(t, t_0, x_0)) \leq 0, \qquad \forall t > t_0,$$

$$D_{\mathbb{K}} \tau(x(t, t_0, x_0)) \geq 1, \qquad \forall t > t_0,$$

hold for any $t_0 \in \mathbb{R}$ *and any solution* $x(\cdot, t_0, x_0)$ *of* (4.13) *as long as*

$$x(t, t_0, x_0) \in \Omega \backslash \{\mathbf{0}\},$$

where Ω *is a neighborhood of the origin (resp.,* $\Omega = \mathbb{R}^n$*).*

The fixed-time stability can be characterized using only one Lyapunov function V provided that settling-time and an escape-time function are continuous at zero and at infinity, respectively.

Corollary 4.9 *Let solutions of the autonomous system* (4.13) *be unique in the forward time and in the backward time for all* $x_0 \neq \mathbf{0}$. *The system* (4.13) *is locally (globally) continuously FxTS if and only if there exist a locally (globally) proper positive definite continuous function* $V : \mathbb{R}^n \mapsto \mathbb{R}$ *and a number* $q > 0$ *such that the inequality*

$$D_K V(x(t, t_0, x_0)) \leq -q(1 + V^2(x(t, t_0, x_0))), \qquad \forall t > t_0 \tag{4.34}$$

hold for any $t_0 \in \mathbb{R}$ *and any solution* $x(\cdot, t_0, x_0)$ *of* (4.13) *as long as*

$$x(t, t_0, x_0) \in \Omega\backslash\{\mathbf{0}\},$$

where Ω *is a neighborhood of the origin (resp.,* $\Omega = \mathbb{R}^n$*). The settling-time function admits the estimate*

$$T(x_0) \leq \frac{\pi}{2q}, \quad \forall x_0 \in U,$$

where $U = \{x \in \mathbb{R}^n : V(x) < r\}$ *with* $r > 0$ *being a maximal number such that* $U \subset \Omega$ *(resp.,* $U = \mathbb{R}^n$*).*

Proof *Sufficiency*. Theorem 4.2 implies that the origin of the system (4.13) is locally (globally) 0-FTS and $t \to V(x(t, t_0, x_0))$ is a decreasing function differentiable almost everywhere as long as $x(t, t_0, x_0)) \in \Omega\backslash\{\mathbf{0}\}$.

Let us prove that the settling-time function is bounded by $0.5\pi/q$. Suppose that there exists a solution x_{t_0,x_0} which exists on $[0, +\infty)$ and does not vanish after the time instant $t = 0.5\pi/q$. For $T(x(t, t_0, x_0)) = \frac{\arctan(V(x(t,t_0,x_0)))}{q} \leq \frac{\pi}{2q}$ using Lemma 4.7, we derive

$$\dot{T}(x(t, t_0, x_0)) \overset{a.e.}{\leq} \frac{\dot{V}(x(t,t_0,x_0))}{q(1+V^2(x(t,t_0,x_0))} \leq -1$$

as long as $x(t, t_0, x_0) \in \Omega\backslash\{\mathbf{0}\}$. Hence, since $t \mapsto T(x(t, t_0, x_0))$ is decreasing then Lemma 4.8 gives

$$T(x(t_0 + 0.5\pi/q, t_0, x_0)) - T(x_0) \leq \int_{t_0}^{t_0+0.5\pi/q} \dot{T}(x(\tau, t_0, x_0))d\tau \leq -0.5\pi/q.$$

Since T is non-negative and $T(x_0) \leq 0.5\pi/q$ by construction then the obtained inequality implies $T(x(t_0 + 0.5\pi/q, t_0, x_0)) = 0$. Since T is locally proper then $x(t_0 + 0.5\pi/q, t_0, x_0) = \mathbf{0}$ and the system is 0-FxTS. Obviously, since $V(x(t, t_0, x_0)) \leq V(x_0)$ then the set U is a positively invariant attraction domain of the system (4.13).

Let us show that the escape time is continuous at infinity. If $t \mapsto x(2t_0 - t, t_0, x_0)$ is a solution in the backward time then

$$\mathcal{D}_K V(x(2t_0 - t, t_0, x_0)) \geq q(1 + V^2(x(2t_0 - t, t_0, x_0))).$$

Hence, we derive

$$\arctan(V(x(2t_0 - t, t_0, x_0))) - \arctan(V(x_0)) \geq q(t_0 - t)$$

and conclude that

- for the case of $U \neq \mathbb{R}^n$ we have $\tau(x_0) \leq \frac{\arctan(r)}{q} - \frac{\arctan(V(x_0))}{q}$;
- for the case of $U = \mathbb{R}^n$ we have $\tau(x_0) \leq \frac{\pi}{2q} - \frac{\arctan(V(x_0))}{q}$.

Since V is continuous and locally (globally) proper on $U \subset \Omega$ then $V(x_0) \to r$ as $x_0 \to \partial U$ (resp., $V(x_0) \to +\infty$ as $x_0 \to \infty$). This means that $\tau(x_0) \to 0$ as $x_0 \to \partial U$ (resp., $x_0 \to \infty$), i.e., the escape-time function is continuous at infinity and the system is $\{0, \infty\}$-FxTS. Applying Propositions 3.2 and 3.4 we conclude that the system (4.13) is continuously FxTS.

Necessity. The continuous FxTS stability means that the settling-time function T is continuous on an open connected set U being a neighborhood of the origin and the escape-time function τ is continuous on $U\backslash\{\mathbf{0}\}$ and at infinity. Moreover, due to uniqueness of solutions, we have

$$T(x_0) + \tau(x_0) \leq T_{\max}$$

for all $x_0 \in U\backslash\{\mathbf{0}\}$. Let us consider the Lyapunov function candidate defined as

$$V(x) = \frac{T(x)}{\tau(x)}, \quad \forall x \in U\backslash\{\mathbf{0}\}$$

and $V(\mathbf{0}) = 0$. The function V is continuous on $U\backslash\{\mathbf{0}\}$. Since U is a neighborhood of $\mathbf{0}$ then there exists $r_0 > 0$ such that $B(r_0) = \{x : |x| \leq r_0\} \subset \operatorname{int} U$. Since under Assumption **A1** (resp., **A2**) there exists $\bar{f} > 0$ such that $|\dot{x}(t, t_0, x_0)| \leq \bar{f}$ for (almost) all $t > t_0$ as long as $x(t, t_0, x_0) \in B(r_0)$ then there exists $c > 0$ such that $\inf_{|x|\leq 0.5r_0} \tau(x) \geq c$. This implies that V is continuous at $\mathbf{0}$, and, consequently, on whole U. On the other hand, since $0 < C := \sup_{x\in U, |x|\leq r_0} T(x) \leq \sup_{x\in U, |x|\leq r} T(x)$ for any $r \geq r_0$ then $V(x) \to +\infty$ as $x \to \partial U$ (resp., as $x \to +\infty$). Taking into account the positiveness of T and τ we conclude V is locally (globally) proper. Moreover, for any solution $x(\cdot, t_0, x_0)$ of the fixed-time stable system we have that the function $t \mapsto T(x(t, t_0, x_0))$ is decreasing and the function $t \mapsto \tau(x(t, t_0, x_0))$ is increasing such that

$$D_{\mathbb{K}} T(x(t, t_0, x_0)) \leq -1 \quad \text{and} \quad D_{\mathbb{K}} \tau(x(t, t_0, x_0)) \geq 1, \quad \forall t \geq t_0$$

as long as $x(t, t_0, x_0) \neq \mathbf{0}$. Hence, using Lemmas 4.3 and 4.7 and the continuity of T, τ we derive

$$\begin{aligned} D_{\mathbb{K}} V(x(t, t_0, x_0)) &\leq \tfrac{-1}{\tau(x(t,t_0,x_0))} - \tfrac{T(x(t,t_0,x_0)))}{\tau^2(x(t,t_0,x_0))} \\ &\leq \tfrac{-1}{T_{\max}} - \tfrac{T^2(x(t,t_0,x_0))}{T_{\max}\tau^2(x(t,t_0,x_0))} \leq -\tfrac{1+V^2(x(t,t_0,x_0))}{T_{\max}} \end{aligned}$$

as long as $x(t, t_0, x_0) \in U\backslash\{\mathbf{0}\}$. The proof is complete. □

Inspired by [32] we provide the following simple sufficient condition of the fixed-time stability.

Corollary 4.10 *If there exists a locally (globally) proper positive definite function* $V : \mathbb{R}^n \mapsto \mathbb{R}$ *such that for some* $q > 0$ *one holds*

$$\dot{V}(t,x) \leq -\left(\rho_1 V^{\alpha} + \rho_2 V^{\beta}\right)^k, \quad \forall x \in \Omega\backslash\{\mathbf{0}\}, \tag{4.35}$$

where $\rho_1 > 0$, $\rho_2 > 0$, $k > 0$, $0 \leq \alpha k < 1$, $\beta k > 1$ *and* $\dot{V}$ *is introduced in Definition 4.5,* $\Omega \subset \mathbb{R}^n$ *is a neighborhood of the origin (resp.,* $\Omega = \mathbb{R}^n$*), then the system* (4.13) *is locally (globally) 0-FxTS and the settling time* $T(x_0)$ *admits the estimate*

$$T(x_0) \leq T_{\max} := \tfrac{1}{\rho_1^k(1-\alpha k)} + \tfrac{1}{\rho_2^k(\beta k-1)}, \quad \forall x_0 \in \mathbb{R}^n. \tag{4.36}$$

Proof For $V(x(t,t_0,x_0)) \leq 1$ we have $D_{\mathbb{K}}V(t,x(t,x_0)) \leq -\rho_1^k V^{\alpha k}(x(t,t_0,x_0))$, so $V(t_1,x_0) \leq 1 \Rightarrow x(t,t_0,x_0) = \mathbf{0}$, $\forall t \geq t_1 + \frac{1}{\rho_1^k(1-\alpha k)}$. Similarly, for $V(x(t,t_0,x_0)) \geq 1$ we have $D_{\mathbb{K}}V(t,t_0,x_0) \leq -\rho_2^k V^{\beta k}(x(t,t_0,x_0))$. Hence, $V(x(t,t_0,x_0)) \leq 1$ for all $t \geq t_0 + \frac{1}{\rho_2^k(\beta k-1)}$. Therefore, we have $x(t,t_0,x_0) = \mathbf{0}$ for all $t \geq T_{\max}$ independently of the initial state x_0. The proof is complete. □

Example 4.13 Inspired by [32] let us consider the sliding-mode control system

$$\begin{cases} \dot{x} = y, \\ \dot{y} = u + d(t), \\ u = -\frac{\alpha+3\alpha x^2+\gamma}{2}\,\mathrm{sign}(s) - \lfloor \beta s + \beta s^3 \rceil^{\frac{1}{2}}, \end{cases}$$

where $x \in \mathbb{R}$, $y \in \mathbb{R}$, $|d(t)| < C$, $\alpha, \beta, C \in \mathbb{R}_+$, $\gamma > 2C$ and

$$s = y + \left\lfloor \lfloor y \rceil^2 + \alpha x + \alpha x^3 \right\rceil^{\frac{1}{2}}.$$

Recall that $\lfloor x \rceil^{\mu} := |x|^{\mu}\,\mathrm{sign}(\mu)$. The original discontinuous systems correspond to the following extended DI:

$$\begin{cases} \dot{x} = y, \\ \dot{y} \in \left\{-\frac{\alpha+3\alpha x^2+\gamma}{2}\right\} \cdot \overline{\mathrm{sign}}(s) \dot{+} \left\{-\lfloor \beta s + \beta s^3 \rceil^{\frac{3}{2}}\right\} \dot{+} [-C, C]. \end{cases}$$

Considering the Lyapunov function candidate $s \mapsto V(s) := |s|^{1/2}$ and calculating its generalized derivatives along trajectories of the DI we derive

$$D_{\mathbb{K}}V(s(t)) \leq -\beta^{\frac{1}{2}}\left(1 + V^4(s(t))\right)^{\frac{1}{2}} \quad \text{for } s(t) \neq 0.$$

Since $\sqrt{2}\left(1+V^4\right)^{\frac{1}{2}} \geq 1+V^2$ then that the surface $s=0$ is fixed-time stable with the following estimate of the settling time:

$$T_s \leq \frac{\pi}{\sqrt{2\beta}}.$$

The sliding motion equation for $s=0$ has the form

$$\dot{x} = -\left\lfloor \frac{\alpha}{2}x + \frac{\alpha}{2}x^3 \right\rceil^{\frac{1}{2}}.$$

This means that the origin of the system is fixed-time stable and the settling-time function T for the original system can be estimated as follows

$$T(x_0, y_0) \leq T_{\max} \leq \frac{\pi}{\sqrt{\alpha}} + \frac{\pi}{\sqrt{2\beta}}, \quad x_0, y_0 \in \mathbb{R}.$$

4.3.5 Asymptotic Stability

A characterization of a uniform asymptotic stability in terms of Lyapunov functions is well known in the literature [1, 4, 18, 33].

Theorem 4.7 (Clarke, Ledyaev & Stern 1998, [18]) *An autonomous system* (4.13) *satisfying Assumption* ***A1*** *is globally uniformly asymptotically stable if and only if there exist a globally proper function* $V \in C^\infty(\mathbb{R}^n; \mathbb{R})$ *and a positive function* $W \in C^\infty(\mathbb{R}^n \backslash \{\mathbf{0}\}; (0, +\infty))$ *:*

$$\dot{V}(x) \leq -W(x), \quad \forall x \neq \mathbf{0}.$$

The above result generalizes Kurzweil's converse Lyapunov theorem [4] to the case of autonomous differential inclusions. Notice that the theorem proves the existence of a C^∞ Lyapunov function for the uniformly asymptotically stable differential inclusion. Lyapunov characterizations of finite-time and fixed-time stability given above are non-smooth, since the corresponding Lyapunov function design uses settling-time and escape-time functions, which are discontinuous in the general case. The problem of characterization of finite-time and nearly fixed-time stability in terms of smooth Lyapunov functions is still opened in the general case. However, such a characterization always exists for homogeneous systems (see Chap. 7) and, consequently, for systems, which are diffeomorphic to homogeneous systems. Moreover, we also show that any asymptotically stable homogeneous system is diffeomorphic to a quadratically stable system. Quadratic stability (i.e., the existence of quadratic Lyapunov function) is a well-known property of stable linear systems.

Example 4.14 *(Quadratic Lyapunov function for linear systems)* Let $A \in \mathbb{R}^{n\times n}$ is a *Hurwitz matrix*[4] then, in the view of Proposition 2.1, the system

$$\dot{x} = Ax, \quad t > 0 \tag{4.37}$$

is globally asymptotically stable. Inspired by [3], let us show that a Lyapunov function for this system is defined by the norm

$$\|x\| = \sqrt{x^\top P x} \tag{4.38}$$

with $P = P^\top \in \mathbb{R}^{n\times n}$ such that

$$PA + A^\top P \prec 0, \quad P \succ 0, \tag{4.39}$$

where the sign $\prec$ (resp., $\succ$) means that the corresponding symmetric matrix is negative (resp., positive) definite.[5]

The *linear matrix inequality* (4.39) is known as *Lyapunov inequality*. This inequality is always feasible [3] and the matrix P can always be defined as

$$P = \int_0^{+\infty} e^{A^\top \tau} Q e^{A\tau} d\tau, \tag{4.40}$$

where $0 \prec Q = Q^\top \in \mathbb{R}^{n\times n}$ is an arbitrary positive definite matrix. Indeed, P is positive definite by construction, but by computing the time derivative of the function $t \mapsto \|x(t)\|$ we derive

$$\frac{d}{dt}\|x(t)\| = \frac{x^\top(t)P\dot{x}(t) + \dot{x}^\top(t)Px(t)}{2\|x(t)\|} = \frac{x^\top(t)PAx(t) + x^\top(t)A^\top Px(t)}{2\|x(t)\|}$$
$$= \frac{x^\top(t)\left(\int_0^{+\infty} e^{A^\top \tau} Q e^{A\tau} A d\tau + \int_0^{+\infty} A^\top e^{A^\top \tau} Q e^{A\tau} d\tau\right)x(t)}{2\|x(t)\|}.$$

Since $e^{A\tau}A = Ae^{At} = \frac{d}{d\tau}e^{A\tau}$ then using integration by parts we obtain

$$\int_0^{+\infty} e^{A^\top \tau} Q e^{A\tau} A d\tau = e^{A^\top \tau} Q e^{A\tau}\Big|_0^{+\infty} - \int_0^{+\infty} A^\top e^{A^\top \tau} Q e^{A\tau} d\tau.$$

[4] A matrix is Hurwitz if all its eigenvalues have negative real parts.

[5] A symmetric matrix is positive (resp., negative) definite if all its eigenvalues are positive (resp., negative) reals.

Since A is Hurwitz then considering the representation (2.10) we conclude

$$e^{A\tau} \to \mathbf{0} \quad \text{as} \quad \tau \to +\infty.$$

Hence, taking into account $e^{A\tau} = I_n$ for $\tau = 0$ we derive

$$\left. e^{A^\top \tau} Q e^{A\tau} \right|_0^{+\infty} = -Q$$

and

$$\frac{d\|x(t)\|}{dt} = \frac{x^\top(t)PAx(t)+x^\top(t)A^\top Px(t)}{2\|x(t)\|} = \frac{-x^\top(t)Qx(t)}{2\|x(t)\|} \le -\frac{\lambda_{\min}(Q)\|x(t)\|}{2}, \quad \forall t > 0$$

independently of the initial state $x(0) \neq \mathbf{0}$. This means that the weighted Euclidean norm with P given by (4.40) is a Lyapunov function of the asymptotically stable linear system (4.37) and

$$PA + A^\top P = -Q \prec 0, \quad P \succ 0. \tag{4.41}$$

Hence, we conclude that the Lyapunov inequality (4.39) is fulfilled as well.

Despite the theoretically proven existence of a C^∞ Lyapunov function for any asymptotically stable system in $\mathbb{R}^n$, the construction of such a smooth Lyapunov function for a concrete nonlinear system is usually a very difficult problem. However, for autonomous systems, the existence of a "non-strict" Lyapunov function could be sufficient due to Barbashin–Krasovskii theorem [34] being a corollary of the more general result known today as *LaSalle invariance principle* [35]. For the differential inclusions, this principle is studied, for example, in [36].

Theorem 4.8 (Barbashin–Krasovskii theorem for differential inclusions) *The autonomous system* (4.13) *is globally uniformly asymptotically stable if there exists a globally proper function* $V \in C^1(\mathbb{R}^n; \mathbb{R}_+)$ *such that*

$$\dot{V}(x) \le 0, \quad \forall x \in \mathbb{R}^n$$

and the set $\Omega = \{x \in \mathbb{R}^n \backslash \{\mathbf{0}\} : V(x) = 0\}$ *does not contain trajectories of the system.*

The latter condition of the Barbashin–Krasovskii theorem means that system trajectories may cross the set Ω, but they cannot stay in Ω even for a small interval of time. This theorem simplifies the qualitative analysis of homogeneous systems, since the uniform asymptotic stability of a homogeneous system implies its finite-time or nearly fixed-time stability dependently of the sign of the homogeneity degree (see Chap. 7 for more details).

4.3.6 ISS Lyapunov Function

Let the system (4.13) have the special structure:

$$\dot{x}(t) = f(x(t), q(t)), \quad t > 0, \quad x(0) = x_0, \tag{4.42}$$

where $x(t)$ is as before, $f : \mathbb{R}^n \times \mathbb{R}^k \mapsto \mathbb{R}^n$ is locally Lipschitz continuous on $\mathbb{R}^n \times \mathbb{R}^k$ and $q \in L^\infty(\mathbb{R}; \mathbb{R}^k)$ is a uniformly essentially bounded Lebesgue measurable function of time.

Definition 4.6 *(ISS Lyapunov function)* A smooth globally proper function $V : \mathbb{R}^n \mapsto [0, +\infty)$ if said to be *ISS Lyapunov function* if there exists $\alpha, \chi \in \mathcal{K}$ such that for all $x \in \mathbb{R}^n$, $q \in \mathbb{R}^k$ one holds

$$|x| \geq \chi(|q|) \quad \Rightarrow \quad \dot{V}(x, q) := \frac{\partial V}{\partial x} f(x, q) \leq -\alpha(|x|).$$

The Input-to-State Stability (ISS) of the considered system can also be characterized in terms of a Lyapunov function.

Theorem 4.9 (Sontag & Wang 1996, [37]) *The system* (4.42) *is ISS if and only if it has an ISS Lyapunov function.*

Notice that the ISS analysis needs regularity (local Lipschitz continuity) of the ISS Lyapunov function. The characterization of the rated ISS (such as finite-time ISS or fixed-time ISS) would also require such a regularity. The existence of the regular Lyapunov function for a finite/fixed-time system is not proven yet in the general case. The converse Lyapunov theorems presented above guarantee the continuity of the Lyapunov function only if the settling/escape-time function is continuous. In the same way, the local Lipschitz continuity of V would, respectively, be guaranteed if the settling/escape-time function is locally Lipschitz continuous on $\mathbb{R}^n \backslash \{\mathbf{0}\}$. Such an assumption is too conservative even for continuous homogeneous systems in $\mathbb{R}^n$ (see Example 3.9). However, in Chap. 7 of this monograph, we show that a characterization of the rated ISS by a regular ISS Lyapunov function is possible for *homogeneous* systems. In Volume II of the monograph, we extend this result to infinite-dimensional systems.

Concluding Remarks

The Lyapunov function method is the main tool for stability analysis of nonlinear systems modeled by ODEs. In this chapter, finite-time and fixed-time stability is characterized by means of locally (globally) proper Lyapunov functions. The smooth, non-smooth and discontinuous Lyapunov functions are studied in order to take into account finite-time stable systems with non-smooth settling-time functions. Despite that the existence of a smooth Lyapunov function is proven for any asymptotically stable ODE, in practice of control systems design, frequently, it is more simple to find a non-smooth Lyapunov function.

Remark 4.2 Most constructions and theorems given above remain valid for stability of a compact set (i.e., practical stability, see Remark 3.1) if the norm (a distance to the origin) is replaced with the Hausdorff distance to the set.

Notice that any level set $\Omega_r = \{x \in \mathbb{R}^n : V(x) = r\}, r > 0$ of the Lyapunov function V is stable. The Lyapunov function, which can be utilized to prove this, is $V_r(x) = \max(V(x), r)$. For more details about the Lyapunov function method for the stability of the sets, we refer the reader to [5, 35, 38].

References

1. Bacciotti A, Rosier L (2001) Lyapunov functions and stability in control theory. Springer, Berlin
2. Polyakov A, Poznyak A (2009) Reaching time estimation for "super-twisting" second order sliding mode controller via Lyapunov function designing. IEEE Trans Automat Control 54(8):1951–1955
3. Lyapunov AM (1992) The general problem of the stability of motion. Taylor & Francis
4. Kurzweil J (1956) On the inversion of Ljapunov's second theorem on stability of motion [in russian]. Czechoslovak Math J 6(2):217–259
5. Zubov VI (1964) Methods of A.M. Lyapunov and their applications. Noordhoff, Leiden, (Translated from V.I. Zubov, Metody A.M. Lyapunova i ih primenenie, Izdatelstvo Leningradskogo universiteta, 1957)
6. Roxin E (1966) On finite stability in control systems. Rendiconti del Circolo Matematico di Palermo 15:273–283
7. Polyakov A, Poznyak A (2009) Lyapunov function design for finite-time convergence analysis: "twisting" controller for second order sliding mode realization. Automatica 45:444–448
8. Natanson IP (1955) Theory of functions of a real variable. Frederick Ungar Publishing Co., New York
9. Szarski J (1965) Differential Inequalities. Polish Scientific Publishers, Warsaw
10. Bruckner AM (1978) Differentiation of real functions. Springer, Berlin
11. Filippov AF (1988) Differential equations with discontinuous right-hand sides. Kluwer Academic Publishers
12. Polyakov A, Fridman L (2014) Stability notions and Lyapunov functions for sliding mode control systems. J Franklin Inst 351(4):1831–1865
13. Clarke F, Ledyaev YS, Stern R, Wolenski P (1995) Qualitative properties of trajectories of control systems: a survey. J Dyn Control Syst 1(1):1–48
14. Aubin JP, Ekeland I (1984) Applied nonlinear analysis. Wiley, New York
15. Clarke F (1990) Optimization and nonsmooth analysis. SIAM, Philadelphia
16. Mordukhovich BS (2006) Variational analysis and generalized differentiation I. Springer, Berlin
17. Moreau JJ, Valadier M (1987) A chain rule involving vector functions of bounded variation. J Funct Anal 74(2):333–345
18. Clarke FH, Ledyaev YuS, Stern RJ (1998) Asymptotic stability and smooth Lyapunov functions. J Differ Eqn 149:69–114
19. Mironchenko A, Wirth F (2019) Non-coercive Lyapunov functions for infinite-dimensional systems. J Differ Eqn 266(11):7038–7072

20. Mironchenko A, Karafyllis I, Krstic M (2019) Monotonicity methods for input-to-state stability of nonlinear parabolic PDEs with boundary disturbances. SIAM J Control Optim 57(1):510–532
21. Coddington EA, Levinson N (1955) Theory of ordinary differential equations. McGraw-Hill, New York
22. Yoshizawa T (1955) On the stability of solutions of a system of differential equations. Mem College Sci Univ Kyoto Ser A: Math 29(1):27–33
23. Zubov VI (1958) On systems of ordinary differential equations with generalized homogeneous right-hand sides. Izvestia vuzov Mathematica (available on-line: http://mi.mathnet.ru/ivm2874) 1:80–88
24. Deimling K (1992) Multivariable differential equations. Walter de Gruyter, Berlin, New York
25. Utkin VI (1992) Sliding modes in control optimization. Springer, Berlin
26. Bhat SP, Bernstein DS (2000) Finite time stability of continuous autonomous systems. SIAM J Control Optim 38(3):751–766
27. Levant A (1993) Sliding order and sliding accuracy in sliding mode control. Int J Control 58(6):1247–1263
28. Moulay E, Perruquetti W (2005) Finite-time stability of differential inclusions. IMA J Math Control Inf Sci 22(4):465–475
29. Moreno J, Osorio M (2012) Strict Lyapunov functions for the super-twisting algorithm. IEEE Trans Automat Control 57:1035–1040
30. Efimov D, Polyakov A (2021) Finite-time stability tools for control and estimation. In: Foundations and trends in systems and control. Now Publishers Inc., Netherlands
31. Nakamura H, Yamashita Y, Nishitani H (2002) Smooth Lyapunov functions for homogeneous differential inclusions. In: Proceedings of the 41st SICE annual conference, pp 1974–1979
32. Polyakov A (2012) Nonlinear feedback design for fixed-time stabilization of linear control systems. IEEE Trans Automat Control 57(8):2106–2110
33. Krassovski N (1963) Stability of motion. Stanford University Press (Translated from Russian: N. Krasovski, Nekotorya Zadachi Teorii Ustoichivosti Dvizhenija, FizMatLit, Moskwa, 1959)
34. Barbashin EA, Krasovskii NN (1952) On the stability of motion as a whole. Doklady Akademii Nauk SSSR (in Russian) 86(3):453–456
35. La Salle J, Lefschetz S (1961) Stability by Liapunov's direct method with applications. Elsevier Science
36. Shevitz D, Paden B (1994) Lyapunov stability theory of nonsmooth systems. IEEE Trans Automat Control 39(9):1910–1914
37. Sontag ED, Wang Y (1996) On characterizations of the input-to-state stability property. Syst Control Lett 24(5):351–359
38. Khalil H (2002) Nonlinear systems. Pearson

Chapter 5
Linear Dilation in Euclidean Space

This chapter introduces linear dilations in Euclidean spaces and studies their properties. Various geometrical structures (such as homogeneous cones, spheres and balls) induced by dilations are considered. Homogeneous norms and homogeneous inner products are studied. Issues of their numerical computation are discussed. Theoretical constructions are supported by examples.

5.1 Generalized Dilations

By definition, the homogeneity is a dilation symmetry (see Introduction for more details). For example, a function $h : \mathbb{R}^n \mapsto \mathbb{R}$ is *standard homogeneous* of degree $\nu \in \mathbb{R}$ if scaling of its argument is equivalent to scaling of its value:

$$h(e^s x) = e^{\nu s} h(x), \quad s \in \mathbb{R}, \quad x \in \mathbb{R}^n,$$

where $e = 2.71828...$ is the Euler's number. Instead of the standard (isotropic) dilation $x \mapsto e^s x$ we may consider a generalized (anisotropic) one

$$x \mapsto \mathbf{d}(s)x,$$

where $\mathbf{d}(s)$ maps $\mathbb{R}^n$ onto $\mathbb{R}^n$ and $s \in \mathbb{R}$ is a parameter of the dilation. To be a dilation, the family of mappings $\mathbf{d}(s)$ must satisfy certain restrictions [1, 2], which, in the case of the Euclidean space $\mathbb{R}^n$, are formalized by the following definition.

A. Polyakov, *Generalized Homogeneity in Systems and Control Volume I*,
Communications and Control Engineering,
https://doi.org/10.1007/978-3-031-76246-8_5

Definition 5.1 A one-parameter family of mappings $\mathbf{d}(s) : \mathbb{R}^n \mapsto \mathbb{R}^n, s \in \mathbb{R}$ is said to be a *dilation* in $\mathbb{R}^n$ if

- **Group property:** $\mathbf{d}(0)x = x$, $\mathbf{d}(s+t)x = \mathbf{d}(s) \circ \mathbf{d}(t)x$, $\forall t, s \in \mathbb{R}$, $\forall x \in \mathbb{R}^n$;
- **Limit property:** $\liminf_{s\to+\infty} \|\mathbf{d}(s)x\| = +\infty$, $\limsup_{s\to-\infty} \|\mathbf{d}(s)x\| = 0$, $\forall x \neq \mathbf{0}$.

Theoretically, the mapping $(s, x) \mapsto \mathbf{d}(s)x$ may be nonlinear and/or discontinuous with respect to one or both arguments.

Example 5.1 If a function $f : \mathbb{R} \mapsto \mathbb{R}$ is additive

$$f(s+t) = f(s) + f(t), \forall s, t \in \mathbb{R}$$

and $f(0) = 0$ then the one-parameter family of linear mappings $\mathbf{d}(s) : \mathbb{R}^2 \mapsto \mathbb{R}^2$ given by the matrix

$$\mathbf{d}(s) = e^s e^{f(s)\left(\begin{smallmatrix} 0 & 1 \\ -1 & 0 \end{smallmatrix}\right)}, \quad s \in \mathbb{R}$$

is a dilation in $\mathbb{R}^2$. Indeed, since for any $\alpha \in \mathbb{R}$ and any $\beta \in \mathbb{R}$ the matrix $\alpha \left(\begin{smallmatrix} 0 & 1 \\ -1 & 0 \end{smallmatrix}\right)$ commutes with the matrix $\beta \left(\begin{smallmatrix} 0 & 1 \\ -1 & 0 \end{smallmatrix}\right)$, then

$$e^{(\alpha+\beta)\left(\begin{smallmatrix} 0 & 1 \\ -1 & 0 \end{smallmatrix}\right)} = e^{\alpha\left(\begin{smallmatrix} 0 & 1 \\ -1 & 0 \end{smallmatrix}\right)} e^{\beta\left(\begin{smallmatrix} 0 & 1 \\ -1 & 0 \end{smallmatrix}\right)}.$$

So, taking into account the additivity of f, we conclude that $\mathbf{d}$ satisfies the group property:

$$\mathbf{d}(s+t) = e^{s+t} e^{f(s+t)\left(\begin{smallmatrix} 0 & 1 \\ -1 & 0 \end{smallmatrix}\right)} = e^s e^t e^{f(s)\left(\begin{smallmatrix} 0 & 1 \\ -1 & 0 \end{smallmatrix}\right)} e^{f(t)\left(\begin{smallmatrix} 0 & 1 \\ -1 & 0 \end{smallmatrix}\right)} = \mathbf{d}(s)\mathbf{d}(t), \forall s, t \in \mathbb{R}$$

and $\mathbf{d}(0) = I_2$. Notice that $e^{f(s)\left(\begin{smallmatrix} 0 & 1 \\ -1 & 0 \end{smallmatrix}\right)} = \left(\begin{smallmatrix} \cos(f(s)) & \sin(f(s)) \\ -\sin(f(s)) & \cos(f(s)) \end{smallmatrix}\right)$ is a rotation matrix, so

$$|\mathbf{d}(s)x| = e^s \sqrt{x^\top \begin{pmatrix} \cos(f(s)) & -\sin(f(s)) \\ \sin(f(s)) & \cos(f(s)) \end{pmatrix} \begin{pmatrix} \cos(f(s)) & \sin(f(s)) \\ -\sin(f(s)) & \cos(f(s)) \end{pmatrix} x} = e^s |x|, \forall x \in \mathbb{R}^2$$

and the limit property is fulfilled as well. Therefore, the considered one-parameter family of linear mappings $\mathbf{d}(s)$ is a dilation in $\mathbb{R}^2$. The additive function f may be discontinuous in a particular case.

Let $\mathcal{B}$ be Hamel's basis in $\mathbb{R}$ over the field of rational numbers $\mathbb{Q}$, which exists due to the axiom of choice. This means that any real number $x \in \mathbb{R}$ can be represented as a linear combination

$$x = q_1 b_1 + \cdots + q_n b_n,$$

where $q_j \in \mathbb{Q}$, $b_j \in \mathcal{B}$, $j = 1, ..., n$ and the number $n \in \mathbb{N}$ of basis elements in the sum depend on $x \in \mathbb{R}$. Let us define the function $f : \mathbb{R} \mapsto \mathbb{R}$ as follows

$$f(x) = q_1 + \cdots + q_n,$$

where $q_i \in \mathbb{Q}$ are rational coordinates of x in the Hamel's basis and $n \in \mathbb{N}$ depends on x. By construction, the function f is additive. However, $f(x) \in \mathbb{Q}$ for all $x \in \mathbb{R}$, so f does not map $\mathbb{R}$ onto an interval in $\mathbb{R}$. Since f is a non-constant function then f is discontinuous. So, the corresponding dilation $\mathbf{d}$ in $\mathbb{R}^2$ is discontinuous too.

The function f is discontinuous at each point of $\mathbb{R}$ (otherwise it will be continuous on $\mathbb{R}$ due to the additivity of f). Moreover, f is unbounded on any nonempty interval from $\mathbb{R}$ (otherwise it will be continuous at zero and, consequently, on $\mathbb{R}$ due to additivity).

Definition 5.2 A dilation $\mathbf{d}$ is said to be

- **Continuous** if the function $s \mapsto \mathbf{d}(s)x$ is continuous for each $x \in \mathbb{R}^n$;
- **Strongly continuous** if the function $(s, x) \mapsto \mathbf{d}(s)x$ is continuous.

For any $x \in \mathbb{R}^n \backslash \{\mathbf{0}\}$, a dilation $\mathbf{d}$ satisfying the above definition generates a continuous curve $\{\mathbf{d}(s)x : s \in \mathbb{R}\}$, which starts at the origin for $s = -\infty$, crosses the point x for $s = 0$ and tends to infinity as $s \to +\infty$ (see Fig. 5.1). Notice that the continuity of dilation is not equivalent to strong continuity.

Example 5.2 Let us consider the family of nonlinear mappings $\mathbf{d}(s) : \mathbb{R}^n \mapsto \mathbb{R}^n$ defined as follows: $\mathbf{d}(s)\,\mathbf{0} = \mathbf{0}$, $\forall s \in \mathbb{R}$ and

$$\mathbf{d}(s)x = e^{s\left(1+\chi_Q\left(\frac{x_1}{\|x\|}\right)\right)}x, \quad s \in \mathbb{R}, \quad x = (x_1, \ldots, x_n)^\top \in \mathbb{R}^n \backslash \{\mathbf{0}\},$$

where $\chi_Q : \mathbb{R} \mapsto \{0\} \cup \{1\}$ is the Dirichlet function (the indicator function for the set of rational numbers). Since $\mathbf{d}(0)x = x$ and

$$\begin{aligned}\mathbf{d}(t) \circ \mathbf{d}(s)x =& e^{t\left(1+\chi_Q\left(\frac{e^{s\left(1+\chi_Q\left(\frac{x_1}{\|x\|}\right)\right)}x_1}{\left\|e^{s\left(1+\chi_Q\left(\frac{x_1}{\|x\|}\right)\right)}x\right\|}\right)\right)} e^{s\left(1+\chi_Q\left(\frac{x_1}{\|x\|}\right)\right)}x \\ =& e^{t\left(1+\chi_Q\left(\frac{x_1}{\|x\|}\right)\right)} e^{s\left(1+\chi_Q\left(\frac{x_1}{\|x\|}\right)\right)}x = \mathbf{d}(t+s)x, \quad \forall x \in \mathbb{R}^n,\end{aligned}$$

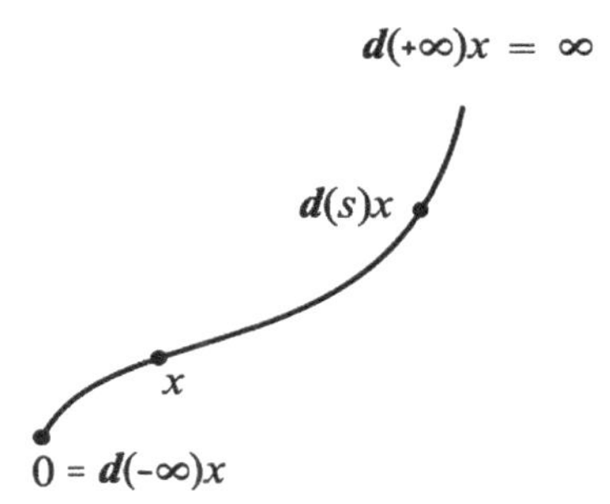

Fig. 5.1 Scaling of a point x by means of a continuous generalized dilation

then $\mathbf{d}$ is a group. The limit property, obviously, holds as well. Therefore, $\mathbf{d}$ is a dilation in $\mathbb{R}^n$. The term $\frac{x_1}{\|x\|}$ is invariant with respect to the dilation $\mathbf{d}$. This implies that $\mathbf{d}$ is continuous (with respect to the parameter $s \in \mathbb{R}$), but it is not strongly continuous due to the discontinuity of the function $x \mapsto \mathbf{d}(s)x$ at any point $x \in \mathbb{R}^n \backslash \{\mathbf{0}\}$.

The dilations presented in Examples 5.1 and 5.2 are rather "exotic". We exclude such dilations from considerations assuming the strong continuity of $\mathbf{d}$.

In [3–5], a strongly continuous dilation $\mathbf{d}$ in $\mathbb{R}^n$ is defined as a flow of a complete C^1 vector field. Namely, let us consider a nonlinear ODE

$$\dot{\phi}(s) = g(\phi(s)), \quad \phi(0) = x, \quad s \in \mathbb{R}, \tag{5.1}$$

where the vector field $g \in C^1(\mathbb{R}^n; \mathbb{R}^n)$ is such that solutions of (5.1) are defined for all $s \in \mathbb{R}$ and all $x \in \mathbb{R}^n$. We assume that the system $\dot{z} = -g(z)$ is globally uniformly asymptotically stable. This means that the system (5.1) is unstable in the forward time such that all its nonzero solutions tend to infinity as time goes to $+\infty$. However, the system (5.1) is globally asymptotically stable in the backward time, so all its solutions tend to zero as time goes to $-\infty$. In this case, a dilation $\mathbf{d}(s) : \mathbb{R}^n \mapsto \mathbb{R}^n$ may be defined as

$$\mathbf{d}(s)x = \phi_x(s), \quad x \in \mathbb{R}^n.$$

Indeed, since ϕ_x is a unique solution of (5.1) then, using semigroup property of solutions(see formula 2.59), we derive

$$\mathbf{d}(0)x = x, \quad \text{and} \quad \mathbf{d}(s+t)x = \mathbf{d}(s) \circ \mathbf{d}(t)x, \forall s, t \in \mathbb{R}, \quad \forall x \in \mathbb{R}^n.$$

Moreover, due to asymptotic stability of (5.1) in the backward time, one holds

$$\|\mathbf{d}(s)x\| \to 0 \text{ as } s \to -\infty,$$

but the instability of (5.1) in the forward time implies that

$$\|\mathbf{d}(s)x\| \to +\infty \text{ as } s \to +\infty \qquad \text{for} \qquad x \neq \mathbf{0}.$$

Such a dilation $\mathbf{d}(s)$ is called *geometric* and the corresponding dilation symmetry is known as *geometric homogeneity* [4, 5]. By construction, any geometric dilation is strongly continuous.

5.1.1 Linear Dilations

In this monograph, we deal only with continuous *linear dilations*, which, in the context of geometric homogeneity, corresponds to a *linear* vector field $g(\phi) = G_{\mathbf{d}}\phi$ in the generative equation (5.1), where $G_{\mathbf{d}} \in \mathbb{R}^{n \times n}$.

Definition 5.3 A dilation $\mathbf{d}$ in $\mathbb{R}^n$ is said to be linear if $\mathbf{d}(s)$ is a linear mapping (i.e., $\mathbf{d}(s) \in \mathbb{R}^{n\times n}$) for any $s \in \mathbb{R}$.

A linear dilation may be discontinuous (see Example 5.1). However, any continuous linear dilation in $\mathbb{R}^n$ is strongly continuous. The following result is a corollary of the well-known representation theorem for continuous linear groups in $\mathbb{R}^n$ (see, e.g., [6, Chap. 1]).

Theorem 5.1 *A continuous dilation* $\mathbf{d}$ *in* $\mathbb{R}^n$ *is linear if and only if there exists an anti-Hurwitz matrix* $G_{\mathbf{d}} \in \mathbb{R}^{n\times n}$ *such that*

$$\mathbf{d}(s) = e^{G_{\mathbf{d}}s} = \sum_{i=0}^{\infty} \frac{s^i G_{\mathbf{d}}^i}{i!}, \quad \forall s \in \mathbb{R}. \tag{5.2}$$

Proof *Sufficiency*. For any anti-Hurwitz matrix $G_{\mathbf{d}} \in \mathbb{R}^{n\times n}$, the family of linear mappings $e^{G_{\mathbf{d}}s} : \mathbb{R}^n \mapsto \mathbb{R}^n$ is a dilation. Indeed, $e^{G_{\mathbf{d}}s}$ satisfies the group property: $e^{0G_{\mathbf{d}}} = I_n$, $e^{(t+s)G_{\mathbf{d}}} = e^{sG_{\mathbf{d}}}e^{tG_{\mathbf{d}}}$, $\forall t, s \in \mathbb{R}$, since $e^{A_1}e^{A_2} = e^{A_1+A_2}$ for all commuting matrices $A_1A_2 = A_2A_1$. If $A \in \mathbb{R}^{n\times n}$ is an anti-Hurwitz matrix then, by Proposition 2.1, the limit property is fulfilled as well.

Necessity. If $\mathbf{d}$ is a continuous linear dilation in $\mathbb{R}^n$ then the matrix-valued function $s \mapsto \mathbf{d}(s)$ is continuous. In this case, we have $\|\mathbf{d}(s) - I_n\| = \|\mathbf{d}(s) - \mathbf{d}(0)\| \to 0$ as $s \to 0$ and

$$\begin{aligned}\left\| I_n - \frac{1}{s}\int_0^s \mathbf{d}(\tau)d\tau \right\| &= \left\| \frac{1}{s}\int_0^s (\mathbf{d}(\tau) - I_n)d\tau \right\| \\ &\le \frac{1}{|s|} \max\left\{ \int_0^{|s|} \|\mathbf{d}(\tau) - I_n\|d\tau, \int_0^{|s|} \|\mathbf{d}(-\tau) - I_n\|d\tau \right\} \\ &\le \max_{\tau\in[0,|s|]} \|\mathbf{d}(\pm\tau) - I_n\| \to 0 \quad \text{as} \quad s \to 0.\end{aligned}$$

Let $\rho \neq 0$ be such that

$$\left\| I_n - \frac{1}{\rho}\int_0^\rho \mathbf{d}(s)ds \right\| < 1 \quad \text{and} \quad \det\left(\frac{1}{\rho}\int_0^\rho \mathbf{d}(s)ds \right) \neq \mathbf{0}.$$

Hence, the matrix $\int_0^\rho \mathbf{d}(s)ds$ is invertible too and

$$
\begin{aligned}
(\mathbf{d}(h) - I_n) \int_0^\rho \mathbf{d}(s)ds &= \int_0^\rho \mathbf{d}(s+h)ds - \int_0^\rho \mathbf{d}(s)ds \\
&= \int_h^{\rho+h} \mathbf{d}(s)ds - \int_0^\rho \mathbf{d}(s)ds \\
&= \int_h^{\rho+h} \mathbf{d}(s)ds - \int_0^\rho \mathbf{d}(s)ds + \int_0^h \mathbf{d}(s)ds - \int_0^h \mathbf{d}(s)ds \\
&= \int_\rho^{\rho+h} \mathbf{d}(s)ds - \int_0^h \mathbf{d}(s)ds \\
&= \int_0^h \mathbf{d}(\rho+s)ds - \int_0^h \mathbf{d}(s)ds = \int_0^h \mathbf{d}(s)ds\,(\mathbf{d}(\rho) - I_n).
\end{aligned}
$$

Hence, we conclude that

$$
h^{-1}(\mathbf{d}(h) - I_n) = h^{-1} \int_0^h \mathbf{d}(s)ds\;(\mathbf{d}(\rho) - I_n) \left(\int_0^\rho \mathbf{d}(s)ds \right)^{-1}.
$$

Since

$$
h^{-1} \int_0^h \mathbf{d}(s)ds = I_n + h^{-1} \int_0^h (\mathbf{d}(s) - I_n)ds
$$

and

$$
\left\| h^{-1} \int_0^h (\mathbf{d}(s) - I_n)ds \right\| \le h^{-1} \int_0^h \|\mathbf{d}(s) - I_n\|\, ds \to 0 \text{ as } h \to 0
$$

then

$$
h^{-1}(\mathbf{d}(h) - I_n) \to G_{\mathbf{d}} := (\mathbf{d}(\rho) - I_n) \left(\int_0^\rho \mathbf{d}(s)ds \right)^{-1} \in \mathbb{R}^{n\times n} \text{ as } h \to 0.
$$

Let us show that $\mathbf{d}(s)$ admits the representation (5.2). Let us fix any $T > 0$. Since $s \mapsto \|\mathbf{d}(s)\|$ and $t \mapsto \|e^{tG_{\mathbf{d}}}\|$ are continuous functions then there exists $C \ge 1$ such that $\|\mathbf{d}(s)\| \le C$ and $\|e^{tG_{\mathbf{d}}}\| \le C$ for all $s, t \in [-T, T]$. Taking into account $\lim_{h\to 0} h^{-1}(e^{G_{\mathbf{d}}h} - I_n) = G_{\mathbf{d}} = \lim_{h\to 0} h^{-1}(\mathbf{d}(s) - I_n)$ we conclude that for any $\varepsilon > 0$ there exists $\delta > 0$ such that

$$
\left\| \frac{\mathbf{d}(h) - e^{hG_{\mathbf{d}}}}{h} \right\| \le \frac{2\varepsilon}{TC^2}, \quad \forall h \in [-\delta, \delta].
$$

Choosing an even $m \in \mathbb{N}$ such that $|s|/m \leq \delta$ we derive

$$\left\|\mathbf{d}(s) - e^{sG_{\mathbf{d}}}\right\| = \left\|\mathbf{d}\left(m\tfrac{s}{m}\right) - e^{m\frac{s}{m}G_{\mathbf{d}}}\right\| = \left\|\mathbf{d}^m\left(\tfrac{s}{m}\right) - \left(e^{\frac{s}{m}G_{\mathbf{d}}}\right)^m\right\|$$

$$= \left\|\mathbf{d}^{m-1}\left(\tfrac{s}{m}\right)\left(\mathbf{d}\left(\tfrac{s}{m}\right) - e^{\frac{s}{m}G_{\mathbf{d}}}\right) + \mathbf{d}^{m-1}\left(\tfrac{s}{m}\right)e^{\frac{s}{m}G_{\mathbf{d}}} - \left(e^{\frac{s}{m}G_{\mathbf{d}}}\right)^m\right\|$$

$$\leq \left\|\mathbf{d}^{m-1}\left(\tfrac{s}{m}\right)\left(\mathbf{d}\left(\tfrac{s}{m}\right) - e^{\frac{s}{m}G_{\mathbf{d}}}\right)\left(e^{\frac{s}{m}G_{\mathbf{d}}}\right)^0 + \mathbf{d}^{m-2}\left(\tfrac{s}{m}\right)\left(\mathbf{d}\left(\tfrac{s}{m}\right) - e^{\frac{s}{m}G_{\mathbf{d}}}\right)\left(e^{\frac{s}{m}G_{\mathbf{d}}}\right)^1 \dots\right\|$$

$$= \left\|\sum_{i=1}^{m} \mathbf{d}^{m-i}\left(\tfrac{s}{m}\right)\left(\mathbf{d}\left(\tfrac{s}{m}\right) - e^{\frac{s}{m}G_{\mathbf{d}}}\right)\left(e^{\frac{s}{m}G_{\mathbf{d}}}\right)^{i-1}\right\|$$

$$\leq \sum_{i=1}^{m} \left\|\mathbf{d}\left((m-i)\tfrac{s}{m}\right)\right\| \left\|\mathbf{d}\left(\tfrac{s}{m}\right) - e^{\frac{s}{m}G_{\mathbf{d}}}\right\| \left\|e^{(i-1)\frac{s}{m}G_{\mathbf{d}}}\right\| \leq mC\frac{\varepsilon\frac{|s|}{m}}{TC^2}C \leq \varepsilon$$

for any $s \in [-T, T]$. Since $\varepsilon > 0$ was arbitrary then $\mathbf{d}(s) = e^{sG_{\mathbf{d}}}$ for all $s \in [-T, T]$. Using the group property we extend this conclusion to $s \in \mathbb{R}$.

Finally, the matrix $G_{\mathbf{d}}$ is anti-Hurwitz. Indeed, otherwise, there exists an eigenvalue $\lambda_* \in \mathbb{C} : \Re(\lambda_*) \leq 0$ and taking any nonzero eigenvector $x_{\lambda_*} \in \mathbb{C}^n\backslash\{\mathbf{0}\}$: $G_{\mathbf{d}}x_{\lambda_*} = \lambda_* x_{\lambda_*}$ we derive

$$e^{sG_{\mathbf{d}}}x_{\lambda_*} = \sum_{i=0}^{+\infty}\frac{s^i G_{\mathbf{d}}^i}{i!}x_{\lambda_*} = \sum_{i=0}^{+\infty}\frac{s^i\lambda_*^i}{i!}x_{\lambda_*} = e^{s\lambda_*}x_{\lambda_*}.$$

If $\Re(\lambda_*) \leq 0$ then $|e^{sG_{\mathbf{d}}}x_{\lambda_*}| = e^{s\Re(\lambda_*)}|x_{\lambda_*}| \leq |x_{\lambda_*}|$ for all $s \geq 0$, where $|z| = \sqrt{z^\top\bar{z}}$ with $z \in \mathbb{C}^n$ is the Euclidean norm in $\mathbb{C}^n$. Taking into account $x_{\lambda_*} \neq \mathbf{0}$, we derive a contradiction to the limit property of the dilation. The proof is complete. □

The anti-Hurwitz matrix $G_{\mathbf{d}} \in \mathbb{R}^{n\times n}$ introduced in Theorem 5.1 is called an (infinitesimal) *generator of linear dilation* $\mathbf{d}$. Usually (see, e.g., [7, 8]), the generator of linear dilation $\mathbf{d}$ is assumed to be diagonalizable. In this monograph, we do not make this assumption allowing the generalized linear dilation to be merged even with rotations. Such dilations naturally appear in the case of consistent finite-dimensional approximations of infinite-dimensional systems (see Volume II of this monograph) and in the optimal control design (see Chap. 13). However, in many other practical cases (see, e.g., Chaps. 9, 10 and 14) it is, indeed, sufficient to consider a diagonalizable linear dilation.

The following dilations in $\mathbb{R}^n$ are rather popular in control literature (see Chap. 1 for more references):

- *Standard dilation* (L. Euler eighteenth century): $\mathbf{d}(s) = e^s I_n$, where I_n is the identity matrix in $\mathbb{R}^{n\times n}$;
- *Weighted dilation* (Zubov 1958 [9]): $\mathbf{d}(s) = \text{diag}\{e^{r_1 s}, ..., e^{r_n s}\} \in \mathbb{R}^n$, where $r_i > 0, i = 1, 2, ..., n$;
- *Geometric dilation* (Khomenuk 1961 [3], Kawski 1991 [2], Rosier 1993 [5]): a flow generated by an unstable C^1 vector field;
- *Linear dilation* (Polyakov 2019, [10]): $\mathbf{d}(s) = e^{sG_{\mathbf{d}}}$, where $G_{\mathbf{d}} \in \mathbb{R}^{n\times n}$ is an anti-Hurwitz matrix.

Obviously, the linear dilation includes standard and weighted dilations as particular cases, but it is less general than the geometric (nonlinear) dilation. Geometrically, differences between various dilations are shown in Fig. 1.3.

Let us introduce the following quantities characterizing a dilation rate:

$$\|\mathbf{d}(s)\| = \sup_{x\in S} \|\mathbf{d}(s)x\| = \sup_{x\neq 0} \tfrac{\|\mathbf{d}(s)x\|}{\|x\|} \text{ and } \lfloor \mathbf{d}(s) \rfloor = \inf_{x\in S} \|\mathbf{d}(s)x\| = \inf_{x\neq 0} \tfrac{\|\mathbf{d}(s)x\|}{\|x\|},$$

where $\|x\|$ is a norm in $\mathbb{R}^n$ and $S = \{x \in \mathbb{R}^n : \|x\| = 1\}$.

Corollary 5.1 *If* $\mathbf{d}$ *be a linear continuous dilation in* $\mathbb{R}^n$ *then*

(1) $\|\mathbf{d}(s+t)\| \leq \|\mathbf{d}(s)\|\|\mathbf{d}(t)\|$ *and* $\lfloor \mathbf{d}(s+t)\rfloor \geq \lfloor \mathbf{d}(s)\rfloor \lfloor \mathbf{d}(t)\rfloor$, $\forall s, t \in \mathbb{R}$;
(2) $\lfloor \mathbf{d}(-s)\rfloor \|\mathbf{d}(s)\| = 1$ *for all* $s \in \mathbb{R}$;
(3) the function $s \mapsto \|\mathbf{d}(s)\|$ *(resp.,* $s \mapsto \lfloor \mathbf{d}(s)\rfloor$*) is continuous;*
(4) there exist $\alpha \geq \beta > 0$ *and* $\exists M_\alpha, M_\beta \geq 1$ *such that*

$$\frac{1}{M_\alpha} e^{\alpha s} \leq \lfloor \mathbf{d}(s)\rfloor \leq \|\mathbf{d}(s)\| \leq M_\beta e^{\beta s}, \forall s \leq 0,$$

$$\frac{1}{M_\beta} e^{\beta s} \leq \lfloor \mathbf{d}(s)\rfloor \leq \|\mathbf{d}(s)\| \leq M_\alpha e^{\alpha s}, \forall s \geq 0.$$

Proof (1) Using the group property and the definitions of $\|\mathbf{d}(\cdot)\|$, $\lfloor \mathbf{d}(\cdot)\rfloor$ we derive

$$\|\mathbf{d}(s+t)x\| = \|\mathbf{d}(s)\mathbf{d}(t)x\| \leq \|\mathbf{d}(s)\|\|\mathbf{d}(t)x\| \leq \|\mathbf{d}(s)\|\|\mathbf{d}(t)\|\|x\|$$

and

$$\|\mathbf{d}(s+t)x\| = \|\mathbf{d}(s)\mathbf{d}(t)x\| \geq \lfloor \mathbf{d}(s)\rfloor \|\mathbf{d}(t)x\| \geq \lfloor \mathbf{d}(s)\rfloor \lfloor \mathbf{d}(t)\rfloor \|x\|$$

for all $x \in \mathbb{R}^n$. Hence, we conclude that the inequalities (1) are fulfilled.

(2) By means of the inequalities $\|x\| = \|\mathbf{d}(s)\mathbf{d}(-s)x\| \leq \|\mathbf{d}(s)\|\|\mathbf{d}(-s)x\|$ and $\|x\| = \|\mathbf{d}(-s)\mathbf{d}(s)x\| \geq \lfloor \mathbf{d}(-s)\rfloor \|\mathbf{d}(s)x\|$ we derive $1 \leq \|\mathbf{d}(s)\|\lfloor \mathbf{d}(-s)\rfloor$ and $1 \geq \lfloor \mathbf{d}(-s)\rfloor \|\mathbf{d}(s)\|$ for all $s \in \mathbb{R}$, i.e., $\|\mathbf{d}(s)\| \cdot \lfloor \mathbf{d}(-s)\rfloor = 1$.

(3) On the one hand, since $\|x\| = \|-x\| = \|(\mathbf{d}(s) - I_n)x + (-\mathbf{d}(s)x)\| \leq \|(\mathbf{d}(s) - I_n)x\| + \|\mathbf{d}(s)x\|$ and $\|\mathbf{d}(s)x\| = \|(\mathbf{d}(s) - I_n)x + x\| \leq \|(\mathbf{d}(s) - I_n)x\| + \|x\|$ then $|\|\mathbf{d}(s)x\| - \|x\|| \leq \|(\mathbf{d}(s) - I_n)x\| \leq \|\mathbf{d}(s) - I_n\|\|x\|$. On the other hand, since

$$\lfloor \mathbf{d}(s) \rfloor - 1 = \inf_{x \neq 0} \frac{\|\mathbf{d}(s)x\|}{\|x\|} - 1 = \inf_{x \neq 0} \frac{\|\mathbf{d}(s)x\| - \|x\|}{\|x\|}$$

then

$$|\lfloor \mathbf{d}(s) \rfloor - 1| = \inf_{x \neq 0} \frac{|\|\mathbf{d}(s)x\| - \|x\||}{\|x\|} \leq \|\mathbf{d}(s) - I_n\|.$$

The matrix-valued function $s \mapsto \mathbf{d}(s)$ (in the view of Proposition 2.1) is continuous and $\|\mathbf{d}(s) - I_n\| \to 0$ as $s \to 0$. Using the group property we conclude the continuity of the function $s \mapsto \lfloor \mathbf{d}(s) \rfloor$ for all $s \in \mathbb{R}$. The continuity of $s \mapsto \|\mathbf{d}(s)\|$ can be proven similarly.

(4) By Proposition 2.1, we have

$$\mathbf{d}(s) = e^{sG_{\mathbf{d}}}x_0 = S \begin{pmatrix} e^{\alpha_1 s} X_{\omega_1}(s) & \mathbf{0} & \dots & \mathbf{0} \\ \mathbf{0} & e^{\alpha_2 s} X_{\omega_2}(s) & \dots & \mathbf{0} \\ \dots & \dots & \dots & \dots \\ \mathbf{0} & \mathbf{0} & \dots & e^{\alpha_k s} X_{\omega_k}(s), \end{pmatrix} S^{-1}, \quad s \in \mathbb{R}$$

where $S \in \mathbb{R}^{n \times n}$ is a real Jordan transformation for $G_{\mathbf{d}}$ and $\alpha_i, \omega_i \geq 0$.

Let $\alpha = \max_i \alpha_i + \varepsilon$ and $\beta = \min_i \alpha_i - \varepsilon$, where $\varepsilon > 0$ is a small number such that $\beta > 0$. Using the polynomial structure of X_{ω_i} (see Proposition 2.1) we conclude

$$\sup_{i=1,\dots,k} \sup_{s \geq 0} |X_{\omega_i}(s)| e^{-\epsilon s} < +\infty$$

and

$$\sup_{i=1,\dots,k} \sup_{s \leq 0} |X_{\omega_i}(s)| e^{\epsilon s} < +\infty.$$

Hence, there exist $M_\alpha, M_\beta \geq 1$ such that

$$\lfloor \mathbf{d}(s) \rfloor \leq \|\mathbf{d}(s)\| \leq M_\alpha e^{\alpha s}, \quad \forall s \geq 0,$$

$$\lfloor \mathbf{d}(s) \rfloor \leq \|\mathbf{d}(s)\| \leq M_\beta e^{\beta s}, \quad \forall s \leq 0.$$

Finally, taking into account $\|z\| = \|\mathbf{d}(-s)\mathbf{d}(s)z\| \leq \|\mathbf{d}(-s)\| \cdot \|\mathbf{d}(s)z\|$, $\forall z \in \mathbb{R}^n$, $\forall s \in \mathbb{R}$, we derive $\lfloor \mathbf{d}(s) \rfloor \geq \frac{1}{\|\mathbf{d}(-s)\|}$ for all $s \in \mathbb{R}$. The proof is complete. □

The linear dilation (5.2) with diagonalizable matrices is studied in the context of the so-called nilpotent Lie groups (see, e.g., [8] and references therein). This monograph, mainly, deals with various applications of generalized homogeneity to stability/robustness/convergence analysis of dynamical systems as well as to control systems design.

5.1.2 Geometric Objects Induced by Dilations

Linear dilation allows an alternative norm topology to be introduced in $\mathbb{R}^n$ by means of the so-called *canonical homogeneous norm* studied in the next section. Let us consider, first, some related geometric objects.

Homogeneous Curve

Given $z \in \mathbb{R}^n$ the set

$$\Gamma_{\mathbf{d}}(z) = \{\mathbf{d}(s)z : s \in \mathbb{R}\} \tag{5.3}$$

is called a **d**-*homogeneous curve*. The homogeneous curve is continuous if the dilation $\mathbf{d}$ is continuous (see Fig. 5.1). Notice that $\mathbf{0} \notin \Gamma_{\mathbf{d}}(z)$ for $z \neq \mathbf{0}$.

Homogeneous Projection

Any element $x \in \mathbb{R}^n \backslash \{\mathbf{0}\}$ can be projected on the unit sphere along the homogeneous curve. For the case of the standard dilation, the corresponding projector $\pi : \mathbb{R}^n \backslash \{\mathbf{0}\} \mapsto S$ is defined as follows

$$\pi(x) = \frac{x}{\|x\|} = \mathbf{d}(-\ln \|x\|)x, \qquad \forall x \neq \mathbf{0},$$

where $\mathbf{d}(s) = e^s I_n$ is the standard dilation and $S = \{x \in \mathbb{R}^n : \|x\| = 1\}$ is a unit sphere in $\mathbb{R}^n$. In fact, the similar projection can be constructed using any continuous dilation.

Proposition 5.1 *If* $\mathbf{d}$ *is a continuous dilation in* $\mathbb{R}^n$ *then for any* $x \in \mathbb{R}^n \backslash \{\mathbf{0}\}$ *there exist* $s_0 \in \mathbb{R}$ *and* $x_0 \in S$ *such that*

$$x_0 = \mathbf{d}(s_0)x.$$

Proof If $x \neq \mathbf{0}$ then, by the limit property, $\|\mathbf{d}(s)x\| \to 0$ as $s \to -\infty$ and $\|\mathbf{d}(s)x\| \to +\infty$ as $s \to +\infty$. Since the function $\|\mathbf{d}(\cdot)x\| : \mathbb{R} \mapsto \mathbb{R}_+$ is continuous then there exists $s_0 \in \mathbb{R}$ such that $\|\mathbf{d}(s_0)x\| = 1$.

The point x_0 defined in the above proposition is called a **d**-*homogeneous projection* of the vector x on the unit sphere S. This projection is obtained as an intersection of the homogeneous curve $\Gamma_{\mathbf{d}}(x)$ with the unit sphere. In the general case, the homogeneous projection is non-unique (see Fig. 5.2) even for a weighted dilation (as shown in the below example).

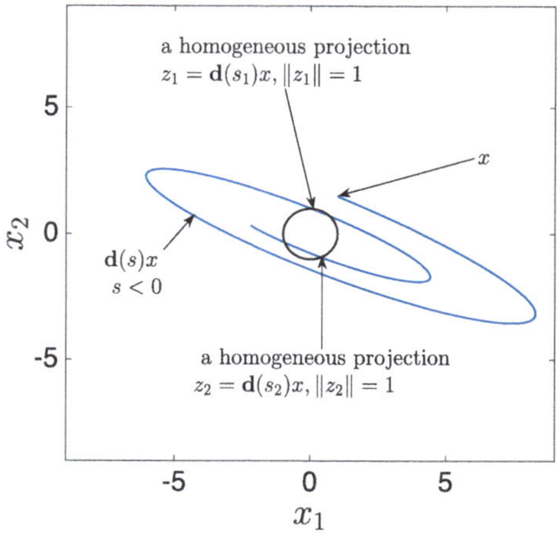

Fig. 5.2 Homogeneous projections

Example 5.3 Let us consider the weighted dilation

$$\mathbf{d}(s) = \begin{pmatrix} e^{\alpha s} & 0 \\ 0 & e^{s} \end{pmatrix}, \quad s \in \mathbb{R}, \quad \alpha > 1 + \frac{2}{\sqrt{5}-1}$$

in the Euclidean space $\mathbb{R}^2$ with the norm

$$\|z\| = \sqrt{z^\top P z}, \quad z \in \mathbb{R}^2, \quad P = P^\top \in \mathbb{R}^{n\times n}, \quad P \succ 0,$$

where

$$P = P_0 + \varepsilon \begin{pmatrix} 0 & 0 \\ 0 & 1 \end{pmatrix} \succ 0, \quad P_0 = \begin{pmatrix} \frac{3-\sqrt{5}}{2} & -1 \\ -1 & \frac{2}{3-\sqrt{5}} \end{pmatrix}, \quad \varepsilon > 0.$$

Let us show that for a sufficiently small $\varepsilon > 0$, the function $\xi : \mathbb{R} \mapsto \mathbb{R}$,

$$\xi(s) = \left\| \mathbf{d}(s) \left(\begin{smallmatrix} 1 \\ 1 \end{smallmatrix}\right) \right\|^2 - 1, \quad s \in \mathbb{R}$$

has at least two zeros: $\xi(s_1) = \xi(s_2) = 0$ and $s_1 \neq s_2$. Let us denote

$$\xi_0(s) = \left(\begin{smallmatrix} 1 \\ 1 \end{smallmatrix}\right)^\top \mathbf{d}^\top(s) P_0 \mathbf{d}(s) \left(\begin{smallmatrix} 1 \\ 1 \end{smallmatrix}\right) - 1.$$

The function $\xi_0 : \mathbb{R} \mapsto \mathbb{R}$ is continuously differentiable on $\mathbb{R}$

$$\xi'(s) = \left(\begin{smallmatrix} 1 \\ 1 \end{smallmatrix}\right)^\top \left(\left(\begin{smallmatrix} \alpha & 0 \\ 0 & 1 \end{smallmatrix}\right)^\top \mathbf{d}^\top(s) P_0 \mathbf{d}(s) + \mathbf{d}^\top(s) P_0 \mathbf{d}(s) \left(\begin{smallmatrix} \alpha & 0 \\ 0 & 1 \end{smallmatrix}\right) \right) \left(\begin{smallmatrix} 1 \\ 1 \end{smallmatrix}\right)$$

and decays in the neighborhood of the point $s = 0$, since

$$\xi_0'(0) = \begin{pmatrix}1\\1\end{pmatrix}^\top \left(\begin{pmatrix}\alpha & 0\\0 & 1\end{pmatrix}^\top P_0 + P_0 \begin{pmatrix}\alpha & 0\\0 & 1\end{pmatrix}\right)\begin{pmatrix}1\\1\end{pmatrix} = 2 + (\alpha - 1)(1 - \sqrt{5}) < 0.$$

Since

$$\xi_0(0) = \begin{pmatrix}1\\1\end{pmatrix}^\top P_0 \begin{pmatrix}1\\1\end{pmatrix} - 1 = \frac{3-\sqrt{5}}{2} + \frac{2}{3-\sqrt{5}} - 3 = 0,$$

$\xi'(0) < 0$ and $\xi_0(s) \to +\infty$ as $s \to +\infty$ then ξ_0 has at least two zeros due to continuity. Moreover, there exists $\delta > 0$ such that

$$\xi_0(s) < 0 \quad \text{for all} \quad 0 < s < \delta.$$

Hence, taking into account

$$\xi(s) = \xi_0(s) + \varepsilon e^{2s}$$

we conclude that ξ also has at least two zeros if $\varepsilon > 0$ is sufficiently small. Therefore, the point $z = \begin{pmatrix}1\\1\end{pmatrix}$ has at least two different homogeneous projections $\mathbf{d}(s_1)z$ and $\mathbf{d}(s_2)z$, $s_1 \neq s_2$ on the unit sphere S.

The uniqueness of homogeneous projection depends on the unit sphere S, i.e., on the norm in $\mathbb{R}^n$. Notice that, for the standard dilation $\mathbf{d}(s) = e^s I_n$, the homogeneous projection is always unique due to the standard homogeneity of the norm: $\|e^s z\| = e^s\|z\|$, $s \in \mathbb{R}$, $z \in \mathbb{R}^n$. The uniqueness of homogeneous projection in the general case is studied in Sect. 5.1.3.

The linear dilation can also be utilized in order to project any nonzero point of a bounded set Ω on the boundary $\partial\Omega$.

Lemma 5.1 *Let* $\mathbf{d}$ *be a continuous dilation in* $\mathbb{R}^n$*. If* $\Omega \subset \mathbb{R}^n$ *is a bounded set then for any* $x \in \Omega\backslash\{\mathbf{0}\}$ *there exists* $s \in [0, +\infty)$ *such that* $\mathbf{d}(s)x \in \partial\Omega$*. Moreover, if* Ω *has nonempty interior and* $x \in \text{int}\,\Omega\backslash\{\mathbf{0}\}$ *then* $s > 0$*.*

Proof If $\mathbf{0} \neq x \in \Omega$ then, due to boundedness of Ω and the limit property of the dilation, there exists $\overline{s}_0 \in (0, +\infty)$ such that $\mathbf{d}(\overline{s}_0)x \notin \Omega$. For $\underline{s}_0 = 0$ we have $\mathbf{d}(\underline{s}_0)x = x \in \Omega$. Let us construct the sequences $\underline{s}_i$ and $\overline{s}_i$ as follows: if $\mathbf{d}\left(\frac{\underline{s}_i+\overline{s}_i}{2}\right)x \in \Omega$ then $\underline{s}_{i+1} = \frac{\underline{s}_i+\overline{s}_i}{2}$ and $\overline{s}_{i+1} = \overline{s}_i$, otherwise, we take $\underline{s}_{i+1} = \underline{s}_i$ and $\overline{s}_{i+1} = \frac{\underline{s}_i+\overline{s}_i}{2}$, where $i = 0, 1, 2, \ldots$. By construction, the sequences $\underline{s}_i$ and $\overline{s}_i$ are monotone and bounded. Consequently, they are convergent. Since $|\overline{s_i} - \underline{s}_i| \to 0$ as $i \to +\infty$ then there exists $s \in [0, +\infty)$ such that

$$\lim_{i\to+\infty} \underline{s}_i = \lim_{i\to+\infty} \overline{s}_i = s.$$

Moreover, the continuity of the dilation $\mathbf{d}$ implies

$$\lim_{i\to+\infty} \mathbf{d}(\underline{s}_i)x = \lim_{i\to+\infty} \mathbf{d}(\overline{s}_i)x = \mathbf{d}(s)x.$$

Therefore, for any neighborhood U_x of the point $\mathbf{d}(s)x$ there exists a sufficiently large index $i^* \geq 0$ such that $\mathbf{d}(\underline{s}_i)x \in U_x$ and $\mathbf{d}(\overline{s}_i)x \in U_x$ for all $i \geq i^*$. Since, by construction, we have $\mathbf{d}(\underline{s}_i)x \in \Omega$ and $\mathbf{d}(\overline{s}_i)x \notin \Omega$ for all i, then, by definition of the boundary point, we conclude that $\mathbf{d}(s)x \in \partial\Omega$. Notice that $s > 0$ if $x \in \text{int}\,\Omega\backslash\{\mathbf{0}\}$. Indeed, otherwise we have $x = \mathbf{d}(0)x \in \partial\Omega$, which contradicts to $x \in \text{int}\,\Omega\backslash\{\mathbf{0}\}$. The proof is complete. □

Homogeneous Spheres and Balls

The set

$$S_{\mathbf{d}}(r, z_0) = \left\{z \in \mathbb{R}^n : \|\mathbf{d}\,(-\ln r)\,(z - z_0)\| = 1\right\}$$

is a **d**-*homogeneous sphere* of the radius $r > 0$ with the center at $z_0 \in \mathbb{R}^n$ *and the set*

$$B_{\mathbf{d}}(r, z_0) = \left\{z \in \mathbb{R}^n : \|\mathbf{d}\,(-\ln r)\,(z - z_0)\| < 1\right\}$$

is an open **d**-*homogeneous ball* of the radius $r > 0$ *with the center at* $z_0 \in \mathbb{R}^n$.

For spheres and balls centered at the origin, we use the notation

$$S_{\mathbf{d}}(r) := S_{\mathbf{d}}(r, \mathbf{0}) \quad \text{and} \quad B_{\mathbf{d}}(r) := B_{\mathbf{d}}(r, \mathbf{0}).$$

The unit homogeneous sphere $S_{\mathbf{d}}(1)$ coincides with the unit sphere S in $\mathbb{R}^n$.

Corollary 5.2 *If* **d** *is a continuous dilation in* $\mathbb{R}^n$ *then*

(1) $B_{\mathbf{d}}(r)\backslash\{\mathbf{0}\} \subset \bigcup\limits_{0<\delta<r} S_{\mathbf{d}}(\delta)$; *(2)* $\mathbb{R}^n\backslash\{\mathbf{0}\} = \bigcup\limits_{r>0} S_{\mathbf{d}}(r)$.

Proof Let $z \in B_{\mathbf{d}}(r)$, $z \neq \mathbf{0}$ and $\xi : (0, +\infty) \mapsto (0, +\infty)$ be defined as follows

$$\xi(\delta) = \|\mathbf{d}(-\ln \delta)z\|.$$

On the one hand, since the dilation **d** is continuous then the function ξ is continuous. On the other hand, we have $\xi(r) < 1$ by definition of $B_{\mathbf{d}}(r)$ and $\xi(\delta) \to +\infty$ as $\delta \to 0$ due to limit property. Hence, by continuity, there exists $\delta \in (0, r)$ such that $\xi(\delta) = 1$. This, by definition, means that $z \in S_{\mathbf{d}}(\delta)$. The second identity is a straightforward corollary of Proposition 5.1. The proof is complete. □

Example 5.4 To illustrate the definition of homogeneous sphere let us consider the dilation

$$\mathbf{d}(s) = e^{0.5\,s}\begin{pmatrix} \cos(s) & \sin(s) \\ -\sin(s) & \cos(s) \end{pmatrix}$$

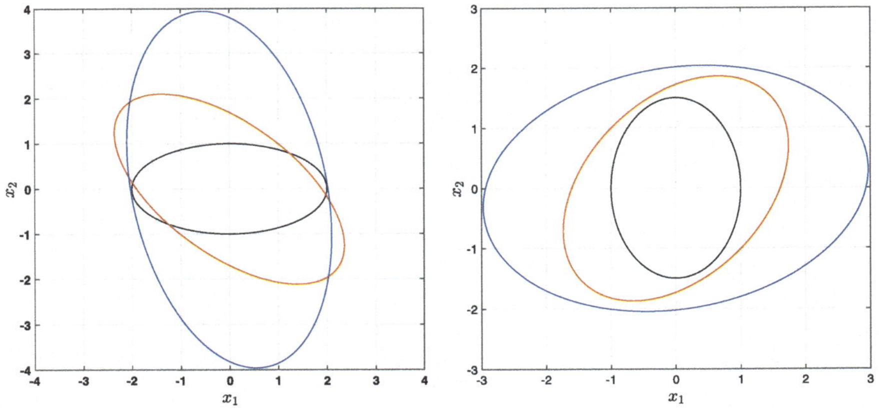

(a) $\|x\| = \sqrt{0.25x_1^2 + x_2^2}$ (b) $\|x\| = \sqrt{x_1^2 + 4x_2^2/9}$

Fig. 5.3 Homogeneous spheres induced by two different norms in $\mathbb{R}^2$: $S_{\mathbf{d}}(1)$—black line; $S_{\mathbf{d}}(2)$—red line; $S_{\mathbf{d}}(3)$—blue line

in $\mathbb{R}^2$ with the weighted Euclidean norm $\|x\| = \sqrt{0.25x_1^2 + x_2^2}$, where $x = (x_1, x_2)^\top \in \mathbb{R}^2$. The homogeneous spheres $S_{\mathbf{d}}(1)$, $S_{\mathbf{d}}(2)$ and $S_{\mathbf{d}}(3)$ are depicted at Fig. 5.3a.

Notice that $S_{\mathbf{d}}(1)$ and $S_{\mathbf{d}}(2)$ have intersections in this case. However, the spheres will not have intersections if we construct them using another norm $\|x\| = \sqrt{x_1^2 + 4x_2^2/9}$ (see, Fig. 5.3b).

If the homogeneous projection is not unique (see Example 5.3 and Fig. 5.2) then the homogeneous sphere $S_{\mathbf{d}}(r)$ with $r \neq 1$ may have intersections with the unit sphere. To guarantee that homogeneous spheres have no intersection, some additional property called *monotonicity* of dilation is imposed below.

Homogeneous Cone

Recall that $\mathcal{D} \subset \mathbb{R}^n$ is a *positive cone* if $\lambda\mathcal{D} \subset \mathcal{D}$, $\forall \lambda > 0$. A positive cone $\mathcal{D}$ is *linear* if $x + y \in \mathcal{D}$, $\forall x, y \in \mathcal{D}$.

Definition 5.4 A nonempty set $\mathcal{D} \subseteq \mathbb{R}^n$ is said to be $\mathbf{d}$-*homogeneous cone* $\mathbb{R}^n$ if $\mathcal{D}$ is invariant with respect to the dilation $\mathbf{d}$:

$$\mathbf{d}(s)z \in \mathcal{D} \text{ for all } z \in \mathcal{D} \text{ and all } s \in \mathbb{R}.$$

Homogeneous cone is a reasonable name for the set $\mathcal{D}$, since any point $z \in \mathcal{D}$ belongs to $\mathcal{D}$ together with the $\mathbf{d}$-homogeneous curve $\Gamma_{\mathbf{d}}(z)$. If $\mathbf{d}$ is the standard dilation (i.e., $\mathbf{d}(s) = e^s I_n$, $s \in \mathbb{R}$), then the homogeneous curve $\Gamma_{\mathbf{d}}(z)$ is a ray initiated

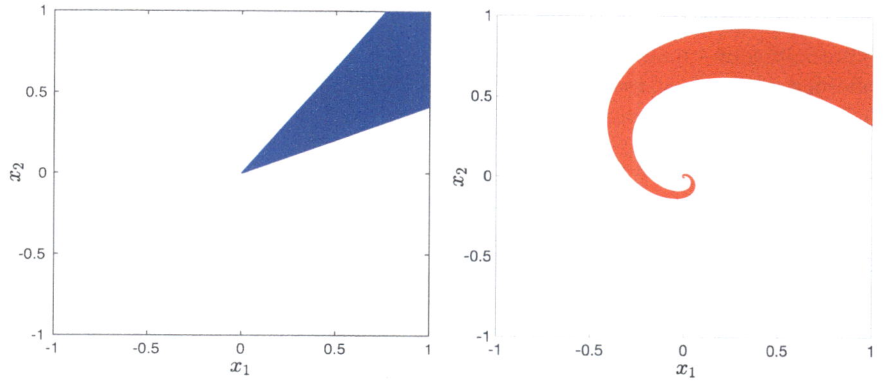

Fig. 5.4 Standard homogeneous cone for $\mathbf{d}(s) = e^s I_2$ (left figure) and generalized homogeneous cone for $\mathbf{d}(s) = e^{s\begin{pmatrix} 1.7 & 2 \\ -2 & 0.7 \end{pmatrix}}$ (right figure)

at the origin and crossing the point z, so $\mathcal{D}$ becomes a conventional positive cone in $\mathbb{R}^n$ in this case. The generalized homogeneous cones in $\mathbb{R}^n$ are usually nonlinear. An exception, for example, is the weighted homogeneous cone of non-negative vectors in $\mathbb{R}^n$. However, such a cone is standard homogeneous too. Geometrically, a difference between a linear standard homogeneous cone and a nonlinear $\mathbf{d}$-homogeneous cone in $\mathbb{R}^2$ is illustrated on Fig. 5.4. The linear dilation $\mathbf{d}$ considered in Example 5.4 is utilized for the plot. Some properties of homogeneous cones are listed below.

- The union $\mathcal{D}_1 \cup \mathcal{D}_2$ of two $\mathbf{d}$-homogeneous cones $\mathcal{D}_1, \mathcal{D}_2 \subset \mathbb{R}^n$ is $\mathbf{d}$-homogeneous cone.
- The intersection $\mathcal{D}_1 \cap \mathcal{D}_2$ of two $\mathbf{d}$-homogeneous cones $\mathcal{D}_1, \mathcal{D}_2 \subset \mathbb{R}^n$ is $\mathbf{d}$-homogeneous cones.
- If $\mathcal{D} \subset \mathbb{R}^n$ is a nonempty closed $\mathbf{d}$-homogeneous cone then $\mathbf{0} \in \mathcal{D}$.

5.1.3 Monotonicity

The monotonicity of dilation plays an important role in further investigations of homogeneous geometrical structures as well as in the analysis and design of homogeneous control systems. This property better characterizes a relation of dilation with a norm topology of a vector space.

Definition 5.5 *(Monotone dilation in* $\mathbb{R}^n$*)* A dilation $\mathbf{d}$ in $\mathbb{R}^n$ is said to be monotone with respect to a norm $\|\cdot\|$ if the function $s \mapsto \|\mathbf{d}(s)x\|$ is strictly increasing for any $x \neq \mathbf{0}$.

Example 5.3 and Figs. 5.2 and 5.3a, illustrate non-monotone dilations in $\mathbb{R}^2$. Figure 5.3b shows that the monotonicity property depends on the norm. The most important property of monotone dilations is the uniqueness of the homogeneous projection on the unit sphere for any $x \in \mathbb{R}^n \setminus \{\mathbf{0}\}$.

Proposition 5.2 (On uniqueness of homogeneous projection in $\mathbb{R}^{\mathbf{n}}$) *If* $\mathbf{d}$ *is a continuous monotone dilation then the homogeneous projection is unique, i.e., for any* $x \neq \mathbf{0}$ *there exists a* ***unique*** *pair* $(s_0, x_0) \in \mathbb{R} \times S$ *such that* $x = \mathbf{d}(s_0)x_0$, *where* $S = \{x \in \mathbb{R}^n : \|x\| = 1\}$ *is the unit sphere.*

Proof The existence of a pair (s_0, x_0) is proven in Proposition 5.1. The uniqueness follows from the strict monotonicity of the function $s \mapsto \|\mathbf{d}(s)x\|$.

For linear dilations, we introduce a more constructive monotonicity property.

Definition 5.6 *(Strictly monotone dilation in* $\mathbb{R}^{\mathbf{n}}$*)* A linear dilation $\mathbf{d}$ in $\mathbb{R}^n$ is said to be strictly monotone with respect to a norm $\|\cdot\|$ if there exists $\beta > 0$ such that

$$\|\mathbf{d}(s)\| \leq e^{\beta s}, \quad \forall s \leq 0. \tag{5.4}$$

Obviously, if a linear continuous dilation $\mathbf{d}$ is strictly monotone then we have $M_\beta = 1$ in Corollary 5.1. Moreover, we can always select $M_\alpha = 1$ and $\alpha = \|G_{\mathbf{d}}\|$, since for $s > 0$ we have

$$\|\mathbf{d}(s)\| = \left\| \sum_{i=0}^{+\infty} \frac{s^i G_{\mathbf{d}}^i}{i!} \right\| \leq \sum_{i=0}^{+\infty} \frac{s^i \|G_{\mathbf{d}}\|^i}{i!} = e^{s\|G_{\mathbf{d}}\|}.$$

The following lemma characterizes families of equivalent dilations in $\mathbb{R}^n$.

Lemma 5.2 *If* $\mathbf{d}$ *is a dilation in* $\mathbb{R}^n$ *then*

- *for any* $\gamma > 0$, *the group* $\mathbf{d}_\gamma(s) = \mathbf{d}(\gamma s)$, $s \in \mathbb{R}$ *is a dilation in* $\mathbb{R}^n$ *too;*
- *(strict) monotonicity of* $\mathbf{d}$ *with respect to a norm* $\|\cdot\|$ *in* $\mathbb{R}^n$ *implies (strict) monotonicity of* $\mathbf{d}_\gamma$ *with respect to the same norm* $\|\cdot\|$ *in* $\mathbb{R}^n$*;*
- *in the case of linear continuous dilation* $\mathbf{d}$, *the generators of* $\mathbf{d}$ *and* $\mathbf{d}_\gamma$ *satisfy the identity*

$$G_{\mathbf{d}_\gamma} = \gamma G_{\mathbf{d}}. \tag{5.5}$$

Proof The group and limit properties of $\mathbf{d}_\gamma$ are obvious. Monotonicity of $\mathbf{d}_\gamma$ follows from monotonicity of $\mathbf{d}$ and the inequality $\gamma > 0$. In the case of linear continuous dilation $\mathbf{d}$, Theorem 5.1 gives $\mathbf{d}_\gamma(s) = \mathbf{d}(\gamma s) = e^{\gamma s G_{\mathbf{d}}}$. Hence, the generator of $\mathbf{d}_\gamma$ is, obviously, $G_{\mathbf{d}_\gamma} = \gamma G_{\mathbf{d}}$. The proof is complete. □

The dilations $\mathbf{d}$ and $\mathbf{d}_\gamma$ considered in the above lemma are called *equivalent* since they generate geometrically identical homogeneous curves in $\mathbb{R}^n$.

Proposition 5.3 (On strict monotonicity of a linear dilation in $\mathbb{R}^n$) *A continuous linear dilation* **d** *in* $\mathbb{R}^n$ *is strictly monotone with respect to the weighted Euclidean norm* $\|x\| = \sqrt{x^\top P x}$, $x \in \mathbb{R}^n$ *with* $0 \prec P = P^\top \in \mathbb{R}^{n\times n}$ ***if and only if*** *the following linear matrix inequality holds*

$$PG_{\mathbf{d}} + G_{\mathbf{d}}^\top P \succ 0, \tag{5.6}$$

where $G_{\mathbf{d}} \in \mathbb{R}^n$ *is the generator of the dilation* **d**. *Moreover, one has*

$$\begin{array}{ll} 2\beta\|\mathbf{d}(s)z\|^2 \leq \frac{d}{ds}\|\mathbf{d}(s)z\|^2 \leq 2\alpha\|\mathbf{d}(s)z\|^2, & \forall s \in \mathbb{R}, \forall z \in \mathbb{R}^n, \\ e^{\alpha s} \leq \lfloor \mathbf{d}(s) \rfloor \leq \|\mathbf{d}(s)\| \leq e^{\beta s}, & \forall s \leq 0, \\ e^{\beta s} \leq \lfloor \mathbf{d}(s) \rfloor \leq \|\mathbf{d}(s)\| \leq e^{\alpha s}, & \forall s \geq 0, \end{array} \tag{5.7}$$

where

$$\alpha = \frac{1}{2}\lambda_{\max}\left(P^{\frac{1}{2}}G_{\mathbf{d}}P^{-\frac{1}{2}} + P^{-\frac{1}{2}}G_{\mathbf{d}}^\top P^{\frac{1}{2}}\right) > 0, \tag{5.8}$$

$$\beta = \frac{1}{2}\lambda_{\min}\left(P^{\frac{1}{2}}G_{\mathbf{d}}P^{-\frac{1}{2}} + P^{-\frac{1}{2}}G_{\mathbf{d}}^\top P^{\frac{1}{2}}\right) > 0. \tag{5.9}$$

Proof *Necessity.* Let **d** be a strictly monotone continuous linear dilation, i.e., $\exists \beta > 0 : \|\mathbf{d}(s)\| \leq e^{\beta s}, \forall s \leq 0$. For any $z \in S$, where $S \subset \mathbb{R}^n$ is the unit sphere for the norm $\|\cdot\|$, we derive

$$1 = \|z\|^2 = \|\mathbf{d}(-s)\mathbf{d}(s)z\|^2 \leq \|\mathbf{d}(-s)\|^2\|\mathbf{d}(s)z\|^2 \leq e^{-2\beta s}\|\mathbf{d}(s)z\|^2, \forall s \geq 0.$$

Hence, we have

$$e^{2\beta s} - 1 \leq z^\top \mathbf{d}^\top(s)P\mathbf{d}(s)z - z^\top Pz = (\mathbf{d}(s)z - z)^\top P\mathbf{d}(s)z + z^\top P(\mathbf{d}(s)z - z).$$

Dividing both sides of the above inequality by s and taking the limit $s \to 0^+$ we obtain

$$0 < 2\beta \leq z^\top G_{\mathbf{d}}^\top Pz + z^\top PG_{\mathbf{d}}z, \quad \forall z \in \mathbb{R}^n\backslash\{\mathbf{0}\}.$$

Sufficiency. Since $\frac{d}{ds}\mathbf{d}(s) = \frac{d}{ds}e^{sG_{\mathbf{d}}} = G_{\mathbf{d}}e^{sG_{\mathbf{d}}}$ then

$$\tfrac{d}{ds}\|\mathbf{d}(s)z\|^2 = \frac{d}{ds}z^\top \mathbf{d}^\top(s)P\mathbf{d}(s)z = z^\top \mathbf{d}^\top(s)\left(PG_{\mathbf{d}} + G_{\mathbf{d}}^\top P\right)\mathbf{d}(s)z$$

and using (5.6) we derive

$$2\beta\|\mathbf{d}(s)z\|^2 \leq \frac{d}{ds}\|\mathbf{d}(s)z\|^2 \leq 2\alpha\|\mathbf{d}(s)z\|^2, \forall s \in \mathbb{R}.$$

The obtained differential inequality yields

$$e^{2\beta s}\|z\| \le e^{2\beta s}\|\mathbf{d}(0)z\|^2 \le \|\mathbf{d}(s)z\|^2 \le e^{2\alpha s}\|\mathbf{d}(0)z\|^2 = e^{2\alpha s}\|z\|^2, \forall s \ge 0$$

and

$$e^{2\alpha s}\|z\| \le e^{2\alpha s}\|\mathbf{d}(0)z\|^2 \le \|\mathbf{d}(s)z\|^2 \le e^{2\beta s}\|\mathbf{d}(0)z\|^2 = e^{2\beta s}\|z\|^2 \quad \forall s \le 0$$

for all $z \in \mathbb{R}^n$. The proof is complete. □

Since the matrix $G_{\mathbf{d}}$ is anti-Hurwitz then the Lyapunov inequality (5.6) is always feasible with respect to a symmetric matrix $P \succ 0$ (see Example 4.14).

> Therefore, *any continuous linear dilation in* $\mathbb{R}^n$ is *strictly monotone with respect to a properly selected weighted Euclidean norm.*

Since in the rest part of the monograph we deal only with continuous linear dilations, for shortness, we omit the words "continuous linear" if the context is clear. In other words, by default, *all generalized dilations considered below are linear and continuous.*

5.2 Generalized Homogeneous Euclidean Space

A *vector space* over the field of reals is a set $\mathbb{V}$ together with two operations (see Appendix A for more details). The first operation is a *vector addition* $\mathbb{V} \times \mathbb{V} \mapsto \mathbb{V}$ denoted $v + w$ for $v, w \in \mathbb{V}$. The second operation is a *multiplication by a scalar* $\mathbb{R} \times \mathbb{V} \mapsto \mathbb{V}$ denoted αv (or $\alpha \cdot v$) for $\alpha \in \mathbb{R}$ and $v \in \mathbb{V}$. For $\mathbb{V} = \mathbb{R}^n$ the multiplication of $v = (v_1, ..., v_n)^\top \in \mathbb{R}^n$ by a scalar $\alpha \in \mathbb{R}$ is traditionally defined as $\alpha v = (\alpha v_1, ..., \alpha v_2)^\top$, which, in fact, is the standard dilation if $\alpha > 0$. In the above section, we have considered some generalizations of the standard dilation. The reasonable question in this case: *Is it possible to construct a vector space using a generalized dilation as a "multiplication by a scalar"?* The positive answer to this question is given below for the case of linear dilations. The topology of the corresponding vector space is defined by means of the so-called canonical homogeneous norm. Moreover, we show that this vector space is, in fact, a Euclidean space $\mathbb{R}^n_{\mathbf{d}} = (\mathbb{R}^n, \langle \cdot, \cdot \rangle_{\mathbf{d}})$ with an inner product $\langle \cdot, \cdot \rangle_{\mathbf{d}}$ defined by means of the linear dilation.

5.2.1 Canonical Homogeneous Norm

By definition, a norm in $\mathbb{R}^n$ is a function $\|\cdot\| : \mathbb{R}^n \mapsto \mathbb{R}_+$ satisfying the following three conditions:

- *Positive Definiteness:* $\|u\| = 0$ if and only if $u = \mathbf{0}$;
- *Dilation+Reflection Symmetry:* $\|\pm e^s u\| = e^s\|u\|, \forall s \in \mathbb{R}, \forall u \in \mathbb{R}^n$;
- *Subadditivity (Triangle Inequality):* $\|u + v\| \leq \|u\| + \|v\|, \forall u, v \in \mathbb{R}^n$.

In the literature, the second condition is frequently called by the homogeneity or the positive homogeneity, but, in fact, it combines two symmetries. One of them is, indeed, the standard homogeneity (i.e., a symmetry with respect to the standard dilation). Omitting the third condition and replacing the standard dilation with a generalized one, the so-called homogeneous norm can be introduced as follows.

Definition 5.7 A continuous function $p : \mathbb{R}^n \mapsto \mathbb{R}_+$ is said to be a **d**-homogeneous norm in $\mathbb{R}^n$ if

- $p(x) = 0$ if and only if $x = \mathbf{0}$;
- $p(\pm\mathbf{d}(s)x) = e^s p(x)$ for $\forall x \in \mathbb{R}^n$, $s \in \mathbb{R}$,

where **d** is a generalized dilation.

For the weighted dilation, the similar definition can be found in [11]. The function p may not satisfy the triangle inequality $p(x + y) \leq p(x) + p(y)$, so, formally, this is not even a semi-norm. However, many authors (see, e.g., [12–15]) call functions satisfying just two above conditions by a "homogeneous norm" (or "dilated norm"). We follow this tradition accepted in control literature, but, basically, we use this name for the so-called *canonical homogeneous norm*, which is, indeed, a norm (in the usual sense) for a homeomorphic Euclidean space $\mathbb{R}^n_{\mathbf{d}}$ introduced below.

Example 5.5 ([2]) For the weighted dilation $\mathbf{d}(s) = \mathrm{diag}\{e^{r_1 s}, e^{r_2 s}, ..., e^{r_n s}\}$, a **d**-homogeneous norm $p : \mathbb{R}^n \mapsto [0, +\infty)$ can be defined as follows

$$p(x) = \sum_{i=1}^{n} |x_i|^{\frac{1}{r_i}}, \quad x = (x_1, x_2, .., x_n)^\top \in \mathbb{R}^n.$$

Indeed, p is continuous, positive definite, $p(\pm\mathbf{d}(s)x) = e^s p(x), \forall x \in \mathbb{R}^n$.

The following method can be suggested for the design of a **d**-homogeneous norm in the case of generalized dilation. Let $\Psi \in C(\mathbb{R}^n; \mathbb{R}^n)$ be such that

$$\Psi(x) = \mathbf{0} \quad \Leftrightarrow \quad x = \mathbf{0} \tag{5.10}$$

and

$$\Psi(\mathbf{d}(s)x) = e^{\bar{r}s}\Psi(x), \quad \forall x \in \mathbb{R}^n, s \in \mathbb{R}, \tag{5.11}$$

where $\bar{r} > 0$. Then the function $\|\cdot\|_{\bar{r}} : \mathbb{R}^n \mapsto \mathbb{R}$ given by

$$\|x\|_{\bar{r}} = \left(\Psi^\top(x) Q \Psi(x)\right)^{\frac{1}{2\bar{r}}}, \quad 0 \prec Q = Q^\top \in \mathbb{R}^{n\times n} \tag{5.12}$$

is a **d**-homogeneous norm in $\mathbb{R}^n$. Indeed, by construction, $\|\cdot\|_{\bar{r}}$ is continuous on $\mathbb{R}^n$ and **d**-homogeneous of the degree 1:

$$\|\mathbf{d}(s)x\|_{\bar{r}} = \left(\Psi^\top(\mathbf{d}(s)x) Q \Psi(\mathbf{d}(s)x)\right)^{\frac{1}{2\bar{r}}} = \left(e^{2\bar{r}}\Psi^\top(x) Q \Psi(x)\right)^{\frac{1}{2\bar{r}}} = e^s \|x\|_{\bar{r}}.$$

Since $\Psi(x) = \mathbf{0} \Leftrightarrow x = \mathbf{0}$ then $\Psi(x) \neq \mathbf{0}$ for $x \neq \mathbf{0}$, but $Q \succ 0$ implies $\|x\|_{\bar{r}} > 0$ for $x \neq \mathbf{0}$. In some cases, the mapping Ψ can be constructed such that the **d**-homogeneous norm $\|\cdot\|_{\bar{r}}$ is *analytic*[1] on $\mathbb{R}^n\backslash\{\mathbf{0}\}$.

Lemma 5.3 (On explicit homogeneous norm for diagonalizable dilation) *Let a linear continuous dilation* **d** *have a diagonalizable generator, i.e.,* $\exists J \in \mathbb{R}^{n\times n}$, $\det(J) \neq 0$ *such that*

$$G_{\mathbf{d}} = J^{-1}\Lambda J, \quad \Lambda = \operatorname{diag}\{r_1, r_2, \ldots, r_n\} \succ 0. \tag{5.13}$$

Then the function $\|\cdot\|_{\bar{r}} : \mathbb{R}^n \mapsto \mathbb{R}$ *defined as follows*

$$\|x\|_{\bar{r}} = \left(\Psi^\top(x) Q \Psi(x)\right)^{\frac{1}{2\bar{r}}}, \quad \Psi(x) = \begin{pmatrix} \lfloor e_1^\top Jx \rceil^{\frac{\bar{r}}{r_1}} \\ \lfloor e_2^\top Jx \rceil^{\frac{\bar{r}}{r_2}} \\ \ldots \\ \lfloor e_n^\top Jx \rceil^{\frac{\bar{r}}{r_n}} \end{pmatrix}, \quad \bar{r} > 0 \tag{5.14}$$

is a **d***-homogeneous norm in* $\mathbb{R}^n$ *and* Ψ *is a homeomorphism on* $\mathbb{R}^n$*, where* $h_i = (0, \ldots, 1, \ldots, 0)^\top \in \mathbb{R}^n$ *is the unit vector and* $0 \prec Q = Q^\top \in \mathbb{R}^{n\times n}$. *Moreover,*

- $\|\cdot\|_{\bar{r}}$ *is continuously differentiable on* $\mathbb{R}^n\backslash\{\mathbf{0}\}$ *provided that* $\bar{r} \geq \max\{r_1, \ldots, r_n\}$;
- $\|\cdot\|_{\bar{r}}$ *is analytic on* $\mathbb{R}^n\backslash\{\mathbf{0}\}$ *provided that* $\bar{r} = p_1 p_2 \ldots p_n$ *and* $r_i = \frac{p_i}{q}$, *where* $p_i, q \in \mathbb{N}$ *are odd natural numbers,* $i = 1, 2, \ldots, n$.

Proof Taking into account $\mathbf{d}(s) = e^{sG_{\mathbf{d}}} = J^{-1}e^{s\Lambda}J$, we derive

$$\Psi(\mathbf{d}(s)x) = \begin{pmatrix} \lfloor e_1^\top J\mathbf{d}(s)x \rceil^{\frac{\bar{r}}{r_1}} \\ \lfloor e_2^\top J\mathbf{d}(s)x \rceil^{\frac{\bar{r}}{r_2}} \\ \ldots \\ \lfloor e_n^\top J\mathbf{d}(s)x \rceil^{\frac{\bar{r}}{r_n}} \end{pmatrix} = \begin{pmatrix} \lfloor e^{r_1 s} e_1^\top Jx \rceil^{\frac{\bar{r}}{r_1}} \\ \lfloor e^{r_2 s} e_2^\top Jx \rceil^{\frac{\bar{r}}{r_2}} \\ \ldots \\ \lfloor e^{r_n s} e_n^\top Jx \rceil^{\frac{\bar{r}}{r_n}} \end{pmatrix} = e^{\bar{r}s}\Psi(x),$$

and

$$\|\mathbf{d}(s)x\|_{\bar{r}} = e^s\|x\|_{\bar{r}}, \quad \forall s \in \mathbb{R}, \forall x \in \mathbb{R}^n.$$

[1] A function $p : \mathbb{R}^n \mapsto \mathbb{R}$ is said to be analytic on an open set $\mathcal{D} \subset \mathbb{R}^n$ if for any $x_0 \in \mathcal{D}$ there exist $a_i \in \mathbb{R}$, $i = 0, 1, 2, \ldots.$ such that $p(x) = \sum_{i=0}^{+\infty} a_i (x - x_0)^i$ and this series is convergent to $p(x)$ is a neighborhood of x_0. Any analytic function belongs to the class C^∞.

On the other hand, since J is invertible then, by construction, Ψ is a homeomorphism on $\mathbb{R}^n$. The inverse of Ψ is given by

$$\Psi^{-1}(y) = J^{-1}\begin{pmatrix} \lfloor y_1 \rceil^{\frac{r_1}{\bar r}} \\ \lfloor y_2 \rceil^{\frac{r_2}{\bar r}} \\ \dots \\ \lfloor y_n \rceil^{\frac{r_n}{\bar r}} \end{pmatrix}, \quad y = (y_1, \dots, y_n)^\top \in \mathbb{R}^n.$$

In this case, $\Psi(x) = \mathbf{0} \Leftrightarrow x = \mathbf{0}$, so $\|x\|_{\bar r}$ is a $\mathbf{d}$-homogeneous norm. Moreover, for $\bar r \geq \max\{r_1, ..., r_n\}$ the function $x \mapsto \lfloor e_i^\top Jx \rceil^{\frac{\bar r}{r_i}}$ is continuously differentiable on $\mathbb{R}^n$. Hence, $\|\cdot\|_{\bar r}$ is continuously differentiable on $\mathbb{R}^n \backslash \{\mathbf{0}\}$.

If $p_i, q \in \mathbb{N}$ are odd numbers then for $\bar r = p_1 p_2 ... p_n$ the number $\frac{\bar r q}{r_i} \in \mathbb{N}$ is odd as well. In this case, $\lfloor \cdot \rceil^{\frac{\bar r q_i}{p_i}} = (\cdot)^{\frac{\bar r q_i}{p_i}}$ and the function $x \mapsto \lfloor h_i^\top Jx \rceil^{\frac{\bar r q_i}{p_i}}$ is a polynomial. Hence, $\|\cdot\|_{\bar r}$ is continuous on $\mathbb{R}^n$ and analytic on $\mathbb{R}^n \backslash \{\mathbf{0}\}$ (as a composition of analytic functions). The proof is complete. □

Geometrically (see [16, Sect. 15] or Fig. 5.5), a conventional norm $\|\cdot\|$ in $\mathbb{R}^n$ can be constructed by means of a standard dilation applied to a convex central-symmetric compact neighborhood $B \subset \mathbb{R}^n$ of the origin (e.g., B can be an ellipsoid/ball in $\mathbb{R}^n$ centered at zero). Indeed, let the boundary of the set B define the unit level set (i.e., we assign $\|x\| = 1, \forall x \in \partial B$), but the boundary of the scaled set $e^s B$ corresponds to the level e^s (i.e., we assign $\|x\| = e^s, \forall x \in \partial(e^s B)$), where $s \in \mathbb{R}$. Let $\|\,\mathbf{0}\,\| = 0$. The function $\|\cdot\|$ constructed in such a way is a norm in $\mathbb{R}^n$. Indeed, since the set B is a convex neighborhood of $\mathbf{0}$ then $\partial(e^s B) \cap \partial(e^t B) = \emptyset$ for $s \neq t$. This means that the function $x \mapsto \|x\|$ is single valued. The positive definiteness comes from the fact that B is a neighborhood of the origin: $\mathbf{0} \in \text{int } B$. The homogeneity is obtained by construction. The reflection symmetry of $\|\cdot\|$ is due to the central symmetry of B. The triangle inequality comes from the convexity of B and, consequently, the convexity of $\|\cdot\|$. A $\mathbf{d}$-homogeneous norm can be designed similarly by means of a linear dilation $\mathbf{d}$ being monotone with respect to a norm $\|\cdot\|$ in $\mathbb{R}^n$. Such a homogeneous norm is called *canonical* [10, 17].

Definition 5.8 *(Canonical homogeneous norm)* Let a linear continuous dilation $\mathbf{d}$ in $\mathbb{R}^n$ be monotone with respect to a norm $\|\cdot\|$. A function $\|\cdot\|_{\mathbf{d}} : \mathbb{R}^n \mapsto [0, +\infty)$ defined as follows: $\|\,\mathbf{0}\,\|_{\mathbf{d}} = 0$ and

$$\|x\|_{\mathbf{d}} = e^{s_x}, \text{ where } s_x \in \mathbb{R} : \|\mathbf{d}(-s_x)x\| = 1, \quad x \neq \mathbf{0} \tag{5.15}$$

is said to be a canonical $\mathbf{d}$-homogeneous norm in $\mathbb{R}^n$.

Below we write shortly "canonical homogeneous norm" if it is clear from the context which dilation $\mathbf{d}$ *has been utilized for its definition.*

The canonical homogeneous norm is defined by means of a homogeneous projection of a vector $x \in \mathbb{R}^n \backslash \{\mathbf{0}\}$ on the unit sphere. The monotonicity of the dilation is required to guarantee that the function $\|\cdot\|_{\mathbf{d}}$ is single valued and continuous at the origin.

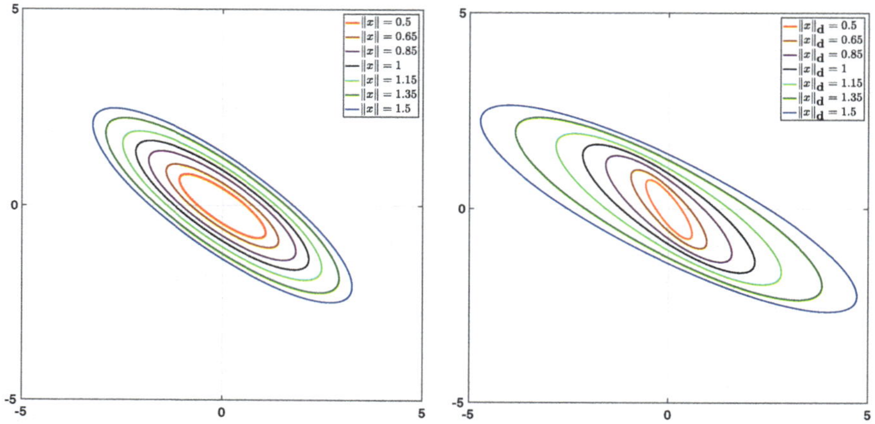

Fig. 5.5 The levels sets of the norm $\|x\| = \sqrt{0.5x_1^2 + 0.4x_1x_2 + x_2^2}$ (the left figure) and the canonical $\mathbf{d}$-homogeneous norm $\|\cdot\|_{\mathbf{d}}$ induced by $\|x\|$ (the right figure)

Obviously, $\|\cdot\|_{\mathbf{d}} = \|\cdot\|$ provided that $\mathbf{d}$ is the standard dilation: $\mathbf{d}(s) = e^s I_n$. In other cases, the canonical homogeneous norm may not coincide with the conventional norm. However, as shown below, the canonical homogeneous norm inherits some properties of the norm. For instance, continuous differentiability of $\|\cdot\|$ on $\mathbb{R}^n\backslash\{\mathbf{0}\}$ implies continuous differentiability of $\|\cdot\|_{\mathbf{d}}$ on $\mathbb{R}^n\backslash\{\mathbf{0}\}$ provided the linear continuous dilation in $\mathbb{R}^n$ is strictly monotone with respect to $\|\cdot\|$.

Figure 5.5 illustrates differences between the weighted Euclidean norm in $\mathbb{R}^2$ and the canonical $\mathbf{d}$-homogeneous norm in $\mathbb{R}^2$ induced by this weighted Euclidean norm using the dilation $\mathbf{d}(s) = \mathrm{diag}\{e^{2s}, e^s\}$, $s \in \mathbb{R}$. The unit level sets are identical for both norms. All other level sets are non-coinciding, since they are obtained by means of a scaling of the unit level set using two different dilations: *standard* and *weighted*.

Examples of canonical homogeneous norms in $\mathbb{R}^n$ for weighted dilations can be discovered in control literature in the form of the so-called controllability function [18, 19] for the integrators chain or as an implicit Lyapunov function [20, 21] for linear controllable system. The generalized homogeneity of these functions was not recognized in the mentioned literature. In this monograph, the canonical homogeneous norm is utilized for the analysis of dynamical systems and control design.

The canonical homogeneous norm $\|\cdot\|_{\mathbf{d}}$ *is defined implicitly*. Indeed, to compute the value $\|x\|_{\mathbf{d}}$ for $x \neq \mathbf{0}$, the algebraic equation $\|\mathbf{d}(-\ln \|x\|_{\mathbf{d}})x\| = 1$ has to be solved with respect to the scalar $\|x\|_{\mathbf{d}} \in (0, +\infty)$. Let us study the existence and uniqueness as well as some basic properties of $\|x\|_{\mathbf{d}}$.

Lemma 5.4 (Properties of canonical homogeneous norm) *If a linear continuous dilation* $\mathbf{d}$ *in* $\mathbb{R}^n$ *is monotone with respect to a norm* $\|\cdot\|$ *then*

(1) $\|\cdot\|_{\mathbf{d}} : \mathbb{R}^n \mapsto \mathbb{R}_+$ *is single valued and continuous on* $\mathbb{R}^n$*;*
(2) $\|\pm\mathbf{d}(s)x\|_{\mathbf{d}} = e^s\|x\|_{\mathbf{d}}, \forall s \in \mathbb{R}, \forall x \in \mathbb{R}^n$*;*
(3) $\|x\|=1 \Leftrightarrow \|x\|_{\mathbf{d}}=1$, $\|x\|<1 \Leftrightarrow \|x\|_{\mathbf{d}}<1$ *and* $\|x\|>1 \Leftrightarrow \|x\|_{\mathbf{d}}>1$*;*
(4) *there exist* $\underline{\sigma}, \overline{\sigma} \in \mathcal{K}_\infty$ *such that*

$$\underline{\sigma}(\|x\|_{\mathbf{d}}) \leq \|x\| \leq \overline{\sigma}(\|x\|_{\mathbf{d}}), \qquad \forall x \in \mathbb{R}^n; \tag{5.16}$$

(5) $\|\cdot\|_{\mathbf{d}}$ *is locally Lipschitz continuous on* $\mathbb{R}^n\backslash\{\mathbf{0}\}$*, moreover, if the linear continuous dilation* $\mathbf{d}$ *is strictly monotone with respect to the norm* $\|\cdot\|$ *then*

$$\left|\|x_1\|_{\mathbf{d}}^{\alpha} - \|x_2\|_{\mathbf{d}}^{\alpha}\right| \leq \|x_1 - x_2\|, \quad \forall x_1, x_2 \in B(1), \tag{5.17}$$

$$\left|\|x_1\|_{\mathbf{d}}^{\beta} - \|x_2\|_{\mathbf{d}}^{\beta}\right| \leq \|x_1 - x_2\|, \quad \forall x_1, x_2 \in \mathbb{R}^n\backslash B(1), \tag{5.18}$$

where $\alpha = \|G_{\mathbf{d}}\|$*,* β *is introduced in Definition 5.6 and* $B(1) = \{x \in \mathbb{R}^n : \|x\| \leq 1\}$ *is the unit ball in* $\mathbb{R}^n$*;*
(6) $\|\cdot\|_{\mathbf{d}} \in C^k(\mathbb{R}^n\backslash\{\mathbf{0}\})$, $1 \leq k \leq \infty$ *provided that the norm* $\|\cdot\| \in C^k(\mathbb{R}^n\backslash\{\mathbf{0}\})$ *and the linear dilation* $\mathbf{d}$ *is strictly monotone with respect to the norm* $\|\cdot\|$*, moreover,*

$$\frac{\partial\|x\|_{\mathbf{d}}}{\partial x} = e^s \left.\frac{\frac{\partial\|\mathbf{d}(-s)x\|}{\partial x}}{\left.\frac{\partial\|z\|}{\partial z}\right|_{z=\mathbf{d}(-s)x} G_{\mathbf{d}}\mathbf{d}(-s)x}\right|_{s=\ln\|x\|_{\mathbf{d}}} \qquad for\, x \neq \mathbf{0}. \tag{5.19}$$

Proof (1) Since the dilation is continuous and monotone then for any $x \in \mathbb{R}^n\backslash\{\mathbf{0}\}$ there exists a unique pair $s_0 \in \mathbb{R}$ and $x_0 \in S$ such that $x = \mathbf{d}(s_0)x_0$ (see, Proposition 5.2). This means that $\|\mathbf{d}(-s_0)x\| = 1$, i.e., $\|x\|_{\mathbf{d}} = e^{s_0}$ is finite, positive and uniquely defined for any $x \in \mathbb{R}^n\backslash\{\mathbf{0}\}$.

Due to the monotonicity of dilation, the locally Lipschitz continuous mapping $(s, x) \mapsto \|\mathbf{d}(s)x\| - 1$ is strictly decreasing on the first argument for any $x \neq \mathbf{0}$. So, the derivative $\frac{\partial\|\mathbf{d}(s)x\|}{\partial s}$ exists almost everywhere and it is negative. Moreover, for any $x \neq \mathbf{0}$, we have $\|\mathbf{d}(s)x\| - 1 \to -1$ as $s \to -\infty$ and $\|\mathbf{d}(s)x\| - 1 \to -1$ as $s \to +\infty$. Hence, there exists a unique $s \in \mathbf{R}$ such that $\|\mathbf{d}(s)x\| - 1 = 0$. This means that $\|\cdot\|_{\mathbf{d}}$ is well defined on $\mathbb{R}^n\backslash\{\mathbf{0}\}$, but the local Lipschitz continuity of $\|\cdot\|_{\mathbf{d}}$ away from the origin follows from Implicit Function theorem for locally Lipschitz mappings (see, e.g., [22, 23] or Appendix D). Let us prove continuity at zero. Since $\|\mathbf{d}(-\ln\|x\|_{\mathbf{d}})x\| = 1$ for $x \in \mathbb{R}^n\backslash\{\mathbf{0}\}$ then $\|\mathbf{d}(-\ln\|x\|_{\mathbf{d}})\| \geq \frac{1}{\|x\|}$. This means that $\|\mathbf{d}(-\ln\|x\|_{\mathbf{d}})\| \to +\infty$ as $\|x\| \to 0$. Due to monotonicity of the dilation $\mathbf{d}$ this is possible if and only if $\ln\|x\|_{\mathbf{d}} \to -\infty$. Therefore, $\|x\|_{\mathbf{d}} \to 0$ as $\|x\| \to 0$.

(2) Let $\|x\|_{\mathbf{d}} = e^{s_x}$, where $s_x \in \mathbb{R} : \|\mathbf{d}(-s_x)x\| = 1$. Since $\|\mathbf{d}(s)x\|_{\mathbf{d}} = e^{\tilde{s}_x}$, where $\tilde{s}_x \in \mathbb{R} : \|\mathbf{d}(-\tilde{s}_x)\mathbf{d}(s)x\| = 1$, then using the group property we derive $\|\mathbf{d}(-\tilde{s}_x +$

$s)x\| = 1$. Due to uniqueness of homogeneous projection, we have $\tilde{s}_x = s + s_x$ and $\|\mathbf{d}(s)x\|_{\mathbf{d}} = e^{\tilde{s}_x} = e^s e^{s_x} = e^s \|x\|_{\mathbf{d}}$ for all $s \in \mathbb{R}$.

(3) If $\|x\| = 1$ then for $s_x = 0$ we have $\|\mathbf{d}(-s_x)x\| = \|x\| = 1$ and $\|x\|_{\mathbf{d}} = e^{s_x} = 1$, and vise versa. The cases $\|x\| > 1$ ($s_x > 0$) and $\|x\| < 1$ ($s_x < 0$) can be analyzed similarly using the monotonicity of dilation.

(4) Since, by definition, $\|\mathbf{d}(-\ln \|x\|_{\mathbf{d}})x\| = 1$ then

$$\lfloor \mathbf{d}(-\ln \|x\|_{\mathbf{d}}) \rfloor \cdot \|x\| \le 1 \le \|\mathbf{d}(-\ln \|x\|_{\mathbf{d}})\| \cdot \|x\|.$$

Using Corollary 5.1 we derive

$$\tfrac{1}{M_\alpha} \|x\|_{\mathbf{d}}^{-\alpha} \|x\| \le 1 \le M_\beta \|x\|_{\mathbf{d}}^{-\beta} \|x\|, \quad \text{if } \|x\| \ge 1,$$

$$\tfrac{1}{M_\beta} \|x\|_{\mathbf{d}}^{-\beta} \|x\| \le 1 \le M_\alpha \|x\|_{\mathbf{d}}^{-\alpha} \|x\|, \quad \text{if } \|x\| \le 1,$$

where $0 < \beta \le \alpha$ and $M_\alpha \ge 1$, $M_\beta \ge 1$. The inequality (5.16) is fulfilled for

$$\overline{\sigma}(\rho) = \max\{M_\alpha \rho^\alpha, M_\beta \rho^\beta\}, \quad \underline{\sigma}(\rho) = \min\{M_\alpha^{-1} \rho^\alpha, M_\beta^{-1} \rho^\beta\}.$$

(5) For $\gamma \in \{\alpha, \beta\}$, by definition of the homogeneous norm, one has

$$|\|u\|_{\mathbf{d}}^\gamma - \|v\|_{\mathbf{d}}^\gamma| = |e^{\gamma s_u} - e^{\gamma s_v}|,$$

where $\|\mathbf{d}(-s_u)u\| = 1$ and $\|\mathbf{d}(-s_v)v\| = 1$. Since for strictly monotone dilations we have

$$e^{\alpha s} \le \lfloor \mathbf{d}(s) \rfloor \le \|\mathbf{d}(s)\| \le e^{\beta s}, \quad \forall s \le 0,$$

$$e^{\beta s} \le \lfloor \mathbf{d}(s) \rfloor \le \|\mathbf{d}(s)\| \le e^{\alpha s}, \quad \forall s \ge 0$$

then for $s_u < s_v$ we derive

$$\begin{aligned} 1 &= \|\mathbf{d}(-s_v)(v - u) + \mathbf{d}(s_u - s_v)\mathbf{d}(-s_u)u\| \le \|\mathbf{d}(-s_v)\| \, \|u - v\| + \|\mathbf{d}(s_u - s_v)\| \\ &\le \|\mathbf{d}(-s_v)\| \, \|u - v\| + e^{\beta(s_u - s_v)} \end{aligned}$$

or, equivalently, $-e^{\beta s_v} \|\mathbf{d}(-s_v)\| \, \|u - v\| \le e^{\beta s_u} - e^{\beta s_v}$. Repeating the same analysis for $s_u > s_v$ we obtain $1 \le \|\mathbf{d}(-s_u)\| \, \|u - v\| + e^{\beta(s_v - s_u)}$ or, equivalently, $e^{\beta s_u} - e^{\beta s_v} \le e^{\beta s_u} \|\mathbf{d}(-s_u)\| \, \|u - v\|$. Therefore,

$$\left| \|u\|_{\mathbf{d}}^\beta - \|v\|_{\mathbf{d}}^\beta \right| = \left| e^{\beta s_u} - e^{\beta s_v} \right| \le \max\left\{ e^{\beta s_v} \|\mathbf{d}(-s_v)\|, e^{\beta s_u} \|\mathbf{d}(-s_u)\| \right\} \|u - v\|.$$

This implies the local Lipschitz continuity of $\|\cdot\|$ on $\mathbb{R}^n \backslash B(1)$. For the strictly monotone dilation $\mathbf{d}$ and $\|u\|_{\mathbf{d}} \ge 1$ and $\|v\|_{\mathbf{d}} \ge 1$ we have $s_u > 0, s_v > 0$ and $e^{\beta s_u} \|\mathbf{d}(-s_u)\| \le 1$, $e^{\beta s_v} \|\mathbf{d}(-s_v)\| \le 1$. Hence,

$$\left|\|u\|_{\mathbf{d}}^{\beta}-\|v\|_{\mathbf{d}}^{\beta}\right|=\left|e^{\beta s_u}-e^{\beta s_v}\right|\leq\|u-v\|,\quad\forall u,v\in\mathbb{R}^n\backslash B(1).$$

Similarly, we derive

$$\left|\|u\|_{\mathbf{d}}^{\alpha}-\|v\|_{\mathbf{d}}^{\alpha}\right|=\left|e^{\alpha s_u}-e^{\alpha s_v}\right|\leq\max\left\{e^{\alpha s_v}\|\mathbf{d}(-s_v)\|,e^{\alpha s_u}\|\mathbf{d}(-s_u)\|\right\}\|u-v\|$$

and $\|u\|_{\mathbf{d}}\leq 1$ and $\|v\|_{\mathbf{d}}\leq 1$ we have $s_u<0, s_v<0$ and $e^{\alpha s_u}\|\mathbf{d}(-s_u)\|\leq 1$, $e^{\alpha s_v}\|\mathbf{d}(-s_v)\|\leq 1$ due to the strict monotonicity of the dilation. Hence,

$$\left|\|u\|_{\mathbf{d}}^{\alpha}-\|v\|_{\mathbf{d}}^{\alpha}\right|=\left|e^{\alpha s_u}-e^{\alpha s_v}\right|\leq\|u-v\|,\quad\forall u,v\in B(1).$$

(6) If the norm $\|\cdot\|$ is smooth outside the origin then $\frac{\partial}{\partial s}\|\mathbf{d}(-s)x\|=-\left.\frac{\partial\|z\|}{\partial z}\right|_{z=\mathbf{d}(-s)x}G_{\mathbf{d}}\mathbf{d}(-s)x$. Taking into account $\frac{\partial}{\partial x}\|x\|_{\mathbf{d}}=e^s\left.\frac{\partial s}{\partial x}\right|_{s=\ln\|x\|_{\mathbf{d}}}$ formula (5.21) is derived from the Implicit Function theorem (see Appendix D) which yields $\frac{\partial s}{\partial x}=-\frac{\frac{\partial\|\mathbf{d}(-s)x\|}{\partial x}}{\frac{\partial\|\mathbf{d}(-s)x\|}{\partial s}}$ for the function $s:\mathbb{R}^n\mapsto\mathbb{R}$ implicitly defined by the equation $\|\mathbf{d}(-s)x\|=1$. The proof is complete. □

A sharp estimate (5.16) can be obtained for strictly monotone dilations.

Corollary 5.3 *Let a continuous linear dilation* $\mathbf{d}$ *in* $\mathbb{R}^n$ *be strictly monotone with respect to the norm* $\|x\|=\sqrt{x^\top Px}$, *where* $0\prec P=P^\top\in\mathbb{R}^{n\times n}$. *Then*

(1) the inequalities (5.17) *and* (5.18) *hold for* $\alpha>0$ *and* $\beta>0$ *defined as in Proposition 5.3;*
(2) the inequalities (5.16) *hold for*

$$\underline{\sigma}(\rho)=\begin{cases}\rho^{\alpha} & \text{if } \rho\leq 1,\\ \rho^{\beta} & \text{if } \rho>1,\end{cases}\qquad\overline{\sigma}(\rho)=\begin{cases}\rho^{\beta} & \text{if } \rho\leq 1,\\ \rho^{\alpha} & \text{if } \rho>1;\end{cases}\tag{5.20}$$

(3) the derivative of $\|\cdot\|_{\mathbf{d}}$ *is given by*

$$\frac{\partial\|x\|_{\mathbf{d}}}{\partial x}=\|x\|_{\mathbf{d}}\frac{x^\top\mathbf{d}^\top(-\ln\|x\|_{\mathbf{d}})P\mathbf{d}(-\ln\|x\|_{\mathbf{d}})}{x^\top\mathbf{d}^\top(-\ln\|x\|_{\mathbf{d}})PG_{\mathbf{d}}\mathbf{d}(-\ln\|x\|_{\mathbf{d}})x},\quad x\neq\mathbf{0},\tag{5.21}$$

moreover, $\|\cdot\|_{\mathbf{d}}^{\lambda}\in C^1(\mathbb{R}^n)$ *provided that* $\lambda>\max_{i=1,2,...,n}\Re(\lambda_i(G_{\mathbf{d}}))$, *where* $\Re(\lambda_i(G_{\mathbf{d}}))>0$ *is the real part of the i-th eigenvalue of* $G_{\mathbf{d}}$;
(4) $\|\cdot\|_{\mathbf{d}}$ *is analytic on* $\mathbb{R}^n\backslash\{\mathbf{0}\}$;
(5) if for some $x,y\in\mathbb{R}^n\backslash\{\mathbf{0}\}$ *the following inequality*

$$x^\top\mathbf{d}^\top(-\ln\|y\|_{\mathbf{d}})P\mathbf{d}(-\ln\|y\|_{\mathbf{d}})x<1\quad(\textit{resp.,}>1)\tag{5.22}$$

holds then $\|x\|_{\mathbf{d}}<\|y\|_{\mathbf{d}}$ *(resp.,* $\|x\|_{\mathbf{d}}>\|y\|_{\mathbf{d}}$*);*

(6) *the function* $V : \mathbb{R}^n \mapsto \mathbb{R}_+$ *defined as* $V(x) = \underline{\sigma}(\|x\|_{\mathbf{d}})$, $x \in \mathbb{R}^n$ *is positive definite, radially unbounded and globally Lipschitz continuous with the Lipschitz constant* 1;

(7) *for any* $\epsilon > 0$ *and any* $x \in \mathbb{R}^n$ *it holds*

$$\frac{\|x\|_{\mathbf{d}}}{\underline{\sigma}^{-1}(\epsilon)} \le \left\| \frac{x}{\epsilon} \right\|_{\mathbf{d}} \le \frac{\|x\|_{\mathbf{d}}}{\overline{\sigma}^{-1}(\epsilon)}, \tag{5.23}$$

where $\overline{\sigma}^{-1}$ *and* $\underline{\sigma}^{-1}$ *are inverse functions to* $\overline{\sigma}$ *and* $\underline{\sigma}$, *respectively.*

Proof (1) For α defined in Proposition 5.3) it holds $\alpha = \|G_{\mathbf{d}}\| = \sup_{x \neq \mathbf{0}} \frac{\|G_{\mathbf{d}} x\|}{\|x\|}$, where $\|x\| = \sqrt{x^\top P x}$. Due to (5.7), the parameter β satisfies Definition 5.6. By Lemma 5.4 the inequalities (5.17) and (5.18) are fulfilled.

(2) The estimates (5.16), (5.20) come from the definition of the canonical homogeneous norm and the estimates (5.7).

(3) Formula (5.21) is a straightforward consequence of (5.19). The obtained formula guarantees the smoothness of the derivative on $\mathbb{R}^n \backslash \{\mathbf{0}\}$. Moreover, since $\|\mathbf{d}(-\ln \|x\|_{\mathbf{d}})x\| = 1$ then, by definition of $\beta > 0$ (see Proposition 5.3) we have

$$x^\top \mathbf{d}^\top(-\ln \|x\|_{\mathbf{d}}) P G_{\mathbf{d}} \mathbf{d}(-\ln \|x\|_{\mathbf{d}})x \ge \beta$$

and using (5.21) we derive

$$\left\| \frac{\partial \|x\|_{\mathbf{d}}^{\lambda}}{\partial x} \right\| \le \lambda \|x\|_{\mathbf{d}}^{\lambda} \frac{\|\mathbf{d}(-\ln \|x\|_{\mathbf{d}})\|}{\beta}.$$

The matrix $\lambda I_n - G_{\mathbf{d}}$ is anti-Hurwitz if $\lambda > \max\limits_{i=1,2,\ldots,n} \Re(\lambda_i(G_{\mathbf{d}}))$. This implies that

$$\|x\|_{\mathbf{d}}^{\lambda} \|\mathbf{d}(-\ln \|x\|_{\mathbf{d}})\| = \|e^{(\lambda I_n - G_{\mathbf{d}}) \ln \|x\|_{\mathbf{d}}}\| \to 0 \quad \text{as} \quad \|x\|_{\mathbf{d}} \to 0,$$

so $\|x\|_{\mathbf{d}}^{\lambda} \in C^1(\mathbb{R}^n)$.

(4) For the norm $\|x\| = \sqrt{x^\top P x}$, denoting $V = -\ln \|x\|_{\mathbf{d}}$, the algebraic equation (5.15) can be rewritten as follows

$$Q(V, x) := x^\top \sum_{i=0}^{+\infty} \frac{(V G_{\mathbf{d}}^\top)^i}{i!} P \sum_{j=0}^{+\infty} \frac{(V G_{\mathbf{d}})^j}{j!} x - 1 = 0.$$

The left-hand side of the above algebraic equation becomes an analytic function with respect to the variables x and V. Hence, by Analytic Implicit Function theorem [24, Theorem 6.1.2], the function $V : \mathbb{R}^n \backslash \{\mathbf{0}\} \mapsto \mathbb{R}$ defined by the algebraic equation is analytic at least on some open ball $B(r_2) \subset \mathbb{R}^n$ with $r_2 > 0$. Since $\ln(\cdot)$ is an analytic function on $(0, +\infty)$ and composition of analytic functions is analytic as well, then the function $\|\cdot\|_{\mathbf{d}}$ is analytic, at least, on $K(r_1, r_2) = \{x \in \mathbb{R}^n : r_1 < \|x\| < r_2\}$ with $r_1 \in (0, r_2)$. Taking into account the identity $\|\mathbf{d}(s)x\|_{\mathbf{d}} = e^s \|x\|_{\mathbf{d}}$, $\forall s \in \mathbb{R}$, we conclude that the canonical homogeneous norm induced by the weighted Euclidean norm is analytic on $\mathbb{R}^n\backslash\{\mathbf{0}\}$.

(5) Given $x \in \mathbb{R}^n\backslash\{\mathbf{0}\}$ let us consider a function $q : \mathbb{R} \mapsto \mathbb{R}$ defined as follows

$$q(s) = x^\top \mathbf{d}^\top(s) P \mathbf{d}(s) x, \quad s \in \mathbb{R}.$$

Since the derivative of q is given by

$$q'(s) = x^\top \mathbf{d}^\top(s)(G_{\mathbf{d}}^\top P + P G_{\mathbf{d}})\mathbf{d}(s)x,$$

then, taking into account the criterion of strict monotonicity (5.6), we conclude that $q'(s) > 0$ for all $s \in \mathbb{R}$. By definition of the canonical homogeneous norm, we have

$$1 = \|\mathbf{d}(-\ln \|x\|_{\mathbf{d}})x\|^2 = x^\top \mathbf{d}^\top(-\ln \|x\|_{\mathbf{d}}) P \mathbf{d}(-\ln \|x\|_{\mathbf{d}})x.$$

In this case, the inequality (5.22) can be rewritten in terms of the function q as follows

$$q(-\ln \|y\|_{\mathbf{d}}) < q(-\ln \|x\|_{\mathbf{d}}).$$

Since the function q is strictly increasing the above inequality is possible if and only if $-\ln \|y\|_{\mathbf{d}} < -\ln \|x\|_{\mathbf{d}}$ or, equivalently, $\|x\|_{\mathbf{d}} < \|y\|_{\mathbf{d}}$. The case of the sign "$>$" in the inequality (5.22) can be treated similarly.

(6) The function V is positive definite and radially unbounded (proper) by construction, but its Lipschitz continuity follows from the property (6) of Lemma 5.4.

(7) For $x = \mathbf{0}$ the inequalities (5.23), obviously, hold. Since for $\epsilon \in (0, 1)$ the first inequality in (5.7) gives $\epsilon \leq \left\lfloor \mathbf{d}\left(\frac{1}{\alpha}\ln \epsilon\right)\right\rfloor$, then, taking into account $\|\mathbf{d}(-\ln \|x\|_{\mathbf{d}})x\| = 1$ for $x \neq \mathbf{0}$, we derive

$$1 \leq \left\lfloor \epsilon^{-1}\mathbf{d}\left(\frac{1}{\alpha}\ln \epsilon\right)\right\rfloor \leq \left\| \epsilon^{-1}\mathbf{d}\left(\frac{1}{\alpha}\ln \epsilon\right)\mathbf{d}(-\ln \|x\|_{\mathbf{d}})x\right\|.$$

Hence, using the property (3) of Lemma 5.4 we obtain

$$1 \leq \left\| \epsilon^{-1}\mathbf{d}\left(\frac{1}{\alpha}\ln \epsilon\right)\mathbf{d}(-\ln \|x\|_{\mathbf{d}})x\right\|_{\mathbf{d}}$$

or, equivalently,

$$\frac{\|x\|_{\mathbf{d}}}{\epsilon^{1/\alpha}} \leq \left\|\frac{x}{\epsilon}\right\|_{\mathbf{d}}.$$

For $\epsilon > 1$ we derive similarly$\epsilon \leq \left\lfloor \mathbf{d}\left(\frac{1}{\beta}\ln \epsilon\right)\right\rfloor$ and

$$1 \leq \left\lfloor \epsilon^{-1} \mathbf{d}\left(\frac{1}{\beta} \ln \epsilon\right) \right\rfloor \leq \left\| \epsilon^{-1} \mathbf{d}\left(\frac{1}{\alpha} \ln \epsilon\right) \mathbf{d}(-\ln \|x\|_{\mathbf{d}}) x \right\|.$$

Hence, we have $\frac{\|x\|_{\mathbf{d}}}{\epsilon^{1/\beta}} \leq \left\|\frac{x}{\epsilon}\right\|_{\mathbf{d}}$. The right inequality in (5.23) can be proven similarly by means of the right inequalities in (5.7). The proof is complete. □

All homogeneous norms are equivalent.

Corollary 5.4 *Let* $\mathbf{d}$ *be a linear continuous dilation in* $\mathbb{R}^n$. *Let* p_1 *and* p_2 *be two* $\mathbf{d}$*-homogeneous norms (see Definition 5.7). Then there exist* $k_1 > 0$, $k_2 \geq k_1$ *such that*

$$k_1 p_1(x) \leq p_2(x) \leq k_2 p_1(x), \quad \forall x \in \mathbb{R}^n.$$

Proof Let $\|\cdot\|_{\mathbf{d}}$ be a canonical $\mathbf{d}$-homogeneous norm induced by a norm $\|\cdot\|$ in $\mathbb{R}^n$. The canonical homogeneous norm in $\mathbb{R}^n$ always exists due to Proposition 5.3. Since p_i is positive definite and continuous then

$$0 < k_{i1} := \inf_{x \in S} p_i(x) \leq k_{i2} := \sup_{x \in S} p_i(x) < +\infty, \quad i = 1, 2,$$

where $S = \{x \in \mathbb{R}^n : \|x\| = 1\}$. Hence, taking into account the condition $p_i(\mathbf{d}(s)x) = e^s p_i(x)$, $\forall s \in \mathbb{R}$, $\forall x \in \mathbb{R}^n$ we derive

$$p_i(x) = \|x\|_{\mathbf{d}} p_i(\mathbf{d}(-\ln \|x\|_{\mathbf{d}})x), \quad \forall x \in \mathbb{R}^n, \quad i = 1, 2.$$

Since $\mathbf{d}(-\ln \|x\|_{\mathbf{d}})x \in S$ then

$$k_{i1} \|x\|_{\mathbf{d}} \leq p_i(x) \leq k_{i2} \|x\|_{\mathbf{d}}, \quad \forall x \in \mathbb{R}^n, \quad i = 1, 2$$

and

$$\frac{k_{21}}{k_{12}} p_1(x) \leq p_2(x) \leq \frac{k_{22}}{k_{11}} p_1(x), \quad \forall x \in \mathbb{R}^n.$$

The proof is complete. □

Remark 5.1 If a linear continuous dilation $\mathbf{d}$ is monotone with respect to a norm $\|\cdot\|$ in $\mathbb{R}^n$ then the homogeneous projector on the unit sphere is given by

$$\pi_{\mathbf{d}}(x) = \mathbf{d}(-\ln \|x\|_{\mathbf{d}})x, \quad x \neq \mathbf{0}. \tag{5.24}$$

Proof Indeed, by definition of $\|\cdot\|_{\mathbf{d}}$, we have $\|\pi_{\mathbf{d}}(x)\| = \|\mathbf{d}(-\ln \|x\|_{\mathbf{d}})x\| = 1$ and $\pi_{\mathbf{d}}(\pi_{\mathbf{d}}(x)) = \mathbf{d}(-\ln \|\pi_{\mathbf{d}}(x)\|_{\mathbf{d}})\pi_{\mathbf{d}}(x) = \mathbf{d}(-\ln 1)\pi_{\mathbf{d}}(x) = \pi_{\mathbf{d}}(x)$.

Formula (5.24) generalizes the definition of the conventional projector on the unit sphere in the case of the standard dilation. Indeed, for $\mathbf{d}(s) = e^s I_n$ we have $\|\cdot\|_{\mathbf{d}} = \|\cdot\|$ and

$$\pi_{\mathbf{d}}(x) = \mathbf{d}(-\ln \|x\|_{\mathbf{d}})x = e^{-\ln \|x\|_{\mathbf{d}}} x = \frac{x}{\|x\|_{\mathbf{d}}} = \frac{x}{\|x\|}.$$

Example 5.6 If

$$\mathbf{d}(s) = e^{sG_{\mathbf{d}}} \quad \text{with} \quad G_{\mathbf{d}} = \begin{pmatrix} 2 & 0 \\ 0 & 1 \end{pmatrix}$$

and the norm of $x = (x_1, x_2)^\top \in \mathbb{R}^2$ is defined as

$$\|x\| = \sqrt{x^\top P x}, \quad P = \begin{pmatrix} p_{11} & p_{12} \\ p_{12} & p_{22} \end{pmatrix} \in \mathbb{R}^{2\times 2} : P \begin{pmatrix} 2 & 0 \\ 0 & 1 \end{pmatrix} + \begin{pmatrix} 2 & 0 \\ 0 & 1 \end{pmatrix} P \succ 0$$

then the Eq. (5.15) becomes

$$x^\top \begin{pmatrix} e^{-2s_x} & 0 \\ 0 & e^{-s_x} \end{pmatrix} P \begin{pmatrix} e^{-2s_x} & 0 \\ 0 & e^{-s_x} \end{pmatrix} x = 1,$$

or, equivalently,

$$V^4 = aV^2 + bV + c,$$

where $V = e^{s_x} = \|x\|_{\mathbf{d}}$ and $a = p_{22}x_2^2, \quad b = 2p_{12}x_1x_2, \quad c = p_{11}x_1^2$. Using the fact that the above quartic equation always has only one real positive root, we solve this equation using Ferrari formulas (see Appendix F) and derive the explicit representation of $\|x\|_{\mathbf{d}}$ in $\mathbb{R}^2$:

$$\|x\|_{\mathbf{d}} = \begin{cases} \sqrt{p_{12}x_2^2} & \text{if} \quad x_1 = 0, \\ \sqrt[4]{p_{12}x_1^2} & \text{if} \quad x_2 = 0, \\ \frac{\sqrt{z}+\sqrt{-z+2a+b/\sqrt{z}}}{2} & \text{if } p_{12}x_1x_2 > 0, \\ \frac{-\sqrt{z}+\sqrt{-z+2a-2b/\sqrt{z}}}{2} & \text{if } p_{12}x_1x_2 < 0, \end{cases} \tag{5.25}$$

where

$$z = \frac{2p_{22}x_2^2 - \sqrt[3]{C_+} - \sqrt[3]{C_-}}{3}, \quad C_\pm = \frac{\Delta_1 \pm \sqrt{\Delta}}{2},$$

$$\Delta = \Delta_1^2 - 4\Delta_0^3, \quad \Delta_1 = 2p_{22}x_2^2(\Delta_0 + 6\tilde{p}x_1^2), \quad \Delta_0 = p_{22}^2x_2^4 - 12p_{11}x_1^2,$$

and $\tilde{p} = 8p_{11} - 9p_{12}^2/p_{22} > 0$.

Canonical homogeneous norms for equivalent linear dilations admit an explicit relation.

Corollary 5.5 *Let* $\mathbf{d}$ *be a linear continuous dilation in* $\mathbb{R}^n$ *and* $\mathbf{d}_\gamma$ *be an equivalent linear continuous dilation in* $\mathbb{R}^n$ *defined in Lemma 5.2 for some* $\gamma > 0$. *If* $\|\cdot\|_{\mathbf{d}}$ *is a canonical homogeneous norm induced by a norm* $\|\cdot\|$ *in* $\mathbb{R}^n$ *then*

$$\|\cdot\|_{\mathbf{d}} = \|\cdot\|_{\mathbf{d}_\gamma}^{\gamma}, \tag{5.26}$$

where $\|\cdot\|_{\mathbf{d}_\gamma}$ *is a canonical homogeneous norm induced by the norm* $\|\cdot\|$.

Proof By Lemma 5.2, the monotonicity of $\mathbf{d}$ with respect to $\|\cdot\|$ implies the monotonicity of $\mathbf{d}_\gamma$ with respect to $\|\cdot\|$. Since $\mathbf{d}_\gamma(s) = \mathbf{d}(\gamma s)$, $s \in \mathbb{R}$ then, by definition of the canonical homogeneous norm, we have

$$1 = \|\mathbf{d}_\gamma(-\ln \|x\|_{\mathbf{d}_\gamma})x\| = \|\mathbf{d}(-\gamma \ln \|x\|_{\mathbf{d}_\gamma})x\| = \|\mathbf{d}(-\ln \|x\|_{\mathbf{d}_\gamma}^{\gamma})x\|$$

and

$$1 = \|\mathbf{d}(-\ln \|x\|_{\mathbf{d}})x\|$$

for any $x \in \mathbb{R}^n \backslash \{\mathbf{0}\}$. Hence, due to uniqueness of the $\mathbf{d}$-homogeneous projection on the unit sphere, we derive $\|x\|_{\mathbf{d}} = \|x\|_{\mathbf{d}_\gamma}^{\gamma}$. The proof is complete. □

The following embedding theorem [25] highlights some similarity of standard and generalized homogeneous cones in the case of a linear continuous dilation *with diagonalizable generator.*

Theorem 5.2 *Let* $\mathbf{d}$ *be a continuous linear dilation in* $\mathbb{R}^n$ *with diagonalizable generator:*

$$\exists \Gamma \in \mathbb{R}^{n\times n} : G_{\mathbf{d}} = \Gamma^{-1}\Lambda\Gamma,\ \Lambda = \mathrm{diag}\{\lambda_1, \lambda_2, ..., \lambda_n\} \succ 0. \tag{5.27}$$

If the canonical homogeneous norm $\|\cdot\|_{\mathbf{d}}$ *is induced by the norm* $\|x\| = \sqrt{x^\top P x}$, $x \in \mathbb{R}^n$ *with a symmetric matrix* $P \in \mathbb{R}^{n\times n}$ *satisfying* (5.6), *then for any* $\delta \in \left(0, \frac{\lambda_{\min}^{1/2}(\Gamma^\top P\Gamma)}{\lambda_{\max}^{1/2}(\Gamma^\top P\Gamma)}\right)$ *and any* $y, z \in \mathbb{R}^n$ *satisfying*

$$|h_i^\top \Gamma^{-1} z| \le \delta |h_i^\top \Gamma^{-1} y|, \quad \forall i = 1 \dots n \tag{5.28}$$

it holds

$$\|z\|_{\mathbf{d}} \le \left(\delta \frac{\lambda_{\max}^{1/2}(\Gamma^\top P\Gamma)}{\lambda_{\min}^{1/2}(\Gamma^\top P\Gamma)}\right)^{\frac{1}{\alpha}} \|y\|_{\mathbf{d}}, \tag{5.29}$$

where $h_i = (0, ..., 1, ..., 0)^\top \in \mathbb{R}^n$ *is the unit vector and* $\alpha > 0$ *is defined in Proposition 5.3.*

Proof Using (5.28) we derive

$$\left.\begin{array}{c} |h_i^\top e^{-\lambda_i s}\Gamma^{-1} z| \le \delta |h_i^\top e^{-\lambda_i s}\Gamma^{-1} y| \\ \forall i = 1, 2, ..., n \end{array}\right\} \Rightarrow |e^{-\Lambda s}\Gamma^{-1} z| \le \delta |e^{-\Lambda s}\Gamma^{-1} y|, \forall s \in \mathbb{R}.$$

By definition of the canonical homogeneous norm, we have

$$1 = \|\mathbf{d}(-s_y)y\|^2 = \|\Gamma e^{-\Lambda s_y}\Gamma^{-1} y\|^2 \ge \lambda_{\min}(\Gamma^\top P\Gamma)|e^{-\Lambda s_y}\Gamma^{-1} y|^2,$$

where $s_y = \ln \|y\|_{\mathbf{d}}$. The above inequality implies that

$$|e^{-\Lambda s_y}\Gamma^{-1}z| \leq \delta|e^{-\Lambda s_y}\Gamma^{-1}y| \leq \frac{\delta}{\lambda_{\min}^{1/2}(\Gamma^\top P\Gamma)}.$$

Taking into account $G_{\mathbf{d}} = \Gamma\Lambda\Gamma^{-1}$, we derive

$$\|\mathbf{d}(-s_y)z\| \leq \lambda_{\max}^{1/2}(\Gamma^\top P\Gamma)|e^{-\Lambda s_y}\Gamma^{-1}z| \leq \delta\frac{\lambda_{\max}^{1/2}(\Gamma^\top P\Gamma)}{\lambda_{\min}^{1/2}(\Gamma^\top P\Gamma)} = \epsilon.$$

Hence, due to $\|x\| \leq 1 \Leftrightarrow \|x\|_{\mathbf{d}} \leq 1, \forall x \in \mathbb{R}^n$ and $s_y = \ln \|y\|_{\mathbf{d}}$ we have

$$\left\|\mathbf{d}(-\ln \|y\|_{\mathbf{d}})\frac{z}{\epsilon}\right\|_{\mathbf{d}} \leq 1.$$

Since $\|\mathbf{d}(s)x\|_{\mathbf{d}} = e^s\|x\|_{\mathbf{d}}, \forall x \in \mathbb{R}^n, \forall s \in \mathbb{R}$ then $\left\|\frac{z}{\epsilon}\right\|_{\mathbf{d}} \leq \|y\|_{\mathbf{d}}$. We complete the proof using (5.23).

Notice that the inequalities (5.28) cannot be relaxed as follows

$$|h_i^\top z| \leq \tilde{\delta}|h_i^\top y|, \quad \forall i = 1\ldots n \tag{5.30}$$

with some small enough $\tilde{\delta} > 0$. Since (5.30) does not imply (5.28) in the general case. Indeed, for $n = 2$ and $\Gamma^{-1} = \left(\begin{smallmatrix}1 & -1\\ 1 & 1\end{smallmatrix}\right)$, $\tilde{\delta} = h_1^\top z = -h_2^\top z \neq 0$ and $1 = h_1^\top y = h_2^\top y \neq 0$ the system of inequalities (5.30) holds but the first inequality ($i = 1$) in (5.28):

$$|h_1^\top\Gamma^{-1}z| \leq \delta|h_1^\top\Gamma^{-1}y|$$

never holds since $|h_1^\top\Gamma^{-1}z| = 2|h_1^\top z| = 2\tilde{\delta} \neq 0$ but $|h_1^\top\Gamma^{-1}y| = 0$.

5.2.2 *Generalized Homogeneous Homeomorphism on* $\mathbb{R}^n$

Formally, the canonical homogeneous norm $\|\cdot\|_{\mathbf{d}}$ is not a norm in the usual sense, but it defines an alternative topology in $\mathbb{R}^n$. Due to the monotonicity of the dilation, it is straightforwardly related to homogeneous spheres and balls. Indeed, for any $r > 0$ the identity $\|u\|_{\mathbf{d}} = r$ is equivalent to $u \in S_{\mathbf{d}}(r)$, but the inequality $\|u\|_{\mathbf{d}} < r$ is equivalent to $u \in B_{\mathbf{d}}(r)$.

Lemma 5.5 (Homeomorphism induced by a linear dilation in $\mathbb{R}^n$) *Let a linear continuous dilation* $\mathbf{d}$ *be monotone with respect to a norm* $\|\cdot\|$. *Let the canonical homogeneous norm* $\|\cdot\|_{\mathbf{d}}$ *be induced by* $\|\cdot\|$. *The mapping* $\Phi : \mathbb{R}^n \mapsto \mathbb{R}^n$ *given by*

$$\Phi(x) = \|x\|_{\mathbf{d}}\mathbf{d}(-\ln \|x\|_{\mathbf{d}})x, \quad x \in \mathbb{R}^n\backslash\{\mathbf{0}\} \tag{5.31}$$

is a homeomorphism in $\mathbb{R}^n$, *its inverse has the form*

$$\Phi^{-1}(z) = \|z\|^{-1}\mathbf{d}(\ln \|z\|)z, \quad z \in \mathbb{R}^n\backslash\{\mathbf{0}\}$$

with $\Phi(\mathbf{0}) = \Phi^{-1}(\mathbf{0}) = \mathbf{0}$ *by continuity. Moreover,* Φ *is a diffeomorphism on* $\mathbb{R}^n\backslash\{\mathbf{0}\}$ *provided that* $\|\cdot\| \in C^1(\mathbb{R}^n\backslash\{\mathbf{0}\})$.

Proof Since

$$\|\Phi(x)\| = \|x\|_{\mathbf{d}}\|\mathbf{d}(-\ln \|x\|_{\mathbf{d}})x\| = \|x\|_{\mathbf{d}}$$

then

$$\Phi^{-1}(\Phi(x)) = \|\Phi(x)\|^{-1}\mathbf{d}(\ln \|\Phi(x)\|)\Phi(x) = \|x\|_{\mathbf{d}}^{-1}\mathbf{d}(\ln \|x\|_{\mathbf{d}})\Phi(x) = x.$$

On the other hand, since

$$\left\|\Phi^{-1}(z)\right\|_{\mathbf{d}} = \left\|\mathbf{d}(\ln \|z\|)\tfrac{z}{\|z\|}\right\|_{\mathbf{d}} = \|z\| \left\|\tfrac{z}{\|z\|}\right\|_{\mathbf{d}} = \|z\|$$

then

$$\Phi(\Phi^{-1}(z)) = \left\|\Phi^{-1}(z)\right\|_{\mathbf{d}}\mathbf{d}\left(-\ln \left\|\Phi^{-1}(z)\right\|_{\mathbf{d}}\right)\Phi^{-1}(z) = \|z\|\mathbf{d}(-\ln \|z\|)\Phi^{-1}(z) = z.$$

Therefore, Φ^{-1} is, indeed, the inverse of Φ.

The continuity Φ at $\mathbf{0}$ follows from the identity $\|\Phi(x)\| = \|x\|_{\mathbf{d}}$. The continuity[2] (differentiability[3]) of Φ at $x \in \mathbb{R}^n\backslash\{\mathbf{0}\}$ follows from continuity (differentiability) of $\|\cdot\|_{\mathbf{d}}$, $\ln(\cdot)$ and $\mathbf{d}(\cdot)$. The proof is complete. □

Now we are ready to introduce a normed vector space homeomorphic to $\mathbb{R}^n$ with the norm $\|\cdot\|_{\mathbf{d}}$.

Theorem 5.3 (Generalized homogeneous normed vector space $\mathbb{R}^n_{\mathbf{d}}$) *Let a linear continuous dilation* $\mathbf{d}$ *in* $\mathbb{R}^n$ *be monotone with respect to a norm* $\|\cdot\|$. *Let the canonical homogeneous norm* $\|\cdot\|_{\mathbf{d}}$ *be induced by* $\|\cdot\|$. *Let an addition of vectors*

$$\tilde{+} : \mathbb{R}^n \times \mathbb{R}^n \mapsto \mathbb{R}^n$$

[2] Homeomorphism is a continuous invertible mapping.

[3] Diffeomorphism is a continuously differentiable invertible mapping.

and a multiplication by scalar

$$\tilde{\cdot} : \mathbb{R} \times \mathbb{R}^n \mapsto \mathbb{R}^n$$

be defined as follows

- $x \tilde{+} y := \Phi^{-1}(\Phi(x) + \Phi(y))$, *where* $x, y \in \mathbb{R}^n$,
- $\lambda \tilde{\cdot} x := \mathrm{sign}(\lambda)\mathbf{d}(\ln|\lambda|)x$, *where* $\lambda \in \mathbb{R}$, $x \in \mathbb{R}^n$,

where $\Phi : \mathbb{R}^n \mapsto \mathbb{R}^n$ *is given by* (5.31). *Then the set* $\mathbb{R}^n$ *together with the operations* $\tilde{+}$ *and* $\tilde{\cdot}$ *is a vector space* $\mathbb{R}^n_{\mathbf{d}}$ *with the norm* $\|\cdot\|_{\mathbf{d}}$.

Proof (1) To prove that $\mathbb{R}^n_{\mathbf{d}}$ is a vector space, we have to show that all required axioms (see Appendix A) hold. Let $u, v, w \in \mathbb{R}^n$ and $a, b \in \mathbb{R}$.

(a) *Associativity of addition*:

$$\begin{aligned} u\tilde{+}(v\tilde{+}w) &= u\tilde{+}\Phi^{-1}(\Phi(v) + \Phi(w)) = \Phi^{-1}(\Phi(u) + \Phi(v) + \Phi(w)) \\ &= \Phi^{-1}(\Phi(u\tilde{+}v) + \Phi(w)) = (u\tilde{+}v)\tilde{+}w. \end{aligned}$$

(b) *Commutativity of addition*: $u\tilde{+}v = \Phi^{-1}(\Phi(u) + \Phi(v)) = v\tilde{+}u$.
(c) *Identity element of addition*: $v\tilde{+}\mathbf{0} = \Phi^{-1}(\Phi(v) + \Phi(\mathbf{0})) = \Phi^{-1}(\Phi(v)) = v$.
(d) *Inverse elements of addition*: since $(-1)\tilde{\cdot}v = -v$ and $\Phi(-v) = -\Phi(v)$ then

$$v\tilde{+}((-1)\tilde{\cdot}v) = \Phi^{-1}(\Phi(v) + \Phi(-v)) = \Phi^{-1}(\Phi(v) - \Phi(v)) = \mathbf{0}.$$

(e) *Compatibility of scalar multiplication with field multiplication*:

$$\begin{aligned} a\tilde{\cdot}(b\tilde{\cdot}v) &= \mathrm{sign}(a)\mathbf{d}(\ln|a|)(b\tilde{\cdot}v) = \mathrm{sign}(a)\mathbf{d}(\ln|a|)\,\mathrm{sign}(b)\mathbf{d}(\ln|b|)v \\ &= \mathrm{sign}(ab)\mathbf{d}(\ln|ab|)v = (ab)\tilde{\cdot}v. \end{aligned}$$

(f) *Identity element of scalar multiplication*: $1\tilde{\cdot}v = \mathrm{sign}(1)\mathbf{d}(\ln 1)v = v$.
(g) *Distributivity of scalar multiplication with respect to vector addition*: since

$$\mathbf{d}(\ln|a|)\Phi^{-1}(w) = \mathbf{d}(\ln|a| + \ln\|w\|)\frac{w}{\|w\|} = \mathbf{d}(\ln|a|\|w\|)\frac{|a|w}{\||a|w\|} = \Phi^{-1}(|a|w)$$

and

$$\begin{aligned} \Phi(\mathbf{d}(\ln|a|)u) &= \|\mathbf{d}(\ln|a|)u\|_{\mathbf{d}}\mathbf{d}(-\ln\|\mathbf{d}(\ln|a|)u\|_{\mathbf{d}})\mathbf{d}(\ln|a|)u \\ &= |a|\|u\|_{\mathbf{d}}\mathbf{d}(-\ln|a| - \ln\|u\|_{\mathbf{d}})\mathbf{d}(\ln|a|)u = |a|\Phi(u) \end{aligned}$$

then

$$\begin{aligned} a\tilde{\cdot}(u\tilde{+}v) &= \mathrm{sign}(a)\mathbf{d}(\ln|a|)\Phi^{-1}(\Phi(u) + \Phi(v)) = \mathrm{sign}(a)\Phi^{-1}(|a|\Phi(u) + |a|\Phi(v)) \\ &= \mathrm{sign}(a)\Phi^{-1}(\Phi(\mathbf{d}(\ln|a|)u) + \Phi(\mathbf{d}(\ln|a|)v)) = a\tilde{\cdot}u\tilde{+}a\tilde{\cdot}v, \end{aligned}$$

where the identities $\Phi(-w) = -\Phi(w)$ and $\Phi^{-1}(-w) = -\Phi^{-1}(w)$ are utilized on the last step.

(h) *Distributivity of scalar multiplication with respect to field addition:*

$$\begin{aligned} a\tilde{\cdot}u\tilde{+}b\tilde{\cdot}u =& \Phi^{-1}(\Phi(\text{sign}(a)\mathbf{d}(\ln|a|)u) + \Phi(\text{sign}(b)\mathbf{d}(\ln|b|)u)) \\ =& \Phi^{-1}(a\Phi(u) + b\Phi(u)) = \Phi^{-1}((a+b)\Phi(u)) \\ =& \,\text{sign}(a+b)\mathbf{d}(\ln|a+b|)u = (a+b)\tilde{\cdot}u. \end{aligned}$$

Therefore, $\mathbb{R}^n_{\mathbf{d}}$ is, indeed, a vector space.

(2) Let us show that the vector space $\mathbb{R}^n_{\mathbf{d}}$ is normed and the function $\|\cdot\|_{\mathbf{d}}$ satisfy all properties of a norm. Positive definiteness

$$\|u\|_{\mathbf{d}} > 0, u \neq \mathbf{0}, \qquad \|\,\mathbf{0}\,\|_{\mathbf{d}} = 0$$

and homogeneity

$$\|a\tilde{\cdot}u\|_{\mathbf{d}} = |a|\|u\|_{\mathbf{d}}, \quad a \in \mathbb{R}, u \in \mathbb{R}_{\mathbf{d}}$$

of $\|\cdot\|_{\mathbf{d}}$ are obvious. Let us prove the sub-additivity (the triangle inequality). Since

$$\|\Phi(w)\| = \|\,\|w\|_{\mathbf{d}}\mathbf{d}(-\ln\|w\|_{\mathbf{d}})w\| = \|w\|_{\mathbf{d}}\|\mathbf{d}(-\ln\|w\|_{\mathbf{d}})w\| = \|w\|_{\mathbf{d}}$$

then

$$\|u\tilde{+}v\|_{\mathbf{d}} = \left\|\Phi^{-1}(\Phi(u)+\Phi(v))\right\|_{\mathbf{d}} = \left\|\mathbf{d}(\ln\|\Phi(u)+\Phi(v)\|)\tfrac{\Phi(u)+\Phi(v)}{\|\Phi(u)+\Phi(v)\|}\right\|_{\mathbf{d}} =$$

$$\|\Phi(u)+\Phi(v)\|\left\|\tfrac{\Phi(u)+\Phi(v)}{\|\Phi(u)+\Phi(v)\|}\right\|_{\mathbf{d}} = \|\Phi(u)+\Phi(v)\| \leq \|\Phi(u)\| + \|\Phi(v)\| = \|u\|_{\mathbf{d}} + \|v\|_{\mathbf{d}}.$$

The proof is complete. □

The above theorem justifies the name "norm" for the function $\|\cdot\|_{\mathbf{d}}$. The **d**-*homogeneous vector space* $\mathbb{R}^n_{\mathbf{d}}$ is not a Euclidean space yet, since it requires an inner product to be introduced. Recall a substruction in vector spaces such as $\mathbb{R}^n_{\mathbf{d}}$ is defined as

$$x\tilde{-}y = x\tilde{+}(-1)\tilde{\cdot}y, \quad x, y \in \mathbb{R}^n_{\mathbf{d}}. \tag{5.32}$$

5.2.3 Canonical Homogeneous Inner Product

Let a linear continuous dilation **d** be monotone with respect to a weighted Euclidean norm $\|x\| = \sqrt{x^\top Px}$ and an inner product in $\mathbb{R}^n$ be defined as

$$\langle x, y\rangle = x^\top Py, \quad x, y \in \mathbb{R}^n,$$

where $0 \prec P = P^\top \in \mathbb{R}^{n\times n}$. In this case, an inner product in the **d**-homogeneous vector space $\mathbb{R}^n_{\mathbf{d}}$ can be introduced as follows

$$\langle x, y\rangle_{\mathbf{d}} = \begin{cases} 0 & \text{if } \|x\|\|y\|=0, \\ \|x\|_{\mathbf{d}}\|y\|_{\mathbf{d}}\langle \mathbf{d}(-\ln\|x\|_{\mathbf{d}})x, \mathbf{d}(-\ln\|y\|_{\mathbf{d}})y\rangle & \text{if } \|x\|\|y\|\neq 0, \end{cases} \tag{5.33}$$

where $\|\cdot\|_{\mathbf{d}}$ is the canonical $\mathbf{d}$-homogeneous norm induced by the norm $\|\cdot\|$.

Theorem 5.4 (On generalized homogeneous Euclidean space $\mathbb{R}^n_{\mathbf{d}}$)
The mapping $\mathbb{R}^n_{\mathbf{d}} \times \mathbb{R}^n_{\mathbf{d}} \mapsto \mathbb{R}$ *given by* (5.33) *is an inner product on* $\mathbb{R}^n_{\mathbf{d}}$.

Proof The *positive definiteness* $\langle x, x\rangle_{\mathbf{d}} = \|x\|^2_{\mathbf{d}} > 0$ for $x \neq \mathbf{0}$ and the conjugate symmetry $\langle x, y\rangle_{\mathbf{d}} = \langle y, x\rangle_{\mathbf{d}}$ are obvious. Let us prove the *linearity*. Since $x\tilde{+}y = \Phi^{-1}(\Phi(x) + \Phi(y))$ then $\|x\tilde{+}y\|_{\mathbf{d}} = \|\Phi(x) + \Phi(y)\|$ and using the definition of Φ^{-1} we obtain

$$\|x\tilde{+}y\|_{\mathbf{d}}\mathbf{d}(-\ln\|x\tilde{+}y\|_{\mathbf{d}})(x\tilde{+}y) =$$

$$\|\Phi(x)+\Phi(y)\|\mathbf{d}(-\ln\|\Phi(x)+\Phi(y)\|)\frac{\mathbf{d}(\ln\|\Phi(x)+\Phi(y)\|)(\Phi(x)+\Phi(y))}{\|\Phi(x)+\Phi(y)\|} = \Phi(x)+\Phi(y).$$

Hence, we derive

$$\begin{aligned}\langle x\tilde{+}y, z\rangle_{\mathbf{d}} &= \|x\tilde{+}y\|_{\mathbf{d}}\|z\|_{\mathbf{d}}\langle \mathbf{d}(-\ln\|x\tilde{+}y\|_{\mathbf{d}})(x\tilde{+}y), \mathbf{d}(-\ln\|z\|_{\mathbf{d}})z\rangle \\ &= \|z\|_{\mathbf{d}}\langle \Phi(x), \mathbf{d}(-\ln\|z\|_{\mathbf{d}})z\rangle + \|z\|_{\mathbf{d}}\langle \Phi(y), \mathbf{d}(-\ln\|z\|_{\mathbf{d}})z\rangle \\ &= \langle x, z\rangle_{\mathbf{d}} + \langle y, z\rangle_{\mathbf{d}},\end{aligned}$$

where the definition of Φ is utilized in the last step. Finally, since $\alpha\tilde{\cdot}x = \text{sign}(\alpha)\mathbf{d}(\ln|\alpha|)x$ then

$$\begin{aligned}\|\alpha\tilde{\cdot}x\|_{\mathbf{d}}\mathbf{d}(-\ln\|\alpha\tilde{\cdot}x\|_{\mathbf{d}})(\alpha\tilde{\cdot}x) &= |\alpha|\|x\|_{\mathbf{d}}\mathbf{d}(-|\alpha| - \ln\|x\|_{\mathbf{d}})\,\text{sign}(\alpha)\mathbf{d}(\ln|\alpha|)x \\ &= \alpha\|x\|_{\mathbf{d}}\mathbf{d}(-\ln\|x\|_{\mathbf{d}})x\end{aligned}$$

and

$$\langle \alpha\tilde{\cdot}x, y\rangle_{\mathbf{d}} = \|\alpha\tilde{\cdot}x\|_{\mathbf{d}}\|y\|_{\mathbf{d}}\langle \mathbf{d}(-\ln\|\alpha\tilde{\cdot}x\|_{\mathbf{d}})(\alpha\tilde{\cdot}x), \mathbf{d}(-\ln\|y\|_{\mathbf{d}})y\rangle = \alpha\langle x, y\rangle_{\mathbf{d}}.$$

The proof is complete. □

Therefore, *any continuous linear dilation* $\mathbf{d}$ in $\mathbb{R}^n$ *induces a generalized homogeneous Euclidean space* $\mathbb{R}^n_{\mathbf{d}}$ *homeomorphic to* $\mathbb{R}^n$.

5.2.4 Convex Cones in Homogeneous Euclidean Spaces

Analysis and design of the so-called safety-critical control systems is related with study of positive invariance of closed sets for dynamical system (see, e.g., [26, 27] or Chaps. 7 and 12). The corresponding analysis may be simplified in the case of convex sets [28, 29]. The generalized homogeneous cones are usually non-convex in $\mathbb{R}^n$ (see Fig. 5.4), but they may be convex in $\mathbb{R}^n_{\mathbf{d}}$. Such convex sets are characterized by the following theorem.

Theorem 5.5 *Let* $\mathbf{d}$ *be a linear continuous dilation being monotone with respect to a norm* $\|\cdot\|$ *in* $\mathbb{R}^n$. *Let a nonempty set* $\mathcal{D} \subset \mathbb{R}^n$ *be a closed* $\mathbf{d}$-*homogeneous cone in* $\mathbb{R}^n$. *The following three claims are equivalent.*

(1) The set $\mathcal{D}$ *is convex in* $\mathbb{R}^n_{\mathbf{d}}$.
(2) The set $\Phi(\mathcal{D})$ *is convex in* $\mathbb{R}^n$.
(3) The following inclusion

$$\frac{\lambda\Phi(x_1) + (1-\lambda)\Phi(x_2)}{\|\lambda\Phi(x_1) + (1-\lambda)\Phi(x_2)\|} \in \mathcal{D} \tag{5.34}$$

holds for any $\lambda \in [0, 1]$ *and any* $x_1, x_2 \in \mathcal{D}$ *such that* $\lambda\Phi(x_1) + (1 - \lambda)\Phi(x_2) \neq \mathbf{0}$.

Proof 1.⇔3. By definition (see Appendix A), the set $\mathcal{D}$ is convex in $\mathbb{R}^n_{\mathbf{d}}$ if

$$\lambda\tilde{\cdot}x_1\tilde{+}(1-\lambda)\tilde{\cdot}x_2 \in \mathcal{D}, \quad \forall\lambda \in [0, 1], \quad \forall x_1, x_2 \in \mathcal{D}$$

or, equivalently,

$$\Phi^{-1}(\Phi(\lambda\tilde{\cdot}x_1) + \Phi((1-\lambda)\tilde{\cdot}x_2)) \in \mathcal{D}, \quad \forall\lambda \in [0, 1], \quad \forall x_1, x_2 \in \mathcal{D}.$$

Since $\mathcal{D}$ is a nonempty closed $\mathbf{d}$-homogeneous cone, then $\mathbf{0} \in \mathcal{D}$ and, taking into account the definition of Φ^{-1}, we conclude that the above condition is equivalent to

$$\frac{\Phi(\lambda\tilde{\cdot}x_1)+\Phi((1-\lambda)\tilde{\cdot}x_2)}{\|\Phi(\lambda\tilde{\cdot}x_1)+\Phi((1-\lambda)\tilde{\cdot}x_2)\|} \in \mathcal{D}, \forall\lambda \in [0, 1], \forall x_1, x_2 \in \mathcal{D} : \Phi(\lambda\tilde{\cdot}x_1) + \Phi((1-\lambda)\tilde{\cdot}x_2) \neq \mathbf{0}.$$

Using the identities $\Phi(\lambda\tilde{\cdot}x_1) = \lambda\Phi(x_1)$ and $\Phi((1-\lambda)\tilde{\cdot}x_2) = (1-\lambda)\Phi(x_2)$, we derive (5.34).

1.⇔2. If $z_1, z_2 \in \Phi(\mathcal{D})$ then there exists $x_1, x_2 \in \mathcal{D}$ such that $z_1 = \Phi(x_1)$, $z_2 = \Phi(x_2)$. In this case, we have

$$\lambda z_1 = \lambda\Phi(x_1) = \Phi(\lambda\tilde{\cdot}x_1)$$

and

$$(1-\lambda)z_2 = (1-\lambda)\Phi(x_2) = \Phi((1-\lambda)\tilde{\cdot}x_2).$$

Therefore, we have

$$\lambda z_1 + (1-\lambda) z_2 = \Phi\left(\underbrace{\Phi^{-1}\left(\Phi(\lambda\tilde{\cdot}x_1) + \Phi((1-\lambda)\tilde{\cdot}x_2)\right)}_{=\lambda\tilde{\cdot}x_1 \tilde{+} (1-\lambda)\tilde{\cdot}x_2}\right).$$

Hence, the convexity of $\mathcal{D}$ in $\mathbb{R}^n_{\mathbf{d}}$ is equivalent to convexity of $\Phi(\mathcal{D})$ in $\mathbb{R}^n$. The proof is complete. □

In the view of the above theorem, the homeomorphism (5.31) may simplify the analysis and design of dynamical systems on homogeneous cones, since, in the new coordinates $z = \Phi(x)$, a non-convex homogeneous set may become convex. Moreover, a nonlinear object (e.g., a cone, a function, an ODE) being **d**-homogeneous in the space $\mathbb{R}^n$ may become linear in the space $\mathbb{R}^n_{\mathbf{d}}$.

Corollary 5.6 *For* $h_i \in \mathbb{R}^n, i = 1, \ldots, p$ *and* Φ *given by* (5.31), *the* **d***-homogeneous cone in the space* $\mathbb{R}^n$

$$\Omega_{\text{lin}} = \{x \in \mathbb{R}^n : h_i^\top \Phi(x) \geq 0, i = 1, \ldots, p\} \tag{5.35}$$

is a linear positive cone in the space $\mathbb{R}^n_{\mathbf{d}}$.

Proof Indeed, since for $x, y \in \Omega_{\text{lin}}$ we have $h_i^\top \Phi(x) \geq 0$, $h_i^\top \Phi(y) \geq 0$ for all $i = 1, \ldots, p$ then using $x \tilde{+} y = \Phi^{-1}(\Phi(x) + \Phi(y))$ we derive $h_i^\top \Phi(x \tilde{+} y) = h_i^\top \Phi(x) + h_i^\top \Phi(y) \geq 0 \quad \Rightarrow \quad x \tilde{+} y \in \Omega_{\text{lin}}$. The proof is complete. □

The set Ω_{lin} introduced in the above corollary is called below by a *linear* **d** *-homogeneous cone*. Linear **d**-homogeneous cones are always convex in $\mathbb{R}^n_{\mathbf{d}}$.

5.3 Computation of Canonical Homogeneous Norm

In this monograph, the canonical homogeneous norm $\|\cdot\|_{\mathbf{d}}$ is utilized in various control algorithms. The norm $\|\cdot\|_{\mathbf{d}}$ is defined implicitly as a positive root of the nonlinear algebraic equation (5.15). Such a root exists and is unique if the dilation **d** is monotone. For practical applications, the canonical homogeneous norm has to be computed somehow. Two approaches are proposed below for this purpose. The first one is based on a computation of $\|\cdot\|_{\mathbf{d}}$ using some numerical method solving the nonlinear algebraic equation. Below we discuss the use of Newton method and bisection method for this purpose. The second approach deals with an approximation of $\|\cdot\|_{\mathbf{d}}$ by an explicit homogeneous norm in $\mathbb{R}^n \setminus \{\mathbf{0}\}$. The so-called Universal Approximation theorem [30] is utilized for this purpose. At first sight, the explicit representation looks more useful for control practice. However, computation of an analytic function (such as $\exp(\cdot)$, $\ln(\cdot)$, $\sqrt{\cdot}$, etc.) in a digital controller with a high enough precision also takes some significant time. In some cases, a controller, which uses the canonical homogeneous norm (see Chap. 9) computed by bisection methods, could be computationally more efficient than an explicit homogeneous controller [31] like $u = -k_n \left(x_n + k_{n-1}\left(x_{n-1} + \cdots\right)^{\lfloor \alpha_{n-1} \rceil}\right)^{\lfloor \alpha_n \rceil}$, where $\alpha_i, k_i > 0$ are control parameters and $x = (x_1, \ldots, x_n)^\top$ is the state vector.

5.3.1 *Numerical Approximation*

Bisection Method

For an arbitrary fixed $x \neq \mathbf{0}$, let us consider the function

$$r \mapsto \|\mathbf{d}(-\ln r)x\|,$$

where $\mathbf{d}$ is a continuous dilation being monotone with respect to the norm $\|\cdot\|$. The considered function is monotone and the limit property implies $\|\mathbf{d}(-\ln r)x\| \to +\infty$ as $r \to 0$ but $\|\mathbf{d}(-\ln r)x\| \to 0$ as $r \to -\infty$ (see Fig. 5.6). In this case, due to the continuity of the dilation, there exists $r^* > 0 : \|\mathbf{d}(-\ln r^*)x\| = 1$ and, by Definition 5.8, we have $r^* = \|x\|_{\mathbf{d}}$. The monotonicity of $\mathbf{d}$ guarantees the uniqueness of $r^* > 0$.

The bisection method for monotone functions can be realized in two stages. In the first stage, the unique root of the function has to be localized, i.e., the upper and lower estimates $\underline{r}$ and $\overline{r}$ of $\|x\|_{\mathbf{d}}$ have to be found. On the second stage, the estimate of $\|x\|_{\mathbf{d}}$ is improved by splitting the interval $[\underline{r}, \overline{r}]$ into two sub-intervals $[\underline{r}, 0.5(\underline{r} + \overline{r})]$ and $[0.5(\underline{r} + \overline{r}), \overline{r}]$, and by selection of the interval, which contains the root of the equation. The latter step is repeated several time in order to reach the required precision of the root estimation. The numerical computation of $\|x\|_{\mathbf{d}}$ is realized by the following algorithm.

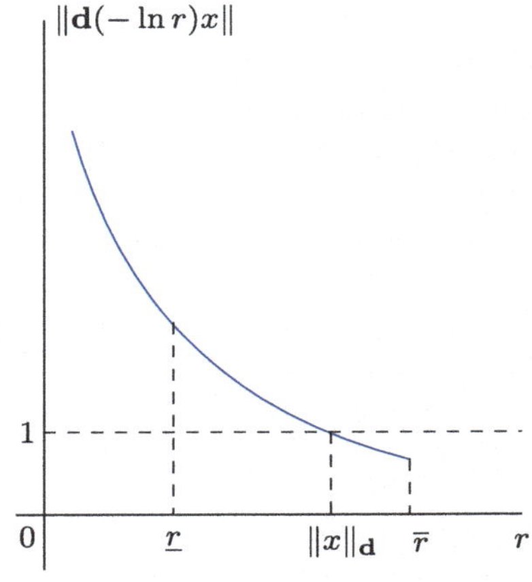

Fig. 5.6 Illustration of the bisection method

Algorithm 5.1 ***(Computation of the canonical homogeneous norm*** $\|\cdot\|_{\mathbf{d}}$***)***

take any $\underline{r} \in [r_{\min}, r_{\max}]$ and any $\overline{r} \in (\underline{r}, r_{\max}]$

while $\|\mathbf{d}(-\ln \overline{r})x\| > 1$ and $\overline{r} < r_{\max}$ **do** $\underline{r} = \overline{r}; \overline{r} = 2\overline{r}$

if $\overline{r} \geq r_{\max}$ **then** $\underline{r} = r_{\max}$; **return** $\underline{r}, \overline{r}$

while $\|\mathbf{d}(-\ln \underline{r})x\| < 1$ and $\underline{r} > r_{\min}$ **do** $\overline{r} = \underline{r}; \underline{r} = \underline{r}/2$

if $\underline{r} \leq r_{\min}$ **then** $\overline{r} = r_{\min}$; **return** $\underline{r}, \overline{r}$

for $j = 1, \ldots, j_{\max}$ **do**
  $c = (\underline{r} + \overline{r})/2$
  if $\|\mathbf{d}(-\ln c)x\| < 1$ **then** $\overline{r} = c$
  else $\underline{r} = c$

return $\underline{r}, \overline{r}$.

A digital device (computer) has a finite machine epsilon $\epsilon > 0$, which characterizes a digital computation error using the floating-point arithmetic. The format of numbers on the computer also assumes that there exist minimal and maximal possible positive numbers $r_{\min} > 0$ and $r_{\max} < +\infty$ such that $r_{\min} < r_{\max}$. Any positive number being less than $r_{\min}$ is considered as zero, but any positive number greater than $r_{\max}$ is treated by the computer as $+\infty$. For example, $r_{\min} \approx 10^{-308}$ and $r_{\max} \approx 10^{308}$ for the double-precision floating-point format of the number, but the machine epsilon is about $\epsilon \approx 10^{-16}$ in this case. By default, the bisection algorithm given above uses these parameters for the localization of the root. In practice, the parameters $r_{\min}$ and $r_{\max}$ can be tuned independently of the floating-point format of the number in order to restrict artificially the admissible values of the canonical homogeneous norm, but $\epsilon > 0$ still can define the required precision of the root estimation.

The parameter $j_{\max}$ in the above algorithm defines a number of steps to perform for improvement of the estimate of the norm $\|x\|_{\mathbf{d}} \in [\underline{r}, \overline{r}]$. The algorithm returns the interval $[\underline{r}, \overline{r}]$ of the length $2^{-j_{\max}}\delta$, where $\delta > 0$ is the initial length of the interval $[\underline{r}, \overline{r}]$ right after localization of the root. The larger $j_{\max}$, the better estimation precision $\epsilon > 0$. Notice that the machine epsilon could be utilized for the minimal length of the interval $[\underline{r}, \overline{r}]$. In this case the number of steps required to reach the estimation of the norm $\|x\|_{\mathbf{d}}$ with the precision $\epsilon > 0$ is given by

$$j_{\max} \geq \log_2 \frac{r_{\max} - r_{\min}}{\epsilon}.$$

For example, if it is known a priori that the values of the canonical homogeneous norm are bounded by $r_{\max} = 10^3$ then the above algorithm with $j_{\max} = 37$ would guarantee the estimation of $\|\cdot\|_{\mathbf{d}}$ with the precision $\epsilon = 10^{-8}$.

Example 5.7 The dilation $\mathbf{d}(s) = e^{sG_{\mathbf{d}}}$ in $\mathbb{R}^2$ with the generator

$$G_{\mathbf{d}} = \begin{pmatrix} 3 & 0 \\ 1 & 1 \end{pmatrix}$$

is strictly monotone with respect to the weighted Euclidean norm $\|x\| = \sqrt{x^\top P x}$ with $x \in \mathbb{R}^2$ and

$$P = \begin{pmatrix} 1 & -1 \\ -1 & 2 \end{pmatrix}.$$

The norm of the vector $x = (1\ -1)^\top$ is

$$\|x\| \approx 2.23606798,$$

but the canonical homogeneous norm computed by the bisection method is

$$\|x\|_{\mathbf{d}} \approx 2.12707008.$$

The norm of the scaled vector $y = \mathbf{d}(\ln 2)x = (8\ \ 1)^\top$ is

$$\|y\| \approx 7.07106781.$$

Due to $\|\mathbf{d}(s)x\|_{\mathbf{d}} = e^s\|x\|_{\mathbf{d}}, \forall x \in \mathbb{R}^n, \forall s \in \mathbb{R}$ we derive $\|y\|_{\mathbf{d}} = 2\|x\|_{\mathbf{d}}$. The computation of $\|y\|_{\mathbf{d}}$ by the bisection method numerically confirms this

$$\|y\|_{\mathbf{d}} \approx 4.25414017.$$

The scaling of the vector x by a scalar parameter, $z = 2x = (2\ -2)$ implies the same scaling of the conventional norm

$$\|z\| = 2\|x\| \approx 4.47213595.$$

Inversely, the $\mathbf{d}$-homogeneous norm is not symmetric with respect to the standard scaling:

$$\|z\|_{\mathbf{d}} = \|2x\|_{\mathbf{d}} \approx 4.24337015 < 2\|x\|_{\mathbf{d}} = 4.25414017.$$

For numerical simulations, the bisection method is implemented with $j_{\max} = 20$ and $\epsilon = 10^{-6}$.

Newton Method

Let us consider the canonical homogeneous norm $\|\cdot\|_{\mathbf{d}}$ induced by the weighted Euclidean norm $\|x\| = \sqrt{x^\top P x}$ with $0 \prec P = P^\top \in \mathbb{R}^{n\times n}$ satisfying the monotonicity condition (5.6). In this case, to find a value of the canonical homogeneous norm for an arbitrary selected $x \in \mathbb{R}^n\backslash\{\mathbf{0}\}$ we need to solve the algebraic equation

$$f(s) = 0 \quad \text{where} \quad f(s) := x^\top \mathbf{d}^\top(-s) P \mathbf{d}(-s) x - 1, s \in \mathbb{R}. \tag{5.36}$$

Notice that $f \in C^\infty(\mathbb{R})$. Since the derivative of the function f is given by

$$f'(s) = -x^\top \mathbf{d}^\top(-s)(PG_{\mathbf{d}} + G_{\mathbf{d}}^\top P)\mathbf{d}(-s)x \tag{5.37}$$

then, due to the monotonicity condition (5.6), the function f is strictly decreasing for $x \neq \mathbf{0}$. Taking into account the limit property of the dilation we conclude that $f(s) \to -1$ as $s \to +\infty$ and $f(s) \to +\infty$ as $s \to -\infty$. So, the function f has a unique root on $\mathbb{R}$, which can be estimated by the Newton method

$$s_{i+1} = s_i - \frac{f(s_i)}{f'(s_i)}, \quad s_0 \in \mathbb{R}. \tag{5.38}$$

Lemma 5.6 *Let a linear continuous dilation* $\mathbf{d}$ *be strictly monotone with respect to the norm* $\|x\| = \sqrt{x^\top P x}$ *with* $0 \prec P = P^\top \in \mathbb{R}^{n\times n}$ *satisfying the monotonicity condition* (5.6). *Let* $\|\cdot\|_{\mathbf{d}}$ *be the canonical homogeneous norm induced by the norm* $\|\cdot\|$. *If*

$$G_{\mathbf{d}}^\top Q + QG_{\mathbf{d}} \succ 0, \tag{5.39}$$

where $Q = G_{\mathbf{d}}^\top P + PG_{\mathbf{d}} \succ 0$, *then, for any* $x \in \mathbb{R}^n\backslash\{\mathbf{0}\}$ *and any* $s_0 \in \mathbb{R}$ *the sequence of numbers* s_i *generated by* (5.38) *is convergent and*

$$e^{s_i} \to \|x\|_{\mathbf{d}} \quad \text{as} \quad i \to +\infty.$$

Proof By Taylor expansion, we have

$$0 = f(s^*) = f(s_i) + f'(s_i)(s^* - s_i) + \frac{f''(\theta)}{2}(s^* - s_i)^2,$$

where s^* is a root of f and $\theta \in \mathbb{R}$ is a number between s_i and s^*. Hence, taking into account

$$f'(s) = -x^\top \mathbf{d}^\top(-s_i) Q \mathbf{d}(-s_i)x < 0, \quad s \in \mathbb{R}$$

we derive

$$-\frac{f(s_i)}{f'(s_i)} + s_i - s^* = \frac{f''(\theta_i)}{2f'(s_i)}(s^* - s_i)^2,$$

or, equivalently,

$$s_{i+1} - s^* = \frac{f''(\theta_i)}{2f'(s_i)}(s_i - s^*)^2, \quad \forall i \geq 0.$$

Hence, taking into account

$$f''(s) = x^\top \mathbf{d}^\top(-s)(QG_\mathbf{d} + G_\mathbf{d}^\top Q)\mathbf{d}(-s)x > 0, \quad \forall s \in \mathbb{R}$$

we conclude that $s_i \leq s^*$ for all $i \geq 1$ and, consequently, $f(s_i) \geq 0$ for all $i \geq 1$ due to monotonicity of f. This implies that $s_{i+1} \geq s_i$ for all $i \geq 1$. Therefore, the sequence s_i is monotone increasing and bounded from above, consequently, it is convergent. Since $f(s) = 0$ if and only if $s = s^*$ then $s_i \to s^*$ as $i + \infty$, i.e., $e^{s_i} \to \|x\|_\mathbf{d}$ as $i \to +\infty$. The proof is complete. □

To avoid computational difficulties in practice, we suggest to implement the Newton method for the function $f(s) = \ln(x^\top \mathbf{d}^\top(-s)P\mathbf{d}(-s)x)$. In this case, the unique solution $s = s^* \in \mathbb{R}$ of the algebraic equation $f(s) = 0$ also corresponds to the value of the canonical homogeneous norm $\|x\|_\mathbf{d} = e^{s^*}$.

Example 5.8 The canonical homogeneous norm of the vector x from Example 5.7 computed by the Newton method is

$$\|x\|_\mathbf{d} \approx 2.12706969.$$

The difference with the estimate obtained by the bisection method is about 10^{-8}. For the numerical simulation, the maximal number of iterations of the Newton method was restricted to 20 and the additional stopping criterion $|\ln(x^\top \mathbf{d}^\top(-s)P\mathbf{d}(-s)x)| \leq 10^{-6}$ is utilized.

5.3.2 *Approximation by an Explicit Homogeneous Norm*

The bisection method discussed above is a fast and efficient tool for the numerical calculation of the canonical homogeneous norm. Most of the control algorithms studied in this monograph can be implemented using the bisection algorithm. However, in some practical applications, an explicit definition of the homogeneous norm may simplify the control system design and analysis (see Chap. 9). A possible explicit $\mathbf{d}$-homogeneous norm $\|\cdot\|_{\tilde{r}}$ is introduced by Lemma 5.3. In the general case, this norm does not coincide with the canonical homogeneous norm $\|\cdot\|_\mathbf{d}$, but, in some cases, it can be utilized for its approximation in view of the following theorem.

Proposition 5.4 (On approximation of $\|\cdot\|_{\mathbf{d}}$ by $\|\cdot\|_{\tilde{r}}$) *Let a linear continuous dilation* $\mathbf{d}$ *in* $\mathbb{R}^n$, *a* $\mathbf{d}$*-homogeneous norm* $\|\cdot\|_{\tilde{r}} \in C(\mathbb{R}^n) \cap C^1(\mathbb{R}^n\backslash\{\mathbf{0}\})$ *be defined as Lemma 5.3 and* $\|\cdot\|_{\mathbf{d}}$ *be the canonical homogeneous norm induced by the weighted Euclidean norm* $\|x\| = \sqrt{x^\top P x}$ *with* $P \succ 0$ *satisfying (5.6). If* $\Delta = \{z_i\}_{i=1}^N$ *is a sufficiently dense grid on the unit sphere* $S = \{x \in \mathbb{R}^n : \|x\| = 1\}$ *such that*

$$\exists \varepsilon \in (0,1) : \inf_i \|z - z_i\| \leq \varepsilon, \quad \forall z \in S, \tag{5.40}$$

then

$$\delta_{\min} \leq \|z\|_{\tilde{r}} \leq \delta_{\max}, \quad \forall z \in S, \tag{5.41}$$

where

$$\delta_{\min} = \inf_i \|z_i\|_{\tilde{r}} - \varepsilon \sup_{y,z\in S, 0\leq\theta<\varepsilon} \left.\frac{\partial \|x\|_{\tilde{r}}}{\partial x}\right|_{x=y+\theta z} z.$$

$$\delta_{\max} = \sup_i \|z_i\|_{\tilde{r}} + \varepsilon \sup_{y,z\in S, 0\leq\theta<\varepsilon} \left.\frac{\partial \|x\|_{\tilde{r}}}{\partial x}\right|_{x=y+\theta z} z.$$

Proof If $z \in S$ and $r \in [0, \varepsilon]$, $0 < \varepsilon < 1$ then $\|z_i + \theta z\| > 0$ for all $\theta \in (0, r)$ and by Mean Value theorem we have

$$\|z_i + rz\|_{\tilde{r}} = \|z_i\|_{\tilde{r}} + r \left.\frac{\partial \|x\|_{\tilde{r}}}{\partial x}\right|_{x=z_i+\theta z} z$$

for some $\theta \in [0, r]$. Since for any $z \in S$, $z \neq z_i$ there exists $i^* \in \{1, 2, ..., N\}$ such that $\|z - z_{i^*}\| \leq \varepsilon$ then for $z^* := \frac{z-z_{i^*}}{\|z-z_{i^*}\|}$ and $r^* := \|z - z_{i^*}\|$ there exists $\theta^* \in [0, r^*]$ such that

$$\|z\|_{\tilde{r}} = \|z_{i^*} + r^* z^*\|_{\tilde{r}} = \|z_{i^*}\|_{\tilde{r}} + r^* \left.\frac{\partial \|x\|_{\tilde{r}}}{\partial x}\right|_{x=z_i+\theta^* z^*} z^*$$

and the estimate (5.41) is fulfilled. The proof is complete. □

Since $|\|x\|_{\tilde{r}} - 1| = |\|x\|_{\tilde{r}} - \|x\|_{\mathbf{d}}|$ if $\|x\| = 1$ then for $\delta_{\min} = \delta_{\max} = 1$ we have $\|x\|_{\tilde{r}} = \|x\|_{\mathbf{d}}, \forall x \in \mathbb{R}^n$ due to homogeneity. Therefore, the parameters $\delta_{\min}$ and $\delta_{\max}$ characterizes a distance between the canonical homogeneous norm $\|\cdot\|_{\mathbf{d}}$ and the explicit homogeneous norm $\|\cdot\|_{\tilde{r}}$.

The above proposition inspires the following scheme for an approximation of the canonical $\mathbf{d}$-homogeneous norm.

Algorithm 5.2 ***(Approximation of*** $\|\cdot\|_{\mathbf{d}}$ ***by*** $\|\cdot\|_{\bar{r}}$ ***)***

Initialization. Select a small $\varepsilon \in (0,1)$ and $\Psi \in C(\mathbb{R}^n;\mathbb{R}^n)$ satisfying conditions (5.10) and (5.11).

Step 1. Generate randomly (uniformly distributed) nonzero vectors $\tilde{z}_i \in [-1,1]^n$, $i=1,2,...,N$ and project them to the unit sphere

$$z_i := \frac{\tilde{z}_i}{\|\tilde{z}_i\|},$$

where $\|\tilde{z}_i\| = \sqrt{z_i^\top P z_i}$.

Step 2. If the inequality (5.40) does not hold for the given $\varepsilon > 0$ then go to **Step 1** taking $N = 2N$.

Step 3. Solve the following semi-definite programming (SDP) problem (see [32] for more details about SDP)

$$\text{minimize} \sum_{i=1}^{N} \left(\Psi^\top(z_i) Q \Psi(z_i)\right)^2 - 2\left(\Psi^\top(z_i) Q \Psi(z_i)\right)$$

subject to $0 \prec Q = Q^\top \in \mathbb{R}^{n\times n}$.

The presented algorithm is aimed at an approximation of the canonical homogeneous norm $\|\cdot\|_{\mathbf{d}}$ by the explicit $\mathbf{d}$-homogeneous norm $\|x\|_{\bar{r}}$ given by (5.12). Indeed, since

$$\left(\Psi^\top(z_i) Q \Psi(z_i)\right)^2 - 2\left(\Psi^\top(z_i) Q \Psi(z_i)\right) = \|z_i\|_{\bar{r}}^{4\bar{r}} - 2\|z_i\|_{\bar{r}}^{2\bar{r}} = \left(\|z_i\|_{\bar{r}}^{2\bar{r}} - 1\right)^2 - 1,$$

then the above optimization problem is equivalent to

$$J_1(Q) := \sum_{i=1}^{N} (\|z_i\|_{\bar{r}}^{2\bar{r}} - 1)^2 \to \min_{0 \prec Q = Q^\top \in \mathbb{R}^{n\times n}}, \tag{5.42}$$

where $z_i \in S$. The closer $J(Q)$ to zero, the more precise estimate of $\|\cdot\|_{\mathbf{d}}$ is. If $\|x\|_{\bar{r}} = 1$ for any $x \in \mathbb{R}^n : x^\top P x = 1$ then $\|x\|_{\mathbf{d}} = \|x\|_{\bar{r}}$ for all $x \in \mathbb{R}^n$ due to homogeneity.

Example 5.9 Let a linear dilation in $\mathbb{R}^2$ be defined as

$$\mathbf{d}(s) = e^{sG_{\mathbf{d}}}, \quad G_{\mathbf{d}} = \begin{pmatrix} 5 & 0 \\ 0 & 4 \end{pmatrix}, \quad s \in \mathbb{R}.$$

This dilation is strictly monotone with respect to the weighted Euclidean norm $\|x\| = \sqrt{x^\top P x}$ with

$$P = \begin{pmatrix} 0.1157 & 0.0194 \\ 0.0194 & 0.1186 \end{pmatrix}.$$

Selecting $\bar{r} = 5$ we derive $\Psi(x) = \begin{pmatrix} x_1 \\ |x_2|^{5/4}\,\mathrm{sign}(x_2) \end{pmatrix}$ and obtain the optimal Q by means of Algorithm 5.2:

$$Q = \begin{pmatrix} 0.1225 & 0.0176 \\ 0.0176 & 0.0720 \end{pmatrix}.$$

The approximation precision

$$\epsilon = \sup_{\|x\|_{\bar{r}}=1} (\|x\|_{\mathbf{d}} - 1)$$

is $\epsilon \approx 0.007$.

For $\bar{r} = 4.5$ we have $\Psi(x) = \begin{pmatrix} |x_1|^{9/10}\,\mathrm{sign}(x_1) \\ |x_2|^{9/8}\,\mathrm{sign}(x_2) \end{pmatrix}$ and the optimal Q is

$$Q = \begin{pmatrix} 0.1455 & 0.0190 \\ 0.0190 & 0.0896 \end{pmatrix}$$

and $\epsilon \approx 0.002$.

The difference between unit level sets of the canonical homogeneous norm $\|x\|_{\mathbf{d}}$ and the homogeneous norm

$$\|x\|_{\bar{r}} = (\Psi^{\top}(x) Q \Psi(x))^{\frac{1}{2\bar{r}}}$$

is illustrated on Fig. 5.7, where the unit level line $\|x\|_{\mathbf{d}} = 1$ is depicted by the red color, while the unit level set $\|x\|_{\bar{r}} = 1$ is plotted by the blue color. The norm $\|\cdot\|_{\bar{r}}$ with $\bar{r} = 5$ is continuously differentiable on $\mathbb{R}^n \backslash \{\mathbf{0}\}$. For $\bar{r} = 4.5$ the norm $\|\cdot\|_{\bar{r}}$ is just continuous on $\mathbb{R}^n$, but it has a better estimation error than the non-smooth norm $\|\cdot\|_{\bar{r}}$ with $\bar{r} = 4.5$.

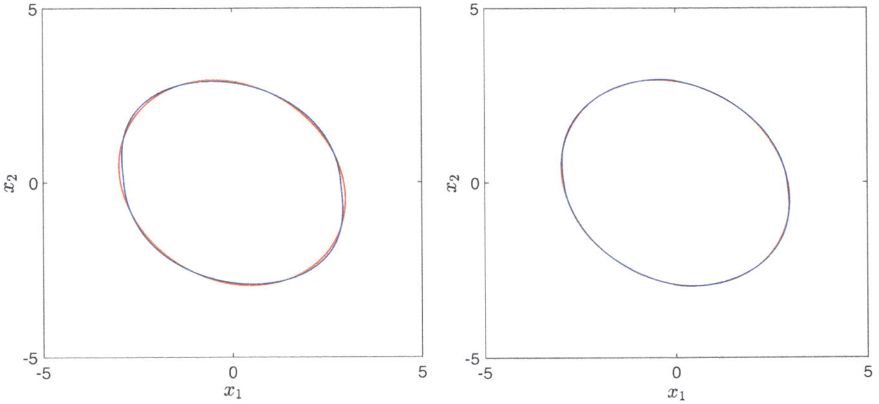

Fig. 5.7 Level lines for $\bar{r} = 5$ (left) and $\bar{r} = 4.5$ (right)

The above example shows an optimization of the parameter $\bar{r}$ could essentially improve the approximation precision. Convexity on the approximation error with respect to the scalar parameter $\bar{r}$ is not obvious, so its selection can be done by a search of a uniform grid.

5.3.3 Arbitrary-Precise Approximation

The approximation precision of $\|x\|_{\mathbf{d}}$ by $\|x\|_{\bar{r}}$ is characterized via the numbers $\delta_{\min}$ and $\delta_{\max}$ from Proposition 5.4. If $\delta_{\min} = \delta_{\max} = 1$ then $\|\cdot\|_{\mathbf{d}} = \|\cdot\|_{\bar{r}}$. However, in the general case, the explicit homogeneous norm $\|\cdot\|_{\bar{r}}$ given by (5.12) may not approximate the canonical homogeneous $\|\cdot\|_{\mathbf{d}}$ with an arbitrary high precision just by tuning of the matrix $Q \succ 0$. The following theorem constructs the required approximation using any explicit homogeneous norm $\|\cdot\|_{\bar{r}}$.

Theorem 5.6 (On arbitrary-precise approximation of $\|\cdot\|_{\mathbf{d}}$) *Let a linear continuous dilation* $\mathbf{d}$ *in* $\mathbb{R}^n$ *be strictly monotone with respect to the weighted Euclidean norm* $\|x\| = \sqrt{x^\top P x}$ *with* $P \succ 0$ *satisfying* (5.6). *Let* $\|\cdot\|_{\mathbf{d}}$ *be a canonical* $\mathbf{d}$*-homogeneous norm induced by* $\|\cdot\|$ *and* $\|\cdot\|_{\bar{r}}$ *be an arbitrary* $\mathbf{d}$*-homogeneous norm.*

Then for any $\epsilon > 0$ *there exist* $N \in \mathbb{N}$, $\beta_i \in \mathbb{R}^n$, $\alpha_i, \gamma_i \in \mathbb{R}$, $i = 1, 2, \ldots, N$ *and* $\xi \in C(\mathbb{R}^n \backslash \{\mathbf{0}\}; \mathbb{R})$ *such that* $|\xi(x)| \leq \epsilon, \forall x \in \mathbb{R}^n \backslash \{\mathbf{0}\}$ *and*

$$\|x\|_{\mathbf{d}} = \|x\|_{\bar{r}} \left(\sum_{i=1}^{N} \frac{\gamma_i}{1+e^{\alpha_i + \beta_i^\top \mathbf{d}(-\ln \|x\|_{\bar{r}})x}} + \xi(x) \right), \quad \forall x \in \mathbb{R}^n \backslash \{\mathbf{0}\}. \tag{5.43}$$

Proof Let us show that the numbers

$$a = \inf_{x \in \mathbb{R}^n \backslash \{\mathbf{0}\}} \|\mathbf{d}(-\ln \|x\|_{\bar{r}})x\| \quad \text{and} \quad b = \sup_{x \in \mathbb{R}^n \backslash \{\mathbf{0}\}} \|\mathbf{d}(-\ln \|x\|_{\bar{r}})x\|$$

are finite and strictly positive. Indeed, since, by definition, the homogeneous norms $\|\cdot\|_{\mathbf{d}}$ and $\|\cdot\|_{\bar{r}}$ are continuous positive definite functions satisfying $\|\mathbf{d}(s)x\|_{\mathbf{d}} = e^s\|x\|_{\mathbf{d}}$ and $\|\mathbf{d}(s)x\|_{\bar{r}} = e^s\|x\|_{\bar{r}}$ then these functions are radially unbounded and

$$0 < \inf_{x:\|x\|=1} \frac{\|x\|_{\mathbf{d}}}{\|x\|_{\bar{r}}} \leq \sup_{x:\|x\|=1} \frac{\|x\|_{\mathbf{d}}}{\|x\|_{\bar{r}}} < +\infty.$$

Since $\|\mathbf{d}(s)x\|_{\mathbf{d}} = e^s\|x\|_{\mathbf{d}}$ and $\|\mathbf{d}(s)x\|_{\bar{r}} = e^s\|x\|_{\bar{r}}$ for all $s \in \mathbb{R}$ and all $x \in \mathbb{R}^n$ then

$$\frac{\|x\|_{\mathbf{d}}}{\|x\|_{\bar{r}}} = \|\mathbf{d}(-\ln \|x\|_{\bar{r}})x\|_{\mathbf{d}}, \quad \forall x \neq \mathbf{0}$$

and

$$0 < \inf_{x \in \mathbb{R}^n \setminus \{\mathbf{0}\}} \|\mathbf{d}(-\ln \|x\|_{\tilde{r}})x\|_{\mathbf{d}} \leq \sup_{x \in \mathbb{R}^n \setminus \{\mathbf{0}\}} \|\mathbf{d}(-\ln \|x\|_{\tilde{r}})x\|_{\mathbf{d}} < +\infty.$$

Using (5.16) we derive $0 < a \leq b < +\infty$.

Let us consider the set

$$\Omega = \{y \in \mathbb{R}^n : a \leq \|y\| \leq b\}$$

and the function

$$y \mapsto h(y) := \|y\|_{\mathbf{d}}.$$

By construction, the set Ω is a compact, $\mathbf{0} \notin \Omega$ and the function h is continuous on Ω. The Universal Approximation theorem (see, e.g., [30] or Chap. 16) proves that for any $\varepsilon > 0$ there exists $N \in \mathbb{N}$, $\beta_i \in \mathbb{R}^n$, $\alpha_i, \gamma_i \in \mathbb{R}$, $i = 1, 2, ..., N$ such that

$$\sup_{y \in \Omega} \left| h(y) - \sum_{i=1}^{N} \frac{\gamma_i}{1+e^{\alpha_i + \beta_i^\top y}} \right| \leq \epsilon.$$

We complete the proof by taking $\xi(x) = h(y) - \sum_{i=1}^{N} \frac{\gamma_i}{1+e^{\alpha_i+\beta_i^\top y}}\Big|_{y=\mathbf{d}(-\ln \|x\|_{\tilde{r}})x}$.

Formula (5.43) inspires the explicit homogeneous norm

$$\|x\|_{\mathbf{d},\epsilon} := \|x\|_{\tilde{r}} \sum_{i=1}^{N} \frac{\gamma_i}{1+e^{\alpha_i+\beta_i^\top \mathbf{d}(-\ln \|x\|_{\tilde{r}})x}}, \quad x \in \mathbb{R}^n \setminus \{\mathbf{0}\}, \tag{5.44}$$

which approximates the canonical homogeneous norm $\|\cdot\|_{\mathbf{d}}$ with an arbitrary precision $\epsilon > 0$ for a sufficiently large $N \in \mathbb{N}$. Notice that any explicit homogeneous norm $\|\cdot\|_{\tilde{r}}$ can be utilized in (5.44) to derive such an approximation. However, if the canonical homogeneous norm is analytic on $\mathbb{R}^n \setminus \{\mathbf{0}\}$, then it is worth to approximate $\|\cdot\|_{\mathbf{d}}$ by an analytic function $\|x\|_{\mathbf{d},\epsilon}$. In the view of Lemma 5.3, such an approximation can always be constructed for a linear continuous dilation with a diagonalizable generator having rational eigenvalues.

In practice [33] the parameters α_i and β_i may be selected randomly to guarantee that the basis functions $y \mapsto \frac{1}{1+e^{\alpha_i+\beta_i^\top y}}$ are linearly independent, where $y \in \mathbb{R}^n$. The parameters γ_i have to be tuned to minimize the approximation error:

$$J_2(\bar{\gamma}) := \left| 1 - \sup_{x^\top P x = 1} \|x\|_{\tilde{r}} \sum_{i=1}^{N} \frac{\gamma_i}{1+e^{\alpha_i+\beta_i^\top \mathbf{d}(-\ln \|x\|_{\tilde{r}})x}} \right| \to \min_{\gamma_i}, \tag{5.45}$$

where $\bar{\gamma} = (\gamma_1, ..., \gamma_N)^\top \in \mathbb{R}^N$. Since $x^\top P x = 1 \Leftrightarrow \|x\|_{\mathbf{d}} = 1$ then the above optimization problem is aimed at minimization of the approximation error. Indeed, if $\|x\|_{\mathbf{d},\epsilon} = 1$ for any $x \in \mathbb{R}^n : x^\top P x = 1$ then $\|x\|_{\mathbf{d}} = \|x\|_{\mathbf{d},\epsilon}$ for all $x \in \mathbb{R}^n$ due to homogeneity. Let us propose a possible algorithm for tuning of the parameters γ_i assuming that the parameter $Q \succ 0$ of the explicit homogeneous norm $\|x\|_{\bar{r}}$ are already tuned (e.g., using Algorithm 5.2).

Algorithm 5.3 ***(Approximation of*** $\|\cdot\|_{\mathbf{d}}$ by $\|\cdot\|_{\mathbf{d},\epsilon}$***)***
Initialization. Select $\epsilon > 0$ and a natural $N \in \mathbb{N}$
Step 1. Generate uniformly distributed random numbers $\alpha_i \in [0, 1]$ and uniformly distributed random vectors $\beta_i \in [0, 1]^n$ with $i = 1, 2, ..., N$.
Step 2. Generate uniformly distributed random points $x_j \in \mathbb{R}^n$ on the unit sphere $x^\top Px = 1$ (as in Algorithm 5.2) with $j = 1, 2, ..., M$.
Step 3. Compute $M \times N$-matrix W with the elements

$$W_{j,i} = \|x_j\|_{\bar{r}} \frac{1}{1 + e^{\alpha_i + \beta_i^\top \mathbf{d}(-\ln \|x_j\|_{\bar{r}}) x_j}}. \tag{5.46}$$

Step 4. Compute $\bar{\gamma}$ by formula

$$\bar{\gamma}^\top = W^+ \begin{pmatrix} 1 \\ 1 \\ \dots \\ 1 \end{pmatrix}, \tag{5.47}$$

where W^+ denotes the Moore pseudo inverse matrix to W.
Step 5. If $J_2(\bar{\gamma}) > \epsilon$ then $N \leftarrow N + 1$ and go to **Step 1**.

Since the matrix W is defined by formula (5.46) then the identity $W\bar{\gamma} = 1$ means that $\|x_j\|_{\mathbf{d},\epsilon} = \|x_j\|_{\mathbf{d}}$ for each $j = 1, 2, ...$ Formula (5.47) defines a least square optimal parameter $\bar{\gamma}$ (see, e.g., [34]), namely, it optimizes $(W\bar{\gamma} - 1)^\top(W\bar{\gamma} - 1) \to \min$. In practice, the condition $J_2(\bar{\gamma}) > \epsilon$ can be checked using Proposition 5.4.

Example 5.10 Let us consider the homogeneous norm $\|x\|_{\bar{r}}$ as in Example 5.9 with $\bar{r} = 5$. Using Algorithm (5.3) with $N = 10$ and $M = 40$ we design the homogeneous norm $\|\cdot\|_{\mathbf{d},\epsilon}$ with parameters

$$\alpha = \left(-0.3108\ -0.0343\ 0.4810\ 0.3202\ -0.3612\ 0.3026\ 0.3577\ 0.1425\ 0.1488\ 0.1608 \right),$$

$$\beta = \begin{pmatrix} 0.0997 & 0.4113 & 0.0557 & 0.3434 & -0.4185 & -0.4357 & 0.4515 & 0.2378 & 0.1478 & 0.0301 \\ -0.4662 & 0.4010 & 0.1654 & 0.3079 & 0.4095 & 0.3874 & -0.1385 & 0.2031 & 0.2659 & -0.3040 \end{pmatrix}$$

$$\gamma = \left(-0.2743\ 0.1192\ -10.6867\ -2.6738\ 1.1245\ -0.2826\ 0.2030\ 0.5774\ 9.7199\ 2.1084 \right).$$

The approximation error on unit sphere for $\|\cdot\|_{\mathbf{d},\epsilon}$ is $\epsilon \approx 0.0034$ that is twice less than the approximation error for $\|\cdot\|_{\bar{r}}$ with $\bar{r} = 5$ (see Fig. 5.8). The number N of basis functions utilized for approximation essentially impacts the precision. For $N = 20$ the approximation error of the unit sphere is $\epsilon \approx 10^{-5}$, but for $N = 30$ we have $\epsilon \approx 10^{-8}$. The plots for the latter cases are not presented since the corresponding spheres are visually indistinguishable with the unit sphere (red line on Fig. 5.8).

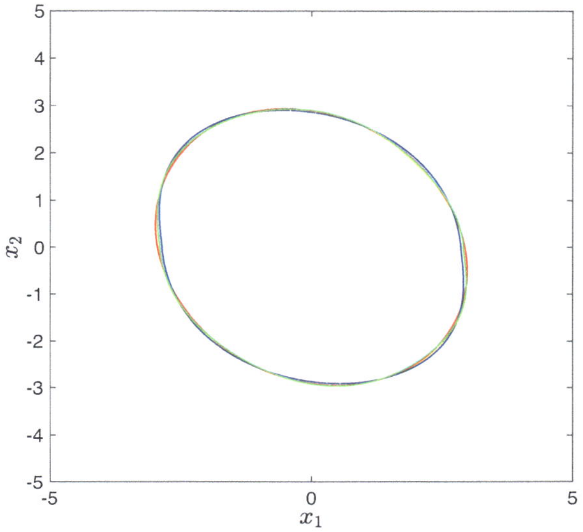

Fig. 5.8 Level lines $\|x\|_{\mathbf{d}} = 1$ (red), $\|x\|_{\tilde{r}} = 1$ (blue) and $\|x\|_{\mathbf{d},\epsilon} = 1$ (green)

Concluding Remarks

Linear dilations in Euclidean spaces and their properties are studied. Any continuous linear dilation in $\mathbb{R}^n$ introduces an alternative topology in $\mathbb{R}^n$ by means of a homogeneous norm. The so-called canonical homogeneous norm is a norm (in the conventional sense) for a Euclidean space homeomorphic to $\mathbb{R}^n$. This norm is the key object introduced in this chapter. Many methods of system analysis and control design developed in this monograph use it. Several algorithms of calculation and approximation of the canonical homogeneous norm with arbitrary high precision are developed. Such an approximation is based on the so-called homogeneous artificial neural network (see Chap. 16), which can be realized in microchips [35]. Therefore, in practice, homogeneous control algorithms developed in this monograph can be efficiently implemented in most engineering systems.

References

1. Husch LS (1970) Topological characterization of the dilation and the translation in frechet spaces. Math Ann 190:1–5
2. Kawski M (1991) Families of dilations and asymptotic stability. In: Analysis of controlled dynamical systems, pp 285–294
3. Khomenuk VV (1961) On systems of ordinary differential equations with generalized homogenous right-hand sides. Izvestia vuzov. Mathematica (in Russian) 3(22):157–164
4. Kawski M (1995) Geometric homogeneity and stabilization. In: Arthur K, David M (eds) Proceedings of IFAC nonlinear control symposium. Lake Tahoe, CA, pp 164–169

5. Rosier L (1993) Etude de quelques problèmes de stabilization. PhD Thesis, Ecole Normale Superieure de Cachan (France)
6. Engel K-J, Nagel R (2000) One-parameter semigroups for linear evolution equations. Springer, Berlin
7. Folland G (1975) Subelliptic estimates and function spaces on nilpotent Lie groups. Arkiv for Matematik 13(1–2):161–207
8. Fischer V, Ruzhansky M (2016) Quantization on nilpotent Lie groups. Springer, Berlin
9. Zubov VI (1958) On systems of ordinary differential equations with generalized homogeneous right-hand sides. Izvestia vuzov. Mathematica (available on-line: http://mi.mathnet.ru/ivm2874) 1:80–88
10. Polyakov A (2019) Sliding mode control design using canonical homogeneous norm. Int J Robust Nonlinear Control 29(3):682–701
11. Hong Y (2001) H_∞ control, stabilization, and input-output stability of nonlinear systems with homogeneous properties. Automatica 37(7):819–829
12. Kawski M (1990) Homogeneous stabilizing feedback laws. Control Theory Adv Technol 6(4):497–516
13. Grüne L (2000) Homogeneous state feedback stabilization of homogeneous systems. SIAM J Control Optim 38(4):1288–1308
14. Bacciotti A, Rosier L (2001) Lyapunov functions and stability in control theory. Springer, Berlin
15. Andrieu V, Praly L, Astolfi A (2008) Homogeneous approximation, recursive observer design, and output feedback. SIAM J Control Optim 47(4):1814–1850
16. Rockafellar RT (1970) Convex analysis. Princeton University Press
17. Polyakov A, Coron J-M, Rosier L (2018) On homogeneous finite-time control for linear evolution equation in Hilbert space. IEEE Trans Automat Control 63(9):3143–3150
18. Korobov VI (1979) A solution of the synthesis problem using controlability function. Doklady Academii Nauk SSSR 248:1051–1063
19. Korobov VI (2007) Method of controlability functions. NIC, Moscow (in Russian)
20. Adamy J, Flemming A (2004) Soft variable-structure controls: a survey. Automatica 40:1821–1844
21. Polyakov A, Efimov D, Perruquetti W (2015) Finite-time and fixed-time stabilization: implicit Lyapunov function approach. Automatica 51(1):332–340
22. Clarke F (1976) On the inverse function theorem. Pac J Math 64(1):97–102
23. Magaril-Il'aev GG (1978) The implicit function theorem for lipschitz map. Russian Math Surv 33(1):209–210
24. Krantz AG, Parks HR (2003) The implicit function theorem: history, theory and appplications. Springer, Berlin
25. Zhou Y, Polyakov A, Zheng G (2025) Finite/fixed-time stabilization of linear systems with states quantization. IEEE Trans Autom Control 70(3):1921–1928
26. Blanchini F (1999) Set invariance in control. Automatica 35:1747–1767
27. Ames AD, Grizzle JW, Tabuada P (2017) Ccontrol barrier function based quadratic programs for safety critical systems. IEEE Trans Automat Control 62:3861–3876
28. Clarke F (1990) Optimization and nonsmooth analysis. SIAM, Philadelphia
29. Clarke F, Ledyaev YS, Stern R, Wolenski P (1995) Qualitative properties of trajectories of control systems: a survey. J Dyn Control Syst 1(1):1–48
30. Cybenko G (1989) Approximation by superpositions of a sigmoidal function. Math Control Signals Syst 4(2):303–3014
31. Levant A (2005) Homogeneity approach to high-order sliding mode design. Automatica 41(5):823–830
32. Boyd S, Ghaoui E, Feron E, Balakrishnan V (1994) Linear matrix inequalities in system and control theory. SIAM, Philadelphia
33. Lewis F, Jagannathan S, Yesildirak A (1998) Neural network control of robot manipulators and non-linear systems. CRC Press

34. Poznyak AS (2008) Advanced mathematical tools for automatic control engineers. Volume 1: deterministic technique. Elsevier
35. Chen T, Du Z, Sun N, Wang J, Wu C, Chen Y, Temam O (2014) Diannao: a small-footprint high-throughput accelerator for ubiquitous machine-learning. ACM SIGARCH Comput Archit News 42(1):269–284

Chapter 6
Finite-Dimensional Homogeneous Mappings

This chapter introduces **d**-homogeneous functions and vector fields in $\mathbb{R}^n$ and studies their properties. In particular, a generalized Euler's homogeneous function theorem is proven for linear dilations; elements of "homogeneous arithmetics" are presented; regularity of **d**-homogeneous mappings is analyzed; homogeneous approximations and homogeneous extensions of non-homogeneous mappings are discussed; a dilation symmetry of tangent cones of homogeneous sets is studied.

6.1 Definitions and Basic Properties

6.1.1 Homogeneous Functions

Generalized homogeneous functions in $\mathbb{R}^n$ (see [1–3]) can be introduced analogously to the Euler's definition (see Chap. 1).

Definition 6.1 *(Homogeneous function)* Let **d** be a dilation in $\mathbb{R}^n$ and $\mathcal{D} \subset \mathbb{R}^n$ be a nonempty **d**-homogeneous cone in $\mathbb{R}^n$. A function $h : \mathcal{D} \subset \mathbb{R}^n \mapsto \mathbb{R}^k$ is said to be **d**-homogeneous of degree $\nu \in \mathbb{R}$ if

$$h(\mathbf{d}(s)x) = e^{\nu s} h(x), \quad \forall s \in \mathbb{R}, \quad \forall x \in \mathcal{D}. \tag{6.1}$$

Since the simplest **d**-homogeneous cone is $\mathcal{D} = \mathbb{R}^n$, the above definition simply repeats the Euler's definition if $\mathbf{d}(s) = e^s I_n$, $s \in \mathbb{R}$ (see Chap. 1). For simplicity, all results given below are proven for scalar-valued functions $\mathbb{R}^n \mapsto \mathbb{R}$, but, obviously, they remain valid for vector-valued functions $\mathbb{R}^n \mapsto \mathbb{R}^k$ in a component-vise manner.

A. Polyakov, *Generalized Homogeneity in Systems and Control Volume I*,
Communications and Control Engineering,
https://doi.org/10.1007/978-3-031-76246-8_6

Example 6.1 • The function $h(x) = x_1 + x_2^2$ is **d**-homogeneous of degree 2 with respect to the standard dilation $\mathbf{d}(s) = \begin{pmatrix} e^{2s} & 0 \\ 0 & e^s \end{pmatrix}$.

• The function $h(x) = x_1^2 + x_2^2$ is **d**-homogeneous of degree 2 with respect to the linear dilation $\mathbf{d}(s) = e^s \begin{pmatrix} \cos(s) & -\sin(s) \\ \sin(s) & \cos(s) \end{pmatrix}$.

In Chap. 5, it is shown that, the so-called canonical homogeneous norm $\|\cdot\|_{\mathbf{d}} \in C(\mathbb{R}^n) \cap C^1(\mathbb{R}^n\backslash\{\mathbf{0}\})$ can always be induced by means a linear continuous dilation $\mathbf{d}$ in $\mathbb{R}^n$ and a weighted Euclidean norm

$$\|x\| = \sqrt{x^\top P x}, \quad x \in \mathbb{R}^n$$

with the positive definite symmetric matrix $P = P^\top \in \mathbb{R}^n$ satisfying the Lyapunov inequality $PG_{\mathbf{d}} + G_{\mathbf{d}}^\top P \succ 0$ (see Corollary 5.3). The above condition means that the dilation $\mathbf{d}$ is strictly monotone with respect to the norm $\|\cdot\|$ (see Proposition 5.3). The norms $\|\cdot\|$ and $\|\cdot\|_{\mathbf{d}}$ are continuously differentiable away from the origin and have the common unit sphere

$$S = \{x \in \mathbb{R}^n : \|x\| = 1\} = \{x \in \mathbb{R}^n : \|x\|_{\mathbf{d}} = 1\}.$$

In this chapter we use only these two norms for analysis of homogeneous functions and vector fields.

Let $\mathcal{H}_{\mathbf{d}}(\mathcal{D})$ denote a set of **d**-homogeneous functions $\mathcal{D} \subset \mathbb{R}^n \mapsto \mathbb{R}$ and let $\deg_{\mathbf{d}}(h) \in \mathbb{R}$ denote the homogeneity degree of $h \in \mathcal{H}_{\mathbf{d}}(\mathcal{D})$.

Any local information about homogeneous function h can be expanded globally using the homogeneity degree of the function. Some properties of locally bounded homogeneous function [4] are summarized in the following proposition.

Proposition 6.1 *Let $h \in \mathcal{H}_{\mathbf{d}}(\mathcal{D})$, $m := \inf_{x \in S\cap\mathcal{D}} |h(x)|$, $M := \sup_{x\in S\cap\mathcal{D}} |h(x)|$.*

- *If $\deg_{\mathbf{d}}(h) > 0$, $M < +\infty$ then h is bounded on any set $B_{\mathbf{d}}(r) \cap \mathcal{D}$ and*

$$\limsup_{x\in\mathcal{D}, \|x\|\to 0} |h(x)| = 0,$$

*where $B_{\mathbf{d}}(r)$ is a **d**-homogeneous ball in $\mathbb{R}^n$ of the radius $r > 0$.*

- *If $\deg_{\mathbf{d}}(h) > 0$, $m > 0$ then h is unbounded on any set $\mathcal{D}\backslash B_{\mathbf{d}}(r)$ and*

$$\liminf_{x\in\mathcal{D}, \|x\|\to+\infty} |h(x)| = +\infty.$$

- *If $\deg_{\mathbf{d}}(h) < 0$, $M < +\infty$ then h is bounded on any set $\mathcal{D}\backslash B_{\mathbf{d}}(r)$ and*

$$\limsup_{x\in\mathcal{D}, \|x\|\to+\infty} |h(x)| = 0.$$

- *If* $\deg_{\mathbf{d}}(h) < 0$, $m > 0$ *then* h *is unbounded on any set* $B_{\mathbf{d}}(r) \cap \mathcal{D}$ *and*

$$\limsup_{x \in \mathcal{D}, \|x\| \to 0} |h(x)| = +\infty.$$

- *If* $\deg_{\mathbf{d}}(h) = 0$ *then* $|h|$ *is uniformly bounded by* $M < +\infty$ *from above and uniformly bounded by* $m \geq 0$ *from below on* $\mathcal{D}\backslash\{\mathbf{0}\}$; *moreover,* $h(x) \equiv \text{const}, \forall x \in \mathcal{D}$ *provided that* h *is continuous at zero.*
- *A non-constant homogeneous function* h *is continuous at zero if and only if* $\deg_{\mathbf{d}}(h) > 0$ *and* $f(\mathbf{0}) = 0$.

Proof Since the identity (6.1) with $s = -\ln \|x\|_{\mathbf{d}}$ yields

$$h(x) = \|x\|_{\mathbf{d}}^{\deg_{\mathbf{d}}(h)} h(\mathbf{d}(-\ln \|x\|_{\mathbf{d}})x)$$

then the claims follow from the inclusion $\mathbf{d}(-\ln \|x\|_{\mathbf{d}})x \in S$ and the properties of the canonical homogeneous norm (see Lemma 5.4). The proof is complete. □

The local boundedness of h on the unit sphere is an important restriction of the above proposition. Indeed, both functions $x \mapsto \frac{x_2^2}{|x_1|}$ for $x_1 \neq 0$ and $x \mapsto \frac{x_2^2}{|x_1|+x_2^2}$ are $\mathbf{d}$-homogeneous of degree 0 with respect to the dilation $\mathbf{d}(s) = \mathrm{diag}(e^{2s}, e^s)$. However, the first function is unbounded on $\mathbb{R}^n$ (and on S), while the second function is globally uniformly bounded on $\mathbb{R}^n$.

Elements of homogeneous arithmetics for $h, g \in \mathcal{H}_{\mathbf{d}}(\mathcal{D})$ are given below:

- the zero function (i.e., $h(x) = 0, \forall x \in \mathcal{D}$) is $\mathbf{d}$-homogeneous of any degree;
- a constant function (i.e., $h(x) = c$, $x \in \mathcal{D}$) is $\mathbf{d}$-homogeneous of zero degree;
- $\alpha h \in \mathcal{H}_{\mathbf{d}}(\mathcal{D})$ and $\deg_{\mathbf{d}}(\alpha h) = \deg_{\mathbf{d}}(h)$ for any $\alpha \in \mathbb{R}$;
- $h + g \in \mathcal{H}_{\mathbf{d}}(\mathcal{D})$ provided that $\deg_{\mathbf{d}}(h) = \deg_{\mathbf{d}}(g)$;
- $h \cdot g \in \mathcal{H}_{\mathbf{d}}(\mathcal{D})$ and $\deg_{\mathbf{d}}(h \cdot g) = \deg_{\mathbf{d}}(h) + \deg_{\mathbf{d}}(g)$;
- $\frac{h}{g} \in \mathcal{H}_{\mathbf{d}}(\mathcal{D}\backslash \ker(g))$ and $\deg_{\mathbf{d}}\left(\frac{h}{g}\right) = \deg_{\mathbf{d}}(h) - \deg_{\mathbf{d}}(g)$, where $\ker(g) = \{x \in \mathbb{R}^n : g(x) = 0\}$ is a $\mathbf{d}$-homogeneous cone.

In the view of Proposition 6.1, the function $\frac{h}{g}$ may belong to $\mathcal{H}_{\mathbf{d}}(\mathcal{D})$ provided that $\ker(g) \subset \ker(h)$, $\sup_{x \in S} |h(x)|/|g(x)| < +\infty$ and $\deg_{\mathbf{d}}(h) \geq \deg_{\mathbf{d}}(g)$.

Homogeneity is inherited somehow by all other objects induced by homogeneous functions. In particular, a derivative of a homogeneous function is homogeneous as well.

Proposition 6.2 *If* $h \in \mathcal{H}_{\mathbf{d}}(\mathcal{D})$ *is differentiable at* $x \in \mathrm{int}\,\mathcal{D}\backslash\{\mathbf{0}\}$ *then* h *is differentiable at any point of the* $\mathbf{d}$*-homogeneous curve* $\Gamma_{\mathbf{d}}(x) \subset \mathrm{int}\,\mathcal{D}$,

$$\left.\frac{\partial h(z)}{\partial z}\right|_{z=\mathbf{d}(s)x} \mathbf{d}(s) = e^{s \deg_{\mathbf{d}}(h)} \frac{\partial h(x)}{\partial x}, \quad \forall s \in \mathbb{R}, \quad \forall x \in \Gamma_{\mathbf{d}}(x). \tag{6.2}$$

Proof Recall that $z \in \Gamma_{\mathbf{d}}(x) \Rightarrow \mathbf{d}(s)z \in \Gamma_{\mathbf{d}}(x), \forall s \in \mathbb{R}$. Since

$$\lim_{\|y\|\to 0} \frac{\left|h(x+y)-h(x)-\frac{\partial h(x)}{\partial x}y\right|}{\|y\|} = 0, \quad y \in \mathbb{R}^n,$$

then, for $z = \mathbf{d}(s)x$, using the $\mathbf{d}$-homogeneity of h we derive

$$\frac{\left|h(\mathbf{d}(s)x+y_s)-h(\mathbf{d}(s)x)-\frac{\partial h(z)}{\partial z}y_s\right|}{\|y_s\|} = \frac{e^{\nu s}\left|h(x+\mathbf{d}(-s)y_s)-h(x)-e^{-\nu s}\frac{\partial h(z)}{\partial z}y_s\right|}{\|y_s\|}$$
$$\leq \frac{e^{\nu s}}{\lfloor \mathbf{d}(s)\rfloor} \frac{\left|h(x+y)-h(x)-e^{-\nu s}\frac{\partial h(z)}{\partial z}\mathbf{d}(s)y\right|}{\|y\|},$$

where $y = \mathbf{d}(-s)y_s$ and $\nu = \deg_{\mathbf{d}}(h)$. Since $\lfloor \mathbf{d}(s)\rfloor > 0$ and $\mathbf{d}(s)$ is an invertible matrix then, for any fixed $s \in \mathbb{R}$, the limit $\|y\| \to 0$ is equivalent to the limit $\|y_s\| \to 0$. Hence, due to the uniqueness of the derivative of a differentiable function, we derive the identity (6.2). The proof is complete. □

An analog of Euler's homogeneous function theorem (see Chap. 1) can be proven for functions being homogeneous with respect to linear dilations in $\mathbb{R}^n$. Let us denote

$$R(x, x^*) = \left\{y = (y_1, ..., y_n)^\top \in \mathbb{R}^n : \min\{x_i, x_i^*\} \leq y_i \leq \max\{x_i, x_i^*\}, i = 1, ..., n\right\},$$

where $x = (x_1, ..., x_n)^\top \in \mathbb{R}^n$ and $x^* = (x_1^*, ..., x_n^*)^\top \in \mathbb{R}^n$

Theorem 6.1 *Let a function $h : \mathbb{R}^n \mapsto \mathbb{R}$ be locally Lipschitz continuous on $\mathbb{R}^n \backslash \{\mathbf{0}\}$ and $h(\mathbf{0}) = 0$. Then the following three claims are equivalent*

(1) $h \in \mathcal{H}_{\mathbf{d}}(\mathbb{R}^n)$ *with* $\nu = \deg_{\mathbf{d}}(h) \in \mathbb{R}$*;*

(2) the identity

$$\frac{\partial h(x)}{\partial x} G_{\mathbf{d}} x \overset{a.e.}{=} \nu h(x) \tag{6.3}$$

holds almost everywhere on $\mathbb{R}^n$, where $x \in \mathbb{R}^n \backslash \{\mathbf{0}\}$ and $G_{\mathbf{d}} \in \mathbb{R}^{n\times n}$ is a generator of the linear dilation $\mathbf{d}$;

(3) the identity

$$\int_{x^*}^{x} (\mathrm{trace}(G_{\mathbf{d}}) + \nu)h(y) - \sum_{i:x_i \neq x_i^*} \frac{h(y)e_i^\top G_{\mathbf{d}} y\Big|_{x_i^*}^{x_i}}{x_i - x_i^*} dy = 0 \tag{6.4}$$

holds for all $x = (x_1, ..., x_i, ..., x_n)^\top$, $x^* = (x_1^*, ..., x_i, ..., x_n^*)^\top$ such that $R(x, x^*) \in \mathbb{R}^n \backslash \{\mathbf{0}\}$, where $\int_{x^*}^{x} = \int_{x_1^*}^{x_1} ... \int_{x_n^*}^{x_n}$, $dy = dy_n...dy_1$, $i = 1, ..., n$, $e_i = (0, ..., 1, ..., 0)^\top$ and

$$h(y)e_i^\top G_{\mathbf{d}} y\Big|_{x_i^*}^{x_i} = h(y_{x_i})e_i^\top G_{\mathbf{d}}(y_{x_i}) - h(y_{x_i^*})e_i^\top G_{\mathbf{d}}(y_{x_i^*})$$

with $y_{x_i} = (y_1, ..., x_i, ..., y_n)^\top$ *and* $y_{x_i^*} = (y_1, ..., x_i^*, ..., y_n)^\top$.

Moreover, if h is differentiable on $\mathbb{R}^n$ *(resp.,* $\mathbb{R}^n\backslash\{\mathbf{0}\}$*), the identity* (6.3) *is defined everywhere on* $\mathbb{R}^n$ *(resp.,* $\mathbb{R}^n\backslash\{\mathbf{0}\}$*).*

Proof (1)⇒(2). Let $h \in \mathcal{H}_{\mathbf{d}}(\mathbb{R}^n)$ and $\nu = \deg_{\mathbf{d}}(h) \in \mathbb{R}$. Due to homogeneity, we have

$$h(x) = h\left(\mathbf{d}(\ln \|x\|_{\mathbf{d}})\mathbf{d}(-\ln \|x\|_{\mathbf{d}})x\right) = \|x\|_{\mathbf{d}}^{\nu} h(\mathbf{d}(-\ln \|x\|_{\mathbf{d}})x), \forall x \neq \mathbf{0}.$$

Since, by Rademacher theorem (see Appendix D), the locally Lipschitz continuous function h is differentiable almost everywhere on $\mathbb{R}^n\backslash\{\mathbf{0}\}$ then for almost all $x \in \mathbb{R}^n\backslash\{\mathbf{0}\}$ using product and chain rules for derivatives we obtain

$$\frac{\partial h(x)}{\partial x} \overset{a.e.}{=} \nu\|x\|_{\mathbf{d}}^{\nu-1} h(\mathbf{d}(-\ln \|x\|_{\mathbf{d}})x)\frac{\partial \|x\|_{\mathbf{d}}}{\partial x} + \|x\|_{\mathbf{d}}^{\nu}\frac{\partial h(z)}{\partial z}\frac{\partial}{\partial x}(\mathbf{d}(-\ln \|x\|_{\mathbf{d}})x),$$

where $z = \mathbf{d}(-\ln \|x\|_{\mathbf{d}})x$. Taking into account the identities $\frac{d}{ds}\mathbf{d}(s) = G_{\mathbf{d}}\mathbf{d}(s) = \mathbf{d}(s)G_{\mathbf{d}}, \forall s \in \mathbb{R}$ we conclude

$$\frac{\partial}{\partial x}(\mathbf{d}(-\ln \|x\|_{\mathbf{d}})x) = \mathbf{d}(-\ln \|x\|_{\mathbf{d}}) - \|x\|_{\mathbf{d}}^{-1} G_{\mathbf{d}}\mathbf{d}(-\ln \|x\|_{\mathbf{d}})x\frac{\partial \|x\|_{\mathbf{d}}}{\partial x}.$$

Finally, since $\frac{\partial \|x\|_{\mathbf{d}}}{\partial x}G_{\mathbf{d}}x = \|x\|_{\mathbf{d}}, \forall x \in \mathbb{R}^n\backslash\{\mathbf{0}\}$ due to formula (5.21), then

$$\frac{\partial\left(\mathbf{d}(-\ln \|x\|_{\mathbf{d}})x\right)}{\partial x}G_{\mathbf{d}}x = \mathbf{0}$$

and the identity (6.3) holds.

(2)⇒(1). For an arbitrary $x \in \mathbb{R}^n\backslash\{\mathbf{0}\}$ let us consider the function $s \mapsto h(\mathbf{d}(s)x)$. Since $\mathbf{d}(s)x \neq \mathbf{0}$ for all $s \in \mathbb{R}$ then the considered function is locally Lipschitz continuous on $\mathbb{R}$ and, consequently, for almost all $x \neq \mathbf{0}$ we have

$$\frac{dh(\mathbf{d}(s)x)}{ds} \overset{a.e}{=} \frac{\partial h(z)}{\partial z}\bigg|_{z=\mathbf{d}(s)x}\frac{d\mathbf{d}(s)x}{ds} \overset{a.e}{=} \frac{\partial h(z)}{\partial z}G_{\mathbf{d}}z\bigg|_{z=\mathbf{d}(s)x}$$

due to $\frac{d}{ds}\mathbf{d}(s) = G_{\mathbf{d}}\mathbf{d}(s), \forall s \in \mathbb{R}$. Using the identity (6.3) we derive

$$\frac{dh(\mathbf{d}(s)x)}{ds} \overset{a.e}{=} \nu h(\mathbf{d}(s)x).$$

Hence, taking into account $h(\mathbf{d}(0)x) = h(x)$ we derive

$$h(\mathbf{d}(s)x) = e^{\nu s} h(x), \forall s \in \mathbb{R}, \quad \forall x \neq \mathbf{0}.$$

Since, by assumption, $h(\mathbf{0}) = 0$ then $h \in \mathcal{H}_{\mathbf{d}}(\mathbb{R}^n)$.

(2)⇔(3). Since we have $\frac{\partial h}{\partial x} G_{\mathbf{d}} x \overset{a.e.}{=} \sum_{i=1}^{n} \frac{\partial h}{\partial x_i} e_i^\top G_{\mathbf{d}} x = h(x)$ then, taking into account local Lipschitz continuity of h on $\mathbb{R}^n \backslash \{\mathbf{0}\}$, we conclude that the identity (6.3) is equivalent

$$\sum_{i=1}^{n} \int_{x_1^*}^{x_1} \cdots \int_{x_n^*}^{x_n} \frac{\partial h}{\partial y_i} e_i^\top G_{\mathbf{d}} y \, dy_1 \ldots dy_n = \nu \int_{x^*}^{x} h(y) \, dy$$

for all $x, x^* \in \mathbb{R}^n$ such that $R(x, x^*) \in \mathbb{R}^n \backslash \{\mathbf{0}\}$. Changing the integration order

$$\int_{x_1^*}^{x_1} \cdots \int_{x_n^*}^{x_n} \tfrac{\partial h}{\partial y_i} e_i^\top G_{\mathbf{d}} y dy = \int_{x_1^*}^{x_1} \cdots \int_{x_{i-1}^*}^{x_{i-1}} \int_{x_{i+1}^*}^{x_{i+1}} \cdots \int_{x_n^*}^{x_n} \int_{x_i^*}^{x_i} \tfrac{\partial h}{\partial y_i} e_i^\top G_{\mathbf{d}} y \, dy_i dy_n dy_{n-1} \ldots$$

and using the integration by parts

$$\int_{x_i^*}^{x_i} \tfrac{\partial h}{\partial y_i} e_i^\top G_{\mathbf{d}} y \, dy_i = h(y) e_i^\top G_{\mathbf{d}} y \Big|_{x_i^*}^{x_i} - \int_{x_i^*}^{x_i} h(y) \tfrac{\partial e_i^\top G_{\mathbf{d}} y}{\partial y_i} \, dy_i$$

we derive

$$\int_{x_1^*}^{x_1} \cdots \int_{x_n^*}^{x_n} \tfrac{\partial h}{\partial y_i} e_i^\top G_{\mathbf{d}} y dy =$$

$$\int_{x_1^*}^{x_1} \cdots \int_{x_{i-1}^*}^{x_{i-1}} \int_{x_{i+1}^*}^{x_{i+1}} \cdots \int_{x_n^*}^{x_n} \left(h(y) e_i^\top G_{\mathbf{d}} y \Big|_{x_i^*}^{x_i} - \int_{x_i^*}^{x_i} h(y) e_i^\top G_{\mathbf{d}} e_i \, dy_i \right) dy_n \ldots dy_{i+1} dy_{i-1} \ldots dy_1.$$

Hence, using the identity $\frac{1}{x_i - x_i^*} \int_{x_i^*}^{x_i} 1 dy_i = 1$ we derive formula (6.4). The proof is complete. □

The identity (6.4) proven by the above theorem is useful for the identification of the generator $G_{\mathbf{d}}$ of a dilation group $\mathbf{d}$ by means of Artificial Neural Networks studied in Chap. 16.

Example 6.2 The function $h(x) = x_1 + x_2^2$ is **d**-homogeneous of degree 2 with respect to the dilation $\mathbf{d}(s)$ generated by $G_{\mathbf{d}} = \begin{pmatrix} 2 & 0 \\ 0 & 1 \end{pmatrix}$, where $s \in \mathbb{R}$ and $x = (x_1, x_2)^\top$. Simple computations show

$$\frac{\partial h}{\partial x} G_{\mathbf{d}} x = (\, 1 \;\, 2x_2 \,) \begin{pmatrix} 2 & 0 \\ 0 & 1 \end{pmatrix} \begin{pmatrix} x_1 \\ x_2 \end{pmatrix} = 2x_1 + 2x_2^2 = 2h(x).$$

The identity (6.4) also holds. Indeed, if $x_1 \neq x_1^*$ and $x_2 \neq x_2^*$ then

$$\int_x^{x^*} (\mu + \operatorname{trace}(G_{\mathbf{d}}))h(y) - \sum_{i: x_i \neq x_i^*} \frac{h(y) e_i^\top G_{\mathbf{d}} y \big|_{x_i^*}^{x_i}}{x_i - x_i^*} \, dy$$

$$= \int_{x_1}^{x_1^*} \int_{x_2}^{x_2^*} 5(y_1 + y_2^2) - \tfrac{(x_1 + y_2^2)2x_1 - (x_1^* + y_2^2)2x_1^*}{x_1 - x_1^*} - \tfrac{(y_1 + x_2^2)x_2 - (y_1 + (x_2^*)^2)x_2^*}{x_2 - x_2^*} \, dy_2 dy_1$$

$$= \int_{x_1}^{x_1^*} \int_{x_2}^{x_2^*} 4y_1 + 3y_2^2 - \left(2(x_1 + x_1^*) + x_2^2 + x_2 x_2^* + (x_2^*)^2\right) \, dy_2 dy_1$$

$$= \tfrac{4(x_1^2 - (x_1^*)^2)(x_2 - x_2^*)}{2} + \tfrac{3(x_1 - x_1^*)(x_2^3 - (x_2^*)^3)}{3}$$
$$- \left(2(x_1 + x_1^*) + x_2^2 + x_2 x_2^* + (x_2^*)^2\right)(x_1 - x_1^*)(x_2 - x_2^*) = 0.$$

The key feature of any homogeneous function is a global expansion of all local properties. For example, a homogeneous function is regular on $\mathbb{R}^n \backslash \{\mathbf{0}\}$ if and only if it is regular on the unit sphere $S = \{x \in \mathbb{R}^n : \|x\| = 1\}$.

Proposition 6.3 (On homogeneous expansion of regularity in $\mathbb{R}^n$)
A function $h \in \mathcal{H}_{\mathbf{d}}(\mathbb{R}^n)$ is uniformly continuous (Lipschitz continuous or differentiable) on any compact from $\mathbb{R}^n \backslash \{\mathbf{0}\}$ if and only if h is uniformly continuous (resp., Lipschitz continuous or differentiable) on S.

Proof The local uniform continuity and the local Lipschitz continuity follow from the local Lipschitz continuity of the canonical homogeneous norm and the estimate[1]

$$\begin{aligned} |h(x_1) - h(x_2)| &= \left| \|x_1\|_{\mathbf{d}}^{\nu} h(v_1) - \|x_2\|_{\mathbf{d}}^{\nu} h(v_2) \right| \\ &\leq |h(v_1)| \left| \|x_1\|_{\mathbf{d}}^{\nu} - \|x_2\|_{\mathbf{d}}^{\nu} \right| + \|x_2\|_{\mathbf{d}}^{\nu} \, |h(v_1) - h(v_2)|, \end{aligned}$$

---

[1] A mapping $f : \mathbb{R}^n \mapsto \mathbb{R}^m$ is said to be

- *uniformly continuous* on $\Omega \subset \mathbb{R}^n$ if there exists the *modulus of continuity* $\omega \in \mathcal{K}$ such that $\|f(x_1) - f(x_2)\|_{\mathbb{R}^m} \leq \omega(\|x_1 - x_2\|_{\mathbb{R}^n})$ for all $x_1, x_2 \in \Omega$;
- *Lipschitz continuous* on $\Omega \subset \mathbb{R}^n$ if there exists a constant $L > 0$ (known as Lipschitz constant) such that $\|f(x_1) - f(x_2)\|_{\mathbb{R}^m} \leq L\|x_1 - x_2\|_{\mathbb{R}^n}$ for all $x_1, x_2 \in \Omega$.

where $v_i = \mathbf{d}(-\ln \|x_i\|_{\mathbf{d}})x_i \in S$, $i = 1, 2$ and $\nu = \deg_{\mathbf{d}}(h)$. The global expansion of the differentiability follows from Proposition 6.2.

We complete this section with the following simple corollary useful for the analysis of homogeneous systems.

Corollary 6.1 *Let $f : \mathbb{R}^n \mapsto \mathbb{R}_+$ be a globally bounded positive function $0 < C_1 \leq f(x) \leq C_2 < +\infty$ and $h_1, h_2 \in \mathcal{H}_{\mathbf{d}}(\mathbb{R}^n)$ be locally bounded **d**-homogeneous functions of the same degree ν. If $\inf_{x\in S} h_i(x) > 0$ then*

$$0 < \frac{C_1 \inf_{x\in S} h_1(x)}{\sup_{x\in S} h_2(x)} \leq \frac{f(x)h_1(x)}{h_2(x)} \leq \frac{C_2 \sup_{x\in S} h_1(x)}{\inf_{x\in S} |h_1(x)|} < +\infty, \quad \forall x \in \mathbb{R}^n\backslash\{\mathbf{0}\},$$

where S is the unit sphere in $\mathbb{R}^n$.

Proof Using the canonical homogeneous norm, the **d**-homogeneous functions h_1 and h_2 can be rewritten as follows

$$h_i(x) = \|x\|_{\mathbf{d}}^{\nu} h_i(\mathbf{d}(-\ln \|x\|_{\mathbf{d}})x), \quad x \in \mathbb{R}^n\backslash\{\mathbf{0}\}.$$

Taking into account $\mathbf{d}(-\ln \|x\|_{\mathbf{d}})x \in S$ we complete the proof taking into account $h_2(x) \geq \|x\|_{\mathbf{d}}^{\nu} \inf_{x\in S} h_2(x) > 0$.

6.1.2 Homogeneous Vector Fields

A vector-valued function $f : \mathcal{D} \subset \mathbb{R}^n \mapsto \mathbb{R}^n$ is a *vector field* if it defines a gradient of some flow in $\mathcal{D}$. In particular, *a function f in the right-hand side of the ODE $\dot{x} = f(x)$ is a vector field*. Solutions of the ODE form a flow.

A dilation symmetry of functions and vector fields (i.e., ODEs) is defined in different ways (see Chap. 1). In this book, we study vector fields which are **d**-homogeneous in the sense of [1, 2, 5, 6] and others.

Definition 6.2 Let **d** be a dilation in $\mathbb{R}^n$ and $\mathcal{D} \subset \mathbb{R}^n$ be a nonempty **d**-homogeneous cone in $\mathbb{R}^n$. A vector field $f : \mathcal{D} \subset \mathbb{R}^n \mapsto \mathbb{R}^n$ is said to be **d**-homogeneous of degree $\mu \in \mathbb{R}$ if

$$f(\mathbf{d}(s)x) = e^{\mu s}\mathbf{d}(s)f(x) \qquad \forall s \in \mathbb{R}, \quad \forall x \in \mathcal{D}. \tag{6.5}$$

We say that an ODE $\dot{x} = f(x)$ is **d**-homogeneous if f is a **d**-homogeneous vector field. The formula (6.5) simply means that *a vector field f commutes with the dilation* **d** *up to the scalar factor $e^{\mu s}$*.

Geometrically, the homogeneity of the vector field f means that any vector $f(x) \in \mathbb{R}^n$ with $x \in \mathcal{D}\backslash\{\mathbf{0}\}$ uniquely defines all other vectors of the field along the homogeneous curve $\Gamma_{\mathbf{d}}(x) \in \mathcal{D}$ as shown in Fig. 6.1.

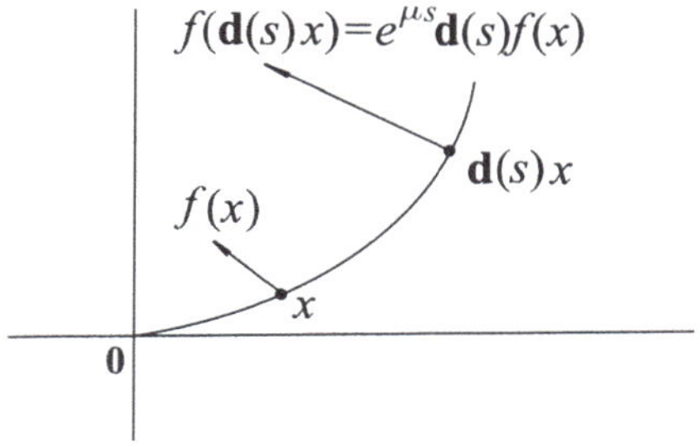

Fig. 6.1 Scaling of the homogeneous vector field along the homogeneous curve

> Let $\mathcal{F}_{\mathbf{d}}(\mathcal{D})$ denote a set of **d**-homogeneous vector fields $\mathcal{D} \subset \mathbb{R}^n \mapsto \mathbb{R}^n$ and let $\deg_{\mathbf{d}}(f) \in \mathbb{R}$ be a homogeneity degree of $f \in \mathcal{F}_{\mathbf{d}}(\mathcal{D})$.

Formally, to avoid collisions in Definitions 6.1 and 6.2 for $n = 1$, a vector field f may be defined as $f : \mathcal{D} \mapsto \mathbb{T}\mathcal{D}$, where $\mathbb{T}\mathcal{D}$ is the tangential space for $\mathcal{D}$. For simplicity, we write $f : \mathcal{D} \subset \mathbb{R}^n \mapsto \mathbb{R}^n$ having in mind that the homogeneity degrees for a scalar function $h \in \mathcal{H}_{\mathbf{d}}(\mathbb{R})$ and for a one-dimensional vector field $f \in \mathcal{F}_{\mathbf{d}}(\mathbb{R})$ (or, equivalently, for a differential equation $\dot{x} = f(x)$) are defined differently.

Example 6.3 Let us consider the three-tank system (Fig. 6.2). A dynamic model of the three-tank system admits the following representation [7]:

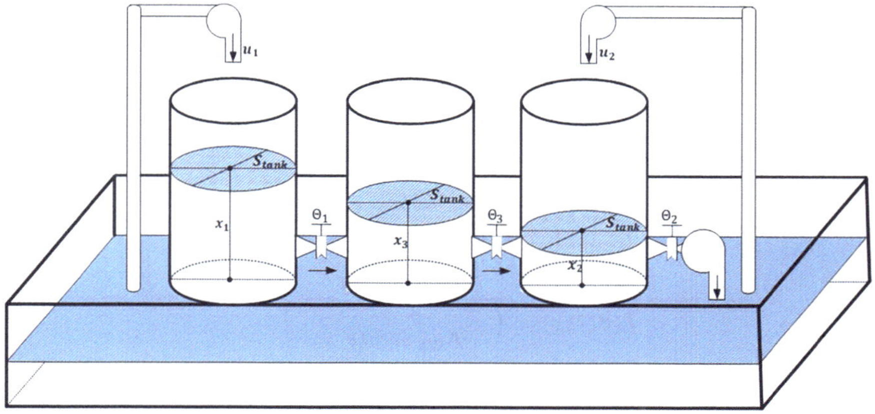

Fig. 6.2 The three-tank system

$$\begin{aligned}\dot{x}_1 &= S_{tank}^{-1}\left[-\theta_1\lceil x_1 - x_3\rfloor^{0.5} + u_1\right],\\ \dot{x}_2 &= S_{tank}^{-1}\left[\theta_3\lceil x_3 - x_2\rfloor^{0.5} - \theta_2\lceil x_2\rfloor^{0.5} + u_2\right],\\ \dot{x}_3 &= S_{tank}^{-1}\left[\theta_1\lceil x_1 - x_3\rfloor^{0.5} - \theta_3\lceil x_3 - x_2\rfloor^{0.5}\right],\end{aligned} \tag{6.6}$$

where x_1, x_2 and x_3 represent the liquid level of each tank respectively, S_{tank}, $[m]$ is the diameter of each tank, the input flows u_1 and u_2 are the control signals and the constant parameters θ_1, θ_2 and θ_3 are coefficients related with the outflow rate according to Torricelli's rule.

The model of the three-tank system is a standard homogeneous $\mathbf{d}(s) = e^s I_3$ of the degree -0.5 for $u_1 = 0, u_2 = 0$.

Example 6.4 Let us consider the differential equation which describes the position dynamics of a controlled rigid body motion on the surface under a dry friction force:

$$\begin{aligned}\dot{r} &= v\\ \dot{v} &= -K_{\text{dry}}\,\text{sign}(v) + u,\end{aligned} \tag{6.7}$$

where $r(t) \in \mathbb{R}^2$ is a position of the body in a planar inertial frame, $v(t) = (v_1(t), v_2(t))^\top \in \mathbb{R}^2$ is the velocity vector, $\text{sign}(v) = (\text{sign}(v_1(t)), \text{sign}(v_2(t)))^\top$ is the vector-valued sign function, $K_{\text{dry}} \in \mathbb{R}^{2\times 2}$ is a diagonal matrix of coefficients of dry frictions, and $u(t) \in \mathbb{R}^2$ is the vector of control inputs. It is assumed that the body has the unit mass. The open-loop ($u = \mathbf{0}$) system is $\mathbf{d}$-homogeneous with respect to the weighted dilation in $\mathbb{R}^4$:

$$\mathbf{d}(s) = \begin{pmatrix} e^{2s} I_2 & \mathbf{0} \\ \mathbf{0} & e^s I_2. \end{pmatrix}. \tag{6.8}$$

Indeed, denoting $x = (x_1^\top, x_2^\top)^\top$, $x_1 = r$, $x_2 = v$ and

$$f(x) = \begin{pmatrix} x_2 \\ -K_{\text{dry}}\,\text{sign}(x_2) \end{pmatrix}$$

we conclude that the open-loop dynamics of the system admits the representation $\dot{x} = f(x)$ and f is the $\mathbf{d}$-homogeneous vector field of degree -1:

$$f(\mathbf{d}(s)x) = \begin{pmatrix} e^s x_2 \\ -K_{\text{dry}}\,\text{sign}(e^s x_2) \end{pmatrix} =$$

$$e^{-s}\begin{pmatrix} e^{2s} x_2 \\ -e^s K_{\text{dry}}\,\text{sign}(x_2) \end{pmatrix} = e^{-s}\mathbf{d}(s) f(x), \quad \forall s \in \mathbb{R}, \quad \forall x \in \mathbb{R}^4.$$

The generator of dilation together with the homogeneity degree of a vector field $f \in \mathcal{F}_{\mathbf{d}}(\mathcal{D})$ characterize some properties of f.

Proposition 6.4 *Let* $\mathbf{d}$ *be linear continuous dilation in* $\mathbb{R}^n$ *and* $\mathcal{D} \subset \mathbb{R}^n$ *be a nonempty* $\mathbf{d}$*-homogeneous cone in* $\mathbb{R}^n$. *Let* $f \in \mathcal{F}_{\mathbf{d}}(\mathcal{D})$,

$$m := \inf_{x \in S \cap \mathcal{D}} \|f(x)\|, \quad M := \sup_{x \in S \cap \mathcal{D}} \|f(x)\|$$

and let parameters $\alpha \geq \beta > 0$ *be defined as in Proposition 5.3.*

(1) If the matrix $\deg_{\mathbf{d}}(f) I_n + G_{\mathbf{d}}$ *is anti-Hurwitz or* $\deg_{\mathbf{d}}(f) + \beta > 0$ *then*

- *for* $M < +\infty$, *the vector field* f *is bounded on any set* $B_{\mathbf{d}}(r) \cap \mathcal{D}$,

$$\sup_{x \in S_{\mathbf{d}}(r) \cap \mathcal{D}} \|f(x)\| \to 0 \quad as \quad r \to 0;$$

- *for* $m > 0$, *the vector field* f *is unbounded on any set* $\mathcal{D} \backslash B_{\mathbf{d}}(r)$,

$$\inf_{x \in S_{\mathbf{d}}(r) \cap \mathcal{D}} \|f(x)\| \to +\infty \quad as \quad r \to +\infty.$$

(2) If the matrix $\deg_{\mathbf{d}}(f) I_n + G_{\mathbf{d}}$ *is Hurwitz or* $\deg_{\mathbf{d}}(f) + \alpha < 0$ *then*

- *for* $M < +\infty$, *the vector field* f *is bounded on any set* $\mathcal{D} \backslash B_{\mathbf{d}}(r)$,

$$\sup_{x \in S_{\mathbf{d}}(r) \cap \mathcal{D}} \|f(x)\| \to 0 \quad as \quad r \to +\infty;$$

- *for* $m > 0$, *the vector field* f *is unbounded on any set* $B_{\mathbf{d}}(r) \cap \mathcal{D}$,

$$\inf_{x \in S_{\mathbf{d}}(r) \cap \mathcal{D}} \|f(x)\| \to +\infty \quad as \quad r \to 0.$$

(3) If $\deg_{\mathbf{d}}(f) I_n + G_{\mathbf{d}} = \mathbf{0}$ *or* $\deg_{\mathbf{d}}(f) + \beta = \deg_{\mathbf{d}}(f) + \alpha = 0$ *then* f *is uniformly bounded on* $\mathcal{D}$.

(4) A non-constant $\mathbf{d}$*-homogeneous vector field* f *with* $f(\mathbf{0}) = \mathbf{0}$ *is continuous at* $\mathbf{0}$ *if and only if the matrix* $\deg_{\mathbf{d}}(f) I_n + G_{\mathbf{d}}$ *is anti-Hurwitz.*

Proof By homogeneity of f we have

$$\begin{aligned} f(x) &= \|x\|_{\mathbf{d}}^{\deg_{\mathbf{d}}(f)} \mathbf{d}(\ln \|x\|_{\mathbf{d}}) f(\mathbf{d}(-\ln \|x\|_{\mathbf{d}})x) \\ &= e^{(\deg_{\mathbf{d}}(f) I_n + G_{\mathbf{d}}) \ln \|x\|_{\mathbf{d}}} f(\mathbf{d}(-\ln \|x\|_{\mathbf{d}})x), \end{aligned}$$

where $\mathbf{d}(-\ln \|x\|_{\mathbf{d}})x \in S$. In this case, the claims (1)–(3) follow from the representations

$$\sup_{x \in S_{\mathbf{d}}(r) \cap \mathcal{D}} \|f(x)\| = \sup_{x \in S \cap \mathcal{D}} \|e^{(\deg_{\mathbf{d}}(f) I_n + G_{\mathbf{d}}) \ln r} f(x)\|,$$

$$\inf_{x \in S_{\mathbf{d}}(r) \cap \mathcal{D}} \|f(x)\| = \inf_{x \in S \cap \mathcal{D}} \|e^{(\deg_{\mathbf{d}}(f) I_n + G_{\mathbf{d}}) \ln r} f(x)\|$$

and Proposition 5.3. Notice that $e^{(\deg_{\mathbf{d}}(f)I_n+G_{\mathbf{d}})s} \to \mathbf{0}$ as $s \to -\infty$ if and only if $\deg_{\mathbf{d}}(f)I_n + G_{\mathbf{d}}$ is anti-Hurwitz. So, if the vector field f is continuous at zero with $f(\mathbf{0}) = \mathbf{0}$ then either $f \equiv \mathbf{0}$ or the matrix $\deg_{\mathbf{d}}(f)I_n + G_{\mathbf{d}}$ is anti-Hurwitz. The proof is complete. □

Most properties of homogeneous vector fields (and the corresponding ODEs) depend mainly on the sign of the homogeneity degree (but not on its value) due to the following corollary.

Corollary 6.2 *Let a vector field f be $\mathbf{d}$-homogeneous of degree $\mu \in \mathbb{R}$. Let $\mathbf{d}_\gamma(s) := \mathbf{d}(\gamma s)$, $s \in \mathbb{R}$ with some $\gamma > 0$. Then f is $\mathbf{d}_\gamma$-homogeneous of the degree $\gamma\mu$.*

Proof Indeed, since $\mathbf{d}(s)$ satisfies the group property and the limit property then $\mathbf{d}_\gamma$ also does. Moreover, the vector field f is $\mathbf{d}$-homogeneous of degree μ, i.e., the identity (6.5) holds for all $s \in \mathbb{R}$ and all $x \in \mathbb{R}^n$. In this case, we have

$$f(\mathbf{d}_\gamma(s)x) = f(\mathbf{d}(\gamma s)x) = e^{\mu\gamma s}\mathbf{d}(\gamma s)f(x) = e^{\mu\gamma s}\mathbf{d}_\gamma(s)f(x), \forall x \in \mathcal{D}, \forall s \in \mathbb{R},$$

i.e., f is $\mathbf{d}_\gamma$-homogeneous of degree $\gamma\mu$. The proof is complete. □

The proven corollary holds for $\mathbf{d}$-homogeneous functions studied above as well as for $\mathbf{d}$-homogeneous set-valued mappings studied below.

Dilation symmetries are inherited by derivatives of homogeneous vector fields.

Proposition 6.5 *If $f \in \mathcal{F}_{\mathbf{d}}(\mathcal{D})$ is differentiable at $x \in \operatorname{int}\mathcal{D}$ then f is differentiable at any point of the $\mathbf{d}$-homogeneous curve $\Gamma_{\mathbf{d}}(x) \subset \operatorname{int}\mathcal{D}$,*

$$\left.\frac{\partial f(z)}{\partial z}\right|_{z=\mathbf{d}(s)x} \mathbf{d}(s) = e^{s\deg_{\mathbf{d}}(f)}\mathbf{d}(s)\frac{\partial f(x)}{\partial x}, \quad s \in \mathbb{R}. \tag{6.9}$$

Proof By definition, we

$$\lim_{\|y\|\to 0} \frac{\left\|f(x+y)-f(x)-\frac{\partial f(x)}{\partial x}y\right\|}{\|y\|} = 0, \quad y \in \mathbb{R}^n.$$

Similarly, f is differentiable in $z = \mathbf{d}(s)x$ if and only if there exists a matrix $\frac{\partial f(z)}{\partial z} \in \mathbb{R}^{n\times n}$ such that

$$\lim_{\|y_s\|\to 0} \frac{\left\|f(\mathbf{d}(s)x+y_s)-f(\mathbf{d}(s)x)-\frac{\partial f(z)}{\partial z}y_s\right\|}{\|y_s\|} = 0, \quad y_s \in \mathbb{R}^n.$$

Using homogeneity we derive

$$\frac{\left\| f(\mathbf{d}(s)x+y_s)-f(\mathbf{d}(s)x)-\frac{\partial f(z)}{\partial z}y_s\right\|}{\|y_s\|} \leq \frac{e^{\mu s}\left\|\mathbf{d}(s)\|\cdot\| f(x+y)-f(x)-e^{-\mu s}\mathbf{d}(-s)\frac{\partial f(z)}{\partial z}\mathbf{d}(s)y\right\|}{\|\mathbf{d}(s)y\|}$$
$$\leq \frac{e^{\mu s}\|\mathbf{d}(s)\|}{\lfloor \mathbf{d}(s)\rfloor}\frac{\left\| f(x+y)-f(x)-e^{-\mu s}\mathbf{d}(-s)\frac{\partial f(z)}{\partial z}\mathbf{d}(s)y\right\|}{\|y\|},$$

where $y = \mathbf{d}(-s)y_s$ and $\mu = \deg_{\mathbf{d}}(f)$. Since $\lfloor \mathbf{d}(s)\rfloor > 0$ and $\mathbf{d}(s)$ is an invertible matrix then $\|y\| \to 0$ is equivalent $\|y_s\| \to 0$. Hence, due to the uniqueness of the derivative, we derive the identity (6.9). The proof is complete. □

An analog of the Euler homogeneous function theorem can be proven for $\mathbf{d}$-homogeneous vector fields in $\mathbb{R}^n$.

Theorem 6.2 *Let a vector field $f : \mathbb{R}^n \mapsto \mathbb{R}^n$ be locally Lipschitz continuous on $\mathbb{R}^n\backslash\{\mathbf{0}\}$ and $f(\mathbf{0}) = \mathbf{0}$. Then the following three claims are equivalent*

(1) $f \in \mathcal{F}_{\mathbf{d}}(\mathbb{R}^n)$ *with* $\mu = \deg_{\mathbf{d}}(f) \in \mathbb{R}$;
(2) the identity

$$\frac{\partial f(x)}{\partial x}G_{\mathbf{d}}x = (\mu I_n + G_{\mathbf{d}})f(x) \tag{6.10}$$

holds almost everywhere on $\mathbb{R}^n$, where $x \in \mathbb{R}^n\backslash\{\mathbf{0}\}$ and $G_{\mathbf{d}} \in \mathbb{R}^{n\times n}$ is a generator of the linear continuous dilation $\mathbf{d}$;
(3) the identity

$$\int_{x^*}^{x} (\text{trace}(G_{\mathbf{d}}) + \mu)f_j(y) + e_j^\top G_{\mathbf{d}} f(y) - \sum_{i:x_i\neq x_i^*} \frac{f_j(y)e_i^\top G_{\mathbf{d}} y\big|_{x_i^*}^{x_i}}{x_i - x_i^*}\, dy = 0 \tag{6.11}$$

holds for all $j = 1, ..., n$, $x = (x_1, ..., x_i, ..., x_n)^\top$, $x^* = (x_1^*, ..., x_i, ..., x_n^*)^\top$ such that $R(x, x^*) \in \mathbb{R}^n\backslash\{\mathbf{0}\}$, where $\int_{x^*}^{x} = \int_{x_1^*}^{x_1} ... \int_{x_n^*}^{x_n}$, $dy = dy_1...dy_n$, $e_i = (0, ..., 1, ..., 0)^\top$, $i = 1, ..., n$,

$$h(y)e_i^\top G_{\mathbf{d}} y\Big|_{x_i^*}^{x_i} = h(y_{x_i})e_i^\top G_{\mathbf{d}}(y_{x_i}) - h(y_{x_i^*})e_i^\top G_{\mathbf{d}}(y_{x_i^*})$$

with $y_{x_i} = (y_1, ..., x_i, ..., y_n)^\top$ and $y_{x_i^*} = (y_1, ..., x_i^*, ..., y_n)^\top$.

Moreover, if f is differentiable on $\mathbb{R}^n$ (resp., $\mathbb{R}^n\backslash\{\mathbf{0}\}$), the identity (6.10) *is defined everywhere on $\mathbb{R}^n$ (resp., $\mathbb{R}^n\backslash\{\mathbf{0}\}$).*

Proof *(1)⇒(2)* Let $f \in \mathcal{F}_{\mathbf{d}}(\mathbb{R}^n)$ and $\mu = \deg_{\mathbf{d}}(f) \in \mathbb{R}$. By Rademacher theorem (see Appendix D), the locally Lipschitz continuous vector field f is differentiable almost everywhere on $\mathbb{R}^n\backslash\{\mathbf{0}\}$. Since, due to homogeneity, we have

$$f(x) = f\left(\mathbf{d}(\ln\|x\|_{\mathbf{d}})\mathbf{d}(-\ln\|x\|_{\mathbf{d}})x\right) = \|x\|^{\mu}\mathbf{d}(\ln\|x\|_{\mathbf{d}})f\left(\mathbf{d}(-\ln\|x\|_{\mathbf{d}})x\right)$$

then for almost all $x \in \mathbb{R}^n \backslash \{\mathbf{0}\}$, using the product and chain rule formulas for derivatives, we derive

$$\begin{aligned}\frac{\partial f(x)}{\partial x} =& \mu \|x\|_{\mathbf{d}}^{\mu-1}\mathbf{d}(\ln\|x\|_{\mathbf{d}}) f(\mathbf{d}(-\ln\|x\|_{\mathbf{d}})x)\frac{\partial\|x\|_{\mathbf{d}}}{\partial x}\\ &+ \|x\|_{\mathbf{d}}^{\mu-1} G_{\mathbf{d}}\mathbf{d}(\ln\|x\|_{\mathbf{d}}) f(\mathbf{d}(-\ln\|x\|_{\mathbf{d}})x)\frac{\partial\|x\|_{\mathbf{d}}}{\partial x}\\ &+ \|x\|_{\mathbf{d}}^{\mu}\mathbf{d}(\ln\|x\|_{\mathbf{d}})\frac{\partial f(z)}{\partial z}\frac{\partial(\mathbf{d}(-\ln\|x\|_{\mathbf{d}})x)}{\partial x},\end{aligned}$$

where $z = \mathbf{d}(-\ln\|x\|_{\mathbf{d}})x$. Since the mapping $s \mapsto \mathbf{d}(s)$ is differentiable and $\frac{d}{ds}\mathbf{d}(s) = G_{\mathbf{d}}\mathbf{d}(s) = \mathbf{d}(s)G_{\mathbf{d}}$ then

$$\frac{\partial(\mathbf{d}(-\ln\|x\|_{\mathbf{d}})x)}{\partial x} = \mathbf{d}(-\ln\|x\|_{\mathbf{d}}) - \frac{1}{\|x\|_{\mathbf{d}}} G_{\mathbf{d}}\mathbf{d}(-\ln\|x\|_{\mathbf{d}})x\frac{\partial\|x\|_{\mathbf{d}}}{\partial x}.$$

Since $\frac{\partial\|x\|_{\mathbf{d}}}{\partial x} G_{\mathbf{d}}x = \|x\|_{\mathbf{d}}$ (see formula (5.21)), then $\frac{\partial(\mathbf{d}(-\ln\|x\|_{\mathbf{d}})x)}{\partial x} G_{\mathbf{d}}x = 0$ and the identity (6.10) holds.

(2)⇒(1) For an arbitrary $x \in \mathbb{R}^n \backslash \{\mathbf{0}\}$, let us consider the vector-valued function $s \mapsto f(\mathbf{d}(s)x)$. Since $\mathbf{d}(s)x \neq \mathbf{0}$ for all $s \in \mathbb{R}$ then the considered function is locally Lipschitz continuous on $\mathbb{R}$ and, consequently,

$$\frac{df(\mathbf{d}(s)x)}{ds} \overset{a.e}{=} \left.\frac{\partial f(z)}{\partial z}\right|_{z=\mathbf{d}(s)x} \frac{d\mathbf{d}(s)x}{ds} \overset{a.e}{=} \left.\frac{\partial f(z)}{\partial z} G_{\mathbf{d}}z\right|_{z=\mathbf{d}(s)x}$$

due to $\frac{d}{ds}\mathbf{d}(s) = G_{\mathbf{d}}\mathbf{d}(s) = \mathbf{d}(s)G_{\mathbf{d}}$, $\forall s \in \mathbb{R}$. Using the identity (6.3) we derive

$$\frac{df(\mathbf{d}(s)x)}{ds} \overset{a.e}{=} \mu(I_n + G_{\mathbf{d}}) f(\mathbf{d}(s)x).$$

Hence, taking into account $f(\mathbf{d}(0)x) = f(x)$ we derive

$$f(\mathbf{d}(s)x) = e^{(\mu I_n + G_{\mathbf{d}})s} f(x) = e^{\mu s}\mathbf{d}(s) f(x), \forall s \in \mathbb{R}, \quad \forall x \neq \mathbf{0}.$$

Since by assumption $f(\mathbf{0}) = \mathbf{0}$ then $f \in \mathcal{F}_{\mathbf{d}}(\mathbb{R}^n)$.

(2)⇔(3) Since the identity (6.10) can be equivalently rewritten

$$\frac{\partial f_j(x)}{\partial x} G_{\mathbf{d}}x = \deg_{\mathbf{d}}(f) f_j(x) + e_j^\top G_{\mathbf{d}} f(x), \quad \forall j = 1, .., n, \quad \forall x \neq \mathbf{0} \tag{6.12}$$

then repeating the proof of Theorem 6.1 we derive the identity (6.11). The proof is complete. □

The representation (6.10) allows rather simple convex embedding of any smooth homogeneous vector field into a linear-like homogeneous differential inclusion to be developed in Sect. 6.3.2. Indeed, if f is non-constant, continuous at zero and $f(\mathbf{0}) = \mathbf{0}$ then the matrix $\deg_{\mathbf{d}}(f)I_n + G_{\mathbf{d}}$ is anti-Hurwitz (see, Proposition 6.4) and, consequently, invertible. In this case, the vector field f admits the representation

$$f(x) = M(x)x, \quad M(x) = (\deg_{\mathbf{d}}(f)I_n + G_{\mathbf{d}})^{-1}\frac{\partial f(x)}{\partial x}G_{\mathbf{d}} \in \mathbb{R}^{n\times n}, \quad x \in \mathbb{R}^n, \tag{6.13}$$

where $M : \mathbb{R}^n \mapsto \mathbb{R}^{n\times n}$ is a homogeneous (with respect to $\mathbf{d}$) matrix-valued function (see, Proposition 6.5). The above representation simplifies the analysis and design of nonlinear homogeneous control systems (see Chap. 7).

Elements of Homogeneous Arithmetics for Vector Fields

The following properties of homogeneous vector fields hold:

- if $h \in \mathcal{H}_{\mathbf{d}}(\mathcal{D}_1)$ and $f \in \mathcal{F}_{\mathbf{d}}(\mathcal{D}_2)$ with $\mathcal{D}_2 \subset \mathcal{D}_1$ then
  - $h \cdot f \in \mathcal{F}_{\mathbf{d}}(\mathcal{D}_2)$ and $\deg_{\mathbf{d}}(h \cdot f) = \deg_{\mathbf{d}}(h) + \deg_{\mathbf{d}}(f)$, where h may be a constant function;
  - $\frac{f}{h} \in \mathcal{F}_{\mathbf{d}}(\mathcal{D}_2 \backslash \ker(h))$ and $\deg_{\mathbf{d}}(h \cdot f) = \deg_{\mathbf{d}}(h) - \deg_{\mathbf{d}}(f)$, where $\ker(h) = \{x \in \mathbb{R}^n : h(x) = 0\}$;
- if $f_1 \in \mathcal{F}_{\mathbf{d}}(\mathcal{D}_1)$, $f_2 \in \mathcal{F}_{\mathbf{d}}(\mathcal{D}_2)$ and $\deg_{\mathbf{d}}(f_1) = \deg_{\mathbf{d}}(f_2)$ then $f_1 + f_2 \in \mathcal{F}_{\mathbf{d}}(\mathcal{D}_1 \cap \mathcal{D}_2)$ and $\deg_{\mathbf{d}}(f_1 + f_2) = \deg_{\mathbf{d}}(f_1) = \deg_{\mathbf{d}}(f_2)$.

The composition of two homogeneous vector fields is homogeneous under an additional restriction to one of the vector fields.

Theorem 6.3 (Composition of homogeneous vector fields in $\mathbb{R}^n$)
If $f_1, f_2 \in \mathcal{F}_{\mathbf{d}}(\mathbb{R}^n)$ then $f_1(f_2) \in \mathcal{F}_{\mathbf{d}}(\mathbb{R}^n)$ provided that $\deg_{\mathbf{d}}(f_2) = 0$ or f_1 is standard homogeneous too, i.e.,

$$\exists \mu \in \mathbb{R} \quad : \quad f_1(e^s x) = e^{\mu s} f_1(x), \quad \forall s \in \mathbb{R}, \forall x \in \mathbb{R}^n.$$

Proof Since $f_i(\mathbf{d}(s)x) = e^{s\deg_{\mathbf{d}}(f_i)}\mathbf{d}(s)f_i(x), \forall x \in \mathbb{R}^n, \forall s \in \mathbb{R}^n$ $i = 1, 2$ then

$$f_1(f_2(\mathbf{d}(s)x)) = f_1(e^{s\deg_{\mathbf{d}}(f_2)}\mathbf{d}(s)f_2(x)) = e^{\deg_{\mathbf{d}}(f_1)}\mathbf{d}(s)f_1(e^{s\deg_{\mathbf{d}}(f_2)}f_2(x)),$$

and the $\mathbf{d}$-homogeneity of the composition $f_1(f_2)$ is preserved if f_1 is standard homogeneous or $\deg_{\mathbf{d}}(f_2) = 0$. The proof is complete. □

Notice that if $f_1(x) = Ax$ with $x \in \mathbb{R}^n$, $A \in \mathbb{R}^{n\times n}$ and $f_1, f_2 \in \mathcal{F}_{\mathbf{d}}(\mathbb{R}^n)$ then the conditions of the above theorem are fulfilled, i.e.,

a composition $f_1(f_2)$ of a nonlinear $\mathbf{d}$-homogeneous vector field f_2 and a linear $\mathbf{d}$-homogeneous vector field f_1 is always $\mathbf{d}$-homogeneous.

An inverse (if it exists) of a homogeneous vector field is homogeneous too.

Theorem 6.4 (Homogeneous inverse mapping in $\mathbb{R}^n$)
If $f \in \mathcal{F}_{\mathbf{d}}(\mathbb{R}^n)$ *is invertible, continuous at* $\mathbf{0}$ *and* $f(\mathbf{0}) = \mathbf{0}$ *then the mapping* f^{-1} *is* $\mathbf{d}_\mu$*-homogeneous of the degree* $-\deg_{\mathbf{d}}(f)$ *with respect to the linear dilation* $\mathbf{d}_\mu(s) = e^{s\deg_{\mathbf{d}}(f)}\mathbf{d}(s)$, $s \in \mathbb{R}$.
Moreover, if, additionally, $f(e^{s\deg_{\mathbf{d}}(f)}x) = e^{s\deg_{\mathbf{d}}(f)}f(x)$ *for all* $x \in \mathbb{R}^n$ *and all* $s \in \mathbb{R}$ *then* f^{-1} *is* $\mathbf{d}$*-homogeneous of the degree* $-\deg_{\mathbf{d}}(f)$.

Proof Let $x \in \mathbb{R}^n\backslash\{\mathbf{0}\}$ and $\mu = \deg_{\mathbf{d}}(f)$. If there exit $y_1 \in \mathbb{R}^n$ and $y_2 \in \mathbb{R}^n$ such that

$$e^{\mu s}\mathbf{d}(s)f(x) = y_1 \quad \text{and} \quad f(\mathbf{d}(s)x) = y_2,$$

where $s \in \mathbb{R}^n$, then

$$x = f^{-1}(e^{-\mu s}\mathbf{d}(-s)y_1) \quad \text{and} \quad x = \mathbf{d}(-s)f^{-1}(y_2).$$

Due to homogeneity we have $e^{\mu s}\mathbf{d}(s)f(x) = f(\mathbf{d}(s)x)$ and $y_1 = y_2$. Hence,

$$f^{-1}(e^{-\mu s}\mathbf{d}(-s)y_1) = \mathbf{d}(-s)f^{-1}(y_1), \quad \forall s \in \mathbb{R}$$

or, equivalently,

$$f^{-1}(\mathbf{d}_\mu(s)y_1) = e^{-\mu s}\mathbf{d}_\mu(s)f^{-1}(y_1), \quad \forall s \in \mathbb{R}.$$

By Proposition 6.4, the matrix $G_{\mathbf{d}} + \deg_{\mathbf{d}}(f)I_n$ is anti-Hurwitz and $\mathbf{d}_\mu$ is a linear continuous dilation in $\mathbb{R}^n$. Moreover, if, additionally, $f(e^{\mu s}x) = e^{\mu s}f(x)$ for $x \in \mathbb{R}^n, s \in \mathbb{R}$ then $f^{-1}(e^{-\mu s}y) = e^{-\mu s}f^{-1}(y)$ for any $y \in \mathbb{R}^n$ and any $s \in \mathbb{R}$, so $f^{-1}(\mathbf{d}(s)x) = e^{\mu s}\mathbf{d}(s)f^{-1}(x)$ for any $x \in \mathbb{R}^n$ and any $s \in \mathbb{R}$. The proof is complete. □

On Regularity of Homogeneous Vector Fields in $\mathbb{R}^n$

A non-homogeneous mapping f may satisfy the Lipschitz condition on any $\mathbf{d}$-homogeneous sphere in $\mathbb{R}^n$ and be nowhere continuous. For example, if $\chi_Q : \mathbb{R} \mapsto \{0, 1\}$ is the Dirichlet function (the indicator function for the set of rational numbers) then the vector field $f : \mathbb{R}^n \mapsto \mathbb{R}^n$ defined as follows $f(\mathbf{0}) = \mathbf{0}$ and

$$f(x) = \left(1 + \chi_Q(|x|)\right)\frac{x}{|x|}, \quad |x| > 0,$$

obviously, satisfies the Lipschitz condition on any sphere of the radius $r > 0$ with the Lipschitz constant $L_r = (1 + \chi_Q(r))/r$. However, f is discontinuous at any point $x \neq \mathbf{0}$ along the ray $\{y \in \mathbb{R}^n : y = \lambda x, \lambda > 0\}$. Notice that f is non-homogeneous.

For homogeneous vector fields such a situation is impossible since f is smooth along any homogeneous curve generated by a linear continuous dilation [8].

> **Theorem 6.5** (On homogeneous expansion of regularity in $\mathbb{R}^{n}$)
> *A* **d***-homogeneous vector field* $f : \mathbb{R}^n \mapsto \mathbb{R}^n$ *is uniformly continuous (Lipschitz continuous or differentiable) on any compact from* $\mathbb{R}^n \backslash \{\mathbf{0}\}$ *if and only if it is uniformly continuous (resp., Lipschitz continuous or differentiable) on the unit sphere* S.

Proof The global differentiability follows from Proposition 6.5. Let us study the analysis of Lipschitz continuity. The uniform continuity of f can be treated analogously. Since *Necessity* is straightforward, we prove only *Sufficiency*. Let $\mu = \deg_{\mathbf{d}}(f)$. Let us denote

$$K_{\mathbf{d}}(r_1, r_2) = \{x \in \mathbb{R}^n : r_1 \leq \|x\|_{\mathbf{d}} \leq r_2\}$$

and show that f satisfy a Lipschitz condition on $K_{\mathbf{d}}(r_1, r_2)$ for any finite $0 < r_1 < r_2$.

First, let us prove that f satisfies Lipschitz condition on any $\mathbf{d}$-homogeneous sphere $S_{\mathbf{d}}(r)$, $r > 0$. Indeed, if $x_i \in S_{\mathbf{d}}(r)$, $i = 1, 2$ then

$$z_i := \mathbf{d}(-\ln(\|x_i\|_{\mathbf{d}}))x_i = \mathbf{d}(-\ln(r))x_i \in S$$

are uniquely defined the homogeneous projections. Hence, if $L > 0$ is a Lipschitz constant of f on S then

$$\begin{aligned}\|f(x_1) - f(x_2)\| =&\|r^{\mu}\mathbf{d}(-\ln r)(f(z_1) - f(z_2))\| \leq r^{\mu}\|\mathbf{d}(-\ln r)\|L\|z_1 - z_2\| \\ =&r^{\mu}\|\mathbf{d}(-\ln r)\|\, L\, \|\mathbf{d}(\ln r)(x_1 - x_2)\| \leq L_r\|x_1 - x_2\|,\end{aligned}$$

where $L_r := r^{\mu}\|\mathbf{d}(-\ln r)\|\, L\, \|\mathbf{d}(\ln r)\| < +\infty$ is the Lipschitz constant for f on $S_{\mathbf{d}}(r)$, $r > 0$ and $x_1, x_2 \in S_{\mathbf{d}}(r)$.

Let $x_i \in K_{\mathbf{d}}(r_1, r_2)$, $i = 1, 2$ and let $r_0 \geq \max\{r_2, r_1^{-1}\}$. In this case,

$$\frac{1}{r_0} \leq \|x_i\| \leq r_0 \quad \text{and} \quad x_i = \mathbf{d}\left(\ln \tfrac{\|x_i\|_{\mathbf{d}}}{r_0}\right) z_i$$

for some $z_i \in S_{\mathbf{d}}(r_0)$. Hence, we derive

$$\begin{aligned} f(x_1) - f(x_2) =& f(\mathbf{d}(\ln \|x_1\|_{\mathbf{d}}/r_0)z_1) - f(\mathbf{d}(\ln \|x_2\|_{\mathbf{d}}/r_0)z_2) \\ =& \tfrac{\|x_1\|_{\mathbf{d}}^{\mu}}{r_0^{\mu}}\mathbf{d}(\ln \|x_1\|_{\mathbf{d}}/r_0)f(z_1) - \tfrac{\|x_2\|_{\mathbf{d}}^{\mu}}{r_0^{\mu}}\mathbf{d}(\ln \|x_2\|_{\mathbf{d}}/r_0)f(z_2) \\ =& \tfrac{\|x_1\|_{\mathbf{d}}^{\mu}}{r_0^{\mu}}\mathbf{d}(\ln \|x_1\|_{\mathbf{d}}/r_0)(f(z_1) - f(z_2)) + \\ & \tfrac{\|x_1\|_{\mathbf{d}}^{\mu} - \|x_2\|_{\mathbf{d}}^{\mu}}{r_0^{\mu}}\mathbf{d}(\ln \|x_1\|_{\mathbf{d}}/r_0)f(z_2) + \\ & \tfrac{\|x_2\|^{\mu}}{r_0^{\mu}}(\mathbf{d}(\ln \|x_1\|_{\mathbf{d}}/r_0) - \mathbf{d}(\ln \|x_2\|_{\mathbf{d}}/r_0)f(z_2). \end{aligned}$$

(a) Let $L_{r_0} > 0$ be a Lipschitz constant for f on the $\mathbf{d}$-homogeneous sphere $S_{\mathbf{d}}(r_0)$. Since $\|x_i\|_{\mathbf{d}} < r_0$ (or, equivalently, $\|\mathbf{d}(-\ln r_0)x_i\| < 1$) and due to strict monotonicity of $\mathbf{d}$ we have $\|\mathbf{d}(\ln \|x_i\|_{\mathbf{d}}/r_0)\| < 1$ then

$$
\begin{aligned}
\frac{\|x_1\|_{\mathbf{d}}^{\mu}\|\mathbf{d}(\ln \|x_1\|_{\mathbf{d}}/r_0)(f(z_1)-f(z_2))\|}{r_0^{\mu}} \leq & L_{r_0}\|z_1 - z_2\| \\
= & L_{r_0}\|\mathbf{d}(\ln r_0/\|x_1\|_{\mathbf{d}})x_1 - \mathbf{d}(\ln r_0/\|x_2\|_{\mathbf{d}})x_2\| \\
\leq & L_{r_0}\|\mathbf{d}(\ln r_0/\|x_1\|_{\mathbf{d}})\| \; \|x_1 - x_2\| \\
& + L_{r_0}\left\|\left(\mathbf{d}\left(\ln \tfrac{r_0}{\|x_1\|_{\mathbf{d}}}\right) - \mathbf{d}\left(\ln \tfrac{r_0}{\|x_2\|_{\mathbf{d}}}\right)\right) x_2\right\| \\
\leq & L_{r_0}\;\|\mathbf{d}\,(2\ln r_0)\| \;\; \|x_1 - x_2\| \\
& + \|x_2\| L_{r_0}\left\|\mathbf{d}\left(\ln \tfrac{r_0}{\|x_1\|_{\mathbf{d}}}\right) - \mathbf{d}\left(\ln \tfrac{r_0}{\|x_2\|_{\mathbf{d}}}\right)\right\|.
\end{aligned}
$$

Since $\mathbf{d}$ is a continuous linear dilation then using Theorem 5.1 we derive $\mathbf{d}(s_1) - \mathbf{d}(s_2) = \int_{s_2}^{s_1} G_{\mathbf{d}}\mathbf{d}(s)ds$ and, consequently,

$$
\begin{aligned}
\left\|\mathbf{d}\left(\ln \tfrac{r_0}{\|x_1\|_{\mathbf{d}}}\right) - \mathbf{d}\left(\ln \tfrac{r_0}{\|x_2\|_{\mathbf{d}}}\right)\right\| \leq & \|G_{\mathbf{d}}\| \sup_{r\in(r_1,r_2)} \left\|\mathbf{d}\left(\ln \tfrac{r_0}{r_1}\right)\right\| \left|\ln \tfrac{r_0}{\|x_1\|_{\mathbf{d}}} - \ln \tfrac{r_0}{\|x_2\|_{\mathbf{d}}}\right| \\
\leq & \|G_{\mathbf{d}}\|\,\|\mathbf{d}(2\ln r_0)\| \;\; |\ln \|x_2\|_{\mathbf{d}} - \ln \|x_1\|_{\mathbf{d}}|\,.
\end{aligned}
$$

By Lemma 5.4 the canonical homogeneous norm $\|\cdot\|_{\mathbf{d}}$ satisfies the Lipschitz condition on any set $K_{\mathbf{d}}(r_1, r_2)$. Hence, from the smoothness of the logarithm function on $(0, +\infty)$, we conclude $\exists \tilde{L}_1 > 0$ such that

$$
\left\|\mathbf{d}\left(\ln \tfrac{r_0}{\|x_1\|_{\mathbf{d}}}\right) - \mathbf{d}\left(\ln \tfrac{r_0}{\|x_2\|_{\mathbf{d}}}\right)\right\| \leq \|G_{\mathbf{d}}\|\,\|\mathbf{d}(2\ln r_0)\| \;\tilde{L}_1\|x_1 - x_2\|
$$

for all $x_1, x_2 \in K(r_1, r_2)$. Therefore, we derive

$$
\left\|\tfrac{\|x_1\|_{\mathbf{d}}^{\mu}}{r_0^{\mu}}\mathbf{d}(\ln \|x_1\|_{\mathbf{d}}/r_0)(f(z_1) - f(z_2))\right\| \leq L_{r_0}^1\|x_1 - x_2\|,
$$

where $L_{r_0}^1 = L_{r_0}\|\mathbf{d}(2\ln r_0)\|(1 + \|G_{\mathbf{d}}\|\tilde{L}_1)$.

(b) Since the canonical homogeneous norm $\|\cdot\|_{\mathbf{d}}$ satisfies the Lipschitz condition on any set $K_{\mathbf{d}}(r_1, r_2)$ then there exists $\tilde{L}_2 > 0$ such that

$$
\begin{aligned}
\left\|\tfrac{\|x_1\|^{\mu}-\|x_2\|^{\mu}}{r_0^{\mu}}\mathbf{d}(\ln \|x_1\|_{\mathbf{d}}/r_0)f(z_2)\right\| \leq & \tilde{L}_2\|x_1 - x_2\|\,\|\mathbf{d}(\ln \|x_1\|_{\mathbf{d}}/r_0)\| \; \|f(z_2)\| \\
\leq & L_{r_0}^2\|x_1 - x_2\|,
\end{aligned}
$$

where $L_{r_0}^2 = \tilde{L}_2 \sup_{z\in S_{\mathbf{d}}(r_0)} \|f(z)\|$ with $r_0 > 0$.

(c) Finally, repeating considerations given in the part (a) of this proof, we conclude

$$\left\| \tfrac{\|x_2\|^{\mu}}{r_0^{\mu}}(\mathbf{d}(\ln \|x_1\|_{\mathbf{d}}/r_0) - \mathbf{d}(\ln \|x_2\|_{\mathbf{d}}/r_0) f(z_2)\right\| \leq L^3_{r_0} \|x_1 - x_2\|,$$

where $L^3_{r_0} = \|G\|_{\mathbf{d}} \tilde{L}_1 \sup_{z \in S_{\mathbf{d}}(r_0)} \|f(s)\|$.

Therefore, f satisfies a Lipschitz condition on $K_{\mathbf{d}}(r_1, r_2)$ for any $0 < r_1 < r_2$. We complete the proof taking into account that any compact from $\mathbb{R}^n \backslash \{\mathbf{0}\}$ can be embedded in $K_{\mathbf{d}}(r_1, r_2)$ under a proper selection of $r_1 > 0$ and $r_2 > r_1$.

The above theorem highlights the following simple but an important fact: *any homogeneous vector field in $\mathbb{R}^n$ is completely characterized by its values on a unit sphere, the homogeneity degree and the dilation group.* Such an equivalence of local and global properties of homogeneous mappings underlies most methods of homogeneous control systems design and analysis developed.

6.1.3 Generalized Homogeneity of Linear Vector Field

Any linear vector field is standard homogeneous of degree 0. Indeed, $A\mathbf{d}(s) = \mathbf{d}(s)A$ if $\mathbf{d}(s) = e^s I_n$. However, it may be generalized homogeneous of nonzero degree as well [8, 9]. Any linear $\mathbf{d}$-homogeneous vector field in the Euclidean space $\mathbb{R}^n$ can be characterized using the generator $G_{\mathbf{d}}$.

Lemma 6.1 (On homogeneous linear vector field in $\mathbb{R}^n$)
The following three claims are equivalent

(1) a vector field $x \mapsto Ax$ with $x \in \mathbb{R}^n$, $A \in \mathbb{R}^{n\times n}$ is $\mathbf{d}$-homogeneous of degree $\mu \neq 0$ with respect to a linear continuous dilation $\mathbf{d}$ in $\mathbb{R}^n$;

(2) there exist an anti-Hurwitz matrix $G_{\mathbf{d}} \in \mathbb{R}^{n\times n}$ and $\mu \neq 0$ such that

$$AG_{\mathbf{d}} = (\mu I_n + G_{\mathbf{d}})A; \tag{6.14}$$

(3) the matrix A is nipotent.

Proof *(1) ⇔ (2).* Since any linear continuous dilation is given by $\mathbf{d}(s) = e^{sG_{\mathbf{d}}}$, $s \in \mathbb{R}$ then, by Definition 6.2, to guarantee homogeneity of the vector field $x \mapsto Ax$, the identity

$$Ae^{sG_{\mathbf{d}}} = e^{\mu s} e^{sG_{\mathbf{d}}} A \tag{6.15}$$

must hold for all $s \in \mathbb{R}$. Notice $e^{\mu s} e^{sG_{\mathbf{d}}} = e^{s(\mu I + G_{\mathbf{d}})}$. Taking into account

$$e^{sG_{\mathbf{d}}} = \sum_{i=0}^{+\infty} \frac{s^n G^n_{\mathbf{d}}}{n!} \quad \text{and} \quad e^{s(\mu I + G_{\mathbf{d}})} = \sum_{i=0}^{+\infty} \frac{s^n(\mu I + G_{\mathbf{d}})^n}{n!}, \quad \forall s \in \mathbb{R}$$

we conclude that the identity (6.15) is fulfilled for all $s \in \mathbb{R}$ if the identity (6.14) holds. Inversely, if the identity (6.15) is fulfilled for all $s \in \mathbb{R}$ then, differentiating both sides on $s \in \mathbb{R}$, we derive

$$AG_{\mathbf{d}}e^{sG_{\mathbf{d}}} = (\mu I_n + G_{\mathbf{d}})e^{s(\mu I_n + G_{\mathbf{d}})}A.$$

Hence, taking $s = 0$ we obtain (6.14).

(2) ⇒ (3). Let $M \in \mathbb{C}^{n\times n}$ be the Jordan transformation for A, i.e.,

$$M^{-1}AM = J = \begin{pmatrix} J_1 & \mathbf{0} & \dots & \mathbf{0} \\ \mathbf{0} & J_2 & \dots & \mathbf{0} \\ \dots & \dots & \dots & \dots \\ \mathbf{0} & \mathbf{0} & \dots & J_k \end{pmatrix},$$

where J_i are the so-called Jordan blocks

$$J_i = \begin{pmatrix} \lambda_i & 1 & 0 & \dots & 0 & 0 \\ 0 & \lambda_i & 1 & \dots & 0 & 0 \\ \dots & \dots & \dots & \dots & \dots & \dots \\ 0 & 0 & 0 & \dots & \lambda_i & 1 \\ 0 & 0 & 0 & \dots & 0 & \lambda_i \end{pmatrix} = \lambda_i I_{n_i} + N_{n_i} \in \mathbb{C}^{n_i}, \quad i = 1, 2, \dots, k,$$

where $\lambda_i \in \mathbb{C}$ denotes an eigenvalue of A, $N_{n_i} = J_i - \lambda_i I_{n_i}$ is an upper diagonal nilpotent matrix and $n_1 + \cdots + n_k = n$. Denoting $\tilde{G}_{\mathbf{d}} = MG_{\mathbf{d}}M^{-1}$ we derive that the identity (6.14) is equivalent to

$$J\tilde{G}_{\mathbf{d}} = (\tilde{G}_{\mathbf{d}} + \mu I_n)J.$$

Let $G_{ij} \in \mathbb{C}^{n_i\times n_j}$ denote the blocks of the matrix $\tilde{G}_{\mathbf{d}} \in \mathbb{C}^{n\times n}$, i.e.,

$$\tilde{G}_{\mathbf{d}} = \begin{pmatrix} G_{11} & G_{12} & \dots & G_{1k} \\ G_{21} & G_{22} & \dots & G_{2k} \\ \dots & \dots & \dots & \dots \\ G_{k1} & G_{k2} & \dots & G_{kk} \end{pmatrix}.$$

In this case, $J\tilde{G}_{\mathbf{d}} = (\tilde{G}_{\mathbf{d}} + \mu I_n)J$ can be rewritten in a block-wise manner as follows

$$J_iG_{ij} = G_{ij}J_j, \quad i \neq j$$

and

$$J_iG_{ii} = G_{ii}J_i + \mu J_i,$$

where $i, j = 1, 2, \dots, k$. If $n_i = 1$ then the latter identity is equivalent to $\lambda_i\mu = 0 \Rightarrow \lambda_i = 0$ for $\mu \neq 0$. For $n_i > 1$ this identity is equivalent to

$$N_{n_i}G_{ii} = G_{ii}N_{n_i} + \mu\lambda_i I_n + \mu N_{n_i}.$$

If g_{kp} with $k, p \in \{1, ..., n_i\}$ are elements of the matrix G_{ii} then the obtained matrix equation implies that

$$g_{21}=\lambda_i\mu,\;\; g_{32}=g_{21}+\lambda_i\mu,\;\; \dots\; g_{n_i\, n_i-1}=g_{n_i-1\, n_i-2}+\lambda_i\mu,\;\; 0=g_{n_i\, n_i-1}+\lambda_i\mu.$$

Hence, $g_{n\,n-1} = (n_i-1)\lambda_i\mu$ and $g_{n\,n-1} = -\lambda_i\mu$. This means that, for a real $\mu \neq 0$, the matrix identity may hold only if $\lambda_i = 0$ (i.e., only if A is nilpotent).

(3) ⇒ (2). If A is nilpotent then all eigenvalues of A are zeros and the Jordan transformation is real: $M \in \mathbb{R}^{n\times n}$. Selecting $G_{ij} = \mathbf{0}$ for $i \neq j$ and $G_{ii} = \mathrm{diag}\{\mu, 2\mu, ..., n_i\mu\}$ for $\mu > 0$ we conclude that the identity $J\tilde{G}_{\mathbf{d}} = (\tilde{G}_{\mathbf{d}} + \mu I_n)J$ holds. For $\mu < 0$ we select $G_{ii} = \mathrm{diag}\{-n_i\mu, ..., -2\mu, -\mu\}$ to fulfill the same identity. In both cases, the matrix $G_{\mathbf{d}} = M^{-1}\tilde{G}_{\mathbf{d}}M$ is anti-Hurwitz. The proof is complete. □

Example 6.5 The integrator chain is the simplest example of a nontrivial linear system that is generalized homogeneous of both negative and positive degrees (dependently of a dilation). Indeed, if

$$\dot{x} = f(x) := Ax, \quad A = \begin{pmatrix} 0 & 1 & 0 & \dots & 0 \\ 0 & 0 & 1 & \dots & 0 \\ \dots & \dots & \dots & \dots & \dots \\ 0 & 0 & 0 & \dots & 1 \\ 0 & 0 & 0 & \dots & 0 \end{pmatrix}, \quad x \in \mathbb{R}^n$$

and

$$\mathbf{d}(s) = \begin{pmatrix} e^{s(1-(n-1)\mu)} & 0 & \dots & 0 & 0 \\ 0 & e^{s(1-(n-2)\mu)} & \dots & 0 & 0 \\ \dots & \dots & \dots & \dots & \dots \\ 0 & 0 & \dots & e^{s(1-\mu)} & 0 \\ 0 & 0 & \dots & 0 & e^{s} \end{pmatrix}, \quad \mu \in \mathbb{R}$$

then

$$A\mathbf{d}(s)x = e^{\mu s}\mathbf{d}(s)A,$$

i.e., $f(\mathbf{d}(s)x) = e^{\mu s}\mathbf{d}(s)f(x)$, $\forall s \in \mathbb{R}$, $\forall x \in \mathbb{R}^n$. Moreover, if $\mu < 1/(n-1)$ then $\mathbf{d}$ is a dilation in $\mathbb{R}^n$.

6.2 Homogeneous Approximations

6.2.1 Homogeneous Approximation on a Sphere

For control systems design and analysis, it is important to know if homogeneity-based methods can be applied to non-homogeneous systems. The so-called homogeneous approximations [1, 3, 10] and homogeneous extensions [11, 12] are appropriate for these purposes.

Definition 6.3 Let $\mathbf{d}$ be a linear continuous dilation in $\mathbb{R}^n$ and $\mathcal{D} \subset \mathbb{R}^n$ be a nonempty $\mathbf{d}$-homogeneous cone in $\mathbb{R}^n$.

- A function $h_r : \mathcal{D} \subset \mathbb{R}^n \mapsto \mathbb{R}^m$ defined as follows: $h(\mathbf{0}) = \mathbf{0}$ and

$$h_r(x) = \tfrac{\|x\|_{\mathbf{d}}^{\nu}}{r^{\nu}} h\left(\mathbf{d}\left(\ln \tfrac{r}{\|x\|_{\mathbf{d}}}\right) x\right), \qquad x \in \mathcal{D}\backslash\{\mathbf{0}\}$$

  is called a $\mathbf{d}$-homogeneous approximation with degree $\mu \in \mathbb{R}$ of a function $h : \mathcal{D} \subset \mathbb{R}^n \mapsto \mathbb{R}^n$ on a $\mathbf{d}$-homogeneous sphere $S_{\mathbf{d}}(r) \cap \mathcal{D}$ of the radius $r > 0$.
- A vector field $f_r : \mathcal{D} \subset \mathbb{R}^n \mapsto \mathbb{R}^n$ defined as follows: $f(\mathbf{0}) = \mathbf{0}$ and

$$f_r(x) = \tfrac{\|x\|_{\mathbf{d}}^{\mu}}{r^{\mu}} \mathbf{d}\left(\ln \tfrac{\|x\|_{\mathbf{d}}}{r}\right) f\left(\mathbf{d}\left(\ln \tfrac{r}{\|x\|_{\mathbf{d}}}\right) x\right), \qquad x \in \mathcal{D}\backslash\{\mathbf{0}\}$$

  is called a $\mathbf{d}$-homogeneous approximation with degree $\mu \in \mathbb{R}$ of a vector field $f : \mathcal{D} \subset \mathbb{R}^n \mapsto \mathbb{R}^n$ on a $\mathbf{d}$-homogeneous sphere $S_{\mathbf{d}}(r) \cap \mathcal{D}$ of the radius $r > 0$.

Since $\mathbf{d}\left(\ln \tfrac{r}{\|x\|_{\mathbf{d}}}\right) = I_n$ for $x \in S_{\mathbf{d}}(r) \cap \mathcal{D}$ then $f_r(x) = f(x)$ if $x \in S_{\mathbf{d}}(r)$, i.e., the vector field f_r coincide with f on the sphere $S_{\mathbf{d}}(r)$. The vector field f_r can be treated as a homogeneous expansion of f from the homogeneous sphere $S_{\mathbf{d}}(r)$ to the whole space $\mathbb{R}^n$. Indeed,

$$\begin{aligned} f_r(\mathbf{d}(s)x) =& \frac{\|\mathbf{d}(s)x\|_{\mathbf{d}}^{\mu}}{r^{\mu}} \mathbf{d}\left(\ln \frac{\|\mathbf{d}(s)x\|_{\mathbf{d}}}{r}\right) f\left(\mathbf{d}\left(\ln \frac{r}{\|\mathbf{d}(s)x\|_{\mathbf{d}}}\right) \mathbf{d}(s)x\right) \\ =& e^{\mu s} \tfrac{\|x\|_{\mathbf{d}}^{\mu}}{r^{\mu}} \mathbf{d}\left(s + \ln \tfrac{\|x\|_{\mathbf{d}}}{r}\right) f\left(\mathbf{d}\left(\ln \tfrac{r}{e^s \|x\|_{\mathbf{d}}}\right) \mathbf{d}(s)x\right) \\ =& e^{\mu s} \mathbf{d}(s) f_r(x), \end{aligned}$$

where $x \in \mathcal{D}\backslash\{\mathbf{0}\}$. Therefore, we have proven the following simple result.

Theorem 6.6 *If f_r (resp., h_r) is a $\mathbf{d}$-homogeneous approximation of a vector field f (resp., a function h) on a sphere $S_{\mathbf{d}}(r), r > 0$ then f_r (resp., h_r) is $\mathbf{d}$-homogeneous vector field (resp., function) of degree $\mu \in \mathbb{R}$.*

If a vector field $f \in \mathcal{F}_{\mathbf{d}}(\mathbb{R}^n)$ is Lipschitz continuous (resp., uniformly continuous or differentiable) on $S_{\mathbf{d}}(r)$ then, by Theorem 6.5, its homogeneous approximation f_r is locally Lipschitz continuous (resp., locally uniformly continuous or differentiable) on $\mathbb{R}^n\backslash\{\mathbf{0}\}$. In this case, due to continuity, global properties of the homogeneous approximation f_r (such as stability or instability) would imply similar local (close to $S_{\mathbf{d}}(r)$) properties of f. The same considerations are valid for homogeneous approximations of functions.

6.2.2 Homogeneity at Zero and at Infinity

Homogeneous approximation on a sphere exists for any nonlinear single-valued mapping and approximates it in a neighborhood of the sphere. A more important (for stability analysis and control design) question is the existence of homogeneous approximations at the origin or at infinity [1, 3]. For example, a scalar function $f : \mathbb{R} \mapsto \mathbb{R}$ given by $f(x) = x^3 + x$ with $x \in \mathbb{R}$ can be approximated by $f_0(x) = x$ close to zero and by $f_\infty = x^3$ "at infinity". Indeed, the linear term dominates the cubic one as $x \to 0$ while the inverse situation happens as $x \to \infty$. Similar considerations can be repeated for vector fields in $\mathbb{R}^n$ and even for certain operators in Banach spaces (see Volume II of this monograph).

Definition 6.4 Let $\mathbf{d}$ be a linear continuous dilation in $\mathbb{R}^n$ and $\mathcal{D}$ be a nonempty $\mathbf{d}$-homogeneous cone in $\mathbb{R}^n$.

- A $\mathbf{d}$-homogeneous function $h_L : \mathcal{D} \subset \mathbb{R}^n \mapsto \mathbb{R}^m$ of degree $\nu \in \mathbb{R}$ is said be $\mathbf{d}$-homogeneous approximation of a function $h : \mathcal{D} \subset \mathbb{R}^n \mapsto \mathbb{R}^m$ at L-limit (with $L = 0$ or $L = \infty$) if

$$\lim_{r \to L} \sup_{x \in S \cap \mathcal{D}} |r^{-\mu} h(\mathbf{d}(\ln r)x) - h_L(x)| = 0,$$

where, as before, S is the unit sphere in $\mathbb{R}^n$. A function h, which admits a $\mathbf{d}$-homogeneous approximation at L-limit, is called locally $\mathbf{d}$-homogeneous at L-limit (with $L = 0$ or $L = \infty$).
- A $\mathbf{d}$-homogeneous vector field $f_L : \mathcal{D} \subset \mathbb{R}^n \mapsto \mathbb{R}^n$ of degree $\mu \in \mathbb{R}$ is said be $\mathbf{d}$-homogeneous approximation of a vector field $f : \mathcal{D} \subset \mathbb{R}^n \mapsto \mathbb{R}^n$ at L-limit (with $L = 0$ or $L = \infty$) if

$$\lim_{r \to L} \sup_{x \in S \cap \mathcal{D}} |r^{-\mu} \mathbf{d}(-\ln r) f(\mathbf{d}(\ln r)x) - f_L(x)| = 0,$$

where, as before, S is the unit sphere in $\mathbb{R}^n$. A vector field f, which admits a $\mathbf{d}$-homogeneous approximation at L-limit, is called locally $\mathbf{d}$-homogeneous at L-limit (with $L = 0$ or $L = \infty$).

Homogeneous approximations at 0 and ∞-limits admit equivalent characterizations presented by the following corollary.

Corollary 6.3 • *A function $h_L \in \mathcal{H}_{\mathbf{d}}(\mathcal{D})$ is a $\mathbf{d}$-homogeneous approximation of a function $h : \mathcal{D} \subset \mathbb{R}^n \mapsto \mathbb{R}$ at L-limit (with $L = 0$ or $L = \infty$) if and only if*

$$\lim_{r \to L} \sup_{x \in S_{\mathbf{d}}(r) \cap \mathcal{D}} |r^{-\deg_{\mathbf{d}}(h_L)}(h(x) - h_L(x))| = 0.$$

- *A vector field $f_L \in \mathcal{F}_{\mathbf{d}}(\mathcal{D})$ is a* **d***-homogeneous approximation of a vector field $f : \mathcal{D} \subset \mathbb{R}^n \mapsto \mathbb{R}^n$ at L-limit (with $L = 0$ or $L = \infty$) if and only if*

$$\lim_{r \to L} \sup_{x \in S_{\mathbf{d}}(r) \cap \mathcal{D}} \|r^{-\deg_{\mathbf{d}}(f_L)} \mathbf{d}(-\ln r)(f(x) - f_L(x))\| = 0.$$

Indeed, since $\mathcal{D}$ is a **d**-homogeneous cone and $u \in S$ is equivalent to $\mathbf{d}(r)u \in S_{\mathbf{d}}(r)$ then

$$\begin{aligned}
&\sup_{u \in S \cap \mathcal{D}} \|r^{-\deg_{\mathbf{d}}(f_L)} \mathbf{d}(-\ln r) f(\mathbf{d}(\ln r)u) - f_L(u)\| \\
&= \sup_{u \in S \cap \mathcal{D}} \|r^{-\deg_{\mathbf{d}}(f_L)} \mathbf{d}(-\ln(r)) f(\mathbf{d}(\ln r)u) - r^{-\mu} \mathbf{d}(-\ln r) f_L(\mathbf{d}(\ln r)u)\| \\
&= \sup_{\tilde{u} \in S_{\mathbf{d}}(r) \cap \mathcal{D}} \|r^{-\deg_{\mathbf{d}}(f_L)} \mathbf{d}(-\ln r)(f(\tilde{u}) - f_L(\tilde{u}))\|.
\end{aligned}$$

The equivalent representation for the functions h and h_L can proven similarly.

A precision of a homogeneous approximation is characterized below.

Theorem 6.7 (Precision of homogeneous approximations)

- *If $h_L \in \mathcal{H}_{\mathbf{d}}(\mathcal{D})$ is a* **d***-homogeneous approximation of a function $h : \mathcal{D} \subset \mathbb{R}^n \mapsto \mathbb{R}$ at L-limit (with $L = 0$ or $L = \infty$) then $\exists \sigma_L \in \mathcal{K}$:*

$$|h_0(x) - h(x)| \le \sigma_0(\|x\|_{\mathbf{d}}) \|x\|_{\mathbf{d}}^{\deg_{\mathbf{d}}(h_0)}, \quad \forall x \in \mathcal{D} : 0 < \|x\| \le 1,$$

or, respectively,

$$|h_\infty(x) - h(x)| \le \sigma_\infty \left(\|x\|_{\mathbf{d}}^{-1}\right) \|x\|_{\mathbf{d}}^{\deg_{\mathbf{d}}(h_\infty)}, \quad \forall x \in \mathcal{D} : \|x\| \ge 1,$$

where the canonical homogeneous norm $\|\cdot\|_{\mathbf{d}}$ is induced by the weighted Euclidean norm $\|\cdot\|$ defined in Proposition 5.3.

- *If $f_L \in \mathcal{F}_{\mathbf{d}}(\mathcal{D})$ is a* **d***-homogeneous approximation of a vector field $f : \mathcal{D} \subset \mathbb{R}^n \mapsto \mathbb{R}^n$ at L-limit (with $L = 0$ or $L = \infty$) then $\exists \sigma_L \in \mathcal{K}$:*

$$\|f_0(x) - f(x)\| \le \sigma_0(\|x\|_{\mathbf{d}}) \|x\|_{\mathbf{d}}^{\deg_{\mathbf{d}}(f_0) + \beta}, \quad \forall x \in \mathcal{D} : 0 < \|x\| \le 1,$$

or, respectively,

$$\|f_\infty(x) - f(x)\| \le \sigma_\infty \left(\|x\|_{\mathbf{d}}^{-1}\right) \|x\|_{\mathbf{d}}^{\deg_{\mathbf{d}}(f_\infty) + \alpha}, \quad \forall x \in \mathcal{D} : \|x\| \ge 1,$$

where the positive parameters $\alpha \ge \beta > 0$ are also defined in Proposition 5.3.

Proof For simplicity, we present the proof for $\mathcal{D} = \mathbb{R}^n$. For $\mathcal{D} \subset \mathbb{R}^n$ the analysis is similar. Let us prove the estimate for the vector fields f and f_L. The analysis for the functions h and h_L is similar.

Case $L = 0$. On the one hand, if f_0 is a **d**-homogeneous approximation of f at 0-limit then the function $\tilde{\sigma}_0 : [0, 1] \mapsto \mathbb{R}$ given by $\tilde{\sigma}_0(0) = 0$ and

$$\tilde{\sigma}_0(\rho) = \sup_{0<r\leq\rho}\sup_{u\in S_{\mathbf{d}}(r)} \|r^{-\mu}\mathbf{d}(-\ln r)(f(u) - f_0(u))\|, \quad \rho \in (0, 1]$$

is continuous at 0, increasing on [0, 1] and bounded on [0, 1]. In this case, there exists $\sigma_0 \in \mathcal{K}$ such that $\tilde{\sigma}_0(\rho) \leq \sigma_0(\rho)$ for all $\rho \in [0, 1]$. For example, we can take $\sigma_0(\rho) = \frac{\rho}{\rho+1} + \hat{\sigma}_0(\rho)$, where the continuous increasing (but probably non-strictly increasing) function $\hat{\sigma} : \mathbb{R}_+ \mapsto \mathbb{R}_+$ is defined as follows

$$\hat{\sigma}_0(\rho) = \begin{cases} \tilde{\sigma}_0(1) + (\rho - 0.5) & \text{for } \rho \geq 0.5, \\ \tilde{\sigma}_0(2^{-i-1})(2 - 2^{i+2}\rho) + \tilde{\sigma}_0(2^{-i})(2^{i+2}\rho - 1) & \text{for } \rho \in [2^{-i-2}, 2^{-i-1}) \end{cases}$$

with $i = 0, 1, \ldots$.

On the other hand, we have

$$\sup_{u\in S_{\mathbf{d}}(r)} \|r^{-\mu}\mathbf{d}(-\ln(r))(f(u) - f_0(u))\| \geq r^{-\mu}\lfloor\mathbf{d}(-\ln r)\rfloor \sup_{u\in S_{\mathbf{d}}(r)} \|f(u) - f_0(u)\|$$

and using the identity $\lfloor\mathbf{d}(-\ln r)\rfloor\,\|\mathbf{d}(\ln r)\| = 1$ (see Corollary 5.1) and the strict monotonicity of $\mathbf{d}$ (see Proposition 5.3), for $r \leq 1$, we derive

$$r^{-(\mu+\beta)}\sup_{x\in S_{\mathbf{d}}(r)} \|f(x) - f_0(x)\| \leq \sup_{x\in S_{\mathbf{d}}(r)} \|r^{-\mu}\mathbf{d}(-\ln(r))(f(x) - f_0(x))\| \leq \sigma_0(r),$$

and

$$\|f(x) - f_0(x)\| \leq \sup_{x\in S_{\mathbf{d}}(\|x\|_{\mathbf{d}})} \|f(x) - f_0(x)\| \leq \sigma_0(\|x\|_{\mathbf{d}})\|x\|_{\mathbf{d}}^{\mu+\beta}.$$

In the **case** $L = \infty$, the function $\tilde{\sigma}_\infty$ can be defined as, $\tilde{\sigma}_\infty(0) = 0$ and

$$\tilde{\sigma}_\infty(\rho) := \sup_{r\geq\rho^{-1}}\sup_{u\in S_{\mathbf{d}}(r)} \|r^{-\mu}\mathbf{d}(-\ln r)(f(u) - f_\infty(u))\|, \quad \rho \in (0, 1)$$

and a function $\sigma_\infty \in \mathcal{K}$ can be constructed similarly to σ_0 such that $\sigma_{\infty(\rho)} \geq \tilde{\sigma}_\infty(\rho)$ for all $\rho \geq 0$. For $r > 1$ we have

$$\begin{aligned} \sup_{u\in S_{\mathbf{d}}(r)} \|r^{-\mu}\mathbf{d}(\ln(r))(f(u) - f_\infty(u))\| &\geq r^{-\mu}\lfloor\mathbf{d}(-\ln r)\rfloor \sup_{u\in S_{\mathbf{d}}(r)} \|f(u) - f_\infty(u)\| \\ &\geq \frac{r^{-\mu}}{\|\mathbf{d}(\ln r)\|}\sup_{u\in S_{\mathbf{d}}(r)} \|f(u) - f_\infty(u)\| \\ &\geq r^{-(\mu+\alpha)}\sup_{u\in S_{\mathbf{d}}(r)} \|f(u) - f_\infty(u)\|. \end{aligned}$$

Hence,

$$\frac{\sup_{x\in S_{\mathbf{d}}(r)} \|f(x) - f_\infty(x)\|}{r^{\mu+\alpha}} \leq \sup_{x\in S_{\mathbf{d}}(r)} \|r^{-\mu}\mathbf{d}(-\ln(r))(f(x) - f_\infty(x))\| \leq \sigma_\infty(r^{-1})$$

and

$$\|f(x) - f_\infty(x)\| \le \sup_{x \in S_{\mathbf{d}}(\|x\|_{\mathbf{d}})} \|f(x) - f_\infty(x)\| \le \frac{\sigma_\infty(\|x\|_{\mathbf{d}}^{-1})}{\|x\|_{\mathbf{d}}^{-\mu-\alpha}}.$$

The proof is complete. □

Example 6.6 Let the vector field $f : \mathbb{R}^n \mapsto \mathbb{R}^n$ is given by $f(x) = (\sin(|x|) + |x|)x$ with $x \in \mathbb{R}^n$ and the dilation is defined as follows $\mathbf{d}(s) = e^s I, s \in \mathbb{R}$, then the $\mathbf{d}$-homogeneous approximation of f at the 0-limit is defined as $f_0(x) = 2|x|x$ and the $\mathbf{d}$-homogeneous approximation of f at the ∞-limit is given by $f_\infty(x) = |x|x$.

In the case of a smooth nonlinear system:

$$\dot{x} = f(x), \quad f \in C^1(\mathbb{R}^n, \mathbb{R}^n), \quad f(\mathbf{0}) = \mathbf{0}$$

the so-called *first-order approximation* (or linear approximation or the first-order Taylor expansion)

$$f(x) = Ax + O(\|x\|^2), \quad A = \left.\frac{\partial f(x)}{\partial x}\right|_{x=0}$$

can be utilized for stability analysis and control design (see, e.g., [13] and references therein). If the system $\dot{x} = Ax$ is asymptotically stable then $\dot{x} = f(x)$ is, at least, locally asymptotically stable.

However, the first-order approximation is not informative in some cases. Indeed, the first-order approximation of the asymptotically stable system $\dot{x} = -x\sin^2(x)$ is the system $\dot{x} = 0$, which is not asymptotically stable. The reasonable question in this case: *is it possible to introduce a homogeneity-based analog of the first-order approximation?* The main advantage of the classical first-order approximations is the possibility to reduce the local stability analysis of a nonlinear system to the same problem for a *linear system* (i.e., just check that a matrix A is Hurwitz). The homogeneous approximations introduced by Definitions 6.3 and 6.4 do not allow this in the general case. That is why, we introduce a special sub-class of $\mathbf{d}$-homogeneous systems which admit a similar "linear" stability analysis locally.

Definition 6.5 A $\mathbf{d}$-homogeneous approximation $\tilde{f} \in \mathcal{F}_{\mathbf{d}}(\mathcal{D})$ is said to be the *first order* $\mathbf{d}$ *-homogeneous approximation* if there exists a matrix $A \in \mathbb{R}^{n\times n}$ such that

$$\tilde{f}(x) = \|x\|_{\mathbf{d}}^{\mu}\mathbf{d}(\ln\|x\|_{\mathbf{d}})A\mathbf{d}(-\ln\|x\|_{\mathbf{d}})x, \quad x \in \mathcal{D}\backslash\{\mathbf{0}\}. \tag{6.16}$$

Homogeneous vector fields admitting the representation (6.16) form an important class of homogeneous models. For example, a vector field of the form (6.16) appears as a Galerkin projection of an infinite-dimensional homogeneous system (see Volume II of this monograph for more details). If the matrix $A \in \mathbb{R}^{n\times n}$ is **d**-homogeneous in the sense of Lemma 6.1 then

$$\|x\|_{\mathbf{d}}^{\mu}\mathbf{d}(\ln \|x\|_{\mathbf{d}})A\mathbf{d}(-\ln \|x\|_{\mathbf{d}})x = Ax.$$

The first-order homogeneous approximation may be designed even for homogeneous vector fields. Indeed, if $f \in \mathcal{F}_{\mathbf{d}}(\mathcal{D}) \cap C^1(\mathbb{R}^n\backslash\{\mathbf{0}\})$ and the matrix $\deg_{\mathbf{d}}(f)I_n + G_{\mathbf{d}}$ is invertible then, due to formula (6.13), we have

$$f(x) = \|x\|_{\mathbf{d}}^{\deg_{\mathbf{d}}(f)}\mathbf{d}(\ln \|x\|_{\mathbf{d}})M(\mathbf{d}(-\ln \|x\|_{\mathbf{d}})x)\mathbf{d}(-\ln \|x\|_{\mathbf{d}})x.$$

To estimate f by $\tilde{f}$ of the form (6.16), the matrix $A \in \mathbb{R}^{n\times n}$ can be defined as follows

$$A \in \arg\min_{\tilde{A}\in\mathbb{R}^{n\times n}} \max_{z\in S} \|M(z) - \tilde{A}\|. \tag{6.17}$$

Since $\mathbf{d}(-\ln \|x\|_{\mathbf{d}})x \in S$ then such a selection of the matrix A minimizes the approximation error

$$\|f(x) - \tilde{f}(x)\| \le \|x\|_{\mathbf{d}}^{\deg_{\mathbf{d}}(f)} \cdot \|\mathbf{d}(\ln \|x\|_{\mathbf{d}})\| \cdot \|M(\mathbf{d}(-\ln \|x\|_{\mathbf{d}})x) - A\|. \tag{6.18}$$

Below we say that a function (resp., a vector field or a dynamical system defined by the corresponding vector field) is locally homogeneous if it has a homogeneous approximation at 0-limit or at ∞-limit.

6.3 Homogeneous Set-Valued Mappings

In this section, we deal with homogeneous set-valued mappings

$$H : \mathbb{R}^n \mapsto 2^{\mathbb{R}^m},$$

which, in particular (for $m = n$), may define set-valued vector fields for dynamical systems governed by differential inclusions in $\mathbb{R}^n$ (see Chap. 2). A homogeneity of set-valued mappings can be introduced in various ways [11, 12, 14, 15]. Similarly to the single-valued case, we may ask

$$H(\mathbf{d}(s)x) = e^{\nu s}H(x), \quad \forall x \in \mathbb{R}^n, \quad \forall s \in \mathbb{R}. \tag{6.19}$$

The following definition proposes possible relaxations of the above identity.

Definition 6.6 Let $\mathbf{d}$ be a dilation in $\mathbb{R}^n$ and $\mathcal{D} \subset \mathbb{R}^n$ be a $\mathbf{d}$-homogeneous cone with nonempty interior. A set-valued mapping $H : \mathcal{D} \subset \mathbb{R}^n \mapsto 2^{\mathbb{R}^m}$ is said to be

- $\mathbf{d}$-homogeneous of degree $\nu \in \mathbb{R}$ if

$$H(\mathbf{d}(s)x) \subset e^{\nu s} H(x), \quad \forall x \in \mathcal{D}, \quad \forall s \in \mathbb{R}; \tag{6.20}$$

- $\mathbf{d}$-*sub*homogeneous of degree $\nu \in \mathbb{R}$ if

$$H(\mathbf{d}(s)x) \subset e^{\nu s} H(x), \quad \forall x \in \mathcal{D}, \quad \forall s < 0; \tag{6.21}$$

- $\mathbf{d}$-*sup*homogeneous of a degree $\nu \in \mathbb{R}$ if

$$H(\mathbf{d}(s)x) \subset e^{\nu s} H(x), \quad \forall x \in \mathcal{D}, \quad \forall s > 0. \tag{6.22}$$

For a set-valued vector field $F : \mathbb{R}^n \mapsto 2^{\mathbb{R}^n}$ we may ask

$$F(\mathbf{d}(s)x) = e^{\mu s}\mathbf{d}(s)F(x) \quad \forall s \in \mathbb{R}, \quad \forall x \in \mathbb{R}^n. \tag{6.23}$$

or relax the above condition as follows.

Definition 6.7 Let $\mathbf{d}$ be a dilation in $\mathbb{R}^n$ and $\mathcal{D} \subset \mathbb{R}^n$ be a $\mathbf{d}$-homogeneous cone with nonempty interior. A set-valued vector field $F : \mathcal{D} \subset \mathbb{R}^n \mapsto 2^{\mathbb{R}^n}$, is said to be

- $\mathbf{d}$-homogeneous of a degree $\mu \in \mathbb{R}$ if

$$F(\mathbf{d}(s)x) \subset e^{\mu s}\mathbf{d}(s)F(x), \quad \forall s \in \mathbb{R}, \quad \forall x \in \mathcal{D}; \tag{6.24}$$

- $\mathbf{d}$-*sub*homogeneous of a degree $\mu \in \mathbb{R}$ if

$$F(\mathbf{d}(s)x) \subset e^{\mu s}\mathbf{d}(s)F(x), \quad \forall s < 0, \quad \forall x \in \mathcal{D}; \tag{6.25}$$

- $\mathbf{d}$-*sup*homogeneous of a degree $\mu \in \mathbb{R}$ if

$$F(\mathbf{d}(s)x) \subset e^{\mu s}\mathbf{d}(s)F(x), \quad \forall s > 0, \quad \forall x \in \mathcal{D}. \tag{6.26}$$

A differential inclusion $\dot{x} \in F(x)$ is said to be (sub/sup)homogeneous if the set-valued vector field F is (sub/sup)homogeneous.

If $\mathbf{d}$ is a linear dilation then for any $\mathbf{d}$-homogeneous set-valued vector field F we have

$$e^{-\mu s}\mathbf{d}(-s)F(\mathbf{d}(s)x) \subset F(x), \quad \forall s \in \mathbb{R}, \quad \forall x \in \mathcal{D}.$$

Since $\mathcal{D}$ is a $\mathbf{d}$-homogeneous cone, then denoting $y = \mathbf{d}(s)x \in \mathcal{D}$ and $\tilde{s} = -s$

$$e^{\mu\tilde{s}}\mathbf{d}(\tilde{s})F(y) \subset F(\mathbf{d}(\tilde{s})y), \quad \forall \tilde{s} \in \mathbb{R}, \quad \forall y \in \mathcal{D}.$$

So, the identity (6.23) holds for any $\mathbf{d}$-homogeneous (in the sense of the above definition) set-valued vector field F. *Sup*/*sub*homogeneity relaxes this identity asking the embeddings of sets $F(\mathbf{d}(s)x)$ and $e^{\mu s}\mathbf{d}(s)F(x)$ only for $s < 0$ (resp., $s > 0$). The same conclusions remain valid for the set-valued mapping H, namely, if H is $\mathbf{d}$-homogeneous (in the sense of Definition 6.6) then the identity (6.19) holds.

Set-valued homogeneous mappings have similar properties to homogeneous function, for example,

- a (geometric) sum of homogeneous set-valued mappings is a homogeneous set-valued mapping provided that their homogeneity degrees are identical;
- a composition of homogeneous set-valued mappings gives a $\mathbf{d}$-homogeneous mapping provided that they satisfy some additional conditions (like in Theorem 6.3);
- local regularity (e.g., upper semi-continuity) of a homogeneous set-valued mapping can be expanded globally.

A set-valued homogeneous vector field appears, for example, as a result of Filippov's regularization (see Chap. 2) applied to a discontinuous homogeneous vector field. The following theorem generalizes the result of [15] to the case of linear dilations.

Theorem 6.8 *If* $f \in \mathcal{F}_{\mathbf{d}}(\mathbb{R}^n)$ *(resp.,* $h \in \mathcal{H}(\mathbb{R}^n)$*) is a piece-wise continuous* $\mathbf{d}$*-homogeneous vector field (resp., function) then Filippov regularization of* f *given by (2.32) is a* $\mathbf{d}$*-homogeneous set-valued vector field (resp., mapping) of degree* $\deg_{\mathbf{d}}(f)$ *(resp.,* $\deg_{\mathbf{d}}(h)$*).*

Proof Let f be a piece-wise continuous vector field and F be a set-valued vector field obtained by means of Filippov regularization (see formula (2.32)). In this case, by Carathéodory lemma (see Appendix A), for any $x \in \mathbb{R}^n$ and any $y \in F(x)$ there exist α_i such that

$$y = \sum_{i=1}^{N} \alpha_i f_i(x), \quad \sum_{i=1}^{N} \alpha_i = 1, \quad \alpha_i \geq 0,$$

where $f_i(x)$ is a limit value of $f(x^*)$ for x^* tending to x under assumption that x^* belongs to the continuity domain of f (see Chap. 2). If x is a discontinuity point of f then the limit is not unique. In the case of piece-wise continuous function, the number of different limits at each point x is finite and bounded by some number $N \in \mathbb{N}$ (see formula (2.32)). The $\mathbf{d}$-homogeneity of f implies the $\mathbf{d}$-homogeneity of f_i. This means that F is $\mathbf{d}$-homogeneous with the same homogeneity degree. The case of $\mathbf{d}$-homogeneous function $h \in \mathcal{H}(\mathbb{R}^n)$ can be analyzed similarly. The proof is complete. □

Example 6.7 Let us consider the so-called super-twisting system (see Example 4.12)

$$\dot{x}_1 = -\alpha\sqrt{|x_1|}\,\mathrm{sign}(x_1) + x_2, \quad \dot{x}_2 = -\beta\,\mathrm{sign}(x), \quad t > 0$$

where $x_1(t), x_2(t) \in \mathbb{R}$, $\alpha > 0$, $\beta > 0$. This differential equation is **d**-homogeneous of degree $\mu = -1$ with respect to the dilation $\mathbf{d}(s) = \left(\begin{smallmatrix} e^{2s} & 0 \\ 0 & e^{s} \end{smallmatrix}\right)$, $s \in \mathbb{R}$.

Applying Filippov regularization (see Remark 2.5) we derive

$$\dot{x}_1 = -\alpha\sqrt{|x_1|}\,\mathrm{sign}(x_1) + x_2, \quad \dot{x}_2 \in -\beta\overline{\mathrm{sign}}(x),$$

where $\overline{\mathrm{sign}}$ is the set-valued sign function defined by formula (0.6). The obtained differential inclusion is **d**-homogeneous of the same degree.

6.3.1 Homogeneous Tangent and Normal Cones

Let Ω be a nonempty closed set in $\mathbb{R}^n$. The (Bouligand) tangent cone $\mathcal{C}_\Omega(x)$ to the set $\Omega \subset \mathbb{R}^n$ at a point $x \in \mathbb{R}^n$ is defined as follows (see, e.g., [16–18])

$$\mathcal{C}_\Omega(x) = \left\{ z \in \mathbb{R}^n : \liminf_{h \to 0^+} \tfrac{\inf_{y \in \Omega} \|x + hz - y\|}{h} = 0 \right\}. \tag{6.27}$$

By this definition, *an infinitesimal increment of $x \in \Omega$ along a vector from $\mathcal{C}_\Omega(x)$ leaves the resulting vector in Ω*. Roughly speaking, the cone $\mathcal{C}_\Omega(x)$ characterizes vectors initiated at $x \in \Omega$ and oriented "inside" the set Ω. Obviously, the set $\mathcal{C}_\Omega(x)$ is a positive cone: $z \in \mathcal{C}_\Omega(x) \Rightarrow \lambda z \in \mathcal{C}_\Omega(x), \forall \lambda > 0$. Moreover, $\mathcal{C}_\Omega(x) = \{\emptyset\}$ for $x \notin \Omega$ and $\mathcal{C}_\Omega(x) = \mathbb{R}^n$ for $x \in \mathrm{int}\,\Omega$. Notice also that C_Ω is independent of the selection of the norm in $\mathbb{R}^n$.

A normal cone (see, e.g., [19, 20] and references therein) defines vectors, which are orthogonal in some sense to the set Ω. An outer *normal cone* to the set $\Omega \subset \mathbb{V}$ at a point $x \in \mathbb{V}$ is defined as follows

$$\mathcal{N}_\Omega(x) = \{z \in \mathbb{V} : \langle z, y \tilde{-} x \rangle \le 0 \text{ for all } y \in \Omega\}, \tag{6.28}$$

where $\mathbb{V}$ is a Euclidean space with an operation $\tilde{+}$ for summation of vectors from $\mathbb{V}$, with an operation $\tilde{\cdot}$ for multiplication of by a scalar from $\mathbb{R}$ and with $\langle \cdot, \cdot \rangle$ being an inner product on $\mathbb{V}$. Recall that, by definition, $y \tilde{-} x = y + (-1) \tilde{\cdot} x$ for all $x, y \in \mathbb{V}$. The normal cone depends on a definition of the inner product. For $\mathbb{V} = \mathbb{R}^n$, the operations of summation and multiplication are defined in the component-wise (conventional) manner, but the inner product can be defined as $\langle x, y \rangle = x^\top P y$ with some positive definite matrix $P \succ 0$. The normal cone depends on the selection of P in this case.

By convention, $\mathcal{N}_\Omega(x) = \emptyset$ for $x \notin \Omega$. One can be shown that $\mathcal{N}_\Omega(x) = \{\mathbf{0}\}$ for $x \in \operatorname{int}\Omega$, so nontrivial normal cones exist only for $x \in \partial\Omega$. A normal cone to a convex set is convex too.

The tangent and normal cones characterize the positive invariance of sets for dynamical systems (see, e.g., [16–18, 21]).

The homogeneity of a mathematical object is expected to be inherited by all other mathematical objects induced by the former one. This means that the homogeneity of the set Ω should imply a dilation symmetry of tangent $\mathcal{C}_\Omega(\cdot)$ and normal $\mathcal{N}_\Omega(\cdot)$ cones considering them as set-valued mappings $\mathbb{R}^n \mapsto 2^{\mathbb{R}^n}$. These mappings are not compact-valued, but they still may be **d**-homogeneous in the sense of Definitions 6.6 and 6.7.

Lemma 6.2 *Let* $\mathbf{d}$ *be a linear continuous dilation in* $\mathbb{R}^n$. *If* $\Omega \subset \mathbb{R}^n$ *is a nonempty closed* **d***-homogeneous cone then* C_Ω *is the* **d***-homogeneous vector field of zero degree:*

$$\mathcal{C}_\Omega(\mathbf{d}(s)x) = \mathbf{d}(s)\mathcal{C}_\Omega(x), \quad \forall x \in \mathbb{R}^n, \quad \forall s \in \mathbb{R}. \tag{6.29}$$

Moreover, one holds $\mathbf{0} \in C_\Omega(\mathbf{0})$.

Proof The set Ω is **d**-homogeneous (i.e., $\mathbf{d}(s)\Omega = \Omega, \forall s \in \mathbb{R}$), by formula (6.27), we have

$$\mathcal{C}_\Omega(\mathbf{d}(s)x) = \left\{z \in \mathbb{R}^n : \liminf_{h\to 0^+} \frac{\inf_{y\in\Omega}\|\mathbf{d}(s)x + hz - y\|}{h} = 0\right\}$$

$$= \left\{\mathbf{d}(s)\tilde{z} \in \mathbb{R}^n : \liminf_{h\to 0^+} \frac{\inf_{\mathbf{d}(s)\tilde{y}\in\Omega}\|\mathbf{d}(s)(x + h\tilde{z} - \tilde{y})\|}{h} = 0\right\}$$

$$= \mathbf{d}(s)\left\{\tilde{z} \in \mathbb{R}^n : \liminf_{h\to 0^+} \frac{\inf_{\tilde{y}\in\Omega}\|\mathbf{d}(s)(x + h\tilde{z} - \tilde{y})\|}{h} = 0\right\}.$$

It is well known [16, 18] that the tangent cone $\mathcal{C}_\Omega(x)$ is independent of the selection of a norm. Notice that $\mathbf{d}(s)$ is invertible for any $s \in \mathbb{R}$, so $\|\cdot\|_s = \|\mathbf{d}(s)\cdot\|$ is a norm in $\mathbb{R}^n$. This means that

$$\mathcal{C}_\Omega(x) = \left\{\tilde{z} \in \mathbb{R}^n : \liminf_{h\to 0^+} \frac{\inf_{\tilde{y}\in\Omega}\|\mathbf{d}(s)(x + h\tilde{z} - \tilde{y})\|}{h} = 0\right\}.$$

Since the **d**-homogeneous cone Ω is nonempty and closed, then $\mathbf{0} \in \Omega$. In this case, for $x = z = \mathbf{0}$ we have $\inf_{y\in\Omega}\|x + hz - y\| = 0$ and

$$\liminf_{h\to 0^+} \frac{\inf_{y\in\Omega}\|x + hz - y\|}{h} = 0,$$

i.e., $\mathbf{0} \in C_\Omega(\mathbf{0})$. The proof is complete. □

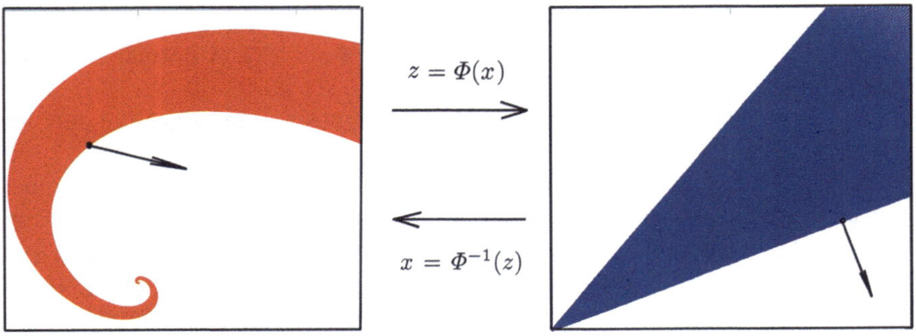

Fig. 6.3 A normal vector in $\mathbb{R}^n_{\mathbf{d}}$

The normal cone introduced above depends on the selection of the inner product in the space. In this monograph, we deal mainly with normal cones in the generalized homogeneous Euclidean space $\mathbb{V} = \mathbb{R}^n_{\mathbf{d}}$ (see Chap. 5):

$$\mathcal{N}_\Omega(x) = \{z \in \mathbb{R}^n_{\mathbf{d}} : \langle z, y \tilde{-} x \rangle_{\mathbf{d}} \leq 0 \text{ for all } y \in \Omega\}, \tag{6.30}$$

where a set Ω in $\mathbb{R}^n$ can be considered as a set Ω in $\mathbb{R}^n_{\mathbf{d}}$ too. Geometrically (see Fig. 6.3), this means that the normal cone is defined for the transformed set $\Phi(\Omega)$, where Φ is a homeomorphism on $\mathbb{R}^n$ introduced in Chap. 5. Such a normal cone is **d**-homogeneous by construction:

$$z \in \mathcal{N}_\Omega(x) \quad \Leftrightarrow \quad \lambda \tilde{\cdot} z \in \mathcal{N}_\Omega(x), \quad \forall \lambda > 0$$

or, equivalently,

$$\mathbf{d}(s)\mathcal{N}_\Omega(x) = \mathcal{N}_\Omega(x), \quad \forall x \in \Omega, \quad \forall s \in \mathbb{R}. \tag{6.31}$$

However, in the case of a **d**-homogeneous set Ω, a symmetry with respect to scaling of the vector x can also be established.

Lemma 6.3 *Let* **d** *be a linear continuous dilation in* $\mathbb{R}^n$. *Let* Ω *be a nonempty closed* **d**-*homogeneous cone in* $\mathbb{R}^n$. *Let* $\mathbb{R}^n_{\mathbf{d}}$ *be a generalized homogeneous Euclidean space with an inner product* $\langle \cdot, \cdot \rangle_{\mathbf{d}}$. *If* $\mathcal{N}_\Omega(x)$ *is a normal cone in* $\mathbb{R}^n_{\mathbf{d}}$ *to the set* Ω *at a point* $x \in \Omega$ *then* $\mathcal{N}_\Omega$ *is* **d**-*homogeneous mapping of the zero degree:*

$$\mathcal{N}_\Omega(\mathbf{d}(s)x) = \mathcal{N}_\Omega(x), \quad \forall x \in \mathbb{R}^n, \quad \forall s \in \mathbb{R}. \tag{6.32}$$

Proof The **d**-homogeneity of Ω in $\mathbb{R}^n$ implies its standard homogeneity in the space $\mathbb{R}^n_{\mathbf{d}}$, i.e., $\lambda \tilde{\cdot} \Omega = \Omega$ for all $\lambda > 0$. In this case, for any $\lambda > 0$ it holds

$$\begin{aligned}
\mathcal{N}_\Omega(\lambda\tilde{\cdot}x) &= \{z \in \mathbb{R}^n_{\mathbf{d}} : \langle z, y\tilde{-}\lambda\tilde{\cdot}x\rangle_{\mathbf{d}} \le 0 \text{ for all } y \in \Omega\} \\
&= \{z \in \mathbb{R}^n_{\mathbf{d}} : \langle z, \lambda\tilde{\cdot}y\tilde{-}\lambda\tilde{\cdot}\tilde{x}\rangle_{\mathbf{d}} \le 0 \text{ for all } \lambda\tilde{\cdot}\tilde{y} \in \Omega\} \\
&= \{z \in \mathbb{R}^n_{\mathbf{d}} : \langle z, \lambda\tilde{\cdot}(y\tilde{-}\tilde{x})\rangle_{\mathbf{d}} \le 0 \text{ for all } \tilde{y} \in \Omega\} \\
&= \{z \in \mathbb{R}^n_{\mathbf{d}} : \lambda\langle z, y\tilde{-}\tilde{x}\rangle_{\mathbf{d}} \le 0 \text{ for all } \tilde{y} \in \Omega\} = \mathcal{N}_\Omega(x),
\end{aligned}$$

where the operations $\tilde{\cdot}$, $\tilde{+}$ and the inner product $\langle\cdot,\cdot\rangle_{\mathbf{d}}$ in $\mathbb{R}^n_{\mathbf{d}}$ are introduced in Chap. 5. Taking into account the definition of $\tilde{\cdot}$ we complete the proof.

Homogeneous tangent and normal cones are utilized in Chaps. 7 and 12 for analysis of invariant sets of homogeneous dynamical systems as well as for safety-critical homogeneous control systems design.

6.3.2 Homogeneous Extensions

The homogeneous approximation f_L at L-limit introduced above is a useful tool for the homogeneity-based analysis of non-homogeneous systems $\dot{x} = f(x)$. However, in many cases, homogeneous approximations may not exist.

Example 6.8 Let $\mathbf{d}$ be a standard dilation in $\mathbb{R}^2$. The continuously differentiable vector field $f : \mathbb{R}^2 \mapsto \mathbb{R}^2$ given by

$$f(x_1, x_2) = \begin{pmatrix} (1 + 0.5\cos(1/x_2))x_2^3 \\ -x_1^3 - x_2^3 \end{pmatrix}$$

does not have a $\mathbf{d}$-homogeneous approximation at 0. Indeed, denoting $f = f_1 + f_2$ with

$$f_1(x) = \begin{pmatrix} x_2^3 \\ -x_1^3 - x_2^3 \end{pmatrix}, \qquad f_2(x) = \begin{pmatrix} 0.5\cos(1/x_2)x_2^3 \\ 0 \end{pmatrix},$$

we conclude that the vector field f_1 is standard homogeneous ($\mathbf{d}(s) = e^s I_2$) of degree $\nu = 2$, so a homogeneous approximation of f_2 at zero must have the same homogeneity degree (otherwise, the sum $f_1 + f_2$ will not be homogeneous). However, for $x_2 \neq 0$ we have

$$r^{-\nu}\mathbf{d}(-\ln r) f_2(\mathbf{d}(\ln r)x) = r^{-\nu+2}\begin{pmatrix} 0.5\cos\left(\frac{1}{rx_2}\right)(x_2)^3 \\ 0 \end{pmatrix} \to ? \text{ as } r \to 0.$$

The above limit exists if and only if $\nu < 2$, so the function $f = f_1 + f_2$ does not have a homogeneous approximation at 0.

The set-valued homogeneous extensions [11] introduced below can be considered as a "robust" alternative to the homogeneous approximation.

Definition 6.8 Let $\mathbf{d}$ be a dilation in $\mathbb{R}^n$ and $\mathcal{D} \subset \mathbb{R}^n$ be a $\mathbf{d}$-homogeneous cone with nonempty interior.

- A set-valued $\mathbf{d}$-homogeneous (resp., *sup*- or *sub*homogeneous) mapping $H : \mathcal{D} \subset \mathbb{R}^n \mapsto 2^{\mathbb{R}^m}$ is said to be a $\mathbf{d}$-homogeneous (resp., *sup*- or *sub*homogeneous) extension of a function $h : \mathcal{D} \subset \mathbb{R}^n \mapsto \mathbb{R}^m$ if $h(x) \subset H(x)$ for any $x \in \mathcal{D}$.
- A set-valued $\mathbf{d}$-homogeneous (resp., *sup*- or *sub*homogeneous) vector field $F : \mathcal{D} \subset \mathbb{R}^n \mapsto 2^{\mathbb{R}^n}$ is said to be a $\mathbf{d}$-homogeneous (resp., *sup*- or *sub*homogeneous) extension of a vector field $f : \mathcal{D} \subset \mathbb{R}^n \mapsto \mathbb{R}^n$ if $f(x) \subset F(x)$ for any $x \in \mathcal{D}$.

In this book, we deal mainly with homogeneous extensions defined as

$$F(x) = \bigcup_{s\in\mathbb{R}} \{e^{-\mu s}\mathbf{d}(-s) f(\mathbf{d}(s)x)\}. \tag{6.33}$$

By construction, the set-valued vector field F given by (6.33) *is* $\mathbf{d}$ *-homogeneous of the degree* μ. If, in the above identity, we restrict $s \le 0$ (or $s \ge 0$) then the set-valued vector field F becomes *sub*homogeneous (resp., *sup*homogeneous).

Example 6.9 The vector field f considered in Example 6.8 has the $\mathbf{d}$-homogeneous extension of the degree $\mu = 2$ given by

$$F(x_1, x_2) = \begin{pmatrix} [0.5, 1.5]x_2^3 \\ -x_1^3 - x_2^3 \end{pmatrix}$$

provided that $\mathbf{d}(s) = e^s I_2$, $s \in \mathbb{R}$.

Formally, the set-valued homogeneous extension (6.33) is defined for any $\mu \in \mathbb{R}$, but this vector field may be unbounded-valued if μ is poorly chosen. For example, if $\mathbf{d}(s) = e^s I$ and $f(x) = -(1 + 0.5\sin(|x|))x$ with $x \in \mathbb{R}$ then the homogeneous extension (6.33) for $\mu \neq 0$ gives $F(x) = -(0, +\infty) \cdot x$ for $\mu \neq 0$ and $F(x) = -\left(\frac{1}{2}, \frac{3}{2}\right) x$ for $\mu = 0$. An analysis of unbounded-valued homogeneous extensions goes out of the scope of this monograph. We refer the reader to [22, 23] for more details about differential inclusions with unbounded-valued right-hand sides.

Homogeneous extensions can also be utilized in order to model uncertainties of control systems [15, 24].

Example 6.10 Let the vector field f be given as follows

$$f(x) = \begin{pmatrix} x_2 \\ x_3 \\ \cdots \\ a(x) + b(x)u(x) \end{pmatrix}, \quad x = (x_1, \ldots, x_n)^\top \in \mathbb{R}^n,$$

where a and b are uncertain functions with known boundaries

$$|a(x)| < C, \quad 0 < b_{\min} < b(x) < b_{\max}, \quad \forall x \in \mathbb{R}^n.$$

Let $u : \mathbb{R}^n \mapsto \mathbb{R}$ be a **d**-homogeneous function (feedback control) of the degree 0 with respect to the dilation $\mathbf{d}(s) = \text{diag}\{e^{ns}, e^{(n-1)s}, ..., e^s\}$. Then the homogeneous extension of f is given by

$$F(x) = \begin{pmatrix} x_2 \\ x_3 \\ \dots \\ [-C, C] + [b_{\min}, b_{\max}]u(x) \end{pmatrix}.$$

The homogeneity degree of F is $\mu = -1$.

The use of homogeneous extension may simplify the stability analysis of essentially nonlinear homogeneous systems. A homogeneous extension of any continuous homogeneous vector field can be constructed using the so-called homogeneous convex embedding procedures [25].

Theorem 6.9 (Convex embedding of continuous homogeneous maps)
Let a linear continuous dilation **d** *in* $\mathbb{R}^n$ *be monotone with respect to a norm* $\|\cdot\|$ *and* $\|\cdot\|_{\mathbf{d}}$ *be the corresponding canonical homogeneous norm. Given* $\varepsilon \in (0, 1)$, *let* $\Delta = \{z_i\}_{i=1}^N$, $z_i \in S$ *be a grid on the unit sphere* $S = \{x \in \mathbb{R}^n : \|x\| = 1\}$ *such that*

$$\inf_i \|z - z_i\| \le \varepsilon, \quad \forall z \in S. \tag{6.34}$$

Let $\chi_i : \mathbb{R}^n \mapsto \{0, 1\}$ *be the indicator function defined as*

$$\chi_i(z) = \begin{cases} 1 & \text{if } \|z - z_i\| \le \varepsilon, \\ 0 & \text{if } \|z - z_i\| > \varepsilon. \end{cases}$$

- *Let* $h \in \mathcal{H}_{\mathbf{d}}(\mathbb{R}^n)$ *be uniformly continuous on the unit sphere* S *with a modulus of continuity* $\omega \in \mathcal{K}$. *Let the set-valued mapping* $H : \mathbb{R}^n \mapsto 2^{\mathbb{R}}$ *be defined as follows*

$$H(x) = \|x\|_{\mathbf{d}}^{\deg_{\mathbf{d}}(f)} H_0(\mathbf{d}(-\ln \|x\|_{\mathbf{d}})x), \quad x \in \mathbb{R}^n, \tag{6.35}$$

where $H_0 : S \mapsto 2^{\mathbb{R}^n}$ *is given by*

$$H_0(z) = \frac{\chi_1(z)\big(h(z_1)+\omega(\varepsilon)[-1,1]\big)+\cdots+\chi_N(z)\big(h(z_n)+\omega(\varepsilon)[-1,1]\big)}{\chi_1(z)+\cdots+\chi_N(z)}, \quad z \in S.$$

Then the set-valued mapping H *is a convex- and compact-valued* **d***-homogeneous extension of the function* h *with degree* $\deg_{\mathbf{d}}(h)$*:*

$$\sup_{y \in H(x)} \|h(x) - y\| \le 2\omega(\varepsilon)\|x\|_{\mathbf{d}}^{\deg_{\mathbf{d}}(h)}, \quad \forall x \neq \mathbf{0}. \tag{6.36}$$

- *Let $f \in \mathcal{F}_{\mathbf{d}}(\mathbb{R}^n)$ be uniformly continuous on the unit sphere S with a modulus of continuity $\omega \in \mathcal{K}$. Let the set-valued vector field $F : \mathbb{R}^n \mapsto 2^{\mathbb{R}^n}$ be defined as follows*

$$F(x) = \|x\|_{\mathbf{d}}^{\deg_{\mathbf{d}}(f)} \mathbf{d}(\ln \|x\|_{\mathbf{d}}) F_0(\mathbf{d}(-\ln \|x\|_{\mathbf{d}})x), \quad x \in \mathbb{R}^n, \tag{6.37}$$

where $F_0 : S \mapsto 2^{\mathbb{R}^n}$ is given by

$$F_0(z) = \frac{\chi_1(z)\big(f(z_1)\dot{+}\overline{B}(\omega(\varepsilon))\big)\dot{+}\cdots\dot{+}\chi_N(z)\big(f(z_n)\dot{+}\overline{B}(\omega(\varepsilon))\big)}{\chi_1(z)+\cdots+\chi_N(z)}, \quad z \in S.$$

Then the set-valued vector field F is a convex- and compact-valued $\mathbf{d}$-homogeneous extension of the vector field f with degree $\deg_{\mathbf{d}}(f)$:

$$\sup_{y \in F(x)} \|f(x) - y\| \leq 2\omega(\varepsilon)\|x\|_{\mathbf{d}}^{\deg_{\mathbf{d}}(f)} \|\mathbf{d}(\ln \|x\|_{\mathbf{d}})\|, \quad \forall x \neq \mathbf{0}. \tag{6.38}$$

Proof Let us prove that F is the homogeneous extension of the vector field $f \in \mathcal{F}_{\mathbf{d}}(\mathbb{R}^n)$. The homogeneous extension H of the function $h \in \mathcal{H}_{\mathbf{d}}(\mathbb{R}^n)$ can be analyzed similarly.

Since the geometric sum of convex (resp., compact) sets is convex (resp., compact) then F_0 is convex- (resp., compact-) valued. Linear transformation of a convex (resp., compact) set gives a compact set. Since the dilation $\mathbf{d}$ is linear then F is compact-(resp., convex-) valued. Using uniform continuity we conclude that if $z \in z_i \dot{+} \overline{B}(\varepsilon)$ (or, equivalently, $\chi_i(z) = 1$) then using uniform continuity we conclude $f(z) \in f(z_i)\dot{+}\overline{B}(\omega(\varepsilon))$. Hence, by construction, we have $f(z) \in F(z)$ for any $z \in S$. Hence, due to $\mathbf{d}$-homogeneity $f(x) \in F(x)$ for any $x \in \mathbb{R}^n \backslash \{\mathbf{0}\}$, i.e., F is, indeed, a $\mathbf{d}$-homogeneous extension of f. Since for $z \in S$ we have

$$\begin{aligned}
\sup_{y \in F(z)} \|f(z) - y\| &= \sup_{y \in F_0(z)} \left\| f(z)\frac{\chi_1(z) + \cdots + \chi_N(z)}{\chi_1(z) + \cdots + \chi_N(z)} - y \right\| \\
&= \sup_{v_i : \|v_i\| \leq 1, i=1,\ldots,N} \left\| \frac{\sum_{i=1}^{n} \chi_i(z)(f(z_i) - f(z) + \omega(\varepsilon)v_i)}{\sum_{i=1}^{n} \chi_i(z)} \right\| \\
&\leq \frac{\sum_{i=1}^{n} \chi_i(z) \sup_{v_i : \|v_i\| \leq 1} \|f(z_i) - f(z) + \omega(\varepsilon)v_i\|}{\sum_{i=1}^{n} \chi_i(z)} \\
&\leq \frac{\sum_{i=1}^{n} \chi_i(z)(\|f(z_i) - f(z)\| + \omega(\varepsilon))}{\sum_{i=1}^{n} \chi_i(z)} \leq 2\omega(\varepsilon)
\end{aligned}$$

then the estimate (6.38) holds for $x \in \mathbb{R}^n\backslash\{\mathbf{0}\}$ due to the representation $f(x) = \|x\|_{\mathbf{d}}^{\deg_{\mathbf{d}}(f)}\mathbf{d}(\ln\|x\|_{\mathbf{d}} f(\mathbf{d}(-\ln\|x\|_{\mathbf{d}})x)$ and the formula (6.37). The proof is complete. □

Notice that the Hausdorff distance (see Chap. 2) between $f(x)$ and $F(x)$ tends to zero as ε tends to zero (see, the estimate (6.38)). This means that any continuous **d**-homogeneous vector field admits an arbitrary-precise approximation on any compact from $\mathbb{R}^n\backslash\{\mathbf{0}\}$ using only a sufficiently dense grid on the unit sphere. Moreover, formula (6.38) is the global estimate of the approximation error on $\mathbb{R}^n$. In Chap. 16, this feature of **d**-homogeneous system is utilized for the construction of homogeneous artificial neural networks.

In the previous section, we have introduced a special class of systems which admit the so-called first order **d**-homogeneous approximations. The first order **d**-homogeneous extensions can be introduced similarly.

Definition 6.9 • A **d**-homogeneous extension $H : \mathcal{D} \subset \mathbb{R}^n \mapsto 2^{\mathbb{R}^m}$ of a function $h : \mathcal{D} \subset \mathbb{R}^n \mapsto \mathbb{R}^m$ is said to be the *first order* **d**-*homogeneous extension* if it admits the representation

$$H(x)=\left\{\|x\|_{\mathbf{d}}^{\deg_{\mathbf{d}}(h)}\tilde{A}\mathbf{d}(-\ln\|x\|_{\mathbf{d}})x \, : \, \tilde{A}\in\Sigma\right\}, \quad x\in\mathcal{D}\backslash\{\mathbf{0}\}, \tag{6.39}$$

where $\Sigma \subset \mathbb{R}^{m\times n}$ is a bounded set.

• A **d**-homogeneous extension $F : \mathcal{D} \subset \mathbb{R}^n \mapsto 2^{\mathbb{R}^n}$ of a vector field $f : \mathcal{D} \subset \mathbb{R}^n \mapsto \mathbb{R}^n$ is said to be the *first order* **d**-*homogeneous extension* if it admits the representation

$$F(x)=\left\{\|x\|_{\mathbf{d}}^{\deg_{\mathbf{d}}(f)}\mathbf{d}(\ln\|x\|_{\mathbf{d}})\tilde{A}\mathbf{d}(-\ln\|x\|_{\mathbf{d}})x \, : \, \tilde{A}\in\Sigma\right\}, \quad x\in\mathcal{D}\backslash\{\mathbf{0}\} \tag{6.40}$$

where $\Sigma \subset \mathbb{R}^{n\times n}$ is a bounded set.

Such a homogeneous extension can be considered a generalization of the so-called linear differential inclusion [26, Chap. 4]. In many practical cases, the set Σ is assumed to be convex and represented as follows

$$\Sigma=\left\{\tilde{A}=\sum_{k=1}^{N}\alpha_k A_k \; : \; \alpha_k\geq 0, \sum_{k=1}^{N}\alpha_k=1\right\}, \quad A_k\in\mathbb{R}^{m\times n}, \quad N\in\mathbb{N}, \tag{6.41}$$

where $m = n$ in the case of a homogeneous extension of a vector field.

Theorem 6.10 (Convex embedding of smooth homogeneous maps)

Let a linear continuous dilation **d** *in* $\mathbb{R}^n$ *be monotone with respect to a norm* $\|\cdot\|$ *and* $\|\cdot\|_{\mathbf{d}}$ *be the corresponding canonical homogeneous norm.*

• *Let* $h \in \mathcal{H}_{\mathbf{d}}(\mathbb{R}^n)$ *be continuously differentiable on the unit sphere* $S = \{x \in \mathbb{R}^n : \|x\| = 1\}$. *Let* $\deg_{\mathbf{d}}(h) \neq 0$ *and the numbers* $a_j \in \mathbb{R}$ *be defined as follows*

$$\underline{a}_j \leq \inf_{z\in S} \frac{\partial h(z)}{\partial z_j} \leq \frac{\partial h(z)}{\partial z_j} \leq \sup_{z\in S} \frac{\partial h(z)}{\partial z_j} \leq \overline{a}_j,$$

where $j \in \{1, 2, ..., n\}$. *Then the function* h *admits the first order* **d**-*homogeneous extension* H *given b* (6.39) *with*

$$\Sigma = \left\{ \tfrac{1}{\deg_{\mathbf{d}}(h)} \left(a_1 \; a_2 \; ... \; a_n \right) G_{\mathbf{d}} : a_j \in [\underline{a}_j, \overline{a}_j] \right\}. \tag{6.42}$$

Moreover, the set Σ *admits the representation* (6.41) *for* $N = 2^n$ *with*

$$A_k = \tfrac{1}{\deg_{\mathbf{d}}(h)} \left(a_1^k \; a_2^k \; ...a_n^k \right) G_{\mathbf{d}} \in \mathbb{R}^{1\times n} \tag{6.43}$$

with the elements $a_j^k \in \{\underline{a}_j, \overline{a}_{ij}\}$, $j = 1, 2, ..., n$.

- *Let* $f \in \mathcal{F}_{\mathbf{d}}(\mathbb{R}^n)$ *be continuously differentiable on the unit sphere* S, *the matrix* $\deg_{\mathbf{d}}(f)I_n + G_{\mathbf{d}}$ *be invertible and the numbers* $a_{ij} \in \mathbb{R}$ *be defined as*

$$\underline{a}_{ij} \leq \inf_{z\in S} \frac{\partial f_i(z)}{\partial z_j} \leq \frac{\partial f_i(z)}{\partial z_j} \leq \sup_{z\in S} \frac{\partial f_i(z)}{\partial z_j} \leq \overline{a}_{ij},$$

where $i, j \in \{1, 2, ..., n\}$. *Then the vector field* f *admits the first order* **d**-*homogeneous extension* F *given by* (6.40) *with*

$$\Sigma = \left\{ (\deg_{\mathbf{d}}(f)I_n + G_{\mathbf{d}})^{-1} \begin{pmatrix} a_{11} & ... & a_{1n} \\ ... & ... & ... \\ a_{n1} & ... & a_{nn} \end{pmatrix} G_{\mathbf{d}} : a_{ij} \in [\underline{a}_{ij}, \overline{a}_{ij}] \right\}. \tag{6.44}$$

Moreover, the set Σ *admits the representation* (6.41) *for* $N = 2^{n^2}$ *and the matrices*

$$A_k = (\deg_{\mathbf{d}}(f)I_n + G_{\mathbf{d}})^{-1} \begin{pmatrix} a_{11}^k & a_{12}^k & ... \\ a_{21}^k & a_{22}^k & ... \\ ... & ... & ... \end{pmatrix} G_{\mathbf{d}} \in \mathbb{R}^{n\times n} \tag{6.45}$$

with the elements $a_{ij}^k \in \{\underline{a}_{ij}, \overline{a}_{ij}\}$, $i, j = 1, 2, ..., n$.

Proof The first result follows from formula (6.3) and from formula (6.2) with $s = -\ln \|x\|_{\mathbf{d}}$. The second result follows from formula (6.13) being the corollary of Euler's theorem for smooth homogeneous vector fields (see Theorem 6.2) and from formula (6.9) with $s = -\ln \|x\|_{\mathbf{d}}$. The proof is complete. □

Using the above theorem, rather simple algorithms of homogeneous stabilizer design are developed in Chap. 15 for nonlinear control systems.

Concluding Remarks

In this chapter, the notions of **d**-homogeneous functions and **d**-homogeneous vector fields are introduced. Some important properties of homogeneous mappings are studied. The **d**-homogeneous approximation and the **d**-homogeneous extension are introduced as possible tools for the analysis of non-homogeneous systems using homogeneity-based methods. They can be utilized as an alternative to a linearization (the first order approximation) of a system for control design and local stability analysis. Special classes of single-valued and multi-valued homogeneous mappings are considered. It is shown that a linear vector field $x \mapsto Ax$ may have a nonzero homogeneity degree if and only if the matrix $A \in \mathbb{R}^{n\times n}$ is nilpotent. The homogeneity of tangent and normal cones for homogeneous sets is proven. Homogeneous convex embedding procedures for continuous and smooth mappings are developed. All these results are utilized in further chapters for control systems design and analysis.

References

1. Zubov VI (1958) On systems of ordinary differential equations with generalized homogeneous right-hand sides. Izvestia vuzov. Mathematica (available on-line: http://mi.mathnet.ru/ivm2874) 1:80–88
2. Kawski M (1991) Families of dilations and asymptotic stability. Analysis of controlled dynamical systems, pp 285–294
3. Andrieu V, Praly L, Astolfi A (2008) Homogeneous approximation, recursive observer design, and output feedback. SIAM J Control Optim 47(4):1814–1850
4. Bhat SP, Bernstein DS (2005) Geometric homogeneity with applications to finite-time stability. Math Control, Signals Syst 17:101–127
5. Khomenuk VV (1961) On systems of ordinary differential equations with generalized homogenous right-hand sides. Izvestia vuzov. Mathematica (in Russian) 3(22):157–164
6. Rosier L (1992) Homogeneous Lyapunov function for homogeneous continuous vector field. Syst Control Lett 19:467–473
7. Zimenko K, Efimov D, Polyakov A, Perruquetti W (2017) A note on delay robustness for homogeneous systems with negative degree. Automatica 79(5):178–184
8. Polyakov A (2019) Sliding mode control design using canonical homogeneous norm. Int J Robust Nonlinear Control 29(3):682–701
9. Zimenko K, Polyakov A, Efimov D, Perruquetti W (2020) Robust feedback stabilization of linear MIMO systems using generalized homogenization. IEEE Trans Automatic Control 65(12):5429–5436
10. Efimov D, Perruquetti W (2010) Oscillations conditions in homogenous systems. In: Proceedings Of IFAC NOLCOS Symposium, pp 1379–1384
11. Polyakov A, Coron J-M, Rosier L (2018) On homogeneous finite-time control for linear evolution equation in Hilbert space. IEEE Trans Automat Control 63(9):3143–3150
12. Braidiz Y, Polyakov A, Efimov D, Perruquetti W (2021) On finite/fixed-time stability analysis based on sup-and sub-homogeneous extensions. Syst Control Lett 150:104893
13. Poznyak A, Polyakov A, Azhmyakov V (2014) Attractive ellipsoids in robust control. Birkhauser
14. Filippov AF (1988) Differential equations with discontinuous right-hand sides. Kluwer Academic Publishers

15. Orlov Y (2005) Finite time stability and robust control synthesis of uncertain switched systems. SIAM J Control Optim 43(4):1253–1271
16. Nagumo M (1942) Uber die lage der integralkurven gewohnlicher differentialgleichungen. Proc Physico-Math Soc Jpn 24:551–559
17. Aubin J-P (1991) Viability theory. Birkhauser
18. Blanchini F (1999) Set invariance in control. Automatica 35:1747–1767
19. Clarke F, Ledyaev YS, Stern R, Wolenski P (1995) Qualitative properties of trajectories of control systems: a survey. J Dyn Control Syst 1(1):1–48
20. Mordukhovich BS (2006) Variational analysis and generalized differentiation I. Springer, Berlin
21. Clarke F (1990) Optim Nonsmooth Anal. SIAM, Philadelphia
22. Ioffe AD (2006) Existence and relaxation theorems for unbounded differential inclusions. J Convex Anal 13:353–362
23. Polovinkin ES (2015) Differential inclusions with unbounded right-hand side and necessary optimality conditions. Proc Steklov Inst Math 291:25–237
24. Levant A (2005) Homogeneity approach to high-order sliding mode design. Automatica 41(5):823–830
25. Zimenko K, Polyakov A, Efimov D (2023) Homogeneous systems stabilization based on convex embedding. Automatica 8(111108)
26. Boyd S, Ghaoui E, Feron E, Balakrishnan V (1994) Linear matrix inequalities in system and control theory. SIAM, Philadelphia

Chapter 7
Analysis of Homogeneous Systems

This chapter introduces a homogeneity-based approach to the stability, regularity and robustness analysis of dynamical systems in $\mathbb{R}^n$. In particular,

- dilation symmetry of solutions of homogeneous systems is established;
- forward/backward completeness is analyzed;
- asymptotic stability is characterized by homogeneous Lyapunov function;
- quadratic stability of homogeneous systems is studied;
- homogeneity-based finite/fixed-time stability analysis is developed;
- dynamics on homogeneous and locally homogeneous cones is analyzed;
- robustness and safety issues of homogeneous systems are studied.

7.1 Dilation Symmetry of Solutions of Differential Equations and Inclusions

In Chap. 1 (Introduction) it is shown that a dilation symmetry (standard homogeneity) of an equation is inherited by its solutions. Namely, if there exists $\nu \in \mathbb{R}$ such that

$$f(e^s x) = e^{(\nu+1)s} f(x), \quad \forall x \in \mathbb{R}^n, \quad \forall s \in \mathbb{R},$$

then for any solution $x(t, x_0)$ of the ODE

$$\dot{x}(t) = f(x(t)), \quad t > 0, \quad x(0) = x_0$$

we have the symmetry of solutions with respect to the dilation of the initial state

$$x(t, e^s x_0) = e^s x(e^{\nu s} t, x_0), \quad t > 0.$$

A. Polyakov, *Generalized Homogeneity in Systems and Control Volume I*,
Communications and Control Engineering,
https://doi.org/10.1007/978-3-031-76246-8_7

This property is crucial for a homogeneity-based analysis of dynamical systems. In particular, it guarantees that the existence (uniqueness, stability, etc.) of solutions for small initial data always implies the existence (resp., uniqueness, stability, etc.) of solutions for large initial data. Below we extend this result to solutions of generalized homogeneous dynamical systems modeled by ordinary differential equations (inclusions) studied in Chap. 2.

7.1.1 Linear Homogeneous Systems

Any linear system

$$\dot{x} = Ax, \quad t \in \mathbb{R}, \quad x(t) \in \mathbb{R}^n, \quad A \in \mathbb{R}^{n\times n}, \quad x(0) = x_0 \tag{7.1}$$

is standard homogeneous and its solution $x(t, x_0) = e^{tA}x_0$ is symmetric with respect to the standard scaling of the initial condition:

$$x(t, e^s x_0) = e^s x(t, x_0).$$

However, if the matrix A is nilpotent then, in the view of Lemma 6.1, the system is $\mathbf{d}$-homogeneous in a generalized sense too.

Lemma 7.1 *Let* $\mathbf{d}$ *be a linear dilation and let a linear vector field* $f : \mathbb{R}^n \mapsto \mathbb{R}^n$ *given by*

$$f(x) = Ax, \quad x \in \mathbb{R}^n, \quad A \in \mathbb{R}^{n\times n}$$

be $\mathbf{d}$*-homogeneous of degree* $\mu \in \mathbb{R}^n$ *then*

$$e^{tA}\mathbf{d}(s) = \mathbf{d}(s)e^{tAe^{\mu s}}, \quad \forall s \in \mathbb{R}, \quad \forall t \in \mathbb{R} \tag{7.2}$$

and

$$\mathbf{d}(s)\int_0^h e^{\tau A}d\tau = e^{\mu s}\int_0^{e^{-\mu s}h} e^{\tau A}d\tau\, \mathbf{d}(s), \quad \forall s \in \mathbb{R}. \tag{7.3}$$

Proof Since the vector field $f(x) = Ax$is $\mathbf{d}$-homogeneous then

$$A\mathbf{d}(s) = e^{\mu s}\mathbf{d}(s)A$$

and

$$e^{tA}\mathbf{d}(s) = \sum_{i=0}^{+\infty}\frac{t^i A^i \mathbf{d}(s)}{i!} = \sum_{i=0}^{+\infty}\frac{t^i e^{\mu s i}\mathbf{d}(s)A^i}{i!} = \mathbf{d}(s)e^{e^{\mu s}tA}.$$

Hence, for all $s \in \mathbb{R}$ and $h \in \mathbb{R}$ we also have

$$\mathbf{d}(s)\int_0^h e^{\tau A}d\tau = \int_0^h e^{e^{-\mu s}\tau A}\mathbf{d}(s)d\tau.$$

Making the change of the integration variable in the above integral we derive (7.3). The proof is complete. □

The formula (7.2) means that the exponential matrix e^{tA} commutes with the linear dilation $\mathbf{d}(s)$ up to certain scaling of the time parameter t. This scaling depends on the homogeneity degree μ of A. Below we show that this degree specifies convergence rate of homogeneous system (see Sect. 7.2.1). Lemma 7.1 proves the identity

$$x(t, \mathbf{d}(s)x_0) = \mathbf{d}(s)x(e^{\mu s}t, x_0), \quad s \in \mathbb{R}, \quad \forall t \in \mathbb{R}, \quad \forall x_0 \in \mathbb{R}^n, \tag{7.4}$$

where $x(\cdot, z)$ denotes a solution of (7.1) with the initial condition $x(0) = z$. Below we show that this identity (7.4) holds for any $\mathbf{d}$-homogeneous system.

7.1.2 Nonlinear Homogeneous Systems

Let us consider the following differential equation

$$\dot{x} = f(x), \quad t \in \mathbb{R}, \tag{7.5}$$

where $f : \mathbb{R}^n \mapsto \mathbb{R}^n$ is a continuous vector field. Inspired by [1–3] we characterize a dilation symmetry of solutions of homogeneous ODEs.

Theorem 7.1 *Let a continuous vector field $f : \mathbb{R}^n \mapsto \mathbb{R}^n$ be $\mathbf{d}$-homogeneous of degree $\mu \in \mathbb{R}$ and a function $t \mapsto x(t, x_0)$ be a classical solution of the system (7.5) with $x(0) = x_0$ defined for $t \in (t_0, t_1)$, where*

$$-\infty \le t_0 \le 0 < t_1 \le +\infty.$$

Then, for any $s \in \mathbb{R}$, the function $t \mapsto x(t, \mathbf{d}(s)x_0)$ given by formula (7.4) for $t \in (e^{-\mu s}t_0, e^{-\mu s}t_1)$ is a solution of the system (7.5) with the initial condition $x(0) = \mathbf{d}(s)x_0$.

Proof Denoting $y(t) = \mathbf{d}(s)x(e^{\mu s}t, x_0)$ and using the formula for the derivative of the composition of functions, we derive

$$\dot{y} = e^{\mu s}\mathbf{d}(s)\left.\frac{\partial x(\tau, x_0)}{\partial \tau}\right|_{\tau = e^{\mu s}t} = e^{\mu s}\mathbf{d}(s)f(x(e^{\mu s}t, x_0)).$$

Due to the $\mathbf{d}$-homogeneity of the vector field f, we have

$$\dot{y}(t) = f(\mathbf{d}(s)x(e^{\mu s}t, x_0)) = f(y(t)),$$

i.e., y is a solution of the system (7.5). Taking into account $y(0) = \mathbf{d}(s)x(0, x_0) = \mathbf{d}(s)x_0$ we complete the proof.

Theorem 7.1 holds for differential equations with discontinuous right-hand sides as well as for differential inclusions like

$$\dot{x} \in F(x), \quad t \in \mathbb{R}, \quad x(t) \in \mathbb{R}^n, \tag{7.6}$$

where the set-valued vector field $F : \mathbb{R}^n \mapsto 2^{\mathbb{R}^n}$ is $\mathbf{d}$-homogeneous of degree $\mu \in \mathbb{R}$. Indeed, on the one hand, if f is a discontinuous vector field then the Filippov regularization (see Remark 2.5) results in a set-valued vector field F, while, by Theorem 6.8, the $\mathbf{d}$-homogeneity of f is inherited by F. The dilation symmetry of solutions of (7.6) can be proven as in Theorem 7.1.

Theorem 7.2 *Let a set-valued vector field* $F : \mathbb{R}^n \mapsto 2^{\mathbb{R}^n}$ *be* $\mathbf{d}$*-homogeneous of degree* $\mu \in \mathbb{R}$ *and a function* $t \mapsto x(t, x_0)$ *be a strong (resp., classical) solution of the system* (7.6) *with* $x(0) = x_0$ *defined for* $t \in (t_0, t_1)$, *where* $-\infty \leq t_0 \leq 0 < t_1 \leq +\infty$. *Then for any* $s \in \mathbb{R}$ *the function* $t \mapsto x(t, \mathbf{d}(s)x_0)$ *given by formula* (7.4) *for* $t \in (e^{-\mu s}t_0, e^{-\mu s}t_1)$ *is a strong (resp., classical) solution of the system* (7.6) *with the scaled initial condition* $x(0) = \mathbf{d}(s)x_0$.

Proof Denoting $y(t) = \mathbf{d}(s)x(e^{\mu s}t, x_0)$ and using the formula for the derivative of the composition of functions, we derive

$$\dot{y} = e^{\mu s}\mathbf{d}(s) \left.\frac{\partial x(\tau, x_0)}{\partial \tau}\right|_{\tau = e^{\mu s}t} \overset{a.e.}{\in} e^{\mu s}\mathbf{d}(s)F(x(e^{\mu s}t, x_0)).$$

Using the $\mathbf{d}$-homogeneity of the set-valued vector field F, we conclude

$$\dot{y}(t) \overset{a.e.}{\in} F(\mathbf{d}(s)x(e^{\mu s}t, x_0)) = F(y(t)),$$

i.e., y is a strong solution of the system (7.6). Taking into account $y(0) = \mathbf{d}(s)x(0, x_0) = \mathbf{d}(s)x_0$ we complete the proof.

Notice that the above theorem is applicable for any (even unbounded) differential inclusion. However, all further results presented in this chapter are proven for differential inclusion (7.6) under the following assumption.

By default, we assume below that the autonomous system (7.6) satisfies Assumption **A1** of Chap. 4 (i.e., F is compact-valued, convex-valued and upper semi-continuous) and $\mathbf{0} \in F(\mathbf{0})$.

For shortness we say that *the system* (7.6) *is* $\mathbf{d}$*-homogeneous* if its right-hand side is a $\mathbf{d}$-homogeneous (possibly set-valued) vector field.

7.1.3 Topological Equivalence of Homogeneous Systems

Any method developed for standard homogeneous systems remains valid for generalized homogeneous systems due to the following result [4, 5].

Theorem 7.3 *Let* $\mathbf{d}$ *be a linear continuous dilation in* $\mathbb{R}^n$. *Let a homeomorphism* $\Phi : \mathbb{R}^n \mapsto \mathbb{R}^n$ *be defined by formula (5.31), namely,* $\Phi(\mathbf{0}) = \mathbf{0}$ *and*

$$\Phi(x) = \|x\|_{\mathbf{d}}\mathbf{d}(-\ln\|x\|_{\mathbf{d}})x, \qquad \Phi^{-1}(z) = \|z\|^{-1}\mathbf{d}(\ln\|z\|)z,$$

where $x, z \in \mathbb{R}^n\backslash\{\mathbf{0}\}$ *and* $\|\cdot\|_{\mathbf{d}}$ *is the canonical homogeneous norm in* $\mathbb{R}^n$ *induced by the weighted norm* $\|z\| = \sqrt{z^\top P z}$ *with* $P = P^\top \in \mathbb{R}^{n\times n}$ *satisfying (5.6). If* $x(\cdot, x_0) : (t_0, t_1) \mapsto \mathbb{R}^n$ *is a strong (classical) solution of the* $\mathbf{d}$*-homogeneous system* (7.6) *such that* $x(t, x_0) \neq \mathbf{0}, \forall t \in (t_0, t_1)$, *where* $-\infty \leq t_0 \leq 0 < t_1 \leq +\infty$, *then the function* $z(\cdot, z_0) : (t_0, t_1) \mapsto \mathbb{R}^n$ *defined as*

$$z(t, z_0) = \Phi(x(t, x_0)), \quad t \in (t_0, t_1) \tag{7.7}$$

is a strong (resp., classical) solution of the standard homogeneous system

$$\dot{z} \in \|z\|^{1+\mu} F_0(z), \quad z(0) = z_0 := \Phi(x_0), \tag{7.8}$$

where $\mu \in \mathbb{R}$ *is the homogeneity degree of* F *and* $F_0(\mathbf{0}) = \mathbf{0}$,

$$F_0(z) = \left(\frac{(I_n - G_{\mathbf{d}})zz^\top P}{z^\top P G_{\mathbf{d}} z} + I_n\right) F\left(\frac{z}{\|z\|}\right), \quad z \neq \mathbf{0}. \tag{7.9}$$

Inversely, if $z(\cdot, z_0) : (t_0, t_1) \mapsto \mathbb{R}^n$ *is a strong (classical) solution of* (7.8) *such that* $z(t) \neq \mathbf{0}, \forall t \in (t_0, t_1)$ *then* $x(t, x_0) = \Phi^{-1}(z(t, z_0)), t \in (t_0, t_1)$ *is a strong (classical) solution of the system* (7.6) *with* $x_0 = \Phi^{-1}(z_0)$.

Proof The change of variable $z = \|x\|_{\mathbf{d}}\mathbf{d}(-\ln\|x\|_{\mathbf{d}})x$ gives

$$\|z\| = \|x\|_{\mathbf{d}}\|\mathbf{d}(-\ln\|x\|_{\mathbf{d}})x\| = \|x\|_{\mathbf{d}}$$

and $x = \mathbf{d}(\ln\|z\|)\frac{z}{\|z\|}$. Moreover, for $x \neq \mathbf{0}$ using formula (5.21) we derive

$$\begin{aligned}
\dot{z} &= \frac{d\|x\|_{\mathbf{d}}}{dt}(I_n - G_{\mathbf{d}})\mathbf{d}(-\ln\|x\|_{\mathbf{d}})x + \|x\|_{\mathbf{d}}\mathbf{d}(-\ln\|x\|_{\mathbf{d}})\dot{x} \\
&= \|x\|_{\mathbf{d}}(I_n - G_{\mathbf{d}})\mathbf{d}(-\ln\|x\|_{\mathbf{d}})x\frac{x^\top\mathbf{d}^\top(-\ln\|x\|_{\mathbf{d}})P\mathbf{d}(-\ln\|x\|_{\mathbf{d}})\dot{x}}{x^\top\mathbf{d}^\top(-\ln\|x\|_{\mathbf{d}})PG_{\mathbf{d}}\mathbf{d}(-\ln\|x\|_{\mathbf{d}})x} \\
&\quad + \|x\|_{\mathbf{d}}\mathbf{d}(-\ln\|x\|_{\mathbf{d}})\dot{x} \\
&= \|x\|_{\mathbf{d}}\left(\frac{(I-G_{\mathbf{d}})\mathbf{d}(-\ln\|x\|_{\mathbf{d}})xx^\top\mathbf{d}^\top(-\ln\|x\|_{\mathbf{d}})P}{x^\top\mathbf{d}^\top(-\ln\|x\|_{\mathbf{d}})PG_{\mathbf{d}}\mathbf{d}(-\ln\|x\|_{\mathbf{d}})x} + I_n\right)\mathbf{d}(-\ln\|x\|_{\mathbf{d}})\dot{x} \\
&= \|z\|\left(\frac{(I_n - G_{\mathbf{d}})z^\top zP}{z^\top PG_{\mathbf{d}}z} + I_n\right)\mathbf{d}(-\ln\|z\|)\dot{x}.
\end{aligned}$$

Due to $\mathbf{d}$-homogeneity of F we have

$$\mathbf{d}(-\ln\|z\|)F(x)=\|z\|^{\mu}F(\mathbf{d}(-\ln\|x\|_{\mathbf{d}})x)=\|z\|_{\mathbf{d}}^{\mu}F\left(\frac{z}{\|z\|}\right).$$

So, if $x(\cdot,x_0)$ is a strong (or classical) solution of (7.6) such that $\|x(t,x_0)\|\neq 0, \forall t\in(t_0,t_1)$ then $z(t,z_0)=\Phi(x(t,x_0)), t\in(t_0,t_1)$ is a strong (or classical) solution of the Cauchy problem (7.8) with $z_0=\Phi(x_0)$.

If F is compact-valued, convex-valued and upper semi-continuous on $\mathbb{R}^n\backslash\{\mathbf{0}\}$ then the set-valued vector field $z\mapsto\|z\|^{1+\mu}\left(\frac{(I_n-G_{\mathbf{d}})z^{\top}zP}{z^{\top}PG_{\mathbf{d}}z}+I_n\right)F\left(\frac{z}{\|z\|}\right)$ is compact-valued, convex-valued and upper semi-continuous on $\mathbb{R}^n\backslash\{\mathbf{0}\}$, as well. Therefore, if the system (7.6) satisfies Assumption **A1** of Chap. 4 then the system (7.8) satisfies Assumption **A1** of Chap. 4, at least, on $\mathbb{R}^n\backslash\{\mathbf{0}\}$. This means that the system (7.8) has strong (or classical) solutions on $\mathbb{R}^n\backslash\{\mathbf{0}\}$. Considering the change of variable $x=\Phi^{-1}(z)=\mathbf{d}(\ln\|z\|)\frac{z}{\|z\|}$ and repeating the above calculations we derive that

$$\begin{aligned}\dot{x}&=\mathbf{d}(\ln\|z\|)(G_{\mathbf{d}}-I_n)\frac{z}{\|z\|^2}\frac{d\|z\|}{dt}+\mathbf{d}(\ln\|z\|)\frac{\dot{z}}{\|z\|}\\&=\mathbf{d}(\ln\|z\|)\left(\frac{(G_{\mathbf{d}}-I_n)zz^{\top}P}{z^{\top}Pz}+I_n\right)\frac{\dot{z}}{\|z\|}.\end{aligned}$$

Since

$$\left(\frac{(G_{\mathbf{d}}-I_n)zz^{\top}P}{z^{\top}Pz}+I_n\right)\left(\frac{(I_n-G_{\mathbf{d}})z^{\top}zP}{z^{\top}PG_{\mathbf{d}}z}+I_n\right)=$$

$$\frac{(G_{\mathbf{d}}-I_n)zz^{\top}P(I_n-G_{\mathbf{d}})z^{\top}zP}{(z^{\top}Pz)(z^{\top}PG_{\mathbf{d}}z)}+\frac{(G_{\mathbf{d}}-I_n)zz^{\top}P}{z^{\top}Pz}+\frac{(I_n-G_{\mathbf{d}})z^{\top}zP}{z^{\top}PG_{\mathbf{d}}z}+I_n=I_n$$

and

$$\|z\|^{\mu}\mathbf{d}(\ln\|z\|)F\left(\frac{z}{\|z\|}\right)=F\left(\mathbf{d}(\ln\|z\|)\frac{z}{\|z\|}\right)=F(x)$$

then for any strong (or classical) solution $z(\cdot,z_0)$ of (7.8) such that $\|z(t,z_0)\|\neq 0, \forall t\in(t_0,t_1)$ we have that $x(t,x_0)=\Phi^{-1}(z(t,z_0))$ is a strong (or classical) solution of the system (7.6) with the initial condition $x_0=\Phi^{-1}(z_0)$. The proof is complete. □

The above theorem proves the equivalence of solutions of standard and generalized homogeneous systems on $\mathbb{R}^n\backslash\{\mathbf{0}\}$. If $\mathbf{0}\in F(\mathbf{0})$ then the system (7.6) has the zero solution. For $\mu\geq -1$, the system (7.8) has the zero solution too. For $\mu>-1$ the right-hand side of the system is continuous at $\mathbf{0}$, so the system (7.8) satisfies Assumption **A1** of Chap. 4 on $\mathbb{R}^n$. Taking into account Corollary 6.2 we make the following conclusion.

Any generalized homogeneous (with respect to linear continuous dilation) system is topologically equivalent[a] *(homeomorphic*[b]*) on* $\mathbb{R}^n$ *to a standard homogeneous system.* Moreover, the transformations Φ and Φ^{-1} are continuously differentiable away from the origin, so the **d**-homogeneous system (7.6) is diffeomorphic[c] on $\mathbb{R}^n\backslash\{\mathbf{0}\}$ to the system (7.8).

[a] Two systems are topologically equivalent if solution sets of the systems coincide up to some continuous transformation of coordinates.

[b] Homeomorphism is a continuous invertible mapping with continuous inverse.

[c] Diffeomorphism is a continuously differentiable invertible mapping with continuously differentiable inverse.

In the view of the proven equivalence, we may omit the word "generalized" when we discuss generalized homogeneous ODEs in $\mathbb{R}^n$.

Example 7.1 The twisting system

$$\dot{x} = f(x) := \begin{pmatrix} x_2 \\ -k_1\,\text{sign}(x_1) - k_2\,\text{sign}(x_2) \end{pmatrix}, \quad t > 0, \quad x = (x_1, x_2)^\top \in \mathbb{R}$$

is **d**-homogeneous of degree -1 with respect to the weighted dilation

$$\mathbf{d}(s) = e^{sG_{\mathbf{d}}}, \quad G_{\mathbf{d}} = \begin{pmatrix} 2 & 0 \\ 0 & 1 \end{pmatrix}$$

in $\mathbb{R}^2$. By Theorem 7.3, this system is equivalent to

$$\dot{z} = \begin{pmatrix} \frac{-z^\top Pf(z/\|z\|)}{z^\top PG_{\mathbf{d}}z} z_1 + \frac{1}{\|z\|} z_2 \\ -k_1\,\text{sign}(z_1) - k_2\,\text{sign}(z_2) \end{pmatrix}, \quad z = (z_1, z_2)^\top \in \mathbb{R}^2,$$

where the matrix $P = P^\top \in \mathbb{R}^{2\times 2}$ satisfies 5.6 and $\|z\| = \sqrt{z^\top Pz}$.

7.1.4 Completeness

Recall that a dynamical system is said to be *forward complete* (resp., *backward complete*) if any solution of this system is defined for all $t \geq t_0$ (resp, for all $t \leq t_0$). The system is *complete* if it is both forward and backward complete.

Corollary 7.1 *The system* (7.6) *is forward (resp., backward) complete if and only if it has forward (resp., backward) complete solutions for all initial conditions from the unit sphere* $S = \{x \in \mathbb{R}^n : \|x\| = 1\}$.

The proof immediately follows from the identity (7.4). Notice that $\|x\|$ is an arbitrary norm in $\mathbb{R}^n$ (e.g., $\|x\| = \sqrt{x^\top Px}$). Therefore, to guarantee the completeness of the homogeneous system, it is sufficient to find a unit sphere, such that any solution initiated on this sphere is defined globally in time.

A degree of a homogeneous system allows an analysis of completeness to be provided without any knowledge of solutions [6].

Corollary 7.2 *If the system* (7.6) *is* **d**-*homogeneous of degree* $\mu \leq 0$ *then it is complete.*

Proof By Theorem 7.3 any generalized homogeneous system (7.6) is topologically equivalent to the standard homogeneous system (7.8), where a set-valued mapping F_0 is uniformly bounded. Hence, using Theorem 2.9 we conclude that for $\mu \leq 0$ the considered standard homogeneous system is complete.

The above result also follows from the remark given below.

Remark 7.1 Let **d** be a linear continuous dilation in $\mathbb{R}^n$ and the canonical homogeneous norm $\|\cdot\|_{\mathbf{d}}$ be induced by the weighted Euclidean norm $\|x\| = \sqrt{x^\top P x}$ with $P \succ 0$ satisfying (5.6). If $f : \mathbb{R}^n \mapsto \mathbb{R}^n$ is a continuous **d**-homogeneous vector field of degree $\mu \in \mathbb{R}$ then there exists $\rho^* > 0$ such that

$$-\rho^* \|x\|_{\mathbf{d}}^{1+\mu} \leq \frac{d\|x\|_{\mathbf{d}}}{dt} \leq \rho^* \|x\|_{\mathbf{d}}^{1+\mu}, \quad \forall t > 0 \tag{7.10}$$

for any solution $t \mapsto x(t)$ of (7.5). Indeed, since using formula (5.21) and the **d**-homogeneity of f we derive

$$\begin{aligned}\frac{d\|x\|_{\mathbf{d}}}{dt} &= \|x\|_{\mathbf{d}} \frac{x^\top \mathbf{d}^\top(-\ln \|x\|_{\mathbf{d}}) P \mathbf{d}(-\ln \|x\|_{\mathbf{d}}) f(x)}{x^\top \mathbf{d}^\top(-\ln \|x\|_{\mathbf{d}}) P G_{\mathbf{d}} \mathbf{d}(-\ln \|x\|_{\mathbf{d}}) x} \\ &= \|x\|_{\mathbf{d}}^{1+\mu} \frac{x^\top \mathbf{d}^\top(-\ln \|x\|_{\mathbf{d}}) P f(\mathbf{d}(-\ln \|x\|_{\mathbf{d}}) x)}{x^\top \mathbf{d}^\top(-\ln \|x\|_{\mathbf{d}}) P G_{\mathbf{d}} \mathbf{d}(-\ln \|x\|_{\mathbf{d}}) x},\end{aligned}$$

then taking into account $\|\mathbf{d}(-\ln \|x\|_{\mathbf{d}}) x\| = 1$ we conclude that the required differential inequality holds for

$$\rho^* = \frac{1}{2\beta} \max_{\|y\|=1} \|f(y)\|, \tag{7.11}$$

where $\beta > 0$ is defined in Proposition 5.3.

Homogeneous systems with positive degrees may have a finite-time blow up in a forward time (e.g., $\dot{x} = x^3$). Moreover, if the homogeneous system with a positive degree is asymptotically stable then, by Theorem 7.5 given below, its nearly fixed-time stable in the forward time, but, in this case, it has a finite-time blow up in the backward time (see, Example 3.10 and Proposition 3.3). However, this does not mean that homogeneous systems with positive degree always have a finite-time blow up in forward or in the backward time. The following example presents a complete homogeneous system with positive degree.

Example 7.2 A finite-dimensional approximation of the non-viscous Burgers equation gives the ODE of the form (see, e.g., [7])

$$\dot{x} = f(x) := W(x)x, \quad t \in \mathbb{R},$$

where $x(t) \in \mathbb{R}^n$ and $W : \mathbb{R}^n \mapsto \mathbb{R}^{n\times n}$ is a skew-symmetric linear matrix-valued function:

$$W^\top(x) = -W(x), \quad W(\alpha x + \beta y) = \alpha W(x) + \beta W(y), \quad \forall \alpha, \beta \in \mathbb{R}, \quad \forall x, y \in \mathbb{R}^n.$$

This system is standard homogeneous of degree 1, i.e., for $\mathbf{d}(s) = e^s I_n$ it holds

$$f(\mathbf{d}(s)x) = W(e^s x)e^s x = e^{2s} W(x)x = e^s \mathbf{d}(s) f(x), \quad \forall x \in \mathbb{R}^n.$$

Using the skew-symmetry of W and the identity

$$z^\top W y = y^\top W^\top z, \quad \forall z, y, \in \mathbb{R}^n, \quad \forall W \in \mathbb{R}^{n\times n}$$

we derive

$$\frac{d}{dt}|x|^2 = 2x^\top W(x)x = x^\top W(x)x + x^\top W^\top(x)x = x^\top W(x)x - x^\top W(x)x = 0,$$

so, $|x(t)| = |x(0)|$ for all $t \in \mathbb{R}$. Therefore, the considered system is a complete standard homogeneous system of positive degree.

7.2 Stability of Homogeneous Systems

The dilation symmetry of solutions (see formula (7.4)) is the key feature of homogeneous systems which, in particular, implies the following corollary.

Corollary 7.3 *If the system* (7.6) *is* **d***-homogeneous and locally Lyapunov (resp., asymptotically) stable then it is globally Lyapunov (resp., asymptotically) stable.*

Proof The local Lyapunov stability implies that all solutions of a system are bounded for all $t \geq 0$ if the initial data is sufficiently small: $\|x_0\| \leq \varepsilon$. In this case, on the one hand, Corollary 7.1 implies that the system is forward complete. On the other hand, the dilation symmetry of solutions means

$$x(t, \mathbf{d}(s)x_0) = \mathbf{d}(s)x(e^{\mu s}t, x_0), \quad \forall t \geq 0,$$

where μ is a homogeneity degree of the system and $s \in \mathbb{R}$ is the group parameter. Then

- uniform (on t) boundedness of $x(t, x_0)$ implies uniform (on t) boundedness of $x(t, \mathbf{d}(s)x_0)$;
- asymptotic convergence of $x(t, x_0)$ to $\mathbf{0}$ as $t \to +\infty$ implies asymptotic convergence of $x(t, \mathbf{d}(s)x_0)$ to $\mathbf{0}$ as $t \to +\infty$

and vice versa. Since $\mathbf{d}$ is a dilation then a global behavior of solutions (for $\|x_0\| < +\infty$) repeats their local behavior (for $\|x_0\| < \varepsilon$). Indeed, if $\|x_0\| < +\infty$ then from the limit property of the dilation (see Chap. 5) we derive that there exists $s_0 \in \mathbb{R}$ such that $\|\mathbf{d}(s_0)x_0\| \leq \varepsilon$. Therefore, local Lyapunov (asymptotic) stability of a $\mathbf{d}$-homogeneous system is equivalent to its global Lyapunov (asymptotic) stability. The proof is complete. □

Similarly to a linear system, the existence of a nonempty nonzero positively invariant compact set implies the Lyapunov stability of a homogeneous system.

Proposition 7.1 *A* $\mathbf{d}$*-homogeneous system is Lyapunov stable if and only if it has a positively invariant*[a] *bounded neighborhood of* $\mathbf{0}$.

[a] A set $\Omega \subset \mathbb{R}^n$ is positively invariant if $x_0 \in \Omega \Rightarrow x(t, x_0) \in \Omega, \forall t \geq 0$.

Proof *Sufficiency.* If Ω is a neighborhood of the origin then $\exists \rho_0 > 0 : \|x_0\|_{\mathbf{d}} \leq \rho_0 \Rightarrow x_0 \in \Omega$. Since Ω is a bounded set then $\exists \rho > 0$ such that $x \in \Omega \Rightarrow \|x\|_{\mathbf{d}} \leq \rho$. Recall that for any $\mathbf{d}$-homogeneous norm we have $\|\mathbf{d}(s)z\|_{\mathbf{d}} = e^s\|z\|_{\mathbf{d}}, \forall s \in \mathbb{R}, \forall z \in \mathbb{R}^n$. If Ω is a positively invariant set then $\|x_0\|_{\mathbf{d}} \leq \rho_0 \Rightarrow \|x(t, x_0)\|_{\mathbf{d}} \leq \rho, \forall t \geq 0$. If $x_0 \in \mathbb{R}^n \backslash \{\mathbf{0}\}$ then taking $s = \ln \rho_0 - \ln \|x_0\|_{\mathbf{d}}$, we derive

$$\|\mathbf{d}(s)x_0\|_{\mathbf{d}} = \|\mathbf{d}(\ln \rho_0)\mathbf{d}(-\ln \|x_0\|_{\mathbf{d}})x_0\|_{\mathbf{d}} = e^{\ln \rho_0}\|\mathbf{d}(-\ln \|x_0\|_{\mathbf{d}})x_0\|_{\mathbf{d}} = \rho_0$$

and $\|x(\tau, \mathbf{d}(s)x_0)\|_{\mathbf{d}} \leq \rho$ for all $\tau \geq 0$. Hence, using the dilation symmetry of solutions (see formula (7.4)), for $s = \ln \rho_0 - \ln \|x_0\|_{\mathbf{d}}$ and any $\forall t \geq 0$ we have

$$\|x(t, x_0)\|_{\mathbf{d}} = \|\mathbf{d}(-s)x(e^{\mu s}t, \mathbf{d}(s)x_0)\|_{\mathbf{d}} = e^{-s}\|x(e^{\mu s}t, \mathbf{d}(s)x_0)\|_{\mathbf{d}} \leq \tfrac{\rho}{e^s} = \tfrac{\rho\|x_0\|_{\mathbf{d}}}{\rho_0}.$$

This inequality means the global Lyapunov stability of the system.

Necessity. If the system is locally Lyapunov stable then by Corollary 7.3 it is globally Lyapunov stable and there exists $\varepsilon \in \mathcal{K}_\infty$ such that

$$\|x(t, x_0)\| \leq \varepsilon(\|x_0\|), \quad \forall t \geq 0, \quad \forall x_0 \in \mathbb{R}^n.$$

Let Ω be a reachability set defined by means of all trajectories of the system initiated in the unit ball:

$$\Omega = \{y \in \mathbb{R}^n : y = x(t, x_0), t \geq 0, x_0 \in \mathbb{R}^n : \|x_0\| \leq 1\}.$$

By construction, the set Ω is positively invariant and it is a neighborhood of the origin, which contains the unit ball $\{x \in \mathbb{R}^n : \|x\| \leq 1\}$. Moreover, Ω is bounded (embedded in the ball of the radius $\varepsilon(1)$) since the system is Lyapunov stable. The proof is complete. □

Inspired by [6], we present the following characterization of the asymptotic stability of homogeneous systems.

Proposition 7.2 *A* $\mathbf{d}$*-homogeneous system* (7.6) *is asymptotically stable if and only if it has a strictly positively invariant*[a] *compact neighborhood of the origin.*

[a] A compact set Ω is strictly positively invariant if $x_0 \in \Omega \Rightarrow x(t, x_0) \in \text{int}\,\Omega, \forall t > 0$.

Proof The *Necessity* follows from Theorem 4.7, since a level set $\Omega = \{x \in \mathbb{R}^n : V(x) \leq 1\}$ of a smooth Lyapunov function V of an asymptotically stable system is a compact strictly positively invariant neighborhood of the origin.

Sufficiency. The Lyapunov stability follows from Proposition 7.1. If Ω is strictly positively invariant set of the system then there exists a limit set $K \subset \text{int}\,\Omega$ such that $x(0) \in \Omega \Rightarrow x(t) \to K$ as $t \to +\infty$. On the one hand, due to homogeneity, the set $\mathbf{d}(s)\Omega$ is a strictly positively invariant compact for the system for any $s \in \mathbb{R}$ and $\mathbf{d}(s)K \subset \text{int}\,\mathbf{d}(s)\Omega$ is the corresponding limit set. On the other hand, since Ω is strictly positively invariant set then $x(t) \in \text{int}\,\Omega$ for all $t > 0$, so, by Lemma 5.1, for any $t > 0$ there exists $s_t < 0$ such that $x(t) \in \partial\mathbf{d}(s_t)\Omega$ (the boundary of the compact $\mathbf{d}(s_t)\Omega$). This means that $x(t) \in \mathbf{d}(s_t)\Omega$ for all $t \geq 0$.

Let us show $s_{t_k} \to -\infty$ as $t_k \to +\infty$. Suppose the contrary: there exists $s_{t_k} < 0$ such that $s_{t_k} \to s^* \in (-\infty, +\infty]$ as $t_k \to +\infty$. On the one hand, since $x(t_k) \in \partial\mathbf{d}(s_{t_k})\Omega \subset \mathbf{d}(s_{t_k})\Omega$ then for any $k \in \mathbb{N}$ we have $x(t) \to \mathbf{d}(s_{t_k})K$ as $t \to +\infty$ and, consequently, $x(t_k) \to \mathbf{d}(s^*)K$ as $t_k \to +\infty$. On the other hand, since $\mathbf{d}(s_{t_k})\Omega \to \mathbf{d}(s^*)\Omega$ as $s_{t_k} \to s^*$ (the limit is understood in Hausdorff topology) then $x(t_k) \to \partial\mathbf{d}(s^*)\Omega$ as $t_k \to +\infty$. This is impossible since $\mathbf{d}(s^*)K \subset \text{int}\,\mathbf{d}(s^*)\Omega$ for any finite $s^* < 0$. Therefore, $s_t \to -\infty$ as $t \to +\infty$. Since Ω is bounded then $\mathbf{d}(s_t)\Omega \to \mathbf{0}$ as $t \to +\infty$ and $x(t) \to \mathbf{0}$ as $t \to +\infty$. The proof is complete. □

As a corollary of the above proposition, one can be shown that asymptotic convergence of solutions of homogeneous system to $\mathbf{0}$ implies stability [1, 6]. Moreover, in the case of the negative homogeneity degree, the finite-time stability can be guaranteed.

Theorem 7.4 *(1) The* $\mathbf{d}$*-homogeneous system* (7.6) *of degree* $\mu < 0$ *is globally uniformly finite-time stable if and only if any trajectory of the system reaches the origin in a finite time being a locally bounded function of the initial state* $x_0 \in \mathbb{R}^n$.

(2) The $\mathbf{d}$*-homogeneous system* (7.6) *is globally uniformly asymptotically stable if and only if the system uniformly asymptotically converges to the origin.*

Proof (1) *Necessity* follows from the definition of the uniform finite-time stability. *Sufficiency*. Let $T(r) > 0$ be a supremum of reaching times over all initial conditions x_0 belonging to the homogeneous sphere of the radius $r > 0$ (i.e., $\|x_0\|_{\mathbf{d}} = r$).

Obviously, $T(0) = 0$. Since the reaching time is a locally bounded function of the initial state x_0 then $\sup_{\rho\in[0,r]} T(\rho) < +\infty$ for any finite $r \geq 0$. Moreover, the **d**-homogeneity of the solutions (see formula (7.4)) implies that

$$T(e^s r) = e^{-\mu s} T(r), \quad \forall s \in \mathbb{R}, \quad \forall r > 0.$$

This means that $T(r) \to 0$ as $r \to \mathbf{0}$, i.e., T is continuous at zero.

Let us show that the system is globally Lyapunov stable. Suppose the contrary, i.e., $\exists \epsilon > 0, \forall \delta \in (0,1), \exists x_0 \in \mathbb{R}^n : \|x_0\|_{\mathbf{d}} = \delta$ and $\exists t^* = t^*(\delta) > 0 : \|x(t^*, x_0)\|_{\mathbf{d}} = \epsilon$, and $\delta < \|x(t, x_0)\|_{\mathbf{d}} < \epsilon$ for all $t \in (0, t^*)$, where $\|\cdot\|_{\mathbf{d}}$ is a **d**-homogeneous norm induced by the weighted dilation $\|x\| = \sqrt{x^\top P x}$. Due to the dilation symmetry of solutions, we can assume $\epsilon = 1$ without loss generality.

Using formula (5.21) we derive

$$\frac{d}{dt}\|x\|_{\mathbf{d}} \leq \sup_{y\in F(x)} \|x\|_{\mathbf{d}} \frac{x^\top \mathbf{d}^\top(-\ln\|x\|_{\mathbf{d}}) P \mathbf{d}(-\ln\|x\|_{\mathbf{d}}) y}{x^\top \mathbf{d}^\top(-\ln\|x\|_{\mathbf{d}}) P G_{\mathbf{d}} \mathbf{d}(-\ln\|x\|_{\mathbf{d}}) x}.$$

Applying **d**-homogeneity of F, we obtain

$$\frac{d}{dt}\|x\|_{\mathbf{d}} \leq \sup_{\tilde{y}\in F(\mathbf{d}(-\ln\|x\|_{\mathbf{d}})x)} \|x\|_{\mathbf{d}}^{1+\mu} \frac{x^\top \mathbf{d}^\top(-\ln\|x\|_{\mathbf{d}}) P \tilde{y}}{x^\top \mathbf{d}^\top(-\ln\|x\|_{\mathbf{d}}) P G_{\mathbf{d}} \mathbf{d}(-\ln\|x\|_{\mathbf{d}}) x},$$

where $\beta = \frac{1}{2}\lambda_{\min}(P^{1/2} G_{\mathbf{d}} P^{-1/2} + P^{-1/2} \bar{G}_{\mathbf{d}}^\top P^{1/2}) > 0$ and

$$M = \sup_{y\in F(x), \|x\|_{\mathbf{d}}=1} |y| < +\infty.$$

Hence, we conclude $\|x_{x_0}(t)\|_{\mathbf{d}}^{-\mu} \leq \|x_0\|_{\mathbf{d}}^{-\mu} + \beta M t$ for all $t \geq 0$ and

$$\|x_{x_0}(t)\|_{\mathbf{d}} < 1 \text{ for all } t \in \left(0, \frac{1-\delta^{-\mu}}{M\beta}\right).$$

This means that $t^* \geq \frac{1-\delta^{-\mu}}{M\beta}$. Since $T(\delta) \to 0$ as $\delta \to 0$ then, for sufficiently small δ, we have the inequality $T(t) < t^*$, which contradicts our supposition.

(2) If the system uniformly asymptotically converges to the origin then for there exists $T \in (0, +\infty)$ such that

$$\|x_0\|_{\mathbf{d}} \leq 1 \Rightarrow \|x(t, x_0)\|_{\mathbf{d}} \leq 1, \quad \forall t \geq T.$$

Let $q = \sup_{\|x_0\|_{\mathbf{d}}\leq 1, t\in[0,T]} \|x(t, x_0)\| < +\infty$. Then

$$\|x_0\|_{\mathbf{d}} \leq 1 \Rightarrow \|x(t, x_0)\|_{\mathbf{d}} \leq q, \quad \forall t \geq 0.$$

In this case, due to the dilation symmetry of solutions (7.4), for any $x_0 \in \mathbb{R}^n \backslash \{\mathbf{0}\}$ we have

$$\|x(t, x_0)\|_{\mathbf{d}} = \|\mathbf{d}(\ln\|x_0\|_{\mathbf{d}}) x(\|x_0\|_{\mathbf{d}}^{\mu} t, \mathbf{d}(-\ln\|x_0\|_{\mathbf{d}}) x_0)\|_{\mathbf{d}}$$

$$= \|x_0\|_{\mathbf{d}} \|x(\|x_0\|_{\mathbf{d}}^{-\mu} t, \mathbf{d}(-\ln \|x_0\|_{\mathbf{d}}) x_0)\|_{\mathbf{d}} \le q \|x_0\|_{\mathbf{d}}, \quad \forall t \ge 0.$$

This means that the system is Lyapunov stable. The proof is complete. □

7.2.1 Convergence Rates

Since 1950s (see, [8, p. 110]) it is known that the homogeneity degree of a uniformly asymptotically stable system specifies its convergence rate. Based on the concept of finite/fixed-time stability discussed in Chap. 3, we provide the following characterization of the convergence rates of homogeneous systems inspired by [9, 10].

Theorem 7.5 *If the system* (7.6) *is* $\mathbf{d}$*-homogeneous of degree* $\mu \in \mathbb{R}$ *and locally uniformly asymptotically stable then*

- *for* $\mu < 0$ *it is globally uniformly finite-time stable and there exists a settling-time function* $T : \mathbb{R}^n \mapsto [0, +\infty)$ *being*
  - $\mathbf{d}$*-homogeneous of degree* $-\mu$*;*
  - *locally bounded on* $\mathbb{R}^n$ *and continuous at* $\mathbf{0}$*;*
  - *continuous on* $\mathbb{R}^n$ *provided that all solutions of the system* (7.6) *are unique in the forward time and depend continuously on the initial data* x_0*;*
- *for* $\mu = 0$ *it is globally uniformly exponentially stable;*
- *for* $\mu > 0$ *it is globally uniformly nearly fixed-time stable and there exists an escape-time function* $\tau : \mathbb{R}^n \backslash \{\mathbf{0}\} \mapsto (0, +\infty)$ *being*
  - $\mathbf{d}$*-homogeneous of degree* $-\mu$*;*
  - *locally bounded on* $\mathbb{R}^n \backslash \{\mathbf{0}\}$ *and continuous at* ∞ *(Definition 3.15);*
  - *continuous on* $\mathbb{R}^n \backslash \{\mathbf{0}\}$ *provided that all solutions of the system* (7.6) *are unique in the backward time and depend continuously on the initial data* x_0.

Proof **I.** Since $\mathbf{d}$ is a dilation then there exists $s > 0$ such that $c := \|\mathbf{d}(s)\| > 1$. By Corollary 7.3 the local uniform asymptotic stability of a homogeneous system is equivalent to its global uniform asymptotic stability. Consequently, there exists $T' > 0$ such that $\|x(t, x_0)\| \le 1/c$ for $t \ge T'$ for all $x_0 \in \mathbb{R}^n : \|x_0\| \le 1$.

Let us introduce the following notation:

- $\Delta T_0 = e^{\mu s} T'$ and $\Delta T_i = e^{\mu s} \Delta T_{i-1}$ for $i = 1, 2, \ldots$,
- $T_0 = 0$ and $T_i = T_{i-1} + \Delta T_{i-1}$ for $i = 1, 2, \ldots$,
- $x_i = x(T_i, x_0)$ for $i = 1, 2, \ldots$,

where $x(\cdot, x_0)$ is a an arbitrary solution of the system (7.6) with $\|x_0\| \le 1$.

Since $x(T_1, x_0) = x_1$ then $x(\tau, x_1) = x(T_1 + \tau, x_0)$, $\tau > 0$ and, by construction, $\|x_1\| \le 1/c$. Using the $\mathbf{d}$-homogeneity we conclude that

$$x(t, \mathbf{d}(s) x_1) = \mathbf{d}(s) x(e^{\nu s} t, x_1) = \mathbf{d}(s) x(T_1 + e^{\mu s} t, x_0), \quad t > 0$$

is also a solution of the system (7.6). Moreover, $x(0, \mathbf{d}(s)x_1) = \mathbf{d}(s)x_1$ and

$$\|x(0, \mathbf{d}(s)x_1)\| \le \|\mathbf{d}(s)\| \, \|x_1\| \le 1.$$

In this case, the uniform asymptotic stability of the system guarantees

$$\|x(T', \mathbf{d}(s)x_1)\| = \|\mathbf{d}(s)x(T_1 + e^{\mu s}T', x_0)\| =$$

$$\|\mathbf{d}(s)x(T_1 + \Delta T_1, x_0)\| = \|\mathbf{d}(s)x(T_2, x_0)\| = \|\mathbf{d}(s)x_2\| \le 1/c$$

and

$$\|\mathbf{d}(2\,s)x_2\| \le \|\mathbf{d}(s)\| \cdot \|\mathbf{d}(s)x_2\| \le 1.$$

Repeating the same consideration we derive $\|\mathbf{d}(is)x_i\| \le 1$ and

$$\|x(T_i, x_0)\| = \|\mathbf{d}(-is)\mathbf{d}(is)x_i\| \le \|\mathbf{d}(-is)\| \to 0 \quad \text{as} \quad i \to \infty$$

due to the limited property of the dilation. Since $\Delta T_i = T'e^{i\mu s}$ then for $\mu < 0$ we have $e^{\mu s} < 1$ and

$$T_i = T' \sum_{j=0}^{i-1} e^{j\mu s} \to \frac{T'}{1 - e^{\mu s}} \quad \text{as} \quad i \to \infty.$$

In other words, $x(t, x_0) \to 0$ as $t \to \frac{T'}{1-e^{\mu s}}$ provided that $\|x_0\| \le 1$. This means that the system (7.6) is locally uniformly finite-time stable (see Definition 3.12), but the homogeneity (7.6) implies its globally uniform finite-time stability.

Let us show now that the settling-time function is $\mathbf{d}$-homogeneous of degree $-\mu > 0$. Let $S(x_0)$ be a set of solutions of our $\mathbf{d}$-homogeneous system (7.6) with the initial condition $x(0) = x_0 \in \mathbb{R}^n$. Let us introduce the following function

$$T(x_0) = \sup_{x(\cdot, x_0) \in S(x_0)} \; \inf_{\tau \ge 0 : x(\tau, x_0) = \mathbf{0}} \tau.$$

The uniform finite-time stability guarantees local boundedness of T on $\mathbb{R}^n$. Lyapunov stability implies that $x(t) = \mathbf{0}$ is the unique solution of the evolution system with the zero initial condition. Hence, by construction, T is the settling-time function. Since $\mathbf{d}$-homogeneity of the system implies the symmetry of solutions $x(\tau, \mathbf{d}(s)x_0) = \mathbf{d}(s)x(e^{\mu s}\tau, x_0)$ then

$$T(\mathbf{d}(s)x_0) = \sup_{x(\cdot, \mathbf{d}(s)x_0) \in S(\mathbf{d}(s)x_0)} \; \inf_{\tau \ge 0 : x(\tau, \mathbf{d}(s)x_0) = \mathbf{0}} \tau =$$

$$\sup_{x(\cdot, x_0) \in S(x_0)} \; \inf_{\tau \ge 0 : x(e^{\mu s}\tau, x_0) = \mathbf{0}} \tau = e^{-\mu s} \sup_{x(\cdot, x_0) \in S(x_0)} \; \inf_{\tau \ge 0 : x(\tau, x_0) = \mathbf{0}} \tau = e^{-\mu s} T(x_0).$$

This means that the function T is $\mathbf{d}$-homogeneous of degree $-\mu$. Finally, since T is locally bounded then

$$\sup_{\|x_0\|_{\mathbf{d}}\leq e^s} T(x_0) = \sup_{\|\mathbf{d}(-s)x_0\|_{\mathbf{d}}\leq 1} T(x_0) = \sup_{\|y_0\|_{\mathbf{d}}\leq 1} T(\mathbf{d}(s)y_0) = e^{-\mu s} \sup_{\|y_0\|_{\mathbf{d}}\leq 1} T(y_0),$$

where $y_0 = \mathbf{d}(-s)x_0$. Taking into account $\mu < 0$ we derive $e^{-\mu s} \to 0$ as $s \to -\infty$ and

$$\sup_{\|x_0\|\leq e^s} T(x_0) \to 0 \quad \text{as} \quad s \to -\infty,$$

i.e., T is continuous at $\mathbf{0}$. The continuity of T on $\mathbb{R}^n$ is follows from Proposition 3.2.

II. Repeating the above considerations for $\mu = 0$ we derive

$$\|x(T_i)\| \leq \|\mathbf{d}(-is)\| \leq \|\mathbf{d}(-s)\|^i, \quad T_i = iT'$$

Due to the limit property of the dilation, the parameter s can always be selected such that $\|\mathbf{d}(s)\| > 1$ and $\|\mathbf{d}(-s)\| < 1$. We have

$$\|x(T_i)\| \leq e^{T_i \ln \frac{\|\mathbf{d}(-s)\|}{T'}}$$

and taking into account uniform asymptotic stability we conclude that the origin is locally exponentially stable, but the homogeneity implies global exponential stability.

III. Let $\mu > 0$ and, as before, the set $S(x_0)$ denote the set of solutions of the system (7.6) with the initial condition $x(0) = x_0$. Let $\|\cdot\|_{\mathbf{d}}$ be a canonical homogeneous norm. Let us consider the functional

$$T(x_0, r) = \sup_{x(\cdot,x_0)\in S(x_0)} \; \inf_{\|x(t,x_0)\|_{\mathbf{d}}\leq r, \forall t\geq\tau} \tau,$$

which estimates the convergence time of solutions (initiated at x_0) into the $\mathbf{d}$-homogeneous ball of the radius $r > 0$. This functional is well defined and locally bounded since the system is assumed to be uniformly asymptotically stable and the canonical homogeneous norm is topologically equivalent to a norm in $\mathbb{R}^n$ (see Theorem 5.3 and Corollary 5.3). Recall that $\|\mathbf{d}(s)x\|_{\mathbf{d}} = e^s\|x\|_{\mathbf{d}}, \forall s \in \mathbb{R}, \forall x \in \mathbb{R}^n$. The $\mathbf{d}$-homogeneity of (7.6) implies the symmetry of solutions (7.4), so for any $s \in \mathbb{R}$ one holds

$$T(\mathbf{d}(s)x_0, r) = \sup_{x(\cdot,x_0)\in S(x_0)} \; \inf_{\|\mathbf{d}(s)x(e^{\mu s}t,x_0)\|_{\mathbf{d}}\leq r, \forall t\geq\tau} \tau =$$

$$\sup_{x(\cdot,x_0)\in S(x_0)} \; \inf_{e^s\cdot\|x(e^{\mu s}t,x_0)\|_{\mathbf{d}}\leq r, \forall t\geq\tau} \tau = \sup_{x(\cdot,x_0)\in S(x_0)} \; \inf_{\|x(\tilde{t},x_0)\|\leq e^{-s}r, \forall\tilde{t}\geq\tilde{\tau}} e^{-\mu s}\tilde{\tau} =$$

$$e^{-\mu s}T(x_0, e^{-s}r).$$

Let us denote $r_i = e^{is_0}r$, $i = 0, 1, 2...$ with $s_0 > 0$ and $r > 0$. Then $r_i < r_{i+1}$ and $r_i \to +\infty$ as $i \to +\infty$ and $r_i \to 0$ as $i \to -\infty$. Let us introduce the numbers

$$T_i = \sup_{\|x_0\| \leq r_i} T(x_0, r_{i-1}), \quad i = 1, 2, \ldots.$$

The number T_i estimates a convergence time of all solutions of the system from the $\mathbf{d}$-homogeneous ball of the radius r_i into the $\mathbf{d}$-homogeneous ball of the radius r_{i-1}. The uniform asymptotic stability of the system (7.6) guarantees that all these numbers are finite. Since $\|x_0\|_{\mathbf{d}} = e^s \|\mathbf{d}(-s)x_0\|_{\mathbf{d}}, \forall s \in \mathbb{R}$ then

$$T_i = \sup_{e^{s_0}\|\mathbf{d}(-s_0)x_0\|_{\mathbf{d}} \leq r_i} T(x_0, r_{i-1}) = \sup_{\|\mathbf{d}(-s_0)x_0\|_{\mathbf{d}} \leq r_{i-1}} T(x_0, r_{i-1}) =$$

$$\sup_{\|\tilde{x}_0\|_{\mathbf{d}} \leq r_{i-1}} T(\mathbf{d}(s_0)\tilde{x}_0, r_{i-1}) = e^{-\mu s_0} \sup_{\|\tilde{x}_0\|_{\mathbf{d}} \leq r_{i-1}} T(\tilde{x}_0, r_{i-1}e^{-s}) = e^{-\mu s_0} T_{i-1} =$$

$$e^{-\mu(i-1)s_0} T_1.$$

Hence, for $x_0 \in B(r_n)$ we derive

$$T(x_0, r) = T_1 \sum_{i=1}^{n} e^{-\mu(i-1)s_0}.$$

Since $r_i \to +\infty$ as $i \to +\infty$ and $\mu > 0$ then $\forall x_0 \in \mathbb{B}$ we have

$$T(x_0, r) \leq T_1 \sum_{i=1}^{+\infty} e^{-\mu(i-1)s_0} = \frac{T_1}{1 - e^{-\mu s_0}} < +\infty, \quad \forall x_0 \in \mathbb{R}^n, \forall r > 0.$$

We complete the proof taking into account that T_1 is independent of x_0. Therefore, the system is nearly fixed-time stable and according to Proposition 3.3 there exists a locally bounded escape-time function $\tau : \mathbb{R}^n \backslash \{\mathbf{0}\} \mapsto \mathbb{R}_+$.

Let us "swap" the zero and infinity points (as in the proof of Proposition (3.4)) and the time direction by introducing an y-system with solutions

$$y(t, y_0) = \frac{\mathbf{d}(-\ln \|x(-t, x_0)\|_{\mathbf{d}})x(-t, x_0)}{\|x(-t, x_0)\|_{\mathbf{d}}}$$

induced by solutions of the x-system obtained in the backward time. Since $\|y(t, y_0)\| = \frac{1}{\|x(-t,x_0)\|_{\mathbf{d}}}$ then

$$x(-t, x_0) = \frac{\mathbf{d}(-\ln \|y(t, y_0)\|)y(t, y_0)}{\|y(t, y_0)\|}.$$

The nearly fixed-time stability of the x-system proven above implies that the y-system is finite-time stable and the setting-time function is given by

$$T(y_0) = \tau\left(\frac{\mathbf{d}(-\ln \|y_0\|)y_0}{\|y_0\|}\right).$$

Moreover, **d**-homogeneity of the x-system with degree μ implies that the y-system is standard homogeneous of degree $-\mu$:

$$y(t, e^s y_0) = e^s y(e^{-\mu s}t, y_0).$$

As shown above, the latter implies that the settling-time function is standard homogeneous of degree μ, i.e., $T(e^s y_0) = e^{\mu s}T(y_0), \forall s \in \mathbb{R}, \forall y_0 \in \mathbb{R}^n$. Consequently, τ is **d**-homogeneous of degree $-\mu$:

$$T(e^s y_0) = \tau\left(\frac{\mathbf{d}(-\ln\|e^s y_0\|)e^s y_0}{\|e^s y_0\|}\right) = \tau\left(\mathbf{d}(-s)\frac{\mathbf{d}(-\ln\|y_0\|)y_0}{\|y_0\|}\right)$$

$$= \tau(\mathbf{d}(-s)x_0) = e^{\mu s}T(y_0) = e^{\mu s}\tau(x_0), \quad \forall s \in \mathbb{R}, \quad \forall x_0 \neq \mathbf{0}.$$

Finally, since the function τ is uniformly bounded on the unit sphere and homogeneous of negative degree $-\mu$ then, by Proposition 6.1, it is continuous at ∞. Continuity of τ on $\mathbb{R}^n \backslash \{\mathbf{0}\}$ is follows from Proposition 3.4. The proof is complete. □

Example 7.3 Let us consider the system

$$\begin{cases} \dot{x}_1 = x_2, \\ \dot{x}_2 = u(x), \end{cases}$$

where $x = (x_1, x_2)^\top \in \mathbb{R}^2$ and

$$u(x) = -k_1\lfloor x_1\rceil^{1+\mu} - k_2\lfloor x_2\rceil^{\frac{2+2\mu}{2+\mu}}, \quad k_1 > 0, \quad k_2 > 0, \quad \mu > -1.$$

Inspired by [10, 12], let us consider the function $V : \mathbb{R}^2 \mapsto \mathbb{R}$ given by

$$V(x) = \frac{1}{2+\mu}k_1|x_1|^{2+\mu} + 0.5x_2^2.$$

The time derivative of V along the trajectories of the considered system is

$$\dot{V}(x) = -k_2|x_2|^{\frac{4+3\mu}{2+\mu}}.$$

Therefore, V is a weak Lyapunov function. On the one hand, the system is asymptotically stable due to Theorem 4.8. On the other hand, the system is **d**-homogeneous of degree μ with respect to the dilation $\mathbf{d}(s) = \begin{pmatrix} e^{2s} & 0 \\ 0 & e^{(2+\mu)s} \end{pmatrix}$, $s \in \mathbb{R}$. According to Theorem 7.5, the system is globally finite-time stable for $\mu \in (-1, 0)$ and globally nearly fixed-time stable for $\mu > 0$.

7.2.2 Homogeneous Lyapunov Function

In Chap. 4, we have discussed a characterization of various stability properties of the system (7.6) in terms of Lyapunov functions. In practice, the design of a Lyapunov function for a nonlinear system is a challenging problem [8, 13, 14]. Any property of the system (such as symmetry) gives some hints for the selection of a Lyapunov function candidate. For example, if a vector field f of an asymptotically stable system $\dot{x} = f(x)$ is central symmetric $f(-x) = -f(x), \forall x \in \mathbb{R}^n$, then the system always admits a symmetric Laypynov function $V(-x) = V(x)$. Indeed, if V is such that $\dot{V}(x) = \frac{\partial V}{\partial x} f(x) < 0$ then $V_{new}(x) = V(x) + V(-x)$, due to central symmetry of f, satisfies $\dot{V}_{new}(x) = \frac{\partial V_{new}}{\partial x} f(x) = \frac{\partial V}{\partial x} f(x) + \frac{\partial V(y)}{\partial y} f(y)\Big|_{y=-x} < 0$. Similarly, a Lyapunov function of a homogeneous system can be transformed into a homogeneous Lyapunov function.

An asymptotic stability of a weighted homogeneous ODE can be characterized in terms of a weighted homogeneous Lyapunov function. The mentioned characterization by a continuous Lyapunov function was originally discovered by Vladimir Zubov in 1958, [1]. A smooth Lyapunov characterization of asymptotically stable homogeneous ODEs was obtained in 1992 by Lionel Rosier, [15]. The Zubov–Rosier theorems for differential equations and inclusions, which are symmetric with respect to linear dilations, were presented in [4, 5], respectively.

Theorem 7.6 *Let an upper semi-continuous vector field $F : \mathbb{R}^n \mapsto \mathbb{R}^n$ be nonempty-valued, compact-valued, convex-valued and* **d***-homogeneous of degree μ. Let $m > 0$ be an arbitrary positive number. The system* (7.6) *is uniformly asymptotically stable if and only if there exists a positive definite homogeneous function $V \in C(\mathbb{R}^n) \cap C^\infty(\mathbb{R}^n \backslash \{\mathbf{0}\})$ of degree m such that*

$$\dot{V}(x) := \sup_{z \in F(x)} \frac{\partial V(x)}{\partial x} z \leq -\rho V^{1+\frac{\mu}{m}}(x), \quad \forall x \in \mathbb{R}^n \backslash \{\mathbf{0}\}, \tag{7.12}$$

where $\rho > 0$ is some number. Moreover, if $m > \alpha$ then $V \in C^1(\mathbb{R}^n) \cap C^\infty(\mathbb{R}^n \backslash \{\mathbf{0}\})$, where $\alpha > 0$ is defined in Proposition 5.3.

Proof *Sufficiency* follows from Theorem 4.7 since positive definite homogeneous function of degree $m > 0$ is always globally proper

$$\underline{k} \|x\|_{\mathbf{d}}^m \leq V(x) = \|x\|_{\mathbf{d}}^\mu V(\mathbf{d}(-\ln \|x\|_{\mathbf{d}})x) \leq \overline{k} \|x\|_{\mathbf{d}}^m,$$

where $0 < \underline{k} = \inf_{y \in S} V(y) \leq \sup_{y \in S} V(y) = \overline{k} < +\infty$.

Necessity. By Corollary 7.3 local stability of a homogeneous system implies global stability. By Theorem 4.7 there exists a globally proper positive definite function $V_0 \in C^\infty(\mathbb{R}^n)$ and a positive function $W_0 \in C^\infty(\mathbb{R}^n \backslash \{\mathbf{0}\}; (0, +\infty))$ such that

$$\dot{V}_0(x) = \sup_{z\in F(x)} \frac{\partial V_0(x)}{\partial x} z \le -W_0(x), \quad \forall x \neq \mathbf{0}.$$

Since V_0 is globally proper then there exists $\underline{\rho}, \overline{\rho} \in \mathcal{K}_\infty$ such that

$$\underline{\rho}(\|x\|_{\mathbf{d}}) \le V_0(x) \le \overline{\rho}(\|x\|_{\mathbf{d}}), \quad \forall x \in \mathbb{R}^n.$$

Let $r > 0$ be an arbitrary number. Let $\overline{v} = \underline{\rho}(r)$ and let $r_0 < r$ be such that $\underline{v} = \overline{\rho}(r_0) < \overline{v}$. Then

$$\underline{v} \le V_0(x) \le \overline{v} \quad \Rightarrow \quad r_0 \le \|x\|_{\mathbf{d}} \le r.$$

Inspired by [15] let us define

$$V(x) = \int_{-\infty}^{+\infty} e^{-ms} a(V_0(\mathbf{d}(s)x))ds,$$

where $a \in C^\infty : a(\sigma) = 0$ for $\sigma \in [0, \underline{v}]$, $a'(\sigma) > 0$ for $\sigma \in (\underline{v}, \overline{v})$, $a(\sigma) = a(\overline{v}) = \text{const}$ for $\sigma \ge \overline{v}$. For example, $a(\sigma) = \int_0^\rho \hat{a}(\tau)d\tau$ with $\hat{a}(\tau) = e^{\frac{-1}{(\overline{v}-\tau)(\tau-\underline{v})}} > 0$ for $\tau \in (\underline{v}, \overline{v})$ and $\hat{a}(\tau) = 0$ otherwise.

The function V is positive definite and $V \in C(\mathbb{R}^n) \cap C^\infty(\mathbb{R}^n\backslash\{\mathbf{0}\})$. Indeed,

- $x = \mathbf{0} \;\Rightarrow\; \mathbf{d}(s)x = \mathbf{0}, \forall s \in \mathbb{R} \;\Rightarrow\; V_0(\mathbf{d}(s)x) = 0, \forall s \in \mathbb{R} \;\Rightarrow\; V(x) = 0$
- $x \neq \mathbf{0} \;\Rightarrow\; 0 < r_0 \le \|\mathbf{d}(s)x\|_{\mathbf{d}} = e^s\|x\|_{\mathbf{d}} \le r \;\Rightarrow\; s \in \left[\ln \frac{r_0}{\|x\|_{\mathbf{d}}}, \ln \frac{r}{\|x\|_{\mathbf{d}}}\right]$.

Hence, we derive

$$\underline{v} \le V_0(\mathbf{d}(s)x) \le \overline{v} \;\Rightarrow\; r_0 \le \|\mathbf{d}(s)x\|_{\mathbf{d}} = e^s\|x\|_{\mathbf{d}} \le r \;\Rightarrow\; s \in \left[\ln \frac{r_0}{\|x\|_{\mathbf{d}}}, \ln \frac{r}{\|x\|_{\mathbf{d}}}\right]$$

and for any $x \neq \mathbf{0}$ we have

$$0 < V(x) = \int_{-\infty}^{\ln \frac{r_0}{\|x\|_{\mathbf{d}}}} e^{-ms} \underbrace{a(\underline{v})}_{=0} ds + \int_{\ln \frac{r_0}{\|x\|_{\mathbf{d}}}}^{\ln \frac{r}{\|x\|_{\mathbf{d}}}} e^{-ms} a(V_0(\mathbf{d}(s)x))ds + \int_{\ln \frac{r}{\|x\|_{\mathbf{d}}}}^{+\infty} e^{-ms} a(\overline{v})ds < +\infty.$$

Making change of integration variable $\tilde{s} = s - \ln \frac{r_0}{\|x\|_{\mathbf{d}}}$ we derive

$$V(x) = \frac{\|x\|_{\mathbf{d}}^m}{r_0^m} \int_0^{+\infty} e^{-m\tilde{s}} a\left(V_0\left(\mathbf{d}\left(\tilde{s} + \ln \frac{r_0}{\|x\|_{\mathbf{d}}}\right) x\right)\right) d\tilde{s}$$

$$= \frac{\|x\|_{\mathbf{d}}^m}{r_0^m} \int_0^{+\infty} e^{-m\tilde{s}} a\left(V_0\left(\mathbf{d}\left(\tilde{s} + \ln r_0\right) \pi(x)\right)\right) d\tilde{s} \le \frac{\|x\|_{\mathbf{d}}^m}{r_0^m} \frac{a(\overline{v})}{m},$$

where $\pi(x) = \mathbf{d}(-\ln \|x\|_{\mathbf{d}})x \in S$. Since $m > 0$ then $\|x\|_{\mathbf{d}} \to 0$ as $x \to 0$ and V is continuous at $\mathbf{0}$. Obviously, by construction, $a, V_0 \in C^{\infty} \Rightarrow V \in C^{\infty}(\mathbb{R}^n\backslash\{\mathbf{0}\})$. Moreover, since $\pi(\mathbf{d}(s)x) = \pi(x), \forall x \neq \mathbf{0}$ and $\|\mathbf{d}(s)x\|_{\mathbf{d}} = e^s\|x\|_{\mathbf{d}}, \forall s \in \mathbb{R}, \forall x \in \mathbb{R}^n$ then V is $\mathbf{d}$-homogeneous of degree m. Using formula (6.2) with $s = -\ln \|x\|_{\mathbf{d}}$ we derive

$$\left|\frac{\partial V(x)}{\partial x}\right| \leq C\|x\|_{\mathbf{d}}^m \|\mathbf{d}(-\ln \|x\|_{\mathbf{d}})\| \leq C\|x\|_{\mathbf{d}}^{m-\alpha},$$

where $C = \max_{z\in S}\left|\frac{\partial V(z)}{\partial z}\right| < +\infty$ and $\alpha > 0$ is defined in Proposition 5.3. Hence, $V \in C^1(\mathbb{R}^n)$ for $m > \alpha$. By construction, the function V is positive definite and

$$\dot{V}(x) = \sup_{z\in F(x)} \tfrac{\partial V(x)}{\partial x} z \leq \int_{-\infty}^{+\infty} e^{-ms} a'(V_0(y)) \sup_{z\in F(x)} \left.\tfrac{\partial V_0(y)}{\partial y}\right|_{y=\mathbf{d}(s)x} \mathbf{d}(s)z ds.$$

Hence, using the $\mathbf{d}$-homogeneity of F we derive

$$\dot{V}(x) = \int_{-\infty}^{+\infty} e^{-ms} a'(V_0(y)) \sup_{z\in e^{-\mu s}\mathbf{d}(-s)F(y)} \left.\tfrac{\partial V_0(y)}{\partial y}\right|_{y=\mathbf{d}(s)x} \mathbf{d}(s)z ds =$$

$$\int_{-\infty}^{+\infty} e^{-(m+\mu)s} a'(V_0(y)) \sup_{\tilde{z}\in F(y)} \left.\tfrac{\partial V_0(y)}{\partial y}\tilde{z}\right|_{y=\mathbf{d}(s)x} ds \leq$$

$$-\int_{-\infty}^{+\infty} e^{-(m+\mu)s} a'(V_0(\mathbf{d}(s)x)) W_0(\mathbf{d}(s)x) ds = -W(x) < 0, \forall x \neq \mathbf{0},$$

where, by construction, $W \in C^{\infty}(\mathbb{R}^n\backslash\{\mathbf{0}\}; (0, +\infty))$ and W is $\mathbf{d}$-homogeneous of the degree $m + \mu$. Hence, there exists $\rho > 0$ such that

$$W(x) = V^{1+\frac{\mu}{m}} W(\mathbf{d}(-\ln V^{1/m}(x))x) \geq \rho V^{1+\frac{\mu}{m}},$$

where $\rho = \inf_{x\neq\mathbf{0}} h(x) > 0$ with $h(x) = W(\mathbf{d}(-\ln V(x))x)$ being positive definite and $\mathbf{d}$-homogeneous of degree 0 ($\Rightarrow h$ is uniformly bounded from below by a positive constant due to Proposition 6.1). The proof is complete. □

Remark 7.2 Asymptotically stable homogeneous ODE may not have a *homogeneous* Lyapunov function which is infinitely smooth at the origin. Indeed, if a Lyapunov function is standard homogeneous of degree m, then, due to formula (6.2), the homogeneity degree of its partial derivative is $m - 1$. In the view of Proposition 6.1, its mth derivative is either discontinuous at the origin or a constant. The latter is possible if and only if the Lyapunov function is a polynomial of degree m, but, as shown in [13], homogeneous systems may not have analytic Lyapunov functions at all.

Example 7.4 Let us consider the twisting system from Example 7.3, which is $\mathbf{d}$-homogeneous of degree -1. Inspired by [10, 16], let us define the Lyapunov function candidate as follows

$$V(x) = \left(k_1|x_1| + \tfrac{x_2^2}{2}\right)^{\frac{3}{2}} + cx_1x_2, \qquad c > 0.$$

The function V is $\mathbf{d}$-homogeneous of degree 3. Indeed,

$$V(\mathbf{d}(s)x) = \left(\alpha|e^{2s}x_1| + \tfrac{(e^s x_2)^2}{2}\right)^{\frac{3}{2}} + ce^{2s}x_1e^s x_2 = e^{3s}V(x).$$

Moreover, using *Young's inequality*[1], it can be shown that V is positive definite for a sufficiently small $c > 0$ and its time derivative along the trajectories of the system admits the following estimate:

$$\dot{V}(x) = -k_2|x_2|\sqrt{k_1|x_1| + 0.5x_2^2} + cx_2^2 - ck_1|x_1| - ck_2x_1\,\mathrm{sign}(x_2) \leq$$

$$\leq -(\tfrac{k_2}{\sqrt{2}} - c)x_2^2 - c(k_1 - k_2)|x_1| \leq -\rho V^{1+\frac{-1}{3}}(x),$$

where the positive parameter $\rho > 0$ depends on $k_1 > k_2 > 0$ and $c > 0$.

For asymptotically stable system, which is homogeneous at 0 and/or ∞-limit, a Lyapunov function can always be designed locally homogeneous [11].

Corollary 7.4 (On Lyapunov function for locally homogeneous system)

- *Let a continuous vector field $f : \mathbb{R}^n \mapsto \mathbb{R}^n$ be $\mathbf{d}_L$-homogeneous of degree $\mu_L \in \mathbb{R}$ at L-limit and $f_L : \mathbb{R}^n \mapsto \mathbb{R}^n$ be a $\mathbf{d}_L$-homogeneous approximation of f, where $L = 0$ or $L = \infty$. If the $\mathbf{d}_L$-homogeneous system*

$$\dot{x} = f_L(x), \quad t > 0$$

is asymptotically stable then for any $m_L > 0$ there exist numbers $\rho_L > 0$, $r_L > 0$ and a positive definite $\mathbf{d}_L$-homogeneous function $V_L \in C(\mathbb{R}^n) \in C^\infty(\mathbb{R}^n\backslash\{\mathbf{0}\})$ of degree m_L such that

$$\tfrac{\partial V_L(x)}{\partial x} f(x) \leq -\rho_L V_L^{1+\frac{\mu_L}{m_L}}(x), \tag{7.13}$$

$\forall x \in \mathbb{R}^n\backslash\{\mathbf{0}\} : \|x\|_{\mathbf{d}} \leq r_0$ *if* $L = 0$ *or* $\forall x \in \mathbb{R}^n : \|x\|_{\mathbf{d}} \geq r_\infty$ *if* $L = \infty$.

- *Let a continuous vector field $f : \mathbb{R}^n \mapsto \mathbb{R}^n$ be $\mathbf{d}_0$-homogeneous of degree $\mu_0 \in \mathbb{R}$ at 0-limit and $\mathbf{d}_\infty$-homogeneous of degree $\mu_\infty \in \mathbb{R}$ at ∞-limit. Let $f_0, f_\infty : \mathbb{R}^n \mapsto \mathbb{R}^n$ be the corresponding homogeneous approximations of f and $m_0 > 0$, $m_\infty > 0$ be arbitrary numbers. If the systems*

[1] $ab \leq \frac{a^p}{p} + \frac{b^q}{q}$, $\forall a, b \geq 0$ if $p, q \geq 1 : \frac{1}{p} + \frac{1}{q} = 1$.

$$\dot{x} = f(x), \quad \dot{x} = f_0(x), \quad \dot{x} = f_\infty(x), \quad t > 0$$

are globally uniformly asymptotically stable, then there exist a globally proper positive definite function $V \in C(\mathbb{R}^n) \cap C^\infty(\mathbb{R}^n \backslash \{\mathbf{0}\})$ and a constant $\rho > 0$ such that

$$\dot{V}(x) \leq -\rho\sigma(V(x)), \quad x \neq \mathbf{0}, \tag{7.14}$$

where

$$\sigma(V) = \begin{cases} V^{1+\frac{\mu_0}{m_0}} & \text{if } V \in (0, 1), \\ V^{1+\frac{\mu_\infty}{m_\infty}} & \text{if } V \geq 1. \end{cases} \tag{7.15}$$

Moreover, for small enough $r_0 > 0$ and large enough $r_\infty > 0$ we have $V(x) = V_\infty(x)$ for all $x : \|x\|_{\mathbf{d}_\infty} \geq 2r_\infty$, where V_L is $\mathbf{d}_L$-homogeneous function of degree $m_L > 0$ satisfying (7.13) *and $L \in \{0, \infty\}$.*

Proof (1) By Theorem 7.6, there exists a $\mathbf{d}_L$-homogeneous Lypapunov function $V_L \in C(\mathbb{R}^n) \cap C^\infty(\mathbb{R}^n \backslash \{\mathbf{0}\})$ of degree $\mu_L > 0$ such that

$$\exists \underline{k}_L, \overline{k}_L > 0 : \quad \underline{k}_L \|x\|_{\mathbf{d}_L}^{m_L} \leq V_L(x) \leq \overline{k}_L \|x\|_{\mathbf{d}_L}^{m_L}, \tag{7.16}$$

$$\frac{\partial V_L(x)}{\partial x} f_L(x) \leq -2\rho_L V_L^{1+\frac{\mu_L}{m_L}}, \quad \forall x \neq \mathbf{0}$$

with some $\rho_L > 0$, where $\|\cdot\|_{\mathbf{d}_L}$ is a $\mathbf{d}_L$-homogeneous norm in $\mathbb{R}^n$. In this case, using $\mathbf{d}_L$-homogeneity of the derivative of V (see Proposition 6.2) we derive

$$\frac{\partial V_L(x)}{\partial x} f(x) = \frac{\partial V_L(x)}{\partial x} f(x) + \frac{\partial V_L(x)}{\partial x}(f(x) - f_L(x))$$

$$\leq -2\rho_L V_L^{1+\frac{\mu_L}{m_L}}(x) + V_L(x)\frac{\partial V(y)}{\partial y}\mathbf{d}_L\left(-\ln V_L^{\frac{1}{m_L}}(x)\right)(f(x) - f_L(x))$$

for all $x \neq \mathbf{0}$, where $y = \mathbf{d}_L\left(-\ln V_L^{\frac{1}{m_L}}(x)\right)x$. Notice that

$$V_L(x)\frac{\partial V(y)}{\partial y}\mathbf{d}_L\left(-\ln V_L^{\frac{1}{m_L}}(x)\right)(f(x) - f_L(x)) = V_L^{1+\frac{\mu_L}{m_L}}(x) \cdot \frac{\partial V(y)}{\partial y}$$

$$\cdot \frac{\|x\|_{\mathbf{d}_L}^{\mu_L}}{V_L^{\frac{\mu_L}{m_L}}(x)} \cdot \mathbf{d}\left(-\ln \frac{V_L^{\frac{1}{m_L}}(x)}{\|x\|_{\mathbf{d}_L}}\right) \cdot \|x\|_{\mathbf{d}_L}^{-\mu_L}\mathbf{d}(-\ln\|x\|_{\mathbf{d}_L})(f(x) - f_L(x).$$

On the one hand, the inequality (7.16) implies that there exists $C_1 > 0$:

$$\left\| \frac{\|x\|_{\mathbf{d}_L}^{\mu_L}}{V_L^{\frac{\mu_L}{m_L}}(x)} \cdot \mathbf{d}\left(-\ln \frac{V_L^{\frac{1}{m_L}}(x)}{\|x\|_{\mathbf{d}_L}}\right) \right\| \leq C_1, \quad \forall x \neq \mathbf{0}.$$

Moreover, since V_L is $\mathbf{d}_L$-homogeneous of degree m_L then

$$V_L(y) = V_L\left(\mathbf{d}_L(-\ln V_L(x))x\right) = e^{-m_L \ln V_L^{\frac{1}{m_L}}(x)} V_L(x) = 1,$$

i.e., the vector $y = \mathbf{d}_L\left(-\ln V_L^{\frac{1}{m_L}}(x)\right)x$ belongs to a compact $\{y \in \mathbb{R}^n : V_L(y) = 1\}$ for any $x \neq \mathbf{0}$. Hence, using $V_L \in C(\mathbb{R}^n)$ we conclude

$$\exists C_2 > 0 \quad : \quad \left| \left.\frac{\partial V(y)}{\partial y}\right|_{y=\mathbf{d}_L\left(-\ln V_L^{\frac{1}{m_L}}(x)\right)x} \right| \leq C_2, \quad \forall x \neq \mathbf{0}.$$

On the other hand, by Corollary 6.3, we have

$$\sup_{\|x\|_{\mathbf{d}_L}=r} \|e^{-\mu_L \ln r}\mathbf{d}_L(-\ln r)(f(x) - f_L(x))\| \to 0 \quad \text{as} \quad r \to L,$$

where $L = 0$ of $L = \infty$. Hence, there exists $r_L > 0$ such that

$$\| \|x\|_{\mathbf{d}_L}^{-\mu_L}\mathbf{d}(-\ln \|x\|_{\mathbf{d}_L})(f(x) - f_L(x))\| \leq \frac{\rho_L}{C_1 C_2}$$

for all $x \in \mathbb{R}^n \backslash \{\mathbf{0}\} : \|x\|_{\mathbf{d}} \leq r_0$ if $L = 0$ and for all $x \in \mathbb{R}^n \backslash \{\mathbf{0}\} : \|x\|_{\mathbf{d}} \geq r_\infty$ if $L = \infty$. Therefore, we derive (7.13).

(2) If f is homogeneous in the bi-limit then, as shown above, there exist $V_L \in C(\mathbb{R}^n) \cap C^\infty(\mathbb{R}^n \backslash \{\mathbf{0}\})$, $\rho_L > 0$ and $r_L > 0$ such that (7.13) holds for both $L = 0$ and $L = \infty$.

By Theorem 4.7 there exist a globally proper positive definite function $V_r \in C^\infty(\mathbb{R}^n)$ and a positive function $W_r \in C^\infty(\mathbb{R}^n \backslash \{\mathbf{0}\}; (0, +\infty))$ such that

$$\dot{V}_r(x) = \tfrac{\partial V_r(x)}{\partial x} f(x) \leq -W_r(x), \quad \forall x \neq \mathbf{0}.$$

Since V_r is globally proper then there exists $\underline{k}, \overline{k} \in \mathcal{K}_\infty$ such that

$$\underline{k}(\|x\|_{\mathbf{d}_0}) \leq V_r(x) \leq \overline{k}(\|x\|_{\mathbf{d}_\infty}), \quad \forall x \in \mathbb{R}^n.$$

Without loss of generality, we assume that $r_0 > 0$ is selected small enough and $r_\infty > 0$ is selected large enough to guarantee $\underline{k}(r_0) < 1 < \overline{k}(r_\infty)$.

Inspired by [11], let us consider the function $V : \mathbb{R}^n \mapsto \mathbb{R}_+$ defined as

$$V = w_\infty \phi_\infty(V_r)V_\infty + (1-\phi_\infty(V_r))\,\phi_0(V_r)V_r + w_0(1-\phi_0(V_r))V_0,$$

where $\phi_0, \phi_\infty \in C^\infty$ are such that

$$\phi_0(\rho)=0 \text{ if } \rho \le \underline{k}\left(\tfrac{r_0}{2}\right),\ \phi_0'(\rho)>0 \text{ if } \rho \in \left(\underline{k}\left(\tfrac{r_0}{2}\right), \underline{k}(r_0)\right),\ \phi_0(\rho)=1 \text{ if } \rho \ge \underline{k}(r_0),$$

$$\phi_\infty(\rho)=0 \text{ if } \rho \le \overline{k}(r_\infty), \phi_\infty'(\rho)>0 \text{ if } \overline{k}(r_\infty)<\rho<\overline{k}(\tfrac{r_\infty}{0.5}), \phi_\infty(\rho)=1 \text{ if } \rho \ge \overline{k}(\tfrac{r_\infty}{0.5}),$$

$$w_0 = \inf_{\underline{k}(0.5r_0)\le V_r(x)\le \overline{k}(2r_\infty)} \frac{V_r(x)}{V_0(x)} \quad \text{and} \quad w_\infty = \sup_{\underline{k}(0.5r_0)\le V_r(x)\le \overline{k}(2r_\infty)} \frac{V_r(x)}{V_\infty(x)}.$$

The function V is positive definite and globally proper by construction. Moreover, $w_\infty \phi_\infty'(V_r)V_\infty \ge \phi_\infty'(V_r)\phi_0(V_r)V_r$ for $\underline{k}(0.5r_0) \le V_r \le \overline{k}(2r_\infty)$, $(1-\phi_\infty(V_r))\,\phi_0'(V_r)V_r \ge w_0\phi_0'(V_r)V_0$ for $\underline{k}(0.5r_0) \le V_r \le \overline{k}(2r_\infty)$, and

$$\dot V = (w_\infty \phi_\infty' V_\infty - \phi_\infty'\phi_0 V_r)\dot V_r + w_\infty\phi_\infty \dot V_\infty$$

$$+ \left((1-\phi_\infty)\,\phi_0' V_r - w_0\phi_0' V_0\right)\dot V_r + (1-\phi_\infty)\,\phi_0 \dot V_r + w_0(1-\phi_0)\dot V_0$$

$$\le w_0(1-\phi_0)\dot V_0 + \phi_0\,(1-\phi_\infty)\,\dot V_r + w_\infty\phi_\infty \dot V_\infty \le$$

$$-\rho_0 w_0(1-\phi_0)V_0^{1+\frac{\mu_0}{m_0}} - \rho_r\phi_0\,(1-\phi_\infty)\,(V^{1+\frac{\mu_0}{m_0}} + V^{1+\frac{\mu_\infty}{m_\infty}}) - \rho_\infty w_\infty \phi_\infty V_\infty^{1+\frac{\mu_\infty}{m_\infty}},$$

where

$$\rho_r = \inf_{\underline{k}(0.5r_0)\le V_r(x)\le \overline{k}(2r_\infty)} \frac{W(x)}{V^{1+\frac{\mu_0}{m_0}}(x)+V^{1+\frac{\mu_L}{m_l}}(x)} > 0.$$

We complete the proof using the identities $V = V_0$ for $V_r(x) \le \underline{k}(0.5r_0)$ and $V = V_\infty$ for $V_r(x) \ge \overline{k}(2r_\infty)$, which hold by construction.

In practice, homogeneous models may appear as local approximations of non-homogeneous physical systems. In this case, it is important to know if a local stability analysis of a system can be based on the analysis of its homogeneous approximation. The following corollary is given without proof, since it is a straightforward consequence of Corollary 7.4.

Corollary 7.5 *Let a continuous vector field* $g : \mathbb{R}^n \mapsto \mathbb{R}^n$ *admit a* $\mathbf{d}_0$-*homogeneous approximation* $g_0 : \mathbb{R}^n \mapsto \mathbb{R}^n$ *of degree* $\mu_0 \in \mathbb{R}$ *at 0-limit and* $\mathbf{d}_\infty$-*homogeneous approximation* $g_\infty : \mathbb{R}^n \mapsto \mathbb{R}^n$ *of degree* $\mu_\infty \in \mathbb{R}$ *at* ∞-*limit.*

(1) If the system $\dot x = g_0(x)$ *is uniformly asymptotically stable then the system* $\dot x = g(x)$ *is*

- *locally uniformly finite-time stable for* $\mu_0 < 0$;
- *locally uniformly exponentially stable for* $\mu_0 = 0$;
- *locally uniformly nearly fixed-time stable for* $\mu_0 > 0$.

(2) If the system $\dot{x} = g_\infty(x)$ is asymptotically stable then there exists $r > 0$ such that the system $\dot{x} = g(x)$ converges into a $\mathbf{d}_\infty$-homogeneous ball $B_{\mathbf{d}}(r) \in \mathbb{R}^n$

- *in a finite time if $\mu_\infty < 0$;*
- *exponentially if $\mu_\infty = 0$;*
- *in a fixed time if $\mu_\infty > 0$.*

(3) If $\mu_0 < 0$, $\mu_\infty > 0$ and the systems $\dot{x} = g(x)$, $\dot{x} = g_0(x)$ and $\dot{x} = g_\infty(x)$ are globally uniformly asymptotically stable, then the origin of the system $\dot{x} = g(x)$ is globally uniformly fixed-time stable.

7.2.3 Quadratic Stability

Theorem 7.6 is useful for further analysis of homogeneous systems. For instance, it simplifies the robustness analysis of homogeneous systems and helps to discover an analog of *quadratic stability*[2] of homogeneous systems. A topological equivalence of any asymptotically stable homogeneous system to a quadratically stable one follows from the following theorem [4, 5].

Theorem 7.7 *Let $\mathbf{d}$ be linear continuous dilation in $\mathbb{R}^n$ and $P \in \mathbb{R}^{n\times n}$ be an arbitrary symmetric positive definite matrix satisfying (5.6). The following five claims are equivalent.*

(1) The $\mathbf{d}$-homogeneous system (7.6) is asymptotically stable.
(2) The origin of the standard homogeneous system (7.8) with

$$\|z\| = \sqrt{z^\top P z}$$

is asymptotically stable.
(3) There exists a $\mathbf{d}$-homogeneous vector field $\Psi : \mathbb{R}^n \mapsto \mathbb{R}^n$ of degree 0 such that $\Psi \in C^\infty(\mathbb{R}^n\backslash\{\mathbf{0}\}; \mathbb{R}^n) \cap C(\mathbb{R}^n; \mathbb{R}^n)$ is diffeomorphism on $\mathbb{R}^n\backslash\{\mathbf{0}\}$, homeomorphism on $\mathbb{R}^n$, $\Psi(\mathbf{0}) = \mathbf{0}$ and

$$x \in \mathbb{R}^n : \Psi^\top(x)P\Psi(x) = 1 \quad \Rightarrow \quad \sup_{y\in F(x)} \frac{\partial(\Psi^\top(x)P\,\Psi(x))}{\partial x} y < 0. \qquad (7.17)$$

Moreover, the function

$$x \mapsto \|\Psi(x)\|_{\mathbf{d}}$$

[2] The ODE $\dot{z} \in \tilde{F}(z)$ with $z(t) \in \mathbb{R}^n$ is said to be quadratically stable if it admits a quadratic Lyapunov function $V(z) = z^\top P z$, $0 \prec P = P^\top \in \mathbb{R}^{n\times n}$, $z \in \mathbb{R}^n$:

$$\sup_{y\in F(x)} z^\top P y < 0, \quad \forall z \neq \mathbf{0}.$$

is a **d***-homogeneous Lyapunov function of the degree* 1 *for the system* (7.6), *where* $\|\cdot\|_{\mathbf{d}}$ *is the canonical homogeneous norm induced by* $\|\cdot\|$.

(4) There exists a mapping

$$\Xi \in C^{\infty}(\mathbb{R}^n \backslash \{\mathbf{0}\}; \mathbb{R}^{n\times n})$$

such that, for any $z = (z_1, \ldots, z_n)^{\top} \in \mathbb{R}^n\backslash\{\mathbf{0}\}$ *and any* $s \in \mathbb{R}$, *it holds*

$$\det(\Xi(z)) \neq 0, \quad \tfrac{\partial \Xi(z)}{\partial z_i} z = 0, \quad \Xi(e^s z) = \Xi(z), \quad i = 1, \ldots, n,$$

and

$$\sup_{y\in F\left(\frac{z}{\sqrt{z^{\top}Pz}}\right)} z^{\top}\Xi^{\top}(z)P\Xi(z)\left(\tfrac{(I_n - G_{\mathbf{d}})zz^{\top}P}{z^{\top}PG_{\mathbf{d}}z} + I_n\right) y < 0, \quad \forall z \neq \mathbf{0}. \tag{7.18}$$

Proof The equivalence (1) $\Leftrightarrow$ (2) follows from Theorem 7.3.

(3)$\Rightarrow$ (1) Notice that $\Psi(x) \neq \mathbf{0}$ for $x \neq \mathbf{0}$, otherwise (i.e., $\exists x^* \neq \mathbf{0} : \Psi(x^*) = \mathbf{0}$), due to homogeneity, we derive that $\Psi(x) = \mathbf{0}$ on a smooth homogeneous curve $\{\mathbf{d}(s)x^*, s \in \mathbb{R}\}$, which starts at the origin and goes to ∞. This contradicts the assumption that Ψ is diffeomorphism (a continuously differentiable *invertible* mapping with a continuously differentiable inverse) on $\mathbb{R}^n\backslash\{\mathbf{0}\}$. Since the homogeneity degree of Ψ is zero then $\|\Psi(\mathbf{d}(s)x)\|_{\mathbf{d}} = \|\mathbf{d}(s)\Psi(x)\|_{\mathbf{d}} = e^s\|\Psi(x)\|_{\mathbf{d}}$ and the function $\|\Psi(\cdot)\|_{\mathbf{d}}$ is **d**-homogeneous of the degree 1, radially unbounded, continuous at the origin and continuously differentiable away from the origin. Using formula (5.19) we derive

$$\frac{\partial\|\Psi(x)\|_{\mathbf{d}}}{\partial x} = e^s \left.\frac{\frac{\partial\|\mathbf{d}(-s)\Psi(x)\|}{\partial x}}{\left.\frac{\partial\|z\|}{\partial z}\right|_{z=\mathbf{d}(-s)\Psi(x)} G_{\mathbf{d}}\mathbf{d}(-s)\Psi(x)}\right|_{s=\ln\|\Psi(x)\|_{\mathbf{d}}} \quad \text{for} \quad \Psi(x) \neq \mathbf{0}.$$

Taking into account $\|\Psi(x)\| = 1 \Leftrightarrow \|\Psi(x)\|_{\mathbf{d}} = 1$ we conclude

$$\left.\left.\frac{\partial\|z\|}{\partial z}\right|_{z=\mathbf{d}(-s)\Psi(x)} G_{\mathbf{d}}\mathbf{d}(-s)\Psi(x)\right|_{s=\ln\|\Psi(x)\|_{\mathbf{d}}} = \Psi^{\top}(x)PG_{\mathbf{d}}\Psi(x) \quad \text{for} \quad \|\Psi(x)\| = 1$$

and

$$\frac{\partial\|\Psi(x)\|_{\mathbf{d}}}{\partial x} = \frac{\frac{\partial\|\Psi(x)\|}{\partial x}}{\Psi^{\top}(x)PG_{\mathbf{d}}\Psi(x)} \quad \text{for} \quad \|\Psi(x)\| = 1.$$

Since P satisfies (5.6) then the inequality $\sup_{y\in F(x)} \left.\frac{\partial\|\Psi(x)\|}{\partial x}y\right|_{x:\|\Psi(x)\|=1} < 0$ implies the inequality $\sup_{y\in F(x)} \left.\frac{\partial\|\Psi(x)\|_{\mathbf{d}}}{\partial x}y\right|_{x:\|\Psi(x)\|_{\mathbf{d}}=1} < 0$. The **d**-homogeneity of Ψ and F yields $\sup_{y\in F(x)} \frac{\partial\|\Psi(x)\|_{\mathbf{d}}}{\partial x}y < 0$ for any $x \in \mathbb{R}^n\backslash\{\mathbf{0}\}$, i.e., $\|\Psi\|_{\mathbf{d}}$ is a Lyapunov function for the system.

(1) $\Rightarrow$ (3) Let $V : \mathbb{R}^n \mapsto \mathbb{R}_+$ be (a continuous on $\mathbb{R}^n$ and smooth on $\mathbb{R}^n\backslash\{\mathbf{0}\}$) $\mathbf{d}$-homogeneous Lyapunov function of the degree 1, which exists due to Zubov–Rosier theorem (see Theorem 7.6). Let us consider the mapping $\Psi : \mathbb{R}^n \mapsto \mathbb{R}^n$ defined as

$$\Psi(x) = \mathbf{d}\left(\ln\left(\frac{V(x)}{\|x\|_{\mathbf{d}}}\right)\right)x.$$

Since the function $x \mapsto \frac{V(x)}{\|x\|_{\mathbf{d}}}$ is $\mathbf{d}$-homogeneous and $\deg_{\mathbf{d}}\left(\frac{V(\cdot)}{\|\cdot\|_{\mathbf{d}}}\right) = 0$ then

$$0 < \inf_{x\in S}\frac{V(x)}{\|x\|_{\mathbf{d}}} = \liminf_{\|x\|\to 0}\frac{V(x)}{\|x\|_{\mathbf{d}}} \leq \limsup_{\|x\|\to 0}\frac{V(x)}{\|x\|_{\mathbf{d}}} = \sup_{x\in S}\frac{V(x)}{\|x\|_{\mathbf{d}}} < +\infty.$$

Hence, we conclude that Ψ is continuous at $\mathbf{0}$ and $\Psi(\mathbf{0}) = \mathbf{0}$. Obviously, $\Psi(\mathbf{d}(s)x) = \mathbf{d}(s)\Psi(x)$ and

$$\|\Psi(x)\|_{\mathbf{d}} = V(x).$$

Hence, $x \mapsto \|\Psi(x)\|_{\mathbf{d}}$ is a $\mathbf{d}$-homogeneous Lyapunov function. Using formula (5.21) we derive (7.17).

The mapping Ψ is a bijection. The inverse mapping $\Psi^{-1} : \mathbb{R}^n \mapsto \mathbb{R}^n$ is given by

$$\Psi^{-1}(y) = \mathbf{d}\left(-\ln\left(\frac{V(y)}{\|y\|_{\mathbf{d}}}\right)\right)y,$$

where $y \in \mathbb{R}^n$. Indeed, $\Psi^{-1}(\Psi(x)) = x$ and $\Psi(\Psi^{-1}(y)) = y$ for all $x, y \in \mathbb{R}^n$. Since Ψ and Ψ^{-1} are continuous at the origin and smooth outside the origin then Ψ is diffeomorphism on $\mathbb{R}^n\backslash\{\mathbf{0}\}$ and homeomorphism on $\mathbb{R}^n$.

(2) $\Rightarrow$ (4)[3] The system (7.8) is $\mathbf{d}_0$-homogeneous of the degree μ with respect to the standard dilation $\mathbf{d}_0(s) = e^s$ with $G_{\mathbf{d}_0} = I_n$. Notice that $P_0G_{\mathbf{d}_0} + G_{\mathbf{d}_0}^\top P_0 = 2P_0 \succ 0$ holds for an arbitrary symmetric positive definite matrix P_0. Taking into account $\|z\|_{\mathbf{d}_0} = \|z\| = \sqrt{z^\top P_0 z}$, we use the already proven claim 3) to obtain the standard homogeneous Lyapunov function V defined as $V(z) = \|\Psi_0(z)\|_{\mathbf{d}_0}^2 = \Psi_0^\top(z)P_0\Psi_0(z)$ for $z \in \mathbb{R}^n$, where $\Psi_0 \in C^\infty(\mathbb{R}^n\backslash\{\mathbf{0}\}; \mathbb{R}^n) \cap C(\mathbb{R}^n; \mathbb{R}^n)$ is a $\mathbf{d}_0$-homogeneous vector field of degree 0. Since Ψ_0 is diffeomorphism then $\det\left(\frac{\partial\Psi_0(z)}{\partial z}\right) \neq 0$ for $z \in \mathbb{R}^n\backslash\{\mathbf{0}\}$. Using the homogeneous function theorem (see, formula (6.10)) we derive $\frac{\partial\Psi_0(z)}{\partial z}z = \Psi_0(z)$ and $\frac{\partial V}{\partial z} = 2z^\top\Xi^\top(z)P_0\Xi(z)$, where $\Xi(z) = \frac{\partial\Psi_0(z)}{\partial z}$. Moreover, $\frac{\partial\Xi(z)}{\partial z_i}z = 0$, since

$$\frac{\partial\Psi_0(z)}{\partial z}z = \Psi_0(z) \;\Leftrightarrow\; \frac{\partial}{\partial z_i}\left(\frac{\partial\Psi_0(z)}{\partial z}z\right) = \frac{\partial\Psi_0(z)}{\partial z_i} \;\Leftrightarrow\; \frac{\partial\Xi(z)}{\partial z_i}z + \frac{\partial\Psi_0(z)}{\partial z_i} = \frac{\partial\Psi_0(z)}{\partial z_i}.$$

---

[3] The proof of the implication 2) $\Rightarrow$ 4) is reproduced from Polyakov A. Sliding mode control design using canonical homogeneous norm. *Int Journal Robust Nonlinear Control*. 2019; 29: 682–701. https://doi.org/10.1002/rnc.4058 ©2018 John Wiley & Sons, Ltd. Reused with permission of John Wiley & Sons.

Finally, since the homogeneity degree of the vector field Ψ_0 is zero then, by Proposition 6.5, $\Xi(e^s z) = \Xi(z)$ for $z \in \mathbb{R}^n \backslash \{\mathbf{0}\}$ and $s \in \mathbb{R}$.

(4) $\Rightarrow$ (2) Let us consider the function $z \mapsto V(z)$ given by $V(z) = z^\top \Xi^\top(z) P \Xi(z) z$. The function V is standard homogeneous of degree 2 and $V \in C(\mathbb{R}^n) \cap C^\infty(\mathbb{R}^n \backslash \{\mathbf{0}\})$. Moreover, V is positive definite and proper due to $\det(\Xi(z)) \neq 0$, $\forall z \neq \mathbf{0}$. Since $\frac{\partial \Xi(z)}{\partial z_i} z = 0$ then $\frac{\partial V(z)}{\partial z} = 2z^\top \Xi^\top(z) P \Xi(z)$ and the condition (7.18) implies that V is the Lyapunov function for the system (7.8). The proof is complete. □

This theorem proves that

> *any asymptotically stable* **d**-*homogeneous system is homeomorphic on* $\mathbb{R}^n$ *and diffeomorphic on* $\mathbb{R}^n \backslash \{\mathbf{0}\}$ *to a* **quadratically stable** *system.*

Indeed, making the change of variables $\xi = \Psi(x)$ in (7.6) we derive

$$\dot{\xi} \in F_{new}(\xi), \quad \xi \neq \mathbf{0},$$

where

$$\tilde{F}_{new}(\xi) = \left. \frac{\partial \Psi(x)}{\partial \xi} F(x) \right|_{x = \Psi^{-1}(\xi)}.$$

The criterion (7.17) implies that $\xi^\top P \dot{\xi} < 0$ if $\xi^\top P \xi = 1$. Hence, the homogeneous norm $\|\cdot\|_{\mathbf{d}}$ is the Lyapunov function of the system. Finally, the change of variable $z = \|\xi\|_{\mathbf{d}} \mathbf{d}(-\ln \|\xi\|_{\mathbf{d}}) \xi$ gives $\|\xi\|_{\mathbf{d}} = \|z\|$, i.e., the transformed (homeomorphic) system

$$\dot{z} \in \tilde{F}(z) := \|z\|^{1+\mu} \left(\frac{(I_n - G_{\mathbf{d}}) z^\top z P}{z^\top P G_{\mathbf{d}} z} + I_n \right) F_{new} \left(\frac{z}{\|z\|} \right)$$

is quadratically stable provided that the system (7.6) is asymptotically stable and vice versa.

> If $\Xi \equiv I_n$ then the inequality (7.18) becomes
>
> $$\sup_{y \in F\left(\frac{z}{\sqrt{z^\top P z}}\right)} z^\top P y < 0, \quad z \neq \mathbf{0}. \tag{7.19}$$
>
> If the above inequality is fulfilled then the canonical homogeneous norm $\|\cdot\|_{\mathbf{d}}$ induced by the weighted Euclidean norm $\|x\| = \sqrt{x^\top P x}$ is a Lyapunov function for the system (7.6) provided that $0 \prec P = P^\top \in \mathbb{R}^{n \times n}$ satisfies (5.6). *A system having a Lyapunov function given by such a canonical homogeneous norm is said to be* ***homogeneously quadratically stable***. Most controllers designed in this monograph are homogeneous quadratic stabilizers (see Chaps. 9, 12, 13 and 15).

In the case of standard dilation, the homogeneous quadratic stability is equivalent to the conventional quadratic stability.

Example 7.5 Let us consider the controlled double integrator

$$\dot{x} = f(x) := \begin{pmatrix} 0 & 1 \\ 0 & 0 \end{pmatrix} x(t) + \begin{pmatrix} 0 \\ 1 \end{pmatrix} u(x), \quad t > 0, \quad x = (x_1, x_2)^\top \in \mathbb{R}^2,$$

where u is the control designed in [17] as follows

$$u(x) = -k_1 \frac{x_1 + k_2 x_2 |x_2|}{|x_1| + k_2 x_2^2}, \quad x \neq \mathbf{0}$$

and $k_1, k_2 \in \mathbb{R}_+$ are control parameters. The vector field f is discontinuous only at zero. The Filippov regularization of f at the origin is given by $K(f)(\mathbf{0}) = (0, [-k_1, k_1])^\top$. The vector field f is $\mathbf{d}$-homogeneous of degree -1 with respect to the weighted dilation $\mathbf{d}$ given by $\mathbf{d}(s) = \left(\begin{smallmatrix} e^{2s} & 0 \\ 0 & e^s \end{smallmatrix}\right)$:

$$f(e^{sG_\mathbf{d}} x) = e^{-s} e^{sG_\mathbf{d}} f(x), \quad \forall x \in \mathbb{R}^2 \backslash \{\mathbf{0}\}, \quad \forall s \in \mathbb{R}, \quad G_\mathbf{d} = \begin{pmatrix} 2 & 0 \\ 0 & 1 \end{pmatrix}.$$

To fulfill the requirements of Theorem 7.7, the parameters $k_1 > 0$ and $k_2 > 0$ can be selected such that the system is homogeneously quadratically stable, i.e., a canonical homogeneous norm $\|\cdot\|_\mathbf{d}$ is a Lyapunov function of the system. For this purpose we check the inequalities (7.19) and (5.6) for the matrix

$$P = \begin{pmatrix} 1+\varepsilon & 1 \\ 1 & 1 \end{pmatrix}.$$

Simple computations show that the inequality (5.6) is fulfilled for any

$$\varepsilon > 1/8.$$

Due to the central symmetry $f(-x) = -f(x)$ we may consider only the case $x_1 \geq 0$. Hence, taking into account

$$x^\top P x = 1 \quad \Leftrightarrow \quad (1+\varepsilon)x_1^2 + 2x_1 x_2 + x_2^2 = 1 \quad \Leftrightarrow \quad x_2 = -x_1 \pm \sqrt{1 - \varepsilon x_1^2}$$

we derive that the inequality (7.19) becomes

$$\left(\begin{smallmatrix} x_1 \\ x_2 \end{smallmatrix}\right)^T \left(\begin{smallmatrix} 1+\varepsilon & 1 \\ 1 & 1 \end{smallmatrix}\right) \left(\begin{smallmatrix} x_2 \\ -k_1 \frac{x_1 + k_2 |x_2| x_2}{|x_1| + \beta x_2^2} \end{smallmatrix}\right) < 0, \quad x_2 = -x_1 \pm \sqrt{1 - \varepsilon x_1^2}, \quad 0 \leq x_1 \leq \frac{1}{\sqrt{\varepsilon}}$$

or, equivalently,

$$q(x_1) < 0 \quad \text{for all} \quad x_1 \in \left[0, \frac{1}{\sqrt{\varepsilon}}\right],$$

where

$$q(x_1)=\max_{x_2\in\left\{-x_1+\sqrt{1-\varepsilon x_1^2},-x_1-\sqrt{1-\varepsilon x_1^2}\right\}}\begin{pmatrix}x_1\\x_2\end{pmatrix}^\top\begin{pmatrix}1+\varepsilon & 1\\ 1 & 1\end{pmatrix}\begin{pmatrix}x_2\\ -k_1\frac{x_1+k_2|x_2|x_2}{|x_1|+k_2x_2^2}\end{pmatrix}.$$

Given $k_1>0, k_2>0$ and $\varepsilon>0$ the above inequality can be checked numerically. For example, it holds if $k_1=1.5$, $k_2=1$ and $\varepsilon=1$. Therefore, the canonical homogeneous norm $\|\cdot\|_{\mathbf{d}}$ induced (see formula (5.15)) by the norm $\|x\|=\sqrt{(1+\varepsilon)x_1^2+2x_1x_2+x_2^2}$ is a Lyapunov function for the considered system with $k_1=1.5$ and $k_2=1$.

The following corollary characterizes a particular class of homogeneously quadratically stable systems.

Corollary 7.6 *Let a linear continuous dilation* $\mathbf{d}$ *be strictly monotone with respect to a weighted Euclidean norm* $\|x\|=\sqrt{x^\top Px}$ *with* $0\prec P=P^\top\in\mathbb{R}^{n\times n}$. *Let the canonical homogeneous norm* $\|\cdot\|_{\mathbf{d}}$ *be induced by* $\|\cdot\|$. *Let a mapping* $A\in C(\mathbb{R}_+;\mathbb{R}^{n\times n})$ *be* $\mathbf{d}$*-homogeneous of degree* $\mu\in\mathbb{R}$ *in the following sense*

$$A(e^s r)\mathbf{d}(s)=e^{\mu s}\mathbf{d}(s)A(r),\quad \forall r\in\mathbb{R}_+, \forall s\in\mathbb{R}.$$

The canonical homogeneous norm $\|\cdot\|_{\mathbf{d}}$ *is a strict Lyapunov function of the system*

$$\dot{x}=A(\|x\|_{\mathbf{d}})x,\quad t>0 \tag{7.20}$$

if and only if

$$A(1)^\top P+PA(1)\prec 0. \tag{7.21}$$

Proof Using formula (5.21) we conclude that the canonical homogeneous norm is a Lyapunov function of the system (7.20) if and only if

$$\frac{x^\top\mathbf{d}^\top(-\ln(\|x\|_{\mathbf{d}})P\mathbf{d}(-\ln\|x\|_{\mathbf{d}})A(\|x\|_{\mathbf{d}}))x}{x^\top\mathbf{d}^\top(-\ln(\|x\|_{\mathbf{d}})PG_{\mathbf{d}}\mathbf{d}(-\ln\|x\|_{\mathbf{d}})x}<0,\quad \forall x\neq\mathbf{0}.$$

Since $P\succ 0$ satisfies the criterion of the strict monotonicity (5.6), then, due to $\mathbf{d}$-homogeneity of A, the above inequality is equivalent to

$$x^\top\mathbf{d}^\top(-\ln(\|x\|_{\mathbf{d}})PA(\|\mathbf{d}(-\ln\|x\|_{\mathbf{d}})x\|_{\mathbf{d}}))\mathbf{d}(-\ln\|x\|_{\mathbf{d}})x<0,\quad \forall x\neq\mathbf{0}.$$

Taking into account $\|\mathbf{d}(-\ln\|x\|_{\mathbf{d}})x\|_{\mathbf{d}}=1$ we complete the proof.

The system of the form (7.20) appears in many chapters of this monograph, for example, as a model of the closed-loop system with implicit homogeneous stabilizer (see Chap. 9).

7.2.4 *Particular Classes of Homogeneous Systems*

Homogeneous ODE with Vector Field Dependent on Initial State

Inspired by [18] let us consider the following system

$$\dot{x}(t) = f(x(t), x(t_0)), \quad t > t_0, \quad x(t_0) = x_0 \in \mathbb{R}^n, \tag{7.22}$$

where the system state is denoted by $x(t) \in \mathbb{R}$ and the vector field $f \in C(\mathbb{R}^n\backslash\{\mathbf{0}\} \times \mathbb{R}^n; \mathbb{R}^n)$ is assumed to be locally bounded on $\mathbb{R}^n \times \mathbb{R}^n$. The system (7.22) is not an ODE and its well-posedness cannot be studied using the results of Chap. 2. Formally, (7.22) can be treated as a functional differential equation or a system with delay, since the dynamics of the system depends on its prehistory (namely, on the initial state). However, the system (7.22) can be rewritten in the ODE form:

$$\begin{cases} \dot{x} = f(x, r), \\ \dot{r} = \mathbf{0}, \end{cases} \quad t > t_0, \quad x(t_0) = r(t_0) = x_0. \tag{7.23}$$

By Peano existence theorem, the continuity of f on $\mathbb{R}^n\backslash\{\mathbf{0}\} \times \mathbb{R}^n$ implies the existence of classical solutions of the system (7.23) on $\mathbb{R}^n\backslash\{\mathbf{0}\} \times \mathbb{R}^n$. The uniqueness of solutions on $\mathbb{R}^n\backslash\{\mathbf{0}\} \times \mathbb{R}^n$ can be guaranteed for f being locally Lipschitz continuous on $\mathbb{R}^n\backslash\{\mathbf{0}\} \times \mathbb{R}^n$. Since f is assumed to be locally bounded $\mathbb{R}^n \times \mathbb{R}^n$, then the classical solution of the ODE (7.23) can be prolonged to the whole $\mathbb{R}^n \times \mathbb{R}^n$ using the Filippov's method (see Chap. 2 or [19]). In the view of Proposition 3.1, the uniqueness of solutions in this case can be guaranteed for locally Lyapunov stable system (7.23). The stability notions for (7.22) are assumed to be introduced in the usual way (as in Chap. 3). Notice that the Lypunov stability of (7.22) is equivalent to the Lyapunov stability of (7.23). However, an asymptotic (resp., a finite/fixed-time) stability of (7.22) corresponds to the partial asymptotic (resp., finite/fixed-time) stability of (7.23) with respect to the variable x.

As shown above, any asymptotically stable homogeneous ODE of negative degree is finite-time stable. The globally homogeneous functional differential equation (7.22) may be fixed-time stable. The following theorem generalizes the result of the paper [18].

Theorem 7.8 *Let* $\mathbf{d}$ *be a linear continuous dilations in* $\mathbb{R}^n$. *Let* $f \in C(\mathbb{R}^n\backslash\{\mathbf{0}\} \times \mathbb{R}^n; \mathbb{R}^n)$ *be locally bounded on* $\mathbb{R}^n \times \mathbb{R}^n$, $f(\mathbf{0}, \cdot) = \mathbf{0}$ *and*

$$f(\mathbf{d}(s)x, \mathbf{d}(s)r) = \mathbf{d}(s)f(x, r), \quad x, r \in \mathbb{R}^n, \quad \forall s \in \mathbb{R}. \tag{7.24}$$

If for any $x_0 \in \mathbb{R}^n$ *the system* (7.22) *has the global-in-time unique solution converging to zero in a finite time then the system* (7.22) *is*

(i) finite-time stable with a discontinuous (at least at $\mathbf{0}$*) settling-time function* $T : \mathbb{R}^n \mapsto [0, +\infty)$*;*

(ii) fixed-time stable provided that the settling-time function T is bounded on a compact set $\tilde{S} \subset \mathbb{R}^n \backslash \{\mathbf{0}\} : \bigcup_{s\in\mathbb{R}} \mathbf{d}(s)\tilde{S} = \mathbb{R}^n \backslash \{\mathbf{0}\}$;

(iii) constant-time stable provided that the settling time is constant on the set $\tilde{S}$ defined above.

Proof First of all, notice that $\tilde{\mathbf{d}}$-homogeneity of the system (7.23) implies the dilation symmetry

$$x(t, \mathbf{d}(s)x_0) = \mathbf{d}(s)x(t, x_0), \quad \forall t \geq 0, \quad \forall s \in \mathbb{R} \tag{7.25}$$

of solutions of the system (7.22), where $x(t, x_0)$ denotes the global-in-time unique solution of (7.22).

Let us show that there exists $C > 0$ such $\sup_{x_0 \in S} \|x(t, x_0)\|_{\mathbf{d}} \leq C$ for all $t \geq 0$, where $S = \{x \in \mathbb{R}^n : \|x\|_{\mathbf{d}} = 1\}$ is the unit sphere in $\mathbb{R}^n$ and $\|\cdot\|_{\mathbf{d}}$ is the canonical homogeneous norm.

Suppose the contrary, i.e., there exists the sequence $(t_i, x_i) \in \mathbb{R}_+ \times S$ such that $\|x(t_i, x_i)\|_{\mathbf{d}} \to +\infty$ as $i \to +\infty$. Since $x_i \in S$ is bounded then, by Bolzano–Weierstrass theorem, there exists a convergent subsequence x_{i_k}, i.e., there exists $x^* \in S$ such that $x_{i_k} \to x^*$ and $\|x(t_{i_k}, x_{i_k})\|_{\mathbf{d}} \to +\infty$ as $i_k \to +\infty$. Recall that $t \mapsto (x(t, x_i), x_i)$ is the solution of the ODE (7.23). In this case the continuous dependence of classical solutions of ODEs on initial conditions and on the right-hand (see Chap. 2, Theorem 2.3) implies that $x(t_{i_k}, x^*) \to +\infty$ as $i_k \to +\infty$. This contradicts the finite-time convergence of all solutions of (7.22) to $\mathbf{0}$.

Therefore, all solutions of (7.22) initiated on the unit sphere are uniformly bounded by a constant $C > 0$ but the dilation symmetry of solutions (7.25) implies the Lyapunov stability of (7.22)

$$\|x_{x_0}(t, x_0)\|_{\mathbf{d}} \leq C\|x_0\|_{\mathbf{d}}, \quad \forall t \geq 0.$$

The finite-time convergence of all solutions to zero yields the finite-time stability, while the dilation symmetry of solutions (7.25) guarantees the symmetry of the settling-time function

$$T(\mathbf{d}(s)x_0) = T(x_0). \tag{7.26}$$

Therefore,[4] the settling-time function of the system (7.22) has a constant value along any homogeneous curve $\Gamma_{\mathbf{d}}(x_0) = \{x \in \mathbb{R}^n : x = \mathbf{d}(s)x_0, s \in \mathbb{R}\}$. In this case, T is always discontinuous at zero since $T(\mathbf{0}) = 0$ and $T(x_0) \neq 0$ for $x_0 \neq \mathbf{0}$. If $\bigcup_{s\in\mathbb{R}} \mathbf{d}(s)\tilde{S} = \mathbb{R}^n \backslash \{\mathbf{0}\}$ then, due to dilation symmetry of T, the boundedness of T on the compact $\tilde{S}$ implies its boundedness on $\mathbb{R}^n$ by the same constant, i.e., the system (7.22) is fixed-time stable in this case. Moreover, if $T(x_0)$ =const for all

[4] Text in this paragraph is reproduced from Polyakov A, Krstic M. Fixed-time stabilization with a prescribed constant settling time by static feedback for delay-free and input delay systems. *Int J Robust Nonlinear Control*. 2024; 34:6393–6415. https://doi.org/10.1002/rnc.7212 ©2024 John Wiley & Sons, Ltd. Reused with permission of John Wiley & Sons.

$x_0 \in \tilde{S}$ then, due to homogeneity, $T(x_0)$ =const for all $x_0 \in \mathbb{R}^n\backslash\{\mathbf{0}\}$. The proof is complete. □

One of the conditions of the above theorem is the finite-time convergence of all solutions of (7.22) to zero. In the view of Theorem 7.5, this can be guaranteed by asymptotic stability and a generalized homogeneity with a negative degree of the system $\dot{x} = f(x, r)$ for each fixed $r \neq \mathbf{0}$. Notice that the dilation group may depend on r in this case.

In Chap. 9 we design a feedback control for a linear plant such that the closed-loop system satisfies Theorem 7.8 with $\tilde{S} = S = \{x \in \mathbb{R}^n : \|x\| = 1\}$ and $\mathbf{d}(s) = e^s I_n$.

Homogeneous Quadratic Stability of Impulsive Systems

Impulsive systems are introduced in Sect. 2.3.1. Let us consider the impulsive system with a linear reset map

$$\begin{cases} \dot{x} = f(t, x) & \text{if } x^- \in \mathcal{C}, \\ x = \Pi x^- & \text{if } x^- \in \mathcal{D}, \end{cases} \quad t > t_0, \quad x(t_0) = x_0, \tag{7.27}$$

where $\mathcal{C} \subset \mathbb{R}^n$ is the set of admissible system states, $x(t) \in \mathcal{C}$ is the state at the time t, $f \in C(\mathbb{R}^{n+1}; \mathbb{R}^n)$ is a vector field governing the continuous dynamics of the system, the matrix $\Pi \in \mathbb{R}^{n\times n}$ defines a linear reset map governing the jump dynamics of the systems, $\mathcal{D} \subset \mathbb{R}^n\backslash\{\mathbf{0}\}$ and x^- is given by (2.55).

There are few studies on *impulsive homogeneous system* and its stability (e.g., [20]), but they require the scaling invariance (**d**-homogeneity) of the vector field f, of the reset map Π and of the jump set $\mathcal{D}$. In this case, the dilation symmetry (7.4) of solutions of (7.27) can be proven in the usual manner (see Theorem 7.2). So, Theorems 7.3, 7.5 and Corollaries 7.1, 7.3 and Propositions 7.1, 7.2 can be straightforwardly extended to the class of impulsive systems.

The following theorem [21] provides a sufficient condition of a finite/fixed-time stability of the Zeno-free impulsive system (7.27).

Theorem 7.9 *Let* $\mathbf{d}$ *be a linear continuous dilation in* $\mathbb{R}^n$. *Let the system* (7.27) *be Zeno-free. Let* $\mathcal{C}$ *be a neighborhood of the origin such that* $\partial\mathcal{C} \subset \mathcal{D}$, $\Pi\mathcal{D} \subset \mathcal{C}$ *and* $\mathbf{0} \notin \mathcal{D}$. *Let the canonical homogeneous norm be induced by the weighted Euclidean norm* $\|x\| = \sqrt{x^\top P x}$ *with* P *satisfying (5.6) If, for some* $\mu \in \mathbb{R}$ *and* $\rho > 0$, *the canonical homogeneous* $\|\cdot\|_{\mathbf{d}}$ *a Lyapunov function of the impulse-free system* $\dot{x} = f(t, x)$ *satisfying*

$$\frac{\partial \|x\|_{\mathbf{d}}}{\partial x} f(t, x) \leq -\rho \|x\|_{\mathbf{d}}^{1+\mu}, \quad \forall x \in \mathcal{C}\backslash\{\mathbf{0}\}, \quad \forall t \in \mathbb{R}, \tag{7.28}$$

and

$$G_{\mathbf{d}}\Pi = \Pi G_{\mathbf{d}}, \tag{7.29}$$

$$y^\top \Pi^\top P \Pi y \leq 1, \quad \forall y \in \pi_{\mathbf{d}}(\mathcal{D}), \tag{7.30}$$

then the closed-loop impulsive system (7.27) *is*

- *uniformly finite-time stable for* $\mu < 0$ *and*

$$x(t) = \mathbf{0}, \quad \forall t \geq t_0 + \tfrac{\|x(t_0)\|_{\mathbf{d}}^{-\mu}}{-\rho\mu}, \quad \forall x(t_0) \in \mathcal{C} \cup \mathcal{D}, \quad \forall t_0 \in \mathbb{R};$$

- *uniformly exponentially stable for* $\mu = 0$ *and*

$$\|x(t)\|_{\mathbf{d}} \leq e^{-\rho(t-t_0)}\|x(t_0)\|_{\mathbf{d}}, \quad \forall t \geq t_0, \quad \forall x(t_0) \in \mathcal{C} \cup \mathcal{D}, \quad \forall t_0 \in \mathbb{R};$$

- *uniformly nearly fixed-time stable for* $\mu > 0$ *and*

$$\|x(t)\|_{\mathbf{d}} \leq r, \quad \forall t \geq t_0 + \tfrac{1}{\rho r^{\mu}}, \quad \forall r > 0, \quad \forall x(t_0) \in \mathcal{C} \cup \mathcal{D}, \quad \forall t_0 \in \mathbb{R}.$$

Proof For any $x(t_0) \in \mathcal{C}$, due to the condition (7.28), the function $t \mapsto V(x(t))$ decreases for $t > t_0$ as long as $x^- \notin \mathcal{D}$. Then, taking into account that any reset for $x^- \in \mathcal{D}$ results in a finite jump, we conclude that the system is forward complete in the Zeno-free case.

On the one hand, since $\mathbf{d}(s) = e^{G_{\mathbf{d}}s} = \sum_{k=0}^{\infty} \frac{s^k G_{\mathbf{d}}^k}{k!}$, then the identity (7.29) implies that

$$\mathbf{d}(s)\Pi = \sum_{k=0}^{\infty} \frac{s^k G_{\mathbf{d}}^k \Pi}{k!} = \sum_{k=0}^{\infty} \frac{s^k \Pi G_{\mathbf{d}}^k}{k!} = \Pi\mathbf{d}(s), \quad \forall s \in \mathbb{R}. \tag{7.31}$$

On the other hand,[5] the Lyapunov function $\|\cdot\|_{\mathbf{d}}$ decreases during the jump if and only if $\|x\|_{\mathbf{d}} \leq \|x^-\|_{\mathbf{d}}$. This is equivalent to $\frac{\|x\|_{\mathbf{d}}}{\|x^-\|_{\mathbf{d}}} = \|\mathbf{d}(-\ln\|x^-\|_{\mathbf{d}})x\|_{\mathbf{d}} \leq 1$. Since $\|y\| \leq 1 \Leftrightarrow \|y\|_{\mathbf{d}} \leq 1$ then $\|\mathbf{d}(-\ln\|x^-\|_{\mathbf{d}})x\|_{\mathbf{d}} \leq 1$ is equivalent to

$$x^\top \mathbf{d}^\top(-\ln\|x^-\|_{\mathbf{d}}) P \mathbf{d}(-\ln\|x^-\|_{\mathbf{d}})x \leq 1. \tag{7.32}$$

Denoting $s^- = \ln\|x^-\|_{\mathbf{d}}$ and using (7.31), we derive

$$\begin{aligned} x^\top\mathbf{d}^\top(-s^-)P\mathbf{d}(-s^-)x &= (\Pi x^-)^\top\mathbf{d}^\top(-s^-)P\mathbf{d}(-s^-)\Pi x^- \\ &= (x^-)^\top\mathbf{d}^\top(-s^-)\Pi^\top P\Pi\mathbf{d}(-s^-)x^-, \quad x^- \in \mathcal{D}. \end{aligned} \tag{7.33}$$

Since $\|\mathbf{d}(-s^-)x^-\|_{\mathbf{d}} = 1$ then $\mathbf{d}(-s^-)x^- \in \pi_{\mathbf{d}}(\mathcal{D})$, where $\pi_{\mathbf{d}}(\mathcal{D})$ is the $\mathbf{d}$-homogeneous projection of the set $\mathcal{D}$ on the unit sphere. It is well defined since $\mathbf{0} \notin \mathcal{D}$. Therefore, the condition (7.30) and the identity (7.33) imply the inequality (7.32), or equivalently, $\|x\|_{\mathbf{d}} \leq \|x^-\|_{\mathbf{d}}$ for any $x^- \in \mathcal{D}$. This means that the function $t \mapsto V(x(t))$ is monotone decreasing along any trajectory $x(t)$ of the system. The sequence t_k of time instances (of jumps) has no accumulation points (by assumption that the system is Zeno-free). In this case, the inequality (7.28) holds almost

[5] Text following Eq. 7.31 until the end of the proof is reprinted from *Automatica*, 163/5, Y. Zhou, A. Polyakov, G. Zheng, Generalized homogeneous rigid body attitude control, 111548, https://doi.org/10.1016/j.automatica.2024.111548. ©2024 Elsevier, with permission from Elsevier.

everywhere along the trajectory of the system and the inequality

$$V(x(t)) \leq V(0) - \rho \int_0^t V^{1+\mu}(x(\tau))d\tau$$

is fulfilled due to monotonicity of $t \mapsto V(x(t))$ (see Lemma 4.8). Hence, we conclude that $(-\mu)^{-1}(V^{-\mu}(x(t)) - V^{-\mu}(x_0)) \leq -\rho t$ for $\mu \neq 0$ and $V(x(t)) \leq e^{-\rho t}V(x_0)$ for $\mu = 0$. The proof is complete. □

In Chap. 15, we investigate a particular case of the impulsive system (7.27) being a common model of a rigid body attitude control system.

7.3 Dynamics on Homogeneous Cones

Control theory frequently deals with a design of dynamical systems satisfying some state constraints. In many cases, the fulfillment of these constraints may be guaranteed by an adjustment of a positively invariant set of the closed-loop system (see, e.g., [22–24]). Positively invariant closed sets of continuous ODEs having unique solutions were first characterized in [25]. Usually, a control design with state constraints considers compact invariant sets, which include the zero equilibrium as an interior point of the safe set (see, for example, [26, 27]). In this case, by Proposition 7.1 (resp., Proposition 7.2), Lyapunov (resp., asymptotic) stability of homogeneous systems follows from (resp., strict) positive invariance of the bounded (resp., compact) set. In this section, we study the case when the equilibrium of a homogeneous system belongs to the boundary of a closed invariant set (e.g., an equilibrium is a vertex of a cone). Such invariant sets naturally appear during the design of the so-called nonovershooting stabilizers and safety filters (see Chap. 12).

7.3.1 Positively Invariant Homogeneous Cones

A positively invariant set $\Omega \subset \mathbb{R}^n$, which ensures restrictions to admissible states of a dynamical system, is also called *safe (or safety) set* (see, e.g., [23] and references therein). The positive invariance of a *closed* set $\Omega \subset \mathbb{R}^n$ for a finite-dimensional dynamical system is well studied in the literature (see, e.g., [25, 26, 28, 29]), where the following criterion has been established.

Let a forward complete system

$$\dot{x} = f(x), \quad t > 0, \quad x(0) = x_0, \quad f \in C(\mathbb{R}^n; \mathbb{R}^n) \tag{7.34}$$

have unique solutions on Ω. *A closed set* Ω *in* $\mathbb{R}^n$ *is positively invariant for the system* (7.34) *if and only if*

$$f(x) \in \mathcal{C}_\Omega(x), \quad \forall x \in \Omega, \tag{7.35}$$

where $\mathcal{C}_\Omega(x) \subset \mathbb{R}^n$ *is the (Bouligand) tangent cone to the set* $\Omega \subset \mathbb{R}^n$ *at the point* $x \in \mathbb{R}^n$ *given by* (6.29). If solutions are not unique then the set Ω is, at least, weakly positively invariant.[a]

[a] A set Ω is positively invariant for a system $\dot{x} = f(x), t \geq 0$ if $\forall x_0 \in \Omega \Rightarrow x(t, x_0) \in \Omega, \forall t \geq 0$ for each solution $x(t, x_0)$ of the system. The set Ω is *weakly* positively invariant if the above implication holds for any $x_0 \in \Omega$ and for, at least, one of solutions $x(t, x_0)$ of the system.

Since $\mathcal{C}_\Omega(x) = \mathbb{R}^n$ for $x \in \operatorname{int}\Omega$, then the condition (7.35) can be relaxed to $x \in \partial\Omega$. This means that, for the analysis of the positive invariance, *uniqueness of solutions is required only in a neighborhood of* $\partial\Omega$.

The tangent cones of homogeneous sets are studied in Sect. 6.3.1. By Lemma 6.2, the mapping $\mathcal{C}_\Omega : \mathbb{R}^n \mapsto 2^{\mathbb{R}^n}$ is **d**-homogeneous provided that Ω is a **d**-homogeneous cone in $\mathbb{R}^n$. As shown in Chap. 6, any **d**-homogeneous mapping is completely defined by its values on the unit sphere and the homogeneity degree. This results in the following corollary discovered in [30].

Corollary 7.7 *Let a linear continuous dilation* **d** *be monotone with respect to a norm* $\|\cdot\|$ *in* $\mathbb{R}^n$. *Let a vector field* $f \in C(\mathbb{R}^n; \mathbb{R}^n)$ *be* **d**-*homogeneous of degree* $\mu \in \mathbb{R}$ *and* $f(\mathbf{0}) = \mathbf{0}$. *Let all solutions of the system* (7.34) *initiated in* Ω *be forward unique and forward complete. A closed* **d**-*homogeneous cone* Ω *in* $\mathbb{R}^n$ *is positively invariant for* (7.34) *if and only if*

$$f(x) \in \mathcal{C}_\Omega(x), \quad \forall x \in \partial\Omega \cap S, \tag{7.36}$$

where $S = \{x \in \mathbb{R}^n : \|x\| = 1\}$ *is the unit sphere. The set* Ω *is, at least, weakly positively invariant if solutions are not forward unique on* Ω.

Proof Due to **d**-homogeneity of f and $\mathcal{C}_\Omega$ (see Lemma 6.2), we have

$$f(x) = \|x\|_{\mathbf{d}}^{-\mu}\mathbf{d}(\ln\|x\|_{\mathbf{d}})f(\mathbf{d}(-\ln\|x\|_{\mathbf{d}})x)$$

and

$$\mathcal{C}_\Omega(x) = \mathbf{d}(\ln\|x\|_{\mathbf{d}})\mathcal{C}_\Omega(\mathbf{d}(-\ln\|x\|_{\mathbf{d}})x)$$

for any $x \neq \mathbf{0}$. In this case, the condition (7.36) yields the inclusion

$$f(\mathbf{d}(-\ln\|x\|_{\mathbf{d}})x) \in \mathcal{C}_\Omega(\mathbf{d}(-\ln\|x\|_{\mathbf{d}})x), \quad \forall x \in \partial\Omega\backslash\{\mathbf{0}\}.$$

or, equivalently,

$$\|x\|_{\mathbf{d}}^{-\mu}\mathbf{d}(-\ln\|x\|_{\mathbf{d}})f(x) \in \mathbf{d}(-\ln\|x\|_{\mathbf{d}})\mathcal{C}_\Omega(x), \quad \forall x \in \partial\Omega\backslash\{\mathbf{0}\}.$$

Since the matrix $\mathbf{d}(s) \in \mathbb{R}^{n\times n}$ is invertible for any $s \in \mathbb{R}$ then the above inclusion means

$$\|x\|_{\mathbf{d}}^{-\mu}f(x) \in \mathcal{C}_\Omega(x), \quad \forall x \in \partial\Omega\backslash\{\mathbf{0}\}.$$

The set $\mathcal{C}_{\Omega}(x)$ is a positive cone (i.e., $y \in \mathcal{C}_{\Omega} \Rightarrow \lambda y \in \mathcal{C}_{\Omega}(x), \forall \lambda > 0$), so we derive the inclusion

$$f(x) \in \mathcal{C}_{\Omega}(x), \quad \forall x \in \partial\Omega \backslash \{\mathbf{0}\}.$$

Taking into account $f(\mathbf{0}) = \mathbf{0}$ and $\mathbf{0} \in \mathcal{C}_{\Omega}(\mathbf{0})$ (see Lemma 6.2) we conclude that the necessary and sufficient condition (7.36) for the set Ω to be positively invariant is fulfilled. The proof is complete. □

For convex sets, the positive invariance admits alternative characterization in terms of outer normal cones (see, e.g., [29] and references therein).

> *Let $F : \mathbb{R}^n \mapsto 2^{\mathbb{R}^n}$ be compact-valued, convex-valued and upper semi-continuous. Let a system*
>
> $$\dot{x} \in F(x), \quad t > 0, \quad x(0) = x_0 \tag{7.37}$$
>
> *be forward complete on a closed convex set $\Omega \subset \mathbb{R}^n$. The set Ω is weakly positively invariant for the system* (7.34) *if and only if*
>
> $$\min_{y \in F(x)} \langle y, z \rangle \leq 0, \quad \forall z \in \mathcal{N}_{\Omega}(x), \quad \forall x \in \partial\Omega, \tag{7.38}$$
>
> *where $\langle \cdot, \cdot \rangle$ is an inner product in $\mathbb{R}^n$* and $\mathcal{N}_{\Omega}(x) \subset \mathbb{R}^n$ *is the normal cone to the set $\Omega \subset \mathbb{R}^n$ at the point $x \in \mathbb{R}^n$ (see formula* (6.28)*). Moreover, the set Ω is positively invariant for the system* (7.37) *if solutions are forward unique on Ω.*

The criterion (7.38) is independent of a selection of an inner product, but

> *an inner product utilized for the definition of the normal cone $\mathcal{N}_{\Omega}$ must be the same as in the condition* (7.38).

The following theorem adapts the above analysis to a generalized homogeneous set Ω, which is convex in a generalized homogeneous Euclidean space $\mathbb{R}^n_{\mathbf{d}}$. Such a generalized *homogeneous convex cone* in $\mathbb{R}^n_{\mathbf{d}}$ is characterized by Theorem 5.5 from Chap. 5.

> **Theorem 7.10** *Let a linear continuous dilation* $\mathbf{d}$ *in $\mathbb{R}^n$ be strictly monotone with respect to a weighted Euclidean norm $\|x\| = \sqrt{x^\top P x}$ with a shape matrix $0 \prec P = P^\top \in \mathbb{R}^{n \times n}$ satisfying (5.6). Let $\mathbb{R}^n_{\mathbf{d}}$ be the corresponding generalized homogeneous Euclidean space with the canonical homogeneous norm $\|\cdot\|_{\mathbf{d}}$ induced by the norm $\|\cdot\|$. Let the closed* $\mathbf{d}$*-homogeneous cone $\Omega \in \mathbb{R}^n$ be convex in $\mathbb{R}^n_{\mathbf{d}}$. Let $F : \mathbb{R}^n \mapsto 2^{\mathbb{R}^n}$ be convex-valued, compact-valued, upper semi-continuous and* $\mathbf{d}$*-homogeneous of degree $\mu \in \mathbb{R}$, $\mathbf{0} \in F(\mathbf{0})$. Let the system* (7.37) *be forward complete. Then the following three claims are equivalent.*
>
> *(1) The set Ω is weakly positively invariant for the system* (7.37).
>
> *(2) The condition*

$$\min_{y\in F(x)} \langle y,\xi\rangle_{\mathbf{d}} \le 0, \quad \forall \xi \in \mathcal{N}_{\Omega}(x), \quad \forall x \in \partial\Omega \cap S \tag{7.39}$$

holds, where $\langle\cdot,\cdot\rangle_{\mathbf{d}}$ *is an inner product in* $\mathbb{R}^n_{\mathbf{d}}$.

(3) The condition

$$\min_{y\in \tilde F(z)} \langle y,\xi\rangle \le 0, \quad \forall \xi \in \mathcal{N}_{\Phi(\Omega)}(z), \quad \forall z \in \partial\Omega \cap S \tag{7.40}$$

holds, where $\langle\cdot,\cdot\rangle$ *is an inner product* $\mathbb{R}^n$,

$$\tilde F(z) = \left(\frac{(I_n - G_{\mathbf{d}})zz^\top P}{z^\top P G_{\mathbf{d}} z} + I_n\right) F\left(\frac{z}{\|z\|}\right) \tag{7.41}$$

and a homeomorphism $\Phi \in C(\mathbb{R}^n,\mathbb{R}^n)$ *is given by (5.31).*

The set Ω *is positively invariant if, additionally, solutions are forward unique on* Ω.

Proof Since F is a $\mathbf{d}$-homogeneous vector field then

$$\dot x \in \|x\|_{\mathbf{d}}^{\mu}\mathbf{d}(\ln\|x\|_{\mathbf{d}})F(\mathbf{d}(-\ln\|x\|_{\mathbf{d}})x),$$

where $\|\cdot\|_{\mathbf{d}}$ is a canonical homogeneous norm. The scalar positive multiplier $\|x\|_{\mathbf{d}}^{\mu}$ has no impact on the analysis of the positive invariance, since it modifies just convergence/transient rates of the system, but it does not change a form of the phase curve. So, without loss of generality, we may assume that $\mu = 0$.

1⇔2 By Lemma 6.3, $\mathcal{N}_{\Omega}(\mathbf{d}(s)x) = \mathcal{N}_{\Omega}(x), \forall s \in \mathbb{R}, \forall x \in \Omega$. Since Ω is a $\mathbf{d}$-homogeneous cone then, using $\|\mathbf{d}(-\ln\|x\|_{\mathbf{d}})x\| = 1, \forall x \neq \mathbf{0}$, we derive

$$\min_{y\in F(x)} \langle y,\xi\rangle_{\mathbf{d}} \le 0, \quad \forall \xi \in \mathcal{N}_{\Omega}(x), \quad \forall x \in \partial\Omega \cap S$$

$$\Leftrightarrow \quad \min_{y\in F(\mathbf{d}(-\ln\|x\|_{\mathbf{d}})x)} \langle y,\xi\rangle_{\mathbf{d}} \le 0, \quad \forall \xi \in \mathcal{N}_{\Omega}(\mathbf{d}(-\ln\|x\|_{\mathbf{d}})x), \quad \forall x \in \partial\Omega\backslash\{\mathbf{0}\}$$

$$\Leftrightarrow \quad \min_{y\in \mathbf{d}(-\ln\|x\|_{\mathbf{d}})F(x)} \langle y,\xi\rangle_{\mathbf{d}} \le 0, \quad \forall \xi \in \mathcal{N}_{\Omega}(x), \quad \forall x \in \partial\Omega\backslash\{\mathbf{0}\}$$

$$\Leftrightarrow \quad \min_{\tilde y\in F(x)} \langle \mathbf{d}(-\ln\|x\|_{\mathbf{d}})\tilde y,\xi\rangle_{\mathbf{d}} \le 0, \quad \forall \xi \in \mathcal{N}_{\Omega}(x), \quad \forall x \in \partial\Omega\backslash\{\mathbf{0}\}$$

$$\Leftrightarrow \quad \|x\|_{\mathbf{d}}^{-1}\min_{\tilde y\in F(x)} \langle \tilde y,\xi\rangle_{\mathbf{d}} \le 0, \quad \forall \xi \in \mathcal{N}_{\Omega}(x), \quad \forall x \in \partial\Omega\backslash\{\mathbf{0}\}.$$

Using $\mathbf{0} \in F(\mathbf{0})$, we conclude that the condition (7.38) is fulfilled.

1⇔3 By Theorem 7.3, the $\mathbf{d}$-homogeneous system (7.37) with $\mu = 0$ is topologically equivalent on $\mathbb{R}^n$ to the standard homogeneous system

$$\dot z \in \|z\|\tilde F(z), \tag{7.42}$$

where $z = \Phi(x)$ and the right-hand side of this system is prolonged to zero by continuity $\|z\|\tilde{F}(z) \to \mathbf{0}$ as $z \to \mathbf{0}$. By Theorem 5.5, the set $\tilde{\Omega} = \Phi(\Omega)$ is convex in $\mathbb{R}^n$. Moreover, since Ω is a $\mathbf{d}$-homogeneous cone then $\tilde{\Omega}$ is standard homogeneous cone. Therefore, the condition (7.40) is, in fact, the condition (7.39) for the standard dilation $\mathbf{d}_0(s) = e^s I_n$, the standard homogeneous system (7.42) and the standard homogeneous cone $\tilde{\Omega}$. Using the equivalence 1⇔2 proven above, we conclude that the condition (7.40) is the criterion of positive invariance of the set $\tilde{\Omega}$ for the standard homogeneous system (7.42). Since Φ is a homeomorphism on $\mathbb{R}^n$ then the positive invariance of $\tilde{\Omega}$ for the system (7.42) is equivalent to the positive invariance of the set $\Omega = \Phi^{-1}(\tilde{\Omega})$ for the system (7.37). The proof is complete. □

Let a safe set be defined as follows

$$\Omega = \{x \in \mathbb{R}^n \ : \ \phi_i(x) \geq 0, \ i = 1, \dots, p\}, \tag{7.43}$$

where $\phi_i \in C(\mathbb{R}^n; \mathbb{R})$, $i = 1, \dots, p$ are known as *barrier functions* [23, 31]. The set Ω is closed due to continuity of ϕ_i. We also assume that Ω is nonempty. In this case,

$$\Omega_i = \{x \in \mathbb{R}^n : \phi_i(x) \geq 0\}, \quad i = 1, ..., p \tag{7.44}$$

are also closed nonempty sets such that $\Omega = \bigcap_{i=1}^{p} \Omega_i$. If Ω_i are convex sets then Ω is convex and (see, e.g., [32, Corollary 3.5])

$$\mathcal{N}_\Omega(x) = \mathcal{N}_{\Omega_1}(x) \dot{+} \cdots \dot{+} \mathcal{N}_{\Omega_i}(x) \dot{+} \cdots \dot{+} \mathcal{N}_{\Omega_p}(x), \tag{7.45}$$

where $\dot{+}$ is the geometric sum of sets.

As an example, let us consider the so-called *linear* $\mathbf{d}$ *-homogeneous cone* (see Corollary 5.6 for the proof of linearity of this cone in $\mathbb{R}^n_{\mathbf{d}}$)

$$\Omega_{\text{lin}} = \{x \in \mathbb{R}^n : \|x\|_{\mathbf{d}} h_i^\top \mathbf{d}(-\ln \|x\|_{\mathbf{d}})x \geq 0, i = 1, \dots, p\}. \tag{7.46}$$

In $\mathbb{R}^n$, this set is usually nonlinear, but it is a linear polyhedral cone in $\mathbb{R}^n_{\mathbf{d}}$. The positive invariance of Ω_{lin} admits the following simple characterization.

Corollary 7.8 *Let a linear continuous dilation* $\mathbf{d}$ *be strictly monotone with respect to a weighted Euclidean norm* $\|x\| = \sqrt{z^\top P z}$ *with a shape matrix* $0 \prec P = P^\top \in \mathbb{R}^{n \times n}$ *satisfying (5.6). Let* $F : \mathbb{R}^n \mapsto 2^{\mathbb{R}^n}$ *be compact-valued, convex-valued, upper semi-continuous and* $\mathbf{d}$*-homogeneous of degree* $\mu \in \mathbb{R}$, $\mathbf{0} \in F(\mathbf{0})$. *Let* $\Omega = \Omega_{\text{lin}}$ *be the linear* $\mathbf{d}$*-homogeneous cone given by* (7.46) *and all solutions of the system* (7.34) *initiated in* Ω_{lin} *be forward complete. Then* Ω_{lin} *is weakly positively invariant for the system* (7.37) *if and only if*

$$\Xi_i \neq \emptyset \quad \Rightarrow \quad \max_{y \in F(x)} h_i^\top y - \frac{h_i^\top G_{\mathbf{d}} x}{x^\top P G_{\mathbf{d}} x} x^\top P y \geq 0, \quad \forall x \in \Xi_i, \tag{7.47}$$

where $i = 1, 2, ..., p$, $\Xi_i := \{z \in \mathbb{R}^n : \|z\| = 1, h_i^\top z = 0, h_j^\top z \geq 0, \forall j \neq i\}$,

$$\partial\Omega_{\rm lin}\cap S=\bigcup_{i=1}^{p}\Xi_i. \tag{7.48}$$

The set Ω is positively invariant if, additionally, solutions are forward unique on Ω.

Proof By Corollary 5.6, the **d**-homogeneous cone $\Omega_{\rm lin}$ in the variables $z=\Phi(x)$ becomes the linear cone:

$$\tilde{\Omega}_{\rm lin}=\bigcap_{i=1}^{p}\tilde{\Omega}_{i,\rm lin},\quad \tilde{\Omega}_{i,\rm lin}=\{z\in\mathbb{R}^n : h_i^\top z\geq 0\}. \tag{7.49}$$

The set $\tilde{\Omega}_{i,\rm lin}$ is convex. Using formula (7.45), we rewrite the criterion (7.40) of positive invariance in the form

$$\min_{y\in\tilde{F}(z)}\sum_{i=1}^{p}\xi_i^\top y\leq 0,\quad \forall\xi_i\in\mathcal{N}_{\tilde{\Omega}_{i,\rm lin}}(x),\quad \forall x\in\partial\tilde{\Omega}_{\rm lin}\cap S,$$

where $\tilde{F}$ is given by (7.41). The normal cone $\mathcal{N}_{\tilde{\Omega}_{i,\rm lin}}(x)$ in $\mathbb{R}^n$ (equipped with the canonical inner product $\langle x,y\rangle=x^\top y$ for $x,y\in\mathbb{R}^n$) is given by

$$\mathcal{N}_{\tilde{\Omega}_{i,\rm lin}}(z)=\begin{cases}\{-\lambda h_i:\forall\lambda\geq 0\} & \text{if}\quad z\in\partial\tilde{\Omega}_{i,\rm lin}\\ \{\mathbf{0}\} & \text{if}\ z\in\operatorname{int}\tilde{\Omega}_{i,\rm lin}\\ \emptyset & \text{if}\quad z\notin\tilde{\Omega}_{i,\rm lin}.\end{cases}$$

Hence, we derive the criterion in the form

$$\max_{y\in\tilde{F}(z)}\sum_{i=1}^{p}\lambda_i\chi_i(z)h_i^\top y\geq 0,\quad \forall\lambda_i\geq 0,\quad \forall z\in\partial\tilde{\Omega}_{\rm lin}\cap S, \tag{7.50}$$

where $\chi_i(z)=1$ if $x\in\partial\tilde{\Omega}_{i,\rm lin}$ and $\chi_i(z)=0$ otherwise. The condition (7.50) is fulfilled if and only if

$$\max_{y\in\tilde{F}(z)}\min_{i=1,\ldots p}\chi_i(z)h_i^\top y\geq 0,\quad \forall z\in\partial\tilde{\Omega}_{\rm lin}\cap S.$$

Indeed, if the above condition holds then, obviously, the condition (7.50) is fulfilled. Inversely, if the above condition does not hold then there exists i^* such that $\max_{y\in\tilde{F}(z)}\chi_{i^*}(z)h_{i^*}^\top y<0$. In this case, taking $\lambda_{i^*}>0$ and $\lambda_j=0$ for $j\neq i^*$, we conclude that the condition (7.50) does not hods as well.

Notice that $\partial\Omega_{\rm lin}\cap S=\partial\tilde{\Omega}_{\rm lin}$ and $\Xi_i=\partial\tilde{\Omega}_{i,\rm lin}\cap\partial\tilde{\Omega}_{\rm lin}\cap S$, so the inclusion (7.48) holds. If $\Xi_i=\emptyset$ then $\partial\tilde{\Omega}_i$ have no common points with $\partial\tilde{\Omega}$ except the origin. Let Θ^* be a set of indexes such that $i\in\Theta^*\Rightarrow\Xi_i\neq\emptyset$, so $i\in[1,\ldots,p]\backslash\Theta^*\Rightarrow$

$\Xi_i = \emptyset$. Therefore, taking into account $\max_{y\in\tilde{F}(z)} \min_{i=1,\dots,p} \chi_i(z)h_i^\top y = \min_{i=1,\dots,p} \max_{y\in\tilde{F}(z)} \chi_i(z)h_i^\top y$, we derive the criterion of the positive invariance in the form

$$\min_{i\in\Theta^*} \max_{y\in\tilde{F}(z)} \chi_i(z)h_i^\top y \geq 0, \quad \forall z \in (\cup_{i\in\Theta^*} \Xi_i)$$

or, equivalently,

$$\max_{y\in\tilde{F}(z)} h_i^\top y \geq 0, \quad \forall z \in \partial\Omega \cap \Xi_i, \quad \forall i \in \Theta^*.$$

Taking into account the inequality $z^\top PG_{\mathbf{d}}z > 0, \forall z \neq \mathbf{0}$ and the identity

$$h_i^\top \tilde{F}(z) = \left(-\frac{h_i^\top G_{\mathbf{d}}z}{z^\top PG_{\mathbf{d}}z} z^\top P + h_i^\top\right) F(z), \quad \forall z \in \partial\tilde{\Omega} \cap \Xi_i,$$

we complete the proof.

Example 7.6 Let us consider the homogeneous system given by

$$\dot{x} = \|x\|_{\mathbf{d}}\mathbf{d}(\ln \|x\|_{\mathbf{d}})A\mathbf{d}(-\ln \|x\|_{\mathbf{d}})x, \tag{7.51}$$

where $A \in \mathbb{R}^{n\times n}$ is a constant matrix, $\|x\|_{\mathbf{d}}$ is induced by the weighted Euclidean norm $\|z\| = \sqrt{z^\top Pz}$ and $\mathbf{d}$ is strictly monotone with respect to $\|\cdot\|$. The transformed system for $z = \Phi(z)$ has the form

$$\dot{z} = \|z\|^{\mu+1} \left(A + \frac{z^\top PAz}{z^\top PG_{\mathbf{d}}z}(G_{\mathbf{d}} - I_n)\right) \frac{z}{\|z\|}. \tag{7.52}$$

Let Ω_{lin} be defined by formula (7.49) with $p = n$ such that the matrix

$$H = \begin{pmatrix} h_1^\top \\ \dots \\ h_n^\top \end{pmatrix}$$

be invertible. Then the condition (7.47) means that the matrix

$$M(z) = H\left(A + \frac{z^\top PAz}{z^\top PG_{\mathbf{d}}z}(I_n - G_{\mathbf{d}})\right) H^{-1} \tag{7.53}$$

is Metzler for all $z \in \partial\Omega_{\text{lin}}$. Since the dilation is assumed to be strictly monotone with respect to the norm $\|z\| = \sqrt{z^\top Pz}$ then $z^\top PG_{\mathbf{d}}z > 0$ for $z \neq \mathbf{0}$ and the above matrix $M(z)$ is Metzler for all $z \neq \mathbf{0}$ if and only if the matrix

$$(z^\top PG_{\mathbf{d}}z)HAH^{-1} - (z^\top PAz)HG_{\mathbf{d}}H^{-1}$$

is Metzler. Linear systems with Metzler matrices are known to be positive (see Lemma 2.1). For the nonlinear case, the uniqueness of solutions is additionally required to ensure that the system (7.52) with the Metzler matrix is positive. Therefore, Corollary 7.8, in some sense, generalizes Lemma 2.1 to homogeneous systems.

7.3.2 Stability on Homogeneous Cone

Let us consider a safe set defined as follows.

Assumption 7.1 Let $\mathbf{d}$ be a linear continuous dilation in $\mathbb{R}^n$. Let the nonzero barrier functions $\phi_i \in C(\mathbb{R}^n; \mathbb{R})$ in (7.43) be $\mathbf{d}$-homogeneous of degrees ν_i, $i = 1, 2, ..., p$. Let the closed set Ω given by (7.43) have nonempty interior.

Obviously that, under this assumption, the set Ω is a $\mathbf{d}$-homogeneous cone. Let us consider the following closed set

$$\Omega_0 = \{x \in \mathbb{R}^n : \phi_i(x) = 0, \ i = 1, 2, \ldots, p\}, \tag{7.54}$$

which belongs to the boundary of Ω.

Assumption 7.2 The closed set Ω_0 given by (7.54) is bounded.

If Ω is a $\mathbf{d}$-homogeneous cone, then Ω_0 is a $\mathbf{d}$-homogeneous cone as well, so *the boundedness of* Ω_0 *means that*

$$\Omega_0 = \{\mathbf{0}\}. \tag{7.55}$$

Indeed, otherwise, the set Ω_0 is unbounded due to homogeneity:

$$\mathbf{d}(s)x \in \Omega_0, \quad \forall x \in \Omega_0, \quad \forall s \in \mathbb{R}$$

but $\|\mathbf{d}(s)x\| \to +\infty$ as $s \to +\infty$ for any $x \neq \mathbf{0}$. Below we study homogeneous systems on homogeneous cones with *bounded and unbounded set* Ω_0.

An asymptotic stability of the system (7.34) on the $\mathbf{d}$-homogeneous cone Ω given by (7.43) is understood in the following sense [30].

Definition 7.1 The system (7.34) is said to be uniformly asymptotically stable on Ω if there exist $\beta_i \in \mathcal{KL}$ such that

$$0 \leq \phi_i(x(t)) \leq \beta_i(|x(0)|, t) \tag{7.56}$$

for all $t \geq 0$, $x(0) \in \Omega$ and $i = 1, \ldots, p$.

If Ω is a homogeneous cone then, by Proposition 6.1, we have

$$\phi_i(z) \to 0 \quad \text{as} \quad |z| \to \mathbf{0},$$

so the above definition introduces an analog of asymptotic stability of the set Ω_0 (see formula (7.54)) belonging to the boundary of the invariant (safe) set Ω. Such a stability is typical for safety-critical control systems (see Chap. 12).

The uniqueness of solutions of (7.34) is important for the analysis of the positive invariance of the closed set Ω. The solutions are unique, for example, if $f \in C^1(\Omega; \mathbb{R}^n)$ or f is locally Lipschitz continuous on Ω. However, in this case, the homogeneous vector field has a non-negative homogeneity degree. To include finite-time stable homogeneous systems into considerations, we assume $f \in C^1(\Omega\backslash\{\mathbf{0}\}; \mathbb{R}^n)$ and we ask for a certain regularity of the vector field at $\mathbf{0}$. Using Lemma 7.7 the Zubov–Rosier theorem (see Theorem 7.6) can be generalized to homogeneous systems on homogeneous cones [30].

Theorem 7.11 *Let a linear continuous dilation* $\mathbf{d}$ *be monotone with respect to a norm* $\|\cdot\|$ *in* $\mathbb{R}^n$*, Assumptions 7.1 and 7.2 be fulfilled and a vector field* $f \in C(\mathbb{R}^n) \cap C^1(\Omega\backslash\{\mathbf{0}\})$ *be* $\mathbf{d}$*-homogeneous of degree* $\mu \in \mathbb{R}$ *such that the zero solution of the system* (7.34) *is unique. The system* (7.34) *is uniformly asymptotically stable on* Ω *if and only if the condition* (7.36) *is fulfilled and there exists a* $\mathbf{d}$*-homogeneous function* $V : \mathbb{R}^n \mapsto \mathbb{R}_+$ *of degree 1 satisfying*

(1) $V \in C(\Omega) \cap C^1(\Omega\backslash\{\mathbf{0}\})$ *and there exist* $0 < k_1 \leq k_2 < \infty$ *such that*

$$k_1 \|x\|_{\mathbf{d}} \leq V(x) \leq k_2 \|x\|_{\mathbf{d}}, \quad \forall x \in \Omega; \tag{7.57}$$

(2) there exists $\rho > 0$ *such that*

$$\dot{V}(x) := \frac{\partial V(x)}{\partial x} f(x) \leq -\rho V^{1+\mu}(x), \quad \forall x \in \Omega\backslash\{\mathbf{0}\}. \tag{7.58}$$

Proof *Sufficiency.* The condition (7.36) guarantees (see Corollary 7.7) that any solution $x(t), t \geq 0$ initiated in Ω belongs to Ω, (i.e., $\phi_i(x(t)) \geq 0$) as long as $|x(t)| < +\infty$. The inequalities (7.57), (7.58), imply that $\exists \tilde{\beta} \in \mathcal{KL}$ (more precisely, $\tilde{\beta} \in \mathfrak{e}\mathcal{KL}$ for $\mu = 0$, $\tilde{\beta} \in \mathfrak{f}\mathcal{KL}$ for $\mu < 0$ and $\tilde{\beta} \in \mathfrak{n}\mathcal{KL}$ for $\mu > 0$) satisfying

$$|x(t)| \leq \tilde{\beta}(|x(0)|, t)$$

for all $t \geq 0$ and any $x(0) \in \Omega$. Since ϕ_i is $\mathbf{d}$-homogeneous of degree $\nu_i > 0$ then

$$0 \leq \phi(x(t)) = \|x(t)\|_{\mathbf{d}}^{\nu_i} \phi(\mathbf{d}(-\ln \|x(t)\|_{\mathbf{d}})x(t)) \leq c_i \|x(t)\|_{\mathbf{d}}^{\nu_i} \tag{7.59}$$

as long as $|x(t)| \neq \mathbf{0}$, where $c_i = \sup_{\|z\|=1} \phi_i(z) < +\infty$ due to continuity of ϕ_i. Using (5.16) and the equivalence of the norms $|\cdot|$ and $\|\cdot\|$ we derive (7.56).

Necessity. The inequality (7.56) guarantees that Ω is positively invariant for the system (7.34), so the inclusion (7.36) holds. Since Assumption 7.2 implies $\Omega_0 = \{\mathbf{0}\}$ then

$$x(t) \to \mathbf{0} \quad \Leftrightarrow \quad x(t) \to \Omega_0$$

and the inequality (7.56) means that the origin is the globally uniformly asymptotically stable equilibrium of the system (7.34) on Ω.

Given $\varrho_0 \in (0, 0.5]$ let a function $a \in C^\infty(\mathbb{R})$ be such that $a(\varrho) = 0$ for $\varrho \in [0, \rho_0]$, $a'(\varrho) > 0$ for $\varrho \in (\varrho_0, 1)$ and $a(\varrho) = 1$ for $\varrho \geq 1$. Inspired by [33], let us define the function

$$V_r(x_0) = \int_0^\infty a(|x(\tau, x_0)|)d\tau,$$

where $x(\cdot, x_0)$ denotes the unique solution of the system (7.34) with the initial condition $x(0) = x_0 \in \Omega$. Since the system (7.34) is uniformly asymptotically stable on Ω, the function V_r is well defined and locally bounded on Ω. Moreover, $V_r(x_0) = 0$ for all $x_0 \in \Omega : |x_0| \leq \varepsilon^{-1}(\rho_0)$, where $\varepsilon \in \mathcal{K}$ is introduced in the definition of Lyapunov stability. Taking into account $g \in C^1(\Omega\backslash\{\mathbf{0}\})$ we conclude that solutions of the system depend smoothly on initial conditions (see Theorem 2.3 or [34]). This means that $V_r \in C(\Omega) \cap C^1(\Omega\backslash\{\mathbf{0}\})$. Using the semigroup identity $x(s + \tau, x_0) = x(s, x(\tau, x_0))$ we derive

$$V_r(x(t+h, x_0)) - V_r(x(t, x_0)) = \int_0^\infty a(|x(\tau + h, x(t, x_0))|)d\tau - \int_0^\infty a(|x(\tau, x(t, x_0))|)d\tau$$

$$= -\int_0^h a(|x(\tau, x(t, x_0))|)d\tau, \quad \forall h > 0$$

and

$$\frac{dV_r(x(t,x_0))}{dt} = \frac{\partial V_r(z)}{\partial z} f(z)\bigg|_{z=x(t,x_0)} = -a(|x(t, x_0)|) \quad \text{if} \quad x(t, x_0) \in \Omega\backslash\{\mathbf{0}\}.$$

Let us introduce the following constants

$$c_1 = \sup_{x_0 \in \Omega : |x_0| \leq \tilde{r}_1} V_r(x_0) \quad \text{and} \quad c_2 = \inf_{x_0 \in \Omega : |x_0| > \tilde{r}_2} V_r(x_0),$$

where $\tilde{r}_1 \in (0, 1)$ and $\tilde{r}_2 > 1$ are selected such that $c_2 > c_1 > 0$. Appropriate numbers $\tilde{r}_1$ and $\tilde{r}_2$ always exist since, by construction,

$$\sup_{x_0 \in \Omega : |x_0| \leq \tilde{r}_1} V_r(x_0) \to 0 \quad \text{as} \quad \tilde{r}_1 \to 0$$

and, due to continuity of V_r,

$$\inf_{x_0 \in \Omega : |x_0| \leq \tilde{r}_2} V_r(x_0) > 0 \text{ for any } \tilde{r}_2 > \varrho_0.$$

Let $b \in C^\infty(\mathbb{R})$ such that $b(\rho) = 0$ for $\rho \in [0, c_1]$, $b'(\rho) > 0$ for $\rho \in (c_1, c_2)$ and $b(\rho) = \text{const}$ for $\rho \geq c_2$. Since Ω is the $\mathbf{d}$-homogeneous cone then the Lyapunov function candidate [15]

$$V(x) = \int_0^{+\infty} e^{-s} b(V_r(\mathbf{d}(s)x))ds$$

is well defined for all $x \in \Omega$. By construction, $V(\mathbf{d}(\tau)x) = e^{\tau}V(x), \forall x \in \Omega$ and $V \in C(\Omega) \cap C^1(\Omega \backslash \{\mathbf{0}\})$. The inequality (5.16) implies that

$$\|\mathbf{d}(s)x\|_{\mathbf{d}} \le r_1 \quad \Rightarrow \quad |\mathbf{d}(s)x| \le \tilde{r}_1,$$

$$\|\mathbf{d}(s)x\|_{\mathbf{d}} \ge r_2 \quad \Rightarrow \quad |\mathbf{d}(s)x| \ge \tilde{r}_2$$

for some $r_2 > 1 > r_1 > 0$. This means that V can be represented as follows

$$V(x) = \int_{\ln \frac{r_1}{\|x\|_{\mathbf{d}}}}^{\ln \frac{r_2}{\|x\|_{\mathbf{d}}}} e^{-s} b(V_r(\mathbf{d}(s)x))ds + \int_{\ln \frac{r_2}{\|x\|_{\mathbf{d}}}}^{+\infty} e^{-s} b(c_2)ds.$$

Hence, we derive the following estimate

$$\tfrac{b(c_2)}{r_2}\|x\|_{\mathbf{d}} \le V(x) \le \tfrac{b(c_2)}{r_1}\|x\|_{\mathbf{d}}, \quad \forall x \in \Omega.$$

Taking into account $\mathbf{d}$-homogeneity of the vector field g, the time derivative of the Lyapunov function candidate V along trajectories of the system can be represented as follows

$$\frac{\partial V}{\partial x} g(x) = \int_0^{+\infty} e^{-s} b'(V_r(\mathbf{d}(s)x)) \left.\tfrac{\partial V_r(z)}{\partial z}\right|_{z=\mathbf{d}(s)x} \mathbf{d}(s) f(x)ds =$$

$$\int_0^{+\infty} e^{-(1+\mu)s} b'(V_r(\mathbf{d}(s)x)) \left.\tfrac{\partial V_r(z)}{\partial z} f(z)\right|_{z=\mathbf{d}(s)x} ds =$$

$$-\int_{\ln \frac{r_1}{\|x\|_{\mathbf{d}}}}^{\ln \frac{r_2}{\|x\|_{\mathbf{d}}}} e^{-(1+\mu)s} b'(V_r(\mathbf{d}(s)x)) a(|\mathbf{d}(s)x|)ds = -W(x).$$

By construction, the function $W \in C(\Omega \backslash \{\mathbf{0}\})$ is positive on $\Omega \backslash \{\mathbf{0}\}$ since

$$b'(V_r(\mathbf{d}(s)x)) > 0 \quad \text{for} \quad \ln \frac{r_1}{\|x\|_{\mathbf{d}}} \le s < \ln \frac{r_2}{\|x\|_{\mathbf{d}}},$$

$$a(|\mathbf{d}(s)x|) = a(1) > 0 \quad \text{for} \quad \ln \frac{1}{\|x\|_{\mathbf{d}}} < s < \ln \frac{r_2}{\|x\|_{\mathbf{d}}}.$$

Hence, $W(x) > 0$ for any $x \in \Omega \backslash \{\mathbf{0}\}$. Simple calculations show that $W(\mathbf{d}(\tau)x) = e^{\tau(\mu+1)}W(x), \forall \tau \in \mathbb{R}, \forall x \in \Omega \backslash \{\mathbf{0}\}$, so

$$W(x) = W(\mathbf{d}(-\ln \|x\|_{\mathbf{d}})x)V^{1+\mu}(x) \ge \rho V^{1+\mu}(x)$$

and

$$\dot{V}(x) \leq -\rho V^{1+\mu}(x), \quad \forall x \in \Omega\backslash\{\mathbf{0}\},$$

where $\rho = \inf_{y\in\mathbb{R}^n:V(y)=1} W(y) > 0$. The proof is complete. □

Due to Corollary 7.8, the condition (7.36) of positive invariance of the set Ω can be replaced with the inequality (7.47) provided that $\Omega = \Omega_{\text{lin}}$ is the linear **d**-homogeneous cone given by (7.46).

7.4 Robust Stability and Invariance

Let us consider a dynamical system modeled by the following ODE

$$\dot{x} = f(x, q), \quad t > t_0, \quad x(t_0) = x_0, \tag{7.60}$$

where $x(t) \in \mathbb{R}^n$ is the system state, $q \in L^\infty(\mathbb{R}; \mathbb{R}^k)$ is the exogenous input, and $f \in C(\mathbb{R}^{n+k}; \mathbb{R}^n)$. The input q can be interpreted as a perturbation or an uncertainty of the system. The considered system always has, at least, a strong (Carathéodory) solution (see Chap. 2). A preservation of a stability property of the system under disturbances or parametric uncertainties is known as *robust stability* [35]. Below we show that homogeneous extensions may simplify an analysis of local/global robust asymptotic stability of the system (7.60).

7.4.1 Homogeneous Domination Approach

Let us consider a system

$$\dot{x} = f_1(x) + f_2(x), \quad t > 0, \quad f_1 : \mathbb{R}^n \mapsto \mathbb{R}^n, \quad f_2 : \mathbb{R}^n \mapsto \mathbb{R}^n. \tag{7.61}$$

The vector fields f_1 and f_2 are assumed to be locally or globally homogeneous. The so-called homogeneous domination has been discovered by Vladimir Zubov in 1958 (see [1, Theorems 6 and 7]) and reinvented about 40 years later (see, e.g., [36]). In particular, it answers the question: *Which vector field f_1 or f_2 dominating close to the origin or close to infinity?* For example, for control systems design and analysis it is important to know if a stable behavior of the homogeneous system $\dot{x} = f_1(x)$ is preserved, at least locally, by the possibly non-homogeneous system (7.61). In the case of homogeneous vector fields f_1 and f_2, the answer to this question can be given by comparison of the homogeneity degrees (see Corollary 7.10).

In this subsection, we extend the homogeneous domination approach to uncertain vector fields $f(x, q)$ which can be embedded into set-valued mappings independent of uncertainty/perturbation q.

Corollary 7.9 (Global homogeneous domination) *Let $F : \mathbb{R}^n \mapsto 2^{\mathbb{R}^n}$ be a* **d**-*homogeneous extension with degree $\mu \in \mathbb{R}$ of the vector field $f : C(\mathbb{R}^{n+k}; \mathbb{R}^n)$ such that $f(x,q) \subset F(x), \forall x \in \mathbb{R}^n, \forall q \in \mathbb{R}^n$.*

If F is compact-valued, convex-valued and upper semi-continuous on $\mathbb{R}^n$ and the homogeneous system $\dot{x} \in F(x)$ is asymptotically stable then the system (7.60) *is*

- *globally uniformly finite-time stable for $\mu < 0$;*
- *globally uniformly exponentially stable for $\mu = 0$;*
- *globally uniformly practically fixed-time stable for $\mu > 0$.*

The proof is straightforward, since the solution set of $\dot{x} = f(x,q)$ is embedded in the solution set of $\dot{x} \in F(x)$.

The above conclusions may be meaningful for unbounded-valued homogeneous extensions. For example, if $f : \mathbb{R}^2 \mapsto \mathbb{R}$ is defined as $f(x,q) = -(1 + |\sin(q)x|)\sqrt{|x|}\,\mathrm{sign}(x)$, $x, q \in \mathbb{R}$ and $\mathbf{d}(s) = e^s$, $\nu = -1/2$, then formula (6.33) gives the set-valued **d**-homogeneous extension

$$F(x) = -\sqrt{|x|}\,\mathrm{sign}(x) \cdot (1, +\infty).$$

The differential equation $\dot{x} = f(x,q)$ has continuous right-hand side so it has well defined (at least locally in time) classical solutions for any $q \in \mathbb{R}$ and strong solutions for $q \in L^\infty(\mathbb{R}; \mathbb{R})$. It is easy to see that all conditions of the latter corollary, except the compactness of $G(x)$, hold and the origin of the system $\dot{x} = f(x,q)$ is globally uniformly finite-time stable.

Example 7.7 Let us consider the non-homogeneous planar system

$$\dot{x}_1 = qx_2 \sin\tfrac{1}{x_1^2+x_2^2} + x_2, \quad \dot{x}_2 = -\tfrac{-k_1x_1-k_2h(x)x_2}{h^2(x)}, \quad h(x) = \sqrt{|x_1| + x_2^2},$$

where $q \in [0, 0.2]$ is an uncertain parameter. Let $\mathbf{d}(s) = \mathrm{diag}\{e^{2s}, e^s\}$ and the canonical homogeneous norm $\|\cdot\|_{\mathbf{d}}$ be induced by the norm $\|x\| = \sqrt{x^\top Px}$ with a positive definite matrix $P = P^\top \in \mathbb{R}^{n\times n}$ satisfying (5.6).

Let us denote

$$b_{min} := \inf_{z^\top Pz=1} h(z) > 0$$

and

$$b_{\max} := \sup_{z^\top Pz=1} h(z) \geq b_{\min}.$$

The right-hand side of this system admits the first order **d**-homogeneous extension of the form

$$f(x) \in \left\{\|x\|_{\mathbf{d}}^{-1}\mathbf{d}(\ln\|x\|_{\mathbf{d}})\tilde{A}\mathbf{d}(-\ln\|x\|_{\mathbf{d}})x : \tilde{A} \in \Sigma\right\},$$

where

$$\Sigma = \left\{\tilde{A} = \begin{pmatrix} 0 & a \\ -\frac{k_1}{b} & -\frac{k_2}{b^2} \end{pmatrix} \in \mathbb{R}^{2\times 2} \;:\; a\in[0.8, 1.2],\;\; b\in[b_{\min}, b_{\max}]\right\}.$$

The sufficient stability condition in this case can be written in terms of the parameterized linear matrix inequality (LMI)

$$P\begin{pmatrix} 0 & \alpha \\ -\frac{k_1}{\beta} & -\frac{k_2}{\beta^2} \end{pmatrix} + \begin{pmatrix} 0 & \alpha \\ -\frac{k_1}{\beta} & -\frac{k_2}{\beta^2} \end{pmatrix}^{\top} P \prec 0, \quad \forall \alpha\in[0.8, 1.2], \;\; \forall \beta\in[\beta_{\min}, \beta_{\max}].$$

This LMI can be solved using a convex embedding procedure [37].

Corollary 7.10 (Local homogeneous domination)

- *Let* $\mathbf{d}_0$ *be a linear continuous dilation in* $\mathbb{R}^n$. *Let* $f \in C(\mathbb{R}^{n+k}; \mathbb{R}^n)$. *Let a nonempty-valued, compact-valued, convex-valued and upper semi-continuous vector field* $F_0 : \mathbb{R}^n \mapsto 2^{\mathbb{R}^n}$ *be* $\mathbf{d}_0$*-homogeneous of degree* $\mu_0 \in \mathbb{R}$. *Let a continuous vector field* $g : \mathbb{R}^n \mapsto \mathbb{R}^n$ *be* $\mathbf{d}_0$*-homogeneous of degree* $\mu > \mu_0$ *at 0-limit. If*

$$f(x, q) \subset F_0(x)\dot{+}\{g(x)\}, \quad \forall x \in \mathbb{R}^n, \quad \forall q \in \mathbb{R}^k$$

  and the system $\dot{x} \in F_0(x)$ *is globally uniformly asymptotically stable then the system* (7.60) *is*

  - *locally uniformly finite-time stable if* $\mu_0 < 0$;
  - *locally uniformly exponentially stable if* $\mu_0 = 0$;
  - *locally uniformly asymptotically stable if* $\mu_0 > 0$.

- *Let* $\mathbf{d}_\infty$ *be a linear continuous dilation in* $\mathbb{R}^n$. *Let a nonempty-valued, compact-valued, convex-valued and upper semi-continuous vector field* $F_\infty : \mathbb{R}^n \mapsto 2^{\mathbb{R}^n}$ *be* $\mathbf{d}_\infty$*-homogeneous of degree* $\mu_\infty \in \mathbb{R}$. *Let a continuous vector field* $g : \mathbb{R}^n \mapsto \mathbb{R}^n$ *be* $\mathbf{d}_\infty$*-homogeneous of degree* $\mu < \mu_\infty$ *at* ∞*-limit. If*

$$f(x, q) \subset F_\infty(x)\dot{+}\{g(x)\}, \quad \forall x \in \mathbb{R}^n, \forall q \in \mathbb{R}^k$$

  and the system $\dot{x} \in F_\infty(x)$ *is globally uniformly asymptotically stable then the system* (7.60) *is*

  - *globally uniformly practically asymptotically stable if* $\mu_\infty < 0$;
  - *globally uniformly practically exponentially stable if* $\mu_\infty = 0$;
  - *globally uniformly practically fixed-time stable if* $\mu_\infty > 0$.

- *If* f *satisfies both above conditions with* $\mu_0 < 0 < \mu_\infty$ *and the system* (7.60) *is globally uniformly asymptotically stable then the system is globally uniformly fixed-time stable.*

Proof If the system $\dot{x} \in F_L(x)$ is asymptotically stable then, by Theorem 7.6, it has a **d**-homogeneous Lyapunov function $V \in C(\mathbb{R}^n) \cap C^\infty(\mathbb{R}^n)$ of degree 1 such that

$$\sup_{y\in F_L(x)} \frac{\partial V}{\partial x} y \le -\rho V^{1+\mu}(x), \quad \forall x \neq \mathbf{0}$$

with some $\rho > 0$. In this case, for the original system we derive

$$\frac{\partial V}{\partial x} f(x,q) \le \sup_{y\in F_0(x)} \frac{\partial V}{\partial x}(y+g(x)) \le -\rho V^{1+\mu_L}(x) + \frac{\partial V}{\partial x} g(x),$$

where $L = 0$ or $L = \infty$. Using the homogeneity of $\frac{\partial V}{\partial x}$ (see, Proposition 6.2) we derive

$$\left|\frac{\partial V}{\partial x} f(x,q)\right| \le V^{1+\mu}(x)\left|\frac{\partial V}{\partial y}\right| \cdot \left|e^{-\mu s}\mathbf{d}_L(-s)g(\mathbf{d}(s)y)\right|,$$

where $s = \ln V(x)$ and $y = \mathbf{d}_L(-\ln V(x))x$. By assumption, g is $\mathbf{d}_L$-homogeneous of degree μ at L-limit, i.e., there exists a $\mathbf{d}$-homogeneous function g_L of degree μ such that

$$\left|e^{-\mu \ln r}\mathbf{d}_L(-\ln r)g(\mathbf{d}(\ln r)y) - g_L(y)\right| \to 0 \text{ as } r \to L$$

uniformly on compacts from $\mathbb{R}^n\backslash\{\mathbf{0}\}$. Since g is locally bounded on $\mathbb{R}^n\backslash\{\mathbf{0}\}$ (due to continuity of g), then, by Theorem 6.7, g_L is locally bounded on $\mathbb{R}^n\backslash\{\mathbf{0}\}$ too. Taking into account that y belongs to the unit level set of the Lyapunov function (being a compact from $\mathbb{R}^n\backslash\{\mathbf{0}\}$), for sufficiently small $r_0 > 0$ if $L = 0$ (resp., sufficiently large r_∞ if $L = \infty$) and for sufficiently large $C > 0$ we have

$$\frac{\partial V}{\partial x} f(x,q) \le -\rho V^{1+\mu_L}(x) + CV^{1+\mu}(x) = -(\rho - CV^{\mu-\mu_L}(x))V^{1+\mu_L}(x),$$

for $V(x) < r_0$ if $L = 0$ and for $V(x) > r_\infty$ if $L = \infty$. Taking into account $\mu > \mu_0$ for $L = 0$ and $\mu < \mu_\infty$for $L = \infty$, we complete the proof.

The above corollary shows that in order to identify, which component of a vector field dominates (close to zero or close to infinity), it is sufficient to compare the homogeneity degrees of their homogeneous approximations/extensions.

Example 7.8 Let us consider the system

$$\dot{x} = f(x,q) := -(2+\sin(q(t)))|x|^{-0.5}x + |x|x, \quad t > 0, \tag{7.62}$$

where $f : \mathbb{R}^n \mapsto \mathbb{R}^n$ and $q \in L^\infty(\mathbb{R};\mathbb{R})$. Obviously, for $\mathbf{d}(s) = e^s I_n$, $s \in \mathbb{R}$ the set-valued vector field $G_0(x) = -[1,3]|x|^{-0.5}x$ is the $\mathbf{d}$-homogeneous of degree $\mu_0 = -0.5$ and the scalar vector field $g(x) = |x|x$ is $\mathbf{d}$-homogeneous of degree $\mu = 1$. Since $V = x^\top x$ is a Lyapunov function for the system

$$\dot{x} \in G_0(x)$$

and

$$\dot{V} \leq -2|x|^{3/2} = -2V^{\frac{3}{4}}$$

then, by Corollary 7.10, the system (7.62) is locally uniformly finite-time stable.

7.4.2 Robustly Positively Invariant Cones

Let us consider the system (7.60) on a homogeneous cone Ω with $q \in \mathcal{Q}$ given by

$$\mathcal{Q} := \{q \in L^{\infty}(\mathbb{R}; \mathbb{R}^k) : q(t) \in Q, \forall t \in \mathbb{R}\}, \tag{7.63}$$

where $Q \subset \mathbb{R}^k$ is a closed set.

Assumption 7.3 Let any solution of the system (7.60) with $q \in \mathcal{Q}$ initiated as $x(t_0) = \mathbf{0}$ may leave the origin only by entering the set $\Omega \backslash \{\mathbf{0}\}$.

As an example of the scalar standard homogeneous system satisfying this assumption, let us consider the differential equation

$$\dot{x} = |q||x|^{1/3}$$

with $\Omega = \{x \in \mathbb{R} : x \geq 0\}$. For any $q \in \mathbb{R}\backslash\{0\}$ and $x(0) = 0$, this ODE has the zero solution. The zero solution is non-unique. For example, $x(t) = (2|q|t/3)^{3/2}$ is a solution too, indeed,

$$\dot{x}(t) = \frac{d}{dt}(2|q|t/3)^{3/2} = (3/2)(2|q|/3)^{3/2}t^{1/2} = |q|(2|q|t/3)^{1/2} = |q|x(t).$$

However, since $|q||x|^{1/3} \geq 0$, then all solutions initiated at 0 remain in Ω.

First, let us study the robustness of a homogeneous system (7.60) on a $\mathbf{d}$-homogeneous cone. The homogeneity of the perturbed system is understood in the following sense.

Assumption 7.4 Let a linear continuous dilation $\mathbf{d}$ in $\mathbb{R}^n$ be monotone with respect to a norm $\|\cdot\|_{\mathbb{R}^n}$, and a linear continuous dilation $\tilde{\mathbf{d}}$ in $\mathbb{R}^k$ be monotone with respect to a norm $\|\cdot\|_{\mathbb{R}^k}$. Let the vector field $\hat{f} \in C(\mathbb{R}^{n+k})$ given by

$$\hat{f}(x, q) = \begin{pmatrix} f(x, q) \\ \mathbf{0} \end{pmatrix}, \quad x \in \mathbb{R}^n, \quad q \in \mathbb{R}^k,$$

be $\hat{\mathbf{d}}$-homogeneous of degree $\mu \in \mathbb{R}$, where $\hat{\mathbf{d}}(s) = \begin{pmatrix} \mathbf{d}(s) & \mathbf{0} \\ \mathbf{0} & \tilde{\mathbf{d}}(s) \end{pmatrix}$, $s \in \mathbb{R}$.

The set Ω may also be positively invariant for the perturbed system (7.60) with $q \in \mathcal{Q}$.

If the solutions of the system (7.60) are forward unique and complete then a closed set Ω is robustly positively invariant[a] (see [26, 28]) if

$$f(x,q) \in \mathcal{C}_\Omega(x), \quad \forall x \in \partial\Omega, \quad \forall q \in Q. \tag{7.64}$$

[a] The set Ω is said to be robustly positively invariant for system (7.60) if $x(t_0) \in \Omega \Rightarrow x(t) \in \Omega$ for any $q \in L^\infty(\mathbb{R};\mathbb{R}^k) : q(t) \in Q \subset \mathbb{R}^k$ and for all $t \geq t_0$.

For homogeneous systems on cones, the analysis of robust positive invariance can be restricted to a sphere [30].

Corollary 7.11 *Let, under Assumptions 7.3 and 7.4, a set $\Omega \subset \mathbb{R}^n$ be a closed* $\mathbf{d}$*-homogeneous cone in $\mathbb{R}^n$. Let any solution of the system* (7.60) *with $q \in L^\infty(\mathbb{R};\mathbb{R}^k)$ be forward complete and forward unique as long as $|x(t)| > 0$. Then the set Ω is robustly positively invariant for the system* (7.60) *with $q \in L^\infty(\mathbb{R};\mathbb{R}^k)$ if and only if*

$$f(x,q) \in \mathcal{C}_\Omega(x), \quad \forall x \in \partial\Omega \cap S, \quad \forall q \in \mathbb{R}^k. \tag{7.65}$$

Proof The $\mathbf{d}$-homogeneity of $\mathcal{C}_\Omega$ (see Lemma 6.2) and the $\hat{\mathbf{d}}$-homogeneity of f (see Assumption 7.4) allows the inclusion (7.65) to be extended globally. Indeed, since $\mathbf{d}(-\ln\|x\|_{\mathbf{d}})x \in S$ then

$$f(\mathbf{d}(-\ln\|x\|_{\mathbf{d}})x, \tilde{\mathbf{d}}(-\ln\|x\|_{\mathbf{d}})q) \in \mathcal{C}_\Omega(\mathbf{d}(-\ln\|x\|_{\mathbf{d}})x), \quad \forall x \in \partial\Omega\backslash\{\mathbf{0}\}, \forall q \in \mathbb{R}^k.$$

Using the homogeneity we derive

$$\|x\|_{\mathbf{d}}^{-\mu}\mathbf{d}(-\ln\|x\|_{\mathbf{d}})f(x,q) \in \mathbf{d}(-\ln\|x\|_{\mathbf{d}})\mathcal{C}_\Omega(x), \quad \forall x \in \partial\Omega\backslash\{\mathbf{0}\}, \quad \forall q \in \mathbb{R}^k.$$

Since $\mathbf{d}(s)$ is a linear invertible transformation then the above inclusion is equivalent to

$$\|x\|^{-\mu} f(x,q) \in \mathcal{C}_\Omega(x) \quad \forall x \in \partial\Omega\backslash\{\mathbf{0}\}, \quad \forall q \in \mathbb{R}^k.$$

Since $\mathcal{C}_\Omega(x)$ is a positive cone then the positive multiplier $\|x\|_{\mathbf{d}}^{-\mu}$ can be omitted. Using Assumption 7.3 and the criterion (7.64), we conclude that the set Ω is positively invariant for the system (7.60). The proof is complete. □

The positive invariance of a closed set $\Omega \subset \mathbb{R}^n$ for a differential inclusion with locally Lipschitz continuous right-hand side can be analyzed by means of normal cones (see, e.g., [29]).

Let $F : \mathbb{R}^n \mapsto 2^{\mathbb{R}^n}$ be compact-valued, convex-valued and locally Lipschitz continuous on $\mathbb{R}^n$. Let a set $\Omega \subset \mathbb{R}^n$ be compact and convex. The following two claims are equivalent

- The set $\Omega \subset \mathbb{R}^n$ is positively invariant for the system (7.37).
- The condition

$$\max_{y \in F(x)} \langle \xi, y \rangle \leq 0, \quad \forall \xi \in \mathcal{N}_\Omega(x), \quad \forall x \in \partial\Omega, \tag{7.66}$$

holds, where $\mathcal{N}_\Omega(x)$ is a normal cone to Ω at the point x and $\langle \cdot, \cdot \rangle$.

For homogeneous DI, the above analysis can be reduced to the unit sphere.

Theorem 7.12 *Let a linear continuous dilation* $\mathbf{d}$ *in* $\mathbb{R}^n$ *be strictly monotone with respect to the weighted Euclidean norm* $\|x\| = \sqrt{x^\top P x}$ *with a shape matrix* $0 \prec P = P^\top \in \mathbb{R}^{n \times n}$ *satisfying (5.6). Let* $\mathbb{R}^n_{\mathbf{d}}$ *be the corresponding generalized homogeneous Euclidean space with the canonical homogeneous norm* $\|\cdot\|_{\mathbf{d}}$ *induced by the norm* $\|\cdot\|$*. Let the closed* $\mathbf{d}$*-homogeneous cone* $\Omega \in \mathbb{R}^n$ *be convex in* $\mathbb{R}^n_{\mathbf{d}}$*. Let* $F : \mathbb{R}^n \mapsto 2^{\mathbb{R}^n}$ *be convex-valued, compact-valued, locally Lipschitz continuous on* $\mathbb{R}^n \backslash \{\mathbf{0}\}$ *and* $\mathbf{d}$*-homogeneous of degree* $\mu \in \mathbb{R}$, $\mathbf{0} \in F(\mathbf{0})$*. Let the system* (7.37) *be forward complete and any solution initiated at zero may leave the origin only by entering the set* $\Omega \backslash \{\mathbf{0}\}$*. Then the following three claims are equivalent.*

(1) The set Ω *is positively invariant for the system* (7.37).
(2) The condition

$$\max_{y \in F(x)} \langle y, \xi \rangle_{\mathbf{d}} \leq 0, \quad \forall \xi \in \mathcal{N}_\Omega(x) \quad \forall x \in \partial\Omega \cap S \tag{7.67}$$

holds, where $\langle \cdot, \cdot \rangle_{\mathbf{d}}$ *is an inner product in* $\mathbb{R}^n_{\mathbf{d}}$.
(3) The condition

$$\max_{y \in \tilde{F}(z)} \langle y, \xi \rangle \leq 0, \quad \forall \xi \in \mathcal{N}_{\Phi(\Omega)}(z), \quad \forall z \in \partial\Omega \cap S \tag{7.68}$$

holds, where $\langle \cdot, \cdot \rangle$ *is an inner product* $\mathbb{R}^n$,

$$\tilde{F}(z) = \left(\frac{(I_n - G_{\mathbf{d}}) z z^\top P}{z^\top P G_{\mathbf{d}} z} + I_n \right) F\left(\frac{z}{\|z\|} \right) \tag{7.69}$$

and a homeomorphism $\Phi \in C(\mathbb{R}^n, \mathbb{R}^n)$ *is given by (5.31).*

Proof Local Lipschitz continuity of F guarantees local Lipschitz continuity of $\tilde{F}$. The rest of the proof repeats the proof of Theorem 7.10 using the criterion (7.66) of strong positive invariance instead of the criterion of weak positive (7.38). The proof is complete. □

If $f \in C(\mathbb{R}^{n+k}, \mathbb{R}^n)$ and $Q \subset \mathbb{R}^k$ is a compact set then

$$f(x, q) \in F(x), \quad \forall x \in \mathbb{R}^n, \quad \forall q \in Q,$$

where $x \mapsto F(x) := \text{co}\{f(x, q) : q \in Q\}$ is compact-valued and convex-valued. Moreover, if f is locally Lipschitz continuous in x then F is locally Lipschitz continuous too (see Lemma 2.2). Therefore, the above theorem can be utilized for an analysis of robust positive invariance of the homogeneous system (7.60) satisfying the condition

$$f(\mathbf{d}(s)x, q) = e^{\mu s}\mathbf{d}(s)f(x, q), \quad s \in \mathbb{R}, \quad \forall x \in \mathbb{R}^n, \quad \forall q \in \mathbb{R}^k. \tag{7.70}$$

The vector field f satisfying this condition fulfills Assumption 7.4 for $\tilde{\mathbf{d}}(s) = I_n$, which is not a dilation in $\mathbb{R}^k$.

In the case of linear $\mathbf{d}$-homogeneous cone $\Omega = \Omega_{\text{lin}}$, the above theorem results in an explicit inequality[30] characterizing its robust positive invariance for the system (7.60) satisfying the condition (7.70).

Corollary 7.12 *Let a linear continuous dilation* $\mathbf{d}$ *be strictly monotone with respect to the weighted Euclidean norm* $\|x\| = \sqrt{z^\top Pz}$ *with the shape matrix* $0 \prec P = P^\top \in \mathbb{R}^{n\times n}$ *satisfying (5.6). Let* Ω_{lin} *be the linear* $\mathbf{d}$*-homogeneous cone given by* (7.46). *Let* $Q \subset \mathbb{R}^k$ *be a compact set. Let a continuous function* $(x, q) \mapsto f(x, q)$ *be locally Lipschitz continuous in* $x \in \mathbb{R}^n\backslash\{\mathbf{0}\}$. *Let the set-valued vector field*

$$x \mapsto F(x) := \text{co}\{f(x, q) : q \in Q\}, \quad x \in \mathbb{R}^n$$

be $\mathbf{d}$*-homogeneous of degree* $\mu \in \mathbb{R}$ *such that* $F(\mathbf{0}) = \mathbf{0}$. *Then, under Assumptions 7.3, the set* Ω_{lin} *is robustly positively invariant for the system* (7.60) *with* $q \in \mathcal{Q}$ *if*

$$\Xi_i \neq \emptyset \quad \Rightarrow \quad \min_{y\in F(x)} h_i^\top y - \frac{h_i^\top G_{\mathbf{d}}x}{x^\top PG_{\mathbf{d}}x}x^\top Py \geq 0, \quad \forall x \in \Xi_i, \tag{7.71}$$

where $i = 1, 2, ..., p$, $\Xi_i := \{z \in \mathbb{R}^n : \|z\| = 1, h_i^\top z = 0, h_j^\top z \geq 0, \forall j \neq i\}$.

Proof By Lemma 2.2, the set-valued vector field F is compact-valued, convex-valued and locally Lipschitz continuous on $\mathbb{R}^n\backslash\{\mathbf{0}\}$.

Repeating the proof of Corollary 7.8 *with the criterion* (7.47), *we conclude that the condition* (7.71) *is a criterion of positive invariance of the linear* $\mathbf{d}$ *-homogeneous cone* Ω_{lin} *for the system* $\dot{x} \in F(x)$ *with compact-valued, convex-valued and locally Lipschitz continuous right-hand side.*

Any a solution x of (7.60) with $q \in \mathcal{Q}$ satisfies the differential inclusion $\dot{x}(t) \overset{a.e.}{=} f(x(t), q(t)) \overset{a.e.}{\in} F(x(t))$. This means the system (7.60) with $q \in \mathcal{Q}$ is robustly positively invariant. The proof is complete. □

7.5 Input-to-State Analysis

7.5.1 Input-to-State Stability

Let us consider the system (7.60). We assume that the unperturbed system ($q = \mathbf{0}$) is asymptotically stable. A bounded perturbation in this case may destroy the asymptotic convergence to the origin implying only a convergence to a zone. In this case, it is important to know if a size of this zone depends continuously on a magnitude of perturbations. *Input-to-State Stability (ISS)* (see [38] or Chap. 3) is a well-known concept of robustness for nonlinear (in particular, homogeneous [11, 39, 40]) systems. Recall that the system (7.60) is said to be *input-to-state stable*, if there exist two functions $\xi \in \mathcal{KL}$ and $\theta \in \mathcal{K}$ such that

$$|x_q(t, t_0, x_0)| \leq \xi(|x_0|, t - t_0) + \theta(\|q\|_{L^\infty(0,t)}), \quad \forall t \geq 0 \tag{7.72}$$

for any solution $x_q(\cdot, t_0, x_0)$ of (7.60) with $q \in L^\infty(\mathbb{R}; \mathbb{R}^k)$, $t_0 \in \mathbb{R}$ and $x_0 \in \mathbb{R}^n$.

The Lyapunov function method can be utilized for the characterization of the input-to-state stability of nonlinear systems (see [38] or Chap. 3). In the case of homogeneity, the construction of ISS Lyapunov is not required for the ISS analysis, since the asymptotic stability of the unperturbed homogeneous system is necessary and sufficient for its ISS. This property was first discovered for standard homogeneous systems by Eugene Ryan in 1995, [39]. The ISS of weighted homogeneous systems was considered in [11, 40, 41]. The case of linear dilations was studied in the first edition of this monograph.

Theorem 7.13 *Let a vector field* $f : \mathbb{R}^n \times \mathbb{R}^k \mapsto \mathbb{R}^n$ *satisfy Assumption 7.4. The system* (7.60) *is ISS if and only if the unperturbed system* $\dot{x} = f(x, \mathbf{0})$ *is asymptotically stable.*

Proof For $q = \mathbf{0}$ the ISS inequality (7.72), obviously, implies the asymptotic stability. Inversely, if the unperturbed system is asymptotically stable then, by Theorem 7.6, there exists a positive definite $\mathbf{d}$-homogeneous Lyapunov function $V \in C(\mathbb{R}^n) \cap C^\infty(\mathbb{R}^n)$ of degree 1 such that

$$\frac{\partial V}{\partial x} f(x, \mathbf{0}) \leq -\rho V^{1+\mu}(x), \quad \forall x \neq \mathbf{0}$$

for some $\rho > 0$. In this case, for the perturbed system we have

$$\frac{\partial V}{\partial x} f(x, q) \leq -\rho V^{1+\mu}(x) + \frac{\partial V}{\partial x}(f(x, q) - f(x, \mathbf{0})), \quad \forall x \neq \mathbf{0}.$$

Using $\mathbf{d}$-homogeneity of $\frac{\partial V}{\partial x}$ and f we derive

$$\begin{aligned}\frac{\partial V}{\partial x}(f(x,q)-f(x,\mathbf{0})) &= \frac{\partial V}{\partial x}\mathbf{d}(\ln V(x))\mathbf{d}(-\ln V(x))(f(x,\tilde{q})-f(x,\mathbf{0}))\\ &= V^{1+\mu}(x)\frac{\partial V}{\partial y}(f(y,\tilde{\mathbf{d}}(-\ln V(x))q)-f(y,\mathbf{0})),\end{aligned}$$

where $y=\mathbf{d}(-\ln V(x))x$. Since y belongs to the unit level set of the $\mathbf{d}$-homogeneous Lyapunov function V then there exists $C_1, C_2, C_3 > 0$ such that

$$0 < C_1 \leq \|y\| \leq C_2 < +\infty \quad \text{and} \quad \left\|\frac{\partial V}{\partial y}\right\| \leq C_3.$$

Since f is assumed to be continuous then, by Heine–Cantor theorem, it is uniformly continuous on any compact set and there exists $\omega \in \mathcal{K}$ such that

$$\|f(y,z)-f(y,\mathbf{0})\| \leq \omega(\|z\|), \quad \forall y\in\mathbb{R}^n : C_1 \leq \|y\| \leq C_2, \quad \forall z\in\mathbb{R}^n : \|z\| \leq \varepsilon,$$

where $\varepsilon > 0$ is any number. Since $\omega \in \mathcal{K}$ then there exists sufficiently small $\varepsilon > 0$ such that

$$\|f(y,z)-f(y,\mathbf{0})\| \leq \frac{\rho}{2C_3}, \quad \forall y \in \mathbb{R}^n : C_1 \leq \|y\| \leq C_2, \quad \forall z \in \mathbb{R}^n : \|z\| \leq \varepsilon.$$

This means that

$$\frac{\partial V}{\partial x}f(x,q) \leq -\frac{\rho}{2}V^{1+\mu}(x) \quad \text{if} \quad \|\tilde{\mathbf{d}}(-\ln V(x))q\| \leq \varepsilon.$$

Let $\|\cdot\|_{\tilde{\mathbf{d}}}$ be a $\tilde{\mathbf{d}}$-homogeneous norm in $\mathbb{R}^k$. Since $\|\tilde{\mathbf{d}}(-\ln V(x))q\| \leq \varepsilon \;\Leftrightarrow\; \|q/\varepsilon\|_{\tilde{\mathbf{d}}} \leq V(x) \leq \bar{k}_2\|x\|_{\mathbf{d}}$, where $\bar{k} > 0$ is defined in the proof of Theorem 7.6, then V is the ISS Lypunov function (see Definition 4.6). We complete the proof applying Theorem 4.9.

Notice that the proof of the above theorem implicitly uses certain dilation symmetry of the perturbed homogeneous systems [42].

Corollary 7.13 *Let $x_q(t,t_0,x_0)$, $t \geq t_0$ denote a solution of the system* (7.60). *Let $x_{q_s}(t,t_0,\mathbf{d}(s)x_0)$, $t \geq t_0$ denote a solution of the system* (7.60) *for $x_0 \in \mathbb{R}^n$ replaced with $\mathbf{d}(s)x_0 \in \mathbb{R}^n$ and $q \in L^\infty(\mathbb{R};\mathbb{R}^k)$ replaced with $q_s \in L^\infty(\mathbb{R};\mathbb{R}^k)$ given by*

$$q_s(t) = \tilde{\mathbf{d}}(s)q(t_0 + e^{\mu s}(t-t_0)), \quad \forall t \in \mathbb{R}. \tag{7.73}$$

Then, under Assumption 7.4, the identity

$$x_{q_s}(t,t_0,\mathbf{d}(s)x_0) = \mathbf{d}(s)x_q(t_0 + e^{\mu s}(t-t_0), t_0, x_0), \quad t \geq t_0 \tag{7.74}$$

holds as long as solutions exist.

Proof Indeed, denoting $\tau = t_0 + e^{\mu s}(t - t_0)$ we obtain

$$\frac{d\mathbf{d}(s)x_q(t_0+e^{\mu s}(t-t_0),t_0,x_0)}{dt} = e^{\mu s}\mathbf{d}(s)\frac{dx_q(\tau,t_0,x_0)}{d\tau} \overset{a.e.}{=} e^{\mu s}\mathbf{d}(s)f(x_q(\tau,t_0,x_0),q(\tau)).$$

Using the homogeneity of f we derive

$$\frac{d\,\mathbf{d}(s)x_q(t_0+e^{\mu s}(t-t_0),x_0)}{dt} \overset{a.e.}{=} f(\mathbf{d}(s)x_q(t_0+e^{\mu s}(t-t_0),x_0),\tilde{\mathbf{d}}(s)q(t_0+e^{\mu s}(t-t_0))).$$

Hence, $\frac{d}{dt}x_{q_s}(t,t_0,\mathbf{d}(s)x_0) \overset{a.e.}{=} f(x_{q_s}(t,t_0,\mathbf{d}(s)x_0),q_s(t))$ and the identity (7.74) holds. The proof is complete. □

The convergence rate of ISS system is defined by the homogeneity degree.

Corollary 7.14 *Let $f : \mathbb{R}^n \times \mathbb{R}^k \mapsto \mathbb{R}^n$ satisfy all conditions of Theorem 7.13. If the system* (7.60) *is ISS then it is*

- *finite-time ISS if $\mu < 0$;*
- *exponentially ISS if $\mu = 0$;*
- *nearly fixed-time ISS if $\mu > 0$.*

Proof The required conclusion follows from the estimate

$$\dot{V}(x) = \frac{\partial V}{\partial x}f(x,q) \leq -\frac{\rho}{2}V^{1+\mu}(x) \quad \text{for} \quad \|q/\varepsilon\|_{\tilde{\mathbf{d}}} \leq V(x)$$

proven in Theorem 7.13. Indeed, the above inequality means that if $\|q/\varepsilon\|_{\tilde{\mathbf{d}}} \leq r$ for $t > 0$ then the system converges to the set $\{x \in \mathbb{R}^n : V(x) \leq r\}$ is a finite time (resp., exponentially or nearly in a fixed time) for $\mu < 0$ (resp., $\mu = 0$ or $\mu > 0$). The proof is complete. □

Example 7.9 Let us consider the system

$$\begin{cases} \dot{x}_1 = x_2, \\ \dot{x}_2 = u(x), \end{cases}$$

where $x = (x_1, x_2)^\top \in \mathbb{R}^2$ and

$$u(x) = -k_1|x_1|^{\frac{1}{3}}\operatorname{sign}(x_1) - k_2|x_1|^{\frac{1}{2}}\operatorname{sign}(x_2), \quad 0 < k_2 < k_1 < +\infty.$$

Inspired by [16], let us consider the function $V : \mathbb{R}^2 \mapsto \mathbb{R}$ given by

$$V(x) = \frac{3}{4}k_1|x_1|^{\frac{4}{3}} + 0.5x_2^2.$$

The time derivative of V along the trajectories of the considered system is

$$\dot{V}(x) = -k_2|x_2|^{\frac{3}{2}}.$$

Therefore, V is a weak Lyapunov function and the system is asymptotically stable due to Theorem 4.8. The closed-loop system is $\mathbf{d}$-homogeneous of degree -1 with respect to the dilation $\mathbf{d}(s) = \begin{pmatrix} e^{3s} & 0 \\ 0 & e^{2s} \end{pmatrix}$, $s \in \mathbb{R}$. According to Theorem 7.5, its origin is globally finite-time stable. Moreover, the vector field of the perturbed system

$$\begin{cases} \dot{x}_1 = x_2, \\ \dot{x}_2 = u(x + q_1) + q_2, \end{cases} \quad q = (q_1, q_2)^\top \in L^\infty(\mathbb{R}; \mathbb{R}^3)$$

satisfies Assumption 7.4 with

$$\tilde{\mathbf{d}}(s) = \begin{pmatrix} \mathbf{d}(s) & 0 \\ 0 & e^s \end{pmatrix} \in \mathbb{R}^{3\times 3}, \quad s \in \mathbb{R}.$$

Since $\tilde{\mathbf{d}}$ is a dilation in $\mathbb{R}^3$ then, by Theorem 7.13, the perturbed system is ISS with respect to q. Moreover, by Corollary 7.14, it is finite-time ISS.

ISS on Cone

ISS of a safety-critical system is understood in the sense of the following definition inspired by [30, 43, 44].

Definition 7.2 Let a set $\Omega \subset \mathbb{R}^n$ be defined by formula (7.43). The system (7.60) is said to be ISS on Ω with $q \in \mathcal{Q}$ if there exist $\beta_i \in \mathcal{KL}$ and $\gamma_i \in \mathcal{K}$ such that

$$0 \le \phi_i(x_q(t, t_0, x_0)) \le \beta_i(|x_0|, t - t_0) + \gamma_i(\|q\|_{L^\infty(t_0,t)}) \tag{7.75}$$

for all $t \ge t_0$, $x_0 \in \Omega$, $q \in \mathcal{Q} \subset L^\infty(\mathbb{R}; \mathbb{R}^k)$ and $i = 1, \ldots, p$.

The definition asks both the positive invariance of Ω (the left inequality in (7.75)) and an ISS of the system (the right inequality in (7.75)).

The ISS on a homogeneous cone Ω is characterized by the following theorem [30] being a generalization of Theorem 7.13 to $\mathbf{d}$-homogeneous systems on cones.

Theorem 7.14 *Let $f(\cdot, \mathbf{0}) \in C^1(\mathbb{R}^n \backslash \{\mathbf{0}\}; \mathbb{R}^n)$ and Assumptions 7.1, 7.2, 7.4, 7.3 be fulfilled. Let a set $\Omega \subset \mathbb{R}^n$ be a closed $\mathbf{d}$-homogeneous cone in $\mathbb{R}^n$. Let any solution of the system (7.60) with $q \in L^\infty(\mathbb{R}, \mathbb{R}^k)$ be forward complete and forward unique as long as $|x(t)| > 0$. The system (7.60) is ISS on Ω with $q \in L^\infty(\mathbb{R}, \mathbb{R}^k)$ if the unperturbed ($q = \mathbf{0}$) system is uniformly asymptotically stable on Ω and the condition (7.65) is fulfiled. Moreover, Assumption 7.2 can be omitted if the unperturbed system (7.60) is asymptotically stable on $\mathbb{R}^n$.*

Proof The robust positive invariance of the set Ω for the system (7.60) follows from Corollary 7.11. If the unperturbed system is uniformly asymptotically stable on Ω (resp., on $\mathbb{R}^n$) then, by Theorem 7.11, there exists a **d**-homogeneous Lyapunov function V on Ω (resp., on $\mathbb{R}^n$) of degree 1. Taking into account **d**-homogeneity of f, V nd $\frac{\partial V}{\partial x}$ we derive

$$\begin{aligned}&\tfrac{\partial V(x)}{\partial x} f(x,q) = \tfrac{\partial V(x)}{\partial x} f(x,0) + \tfrac{\partial V(x)}{\partial x}(f(x,q) - f(x,0)) \leq \\ &-\rho V^{1+\mu}(x) + V^{1+\mu}(x)\tfrac{\partial V(z)}{\partial z}\Big(f(z,\tilde{\mathbf{d}}(-\ln V(x))q) - f(z,\mathbf{0})\Big)\Big|_{z=\mathbf{d}(-\ln V(x))x} \leq \\ &-V^{1+\mu}(x)\left(\rho + \tfrac{\partial V(z)}{\partial z}\left(f(z,\tilde{\mathbf{d}}(-\ln \|x\|_{\mathbf{d}})q) - f(z,\mathbf{0})\right)\right)\Big|_{z=\mathbf{d}(-\ln V(x))x},\end{aligned}$$

where $\rho > 0$ is defined in Theorem 7.11. Since $f \in C(\mathbb{R}^{n+k})$ then, by Heine–Cantor theorem, f is uniformly continuous on any compact. Hence, taking into account $0 < r_1 \leq |z| = |\mathbf{d}(-\ln V(x))x| \leq r_2$ for some $r_2 > r_1 > 0$ we conclude that $f(z,\tilde{q}) \to f(z,\mathbf{0})$ as $\|\tilde{q}\| \to 0$ uniformly on z from the compact $\{z \in \mathbb{R}^n : r_1 \leq |z| \leq r_2\}$, and there exists $\delta > 0$ such that

$$\|\tilde{q}\| \leq \delta \quad \Rightarrow \quad \tfrac{\partial V(z)}{\partial z}(f(z,\tilde{q}) - f(z,\mathbf{0})) \leq \tfrac{\rho}{2}$$

for all $z \in \{z \in \mathbb{R}^n : r_1 \leq |z| \leq r_2\}$. Therefore, we derive

$$\|\tilde{\mathbf{d}}(-\ln V(x))q\| \leq \delta \quad \Rightarrow \quad \dot{V}(x) \leq -\tfrac{\rho V^{1+\mu}}{2}.$$

The inequality $\|\tilde{\mathbf{d}}(-\ln \|x\|_{\mathbf{d}})q\| \leq \delta$ is equivalent to $\|q/\delta\|_{\tilde{\mathbf{d}}} \leq V(x)$, so V is an ISS Lyapunov function (see [45] or Chap. 4) for the system (7.60) on Ω (resp., on $\mathbb{R}^n$). Using the inequality (7.59) we complete the proof.

The following straightforward corollary of the above theorem can be derived using the condition of robust positive invariance for the linear **d**-homogeneous cone $\Omega = \Omega_{\text{lin}}$.

Corollary 7.15 *Let a linear continuous dilation* **d** *be strictly monotone with respect to the weighted Euclidean norm* $\|x\| = \sqrt{z^\top P z}$ *with the shape matrix* $0 \prec P = P^\top \in \mathbb{R}^{n\times n}$ *satisfying (5.6). Let* Ω_{lin} *be the linear* **d**-*homogeneous cone given by* (7.46). *Let* $Q \subset \mathbb{R}^k$ *be a compact set. Let* $f \in C(\mathbb{R}^{n+k}, \mathbb{R}^n)$. *Let the set-valued vector field*

$$x \mapsto F(x) := \text{co}\{f(x,q) : q \in Q\}, x \in \mathbb{R}^n$$

be locally Lipschitz continuous and **d**-*homogeneous of degree* $\mu \in \mathbb{R}$. *Then, under Assumptions 7.1, 7.2, 7.4, 7.3, the system* (7.60) *is ISS on* Ω *with* $q \in Q$ *if the unperturbed system (*$q = \mathbf{0}$*) is asymptotically stable on* Ω *and the condition* (7.71) *is fulfilled. Moreover, Assumption 7.2 can be omitted if the unperturbed system* (7.60) *is asymptotically stable on* $\mathbb{R}^n$.

ISS of Locally Homogeneous Systems

The local ISS can be easily analyzed by mixing the local homogeneity with the homogeneous domination approach. Moreover, the homogeneity in the bi-limit can be utilized for the global ISS analysis [11] as well as for fixed-time ISS analysis [46].

Corollary 7.16 *Let $\mathbf{d}_L$ and $\tilde{\mathbf{d}}_L$ be linear dilations in $\mathbb{R}^n$ in and $\mathbb{R}^k$, respectively, where $L = 0$ or $L = \infty$. Let a continuous vector field*

$$\begin{pmatrix} x \\ q \end{pmatrix} \in \mathbb{R}^{n+k} \mapsto \hat{f}(x, q) := \begin{pmatrix} f(x, q) \\ \mathbf{0} \end{pmatrix} \in \mathbb{R}^{n+k}, \quad x \in \mathbb{R}^n, \quad q \in \mathbb{R}^k$$

be locally $\hat{\mathbf{d}}_0$-homogeneous of degree μ_0 at 0-limit and $\hat{\mathbf{d}}_\infty$-homogeneous of degree μ_∞ at ∞-limit, where

$$\hat{\mathbf{d}}_0 = \begin{pmatrix} \mathbf{d}_0 & \mathbf{0} \\ \mathbf{0} & \tilde{\mathbf{d}}_0 \end{pmatrix}, \quad \hat{\mathbf{d}}_\infty = \begin{pmatrix} \mathbf{d}^\infty & \mathbf{0} \\ \mathbf{0} & \tilde{\mathbf{d}}_\infty \end{pmatrix},$$

and let $\hat{f}_0 = \left(\begin{smallmatrix} f_0 \\ \mathbf{0} \end{smallmatrix}\right) : \mathbb{R}^{n+k} \mapsto \mathbb{R}^{n+k}$ and $\hat{f}_\infty = \left(\begin{smallmatrix} f_\infty \\ \mathbf{0} \end{smallmatrix}\right) : \mathbb{R}^{n+k} \mapsto \mathbb{R}^{n+k}$ be the corresponding homogeneous approximations. If the systems

$$\dot{x} = f_0(x, \mathbf{0}), \quad \dot{x} = f(x, \mathbf{0}), \quad \dot{x} = f_\infty(x, \mathbf{0}), \quad t > 0$$

are globally asymptotically stable then the system (7.60) is ISS. Moreover, it is

- *finite-time ISS if $\mu_0 < 0$;*
- *nearly fixed-time ISS if $\mu_\infty > 0$;*
- *fixed-time ISS provided that $\mu_0 < 0$ and $\mu_\infty > 0$.*

Proof By Corollary 7.4 there exists a Lyapunov function V such that

$$\frac{\partial V(x)}{\partial x} f(x, \mathbf{0}) \leq -\rho\sigma(V), \quad \forall x \neq \mathbf{0},$$

where $\rho > 0$ and $\sigma \in \mathcal{K}_\infty$ has the form (7.15) with $m_0 > \max\{0, -\mu_0\}$ and $m_\infty > \max\{0, -\mu_\infty\}$. Moreover,

$$\frac{\partial V(x)}{\partial x} f(x, q) = \frac{\partial V(x)}{\partial x} f(x, \mathbf{0}) + \frac{\partial V(x)}{\partial x}(f(x, q) - f(x, \mathbf{0}))$$

$$\leq -\rho\sigma(V) + \frac{\partial V(x)}{\partial x}(f(x, q) - f(x, \mathbf{0})).$$

Let us consider the compact set

$$K(r) = \{x \in \mathbb{R}^n : \frac{1}{r} \leq \|x\|_{\mathbf{d}^0} \text{ and } \|x\|_{\mathbf{d}^\infty} \leq r\},$$

where $r > 1$ and $\|x\|_{\mathbf{d}_0}$ is a canonical $\mathbf{d}_0$-homogeneous norm in $\mathbb{R}^n$ and $\|\cdot\|_{\mathbf{d}_\infty}$ is a canonical $\mathbf{d}_\infty$-homogeneous norm in $\mathbb{R}^n$. For sufficiently large $r > 0$, the set K_r is nonempty. Since f is continuous then it is uniformly continuous on the compact set

$$\Theta_r = \{(x, q) : x \in K_r, \|q\| \leq V(x)\}$$

and there exists $\omega_r \in \mathcal{K}_\infty$ such that

$$\left\| \frac{\partial V(x)}{\partial x}(f(x,q) - f(x,\mathbf{0})) \right\| \leq \omega_r(\|q\|), \quad \forall (x,q) \in \Theta_r.$$

Therefore, we derive

$$\dot{V} \leq -\rho\sigma(V) + \omega_r(\|q\|) \quad \text{if} \quad x \in K_r, \quad \|q\| \leq V(x).$$

For sufficiently large $r > 1$, by Corollary 7.4, we have $V(x) = V_0(x)$ for x : $\|x\|_{\mathbf{d}_0} \leq \frac{1}{r}$ and $V(x) = V_\infty(x)$ for $x : \|x\|_{\mathbf{d}^\infty} \geq r$, where V_L is a $\mathbf{d}_L$-homogeneous function of degree m_L satisfying (7.13) with $L \in \{0\} \cup \{\infty\}$. Due to local homogeneity of V, we have

$$\|x\|_{\mathbf{d}_x^0} \leq \frac{1}{r} \quad \Rightarrow \quad \dot{V} \leq -\rho V^{1+\frac{\mu_0}{m_L}} + V\frac{\partial V}{\partial y}\mathbf{d}_x^0\left(-\ln V^{\frac{1}{m_0}}\right)(f(x,q) - f(x,\mathbf{0})),$$

where $y = \mathbf{d}_0\left(-\ln V^{\frac{1}{m_0}}\right)x$ belongs to a compact set for all $0 < \|x\|_{\mathbf{d}_0} \leq \frac{1}{r}$. Since f_0 is $\hat{\mathbf{d}}_0$-homogeneous approximation of f at 0-limit, then

$$\sup_{\|x\|_{\mathbf{d}_0}+\|q\|_{\tilde{\mathbf{d}}_0}=p} \|e^{-\mu_0 p}\mathbf{d}_0(-\ln p)(f(x,q) - f(x,\mathbf{0})) - (f_0(x,q) - f_0(x,\mathbf{0}))\| \to 0$$

as $p \to 0$. Hence, repeating the proof of Corollary 7.4, we conclude that, for sufficiently large $r > 1$, there exists $c_0 > 0$ such that the inequalities $\|q\|_{\tilde{\mathbf{d}}_0} \leq \|x\|_{\mathbf{d}_0} \leq \frac{1}{r}$ imply

$$\dot{V} \leq -V^{1+\frac{\mu_0}{m_0}}\left(\rho - c_0\left|\|x\|_{\mathbf{d}_0}^{-\mu_0}\mathbf{d}_0(-\ln\|x\|_{\mathbf{d}_0})(f_0(x,q) - f_0(x,\mathbf{0}))\right|\right).$$

Similarly, for sufficiently large $r > 0$, there exists $c_\infty > 0$ such that the inequalities $\|x\|_{\mathbf{d}_\infty} \geq \max\{r, \|q\|_{\tilde{\mathbf{d}}_\infty}\}$, imply

$$\dot{V} \leq -V^{1+\frac{\mu_\infty}{m_\infty}}\left(\rho - c_\infty\left|\|x\|_{\mathbf{d}_\infty}^{-\mu_\infty}\mathbf{d}_\infty(-\ln\|x\|_{\mathbf{d}_\infty})(f_\infty(x,q) - f_\infty(x,\mathbf{0}))\right|\right).$$

Repeating the proof of Theorem 7.13 we conclude that there exists $\omega \in \mathcal{K}_\infty$ such that

$$\dot{V}(x) \leq -\sigma(V(x)) + \omega(\|q\|), \quad \forall x \neq \mathbf{0}.$$

The proof is complete. □

ISS with Homogeneous Domination

The ISS of systems, which do not have homogeneous approximations, can be analyzed using homogeneous domination. Let us consider the system

$$\dot{x} = f(x, q) + \delta g(x, q), \quad t > 0, \tag{7.76}$$

where x, q, f are as before, $\delta \in L^{\infty}(\mathbb{R}; \mathbb{R})$, $f \in C(\mathbb{R}^{n+k}; \mathbb{R}^n)$ is a locally or globally homogeneous vector field (in the sense of Theorem 7.13 or Corollary 7.16), but $g : \mathbb{R}^{n+k} \mapsto \mathbb{R}^n$ is probably neither globally nor locally homogeneous vector field. If for $\delta = \mathbf{0}$ the homogeneous system (7.76) is ISS then, for small enough δ, the system (7.76) should also be ISS provided that f homogeneously dominates g. Inspired by [47], the rigorous analysis of this issue is given by the following corollary.

Corollary 7.17 *Let f satisfy Assumption 7.4. Let $g : \mathbb{R}^{n+k} \mapsto \mathbb{R}^n$ be locally bounded measurable vector field and a vector field $\tilde{g} : \mathbb{R}^{n+k} \mapsto \mathbb{R}^{n+k}$ be defined as $\hat{g} = \binom{g}{\mathbf{0}}$. Let $\hat{G} : \mathbb{R}^{n \times k} \mapsto 2^{\mathbb{R}^{n+k}}$ be a $\hat{\mathbf{d}}$-homogeneous extension of $\hat{g}$ with degree μ such that*

$$K[\hat{g}](x, q) \subset \hat{G}(x, q), \quad \forall x \in \mathbb{R}^n, \quad \forall q \in \mathbb{R}^k,$$

where $K[\hat{g}]$ is a Filippov regularization of $\hat{g}$.

If $\hat{G}$ is compact-valued, convex-valued and upper semi-continuous on $\mathbb{R}^{n+k}$ then the system (7.76) is ISS with respect to q provided that $\|\delta\|_{\infty}$ is small enough.

Proof If g is discontinuous then a solution of (7.76) is understood in the sense of Filippov, i.e., as a strong solution of the differential inclusion

$$\dot{x} \in f(x, q) + \delta K[g](x, q), \quad t > 0. \tag{7.77}$$

For $\delta = \mathbf{0}$, as shown in the proof of Theorem 7.13, there exists a $\mathbf{d}$-homogeneous ISS Lyapunov function of degree 1:

$$\|\tilde{\mathbf{d}}(-\ln V)q\| \leq C \quad \Rightarrow \quad \frac{\partial V}{\partial x} f(x, q) \leq -\frac{\rho}{2} V^{1+\mu}.$$

Since $\hat{G}$ is compact-valued and upper semi-continuous in $\mathbb{R}^{n \times k}$ then it is locally bounded on any compact from $\mathbb{R}^{n \times k}$ and

$$M_g := \sup_{x \in \Omega, \|q\| \leq C} \; \sup_{y \in \tilde{G}(x,y)} \|y\| < +\infty,$$

where the compact set $\Omega = \{x \in \mathbb{R}^n : V(x) = 1\}$ is the unit level set of the Lyapunov function V. Since $K[\hat{g}](x, q) \subset \hat{G}(x, q)$ then

$$\sup_{y\in K[g](x,q)} \frac{\partial V}{\partial x} y \leq \sup_{\binom{y}{\mathbf{0}}\in \hat{G}(x,q)} V^{1+\mu} \left.\frac{\partial V(z)}{\partial z}\right|_{z=\mathbf{d}(-\ln V)x} V^{-\mu}\mathbf{d}_x(-\ln V)y$$

$$= \sup_{\binom{\tilde{y}}{\mathbf{0}}\in \hat{G}(\mathbf{d}(-\ln V(x))x,\tilde{\mathbf{d}}(-\ln V(x))q)} V^{1+\mu} \left.\frac{\partial V(z)}{\partial z}\right|_{z=\mathbf{d}(-\ln V)x} \cdot \tilde{y},$$

where the $\hat{\mathbf{d}}$-homogeneity of $\hat{G}$ is utilized on the last step. Since $\mathbf{d}(-\ln V)x \in \Omega$ then

$$\|\tilde{\mathbf{d}}(-\ln V)q\| \leq C \Rightarrow \sup_{y\in K[g](x,q)} \frac{\partial V}{\partial x} y \leq V^{1+\mu} M_v M_q,$$

where $M_v = \sup_{z\in\Omega} \left|\frac{\partial V(z)}{\partial z}\right|$ is a finite number due to compactness of Ω and continuity of the derivative of V on Ω. Therefore, for $\|\delta\|_\infty \leq \frac{\rho}{4M_v M_g}$ we have

$$\|\tilde{\mathbf{d}}(-\ln V)q\| \leq C \Rightarrow \sup_{y\in K[g](x,q)} \frac{\partial V}{\partial x}(f(x,q)+\delta y) \leq -\frac{\rho}{4} V^{1+\mu},$$

i.e., V is an ISS Lyapunov function of the system (7.77). The proof is complete. □

7.5.2 *Input-to-State Safety*

The ISS analysis on cones assumes that the set Ω is robustly positively invariant for all $q \in L^\infty$ (see the left inequality in (7.75)). This condition is very restrictive in many practical cases. The Input-to-State Safety introduced in [48] relaxes the invariance of Ω allowing a continuous dependence of the invariant set on the magnitude of the perturbation.

> **Definition 7.3** Let the set $\Omega \subset \mathbb{R}^n$ be defined by (7.43). The system (7.60) is said to be input-to-state safe (ISSf) on Ω if there exist $\tilde{\gamma}_i \in \mathcal{K}$:
>
> $$-\tilde{\gamma}_i\left(\|q\|_{L^\infty_{(t_0,t)}}\right) \leq \phi_i(x_q(t,t_0,x_0)) \tag{7.78}$$
>
> for all $t \geq t_0$, $x_0 \in \Omega$, $q \in \mathcal{Q} \subset L^\infty(\mathbb{R};\mathbb{R}^k)$ and $i = 1,\ldots,p$, where $x_q(\cdot,t_0,x_0)$ denotes a solution with $x(t_0) = x_0$ and the input q.

The inequality (7.78) in the ISSf definition simply means that a "magnitude of violation of the safety constraint" $\phi_i(x) \geq 0$ by the perturbed system depends continuously on the magnitude of the perturbation. In the case of a homogeneous system and a homogeneous cone Ω, such "a violation of the safety constraint" should be homogeneous in some sense.

ISSf on Globally Homogeneous Cones

To characterize the perturbed invariant set of the system (7.60), let us consider the family of the sets

$$\Omega(r) = \{x \in \mathbb{R}^n : \phi_i(x) + r_i^{\nu_i} \geq 0, i = 1, 2, \ldots, p\} \tag{7.79}$$

parameterized by $r = (r_1, \ldots, r_p)^\top \in \mathbb{R}^n_+$, where $\nu_i > 0$ is the **d**-homogeneity degree of ϕ_i.

Lemma 7.2 *Under Assumption 7.1,*

1. $\Omega(0) = \Omega$ *and the inclusions* $\Omega \subset \Omega(r)$ *holds for any* $r \in \mathbb{R}^p_+$*;*
2. *if* $r, \tilde{r} \in \mathbb{R}^p_+$ *are such that* $r \leq \tilde{r}$ *then*

$$\Omega(r) \subset \Omega(\tilde{r}) \quad \textit{and} \quad \partial\Omega(r) \subset \mathbb{R}^n \backslash \operatorname{int} \Omega,$$

where the order relation "$\leq$" for the vectors $r, \tilde{r} \in \mathbb{R}^p_+$ *is understood in the component-wise manner;*

3. *if* $r \in \mathbb{R}^p_+ : r_i > 0, i = 1, \ldots, p$ *then* $\Omega \subset \operatorname{int} \Omega(r)$ *and*

$$\mathcal{C}_{\Omega(r)}(\mathbf{d}(s)x) = \mathbb{R}^n, \quad \forall s > 0, \quad \forall x \in \Omega(r); \tag{7.80}$$

4. *for any* $r \in \mathbb{R}^p_+$*, the tangent cones* $\mathcal{C}_{\Omega(r)}$ *are* **d***-homogeneous as follows:*

$$\mathcal{C}_{\Omega(r)}(\mathbf{d}(s)x) = \mathbf{d}(s)\mathcal{C}_{\Omega(e^{-s}r)}(x), \quad \forall x \in \mathbb{R}^n, \quad \forall s \in \mathbb{R}. \tag{7.81}$$

Proof (1) Indeed, since

$$x \in \Omega \quad \Rightarrow \quad \phi_i(x) \geq 0 \quad \Rightarrow \quad \phi_i(x) + r_i^{\nu_i} \geq 0$$

then $\Omega \subset \Omega(r)$ for any $r \in \mathbb{R}^n_+$.

(2) Similarly, if $r \leq \tilde{r}$ then

$$\phi_i(x) + r_i^{\nu_i} \geq 0 \quad \Rightarrow \quad \phi_i(x) + \tilde{r}_i^{\nu_i} \geq 0$$

and $x \in \Omega(r) \Rightarrow x \in \Omega(\tilde{r})$.

(3) If $r_i > 0, \forall i = 1, 2, \ldots, p$ and $x \in \Omega$ then the continuity of ϕ_i implies that there exists $\rho > 0$ such that $\phi_i(z) + r_i^{\nu_i} \geq 0$ for all $z \in x \dot{+} B(\rho)$ and for all $i = 1, 2, \ldots, p$. This means that $\Omega \subset \operatorname{int} \Omega(r)$.

If $r_i > 0, \forall i = 1, 2, \ldots, p$ and $x \in \Omega(r)$ then $\phi_i(x) + (r_i)^{\nu_i} \geq 0$ and

$$\phi_i(\mathbf{d}(s)x) + (r_i)^{\nu_i} = e^{\nu_i s}\left(\phi_i(\mathbf{d}(s)x) + (e^{-s}r_i)^{\nu_i}\right) > 0, \quad \forall s > 0, \quad \forall i = 1, 2, \ldots, p.$$

This means that $\mathbf{d}(s)x \in \operatorname{int} \Omega$ and $\mathcal{C}_{\Omega(r)}(\mathbf{d}(s)x) = \mathbb{R}^n$ for all $x \in \Omega(r)$ and $\forall s > 0$. This implies (7.80).

(4) The proof of the identity (7.81) repeats the proof of Lemma 6.2.

The ISS on Ω is studied under Assumption 7.3. The ISSf analysis presented below requires a more conservative assumption asking the forward uniqueness of the perturbed system up to the boundary of the set Ω.

Assumption 7.5 For any $q \in L^\infty(\mathbb{R}; \mathbb{R}^k)$ and any $x_0 \in \mathbb{R}^n \backslash \Omega$, a solution of the system (7.60) is forward unique, at least, locally in time.

Due to the semigroup property of solutions, under the above assumption, any solution of system (7.60) initiated away from Ω is unique, at least, until the solution exists and stays away from Ω.

Let us introduce the following notation

$$\|q\|_{\tilde{\mathbf{d}},L^\infty(t_0,t)} = ess \sup_{\tau\in(t_0,t)} \|q(\tau)\|_{\tilde{\mathbf{d}}}. \tag{7.82}$$

The theorem given below [30] presents a sufficient condition of ISSf of the homogeneous system (7.60) on a homogeneous Ω.

Theorem 7.15 *Under Assumptions 7.1, 7.4 and 7.5, the system* (7.60) *is ISSf on Ω if there exists a vector $r \in \mathbb{R}^p_+$ and a constant $q_r > 0$:*

$$f(x,q) \in C_{\Omega(r)}(x), \quad \forall x \in \partial\Omega(r), \quad \forall q \in \mathbb{R}^k : \|q\|_{\mathbb{R}^k} \le q_r. \tag{7.83}$$

Proof Since $\partial\Omega(r) \subset \mathbb{R}^n \backslash \Omega$, then, under Assumption 7.5, the set $\Omega(r)$ is positively invariant for the system (7.60) with any $\|q\|_{L^\infty} \le q_r$ due to the condition (7.83).

Let $t_0 < t_1 < +\infty$ be arbitrary instances of time such that $0 < \|q\|_{\tilde{\mathbf{d}},L^\infty(t_0,t_1)} < +\infty$ and q_s be given by formula (7.73) with $s = -\ln \|q/q_r\|_{\tilde{\mathbf{d}},L^\infty(t_0,t_1)}$. Then

$$\left\|\frac{q_s}{q_r}\right\|_{\mathbf{d},L^\infty(t_0,t_1')} = ess \sup_{\tau\in(t_0,t_1')} \left\|\frac{\tilde{\mathbf{d}}(s)q(t_0+e^{\mu s}(\tau-t_0))}{q_r}\right\|_{\tilde{\mathbf{d}}} = e^s \left\|\frac{q}{q_r}\right\|_{\tilde{\mathbf{d}},L^\infty(t_0,t_1)} = 1$$

(or, equivalently, $\|q_s\|_{L^\infty(t_0,t_1')} = q_r$), where $t_1' = t_0 + e^{-\mu s}(t_1 - t_0)$. Hence, due to positive invariance of $\Omega(r)$ for any $\|q_s\|_{L^\infty} \le q_r$, we conclude that any solution $x_{q_s}(\cdot, t_0, z)$ of the system (7.60) with the perturbation q_s and the initial condition $x(t_0) = z \in \Omega$ satisfies

$$\phi_i(x_{q_s}(t, t_0, z)) + r_i^{\nu_i} \ge 0$$

for all $i = 1, \ldots, p$ and all $t \in [t_0, t_1']$. Taking into account the dilation symmetry of solutions of perturbed homogeneous system (see formula (7.74)) we derive

$$\phi_i(\mathbf{d}(s)x_q(t_0 + e^{\mu s}(t - t_0)), t_0, \mathbf{d}(-s)z) + r_i^{\nu_i} \ge 0, \quad i = 1, 2, ..., p$$

for all $t \in [t_0, t_1']$ and for all $z \in \Omega$, where x_q is a solutions of the system (7.60) with the perturbation q and the initial condition $x(t_0) = \mathbf{d}(-s)z$. Since $s = -\ln \|q/$

$q_r\|_{\mathbf{d},L^\infty(t_0,t_1)}$ and $t_1' = t_0 + e^{-\mu s}(t_1 - t_0)$ then, taking into account that $\mathbf{d}(-s)z \in \Omega$ for all $z \in \Omega$, the obtained inequality and the **d**-homogeneity of ϕ_i gives

$$\phi_i(x_q(t_1, t_0, x_0)) + (r_i\|q/q_r\|_{\mathbf{d},L^\infty(t_0,t_1)})^{\nu_i} \geq 0, \quad \forall x_0 \in \Omega, \quad i = 1, 2, ..., p,$$

where $t_0 < t_1$ are arbitrary selected instances of time. This means the ISSf of the system (7.60). The proof is complete. □

The following lemma shows that the final inequality obtained in the proof of the above theorem is necessary for ISSf of a homogeneous system (7.60).

Lemma 7.3 *Let Assumptions 7.1, 7.4 and 7.5 be fulfilled. If the system* (7.60) *is ISSf on* Ω *then there exists a vector* $r \in \mathbb{R}^p_+$ *such that*

$$\phi_i(x_q(t, t_0, x_0)) + (r_i\|q\|_{\tilde{\mathbf{d}},L^\infty(t_0,t)})^{\nu_i} \geq 0 \tag{7.84}$$

for all $t \geq t_0$, $x_0 \in \Omega$, $q \in L^\infty(\mathbb{R}; \mathbb{R}^k)$, $i = 1, \ldots, p$.

Proof By Lemma 5.4, the canonical homogeneous norm is equivalent to the conventional norm in the sense of the inequalities (5.16). In this case, if the system (7.60) is ISSf on Ω then there exists $\tilde{\gamma}_i \in \mathcal{K}$ such that

$$-\tilde{\gamma}_i(\|q\|_{\tilde{\mathbf{d}},L^\infty(t_0,t)}) \leq \phi_i(x_q(t, t_0, x_0))$$

for all $t \geq t_0$, $x_0 \in \Omega$, $q \in L^\infty(\mathbb{R}; \mathbb{R}^k)$ and $i = 1, \ldots, p$. Let us consider q_s defined in the proof of Theorem 7.15. Since $-\tilde{\gamma}_i(\|q_s\|_{\tilde{\mathbf{d}},L^\infty(t_0,t)}) \leq \phi_i(x_{q_s}(t, t_0, \mathbf{d}(s)x_0))$ for all $t \geq t_0$ then using (7.74), **d**-homogeneity of ϕ_i and $\tilde{\mathbf{d}}$-homogeneity of $\|\cdot\|_{\tilde{\mathbf{d}}}$ we derive

$$-\tilde{\gamma}_i(e^s\|q\|_{\tilde{\mathbf{d}},L^\infty(t_0,t)}) \leq e^{\nu_i s}\phi_i(x_q(t, t_0, x_0)), \forall s \in \mathbb{R}, \forall t \geq t_0,$$

where x_{q_s} is defined in Corollary 7.13. Hence, taking

$$\gamma_i(\sigma) = \inf_{s\in\mathbb{R}} e^{-\nu_i s}\tilde{\gamma}_i(e^s\sigma), \quad \sigma \geq 0$$

we derive $-\gamma_i(\|q\|_{\tilde{\mathbf{d}},L^\infty(t_0,t)}) \leq \phi_i(x_q(t, t_0, x_0)), \forall t \geq t_0$. By construction, γ_i is a standard homogeneous function of the positive degree ν_i such that $0 \leq \gamma_i(\sigma) \leq \tilde{\gamma}_i(\sigma), \forall \sigma \geq 0$. Hence, by Proposition 6.1, the function γ_i is continuous at zero. Since $\gamma(\sigma) = \sigma^{\nu_i}\gamma_i(1), \forall \sigma \neq 0$ then the inequality (7.84) is fulfilled for $r_i^{\nu_i} = \gamma_i(1)$. The proof is complete. □

Notice that if the inequalities (7.84) hold for all $x_0 \in \Omega(r)$ then the set $\Omega(r)$ is positively invariant for the system (7.60) with $\|q\|_{L^\infty} \leq 1$. By Theorem 7.15, this means the system is ISSf, so the necessary condition (7.84) is rather close to the sufficient ISSf condition (7.83). Moreover, if $\|q\|_{L^\infty} \leq r_0$ then the obtained condition means that the set $\Omega(q_0 r)$ is robustly positively invariant set of the system. Due to Lemma 7.2, we have "$\Omega(q_0 r) \to \Omega(0) = \Omega$" as $q_0 \to 0$ in some sense.

Obviously, the positive invariance of the set of $\Omega = \Omega(0)$ for the unperturbed system is necessary for ISSf of (7.60) on Ω. The following lemma [30] shows that, in the case of linear homogeneous cone, the strict positive invariance of a set $\Omega(r)$ for the unperturbed system is sufficient for ISSf of (7.60) on Ω.

Corollary 7.18 *Let* $\mathbf{d}$ *be strictly monotone with respect to a weighted Euclidean norm* $\|x\| = \sqrt{x^\top P x}$ *with* $P \succ 0$ *satisfying (5.6). Let* $f \in C(\mathbb{R}^n \times \mathbb{R}^k, \mathbb{R}^n)$ *be locally Lipschitz continuous in the first argument on* $\mathbb{R}^n \backslash \{\mathbf{0}\}$*. Let* $\Omega = \Omega_{\text{lin}}$ *be given by* (7.46) *with* $h_i \neq \mathbf{0}$, $i = 1, \ldots, p$*. Under Assumptions 7.1, 7.4 and 7.5, the system* (7.60) *is ISSf on* Ω *if*

$$\exists r = \begin{pmatrix} r_1 \\ \cdots \\ r_p \end{pmatrix} > \mathbf{0} : \inf_{z \in \Xi_i(r)} h_i^\top f\left(\tfrac{z}{\|z\|}, \mathbf{0}\right) - \frac{h_i^\top (G_{\mathbf{d}} - I_n) z z^\top P f\left(\frac{z}{\|z\|}, \mathbf{0}\right)}{z^\top P G_{\mathbf{d}} z} > 0, \tag{7.85}$$

where $i = 1, \ldots, p$ *and*

$$\Xi_i(r) := \{z \in \mathbb{R}^n : h_i^\top z + r_i = 0, h_j^\top z + r_j \geq 0, \forall j \neq i\}. \tag{7.86}$$

Proof Since $\phi_i(x) = \|x\|_{\mathbf{d}} h_i^\top \mathbf{d}(-\ln \|x\|_{\mathbf{d}})x$ then, for $\Omega = \Omega_{\text{lin}}$, the set $\Omega(r) = \Omega_{\text{lin}}(r)$ is given

$$\Omega_{\text{lin}}(r) := \{x \in \mathbb{R}^n : h_i^\top \mathbf{d}(-\ln \|x\|_{\mathbf{d}})x + r_i \geq 0, i = 1, 2, \ldots, p\}.$$

In the coordinates $z = \Phi(x)$, we have

$$\tilde{\Omega}_{\text{lin}}(r) = \bigcap_{i=1}^{p} \tilde{\Omega}_{i,\text{lin}}(r), \quad \tilde{\Omega}_{i,\text{lin}}(r) = \{z \in \mathbb{R}^n : h_i^\top z + r_i \geq 0\}.$$

Due to Theorem 7.15, to prove the corollary, it is sufficient to show the set $\Omega_{\text{lin}}(r)$ is robustly positively invariant for the system (7.60) with $\|q\|_{L^\infty} \leq q_0$ for some $q_0 > 0$ and for $r \in \mathbb{R}^p_+$ satisfying (7.85).

In the variables $z = \Phi(x)$, the system (7.60) has the form

$$\dot{z} = \|z\|^{1+\mu} \left(\frac{I_n - G_{\mathbf{d}} z z^\top P}{z^\top P G_{\mathbf{d}} z} + I_n \right) f\left(\frac{z}{\|z\|}, \tilde{\mathbf{d}}(-\ln \|z\|) q \right). \tag{7.87}$$

Therefore, to complete the proof we need to show there exists $q_0 > 0$ such that the set $\tilde{\Omega}_{\text{lin}}(r)$ is positively invariant for the system (7.87) with $\|q\|_{L^\infty} \leq q_0$.

Since $r \in \mathbb{R}^p_+$, $r_i > 0, \forall i = 1, \ldots, p$ then, by Lemma 7.2, $\mathbf{0} \in \operatorname{int} \Omega(r)$, so there exists $\rho_0 > 0$ such that for any $i \in \{1, \ldots, p\}$ we have

$$z \in \Xi_i(r) \Rightarrow \|z\| \geq \rho_0.$$

Let us introduce the set

$$Q = \left\{ \tilde{\mathbf{d}}(-\ln \rho) q : \rho \geq \rho_0, |q| \leq q_0 \right\}.$$

Since $\tilde{\mathbf{d}}$ is a linear continuous dilation in $\mathbb{R}^k$ then $|\mathbf{d}(-\ln \rho)q| \to 0$ as $\rho \to +\infty$ and Q is a compact in $\mathbb{R}^k$.

By construction, the set Q is proportional to q_0, so $Q \to \{\mathbf{0}\}$ as $q_0 \to 0$ (in the Hausdorff topology). In this case, due to continuity of f, the condition (7.85) guarantees that there exists $q_0 > 0$ such that

$$\min_{y \in f\left(\frac{z}{\|z\|}, Q\right)} \frac{h_i^\top (I_n - G_{\mathbf{d}}) z z^\top P y}{z^\top P G_{\mathbf{d}} z} + h_i^\top y \geq 0, \quad \forall z \in \Xi_i \tag{7.88}$$

for any $i = 1, \ldots, p$.

Let $\sigma \in C^\infty(\mathbb{R}_+; \mathbb{R}_+)$ such that $\sigma(s) = 0$ for $s \in [0, \alpha]$ and $\sigma(s) = 1$ for $s \geq 2\alpha/3$ with $\alpha \in (0, \rho_0)$. Let us consider the system

$$\dot{z} \in F(z) := \|z\|^{1+\mu} \operatorname{co} \left\{ \left(\sigma(\|z\|) \frac{(I_n - G_{\mathbf{d}}) z z^\top P}{z^\top P G_{\mathbf{d}} z} + I_n \right) f \left(\sigma(\|z\|) \frac{z}{\|z\|}, \tilde{Q} \right) \right\}. \tag{7.89}$$

By construction, F is compact-valued, convex-valued and locally Lipschitz continuous (see Lemma 2.2). Moreover, any solution $z(t)$ of the system (7.87) belongs to the solution set of the system (7.89) as long as $\frac{2}{3\alpha} < \|z(t)\| < +\infty$.

By Carathéodory lemma (see Appendix A), any vector $y \in F(z)$ can be represented as a convex combination

$$y = \sum_{i=0}^{n} \alpha y_i, \quad \alpha_i \geq 0 : \sum_{i=0}^{n} \alpha_i = 1,$$

where $y_i = \|z\|^{1+\mu} \left(\frac{I_n - G_{\mathbf{d}} z z^\top P}{z^\top P G_{\mathbf{d}} z} + I_n \right) f \left(\frac{z}{\|z\|}, q_i \right)$ with $q_i \in Q$. This means that

$$h_i^\top y = \sum_{i=0}^{n} \frac{h_i^\top (I_n - G_{\mathbf{d}}) z z^\top P f\left(\frac{z}{\|z\|}, q_i\right)}{z^\top P G_{\mathbf{d}} z} + h_i^\top f\left(\frac{z}{\|z\|}, q_i\right) \geq 0, \quad \forall z \in \Xi_i(r)$$

due to the condition (7.88). Therefore, we have

$$\min_{y \in F(z)} h_i^\top y \geq 0, \quad \forall z \in \Xi_i(r).$$

Since $\partial \tilde{\Omega}_{\text{lin}}(r) = \bigcup_{i=1}^{p} \Xi_i(r)$ then, by the criterion (7.66), the obtained condition guarantees the positive invariance of the set $\tilde{\Omega}_{\text{lin}}(r)$ for the system (7.89). Hence, $\tilde{\Omega}_{\text{lin}}(r)$ is robustly positively invariant for the system (7.87) with $\|q\|_{L^\infty} \leq q_0$ and $\Omega_{\text{lin}}(r)$ s robustly positively invariant for the system (7.87) with $\|q\|_{L^\infty} \leq q_0$ for the original system (7.60). Applying Theorem (7.15) we complete the proof.

One can be shown that the strict inequality (7.85), in fact, implies a strict positive invariance of the set $\Omega_{\text{lin}}(r)$ for the unperturbed ($q = \mathbf{0}$) system (7.60).

> Therefore, *the ISSf analysis of a homogeneous system on a linear homogeneous cone can be reduced to an analysis of the unperturbed system.*

7.5.3 Input-to-State Safety and Stability

Robustness analysis of the so-called nonovershooting stabilizers (see Chap. 12) deals with ISS on Ω and ISSf on Ω, simultaneously. The corresponding property is introduced by the following definition [30].

Definition 7.4 The system (7.60) is said to be Input-to-State Safe & Stable (ISSfS) on Ω if there exist $\beta_i \in \mathcal{KL}$ and $\gamma_i, \tilde{\gamma}_i \in \mathcal{K}$ such that

$$-\tilde{\gamma}_i(\|q\|_{L^\infty(t_0,t)}) \leq \phi_i(x(t, t_0, x_0)) \leq \beta_i(|x_0|, t - t_0) + \gamma_i(\|q\|_{L^\infty(t_0,t)}) \quad (7.90)$$

for all $t \geq t_0$, $x_0 \in \Omega$, $q \in L^\infty(\mathbb{R}; \mathbb{R}^k)$ and $i = 1, \ldots, p$.

The following theorem [30] characterizes the robustness of both safety and stability properties of the homogeneous system (7.60) on the homogeneous cone Ω in the sense of the above definition.

Theorem 7.16 *Under Assumptions 7.1, 7.4, 7.5, the system* (7.60) *is ISSfS on Ω if the unperturbed ($q = \mathbf{0}$) system is asymptotically stable on $\mathbb{R}^n$ and the condition* (7.83) *is fulfilled for some $r \in \mathbb{R}^n_+$ and some $q_r > 0$.*

Proof On the one hand, under Assumptions 7.1, 7.4, 7.5, the condition (7.83) implies ISSf of the system (7.60) on Ω provided that the system is forward complete. By Theorem 7.13, under Assumption 7.4, the asymptotic stability of the unperturbed ($q = \mathbf{0}$) system (7.60) on $\mathbb{R}^n$ implies its ISS on $\mathbb{R}^n$. Consequently, the perturbed system is forward complete. Finally, using the estimates (7.59) and (5.16) we complete the proof.

Remark 7.3 For the linear $\mathbf{d}$-homogeneous cone $\Omega = \Omega_{\text{lin}}$, by Corollary 7.18, the condition (7.83) can be replaced with the inequality (7.85), which is independent of q. This means that similarly to ISS analysis on $\mathbb{R}^n$, the ISSfS analysis on Ω_{lin} can be reduced to an investigation of the unperturbed ($q = \mathbf{0}$) homogeneous system. However, in addition to asymptotic stability, the strict positive invariance of a set $\Omega_{\text{lin}}(r), r > \mathbf{0}$ for the unperturbed system is required to guarantee ISSfS (see, (7.85)).

Examples of ISS, ISSf and ISSfS analysis of homogeneous systems on homogeneous cones can be found in Chap. 12.

Concluding Remarks

In this chapter, several important qualitative properties of homogeneous systems are established. In particular, it is shown that any local property of a homogeneous system holds globally; all homogeneous systems of a non-positive degree are complete; the homogeneity degree specifies the convergence rate (finite-time, exponential and nearly fixed-time) of a stable homogeneous system. Homogeneous systems are robust (in the ISS sense) with respect to a rather large class of perturbations. This conclusion remains valid for homogeneous systems on cones, which can be treated as models of control systems with conic state constraints.

References

1. Zubov VI (1958) On systems of ordinary differential equations with generalized homogeneous right-hand sides. Izvestia vuzov. Math 1:80–88
2. Khomenuk VV (1961) On systems of ordinary differential equations with generalized homogenous right-hand sides. Izvestia vuzov. Math (in Russian) 3(22):157–164
3. Kawski M (1991) Families of dilations and asymptotic stability. In: Analysis of controlled dynamical systems, pp 285–294
4. Polyakov A (2019) Sliding mode control design using canonical homogeneous norm. Int J Robust Nonlinear Control 29(3):682–701
5. Zimenko K, Polyakov A, Efimov D (2022) Robust feedback stabilisation of homogeneous differential inclusions. Int J Control 95(1):33–41
6. Bhat SP, Bernstein DS (2005) Geometric homogeneity with applications to finite-time stability. Math Control, Signals Syst 17:101–127
7. Zhuk S, Polyakov A (2020) Homogeneous observer design for finite-dimensional projections of homogeneous pdes. In: Conference on Decision and Control
8. Zubov VI (1964) Methods of A.M. Lyapunov and Their Applications. Noordhoff, Leiden, (Translated from V.I. Zubov, Metody A.M. Lyapunova i ih primenenie, Izdatelstvo Leningradskogo universiteta, 1957) edition
9. Nakamura H, Yamashita Y, Nishitani H (2002) Smooth Lyapunov functions for homogeneous differential inclusions. In: Proceedings of the 41st SICE annual conference, pp 1974–1979
10. Orlov Y (2005) Finite time stability and robust control synthesis of uncertain switched systems. SIAM J Control Optim 43(4):1253–1271
11. Andrieu VA, Praly L, Astolfi A (2008) Homogeneous approximation, recursive observer design, and output feedback. SIAM J Control Optim 47(4):1814–1850
12. Pyatnitskii ES (1999) Control of mechanical systems under uncertainty conditions in the absence of quantitative information on the current state. Autom Remote Control 60(5):739–743
13. Bacciotti A, Rosier L (2001) Lyapunov functions and stability in control theory. Springer
14. Polyakov A, Poznyak A (2009) Lyapunov function design for finite-time convergence analysis: twisting controller for second order sliding mode realization. Automatica 45:444–448
15. Rosier L (1992) Homogeneous Lyapunov function for homogeneous continuous vector field. Syst Control Lett 19:467–473
16. Polyakov A, Orlov Y, Oza H, Spurgeon S (2015) Robust finite-time stabilization and observation of a planar system revisited. In: Conference on Decision and Control
17. Levant A (2005) Quasi-continuous high-order sliding-mode controllers. IEEE Trans Autom Control 50(11):1812–1816

18. Polyakov A, Krstic M (2024) Fixed-time stabilization with a prescribed constant settling time by static feedback for delay-free and input delay systems. Int J Robust Nonlinear Control 34(10):6393–6415
19. Filippov AF (1988) Differential equations with discontinuous right-hand sides. Kluwer Academic Publishers
20. Tuna S, Teel A (2006) Homogeneous hybrid systems and a converse Lyapunov theorem. In: Conference on Decision and Control, pp 6235–6240
21. Zhou Y, Polyakov A, Zheng G (2024) Generalized homogeneous rigid-body attitude control. Automatica 163(111548)
22. Blanchini F, Miani S (2016) Set-theoretic methods in control. Birkhauser
23. Ames AD, Grizzle JW, Tabuada P (2017) Control barrier function based quadratic programs for safety critical systems. IEEE Trans Autom Control 62:3861–3876
24. Abel I, Steeves D, Krstic M, Jankovic M (2022) Prescribed-time safety design for a chain of integrators. In: American control conference
25. Nagumo M (1942) Uber die lage der integralkurven gewohnlicher differentialgleichungen. Proc Phys-Math Soc Jpn 24:551–559
26. Blanchini F (1999) Set invariance in control. Automatica 35:1747–1767
27. Poznyak A, Polyakov A, Azhmyakov V (2014) Attractive ellipsoids in robust control. Birkhauser
28. Aubin JP (1991) Viability theory. Birkhauser
29. Clarke F, Ledyaev YS, Stern R, Wolenski P (1995) Qualitative properties of trajectories of control systems: a survey. J Dyn Control Syst 1(1):1–48
30. Polyakov A, Krstic M (2023) Homogeneous control systems on cones and nonovershooting finite-time stabilizers. arXiv:2305.08951
31. Jankovic M (2018) Robust control barrier functions for constrained stabilization of nonlinear systems. Automatica 96:359–367
32. Mordukhovich BS (2006) Variational analysis and generalized differentiation I. Springer
33. Massera JL (1949) On lyapunovff's conditions of stability. Ann Math 50:705–721
34. Coddington EA, Levinson N (1955) Theory of ordinary differential equations. McGraw-Hill, New York
35. Kharitonov V (1998) Robust stability analysis of time delay systems: a survey. IFAC Proc 31(18):1–12
36. Qian C (2005) A homogeneous domination approach for global output feedback stabilization of a class nonlinear systems. In: American control conference, pp 4708–4715
37. Boyd S, Ghaoui E, Feron E, Balakrishnan V (1994) Linear matrix inequalities in system and control theory. SIAM, Philadelphia
38. Sontag ED (2007) Input to state stability: basic concepts and results. In: Nistri PP, Stefani G (eds) Nonlinear and optimal control theory. Springer, Berlin, pp 163–220
39. Ryan EP (1995) Universal stabilization of a class of nonlinear systems with homogeneous vector fields. Syst Control Lett 26:177–184
40. Bernuau E, Polyakov A, Efimov D, Perruquetti W (2013) Verification of ISS, iISS and IOSS properties applying weighted homogeneity. Syst Control Lett 62(12):1159–1167
41. Hong Y (2001) H_{∞} control, stabilization, and input-output stability of nonlinear systems with homogeneous properties. Automatica 37(7):819–829
42. Polyakov A (2021) Input-to-state stability of homogeneous infinite dimensional systems with locally lipschitz nonlinearities. Automatica 129:109615
43. Sontag ED (1989) Smooth stabilization implies coprime factorization. IEEE Trans Autom Control 34:435–443
44. Magni L, Raimondo DM, Scattolini R (2006) Regional input-to-state stability for nonlinear model predictive control. IEEE Trans Autom Control 51(9):1548–1553
45. Sontag ED, Wang Y (1996) On characterizations of the input-to-state stability property. Syst Control Lett 24(5):351–359
46. Lopez-Ramirez F, Efimov D, Polyakov A, Perruquetti W (2020) Finite-time and fixed-time input-to-state stability: explicit and implicit approaches. Syst Control Lett 144:104775

47. Braidiz Y, Polyakov A, Efimov D, Perruquetti W (2021) On finite/fixed-time stability analysis based on sup-and sub-homogeneous extensions. Syst Control Lett 150:104893
48. Kolathaya S, Ames AD (2019) Input-to-state safety with control barrier functions. IEEE Control Syst Lett 3(1):108–113

Chapter 8
Discretization of Homogeneous Systems

A discretization of continuous-time models is an important issue for numerical simulations of dynamical systems as well as for digital implementation of control laws. Continuous-time homogeneous systems have specific features such as finite/fixed-time convergence, which may be destroyed by improper discretization. This chapter studies the problem of the so-called consistent discretization asymptotically stable homogeneous systems, which preserve the convergence rate of the continuous-time system in its discrete-time counterpart. The classical Euler methods are shown to be inconsistent in the above sense. However, their combinations with a special coordinate transformation provide the desired discretization algorithms.

8.1 Euler Methods

The Euler methods are the most famous discretization algorithms [1] for the nonlinear ODE

$$\dot{x} = f(x), \quad t > 0, \quad x(0) = x_0, \tag{8.1}$$

where, for simplicity, we assume that f is locally bounded piece-wise continuous vector field (see Chap. 2) such that $f(\mathbf{0}) = \mathbf{0}$. Since

$$\dot{x}(t) = \lim_{h\to 0} \frac{x(t+h) - x(t)}{h} \approx \frac{x(t+h) - x(t)}{h}$$

for a sufficiently small $h > 0$ then the simplest discrete-time approximation of the system (8.1) can be given by

$$x_{i+1} = x_i + hf(x_i), \quad i = 0, 1, \ldots \tag{8.2}$$

A. Polyakov, *Generalized Homogeneity in Systems and Control Volume I*,
Communications and Control Engineering,
https://doi.org/10.1007/978-3-031-76246-8_8

Under some restrictions on f, one can be shown that a solution of (8.1) is approximated by the solution of (8.2), namely, $x(ih) \approx x_i$ in the following sense. Considering

$$\hat{x}_h(t) = x_i + \frac{x_{i+1} - x_i}{h}(t - ih), \quad t \in [ih, (i+1)h), i = 0, 1, 2, ...,$$

one can be shown that $\hat{x}_h \to x$ as $h \to 0$ uniformly on compact intervals of time (see, e.g., the proof of Theorem 2.8 for more details). The discrete-time equation (8.2) is known as *explicit Euler discretization* of (8.1). If f is a locally bounded function then the discrete-time equation (8.2) has a solution x_i defined for all i.

Similar considerations introduce the *implicit Euler discretization* as follows

$$x_{i+1} = x_i + hf(x_{i+1}), \quad i = 0, 1, \ldots \tag{8.3}$$

In the latter case, to find an approximation of the system state at the next time instant $t = (i+1)h$, the nonlinear algebraic equation (8.3) has to be solved with respect to the variable x_{i+1}. We say that *the implicit Euler discretization of* (8.1) *exists if the algebraic equation* (8.3) *has a solution with respect to* x_{i+1} for any $x_i \in \mathbb{R}^n$. Most claims presented in this section are proven under assumption that the implicit Euler discretization of (8.1) exists. However, the existence of the implicit Euler discretization is a nontrivial problem in the general case (see, e.g., [2]). For $f \in C^1$ the simple sufficient condition is given by the Hadamard Global Inverse Function theorem [3]:

$$\det\left(I_n - h\frac{\partial f(x)}{\partial x}\right) \neq \mathbf{0}, \forall x \in \mathbb{R}^n \quad \text{and} \quad \liminf_{\|x\| \to +\infty} \|f(x)\| = +\infty. \tag{8.4}$$

For some classes of nonlinear systems the so-called *semi-implicit Euler discretization* can be constructed. For example, if

$$f(x) = A(x)x, \quad x \in \mathbb{R}^n,$$

where $A : \mathbb{R}^n \mapsto \mathbb{R}^{n \times n}$ is a matrix-valued function, then the following discrete-time approximation of the continuous-time system (8.1) can be proposed

$$x_{i+1} = x_i + hA(x_i)x_{i+1}, \quad i = 0, 1, 2, \ldots \tag{8.5}$$

Such a discretization is called semi-implicit, since, the terms $A(x)$ and x in the product $A(x)x$ are discretized by mixing explicit and implicit approaches, respectively. On the one hand, the vector x_{i+1} is still defined implicitly and the algebraic equation (8.5) has to be resolved with respect to x_{i+1}. On the other hand, this equation is linear, so x_{i+1} admits the explicit representation

$$x_{i+1} = (I_n - hA(x_i))^{-1}x_i$$

provided that the matrix $(I_n - hA(x_i))^{-1}$ is invertible. Despite that the discretization (8.5) is still implicit, its existence can be easily proven in some cases. Indeed, if the matrix $A(x)$ is Hurwitz for any $x \neq \mathbf{0}$ then the matrix $-hA(x_i)$ is anti-Hurwitz for any $x \neq \mathbf{0}$. This implies that $I_n - hA(x_i)$ is invertible and the spectral radius of the matrix $\Delta_i = (I_n - hA(x_i))^{-1}$ is less than 1 for any $i = 0, 1, 2, \ldots$. Latter implies stability of the semi-implicit discretization provided that the matrices Δ_i commute.

8.1.1 Homogeneity of Euler Discretizations

The symmetry of solutions with respect to a dilation of the initial state x_0 can also be discovered for the discretized systems [4].

Theorem 8.1 *Let* $\mathbf{d}$ *be a linear continuous dilation in* $\mathbb{R}^n$. *Let* f *be a* $\mathbf{d}$-*homogeneous continuous vector field of degree* $\mu \neq 0$. *If the sequence* $x_i, i = 0, 1, \ldots$ *is a solution of* (8.2) *(resp.,* (8.3)*) then the sequence*

$$y_i = \mathbf{d}(s)x_i, \quad s \in \mathbb{R} \tag{8.6}$$

is a solution of (8.2) *(resp.,* (8.3)*) with the scaled sampling step* $e^{-\mu s}h$.

Proof Since the sequence x_i satisfies (8.2) then multiplying both sides on $\mathbf{d}(s)$ we derive

$$\mathbf{d}(s)x_{i+1} = \mathbf{d}(s)x_i + h\mathbf{d}(s)f(x_i).$$

Denoting $y_i = \mathbf{d}(s)x_i$, due to homogeneity $f(\mathbf{d}(s)x_i) = e^{\mu s}\mathbf{d}(s)f(x_i)$, we obtain

$$y_{i+1} = y_i + h\mathbf{d}(s)f(x_i) = y_i + he^{-\mu s}f(y_i).$$

The case of the implicit Euler discretization (8.3) can be treated similarly. □

The same symmetry can be proven for the semi-implicit discretization of certain homogeneous ODE.

Theorem 8.2 *Let* $\mathbf{d}$ *be a linear continuous dilation in* $\mathbb{R}^n$. *Let* $f : \mathbb{R}^n \mapsto \mathbb{R}^n$ *be a* $\mathbf{d}$-*homogeneous continuous vector field of degree* $\mu \neq 0$ *such that*

$$f(x) = A(\|x\|_{\mathbf{d}})x, \quad \forall x \in \mathbb{R}^n, \tag{8.7}$$

where $A \in C((0, +\infty); \mathbb{R}^{n\times n})$ *and* $\|\cdot\|_{\mathbf{d}}$ *is a canonical homogeneous norm induced by the weighted Euclidean norm* $\|x\| = \sqrt{x^\top Px}$ *with* $P \in \mathbb{R}^{n\times n}$ *satisfying (5.6). If the sequence* $x_i, i = 0, 1, \ldots$ *is a solution of* (8.5) *then the sequence* y_i *given by* (8.6) *is a solution of* (8.5) *with the scaled sampling step* $e^{-\mu s}h$.

Proof First of all, let us prove that the $\mathbf{d}$-homogeneity of f implies that A is $\mathbf{d}$-homogeneous in the following sense

$$A(e^s r)\mathbf{d}(s) = e^{\mu s}\mathbf{d}(s)A(r), \quad \forall r > 0, \forall s \in \mathbb{R}. \tag{8.8}$$

Indeed, since $f(\mathbf{d}(s)x) = e^{\mu s}\mathbf{d}(s)f(x)$ for all $x \in \mathbb{R}^n$ and all $s \in \mathbb{R}$ then, taking into account $\|\mathbf{d}(s)x\|_{\mathbf{d}} = e^s\|x\|_{\mathbf{d}}$, we derive

$$A(e^s\|x\|_{\mathbf{d}})\mathbf{d}(s)x = e^{\mu s}\mathbf{d}(s)A(\|x\|_{\mathbf{d}})x, \quad \forall x \in \mathbb{R}^n, \forall s \in \mathbb{R}.$$

Using the equivalence $\|x\|_{\mathbf{d}} = 1 \;\Leftrightarrow\; \|x\| = 1$, we obtain

$$A(e^s)\mathbf{d}(s)x = e^{\mu s}\mathbf{d}(s)A(1)x, \quad \forall x \in S, \forall s \in \mathbb{R},$$

where $S = \{x \in \mathbb{R}^n : \|x\| = 1\}$ is the unit sphere. The kernel of the matrix $A(e^s)\mathbf{d}(s) - e^{\mu s}\mathbf{d}(s)A(1)$ contains the unit sphere S if and only if this matrix is zero. Therefore, we derive the matrix identity

$$A(e^s)\mathbf{d}(s) = e^{\mu s}\mathbf{d}(s)A(1), \quad \forall s \in \mathbb{R},$$

and

$$A(r) = e^{\mu \ln r}\mathbf{d}(\ln r)A(1)\mathbf{d}(-\ln r), \quad \forall r > 0.$$

This implies (8.8) in the view of the group property of the dilation $\mathbf{d}$.

Let us show now that the sequence y_i given by (8.6) is a solution of (8.5) with the scaled sampling step $e^{-\mu s}h$. Indeed, multiplying both sides on $\mathbf{d}(s)$ we derive

$$\mathbf{d}(s)x_{i+1} = \mathbf{d}(s)x_i + h\mathbf{d}(s)A(\|x_i\|_{\mathbf{d}})x_{i+1}.$$

Denoting $y_i = \mathbf{d}(s)x_i$, due to homogeneity of A (see formula (8.8)), we obtain

$$y_{i+1} = y_i + he^{-\mu s}A(e^s\|x_i\|_{\mathbf{d}})\mathbf{d}(s)x_{i+1} = y_i + he^{-\mu s}A(\|y_i\|_{\mathbf{d}})y_{i+1},$$

where the identity $\|\mathbf{d}(s)x_i\|_{\mathbf{d}} = e^s\|x_i\|_{\mathbf{d}}$ is utilized on the last step. The proof is complete. □

Theoretically, all Euler discretizations have a similar local-in-time approximation precision $O(h)$ of the system trajectory (8.1) (see, e.g., [1]). However, they have totally different global-in-time properties (such as stability) especially in the case of homogeneous systems.

8.1.2 Stability Analysis

The symmetry of solutions discovered in Theorems 8.1, 8.2 requires the simultaneous scaling of the initial state and the sampling period $h > 0$. Therefore, if the Euler discretization has some global property (e.g., global stability) for some sampling period then it has the same property for any sampling period. However, such a dilation symmetry does not imply an equivalence of local and global stability properties of the discrete-time system as it was established for the continuous-time case.

Example 8.1 Let us consider the standard homogeneous globally asymptotically stable

$$\dot{x}(t) = -x^3(t), \quad t > 0, \quad x(0) = x_0 \in \mathbb{R}.$$

The explicit Euler discretization gives

$$x_{i+1} = x_i - hx_i^3.$$

For $|x_i| > \sqrt{\frac{2}{h}}$ we have $|x_{i+1}| > |x_i|$ and the discrete-time system diverges (hyper exponentially). Therefore, the explicit Eulers discretization of the simplest scalar globally stable homogeneous system is not globally stable.

The local (close to zero or close to infinity) instability of the explicit Euler discretization of any globally asymptotically stable $\mathbf{d}$-homogeneous system of nonzero degree can be proven even in a more general case [5].

Theorem 8.3 *Let* $\mathbf{d}$ *be a linear continuous dilation in* $\mathbb{R}^n$. *Let* $f : \mathbb{R}^n \mapsto \mathbb{R}^n$ *be a* $\mathbf{d}$*-homogeneous continuous vector field of degree* $\mu \neq 0$ *and the system* (8.1) *be globally asymptotically stable. Then the explicit Euler discretization* (8.2) *of the system* (8.1) *is locally asymptotically stable if* $\mu > 0$ *and globally practically asymptotically stable if* $\mu < 0$.

Proof Since the $\mathbf{d}$-homogeneous system (8.1) is globally asymptotically stable then, by Zubov–Rosier theorem (see Theorem 7.6), it has a $\mathbf{d}$-homogeneous Lyapunov function $V \in C(\mathbb{R}^n) \cap C^\infty(\mathbb{R}^n \backslash \{\mathbf{0}\})$ of degree 1. Let $s_i = \ln V(x_i)$ with $x_i \neq \mathbf{0}$. In this case, we have

$$V(x_{i+1}) - V(x_i) = e^{s_i}(V(\mathbf{d}(-s_i)x_{i+1}) - V(\mathbf{d}(-s_i)x_i)) =$$

$$e^{s_i}(V(\mathbf{d}(-s_i)x_i + h\mathbf{d}(-s_i)f(x)) - V(\mathbf{d}(-s_i)x_i)) =$$

$$e^{s_i}\left(V\left(\mathbf{d}(-s)x_i + he^{\mu s_i}f(\mathbf{d}(-s_i)x)\right) - V(\mathbf{d}(-s_i)x_i\right) =$$

$$e^{s_i}\left(V\left(z_i + he^{\mu s_i}f(z_i)\right) - V(z_i)\right),$$

where $z_i = \mathbf{d}(-s_i)x_i \in \partial\Omega(1)$ is the $\mathbf{d}$-homogeneous projection of the vector $x_i \in \mathbb{R}^n\backslash\{\mathbf{0}\}$ on the unit level set $\partial\Omega(1) = \{x \in \mathbb{R}^n : V(x) = 1\}$ of the Lyapunov function V. Due to homogeneity of V, such a projection is unique.

Taking into account that $V(z_i) = 1$ and V is positive definite, we conclude that

$$\exists\varepsilon \in (0,1) : e^{\mu s_i} < \varepsilon \quad \Rightarrow \quad 0.5 \le V(z_i + he^{\mu s_i} f(z_i)) \le 1.5.$$

In this case, using Taylor expansion we derive

$$V(x_{i+1}) - V(x_i) = V(x_i)\left(he^{\mu s_i}\frac{\partial V(z_i)}{\partial z_i} f(z_i) + O(he^{\mu s_i}, z_i)he^{\mu s_i}\right),$$

where $O(he^{\mu s_i}, z_i) \to 0$ as $he^{\mu s_i} \to 0$ uniformly on $z_i \in \Omega_\varepsilon = \{z \in \mathbb{R}^n : 0.5 \le V(z) \le 1.5\}$. Since V is a Lyapunov function then

$$a = -\sup_{z\in S}\frac{\partial V(z)}{\partial z} f(z). \tag{8.9}$$

Therefore, for sufficiently small $\varepsilon > 0$ we have

$$V^{\mu}(x_i) < \varepsilon \quad \Rightarrow \quad V(x_{i+1}) \le V(x_i)(1 - 0.5ahe^{\mu s_i}) = V(x_i)\left(1 - 0.5ahV^{\mu}(x_i)\right).$$

Therefore, we derive that the $\mathbf{d}$-homogeneous Lyapunov function V is a local Lyapunov function for the explicit Euler discretization. For $\mu < 0$, this yields the global practical asymptotic stability, but, for $\mu > 0$, the local asymptotic stability is guaranteed. The proof is complete. □

The implicit Euler discretization for homogeneous systems has better stability properties [4]. Its global stability can be proven under some mild conditions.

Theorem 8.4 *Let* $\mathbf{d}$ *be a linear continuous dilation in* $\mathbb{R}^n$. *Let* $f : \mathbb{R}^n \mapsto \mathbb{R}^n$ *be a* $\mathbf{d}$*-homogeneous continuous vector field such that the implicit Euler discretization* (8.3) *exists for some* $h > 0$. *If a* $\mathbf{d}$*-homogeneous Lyapunov function* $V \in C(\mathbb{R}^n) \cap C^\infty(\mathbb{R}^n\backslash\{\mathbf{0}\})$ *of the globally asymptotically stable system* (8.1) *has a convex level set, i.e.,*

$$\exists r > 0 \quad : \quad \Omega(r) = \overline{\text{co}}\,\partial\Omega(r), \quad \text{where } \Omega(r) = \{x \in \mathbb{R}^n : V(x) \le r\}, \tag{8.10}$$

then the implicit Euler discretization (8.2) *of the system* (8.1) *is globally asymptotically stable.*

Proof First of all notice that, since linear transformation of any convex set gives a convex set, then due to $\mathbf{d}$-homogeneity of V and the linearity of the dilation $\mathbf{d}$, the level sets $\Omega(r)$ are convex for all $r > 0$. Moreover, due to Theorem 8.1, the existence of the implicit Euler discretization for some $h > 0$ implies the existence of the implicit Euler discretization for any $h > 0$.

Let $r > 0$ be an arbitrary number. Due to convexity of $\Omega(r)$, for any $x \in \partial\Omega(r)$ and any $z \in \Omega(r) : z \neq x$ there exists $y \in \partial\Omega(r)$ and $\alpha > 0$ such that $z = x + \alpha(x - y)$.

Let us denote $r_i = V(x_i)$ and show that $x_i \notin \Omega(r_{i+1})$. Suppose the contrary, i.e., $x_i \in \Omega(r_{i+1})$. Since the system (8.1) is globally asymptotically stable $f(x) = \mathbf{0} \Leftrightarrow x = \mathbf{0}$. This implies that $x_{i+1} \neq x_i$ provided that $h > 0$ and $x_i \neq \mathbf{0}$. Then, there exists $\alpha > 0$ and $y \in \Omega(r_{i+1})$ such that $x_i = x_{i+1} + \alpha(x_{i+1} - y)$. Hence, we have $hf(x_{i+1}) = \alpha(x_{i+1} - y)$, i.e., the vectors $f(x_{i+1})$ and $x_{i+1} - y$ are co-directional. This is impossible, since the vector $f(x_i)$ initiated at the point x_{i+1} is always oriented inside the level set $\Omega(r_{i+1})$ (due $\frac{\partial V(x)}{\partial x} f(x) < 0$ for all $x \neq \mathbf{0}$), but the vector $\alpha(x_{i+1} - y)$ initiated at x_{i+1} is oriented outside of the level set $\Omega_{r_{i+1}}$ (due to convexity of $\Omega(r_{i+1})$). We obtain the contradiction.

Since $x_i \notin \Omega(x_{i+1})$ then $V(x_i) > V(x_{i+1}) \geq 0$ for any $i \geq 0$. This implies that the sequence $r_i = V(x_i)$ is convergent. Let us show that $r_i \to 0$ as $i \to +\infty$. Suppose the contrary, i.e., there exists $r^* = \lim_{i+\infty} r_i > 0$. Hence, for any $\epsilon \in (0, r^*)$ there exists a sufficiently large i_ϵ such that $0 < r_i - r^* \leq \epsilon$ for all $i \geq i_\epsilon$. Hence, taking into account $r_i = V(x_i)$, we conclude that $x_i \in \partial\Omega^*_\epsilon = \Omega(r^*) \dot{+} \gamma(\varepsilon)\overline{B}$, where $\overline{B} = \{x \in \mathbb{R}^n : \|x\| \leq 1\}$ and $\gamma \in \mathcal{K}$. Since $\gamma(\epsilon) \to 0$ as $\epsilon \to 0$ then the set Ω^*_ϵ is a compact from $\mathbb{R}^n \backslash \{\mathbf{0}\}$.

On the one hand, since V is infinitely smooth on Ω^*_ϵ, then, using Taylor expansion, we derive

$$V(x_i) = V(x_{i+1} - hf(x_i)) = V(x_i) - h\frac{\partial V(x_{i+1})}{\partial x_{i+1}} f(x_{i+1}) + O(h, x_{i+1})h,$$

where $O(h, x_i) \to 0$ as $h \to 0$ uniformly on $x_{i+1} \in \Omega^*_\epsilon$. Therefore, for a sufficiently small $h > 0$, we have

$$V(x_{i+1}) - V(x_i) < -\delta \quad \text{for all} \quad i \geq i_\epsilon,$$

where

$$\delta = -0.5h \sup_{x \in \Omega^*_\epsilon} \frac{\partial V(x)}{\partial x} f(x) > 0.$$

This means that $r_i < r^*$ for $i > i_\epsilon + \frac{\epsilon}{\delta}$. We obtain the contradiction. The proof is complete. □

The condition of convexity of the level sets of the Lyapunov function may be too conservative for some homogeneous systems [6]. It can be slightly relaxed [4] as follows

$$(x - y)^\top f(x) \neq \sqrt{(x - y)^\top (x - y)} \sqrt{f^\top(x) f(x)}, \quad \forall x, y \in \partial\Omega(1) : x \neq y. \quad (8.11)$$

The above conditions explicitly ask that the vectors $f(x)$ and $x - y$ cannot be co-directional. This is the main property of the stable homogeneous vector field f, which is required in the proof of Theorem 8.4. The convexity of the level set $\Omega(1)$ is just a sufficient condition to fulfill (8.11).

The stability of the semi-implicit Euler discretization is studied for a class of homogeneously quadratically stable systems (see Corollary 7.6), which appear in many chapters of this monograph, for example, as a finite-dimensional approximation of an infinite-dimensional homogeneous system (see Volume II of this monograph).

Theorem 8.5 *Let* $\mathbf{d}$ *be a linear continuous dilation in* $\mathbb{R}^n$. *Let* f *be a* $\mathbf{d}$-*homogeneous continuous vector field of degree* $\mu \neq 0$ *satisfying the conditions of Theorem 8.2. If*

$$PA(1) + A^\top(1)P \prec 0, \quad P \succ 0 \tag{8.12}$$

then the semi-implicit Euler discretization (8.5) *of the system* (8.1) *is globally asymptotically stable.*

Proof The inequality (8.12) implies the matrix $A(1)$ is Hurwitz, but, taking into account $\mathbf{d}(s)\mathbf{d}(-s) = I_n$, we conclude that the identity (8.8) guarantees that $A(r)$ is Hurwitz for any $r > 0$. So, the semi-implicit Euler discretization exists for any $h > 0$. Moreover, we have

$$(I_n - hA(\|x\|_\mathbf{d}))^{-1}x = (I_n - he^{-\mu s}\mathbf{d}(-s)A(e^s\|x\|_\mathbf{d})\mathbf{d}(s))^{-1}x =$$

$$\left(\mathbf{d}(-s)\left[I_n - he^{-\mu s}A(e^s\|x\|_\mathbf{d})\right]\mathbf{d}(s)\right)^{-1}x =$$

$$\mathbf{d}(-s)\left(I_n - he^{-\mu s}A(e^s\|x\|_\mathbf{d})\right)^{-1}\mathbf{d}(s)x$$

for any $x \in \mathbb{R}^n \backslash \{\mathbf{0}\}$ and any $s \in \mathbb{R}$. Taking $s = -\ln\|x_i\|_\mathbf{d}$ we derive

$$x_{i+1} = (I_n - hA(\|x_i\|_\mathbf{d}))^{-1}x_i = \mathbf{d}(\ln\|x_i\|_\mathbf{d})(I_n - h\|x_i\|_\mathbf{d}^\mu A(1))^{-1}\mathbf{d}(-\ln\|x_i\|_\mathbf{d})x_i.$$

Let us show that

$$x_{i+1}^\top \mathbf{d}^\top(-\ln\|x_i\|_\mathbf{d}|)P\mathbf{d}(-\ln\|x_i\|_\mathbf{d})x_{i+1} < 1, \forall x_i \neq \mathbf{0}. \tag{8.13}$$

Since $1 = \|\mathbf{d}(-\ln\|x_i\|_\mathbf{d})x_i\|^2 = x_i^\top \mathbf{d}^\top(-\ln\|x_i\|_\mathbf{d}|)P\mathbf{d}(-\ln\|x_i\|_i)x_i$ then, to prove the required inequality, it is sufficient to show that the following matrix inequality

$$(I_n - h\|x_i\|_\mathbf{d}^\mu A(1))^{-\top}P(I_n - h\|x_i\|_\mathbf{d}^\mu A(1))^{-1} \prec P$$

holds for any $x_i \neq \mathbf{0}$. This matrix inequality is equivalent to

$$P \prec (I_n - h\|x_i\|_\mathbf{d}^\mu A(1))^\top P(I_n - h\|x_i\|_\mathbf{d}^\mu A(1)). \tag{8.14}$$

Since $(I_n - h\|x_i\|_\mathbf{d}^\mu A(1))^\top P(I_n - h\|x_i\|_\mathbf{d}^\mu A(1)) = P - h\|x_i\|_\mathbf{d}^\mu(PA(1) + A(1)^\top P)$ $+ h^2\|x\|_\mathbf{d}^{2\mu}A^\top(1)PA(1)$, then the matrix inequality is fulfilled due to (8.12).

By Corollary 5.3, the inequality (8.13) implies $\|x_{i+1}\|_\mathbf{d} < \|x_i\|_\mathbf{d}$, i.e., the canonical homogeneous norm is a Lyapunov function $V(x) = \|x\|_\mathbf{d}$ of the semi-implicit Euler discretization too[1] The proof of convergence of the sequence $V(x_i)$ to 0 literally repeats the similar proof for the implicit Euler discretization (see Theorem 8.4). □

Examples of explicit and implicit Euler discretization of homogeneous systems are given below, where, in particular, their inconsistency in the sense of the convergence rate is demonstrated.

8.2 Consistent Discretization

We start this section with two motivating examples which, on the one hand, demonstrate some inconsistency of the explicit and implicit Euler methods for homogeneous systems of nonzero degree and, on the other hand, explain the basic ideas of the so-called *consistent discretization* of a homogeneous system to be developed in this section.

Finite-Time Stable Implicit Discretization

Inspired by [7] let us consider the following homogeneous systems:

$$\dot{x} = -2\sqrt{|x|}\overline{\text{sign}}(x), \qquad \overset{y=\sqrt{|x|}\text{sign}(x)}{\Leftrightarrow} \qquad \dot{y} \in -\overline{\text{sign}}(y),$$

where

$$\overline{\text{sign}}(\rho) = \begin{cases} 1 & \text{if } \rho > 0, \\ [-1, 1] & \text{if } \rho = 0, \\ -1 & \text{if } \rho < 0. \end{cases}$$

These systems are standard homogeneous, finite-time stable and topologically equivalent (homeomorphic on $\mathbb{R}$ and diffeomorphic on $\mathbb{R}\backslash\{\mathbf{0}\}$). Indeed, if $x(\cdot, x_0)$ is the solution of the first system with $x(0) = x_0 \in \mathbb{R}$ then $y(\cdot, y_0) = \sqrt{|x(\cdot, x_0)|}\,\text{sign}(x(\cdot, x_0))$ is the solution of the second system with $y(0) = y_0 = \sqrt{|x_0|}\,\text{sign}(x_0)$ and vice versa.

The implicit Euler discretizations of these systems are given by:

$$x_{i+1} = x_i - 2h\sqrt{|x_{i+1}|}\,\text{sign}(x_{i+1}), \qquad \not\Leftrightarrow \qquad \begin{array}{l} y_{i+1} = y_i + h\tilde{u}_i, \\ \tilde{u}_i \in -\overline{\text{sign}}(y_{i+1}), \end{array}$$

where $h > 0$ is the sampling period, $x_i = x(ih, x_0)$, $y_i = y(ih, y_0)$ for $t \in [ih, (i+1)h)$, $\tilde{u}_i$ is an auxiliary variable, $i = 0, 1, 2...$ and $\overline{\text{sign}}$ is the set-valued sign function (see formula (0.6)). We refer the reader to [8] for more details about the implicit discretization of the discontinuous system.

[1] In Corollary 7.6, it was shown that the canonical homogeneous norm is a Lyapunov function of the continuous-time system (8.1), (8.7).

The above difference equations admit explicit representations

$$x_{i+1} = \left(\sqrt{h^2 + |x_i|} - h\right)^2 \mathrm{sign}(x_i), \quad \not\Leftrightarrow \quad y_{i+1} = \begin{cases} y_i - h\ \mathrm{sign}(y_i) & \text{if } |y_i| > h, \\ 0 & \text{if } |y_i| \leq h. \end{cases}$$

The discretization destroys the topological equivalence between systems. Indeed, the discrete-time approximation of the first system is just asymptotically stable ($x_0 \neq 0 \Rightarrow x_i \neq 0, \forall i$), but the discretization of the second equation remains finite-time stable ($\exists i^* = i^*(y_0) : y_i = 0, \forall i \geq i^*$). Therefore, a continuous invertible coordinate transformation, which transforms a solution set of the first discrete-time system to a solution set of the second one, does not exist anymore.

The discrete-time approximation of the first (continuous) homogeneous system is inconsistent with its continuous-time counterpart in the context of the convergence rate. A reasonable way to discretize it consistently is to use the equivalence with the second system. Using solutions of the consistently discretized (second) system we can recover the finite-time convergent solutions of the first dynamics by means of the posterior coordinate transformation $\hat{x}_i = y_i^2 \, \mathrm{sign}(y_i)$. The suggested approach gives the following approximation

$$\hat{x}_{i+1} = \begin{cases} \left(\sqrt{|\hat{x}_i|} - h\right)^2 \mathrm{sign}(\hat{x}_i) & \text{if } |\hat{x}_i| > h^2, \\ 0 & \text{if } |\hat{x}_i| \leq h^2, \end{cases}, \quad \tilde{x}_0 = x_0$$

which is, obviously, finite-time stable.

Below we show that any asymptotically stable continuous homogeneous system of a degree $\mu < 0$ admits *a consistent discrete-time approximation* that preserves the finite-time convergence of all trajectories of the system to the origin.

Nearly Fixed-Time Stable Semi-implicit Discretization

Again inspired by [7] let us consider the scalar system

$$\dot{x} = -|x|x, \tag{8.15}$$

which is globally uniformly nearly fixed-time stable: $|x(t, x_0)| < \varepsilon$ for $t > \frac{1}{\varepsilon}$ independently of the initial state x_0. The *explicit* Euler method applied to the system (8.15) gives the discrete-time system

$$x_{i+1} = x_i - h|x_i|x_i,$$

which has blowing-up solutions if $h > 2/|x_0|$, i.e., the discrete-time approximation by the explicit Euler method is not globally stable.

The *implicit* Euler discretization yields the globally asymptotically stable system

$$x_{i+1} = \frac{\sqrt{1+4h|x_i|}-1}{2h} \, \mathrm{sign}(x_i),$$

which does not preserve nearly fixed-time stability. Thus it is also inconsistent with the original continuous-time model in the context of the decay rate. However, using a *semi-implicit* Euler discretization, we derive

$$\tfrac{x_{i+1}-x_i}{h} = -|x_i|x_{i+1}$$

or, equivalently,

$$x_{i+1} = \tfrac{x_i}{1+h|x_i|}.$$

It is easy see that $|x_1| \leq (h)^{-1}$ *independently of* x_0, but

$$|x_2| = \tfrac{1}{|x_1|^{-1}+h} \leq \tfrac{1}{h+h} = (2h)^{-1}$$

$$|x_3| = \tfrac{1}{|x_2|^{-1}+h} \leq \tfrac{1}{2h+h} = (3h)^{-1},$$

$$\ldots$$

$$|x_i| \leq (ih)^{-1}.$$

Therefore, the discrete-time approximation by the semi-implicit Euler method preserves the nearly fixed-time stability of the original continuous-time system. Below we extend this approach to asymptotically stable homogeneous systems with positive degrees.

8.2.1 Definition of Consistent Discretization

Let us consider the differential inclusion

$$\dot{x} \in F(x), \quad t > 0, \quad x(0) = x_0, \tag{8.16}$$

where $x(t) \in \mathbb{R}^n$ is the system state and $F : \mathbb{R}^n \mapsto 2^{\mathbb{R}^n}$ is a (possibly) set-valued vector field. In particular, F may be single-valued vector field $F(x) = \{f(x)\}$ with $f \in C(\mathbb{R}^n; \mathbb{R}^n)$ or $F(x) = K[f](x)$ may be a set-valued regularization (see Chap. 2 or [9]) of a locally bounded measurable vector field f. Inspired by [7] we introduce the following definition.

Definition 8.1 Let $Q : \mathbb{R}_+ \times \mathbb{R}^n \times \mathbb{R}^n \mapsto 2^{\mathbb{R}^n}$ be a (possibly) set-valued mapping. The discrete-time inclusion

$$\mathbf{0} \in Q(h, x_i, x_{i+1}), \quad h > 0, \quad i = 0, 1, 2, \ldots \tag{8.17}$$

is said to be a *consistent discretization* (discrete-time approximation) of the globally uniformly finite-time stable system (8.16) if

1. for any $\tilde{x} \in \mathbb{R}^n$ and any $h > 0$, there exists $\tilde{x}_h \in \mathbb{R}^n$:
$$\mathbf{0} \in Q(h, \tilde{x}, \tilde{x}_h), \tag{8.18}$$
and $\tilde{x}_h = \mathbf{0}$ is the unique solution to $\mathbf{0} \in Q(h, \mathbf{0}, \tilde{x}_h)$.
2. for any $h > 0$ each sequence
$$\{x_i\}_{i=0}^{+\infty} \tag{8.19}$$
generated by (8.17) converges to zero in a finite number of steps, i.e., for any $x_0 \in \mathbb{R}^n \backslash \{\mathbf{0}\}$ there exists $i^* > 0$ such that
$$x_i = \mathbf{0} \qquad \text{for} \quad i \geq i^*$$
and $x_{i^*-1} \neq \mathbf{0}$.
3. for any $x_0 \in \mathbb{R}^n$ and for any strong solution $\phi(t, x_0), t \in [0, T]$ of the continuous-time system (8.16), the family of continuous functions
$$\phi_h(t, x_0) = x_i + (x_{i+1} - x_i)\tfrac{t-ih}{h}, \quad t \in [ih, (i+1)h), \quad i = 0, 1, 2, \ldots \tag{8.20}$$
with $h > 0$ and $\{x_i\}_{i=0}^{+\infty}$ being solutions of the discrete-time system (8.17), contains a sequence ϕ_{h_k} such that
$$\sup_{t\in[0,T]} |\phi(t, x_0) - \phi_{h_k}(t, x_0)| \to 0 \quad \text{as } h_k \to 0, \tag{8.21}$$
and, inversely, if a sequence of functions $\{\phi^{h_k}\}$ generated by solutions of the discrete-time system (8.17) by formula (8.20) converges to a continuous function ϕ as $h_k \to 0$ then this function is a strong solution of (8.16).

The latter condition in the definition guarantees that the discrete-time model (8.17) is, indeed, an approximation of the continuous-time system (8.16). Below we use (implicit and semi-implicit) Euler approximates to construct consistent discretization (see Chap. 2 for more details about Euler approximates).

Definition 8.2 Let $Q : \mathbb{R}_+ \times \mathbb{R}^n \times \mathbb{R}^n \mapsto 2^{\mathbb{R}^n}$ be a (possibly) set-valued mapping. The equation

$$\mathbf{0} \in Q(h, x_i, x_{i+1}), \quad i = 0, 1, 2, ... \tag{8.22}$$

is said to be a consistent discretization (discrete-time approximation) of the globally nearly fixed-time stable system (8.16) if it satisfies the conditions 1 and 3 of Definition 8.1 and for any $r > 0$ and any $h > 0$ there exists $N(r, h) > 0$ such that any sequence

$$\{x_i\}_{i=0}^{+\infty} \tag{8.23}$$

generated by Eq. (8.22) satisfies

$$\|x_i\| \leq r \qquad \text{for} \quad i \geq N(r, h)$$

independently of $x_0 \in \mathbb{R}^n$.

8.2.2 *Finite-Time Stable Implicit Discretization*

First, we design a consistent discretization of the $\mathbf{d}$-homogeneous differential inclusion (8.16) with degree $\mu = -1$. The coordinate transformation (5.31) is utilized for this purpose. The following theorem generalizes the result of the paper [7] to differential inclusions.

Theorem 8.6 *Let a set-valued vector field $F : \mathbb{R}^n \mapsto \mathbb{R}^n$ be compact-valued, convex-valued, upper semi-continuous and $\mathbf{d}$-homogeneous of the degree -1. Let $G_{\mathbf{d}} \in \mathbb{R}^{n \times n}$ be the generator of the dilation $\mathbf{d}$ and a symmetric matrix $P \in \mathbb{R}^{n \times n}$ satisfies (5.6).*

If the condition (7.18) holds with $\Xi \equiv I_n$, then the mapping $Q : \mathbb{R}_+ \times \mathbb{R}^n \times \mathbb{R}^n \mapsto 2^{\mathbb{R}^n}$ given by

$$\begin{aligned} Q(h, x_i, x_{i+1}) &= \tilde{Q}(h, \Phi(x_i), \Phi(x_{i+1})), \\ \Phi(x) &= \|x\|_{\mathbf{d}} \mathbf{d}(-\ln \|x\|_{\mathbf{d}})x \end{aligned} \tag{8.24}$$

where $h > 0$ and

$$\tilde{Q}(h, y_i, y_{i+1}) = h\tilde{F}(y_{i+1})\dot{+}\{y_i - y_{i+1}\}, \tag{8.25}$$

$$\tilde{F}(y):=\left(\frac{(I-G_{\mathbf{d}})yy^{\top}P}{y^{\top}G_{\mathbf{d}}Py}+I_n\right)F\left(\frac{y}{\sqrt{y^{\top}Py}}\right), \qquad y\in\mathbb{R}^n\backslash\{\mathbf{0}\},$$

$$\tilde{F}(\mathbf{0})=\bigcap_{\varepsilon>0}\mathrm{co}\,\tilde{F}(B(\varepsilon)\backslash\{\mathbf{0}\})$$

defines a consistent (in the sense of Definition 8.1) discrete-time approximation of the system (8.16).

Proof By Theorem 7.3 the system

$$\dot{y} \in \tilde{F}(y) \tag{8.26}$$

is topologically equivalent (homeomorphic on $\mathbb{R}^n$ and diffeomorphic on $\mathbb{R}^n\backslash\{\mathbf{0}\}$) to the system (8.16) with the change of coordinates given by

$$y = \Phi(x), \qquad x = \Phi^{-1}(y) = \mathbf{d}(\ln\|y\|)\frac{y}{\|y\|},$$

where, as before, $\|y\| = \sqrt{y^{\top}Py}$. Therefore, we need to show that the inclusion $\mathbf{0} \in \tilde{Q}(h, y_i, y_{i+1})$ is a defines a consistent discrete-time approximation of (8.26). Due to fulfillment of the condition (7.18), the origin of this system is globally finite-time stable and $\mathbf{0} \in \tilde{F}(\mathbf{0})$.

(1) Let us show that for any $\tilde{y} \in \mathbb{R}^n$ and any $h > 0$, there exists $\tilde{y}_h \in \mathbb{R}^n$ such that $\mathbf{0} \in \tilde{Q}(h, \tilde{y}, \tilde{y}_h)$. Let us show that the inclusion $\mathbf{0} \in \tilde{Q}(h, \mathbf{0}, \tilde{y}_h)$ has the unique solution $\tilde{y}_h = \mathbf{0}$. Suppose the contrary, i.e., $\tilde{y}_h \neq \mathbf{0}$. In this case, $\tilde{y}_h \in h\tilde{F}(\tilde{y}_h)$ and

$$0 < \tilde{y}_h^{\top}P\tilde{y}_h \leq \sup_{q\in\tilde{F}(\tilde{y}_h)} h\tilde{y}_h^{\top}Pq < 0$$

due to (7.18) with $\Xi = I_n$, i.e., we obtain the contradiction.

Since $\tilde{F}(e^s x) = \tilde{F}(x)$, $\forall x \neq \mathbf{0}$, $\forall s \in \mathbb{R}$ and, by assumption, $x = \mathbf{0}$ is the unique equilibrium of globally asymptotically stable system (8.16). Moreover, by construction, $\tilde{F}(x) \subset \tilde{F}(\mathbf{0})$ for all $x \neq \mathbf{0}$, so $\tilde{F}$ is compact-valued, convex-valued and upper semi-continuous on $\mathbb{R}^n$. Therefore, for any $\tilde{y} \in \mathbb{R}^n$ the set-valued mapping $h\tilde{F}(\tilde{y} + \cdot) : \Omega \mapsto \Omega$ has a closed graph, where $\Omega = h\tilde{F}(\mathbf{0})$. Since Ω is a convex and compact set, then, according to Kakutani fixed-point theorem (see [10] or Appendix C), $\exists\Delta^* \in \Omega$ such that $\Delta^* \in h\tilde{F}(\tilde{y} + \Delta^*)$, and taking $\tilde{y}_h = \tilde{y} + \Delta^*$ we derive $\mathbf{0} \in \tilde{Q}(h, \tilde{y}, \tilde{y}_h)$.

(2) Let us show that the sequence $\{y_i\}_{i=0}^{+\infty}$ generated by $\mathbf{0} \in \tilde{Q}(h, y_i, y_{i+1})$ converges to zero in a finite number of steps. Since the stability condition (7.18) holds for $\Xi = I_n$, then $V(z) = z^{\top}Pz$ is a positive definite Lyapunov function for the system

(8.26). Let us show that there exists i^* such that $y_{i^*} = \mathbf{0}$. Suppose that $y_i \neq \mathbf{0}$ for all $i > 0$. In this case, we derive $y_{i-1} = y_i - hq_i$ with $q_i \in \tilde{F}(y_i)$ and

$$V(y_{i-1}) = (y_i - hq_i)^\top P(y_i - q_i) =$$

$$V(y_i) - 2h \underbrace{y_i^\top P q_i}_{<0} + h^2 \underbrace{q_i^\top P q_i}_{\geq c} > V(y_i) + h^2 c,$$

where the constant

$$c = \inf_{q \in \tilde{F}(x), x \neq \mathbf{0}} q^\top P q > 0$$

is independent of $y_i \neq \mathbf{0}$ due to standard homogeneity of $\tilde{F}$. This means that, after a finite number of steps, we have $V(y_i) < 0$, but this is impossible due to the positive definiteness of the Lyapunov function V. Therefore, there exists i^* such that $y_{i^*} = \mathbf{0}$.

(3) If $\{y_i\}_{i=0}^{+\infty}$ is generated $\mathbf{0} \in \tilde{Q}(h, y_i, y_{i+1})$ then

$$\tilde{\phi}_h(t, y_0) = y_i + (y_{i+1} - y_i)\tfrac{t-ih}{h}, \quad t \in [ih, (i+1)h), \quad i = 0, 1, 2, ...$$

are Euler approximates of strong solutions of the system (8.26) (see the proof of Theorem 2.8). The condition 3 of Definition 8.1 is fulfilled. The proof is complete. □

Formally, the proof is based on the fact that the system $\dot{y} = \tilde{f}(y)$ admits a quadratic Lyapunov function ($\Leftrightarrow$ the condition (7.18) with Ξ =const). However, Theorem 7.7 shows that any stable homogeneous system is equivalent to a quadratically stable one. If F in Theorem 8.6 is replaced with

$$F^{new}(x) = \tfrac{\partial \Psi(\xi)}{\partial \xi} F(\xi)\Big|_{\xi = \Psi^{-1}(x)}, \quad x \in \mathbb{R}^n, \quad \xi \in \mathbb{R}^n,$$

where $\Psi \in \mathbb{F}_{\mathbf{d}}(\mathbb{R}^n)$ is a diffeomorphism on $\mathbb{R}^n \backslash \{\mathbf{0}\}$ introduced in Theorem 7.7, then the condition $\Xi = I_n$ is fulfilled for F^{new}. According to Theorem 7.6 a $\mathbf{d}$-homogeneous Lyapunov function $V \in C(\mathbb{R}^n) \cap C^\infty(\mathbb{R}^n \backslash \{\mathbf{0}\})$ with the degree 1 can always be found for any asymptotically stable system with a $\mathbf{d}$-homogeneous vector field $f : \mathbb{R}^n \mapsto \mathbb{R}^n$. In this case, the required transformation Ψ can be defined as follows (see Page 258 for more details)

$$\Psi(\xi) = \mathbf{d}\left(\ln \tfrac{V(\xi)}{\|\xi\|_{\mathbf{d}}}\right) \xi. \tag{8.27}$$

Therefore, taking into account Corollary 6.2, we complete the proof of the following theorem.

$$\begin{array}{ccccc}
\dot{\xi}(t) \in F(\xi(t)) & \overset{x=\Psi(\xi)}{\Longleftrightarrow} & \dot{x}(t) \in F_{new}(x(t)) & \overset{y=\Phi(x)}{\Longleftrightarrow} & \dot{y}(t) \in \tilde{F}(y(t)) \\
\uparrow & & \uparrow & & \uparrow \\
\text{consistent} & & \text{consistent} & & \text{consistent} \\
\text{approximation} & & \text{approximation} & & \text{approximation} \\
\downarrow & & \downarrow & & \downarrow \\
\mathbf{0} \in Q(h, \Psi(\xi_k), \Psi(\xi_{k+1})) & \overset{\xi_k=\Psi^{-1}(x_k)}{\Longleftrightarrow} & \mathbf{0} \in Q(h, x_k, x_{k+1}) & \overset{x_k=\Phi^{-1}(y_k)}{\Longleftrightarrow} & \mathbf{0} \in \tilde{Q}(h, y_k, y_{k+1})
\end{array}$$

Fig. 8.1 The scheme of the consistent implicit discretization of $\dot{\xi} \in F(\xi)$

Theorem 8.7 *A* **d***-homogeneous finite-time stable system*

$$\dot{\xi} \in F(\xi), \quad t > 0$$

admits a consistent implicit approximation provided that $F : \mathbb{R}^n \mapsto 2^{\mathbb{R}^n}$ *is compact-valued, convex-valued and upper semi-continuous.*

If we know a homogeneous Lyapunov function for a finite-time stable homogeneous system then, to design a consistent (in the sense of Definition 8.1) implicit discretization scheme we have to

(1) modify the dilation group **d** (using Corollary 6.2) in order to have the homogeneity degree -1 of the vector field;
(2) apply the coordinate transformation Ψ given by (8.27) in order to obtain a system which satisfies the condition (7.18) with $\Xi = I_n$;
(3) use Theorem 8.6.

The suggested discretization approach is shown schematically in Fig. 8.1.

Example 8.2 Inspired by [11] let[2] us consider the system

$$\dot{x}(t) = f(x(t)) = \begin{pmatrix} 0 & 1 \\ 0 & 0 \end{pmatrix} x(t) + \begin{pmatrix} 0 \\ 1 \end{pmatrix} u(x(t)), \quad x = (x_1, x_2)^\top \in \mathbb{R}^2,$$

where u is the so-called control $u(x) \in -\alpha \frac{x_1 + \beta x_2 |x_2|}{|x_1| + \beta x_2^2}$ proposed in [12] with constant parameters $\alpha, \beta \in \mathbb{R}_+$ to be defined below. The vector field $f(\cdot)$ has a unique discontinuity at zero, so, according to the Filippov method, the considered ODE such be represented in the form (8.16) with $F(x) = \{f(x)\}$ for $x \neq \mathbf{0}$ and $F(\mathbf{0}) = (0, [-\alpha, \alpha])^T$. In Example 7.5, it is shown that the closed-loop system is **d**-homogeneous of degree -1 with respect to the linear dilation generated by $G_{\mathbf{d}} = \left(\begin{smallmatrix} 2 & 0 \\ 0 & 1 \end{smallmatrix}\right)$, and the canonical

---

[2] This example is reproduced from Brogliato B, Polyakov A. Digital implementation of sliding mode control via the implicit method: A tutorial. *Int J Robust Nonlinear Control*. 2021; 31:3528–3586. https://doi.org/10.1002/rnc.5121 ©2020 John Wiley & Sons, Ltd. Reused with permission of John Wiley & Sons.

homogeneous norm $\|\cdot\|_{\mathbf{d}}$ induced (see formula (5.15)) by the norm $\|x\| = \sqrt{x^\top Px}$ with $P = \left(\begin{smallmatrix}2 & 1\\ 1 & 1\end{smallmatrix}\right)$ is a Lyapunov function with the considered system for $\alpha = 1.5$ and $\beta = 1$.

According to the scheme presented above (see Fig. 8.1), a computational algorithm first needs to be developed for the transformed system (8.25).

Let us denote $y = q\tilde{z}$, where $q = \|y\| = \sqrt{y^\top Py}$ and $\tilde{z} = (\tilde{z}_1, \tilde{z}^2)^\top \in S = \{z \in \mathbb{R}^2 : \|z\| = 1\}$. In this case, we have

$$\tilde{f}(y) = \left(\frac{(I - G_{\mathbf{d}})yy^\top P}{y^\top PG_{\mathbf{d}}y} + I_2\right) f\left(\frac{y}{\|y\|}\right) = \begin{pmatrix} -\tilde{z}_1 \frac{\tilde{z}^\top Pf(\tilde{z})}{\tilde{z}^\top PG_{\mathbf{d}}\tilde{z}} + \tilde{z}_2 \\ \\ -\alpha \frac{\tilde{z}_1 + \beta|\tilde{z}_2|\tilde{z}_2}{|\tilde{z}_1| + \beta\tilde{z}_2^2} \end{pmatrix} = \tilde{f}(\tilde{z}).$$

Let $q_k = \|y_k\|$ and $z_k = y_k/q_k$. Hence, for $-\frac{y_k}{h} \in \tilde{F}(\mathbf{0}) = \mathrm{co}\bigcup_{\tilde{z}\in S} \tilde{f}(\tilde{z})$, the inclusion $\mathbf{0} \in \tilde{Q}(h, y_k, y_{k+1})$ has the unique solution $q_{k+1} = 0$ and $y_{k+1} = \mathbf{0}$, but if $-\frac{y_k}{h} \notin \tilde{F}(\mathbf{0})$, then the inclusion becomes the algebraic equation

$$\begin{aligned} &q_{k+1}z_{k+1} - y_k = h\tilde{f}(z_k), \\ &q_{k+1} > 0, \quad z_{k+1}^\top Pz_{k+1} = 1. \end{aligned}$$

Simple computations show that

$$q_{k+1} = z_{k+1}^\top Py_k + hz_{k+1}Pf(z_{k+1})$$

and

$$e_2^\top \tilde{f}(z_{k+1}) = e_2^\top f(z_{k+1}),$$

where $e_2 = (0, 1)^\top \in \mathbb{R}^2$. Therefore, we derive the system of two equations

$$(hz_{k+1}^\top Pf(z_{k+1}) + z_{k+1}^\top Py_k)e_2^\top z_{k+1} - e_2^\top y_k - he_2^\top f(z_{k+1}) = 0, \quad z_{k+1}^\top Pz_{k+1} = 1$$

with respect to two variables $(a, b)^T \triangleq z_{k+1} \in \mathbb{R}^2$. Notice that multiplying the first equation by $|a| + \beta b^2$ yields a polynomial-like equation:

$$m(a, b) = 0,$$

where $m(a, b) = (|a| + \beta b^2)\{(a \;\; b)Py_k + h(2a + b)b\}\, b - (|a| + \beta b^2)e_2^\top y_k + \alpha h(a + \beta b|b|)(1 - (a + b)b)$. Since

$$z_{k+1}^\top Pz_{k+1} = 1 \quad \Leftrightarrow \quad b = -a \pm \sqrt{1 - \varepsilon a^2}, \quad -\frac{1}{\sqrt{\varepsilon}} \le a \le \frac{1}{\sqrt{\varepsilon}},$$

then to compute z_{k+1} and q_{k+1}, it is necessary and sufficient to find a real root of one of the two scalar equations:

$$\begin{aligned} &(a)\ \ m_1(a, -a+\sqrt{1-\varepsilon^2 a})=0, \quad -\tfrac{1}{\sqrt{\varepsilon}}\leq a\leq \tfrac{1}{\sqrt{\varepsilon}} \\ &(b)\ \ m_2(a, -a-\sqrt{1-\varepsilon^2 a})=0, \quad -\tfrac{1}{\sqrt{\varepsilon}}\leq a\leq \tfrac{1}{\sqrt{\varepsilon}}. \end{aligned} \tag{8.28}$$

Theorem 8.6 implies that such a real root always exists at least for one of these two equations.

8.2.3 Nearly Fixed-Time Stable Semi-implicit Discretization

Recall that a positive homogeneity degree corresponds to the case of nearly fixed-time stability. In view of Corollary 6.2 we can restrict ourselves to the case of the homogeneity degree 1 without loss of generality. Notice also that the homogeneous vector field $f:\mathbb{R}^n\mapsto\mathbb{R}^n$ with $f(\mathbf{0})=\mathbf{0}$ and with positive degree is continuous at the origin (see Proposition 6.4). Inspired by [7] we prove the following result.

Theorem 8.8 *Let a vector field $f:\mathbb{R}^n\mapsto\mathbb{R}^n$ be uniformly continuous on a unit sphere $S=\{z\in\mathbb{R}^n : \|z\|=1\}$, $\mathbf{d}$-homogeneous of the degree 1 and*

$$\tilde f(z)=\left(\tfrac{(I-G_{\mathbf{d}})zz^\top P}{z^\top G_{\mathbf{d}}Pz}+I_n\right)f\left(\tfrac{z}{\|z\|}\right),\quad z\in\mathbb{R}^n\backslash\{\mathbf{0}\},$$

where $G_{\mathbf{d}}$ is the generator of the dilation $\mathbf{d}$, $\|z\|=\sqrt{z^\top Pz}$ and the positive definite matrix $P\in\mathbb{R}^{n\times n}$ satisfies (5.6). If the condition (7.18) holds with $\Xi=I_n$ and the mapping

$$y\mapsto Q_\alpha(y):=y-\alpha\|y\|\tilde f(y),\quad y\in\mathbb{R}^n \tag{8.29}$$

be a homeomorphism on $\mathbb{R}^n$ for any $\alpha>0$, then the function $Q:\mathbb{R}_+\times\mathbb{R}^n\times\mathbb{R}^n\mapsto\mathbb{R}^n$ given by

$$Q(h,x_i,x_{i+1})=\tilde Q(h,\Phi(x_i),\Phi(x_{i+1})), \tag{8.30}$$

where $h>0$, $\Phi(x)=\|x\|_{\mathbf{d}}\mathbf{d}(-\ln\|x\|_{\mathbf{d}})x$ for $x\in\mathbb{R}^n$, and

$$\tilde Q(h,y_i,y_{i+1})=y_{i+1}-y_i-h\|y_i\|\,\|y_{i+1}\|\,\tilde f(y_{i+1}) \tag{8.31}$$

defines a consistent discrete-time approximation of the nearly fixed-time stable system (8.16) *with $F(x)=\{f(x)\}$ in the sense of Definition 8.2.*

Proof (1) Since f has the homogeneity degree 1, then the system

$$\dot y=\|y\|^2\tilde f(y) \tag{8.32}$$

is equivalent to (8.16) with the change of coordinates (see Theorem 7.7)

$$y = \Phi(x), \qquad x = \Phi^{-1}(y) = \mathbf{d}(\ln \|y\|)\frac{y}{\|y\|}.$$

Hence, $\tilde{q}$ in (8.30), (8.31) corresponds to the semi-implicit discretization

$$\frac{y_{i+1} - y_i}{h} = \|y_i\| \, \|y_{i+1}\| \, \tilde{f}(y_{i+1}), \quad i = 0, 1, 2, \ldots \tag{8.33}$$

Notice that for $y_i = \mathbf{0}$ the presented discrete-time model gives the unique solution $y_{i+1} = \mathbf{0}$. From the above equation, we derive

$$\tfrac{y_{i+1}}{\|y_i\|} - h\|y_i\| \left\| \tfrac{y_{i+1}}{\|y_i\|} \right\| \tilde{f}\left(\tfrac{y_{i+1}}{\|y_i\|}\right) = \tfrac{y_i}{\|y_i\|} \tag{8.34}$$

for any $y_i \neq \mathbf{0}$. The condition (8.29) implies that

$$S \subset Q_\alpha(\mathbb{R}^n) \quad \text{for any} \quad \alpha \in (0, +\infty), \tag{8.35}$$

where $S = \{x \in \mathbb{R}^n : \|x\| = 1\}$.

Since the stability criterion (7.18) holds for $\Xi = I_n$, then

$$V(y) = y^\top P y, \quad y \in \mathbb{R}^n$$

defines a Lyapunov function for the system (8.32), so $y^\top P \tilde{f}(y) < 0$ for $y \neq \mathbf{0}$.

(2) Let us show that any sequence $\{y_i\}_{i=0}^{+\infty}$ generated by $\tilde{q}(h, y_i, y_{i+1}) = 0$ satisfies the practical fixed-time convergence property. Using $y^\top P \tilde{f}(y) < 0$ for $y \neq \mathbf{0}$, and we derive

$$\begin{aligned} y_{i+1}^\top P y_{i+1} &= y_{i+1}^\top P y_i + h\|y_i\| \, \|y_{i+1}\| \, y_{i+1}^\top P \tilde{f}(y_{i+1}) \\ &= y_{i+1}^\top P y_i + h\|y_i\| \, \|y_{i+1}\|^2 \left(\tfrac{y_{i+1}}{\|y_{i+1}\|}\right)^\top P \tilde{f}\left(\tfrac{y_{i+1}}{\|y_{i+1}\|}\right) \\ &\leq y_{i+1}^\top P y_i - ch\|y_i\| \, \|y_{i+1}\|^2, \end{aligned}$$

where

$$c = \inf_{z \in S} -z^\top P \tilde{f}(z) > 0.$$

Hence, using the Cauchy–Schwarz inequality we conclude

$$\|y_{i+1}\|^2 \leq \|y_{i+1}\| \, \|y_i\| - ch\|y_i\| \, \|y_{i+1}\|^2,$$

or, equivalently,

$$\|y_{i+1}\| \leq \frac{\|y_i\|}{1 + ch\|y_i\|} = \frac{1}{\|y_i\|^{-1} + ch},$$

where $\|y\| = \sqrt{y^\top P y}$. The latter inequality implies that

$$\|y_i\| \le \frac{1}{ich} \quad \text{for} \quad i = 1, 2, \ldots$$

independently of the initial state y_0.

(3) Since the discrete-time equation (8.33) generates Euler approximates

$$\tilde{\phi}_h(t, y_0) = y_i + (y_{i+1} - y_i)\tfrac{t-ih}{h}, \quad t \in [ih, (i+1)h), \quad i = 0, 1, 2, \ldots$$

for solutions of the ODE (8.32). The proof is complete. □

Notice that the fixed-time stable discretization suggested in [7] uses the inclusion (8.35) as a sufficient condition for the existence of a solution of Eq. (8.34). The condition (8.29) given in Theorem 8.8 is more constructive, since, for $f \in C^1(\mathbb{R}^n, \mathbb{R}^n)$, due to homogeneity and Hadamard Global Inverse Function theorem [3], it is equivalent to the algebraic criterion

$$\|v - \alpha \tilde{f}(v)\| \neq \mathbf{0}, \text{ and } \det\left(\alpha \frac{\partial \tilde{f}(v)}{\partial v} - I_n\right) \neq \mathbf{0}, \quad \forall v \in S, \quad \forall \alpha > 0, \tag{8.36}$$

which, in particular, is fulfilled if all eigenvalues of the matrix $\frac{\partial \tilde{f}(v)}{\partial v}$ are placed in the left complex half-plane for any $v \in S$.

Example 8.3 Inspired by [7], let us consider the scalar sliding-mode control system

$$\dot{x} = u, \quad x \in \mathbb{R}$$

with

$$u = \begin{cases} -|x|x & \text{if } |x| > 1, \\ -\operatorname{sign}(x) & \text{if } |x| \le 1. \end{cases}$$

The control is designed by means of the combination of two homogeneous systems with positive and negative degrees. The closed-loop system is globally fixed-time stable, i.e., it is finite-time stable with the globally bounded settling time $T(x_0) \le T^{\max}$. We have $T^{\max} = 2$ for the considered scalar control system.

Theorems 8.8 and 8.6 can be directly applied to subsystems

$$\dot{x} = -x|x|, |x| > 1 \qquad \text{and} \qquad \dot{x} = -\operatorname{sign}(x), |x| \le 1,$$

which are $\mathbf{d}$-homogeneous of degree -1 and 1, respectively, with $\mathbf{d}(s) = e^{G_{\mathbf{d}} s}$, $s \in \mathbb{R}$ and $G_{\mathbf{d}} = 1$. Applying the consistent discretization schemes to these subsystems we derive

$$x_{i+1} = \begin{cases} \frac{x_i}{1+h|x_i|} & \text{if } |x_i| \geq 1, \\ x_i - h\,\text{sign}(x_i) & \text{if } h < |x_i| < 1, \\ 0 & \text{if } |x_i| \leq h \end{cases} \quad \begin{matrix} \text{– the semi-implicit method} \\ \text{– the implicit method} \end{matrix}$$

where $h > 0$ is the sampling period, $i = 0, 1, \ldots$. The obtained discretized model is fixed-time stable, i.e., its trajectories converge to zero in $N(x_0)$ samplings steps and $N(x_0)$ is globally bounded. Since $N_1(x_0)$ steps ($N_1 \leq 1/h$) are needed to guarantee $|x(th)| = |x_i| \leq 1$ for all $i \geq N_1$ provided that $|x_0| > 1$, and $N_2(x_0)$ steps ($N_2 \leq 1/h$) are needed to reach the origin if $|x_0| \leq 1$, then $N \leq N_1 + N_2$ and

$$N(x_0) \leq \frac{2}{h}.$$

Notice that if the control law is implemented in a sampled way as

$$u(t) = u_i, \quad t \in [ih, (i+1)h),$$

then

$$x(t) = x_i + (t - h)u_i \quad t \in [ih, (i+1)h).$$

In order to guarantee $x((i+1)h) = x_{i+1}$ the control values u_i should be selected as follows

$$u_i = \begin{cases} -\frac{|x_i|x_i}{1+h|x_i|} & \text{if } |x_i| \geq 1, \\ -\text{sign}(x_i) & \text{if } h < |x_i| < 1, \\ -\frac{x_i}{h} & \text{if } |x_i| \leq h, \end{cases} \quad \begin{matrix} \text{– the semi-implicit method} \\ \text{– the implicit method} \end{matrix}$$

which correspond to the semi-implicit $u_i = -|x_i|x_{i+1}$ and the implicit $u_i \in -\text{sign}(x_{i+1})$ discretizations of the control law, respectively. Therefore, the closed-loop system with the sampled control is also fixed-time stable, i.e., $x(t) = 0$ for $t \geq hN(x_0)$. This perfectly corresponds to the continuous-time case. Since $hN(x_0) \to T(x_0)$ as $h \to 0^+$, we conclude that the same settling time estimate $hN(x_0) \leq T_{\max} = 2$ holds even for the system with the sampled control actions.

Concluding Remarks

The homogeneity is shown to be useful for the analysis and design of a discretization (discrete-time approximation) for a continuous-time system. The concept of consistent discretization preserving the finite/fixed-time convergence rate of the continuous-time system is introduced. Euler's methods are shown to be inconsistent in the above sense. Consistent discretization is developed for homogeneous systems of nonzero degrees.

References

1. Butcher JC (2016) Numerical methods for ordinary differential equations. Wiley
2. Berger T, Haller F (2022) On an extension of a global implicit function theorem. Comptes Rendus. Mathématique 360:439–450
3. Ruzhansky M, Sugimoto M (2015) On global inversion of homogeneous maps. Bull Math Sci 5:13–18
4. Efimov D, Polyakov A, Levant A, Perruquetti W (2017) Realization and discretization of asymptotically stable homogeneous systems. IEEE Trans Autom Control 62(11):5962–5969
5. Sanchez T, Polyakov A, Efimov D (2021) Lyapunov-based consistent discretisation of stable homogeneous systems. Int J Robust Nonlinear Control 31(9):3587–3605
6. Polyakov A, Poznyak A (2009) Reaching time estimation for "super-twisting" second order sliding mode controller via Lyapunov function designing. IEEE Trans Autom Control 54(8):1951–1955
7. Polyakov A, Efimov D, Brogliato B (2019) Consistent discretization of finite-time and fixed-time stable systems. SIAM J Control Optim 57(1):78–103
8. Acary V, Brogliato B, Orlov Y (2012) Chattering-free digital sliding-mode control with state observer and disturbance rejection. IEEE Trans Autom Control 57(5):1087–1101
9. Filippov AF (1988) Differential equations with discontinuous right-hand sides. Kluwer Academic Publishers
10. Kakutani S (1941) A generalization of Brouwer's fixed point theorem. Duke Math J 8(3):457--459
11. Brogliato B, Polyakov A (2021) Digital implementation of sliding-mode control via the implicit method: a tutorial. Int J Robust Nonlinear Control 31:3528–3586
12. Levant A (2005) Quasi-continuous high-order sliding-mode controllers. IEEE Trans Autom Control 50(11):1812–1816

Chapter 9
Homogeneous Stabilization of Linear Plant

This chapter is devoted to **d**-homogeneous stabilization of linear time-invariant (LTI) systems.

Definition 9.1 A system

$$\dot{x} = f(x, u), \quad t > 0, \quad f \in C(\mathbb{R}^n \times \mathbb{R}^m; \mathbb{R}^n)$$

is said to be continuously **d**-homogeneously stabilizable with degree $\mu \in \mathbb{R}$ if there exists a feedback law $u \in C(\mathbb{R}^m; \mathbb{R}^n)$ such that the closed-loop system $\dot{x} = f(x, u(x))$ is globally uniformly asymptotically stable and **d**-homogeneous of the degree μ, where **d** is a dilation in $\mathbb{R}^n$.

We present a criterion of **d**-homogeneous stabilizability with nonzero degree for LTI plant, where **d** is a linear continuous dilation in $\mathbb{R}^n$. We design various finite-time and fixed-time homogeneous stabilizers. We study their robustness with respect to measurement noise and additive perturbations. We also develop generalized homogeneous sliding-mode control algorithms rejecting matched perturbations.

9.1 Homogeneous Stabilizers

Let us consider a control system modeled by the following ODE:

$$\dot{x} = Ax + Bu, \quad t > 0, \tag{9.1}$$

$$x(0) = x_0, \tag{9.2}$$

A. Polyakov, *Generalized Homogeneity in Systems and Control Volume I*,
Communications and Control Engineering,
https://doi.org/10.1007/978-3-031-76246-8_9

where $x(t) \in \mathbb{R}^n$ is the system state, $u(t) \in \mathbb{R}^m$ is a control to be designed, $A \in \mathbb{R}^{n\times n}$ and $B \in \mathbb{R}^{n\times m}$ are system matrices.

Linear feedback law $u = Kx$ is the conventional controller [1] for asymptotic (exponential) stabilization of the linear plant (9.1). Nonlinear algorithms are utilized for linear plants if some additional restrictions, such as a uniform boundedness of the control input [2], time optimality [3] or an advanced robustness/convergence property [4] of the control system, have to be fulfilled. In this section, we solve the problem of homogeneous stabilization of the linear plant in $\mathbb{R}^n$. In particular, we show that the linear system can be homogeneously stabilized with a degree $\mu \neq 0$ if and only if the pair $\{A, B\}$ is *controllable* or, equivalently, if

$$\mathrm{rank}(B, AB, ..., A^{n-1}B) = n. \tag{9.3}$$

We refer the reader to [1, 5] for more details about controllability of linear plants. Dependently of the homogeneity degree μ, the obtained feedback law is a finite-time or a nearly fixed-time stabilizer of the system. We also characterize a class of additive perturbations and system uncertainties which can be rejected by the homogeneous controller. Homogeneous controller with integral action as well as globally fixed-time stabilizing homogeneous feedback laws is presented as well.

9.1.1 Implicit Homogeneous Stabilizer

The homogeneous control algorithm presented in this section is a result of a long-term evolution of several design schemes. At the first glance, it does not look like any known before control law. However, under a particular selection of a plant model, a dilation and a homogeneity degree, it coincides with an algorithm developed in 1970s by Valeriy Korobov [6] in the context of the so-called *controllability function* method [7]. Korobov's controllability functions for linear plants were partially rediscovered in 1991 by Jürgen Adamy [8, 9], who called them by *implicit Lyapunov functions*. In [10], the latter method was applied for the finite/fixed-time stabilization of the integrator chain. All mentioned papers implicitly use a dilation symmetry for the control design, but none of them claims this explicitly. It seems a bridging of Korobov's and Adamy's results with the homogeneity theory was first done in 2016 (see [11]). However, it is worth noticing that Korobov's algorithm for the integrator chain [6] was reinvented by Laurent Praly in 1997 (see [12]) based on the so-called generalized homogeneous polar coordinates. Further generalizations of the mentioned design schemes were proposed in [13–16] and in the first edition of this monograph.

The idea of homogeneity-based control design for LTI system can be explained using the following planar example:

$$\begin{aligned} \dot{x}_1 &= x_2, \\ \dot{x}_2 &= a_1x_1 + a_2x_2 + u, \end{aligned}$$

where $a_1, a_2 \in \mathbb{R}$ are parameters, $x_1(t), x_2(t) \in \mathbb{R}$ are components of the state vector $x(t) = (x_1(t), x_2(t))^\top \in \mathbb{R}^2$ and $u(t) \in \mathbb{R}$ is the control input. The linear part of the system may be **d**-homogeneous of non-negative degree if and only if the matrix of the system is nilpotent (see Lemma 6.1). Therefore, a feedback law u, which homogeneously stabilizes the system (in the sense of Definition 9.1), should have the form:

$$u = -a_1 x_1 - a_2 x_2 + u_{\text{hom}}(x),$$

where u_{hom} is a homogeneous component of u to be designed. In this case, the closed-loop system becomes

$$\dot{x} = f(x) := A_0 x + B u_{\text{hom}}(x), \quad A_0 = \begin{pmatrix} 0 & 1 \\ 0 & 0 \end{pmatrix}, \quad B = \begin{pmatrix} 0 \\ 1 \end{pmatrix},$$

where the matrix A_0 is nilpotent. So, the linear part of the system is **d**-homogeneous of degree μ with respect to the dilation:

$$\mathbf{d}(s) = \begin{pmatrix} e^{s(1-\mu)} & 0 \\ 0 & e^{s} \end{pmatrix},$$

where $\mu < 1$. Indeed, simple calculations give $A_0\mathbf{d}(s) = e^{\mu s}\mathbf{d}(s)A_0, \forall s \in \mathbb{R}$. If $\|\cdot\|_{\mathbf{d}}$ is an arbitrary (explicit or implicit) **d**-homogeneous norm then selecting

$$u_{\text{hom}} = \|x\|_{\mathbf{d}}^{2\mu} k_1 x_1 + \|x\|_{\mathbf{d}}^{\mu} k_2 x_2, \quad K = (k_1, k_2) \in \mathbb{R}^{1\times 2}$$

we derive

$$f(\mathbf{d}(s)x) = e^{\mu s}\mathbf{d}(s) f(x), \quad \forall s \in \mathbb{R}, \forall x \neq \mathbf{0},$$

i.e., the closed-loop system is **d**-homogeneous of degree $\mu < 1$. Therefore, to complete the design of a **d**-homogeneous control, we just need to select a **d**-homogeneous norm $\|\cdot\|_{\mathbf{d}}$ and the parameters k_1, k_2 such that the closed-loop system is globally asymptotically stable. Notice that we have

$$u_{\text{hom}} = Kx \quad \text{for} \quad \|x\|_{\mathbf{d}} = 1,$$

i.e., the homogeneous control coincides with the linear feedback on the unit homogeneous sphere $S_{\mathbf{d}}(1)$. If the function $x \mapsto \|x\|_{\mathbf{d}}$ is continuous differentiable away from the origin then the inequality

$$\frac{\partial \|x\|_{\mathbf{d}}}{\partial x}(A_0 + BK)x < 0 \quad \text{for} \quad \|x\|_{\mathbf{d}} = 1 \tag{9.4}$$

guarantees that the set $\{x \in \mathbb{R}^n : \|x\|_{\mathbf{d}} \leq 1\}$ is strictly positively invariant. In this case, the homogeneity implies global uniform asymptotic stability of the closed-loop system (see Proposition 7.2). Moreover, $x \mapsto \|x\|_{\mathbf{d}}$ is a strict Lyapunov function.

The only difficulty of such a design is a selection of a homogeneous norm $\|\cdot\|_{\mathbf{d}}$ such that the inequality (9.4) becomes a simple algebraic condition for selection of parameters k_1, k_2. To represent the mentioned condition in terms of Linear Matrix Inequalities (LMIs), the theorem given below uses the canonical homogeneous norm $\|\cdot\|_{\mathbf{d}}$ induced by the weighted Euclidean norm (see Chap. 5).

Theorem 9.1 *The system* (9.1) *is continuously* $\mathbf{d}$*-homogeneously stabilizable with degree* $\mu \neq 0$ *if and only if the pair* $\{A, B\}$ *is controllable. Moreover, for any controllable pair* $\{A, B\}$*, one holds*

(1) the linear algebraic equation

$$AG_0 - G_0A + BY_0 = A, \quad G_0B = \mathbf{0} \tag{9.5}$$

has a solution $Y_0 \in \mathbb{R}^{m\times n}$, $G_0 \in \mathbb{R}^{n\times n}$ *and for any solution of* (9.5)

- *the matrix*

$$G_{\mathbf{d}} = I_n + \mu G_0 \tag{9.6}$$

is anti-Hurwitz for $\mu < 1/(\tilde{n}-1)$*, where* $\tilde{n} \in \mathbb{N}$ *is a minimal number such that* $\operatorname{rank}(B, AB, ..., A^{\tilde{n}-1}B) = n$*;*

- *the matrix* $G_0 - I_n$ *is invertible and the matrix*

$$A_0 = A + BY_0(G_0 - I_n)^{-1} \tag{9.7}$$

satisfies the identity

$$A_0G_{\mathbf{d}} = (G_{\mathbf{d}} + \mu I_n)A_0 \quad \text{and} \quad G_{\mathbf{d}}B = B; \tag{9.8}$$

(2) the linear algebraic system

$$\begin{array}{c} A_0X + XA_0^\top + BY + Y^\top B^\top + \rho(G_{\mathbf{d}}X + XG_{\mathbf{d}}^\top) = \mathbf{0}, \\ G_{\mathbf{d}}X + XG_{\mathbf{d}}^\top \succ 0, \quad X = X^\top \succ 0 \end{array} \tag{9.9}$$

has a solution $X \in \mathbb{R}^{n\times n}$, $Y \in \mathbb{R}^{m\times n}$ *for any* $\rho \in \mathbb{R}_+$*;*

(3) the canonical homogeneous norm $\|\cdot\|_{\mathbf{d}}$ *induced by* $\|x\| = \sqrt{x^\top X^{-1}x}$ *is a Lyapunov function of the closed-loop system* (9.1), (9.2) *with the feedback control*

$$u(x) = K_0x + \|x\|_{\mathbf{d}}^{1+\mu}K\mathbf{d}(-\ln\|x\|_{\mathbf{d}})x, \tag{9.10}$$

where $\mathbf{d}(s) = e^{sG_{\mathbf{d}}}$, $s \in \mathbb{R}$ *is a linear dilation generated by* $G_{\mathbf{d}}$ *and*

$$K_0 = Y_0(G_0 - I_n)^{-1}, \qquad K = YX^{-1}; \tag{9.11}$$

moreover,

$$\tfrac{d}{dt}\|x\|_{\mathbf{d}} = \tfrac{\partial\|x\|_{\mathbf{d}}}{\partial x}(Ax + Bu(x)) = -\rho\|x\|_{\mathbf{d}}^{1+\mu}, \quad \forall x \neq \mathbf{0}; \tag{9.12}$$

(4) *u is analytic on* $\mathbb{R}^n\backslash\{\mathbf{0}\}$ *for* $\mu \in [-1, 1/(\tilde{n}-1))$ *and* $u \in C(\mathbb{R}^n;\mathbb{R}^m)$ *for* $\mu \in (-1, 1/(\tilde{n}-1))$,

$$|u(x)| \leq |K_0 x| + \tilde{k}\|x\|_{\mathbf{d}}^{1+\mu}, \quad \forall x \in \mathbb{R}^n\backslash\{\mathbf{0}\}, \tag{9.13}$$

where $\tilde{k} = \sqrt{\lambda_{\max}\left(X^{1/2}K^\top K X^{1/2}\right)}$;

(5) *the closed-loop system* (9.1), (9.2), (9.10) *is* **d**-*homogeneous of degree* μ *and*

- *globally uniformly finite-time stable for* $\mu \in [-1, 0)$ *such that*

$$x(t) = \mathbf{0}, \quad \forall t \geq T(x_0) := \frac{\|x_0\|_{\mathbf{d}}^{-\mu}}{-\mu\rho}, \tag{9.14}$$

but $x(t) \neq \mathbf{0}$ *for all* $t \in [0, T(x_0))$ *provided that* $x(0) \neq \mathbf{0}$;

- *globally uniformly exponentially stable for* $\mu = 0$ *such that*

$$\|x(t)\|_{\mathbf{d}} = e^{-\rho t}\|x_0\|_{\mathbf{d}}, \quad \forall t \geq 0; \tag{9.15}$$

- *globally uniformly nearly fixed-time stable for* $\mu \in (0, 1/(\tilde{n}-1))$ *such that*

$$\forall r > 0, \quad \|x(t)\|_{\mathbf{d}} \leq r, \quad \forall t \geq \frac{1}{r^\mu \rho} \tag{9.16}$$

independently of $x_0 \in \mathbb{R}^n$, *moreover, if* $\|x_0\|_{\mathbf{d}} > r$ *then* $\|x(t)\|_{\mathbf{d}} > r$ *for all* $t \in \left[0, \frac{1}{r^\mu\rho}\right)$.

Let us make a few remarks prior giving the proof of the theorem.

The main advantage of the proposed control algorithm is simplicity of the design even in the case of the multi-input system. Indeed, to find control parameters, first, we define a dilation in $\mathbb{R}^n$ by solving the system of linear algebraic equations (9.5). The generator of the dilation is given by the formula (9.6). Moreover, a solution of (9.5) also defines the linear component $K_0 x$ of the feedback law, which "homogenizes" the linear part of the system. Namely, the gain K_0 is selected such that the matrix $A_0 = A + BK_0$ is nilpotent and satisfies the identity $A_0 G_{\mathbf{d}} = (G_{\mathbf{d}} + \mu I_n)A_0$. This identity is shown to be necessary and sufficient for **d**-homogeneity of the linear vector field $x \mapsto A_0 x$ with degree $\mu \neq 0$ (see Lemma 6.1). The homogeneous component $\|x\|_{\mathbf{d}}^{1+\mu} K\mathbf{d}(-\ln\|x\|_{\mathbf{d}})x$ of the feedback law u is designed by solving the linear algebraic system (9.9), which is very similar to the LMI for linear feedback design [17]. Moreover, for $\mu = 0$ we have $G_{\mathbf{d}} = I_n, G_{\mathbf{d}}X + XG_{\mathbf{d}}^\top = 2X$ and the design of the homogeneous feedback design simply repeats the linear feedback design resulting in $u = (K_0 + K)x$.

The canonical homogeneous norm $\|\cdot\|_{\mathbf{d}}$ is well defined due to monotonicity condition $G_{\mathbf{d}}X + XG_{\mathbf{d}}^\top \succ 0$ (see Proposition 5.3) and utilized for both control design and analysis of the system. The identity (9.12) means that $\|\cdot\|_{\mathbf{d}} \in C(\mathbb{R}^n) \cap C^\infty(\mathbb{R}^n\backslash\{\mathbf{0}\})$ is a **d**-homogeneous Lyapunov function of the closed-loop system. Moreover, it defines the settling time T of the system. Indeed, if $\mu < 0$ then the closed-loop system (9.1), (9.2), (9.10) is finite-time stable:

$$x_{x_0}(t) = \mathbf{0}, \quad \forall t \geq \frac{\|x_0\|_{\mathbf{d}}^{-\mu}}{-\mu\rho}, \quad \forall x_0 \in \mathbb{R}^n$$

but if $\mu > 0$ then the closed-loop system is nearly fixed-time stable:

$$\|x_{x_0}(t)\|_{\mathbf{d}} \leq r, \quad \forall t \geq \frac{1}{\rho r^{\mu}\mu}, \quad \forall r > 0, \quad \forall x_0 \in \mathbb{R}^n.$$

The convergence time in both cases can be adjusted by tuning the parameter $\rho > 0$ and the homogeneity degree $\mu \in [-1, 1/(\tilde{n} - 1))$. Moreover, *the settling-time estimates are exact ("non-improvable")* due to the identity (9.12). For $\mu = -1$ and $\rho = 1$ the canonical homogeneous norm is the settling-time function of the finite-time stable closed-loop system.

The term $\pi_{\mathbf{d}}(x) = \mathbf{d}(-\ln \|x\|_{\mathbf{d}})x$ in u is the $\mathbf{d}$-homogeneous projector of x to the unit sphere (see the formula (5.24)), which is well defined for $x \neq \mathbf{0}$ such that $\|\pi_{\mathbf{d}}(x)\| = 1$. For $\mu > -1$ we have $u \to \mathbf{0}$ as $x \to \mathbf{0}$, so u can be continuously prolonged to zero assigning $u(\mathbf{0}) = \mathbf{0}$. If $\mu = -1$ then u is discontinuous at $\mathbf{0}$ and locally bounded in the view of the estimate (9.13). So, the closed-loop system with $\mu = -1$ has solutions in the sense of Filippov (see Sect. 2.1) in this case. In all cases solutions are unique. The homogeneous control with $\mu = -1$ can be interpreted as a version of the sliding-mode algorithm (see Sect. 9.2 for more details). For $\mu < -1$ the control u is unbounded at $\mathbf{0}$. So, the practically feasible stabilizer (9.10) should have the homogeneity degree $\mu \in [-1, (\tilde{n} - 1)^{-1})$, but for theoretical and numerical investigations of the closed-loop system on $\mathbb{R}^n \backslash \{\mathbf{0}\}$ we can select $\mu < -1$.

The proof of Theorem 9.1 is constructive and uses some supporting results. The following two lemmas are inspired by [18, 19], while the further constructions are based on a methodology proposed in [11] for weighted homogeneous control design.

Lemma 9.1 *If the pair $\{A, B\}$ is controllable then there exists an orthogonal matrix $T \in \mathbb{R}^{n\times n}$ such that*

$$TAT^{\top} = \begin{pmatrix} A_{11} & A_{12} & \mathbf{0} & \dots & \mathbf{0} \\ A_{21} & A_{22} & A_{23} & \dots & \mathbf{0} \\ \dots & \dots & \dots & \dots & \dots \\ A_{\tilde{n}-1\,1} & A_{\tilde{n}-1\,2} & A_{\tilde{n}-1\,3} & \dots & A_{\tilde{n}-1\,\tilde{n}} \\ A_{\tilde{n}\,1} & A_{\tilde{n}\,2} & A_{\tilde{n}\,3} & \dots & A_{\tilde{n}\,\tilde{n}} \end{pmatrix}, \quad TB = \begin{pmatrix} \mathbf{0} \\ \mathbf{0} \\ \dots \\ \mathbf{0} \\ A_{\tilde{n},\tilde{n}+1} \end{pmatrix}, \tag{9.17}$$

where A_{ij}—possibly nonzero blocks, $i = 1, ..., \tilde{n}$, $j = 1, i+1$ $A_{ii+1} \in \mathbb{R}^{n_i\times n_{i+1}}$—full row rank matrices, $n_i \in \mathbb{N}$ such that

$$\sum_{i=1}^{\tilde{n}} n_i = n,$$

where $\tilde{n} \in \mathbb{N}$ is a minimal number such that $\text{rank}(B, AB, ..., A^{\tilde{n}-1}B) = n$.

Proof (1) Let $\mathrm{rown}(W)$ and $\mathrm{coln}(W)$ denote the number of rows and columns of a matrix W, respectively. Let $\ker(W)$ and $\mathrm{range}(W)$ be the null space and the column space of W, respectively. Finally, let $\dim(X)$ be a dimension of a finite-dimensional space X and $\mathrm{null}(W)$ denote a matrix, which columns define an orthonormal basis in $\ker(W)$.

Let the matrices T_i be defined by the following algorithm:
Initialization: $A_0 = A$, $B_0 = B$, $T_0 = I_n$, $k = 0$.
Loop: While $\mathrm{rank}(B_k) < \mathrm{rown}(A_k)$ do

$$A_{k+1} = B_k^{\perp} A_k \left(B_k^{\perp}\right)^{\top}, \quad B_{k+1} = B_k^{\perp} A_k \hat{B}_k, \quad T_{k+1} = \begin{pmatrix} B_k^{\perp} \\ \hat{B}_k \end{pmatrix}, \quad k = k+1,$$

where $B_k^{\perp} := \left(\mathrm{null}(B_k^{\top})\right)^{\top}$, $\hat{B}_k := \left(\mathrm{null}\left(B_k^{\perp}\right)\right)^{\top}$.

Let us denote r_0=$\mathrm{rank}(B_0)$. Controllability of $\{A_0, B_0\}$ implies $r_0 > 0$. Notice that if $r_0 = n$ then the above algorithm stops after initialization showing that the original system does not require transformation.

Let us consider the case $r_0 < n$. In this case, $\dim(\mathrm{range}(B_0))$=$r_0$, $\dim(\ker(B_0^{\top}))$= $n - r_0$ and $\mathbb{R}^n$=$\mathrm{range}(B_0) \oplus \ker(B_0^{\top})$. Columns of the matrix $\mathrm{null}(B_0^{\top}) \in \mathbb{R}^{n\times(n-r_0)}$ by definition form an orthonormal basis in $\ker(B_0^{\top}) \subset \mathbb{R}^n$, so $\ker(B_0^{\top})$=$\mathrm{range}(\mathrm{null}(B_0^{\top}))$. Since $\mathrm{range}(B_0)$ is an orthogonal complement to $\ker(B_0^{\top})$ then $\mathrm{range}(B_0)$=$\ker(\mathrm{null}(B_0^{\top})^{\top})$ and columns of the matrix $(B_0^{\perp})^{\top}$=$\mathrm{null}(B_0^{\top})$ together with columns of the matrix $(\tilde{B}_0)^{\top}$=$\mathrm{null}(\mathrm{null}(B_0^{\top})^{\top})$ define the orthonormal basis in $\mathbb{R}^n$. So, the block matrix T_1 is square and orthogonal. Since $B_0^{\top}\,\mathrm{null}(B_0^{\top})$=0 then the orthogonal transformation T_1 gives

$$T_1 A T_1^{\top} = \begin{pmatrix} \tilde{A}_{11} & \tilde{A}_{12} \\ \tilde{A}_{12} & \tilde{A}_{22} \end{pmatrix} \quad T_1 B = \begin{pmatrix} \mathbf{0} \\ \hat{B}_0 B_0 \end{pmatrix},$$

where $\tilde{A}_{11} \in \mathbb{R}^{(n-r_0)\times(n-r_0)}$, $\tilde{A}_{12} \in \mathbb{R}^{(n-r_0)\times r_0}$, $\tilde{A}_{21} \in \mathbb{R}^{r_0\times(n-r_0)}$, $\tilde{A}_{22} \in \mathbb{R}^{r_0\times r_0}$ are the blocks of the matrix $T_1 A T_1^{T}$, $\hat{B}_0 B_0 \in \mathbb{R}^{r_0\times m}$. Obviously that $\tilde{A}_{11} = A_1$ and $\tilde{A}_{12} = B_1$, where matrices A_1 and B_1 are generated by **Loop** of the algorithm.

The controllability of the pair (A_0, B_0) implies the controllability of the pair (A_1, B_1), so we can repeat **Loop** of the algorithm for A_1 and B_1.

Since for each recursive step we have $\mathrm{rown}(A_{i+1}) = \mathrm{rown}(B_{i+1}) = \mathrm{rown}(A_i) - \mathrm{rank}(B_i)$ and $\mathrm{rank}(B_i) > 0$ then the algorithm terminates in a finite number of steps $\tilde{n} \leq n$ and the orthogonal transformation

$$T = \begin{pmatrix} T_{\tilde{n}} & 0 \\ 0 & I_{w_{\tilde{n}}} \end{pmatrix} \begin{pmatrix} T_{\tilde{n}-1} & \mathbf{0} \\ \mathbf{0} & I_{w_{\tilde{n}-1}} \end{pmatrix} \cdots \begin{pmatrix} T_2 & \mathbf{0} \\ \mathbf{0} & I_{w_2} \end{pmatrix} T_1,$$
where $w_i := n - \mathrm{rown}(T_i)$

gives Eq. (9.17), where $A_{\tilde{n}\,\tilde{n}+1} := \hat{B}_0 B_0$, $A_{i\,j} \in \mathbb{R}^{n_i \times n_j}$, $i > j$, $n_i \leq n_{i+1}$, $i = 1, 2, ..., \tilde{n}$, $n_1 + ... + n_{\tilde{n}} = n$ and

$$\mathrm{rank}(A_{i\,i+1}) = \mathrm{rank}(\hat{B}_{\tilde{n}-i} B_{\tilde{n}-i}) = \mathrm{rank}(T_{\tilde{n}-i} B_{\tilde{n}-i}) = \mathrm{rank}(B_{\tilde{n}-i}) = n_i > 0.$$

Hence, we derive

$$\mathrm{rank}(B, AB, ..., A^{\tilde{n}-1}B) = \mathrm{rank}(TB, TAT^{\top}TB, ..., (TAT^{\top})^{\tilde{n}-1}TB) =$$

$$\mathrm{rank}\begin{pmatrix} \mathbf{0} & \mathbf{0} & ... & A_{12}...A_{\tilde{n}\,\tilde{n}+1} \\ ... & ... & ... & ... \\ \mathbf{0} & A_{\tilde{n}-1\,\tilde{n}}A_{\tilde{n}\,\tilde{n}+1} & ... & * \\ A_{\tilde{n}\,\tilde{n}+1} & * & ... & * \end{pmatrix} = n_{\tilde{n}} + \cdots + n_1 = n,$$

i.e., $\tilde{n}$ is a minimal number such that $\mathrm{rank}(B, AB, ..., A^{\tilde{n}-1}B) = n$. The proof of lemma is complete. □

A controllable system can be transformed to a canonical block controllability form by means of an invertible transformation.

Lemma 9.2 *If the pair $\{A, B\}$ is controllable then there exists an invertible matrix $M \in \mathbb{R}^{n \times n}$ such that*

$$MAM^{-1} = \begin{pmatrix} \mathbf{0} & A_{12} & \mathbf{0} & ... & \mathbf{0} & \mathbf{0} \\ \mathbf{0} & \mathbf{0} & A_{23} & ... & \mathbf{0} & \mathbf{0} \\ ... & ... & ... & ... & ... & ... \\ \mathbf{0} & \mathbf{0} & \mathbf{0} & ... & \mathbf{0} & A_{\tilde{n}-1,\tilde{n}} \\ A_{\tilde{n}1} & A_{\tilde{n}2} & A_{\tilde{n}3} & ... & A_{\tilde{n},\tilde{n}-1} & A_{\tilde{n},\tilde{n}} \end{pmatrix}, \quad M^{-1}B = \begin{pmatrix} \mathbf{0} \\ \mathbf{0} \\ ... \\ \mathbf{0} \\ A_{\tilde{n},\tilde{n}+1} \end{pmatrix}, \tag{9.18}$$

where A_{ii+1}, n_i and $\tilde{n}$ are defined in Lemma 9.1.

Proof Let T be an orthogonal transformation defined in Lemma 9.1. Then $A' = TAT^{\top}$ and $B' = TB$ have the form (9.17).

Since $\mathrm{rank}(A_{i\,i+1}) = n_i = \mathrm{rown}(A_{i\,i+1})$ then $A_{i\,i+1}A^T_{i\,i+1}$ is invertible and $A^+_{i\,i+1} = A^T_{i\,i+1}(A_{i\,i+1}A^T_{i\,i+1})^{-1}$ is the right inverse matrix of $A_{i\,i+1}$ such that

$$A^+_{i\,i+1}A_{i\,i+1} = I_{n_i}.$$

Let us introduce the linear (back-stepping) transformation $z = \Phi y$, $z = (z_1, ..., z_k)^T$, $z_i \in \mathbb{R}^{n_i}$, $y = (y_1, ..., y_k)^T$, $y_i \in \mathbb{R}^{n_i}$ by the formulas:

$$z_i = y_i + \varphi_i, \quad i = 1, 2, ..., k,$$

$$\varphi_1 = 0, \quad \varphi_{i+1} = A^+_{i\,i+1}\left(\sum_{j=1}^{i} A_{ij}y_j + \sum_{r=1}^{i} \frac{\partial \varphi_i}{\partial y_r} \sum_{j=1}^{r+1} A_{rj}y_j\right).$$

Since the dynamics of y is given by the system

$$\begin{aligned}\dot{y}_1 &= A_{11}y_1 + A_{12}y_2,\\ \dot{y}_2 &= A_{21}y_1 + A_{22}y_2 + A_{23}y_3,\\ \dot{y}_3 &= \dots\\ &\dots\end{aligned}$$

then simple calculations show $\dot{\varphi}_i = \sum_{r=1}^{i} \frac{\partial \varphi_i}{\partial y_r} \sum_{j=1}^{r+1} A_{rj} y_j$ and

$$\begin{aligned}z_1 &= y_1\\ z_2 &= y_2 + A_{12}^{+} A_{11} y_1\\ z_3 &= y_3 + A_{23}^{+}(A_{21}y_1 + A_{22}y_2 + A_{12}^{+}A_{11}\dot{y}_1)\\ &\dots\end{aligned}$$

Hence, we derive the dynamics of z in the form:

$$\begin{aligned}\dot{z}_1 &= A_{11}y_1 + A_{12}y_2 = A_{12}z_2\\ \dot{z}_2 &= \dot{y}_2 + A_{12}^{+}A_{11}\dot{y}_1 = A_{23}z_3\\ \dot{z}_3 &= \dots\\ &\dots\end{aligned}$$

Therefore, the formula (9.18) holds for $M = \Phi T$. The presented coordinate transformation is linear and nonsingular. The inverse transformation $y = \Phi^{-1}s$ is defined as follows:

$$y_i = z_i + \psi_i, \quad i = 1, 2, \dots, k,$$

$$\psi_1 = 0, \quad \psi_{i+1} = A_{ii+1}^{+}\left(\sum_{k=1}^{i} \frac{\partial \psi_i}{\partial z_k} A_{ii+1} z_{k+1} - \sum_{j=1}^{i} A_{ij}(z_j + \psi_j)\right).$$

The proof of lemma is complete. □

Proof of Theorem 9.1.

I. *Sufficiency*. First we transform the system to the block controllability form (9.18):

$$\tilde{A} = M^{-1}AM, \quad \tilde{B} = MB.$$

(1) In this case, the Eq. (9.5) becomes

$$\tilde{A}\tilde{G}_0 + \tilde{B}\tilde{Y}_0 = (I_n + \tilde{G}_0)\tilde{A}, \quad \tilde{G}_0\tilde{B} = \mathbf{0},$$

where $\tilde{Y}_0 = Y_0 M$ and $\tilde{G}_0 = M G_0 M^{-1}$. Denoting

$$\tilde{G}_0 = \begin{pmatrix} G_{11} & G_{12} & G_{13} & \dots & G_{1\tilde{n}-1} & G_{1\tilde{n}} \\ G_{21} & G_{22} & G_{23} & \dots & G_{2\tilde{n}-1} & G_{2\tilde{n}} \\ \dots & \dots & \dots & \dots & \dots & \dots \\ G_{\tilde{n}-1\,1} & G_{\tilde{n}-1\,2} & G_{\tilde{n}-1\,3} & \dots & G_{\tilde{n}-1\,\tilde{n}-1} & G_{\tilde{n}-1\tilde{n}} \\ G_{\tilde{n}1} & G_{\tilde{n}2} & G_{\tilde{n}3} & \dots & G_{\tilde{n}\tilde{n}-1} & G_{\tilde{n}\tilde{n}} \end{pmatrix},$$

we derive

$$\tilde{G}_0\tilde{B} = \tilde{G}_0 \begin{pmatrix} \mathbf{0} \\ \mathbf{0} \\ \dots \\ \mathbf{0} \\ A_{\tilde{n},\tilde{n}+1} \end{pmatrix} = \begin{pmatrix} G_{1\tilde{n}}A_{\tilde{n}\tilde{n}+1} \\ G_{2\tilde{n}}A_{\tilde{n}\tilde{n}+1} \\ \dots \\ G_{\tilde{n}-1\tilde{n}}A_{\tilde{n}\tilde{n}+1} \\ G_{\tilde{n}\tilde{n}}A_{\tilde{n}\tilde{n}+1} \end{pmatrix} = \mathbf{0} \quad \Rightarrow \quad G_{i\tilde{n}} = \mathbf{0},\, i = 1, \dots, \tilde{n}$$

(since $A_{\tilde{n}\tilde{n}+1}$ has the full row rank). In this case, for $\tilde{A}\tilde{G}_0$ we have

$$\begin{pmatrix} \mathbf{0} & A_{12} & \mathbf{0} & \dots & \mathbf{0} & \mathbf{0} \\ \mathbf{0} & \mathbf{0} & A_{23} & \dots & \mathbf{0} & \mathbf{0} \\ \dots & \dots & \dots & \dots & \dots & \dots \\ \mathbf{0} & \mathbf{0} & \mathbf{0} & \dots & \mathbf{0} & A_{\tilde{n}-1,\tilde{n}} \\ A_{\tilde{n}1} & A_{\tilde{n}2} & A_{\tilde{n}3} & \dots & A_{\tilde{n},\tilde{n}-1} & A_{\tilde{n},\tilde{n}} \end{pmatrix}\tilde{G}_0 = \begin{pmatrix} A_{12}G_{21} & A_{12}G_{22} & \dots & A_{12}G_{2\tilde{n}-1} & \mathbf{0} \\ A_{23}G_{31} & A_{23}G_{32} & \dots & A_{23}G_{3\tilde{n}-1} & \mathbf{0} \\ \dots & \dots & \dots & \dots & \dots \\ A_{\tilde{n}-1\tilde{n}}G_{\tilde{n}1} & A_{\tilde{n}-1\tilde{n}}G_{\tilde{n}2} & \dots & A_{\tilde{n}-1\tilde{n}}G_{\tilde{n}\tilde{n}-1} & \mathbf{0} \\ \sum_{i=1}^{\tilde{n}} A_{\tilde{n}i}G_{i1} & \sum_{i=1}^{\tilde{n}} A_{\tilde{n}i}G_{i2} & \dots & \sum_{i=1}^{\tilde{n}} A_{\tilde{n}i}G_{i\tilde{n}-1} & \mathbf{0} \end{pmatrix}$$

and

$$\tilde{B}\tilde{Y}_0 = \begin{pmatrix} \mathbf{0} & \mathbf{0} & \dots & \mathbf{0} & \mathbf{0} \\ \mathbf{0} & \mathbf{0} & \dots & \mathbf{0} & \mathbf{0} \\ \dots & \dots & \dots & \dots & \dots \\ \mathbf{0} & \mathbf{0} & \dots & \mathbf{0} & \mathbf{0} \\ * & * & * & * & * \end{pmatrix},\; \tilde{G}_0\tilde{A} = \begin{pmatrix} \mathbf{0} & G_{11}A_{12} & \dots & G_{1\tilde{n}-2}A_{\tilde{n}-2\tilde{n}-1} & G_{1\tilde{n}-1}A_{\tilde{n}-1\tilde{n}} \\ \mathbf{0} & G_{21}A_{12} & \dots & G_{2\tilde{n}-2}A_{\tilde{n}-2\tilde{n}-1} & G_{2\tilde{n}-1}A_{\tilde{n}-1\tilde{n}} \\ \dots & \dots & \dots & \dots & \dots \\ \mathbf{0} & G_{\tilde{n}-1\,1}A_{12} & \dots & G_{\tilde{n}-1\tilde{n}-2}A_{\tilde{n}-2\tilde{n}-1} & G_{\tilde{n}-1\tilde{n}-1}A_{\tilde{n}-1\tilde{n}} \\ \mathbf{0} & G_{\tilde{n}1}A_{12} & \dots & G_{\tilde{n}-2\tilde{n}}A_{\tilde{n}-2\tilde{n}-1} & G_{\tilde{n}\tilde{n}-1}A_{\tilde{n}-1\tilde{n}} \end{pmatrix},$$

where $*$ are possibly nonzero blocks. Due to the structure $\tilde{B}$, the nonzero blocks in $\tilde{B}\tilde{Y}_0$ can be arbitrary selected by tuning $\tilde{Y}_0$.

In this case the identity $\tilde{A}\tilde{G}_0 + \tilde{B}\tilde{Y}_0 = (I_n + \tilde{G}_0)\tilde{A}$ can be rewritten as

$$\begin{pmatrix} A_{12}G_{21} & A_{12}G_{22} & \dots & A_{12}G_{2\tilde{n}-1} & \mathbf{0} \\ A_{23}G_{31} & A_{23}G_{32} & \dots & A_{23}G_{3\tilde{n}-1} & \mathbf{0} \\ \dots & \dots & \dots & \dots & \dots \\ A_{\tilde{n}-1\tilde{n}}G_{\tilde{n}1} & A_{\tilde{n}-1\tilde{n}}G_{\tilde{n}2} & \dots & A_{\tilde{n}-1\tilde{n}}G_{\tilde{n}\tilde{n}-1} & \mathbf{0} \\ * & * & \dots & * & * \end{pmatrix}$$

$$= \begin{pmatrix} \mathbf{0} & * & \dots & G_{1\tilde{n}-2}A_{\tilde{n}-2\tilde{n}-1} & G_{1\tilde{n}-1}A_{\tilde{n}-1\tilde{n}} \\ \mathbf{0} & * & \dots & G_{2\tilde{n}-2}A_{\tilde{n}-2\tilde{n}-1} + A_{\tilde{n}-2\tilde{n}-1} & G_{2\tilde{n}-1}A_{\tilde{n}-1\tilde{n}} \\ \dots & \dots & \dots & \dots & \dots \\ \mathbf{0} & * & \dots & G_{\tilde{n}-1\tilde{n}-2}A_{\tilde{n}-2\tilde{n}-1} & G_{\tilde{n}-1\tilde{n}-1}A_{\tilde{n}-1\tilde{n}} + A_{\tilde{n}-1\tilde{n}} \\ \mathbf{0} & * & \dots & * & \end{pmatrix}.$$

Hence, taking into account that $A_{\tilde{n}-1\,\tilde{n}}$ has the full row rank, we derive $G_{1\tilde{n}-1} = \mathbf{0}$,... $G_{\tilde{n}-2\tilde{n}-1} = \mathbf{0}$, $G_{\tilde{n}-1\tilde{n}-1} = -I_{n_{\tilde{n}-1}}$.

Repeating the above considerations we conclude that if a solution (G_0, Y_0) of the system (9.5) exists then the transformed matrix G_0 has the form:

$$\tilde{G}_0 = \begin{pmatrix} -(\tilde{n}-1)I_{n_1} & \mathbf{0} & \dots & \mathbf{0} & \mathbf{0} \\ * & -(\tilde{n}-2)I_{n_2} & \dots & \mathbf{0} & \mathbf{0} \\ \dots & \dots & \dots & \dots & \dots \\ * & * & \dots & -I_{n_{\tilde{n}-1}} & \mathbf{0} \\ * & * & \dots & * & \mathbf{0} \end{pmatrix}. \tag{9.19}$$

If, in particular, we take

$$\tilde{G}_0 = \mathrm{diag}\{-(\tilde{n}-1)I_{n_1}, -(\tilde{n}-2)I_{n_2}, -I_{n_{\tilde{n}-1}}, \mathbf{0}\} \tag{9.20}$$

then

$$\tilde{A}\tilde{G}_0 = \begin{pmatrix} \mathbf{0} & -(\tilde{n}-2)A_{12} & \mathbf{0} & \dots & \mathbf{0} & \mathbf{0} \\ \mathbf{0} & \mathbf{0} & -(\tilde{n}-3)A_{23} & \dots & \mathbf{0} & \mathbf{0} \\ \dots & \dots & \dots & \dots & \dots & \dots \\ \mathbf{0} & \mathbf{0} & \mathbf{0} & \dots & \mathbf{0} & \mathbf{0} \\ -(\tilde{n}-1)G_{11} & -(\tilde{n}-2)A_{\tilde{n}2} & -(\tilde{n}-3)A_{\tilde{n}3} & \dots & -A_{\tilde{n}\tilde{n}-1} & \mathbf{0} \end{pmatrix},$$

$$\tilde{G}_0\tilde{A} = \begin{pmatrix} \mathbf{0} & -(\tilde{n}-1)A_{12} & \mathbf{0} & \dots & \mathbf{0} & \mathbf{0} \\ \mathbf{0} & \mathbf{0} & -(\tilde{n}-2)A_{23} & \dots & \mathbf{0} & \mathbf{0} \\ \dots & \dots & \dots & \dots & \dots & \dots \\ \mathbf{0} & \mathbf{0} & \mathbf{0} & \dots & \mathbf{0} & -A_{\tilde{n}-1\tilde{n}} \\ \mathbf{0} & \mathbf{0} & \mathbf{0} & \dots & \mathbf{0} & \mathbf{0} \end{pmatrix},$$

so the identity $\tilde{A}\tilde{G}_0 + \tilde{B}\tilde{Y}_0 = (I_n + \tilde{G}_0)\tilde{A}$ holds for

$$\tilde{Y}_0 = A^{+}_{\tilde{n},\tilde{n}+1}\left(\tilde{n}A_{\tilde{n}1} \;\; (\tilde{n}-1)A_{\tilde{n}2} \;\; \dots \;\; 2A_{\tilde{n}-1,\tilde{n}} \;\; A_{\tilde{n},\tilde{n}}\right). \tag{9.21}$$

This means that

- algebraic equation (9.5) always has a solution $\{G_0, Y_0\}$ provided that the pair $\{A, B\}$ is controllable;
- for any solution (G_0, Y_0) of (9.5), the transformed matrix $\tilde{G}_0$ always has the form (9.19) and, consequently, the matrix $I_n - G_0$ is always invertible, the matrix $G_{\mathbf{d}} = I_n + \mu G_0$ is anti-Hurwitz for any $\mu < 1/(\tilde{n}-1)$;
- *at least one solution (G_0, Y_0) is such that G_0 is diagonalizable.*

Moreover, for $\mu \neq 0$, $\tilde{K}_0 = \tilde{Y}_0(\tilde{G}_0 - I_n)^{-1}$ and $\tilde{A}_0 = \tilde{A} + \tilde{B}\tilde{K}_0$ we have

$$\begin{aligned} & \tilde{A}\tilde{G}_0 - \tilde{G}_0\tilde{A} + \tilde{B}\tilde{K}_0(\tilde{G}_0 - I_n) = \tilde{A}, \quad \tilde{G}_0\tilde{B} = \mathbf{0} \\ \Rightarrow \quad & (\tilde{A} + \tilde{B}\tilde{K}_0)\tilde{G}_0 = \tilde{G}_0\tilde{A} + \tilde{G}_0\tilde{B}\tilde{K}_0 + \tilde{A} + \tilde{B}\tilde{K}_0 \\ \Rightarrow \quad & \mu\tilde{A}_0\tilde{G}_0 = \mu(I_n + \tilde{G}_0)\tilde{A}_0 \\ \Rightarrow \quad & \tilde{A}_0\tilde{G}_{\mathbf{d}} = (\mu I_n + \tilde{G}_{\mathbf{d}})\tilde{A}_0, \end{aligned}$$

i.e., $\tilde{A}_0$ is $\tilde{\mathbf{d}}$-homogeneous of degree μ (see Lemma 6.1) with $\tilde{\mathbf{d}}(s) = e^{s\tilde{G}_{\mathbf{d}}}$, $s \in \mathbb{R}$.

(2) Let us prove feasibility of the transformed algebraic system (9.9), where $A_0, B, X, Y, G_{\mathbf{d}}$ are replaced with $\tilde{A}_0, \tilde{B}, \tilde{X} = MXM^\top, \tilde{Y} = YM^\top$ and $\tilde{G}_{\mathbf{d}} = I_n + \mu\tilde{G}_0$. As shown above the matrix $\tilde{G}_0$ is lower triangular. Taking into account that the inverse of the lower triangular matrix $\tilde{G}_0 - I_n$ is lower triangular too, we derive

$$\tilde{A}_0 = \tilde{A} + \tilde{B}K_0 = \tilde{A} + \tilde{B}\tilde{Y}_0(\tilde{G}_0 - I_n)^{-1} = \tilde{G}_0\tilde{A}(\tilde{G}_0 - I_n)^{-1} =$$

$$\begin{pmatrix} \mathbf{0} & A_{12} & \mathbf{0} & \dots & \mathbf{0} & \mathbf{0} \\ \mathbf{0} & \mathbf{0} & A_{13} & \dots & \mathbf{0} & \mathbf{0} \\ \dots & \dots & \dots & \dots & \dots & \dots \\ \mathbf{0} & \mathbf{0} & \mathbf{0} & \dots & \mathbf{0} & A_{\tilde{n}-1,\tilde{n}} \\ \mathbf{0} & \mathbf{0} & \mathbf{0} & \dots & \mathbf{0} & \mathbf{0} \end{pmatrix} + D,$$

where $D \in \mathbb{R}^{n\times n}$ is a matrix having the lower triangular block form:

$$D = \begin{pmatrix} * & \mathbf{0} & \mathbf{0} & \dots & \mathbf{0} & \mathbf{0} \\ * & * & \mathbf{0} & \dots & \mathbf{0} & \mathbf{0} \\ \dots & \dots & \dots & \dots & \dots & \dots \\ * & * & * & \dots & * & \mathbf{0} \\ * & * & * & \dots & * & * \end{pmatrix}.$$

Denoting

$$\tilde{X} = \begin{pmatrix} X_{11} & X_{12} & \dots & X_{1\tilde{n}} \\ X_{12}^\top & X_{22} & \dots & X_{2\tilde{n}} \\ \dots & \dots & \dots & \dots \\ X_{1\tilde{n}}^\top & X_{2\tilde{n}}^\top & \dots & X_{\tilde{n},\tilde{n}} \end{pmatrix}, \qquad \tilde{Y} = (\, Y_1 \; Y_2 \; \dots \; Y_{\tilde{n}} \,),$$

$$D + \tilde{G}_{\mathbf{d}} = \begin{pmatrix} D_{11} & \mathbf{0} & \mathbf{0} & \dots & \mathbf{0} & \mathbf{0} \\ D_{21} & D_{22} & \mathbf{0} & \dots & \mathbf{0} & \mathbf{0} \\ \dots & \dots & \dots & \dots & \dots & \dots \\ D_{n-1\,1} & D_{n-1\,2} & D_{n-1\,3} & \dots & D_{n-1\,n-1} & \mathbf{0} \\ D_{n\,1} & D_{n\,2} & D_{n\,3} & \dots & D_{n\,n-1} & D_{n\,n} \end{pmatrix}$$

the linear algebraic equation in (9.9) can be rewritten in the block form:

$$A_{i\,i+1}X_{i\,i+1}^\top + X_{i\,i+1}A_{i\,i+1}^\top + \sum_{k=1}^{i} D_{ik}X_{ki} + X_{ik}^\top D_{ik}^\top = 0, \qquad (9.22)$$

$$A_{i\,i+1}X_{i+1\,j} + X_{i\,j+1}A_{j\,j+1}^\top + \sum_{k=i+1}^{j} X_{ik}D_{jk}^\top + \sum_{k=1}^{i} D_{ik}X_{kj} + X_{ki}^\top D_{jk}^\top = 0, \qquad (9.23)$$

$$A_{i\,i+1}X_{i+1\,\tilde{n}} + Y_i^\top A_{\tilde{n},\tilde{n}+1}^\top + \sum_{k=i+1}^{\tilde{n}} X_{ik}D_{\tilde{n}k}^\top + \sum_{k=1}^{i} D_{ik}X_{k\tilde{n}} + X_{ki}^\top D_{\tilde{n}k}^\top = 0, \qquad (9.24)$$

$$Y_{\tilde{n}}^\top A_{\tilde{n},\tilde{n}+1}^\top + A_{\tilde{n},\tilde{n}+1}Y_{\tilde{n}} + \sum_{k=1}^{\tilde{n}} D_{\tilde{n}k}X_{k\tilde{n}} + X_{k\tilde{n}}^\top D_{\tilde{n}k}^\top = 0, \qquad (9.25)$$

where $i = 1, 2, \dots, \tilde{n} - 1$ and $i < j < \tilde{n}$. Let us prove the feasibility of the above block system with

$$\tilde{X} \succ 0, \qquad Z := \tilde{G}_{\mathbf{d}}\tilde{X} + \tilde{X}\tilde{G}_{\mathbf{d}}^\top \succ 0.$$

For $i_1 \le i_2$, $j_1 \le j_2$ let us introduce the following notation:

$$X_{(i_1:i_2;\, j_1:j_2)} = \begin{pmatrix} X_{i_1\,i_1} & X_{i_1\,i_1+1} & \dots & X_{i_1\,i_2} \\ X_{i_1\,i_1+1}^\top & X_{i_1+1\,i_1+1} & \dots & X_{i_1+1\,i_2} \\ \dots & \dots & \dots & \dots \\ X_{i_1\,i_2}^\top & X_{i_1+1\,i_2}^\top & \dots & X_{i_2\,i_2} \end{pmatrix},$$

$$X_i = X_{(1:i;\,1:i)}, \quad \tilde{X}_i = X_{(1:i-1;\,i:i)}, \quad G_i = (\tilde{G}_{\mathbf{d}})_{(1:i;\,1:i)} \quad \tilde{G}_i = (\tilde{G}_{\mathbf{d}})_{(i:i;\,1:i-1)},$$

$$Z_1 = X_{11}G_{11}^\top + G_{11}X_{11}, \quad Z_i = \begin{pmatrix} Z_{i-1} & \tilde{X}_i G_{ii}^\top + G_i\tilde{X}_i + X_{i-1}\tilde{G}_i^\top \\ * & G_{ii}X_{ii} + X_{ii}G_{ii}^\top + \tilde{G}_i\tilde{X}_i + \tilde{X}_i^\top\tilde{G}_i^\top \end{pmatrix},$$

where $*$ denotes a symmetric term and $G_{ii} = (1 + \mu(\tilde{n} - i))I_{n_i} \succ 0$ denotes the i-th diagonal block of the lower triangular matrix $\tilde{G}_{\mathbf{d}}$.

Let us construct $\tilde{X} = X_{\tilde{n}}$ satisfying (9.9) by induction:

$\underline{i = 1}$. Let us select $X_{11} = \alpha_1 I_{n_1}$ with $\alpha_1 > 0$ being an arbitrary number. In this case, we have $Z_1 \succ 0$ and the formula (9.22) implies

$$X_{12} = -\frac{1}{2}(D_{11}X_{11} + X_{11}D_{11}^{\top})(A_{12}A_{12}^{\top})^{-1}A_{12},$$

where the matrix $(A_{12}A_{12}^{\top})^{-1}$ is invertible, since A_{12} is a full row rank matrix.

$\underline{i = 2}$. Since $X_{11} \succ 0$ and $\tilde{Z}_1 \succ 0$ then taking $X_{22} = \alpha_2 I_{n_2}$ with a sufficiently large $\alpha_2 > 0$ we obtain

$$X_2 = \begin{pmatrix} X_1 & \tilde{X}_1 \\ * & X_{22} \end{pmatrix} \succ 0, \qquad Z_2 = \begin{pmatrix} Z_1 & \tilde{X}_2 G_{22}^{\top} + G_2 \tilde{X}_2 + X_1 \tilde{G}_2^{\top} \\ * & G_{22}X_{22} + X_{22}G_{22}^{\top} + \tilde{G}_i \tilde{X}_2 + \tilde{X}_2^{\top} \tilde{G}_2^{\top} \end{pmatrix} \succ 0.$$

Using (9.22) we derive

$$X_{23} = -\tfrac{1}{2}\left(\sum_{k=1}^{2} D_{2k}X_{k2} + X_{2k}^{\top} D_{2k}^{\top}\right)\left(A_{23}A_{23}^{\top}\right)^{-1} A_{23},$$

but the formula (9.23) gives

$$X_{13} = -\left(A_{12}X_{22} + \sum_{k=2}^{2} X_{1k}D_{3k}^{\top} + \sum_{k=1}^{1} D_{1k}X_{k2} + X_{k1}^{\top} D_{2k}^{\top}\right)\left(A_{23}A_{23}^{\top}\right)^{-1} A_{23}.$$

Hence, we derive

$$\tilde{X}_2 = \begin{pmatrix} X_{13} \\ X_{23} \end{pmatrix}$$

$\underline{i = 3.}$ Since $X_2 \succ 0$ and $Z_2 \succ 0$ then taking $X_{33} = \alpha_3 I_{n_3}$ with a sufficiently large $\alpha_3 > 0$ we obtain

$$X_3 = \begin{pmatrix} X_2 & \tilde{X}_2 \\ * & X_{33} \end{pmatrix} \succ 0, \quad Z_3 = \begin{pmatrix} Z_2 & \tilde{X}_3 G_{33}^{\top} + G_3 \tilde{X}_3 + X_2 \tilde{G}_3^{\top} \\ * & G_{33}X_{33} + X_{33}G_{33}^{\top} + \tilde{G}_3 \tilde{X}_3 + \tilde{X}_3^{\top} \tilde{G}_3^{\top} \end{pmatrix} \succ 0.$$

Using (9.22) we derive X_{34}, but the formula (9.23) gives X_{14}, X_{24}, so the matrix

$$\tilde{X}_3 = \begin{pmatrix} X_{14} \\ X_{24} \\ X_{34} \end{pmatrix}$$

can be utilized on the step $\underline{i = 4}$ to define $X_{44} = \alpha_4 I_{n_4}$ such that $X_4 \succ 0$ and $Z_4 \succ 0$.

$\underline{i = \ldots}$

$\underline{i = \tilde{n}}$. Since $X_{\tilde{n}-1} \succ 0$ and $Z_{\tilde{n}-1} \succ 0$ then taking $X_{\tilde{n}\tilde{n}} = \alpha_{\tilde{n}} I_{n_{\tilde{n}}}$ with a sufficiently large $\alpha_{\tilde{n}} > 0$ we obtain

$$X=\begin{pmatrix} X_{\tilde{n}-1} & \tilde{X}_{n-1} \\ * & X_{\tilde{n}\tilde{n}} \end{pmatrix} \succ 0, \quad Z=\begin{pmatrix} Z_{\tilde{n}-1} & \tilde{X}_{\tilde{n}}G_{\tilde{n}\tilde{n}}^{\top}+G_{\tilde{n}}\tilde{X}_{\tilde{n}}+X_{\tilde{n}-1}\tilde{G}_{\tilde{n}}^{\top} \\ * & G_{\tilde{n}\tilde{n}}X_{\tilde{n}\tilde{n}}+X_{\tilde{n}\tilde{n}}G_{\tilde{n}\tilde{n}}^{\top}+\tilde{G}_{\tilde{n}}\tilde{X}_{\tilde{n}}+\tilde{X}_{\tilde{n}}^{\top}\tilde{G}_{\tilde{n}}^{\top} \end{pmatrix}, \succ 0.$$

Taking

$$Y_i^{\top}=-\left(A_{i\,i+1}X_{i+1\,\tilde{n}}+\sum_{k=i+1}^{\tilde{n}}X_{ik}D_{\tilde{n}k}^{\top}+\sum_{k=1}^{i}D_{ik}X_{k\tilde{n}}+X_{ki}^{\top}D_{\tilde{n}k}^{\top}\right)A_{\tilde{n}\tilde{n}+1}^{+}$$

with $A_{\tilde{n}\tilde{n}+1}^{+}=(A_{\tilde{n}\tilde{n}+1}A_{\tilde{n}\tilde{n}+1}^{\top})^{-1}A_{\tilde{n}\tilde{n}+1}$ and $i=1,..,\tilde{n}-1$ and

$$Y_{\tilde{n}}^{\top}=-\frac{1}{2}\left(\sum_{k=1}^{\tilde{n}}D_{\tilde{n}k}X_{k\tilde{n}}+X_{k\tilde{n}}^{\top}D_{\tilde{n}k}^{\top}\right)A_{\tilde{n}\tilde{n}+1}^{+}$$

we derive that the system of Eqs. (9.22)–(9.25) is fulfilled with $\tilde{X}\succ 0$ and $Z\succ 0$. Therefore, a solution (X,Y) of the linear algebraic systems (9.9) always exists and can be constructed by means of the proposed algorithm.

(3) If the pair (X,Y) is a solution of (9.9) then the dilation $\mathbf{d}(s)=e^{sG_{\mathbf{d}}}$ is monotone with respect to the weighted Euclidean norm $\|x\|=\sqrt{x^{\top}X^{-1}x}$. In this case, the canonical homogeneous norm $\|\cdot\|_{\mathbf{d}}$ induced by this weighted Euclidean norm is well defined, continuous of $\mathbb{R}^n$ and infinitely smooth on $\mathbb{R}^n\backslash\{\mathbf{0}\}$ (see Corollary 5.3). Let us show that $\|\cdot\|_{\mathbf{d}}$ is a Lyapunov function for the closed-loop system. Indeed, it is globally proper and, using the formula (5.21) we derive

$$\frac{d}{dt}\|x\|_{\mathbf{d}}=\frac{\partial\|x\|_{\mathbf{d}}}{\partial x}\dot{x}=\|x\|_{\mathbf{d}}\frac{\pi_{\mathbf{d}}^{\top}(x)X^{-1}\mathbf{d}(-\ln\|x\|_{\mathbf{d}})(Ax+Bu(x))}{\pi_{\mathbf{d}}^{\top}(x)X^{-1}G_{\mathbf{d}}\pi_{\mathbf{d}}(x)}=$$

$$=\|x\|_{\mathbf{d}}\frac{\pi_{\mathbf{d}}^{\top}(x)X^{-1}\mathbf{d}(-\ln\|x\|_{\mathbf{d}})(A_0x+\|x\|_{\mathbf{d}}^{1+\mu}BK\mathbf{d}(-\ln\|x\|_{\mathbf{d}})x}{\pi_{\mathbf{d}}^{\top}(x)X^{-1}G_{\mathbf{d}}\pi_{\mathbf{d}}(x)},\quad x\neq\mathbf{0},$$

where $\pi_{\mathbf{d}}(x)=\mathbf{d}(-\ln\|x\|_{\mathbf{d}})x$ is $\mathbf{d}$-homogeneous projector (see Chap. 5).

By construction, we have $A_0G_{\mathbf{d}}=(\mu I_n+G_{\mathbf{d}}), G_{\mathbf{d}}B=B$ this means (see Lemma 6.1):

$$A_0\mathbf{d}(s)=e^{\mu s}\mathbf{d}(s)A_0 \quad\text{and}\quad \mathbf{d}(s)B=\sum_{i=0}^{\infty}\frac{s^iG_{\mathbf{d}}^iB}{i!}=e^sB \quad\text{for all } s\in\mathbb{R}.$$

Hence, we conclude

$$\frac{d}{dt}\|x\|_{\mathbf{d}}=\|x\|_{\mathbf{d}}^{1+\mu}\frac{\pi_{\mathbf{d}}^{\top}(x)X^{-1}(A_0+BK)\overbrace{\mathbf{d}(-\ln\|x\|_{\mathbf{d}})x}^{=\pi_{\mathbf{d}}(x)}}{\pi^{\top}{}_{\mathbf{d}}(x)X^{-1}G_{\mathbf{d}}\pi_{\mathbf{d}}(x)}=$$

$$=\|x\|_{\mathbf{d}}^{1+\mu}\frac{\pi_{\mathbf{d}}^{\top}(x)X^{-1}\{A_0X+BY+XA_0^{\top}+Y^{\top}B^{\top}\}X^{-1}\pi_{\mathbf{d}}(x)}{\pi_{\mathbf{d}}^{\top}(x)X^{-1}\{G_{\mathbf{d}}X+XG_{\mathbf{d}}^{\top}\}X^{-1}\pi_{\mathbf{d}}(x)}=-\rho\|x\|_{\mathbf{d}}^{1+\mu},\quad x\neq\mathbf{0},$$

where the identity (9.9) is utilized on the last step. Therefore, $\|x\|_{\mathbf{d}}$ is a Lyapunov function of the system and the identity (9.12) holds, but the estimates (9.14), (9.15),(9.16) follow from the identity (9.12), which, for $\mu \neq 0$, gives

$$\|x(t)\|_{\mathbf{d}}^{-\mu} = \|x_0\|_{\mathbf{d}}^{-\mu} + \mu\rho t, \quad \forall t \geq 0 : \|x_0\|_{\mathbf{d}}^{-\mu} + \mu\rho t \geq 0. \tag{9.26}$$

(4) Since, by Corollary 5.3, the canonical homogeneous norm $\|\cdot\|_{\mathbf{d}}$ is analytic on $\mathbb{R}^n\backslash\{\mathbf{0}\}$ then u is analytic on $\mathbb{R}^n\backslash\{\mathbf{0}\}$ as a composition of analytic functions. Moreover,

$$\|\mathbf{d}(-\ln\|x\|_{\mathbf{d}})x\| = \|\pi_{\mathbf{d}}(x)\| = \sqrt{\pi_{\mathbf{d}}^{\top}(x)X^{-1}\pi_{\mathbf{d}}(x)} = 1$$

and for $x \neq \mathbf{0}$ we have

$$\begin{aligned}
|u(x)| &\leq |K_0x| + \|x\|_{\mathbf{d}}^{\mu+1}|K\pi_{\mathbf{d}}(x)| \\
&= |K_0x| + \|x\|_{\mathbf{d}}^{\mu+1}\sqrt{\pi_{\mathbf{d}}^{\top}(x)K^{\top}K\pi_{\mathbf{d}}(x)} \\
&= |K_0x| + \|x\|_{\mathbf{d}}^{\mu+1}\sqrt{\pi_{\mathbf{d}}^{\top}(x)X^{-1/2}X^{1/2}K^{\top}KX^{1/2}X^{-1/2}\pi_{\mathbf{d}}(x)} \\
&\leq |K_0x| + \lambda_{\max}^{1/2}(X^{1/2}K^{\top}KX^{1/2})\|x\|_{\mathbf{d}}^{\mu+1}\underbrace{\sqrt{\pi_{\mathbf{d}}^{\top}(x)X^{-1}\pi_{\mathbf{d}}(x)}}_{=1}.
\end{aligned}$$

(5) Denoting $f(x) = Ax + Bu(x)$ we derive

$$f(\mathbf{d}(s)x) = A_0\mathbf{d}(s)x + \|\mathbf{d}(s)x\|_{\mathbf{d}}^{\mu+1}BK\mathbf{d}(-\ln\|\mathbf{d}(s)x\|_{\mathbf{d}})\mathbf{d}(s)x =$$

$$e^{\mu s}\mathbf{d}(s)A_0x + e^{\mu s}\|x\|_{\mathbf{d}}^{\mu+1}e^{s}BK\mathbf{d}(-\ln\|x\|_{\mathbf{d}})x = e^{\mu s}\mathbf{d}(s)f(x),$$

where the identity $e^s B = \mathbf{d}(s)B$ is utilized on the last step. Therefore, under condition of controllability of the pair $\{A, B\}$, a continuous homogeneous stabilizer can always be designed in the form (9.10) for any $\mu \in (-1, 1/(\tilde{n}-1))$. For $\mu = -1$, the controller is discontinuous at the origin but locally bounded on $\mathbb{R}^n$. Taking into account Corollary 6.2, we complete the proof of sufficiency.

II. *Necessity*. Suppose the contrary, i.e., the system is homogeneously stabilizable with degree $\mu \neq 0$, but the pair $\{A, B\}$ is not controllable. In this case, our algorithm for construction of T (see Lemma 9.1) will have $r_i = 0$ after some number of steps $\tilde{m}$ and the coordinate transformation $x = T^{\top}y$ will give

$$\dot{y} = TAT^{\top}y + TBu = \begin{pmatrix} A_{\tilde{m}} & \mathbf{0} \\ * & * \end{pmatrix} y + \begin{pmatrix} \mathbf{0} \\ * \end{pmatrix} u.$$

So, the uncontrollable part of the system state may have only an exponential converge rate, but, by Theorem 7.5, any asymptotically stable homogeneous system with $\mu < 0$ (resp., $\mu > 0$) is finite-time stable (resp., nearly fixed-time stable). Therefore, we obtain a contradiction. The proof of Theorem 9.1 is complete.

In the proof of Theorem 9.1, it is shown that the generator of the dilation $G_{\mathbf{d}}$ can always be defined as a diagonalizable matrix. In many practical cases such a selection of $G_{\mathbf{d}}$ is sufficient for homogeneous stabilization of LTI plant. Moreover, in the case of the single-input controllable LTI system ($m = 1$), the generator $G_{\mathbf{d}}$ of the linear dilation $\mathbf{d}$ for homogeneous stabilization is always diagonalizable.

Example 9.1 Let us design a $\mathbf{d}$-homogeneous stabilizer for the controlled linear oscillator

$$\dot{x} = \begin{pmatrix} 0 & 1 \\ -1 & 0 \end{pmatrix} x + \begin{pmatrix} 0 \\ 1 \end{pmatrix} u, \quad t > 0, \quad x = (x_1, x_2)^\top \in \mathbb{R}^2, \quad u \in \mathbb{R}. \tag{9.27}$$

Solving the Eq. (9.5) we derive

$$G_0 = \begin{pmatrix} -1 & 0 \\ 0 & 0 \end{pmatrix}, \quad Y_0 = \begin{pmatrix} 2 & 0 \end{pmatrix}.$$

Hence, taking the homogeneity degree $\mu = -0.5$, we define the generator of the dilation

$$G_{\mathbf{d}} = \begin{pmatrix} 1.5 & 0 \\ 0 & 1 \end{pmatrix}.$$

The gain of the linear component $K_0 x$ of the stabilizer (9.10) is given by

$$K_0 = Y_0 (I_2 - G_0)^{-1} = \begin{pmatrix} 1 & 0 \end{pmatrix}.$$

Solving the algebraic system (9.9) with $\rho = 1$ we derive

$$X = \begin{pmatrix} 1 & -1.5 \\ -1.5 & 3.3437 \end{pmatrix}, \quad K = Y X^{-1} = \begin{pmatrix} -3.3438 & -2.5 \end{pmatrix}.$$

The simulation results of the closed-loop system with $x(0) = \begin{pmatrix} 1 & 0 \end{pmatrix}^\top$ are given in Fig. 9.1 The closed-loop system has been simulated in MATLAB® by means of the zero-order-hold method and a consistent discretization of homogeneous control (see Chap. 10). The sampling period is $h = 0.001$. *Homogeneous Control Systems Toolbox for MATLAB®*[1] has been utilized for most simulations in this monograph.

[1] https://gitlab.inria.fr/polyakov/hcs-toolbox-for-matlab.

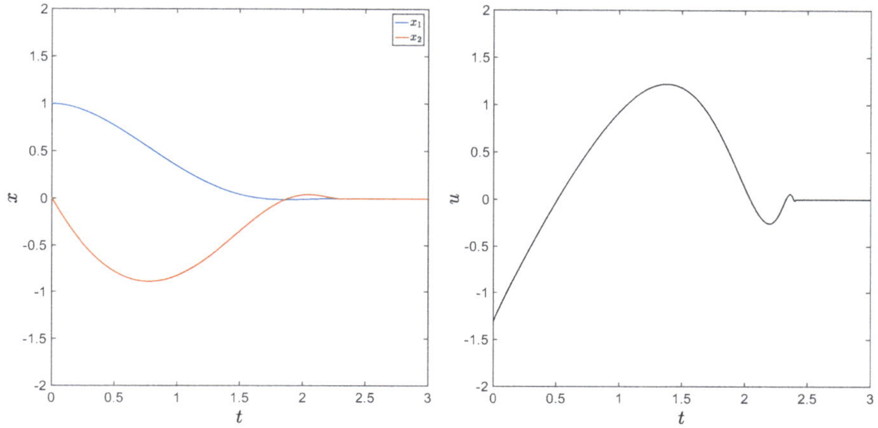

Fig. 9.1 Simulation results for homogeneously stabilized linear oscillator

The obtained closed-loop homogeneous system has the negative homogeneity degree $\mu = -0.5$. By Theorem 9.1, it is finite-time stable and its settling time is given by $2\sqrt{\|x(0)\|_{\mathbf{d}}} \approx 2.4094$. The settling time in the numerical simulation is identified as the first instant of time when the norm of the state becomes less than the machine precision (also known as "machine epsilon"): $\varepsilon = 2^{-54} \approx 1.11 \cdot 10^{-16}$ in the case of double-precision floating point format of the number. The settling time computed by the numerical simulations is 2.41. This perfectly fits the theoretical estimate (up to the discretization period $h = 0.001$).

For $\mu = 0$ the homogeneous controller (9.10) is linear $u = (K_0 + K)x$, so the solution of the closed-loop system can be represented in an explicit form using the matrix exponent: $x(t) = e^{t(A+B(K_0+K))}x_0$. For $\mu \neq 0$ the implicit homogeneous controller (9.10) is essentially nonlinear. However, as discovered in [20], the unique solution of the closed-loop system admits the explicit representation even in this case.[2]

Corollary 9.1 (Explicit representation of solutions) *Under conditions of Theorem 9.1 with* $\mu \neq 0$*, a solution of the closed-loop system* (9.1), (9.10) *is unique and*

$$x(t+\tau) = Q_\tau(\|x(t)\|_{\mathbf{d}})x(t), \tag{9.28}$$

[2] This corollary is reprinted from *Automatica*, 155/9, A. Polyakov, D. Efimov, Z. Ping, Consistent discretization of homogeneous finite/fixed-time controllers for LTI systems, 111118. https://doi.org/10.1016/j.automatica.2023.111118. ©2023, with permission from Elsevier.

where $\tau, t \geq 0$, $Q_\tau(0) = \mathbf{0}$ *and for* $r > 0$ *one has*

$$Q_\tau(r) = \begin{cases} e^{G_{\mathbf{d}} \ln r} \hat{Q}\left(\frac{\ln(1+\mu\rho\tau r^\mu)}{\rho\mu}\right) e^{-G_{\mathbf{d}} \ln r} & \text{if } \frac{1}{r^\mu} > -\mu\rho\tau, \\ \mathbf{0} & \text{if } \frac{1}{r^\mu} \leq -\mu\rho\tau, \end{cases} \tag{9.29}$$

$$\hat{Q}(\hat{s}) = e^{-\rho G_{\mathbf{d}} \hat{s}} e^{(A+B(K_0+K)+\rho G_{\mathbf{d}})\hat{s}}, \quad \hat{s} \geq 0. \tag{9.30}$$

Proof The uniqueness of solutions follows from smoothness of u on $\mathbb{R}^n \backslash \{\mathbf{0}\}$ and stability of the closed-loop system. Denoting $y = \|x\|\mathbf{d}(-\|x\|_{\mathbf{d}})x$, we derive $\|y\| = \|x\|_{\mathbf{d}} \|\mathbf{d}(-\ln \|x\|_{\mathbf{d}})x\| = \|x\|_{\mathbf{d}}$. Due to Theorem 7.3 the closed-loop system (9.1), (9.10) is topologically equivalent (homeomorphic on $\mathbb{R}^n$ and diffeomorphic on $\mathbb{R}^n \backslash \{\mathbf{0}\}$) to the standard homogeneous system:

$$\dot{y} = \|y\|^{1+\mu} \left(\frac{(I_n - G_{\mathbf{d}})yy^\top P}{y^\top P G_{\mathbf{d}} y} + I_n \right) (A + B(K_0 + K)) \frac{y}{\|y\|}$$

$$= \|y\|^\mu (A_0 + BK + \rho(G_{\mathbf{d}} - I_n))y,$$

where $P = X^{-1}$ and the identity from (9.9) is utilized on the last step. In this case, using the same identity, we conclude

$$\frac{d}{dt}\|y(t)\| = \frac{y^\top P \dot{y}(t)}{\|y(t)\|} = \|y\|^{\mu-1} y^\top P \, (A_0 + BK + \rho(G_d - I_n))y$$

$$= \|y\|^{\mu-1} \left(\frac{y^\top \{P(A_0+BK+\rho G_{\mathbf{d}}) + (A_0+BK+\rho G_{\mathbf{d}})^\top P\} y}{2} - \rho y^\top P y \right)$$

$$= -\rho \|y(t)\|^{\mu+1}$$

and $\|y(t+\tau)\|^{-\mu} = \|y(t)\|^{-\mu} + \mu\rho\tau$ for $\|y(t)\|^{-\mu} + \mu\rho\tau \geq 0$. Obviously, $\|y(t + \tau)\| = 0$ if $\|y(t)\|^{-\mu} + \mu\rho\tau \leq 0$. This case corresponds to the negative homogeneity degree $\mu < 0$ and the finite-time stability of the closed-loop system. Hence, denoting $\tilde{A} = A_0 + BK + \rho G_{\mathbf{d}}$ we obtain

$$\begin{aligned} y(t+\tau) &= e^{(\tilde{A}-\rho I_n)\int_0^\tau \|y(t+\sigma)\|^\mu d\sigma} y(t) \\ &= e^{\tilde{A}-\rho I_n)) \int_0^\tau \frac{1}{\|y(t)\|^{-\mu}+\mu\rho\sigma} d\sigma} y(t) \\ &= e^{(\tilde{A}-\rho I_n)\frac{1}{\mu\rho} \ln \frac{\|y(t)\|^{-\mu}+\mu\rho\tau}{\|y(t)\|^{-\mu}}} y(t) \\ &= e^{(\tilde{A}-\rho I_n)\frac{1}{\mu\rho} \ln(1+\mu\rho\tau\|y(t)\|^\mu)} y(t). \end{aligned}$$

Since $\|y(t)\| = \|x(t)\|_{\mathbf{d}}$ then returning to the original coordinates we derive

$$\begin{aligned}
x(t+\tau) =&\mathbf{d}(\ln \|y(t+\tau\|)\tfrac{y(t+\tau)}{\|y(t+\tau)\|}\\
=&\mathbf{d}(\ln(\|y(t)\|^{-\mu}+\mu\rho\tau)^{\frac{1}{-\mu}})\cdot e^{(A+BK_{lin}+\rho(G_{\mathbf{d}}-I_n))\frac{1}{\rho\mu}\ln(1+\mu\rho\tau\|y(t)\|^{\mu})}\\
&\cdot\frac{y(t)}{(\|y(t)\|^{-\mu}+\mu\rho\tau)^{\frac{1}{-\mu}}}\\
=&\mathbf{d}(\ln(\|x(t)\|_{\mathbf{d}}^{-\mu}+\mu\rho\tau)^{\frac{1}{-\mu}})\cdot e^{(A+BK_{lin}+\rho G_{\mathbf{d}})\frac{1}{\rho\mu}\ln(1+\mu\rho\tau\|x(t)\|_{\mathbf{d}}^{\mu})}\\
&\cdot\mathbf{d}(-\ln\|x(t)\|_{\mathbf{d}})x(t)\\
=&Q_\tau(\|x(t)\|_{\mathbf{d}})x(t)
\end{aligned}$$

for all $t \geq 0$ and all $\tau \geq 0$. The proof is complete. □

The above corollary is useful for digitization of homogeneous controller, more precisely, for a design of its *consistent discretization* which preserves the finite/fixed-time convergence rate of a continuous-time system with a homogeneous controller in its sampled-time counterpart (see Chap. 10). Moreover, in the proof it is shown that, under conditions of Theorem 9.1, *the closed-loop system* (9.1), (9.10) *is a linear system in the generalized homogeneous space* $\mathbb{R}^n_{\mathbf{d}}$. Indeed, the change of variable $y = \Phi(y)$ gives

$$\dot{y} = \|y\|^{\mu}(A + B(K_0 + K) + \rho(G_{\mathbf{d}} - I_n))y. \tag{9.31}$$

The multiplier $\|y\|^{\mu}$ in the above equation just modifies the convergence rate close to zero or close to infinity dependently of μ. A state-dependent change of time $\tau = \int_0^t \|y(s)\|^{\mu}ds$ makes the above system linear $\dot{y}(\tau) = (A + B(K_0 + K) + \rho(G_{\mathbf{d}}) - I_n)y(\tau)$, $\tau > 0$. *A similarity of the homogeneous control system to a linear control system is implicitly utilized almost everywhere in the monograph.*

Remark 9.1 The algebraic system (9.9) can be replaced with the LMI

$$A_0X + A_0^{\top}X + BY + Y^{\top}B^{\top} + \rho(G_{\mathbf{d}}X + XG_{\mathbf{d}}^{\top}) \preceq 0,\; G_{\mathbf{d}}X + XG_{\mathbf{d}}^{\top} \succ 0,\; X \succ 0$$

implying the inequality

$$\frac{d}{dt}\|x\|_{\mathbf{d}} \leq -\rho\|x\|_{\mathbf{d}}^{1+\mu}$$

instead of the identity (9.12).

9.1.2 Explicit Homogeneous Stabilizer

An implementation of the control law (9.10) requires a computation of the canonical homogeneous norm $\|\cdot\|_{\mathbf{d}}$, which is defined implicitly (see Definition 5.8). In Chap. 5, various computational schemes are developed for this purpose. By Theorem 5.6, any canonical homogeneous norm can be approximated with an arbitrary high precision by means of a **d**-homogeneous norm defined explicitly. The reasonable question:

> *Is it possible to stabilize the system* (9.1) *replacing* $\|\cdot\|_{\mathbf{d}}$ *with an explicitly defined homogeneous norm in the controller* (9.10)*?*

The following theorem positively answers to this question. This possibility was originally proven in [14] just for the case $\mu = -1$. The similar design for the weighted homogeneous integrator chain has been proposed in [21]. Below we present the proof for the general case.

Theorem 9.2 *Let* μ, $G_{\mathbf{d}}$, K_0 *and* A_0 *be defined as in Theorem 9.1. Let the system of linear matrix inequalities*

$$\begin{pmatrix} (\tilde{A}X+X\tilde{A}^{\top})+BY+Y^{\top}B^{\top}-X & BY & X\mathbf{d}\left(\ln\frac{\delta_{\max}}{\delta_{\min}}\right) \\ Y^{\top}B^{\top} & -\frac{G_{\mathbf{d}}^{-1}X+XG_{\mathbf{d}}^{-\top}}{\omega} & \mathbf{0} \\ \mathbf{d}\left(\ln\frac{\delta_{\max}}{\delta_{\min}}\right)X & \mathbf{0} & -X \end{pmatrix} \prec 0 \tag{9.32}$$

be feasible with respect to the variable $X \in \mathbb{R}^{n\times n}$, $Y \in \mathbb{R}^{m\times n}$ *for some* $0 < \delta_{\min} \leq \delta_{\max}$ *and* $\rho > 0$, *where*

$$\tilde{A} = \left(1 + (1+\mu)\ln\left(\frac{\delta_{\max}}{\delta_{\min}}\right)\right) A_0 + \rho\left(\frac{\delta_{\max}}{\delta_{\min}}\right)^{1+\mu} G_{\mathbf{d}} \tag{9.33}$$

and

$$\omega = \int_{\delta_{\min}}^{\delta_{\max}} \sigma^{2\mu+1} d\sigma > 0. \tag{9.34}$$

Let $\|\cdot\|_{\mathbf{d}}$ *be the canonical homogeneous norm induced by the norm* $\|x\| = \sqrt{x^{\top}Px}$ *with* $P = X^{-1}$. *If a function* $p : \mathbb{R}^n \mapsto \mathbb{R}$ *is continuous, positive definite,* **d***-homogeneous of the degree* 1 *and satisfies the inequalities*

$$0 < \delta_{\min} \leq p(y) \leq \delta_{\max}, \quad \forall y \in S := \{z \in \mathbb{R}^n : z^{\top}X^{-1}z = 1\}, \tag{9.35}$$

then the closed-loop system (9.1) *with the control*

$$u = K_0x + p^{1+\mu}(x)K\mathbf{d}(-\ln p(x))x, \quad K = YX^{-1}\mathbf{d}(\ln\delta_{\max}) \tag{9.36}$$

is **d***-homogeneous of the degree* $\mu \in \mathbb{R}$ *and globally asymptotically stable. Moreover, the canonical homogeneous norm is a Lyapunov function of the closed-loop system* (9.1), (9.36)*:*

$$\tfrac{d}{dt}\|x\|_{\mathbf{d}} \leq -\rho\|x\|_{\mathbf{d}}^{1+\mu}, \quad \forall x \neq \mathbf{0}, \tag{9.37}$$

and the system (9.1), (9.36) *is globally uniformly*

- *finite-time stable for* $\mu \in [-1, 0)$ *such that the estimate* (9.14) *is fulfilled;*
- *exponentially stable for* $\mu = 0$ *such that*

$$\|x(t)\|_{\mathbf{d}} \leq e^{-\rho t}\|x_0\|_{\mathbf{d}}, \quad \forall t \geq 0;$$

- *nearly fixed-time stable for* $\mu \in (0, 1/(\tilde{n}-1))$ *and the estimate* (9.14) *is fulfilled.*

Proof (1) Since $\omega > 0$ then for any solution (X, Y) of the linear matrix inequality (9.32) we have $G_{\mathbf{d}}^{-1}X + XG_{\mathbf{d}}^{-\top} \succ 0$ or, equivalently, $PG_{\mathbf{d}}^{-1} + G_{\mathbf{d}}^{-\top}P \succ 0$ and $G_{\mathbf{d}}^{\top}P + PG_{\mathbf{d}} \succ 0$, where $P = X^{-1} \succ 0$. Let us show that

$$2z^{\top}P\left(A_0 + \delta^{1+\mu}BK\mathbf{d}(-\ln\delta)\right)z < -\rho z^{\top}(PG_{\mathbf{d}} + G_{\mathbf{d}}P)z, \tag{9.38}$$

$$\forall z \in \mathbb{R}^n\backslash\{\mathbf{0}\}, \quad \forall \delta \in [\delta_{\min}, \delta_{\max}].$$

Let us fix an arbitrary $z \in \mathbb{R}^n\backslash\{\mathbf{0}\}$ and consider the smooth function $q : [\delta_{\min}, \delta_{\max}] \mapsto \mathbb{R}$ defined as

$$q(\delta) = z^T P\left(A_0 + \delta^{1+\mu}BK\mathbf{d}(-\ln\delta)\right)z.$$

To estimate its derivative $q'(\delta)$ we use the inequality

$$2y^{\top}v \leq y^{\top}y + v^{\top}v,$$

which, obviously, holds for all $y, v \in \mathbb{R}^n$ and, in particular, for

$$y = \delta^{0.5+0.5\mu}\Gamma^{-\frac{1}{2}}\mathbf{d}(-\ln\delta_{\max})K^{\top}B^{\top}Pz$$

and

$$v = \delta^{-0.5-0.5\mu}\Gamma^{\frac{1}{2}}G_{\mathbf{d}}\mathbf{d}\left(-\ln\frac{\delta}{\delta_{\max}}\right)z,$$

where $\Gamma = G_{\mathbf{d}}^{-\top}P + PG_{\mathbf{d}}^{-1}$. Since $\frac{d}{ds}\mathbf{d}(s) = G_{\mathbf{d}}\mathbf{d}(s) = \mathbf{d}(s)G_{\mathbf{d}}$ for $s \in \mathbb{R}$ then

$$(1+\mu)\tfrac{q(\delta)-z^{\top}PA_0z}{\delta} - \tfrac{dq}{d\delta} = \delta^{\mu}z^T P\hat{B}KG_{\mathbf{d}}\mathbf{d}(-\ln\delta)z = \delta^{\mu}y^{\top}v \leq$$

$$\tfrac{\delta^{2\mu+1}z^{\top}P\hat{B}K\mathbf{d}^{\top}(\ln\delta_{\max}^{-1})\Gamma^{-1}\mathbf{d}(\ln\delta_{\max}^{-1})K^{\top}\hat{B}^{\top}Pz + \frac{1}{\delta}z^{\top}\mathbf{d}^{\top}\left(\ln\frac{\delta_{\max}}{\delta}\right)(PG_{\mathbf{d}}+G_{\mathbf{d}}P)\mathbf{d}\left(\ln\frac{\delta_{\max}}{\delta}\right)z}{2} =$$

$$\tfrac{\delta^{2\mu+1}z^T P\hat{B}K\mathbf{d}^{\top}(\ln\delta_{\max}^{-1})\Gamma^{-1}\mathbf{d}(\ln\delta_{\max}^{-1})K^{\top}\hat{B}^{\top}Pz - \frac{d}{d\delta}\left(z^{\top}\mathbf{d}^{\top}\left(\ln\frac{\delta_{\max}}{\delta}\right)P\mathbf{d}\left(\ln\frac{\delta_{\max}}{\delta}\right)z\right)}{2}.$$

Using the Newton–Leibniz formula, we derive

$$q(\delta) = q(\delta_{\max}) - \int_{\delta}^{\delta_{\max}} q'(s)ds \leq z^{\top}P\left(A_0 + \delta_{\max}^{1+\mu}\hat{B}K\mathbf{d}(-\ln\delta_{\max})\right)z$$

$$+ (1+\mu)z^{\top}PA_0 z\ln\tfrac{\delta_{\max}}{\delta_{\min}} + \tfrac{z^{\top}P\hat{B}K\mathbf{d}^{\top}(\ln\delta_{\max}^{-1})\Gamma^{-1}\mathbf{d}(\ln\delta_{\max}^{-1})K^{\top}\hat{B}^{\top}Pz}{2}\int_{\delta_{\min}}^{\delta_{\max}}\sigma^{2\mu+1}d\sigma$$

$$+ \frac{z^{\top}\mathbf{d}^{\top}\left(\ln\frac{\delta_{\max}}{\delta_{\min}}\right)P\mathbf{d}\left(\ln\frac{\delta_{\max}}{\delta_{\min}}\right)z}{2} - \frac{z^{\top}Pz}{2} - \int_{\delta}^{\delta_{\max}}(1+\mu)\tfrac{q(\sigma)}{\sigma}d\sigma$$

for all $\delta \in [\delta_{\min}, \delta_{\max}]$. Hence, using Grönwall–Bellman inequality we derive

$$q(\delta) \leq \frac{z^{\top}Wz}{2}e^{-(1+\mu)\int_{\delta_{\min}}^{\delta_{\max}}\frac{d\sigma}{\sigma}} = \frac{z^{\top}Wz}{2}\left(\tfrac{\delta_{\min}}{\delta_{\max}}\right)^{1+\mu}, \quad \forall\delta\in[\delta_{\min},\delta_{\max}],$$

where

$$W = P\left(A_0 + \delta_{\max}^{1+\mu}\hat{B}K\mathbf{d}(\ln\delta_{\max}^{-1})\right) + \left(A_0 + \delta_{\max}^{1+\mu}\hat{B}K\mathbf{d}(\ln\delta_{\max}^{-1})\right)^{\top}P$$

$$+ \omega P\hat{B}K\mathbf{d}^{\top}(\ln\delta_{\max}^{-1})\Gamma^{-1}\mathbf{d}(\ln\delta_{\max}^{-1})K^{\top}\hat{B}^{\top}P$$

$$+ \mathbf{d}^{\top}\left(\ln\tfrac{\delta_{\max}}{\delta_{\min}}\right)P\mathbf{d}\left(\ln\tfrac{\delta_{\max}}{\delta_{\min}}\right) - P + (1+\mu)PA_0\ln\tfrac{\delta_{\max}}{\delta_{\min}}.$$

Therefore, the matrix inequality

$$\left(\tfrac{\delta_{\min}}{\delta_{\max}}\right)^{1+\mu}W \prec -\rho(PG_{\mathbf{d}} + G_{\mathbf{d}}P)$$

implies (9.38). Applying *Schur complement*[3] we derive that the LMI (9.32) is equivalent to the system of matrix inequalities

$$W + \rho\left(\tfrac{\delta_{\max}}{\delta_{\min}}\right)^{1+\mu}(PG_{\mathbf{d}} + G_{\mathbf{d}}P) \prec 0, \quad P \succ 0, \quad PG_{\mathbf{d}} + G_{\mathbf{d}}^{\top}P \succ 0$$

with $X = P^{-1}$ and $Y = K\mathbf{d}(-\ln\delta_{\max})P^{-1}$.

(2) Since p and $\|\cdot\|_{\mathbf{d}}$ are $\mathbf{d}$-homogeneous of the degree 1 and $\|x\| = 1 \Leftrightarrow \|x\|_{\mathbf{d}} = 1$ then

$$0 < \delta_{\min} \leq \frac{p(x)}{\|x\|_{\mathbf{d}}} < \delta_{\max}, \quad \forall x \neq \mathbf{0}.$$

Let us compute the time derivative of the canonical homogeneous norm along the trajectories of the system (9.1), (9.36)

---

[3] $\begin{pmatrix} X_{11} & X_{12} \\ X_{12}^{\top} & X_{22}\end{pmatrix} \prec 0 \;\Leftrightarrow\; X_{22} \prec 0,\ X_{11} - X_{12}X_{22}^{-1}X_{12}^{\top} \prec 0 \;\Leftrightarrow\; X_{11} \prec 0,\ X_{22} - X_{12}^{\top}X_{11}^{-1}X_{12} \prec 0.$

$$\frac{d\|x\|_{\mathbf{d}}}{dt} = \|x\|_{\mathbf{d}} \frac{x^\top \mathbf{d}^\top(-\ln\|x\|_{\mathbf{d}})X^{-1}\mathbf{d}(-\ln\|x\|_{\mathbf{d}})(A_0x+p^{1+\mu}(x)BK\mathbf{d}(-\ln p(x))x}{x^\top\mathbf{d}^\top(-\ln\|x\|_{\mathbf{d}})X^{-1}G_{\mathbf{d}}\mathbf{d}(-\ln\|x\|_{\mathbf{d}})x}$$

$$= \|x\|_{\mathbf{d}}^{1+\mu} \frac{x^\top\mathbf{d}^\top(-\ln\|x\|_{\mathbf{d}})X^{-1}\left(A_0+\left(\frac{p(x)}{\|x\|_{\mathbf{d}}}\right)^{1+\mu}BK\mathbf{d}\left(-\ln\frac{p(x)}{\|x\|_{\mathbf{d}}}\right)\right)\mathbf{d}(-\ln\|x\|_{\mathbf{d}})x}{x^\top\mathbf{d}^\top(-\ln\|x\|_{\mathbf{d}})X^{-1}G_{\mathbf{d}}\mathbf{d}(-\ln\|x\|_{\mathbf{d}})x}$$

$$\leq -\rho\|x\|_{\mathbf{d}}^{1+\mu}, \quad \forall x \neq \mathbf{0},$$

where the inequality (9.38) is utilized on the last step. The proof is complete. □

Notice that, for $\delta_{\min} \to 1$ and $\delta_{\max} \to 1$, we have $\ln\frac{\delta_{\max}}{\delta_{\min}} \to 0$ and $\omega \to 0$, so, using Schur complement, we conclude that, in the limit case, the LMI (9.32) is equivalent to

$$A_0X + A_0^\top X + BY + Y^\top B^\top + \rho(G_{\mathbf{d}}X + XG_{\mathbf{d}}^\top) \prec 0, \;\; G_{\mathbf{d}}X + XG_{\mathbf{d}}^\top \succ 0, \;\; X \succ 0.$$

This LMI is feasible in the view of Theorem 9.1, so the LMI (9.32) is feasible at least for $\delta_{\min} < 1 < \delta_{\max}$ being close to 1. In this case, the function p in Theorem 9.2 is a $\mathbf{d}$-homogeneous norm approximating the canonical homogeneous norm $\|\cdot\|_{\mathbf{d}}$ with the precision $\max\{1-\delta_{\min}, \delta_{\max}-1\}$ on the unit sphere.

In Sect. 5.3, two explicit $\mathbf{d}$-homogeneous norms $\|\cdot\|_{\tilde{r}}$ and $\|\cdot\|_{\mathbf{d},\epsilon}$ given by the formulas (5.12) and (5.43), respectively, have been introduced in order to approximate the canonical homogeneous norm $\|\cdot\|_{\mathbf{d}}$. Each of them can be utilized for the design of the explicit homogeneous control (9.36). According to Theorem 5.6, the norm $p(x) = \|x\|_{\mathbf{d},\epsilon}$ can approximate $\|x\|_{\mathbf{d}}$ with an arbitrary high precision, so for any $0 < \delta_{\min} < 1 < \delta_{\max} < +\infty$ the condition (9.35) can always be fulfilled under proper tuning of parameters of the norm $p(x) = \|x\|_{\mathbf{d},\epsilon}$. In this case, the explicit homogeneous control (9.36) with $p(x) = \|x\|_{\mathbf{d},\epsilon}$ approximates the implicit homogeneous control (9.10) with an arbitrary high precision and $p(x) = \|x\|_{\mathbf{d},\epsilon}$ satisfies Theorem 9.2 with $\delta_{\min} < 1 < \delta_{\max}$ being selected arbitrarily close to 1. The following algorithm can be suggested for the design of the explicit homogeneous control:

Algorithm 9.1 ***(Design of explicit homogeneous feedback)***

(0) *Select a desired homogeneity degree μ, a desired rate $\rho > 0$ and some $0 < \delta_{\min} < 1 < \delta_{\max}$ close to* 1.
(1) *Solve the LMI* (9.32).
(2) *If the LMI* (9.32) *is infeasible then take $\delta_{\min} = 0.5(\delta_{\min}+1)$, $\delta_{\max} = 0.5(\delta_{\max}+1)$ and go to Step 1.*
(3) *Compute parameters of an explicit homogeneous norm $p(x) = \|x\|_{\mathbf{d},\epsilon}$ given by* (5.43) *using Algorithm 5.2 for $P = X^{-1} \succ 0$, where X is a solution of the LMI* (9.32).

Due to Theorem 5.6 and the feasibility of the LMI (9.32) for $\delta_{\min}, \delta_{\max}$ selected sufficiently close to 1, the explicit homogeneous stabilizer (9.36) can always be constructed using the above algorithm.

The **d**-homogeneous norm $\|\cdot\|_{\bar{r}}$ given by (5.12) has much more simple representation, but it cannot guarantee an arbitrary high approximation precision of $\|\cdot\|_{\mathbf{d}}$, so the successful design of the explicit homogeneous control using the **d**-homogeneous norm $\|\cdot\|_{\bar{r}}$ cannot always be provided. However, a possibility of such a design in particular cases can be demonstrated on the numerical example.

Example 9.2 To illustrate the above theorem we consider the system

$$\dot{x} = \begin{pmatrix} 0 & 1 \\ 0 & 0 \end{pmatrix} x + \begin{pmatrix} 0 \\ 1 \end{pmatrix} u(x), \quad u : \mathbb{R}^2 \mapsto \mathbb{R}. \tag{9.39}$$

For the homogeneity degree $\mu = -1/4$, the corresponding weighted dilation satisfying Theorem 9.10 is given by

$$\mathbf{d}(s) = e^{sG_{\mathbf{d}}}, \quad G_{\mathbf{d}} = \begin{pmatrix} 5/4 & 0 \\ 0 & 1 \end{pmatrix}, \quad s \in \mathbb{R}.$$

For $\delta_{\min} = 0.99$, $\delta_{\max} = 1.01$ we solve the linear matrix inequality (9.32) and obtain

$$P = X^{-1} = \begin{pmatrix} 0.1157 & 0.0194 \\ 0.0194 & 0.1186 \end{pmatrix}$$

and

$$K = YX^{-1}\mathbf{d}(\ln \delta_{\max}) = -\begin{pmatrix} 1.1096 & 0.6944 \end{pmatrix}.$$

By Lemma 5.2, the linear continuous dilation **d** is equivalent to the linear continuous dilation $\mathbf{d}_\gamma(s) = \mathbf{d}(4s)$ studied in Example 5.9, where an approximation of $\|\cdot\|_{\mathbf{d}}$ by an explicit homogeneous norm is designed. By Corollary 5.5, we have $\|\cdot\|_{\mathbf{d}} = \|\cdot\|^4_{\mathbf{d}_\gamma}$. Therefore, the canonical homogeneous norm induced by $\|x\| = \sqrt{x^\top Px}$ can be approximated by the explicit norm $\|x\|_{\bar{r}}$ given by (5.12) with $\bar{r} = 5/4$, $\Psi(x) = \begin{pmatrix} x_1 \\ |x_2|^{5/4}\,\mathrm{sign}(x_2) \end{pmatrix}$ and

$$Q = \begin{pmatrix} 0.0720 & 0.0176 \\ 0.0176 & 0.1225 \end{pmatrix}.$$

One can be checked numerically that $|\|x\|_{\bar{r}} - \|x\|_{\mathbf{d}}| \leq 0.01$ for $x \in S$, so all conditions of Theorem 9.2 are fulfilled.

The results of the numerical simulation for the considered system with the controller (9.38) are depicted in Fig. 9.2.

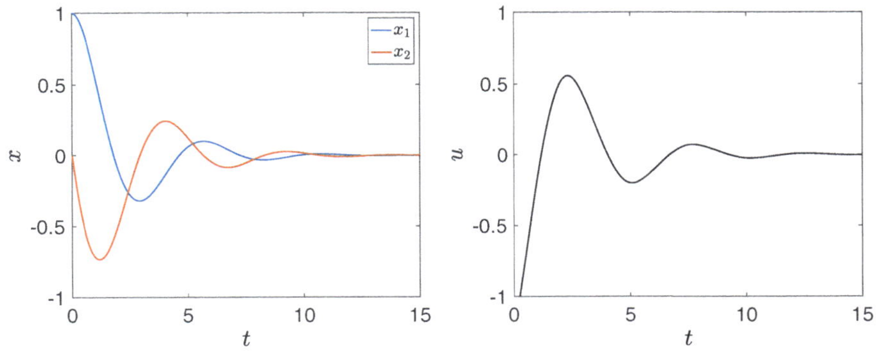

Fig. 9.2 Evolution of the state of the system (9.39) with the explicit homogeneous feedback (9.36), (5.12)

9.1.3 Robustness Analysis

Homogeneous Domination Approach

Usually, in practice, the linear model like (9.1) is a comfortable approximation of a more realistic model of the plant. In this case, it is important to characterize a class of perturbations and uncertainties which can be rejected by the given feedback law. Let us consider the nonlinear system

$$\dot{x} = Ax + Bu + g(t, x), \quad t \geq t_0, \quad x(t_0) = x_0 \in \mathbb{R}^n, \tag{9.40}$$

where x, A, B, u are as before, but an unknown function $g : \mathbb{R} \times \mathbb{R}^n \mapsto \mathbb{R}^n$ models uncertainties and disturbances of the system.

Corollary 9.2 *Let u be defined as in Theorem 9.1 (resp., as in Theorem 9.2) and $g : \mathbb{R} \times \mathbb{R}^n \mapsto \mathbb{R}^n$ be locally bounded measurable function. If g satisfies the inequality*

$$\frac{x^\top \mathbf{d}^\top(-\ln\|x\|_{\mathbf{d}})X^{-1}\mathbf{d}(-\ln\|x\|_{\mathbf{d}})g(t,x)}{x^\top \mathbf{d}^\top(-\ln\|x\|_{\mathbf{d}})X^{-1}G_{\mathbf{d}}\mathbf{d}(-\ln\|x\|_{\mathbf{d}})x} \leq \kappa\|x\|_{\mathbf{d}}^{\mu}, \quad \forall x \in \mathbb{R}^n, \quad \forall t \in \mathbb{R}, \tag{9.41}$$

for some $\kappa \in (0, \rho)$ then

$$\frac{d}{dt}\|x\|_{\mathbf{d}} \leq -(\rho - \kappa)\|x\|_{\mathbf{d}}^{1+\mu}$$

and the system (9.40) *with the control* (9.10) *(resp.,* (9.15)*) is*

- *finite-time stable for* $\mu < 0$ *and*

$$x_{x_0}(t) = \mathbf{0}, \quad \forall t \geq t_0 + \frac{\|x_0\|_{\mathbf{d}}^{-\mu}}{-\mu(\rho-\kappa)}, \quad \forall x_0 \in \mathbb{R}^n; \tag{9.42}$$

- *exponentially stable for* $\mu = 0$ *and*

$$\|x(t)\| \leq e^{-(\rho-\kappa)(t-t_0)}\|x_0\|, \quad \forall t \geq t_0;$$

- *nearly fixed-time stable for* $\mu > 0$ *and*

$$\|x_{x_0}(t)\|_{\mathbf{d}} \leq r, \quad \forall t \geq t_0 + \frac{1}{(\rho-\kappa)r^{\mu}\mu}, \quad \forall r > 0, \quad \forall x_0 \in \mathbb{R}^n. \tag{9.43}$$

Proof First, let us consider the case of continuous g. Calculating the time derivative of $\|\cdot\|_{\mathbf{d}}$ along the trajectories of the closed-loop system (9.40), (9.10), we derive

$$\frac{d\|x\|_{\mathbf{d}}}{dt} = -\rho\|x\|_{\mathbf{d}}^{1+\mu} + \|x\|_{\mathbf{d}} \frac{x^\top \mathbf{d}^\top(-\ln\|x\|_{\mathbf{d}})X^{-1}\mathbf{d}(-\ln\|x\|_{\mathbf{d}})g(t,x)}{x^\top \mathbf{d}^\top(-\ln\|x\|_{\mathbf{d}})X^{-1}G_{\mathbf{d}}\mathbf{d}(-\ln\|x\|_{\mathbf{d}})x} \leq (\kappa-\rho)\|x\|_{\mathbf{d}}^{1+\mu}.$$

If g is locally bounded measurable function, then the closed-loop system has solutions in the sense of Filippov (see Chap. 2). In this case, the above inequality holds for almost all $t \in \mathbb{R}$ if g is replaced with $\tilde{g}(t,x) \in K[g]$, where $K[g]$ is the Filippov regularization of g (see Chap. 2). Taking into account $\kappa \in (0,\rho)$ we complete the proof using Corollary 4.6 if $\mu < 0$ or Corollary 4.8 if $\mu > 0$. The case $\mu = 0$ can be treated by means of the conventional Lyapunov theorems for ODEs and DIs, since $\|x\|_{\mathbf{d}} = \|x\|$ in this case. The proof is complete. □

A more simple characterization of g can be obtained in the case of *matched perturbations*:

$$g(t,x) = B\gamma(t,x). \tag{9.44}$$

Indeed, in this case, the inequality (9.41) becomes

$$\frac{\pi_{\mathbf{d}}^\top(x)X^{-1}B\gamma(t,x)}{\pi_{\mathbf{d}}^\top(x)X^{-1}G_{\mathbf{d}}\pi_{\mathbf{d}}(x)} \leq \kappa\|x\|_{\mathbf{d}}^{1+\mu},$$

where $\pi_{\mathbf{d}}(x) = \mathbf{d}(-\ln\|x\|_{\mathbf{d}})x$ and the identity $\mathbf{d}(s)B = e^s B$ is utilized. Since $\|\pi_{\mathbf{d}}(x)\|^2 = \pi_{\mathbf{d}}^\top(x)X^{-1}\pi_{\mathbf{d}}(x) = 1$ then, using Cauchy–Schwarz inequality[4] we derive

$$\frac{\pi_{\mathbf{d}}^\top(x)X^{-1}B\gamma(t,x)}{\pi_{\mathbf{d}}^\top(x)X^{-1}G_{\mathbf{d}}\pi_{\mathbf{d}}(x)} = \frac{\overbrace{(X^{-1/2}\pi_{\mathbf{d}}(x))^\top}^{y^\top}\overbrace{X^{-1/2}B\gamma(t,x)}^{z}}{0.5\pi_{\mathbf{d}}^\top(x)(X^{-1}G_{\mathbf{d}}+G_{\mathbf{d}}^\top X^{-1})\pi_{\mathbf{d}}(x)} \leq \frac{2\|\pi_{\mathbf{d}}(x)\|^2\|B\gamma(t,x)\|}{\beta\|\pi_{\mathbf{d}}(x)\|^2},$$

[4] $|y^\top z| \leq |y|\cdot|z|$.

where $\beta=\lambda_{\min}(X^{-1/2}G_{\mathbf{d}}X^{1/2}+X^{1/2}G_{\mathbf{d}}^{\top}X^{-1/2})$. Therefore, if

$$\|B\gamma(t,x)\| \leq \frac{\beta\kappa}{2}\|x\|_{\mathbf{d}}^{1+\mu} \tag{9.45}$$

then (9.41) holds.

For $\mu=-1$ we have $\|B\gamma(t,x)\| \leq 0.5\beta\kappa$, i.e., the discontinuous homogeneous controller (9.10) can reject bounded matched perturbations. We use this property in Sect. 9.2 in order to design generalized homogeneous sliding-mode controllers.

Example 9.3 Let us consider the system studied in Example 9.1 with the additive perturbation:

$$g(t,x)=\alpha\begin{pmatrix}0\\1\end{pmatrix}\sin(10t)\sqrt{\|x\|_{\mathbf{d}}}, \quad \alpha\in\mathbb{R}. \tag{9.46}$$

In the view of the estimate (9.45), the closed-loop system with the considered perturbation function g remains stable for α being sufficiently close to 0. The simulation results for $\alpha=0.5$ are depicted on Fig. 9.3.

ISS Analysis

In Chap. 7, we have proven that the robustness in the sense of the Input-to-State Stability (ISS) of a homogeneous system immediately follows from asymptotic stability of the unperturbed system provided that the perturbations are involved into the system in a homogeneous way. Let us study the ISS of the closed-loop system

$$\dot{x}=Ax+Bu(x+q_1)+q_2 \tag{9.47}$$

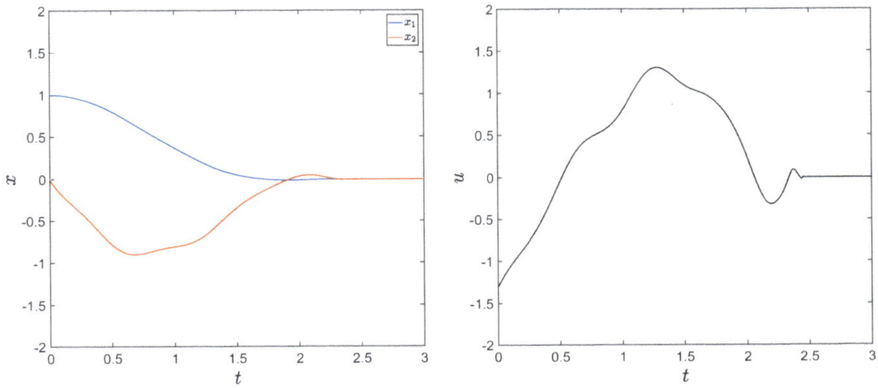

Fig. 9.3 Simulation results of the homogeneously stabilized harmonic oscillator with the perturbation (9.46)

with respect to the measurement noise $q_1 \in L^\infty(\mathbb{R};\mathbb{R}^n)$ and additive perturbation $q_2 \in L^\infty(\mathbb{R};\mathbb{R}^n)$.

Corollary 9.3 *Let u be defined as in Theorem 9.1 or as in Theorem 9.2. The system (9.47) is*

- *ISS with respect to $q = \left(q_1^\top, q_2^\top\right)^\top$ provided that $\mu \in (-1, (\tilde{n}-1)^{-1})$;*
- *ISS with respect to q_1 provided that $\mu = -1$, $q_2 = \mathbf{0}$ and $K_0 = \mathbf{0}$.*

Proof Notice that

$$Ax + Bu(x+q_1) + q_2 = A_0 x + BK_0 q_1 + Bu_{\mathrm{h}}(x_1 + q_1),$$

where

$$u_{\mathrm{h}}(z) = p^{1+\mu}(z) K \mathbf{d}(-\ln p(z)) z, \quad z \in \mathbb{R}^n$$

and $p(z) = \|z\|_{\mathbf{d}}$ as in Theorem 9.1 or $p \in C(\mathbb{R}^n;\mathbb{R}_+)$ as in Theorem 9.2. For $q_1, q_2, q_3 \in \mathbb{R}^n$ let us consider the vector field

$$f(x, \tilde{q}) = A_0 x + q_2 + BK_0 q_3 + Bu_{\mathrm{h}}(x + q_1)$$

and let a one-parameter group $\mathbf{d}_q$ of matrices $\mathbb{R}^{3n\times 3n}$ be defined as

$$\mathbf{d}_q(s) = \begin{pmatrix} \mathbf{d}(s) & \mathbf{0} & \mathbf{0} \\ \mathbf{0} & e^{\mu s}\mathbf{d}(s) & \mathbf{0} \\ \mathbf{0} & \mathbf{0} & e^{s(1+\mu)} I_n \end{pmatrix},$$

where the linear dilation $\mathbf{d}$ is defined in Theorem 9.1. Since $\mathbf{d}(s)B = e^s B$, $A_0\mathbf{d}(s) = e^{\mu s}\mathbf{d}(s)A_0$ and $\|\mathbf{d}(s)z\|_{\mathbf{d}} = e^s\|z\|_{\mathbf{d}}$, $\forall z \in \mathbb{R}^n$ then

$$\begin{aligned} f(\mathbf{d}(s)x, \mathbf{d}_q(s)\tilde{q}) =& A_0\mathbf{d}(s)x + e^{\mu s}\mathbf{d}(s)q_2 + e^{\mu s}\mathbf{d}(s)BK_0q_3 + Bu_{\mathrm{h}}(\mathbf{d}(s)(x+q_1)) \\ =& e^{\mu s}\mathbf{d}(s) f(x,\tilde{q}), \quad \forall s \in \mathbb{R}, \quad \forall x \in \mathbb{R}^n, \quad \forall \tilde{q} \in \mathbb{R}^{3n}. \end{aligned}$$

Taking into account that $\mathbf{d}_q$ is a linear dilation in $\mathbb{R}^n$ for $\mu > -1$, we complete the proof of this corollary using Theorem 7.13. The same conclusion remains valid if $\mu = -1$, $K_0 = \mathbf{0}$ and $q_2 = q_3 = \mathbf{0}$. In this case, the ISS can be proven with respect to q_1 by taking small enough $\varepsilon > 0$ in the proof of Theorem 7.13. The proof is complete. □

Example 9.4 By the above corollary, the system studied in Example 9.1 is ISS with respect to the measurement noise. The simulation results for the case of uniformly distributed random noise of the magnitude 0.01 are presented in Fig. 9.4. The sampling period is $h = 0.001$.

By Theorem 7.13, ISS of the closed-loop system can also be guaranteed in a more general case, when, for example, the additive perturbation is state dependent.

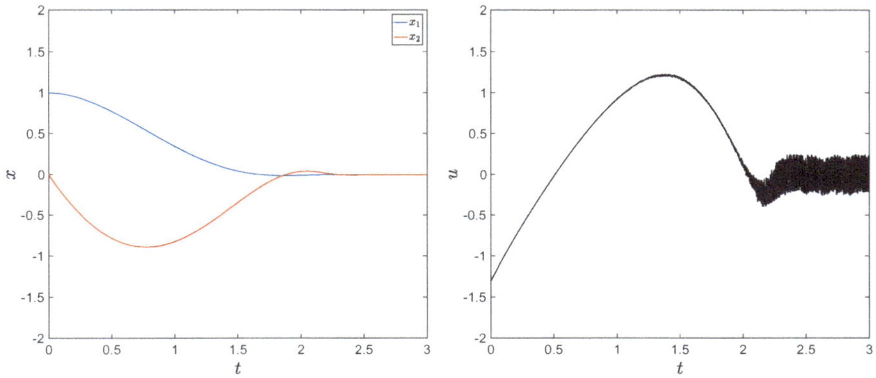

Fig. 9.4 Simulation results of the homogeneously stabilized harmonic oscillator with the measurement noise

Example 9.5 Let us consider the additive perturbation

$$g(t,x) = \begin{pmatrix} 0 \\ 1 \end{pmatrix} q_2(t)\|x\|_{\mathbf{d}}^{\frac{1}{12}}, \quad \alpha \in \mathbb{R} \tag{9.48}$$

for the system studied in Example 9.1. It can be shown that the inequality (9.45) is not fulfilled (close to zero) for any $q_2 \neq \mathbf{0}$ (despite that g vanishes as $x \to \mathbf{0}$). So, the closed-loop system (9.40) cannot be globally uniformly asymptotically stable but it is ISS in the view of Theorem 7.13 (with the dilation $\tilde{\mathbf{d}}(s) = e^{\frac{11}{12}s}$ for the variable q_2). The simulation results for $q_2(t) = 0.5\sin(10t)$ are given in Fig. 9.5.

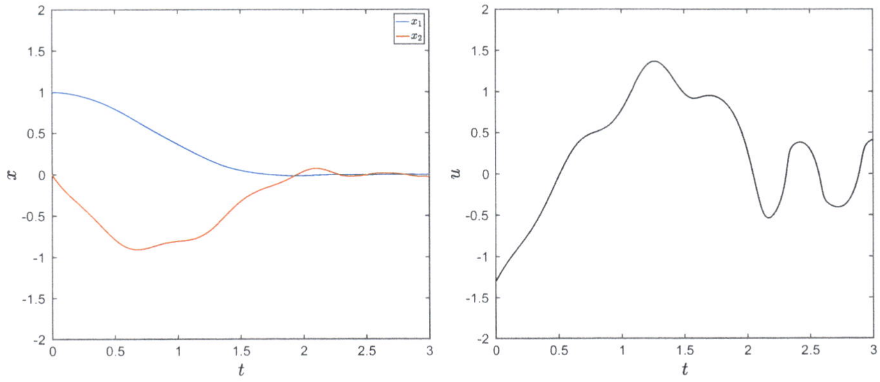

Fig. 9.5 Simulation results of the homogeneously stabilized harmonic oscillator with the perturbation (9.48)

9.1.4 Homogeneous Control for Cascade Systems

Let us consider the system (9.1) having the following structure:

$$A = \begin{pmatrix} A_1 & * & \dots & * & * \\ \mathbf{0} & A_2 & \dots & * & * \\ \dots & \dots & \dots & \dots & \dots \\ \mathbf{0} & \mathbf{0} & \dots\dots & A_{m-1} & * \\ \mathbf{0} & \mathbf{0} & \dots\dots & \mathbf{0} & A_m \end{pmatrix}, \quad B = \begin{pmatrix} B_1 & \mathbf{0} & \dots & \mathbf{0} & \mathbf{0} \\ \mathbf{0} & B_2 & \dots & \mathbf{0} & \mathbf{0} \\ \dots & \dots & \dots & \dots & \dots \\ \mathbf{0} & \mathbf{0} & \dots & B_{m-1} & \mathbf{0} \\ \mathbf{0} & \mathbf{0} & \dots & \mathbf{0} & B_m, \end{pmatrix}, \tag{9.49}$$

where $A_i \in \mathbb{R}^{n_i \times n_i}$, $B_i \in \mathbb{R}^{n_i \times m_i}$, $n_1 + n_2 + \dots + n_m = n$, $m_1 + m_2 + \dots m_m = \tilde{m}$ and $*$ is a (possibly) nonzero block of the matrix A. For each $i = 1, \dots, m$, the pair $\{A_i, B_i\}$ is assumed to be controllable. The model (9.1), (9.49) can be treated as a system of interconnected subsystems with state vectors $x_i \in \mathbb{R}^{n_i}$ and control inputs $u_i \in \mathbb{R}^{m_i}$. The homogeneous control design for such a cascade system was first developed in [20].

Corollary 9.4 (Cascade homogeneous control system) *Let* $G_{\mathbf{d}_i} \in \mathbb{R}^{n_i \times n_i}$, $K_i \in \mathbb{R}^{m_i \times n_i}$, $P_i \in \mathbb{R}^{n_i \times n_i}$ *and the control* $u_i = u_i(x_i)$ *with* $x_i \in \mathbb{R}^{n_i}$ *be defined by Theorem 9.1 for the pairs* $\{A_i, B_i\}$ *and some* $\mu_i \in [-1, (\tilde{n}_i - 1)^{-1})$, $\rho_i > 0$, *respectively. If*

$$-\mu_i < \beta_i = 0.5\lambda_{\min}(P_i^{1/2} G_{\mathbf{d}_i} P^{-1/2} + P_i^{-1/2} G_{\mathbf{d}_i}^\top P^{1/2}), \quad i = 1, 2, \dots, n$$

then the system (10.20) with the control $u = (u_1, \dots, u_m)$ *is globally uniformly asymptotically stable. Moreover, it is globally uniformly finite-time (nearly fixed time) stable if* $\max \mu_i \in [-1, 0)$ *(resp.,* $\min \mu_i \in (0, (\tilde{n}_i - 1)^{-1})$*).*

Proof In Theorem 9.1, the norm in $\mathbb{R}^{n_i}$ is defined as $\|z\| = \sqrt{z^\top P_i z}$, $z \in \mathbb{R}^{n_i}$ with $P_i \succ 0$ satisfying $P_i G_{\mathbf{d}_i} + G_{\mathbf{d}_i}^\top P_i \succ 0$. By definition of the canonical homogeneous norm, we have

$$x_i^\top \mathbf{d}_i^\top(-\|x_i\|_{\mathbf{d}_i}) P_i \mathbf{d}_i(-\|x_i\|_{\mathbf{d}_i}) x_i = 1,$$

or, equivalently, $\|\mathbf{d}_i(-\|x_i\|_{\mathbf{d}_i})x_i\| = 1$. By the formula (5.21), we derive

$$\tfrac{d}{dt}\|x_i\|_{\mathbf{d}_i} = -\rho_i \|x_i\|_{\mathbf{d}_i}^{1+\mu_i} + \|x_i\|_{\mathbf{d}_i} \frac{x_i^\top \mathbf{d}_i^\top(-\ln\|x_i\|_{\mathbf{d}_i}) P_i \mathbf{d}_i(-\ln\|x_i\|_{\mathbf{d}_i}) \sum_{j=i+1}^m A_{ij} x_j}{x_i^\top \mathbf{d}_i^\top(-\ln\|x_i\|_{\mathbf{d}_i}) P_i G_{\mathbf{d}_i} \mathbf{d}_i(-\ln\|x_i\|_{\mathbf{d}_i}) x_i},$$

where A_{ij} are over diagonal blocks of the matrix A and[5]

$$x_i^\top \mathbf{d}_i^\top(-\ln\|x_i\|_{\mathbf{d}_i}) P_i G_{\mathbf{d}_i} \mathbf{d}_i(-\ln\|x_i\|_{\mathbf{d}_i}) x_i \ge \beta_i > 0.$$

[5] The passage of text between the footnote marker and the end of the proof is reprinted from *Automatica*, 155/9, A. Polyakov, D. Efimov, Z. Ping, Consistent discretization of homogeneous finite/fixed-time controllers for LTI systems, 111118. https://doi.org/10.1016/j.automatica.2023.111118. ©2023, with permission from Elsevier.

Applying the Cauchy–Schwarz inequality, we obtain

$$\frac{x_i^\top \mathbf{d}_i^\top(-\ln\|x_i\|_{\mathbf{d}_i})P_i\mathbf{d}_i(-\ln\|x_i\|_{\mathbf{d}_i})\sum_{j=i+1}^m A_{ij}x_j}{x_i^\top \mathbf{d}_i^\top(-\ln\|x_i\|_{\mathbf{d}_i})P_iG_{\mathbf{d}_i}\mathbf{d}_i(-\ln\|x_i\|_{\mathbf{d}_i})x_i} \leq$$

$$\frac{1}{\beta_i}\left\|\mathbf{d}_i(-\ln\|x_i\|_{\mathbf{d}_i})\Sigma_{j=i+1}^m A_{ij}x_j\right\| \leq \frac{\left\|\Sigma_{j=i+1}^m A_{ij}x_j\right\|}{\beta_i\sigma_i(\|x_i\|_{\mathbf{d}_i})},$$

where $\sigma_i = \underline{\sigma}$ is given by Corollary 5.3 with $\alpha = \alpha_i = 0.5\lambda_{\max}(P_i^{1/2}G_{\mathbf{d}_i}P_i^{-1/2} + P_i^{-1/2}G_{\mathbf{d}_i}^\top P_i^{1/2})$ and $\beta = \beta_i$. Notice that, by Corollary 5.3, the function $x_i \mapsto V_i(x_i) = \sigma_i(\|x_i\|_{\mathbf{d}_i})$ is positive definite, radially unbounded and globally Lipschitz continuous with the Lipschitz constant 1. Then for $\|x_i\|_{\mathbf{d}_i} > 1$ we have

$$\frac{dV_i}{dt} = \beta_i\frac{V_i\cdot\frac{d}{dt}\|x_i\|_{\mathbf{d}}}{\|x_i\|_{\mathbf{d}}} \leq -\beta_i\rho_i V_i^{1+\frac{\mu_i}{\beta_i}} + \left\|\sum_{j=i+1}^m A_{ij}x_j\right\|.$$

For $0 < \|x_i\|_{\mathbf{d}} < 1$, we derive

$$\frac{dV_i}{dt} = \alpha_i\frac{V_i\cdot\frac{d}{dt}\|x_i\|_{\mathbf{d}}}{\|x_i\|_{\mathbf{d}}} \leq -\alpha_i\rho_i V_i^{1+\frac{\mu_i}{\alpha_i}} + \frac{\alpha_i}{\beta_i}\left\|\sum_{j=i+1}^m A_{ij}x_j\right\|.$$

Using Clarke's gradient (see Appendix D) for $\|x_i\|_{\mathbf{d}_i} = 1$, we derive

$$\begin{aligned}\frac{dV_i}{dt} \leq &-\lambda_i(x_i)\alpha_i\rho_i V_i^{1+\frac{\mu_i}{\alpha_i}} - (1-\lambda_i(x_i))\beta_i\rho_i V_i^{1+\frac{\mu_i}{\beta_i}}\\ &+\left(\lambda_i(x_i)\frac{\alpha_i}{\beta_i} + 1 - \lambda_i(x_i)\right)\left\|\Sigma_{j=i+1}^m A_{ij}x_j\right\|\end{aligned}$$

with some $\lambda(x_i) \in [0, 1]$. This means that $x_i \mapsto V_i(x_i)$ is an ISS Lyapunov function [22] of the i-th subsystem with respect to the input $\sum_{j=i+1}^m A_{ij}x_j$ provided that $\mu_i > -\beta_i$. Since the m-th subsystem is globally uniformly asymptotically stable then using the cascade structure and the ISS property of each subsystem, we conclude that the whole system is globally uniformly asymptotically stable as well (see, e.g., [23] for more details about ISS analysis of cascade systems). Moreover, the obtained estimate of $\frac{d}{dt}V_i$ implies that the finite-time, exponential and nearly fixed-time convergence rates are preserved dependently of the sign of μ_i for all $i = 1, \ldots, m$. The proof is complete. □

Example 9.6 Let us consider the system (9.1) with

$$A = \begin{pmatrix} A_1 & A_{12} \\ \mathbf{0} & A_2 \end{pmatrix},\ B = \begin{pmatrix} B_1 & \mathbf{0} \\ \mathbf{0} & B_2 \end{pmatrix},$$

where$A_1 = \begin{pmatrix} \mathbf{0} & I_2 \\ 0 & \mathbf{0} \end{pmatrix}$, $A_{12} = \begin{pmatrix} 1 & 0 \\ 0 & 0 \\ 0 & 0 \end{pmatrix}$, $A_2 = \begin{pmatrix} 0 & 1 \\ 0 & 0 \end{pmatrix}$, $B_1 = \begin{bmatrix} 0 \\ 0 \\ 1 \end{bmatrix}$, $B_2 = \begin{pmatrix} 0 \\ 1 \end{pmatrix}$.

For $\mu_1 = -0.5$, $\rho_1 = 1$, and the pair $\{A_1, B_1\}$ by Theorem 9.1, we derive a finite-time homogeneous control (9.10) with parameters

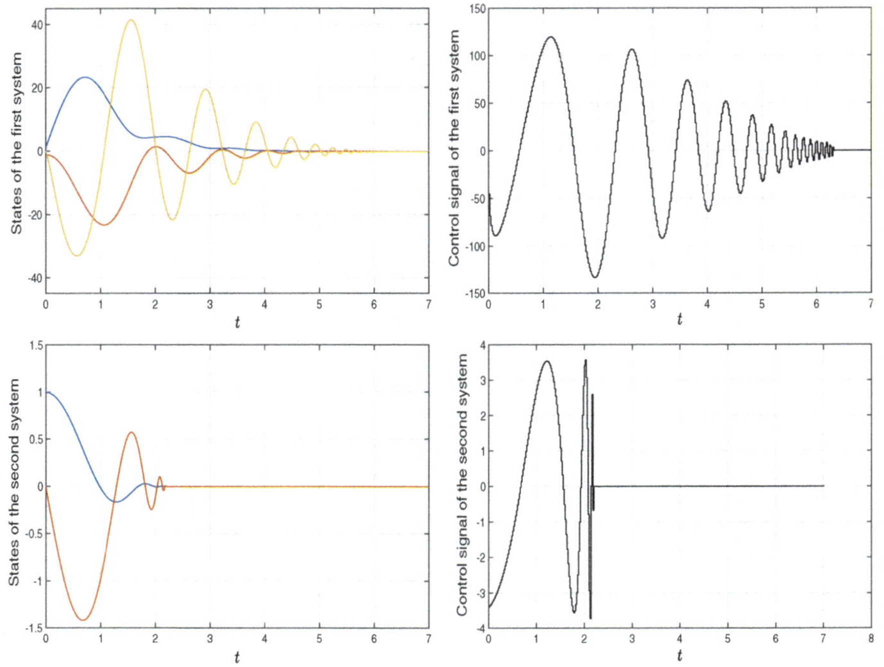

Fig. 9.6 The simulation results for the cascade system (9.1) with consistently discretized finite-time controllers with $n_1 = 3, n_2 = 2, \mu_1 = -0.5, \rho_1 = 1, \mu_2 = -1, \rho_2 = 0.5$

$$G_{\mathbf{d}_1} = \begin{pmatrix} 2 & 0 & 0 \\ 0 & 1.5 & 0 \\ 0 & 0 & 1 \end{pmatrix}, X_1 = \begin{pmatrix} 1.0000 & -2.0000 & 1.6667 \\ -2.0000 & 5.3333 & -8.0000 \\ 1.6667 & -8.0000 & 101.2500 \end{pmatrix}, K_1 = [\, -133.8750 \; -72.1875 \; -4.5000 \,].$$

For $\mu_2 = -1, \rho_2 = 0.5$ we derive $G_{\mathbf{d}_2} = \begin{pmatrix} 2 & 0 \\ 0 & 1 \end{pmatrix}$, $X_2 = \begin{pmatrix} 1.0000 & -1.0000 \\ -1.0000 & 3.9375 \end{pmatrix}$, $K = (\, -3.9375 \; -1.5000 \,)$. The numerical simulation of the cascade control system has been done using the method of consistent discretization (see Chap. 10) with the sampling period $h = 0.02$. The simulation results are depicted in Fig. 9.6.

9.1.5 Locally Homogeneous Fixed-Time Stabilizer

The following corollary of Theorem 9.1 introduces a simple scheme of the fixed-time stabilization for the linear plant. It combines finite-time and nearly fixed-time homogeneous stabilizers in order to design a globally fixed-time stable system, which is locally homogeneous of negative degree at 0-limit and locally homogeneous of positive degree at ∞-limit. The following theorem generalizes the results of the papers [10, 11] to the case of linear dilations.

Corollary 9.5 *Let $K_0 \in \mathbb{R}^{m\times n}$ and $G_0, A_0 \in \mathbb{R}^{n\times n}$ be defined as in Theorem 9.1 and*

$$G_{\mathbf{d}_1} = I_n + \mu_1 G_0, \qquad G_{\mathbf{d}_2} = I_n + \mu_2 G_0 \tag{9.50}$$

for some $\mu_1 \in [-1, 0)$ and some $\mu_2 \in (0, (\tilde{n}-1)^{-1})$. If $X \in \mathbb{R}^{n\times n}$ and $Y \in \mathbb{R}^{m\times n}$ satisfy the following system of linear matrix inequalities

$$\begin{cases} XA_0^\top + A_0X + Y^\top B^\top + BY + \rho_i(XG_{\mathbf{d}}^\top + G_{\mathbf{d}}X) \preceq \mathbf{0}, \\ XG_{\mathbf{d}_i}^\top + G_{\mathbf{d}_i}X \succ 0, \quad X \succ 0, \quad i = 1, 2 \end{cases} \tag{9.51}$$

for some $\rho_1 > 0$ and $\rho_2 > 0$, then the control

$$u = K_0x + YX^{-1}\begin{cases} \|x\|_{\mathbf{d}_1}^{1+\mu_1}\mathbf{d}_1(-\ln\|x\|_{\mathbf{d}_1})x & \text{if } \|x\| \le 1, \\ \|x\|_{\mathbf{d}_2}^{1+\mu_2}\mathbf{d}_2(-\ln\|x\|_{\mathbf{d}_2})x & \text{if } \|x\| > 1 \end{cases} \tag{9.52}$$

stabilizes the system (9.1), (9.2) to zero in the fixed time $T_{\max} > 0$ independently of the initial condition:

$$T(x_0) \le T_{\max} := \frac{1}{-\mu_1\rho_1} + \frac{1}{\mu_2\rho_2}, \quad \forall x_0 \in \mathbb{R}^n, \tag{9.53}$$

where $T : \mathbb{R}^n \mapsto \mathbb{R}$ is a settling-time function of the closed-loop system and the canonical homogeneous norms $\|\cdot\|_{\mathbf{d}_1}$ and $\|\cdot\|_{\mathbf{d}_2}$ are induced by the norm $\|x\| = \sqrt{x^\top X^{-1}x}$.

Moreover, the system (9.1), (9.54) is $\mathbf{d}_1$-homogeneous at 0-limit of the negative degree $\mu_1 \in [-1, 0)$ and $\mathbf{d}_2$-homogeneous at ∞-limit of the positive degree $\mu_2 \in (0, (\tilde{n}-1)^{-1})$.

Proof In the view of definition of the canonical homogeneous norm, we have $\|x\| \le 1 \Leftrightarrow \|x\|_{\mathbf{d}_i} \le 1$ for $i = 1, 2$, then, by construction, u is a continuous on the unit sphere $S = \{x : \|x\| = 1\}$. Moreover,

$$\frac{d}{dt}\|x\|_{\mathbf{d}_1} \le -\rho_1\|x\|_{\mathbf{d}_1}^{1+\mu_1}, \quad \forall x \neq 0 : \|x\| < 1,$$

$$\frac{d}{dt}\|x\|_{\mathbf{d}_2} \le -\rho_2\|x\|_{\mathbf{d}_2}^{1+\mu_2}, \quad \forall x : \|x\| > 1.$$

Taking $V = \max\{\|x\|_{\mathbf{d}_1}, \|x\|_{\mathbf{d}_2}\}$ we derive the locally Lipschitz continuous Lyapunov function satisfying

$$\dot{V}(x) \le \begin{cases} -\rho_1 V^{1+\mu_1} & \text{if } \|x\| < 1, \\ -\rho_2 V^{1+\mu_2} & \text{if } \|x\| > 1, \\ -\min\{\rho_1, \rho_2\} & \text{if } \|x\| = 1, \end{cases}$$

where $\dot{V}(x)$ is defined by the formula (4.19). By Corollary 4.10 the system is fixed-time stable. Moreover, using (9.16) we conclude that the state of the closed-loop system belongs to unit ball $\{x : \|x\| \leq 1\}$ for all $t > T_1 = \frac{1}{\mu_2 \rho}$. Being a level set of the Lyapunov function V, this unit ball is an invariant set of the closed-loop system. Hence, using the estimate (9.14) we derive $x(t) = \mathbf{0}$ for all $t \geq T_1 + \frac{\|x(T_1)\|_{\mathbf{d}_1}^{-\mu_1}}{-\mu_1 \rho_1}$. Therefore, the estimate (9.53) holds. By construction, the closed-loop system is locally homogeneous. The proof is complete. □

Let us study the feasibility of the LMI (9.51). Since $XG_{\mathbf{d}_i}^\top + G_{\mathbf{d}_i}X \to 2X$ as $\mu_i \to 0$ then, for $\mu = 0$, the LMI (9.51) becomes

$$XA_0^\top + A_0X + Y^\top B^\top + BY + 2\rho_i X \preceq \mathbf{0}, \quad X \succ 0.$$

The above LMI is feasible due to controllability of the pair $\{A_0, B\}$ (see, e.g., [17]). Hence, the LMI (9.51) is feasible at least for μ_i close to 0.

Remark 9.2 If, under conditions of Corollary 9.5, we take $\mu_2 = 0$ then the control (9.54) becomes a finite-time stabilizer

$$u = \begin{cases} K_0x + \|x\|_{\mathbf{d}_1}^{1+\mu_1} YX^{-1}\mathbf{d}_1(-\ln \|x\|_{\mathbf{d}_1})x & \text{if } \|x\| \leq 1, \\ (K_0 + YX^{-1})x & \text{if } \|x\| > 1, \end{cases} \tag{9.54}$$

of the system (9.1), (9.2) with the settling-time estimate

$$T(x_0) \leq \begin{cases} \frac{\|x_0\|_{\mathbf{d}_1}^{-\mu_1}}{-\mu_1 \rho_1} & \text{if } \|x_0\| \leq 1, \\ \frac{1}{-\mu_1 \rho_1} + \frac{\ln \|x_0\|}{\rho_2} & \text{if } \|x_0\| \geq 1. \end{cases}$$

Example 9.7 Let us consider the system (9.1) with

$$A = \begin{pmatrix} 0.7000 & 0.6500 & 0.3750 & 2.3250 & 0.7000 \\ 1.7000 & 1.7250 & 1.4750 & 1.2250 & 2.4500 \\ 2.2500 & 0.3250 & 0.1750 & 1.6250 & 0.1000 \\ 2.2750 & 0.3000 & 2.0500 & 2.2250 & 0.8250 \\ 1.8750 & 0.4750 & 1.8000 & 1.3500 & 2.4250 \end{pmatrix} \text{ and } B = \begin{pmatrix} 0.2500 & 0.8250 \\ 0.5000 & 2.2500 \\ 0.7500 & 1.2500 \\ 1.0000 & 1.5500 \\ 1.2500 & 1.4500 \end{pmatrix}.$$

We design the fixed-time stabilizing feedback in the form (9.54). Solving (9.5) we derive the gain of the homogenizing linear feedback

$$K_0 = \begin{pmatrix} 0.3594 & -0.8229 & 2.6280 & 6.1596 & -5.6135 \\ -1.6846 & 0.3058 & -2.9129 & -5.1833 & 2.9895 \end{pmatrix}$$

and for $\mu = -0.0575$ and $\mu_2 = 0.0455$ we have the generators of the dilations

$$G_{\mathbf{d}_1} = \begin{pmatrix} 1.0599 & 0.0131 & 0.0681 & -0.2045 & 0.1055 \\ -0.0054 & 1.1005 & 0.2126 & -0.6300 & 0.3373 \\ 0.0149 & 0.0021 & 1.0834 & -0.1105 & 0.0346 \\ 0.0043 & -0.0041 & -0.0099 & 1.0213 & -0.0103 \\ 0.0263 & -0.0163 & -0.0532 & 0.0853 & 0.9649 \end{pmatrix},$$

$$G_{\mathbf{d}_2} = \begin{pmatrix} 0.9526 & -0.0103 & -0.0539 & 0.1618 & -0.0835 \\ 0.0043 & 0.9205 & -0.1682 & 0.4984 & -0.2668 \\ -0.0118 & -0.0016 & 0.9340 & 0.0874 & -0.0273 \\ -0.0034 & 0.0033 & 0.0078 & 0.9831 & 0.0082 \\ -0.0208 & 0.0129 & 0.0421 & -0.0675 & 1.0278 \end{pmatrix}.$$

Solving (9.51) for $\rho_1 = \rho_2 = 0.6132$ we derive the gain of the homogeneous feedback

$$K = \begin{pmatrix} 1528.7718 & -3346.9388 & -50.2227 & 15541.4682 & -11206.0302 \\ -393.2500 & 850.4500 & 13.5300 & -3945.1840 & 2844.4810 \end{pmatrix}$$

and the shape matrix for the weighted Euclidean norm

$$X^{-1} = \begin{pmatrix} 4.8801 & -9.5997 & -0.2888 & 44.1114 & -31.7613 \\ -9.5997 & 20.4638 & 0.4640 & -94.8473 & 68.3154 \\ -0.2888 & 0.4640 & 0.2248 & -2.3103 & 1.5578 \\ 44.1114 & -94.8473 & -2.3103 & 440.3926 & -317.0759 \\ -31.7613 & 68.3154 & 1.5578 & -317.0759 & 228.3587 \end{pmatrix}.$$

A simulation for the closed-loop system with the fixed-time control has been done by means of the semi-implicit discretization (see Chap. 10) with sampling period $h = 0.01$. The results are depicted in Fig. 9.7.

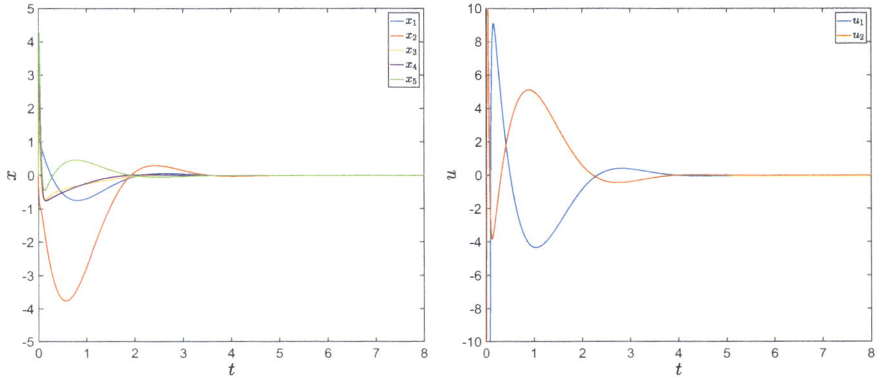

Fig. 9.7 Simulation results for the fixed-time controller (9.54)

Notice that the formula (9.53) is a rather conservative global estimate of the settling-time function. For example, it gives $T_{\max} = 64.2198$ for the above example. Below we design a homogeneous stabilizer allowing us to assign an exact constant settling time for the closed-loop system for all initial states $x_0 \in \mathbb{R}^n$.

9.1.6 Globally Homogeneous Fixed-Time Stabilizer

In this subsection, we consider the problem of fixed-time stabilization with a constant settling time $T > 0$. This means that the settling-time function of the closed-loop system must be equal to a prescribed constant $T > 0$ on whole $\mathbb{R}^n\backslash\{\mathbf{0}\}$. In Chap. 3, such a property was called *constant-time stability*.

Example 9.8 As a possible motivating example, let us consider a very particular rendezvous problem, when two objects (the object q and the object y) have to meet in a given point at a desired instant of time (see Fig. 9.8). We assume that the object y may assign a new meeting time or modify a meeting point, but there is no good communication channel between objects. So, the object q may receive the information about the meeting point/time at some unknown instances of time: t_1, t_2, etc. Moreover, it is prohibited for the object q to reach the meeting point earlier than the object y (otherwise the object y will definitely modify the meeting point).

Let $x(t) = q(t) - r_1$ for the meeting point r_1. Let us consider the simplest possible model of the regulation problem:

$$\dot{x} = u, \quad t > t_1, \quad x(t_1) \neq 0, \tag{9.55}$$

where $x(t) \in \mathbb{R}$ is the system state and $u(t) \in \mathbb{R}$ is a control input. The control aim is to regulate the state x to zero exactly at the prescribed time $t = t_1 + T$ avoiding

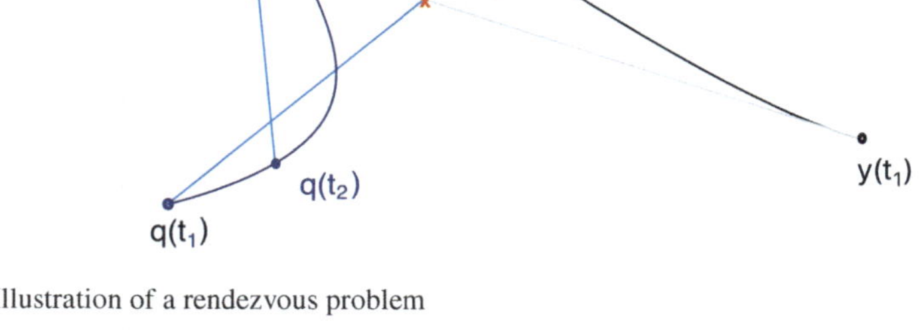

Fig. 9.8 Illustration of a rendezvous problem

crossing the zero in previous time instances. The following three solutions can be proposed for this problem.

1. The simplest prescribed-time regulator is designed in [24]:

$$u(t) = -\frac{x(t)}{T + t_1 - t}. \tag{9.56}$$

The solution of the closed loop is

$$x(t) = (1 - (t - t_1)/T)x(t_1), \quad t \in [t_1, t_1 + T), \tag{9.57}$$

i.e., $x(t) \neq 0$ for $t \in [t_1, t_1 + T)$ and $x(t) \to 0$ as $t \to t_1 + T$. However, this controller has the time-varying gain, which tends to infinity as $t \to t_1 + T$. This implies high sensitivity of this controller with respect to measurement noise $\hat{x}(t) = x(t) + \eta(t)$, since the noise $\eta \in L^\infty$ is amplified by the infinite gain $1/(t_1 + T - t)$ as $t \to t_1 + T$. We refer the reader to [25] and [26] for more details about robustness analysis of this time-varying controller.
2. Another possible solution to the considered regulation problem is the initial-state-dependent constant-time stabilizer [27]:

$$u(t) = -\frac{|x(t_1)|}{T} \operatorname{sign}(x(t)). \tag{9.58}$$

The closed-loop system has the same solution (9.57). The robustness properties of the controller with respect to measurement noises are improved, but the robustness of this controller with respect to additive perturbations is unsatisfactory. Indeed, since the control gain tends to zero as $|x(t_1)|$ tends to 0 than an additive perturbation of the magnitude greater than $|x(t_1)|/T$ can make system unstable.
3. To improve the robustness of the above constant-time stabilizer, it can be modified as follows [28]:

$$u(t) = -\frac{\max\{|x(t_1)|, |x(t)|\}}{T} \operatorname{sign}(x(t)). \tag{9.59}$$

The unperturbed closed-loop system has the same solution (9.57), but the robustness of the control system with respect to measurement noises and additive perturbations is essentially improved. Indeed, if, due to perturbations, we have $|x(t_1)| < |x(t)|$ then $u = -x(t)/T$ is a linear feedback. So, the controller is robust (in the ISS-like sense) with respect to additive perturbations and measurement noises.

Using linear homogeneity, below we generalize the proposed initial-state-dependent feedback design to the case of linear MIMO systems.

Inspired by the above example and the papers [27, 28], let us consider the system (9.1), (9.2) and modify the homogeneous feedback (9.10) as follows:

$$u(x,x_0):=\begin{cases} K_0x+T^{1+\mu}\left\|\frac{x}{\|x_0\|}\right\|_{\mathbf{d}}^{1+\mu}K\mathbf{d}(-\ln T)\mathbf{d}\left(-\ln\left\|\frac{x}{\|x_0\|}\right\|_{\mathbf{d}}\right)x & \text{if } \|x\|<\|x_0\|,\\ (K_0+T^{1+\mu}K\mathbf{d}(-\ln T))x & \text{otherwise.}\end{cases} \tag{9.60}$$

Since $\|x\|\geq\|x_0\| \Leftrightarrow \left\|\frac{x}{\|x_0\|}\right\|_{\mathbf{d}}\geq 1$ then, for $x\neq\mathbf{0}$, we have

$$u(x,x_0)\to u_{\mathrm{lin}}(x):=(K_0+T^{1+\mu}K\mathbf{d}(-\ln T))x \quad \text{as} \quad x_0\to\mathbf{0}. \tag{9.61}$$

Moreover, $u(x,x_0)=u_{\mathrm{lin}}(x)$ for $\|x\|\geq\|x_0\|$, so, the above limit is uniform on any compact from $\mathbb{R}^n\backslash\{\mathbf{0}\}$ and $u\in C(\mathbb{R}^n\backslash\{\mathbf{0}\}\times\mathbb{R}^n;\mathbb{R}^m)$ is locally bounded on $\mathbb{R}^n\times\mathbb{R}^n$. Notice that, for $n=m=1$, $K_0=0$, $K=1$, $\mu=-1$ and $\mathbf{d}(s)=e^s$, the control law (9.60) becomes the scalar feedback (9.59). The following theorem generalizes the main result of [28] to the case $\mu\in[-1,0)$.

Theorem 9.3 *Given $T>0$, let the parameters $G_{\mathbf{d}}$, K_0, K and X be defined as in Theorem 9.1 for $\mu\in[-1,0)$, $\rho=\frac{1}{-\mu T^{1+\mu}}$. and let $\|\cdot\|_{\mathbf{d}}$ be the canonical homogeneous norm induced by the norm:*

$$\|x\|=\sqrt{x^\top\mathbf{d}^\top(-\ln T)X^{-1}\mathbf{d}(-\ln T)x}. \tag{9.62}$$

Then

- *the closed-loop system* (9.1), (9.2), (9.60) *is constant-time stable with the settling-time constant $T>0$ and it has unique solutions for any $x_0\in\mathbb{R}^n$;*
- *all solutions of the perturbed closed-loop system*

$$\begin{array}{l}\dot{x}\overset{a.e.}{=}Ax+Bu(x+q_1,x_0+q_0)+q_2,\quad t>t_0,\quad x(t_0)=x_0,\\ q=(q_0,q_1,q_2)\in\mathbb{R}^n\times L^\infty(\mathbb{R};\mathbb{R}^k)\times L^\infty(\mathbb{R};\mathbb{R}^k)\end{array} \tag{9.63}$$

admit the ISS-like estimate

$$\frac{\|x(t)\|}{\|x_0+q_0\|}\leq\beta\left(\frac{\|x_0\|}{\|x_0+q_0\|},t-t_0\right)+\gamma_1\left(\frac{\|q_1\|_{L^\infty(t_0,t)}}{\|x_0+q_0\|}\right)+\gamma_2\left(\frac{\|q_2\|_{L^\infty(t_0,t)}}{\|x_0+q_0\|}\right) \tag{9.64}$$

for $x_0+q_0\neq\mathbf{0}$, where $\beta\in\mathcal{KL}$ and $\gamma_1,\gamma_2\in\mathcal{K}$ are independent of x_0 and q, but depend on $T>0$; moreover, there exists $C_i>0$ dependent of $T>0$ such that

$$\begin{array}{ll}\|x_0+q_0\|\gamma_i\left(\frac{\|q_i\|_{L^\infty(t_0,t)}}{\|x_0+q_0\|}\right)\to C_i\,\|q_i\|_{L^\infty(t_0,t)} & \text{as } \|x_0+q_0\|\to 0,\\ \|x_0+q_0\|\beta\left(\frac{\|x_0\|}{\|x_0+q_0\|},t-t_0\right)\to\|x_0\|e^{-\rho_0(t-t_0)} & \text{as } \|x_0+q_0\|\to 0,\end{array} \tag{9.65}$$

where $i=1,2$ and $\rho_0=\frac{1}{-2\mu T}\lambda_{\min}(X^{-1}G_{\mathbf{d}}+G_{\mathbf{d}}^\top X^{-1})$;

- *the system* (9.63), (9.60) *is fixed-time stable with the settling-time estimate* $T_{\max} = \frac{T}{1-\kappa}$ *provided that* $q_1 = \mathbf{0}$ *and*

$$\sup_{t\in\mathbb{R}} \left\| \mathbf{d}\left(-\ln\left\|\frac{x}{\|x_0\|}\right\|_{\mathbf{d}}\right)\frac{q_2(t)}{\|x_0\|}\right\| \leq \kappa\frac{\lambda_{\min}\left(X^{-\frac{1}{2}}G_{\mathbf{d}}X^{\frac{1}{2}}+X^{\frac{1}{2}}G_{\mathbf{d}}^{\top}X^{-\frac{1}{2}}\right)}{-2\mu T}\left\|\frac{x}{\|x_0\|}\right\|_{\mathbf{d}}^{\mu} \tag{9.66}$$

with some $\kappa \in (0, 1)$.

Proof (1) *Well-posedness.* Since $\|\cdot\|_{\mathbf{d}} \in C^{\infty}(\mathbb{R}^n\backslash\{\mathbf{0}\})$ then the control law (9.60) is locally Lipschitz continuous on $\mathbb{R}^n\backslash\{\mathbf{0}\} \times \mathbb{R}^n$. So, for any $x_0 \neq \mathbf{0}$ the closed-loop system has a unique classical solution as long as $0 < \|x(t)\| < +\infty$. For $\mu \in (-1, 0)$, the control is continuous at $x = \mathbf{0}$, so the (probably non-unique) classical solution exists as long as $\|x(t)\| < +\infty$. For $\mu = -1$, control law (9.60) is discontinuous at on the first argument at $x = \mathbf{0}$. According to Filippov's method (see Chap. 2 or [29]), to define a solution of the closed-loop system, the discontinuous feedback has to be regularized at $x = \mathbf{0}$. Since, by definition of the canonical homogeneous norm, we have

$$\left\|\mathbf{d}\left(-\ln\left\|\frac{x}{\|x_0\|}\right\|_{\mathbf{d}}\right)x\right\| = \|x_0\| \tag{9.67}$$

then the Filippov regularization of u at $x = \mathbf{0}$ is given by

$$K[u](\mathbf{0}, x_0) = \overline{\text{co}}\bigcap_{\varepsilon>0}\bigcap_{\mu(N)=0} u(B(\epsilon)\backslash N, x_0) = \|x_0\|K\mathbf{d}(-\ln T)\overline{B}(1),$$

where $\overline{co}\Omega$ is the convex closure of a set Ω (see Appendix A), $\overline{B}(r) = \{x \in \mathbb{R}^n : \|x\| \leq r\}$ is the closed ball of radius $r > 0$ and $\mu(N) = 0$ means that a set $N \subset \mathbb{R}^n$ has the measure zero in $\mathbb{R}^n$. The right-hand side of the regularized closed-loop system (9.1), (9.2), (9.60) becomes a differential inclusion with an upper semi-continuous right-hand side, which is single-valued at $x \neq \mathbf{0}$ and set valued, compact and convex valued at $x = \mathbf{0}$. By Theorem 2.53, the system has a Filippov solution for any $x_0 \neq \mathbf{0}$ as long as $\|x(t)\| < +\infty$. For $x_0 = \mathbf{0}$, the closed-loop system becomes linear

$$\dot{x} = (A + B(K_0 + K\mathbf{d}(-\ln T)))x, \quad t > 0, \quad x(0) = \mathbf{0},$$

so it has the unique zero solution too.

(2) *Constant-time stability.* For any constant $r > 0$, using the formula (5.21) we derive

$$\frac{d}{dt}\left\|\frac{x}{r}\right\|_{\mathbf{d}} = \left\|\frac{x}{r}\right\|_{\mathbf{d}} \frac{(x/r)^{\top}\mathbf{d}^{\top}(-\ln\|x/r\|_{\mathbf{d}})\mathbf{d}^{\top}(-\ln T)X^{-1}\mathbf{d}(-\ln T)\mathbf{d}(-\ln\|x/r\|_{\mathbf{d}})(\dot{x}/r)}{(x/r)^{\top}\mathbf{d}^{\top}(-\ln\|x/r\|_{\mathbf{d}})\mathbf{d}^{\top}(-\ln T)X^{-1}G_{\mathbf{d}}\mathbf{d}(-\ln T)\mathbf{d}(-\ln\|x/r\|_{\mathbf{d}})(x/r)}. \tag{9.68}$$

The identity (9.8) implies that

$$A_0\mathbf{d}(s) = e^{\mu s}\mathbf{d}(s)A_0 \quad \text{and} \quad \mathbf{d}(s)B = e^{s}B. \tag{9.69}$$

Then for $\|x\| \leq \|x_0\|$ we have

$$\mathbf{d}(-\ln T)\mathbf{d}\left(-\ln\left\|\tfrac{x}{x_0}\right\|_{\mathbf{d}}\right)\left(A_0 x + T^{1+\mu}\left\|\tfrac{x}{x_0}\right\|_{\mathbf{d}}^{1+\mu} BK\mathbf{d}\left(-\ln\|\tfrac{x}{x_0}\|_{\mathbf{d}}\right)x\right)$$

$$= T^{\mu}\left\|\tfrac{x}{x_0}\right\|_{\mathbf{d}}^{\mu}(A_0 x + BK)\,\mathbf{d}(-\ln T)\mathbf{d}\left(-\ln\|\tfrac{x}{x_0}\|_{\mathbf{d}}\right).$$

For any solution of the closed-loop system (9.1), (9.2), (9.60) with $x_0 \neq \mathbf{0}$ the following identity

$$\frac{d\left\|\frac{x}{\|x_0\|}\right\|_{\mathbf{d}}}{dt} = \frac{\left\|\frac{x}{x_0}\right\|_{\mathbf{d}}^{1+\mu} x^\top \mathbf{d}^\top\left(-\ln\left\|\frac{x}{x_0}\right\|_{\mathbf{d}}\right)\mathbf{d}^\top(-\ln T)X^{-1}(A_0+BK)\mathbf{d}(-\ln T)\mathbf{d}\left(-\ln\left\|\frac{x}{x_0}\right\|_{\mathbf{d}}\right)x}{T^{-\mu}x^\top\mathbf{d}^\top\left(-\ln\left\|\frac{x}{x_0}\right\|_{\mathbf{d}}\right)\mathbf{d}^\top(-\ln T)X^{-1}G_{\mathbf{d}}\mathbf{d}(-\ln T)\left(-\ln\left\|\frac{x}{x_0}\right\|_{\mathbf{d}}\right)x} \tag{9.70}$$

holds. Using (9.9), we derive

$$\tfrac{d}{dt}\left\|\tfrac{x}{\|x_0\|}\right\|_{\mathbf{d}} = -\frac{\rho}{T^{-\mu}}\left\|\tfrac{x}{\|x_0\|}\right\|_{\mathbf{d}}^{1+\mu} \quad \text{for} \quad 0 \neq \|x\| \leq \|x_0\|. \tag{9.71}$$

For $t = 0$ and $x(0) = x_0 \neq \mathbf{0}$, we have $\|x(0)/\|x_0\|\|_{\mathbf{d}} = 1$. Since the derivative of the function $t \mapsto \|x(t)/\|x_0\|\|_{\mathbf{d}}$ is negative at $t = 0$ then $\|x(t)/\|x_0\|\|_{\mathbf{d}} \leq 1$ or, equivalently, $\|x(t)\| \leq \|x_0\|$ for all $t \geq 0$. So, the closed-loop system is Lyapunov stable. Moreover, $x(t) = \mathbf{0}$ for $t \geq \frac{T^{-\mu}}{-\mu\rho} = T$ and $x(t) \neq \mathbf{0}$ for all $t \in [0, T)$ if $x_0 \neq \mathbf{0}$, i.e., the system is constant-time stable with the settling-time constant $T > 0$.

(3) *Robustness analysis.* Making the change of variables $z = x/\|\tilde{x}_0\|$ with $\tilde{x}_0 = x_0 + q_0 \neq \mathbf{0}$ we derive

$$\begin{cases} \dot{z} \overset{a.e.}{=} A_0 z + r^{1+\mu} BK_T \mathbf{d}(-\ln r)\left(z + q_1^{x_0}\right) + q_2^{x_0}, \\ z(0) = z_0, \end{cases} \quad r := \min\{\left\|z + q_1^{x_0}\right\|_{\mathbf{d}}, 1\}, \tag{9.72}$$

where $q_{1,x_0} = \frac{q_1}{\|\tilde{x}_0\|}$, $q_{2,x_0} = \frac{q_2}{\|\tilde{x}_0\|} + BK_0 q_1^{x_0}$, $K_T = T^{1+\mu}K\mathbf{d}(-\ln T)$ and $z_0 = \frac{x_0}{\|\tilde{x}_0\|}$. The system (9.72) with $q_{1,x_0} = q_{2,x_0} = \mathbf{0}$ is $\mathbf{d}$-homogeneous in the bi-limit with zero degree in ∞-limit and negative degree in 0-limit. The unperturbed system (9.72) as well as its homogeneous approximations are globally asymptotically stable. By Corollary 7.16, this implies ISS with respect to $q_1^{x_0}$ and $q_2^{x_0}$. Hence, we derive the estimate (9.64). By construction, the stabilizer $u(x + q_1, x_0 + q_0)$ converges (uniformly on compacts from $\mathbb{R}^n$) to the linear feedback $u_{\text{lin}}(x + q_1) = (K_0 + K_T)(x + q_1)$ as $x_0 + q_0 \to \mathbf{0}$. Moreover, for $\|x + q_1\| \geq \|x_0 + q_0\|$ we have $u(x + q_1, x_0 + q_0) = u_{\text{lin}}(x + q_1)$ and

$$\frac{d\|x\|}{dt} \overset{a.e.}{=} \frac{x^\top\mathbf{d}^\top(-\ln T)X^{-1}\mathbf{d}(-\ln T)(Ax+q_2+B(K_0+K_T)(x+q_1)}{\|x\|}$$

$$= \frac{x^\top\mathbf{d}^\top(-\ln T)X^{-1}(A_0+BK)\mathbf{d}(-\ln T)x}{T^{-\mu}\|x\|}$$

$$+ \frac{x^\top\mathbf{d}^\top(-\ln T)X^{-1}\mathbf{d}(-\ln T)q_2 + x^\top\mathbf{d}^\top(-\ln T)X^{-1}\mathbf{d}(-\ln T)B(K_0+K_T)q_1}{\|x\|}$$

where the homogeneity identities

$$\mathbf{d}(-\ln T)(A+BK_0)=e^{\mu\ln T}(A+BK_0)\mathbf{d}(-\ln T),\quad \mathbf{d}(-\ln T)B=e^{-\ln T}B$$

have been utilized on the last step. Using the identity (9.9) and the Cauchy–Schwarz inequality, we derive

$$\tfrac{d\|x\|}{dt}\overset{a.e.}{\leq}-\rho_0\|x\|+\|B(K_0+K_T)q_1\|+\|q_2\| \quad \text{if} \quad \|x+q_1\|\geq\|x_0+q_0\|.$$

Therefore, the property (9.65) holds.

Finally, assuming $q_1=\mathbf{0}$ we have $q_1^{x_0}=\mathbf{0}$ and $q_2^{x_0}=\frac{q_2}{\|x_0\|}$. In this case, repeating the above considerations for $0<\|z\|\leq 1$ we derive

$$\tfrac{d\|z\|_{\mathbf{d}}}{dt}\overset{a.e.}{=}-\tfrac{\rho\|z\|_{\mathbf{d}}^{1+\mu}}{T^{-\mu}}+\tfrac{\|z\|_{\mathbf{d}}z^{\top}\mathbf{d}^{\top}(-\ln\|z\|_{\mathbf{d}})\mathbf{d}^{\top}(-\ln T)X^{-1}\mathbf{d}(-\ln T)\mathbf{d}(-\ln\|z\|_{\mathbf{d}})\frac{q_2}{\|x_0\|}}{z^{\top}\mathbf{d}^{\top}(-\ln\|z\|_{\mathbf{d}})\mathbf{d}^{\top}(-\ln T)X^{-1}G_{\mathbf{d}}\mathbf{d}(-\ln T)\mathbf{d}(-\ln\|z\|_{\mathbf{d}})z}. \tag{9.73}$$

Since $\|\mathbf{d}(-\ln\|z\|_{\mathbf{d}})z\|=1$ and, equivalently,

$$z^{\top}\mathbf{d}^{\top}(-\ln\|z\|_{\mathbf{d}})\mathbf{d}^{\top}(-\ln T)X^{-1}\mathbf{d}(-\ln T)\mathbf{d}(-\ln\|z\|_{\mathbf{d}})z=1 \tag{9.74}$$

then using the Cauchy–Schwarz inequality we derive

$$z^{\top}\mathbf{d}^{\top}(-\ln\|z\|_{\mathbf{d}})\mathbf{d}^{\top}(-\ln T)X^{-1}\mathbf{d}(-\ln T)\mathbf{d}(-\ln\|z\|_{\mathbf{d}})q_2\leq\|\mathbf{d}(-\ln\|z\|_{\mathbf{d}})q_2\|.$$

Hence, we have

$$\tfrac{d\|z\|_{\mathbf{d}}}{dt}\overset{a.e.}{\leq}-\tfrac{\rho}{T^{-\mu}}\|z\|_{\mathbf{d}}^{1+\mu}+\frac{2\left\|\mathbf{d}(-\ln\|z\|_{\mathbf{d}})\frac{q_2(t)}{\|x_0\|}\right\|}{\lambda_{\min}\left(X^{-\frac{1}{2}}G_{\mathbf{d}}X^{\frac{1}{2}}+X^{\frac{1}{2}}G_{\mathbf{d}}^{\top}X^{-\frac{1}{2}}\right)}\leq-(1-\kappa)\tfrac{\rho}{T^{-\mu}}\|z\|_{\mathbf{d}}^{1+\mu}. \tag{9.75}$$

Taking into account $\|z(0)\|_{\mathbf{d}}=\left\|\frac{x(0)}{\|x_0\|}\right\|_{\mathbf{d}}=1$ we conclude $\|z(t)\|\leq 1$ for all $t\geq 0$ and $z(t)=\mathbf{0}$ for $t\geq T_{\max}$. The proof is complete. □

Remark 9.3 If the control parameters K and X are defined as in Remark 9.1 then the feedback (9.60) is a fixed-time stabilizer with an upper estimate of the settling time prescribed by the constant $T>0$.

If $q_2=B\gamma$ then, for $\mu=-1$, the condition (9.66) becomes

$$\|B\gamma\|\leq\kappa\frac{\|x_0\|\lambda_{\min}\left(X^{-\frac{1}{2}}G_{\mathbf{d}}X^{\frac{1}{2}}+X^{\frac{1}{2}}G_{\mathbf{d}}^{\top}X^{-\frac{1}{2}}\right)}{-2\mu T}, \tag{9.76}$$

i.e., the control law (9.60) with $\mu=-1$ rejects bounded matched perturbations of a certain magnitude proportional to $\|x_0\|$.

Example 9.9 Let us consider again the controlled harmonic oscillator studied in Example 9.1 and design the constant-time stabilizer (9.60) with $\mu = -1$. The parameters of the stabilizer

$$K_0 = (1\ \ 0), \quad G_{\mathbf{d}} = \begin{pmatrix} 2 & 0 \\ 0 & 1 \end{pmatrix}, \quad K = (-5.5\ \ -3), \quad X = \begin{pmatrix} 1 & -2 \\ -2 & 5.5 \end{pmatrix} \tag{9.77}$$

are defined according to Theorem 9.3.

The simulation has been done in MATLAB® using the zero-order-hold method and the consistent discretization of the homogeneous controller (9.60) (see Chap. 10). The sampling period is $h = 0.025$. The simulation results illustrate the constant-time stability of the closed-loop system with the settling time $T = 1.5$. Indeed, independently of the selected initial condition (see Figs. 9.9 and 9.10) the state of

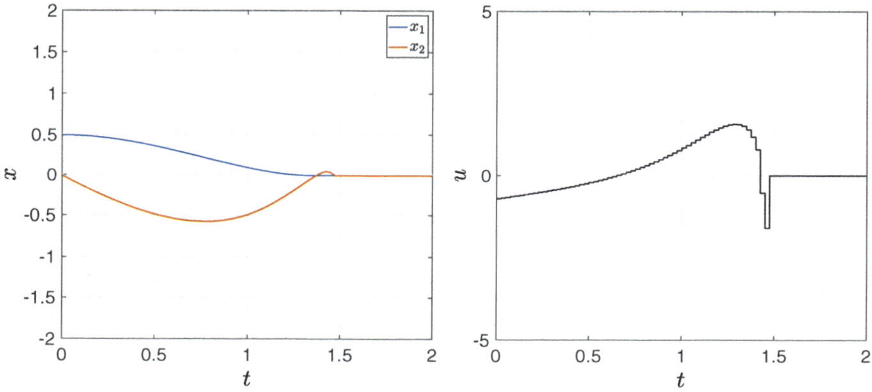

Fig. 9.9 Stabilization at the prescribed time $T = 1.5$ for $x_0 = (0.5\ \ 0)^\top$

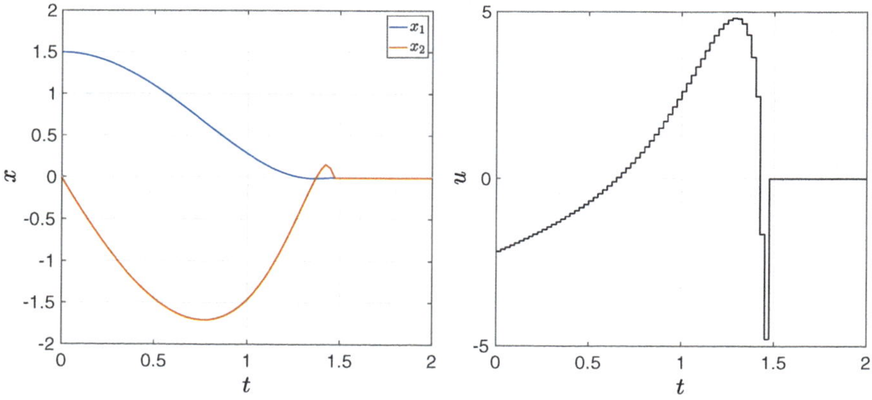

Fig. 9.10 Stabilization at the prescribed time $T = 1.5$ for $x_0 = (1.5\ \ 0)^\top$

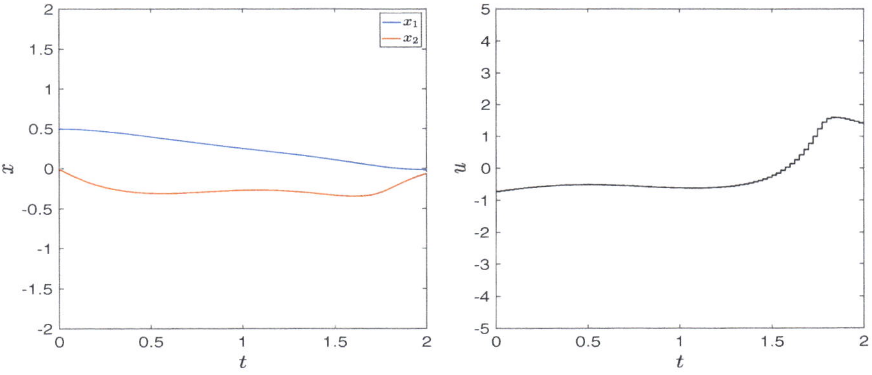

Fig. 9.11 The case of matched additive disturbance $B\sin(2t)$ for $x_0 = (0.5\ \ 0)^\top$

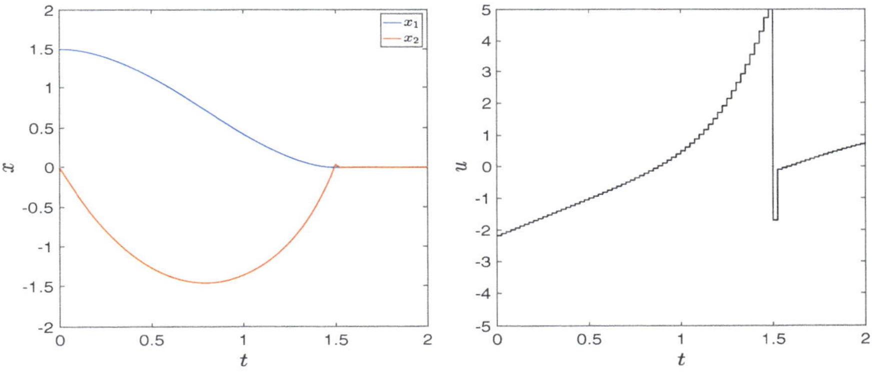

Fig. 9.12 The case of matched additive disturbance $B\sin(2t)$ for $x_0 = (1.5\ \ 0)^\top$

the closed-loop system converges to zero with the precision of the machine epsilon ($\approx 10^{-16}$) at the time instant 1.475, which perfectly corresponds to the prescribed settling time $T = 1.5$ (up to the sampling period).

By the formula (9.66), the considered constant-time stabilizer should reject some bounded matched perturbations if $\mu = -1$. However, this formula highlights that, to guarantee a fixed-time stability, the magnitude of the perturbation must depend on x_0. To study robustness properties of the closed-loop system, the simulations have been done, first, for the system with matched additive perturbation $B\sin(2t)$. In the case of constant-time stabilizer, a perturbation cannot be rejected by the prescribed-time controller (9.60) if the initial state is too small (see Fig. 9.11). The larger the initial condition, the larger matched perturbation can be rejected (see Fig. 9.12).

The ISS with respect to noisy measurements is quite opposite to the case of ISS with respect to additive perturbations in the sense that the smaller the initial deviation from the equilibrium, the less sensitive is the closed-loop system with

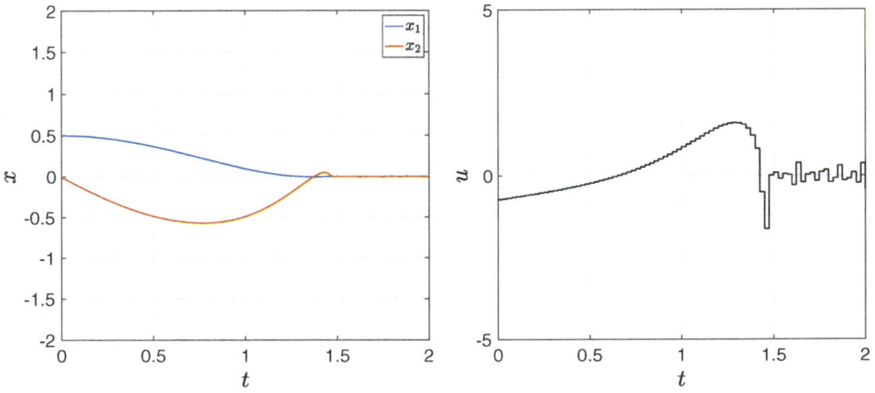

Fig. 9.13 The case of noisy measurements for $x_0 = (0.5\ \ 0)^\top$

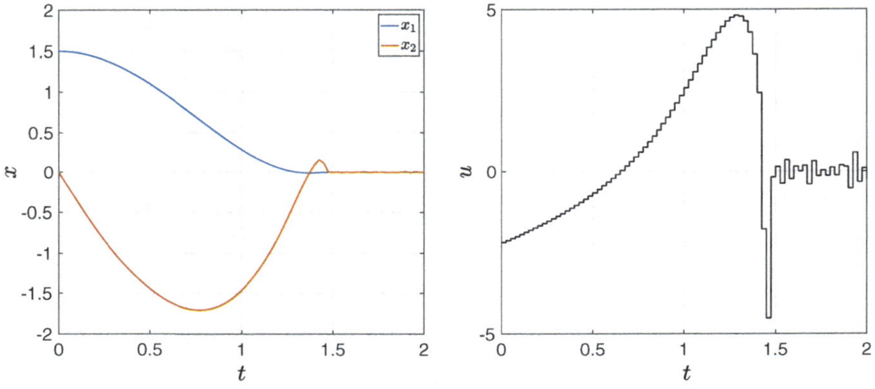

Fig. 9.14 The case of noisy measurements for $x_0 = (1.5\ \ 0)^\top$

respect to measurement noise (see Figs. 9.13 and 9.14). The numerical simulations for this case have been done by adding a noise η of the magnitude 0.001 to the state measurements $\hat{x} = x + \eta$. The noise is simulated by MATLAB® as a uniformly distributed (pseudo-)random variable $\eta \in [-0.001, 0.001]$.

9.1.7 Homogeneous Control with Integral Action

The linear control theory uses an integral feedback in order to improve the robustness properties of a proportional law with respect to a constant or "slowly varying" perturbation. A homogeneous integral action can be added to the implicit homogeneous feedback by the same reason. The following theorem generalizes the result of [30] to the case of linear dilations in $\mathbb{R}^n$.

Theorem 9.4 *Let* $K_0, K \in \mathbb{R}^{m\times n}$, $A_0, G_{\mathbf{d}}, X, \in \mathbb{R}^{n\times n}$, $X \in \mathbb{R}^{n\times n}$, $\tilde{n} \in \mathbb{N}$ *be defined as in Theorem 9.1 for with* $\mu \in [-0.5, (\tilde{n}-1)^{-1})$.

Then for any positive definite matrix $Q \in \mathbb{R}^{m\times m}$ *and any constant vector* $p \in \mathbb{R}^m$ *the control law*

$$u = K_0 x + u_{\text{hom}}(x) + \int_0^t u_{\text{int}}(x(s))ds, \tag{9.78}$$

$$u_{\text{hom}}(x) = \|x\|_{\mathbf{d}}^{1+\mu} Y X^{-1}\mathbf{d}(-\ln \|x\|_{\mathbf{d}})x,$$

$$u_{\text{int}}(x) = -\|x\|_{\mathbf{d}}^{1+2\mu} \frac{QB^\top P\mathbf{d}(-\ln \|x\|_{\mathbf{d}})x}{x^\top \mathbf{d}^\top(-\ln \|x\|_{\mathbf{d}})PG_{\mathbf{d}}\mathbf{d}(-\ln \|x\|_{\mathbf{d}})x}$$

stabilizes the origin of the system

$$\dot{x} = Ax + B(u+p) \tag{9.79}$$

in a finite time if $\mu \in [-0.5, 0)$, *exponentially if* $\mu = 0$, *and in a nearly fixed time if* $\mu \in (0, (\tilde{n}-1)^{-1})$, *where, as before, the canonical homogeneous norm* $\|\cdot\|_{\mathbf{d}}$ *is induced by the norm* $\|x\| = \sqrt{x^\top X^{-1} x}$.

Proof Let us introduce the following virtual variable:

$$x_{n+1} = p + \int_0^t u_{\text{int}}(x(s))ds.$$

In this case, the closed-loop system becomes

$$\dot{x} = Ax + B(K_0 x + u_{\text{hom}}(x) + x_{n+1}), \quad \dot{x}_{n+1} = u_{\text{int}}(x). \tag{9.80}$$

Since $\|\mathbf{d}(-\ln \|x\|_{\mathbf{d}})x\| = 1$ then u_{int} is globally bounded and discontinuous at $x = \mathbf{0}$ if $\mu = -0.5$. In all other cases, the considered system has the continuous right-hand side.

Let us show that the obtained system is globally asymptotically stable. Inspired by [30, 31] let us consider the following Lyapunov function candidate:

$$V = \frac{1}{2+2\mu}\|x\|_{\mathbf{d}}^{2+2\mu} + \frac{1}{2}x_{n+1}^\top Q^{-1} x_{n+1}.$$

Calculating the time derivative of V along the trajectories of the closed-loop system we derive

$$\dot{V} = \|x\|_{\mathbf{d}}^{1+2\mu}\frac{\partial \|x\|_{\mathbf{d}}}{\partial x}\dot{x} + x_{n+1}^{\top}Q^{-1}\dot{x}_{n+1}$$
$$= -\|x\|_{\mathbf{d}}^{2+3\mu} + \|x\|_{\mathbf{d}}^{1+2\mu}\frac{\partial \|x\|_{\mathbf{d}}}{\partial x}Bx_{n+1} - \frac{\|x\|^{1+2\mu}x_{n+1}^{\top}B^{\top}P\mathbf{d}(-\ln\|x\|_{\mathbf{d}})x}{x^{\top}\mathbf{d}^{\top}(-\ln\|x\|_{\mathbf{d}})PG_{\mathbf{d}}\mathbf{d}(-\ln\|x\|_{\mathbf{d}})x}$$
$$= -\|x\|_{\mathbf{d}}^{2+3\mu},$$

where the formula (5.21) and the identity $e^{sG_{\mathbf{d}}}B = e^{s}B$ are utilized on the last step.

The origin $(x, x_{n+1}) = \mathbf{0}$ is the unique equilibrium of the system (9.80) and the hyperplane $\{(x, x_{n+1}) \in \mathbb{R}^{n+m} : x = \mathbf{0}\}$ does not contain nonzero trajectories of this system. Then its origin is globally asymptotically stable (in the view of LaSalle invariance principle, see Chap. 4).

Finally, the system (9.80) is $\tilde{\mathbf{d}}$-homogeneous of the degree μ with respect to the dilation:

$$\tilde{\mathbf{d}}(s) = \begin{pmatrix} e^{sG_{\mathbf{d}}} & \mathbf{0} \\ \mathbf{0} & e^{s(1+\mu)}I_m \end{pmatrix}.$$

We complete the proof using Theorem 7.5. □

Remark 9.4 Since the scalar-valued function

$$x \mapsto x^{\top}\mathbf{d}^{\top}(-\ln\|x\|_{\mathbf{d}})PG_{\mathbf{d}}\mathbf{d}(-\ln\|x\|_{\mathbf{d}})x$$

is $\mathbf{d}$-homogeneous of the degree 0 and uniformly bounded from above and from below then its replacement in the definition of u_{int} with a constant does not destroy the homogeneity properties of the system. Therefore, for practical reasons, the simplified integral term

$$u_{int}(x) = -\|x\|_{\mathbf{d}}^{1+2\mu}QB^{\top}P\mathbf{d}(-\ln\|x\|_{\mathbf{d}})x$$

can be utilized in (9.78) provided that the stability is preserved. An LMI-based sufficient stability condition for the homogeneous control with the simplified integral term u_{int} is given in [32].

Example 9.10 Let us consider the harmonic oscillator studied in Example 9.1 with the constant additive perturbation Bp. Let the homogeneous feedback be designed as in Example 9.1 with $\mu = -0.5$, but the integral action is defined as in Theorem 9.4 with $Q = 1$. In this case, the closed-loop system (9.79), (9.78) is globally finite-time stable for any $p \neq 0$. For numerical simulation, we consider the constant perturbation $p = 1$ and use the explicit discretization of the control with the step size $h = 0.001$. The designed homogeneous control with integral action rejects the bounded unknown disturbance p (see Fig. 9.15) and stabilizes the state of the system to zero in a finite time.

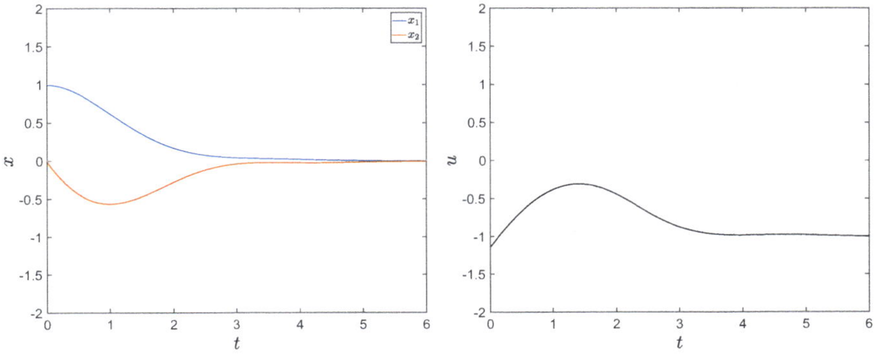

Fig. 9.15 Simulation results of the harmonic oscillator stabilized by a homogeneous control with integral action

9.2 Homogeneous Sliding-Mode Control

Sliding-mode control (SMC) is a nonlinear method that forces the system to slide along a manifold which is selected such that the system in such a sliding mode has a desired dynamics (see [33] and references therein). Traditionally, to enforce the sliding motion on a manifold, a discontinuous function, which can switch the system dynamics based on the current position in the state space, is utilized as SMC (see [4, 34]). The modern SMC theory [35, 36] admits any discontinuous feedback law, which enforces the sliding motion on the desired manifold in a finite time. The weighted homogeneity is rather popular for SMC design [37, 38], since it simplifies analysis and design of the so-called High-Order Sliding-Mode (HOSM) algorithms [35]. In this section, we show that the linear homogeneity can also be utilized for the SMC design especially in the case of multi-input system.

9.2.1 Sliding-Mode Control Principles

In the previous section, the plant is assumed to be modeled by a linear time-invariant (LTI) system without uncertainties. The SMC is a rather famous control methodology, which is aimed at rejection of disturbances and uncertainties in the system dynamics. However, some a priori information about a class of perturbations to be rejected is required for the SMC design. Below we consider the system, which have two kinds of perturbations. The so-called matched perturbations are assumed to be supported with some a priori information (e.g., about maximal magnitude), which can be utilized for the control design. The SMC must reject the matched perturbations satisfying certain assumption. The unstructured (possibly mismatched) perturbations are assumed to be completely unknown. A class of such perturbations rejected by the SMC is going to be characterized a posteriori and depends on the designed SMC algorithm.

Let us consider the control system modeled by the following ODE:

$$\dot{x} = Ax + B(u + \gamma(t, x)) + g(t, x), \quad t > t_0, \quad x(t_0) = x_0, \tag{9.81}$$

where $x(t) \in \mathbb{R}^n$ is the system state, $u(t) \in \mathbb{R}^m$ is the control input, the matrices $A \in \mathbb{R}^{n\times n}$ and $B \in \mathbb{R}^{n\times m}$ are known matrices, $\gamma : \mathbb{R} \times \mathbb{R}^n \mapsto \mathbb{R}^m$ is an unknown locally integrable function, which describes the matched system uncertainties/disturbances satisfying certain assumption. A locally bounded measurable function $g : \mathbb{R} \times \mathbb{R}^n \mapsto \mathbb{R}^k$ models unstructured perturbations, which are not supported with any a priori information utilizable for the design. Sliding-mode control for the LTI system follows the rather simple principles [4]. First, a linear surface

$$Cx = \mathbf{0}, \quad C \in \mathbb{R}^{p\times n}, \quad \operatorname{rank} C = p \tag{9.82}$$

in the state space $\mathbb{R}^n$ has to be selected such that the system on this surface would have a desired (e.g., stable) dynamics. Next, SMC, that steers the system in a finite time to the selected surface, has to be designed. The so-called *sliding variable*

$$\sigma = Cx \tag{9.83}$$

is governed by the differential equation:

$$\dot{\sigma} = CAx + CB(u + \gamma) + Cg. \tag{9.84}$$

Traditionally [4], it is assumed that an estimate of the matched perturbation γ is available for the SMC design.

Assumption 9.1 There exists a known constant $\overline{\gamma} \geq 0$:

$$|\gamma(t, x)| \leq \overline{\gamma}, \quad \forall t \in \mathbb{R}, \quad \forall x \in \mathbb{R}^n. \tag{9.85}$$

The *nominal system* for SMC design corresponds to the case $g = \mathbf{0}$ in (9.81) and in (9.84), respectively. In terms of stability theory, the SMC design problem can be formulated as follows: *given* $\overline{\gamma} > 0$, *we need to stabilize the nominal system* (9.84) *to zero in a finite-time uniformly on* γ *satisfying* (9.85).

The SMC design methodology [4, 33, 34] is developed under the conventional assumption:

$\mathcal{H}$: $p = m$ and the matrix CB is invertible.

One of popular SMC algorithms (see [4, 34, 39]) is the unit control :

$$u = \hat{K}x + K\frac{Cx}{|Cx|}, \tag{9.86}$$

where the matrices $\hat{K} \in \mathbb{R}^{m\times n}$ and $K \in \mathbb{R}^{m\times m}$ are selected as follows

$$\hat{K} = -(CB)^{-1}CA \quad \text{and} \quad K = -(\rho + |CB|\overline{\gamma})(CB)^{-1} \quad \text{with} \quad \rho > 0$$

in order to guarantee (see below) that any trajectory of the nominal closed-loop system (9.84) reaches the surface $Cx = \mathbf{0}$ in a finite time and slides on it after then. In this case, the class of the unstructured perturbations g, which can be rejected using the unit control, is characterized by the inequality:

$$\sup_{(t,x)\in\mathbb{R}\times\mathbb{R}^n} |Cg(t,x)| < \rho.$$

The assumption $\mathcal{H}$ is rather restrictive. In this section, we use generalized homogeneity in order to eliminate the restriction $p = m$. We develop a homogeneous unit control ensuring sliding mode on the linear surface (9.82) with $p > m$ under the following assumption inspired by [40].

Assumption 9.2 There exists $\tilde{K} \in \mathbb{R}^{m\times n}$ such that

$$C(A + B\tilde{K})\left(I_n - C^{\top}(CC^{\top})^{-1}C\right) = \mathbf{0} \tag{9.87}$$

and the pair $\{A_\sigma, B_\sigma\}$ is controllable, where

$$A_\sigma = C(A+B\tilde{K})C^{\top}(CC^{\top})^{-1}, \quad B_\sigma = CB. \tag{9.88}$$

Below we show that Assumption 9.2 simply asks for the controllability of the nominal dynamics of the sliding variable $\sigma = Cx$, such that an influence of any other uncontrollable component of the system state can be rejected by the control input. We introduce a homogeneous unit control, which enforces the sliding mode on the surface $\sigma = 0$ under the classical Assumption 9.1 and the relaxed Assumption 9.2.

The controllability of the pair $\{A_\sigma, B_\sigma\}$ is necessary for a global finite-time stabilization of the sliding variable σ. This means that the matrix CB must be necessarily nonzero. To fulfill this condition, the original sliding surface $Cx = 0$ can be extended artificially [41]. Namely, if $CA^jB = \mathbf{0}$, $j = 0, 1, ..., r$, but $CA^{r+1}B \neq \mathbf{0}$ then the augmented sliding variable can be defined as

$$\sigma_{new} = \begin{pmatrix} C \\ CA \\ \cdots \\ CA^r \end{pmatrix} x.$$

The sliding motion on the surface $\sigma_{new} = \mathbf{0}$ implies $\sigma = 0$. The parameter $r \leq n$ should be selected such that $\{A_{\sigma_{new}}, B_{\sigma_{new}}\}$ is controllable. If this par is not controllable for any $r \leq n$ then the homogeneous unit control cannot be designed.

9.2.2 Homogeneous Unit Control

In the nominal case ($g = \mathbf{0}$), the dynamics of the sliding variable σ of the closed-loop system (9.81) with the *classical unit control* (9.86) is governed by

$$\dot{\sigma} = (CA + CB\hat{K})x + CBK\frac{\sigma}{|\sigma|} + CB\gamma. \tag{9.89}$$

Under restriction $\mathcal{H}$ we have $p = m$ and $\det(CB) \neq 0$. The matrix $\hat{K}$ is traditionally selected [4] as

$$\hat{K} = -(CB)^{-1}CA$$

in order to suppress the linear dynamics in (9.89) and to obtain the system

$$\dot{\sigma} = CBK\frac{\sigma}{|\sigma|} + CB\gamma, \tag{9.90}$$

being standard homogeneous of degree $\mu = -1$ in the unperturbed case (i.e., for $\gamma = \mathbf{0}$). To stabilize the system (9.90) under Assumption 9.1, the gain K can be selected as follows:

$$K = -(\rho + |CB|\overline{\gamma})(CB)^{-1}, \quad \rho > 0.$$

Indeed, in this case, we have

$$\frac{1}{2}\frac{d}{dt}\sigma^2 = \sigma^\top\dot{\sigma} = -(\rho + |CB|\overline{\gamma})|\sigma| + \sigma^\top CB\gamma \leq -\rho|\sigma| < 0.$$

This classical approach to the unit control design can be explained by the homogeneity theory as follows:

- the linear term homogenizes the dynamics of the sliding variable $\sigma = Cx$;
- the nonlinear term is a finite-time homogeneous stabilizer, which rejects matched perturbations of a known magnitude.

We follow this homogeneity-based interpretation in order to design a unit control for the case $p > m$. By Lemma 6.1, the linear vector field

$$\sigma \mapsto \tilde{A}\sigma, \quad \tilde{A} \in \mathbb{R}^{p\times p}$$

is standard homogeneous of nonzero degree if and only if $\tilde{A} = \mathbf{0}$. This explains the selection of the linear gain $\hat{K} = -(CB)^{-1}CA$ given above. The generalized homogeneity relaxes this restriction, since, by Lemma 6.1, a linear vector field $\sigma \mapsto \tilde{A}\sigma$ may be $\mathbf{d}$-homogeneous of nonzero degree if the (possibly nonzero) matrix $\tilde{A}$ is nilpotent. This motivates us to design a similar two-component feedback with a linear term homogenizing the unperturbed system and a nonlinear term stabilizing the homogeneous system to zero in a finite time.

The feedback law (9.86) is called the unit control since $\frac{\sigma}{|\sigma|}$ is the unit vector defined by mean of the (standard homogeneous) projection of the sliding variable σ on the unit Euclidean sphere. For a linear continuous dilation $\mathbf{d}$, such a projector is given by (see Chap. 5)

$$\pi_{\mathbf{d}}(\sigma) = \mathbf{d}(-\ln \|\sigma\|_{\mathbf{d}})\sigma,$$

where $\|\cdot\|_{\mathbf{d}}$ is a $\mathbf{d}$-homogeneous norm in $\mathbb{R}^m$. So, the *homogeneous unit control* can be defined in the form:

$$u = \hat{K}x + K\mathbf{d}(-\ln \|Cx\|_{\mathbf{d}})Cx, \quad \hat{K} \in \mathbb{R}^{m\times n}, \quad K \in \mathbb{R}^{m\times p}. \tag{9.91}$$

Indeed, if $\tilde{K}$ satisfies Assumption 9.2 then

$$\begin{aligned} CA+CB\tilde{K} &= \overbrace{C(A+B\tilde{K})(I_n - C^\top(CC^\top)^{-1}C)}^{=\mathbf{0}} \\ &\quad + C(A+B\tilde{K})C^\top(CC^\top)^{-1}C = A_\sigma C, \end{aligned}$$

where $A_\sigma = C(A+B\tilde{K})C^\top(CC^\top)^{-1}$. In this case, the dynamics of the sliding variable in the nominal case is given by

$$\dot{\sigma} = A_\sigma \sigma + B_\sigma(u - \tilde{K}x + \gamma). \tag{9.92}$$

In the view of Corollary 9.2, to stabilize the system (9.92) in a finite time and reject, simultaneously, the bounded perturbation γ, the control u can be selected in the form (9.91) with

$$\hat{K} = \tilde{K} + K_0 C, \tag{9.93}$$

where the matrix $\tilde{K} \in \mathbb{R}^{m\times n}$ is defined in Assumption 9.2 and the matrix $K_0 \in \mathbb{R}^{m\times p}$ is required for $\mathbf{d}$-homogenization of a linear component of the dynamics of σ (see Theorem 9.1). The feedback law (9.91) is a unit control as well. However, its design is based on the concept of a generalized (non-standard) homogeneity, so, in order to distinguish the conventional unit control and the suggested algorithms, the feedback law (9.91) was called *the homogeneous unit control* [40].

For $g \neq \mathbf{0}$ the sliding variable $\sigma = Cx$ of the closed-loop system (9.81), (9.91) satisfies the equation:

$$\dot{\sigma} = (A_\sigma + B_\sigma K_0)\sigma + B_\sigma K\mathbf{d}(-\ln \|\sigma\|_{\mathbf{d}})\sigma + B_\sigma\gamma + Cg, \tag{9.94}$$

where $B_\sigma = CB$. The following theorem refines the result of [40] and presents a characterization of unstructured perturbations g to be rejected by the homogeneous unit control.

Theorem 9.5 *If Assumptions 9.1 and 9.2 are fulfilled for* $\overline{\gamma} \geq 0$ *and* $C \in \mathbb{R}^{p\times m}$ *then*

(1) *the linear algebraic equation*

$$A_\sigma G_0 - G_0 A_\sigma + B_\sigma Y_0 = A_\sigma, \quad G_0 B_\sigma = \mathbf{0} \tag{9.95}$$

always has a solution $Y_0 \in \mathbb{R}^{m\times p}$, $G_0 \in \mathbb{R}^{p\times p}$*; moreover, for any solution of* (9.5)*, the matrix*

$$G_{\mathbf{d}} = I_p - G_0 \tag{9.96}$$

is anti-Hurwitz, the matrix

$$A_0 = A_\sigma - B_\sigma Y_0 G_{\mathbf{d}}^{-1} \tag{9.97}$$

is nilpotent and satisfies the identity

$$A_0 G_{\mathbf{d}} = (G_{\mathbf{d}} - I_p)A_0, \quad G_{\mathbf{d}} B_\sigma = B_\sigma; \tag{9.98}$$

(2) *the linear algebraic system*

$$\begin{aligned}(A_0 + \rho G_{\mathbf{d}})X + X(A_0 + \rho G_{\mathbf{d}})^\top + B_\sigma Y + Y^\top B_\sigma^\top + \overline{\gamma} X = -\overline{\gamma} B_\sigma B_\sigma^\top,\\ G_{\mathbf{d}} X + X G_{\mathbf{d}}^\top \succ 0, \quad X = X^\top \succ 0\end{aligned} \tag{9.99}$$

has a solution $X \in \mathbb{R}^{p\times p}$, $Y \in \mathbb{R}^{m\times n}$ *for any* $\rho > 0$*;*

(3) *the closed system* (9.94) *with the control* (9.91) *is globally uniformly finite-time stable and*

$$\sigma(t) = \mathbf{0}, \quad \forall t \geq \tfrac{\|\sigma(0)\|_{\mathbf{d}}}{\rho-\kappa},$$

provided that $\mathbf{d}(s) = e^{sG_{\mathbf{d}}}$ *with* $s \in \mathbb{R}$,

$$K_0 = -Y_0 G_{\mathbf{d}}^{-1}, \quad K = Y X^{-1}, \tag{9.100}$$

$\|\cdot\|_{\mathbf{d}}$ *is the canonical homogeneous norm induced by the weighted Euclidean norm* $\|\sigma\| = \sqrt{\sigma^\top X^{-1}\sigma}$ *and*

$$\sup_{(t,x)\in\mathbb{R}\times\mathbb{R}^n} \frac{\|Cx\|_{\mathbf{d}} x^\top C^\top \mathbf{d}^\top(-\ln\|Cx\|_{\mathbf{d}})X^{-1}\mathbf{d}(-\ln\|Cx\|_{\mathbf{d}})Cg(t,x)}{x^\top C^\top \mathbf{d}^\top(-\ln\|Cx\|_{\mathbf{d}})X^{-1}G_{\mathbf{d}}\mathbf{d}(-\ln\|Cx\|_{\mathbf{d}})Cx} \leq \kappa \in [0,\rho). \tag{9.101}$$

Proof The feasibility of algebraic equations (9.95) and (9.99) is proven in Theorem 9.1. Indeed, introducing the new variable $\tilde{Y} = Y + 0.5\overline{\gamma} B_\sigma^\top$ and the new dilation with the generator $\tilde{G}_{\mathbf{d}} = G_{\mathbf{d}} + \frac{\overline{\gamma}}{\rho} I_p$ we derive (9.5) and (9.9), respectively.

Let us prove the finite-time stability of (9.94). Inspired by [40] let us consider the canonical homogeneous norm $\|\sigma\|_{\mathbf{d}}$ as a Lyapunov function candidate. Using (5.21) we derive

$$\frac{d\|\sigma(t)\|_{\mathbf{d}}}{dt} = \|\sigma\|_{\mathbf{d}} \frac{\sigma^\top \mathbf{d}^\top(-\ln\|\sigma\|_{\mathbf{d}})X^{-1}\mathbf{d}(-\ln\|\sigma\|_{\mathbf{d}})\dot{\sigma}}{\sigma^\top \mathbf{d}^\top(-\ln\|\sigma\|_{\mathbf{d}}X^{-1}G_{\mathbf{d}}\mathbf{d}(-\ln\|\sigma\|_{\mathbf{d}})\sigma}.$$

By Lemma 6.1 the identity (9.98) implies that

$$A_0\mathbf{d}(s) = e^{-s}\mathbf{d}(s)A_0, \quad \mathbf{d}(s)B = e^s B, \quad \forall s \in \mathbb{R}.$$

Hence, we conclude

$$\begin{aligned}\frac{d\|\sigma(t)\|_{\mathbf{d}}}{dt} =& \frac{\sigma^\top \mathbf{d}^\top(-\ln\|\sigma\|_{\mathbf{d}})X^{-1}(A_0+B_\sigma K)\mathbf{d}(-\ln\|\sigma\|_{\mathbf{d}})\sigma}{\sigma^\top \mathbf{d}^\top(-\ln\|\sigma\|_{\mathbf{d}})X^{-1}G_{\mathbf{d}}\mathbf{d}(-\ln\|\sigma\|_{\mathbf{d}})\sigma}\\ &+ \frac{\sigma^\top \mathbf{d}^\top(-\ln\|\sigma\|_{\mathbf{d}})X^{-1}B_\sigma\gamma}{\sigma^\top \mathbf{d}^\top(-\ln\|\sigma\|_{\mathbf{d}})X^{-1}G_{\mathbf{d}}\mathbf{d}(-\ln\|\sigma\|_{\mathbf{d}})\sigma}\\ &+ \frac{\|\sigma\|_{\mathbf{d}}\sigma^\top \mathbf{d}^\top(-\ln\|\sigma\|_{\mathbf{d}})X^{-1}\mathbf{d}(-\ln\|\sigma\|_{\mathbf{d}})Cg(t,x)}{\sigma^\top \mathbf{d}^\top(-\ln\|\sigma\|_{\mathbf{d}})X^{-1}G_{\mathbf{d}}\mathbf{d}(-\ln\|\sigma\|_{\mathbf{d}})\sigma}.\end{aligned}$$

Since $X^{-1}(A_0 + B_\sigma K) = X^{-1}(A_0X + B_\sigma Y)X^{-1}$ then using the identity (9.99) we derive

$$\begin{aligned}\frac{d\|\sigma(t)\|_{\mathbf{d}}}{dt} =& -\rho\frac{\sigma^\top \mathbf{d}^\top(-\ln\|\sigma\|_{\mathbf{d}})(X^{-1}G_{\mathbf{d}}+G_{\mathbf{d}}^\top X^{-1})\mathbf{d}(-\ln\|\sigma\|_{\mathbf{d}})\sigma}{\sigma^\top \mathbf{d}^\top(-\ln\|\sigma\|_{\mathbf{d}})(X^{-1}G_{\mathbf{d}}+G_{\mathbf{d}}^\top X^{-1})\mathbf{d}(-\ln\|\sigma\|_{\mathbf{d}})\sigma}\\ &- \frac{\overline{\gamma}\sigma^\top \mathbf{d}^\top(-\ln\|\sigma\|_{\mathbf{d}})X^{-1}\mathbf{d}(-\ln\|\sigma\|_{\mathbf{d}})\sigma}{\sigma^\top \mathbf{d}^\top(-\ln\|\sigma\|_{\mathbf{d}})(X^{-1}G_{\mathbf{d}}+G_{\mathbf{d}}^\top X^{-1})\mathbf{d}(-\ln\|\sigma\|_{\mathbf{d}})\sigma}\\ &- \frac{\overline{\gamma}\sigma^\top \mathbf{d}^\top(-\ln\|\sigma\|_{\mathbf{d}})X^{-1}B_\sigma B_\sigma^\top X^{-1}\mathbf{d}(-\ln\|\sigma\|_{\mathbf{d}})\sigma}{\sigma^\top \mathbf{d}^\top(-\ln\|\sigma\|_{\mathbf{d}})(X^{-1}G_{\mathbf{d}}+G_{\mathbf{d}}^\top X^{-1})\mathbf{d}(-\ln\|\sigma\|_{\mathbf{d}})\sigma}\\ &+ \frac{2\sigma^\top \mathbf{d}^\top(-\ln\|\sigma\|_{\mathbf{d}})X^{-1}B_\sigma\gamma}{\sigma^\top \mathbf{d}^\top(-\ln\|\sigma\|_{\mathbf{d}})(X^{-1}G_{\mathbf{d}}+G_{\mathbf{d}}^\top X^{-1})\mathbf{d}(-\ln\|\sigma\|_{\mathbf{d}})\sigma}\\ &+ \frac{\|\sigma\|_{\mathbf{d}}\sigma^\top \mathbf{d}^\top(-\ln\|\sigma\|_{\mathbf{d}})X^{-1}\mathbf{d}(-\ln\|\sigma\|_{\mathbf{d}})Cg(t,x)}{\sigma^\top \mathbf{d}^\top(-\ln\|\sigma\|_{\mathbf{d}})X^{-1}G_{\mathbf{d}}\mathbf{d}(-\ln\|\sigma\|_{\mathbf{d}})\sigma}.\end{aligned}$$

For $\overline{\gamma} > 0$ and $\xi = \begin{pmatrix} \frac{\gamma}{\sqrt{\overline{\gamma}}} \\ \sqrt{\overline{\gamma}}B_\sigma^\top X^{-1}\mathbf{d}(-\ln\|\sigma\|_{\mathbf{d}})\sigma \end{pmatrix}$ we have

$$0 \le \xi^\top \begin{pmatrix} I_m & -I_m \\ -I_m & I_m \end{pmatrix} \xi = \frac{|\gamma|^2}{\overline{\gamma}} - 2\sigma^\top \mathbf{d}^\top(-\ln\|\sigma\|_{\mathbf{d}})X^{-1}B_\sigma\gamma$$

$$+ \overline{\gamma}\sigma^\top \mathbf{d}^\top(-\ln\|\sigma\|_{\mathbf{d}})X^{-1}B_\sigma B_\sigma^\top X^{-1}\mathbf{d}(-\ln\|\sigma\|_{\mathbf{d}})\sigma.$$

Therefore, using the identity

$$\sigma^\top \mathbf{d}^\top(-\ln\|\sigma\|_{\mathbf{d}})X^{-1}\mathbf{d}(-\ln\|\sigma\|_{\mathbf{d}})\sigma = 1$$

(see the definition of $\|\cdot\|_{\mathbf{d}}$ in Chap. 5) we derive

$$
\begin{aligned}
\frac{d\|\sigma(t)\|_{\mathbf{d}}}{dt} &= -\rho - \frac{\overline{\gamma}-\frac{|\gamma|^2}{\overline{\gamma}}}{\sigma^\top \mathbf{d}^\top(-\ln\|\sigma\|_{\mathbf{d}})(X^{-1}G_{\mathbf{d}}+G_{\mathbf{d}}^\top X^{-1})\mathbf{d}(-\ln\|\sigma\|_{\mathbf{d}})\sigma} \\
&\quad - \frac{\xi^\top \begin{pmatrix} I_m & -I_m \\ -I_m & I_m \end{pmatrix}\xi}{\sigma^\top \mathbf{d}^\top(-\ln\|\sigma\|_{\mathbf{d}})(X^{-1}G_{\mathbf{d}}+G_{\mathbf{d}}^\top X^{-1})\mathbf{d}(-\ln\|\sigma\|_{\mathbf{d}})\sigma} \\
&\quad + \frac{\|\sigma\|_{\mathbf{d}}\sigma^\top \mathbf{d}^\top(-\ln\|\sigma\|_{\mathbf{d}})X^{-1}\mathbf{d}(-\ln\|\sigma\|_{\mathbf{d}})Cg(t,x)}{\sigma^\top \mathbf{d}^\top(-\ln\|\sigma\|_{\mathbf{d}})X^{-1}G_{\mathbf{d}}\mathbf{d}(-\ln\|\sigma\|_{\mathbf{d}})\sigma} \\
&\leq -\rho+\kappa, \quad \text{for} \quad \sigma \neq \mathbf{0},
\end{aligned}
$$

so $\sigma(t) = 0$ for $t \geq \|\sigma(0)\|_{\mathbf{d}}/(\rho - \kappa)$. The proof is complete. □

The homogeneous unit control (9.91) is discontinuous only on the sliding surface $Cx = 0$ and rejects bounded matched perturbations $|\gamma(t,x)| \leq \overline{\gamma}$. It also rejects some unstructured perturbations g satisfying (9.101). Inspired by Theorems 9.5 and [40], the following *algorithm* can be proposed.

Algorithm 9.2 ***(Homogeneous unit control design)***

- ***Step 1.*** *Given $C \in \mathbb{R}^{p\times m}$ solve the linear algebraic equation* (9.87) *with respect to the variable $\tilde{K} \in \mathbb{R}^{m\times n}$.*
- ***Step 2.*** *Check the controllability of the pair $\{A_\sigma, B_\sigma\}$, where $A_\sigma \in \mathbb{R}^{p\times p}$ and $B_\sigma \in \mathbb{R}^{p\times n}$ are given by* (9.88).
- ***Step 3.*** *Given $\overline{\gamma} \geq 0$ solve the linear algebraic equation* (9.95) *with respect to $Y_0 \in \mathbb{R}^{m\times p}$ and $G_0 \in \mathbb{R}^{p\times p}$ for some $\rho > 0$.*
- ***Step 4.*** *Solve the linear matrix inequality* (9.99) *with respect to $Y \in \mathbb{R}^{m\times p}$ and $X \in \mathbb{R}^{p\times p}$.*
- ***Step 5.*** *Compute the parameters $K_0 \in \mathbb{R}^{m\times p}$ and $K \in \mathbb{R}^{m\times p}$ of the homogeneous unit control* (9.91) *by the formulas* (9.100), (9.96).
- ***Step 6.*** *Repeat the design with a larger $\rho > 0$ if it is necessary to reject a larger set of unstructured perturbations g.*

If rank $B_\sigma = m$ then using the equivalent control method (see [4] or Chap. 2) for $t \geq \|Cx(0)\|_{\mathbf{d}}$ we derive

$$
\sigma(t) = \dot{\sigma}(t) = \mathbf{0} \quad \Rightarrow \quad u = u^{eq} := -\gamma + (B_\sigma^\top B_\sigma)^{-1}Cg
$$

and the system dynamics in the sliding mode is governed by the differential algebraic equation:

$$
\dot{x}(t) = (A + B\tilde{K})x(t) + (I_p - B(B_\sigma^\top B_\sigma)^{-1}C)g(t,x), \quad Cx(t) = \mathbf{0}, \tag{9.102}
$$

where $\tilde{K} \in \mathbb{R}^{m\times n}$ is defined in Assumption 9.2. If the control aim is to stabilize the system (9.81) then the sliding surface $Cx = 0$ has to be selected such that the system (9.102) is globally asymptotically stable. For $g = \mathbf{0}$ the simplest condition to C in this case is given by the following matrix inequality:

$$
P(A+B\tilde{K})+(A+B\tilde{K})^\top P-\beta C^\top C \prec 0, \quad P \succ 0, \quad \beta > 0. \tag{9.103}
$$

Indeed, considering the Lyapunov function candidate $V = x^\top Px$, we derive the following estimate for the time derivative of V along (9.102):

$$\begin{aligned}\dot{V} &= 2x^\top P(A + B\tilde{K})x \\ &= x^\top (P(A+B\tilde{K})+(A+B\tilde{K})^\top P-\beta C^\top C)x < 0,\end{aligned}$$

where the identity $Cx = 0$ and the inequality (9.103) are utilized on the last step. Therefore, to design a stabilizing homogeneous unit control (9.91) for the nominal system (9.81) with $g = \mathbf{0}$ it is sufficient to find a matrix C satisfying Assumption 9.2 and the matrix inequality (9.103). The unknown perturbation g does not destroy asymptotic stability of the closed-loop system if g satisfies (9.101) and

$$\sup_{t\in\mathbb{R}} x^\top P(I_n - B(B_\sigma^\top B_\sigma)^{-1}C)g(t,x) < -\tfrac{x^\top(P(A+B\tilde{K})+(A+B\tilde{K})^\top P-\beta C^\top C)x}{2},$$

for all $x \neq \mathbf{0}$. The latter restriction implies $\dot{V} < 0$ in the perturbed case $g \neq \mathbf{0}$. More advanced methods of sliding surface design can be found in [4, 34–36].

Remark 9.5 ***(On chattering attenuation)*** Following the conventional approach to the chattering reduction (see, e.g., [4]), the discontinuous term in (9.91) can be "smoothed" using the saturation function:

$$u_\alpha = \hat{K}x + K\mathbf{d}(-\ln\max\{\epsilon, \|Cx\|_{\mathbf{d}}\})Cx, \quad \epsilon \geq 0. \tag{9.104}$$

This control law coincides with (9.91) for $\epsilon = 0$. On the one hand, the parameter $\epsilon > 0$ defines a set into which the sliding variable σ converges. Indeed, in the view of Theorem 9.5 for the closed-loop system (9.81), (9.104), we have

$$\tfrac{d\|\sigma(t)\|_{\mathbf{d}}}{dt} \leq -\rho + \kappa < 0 \quad \text{if} \quad \|\sigma(t)\|_{\mathbf{d}} \geq \epsilon.$$

This means that the sliding variable σ converges to the homogeneous ball $\|\sigma\|_{\mathbf{d}} \leq \epsilon$ in a finite time. On the other hand, the nonlinear term in (9.104) becomes linear $K\mathbf{d}(-\ln\epsilon)Cx$ for $\|x\|_{\mathbf{d}} \leq \epsilon$, so the feedback law u_α is continuous on $\mathbb{R}^n$ if $\epsilon > 0$. In practice, the parameter ϵ can be utilized for an attenuation of the chattering caused by discontinuity of the SMC (see, e.g., [4] for mode details about chattering attenuation).

If $\gamma = \mathbf{0}$, $g = \mathbf{0}$ and $\rho = 1$ then the closed-loop system (9.81), (9.91) reaches the sliding surface $Cx = 0$ *exactly* at the time instant $T(x_0) = \|Cx_0\|_{\mathbf{d}}$. Indeed, repeating the proof of Theorem 9.5 we derive

$$\tfrac{d\|\sigma(t)\|_{\mathbf{d}}}{dt} = -1, \quad t > 0,$$

i.e., $\sigma(t) \neq \mathbf{0}$ for all $t \in [0, \|\sigma(0)\|_{\mathbf{d}})$ and $\sigma(t) = \mathbf{0}$ for all $t \geq \|\sigma(0)\|_{\mathbf{d}}$. The reaching time can be tuned by the parameter $\rho > 0$. The larger the ρ, the faster is the reaching of the sliding surface (see the estimate of the reaching time in Theorem 9.5). Theoretically, the reaching time may be adjusted arbitrarily small. However, a selection of a large ρ may lead to a large feedback gain and a large chattering in practice. Therefore, in practice, ρ could be one more tuning parameter for the chattering attenuation.

Example 9.11 Let us consider the system (9.81) with

$$A=\begin{pmatrix} 0 & 1 & 0 & 0 & 0\\ -1 & 0 & 0 & 0 & 0\\ 0 & 0 & 1 & -1 & 1\\ 0 & 1 & 0 & 2 & 1\\ 1 & 0 & 1 & -1 & 3\end{pmatrix},\ B=\begin{pmatrix} 0 & 0\\ 0 & 0\\ 0 & 0\\ 1 & 0\\ 0 & 1\end{pmatrix},\ C=\begin{pmatrix} 0 & 0 & 1 & 0 & 0\\ 0 & 0 & 0 & 1 & 0\\ 0 & 0 & 0 & 0 & 1\end{pmatrix},\ g(t,x)=\begin{pmatrix} 0\\ 0\\ 0.1|Cx|\\ 0\\ 0\end{pmatrix}.$$

The pair $\{A, B\}$ is not controllable: the first two components of the state vector cannot be affected by the control. However, the selected matrix C means that the sliding mode needs to be enforced in the last three components of the state vector simultaneously. Since the dimensions of C and B are inconsistent ($p \neq m$) then we cannot use the conventional scheme of the unit control design.

Let us design the homogeneous unit control according to Theorem 9.5. Using (9.87) we derive $\tilde{K}=\begin{pmatrix} 0 & -1 & 0 & 0 & 0\\ -1 & 0 & 0 & 0 & 0\end{pmatrix}$, so $A_\sigma=\begin{pmatrix} 1 & -1 & 1\\ 0 & 2 & 1\\ 1 & -1 & 3\end{pmatrix}$, $B_\sigma=\begin{pmatrix} 0 & 0\\ 1 & 0\\ 0 & 1\end{pmatrix}$. The pair $\{A_\sigma, B_\sigma\}$ is controllable, so Assumption 9.2 is fulfilled. A solution of (9.95) is given by $G_0=\begin{pmatrix} -1 & 0 & 0\\ -0.5 & 0 & 0\\ 0.5 & 0 & 0\end{pmatrix}$, $Y_0=\begin{pmatrix} -0.5 & -2.5 & -0.5\\ -1.5 & 1.5 & -3.5\end{pmatrix}$, so the generator of dilation $\mathbf{d}$ is given by

$$G_{\mathbf{d}}=\begin{pmatrix} 2 & 0 & 0\\ 0.5 & 1 & 0\\ -0.5 & 0 & 1\end{pmatrix},\ \text{and } K_0=\begin{pmatrix} 0 & 0 & 0.5000 & -2.5000 & -0.5000\\ 0 & 0 & -1.5000 & 1.5000 & -3.5000\end{pmatrix}.$$

Solving the linear system (9.99) for $\overline{\gamma}=0.5$ and $\rho=1$ we derive

$$K=\begin{pmatrix} 5.8379 & -2.5103 & 1.1938\\ -5.8379 & 1.1938 & -2.5103\end{pmatrix},\ P=\begin{pmatrix} 5.3112 & -1.3265 & 1.3265\\ -1.3265 & 0.5413 & -0.2751\\ 1.3265 & -0.2751 & 0.5413\end{pmatrix}.$$

According to Theorem 9.5, the homogeneous unit control rejects bounded matched disturbances γ of the magnitude $\overline{\gamma}=0.5$. Moreover, it rejects some mismatched perturbation g provided that g satisfies the condition (9.101). The numerical simulations for the closed-loop system with $\gamma(t)=0.5(\sin(10t), \cos(10t))^\top$ are presented in Fig. 9.16. The simulation is made using explicit Euler method which results in the numerical chattering, which can be observed on the figure. The step of discretization is 0.01.

To demonstrate the efficiency of the chattering attenuation scheme suggested in Remark 9.5, we present numerical simulations for the homogeneous unit control with the saturated canonical homogeneous norm (see the formula (9.104)). The simulation results for the system with the same perturbations and the saturation parameter $\epsilon=0.005$ are depicted in Fig. 9.17.

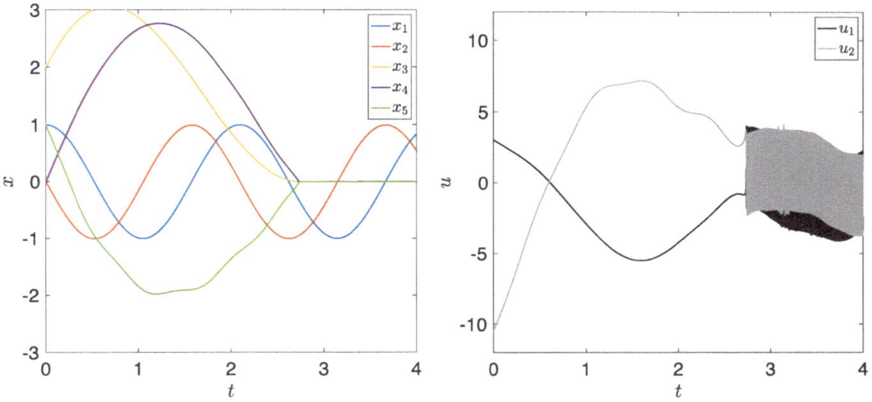

Fig. 9.16 The system with homogeneous unit control

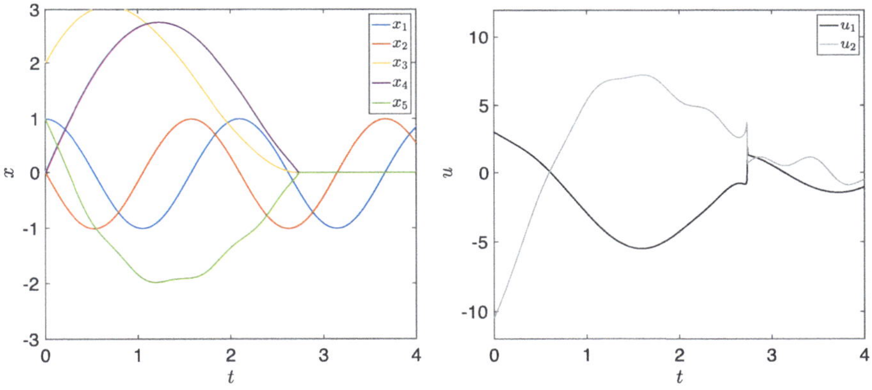

Fig. 9.17 The system with homogeneous unit control and chattering attenuation $\epsilon = 0.005$

9.2.3 Homogeneous Super-Twisting Control

By Assumption 9.1 the maximal magnitude of the matched perturbation γ has to be known in order to design the homogeneous unit control. In this section, we relax this assumption allowing the magnitude of γ to be unknown. However, another a priori information is required in this case.

Assumption 9.3 Let the unknown function γ admit the representation:

$$\gamma(t, x) = \gamma_0(t) + \tilde{\gamma}(t, x), \quad \forall t \geq 0, \forall x \in \mathbb{R}^n, \tag{9.105}$$

$$|\tilde{\gamma}(t, x)| \leq \overline{\gamma} \min\{|Cx|, 1\}, \quad \forall t \geq 0, \forall x \in \mathbb{R}^n, \tag{9.106}$$

where $\overline{\gamma}$ is a known scalar and $\gamma_0 : \mathbb{R} \mapsto \mathbb{R}^m$ is an unknown slowly varying locally Lipschitz continuous function

$$|\dot{\gamma}_0(t)| \overset{a.e.}{\leq} \overline{\gamma}_0, \quad t \geq 0 \tag{9.107}$$

with a small enough $\overline{\gamma}_0 \geq 0$.

Notice that for $\overline{\gamma} = \overline{\gamma_0} = 0$ the unknown function γ satisfying the above assumption becomes an unknown constant. The linear control theory uses the integral term in PID (proportional–integral–derivative) controller in order to reject an unknown matched constant perturbation. The sliding-mode control theory (see, e.g., [42, 43]) follows the same idea in order to enforce the sliding motion on the surface $Cx = 0$ in the case of matched Lipschitz continuous (slowly varying) perturbations of unknown magnitude. Inspired by [40] we define the homogeneous unit control law with integral action as follows:

$$\begin{aligned} u &= \hat{K}x + \sqrt{\|\sigma\|_{\mathbf{d}}} K \mathbf{d}(-\ln \|\sigma\|_{\mathbf{d}})\sigma + \textstyle\int_0^t u_{\mathrm{i}} \, d\tau, \\ u_{\mathrm{i}} &= \frac{K_{\mathrm{i}} \mathbf{d}(-\ln \|\sigma\|_{\mathbf{d}})\sigma}{q(\sigma)}, \quad \sigma = Cx. \quad K_{\mathrm{i}} \in \mathbb{R}^{m \times p}, \end{aligned} \tag{9.108}$$

where $\hat{K} = \tilde{K} + K_0 C$, $\mathbf{d}(s) = e^{sG_{\mathbf{d}}}$, $G_{\mathbf{d}} = I_p - 0.5G_0$, $K = YX^{-1}$, $P = X^{-1}$, K_0 and G_0 are defined as in Theorem 9.5,

$$q(\sigma) = \sigma^\top \mathbf{d}^\top(-\ln \|\sigma\|_{\mathbf{d}})(PG_{\mathbf{d}} + G_{\mathbf{d}}^\top P)\mathbf{d}(-\ln \|\sigma\|_{\mathbf{d}})\sigma.$$

In this case, under Assumptions 9.2 and 9.3, the dynamics of the sliding variable admits the representation

$$\begin{aligned} \dot{\sigma} &= A_0\sigma + B_\sigma \left(\|\sigma\|_{\mathbf{d}}^{\frac{1}{2}} K \mathbf{d}(-\ln \|\sigma\|_{\mathbf{d}})\sigma + \eta + \tilde{\gamma} \right) + Cg(t, x), \\ \dot{\eta} &= u_{\mathrm{i}}(\sigma) + \dot{\gamma}_0, \end{aligned} \tag{9.109}$$

where the nilpotent matrix A_0 is defined in Theorem 9.5 and

$$\eta = \gamma_0 + \int_0^t u_{\mathrm{i}} \, d\tau.$$

If $m = 1$, $A_\sigma \in \mathbb{R}$, $B_\sigma \in \mathbb{R}\backslash\{0\}$ and $\sigma \in \mathbb{R}$ then $\hat{K} = -A_\sigma / B_\sigma$, $P > 0$, $K, K_{\mathrm{i}} \in \mathbb{R}$, $G_{\mathbf{d}} = 1$, $\mathbf{d}(s) = e^s$, $\|\sigma\|_{\mathbf{d}} = |\sigma|$ and $\mathbf{d}(-\ln \|\sigma\|_{\mathbf{d}})\sigma = \mathrm{sign}(\sigma)$. Therefore, for the *singe-input single-output (SISO)* system, the control law (9.108) coincides with the so-called super-twisting algorithm [42]:

$$u = \hat{K}x + K\sqrt{|\sigma|}\,\mathrm{sign}(\sigma) + K_{\mathrm{i}} \int_0^t \mathrm{sign}(\sigma(\tau))d\tau. \tag{9.110}$$

The homogeneous control with integral action given by (9.108) *is a multi-input multi-output (MIMO) version of the super-twisting algorithm.*

The following theorem refines the result of [40] and presents the characterization of unstructured perturbations g to be rejected by the homogeneous super-twisting MIMO control.

Theorem 9.6 *Let Assumptions 9.2, 9.3 be fulfilled for $C \in \mathbb{R}^{p\times n}$, $\overline{\gamma}>0$.*

- *If the parameters K_0, G_0 are defined as in Theorem 9.5 and $G_{\mathbf{d}} = I_p - 0.5G_0$ then the system of linear matrix equations/inequalities*

$$\begin{gathered}(A_0 + \rho G_{\mathbf{d}})X + X(A_0 + \rho G_{\mathbf{d}})^\top + B_\sigma Y + Y^\top B_\sigma^\top + \overline{\gamma}X = -\overline{\gamma}B_\sigma B_\sigma^\top,\\ G_{\mathbf{d}}X + XG_{\mathbf{d}}^\top \succ X, \quad I_n \succeq X \succ 0\end{gathered} \tag{9.111}$$

is feasible with respect to $X \in \mathbb{R}^{n\times n}$, $Y \in \mathbb{R}^{m\times n}$ for any $\bar{\gamma} \geq 0$ and $\rho > 0$.
- *The unperturbed system* (9.109) *is $\tilde{\mathbf{d}}$-homogeneous of degree $-1/2$ with respect to the dilation in $\mathbb{R}^{p+m}$ defined as follows:*

$$\tilde{\mathbf{d}}(s) = \begin{pmatrix} \mathbf{d}(s) & \mathbf{0} \\ \mathbf{0} & e^{\frac{1}{2}s}I_m \end{pmatrix}, \quad s \in \mathbb{R}.$$

- *For $K = YX^{-1}$, $K_{\mathrm{i}} = -QB_\sigma^\top X^{-1}$, $0 \prec Q = Q^\top \in \mathbb{R}^{m\times m}$, the system* (9.109) *is globally uniformly finite-time stable provided that $\overline{\gamma}_0 \geq 0$ is small enough and there exists $\kappa \in [0, \rho)$ such that g satisfies the inequality:*

$$\left\| \frac{Cg(t,x)}{\beta} \right\|_{\hat{\mathbf{d}}} \leq \|Cx\|_{\mathbf{d}}, \quad \forall (t,x) \in \mathbb{R} \times \mathbb{R}^n, \tag{9.112}$$

where β is defined for $\mathbf{d}$ in Corollary 5.3 and $\|\cdot\|_{\hat{\mathbf{d}}}$ is the canonical homogeneous norm induced by the norm $\|\cdot\|$ and the dilation $\hat{\mathbf{d}}(s) = e^{0.5s(I_p - G_0)}$, $s \in \mathbb{R}$.

Proof (1) *Construction of a homogeneous extension.* Denoting $\tilde{G}_{\mathbf{d}} = I_p - G_0$ and $Y = Y_1 + 0.5\rho B_\sigma^\top$ and the system (9.111) without $X \succeq I_n$ can be rewritten as

$$\begin{gathered}(A_0 + \tfrac{1}{2}\rho\tilde{G}_{\mathbf{d}})X + X(A_0 + \tfrac{1}{2}\rho G_{\mathbf{d}})^\top + B_\sigma Y_1 + Y_1^\top B_\sigma^\top + (\rho + \overline{\gamma})X = -(\rho + \overline{\gamma})B_\sigma B_\sigma^\top,\\ 0.5(\tilde{G}_{\mathbf{d}}X + X\tilde{G}_{\mathbf{d}}^\top) \succ 0, \quad X \succ 0\end{gathered}.$$

The above system of LMIs is feasible (see Theorem 9.5) for any $\overline{\gamma} \geq 0$ and any $\rho > 0$. Making the change of variable $Y_1 = \tilde{Y} - 0.5(\overline{\gamma} + \rho)B_\sigma^\top$ we derive the equivalent feasible system of LMIs:

$$\begin{gathered}(A_0 + \tfrac{1}{2}\rho\tilde{G}_{\mathbf{d}})X + X(A_0 + \tfrac{1}{2}\rho\tilde{G}_{\mathbf{d}})^\top + B_\sigma\tilde{Y} + \tilde{Y}^\top B_\sigma^\top + (\rho + \overline{\gamma})X = 0,\\ \tilde{G}_{\mathbf{d}}X + X\tilde{G}_{\mathbf{d}}^\top \succ 0, \quad X \succ 0\end{gathered}.$$

If the pair X and $\tilde{Y}$ is a solution of the above LMI then for any $\lambda > 0$ the pair λX and $\lambda\tilde{Y}$ is also a solution of the latter system of LMIs as well. Hence, for a sufficiently small $\lambda > 0$ we derive $I_n \succeq \lambda X$. Therefore, the system (9.111) is feasible too.

For the selected matrix $G_{\mathbf{d}}$, we have

$$A_0 G_{\mathbf{d}} = (-0.5 I_n + G_{\mathbf{d}})A_0,\ G_{\mathbf{d}} B_\sigma = B_\sigma,\ A_0\mathbf{d}(s) = e^{-\frac{1}{2}s}\mathbf{d}(s)A_0,\ \mathbf{d}(s)B_\sigma = e^s B_\sigma,$$

and

$$\begin{aligned} f(\tilde{\mathbf{d}}(s)\zeta) &= \begin{pmatrix} A_0\mathbf{d}(s)\sigma + B_\sigma \|\mathbf{d}(s)\sigma\|_{\mathbf{d}}^{\frac{1}{2}} K\mathbf{d}(-\ln \|\mathbf{d}(s)\sigma\|_{\mathbf{d}})\mathbf{d}(s)\sigma + B_\sigma e^{\frac{1}{2}s}\eta \\ u_i(\mathbf{d}(s)\sigma) \end{pmatrix} \\ &= \begin{pmatrix} e^{-\frac{1}{2}s}\mathbf{d}(s)\left(A_0\sigma + B_\sigma \|\sigma\|_{\mathbf{d}}^{\frac{1}{2}} K\mathbf{d}(-\ln \|\sigma\|_{\mathbf{d}})\sigma + B_\sigma\eta\right) \\ u_i(\sigma) \end{pmatrix} \\ &= e^{-\frac{1}{2}s}\tilde{\mathbf{d}}(s) f(\zeta), \end{aligned}$$

where

$$f(\zeta) = \begin{pmatrix} A_0\sigma + B_\sigma \|\sigma\|_{\mathbf{d}}^{\frac{1}{2}} K\mathbf{d}(-\ln \|\sigma\|_{\mathbf{d}})\sigma + B_\sigma\eta \\ u_i(\sigma) \end{pmatrix}, \quad \zeta = \begin{pmatrix} \sigma \\ \eta \end{pmatrix}.$$

Therefore, the unperturbed system (9.109) is $\tilde{\mathbf{d}}$-homogeneous of degree $\mu = -0.5$.

Notice that the LMI $G_{\mathbf{d}}X + XG_{\mathbf{d}}^\top \succ X$ implies the monotonicity of the dilation $\hat{\mathbf{d}}$, so the inequality (9.112)is well defined.

Let us introduce the following differential inclusion:

$$\dot{\zeta} \in F(\zeta) := f(\zeta) \dot{+} \begin{pmatrix} B_\sigma \Gamma(\sigma) \dot{+} G_\kappa(\sigma) \\ \bar{B}(\overline{\gamma}_0) \end{pmatrix}, \quad t > 0, \quad \zeta = \begin{pmatrix} \sigma \\ \eta \end{pmatrix}, \tag{9.113}$$

where

$$\Gamma(\sigma) = \bar{B}(\bar{\gamma}\sqrt{\|\sigma\|_{\mathbf{d}}}),$$

$$G_\kappa(\sigma) = \left\{ z \in \mathbb{R}^p : \left\| \frac{z}{\kappa\beta} \right\|_{\hat{\mathbf{d}}} \le \|\sigma\|_{\mathbf{d}} \right\}, \quad G_\kappa(\mathbf{0}) = \{\mathbf{0}\},$$

and $\bar{B}(r) = \{y \in \mathbb{R}^m : |y| \le r\}$ is a ball in $\mathbb{R}^m$ of the radius $r > 0$. Since the mapping G_κ is compact valued and upper semi-continuous due to continuity of the canonical homogeneous norms $\|\cdot\|_{\hat{\mathbf{d}}}$ and $\|\cdot\|_{\mathbf{d}}$. Moreover, taking into account the equivalence (see Lemma 5.4)

$$\left\| \frac{z}{\kappa\beta} \right\|_{\hat{\mathbf{d}}} \le \|\sigma\|_{\mathbf{d}} \quad \Leftrightarrow \quad \left\| \hat{\mathbf{d}}(-\ln \|\sigma\|_{\mathbf{d}}) \frac{z}{\kappa\beta} \right\| \le 1$$

we conclude that G_κ is convex valued. Being the geometric sum of convex-valued, compact-valued and upper semi-continuous mappings, the set-valued vector field F is convex valued, compact valued and upper semi-continuous too. By Theorem 2.8, the system (9.113) has strong solutions.

By construction, F is $\tilde{\mathbf{d}}$-homogeneous of degree $\mu = -0.5$:

$$F(\mathbf{d}(s)\zeta) = e^{-0.5s}\tilde{\mathbf{d}}(s)F(\zeta), \quad \forall \zeta \in \mathbb{R}^{m+p}, \quad \forall s \in \mathbb{R}.$$

(2) *Finite-time stability of the* $\tilde{\mathbf{d}}$*-homogeneous inclusion.* First, let us show that the system (9.113) is finite-time stable for $\overline{\gamma}_0 = 0$ and for any $\kappa \in [0, \rho)$. Let us consider the Lyapunov function candidate

$$V = \|\sigma\|_{\mathbf{d}} + \eta^\top Q^{-1}\eta.$$

For $\overline{\gamma}_0 = 0$ using $\mathbf{d}$-homogeneity of F, we derive

$$\begin{aligned}
\dot{V}(\zeta) := & \sup_{y\in\Gamma(\sigma), z\in G_\kappa(\sigma)} \tfrac{\partial V(\zeta)}{\partial \zeta}\left(f(\zeta) + \begin{pmatrix} B_\sigma y + z \\ \mathbf{0}\end{pmatrix}\right) \\
& \sqrt{\|\sigma\|_{\mathbf{d}}}\frac{\sigma^\top \mathbf{d}^\top(-\ln\|\sigma\|_{\mathbf{d}})X^{-1}(A_0+B_\sigma K)\mathbf{d}(-\ln\|\sigma\|_{\mathbf{d}})\sigma}{0.5q(\sigma)} \\
& + \frac{\sigma^\top \mathbf{d}^\top(-\ln\|\sigma\|_{\mathbf{d}})X^{-1}B_\sigma(y+\eta)}{0.5q(\sigma)} + 2\eta^\top Q^{-1}u_{\mathrm{i}}(\sigma) \\
& + \frac{\|\sigma\|_{\mathbf{d}}\pi_{\mathbf{d}}^\top(\sigma)X^{-1}\mathbf{d}(-\ln\|\sigma\|_{\mathbf{d}})z}{\pi_{\mathbf{d}}{}^\top(\sigma)X^{-1}G_{\mathbf{d}}\pi_{\mathbf{d}}(\sigma)} \\
= & -\rho\|\sigma\|_{\mathbf{d}}^{\frac{1}{2}} + \frac{2\pi_{\mathbf{d}}^\top(\sigma)PB_\sigma\frac{y}{\sqrt{\|\sigma\|_{\mathbf{d}}}} - \bar{\gamma} - \bar{\gamma}\pi_{\mathbf{d}}^\top(\sigma)X^{-1}B_\sigma B_\sigma^\top X^{-1}\pi_{\mathbf{d}}(\sigma)}{q(\sigma)} \\
& + \frac{\|\sigma\|_{\mathbf{d}}\pi_{\mathbf{d}}^\top(\sigma)X^{-1}\mathbf{d}(-\ln\|\sigma\|_{\mathbf{d}})z}{0.5q(\sigma)},
\end{aligned}$$

where $\pi_{\mathbf{d}}(\sigma) = \mathbf{d}(-\ln\|\sigma\|_{\mathbf{d}})\sigma$ is a $\mathbf{d}$-homogeneous projection of σ on the unit sphere $\|\sigma\| = 1$. Notice that $\hat{\mathbf{d}}(s) = e^{0.5s}\mathbf{d}(s)$ and

$$z \in G_\kappa(\sigma) \quad \Rightarrow \quad \frac{\left\|\hat{\mathbf{d}}(-\ln\|\sigma\|_{\mathbf{d}})z\right\|}{\beta} \le \kappa \quad \Rightarrow \quad \frac{\sqrt{\|\sigma\|_{\mathbf{d}}}\pi_{\mathbf{d}}^\top(\sigma)X^{-1}\mathbf{d}(-\ln\|\sigma\|_{\mathbf{d}})z}{0.5q(\sigma)} \le \kappa.$$

Since $y \in \Gamma(\sigma)$ then $|y| \le \overline{\gamma}\sqrt{\|\sigma\|_{\mathbf{d}}}$ then repeating the proof of Theorem 9.5 we derive

$$\dot{V} \le -(\rho - \kappa)\|\sigma\|_{\mathbf{d}}^{0.5}. \tag{9.114}$$

Let us consider the set $\Omega = \{\zeta \in \mathbb{R}^{p+m} : V(\zeta) \le 1\}$ and let $\zeta(t_0) = \begin{pmatrix}\sigma(t_0)\\ \eta(t_0)\end{pmatrix} \in \partial\Omega$. If $\sigma(t_0) \neq \mathbf{0}$ then $F(\zeta(t_0)) = \begin{pmatrix}\eta \\ K[u_{\mathrm{i}}](\mathbf{0})\end{pmatrix}$ and for $\eta(t_0) \neq \mathbf{0}$ we derive $\sigma(t) \neq \mathbf{0}$ for all $t \in (t_0, t_0 + \delta)$, where $\delta > 0$ is small enough. If $\sigma(t_0) \neq \mathbf{0}$ then the inequality (9.114) implies $\zeta(t) \in \operatorname{int}\Omega$ for $t > t_0$. This means that Ω is a strictly positively invariant set of the system (9.113) with $\bar{\gamma}_0 = 0$ and $\kappa \in [0, \rho)$. By Proposition 7.2, this implies asymptotic stability of the $\mathbf{d}$-homogeneous system (9.113), but the negative homogeneity degree guarantees the finite-time stability (see Theorem 7.5).

Let us show that the same conclusion remains valid for a small enough $\bar{\gamma}_0 > 0$. Due to asymptotic stability of the system (9.113) for $\bar{\gamma}_0 = 0$ and $\kappa \in [0, \rho)$, by Zubov--Rosier theorem (see Theorem 7.6), there exists a $\mathbf{d}$-homogeneous Lyapunov function $V_1 \in C(\mathbb{R}^{p+m}) \cap C^\infty(\mathbb{R}^{p+m}\backslash\{\mathbf{0}\})$ of degree 1 and a constant $c_0 > 0$:

$$\sup_{y\in\Gamma(\sigma), z\in G_\kappa(\sigma)} \frac{\partial V_1(\zeta)}{\partial\zeta}\left(f(\zeta) + \begin{pmatrix} B_\sigma y + z \\ \mathbf{0}\end{pmatrix}\right) \leq -c_0, \quad \forall \zeta \neq \mathbf{0}.$$

If $\overline{\gamma}_0 \neq \mathbf{0}$ for the same Lyapunov function we have

$$\dot{V}_1(\zeta) = \sup_{q\in F(\zeta)} \frac{\partial V_1(\zeta)}{\partial\zeta} q \leq -c_0 + \sup_{y\in\bar{B}(\overline{\gamma}_0)} \frac{\partial V_1(\zeta)}{\partial\zeta}\left(\begin{smallmatrix}\mathbf{0}\\ y\end{smallmatrix}\right) \leq -c_0 + \overline{\gamma}_0 \left|\frac{\partial V_1(\zeta)}{\partial\zeta}\right|.$$

Since V is a $\mathbf{d}$-homogeneous function of degree 1 then $\zeta \mapsto \left|\frac{\partial V_1(\zeta)}{\partial\zeta}\right|$ is a $\mathbf{d}$ homogeneous function of degree 0. Since $V_1 \in C^\infty(\mathbb{R}^{p+m}\backslash\{\mathbf{0}\})$ then $\left|\frac{\partial V_1(\zeta)}{\partial\zeta}\right|$ is bounded on the unit sphere and, by Proposition 6.1, it is uniformly bounded on $\mathbb{R}^n\backslash\{\mathbf{0}\}$. Therefore, the system (9.113) remains finite-time stable provided that $\overline{\gamma}_0$ is sufficiently small.

(3) *Embedding of the system* (9.109) *into the homogeneous system* (9.113).

Notice that $\sigma^\top X^{-1}\sigma \geq \sigma^\top\sigma$ due to the LMI (9.111). For $|\sigma| \geq 1$ we have $\|\sigma\| \geq 1$ and $\|\sigma\|_{\mathbf{d}} \geq 1$. By Corollary 5.3, we have

$$|\sigma| \leq \|\sigma\| \leq \overline{\sigma}(\|\sigma\|_{\mathbf{d}}),$$

where

$$\underline{\sigma}(\rho) = \begin{cases} \rho^\beta & \text{if } \rho \leq 1, \\ \rho^\alpha & \text{if } \rho > 1, \end{cases}$$

$\alpha = \lambda_{\max}(X^{-1/2}G_{\mathbf{d}}X^{1/2} + X^{1/2}G_{\mathbf{d}}^\top X^{-1/2})$, $\beta = \lambda_{\min}(X^{-1/2}G_{\mathbf{d}}X^{1/2} + X^{1/2}G_{\mathbf{d}}^\top X^{-1/2})$. The inequality $G_{\mathbf{d}}X + XG_{\mathbf{d}}^\top \succ X$ (see (9.111)) guarantees that $\alpha \geq \beta = \lambda_{\min}(X^{-1/2}G_{\mathbf{d}}X^{1/2} + X^{1/2}G_{\mathbf{d}}^\top X^{-1/2}) \geq 0.5$. So, for $|\sigma| \leq 1$ we have

$$|\sigma|^2 \leq |\sigma|^{1/\beta} \leq \|\sigma\|^{1/\beta} \leq \|\sigma\|_{\mathbf{d}}.$$

Therefore, we derive

$$\min\{|\sigma|^2, 1\} \leq \|\sigma\|_{\mathbf{d}}, \quad \forall \sigma \in \mathbb{R}^p.$$

This means that under Assumptions 9.2 and 9.3 the system (9.113) is $\tilde{\mathbf{d}}$-homogeneous extension of the system (9.109). So, all solutions of the system (9.109) belong to the solution set of the finite-time stable system (9.113). The proof is complete. □

According to the proven theorem, the homogeneous control (9.108) with integral action enforces a sliding mode in the system (9.81) on the surface $Cx = 0$ in a finite

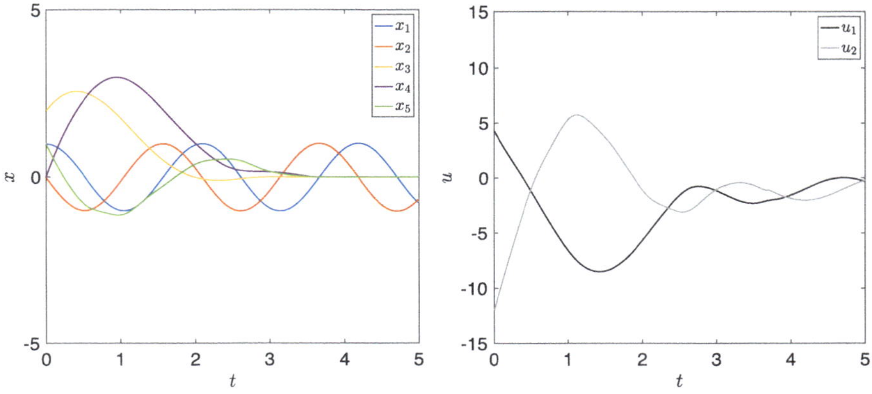

Fig. 9.18 The system with homogeneous SMC with integral action

time provided that the uncertain function γ satisfies Assumption 9.3. The known parameter $\overline{\gamma}$ is utilized for the tuning of the controller.

Example 9.12 Let us consider the system studied in Example 9.11 and select the parameters of the homogeneous super-twisting control according to Theorem 9.6:

$$K = \begin{pmatrix} 4.4043 & -2.2933 & 0.9462 \\ -4.4043 & 0.9462 & -2.2933 \end{pmatrix}, \; K_{\rm i} = \begin{pmatrix} 1.3174 & -0.6732 & 0.2849 \\ -1.3174 & 0.2849 & -0.6732 \end{pmatrix}$$

with $\bar{\gamma} = 0.5$. The simulation for homogeneous super-twisting control with $\gamma_0 = 1$ and $\tilde{\gamma} = |x_2|0.5(\sin(10t), \cos(10t))^\top$ and $\epsilon = 0.1$ is depicted in Fig. 9.18. The homogeneous super-twisting control does not have a "visible" chattering of the control input, which is typical for the conventional SMC without chattering attenuation (see Fig. 9.16). This feature of the conventional SISO super-twisting algorithm [37, 42] is inherited by its generalized homogeneous MIMO version (9.108).

Concluding Remarks

Various explicit and implicit homogeneous controllers for controllable linear plants are designed. The key feature of the presented controllers is the simplicity of their design and tuning. The feedback gains can always be selected by solving the linear matrix equations and inequalities. The settling time of the closed-loop system can be prescribed exactly by an arbitrary selected constant. The controllers reject a class of "conic" (in a generalized homogeneous sense) additive perturbations. The robustness (input-to-state stability) of the closed-loop control systems with respect to measurement noise and additive perturbations is proven as well. The obtained generalized homogeneous feedback can be treated as a linear feedback with the gain dependent on homogeneous norm. For the case of standard homogeneity, such a feedback with the degree $\mu = -1$ corresponds to the unit sliding-mode control. A generalized homogeneous unit sliding-mode control is introduced in this chapter. A homogeneous sliding-mode control with integral action is designed as well. It can be treated as a generalized homogeneous version of the super-twisting sliding-mode algorithm in the MIMO case.

References

1. Wonham WM (1985) Linear multivariable control: a geometric approach. Springer
2. Tarbouriech S, Garcia G (eds) (1997) Control of uncertain systems with bounded inputs. Lecture notes in control and information sciences. Springer
3. Chernous'ko FL, Ananievski IM, Reshmin SA (2008) Control of nonlinear dynamical systems. Springer
4. Utkin VI (1992) Sliding modes in control optimization. Springer, Berlin
5. Kalman R (1960) On the general theory of control systems. In: Proceedings of the first international congress of the IFAC, Moscow
6. Korobov VI (1979) A solution of the synthesis problem using controlability function. Doklady Academii Nauk SSSR 248:1051–1063
7. Korobov VI (2007) Method of controlability functions. NIC, Moscow (in Russian)
8. Adamy J (1991) Strukturvariable Regelungen mittels impliziter Ljapunov-Funktionen. PhD thesis, University of Dortmund, Germany
9. Adamy J, Flemming A (2004) Soft variable-structure controls: a survey. Automatica 40:1821–1844
10. Polyakov A, Efimov D, Perruquetti W (2015) Finite-time and fixed-time stabilization: implicit lyapunov function approach. Automatica 51(1):332–340
11. Polyakov A, Efimov D, Perruquetti W (2016) Robust stabilization of MIMO systems in finite/fixed time. Int J Robust Nonlinear Control 26(1):69–90
12. Praly L (1997) Generalized weighted homogeneity and state dependent time scale for linear controllable systems. In: Proceedings of IEEE conference on decision and control, San Diego, USA, pp 4342–4347
13. Polyakov A, Coron J-M, Rosier L (2018) On homogeneous finite-time control for linear evolution equation in Hilbert space. IEEE Trans Autom Control 63(9):3143–3150
14. Polyakov A (2019) Sliding mode control design using canonical homogeneous norm. Int J Robust Nonlinear Control 29(3):682–701
15. Zimenko K, Polyakov A, Efimov D, Perruquetti W (2020) Robust feedback stabilization of linear mimo systems using generalized homogenization. IEEE Trans Autom Control 65(12):5429–5436
16. Nekhoroshikh A, Efimov D, Polyakov A, Perruquetti W, Furtat I (2021) Finite-time stabilization under state constraints. In: Conference on decision and control
17. Boyd S, Ghaoui E, Feron E, Balakrishnan V (1994) Linear matrix inequalities in system and control theory. SIAM, Philadelphia
18. Polyakov A (2012) Nonlinear feedback design for fixed-time stabilization of linear control systems. IEEE Trans Autom Control 57(8):2106–2110
19. Misrikhanov M, Ryabchenko V (2011) Pole placement for controlling a large scale power system. Autom Remote Control 72(10):2123–2146
20. Polyakov A, Efimov D, Ping X (2023) Consistent discretization of homogeneous finite/fixed-time controllers for LTI systems. Automatica 155:111118
21. Zimenko K, Polyakov A, Efimov D, Perruquetti W (2020) On simple scheme of finite/fixed-time control design. Int J Control 93(6):1353–1361

22. Sontag ED, Wang Y (1996) On characterizations of the input-to-state stability property. Syst Control Lett 24(5):351–359
23. Chaillet A, Angeli D, Ito H (2014) Combining iiss and iss with respect to small inputs: the strong iiss property. IEEE Trans Autom Control 59(9):2518–2524
24. Song Y, Wang Y, Holloway J, Krstic M (2017) Time-varying feedback for regulation of normal-form nonlinear systems in prescribed finite time. Automatica 83:243–251
25. Polyakov A, Efimov D, Ping X (2021) On energetically optimal finite-time stabilization. In: Conference on decision and control
26. Aldana-Lopez R, Seeber R, Haimovich H, Gomez-Gutierrez D (2023) On inherent limitations in robustness and performance for a class of prescribed-time algorithms. Automatica 158:111284
27. Polyakov A, Krstic M (2023) Finite-and fixed-time nonovershooting stabilizers and safety filters by homogeneous feedback. IEEE Trans Autom Control
28. Polyakov A, Krstic M (2024) Fixed-time stabilization with a prescribed constant settling time by static feedback for delay-free and input delay systems. Int J Robust Nonlinear Control 34(10):6393–6415
29. Filippov AF (1988) Differential equations with discontinuous right-hand sides. Kluwer Academic Publishers
30. MercadoUribe A, Moreno, JA Polyakov A, Efimov D (2021) Multiple input multiple output homogeneous integral control design using the implicit Lyapunov function approach. Int J Robust and Nonlinear Control
31. Polyakov A (2020) Generalized homogeneity in systems and control. Springer
32. Zhou Y, Polyakov A, Zheng G (2023) Generalized homogeneous control with integral action. Int J Robust Nonlinear Control 33(8):4345–4366
33. Utkin VI (1977) Variable structure systems with sliding modes. IEEE Trans Autom Control AC-22:212–222
34. Edwards C, Spurgeon S (1998) Sliding mode control: theory and applications. Taylor and Francis
35. Shtessel Y, Edwards C, Fridman L, Levant A (2014) Sliding mode control and observation. Birkhauser
36. Utkin V, Poznyak A, Orlov Y, Polyakov A (2020) Road map for sliding mode control design. Springer
37. Levant A (2003) Higher-order sliding modes, differentiation and output-feedback control. Int J Control 76(9–10):924–941
38. Orlov Y (2005) Finite time stability and robust control synthesis of uncertain switched systems. SIAM J Control Optim 43(4):1253–1271
39. Gutman S, Leitmann G (1976) Stabilizing feedback control for dynamical systems with bounded uncertainties. In: IEEE conference on decision and control, pp 94–99
40. Polyakov A (2024) Homogeneous unit sliding mode control. IEEE Trans Autom Control 69(2):1157–1163
41. Levant A (2005) Homogeneity approach to high-order sliding mode design. Automatica 41(5):823–830
42. Levant A (1993) Sliding order and sliding accuracy in sliding mode control. Int J Control 58(6):1247–1263
43. Moreno J, Osorio M (2012) Strict Lyapunov functions for the super-twisting algorithm. IEEE Trans Autom Control 57:1035–1040

Chapter 10
Digitization of Homogeneous Control

Digitization is the creation of digital signals from its analog originals. Considering a continuous signal as a function of time, the *digitization of the time domain* is known as sampling, but the *digitization of the co-domain* is referred to as quantization. Control laws developed for a continuous-time models require a digitization in order to be implemented in computer-based controllers. This chapter studies various issues related with digitization of homogeneous control algorithms. In particular, a consistent discretization (sampled-time implementation) of the implicit homogeneous control law is developed in such a way that the finite/fixed-time convergence rate of the original continuous-time system is preserved. Various issues of homogeneous control design with quantization are discussed. Finite/fixed-time stabilization by means of a quantized homogeneous feedback is proven.

10.1 Realization by Switched Linear Feedback

In this chapter, we deal with LTI model

$$\dot{x} = Ax + Bu, \quad t > 0, \quad x(t) \in \mathbb{R}^n, \quad u(t) \in \mathbb{R}^m, \quad x(0) = x_0 \tag{10.1}$$

stabilized by the implicit homogeneous controller (see Chap. 9)

$$u = \hat{K}(\|x\|_{\mathbf{d}})x, \quad \hat{K}(r) = K_0 + r^{1+\mu} K\mathbf{d}(-r), \quad r > 0. \tag{10.2}$$

For any fixed $\|x\|_{\mathbf{d}} = r$, such a homogeneous controller becomes the static linear feedback. In the first section of this chapter, we consider the simplest digitization of the homogeneous controller, where the linear feedback in the control system is

A. Polyakov, *Generalized Homogeneity in Systems and Control Volume I*,
Communications and Control Engineering,
https://doi.org/10.1007/978-3-031-76246-8_10

assumed to be realized by an analog controller, but its gain $\hat{K}(r)$ can be assigned by a digital device. In other words, the homogeneous controller is realized by a switched linear feedback obtained via digitization (sampling or quantization) of the continuous scalar signal $t \mapsto \|x(t)\|_{\mathbf{d}}$ in (10.2).

10.1.1 Norm-Quantized Homogeneous Control

First, we study the digitization of the co-domain of the signal $t \mapsto \|x(t)\|_{\mathbf{d}}$. This means that the values of $\|x(t)\|_{\mathbf{d}}$ in (10.2) are assumed to be quantized, so the gain $\hat{K}(r)$ may take values $\hat{K}(r_i)$, where $r_i > 0$ belongs to a *discrete set*

$$\mathcal{Q} = \{r_0, r_{\pm 1}, r_{\pm 2}, \ldots\}.$$

This situation may happen in a real control system having an analog linear feedback, which gain $\hat{K}(r)$ can be modified by a *supervisor* (an external digital device) connected with the controller by a network. The supervisor can measure the whole state vector of the system and compute/estimate the value of the canonical homogeneous norm (e.g., using the algorithms developed in Chap. 5), but, to optimize, the network transmission rate, the supervisor can send to the controller only the integer number i, which corresponds to the static linear feedback law $\hat{K}(r_i)x$ to be applied. A possible mathematical description of such a piece-wise linear realization of the homogeneous control law is given by

$$u_{\sigma(t)}(x(t)) = r_{\sigma(t)}^{1+\mu} K\mathbf{d}(-\ln r_{\sigma(t)})x(t), \tag{10.3}$$

with a state-dependent switching function σ satisfying $\sigma(0) \in \mathbb{Z}$,

$$\sigma(t) = S(\sigma(t+0^-), x(t+0^-)), \quad S : \mathbb{Z} \times \mathbb{R}^n \mapsto \mathbb{Z},$$

$$S(\sigma, x) := \begin{cases} i-1 \text{ if } & \sigma = i \text{ and } \|x\|_{\mathbf{d}} \geq r_{i-1}, \\ \quad i \quad \text{if } \sigma = i \text{ and } r_{i+1} < \|x\|_{\mathbf{d}} < r_{i-1}, \\ i+1 \text{ if } & \sigma = i \text{ and } r_{i+1} \leq \|x\|_{\mathbf{d}}, \end{cases} \tag{10.4}$$

where $r_i \in \mathbb{R}_+ : r_{i+1} < r_i$, $i \in \mathbb{Z}$ and $\sigma(t+0^-) := \lim_{\tau \to 0^-} \sigma(t+\tau)$ (resp., $x(t+0^-) := \lim_{\tau \to 0^-} x(t+\tau)$). The switching variable σ has its own evolution equation given by (10.4), which represents the behavior of the supervisor. So, the pair $(x(t), \sigma(t))$ can be treated as an augmented state of the closed-loop switched system. We refer the reader to [1] for more details about switched systems.

By definition, a solution of the system (10.1), (10.3), (10.4) is a pair $(x(t), \sigma(t)) \in \mathbb{R}^n \times \mathbb{Z}$ defined on $[0, T)$ such that a right-continuous[1] function σ satisfies (10.4)

[1] A function $\sigma : [0, T) \mapsto \mathbb{R}$ is said to be right continuous if for any $t \in [0, T)$ and any $\varepsilon > 0$ there exists $\delta > 0$ such that $\tilde{t} \in [t, t+\delta) \cap [0, T) \Rightarrow |\sigma(\tilde{t}) - \sigma(t)| < \varepsilon$.

for all $t \in [0, T)$ and an absolutely continuous function x is a strong solution of the system (see Chap. 2), where $T \leq +\infty$. Notice that for any locally measurable function $\sigma : [0, T) \mapsto \mathbb{Z}$ the system (10.1), (10.3) is linear and time varying. It, indeed, has a unique strong (Carathéodory) solution.

To preserve the stability properties of the original homogeneous control system, the quantization levels r_i have to be properly selected.

Corollary 10.1 *Let the homogeneous control* (10.2) *be defined according to Theorem 9.1 for some $\mu < 0$. If the linear switching control* (10.3), (10.4) *has the parameters*

$$r_0 = 1, \quad r_{i+1} = e^{-q_i} r_i, \quad i \in \mathbb{Z}$$

and $q_i > 0$ such that

$$\sum_{i=0}^{+\infty} q_i = +\infty \quad \text{and} \quad \lim_{i \to +\infty} q_{i+1} e^{\mu q_i} / q_i < 1,$$

then

- *for any $x_0 \in \mathbb{R}^n \backslash \{\mathbf{0}\}$ the closed-loop system* (10.1), (10.3), (10.4) *has a unique solution $(x(t), \sigma(t)) \in \mathbb{R}^n \times \mathbb{Z}$ defined on $[0, T(x_0))$ such that*

$$\|x(t)\| \to 0 \text{ as } t \to T(x_0),$$

where the settling-time function $T : \mathbb{R}^n \mapsto \mathbb{R}$ admits the estimate

$$T(x_0) \leq \frac{\alpha}{\beta \rho} \sum_{j=i_0}^{\infty} r_j^{-\mu} q_j < +\infty$$

with

$$i_0 \in \mathbb{N} \quad : \quad \|x_0\|_{\mathbf{d}} \in (r_{i_0+1}, r_{i_0}];$$

- *the time instants $t_j : \|x(t_j)\|_{\mathbf{d}} = r_j$ are isolated and the set $\{t_j\}_{j=i_0+1}^{+\infty}$ has the unique accumulation point $T(x_0)$, i.e., $t_j \to T(x_0)$ as $j \to +\infty$.*

Proof Let us denote $\mathcal{K}(a, b) = \{x \in \mathbb{R}^n : a < \|x\|_{\mathbf{d}} \leq b\}$ and $K_i = K_0 + r_i^{1+\mu} K\mathbf{d}(-\ln r_i)$, where $\|x\|_{\mathbf{d}}$ is the canonical homogeneous norm induced by the norm $\|x\| = \sqrt{x^\top X^{-1} x}$ and $X \succ 0$ is defined in Theorem 9.1. For any initial condition $x(t_i) = x_i \in \mathbb{R}^n$ the equation

$$\dot{x}(t) = (A + BK_i)x(t), \quad t > t_i \tag{10.5}$$

has a unique classical solution $x_{i,x_i}(t) \in \mathbb{R}^n$ defined for all $t \geq t_i$. By construction we have $r_{i+1} < r_i$, $i \in \mathbb{Z}$. Since $\|x_{i,x_i}(\cdot)\|_{\mathbf{d}} : \mathbb{R} \mapsto \mathbb{R}_+$ is continuous then for $x(t_i) =$

$x_i \in \mathcal{K}(r_{i+1}, r_{i-1})$ we have $\sigma(t_i) = i$ and the closed-loop system (10.1), (10.3), (10.4) has a unique classical solution x defined at least on $[t_i, t_{i+1})$ and $x(t) = x_{i,x_i}(t)$ for $t \in [t_i, t_{i+1})$, where $t_{i+1} = +\infty$ or $\lim\limits_{t\to t_{i+1}} \|x(t)\|_{\mathbf{d}} = r_{i-1}$ or $\lim\limits_{t\to t_{i+1}} \|x(t)\|_{\mathbf{d}} = r_{i+1}$. Let us show that the latter case with $t_{i+1} < +\infty$ is only possible. Indeed, since

$$\begin{gathered}
\tfrac{d}{dt}\|\mathbf{d}(-\ln r_i)x(t)\|^2 = 2x^\top(t)\mathbf{d}^\top(-\ln r_i)X^{-1}\mathbf{d}(-\ln r_i)\dot{x}(t) = \\
x^\top(t)\mathbf{d}^\top(-\ln r_i)X^{-1}\mathbf{d}(-\ln r_i)\left(A_0 + r_i^{1+\mu}BK\mathbf{d}(-\ln r_i)\right)x(t) = \\
2r_i^\mu x^\top(t)\mathbf{d}^\top(-\ln r_i)X^{-1}(A_0 + BK)\mathbf{d}(-\ln r_i)x(t) = \\
-2\rho r_i^\mu x^\top(t)\mathbf{d}^\top(-\ln r_i)X^{-1}G_{\mathbf{d}}\mathbf{d}(-\ln r_i)x(t) \le -2\rho r_i^\mu \beta\|\mathbf{d}(-\ln r_i)x(t)\|^2,
\end{gathered}$$

where $A_0 = A + BK_0$, the parameter $\beta > 0$ is defined in Proposition 5.3, then the locally Lipschitz continuous function $t \to \|\mathbf{d}(-\ln r_i)x(t)\|$ decays exponentially. Taking into account $\|\mathbf{d}(-\ln r_i)x\|^2 = 1$ if $\|x\|_{\mathbf{d}} = r_i$ and $\|\mathbf{d}(-\ln r_i)x\|^2 < 1$ if $\|x\|_{\mathbf{d}} < r_i$, we derive $\|x(t)\|_{\mathbf{d}} < r_i$ for $t > t_i$ and $\lim_{t\to t_{i+1}} \|x(t)\|_{\mathbf{d}} = r_{i+1}$. Moreover, due to exponential decay rate the time instant $t_{i+1} : \|x(t_{i+1})\|_{\mathbf{d}} = r_{i+1}$ admits the estimate

$$t_{i+1} \le t_i + r_i^{-\mu}\beta^{-1}\rho^{-1}\ln\left(\frac{\|\mathbf{d}(-\ln r_i)x(t_i)\|}{\varepsilon_i}\right) \le t_i + r_i^{-\mu}\beta^{-1}\rho^{-1}\ln\varepsilon_i^{-1},$$

where $\varepsilon_i \in (0, +\infty)$ is such that

$$\|\mathbf{d}(-\ln r_i)x\| \le \varepsilon_i \quad \Rightarrow \quad \|x\|_{\mathbf{d}} \le r_{i+1}, \quad \forall x \in \mathbb{R}^n.$$

Repeating the same considerations for $[t_{i+1}, t_{i+2})$, for $[t_{i+2}, t_{i+3})$, …and taking into account uniqueness and continuity of solutions between switchings we derive that for any initial condition $x_0 \in \mathbb{R}^n$ and $\sigma(0) = i_0$ such that $r_{i_0+1} < \|x_0\|_{\mathbf{d}} < r_{i_0-1}$ the closed-loop system (10.1), (10.3), (10.4) has a unique solution $(x(t), \sigma(t)) \in \mathbb{R}^n \times \mathbb{Z}$ defined on $[0, T(x_0))$ and the continuous trajectory $t \mapsto x(t)$ may not have a time derivative only at time instances $t_i : \|x(t_i)\|_{\mathbf{d}} = r_i$, where

$$T(x_0) = \sum_{j=i_0}^{+\infty} t_{j+1} - t_j \le \sum_{j=i_0}^{+\infty} r_j^{-\mu}\beta^{-1}\rho^{-1}\ln\varepsilon_j^{-1}$$

is the unique accumulation point of the set $\{t_i\}$. In addition, $\sigma(t) = \text{const}$ for $t \in [t_i, t_{i+1})$. Since $\sum_{i=0}^{+\infty} q_i = +\infty$ then $r_i = \frac{1}{e^{\sum_{j=0}^{i-1} q_j}} \to 0$ as $i \to +\infty$ and $\|x(t)\| \to 0$ as $t \to T(x_0)$.

Let us show that

$$\|\mathbf{d}(-\ln r_i)x\| \le e^{-\alpha q_i} \Rightarrow \|\mathbf{d}(-\ln r_{i+1})x\|_{\mathbf{d}} \le 1.$$

Indeed, using the $\mathbf{d}$-homogeneity of $\|\cdot\|_{\mathbf{d}}$ we derive

$$\|\mathbf{d}(-\ln r_{i+1})x\|_{\mathbf{d}}^{\alpha} = \|\mathbf{d}(q_i)\mathbf{d}(-\ln r_i)x\|_{\mathbf{d}}^{\alpha} =$$

$$\left(e^{q_i}\|\mathbf{d}(-\ln r_i)x\|_{\mathbf{d}}\right)^{\alpha} = e^{\alpha q_i}\|\mathbf{d}(-\ln r_i)x\|_{\mathbf{d}}^{\alpha}.$$

Therefore, taking $\varepsilon_i = e^{-\alpha q_i}$ we derive

$$t_{i+1} \le t_i + \alpha\rho^{-1}\beta^{-1}r_i^{-\mu}q_i$$

and the settling time $T(x_0)$ is finite since the ratio test

$$\lim_{i\to+\infty}\frac{r_{i+1}^{-\mu}q_{i+1}}{r_i^{-\mu}q_i} = \lim_{i\to+\infty} q_{i+1}e^{\mu q_i}/q_i < 1$$

implies that the series, which estimates $T(x_0)$, is convergent. The proof is complete. □

10.1.2 Norm-Sampled Homogeneous Control

Let us study the robustness of the homogeneous feedback (10.2) in the case of the digitization of the domain of the signal $t \mapsto \|x(t)\|_{\mathbf{d}}$. In other words, we assume that the gain $\hat{K}(\|x(t)\|_{\mathbf{d}})$ can be changed only in sampled instances of time:

$$u(t) = \hat{K}(\|x(t_i)\|_{\mathbf{d}})x(t), \quad t \in [t_i, t_{i+1}), \tag{10.6}$$

where $t_i < t_{i+1}$ with $i = 0, 1, \ldots$ and $t_0 = 0$. The control law (10.6) can be interpreted as a piece-wise linear feedback with both time- and state-dependent switching law.

In this subsection, we study the perturbed system

$$\dot{x} = Ax + Bu + g(t, x), \quad t \ge t_0, \quad x(t_0) = x_0 \in \mathbb{R}^n, \tag{10.7}$$

where x, A, B, u are as before, but the function $g : \mathbb{R} \times \mathbb{R}^n \mapsto \mathbb{R}^n$ models uncertainties and disturbances of the system.

Theorem 10.1 *Let the homogeneous control* (10.2) *be defined according to Theorem 9.1 and all conditions of Corollary 9.2 be fulfilled. For any fixed $r > 0$, the closed* $\mathbf{d}$*-homogeneous ball $\overline{B}_{\mathbf{d}}(r)$ is a strictly positively invariant*[a] *compact set of the perturbed closed-loop system* (10.7) *with the linear control*

$$u_r(x) = K_0x + r^{1+\mu}K\mathbf{d}(-\ln r)x. \tag{10.8}$$

[a] A compact set $\Omega \subset \mathbb{R}^n$ with nonempty interior is said to be strictly positively invariant for a dynamical system if $x(t_0) \in \Omega \Rightarrow x(t) \in \text{int}\,\Omega$, $t > t_0$, where x denotes a trajectory of the dynamical system and int Ω is the interior of Ω.

Proof Let us denote

$$K_r = r^{1+\mu} K\mathbf{d}(-\ln r), \quad P_r = \mathbf{d}^\top(\ln r)P\mathbf{d}(\ln r), \quad \rho_r = r^\mu \rho, \quad \kappa_r = r^\mu \kappa.$$

In this case, multiplying (9.9) on $\mathbf{d}^\top(-\ln r)$ from the left and on $\mathbf{d}(-\ln r)$ from the right, we derive

$$\mathbf{d}^\top(-\ln r)(A_0 + BK)^\top \mathbf{d}^\top(\ln r)P_r + P_r\mathbf{d}(\ln r)(A_0 + BK)\mathbf{d}(-\ln r) +$$

$$\rho(\mathbf{d}^\top(-\ln r)G_{\mathbf{d}}^\top \mathbf{d}^\top(\ln r)P_r + P_r\mathbf{d}(\ln r)G_{\mathbf{d}}\mathbf{d}(-\ln r)) = \mathbf{0}\,.$$

Taking into account $G_{\mathbf{d}}\mathbf{d}(-\ln r) = \mathbf{d}(-\ln r)G_{\mathbf{d}}$, $A_0\mathbf{d}(-\ln r) = r^{-\mu}\mathbf{d}(-\ln r)A_0$ and $\mathbf{d}(\ln r)B = rB$ we derive

$$r^{-\mu}[(A_0 + BK_r)^\top P_r + P_r(A_0 + BK_r)] + \rho(G_{\mathbf{d}}^\top P_r + P_r G_{\mathbf{d}}) = \mathbf{0}$$

or, equivalently,

$$(A_0 + BK_r)^\top P_r + P_r(A_0 + BK_r) + \rho_r(G_{\mathbf{d}}^\top P_r + P_r G_{\mathbf{d}}) = 0.$$

Hence, the time derivative of the Lyapunov function candidate $V(x) = x^T P_r x$ with $x \in \mathbb{R}^n$ along a trajectory of the closed-loop linear system (10.7), (10.8) is given by

$$\dot{V}(x(t)) = x^\top(t)[P_r(A_0 + BK_r) + (A_0 + BK_r)^\top P_r + \rho_r(G_{\mathbf{d}}^\top P_r + P_r G_{\mathbf{d}})]x(t) +$$

$$2\tilde{g}^\top(t)P_r x(t) - \rho_r x^\top(t)(P_r G_{\mathbf{d}} + G_{\mathbf{d}}^\top P_r)x(t),$$

where $\tilde{g}(t) \overset{a.e.}{\in} K[g](t, x(t))$ and $K[g](t, x)$ is a Filippov regularization of a (possibly discontinuous) perturbation $g(t, x)$ (see Chap. 2 or the book [2]). For $\|x(t)\|_{\mathbf{d}} = r$ using (9.41), we derive

$$2\tilde{g}^\top(t)P_r x(t) \le \kappa_r x^\top(t)(P_r G_{\mathbf{d}} + G_{\mathbf{d}}^\top P_r)x(t).$$

Hence, we conclude that $\dot{V}(x) \le -(\rho_r - \kappa_r)x^\top(P_r G_{\mathbf{d}} + G_{\mathbf{d}}^\top P_r)x < 0$ if $\|x\|_{\mathbf{d}} = r$ (or, equivalently, if $V(x) = r$). This guarantees that $B_{\mathbf{d}}(r)$ is a strictly positively invariant set of the closed-loop linear system. The proof is complete. □

Now we assume that the value $\|x(t)\|_{\mathbf{d}}$ in the implicit homogeneous controller can be changed only in some sampled instances of time. Theorem [3] shows that the corresponding linear switched feedback robustly stabilizes the perturbed linear system.

Corollary 10.2 *Let the homogeneous control* (10.2) *be defined according to Theorem 9.1, all conditions of Corollary 9.2 be fulfilled and* $\{t_i\}_{i=0}^{+\infty}$ *be an arbitrary sequence of time instances such that*

$$0 = t_0 < t_1 < t_2 < \dots \quad \textit{and} \quad \lim_{i\to+\infty} t_i = +\infty.$$

The closed-loop system (10.7), (10.6) *is globally asymptotically stable.*

Proof **I.** Let us show that the sequence $\{\|x(t_i)\|_{\mathbf{d}}\}_{i=1}^{+\infty}$ is monotone decreasing along any solution of the closed-loop system. Indeed, let us define the quadratic positive definite function $V_i : \mathbb{R}^n \mapsto \mathbb{R}_+$ given by $V_i(x) := x^T P_i x$, where $P_i := \mathbf{d}^\top(-\ln\|x(t_i)\|_{\mathbf{d}})P\mathbf{d}(-\ln\|x(t_i)\|_{\mathbf{d}}) > 0$ and $P = X^{-1}$.

On the time interval $[t_i, t_{i+1})$ we have $u(x) = K_i x$, where

$$K_i := K_0 + \|x(t_i)\|_{\mathbf{d}}^{1+\mu} K\mathbf{d}(-\ln\|x(t_i)\|_{\mathbf{d}}).$$

Repeating the proof of Theorem 10.1, we derive

$$\dot{V}_i(x(t)) \le -(\rho - \kappa)\|x(t_i)\|_{\mathbf{d}}^{\mu} x(t)^T(P_i G_{\mathbf{d}} + G_{\mathbf{d}}^\top P_i)x(t) < 0$$

for $t \in [t_i, t_{i+1})$. From the definition of the canonical homogeneous norm $\|\cdot\|_{\mathbf{d}}$ we derive $V_i(x(t_i)) = 1$. In this case, for all $t \in (t_i, t_{i+1}]$ we have

$$V_i(x(t)) - 1 = x^\top(t)\mathbf{d}^\top(-\ln\|x(t_i)\|_{\mathbf{d}})P(-\ln\|x(t_i)\|_{\mathbf{d}})x(t) - 1 < 0 = \\ x^\top(t)\mathbf{d}^\top(-\ln\|x(t)\|_{\mathbf{d}})P(-\ln\|x(t)\|_{\mathbf{d}})x(t) - 1.$$

This implies $\|x(t)\|_{\mathbf{d}} < \|x(t_i)\|_{\mathbf{d}}$ for all $t \in (t_i, t_{i+1}]$, i.e., the sequence $\{\|x(t_i)\|_{\mathbf{d}}\}_{i=1}^{+\infty}$ is monotone decreasing and $x(t) \in B_{\mathbf{d}}(\|x(t_i)\|_{\mathbf{d}})$ for all $t \ge t_i$. Moreover, $V(x(t)) \le V(x(0))$ for all $t \ge 0$, i.e., the origin of the closed-loop system is *Lyapunov stable*.

II. Since the canonical homogeneous norm $\|\cdot\|_{\mathbf{d}}$ is positive definite then the monotone decreasing sequence $\{\|x(t_i)\|_{\mathbf{d}}\}_{i=1}^{\infty}$ converges to some limit. Let us show now that this limit is zero. Suppose the contrary, i.e., $\lim_{i\to+\infty}\|x(t_i)\|_{\mathbf{d}} = r_* > 0$ or, equivalently, $\forall \varepsilon > 0, \exists N = N(\varepsilon) : r_* \le \|x(t_i)\|_{\mathbf{d}} < r_* + \varepsilon,\ \forall i \ge N$. The control function $(r, x) \mapsto K(r)x$ is continuous on $(0, +\infty) \times \mathbb{R}^n\backslash\{\mathbf{0}\}$, so

$$\left\| \|x(t_i)\|_{\mathbf{d}}^{\mu+1} K\mathbf{d}(-\ln\|x(t_i)\|_{\mathbf{d}})x - r_*^{\mu+1}K\mathbf{d}(-\ln r_*)x \right\| \le \sigma(\varepsilon)\|x\|, \quad \forall i \ge N,$$

where $\sigma(\cdot) \in \mathcal{K}$. Therefore, for $t > t_N$ the closed-loop system can be presented in the form:

$$\dot{x}(t) = (A + B(K_* + \Delta(t,\varepsilon))x + \tilde{g}(t), \tag{10.9}$$

where $K_* = K_0 + r_*^{1+\mu}K\mathbf{d}(-\ln r_*)$, $\tilde{g}(t) \in K[g](t, x(t))$ and $\Delta(t,\varepsilon) \in \mathbb{R}^{m\times n}$ is such that $\|\Delta(t,\varepsilon)\| \le \sigma(\varepsilon)$.

Let us consider the quadratic positive definite Lyapunov function candidate $V_*(x) = x^T P_* x$, where $P_* = \mathbf{d}^\top(-\ln r_*) P \mathbf{d}(-\ln r_*)$. For $t > t_N$ we have

$$\dot{V}_*(x(t)) \leq -(\rho - \kappa) r_*^\mu x(t)^\top (P_* G_\mathbf{d} + G_\mathbf{d}^\top P_*) x(t) + \\ x^\top(t)(P_* B \Delta + \Delta^T B^\top P_*) x(t).$$

Notice that $P_* G_\mathbf{d} + G_\mathbf{d}^\top P_* \succ 0 \Leftrightarrow P G_\mathbf{d} + G_\mathbf{d}^\top P \succ 0$. Taking into account $\sigma \in \mathcal{K}$, we conclude that

$$\lambda_{\max}(P_* B \Delta + \Delta^T B^\top P_*) \to 0 \quad \text{as} \quad \varepsilon \to 0.$$

Since $\rho - \kappa > 0$ (see Corollary 9.2) then for sufficiently small $\varepsilon > 0$ (i.e., for sufficiently large t_N) we have

$$\dot{V}_*(x(t)) \leq -\frac{\rho - \kappa}{2} r_*^\mu x(t)^\top (P_* G_\mathbf{d} + G_\mathbf{d}^\top P_*) x(t),$$

and the solution of the closed-loop system decays exponential to zero implying the existence of an instant of time $t_* > t_N$ such that $\|x(t_*)\|_\mathbf{d} < r_*$. This contradicts our supposition, so $\lim_{i\to\infty} \|x(t_i)\|_\mathbf{d} = 0$. Hence, using the Lyapunov stability proven above, we conclude the global asymptotic stability of the closed-loop system with the switched homogeneous control (10.6). The proof is complete. □

According to the proven corollary, the proposed sampled-time realization of the implicit homogeneous controller guarantees an asymptotic stabilization of the closed-loop system *independently of the dwell time* (a time between to sampling instants). Such property is rather usual for sampled and switched linear control systems with additive disturbances [1]. However, without an information about the dwell time, we cannot estimate the convergence rate of this system. Obviously, if the dwell time tends to zero the convergence rate tends to the rate of the original continuous system.

Some advanced schemes for a discrete-time approximation of homogeneous control systems are developed in the next section. They preserve the convergence rate (e.g., a finite-time or a fixed-time stability) of the original continuous-time homogeneous system in its discrete-time counterpart. However, these algorithms still need online computation of the canonical homogeneous (or its discrete-time analog). Fortunately, rather simple numerical procedures can be utilized for this purpose (see Chap. 5). Moreover, the canonical homogeneous norm can be estimated online together with piece-wise sampled-time implementation of the implicit homogeneous control.

Let $0 = t_0 < t_1 < t_2 < \ldots$ be an arbitrary sequence of time instants such that $t_i \to +\infty$ as $i \to +\infty$. Inspired by [3, 4], let us propose the following algorithm:

Algorithm 10.1 *(Sampled estimation of homogeneous norm)*

Initialization *at* $t = t_0$*:* $\underline{V}(t_0) = a$*;* $\overline{V}(t_0) = b$*;* $N_{\max} \in \mathbb{N}$*;*

Step *at* $t = t_{i+1}$*:* $\underline{V}(t_{i+1}) = \underline{V}(t_i)$*;* $\overline{V}(t_{i+1}) = \overline{V}(t_i)$*;*

if $x^\top(t_{i+1})\mathbf{d}^\top(-\ln \overline{V}(t_{i+1}))P\mathbf{d}(-\ln \overline{V}(t_{i+1}))x(t_{i+1}) > 1$ *then*

$\underline{V}(t_{i+1}) = \overline{V}(t_{i+1})$*;* $\overline{V}(t_{i+1}) = \min(b, 2\overline{V}(t_{i+1}))$*;*

elseif $x^\top(t_{i+1})\mathbf{d}^\top(-\ln \underline{V}(t_{i+1}))P\mathbf{d}(-\ln \underline{V}(t_{i+1}))x(t_{i+1}) > 1$ *then*

$\overline{V}(t_{i+1}) = \underline{V}(t_{i+1})$*;* $\underline{V}(t_{i+1}) = \max(0.5\underline{V}(t_{i+1}), a)$*;*

else

for $i = 1 : N_{\max}$

$V = \frac{\underline{V}(t_{i+1})+\overline{V}(t_{i+1})}{2}$*;*

if $x^\top(t_{i+1})\mathbf{d}^\top(-\ln V)P\mathbf{d}(-\ln V)x(t_{i+1}) < 1$ *then*

$\overline{V}(t_{i+1}) = V$*;*

else $\underline{V}(t_{i+1}) = V$*;*

endif;

endfor;

endif;

$\|x(t_{i+1})\|_{\mathbf{d}} \approx \overline{V}(t_{i+1})$*;*

Let $x(t_i) \in \mathbb{R}^n \backslash \{\mathbf{0}\}$ be a given vector and $a = 0, b = +\infty$. If the system is slow enough (or sampling is fast enough) then $\|x(t_i)\|_{\mathbf{d}} \approx \|x(t_{i-1})\|_{\mathbf{d}}$. So, applying **Step** of the algorithm at each sampled instant of time $t = t_i$ we realize

(1) a localization of the unique positive root of the algebraic equation $\|\mathbf{d}(-\ln V(t_i))x(t_i)\| = 1$ with respect to $V(t_i) > 0$, i.e., we have $V(t_i) \in [\underline{V}(t_i), \overline{V}(t_i)]$ after a finite number of time samplings t_i;

(2) improvement of the obtained localization by means of the bisection method, i.e., $(\overline{V}(t_i) - \underline{V}(t_i)) \to 0$ as time grows.

In between two sampled time instances, the control is realized by means of the linear static feedback

$$u(t) = K_i x(t), \quad \forall t \in [t_i, t_{i+1}), \tag{10.10}$$

with the gain

$$K_i = (K_0 + \overline{V}^{\mu}(t_i)K\mathbf{d}(-\ln \overline{V}(t_i))).$$

At time $t = t_{i+1}$ the gain has to be recomputed. To obtain K_{i+1} the estimate $\overline{V}(t_{i+1})$ of $\|x(t_{i+1})\|_{\mathbf{d}}$ is obtained by means of **Step** of Algorithm 10.1.

Such an application of the homogeneous control law allows us to estimate the canonical homogeneous norm $\|x(t_i)\|_{\mathbf{d}}$ online with rather high precision at each step provided that $N_{\max}$ large enough. However, the large $N_{\max}$ would require a high computational capability of a digital device. If the computational power is very limited, then the step of bisection method can be realized just once at each sampled instant of time, i.e., we may select $N_{\max} = 1$. The stability of the closed-loop system may be guaranteed even in this case. Indeed, Theorem 10.1 proves that the $\mathbf{d}$-homogeneous ball $\bar{B}_{\mathbf{d}}(\overline{V}(t_i))$ is a strictly positively invariant set of the closed-loop system with the control $u(x) = (K_0 + \overline{V}^{\mu}(t_i)K\mathbf{d}(-\ln \overline{V}(t_i)))x$. If the root of the equation $\|\mathbf{d}(-\ln V)x(t_i)\| = 1$ is localized (i.e., if $\|x(t_i)\|_{\mathbf{d}} \leq \overline{V}(t_i)$), then Algorithm 10.1 always selects an upper estimate $\overline{V}(t_i)$ of $\|x(t_i)\|_{\mathbf{d}}$ to guarantee $x(t_i) \in \bar{B}_{\mathbf{d}}(\overline{V}(t_i))$. This means that $\|x(t)\|_{\mathbf{d}}$ never leaves the ball $\bar{B}_{\mathbf{d}}(\overline{V})$ even when $x(t)$ and $\overline{V}(t_i)$ vary in time.

The parameters a and b define lower and upper admissible values for $\|x\|_{\mathbf{d}}$. These restrictions are caused by practical issues. For instance, the parameter a cannot be selected arbitrary small due to a finite numerical precision of the digital device and due to measurement errors.

The suggested algorithm of digital implementation of homogeneous feedback is successfully validated on real experiments with electro-mechanical systems (see Chap. 11).

10.2 Sampled Homogeneous Control

In this section, we study the digitization of the domain of the control signal $t \mapsto u(t)$, i.e., we assume that the control signal is generated by a digital device such that

$$u(t) = u_i, \quad t \in [t_i, t_{i+1}), \tag{10.11}$$

where $u_i \in \mathbb{R}^m$, $t_i = ih$, $i = 0, 1, 2, \ldots$ and $h > 0$ is sampling period. If a feedback law is already designed for a continuous-time system, then it can be discretized somehow in order to generate the sequence $u_i \in \mathbb{R}^m$. The process of transformation of a continuous-time control law into a digital one is known as *control discretization* (see, e.g., [5, 6] and references therein).

10.2.1 Euler Discretization

The simplest discretization of a continuous-time control law $u(t) = \tilde{u}(x(t))$ is

$$u_i = \tilde{u}(x_i), \quad x_i = x(t_i) \quad i = 0, 1, 2, \ldots \tag{10.12}$$

Such a discretization of a feedback law $\tilde{u} : \mathbb{R}^n \mapsto \mathbb{R}^m$ is known as *explicit* [5]. The name "explicit discretization" is inspired by the explicit Euler method being well known for nonlinear ODEs (see Chap. 8). The explicit discretization exists for any control law. In particular, for the homogeneous control (10.2) it is given by

$$u(t) = u_i := \hat{K}(\|x_i\|_{\mathbf{d}})x_i, \quad t \in [t_i, t_{i+1}). \tag{10.13}$$

In Chap. 8, it is shown that explicit Euler discretization destroys the asymptotic stability properties of a continuous-time homogeneous ODE. Just a local practical stability can be guaranteed in this case. The same conclusion can be made in the case of the homogeneous control discretization [7].

Inspired by the Euler methods, the so-called *semi-implicit discretization* of the homogeneous control (10.2) can be designed as follows:

$$u(t) = u_i := \hat{K}(\|x_i\|_{\mathbf{d}})\tilde{x}_{i+1}, \quad t \in [t_i, t_{i+1}), \tag{10.14}$$

where $\tilde{x}_{i+1} \in \mathbb{R}^n$ is a solution of the equation

$$\tilde{x}_{i+1} = x_i + hA\tilde{x}_{i+1} + hB\hat{K}(\|x_i\|_{\mathbf{d}})\tilde{x}_{i+1} \tag{10.15}$$

being the semi-implicit Euler discretization (see Chap. 8) of the ODE $\dot{x} = Ax + B\hat{K}(\|x\|_{\mathbf{d}})x$.

Corollary 10.3 (Existence of semi-implicit discretization) *Let the homogeneous control* (10.2) *be defined according to Theorem 9.1 for $\mu > -1$. Then for any $\tilde{x}_i \in \mathbb{R}^n \backslash \{\mathbf{0}\}$ the algebraic equation*

$$\tilde{x}_{i+1} = \tilde{x}_i + hA\tilde{x}_{i+1} + hB\hat{K}(\|\tilde{x}_i\|_{\mathbf{d}})\tilde{x}_{i+1} \tag{10.16}$$

has a unique solution with respect to $\tilde{x}_{i+1} \in \mathbb{R}^n$. Moreover, the discrete-time system (10.16) *is globally asymptotically stable.*

Proof The closed-loop system (10.1), (10.2) has the form:

$$\dot{x} = \tilde{A}(\|x\|_{\mathbf{d}})x, \quad \tilde{A}(r) = A + B\hat{K}(r), \quad r > 0.$$

Since $A(1) = A + B(K_0 + K)$ then, due to the LMI (9.9), all conditions of Theorem 8.5 are fulfilled for $\mu > -1$, so the semi-implicit Euler discretization of the ODE $\dot{x} = \tilde{A}(\|x\|_{\mathbf{d}})x$ is well defined and globally asymptotically stable. The proof is complete. □

The *implicit discretization* of the homogeneous control (10.2) is given by

$$u(t) = u_i := \hat{K}(\|\tilde{x}_{i+1}\|_{\mathbf{d}})\tilde{x}_{i+1}, \quad t \in [t_i, t_{i+1}), \tag{10.17}$$

where $\tilde{x}_{i+1} \in \mathbb{R}^n$ is a solution of the equation

$$\tilde{x}_{i+1} = x_i + hA\tilde{x}_{i+1} + hB\hat{K}(\|\tilde{x}_{i+1}\|_{\mathbf{d}})\tilde{x}_{i+1} \tag{10.18}$$

being the implicit Euler discretization (see Chap. 8) of the ODE $\dot{x} = Ax + B\hat{K}(\|x\|_{\mathbf{d}})x$.

Corollary 10.4 (Existence of implicit discretization) *Let the homogeneous control* (10.2) *be defined according to Theorem 9.1 for* $\mu > -1$. *Then for any* $\tilde{x}_i \in \mathbb{R}^n \backslash \{\mathbf{0}\}$ *the algebraic equation*

$$\tilde{x}_{i+1} = \tilde{x}_i + hA\tilde{x}_{i+1} + hB\hat{K}(\|\tilde{x}_{i+1}\|_{\mathbf{d}})\tilde{x}_{i+1} \tag{10.19}$$

has a solution with respect to $\tilde{x}_{i+1} \in \mathbb{R}^n$. *Moreover, the discrete-time system* (10.19) *is globally asymptotically stable.*

Proof Notice that if $\tilde{x}_i \neq \mathbf{0}$ then, due to continuity of the feedback $\hat{K}(\|x\|_{\mathbf{d}})x$ for $\mu > -1$, a solution of (10.19) (if it exists) is nonzero too. Let us denote

$$q = \|\tilde{x}_{i+1}\|_{\mathbf{d}} \quad \text{and} \quad z = \mathbf{d}(-\ln \|\tilde{x}_{i+1}\|_{\mathbf{d}})\tilde{x}_{i+1}.$$

In this case, Equation (10.19) can be rewritten as follows:

$$\mathbf{d}(\ln q)z = \tilde{x}_i + hA_0\mathbf{d}(\ln q)z + hq^{1+\mu}BKz,$$

where $A_0 = A + BK_0$. Using the $\mathbf{d}$-homogeneity of A_0, namely, $A_0\mathbf{d}(s) = e^{\mu s}\mathbf{d}(s)$ A_0, $\forall s \in \mathbb{R}$ and the $\mathbf{d}$-homogeneity of B, namely, $\mathbf{d}(s)B = e^s B$, $\forall s \in \mathbb{R}$ we derive

$$\mathbf{d}(\ln q)z = \tilde{x}_i + hq^{\mu}\mathbf{d}(\ln q)(A_0 + BK)z.$$

Hence, we derive

$$(I_n - hq^{\mu}(A_0 + BK))z = \mathbf{d}(-\ln q)\tilde{x}_i.$$

Due to LMI (9.9), the matrix $A_0 + BK$ is Hurwitz , so the matrix $I_n - hq^{\mu}(A_0 + BK)$ is anti-Hurwitz and consequently invertible. Taking into account $z^\top Pz = 1$ we derive the following algebraic equation:

$$\tilde{x}_i^\top \mathbf{d}^\top(-\ln q)(I_n - hq^{\mu}(A_0 + BK))^{-\top}P(I_n - hq^{\mu}(A_0 + BK))^{-1}\mathbf{d}(-\ln q)\tilde{x}_i = 1$$

with respect to the scalar variable $q \in (0, +\infty)$. Let us consider the function $\xi :$ $(0, +\infty) \mapsto (0, +\infty)$ defined as follows:

$$\xi(q) = \tilde{x}_i^\top \mathbf{d}^\top(-\ln q)(I_n - hq^{\mu}(A_0 + BK))^{-\top}P(I_n - hq^{\mu}(A_0 + BK))^{-1}\mathbf{d}(-\ln q)\tilde{x}_i.$$

On the one hand, $\xi(\|\tilde{x}_i\|_{\mathbf{d}}) < 1$ due to LMI (8.14). On the other hand, we have $\xi(q) \to +\infty$ as $q \to 0$. Taking into account the continuity of the function ξ we conclude that there exists $q_* \in (0, \|\tilde{x}_i\|_{\mathbf{d}})$ such that $\xi(q_*) = 1$. In this case,

$$\tilde{x}_{i+1} = \mathbf{d}(\ln q_*)(I_n - hq_*^{\mu}(A_0 + BK))^{-1}\mathbf{d}(-\ln q_*)\tilde{x}_i$$

is a solution of (10.19). Since $q_* = \|x_{i+1}\|_{\mathbf{d}} < \|x_i\|_{\mathbf{d}}$ then the canonical homogeneous norm is a Lyapunov function of the discrete-time system (10.19). The proof is complete. □

Therefore, the semi-implicit and implicit Euler discretizations of the homogeneous ODE $\dot{x} = Ax + BK(\|x\|_{\mathbf{d}})x$ are well defined and globally asymptotically stable. However, in the general case, this does not guarantee an asymptotic stability of the system (10.1) with the sampled-time control (10.17) (resp., (10.14)) since $x(t_{i+1}) \neq \tilde{x}_{i+1}$. Moreover, the Euler discretization usually destroys finite/fixed-time convergence of the continuous homogeneous system (see Chap. 10). To overcome these drawbacks of Euler methods, we develop a special (consistent) discretization of the homogeneous controller (10.2).

10.2.2 Exact Sample-and-Hold Discretization

Let us consider a linear control system

$$\dot{x}(t) = Ax(t) + Bu(t), \quad t > 0, \quad x(0) = x_0 \in \mathbb{R}^n, \tag{10.20}$$

where $x(t) \in \mathbb{R}^n$ is the system state, $u(t) \in \mathbb{R}^m$ is the control input, and $A \in \mathbb{R}^{n\times n}$, $B \in \mathbb{R}^{n\times m}$ are known matrices. Let the system (10.20) be regulated by the sampled-time control

$$u(t) = u(t_k) \quad \text{for} \quad t \in [t_k, t_{k+1}). \tag{10.21}$$

Using the Cauchy formula (2.13) the system can be discretized as follows:

$$x_{k+1} = A_h x_k + B_h u(t_k), \quad k \in \mathbb{N}, \tag{10.22}$$

$$x_k = x(t_k), \quad t_k = kh, \quad A_h = e^{hA} \quad \text{and} \quad B_h = \int_0^h e^{sA} B ds.$$

Contrary to the Euler method, the Cauchy formula provides an exact discretization in the sense that any solution of the continuous-time system (10.20), (10.21) coincides with the solution of the discrete-time system (10.22) at the sampled instances of time. For more details about discretization of linear control systems, we refer the reader to [8].

Since $x_{k+2} = A_h x_{k+1} + B_h u(t_{k+1}) = A_h(A_h x_k + B_h u(t_k)) + B_h u(t_{k+1}) = B_h u(t_{k+1}) + A_h B_h u(t_k)) + A_h^2 x_k$, then, considering recursively n steps, we come up with the following formula:

$$x_{k+n} = B_h u(t_{k+n-1}) + \cdots + A_h^{n-1} B_h u(t_k) + A_h^n x_k. \tag{10.23}$$

The controllability of the pair $\{A, B\}$ implies the controllability of the pair $\{A_h, B_h\}$ (see, e.g., [9] or the lemma below):

$$\text{rank}(B_h, A_h B_h, \ldots, A_h^{n-1} B_h) = n.$$

Notice that taking $k = ni, i = 0, 1, \ldots$ the discrete-time system (10.23) can be treated as a system with maximal number of inputs (or an over-regulated system) with the respect to the augmented control input $v_i = \begin{pmatrix} u(t_{ni+n-1}) \\ \ldots \\ u(t_{ni}) \end{pmatrix}$. Below we follow this interpretation in order to design exact discretization of the homogeneous control (10.2).

Single-Input Case

First, let us study the single-input system ($m = 1$). In this case, Eq. (10.23) can be rewritten as

$$x_{k+n} = W_h \begin{pmatrix} u(t_{k+n-1}) \\ u(t_{k+n-2}) \\ \ldots \\ u(t_k) \end{pmatrix} + A_h^n x_k,$$

where the matrix

$$W_h = (B_h, A_h B_h, \ldots, A_h^{n-1} B_h) \tag{10.24}$$

is square and invertible due to controllability of the pair $\{A_h, B_h\}$.

Lemma 10.1 *Let $m = 1$. Let the gain $K_0 \in \mathbb{R}^{m\times n}$ and the matrix $G_0 \in \mathbb{R}^{n\times m}$ be defined as in Theorem 9.1. Let*

$$A_0 = A + BK_0, \quad A_h^0 = e^{hA_0} \quad \textit{and} \quad B_h^0 = \int_0^h e^{sA_0} B ds, \tag{10.25}$$

$$W_h^0 = (B_0^h, A_h^0 B_h^0, \ldots, (A_h^0)^{n-1} B_h^0). \tag{10.26}$$

Then the matrices W_h and W_h^0 are invertible and

$$W_h^0 = \hat{\mathbf{d}}(\ln h) \int_0^1 e^{\tau A_0} d\tau \left(B, e^{A_0} B, \ \ldots, \ e^{(n-1)A_0} B\right), \tag{10.27}$$

$$(W_h^0)^{-1}\hat{\mathbf{d}}(\ln h)\int_0^n e^{\tau A_0}d\tau B = \begin{pmatrix}1\\ \dots\\ 1\end{pmatrix}, \tag{10.28}$$

$$\hat{\mathbf{d}}(-\ln h)e^{sA}\hat{\mathbf{d}}(\ln h) = e^{s/h(A_0-BK_0\hat{\mathbf{d}}(\ln h))}, \tag{10.29}$$

$$W_h = \hat{\mathbf{d}}(\ln h)\int_0^1 e^{s(A_0-R(h))}ds\,(B, e^{A_0-R(h)}B, \dots, e^{(n-1)(A_0-R(h))}B), \tag{10.30}$$

where $\hat{\mathbf{d}}(s) = e^{s(I_n-G_0)}$, $s\in\mathbb{R}$ *is the linear continuous dilation in* $\mathbb{R}^n$ *and* $R(h) = BK\hat{\mathbf{d}}(\ln h)$.

Proof (1) Since e^{sA_0} and $\int_0^h e^{\tau A_0}d\tau$ commute for any $s\in\mathbb{R}$ then taking into account $(A_h^0)^i = e^{ihA_0}$ we derive

$$W_h^0 = \left(B_h^0, A_h^0 B, (A_h^0)^2 B_h^0, \dots, (A_h^0)^{n-1}B_h^0\right) =$$

$$\left(\int_0^h e^{sA_0}ds\right)\cdot(B\ \ e^{hA_0}B\ \ e^{2hA_0}B\ \ \dots\ \ e^{(n-1)hA_0}B).$$

The matrix A_0 is nilpotent and $e^{khA_0} = \sum_{i=0}^{n-1}\frac{(kh)^i A_0^i}{i!}$ for any $kh\in\mathbb{R}$. Then

$$\mathrm{rank}(B\ \ e^{hA_0}B\ \ e^{2hA_0}B\ \ \dots\ \ e^{(n-1)hA}B) = \mathrm{rank}(B\ \ A_0B\ \ A_0^2B\ \ \dots\ \ A_0^{n-1}B).$$

Hence, the controllability of the pair (A, B) and the invertibility of the matrix $\int_0^h e^{sA_0}ds$ imply the invertibility of the matrix W_h^0. The invertibility of W_h can be proven similarly.

By Theorem 9.1, the group $\hat{\mathbf{d}}(s) = e^{s(I_n-G_0)}$, $s\in\mathbb{R}$ is a linear continuous dilation $\mathbb{R}^n$, but the homogeneity identities

$$A_0 + A_0G_0 = G_0A_0, \quad G_0B = \mathbf{0}$$

imply that A_0 is $\hat{\mathbf{d}}$-homogeneous of degree -1 and

$$hB = e^{\ln h(I_n-G_0)}B \ \text{ and } \ e^{ihA_0}\hat{\mathbf{d}}(\ln h) = \hat{\mathbf{d}}(\ln h)e^{iA_0}, \forall i\in\mathbb{N}, \forall h>0.$$

Hence, we have

$$W_h^0 = \left(\frac{1}{h}\int_0^h e^{sA_0}ds\right)\hat{\mathbf{d}}(\ln h)(B, e^{A_0}B, e^{2A_0}B, \ldots, e^{(n-1)A_0}B)$$
$$= \hat{\mathbf{d}}(\ln h)\int_0^1 e^{sA_0}ds(B, e^{A_0}B, e^{2A_0}B, \ldots, e^{(n-1)A_0}B),$$

where the identity (7.3) is utilized on the last step. Therefore, the formula (10.27) holds. Since

$$\int_0^n e^{\tau A_0}d\tau = \sum_{i=0}^{n-1}\int_i^{i+1} e^{\tau A_0}d\tau = \int_0^1 e^{\tau A_0}d\tau \sum_{i=0}^{n-1} e^{iA_0}$$

then

$$(W_h^0)^{-1}\hat{\mathbf{d}}(\ln h)\int_0^n e^{\tau n A_0}d\tau B = (W_h^0)^{-1}\hat{\mathbf{d}}(\ln h)\int_0^1 e^{\tau A_0}d\tau \sum_{i=0}^{n-1} e^{iA_0}B$$
$$= (B, e^{A_0}B, \ldots, e^{(n-1)A_0}B)^{-1}\sum_{i=0}^{n-1} e^{iA_0}B = \sum_{j=1}^{n} e_j,$$

where $e_j = (0, \ldots, 0, 1, 0, \ldots, 0)^\top \in \mathbb{R}^n$ is the identity vector.

(2) Due to the identity $\hat{\mathbf{d}}(s)B = e^s B, \forall s \in \mathbb{R}$, the matrix W_h admits the representation

$$W_h = \left(\int_0^h e^{sA}ds\right)\cdot(B, e^{hA}B, e^{2hA}B, \ldots, e^{(n-1)hA}B)$$
$$= \left(\frac{1}{h}\int_0^h e^{sA}ds\right)\hat{\mathbf{d}}(\ln h)(B, R_1(h)B, R_2(h)B, .., R_{n-1}(h)B),$$

where $R_i(h) = \hat{\mathbf{d}}(-\ln h)e^{ihA}\hat{\mathbf{d}}(\ln h)$. Since

$$\hat{\mathbf{d}}(-\ln h)A = h^{-1}A_0\hat{\mathbf{d}}(\ln h) - h^{-1}BK_0 = h^{-1}\left(A_0 - BK_0\hat{\mathbf{d}}(\ln h)\right)\hat{\mathbf{d}}(-\ln h),$$

then

$$\hat{\mathbf{d}}(-\ln h)e^{sA}\hat{\mathbf{d}}(\ln h) = \hat{\mathbf{d}}(-\ln h)\sum_{k=0}^{+\infty}\frac{s^k}{k!}(A_0 - BK_0)^k\hat{\mathbf{d}}(\ln h)$$
$$= \sum_{k=0}^{+\infty}\frac{(s/h)^k}{k!}(A_0 - BK_0\hat{\mathbf{d}}(\ln h))^k = e^{s/h(A_0-BK_0\hat{\mathbf{d}}(\ln h))}.$$

Hence, we derive $R_i(h) = e^{i(A_0-BK_0\hat{\mathbf{d}}(\ln h))}$ and

$$\frac{1}{h}\int_0^h e^{sA}ds\;\hat{\mathbf{d}}(\ln h) = \hat{\mathbf{d}}(\ln h)\int_0^1 e^{s(A_0-BK\hat{\mathbf{d}}(\ln h))}ds.$$

The proof is complete. □

Let $\tilde{x} \in \mathbb{R}^n$ be an arbitrary selected vector. To guarantee $x_{k+n} = \tilde{x}$ we can always select the control inputs

$$\begin{bmatrix} u(t_{k+n-1}) \\ u(t_{k+n-2}) \\ \dots \\ u(t_k) \end{bmatrix} = W_h^{-1}(\tilde{x} - A_h^n x_k), \quad \tilde{x} \in \mathbb{R}^n. \tag{10.31}$$

By Corollary 9.1, the exact solution $x(t)$ of the continuous-time closed-loop system (10.1), (10.2) admits the explicit representation (9.28). This means that taking $\tilde{x} = x(t_{n(k+1)})$ in (10.31), the discrete-time system (10.22) would track some exact continuous-time trajectories x at the sampled instances of time t_{in}, $i = 1, 2, \dots$ as shown in Fig. 10.1.

Inspired by [10] the discretization algorithm for the homogeneous control law can be designed as follows.

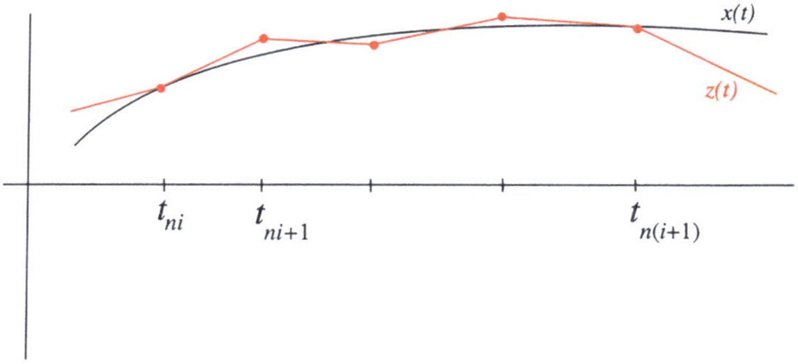

Fig. 10.1 A discrete-time tracking of a continuous-time trajectory at time instances t_{ni}, where $z(t) = x_k + \frac{t-t_k}{t_{k+1}-t_k}(x_{k+1} - x_k)$ for $t \in [t_k, t_{k+1})$

Let the parameters of the homogeneous stabilizer (10.2) be designed as in Theorem 9.1 for $\mu \neq 0$. By Corollary 9.1 the unique solution of the continuous-time closed-loop system (10.1), (10.2) satisfies

$$x(\tau + t) = Q_\tau(\|x(t)\|_{\mathbf{d}})x(t), \quad \forall t, \tau \geq 0,$$

where $Q_\tau : \mathbb{R}_+ \mapsto \mathbb{R}^{n\times n}$ is given by the formula (9.29). To track this solution in the time instances t_{ni} the sampled-time control can be defined as follows:

$$\begin{bmatrix} u(t_{k+n-1}) \\ u(t_{k+n-2}) \\ \dots \\ u(t_k) \end{bmatrix} = W_h^{-1}\left(Q_{nh}(\|x_k\|_{\mathbf{d}}) - A_h^n\right)x_k, \tag{10.32}$$

where $k = 0, n, 2n, 3n, \dots$. As shown in [10], the single-input closed-loop system (10.22), (10.32) preserves stability properties of the original continuous-time system. Below we generalize this result to systems with multiple inputs.

Multi-input Case

The multi-input system (10.20) admits the representation (10.22) with $x_k = x(t_k) \in \mathbb{R}^n$ and $A_h = e^{hA} \in \mathbb{R}^{n\times n}$ as before, but $B_h = \int_0^h e^{sA}Bds \in \mathbb{R}^{n\times m}$ and $u_k \in \mathbb{R}^m$. Similar to the single-input case we derive

$$x_{k+\tilde{n}} = \overline{W}_h \begin{bmatrix} u(t_{k+\tilde{n}-1}) \\ u(t_{k+\tilde{n}-2}) \\ \dots \\ u(t_k) \end{bmatrix} + A_h^{\tilde{n}}x_k, \tag{10.33}$$

where the number $\tilde{n} \in \{1, 2, \dots, n\}$ is such that $\operatorname{rank}[B, AB, \dots, A^{\tilde{n}-1}B] = n$ (see Theorem 9.1) and the matrix

$$\overline{W}_h = [B_h, A_hB_h, \dots, A_h^{\tilde{n}-1}B_h] \tag{10.34}$$

has the full row rank n due to controllability of $\{A_h, B_h\}$. In this case, the discrete-time system (10.22) with the control

$$\begin{bmatrix} u(t_{k+\tilde{n}-1}) \\ u(t_{k+\tilde{n}-2}) \\ \dots \\ u(t_k) \end{bmatrix} = \overline{W}_h^\top(\overline{W}_h\overline{W}_h^\top)^{-1}\left(Q_{\tilde{n}h}(\|x_k\|_{\mathbf{d}}) - A_h^{\tilde{n}}\right)x_k, \tag{10.35}$$

tracks any trajectory of the continuous-time system (10.20), (10.2) at time instances $t = t_{k\tilde{n}}$, $k \in \mathbb{N}$.The formula (10.35) becomes the formula (10.32) for the single-input case $m = 1$.

Theorem 10.2 *The system* (10.20) *with the sampled-time control* (10.32) *is globally uniformly finite-time (nearly fixed-time) stable if* $\mu < 0$ *(*$\mu > 0$*).*

Proof The closed-loop system (10.33), (10.35) has the form:

$$x_{k+\tilde{n}} = Q_{\tilde{n}h}(\|x_k\|_{\mathbf{d}})x_k. \tag{10.36}$$

By Corollary 9.1 this means that the states of the discrete-time system (10.22) with the control (10.32) coincide with the states of the original continuous-time system (10.20), (10.2) at time instances $t_{k\tilde{n}}$, $k = 0, 1, \ldots$. In this case, the trajectories of the discrete-time system converge to zero if $\mu < 0$ or into any neighborhood of the origin in a fixed time independently of the initial state if $\mu > 0$. To complete the proof we just need to show the global Lyapunov stability of the system.

In the view of Theorem 9.1 and Corollary 9.1, we have

$$\|Q_\tau(\|x\|_{\mathbf{d}})x\|_{\mathbf{d}}^{-\mu} = \|x\|_{\mathbf{d}}^{-\mu} + \mu\rho\tau,$$

for all $x \in \mathbb{R}^n$ and $\tau > 0$ such that $\|x\|_{\mathbf{d}}^{-\mu} + \mu\rho\tau \geq 0$. If $\|x\|_{\mathbf{d}}^{-\mu} + \mu\rho < 0$ then $Q_\tau(\|x\|_{\mathbf{d}})x = \mathbf{0}$ otherwise. Hence, we conclude

$$\|Q_\tau(\|x\|_{\mathbf{d}})x\|_{\mathbf{d}} = (\|x\|_{\mathbf{d}}^{-\mu} + \mu\rho\tau)^{-1/\mu} \leq \|x\|_{\mathbf{d}}, \forall x \in \mathbb{R}^n.$$

This means that $\|Q_{nh}(\|x_k\|_{\mathbf{d}})x_k\|_{\mathbf{d}} \leq \|x_k\|_{\mathbf{d}}$, i.e., the system (10.36) is *globally uniformly Lyapunov stable*.[2] Moreover, since

$$|u(t_{k+j})| \leq \left\| \overline{W}_h^\top (\overline{W}_h \overline{W}_h^\top)^{-1} \right\| (\|Q_\tau(\|x_k\|_{\mathbf{d}})x_k\| + \|A_h^{\tilde{n}}\| \|x_k\|)$$

then, by Corollary 5.3, there exists $\xi \in \mathcal{K}_\infty$ such that

$$|u(t_{k+j})| \leq \xi(\|x_k\|_{\mathbf{d}}), \quad \forall j = 0, 1, \ldots, \tilde{n} - 1, \quad \forall k \in \mathbb{N}.$$

Hence, due to the formula (10.22), there exists $\sigma \in \mathcal{K}_\infty$ such that

$$\|x_{k+j+1}\|_{\mathbf{d}} \leq \sigma(\|x_k\|_{\mathbf{d}}), \quad \forall j = 0, 1, \ldots, \tilde{n} - 1, \quad \forall k \in \mathbb{N}.$$

This means that the global Lyapunov stability of the system (10.36) implies the global uniform Lyapunov stability of the system (10.20) with the sampled-time control (10.35). The proof is complete. □

---

[2] A discrete-time system $x_{k+1} = f(k, x_k)$ is globally uniformly Lyapunov stable if there exists $\varepsilon \in \mathcal{K}$ such that $\|x_k\| \leq \varepsilon(\|x_0\|)$ for all $k \in \mathbb{N}$..

Robustness Analysis

In Chap. 7, we have shown that continuous-time homogeneous systems are Input-to-State Stable (ISS) with respect to sufficiently large class of perturbations. Let us investigate the robustness of the system with discretized control (10.34) with respect to additive perturbations. Recall [11] that a system

$$\dot{x} = f(t, x, q), \quad t > t_0 \tag{10.37}$$

is practically ISS with respect to $q \in L^\infty(\mathbb{R}; \mathbb{R}^l)$ if there exist, $c > 0$, $\xi \in \mathcal{KL}$ and $\gamma \in \mathcal{K}$ such that

$$\|x(t)\| \le c + \xi(\|x_0\|, t - t_0) + \gamma(\|q\|_{L^\infty((t_0,t),\mathbb{R}^l)}). \tag{10.38}$$

If $c = 0$ then the system is ISS (see Chap. 3). The local ISS corresponds to the restricted set of initial conditions $\|x_0\| \le r_x$ and the restricted maximal magnitude of the input $\|q\| \le r_q$.

Let us consider the linear plant with the sampled control:

$$\begin{aligned} \dot{x}(t) &= Ax(t) + Bu(t) + q(t), & t > 0, \\ u(t) &= u(t_k), & t \in [t_k, t_{k+1}), \end{aligned} \tag{10.39}$$

where $u(t_k) \in \mathbb{R}$ is given by (10.32) and $q \in L^\infty(\mathbb{R}; \mathbb{R}^n)$ is an exogenous input.

Theorem 10.3 *The system* (10.39) *is (1) locally ISS; (2) practically fixed-time stable if* $\mu > 0$*; (3) ISS if* $\mu > -\beta$*, where* $\beta > 0$ *is defined in Corollary 5.3.*

Proof Let us rewrite the system (10.39) in the discrete time as follows:

$$x^q_{k+n} = Q_{nh}(\|x^q_k\|_{\mathbf{d}})x^q_k + \sum_{i=0}^{n-1} A_h^{n-i}\tilde{q}_{k+i},$$

where $\tilde{q}_k = \int_0^h e^{A(h-\tau)}q(t_k + \tau)d\tau$ is the sampled-time realization of the exogenous perturbation. Since $q \in L^\infty$ then $\{\tilde{q}_k\} \in \ell^\infty$ for any $h > 0$.

By Theorem 9.1 and Corollary 9.1, we have

$$\|Q_{nh}(\|x\|_{\mathbf{d}})x\|_{\mathbf{d}}^{-\mu} = \|x\|_{\mathbf{d}}^{-\mu} + \mu\rho nh, \quad \forall x \in \mathbb{R}^n : \|x\|_{\mathbf{d}}^{-\mu} + \mu\rho nh \ge 0$$

then

$$\|Q_{nh}(\|x^q_k + \hat{q}_k\|_{\mathbf{d}})(x^q_k + \hat{q}_k)\|_{\mathbf{d}}^{-\mu} = \|x^q_k + \hat{q}_k\|_{\mathbf{d}}^{-\mu} + \mu\rho nh$$

and

$$\left\|x^q_{k+n} - \bar{q}_k\right\|_{\mathbf{d}} = \left(\|x^q_k\|_{\mathbf{d}}^{-\mu} + \mu\rho nh\right)^{-\frac{1}{\mu}},$$

where $\bar{q}_k = \sum_{i=0}^{n-1} A_h^{n-i} \tilde{q}_{k+i}$. Notice, since $\bar{q}_k$ is uniformly bounded, then for $\mu > 0$ it guarantees a practical fixed-time stability:

$$\left\| x_{k+n}^q - \bar{q}_k \right\|_{\mathbf{d}} \leq \frac{1}{(\mu\rho n h)^{\frac{1}{\mu}}}, \quad \forall k \in \mathbb{N}, \quad \forall x_0 \in \mathbb{R}^n.$$

Let V be defined as in Corollary 5.3. Since V is globally Lipschitz continuous with the Lipschitz constant 1, then

$$\begin{aligned} V(x_{k+n}^q) - V(x_k^q) &= V(x_{k+n}^q) - V(x_{k+n}^q - \bar{q}_k) + V(x_{k+n}^q - \bar{q}_k) - V(x_k^q) \\ &\leq \|\bar{q}_k\| + V(x_{k+n}^q - \bar{q}_k) - V(x_k^q) = \|\bar{q}_k\| + \underline{\sigma}(\|x_k^q - \bar{q}_k\|_{\mathbf{d}}) - V(x_k^q) \\ &= \|\bar{q}_k\| + \underline{\sigma}\left(\left(\|x_k^q\|_{\mathbf{d}}^{-\mu} + \mu\rho n h \right)^{-\frac{1}{\mu}} \right) - V(x_k^q) = \|\bar{q}_k\| - W(V(x_k^q)), \end{aligned}$$

where

$$W(V) = V - \underline{\sigma}\left(\left(\left(\underline{\sigma}^{-1}(V) \right)^{-\mu} + \mu\rho n h \right)^{-1/\mu} \right).$$

For $\mu > -\beta$, we have $W \in \mathcal{K}_\infty$ and V is an ISS Lyapunov function of the discrete-time system [12]. Hence, we conclude the ISS of the original system (10.39). The proof is complete. □

Since $u(t_{k+i-1})$ depends only on $x_k = x(t_k)$ with $k = \tilde{n}i$, $i = 0, 1, \ldots$, then the discretization (10.32) of the homogeneous control (10.2) could be useful, for example, if the control sampling is a least $\tilde{n}$ times faster than a measurement sampling. In other cases, the sampled control law (10.32) is a mixture of feedforward and feedback algorithms, where the state measurements $x(t_{k+i-1})$ for $i = 2, \ldots, \tilde{n} - 1$ are simply omitted during the control implementation. This could badly impact to a robustness and to a precision of the sampled-time controller. To avoid this drawback, we introduce a discretization, which provides a static feedback preserving the finite/fixed-time stability property of the original continuous-time system.

10.2.3 Consistent Discretization

In this section, we introduce the so-called consistent discretization [10, 13] of the homogeneous control (10.2) and study its robustness with respect to additive perturbations and measurement noises.

Definition 10.1 Let the system (10.20) with a feedback control $u = \tilde{u}(x)$, $\tilde{u} \in C(\mathbb{R}^n\backslash\{\mathbf{0}\}; \mathbb{R}^m)$ be globally uniformly finite-time (nearly fixed-time) stable. A family of functions $\tilde{u}_h : \mathbb{R}^n \mapsto \mathbb{R}^m$ parameterized by a scalar $h > 0$ is said to be a *consistent discretization* of $\tilde{u}$ if

- **Consistency of Stability:** The system (10.20) with

$$u(t) = \tilde{u}_h(x(t_i)), \quad t \in [t_i, t_{i+1}), \quad t_i = ih, i \in \mathbb{N} \tag{10.40}$$

is globally uniformly finite-time (resp., nearly fixed-time) stable for any $h > 0$.
- **Control Approximation:** $\forall r_1 > 0, \forall r_2 > r_1, \exists\, \omega_r \in \mathcal{K}$:

$$\sup_{r_1 \le \|x\| \le r_2} \|\tilde{u}_h(x) - \tilde{u}(x)\| \le \omega_r(h), \quad \forall h > 0. \tag{10.41}$$

If the above properties are fulfilled just for sufficiently small h (i.e., $\forall h \in (0, h_{\max})$ with $0 < h_{\max} < +\infty$) then the discretization $\tilde{u}_h$ is called *conditionally consistent*.

The first condition in Definition 10.1 asks the sampled-time control system (10.20), (10.40) to preserve the stability property of the original continuous-time system. The second condition guarantees that the control $\tilde{u}_h$ is, indeed, an approximation of $\tilde{u}$ means that $\tilde{u}_h(x) \to \tilde{u}(x)$ as $h \to 0^+$ uniformly on compacts from $\mathbb{R}^n\backslash\{\mathbf{0}\}$. The origin is excluded since a finite-time stabilizing feedback is always non-smooth or even discontinuous at zero.

Single-Input System

Let us design, first, a consistent discretization for a generalized homogeneous controllable LTI plant with a single input $m = 1$. Inspired by [10] let us define the static feedback law:

$$\tilde{u}_h(x_k) = \tilde{K}_h(\|x_k\|_{\mathbf{d}})x_k, \quad \tilde{K}_h(\|x_k\|_{\mathbf{d}}) = e_n^\top W_h^{-1}\left(Q_{nh}(\|x_k\|_{\mathbf{d}}) - A_h^n\right), \tag{10.42}$$

which is obtained from (10.32) selecting only $u(t_k)$ at each sampled instant of time. The unit vector $e_n = (0, \ldots, 0, 1)^\top \in \mathbb{R}^n$ is utilized in order to make such a selection. To prove that the feedback law (10.42) defines a *consistent discretization* of the homogeneous controller (10.2) in the sense of Definition (10.1), first, we analyze the stability properties of the feedback (10.42) and, next, we prove that it approximates the feedback (10.2). The consistency of the discretization (10.42) was originally proven in [10] for a system with nilpotent matrix A. Below we generalize this result to the system with an arbitrary matrix A.

For $m = 1$ the matrix W_h is invertible. Let us denote

$$L_h = B_h e_n^\top W_h^{-1}, \quad F_h = A_h - L_h A_h^n, \quad h > 0, \tag{10.43}$$

and rewrite the closed-loop system (10.22), (10.42) in the form:

$$x_{k+1} = F_h x_k + L_h Q_{nh}(\|x_k\|_{\mathbf{d}})x_k, \quad k \in \mathbb{N}. \tag{10.44}$$

As shown below, the matrix F_h is nilpotent ($F_h^n = \mathbf{0}$) even if the matrix A is not. This is the key feature of the proposed control discretization, which, together with properties of $Q_{nh}(\|x_k\|_{\mathbf{d}})x_k$, allows the controller (10.42) to preserve stability properties of the original system in its sampled-time counterpart.

> **Lemma 10.2** *Let u be a homogeneous control* (10.2) *designed by Theorem 9.1. Then the matrix F_h is nilpotent and the closed-loop discrete-time system* (10.20), (10.40), (10.42) *is*
>
> *(1) locally uniformly finite-time stable for $\mu < 0$;*
> *(2) globally practically fixed-time stable for $\mu > 0$.*

Proof Let us show that the matrix F_h is nilpotent. Notice that F_h can be rewritten as follows:

$$\begin{aligned}
F_h &= \left(I_n - [\mathbf{0} \;\; \mathbf{0} \;\; \ldots \;\; \mathbf{0} \;\; B_h]W_h^{-1}A_h^{n-1}\right)A_h \\
&= \left(A_h^{1-n}W_hW_h^{-1}A_h^{n-1} - [\mathbf{0} \;\; \mathbf{0} \;\; \ldots \;\; \mathbf{0} \;\; B_h]W_h^{-1}A_h^{n-1}\right)A_h \\
&= A_h^{1-n}(W_h - [\mathbf{0} \;\; \mathbf{0} \;\; \ldots \;\; \mathbf{0} \;\; A_h^{n-1}B_h])W_h^{-1}A_h^n \\
&= A_h^{1-n}[B_h \;\; A_hB_h \;\; \ldots \;\; A_h^{n-2}B_h \;\; \mathbf{0}]W_h^{-1}A_h^n.
\end{aligned}$$

Since $W_h^{-1}[A_hB_h \;\; \ldots \;\; A_h^{n-1}B_h \;\; \mathbf{0}] = [e_2 \;\; \ldots \;\; e_n \;\; \mathbf{0}]$ then

$$\begin{aligned}
F_h^2 &= A_h^{1-n}[B_h \; A_hB_h \; \ldots \; A_h^{n-2}B_h \; \mathbf{0}][e_2 \; e_3 \; \ldots \; e_n \; \mathbf{0}]W_h^{-1}A_h^n \\
&= A_h^{1-n}[A_hB_h \; \ldots \; A_h^{n-2}B_h \; \mathbf{0} \; \mathbf{0}]W_h^{-1}A_h^n.
\end{aligned}$$

Using the above identities we derive

$$\begin{aligned}
F_h^3 &= A_h^{1-n}[A_h^2B_h \; \ldots \; A_h^{n-2}B_h \; \mathbf{0} \; \mathbf{0}]W_h^{-1}A_h^nF_h \\
&= A_h^{1-n}[A_h^2B_h \; \ldots \; A_h^{n-2}B_h \; \mathbf{0} \; \mathbf{0}][e_2 \; e_3 \; \ldots \; e_n \; \mathbf{0}]W_h^{-1}A_h^n \\
&= A_h^{1-n}[A_h^3B_h \; \ldots \; A_h^{n-2}B_h \; \mathbf{0} \; \mathbf{0} \; \mathbf{0}]W_h^{-1}A_h^n.
\end{aligned}$$

After n steps we come up with the identity

$$F_h^n = A_h^{1-n}[\mathbf{0} \; \ldots \; \mathbf{0}]W_h^{-1}A_h^n = \mathbf{0}.$$

(1) Local Finite-Time Stability for $\mu < 0$.

By Corollary 9.1, for $\mu < 0$ we have $Q(\|x\|_{\mathbf{d}}) = \mathbf{0}$ if $\|x\|_{\mathbf{d}} \leq (-\mu\rho nh)^{-1/\mu}$. This means that close to the origin the closed-loop system (10.22), (10.42) becomes linear

$$x_{k+1} = F_h x_k$$

for $\|x_k\|_{\mathbf{d}} \leq (-\mu\rho nh)^{-1/\mu}$, but the nilpotence of F_h implies $x_{k+n} = \mathbf{0}$ and the local finite-time stability of the system (10.20), (10.40), (10.42).

(2) Practical Fixed-time Stability for $\mu > 0$ By Theorem 9.1 the canonical homogeneous norm is a Lyapunov function of the system and

$$\frac{d}{dt}\|x(t)\|_{\mathbf{d}} = -\rho\|x(t)\|_{\mathbf{d}}^{1+\mu}.$$

This implies the identity $\|x(t+nh)\|_{\mathbf{d}}^{-\mu} = \|x(t)\|_{\mathbf{d}}^{-\mu} + \mu\rho nh$, and for $\mu > 0$ we have $\|x(t+nh)\|_{\mathbf{d}} < (\mu\rho nh)^{-1/\mu}$ independently of $x(t)$. By Corollary 9.1, we have $x(t+nh) = Q_{nh}(\|x(t)\|_{\mathbf{d}})x(t)$, so

$$\|Q_{nh}(\|\tilde{x}\|_{\mathbf{d}})\tilde{x}\|_{\mathbf{d}} < (\mu\rho nh)^{-1/\mu}, \quad \forall \tilde{x} \in \mathbb{R}^n.$$

For any $x_0 \in \mathbb{R}^n$ the solution x_k, $k = 0, 1, 2, \ldots$ of the discrete-time system (10.22), (10.42) satisfies

$$\begin{aligned}
x_1 &= F_h x_0 + L_h y_1, \\
x_2 &= F_h^2 x_0 + F_h L_h y_1 + L_h y_2, \\
&\ldots \\
x_n &= F_h^n x_0 + F_h^{n-1} L_h y_1 + F_h^{n-2} L_h y_2 + \cdots + L_h y_n,
\end{aligned}$$

where $y_i = Q_{nh}(\|x_{i-1}\|_{\mathbf{d}})x_{i-1}$, $i = 1, 2, \ldots$. Since the matrix F_h is nilpotent then $F_h^n = \mathbf{0}$ and

$$x_k = F_h^{n-1} L_h y_{k-n+1} + F_h^{n-2} L_h y_{k-n+2} + \cdots + L_h y_k, \forall k \geq n.$$

Taking into account, the inequality

$$\|y_i\|_{\mathbf{d}} = \|Q_{nh}(\|x_{i-1}\|_{\mathbf{d}})x_{i-1}\| < (\mu\rho nh)^{-1/\mu}$$

we conclude that there exists $C > 0$

$$\|x_k\| \leq C, \quad \forall k \geq n, \quad \forall x_0 \in \mathbb{R}^n,$$

i.e., the system converges to the ball $B(C)$ in a fixed time. The Lyapunov stability for the closed-loop system (10.20), (10.40), (10.42) can be analyzed similar to the proof of Theorem 10.2. The proof is complete. □

The above lemma proves that the discretization (10.42) of the controller (10.2), indeed, preserves a stability property of the original system at least locally (close to zero for $\mu < 0$ and close to infinity for $\mu > 0$). The sufficient condition of the global stability is given by the following corollary.

Corollary 10.5 *Let* $x_0 \in \mathbb{R}^n$, $x_i = F_h x_{i-1} + L_h y_i$, $y_i = Q_{nh}(\|x_{i-1}\|_{\mathbf{d}})x_{i-1}$ *and* $i = 1, 2, \ldots, n$. *Let there exists* $\tilde{\rho} > 0$ *such that*

$$\|F_h^{n-1} L_h y_1 + F_h^{n-2} L_h y_2 + \cdots + L_h y_n\|_{\mathbf{d}} \leq \gamma(\|x_0\|_{\mathbf{d}})\|x_0\|_{\mathbf{d}}, \forall x_0 \in \mathbb{R}^n, \tag{10.45}$$

where

$$\gamma(r) = \max\left\{0, (1 + \mu\tilde{\rho}r^{\mu})^{-1/\mu}\right\}. \tag{10.46}$$

Then closed-loop system (10.22), (10.42) *is*

- *globally uniformly finite-time stable for* $\mu < 0$;
- *globally uniformly nearly fixed-time stable for* $\mu > 0$.

Proof In the proof of Lemma 10.2 it is shown that

$$x_n = F_h^{n-1} L_h y_1 + F_h^{n-2} L_h y_2 + \cdots + L_h y_n.$$

In this case the condition (10.45) means that

$$\|x_{k+n}\|_{\mathbf{d}} \leq \gamma(\|x_k\|_{\mathbf{d}})\|x_k\|_{\mathbf{d}}, \quad \forall k \geq 0,$$

i.e., the function $x \mapsto V(x) := \|x\|_{\mathbf{d}}$ is a Lyapunov function for the discrete-time system:

$$V(x_{in}) \leq V(x_0) \prod_{j=0}^{i-1} \underbrace{\gamma(\|x_{nj}\|_{\mathbf{d}})}_{<1} \to 0 \quad \text{as} \quad i \to +\infty.$$

This implies global asymptotic stability of the closed-loop system (10.20), (10.40), (10.42), but Lemma 10.2 yields its finite-time stability for $\mu < 0$ and nearly fixed-time stability for $\mu > 0$. The proof is complete. □

Using Corollary 9.1, for the system (10.20) with the continuous-time homogeneous controller (10.2) we have

$$\|x(t)\|_{\mathbf{d}} = \max\left\{0, (1 + \mu\rho t\|x_0\|_{\mathbf{d}}^{\mu})^{-1/\mu}\right\} \|x_0\|_{\mathbf{d}}. \tag{10.47}$$

Therefore, the sufficient stability condition (10.45), which simply means

$$\|x_n\|_{\mathbf{d}} \leq \max\left\{0, (1 + \mu\tilde{\rho}\|x_0\|_{\mathbf{d}}^{\mu})^{-1/\mu}\right\} \|x_0\|_{\mathbf{d}}, \tag{10.48}$$

seems to be very close to a necessary condition. In the view of the theorem given below, we conjecture that

> *the condition* (10.48) *holds uniformly on* x_0 *from a compact in* $\mathbb{R}^n\backslash\{\mathbf{0}\}$, *at least, for small enough* $\tilde{\rho} > 0$ *and small enough* $h > 0$.

Let us show the discrete-time control (10.42) approximates the original feedback as $h \to 0^+$.

> **Theorem 10.4** *Let* u *be a homogeneous control* (10.2) *designed by Theorem 9.1 for* $m = 1$ *and* $\tilde{u}_h$ *be defined by the formula* (10.42). *Then* $\tilde{u}_h(x) \to u(x)$ *as* $h \to 0^+$ *uniformly on compacts from* $\mathbb{R}^n\backslash\{\mathbf{0}\}$.

Proof Let us denote $s_h = \ln(1+\mu\rho n h r^{\mu})^{1/(\rho\mu)}$ with $r = \|x\|_{\mathbf{d}}$ and show

$$\tilde{K}_h(r) = e_n^\top W_h^{-1}\left(Q_{nh}(r) - A_h^n\right) \;\to\; K_0 + r^{1+\mu}K\mathbf{d}(-\ln r) \quad \text{as} \quad h \to 0^+,$$

where K_0 and K are defined in Theorem 9.1.

First, we show that

$$e_n^\top W_h^{-1}\left(Q_{nh}(r) - e^{nhA_0}\right) \;\to\; r^{1+\mu}K\mathbf{d}(-\ln r) \quad \text{as} \quad h \to 0^+,$$

next, we demonstrate

$$e_n^\top W_h^{-1}\left(e^{nhA_0} - A_h^n\right) \;\to\; K_0 \quad \text{as} \quad h \to 0^+,$$

where $A_0 = A + BK_0$.

(1) Let $\hat{\mathbf{d}}(s) = e^{\ln h(I_n - G_0)}$, $s \in \mathbb{R}$, where K_0 and G_0 are defined in Theorem 9.1. By construction, the matrix A_0 is nilpotent, $\mathbf{d}$-homogeneous of degree μ and $\hat{\mathbf{d}}$-homogeneous of the degree -1, i.e., $\hat{\mathbf{d}}(s)A_0 = e^s A_0\hat{\mathbf{d}}(s)$, $\forall s \in \mathbb{R}$ (see the proof of Theorem 9.1. Then

$$\begin{aligned}
\hat{\mathbf{d}}(-\ln h)e^{(A_0+BK+\rho G_{\mathbf{d}})s_h} &= \sum_{i=0}^{\infty}\frac{s_h^i\hat{\mathbf{d}}(-\ln h)(A_0 + BK + \rho G_{\mathbf{d}})^i}{i!}\\
&= \sum_{i=0}^{\infty}\frac{s_h^i(A_0 + BK\hat{\mathbf{d}}(\ln h) + \rho h G_{\mathbf{d}})^i\hat{\mathbf{d}}(-\ln h)}{h^i i!}\\
&= e^{\frac{s_h(A+\rho h G_{\mathbf{d}})}{h}}\hat{\mathbf{d}}(-\ln h) + \sum_{i=1}^{n-1}\frac{s_h^i}{h^i i!}A_0^{i-1}BK + O(h).
\end{aligned}$$

Indeed, for $i = 2$ we have

$$(A_0 + BK\hat{\mathbf{d}}(\ln h) + \rho h G_{\mathbf{d}})^2\hat{\mathbf{d}}(-\ln h) = (A_0 + \rho h G_{\mathbf{d}})^2\hat{\mathbf{d}}(-\ln h) + A_0BK + O(h),$$

for $i = 3$ we derive

$$(A_0 + BK\hat{\mathbf{d}}(\ln h) + \rho h G_{\mathbf{d}})^3\hat{\mathbf{d}}(-\ln h) =$$

$$(A_0 + BK\hat{\mathbf{d}}(\ln h) + \rho h G_{\mathbf{d}})\left\{(A_0 + \rho h G_{\mathbf{d}})^2\hat{\mathbf{d}}(-\ln h) + A_0 BK + O(h)\right\} =$$

$$(A_0 + \rho h G_{\mathbf{d}})^3\hat{\mathbf{d}}(-\ln h) + A_0^2 BK + O(h)$$

and, by induction, we conclude

$$(A_0 + BK\hat{\mathbf{d}}(\ln h) + \rho h G_{\mathbf{d}})^i\hat{\mathbf{d}}(-\ln h) = (A_0 + \rho h G_{\mathbf{d}})^i\hat{\mathbf{d}}(-\ln h) + A_0^{i-1}BK + O(h).$$

Since A_0 is nilpotent then for $i \geq n+1$ we have $A_0^{i-1} = \mathbf{0}$ and

$$(A_0 + BK\mathbf{d}_*(\ln h) + \rho h G_{\mathbf{d}})^i\hat{\mathbf{d}}(-\ln h) = (A_0 + \rho h G_{\mathbf{d}})^i\hat{\mathbf{d}}(-\ln h) + O(h).$$

Since A_0 is $\mathbf{d}$-homogeneous then, by Lemma 7.1, we have

$$\mathbf{d}(-s)e^{\tau A_0}\mathbf{d}(s) = e^{e^{s\mu}\tau A_0} \quad \forall s, \tau \in \mathbb{R}$$

and

$$\begin{aligned} Q_{nh}(r) - e^{nhA_0} &= \mathbf{d}(\ln r)\hat{Q}(s_h)\mathbf{d}(-\ln r) - e^{nhA_0} \\ &= \mathbf{d}(\ln r)(\hat{Q}(s_h) - e^{nhr^{\mu}A_0})\mathbf{d}(-\ln r), \end{aligned}$$

where $\hat{Q}(s_h) = e^{-\rho G_{\mathbf{d}} s_h}e^{(A_0+BK+\rho G_{\mathbf{d}})s_h}$ is given by the formula (9.30).

Below we use the notation ω for any function (without specification of co-domain) which admits the estimate $\|w(r)\| \leq \tilde{\omega}(r)$, $\forall r \geq 0$ by some class $\mathcal{K}$ function $\tilde{\omega}$. Since $\hat{\mathbf{d}}(s)$ commutes with $\mathbf{d}(\tau)$, $\forall s, \tau \in \mathbb{R}$ then using the identities $A_0\mathbf{d}(s) = e^{\mu s}\mathbf{d}(s)A_0$ and $\mathbf{d}(s)B = e^s B$ and the estimate of $\hat{\mathbf{d}}(-\ln h)e^{(A_0+BK+\rho G_{\mathbf{d}})s_h}$ obtained above, we derive

$$\begin{gathered} \hat{\mathbf{d}}(-\ln h)\left(e^{(A_0+BK+\rho G_{\mathbf{d}})s_h} - e^{\rho s_h G_{\mathbf{d}}}e^{nhr^{\mu}A_0}\right) = \\ \hat{\mathbf{d}}(-\ln h)e^{(A_0+BK+\rho G_{\mathbf{d}})s_h} - e^{\rho s_h G_{\mathbf{d}}}e^{nr^{\mu}A_0}\hat{\mathbf{d}}(-\ln h) = \\ \sum_{i=0}^{n-1}\frac{s_h^i A_0^{i-1}BK}{h^i i!} + \left(e^{\frac{s_h(A_0+\rho h G_{\mathbf{d}})}{h}} - e^{\rho s_h G_{\mathbf{d}}}e^{nr^{\mu}A_0}\right)\hat{\mathbf{d}}(-\ln h) + \omega(h). \end{gathered}$$

If $Z_1Z_2 - Z_2Z_1 = qZ_2$ and $e^{Z_1}e^{Z_2} = e^{Z_1 + \frac{q}{1-e^{-q}}Z_2}$ (see [14, page 136]). Taking $Z_1 = \rho s_h G_{\mathbf{d}}$, $Z_2 = nr^{\mu}A_0$, $q = -\mu\rho s_h$ we derive

$$e^{\rho s_h G_{\mathbf{d}}}e^{nr^{\mu}A_0} = e^{\rho s_h G_{\mathbf{d}} + \frac{s_h}{h}A_0}.$$

Therefore, since

$$s_h = \frac{\ln(1 + \mu\rho nhr^{\mu})}{\rho\mu} = \frac{\mu\rho nhr^{\mu} - \frac{(\mu\rho 2hr^{\mu})^2}{2} + h^2\omega(h)}{\rho\mu} = hnr^{\mu} + h\omega(h)$$

then

$$\hat{\mathbf{d}}(-\ln h)\left(e^{(A_0+BK+\rho G_{\mathbf{d}})s_h}-e^{\rho s_h G_{\mathbf{d}}}e^{nhr^{\mu}A_0}\right)=\sum_{i=0}^{n-1}\frac{s_h^i A_0^{i-1}BK}{h^i i!}+\omega(h)=\sum_{i=0}^{n-1}\frac{(nr^{\mu})^i A_0^{i-1}BK}{i!}+\omega(h)=\int_0^{nr^{\mu}} e^{\tau A_0}d\tau BK+\omega(h).$$

Moreover, since $\mathbf{d}(s)=e^{sG_{\mathbf{d}}}$ then

$$\hat{\mathbf{d}}(-\ln h)(Q_{nh}(r)-e^{hnA_0})=\hat{\mathbf{d}}(-\ln h)\mathbf{d}(\ln r)(\hat{Q}(s_h)-e^{nhr^{\mu}A_0})\mathbf{d}(-\ln r)=$$

$$\mathbf{d}(\ln r)e^{-\rho s_h G_{\mathbf{d}}}\hat{\mathbf{d}}(-\ln h)\left(e^{(A_0+BK+\rho G_{\mathbf{d}})s_h}-e^{\rho s_h G_{\mathbf{d}}}e^{nhr^{\mu}A_0}\right)\mathbf{d}(-\ln r)=$$

$$\mathbf{d}(\ln r)e^{-\rho s_h G_{\mathbf{d}}}\int_0^{nr^{\mu}} e^{\tau A_0}d\tau BK\mathbf{d}(-\ln r)+\omega(h)=$$

$$\int_0^{nr^{\mu}} \mathbf{d}(\ln r-\rho s_h)e^{\tau A_0}d\tau BK\mathbf{d}(-\ln r)+\omega(h)=$$

$$\int_0^{nr^{\mu}} e^{e^{-\mu(\ln r-\rho s_h)}\tau A_0}d\tau\mathbf{d}(\ln r-\rho s_h)BK\mathbf{d}(-\ln r)+\omega(h)=$$

$$\int_0^{nr^{\mu}} e^{\frac{1+\mu\rho nhr^{\mu}}{r^{\mu}}\tau A_0}d\tau\frac{r}{(1+\mu\rho nhr^{\mu})^{1/\mu}}BK\mathbf{d}(-\ln r)+\omega(h)=$$

$$\int_0^{nr^{\mu}} e^{\frac{1}{r^{\mu}}\tau A_0}d\tau BrK\mathbf{d}(-\ln r)+\omega(h)=\int_0^{n} e^{\tau A_0}d\tau Br^{1+\mu}K\mathbf{d}(-\ln r)+\omega(h),$$

where the estimates $e^{\frac{1+\mu\rho nhr^{\mu}}{r^{\mu}}\tau A_0}=e^{\frac{1}{r^{\mu}}\tau A_0}+\omega(h)$ and $\frac{r}{(1+\mu\rho nhr^{\mu})^{1/\mu}}=r+\omega(h)$ are utilized on the last step. Therefore, using Lemma 10.1 we conclude

$$e_n^{\top}W_h^{-1}(Q_{nh}(s_h)-e^{nhA_0})=$$

$$e_n^{\top}W_h^{-1}\hat{\mathbf{d}}(\ln h)\left(\int_0^{n} e^{\tau A_0}d\tau Br^{1+\mu}K\mathbf{d}(-\ln r)+\omega(h)\right)=$$

$$e_n^{\top}W_h^{-1}W_h^0 r^{1+\mu}K\mathbf{d}(-\ln r)+\omega(h),$$

where W_h^0 is given by (10.26). Using Lemma 10.1 we conclude $W_h^{-1}W_h^0 \to I_n$ as $h \to 0^+$.

(2) Since

$$e^{hnA} - I_n = \int_0^{nh} \frac{d}{d\tau} e^{\tau A} d\tau = \int_0^{nh} e^{\tau A} d\tau A$$

$$e^{hnA_0} - I_n = \int_0^{nh} \frac{d}{d\tau} e^{\tau A_0} d\tau = \int_0^{nh} e^{\tau A_0} d\tau A_0$$

then

$$\hat{\mathbf{d}}(-\ln h)\left(e^{nhA_0} - A_h^n\right) = \hat{\mathbf{d}}(-\ln h)\int_0^{nh} e^{\tau A_0} d\tau A_0 - \hat{\mathbf{d}}(-\ln h)\int_0^{nh} e^{\tau A} d\tau A$$

$$= \hat{\mathbf{d}}(-\ln h)\left(\int_0^{nh} e^{\tau A_0} d\tau - \int_0^{nh} e^{\tau A} d\tau\right) A_0 + \hat{\mathbf{d}}(-\ln h)\int_0^{nh} e^{\tau A} d\tau B K_0.$$

Using the identity (10.29) we derive

$$\hat{\mathbf{d}}(-\ln h)\left(\int_0^{nh} e^{\tau A_0} d\tau - \int_0^{nh} e^{\tau A} d\tau\right) A_0 =$$

$$h\left(\int_0^{n} e^{\tau A_0} d\tau - \int_0^{n} e^{\tau (A_0 - BK_0\hat{\mathbf{d}}(\ln h))} d\tau\right) \hat{\mathbf{d}}(-\ln h) A_0 =$$

$$h\int_0^{n} \left(e^{\tau A_0} - e^{\tau (A_0 - BK_0\hat{\mathbf{d}}(\ln h))}\right) d\tau \hat{\mathbf{d}}(-\ln h) A_0 =$$

$$\sum_{i=0}^{\infty} h \int_0^{n} \frac{\tau^i}{i!} d\tau \left(A_0^i - (A_0 - BK_0\hat{\mathbf{d}}(\ln h))^i\right) \hat{\mathbf{d}}(-\ln h) A_0 =$$

$$\sum_{i=0}^{\infty} \frac{n^{i+1}}{(i+1)!} h \left(A_0^i - (A_0 - BK_0\hat{\mathbf{d}}(\ln h))^i\right) \hat{\mathbf{d}}(-\ln h) A_0.$$

For $i = 0$ we have

$$h\left(A_0^i - (A_0 - BK_0\hat{\mathbf{d}}(\ln h))^i\right)\hat{\mathbf{d}}(-\ln h) = I_n - I_n = \mathbf{0}\,.$$

For $i = 1$ we derive

$$h\left(A_0^i - (A_0 - BK_0\hat{\mathbf{d}}(\ln h))^i\right)\hat{\mathbf{d}}(-\ln h) = hBK_0.$$

For $i = 2$ we obtain

$$h\left(A_0^i - (A_0 - BK_0\hat{\mathbf{d}}(\ln h))^i\right)\hat{\mathbf{d}}(-\ln h)$$

$$= h\left(A_0BK_0 + BK_0\mathbf{d}(\ln h)A_0\mathbf{d}(-\ln h) - h(BK_0)^2\right)$$

$$= h\left(A_0BK_0 + BK_0\mathbf{d}(\ln h)A_0\mathbf{d}(-\ln h) - h(BK_0)^2\right)$$

$$= h\left(A_0BK_0 + hBK_0A_0 - h(BK_0)^2\right),$$

etc. Therefore, we have

$$\hat{\mathbf{d}}(-\ln h)\left(\int_0^{nh} e^{\tau A_0}d\tau - \int_0^{nh} e^{\tau A}d\tau\right)A_0 = O(h).$$

Finally, using the identity (10.29) we conclude

$$\hat{\mathbf{d}}(-\ln h)\int_0^{nh} e^{\tau A}d\tau BK_0 = h\int_0^{n} e^{\tau(A-BK_0\hat{\mathbf{d}}(\ln h))}d\tau\hat{\mathbf{d}}(-\ln h)BK_0$$

$$= \int_0^{n} e^{\tau(A-BK_0\hat{\mathbf{d}}(\ln h))}d\tau BK_0 \to \int_0^{n} e^{\tau A}d\tau\ BK_0 \quad \text{as} \quad h \to 0^+.$$

Using Lemma 10.1, we derive

$$e_n^\top(W_h^0)^{-1}\left(e^{nhA_0} - A_h^n\right) \to K_0 \quad \text{as} \quad h \to 0^+.$$

Taking into account $W_h^{-1}W_h^0 \to I_n$ as $h \to 0^+$, we complete the proof. □

The proven theorem, in particular, implies that for a sufficiently small $h > 0$ the system (10.22), (10.42) behaves similar to the continuous-time system (10.20), (10.2) at least on compacts from $\mathbb{R}^n\backslash\{\mathbf{0}\}$.

Let us consider the system

$$\begin{aligned} \dot{x}(t) &= Ax(t) + Bu(t) + q_p(t), \quad t > 0, \\ u(t) &= \tilde{u}_h(x(t_k) + q_m(t_k)), \qquad t \in [t_k, t_{k+1}), \end{aligned} \tag{10.49}$$

where $\tilde{u}_h$ given by (10.42) is a discretization of a control (10.2) designed by Theorem 9.1 for $m = 1$ and the exogenous input $q = (q_p^\top, q_m^\top)^\top \in L^\infty(\mathbb{R}; \mathbb{R}^{2n})$ is such that

$$\max_{k \in \mathbb{N}: 0 \le t_k \le t} |q_m(t_k)| \le \|q_m\|_{L^\infty_{(0,t)}}, \forall t > 0. \tag{10.50}$$

This condition simply means the sampled-time model of the measurement noise is consistent with the continuous-time model of the measurement noise $q_m \in L^\infty(\mathbb{R}; \mathbb{R})$. So, we can use the conventional definition of ISS for robustness analysis of the system (10.49).

Here q_p models an external perturbation and q_m is a measurement noise. The ISS for the system with the consistently discretized of homogeneous controller is established by the following theorem.

Theorem 10.5 *If the condition* (10.45) *is fulfilled then the system* (10.49) *is globally ISS for* $\mu > -\beta$.

Proof In a discrete time, the system (9.63) can be rewritten as

$$x_{k+1}^q = A_h x_k^q + B_h \tilde{K}_h(\|x_k^q + \hat{q}_k\|)(x_k^q + \hat{q}_k) + \tilde{q}_k, \tag{10.51}$$

where $\{\hat{q}_k\}, \{\tilde{q}_k\} \in \ell^\infty$ for any $h > 0$, $\hat{q}_k = q_m(t_k)$ is the sampled-time realization of the measurement noise and $\tilde{q}_k = \int_0^h e^{A(h-\tau)} q_p(t_k + \tau) d\tau$ is the sampled-time realization of the external perturbation.

In this case, we have

$$x_{k+1}^q = F_h x_k^q + L_h y_k^q + q_k^*,$$

where

$$y_k^q = Q_{nh}(\|x_k^q + \hat{q}_k\|_{\mathbf{d}})(x_k^q + \hat{q}_k), \quad \hat{q}_k^* = \tilde{q}_k + L_h \hat{q}_k.$$

Hence, we derive

$$x_{k+n}^q = F_h^{n-1}(L_h y_k^q + q_k^*) + F_h^{n-2}(L_h y_{k+1}^q + q_{k+1}^*) + \cdots + L_h y_{k+n-1} + q_{k+n-1}^*,$$

but the condition (10.45) gives

$$\|x_{k+n}^q - \bar{q}_k\|_{\mathbf{d}} \le \gamma(\|x_k^q + \hat{q}_k\|_{\mathbf{d}})\|x_k^q + \hat{q}_k\|_{\mathbf{d}}, \quad \forall x_k^q \in \mathbb{R}^n,$$

where $\bar{q}_k = F_h^{n-1}q_k^* + F_h^{n-2}q_{k+1}^* + \cdots + q_{k+n-1}^*$. By Corollary 5.3, the function

$$x \mapsto V(x) := \underline{\sigma}(\|x\|_{\mathbf{d}})$$

is radially unbounded positive definite and globally Lipschitz continuous with the Lipschitz constant 1. Hence, we derive

$$V(x_{k+n}^q) = V(x_{k+n}^q - \bar{q}_k) + V(x_{k+n}^q) - V(x_{k+n}^q - \bar{q}_k) \leq \underline{\sigma}(\|x_{k+n}^q - \bar{q}_k\|_{\mathbf{d}}) + \|\bar{q}_k\|$$
$$\leq \underline{\sigma}(\gamma(\|x_k^q + \hat{q}_k\|_{\mathbf{d}})\|x_k^q + \hat{q}_k\|_{\mathbf{d}}) + \|\bar{q}_k\|.$$

Introducing the function

$$r \mapsto \xi(r) := \underline{\sigma}(r) - \underline{\sigma}(r\gamma(r))$$

we derive

$$V(x_{k+n}^q) \leq V(x_k^q + \hat{q}_k) - \xi(\|x_k^q + \hat{q}_k\|_{\mathbf{d}}) + \|\bar{q}_k\|.$$

Denoting $z_k^q = x_k^q + \hat{q}_k$ and using the global Lipschitz continuity of V we derive

$$V(z_{k+n}^q) = V(z_{k+n}^q - \hat{q}_{k+n}) + V(z_{k+n}^q) - V(z_{k+n}^q - \hat{q}_{k+n})$$
$$\leq V(z_k^q) - \xi(\|z_k^q\|_{\mathbf{d}}) + \|\bar{q}_k\| + \|\hat{q}_{k+n}\|.$$

By construction $\xi \in C(\mathbb{R}_+)$ and ξ is positive definite due to $\gamma(r) < 1, \forall r > 0$. Taking into account the representation (5.23), for large $r > 0$, we derive

$$\underline{\sigma}(r) - \underline{\sigma}(r\gamma(r)) = r^\beta - \gamma^\beta(r)r^\beta = r^\beta - (1 + \mu\tilde{\rho}r^\mu)^{-\frac{\beta}{\mu}}r^\beta.$$

Hence, for $\mu > -\beta$ we have $\xi(r) \to +\infty$ as $r \to +\infty$ and the function V is a global ISS Lyapunov function (see [12]) of the discrete-time system with respect to the virtual variable $z_k = x_k^q + \hat{q}_k$. This implies that the global ISS of the system (10.49). The proof is complete. □

Multi-input System

The MIMO analog of the static feedback controller (10.42) is given by

$$\tilde{u}_h(x_k) = \tilde{K}_h(\|x_k\|_{\mathbf{d}})x_k, \tag{10.52}$$

where $\tilde{K}_h(\|x_k\|_{\mathbf{d}}) = (A_h^{n-1}B_h)^\top \left(\overline{W}_h\overline{W}_h^\top\right)^{-1}\left(Q_{nh}(\|x_k\|_{\mathbf{d}}) - A_h^n\right)$ and $\overline{W}_h$ is given by 10.34.

Remark 10.1 The above analysis of the consistency of the control discretization does not take into account a finite numerical precision of any real digital device (computer). The computation errors may essentially impact at the results of consistent discretization if the system is stiff, poorly conditioned or weakly controllable. In these cases, the exact convergence of the state to zero cannot be observed in simulations. However, for regular systems with consistently discretized homogeneous finite-time control, the state converges to machine zero exactly (up to the sampling period) at the time instant estimated theoretically in Theorem 9.1.

Remark 10.2 The consistent discretization is explicit but its application requires some time for computation of $Q_{nh}(\|x_k\|_{\mathbf{d}})$. So, the consistent (as well as implicit) discretization is applicable only in digital controllers having a sufficient computational power. The consistent discretization may be efficiently parallelized in order to decrease the computation time in the real-time applications. If the computational power is very limited then the real-time application of the homogeneous control can be based on the explicit discretization with online estimation of $\|x_k\|_{\mathbf{d}}$ (see Algorithm 10.1 and its discussion). Examples of real physical systems with homogeneous controllers are presented in Chap. 11.

10.3 Quantized Homogeneous Control

In this section, we study the digitization of the co-domain of the analog signal $t \mapsto x(t)$ being the real-time state measurements. Even if $\tilde{u} : \mathbb{R}^n \mapsto \mathbb{R}^m$ is a continuous feedback law for the system (10.1), the quantization of the signal $t \mapsto x(t)$ implies a quantization of the signal $t \mapsto \tilde{u}(x(t))$.

Mathematically, the state quantization involves partitioning the state space into disjoint sets [15]. We define a quantization of $\mathbb{R}^n$ following ideas of the Voronoi diagram [16].

Let $\mathcal{E}$ be a subset of $\mathbb{R}^n$ and Ξ be a subset of $\mathbb{N}$. Let $\mathcal{S}_i \subseteq \mathcal{E}$ with $i \in \Xi$ be nonempty disjoint subsets covering $\mathcal{E}$: $\mathcal{S}_i \cap \mathcal{S}_j = \emptyset$ if $i \neq j$ and $\bigcup_{i\in\Xi} \mathcal{S}_i = \mathcal{E}$. Let $\mathcal{Q}_{\mathcal{E}} = \{q_i\}_{i\in\Xi}$ such that $q_i \in \mathcal{S}_i$ for any $i \in \Theta$.

Definition 10.2 A function $\mathsf{q} : \mathcal{E} \mapsto \mathcal{Q}_{\mathcal{E}}$ is said to be a quantization function (**quantizer**) on $\mathcal{E}$ if $\mathsf{q}(x) = q_i$ for all $x \in \mathcal{S}_i$ and all $i \in \Xi$.

The set $\mathcal{Q}_{\mathcal{E}}$ represents a collection of all quantization values. Each disjoint set $\mathcal{S}_i$ is known as quantization cell, but the vector $\mathsf{q}_i \in \mathcal{Q}_{\mathcal{E}}$ is the quantization seed of $\mathcal{S}_i$ which represents all states within the quantization cell.

The control system with quantized state measurements has the form:

$$\dot{x} = Ax + B\tilde{u}(\mathsf{q}(x)), \tag{10.53}$$

where $\tilde{u}(\cdot)$ is a state feedback law which uses the quantized measurements $q(x)$ of the state x. Such a closed-loop system is discontinuous even if $\tilde{u} \in C(\mathbb{R}^n; \mathbb{R}^m)$. Its solutions are understood in the sense of Filippov (see Chap. 2 or [2]). Below we consider various quantization functions defined on $\mathbb{R}^n$:

$$\mathfrak{q} : \mathbb{R}^n \mapsto \mathcal{Q}_{\mathbb{R}^n}.$$

10.3.1 Quantization Functions

First, let us recall briefly the two well-known quantization algorithms [15].

Uniform Quantizer

Uniform quantizer on $\mathbb{R}$ has a quantization step size equal to some value $\Delta \in \mathbb{R}_+$ (see Fig. 10.2, left side). It can be expressed as

$$q(\phi) = \Delta \cdot \left\lfloor \frac{\phi}{\Delta} + \frac{1}{2} \right\rfloor, \quad \phi \in \mathbb{R},$$

where the notation $\lfloor \cdot \rfloor$ denotes the floor function (the greatest integer less than or equal to the argument). The quantization error is bounded by the quantization step:

$$|q(\phi) - \phi| \leq \Delta. \tag{10.54}$$

The uniform quantizer for the vector $x \in \mathbb{R}^n$ can be defined in a component-wise manner

$$\mathfrak{q}(x) = (q(x_1), q(x_2), \ldots, q(x_n))^\top, \quad x = (x_1, x_2, \ldots, x_n)^\top \in \mathbb{R}^n. \tag{10.55}$$

Logarithmic Quantizer

Logarithmic quantizer on $\mathbb{R}$ is described by

$$q(\phi) = \begin{cases} \nu^i, \ \dfrac{1}{1+\delta}\nu^i < \phi \leq \dfrac{1}{1-\delta}\nu^i, i = 0, \pm 1, \pm 2, \ldots, \\ 0, \quad \phi = 0, \\ -q(-\phi), \ \phi < 0, \end{cases} \tag{10.56}$$

where $\nu \in (0, 1)$ represents the quantization density and $\delta = (1 - \nu)/(1 + \nu)$. A small ν (resp., large δ) implies a coarse quantization, but a large ν (resp., a small δ)

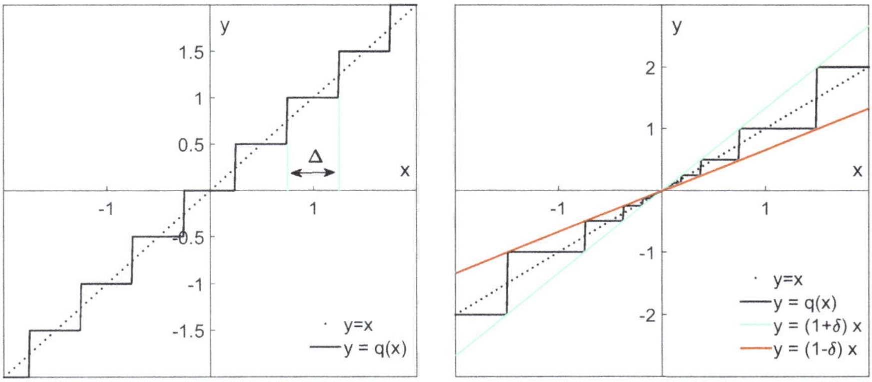

Fig. 10.2 Uniform (left) and logarithmic (right) quantizers

means a dense quantization. For the logarithmic quantizer, the error is sector bounded [17]:

$$|q(\phi) - \phi| \leq \delta|\phi|, \ \delta \in (0, 1) \tag{10.57}$$

and vanishing as ϕ goes to zero (see Fig. 10.2, right side). The parameter $\delta > 0$ can be utilized for tuning of the quantization error: the smaller the δ, the less the error is.

Below we show that homogeneous control can preserve finite-time and fixed-time stabilization property of the closed-loop system in the case of certain logarithmic quantization of state measurements as well as in the case of the so-called *homogeneous quantization*.

Homogeneous Quantizer

Let us represent the vector $x \in \mathbb{R}^n\backslash\{\mathbf{0}\}$ in the *polar homogeneous coordinates*

$$x = \mathbf{d}(\ln r)z, \quad r > 0, \quad z \in S = \{x \in \mathbb{R}^n : \|x\| = 1\}, \tag{10.58}$$

where $\mathbf{d}$ is a linear continuous dilation in $\mathbb{R}^n$ being monotone with respect to a norm $\|x\|$ in $\mathbb{R}^n$. In this case (see Chap. 5), the scalar $r > 0$ corresponds to the canonical homogeneous norm of the vector $x \in \mathbb{R}^n$,

$$r = \|x\|_{\mathbf{d}},$$

induced by the norm $\|x\|$, but $z \in S$ is defined by the $\mathbf{d}$-homogeneous projection of the vector $x \in \mathbb{R}^n\backslash\{\mathbf{0}\}$ on the unit sphere (see Chap. 5):

$$z = \pi_{\mathbf{d}}(x) := \mathbf{d}(-\ln \|x\|_{\mathbf{d}})x.$$

To design a homogeneous quantizer on $\mathbb{R}^n$, we suggest to use the representation of $x \in \mathbb{R}^n$ in the homogeneous polar coordinates. The logarithmic quantizer q on $\mathbb{R}$ can be utilized for the polar radius $r > 0$, but a uniform quantizer on the unit sphere S can be designed for the vector $z \in S$ using the Voronoi diagram [16].

Let $\Xi = \{1, 2, \ldots, N\}$ and

$$\mathcal{Q}_S = \{q_1, \ldots, q_N\} \subset S,$$

where $q_i \in S$ are disjoint points distributed on the unit sphere $S \subset \mathbb{R}^n$. Such a grid on the unit sphere S can be constructed, for example, using Step 1 of Algorithm 5.2.

Let us define $\mathcal{S}_i \subset S$ as follows:

$$\mathcal{S}_i = \{x \in S : i = \min \operatorname*{argmin}_{j \in \Xi} \|x - q_j\|\}, \quad i \in \Xi. \tag{10.59}$$

The set $\mathcal{S}_i$ defined in such a way contains points of the unit sphere S, which are closer to q_i than to any other point from $\mathcal{Q}_S$. If the distances from x to several points $q_{j_1}, \ldots, q_{j_k} \in \mathcal{Q}_S$ are minimal and identical, then the minimal index is selected (i.e., $x \in \mathcal{S}_{j_*}$ with $j_* = \min\{j_1, \ldots, j_k\}$). So, the sets $\mathcal{S}_i$ are disjoint and cover the whole unit sphere S.

Let $\mathfrak{q}_\pi : S \mapsto \mathcal{Q}_S$ be the quantization function (see Definition 10.2) constructed by means of the sets $\mathcal{Q}_S$ and $\mathcal{S}_i$ with $i = 1, 2, \ldots, N$. The parameter

$$\gamma = \sup_{x \in S} \inf_{i \in \Xi} \|x - q_i\| \tag{10.60}$$

characterizes the quantization error of $\mathfrak{q}_\pi$. The parameter $\gamma > 0$ can be tuned arbitrarily small by a construction of a dense enough grid $\mathcal{Q}_S$ on S. Inspired by [18] let us define the following quantizers.

Definition 10.3 Let $\mathbf{d}$ be a linear continuous dilation in $\mathbb{R}^n$ being monotone with respect to a norm $\|\cdot\|$ in $\mathbb{R}^n$. Let $q : \mathbb{R} \mapsto \mathbb{R}$ be a logarithmic quantizer on $\mathbb{R}$, $q_\pi : S \mapsto \mathcal{Q}_S$ be a uniform quantizer on the unit sphere S, $\pi_{\mathbf{d}} : \mathbb{R}^n \backslash \{\mathbf{0}\} \mapsto S$ be a $\mathbf{d}$-homogeneous projector on S and $\|\cdot\|_{\mathbf{d}}$ be a $\mathbf{d}$-homogeneous norm in $\mathbb{R}^n$.

A function $\mathfrak{q}_{\mathbf{d}} : \mathbb{R}^n \mapsto \mathcal{Q}_{\mathbf{d}}$ is said to be

- $\mathbf{d}$-*homogeneous logarithmic quantizer* on $\mathbb{R}^n$ if $\mathfrak{q}_{\mathbf{d}}(\mathbf{0}) = \mathbf{0}$,

$$\mathfrak{q}_{\mathbf{d}}(x) = \mathbf{d}(\ln q(\|x\|_{\mathbf{d}}))\mathfrak{q}_\pi(\pi_{\mathbf{d}}(x)), \tag{10.61}$$

and

$$\mathcal{Q}_{\mathbf{d}} = \bigcup_{r \in q(\mathbb{R}_+)} \mathbf{d}(\ln r)\mathcal{Q}_S; \tag{10.62}$$

- **d**-*homogeneous spherical quantizer* on $\mathbb{R}^n$ if $\mathfrak{q}_{\mathbf{d}}(\mathbf{0}) = \mathbf{0}$,

$$\mathfrak{q}_{\mathbf{d}}(x) = \mathfrak{q}_{\pi}(\pi_{\mathbf{d}}(x)), \quad \forall x \neq \mathbf{0} \tag{10.63}$$

and $\mathcal{Q}_{\mathbf{d}} = \mathcal{Q}_S$.

The uniform spherical quantizers q_π are well known in control literature (see, e.g., [19–21] and references therein). The homogeneous spherical quantizer (10.63) conceptually differs from spherical quantizers considered in the mentioned literature. It uses a uniform quantizer $\mathfrak{q}_\pi$ on the unit sphere S to quantize whole $\mathbb{R}^n$. It has a *finite number* N of quantization seeds q_i distributed on the unit sphere, which correspond to N unbounded quantization cells being homogeneous cones in $\mathbb{R}^n$. By construction, the *homogeneous spherical quantizer* (10.63) *is* **d**-*homogeneous function of degree zero*:

$$q_{\mathbf{d}}(\mathbf{d}(s)x) = q_{\mathbf{d}}(x), \quad \forall x \in \mathbb{R}^n, \quad \forall s \in \mathbb{R},$$

since $\pi_{\mathbf{d}}(\mathbf{d}(s)x) = \pi_{\mathbf{d}}(x), \forall x \neq \mathbf{0}, \forall s \in \mathbb{R}$. The number $\gamma > 0$ defined by (10.60) is the only tuning parameter of this quantizer.

The homogeneous logarithmic quantizer combines the scalar logarithmic quantizer q on $\mathbb{R}_+$ with the homogeneous spherical quantizer. This quantizer has two tuning parameters: $\gamma > 0$ for $\mathfrak{q}_\pi$ and $\delta = \frac{1-\nu}{1+\nu} \in (0, 1)$ for q. A "discrete" dilation symmetry of the homogeneous logarithmic quantizer is proven by the following lemma.

Lemma 10.3 *The* **d**-*homogeneous logarithmic quantizer* (10.61) *satisfies*

$$\mathfrak{q}_{\mathbf{d}}(\mathbf{d}(s)x) = \mathbf{d}(s)\mathfrak{q}_{\mathbf{d}}(x), \quad \forall x \in \mathbb{R}^n, \quad \forall s = k \ln \nu \ : \ k \in \mathbb{Z}, \tag{10.64}$$

where $\nu \in (0, 1)$ *is the parameter of the logarithmic quantizer* (10.56).

Proof On the one hand, for the logarithmic quantizer q we derive

$$\ln q(e^s \|x\|_{\mathbf{d}}) = i \ln \nu \quad \text{if} \quad \ln \frac{1}{1+\delta} + i \ln \nu < s + \ln \|x\|_{\mathbf{d}} \leq \ln \frac{1}{1-\delta} + i \ln \nu,$$

where $i \in \mathbb{Z}$. On the other hand, we also obtain

$$\ln \left(e^s q(\|x\|_{\mathbf{d}})\right) = s + j \ln \nu \quad \text{if} \quad \ln \frac{1}{1+\delta} + j \ln \nu < \ln \|x\|_{\mathbf{d}} \leq \ln \frac{1}{1-\delta} + j \ln \nu,$$

where $j \in \mathbb{Z}$. Hence, for any $k \in \mathbb{Z}$ we have

$$\ln q(e^s \|x\|_{\mathbf{d}}) = \ln \left(e^s q(\|x\|_{\mathbf{d}})\right) \text{ if } s = k \ln \nu.$$

Using the dilation symmetry of the homogeneous norm $\|\mathbf{d}(s)x\|_{\mathbf{d}} = \|x\|_{\mathbf{d}}$ we derive

$$\mathfrak{q}_{\mathbf{d}}(\mathbf{d}(s)x) = \mathbf{d}(\ln q(\|\mathbf{d}(s)x\|_{\mathbf{d}}))\mathfrak{q}_{\pi}(\pi_{\mathbf{d}}(\mathbf{d}(s)x)) = \mathbf{d}(\ln q(e^{s}\|x\|_{\mathbf{d}}))\mathfrak{q}_{\pi}(\pi_{\mathbf{d}}(x))$$

$$= \mathbf{d}(s + \ln q(\|x\|_{\mathbf{d}}))\mathfrak{q}_{\pi}(\pi_{\mathbf{d}}(x)) = \mathbf{d}(s)\mathbf{d}(\ln q(\|x\|_{\mathbf{d}}))\mathfrak{q}_{\pi}(\pi_{\mathbf{d}}(x)) = \mathbf{d}(s)\mathfrak{q}(x)$$

for all $x \in \mathbb{R}^n$ and $\forall s = k \ln \nu \ : \ k \in \mathbb{Z}$. The proof is complete. □

The quantization error of the **d**-homogeneous logarithmic quantizer admits the conic estimate in terms of the canonical homogeneous norm.

Lemma 10.4 *For any $\epsilon > 0$ there exists $\gamma \in (0, 1)$ and $\delta \in (0, 1)$ such that the **d**-homogeneous logarithmic quantizer $\mathfrak{q}_{\mathbf{d}}$ given by* (10.61) *on $\mathbb{R}^n$ satisfies the inequality:*

$$\|\mathfrak{q}_{\mathbf{d}}(x) - x\|_{\mathbf{d}} \le \epsilon\|x\|_{\mathbf{d}}, \quad \forall x \in \mathbb{R}^n, \tag{10.65}$$

where $\gamma \in (0, 1)$ is the quantization error of $\mathfrak{q}_{\pi}$ given by (10.60) *and $\delta = \frac{1-\nu}{1+\nu} \in (0, 1)$ is the parameter of the logarithmic quantizer* (10.56).

Proof The formula of the logarithmic quantization (10.56) gives

$$\frac{1}{1+\delta} \le \frac{\|x\|_{\mathbf{d}}}{q(\|x\|_{\mathbf{d}})} \le \frac{1}{1-\delta} \tag{10.66}$$

for all $x \neq \mathbf{0}$. Using the dilation symmetry of the canonical homogeneous norm we derive

$$\|\mathfrak{q}_{\mathbf{d}}(x) - x\|_{\mathbf{d}} = \|\mathbf{d}\,(\ln q(\|x\|_{\mathbf{d}}))\,\mathfrak{q}_{\pi}(\mathbf{d}(-\ln\|x\|_{\mathbf{d}})x) - x\|_{\mathbf{d}}$$

$$= q(\|x\|_{\mathbf{d}})\,\|\mathfrak{q}_{\pi}(\mathbf{d}(-\ln\|x\|_{\mathbf{d}})x) - \mathbf{d}\,(-\ln q(\|x\|_{\mathbf{d}}))\,x\|_{\mathbf{d}}\,.$$

Since $\|\mathbf{d}\,(-\ln q(\|x\|_{\mathbf{d}}))\,x\|_{\mathbf{d}} = \frac{\|x\|_{\mathbf{d}}}{q(\|x\|_{\mathbf{d}})}$ then due to (10.66) we have

$$\mathbf{d}\,(-\ln q(\|x\|_{\mathbf{d}}))\,x \in \mathcal{C} = \left\{z \in \mathbb{R}^n : \frac{1}{1+\delta} \le \|z\|_{\mathbf{d}} \le \frac{1}{1-\delta}\right\}.$$

On the other hand, by construction, the vector $\mathfrak{q}_{\pi}(\mathbf{d}(-\ln\|x\|_{\mathbf{d}})x)$ belongs to the unit sphere $S \subset \mathcal{C}$. Since $\mathbf{0} \notin \mathcal{C}$, then, by Lemma 5.4, the canonical homogeneous norm satisfies the Lipschitz condition on the compact $\mathcal{C}$, i.e., $\exists L > 0$ such that

$$\|z_1 - z_2\|_{\mathbf{d}} \le L\|z_1 - z_2\|, \quad \forall z_1, z_2 \in \mathcal{C}. \tag{10.67}$$

Hence, we derive

$$\|\mathfrak{q}_{\mathbf{d}}(x) - x\|_{\mathbf{d}} \le (1+\delta)L\|x\|_{\mathbf{d}}\|\mathfrak{q}_{\pi}(\mathbf{d}(-\ln\|x\|_{\mathbf{d}})x) - \mathbf{d}(-\ln q(\|x\|_{\mathbf{d}}))\,x\|$$

$$\le (1+\delta)L\|x\|_{\mathbf{d}}\|\mathfrak{q}_{\pi}(\mathbf{d}(-\ln\|x\|_{\mathbf{d}})x) - \mathbf{d}(-\ln\|x\|_{\mathbf{d}})\,x\|$$

$$+ (1+\delta)L\|x\|_{\mathbf{d}}\|\mathbf{d}(-\ln\|x\|_{\mathbf{d}})x - \mathbf{d}(-\ln q(\|x\|_{\mathbf{d}}))\,x\|$$

$$\le (1+\delta)L\|x\|_{\mathbf{d}}\left(\gamma + \left\|\left(I_n - \mathbf{d}\left(\ln\frac{\|x\|_{\mathbf{d}}}{q(\|x\|_{\mathbf{d}})}\right)\right)\mathbf{d}(-\ln\|x\|_{\mathbf{d}})x\right\|\right)$$

$$\le (1+\delta)L\|x\|_{\mathbf{d}}\left(\gamma + \left\|I_n - \mathbf{d}\left(\ln\frac{\|x\|_{\mathbf{d}}}{q(\|x\|_{\mathbf{d}})}\right)\right\|\right).$$

Using (10.66) we conclude that

$$\sup_{x\in\mathbb{R}^n}\left\|I_n - \mathbf{d}\left(\ln\frac{\|x\|_{\mathbf{d}}}{q(\|x\|_{\mathbf{d}})}\right)\right\| \to 0 \text{ as } \delta\to 0. \tag{10.68}$$

The proof is complete. □

10.3.2 Quantized Stabilizers

Let us consider the system (10.1) with the homogeneous feedback (10.2) in the quantized form:

$$u = K_0\mathfrak{q}(x) + \|\mathfrak{q}(x)\|_{\mathbf{d}}^{1+\mu}K\mathbf{d}(-\ln\|\mathfrak{q}(x)\|_{\mathbf{d}})\mathfrak{q}(x), \tag{10.69}$$

where $\mathfrak{q} : \mathbb{R}^n \mapsto \mathcal{Q}$ is a quantization function (see Definition 10.2). We assume that the non-quantized feedback (10.2) is a homogeneous stabilizer for the system (10.1). The following theorem shows that, for logarithmic-like quantizers, a global stability can be preserved for $K_0 = \mathbf{0}$, while, in the case $K_0 \ne \mathbf{0}$, just a local stability (close to zero or close to infinity) can be guaranteed (dependently of the sign of the homogeneity degree). Recall that the condition $K_0 = \mathbf{0}$ means that the linear plant is already homogeneous and it does not require any homogenization by the linear feedback K_0x (see Chap. 9 for more details). The following theorem is inspired by [22, 23].

Theorem 10.6 *Let* $\|\cdot\|$, $\|\cdot\|_{\mathbf{d}}$ *and all parameters of the homogeneous feedback* (10.2) *be designed as in Theorem 9.1.*

If $K_0 = \mathbf{0}$ *and the quantization error satisfies*

$$\|\mathsf{q}(x) - x\|_{\mathbf{d}} \le \epsilon \|x\|_{\mathbf{d}}, \quad \forall x \in \mathbb{R}^n, \tag{10.70}$$

then, for a sufficiently small $\epsilon > 0$, *the system* (10.1), (10.69) *is globally uniformly*

- *finite-time stable for* $\mu \in (-1, 0)$;
- *exponentially stable for* $\mu = 0$;
- *nearly fixed-time stable for* $\mu > (0, (\tilde{n} - 1)^{-1})$.

If $K_0 \neq \mathbf{0}$, *the quantization error satisfies* (10.70) *and*

$$\|\mathsf{q}(x) - x\| \le \kappa \|x\|, \quad \forall x \in \mathbb{R}^n, \tag{10.71}$$

then, for a sufficiently small $\epsilon > 0$, $\kappa > 0$, *the system* (10.1), (10.69) *is*

- *locally uniformly finite-time stable if* $\mu \in (-1, \min\{\beta - 1, 0\})$;
- *globally uniformly practically fixed-time stable if* $\mu > \min\{\alpha - 1, 0\}$,

where $\alpha = 0.5\lambda_{\max}(P^{1/2}G_{\mathbf{d}}P^{-1/2} + P^{1/2}G_{\mathbf{d}}^{\top}P^{-1/2})$ *and* $\beta = 0.5\lambda_{\min}(P^{1/2}G_{\mathbf{d}}P^{-1/2} + P^{1/2}G_{\mathbf{d}}^{\top}P^{-1/2}) > 0$.

Proof The inequality (10.70) implies that $\mathsf{q}(\mathbf{0}) = \mathbf{0}$. The canonical homogeneous norm $\|x\|_{\mathbf{d}}$ serves as a Lyapunov function for the quantization-free system (10.1), (10.2). Let us calculate its derivative along the trajectories of the quantized system (10.1), (10.69):

$$\begin{aligned}\frac{d}{dt}\|x\|_{\mathbf{d}} &= \|x\|_{\mathbf{d}} \frac{x^{\top}\mathbf{d}(-\ln\|x\|_{\mathbf{d}})P\mathbf{d}(-\ln\|x\|_{\mathbf{d}})\,(Ax + Bu(\mathsf{q}(x)))}{x^{\top}\mathbf{d}^{\top}(-\ln\|x\|_{\mathbf{d}})PG_{\mathbf{d}}\mathbf{d}(-\ln\|x\|_{\mathbf{d}})x}\\ &= \|x\|_{\mathbf{d}} \frac{x^{\top}\mathbf{d}^{\top}(-\ln\|x\|_{\mathbf{d}})P\mathbf{d}(-\ln\|x\|_{\mathbf{d}})\,(Ax + Bu(x) + Bu(\mathsf{q}(x)) - Bu(x))}{x^{\top}\mathbf{d}^{\top}(-\ln\|x\|_{\mathbf{d}})PG_{\mathbf{d}}\mathbf{d}(-\ln\|x\|_{\mathbf{d}})x}.\end{aligned} \tag{10.72}$$

Using the dilation symmetry $A_0\mathbf{d}(s) = e^{\mu s}\mathbf{d}(s)A_0$, $\mathbf{d}(s)B = e^{s}B, \forall s \in \mathbb{R}$, where $A_0 = A + BK_0$, we derive

$$\frac{d}{dt}\|x\|_{\mathbf{d}} = -\rho\|x\|_{\mathbf{d}}^{1+\mu} + \frac{x^{\top}\mathbf{d}(-\ln\|x\|_{\mathbf{d}})PB\,(u(\mathsf{q}(x)) - u(x))}{x^{\top}\mathbf{d}^{\top}(-\ln\|x\|_{\mathbf{d}})PG_{\mathbf{d}}\mathbf{d}(-\ln\|x\|_{\mathbf{d}})x}. \tag{10.73}$$

The control $u(x)$ consists of two parts: the linear part K_0x and the homogeneous part

$$\tilde{u}(x) = \|x\|_{\mathbf{d}}^{1+\mu}K\mathbf{d}(-\ln\|x\|_{\mathbf{d}})x.$$

Therefore, we derive

$$\begin{aligned}\frac{d\|x\|_{\mathbf{d}}}{dt} = & -\rho\|x\|_{\mathbf{d}}^{1+\mu} + \frac{x^{\top}\mathbf{d}(-\ln\|x\|_{\mathbf{d}})PBK_0(\mathfrak{q}_x - x)}{x^{\top}\mathbf{d}^{\top}(-\ln\|x\|_{\mathbf{d}})PG_{\mathbf{d}}\mathbf{d}(-\ln\|x\|_{\mathbf{d}})x} \\ & + \frac{x^{\top}\mathbf{d}^{\top}(-\ln\|x\|_{\mathbf{d}})PB\left(\tilde{u}(\mathfrak{q}(x)) - \tilde{u}(x)\right)}{x^{\top}\mathbf{d}^{\top}(-\ln\|x\|_{\mathbf{d}})PG_{\mathbf{d}}\mathbf{d}(-\ln\|x\|_{\mathbf{d}})x}.\end{aligned} \tag{10.74}$$

By definition of canonical homogeneous norm we have $\|\mathbf{d}(-s_x)x\| = 1$. Using Cauchy–Schwarz inequality we conclude

$$\frac{d}{dt}\|x\|_{\mathbf{d}} \leq -\rho\|x\|_{\mathbf{d}}^{1+\mu} + c_1\|\mathfrak{q}(x) - x\| + \frac{x^{\top}\mathbf{d}(-\ln\|x\|_{\mathbf{d}})PB\left(\tilde{u}(\mathfrak{q}(x)) - \tilde{u}(x)\right)}{x^{\top}\mathbf{d}^{\top}(-\ln\|x\|_{\mathbf{d}})PG_{\mathbf{d}}\mathbf{d}(-\ln\|x\|_{\mathbf{d}})x}, \tag{10.75}$$

where $c_1 = \frac{2\sqrt{\lambda_{\max}(P^{-1/2}K_0^{\top}B^{\top}PBK_0P^{-1/2})}}{\lambda_{\min}(P^{1/2}G_{\mathbf{d}}P^{-1/2}+P^{-1/2}G_{\mathbf{d}}^{\top}P^{1/2})}$. Since

$$B\tilde{u}(\mathbf{d}(s)z) = e^{(1+\mu)s}B\tilde{u}(z), \quad \forall s \in \mathbb{R}, \quad \forall z \in \mathbb{R}^n$$

then

$$\begin{aligned}\frac{d}{dt}\|x\|_{\mathbf{d}} \leq & -\rho\|x\|_{\mathbf{d}}^{1+\mu} + c_1\|\mathfrak{q}(x) - x\| \\ & + c_2\|x\|_{\mathbf{d}}^{1+\mu}\left|\tilde{u}\left(\mathbf{d}(-\ln\|x\|_{\mathbf{d}})\mathfrak{q}(x)\right) - \tilde{u}(\mathbf{d}(-\ln\|x\|_{\mathbf{d}})x)\right|,\end{aligned} \tag{10.76}$$

where $c_2 = \frac{2\sqrt{\lambda_{\max}(B^{\top}PB)}}{\lambda_{\min}(P^{1/2}G_{\mathbf{d}}P^{-1/2}+P^{-1/2}G_{\mathbf{d}}^{\top}P^{1/2})}$. The condition (10.70) implies that

$$\begin{aligned}& e^{-\ln\epsilon}e^{-\ln\|x\|_{\mathbf{d}}}\|\mathfrak{q}(x) - x\|_{\mathbf{d}} \leq 1 \\ \Rightarrow\ & \|\mathbf{d}(-\ln\epsilon)\mathbf{d}(-\ln\|x\|_{\mathbf{d}})(\mathfrak{q}(x) - x)\|_{\mathbf{d}} \leq 1 \\ \Rightarrow\ & \|\mathbf{d}(-\ln\epsilon)\mathbf{d}(-\ln\|x\|_{\mathbf{d}})(\mathfrak{q}(x) - x)\| \leq 1 \\ \Rightarrow\ & \lfloor\mathbf{d}(-\ln\epsilon)\rfloor \cdot \|\mathbf{d}(-\ln\|x\|_{\mathbf{d}})(\mathfrak{q}(x) - x)\| \leq 1.\end{aligned} \tag{10.77}$$

By Proposition 5.3, there exists $\alpha \in \mathcal{K}_{\infty} : \frac{1}{\lfloor\mathbf{d}(-\ln\epsilon)\rfloor} \leq \alpha(\epsilon)$ leading to

$$\|\mathbf{d}(-\ln\|x\|_{\mathbf{d}})(\mathfrak{q}_x - x)\| \leq \alpha(\epsilon)$$

and using the identities $\|\mathbf{d}(-\ln\|x\|_{\mathbf{d}})x\| = 1$,

$$\|\mathbf{d}(-\ln\|x\|_{\mathbf{d}})\mathfrak{q}(x)\| = \|\mathbf{d}(-\ln\|x\|_{\mathbf{d}})x + \mathbf{d}(-\ln\|x\|_{\mathbf{d}})(\mathfrak{q}(x) - x)\|$$

we derive

$$1 - \alpha(\epsilon) \leq \|\mathbf{d}(-\ln\|x\|_{\mathbf{d}})\mathfrak{q}(x)\| \leq 1 + \alpha(\epsilon)$$

for all $x \neq \mathbf{0}$ with a sufficiently small $\epsilon > 0$. Hence, taking into account that $\mathbf{d}(-\ln\|x\|_{\mathbf{d}})x \in S$ for any $x \neq \mathbf{0}$, we conclude that for a sufficiently small $\epsilon > 0$ there exists a compact set $\Omega \subset \mathbb{R}^n$ such that $\mathbf{d}(-\ln\|x\|_{\mathbf{d}})\mathfrak{q}(x) \in \Omega$ and $\mathbf{d}(-\ln\|x\|_{\mathbf{d}})$

$x \in \Omega$ for all $x \neq \mathbf{0}$. For $\mu > -1$ the feedback law $\tilde{u}$ is continuous on $\mathbb{R}^n$, then, by Heine–Cantor theorem (see Appendix D), it is uniformly continuous on any compact form $\mathbb{R}^n$. Therefore, there exists a function $\gamma \in \mathcal{K}$ such that $|\tilde{u}(z_1) - \tilde{u}(z_2)| \leq \gamma(\|z_1 - z_2\|), \forall z_1, z_2 \in \Omega$, so

$$|\tilde{u}(\mathbf{d}(-\ln \|x\|_{\mathbf{d}})\mathrm{q}(x)) - \tilde{u}(\mathbf{d}(-\ln \|x\|_{\mathbf{d}})x)| \leq \gamma\left(\|\mathbf{d}(-\ln \|x\|_{\mathbf{d}})(\mathrm{q}(x) - x)\|\right) \tag{10.78}$$

for all $x \neq \mathbf{0}$ Then, using the above inequality and the inequality (10.77), we derive

$$\frac{d}{dt}\|x\|_{\mathbf{d}} \leq -(\rho - c_2\overline{\gamma}(\epsilon))\|x\|_{\mathbf{d}}^{1+\mu} + c_1\|\mathrm{q}_x - x\|, \tag{10.79}$$

where $\overline{\gamma} = \gamma \circ \alpha \in \mathcal{K}$.

For $K_0 = \mathbf{0}$ we have $c_1 = 0$ and

$$\frac{d}{dt}\|x\|_{\mathbf{d}} \leq -(\rho - c_2\overline{\gamma}(\epsilon))\|x\|_{\mathbf{d}}^{1+\mu}, \tag{10.80}$$

and if $0 < \epsilon < \overline{\gamma}^{-1}\left(\frac{\rho}{c_2}\right)$, where $\overline{\gamma}^{-1}$ is an inverse function of $\overline{\gamma}$, then the system is globally the finite-time stable for $\mu \in (-1, 0)$, exponentially stable for $\mu = 0$ and nearly fixed-time stable for $\mu \in (0, (\tilde{n} - 1)^{-1})$.

If $K_0 \neq \mathbf{0}$ (i.e., $c_1 \neq 0$), then the relation between the Euclidean norm and the canonical homogeneous norm (see Lemma 5.4 and Corollary 5.3) yields

$$\begin{aligned} \|\mathrm{q}(x) - x\| &\leq \kappa\|x\|_{\mathbf{d}}^{\beta} \quad \text{for} \quad \|x\|_{\mathbf{d}} \leq 1, \\ \|\mathrm{q}(x) - x\| &\leq \kappa\|x\|_{\mathbf{d}}^{\alpha} \quad \text{for} \quad \|x\|_{\mathbf{d}} \geq 1. \end{aligned} \tag{10.81}$$

Using this estimates and the condition (10.79), we derive

$$\frac{d\|x\|_{\mathbf{d}}}{dt} \leq \begin{cases} -\left[\rho - c_2\overline{\gamma}(\epsilon) - c_1\kappa\|x\|_{\mathbf{d}}^{\beta-(1+\mu)}\right]\|x\|_{\mathbf{d}}^{1+\mu}, & \|x\|_{\mathbf{d}} \leq 1, \\ -\left[\rho - c_2\overline{\gamma}(\epsilon) - c_1\kappa\|x\|_{\mathbf{d}}^{\alpha-(1+\mu)}\right]\|x\|_{\mathbf{d}}^{1+\mu}, & \|x\|_{\mathbf{d}} \geq 1 \end{cases} \tag{10.82}$$

for some $\kappa > 0$. Let $0 < \epsilon < \overline{\gamma}^{-1}(\frac{\rho}{c_2})$. If $\mu > \max\{\alpha - 1, \beta - 1\} = \alpha - 1$ we have $\|x\|_{\mathbf{d}}^{\alpha-(1+\mu)} \to 0$ and $\|x\|_{\mathbf{d}}^{\beta-(1+\mu)} \to 0$ as $\|x\|_{\mathbf{d}} \to \infty$. Thus, the system is globally practically fixed-time stable due to $\mu > 0$. If $\mu \leq \min\{\alpha - 1, \beta - 1\} = \beta - 1$ then $\|x\|_{\mathbf{d}}^{\alpha-(1+\mu)} \to 0$ and $\|x\|_{\mathbf{d}}^{\beta-(1+\mu)} \to 0$ as $\|x\|_{\mathbf{d}} \to 0$. Hence, the system is locally finite-time stable if $\mu \in (-1, \min\{\beta - 1, 0\})$. The proof is complete. □

The proven theorem characterizes quantizers which preserve the convergence properties of the homogeneous system.

Due to Lemma 10.4, the quantizer (10.61) satisfies the condition (10.70) of Theorem 10.6. Therefore, *the control system* (10.1),(10.69) *with the* **d**-*homogeneous logarithmic quantizer* (10.61) *preserves the stability property of the original homogeneous control system* (10.1), (10.2) *provided that* $K_0 = \mathbf{0}$ *and quantizer's parameters* $\epsilon > 0, \kappa > 0$ *are small enough.*

The uniform quantizer does not satisfy the condition (10.70) implying a chattering of the system with a homogeneous control [24]. However, the logarithmic quantizer combined with a linear coordinate transformation may satisfy the condition (10.70) in many cases.

Stabilization Using Conventional Logarithmic Quantization

Let us design a logarithmic quantizer[25] satisfying conditions of Theorem 10.6.

Corollary 10.6 *Let* $G_{\mathbf{d}} \in \mathbb{R}^{n\times n}$ *be a diagonalizable generator of the linear continuous dilation* **d** *in* $\mathbb{R}^n$*:*

$$\exists \Gamma \in \mathbb{R}^{n\times n}: \quad G_{\mathbf{d}} = \Gamma \Lambda \Gamma^{-1}, \quad \Lambda = \mathrm{diag}(\lambda_1, \lambda_2, \ldots, \lambda_n) \succ 0$$

and the canonical homogeneous norm $\|x\|_{\mathbf{d}}$ *be induced by the weighted Euclidean norm* $\|x\| = \sqrt{x^\top P x}$*, where* $x \in \mathbb{R}^n$ *and* $0 \prec P = P^\top \in \mathbb{R}^{n\times n}$ *satisfies (5.6). Let* q *be a logarithmic quantizer on* $\mathbb{R}$ *with a parameter* $\delta > 0$*. If a quantizer* $\mathfrak{q}$ *on* $\mathbb{R}^n$ *is defined as follows:*

$$\mathfrak{q}(x) = \Gamma[q(\xi_{x,1}), q(\xi_{x,2}), \cdots, q(\xi_{x,n})]^\top, \tag{10.83}$$

where $x \in \mathbb{R}^n$ *and* $[\xi_{x,1}, \xi_{x,2}, \cdots, \xi_{x,n}]^\top = \Gamma^{-1}x$*, then for any* $\epsilon > 0$ *and any* $\kappa > 0$ *there exists* $\delta > 0$ *such that the inequalities* (10.70), (10.71) *are fulfilled simultaneously.*

Proof On the one hand, from the formula of quantization function $\mathfrak{q}$, we have

$$\Gamma^{-1}(\mathfrak{q}(x) - x) = [q(\xi_{x,1}) - \xi_{x,1}, q(\xi_{x,2}) - \xi_{x,2}, \cdots, q(\xi_{x,n}) - \xi_{x,n}]^\top,$$

$\Gamma^{-1}x = [\xi_{x,1}, \xi_{x,2}, \cdots, \xi_{x,n}]^\top$. Then, since $G_{\mathbf{d}}$ is diagonalizable, according to the quantization error of the logarithmic quantizer (10.57) and Theorem 5.2, we conclude that for any given $\epsilon > 0$, there exists a sufficiently small $\delta > 0$ such that the inequality $|q(\xi_{x,i}) - \xi_{x,i}| \le \delta|\xi_{x,i}|, \forall i = 1, 2, \cdots, n$ implies the inequality (10.70).

On the other hand, since Γ is invertible then $\Gamma^{-\top}\Gamma^{-1}$ is positive definite, the inequality (10.57) also implies that

$$
\begin{aligned}
|\mathfrak{q}(x)-x| &= \left|\Gamma\left([q(\xi_{x,1}),q(\xi_{x,2}),\cdots,q(\xi_{x,n})]^\top-\xi_x\right)\right| \\
&\leq \lambda_{\max}^{1/2}(\Gamma^\top\Gamma)\left(\sum_{i=1}^{n}|q(\xi_{x,i})-\xi_{x,i}|^2\right)^{1/2} \\
&\leq \lambda_{\max}^{1/2}(\Gamma^\top\Gamma)\left(\sum_{i=1}^{n}|\delta\xi_{x,i}|^2\right)^{1/2} \\
&\leq \delta\lambda_{\max}^{1/2}(\Gamma^\top\Gamma)\lambda_{\max}^{1/2}(\Gamma^{-\top}\Gamma^{-1})|x|.
\end{aligned}
$$

The above inequality yields (10.71) with

$$
\kappa=\delta\lambda_{\max}^{1/2}(\Gamma^\top\Gamma)\lambda_{\max}^{1/2}(\Gamma^{-\top}\Gamma^{-1})\frac{\sqrt{\lambda_{\min}(P)}}{\sqrt{\lambda_{\max}(P)}}. \tag{10.84}
$$

The proof is complete. □

Recall (see the proof of Theorem 9.1) that the generator of the dilation $\mathbf{d}$ can always be selected diagonizable during the design of the control (10.2).

Therefore, *the control system* (10.1), (10.69) *with the logarithmic quantizer* (10.83) *with a sufficiently small* $\delta>0$ *preserves the stability property of the original homogeneous control system* (10.1), (10.2) *locally if* $K_0\neq\mathbf{0}$ *or globally if* $K_0=\mathbf{0}$.

For the controlled chain of integrators:

$$
A=\begin{pmatrix} 0 & 1 & 0 & \dots & 0 & 0\\ 0 & 0 & 1 & \dots & 0 & 0\\ \dots & \dots & \dots & \dots & \dots & \dots\\ 0 & 0 & 0 & \dots & 0 & 1\\ 0 & 0 & 0 & \dots & 0 & 0 \end{pmatrix},\quad B=\begin{pmatrix}0\\0\\ \dots\\0\\1\end{pmatrix}
$$

the linear continuous dilation $\mathbf{d}$ always has a diagonal generator, so $\Gamma=I_n$ and all conditions of Theorem 10.6 are fulfilled for the component-wise logarithmic quantizer: $\mathfrak{q}(x)=[q(x_1),\dots,q(x_n)]^\top$, $x=(x_1,\dots,x_n)^\top\in\mathbb{R}^n$.

Example 10.1 ([23]) Let us consider the LTI system (10.1) with parameters

$$
A=\begin{bmatrix}0\,1\,0\\0\,0\,1\\0\,0\,0\end{bmatrix},\quad B=\begin{bmatrix}0\,0\\0\,0\\1\,2\end{bmatrix}.
$$

The matrix A is nilpotent and the pair (A,B) is controllable. The initial state is $x_0=[20,20,10]^\top$.

The matrix A is nilpotent and $K_0=\mathbf{0}$ the control parameters are obtained by solving the LMIs (9.9) for $\mu=-0.2$, $\rho=2$:

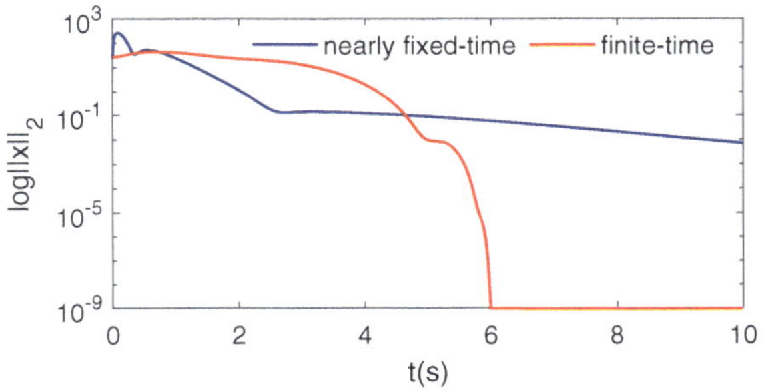

Fig. 10.3 Trajectories of the closed-loop system (10.1), (10.69) with $\mu = -0.2$ and $\mu = 0.2$ and the logarithmic quantizer for $\nu = 0.9$

$$K = -\begin{pmatrix} 5.65 & 4.51 & 1.44 \\ 11.30 & 9.03 & 2.88 \end{pmatrix}, \quad G_{\mathbf{d}} = \begin{pmatrix} 1.4 & 0 & 0 \\ 0 & 1.2 & 0 \\ 0 & 0 & 1 \end{pmatrix}, \quad P = \begin{pmatrix} 70.88 & 39.18 & 7.02 \\ 39.18 & 22.32 & 4.11 \\ 7.02 & 4.11 & 0.79 \end{pmatrix}.$$

For $\mu = 0.2$, $\rho = 2$ the parameters are

$$K = -\begin{pmatrix} 0.87 & 1.68 & 0.96 \\ 1.73 & 3.23 & 1.92 \end{pmatrix}, \quad G_{\mathbf{d}} = \begin{pmatrix} 0.6 & 0 & 0 \\ 0 & 0.8 & 0 \\ 0 & 0 & 1 \end{pmatrix}, \quad P = \begin{pmatrix} 7.61 & 7.08 & 2.11 \\ 7.08 & 7.67 & 2.39 \\ 2.11 & 2.39 & 0.86 \end{pmatrix}.$$

Recall that $\mu > 0$ corresponds to the case of nearly fixed-time stabilization, while the homogeneous controller with $\mu < 0$ stabilizes the system in a finite time. The simulation results of the closed-loop system (10.1), (10.69) are presented in Fig. 10.3 using the explicit Euler discretization. The obtained trajectories are typical for finite-time and nearly fixed-time convergence. The simulation was conducted with the parameter $\nu = 0.9$ of the logarithmic quantizer (see the formula (10.56)).

The global fixed-time stabilizer can be designed by a combination homogeneous stabilizers with negative and positive degrees (see Chap. 9). The fixed-time stability of such a closed-loop system is preserved under the logarithmic quantization [23].

Corollary 10.7 *Let G_0, Y_0, $\tilde{n} \in \mathbb{N}$ and $K_0 = Y_0(G_0 - I_n)^{-1}$, $A_0 = A + BK_0$ be defined as in Theorem 9.1. Let G_0 be diagonalizable:*

$$\exists \Gamma \in \mathbb{R}^{n\times n} : \quad G_0 = \Gamma \Lambda_0 \Gamma^{-1}, \quad \Lambda_0 = \mathrm{diag}(\tilde{\lambda}_1, \tilde{\lambda}_2, \ldots, \tilde{\lambda}_n) \succ 0 \tag{10.85}$$

and the logarithmic quantizer $\mathfrak{q}$ be defined by the formula (10.83) with a parameter $\delta > 0$. Let $X \in \mathbb{R}^{n\times n}$, $Y \in \mathbb{R}^{m\times n}$ satisfy the following algebraic system:

$$\begin{cases} XA_0^\top + A_0 X + Y^\top B^\top + BY \prec 0, \quad X \succ 0, \\ XG_{\mathbf{d}_1}^\top + G_{\mathbf{d}_1} X \succeq 2(1+\mu_1)X, \\ 2(1+\mu_2)X \succeq XG_{\mathbf{d}_2}^\top + G_{\mathbf{d}_2} X \succ 0 \end{cases} \tag{10.86}$$

Let us define a feedback control with quantized states as follows:

$$u = \begin{cases} K_0\mathfrak{q}(x) + \|\mathfrak{q}(x)\|_{\mathbf{d}_1}^{1+\mu_1} K\mathbf{d}_1(-\ln\|\mathfrak{q}(x)\|_{\mathbf{d}_1})\mathfrak{q}(x) & \text{if } \|\mathfrak{q}(x)\| \le 1, \\ K_0\mathfrak{q}(x) + \|\mathfrak{q}(x)\|_{\mathbf{d}_2}^{1+\mu_2} K\mathbf{d}_2(-\ln\|\mathfrak{q}(x)\|_{\mathbf{d}_2})\mathfrak{q}(x) & \text{if } \|\mathfrak{q}(x)\| > 1, \end{cases} \tag{10.87}$$

where $\mathbf{d}_i(\cdot), i = 1, 2$ *are generated by* $G_{\mathbf{d}_i} = \mu_i G_0 + I_n$ *with* $-1 < \mu_1 < 0$ *and* $0 \le \mu_2 < 1/(\tilde{n} - 1)$, *the canonical homogeneous norms* $\|\cdot\|_{\mathbf{d}_i}$ *are induced by the norm* $\|x\| = \sqrt{x^\top Px}$ *with* $P = X^{-1}$. *Then, for sufficiently small* $0 < \delta < 1$, *the system* (10.1), (10.87) *is*

- *globally uniformly finite-time stable if* $\mu_2 = 0$;
- *globally uniformly fixed-time stable if* $\mu_2 \in (0, (\tilde{n} - 1)^{-1})$.

Proof Let us consider the candidate to a Lyapunov function:

$$V = \begin{cases} \|x\|_{\mathbf{d}_1}, & \|x\| \le 1, \\ \|x\|_{\mathbf{d}_2}, & \|x\| \ge 1. \end{cases} \tag{10.88}$$

Notice that $V \in C(\mathbb{R}^n) \cap C^1(\mathbb{R}^n\backslash S\backslash\{\mathbf{0}\})$ and locally Lipschitz continuous on $\mathbb{R}^n\backslash\{\mathbf{0}\}$, where $S = \{x \in \mathbb{R}^n : \|x\| = 1\}$. For sufficiently small $\delta \in (0, 1)$ we have

$$(1 - \kappa)\|x\| \le \|\mathfrak{q}(x)\| \le (1 + \kappa)\|x\|, \tag{10.89}$$

where $\kappa = \kappa(\delta)$ has the form as (10.84), then

$$\frac{1}{(1-\kappa)} \le \|x\| \Rightarrow 1 \le \|\mathfrak{q}(x)\|, \quad \|x\| \le \frac{1}{(1+\kappa)} \Rightarrow \|\mathfrak{q}(x)\| \le 1. \tag{10.90}$$

Let $\Omega_1 := \left\{x \in \mathbb{R}^n : \|x\| \le \frac{1}{(1+\kappa)}\right\}$, $\Omega_2 := \left\{x \in \mathbb{R}^n : \frac{1}{(1-\kappa)} \le \|x\|\right\}$, $\Omega_3 := \left\{x \in \mathbb{R}^n : \frac{1}{(1+\kappa)} \le \|x\| \le \frac{1}{(1-\kappa)}\right\}$. By Corollary 10.6, the logarithmic quantizer $\mathfrak{q}(x)$ satisfies the conditions (10.70) and (10.71) for a sufficiently small $0 < \delta < 1$. Thus according to Theorem 10.6 and (10.82), the derivative of V admits the estimate

$$\dot{V} \le \begin{cases} -\left[\rho_1 - c_{2,\mu_1}\overline{\gamma}_1(\epsilon) - c_{1,\mu_1}\kappa\|x\|_{\mathbf{d}_1}^{\beta_1-(1+\mu_1)}\right]\|x\|_{\mathbf{d}_1}^{1+\mu_1}, & x \in \Omega_1 \\ -\left[\rho_2 - c_{2,\mu_2}\overline{\gamma}_2(\epsilon) - c_{1,\mu_2}\kappa\|x\|_{\mathbf{d}_2}^{\alpha_2-(1+\mu_2)}\right]\|x\|_{\mathbf{d}_2}^{1+\mu_2}, & x \in \Omega_2, \end{cases} \tag{10.91}$$

where

$$c_{1,\mu_i} = \frac{\lambda_{\max}^{1/2}(P^{-1/2}K_0^\top B^\top PBK_0P^{-1/2})}{\beta_i}, \quad c_{2,\mu_i} = \frac{\lambda_{\max}^{1/2}(B^\top PB)}{\beta_i},$$

$$\rho_i = \frac{\lambda_{max}(XA_0^\top + A_0X + Y^\top B^\top + BY)}{\beta_i},$$

$$\alpha_i = \tfrac{1}{2}\lambda_{\max}(P^{\frac{1}{2}}G_{\mathrm{d}_i}P^{-\frac{1}{2}} + P^{-\frac{1}{2}}G_{\mathrm{d}_i}^{\top}P^{\frac{1}{2}}),$$

$$\beta_i = \frac{1}{2}\lambda_{\min}(P^{1/2}G_{\mathbf{d}_i}P^{-1/2}+P^{-1/2}G_{\mathbf{d}_i}^{\top}P^{1/2}), \quad i = 1, 2.$$

The functions $\overline{\gamma}_1, \overline{\gamma}_2 \in \mathcal{K}$ are obtained in the proof of Theorem 10.6 using (10.77), (10.78) with $\mathbf{d}_1$ and $\mathbf{d}_2$, respectively.

The LMIs guarantee $\alpha_2 - 1 \le \mu_2$ and $\mu_1 \le \beta_1 - 1$. Then, we have $\|x\|_{\mathbf{d}_1}^{\alpha_2-(1+\mu_2)}$ monotonically decreases to zero as $\|x\|_{\mathbf{d}_2}$ increases to infinity, and $\|x\|_{\mathbf{d}_1}^{\beta_1-(1+\mu_1)}$ monotonically decreases to zero as $\|x\|_{\mathbf{d}_1}$ decreases to zero. Thus, we conclude that for a sufficiently small δ, such that

$$\dot{V} < 0, \quad \forall x \in \Omega_1 \cup \Omega_2.$$

On the other hand, for $\frac{1}{1+\kappa} \le \|x\| \le \frac{1}{1-\kappa}$, the quantization error is

$$\|\mathfrak{q}_x - x\| \le \frac{\kappa}{1-\kappa}.$$

Using (10.79), for a sufficiently small $\delta > 0$, we have small enough $\kappa > 0$ and $\epsilon > 0$ such that we derive

$$\dot{V} \le \max_{i=1,2}\left\{-\frac{\rho_i - c_{2,\mu_i}\overline{\gamma}_i(\epsilon)}{(1+\kappa)^{1+\mu_i}} + c_{1,\mu_i}\frac{\kappa}{1-\kappa}\right\} < 0, \quad \forall x \in \Omega_3. \tag{10.92}$$

Therefore, V is a global Lyapunov function and the system is globally asymptotically stable. Since $\mu_1 < 0$ then, using Theorem 10.6, we conclude that the system is globally finite-time stable if $\mu_2 = 0$ and globally fixed-time stable if $\mu_2 > 0$. The proof is complete. □

Example 10.2 ([23]) Let us consider the LTI system (10.1) with the parameters

$$A_2 = \begin{pmatrix} 1 & 1 & 0 \\ 0 & 0 & -0.5 \\ 0.2 & 0 & 0.5 \end{pmatrix}, \quad B_2 = \begin{pmatrix} 0 & 0 \\ 0 & 1 \\ 1 & 2 \end{pmatrix}.$$

The pair $\{A, B\}$ is controllable but the matrix A is not nilpotent. We design a global stabilizer (10.87) by Corollary 10.7. The control parameters are obtained by solving LMIs (10.86) with $\mu_1 = -0.5$, $\mu_2 = 0.2$ for the global fixed-time stabilization and with $\mu_1 = -0.5$, $\mu_2 = 0$ for the global finite-time stabilization:

$$K = \begin{pmatrix} 10.09 & 2.66 & -0.61 \\ -5.05 & -1.33 & -0.0001 \end{pmatrix}, \quad P = \begin{pmatrix} 0.4959 & 0.1629 & 0 \\ 0.1629 & 0.1179 & 0 \\ 0 & 0 & 0.1131 \end{pmatrix},$$

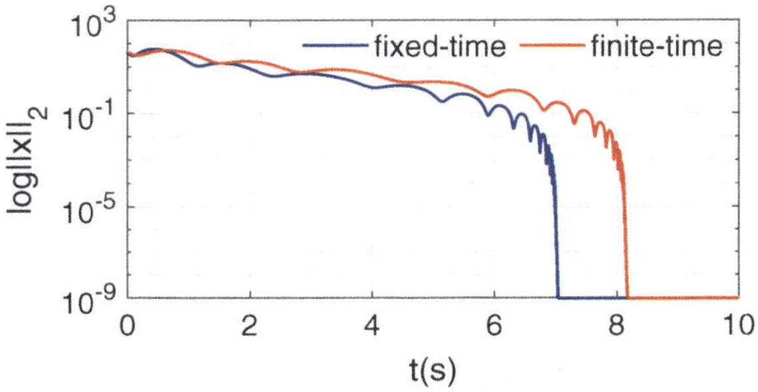

Fig. 10.4 Trajectories of the closed-loop system (10.1), (10.87)

$$K_0 = \begin{pmatrix} 1.8 & 2 & -1.5 \\ -1 & -1 & 0.5 \end{pmatrix}, \quad \Gamma = \begin{pmatrix} 0 & 0 & 0.7071 \\ 0 & 1 & -0.7071 \\ 1 & 0 & 0.0010 \end{pmatrix},$$

$$G_{\mathbf{d}_1} = \begin{pmatrix} 1.5 & 0 & 0 \\ -0.5 & 1 & 0 \\ 0.0007 & 0 & 1 \end{pmatrix}, G_{\mathbf{d}_2} = \begin{pmatrix} 0.8 & 0 & 0 \\ 0.2 & 1 & 0 \\ -0.0003 & 0 & 1 \end{pmatrix}.$$

The logarithmic quantizer in $\mathbb{R}$ is designed for $\delta = 1/19$. The quantizer $\mathfrak{q}$ in $\mathbb{R}^n$ is defined by the formula (10.83). The simulation results of the closed-loop systems with the quantized controller(10.87) are presented in Fig. 10.4. To illustrate the quantization effect, Fig. 10.5 shows the evolution of the quantized measurements.

Finite-Time Stabilization with Limited Data Transmission

The homogeneous controller (10.2) with the **d**-homogeneous spherical quantizer (10.63) ensures the global finite-time stabilization of homogeneous linear plant [18]. If the plant is not homogeneous, then the finite-time stabilization is, at least, local.

Corollary 10.8 *Let all parameters of the homogeneous feedback* (10.2) *be designed by Theorem 9.1. If $K_0 = \mathbf{0}$ then, for a sufficiently small $\gamma > 0$ given by* (10.60)*, the closed-loop system* (10.1), (10.69) *with the* **d***-homogeneous spherical quantizer* (10.63) *is globally uniformly finite-time stable, where $\gamma > 0$ is the tuning parameter of the quantizer. If $K_0 \neq \mathbf{0}$ then the closed-loop system* (10.1), (10.69) *is locally finite-time stable.*

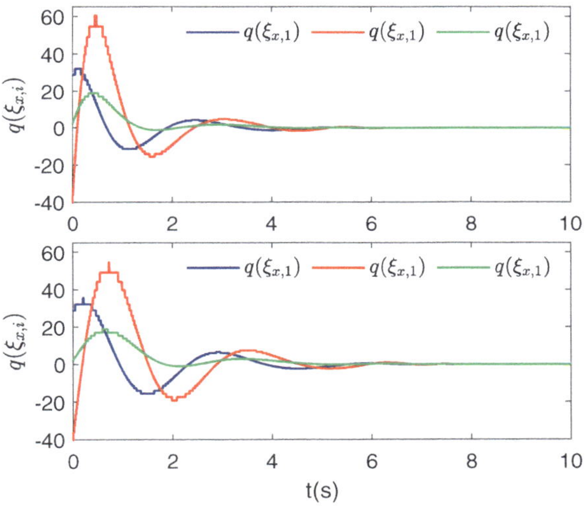

Fig. 10.5 *Quantized measurements of the vector $\Gamma^{-1}x$ with a logarithmic quantizer*

Proof If $K_0 = \mathbf{0}$ the resulting feedback law becomes a unit control (see Chap. 9):

$$u(x) = \|x\|_{\mathbf{d}}^{1+\mu} K\mathbf{d}(-\ln \|x\|_{\mathbf{d}})x. \tag{10.93}$$

In the case of the $\mathbf{d}$-homogeneous finite quantizer (10.63) we have $\|\mathfrak{q}_{\mathbf{d}}(x)\|_{\mathbf{d}} = 1$ and

$$u(\mathfrak{q}_{\mathbf{d}}(x)) = K\mathfrak{q}_{\pi}(\mathbf{d}(-\ln \|x\|_{\mathbf{d}})x), \tag{10.94}$$

where $q_\pi : S \mapsto \mathcal{Q}_S$ is a uniform quantizer on the unit sphere $S = \{x \in \mathbb{R}^n : \|x\| = 1\}$ with the tuning parameter $\gamma > 0$ given by (10.60). Considering the canonical homogeneous norm $\|\cdot\|_{\mathbf{d}}$ as a Lyapunov function for the closed-loop system we derive

$$\begin{aligned}
\frac{d}{dt}\|x\|_{\mathbf{d}} &= \|x\|_{\mathbf{d}} \frac{x^\top \mathbf{d}(-s_x)P\mathbf{d}(-s_x)\left(Ax + BK\mathfrak{q}_{\pi}(\mathbf{d}(-s_x)x)\right)}{x^\top \mathbf{d}(-s_x)PG_{\mathbf{d}}\mathbf{d}(-s_x)x} \\
&= \|x\|_{\mathbf{d}} \frac{x^\top \mathbf{d}(-s_x)P\mathbf{d}(-s_x)\left(Ax + BK\mathbf{d}(-s_x)x + BK\mathfrak{q}_{\pi}(\mathbf{d}(-s_x)x) - BK\mathbf{d}(-s_x)x\right)}{x^\top \mathbf{d}(-s_x)PG_{\mathbf{d}}\mathbf{d}(-s_x)x} \\
&= -\rho + \frac{x^\top \mathbf{d}(-s_x)PBK(\mathfrak{q}_{\pi}(\mathbf{d}(-s_x)x) - \mathbf{d}(-s_x)x)}{x^\top \mathbf{d}(-s_x)PG_{\mathbf{d}}\mathbf{d}(-s_x)x} \\
&\le -\rho + \frac{\|BK\| \cdot \|\mathfrak{q}_{\pi}(\mathbf{d}(-s_x)x) - \mathbf{d}(-s_x)x\|}{2\beta},
\end{aligned}$$

where $s_x = \ln \|x\|_{\mathbf{d}}$ and $\beta = 0.5\lambda_{\min}(P^{1/2}G_{\mathbf{d}}P^{-1/2} + P^{1/2}G_{\mathbf{d}}^\top P^{-1/2})$. We complete the proof taking into account $\|\mathfrak{q}_{\pi}(\mathbf{d}(-s_x)x) - \mathbf{d}(-s_x)x\| \le \gamma$.

Local stability for $K_0 \neq \mathbf{0}$ can be proven analogously.

Notice that, for $K_0 = \mathbf{0}$, the $\mathbf{d}$-homogeneity of the quantizer (10.63) implies the $\mathbf{d}$-homogeneity of the closed-loop system (10.1) with the quantized unit control (10.94).

> Therefore, *the generalized homogeneous LTI system* (10.1) *can be stabilized globally in a finite time by means of a uniformly bounded homogeneous feedback* (10.94) *with quantized state measurements having only a finite number of quantization seeds distributed on a unit sphere.*

Notice also that Corollary 10.8 remains valid for perturbed system (9.40) satisfying Corollary 9.2 with $\mu = -1$.

Concluding Remarks

In this chapter, several issues of digitization of homogeneous control are studied. Realization of a homogeneous feedback by a linear switched controller is discussed. A consistent discretization (sampled-time implementation) of the homogeneous control for LTI plant is presented. It preservers stability and finite/fixed-time convergence rates of the original continuous-time system in its sampled-time counterpart. It is shown that the finite/fixed-time stability of LTI system with a (locally) homogeneous controller can be preserved in the case of logarithmic quantization of the state measurements. A new type of state quantization appropriate for homogeneous control systems is introduced. Such a quantization allows, in particular, a homogeneous finite-time stabilization of LTI system using only a finite number of quantization seeds for the whole state space $\mathbb{R}^n$.

References

1. Liberzon D (2003) Switchings in systems and control. Birkhauser, Boston
2. Filippov AF (1988) Differential equations with discontinuous right-hand sides. Kluwer Academic Publishers
3. Polyakov A, Efimov D, Perruquetti W (2016) Robust stabilization of MIMO systems in finite/fixed time. Int J Robust Nonlinear Control 26(1):69–90
4. Polyakov A, Efimov D, Perruquetti W (2015) Finite-time and fixed-time stabilization: implicit Lyapunov function approach. Automatica 51(1):332–340
5. Acary V, Brogliato B (2010) Implicit Euler numerical scheme and chattering-free implementation of sliding mode systems. Syst Control Lett 59(5):284–293
6. Brogliato B, Polyakov A (2021) Digital implementation of sliding-mode control via the implicit method: a tutorial. Int J Robust Nonlinear Control 31:3528–3586
7. Josse F, Bernuau E, Moulay E, Coirault P (2021) Robustness of sampled-data homogeneous systems. Automatica 123:109345
8. DeCarlo R (1989) Linear systems: a state variable approach with numerical implementation. Prentice Hall, NJ
9. Kalman R (1960) On the general theory of control systems. In: Proceedings of the first international congress of the IFAC. Moscow

10. Polyakov A, Efimov D, Ping X (2023) Consistent discretization of homogeneous finite/fixed-time controllers for LTI systems. Automatica 155:111118
11. Sontag ED (1989) Smooth stabilization implies coprime factorization. IEEE Trans Autom Control 34:435–443
12. Jiang Z-P, Wang Y (2001) Input-to-state stability for discrete-time nonlinear systems. Automatica 37(6):856–869
13. Polyakov A, Efimov D, Brogliato B (2019) Consistent discretization of finite-time and fixed-time stable systems. SIAM J Control Optimization 57(1):78–103
14. Hall BC (2015) Lie groups, Lie algebras, and representations. Springer
15. Nesic D, Liberzon D (2009) A unified framework for design and analysis of networked and quantized control systems. IEEE Trans Autom Control 54(4):732–747
16. Boots B, Sugihara K, Chiu SN, Okabe A (2009) Spatial tessellations: concepts and applications of Voronoi diagrams. Wiley
17. Fu M, Xie L (2005) The sector bound approach to quantized feedback control. IEEE Trans Autom Control 50(11):1698–1711
18. Zhou Y, Polyakov A, Zheng G (2025) Robust finite-time stabilization of linear systems with limited state quantization. Automatica 171(1):111967
19. Haimovich H, Seron MM, Goodwin GC (2006) Geometric characterization of multivariable quadratically stabilizing quantizers. Int J Control 79(8):845–857
20. Bullo F, Liberzon D (2006) Quantized control via locational optimization. IEEE Trans Autom Control 51(1):2–13
21. Wang J (2021) Quantized feedback control based on spherical polar coordinate quantizer. IEEE Trans Autom Control 66(12):6077–6084
22. Zhou Y, Polyakov A, Zheng G (2023) Homogeneous finite/fixed-time stabilization with quantization. In: IEEE conference on decision and control
23. Zhou Y, Polyakov A, Zheng G (2025) Finite/fixed-time stabilization of linear systems with states quantization. IEEE Trans Autom Control 70(3):1921–1928
24. Bernuau E, Moulay E, Coirault P (2017) Stability of homogeneous nonlinear systems with sampled-data inputs. Automatica 85(11):349–355
25. Zhou Y, Polyakov A, Zheng G (2024) Generalized homogeneous rigid-body attitude control. Automatica 163(111548)

Chapter 11
Upgrading Linear Controller to Homogeneous Feedback

The Proportional–Integral–Derivative (PID) regulator is the most popular linear control algorithm utilized in real applications. From theoretical point of view, we have shown that homogeneous systems may have better robustness properties and faster convergence rates than linear systems. This chapter is aimed at practical demonstration of the announced theoretical advantages. More precisely, we study the following question:

> *Is it possible to "upgrade" an existing linear (e.g., PID) controller to a* **d***-homogeneous one with an improved regulation quality (e.g., with faster convergence rates, smaller overshoots and better robustness properties of the closed-loop system)?*

A scheme of the upgrade must prevent a possible degradation of the regulation precision allowing only its improvement.

11.1 Upgrading Linear State Feedback

In this chapter, we develop a model-based methodology for an upgrading linear controllers. This means that we assume that an approximate linear model of the control system is known and some linear stabilizing feedback is already designed. First, we develop a simple way of transformation of the linear stabilizer to a homogeneous one; next, we introduce a general scheme of upgrading linear controllers to nonlinear one in such a way that the quality of the nonlinear control system cannot be worse than the quality of the linear control system.

A. Polyakov, *Generalized Homogeneity in Systems and Control Volume I*,
Communications and Control Engineering,
https://doi.org/10.1007/978-3-031-76246-8_11

11.1.1 Design of Homogeneous State Feedback Using an Existing Linear Stabilizer

Let us consider the linearized model of a control system

$$\dot{x} = Ax + Bu, \quad t > 0, \tag{11.1}$$

where $x(t) \in \mathbb{R}^n$ is the system state, $u(t) \in \mathbb{R}^n$ is a control input, $A \in \mathbb{R}^{n\times n}$, $B \in \mathbb{R}^{n\times m}$ are known system matrices.

Assumption 11.1 Let us assume that a linear control

$$u_{\text{lin}}(x) = K_{\text{lin}}x, \quad K_{\text{lin}} \in \mathbb{R}^{m\times n}, \quad x \in \mathbb{R}^n \tag{11.2}$$

for the control system modeled by (11.1) is already designed such that the matrix $A + BK_{\text{lin}}$ is Hurwitz.

Let some (not yet selected) linear dilation $\mathbf{d}(s) = e^{sG_{\mathbf{d}}}$ in $\mathbb{R}^n$ be strictly monotone with respect to a weighted Euclidean norm $\|x\| = \sqrt{x^\top Px}$ with some $0 \prec P = P^\top \in \mathbb{R}^{n\times n}$ satisfying (5.6). Let $\|\cdot\|_{\mathbf{d}}$ be the canonical homogeneous norm induced by $\|\cdot\|$. The homogeneous control (9.10)

$$u(x) = K_0x + \|x\|_{\mathbf{d}}^{1+\mu} K\mathbf{d}(-\ln \|x\|_{\mathbf{d}})x$$

becomes linear for $\|x\|_{\mathbf{d}} = 1$:

$$u(x) = (K_0 + K)x \quad \text{if} \quad \|x\|_{\mathbf{d}} = 1.$$

Recall that $\|x\|_{\mathbf{d}} = 1 \Leftrightarrow \|x\| = 1$. Therefore, the homogeneous controller coincides with a linear stabilizing feedback on the unit sphere. The dilation $\mathbf{d}$ is just utilized to scale the gain K dependently on the canonical homogeneous norm. Notice that, according to the control design suggested by Theorem 9.1, the gain K_0 and the generator of the dilation $G_{\mathbf{d}}$ are selected independently of K, so we can take $K = K_{\text{lin}} - K_0$ and define the homogeneous feedback upgraded from the linear controller as follows:

$$u_\mu(x) = K_0x + \|x\|_{\mathbf{d}}^{1+\mu}(K_{\text{lin}} - K_0)\mathbf{d}(-\ln \|x\|_{\mathbf{d}})x. \tag{11.3}$$

With such a definition the homogeneous feedback coincides with the given linear controller on the unit sphere:

$$u_\mu(x) = u_{\text{lin}}(x) \quad \text{if} \quad \|x\| = 1.$$

Therefore, to complete the design of the homogeneous feedback (11.3) we just need to define a positive definite symmetric matrix P such that the unit ball $\{x \in \mathbb{R}^n : \|x\| \leq 1\}$ is the strictly positively invariant set of the linear dilation $\mathbf{d}$ is monotone with respect to the $\|\cdot\|$. The following corollary proves that this is always possible if the pair $\{A, B\}$ is controllable and the requested homogeneity degree μ of the closed-loop system satisfies certain restrictions.

Corollary 11.1 *Let the pair $\{A, B\}$ be controllable. Let $K_0 \in \mathbb{R}^{m\times n}$, $A_0 \in \mathbb{R}^{n\times n}$, $G_0 \in \mathbb{R}^{n\times n}$ and $\tilde{n} \in \mathbb{N}$ be defined as in Theorem 9.1.*

If Assumption 11.1 is fulfilled then the linear matrix inequality (LMI)

$$(A + BK_{\text{lin}})^\top P + P(A + BK_{\text{lin}}) + 2\rho P \prec 0, \quad P \succ 0 \tag{11.4}$$

is always feasible with respect to $P = P^\top \in \mathbb{R}^{n\times n}$ for a sufficiently small $\rho > 0$. Moreover, denoting $W = (A + BK_{\text{lin}})^\top P + P(A + BK_{\text{lin}}) + 2\rho P$ and $Q = PG_0 + G_0^\top P$, for any solution P of the LMI (11.4), any $\delta \in (0, 1)$ and for any $\mu \in [\mu_{\min}, \mu_{\max}]$ satisfying

$$\mu_{\min} = \max\left\{-1, \frac{-2\delta}{\max\left\{0, \lambda_{\max}\left(P^{-\frac{1}{2}}QP^{-\frac{1}{2}}\right)\right\}}, \frac{\lambda_{\max}(W)}{\rho\max\{0, -\lambda_{\min}(Q)\}}\right\} < 0, \tag{11.5}$$

$$\mu_{\max} = \min\left\{\frac{1}{\tilde{n}}, \frac{2\delta}{\max\left\{0, -\lambda_{\min}\left(P^{-\frac{1}{2}}QP^{-\frac{1}{2}}\right)\right\}}, \frac{-\lambda_{\max}(W)}{\rho\max\{0, \lambda_{\max}(Q)\}}\right\} > 0, \tag{11.6}$$

(1) the matrix $G_{\mathbf{d}} = I_n + \mu G_0$ is anti-Hurwitz and $PG_{\mathbf{d}} + G_{\mathbf{d}}^\top P \succ 0$;
(2) the closed-loop system (11.1), (11.3) is globally uniformly asymptotically stable and

$$\frac{d}{dt}\|x\|_{\mathbf{d}} \le -\rho\|x\|_{\mathbf{d}}^{1+\mu}, \quad \forall x \neq \mathbf{0}. \tag{11.7}$$

Proof Since, by Assumption 11.1, the matrix $A + BK_{\text{lin}}$ is Hurwitz then, for a sufficiently small $\rho > 0$, the matrix $A + BK_{\text{lin}} + \rho I_n$ is Hurwitz as well, so the LMI (11.4) is the Lyapunov inequality being always feasible (see [1] or Example 4.14). In this case, we have

$$\lambda_{\max}(W) = \lambda_{\max}((A + BK_{\text{lin}})^\top P + P(A + BK_{\text{lin}}) + 2\rho P) < 0.$$

Hence, by construction, $-1 \le \mu_{\min} < 0$ and $0 < \mu_{\max} \le \frac{1}{\tilde{n}}$, where $\tilde{n} \in \mathbb{N}$ is defined in the formulation of Theorem 9.1.

(1) Let us show that $PG_{\mathbf{d}} + G_{\mathbf{d}}^\top P \succ 0$. Indeed, if $\lambda_{\max}\left(P^{-\frac{1}{2}}QP^{-\frac{1}{2}}\right) \le 0$ then $\mu_{\min}P^{-\frac{1}{2}}QP^{-\frac{1}{2}} \succeq 0$ and

$$2I_n + \mu_{\min}P^{-\frac{1}{2}}QP^{-\frac{1}{2}} \succ 0. \tag{11.8}$$

If $\lambda_{\max}\left(P^{-\frac{1}{2}}QP^{-\frac{1}{2}}\right) > 0$ then using the definition of $\mu_{\min}$ we derive

$$\mu_{\min}\lambda_{\max}\left(P^{-\frac{1}{2}}QP^{-\frac{1}{2}}\right) \geq -2\delta \quad \Rightarrow \quad 2\delta I_n + \mu_{\min}P^{-\frac{1}{2}}QP^{-\frac{1}{2}} \succeq 0,$$

so the inequality (11.8) is fulfilled again.

Similarly, if $\lambda_{\min}\left(P^{-\frac{1}{2}}QP^{-\frac{1}{2}}\right) \geq 0$ then $\mu_{\max}\lambda_{\min}\left(P^{-\frac{1}{2}}QP^{-\frac{1}{2}}\right) \geq 0$ and

$$2I_n + \mu_{\max}P^{-\frac{1}{2}}QP^{-\frac{1}{2}} \succ 0. \tag{11.9}$$

If $\lambda_{\min}\left(P^{-\frac{1}{2}}QP^{-\frac{1}{2}}\right) < 0$ then using the definition of $\mu_{\max}$ we derive

$$-\mu_{\max}\lambda_{\min}\left(P^{-\frac{1}{2}}QP^{-\frac{1}{2}}\right) \leq 2\delta \quad \Rightarrow \quad 2\delta I_n + \mu_{\max}P^{-\frac{1}{2}}QP^{-\frac{1}{2}} \succeq 0,$$

so the inequality (11.9) is fulfilled again.

Using the convex embedding $\mu = \lambda\mu_{\min} + (1-\lambda)\mu_{\max}$ for some $\lambda \in [0, 1]$, from the LMIs (11.8), (11.9) we derive the LMI

$$2I_n + \mu P^{-\frac{1}{2}}(PG_0 + G_0^\top P)P^{-\frac{1}{2}} \succ 0. \tag{11.10}$$

Due to $G_{\mathbf{d}} = I_n + \mu G_0$, this LMI is equivalent to $PG_{\mathbf{d}} + G_{\mathbf{d}}P \succ 0$.

(2) If $\lambda_{\min}(PG_0+G_0^\top P)\geq 0$ then $PG_0+G_0^\top P \succeq 0$, $\mu_{\min}(PG_0+G_0^\top P) \preceq 0$ and

$$(A + BK_{\text{lin}})^\top P + P(A + BK_{\text{lin}}) + 2\rho P + \mu_{\min}\rho(PG_0 + G_0^\top P) \preceq 0. \tag{11.11}$$

If $\lambda_{\min}(PG_0 + G_0^\top P) < 0$ then using the definition of $\mu_{\min}$ we derive

$$-\mu_{\min}\rho\lambda_{\min}(PG_0 + G_0^\top P) \geq \lambda_{\max}((A + BK_{\text{lin}})^\top P + P(A + BK_{\text{lin}}) + 2\rho P).$$

Hence, taking into account $-\mu_{\min}\rho(PG_0 + G_0^\top P) \succeq -\mu_{\min}\rho\lambda_{\min}(PG_0 + G_0^\top P)I_n$ we conclude that the inequality (11.11) is fulfilled again.

If $\lambda_{\max}(PG_0+G_0^\top P)\leq 0$ then $PG_0+G_0^\top P \preceq 0$, $\mu_{\max}(PG_0+G_0^\top P) \preceq 0$ and

$$(A + BK_{\text{lin}})^\top P + P(A + BK_{\text{lin}}) + 2\rho P + \mu_{\max}\rho(PG_0 + G_0^\top P) \preceq 0. \tag{11.12}$$

If $\lambda_{\max}(PG_0 + G_0^\top P) > 0$ then using the definition of $\mu_{\max}$ we derive

$$\mu_{\max}\rho\lambda_{\max}(PG_0 + G_0^\top P) \leq -\lambda_{\max}((A + BK_{\text{lin}})^\top P + P(A + BK_{\text{lin}}) + 2\rho P).$$

Hence, taking into account $\mu_{\max}\rho\lambda_{\max}(PG_0 + G_0^\top P)I_n \succeq \mu_{\max}\rho(PG_0 + G_0^\top P)$ we conclude that the inequality (11.12) is fulfilled again. Using the convex embedding $\mu = \lambda\mu_{\min} + (1-\lambda)\mu_{\max}$ for some $\lambda \in [0, 1]$, from the LMIs (11.11), (11.12) we derive that the LMI

$$(A + BK_{\text{lin}})^\top P + P(A + BK_{\text{lin}}) + 2\rho P + \mu\rho(PG_0 + G_0^\top P) \preceq 0 \qquad (11.13)$$

holds for any $\mu \in [\mu_{\min}, \mu_{\max}]$.

Taking into account $G_{\mathbf{d}} = I_n + \mu G_0$ and $A + BK_{\text{lin}} = A_0 + BK$ for $K = K_{\text{lin}} - K_0$ we derive that the following LMI

$$(A_0 + BK)^\top P + P(A_0 + BK) + \rho(PG_{\mathbf{d}} + G_{\mathbf{d}}^\top P) \preceq 0$$

is fulfilled. Hence, repeating the proof of Theorem 9.1 we conclude

$$\frac{d\|x\|_{\mathbf{d}}}{dt} = \|x\|_{\mathbf{d}}^{1+\mu} \frac{x^\top \mathbf{d}^\top(-\ln\|x\|_{\mathbf{d}})((A_0 + BK)^\top P + P(A_0 + BK))\mathbf{d}(-\ln\|x\|_{\mathbf{d}})x}{x^\top \mathbf{d}(-\ln\|x\|_{\mathbf{d}})(G_{\mathbf{d}}^\top P + PG_{\mathbf{d}})\mathbf{d}(-\ln\|x\|_{\mathbf{d}})x} \leq -\rho\|x\|_{\mathbf{d}}^{1+\mu}.$$

The proof is complete. □

The above corollary in particular proves that the linear controller can always be upgraded to homogeneous provided that the pair $\{A, B\}$ is controllable. Indeed, in this case, the interval $[\mu_{\min}, \mu_{\max}]$ is always non-empty and a finite-time (or a nearly fixed-time) controller can be designed using the gains of the linear controller.

Example 11.1 Let us consider the linear model of a control system with the parameters:

$$A = \begin{pmatrix} 0 & 0 & 1.0000 & 0 \\ 0 & 0 & 0 & 1.0000 \\ 0 & 152.0057 & -12.2542 & -0.5005 \\ 0 & 264.3080 & -12.1117 & -0.8702 \end{pmatrix}, \quad B = \begin{pmatrix} 0 \\ 0 \\ 50.6372 \\ 50.0484 \end{pmatrix}.$$

Let the linear feedback (11.2) has the gain

$$K_{\text{lin}} = \begin{pmatrix} 2 & -35 & 1.5 & -3 \end{pmatrix}.$$

Since $A + BK_{lin}$ is Hurwitz and $\{A, B\}$ is controllable then we can apply Corollary (11.1) for upgrading the linear feedback. We derive

$$K_0 = \begin{pmatrix} 0 & -5.2811 & 0.2420 & 0.0174 \end{pmatrix}, \quad G_0 = \begin{pmatrix} -3 & 2.0248 & -0.0033 & 0.0033 \\ 0 & -1 & 0 & 0 \\ 0 & 0.3800 & -2 & 2.0235 \\ 0 & 0 & 0 & 0 \end{pmatrix},$$

$$P = \begin{pmatrix} 3.1581 & -7.2001 & 0.5908 & -0.6077 \\ -7.2001 & 96.7019 & -6.8570 & 7.4704 \\ 0.5908 & -6.8570 & 0.5539 & -0.5856 \\ -0.6077 & 7.4704 & -0.5856 & 0.6472 \end{pmatrix}, \quad \mu_{\min} = -1, \quad \mu_{\max} = 0.1607.$$

For any $\mu \in [\mu_{\min}, \mu_{\max}]$ the controller (11.3) is a **d**-homogeneous stabilizer of degree μ for the linear plant (11.1) with the parameters A and B given above.

11.1.2 Practical Tuning of Control Parameters

For practice, the upgrade should be based on the linear controller, which is already well tuned for the concrete real control system. Any model of control system (e.g., the linear system (11.1)) is just an approximation of reality. The control system may have many elements, which are not included in the model (such as nonlinearities, uncertainties, delays, and measurement noise). The procedure of an upgrade should be developed in such a way that a degradation of the control quality could be prevented.

In this section, we assume that the quality of the algorithm is estimated on a series of the experiments (physical or numerical) which allow to evaluate quantitatively the regulation quality.

> Let $Q_i(u) \in \mathbb{R}^n$, $i = 1, 2, ..p$ denote the *quality indices* of the closed-loop system with the feedback control u. Assume that a smaller Q_i corresponds to a better quality of the regulation.

We present two algorithms for upgrading the linear feedback. The first one proposes a global upgrade, in the sense that the linear controller is replaced with the homogeneous one on the whole state space. The second algorithm is local. It replaces locally in the state space the linear feedback with a homogeneous stabilizer. The mentioned upgrades have different procedures for control parameter's tuning.

For $\mu = 0$ the control u_μ designed in Corollary 11.1 coincides with u_{lin} on the whole state space. In this case, the indexes Q_i of the linear controller coincide with the indexes of the homogeneous controller with $\mu = 0$, i.e.,

$$Q_i(u_{\text{lin}}) = Q_i(u_0), i = 1, 2, \ldots, p.$$

This inspires the following procedure for upgrading of the linear control.

> **Algorithm 11.1** ***(Global upgrade of linear feedback)***
>
> *(1) Define parameters $K_0 \in \mathbb{R}^{m\times n}$ and $G_0 \in \mathbb{R}^{n\times n}$ as in Theorem 9.1.*
> *(2) Initiate $j = 0$, $\mu_0 = 0$ and $\epsilon \neq 0$ close to 0.*
> *(3) Evaluate $Q_i(u_{\mu_0}), i = 1, 2, \ldots, p$—the indexes of the linear controller.*
> *(4) Take $\mu_{j+1} = \mu_j + \epsilon$, define the homogeneous control $u_{\mu_{j+1}}$ by Corollary 11.1 and evaluate $Q_i(u_{\mu_{j+1}}), i = 1, 2, \ldots, p$.*
> *(5) If $\max_i(Q_i(u_{\mu_{j+1}}) - Q_i(u_{\mu_j})) < 0$ and $\mu_{j+1} \in [\mu_{\min}, \mu_{\max}]$ then take $j = j + 1$ and go to (4)*
> *else Stop (no more improvement of the control quality).*

Dependently of the sign of ϵ, the above algorithm is aimed at the upgrade of the linear controller to a homogeneous stabilizer of negative degree for $\epsilon < 0$ or of positive degree if $\epsilon > 0$. It starts with the controller having the zero homogeneity degree $\mu_0 = 0$ which corresponds to the original linear controller. Next, the degree is decreased (if $\epsilon < 0$) or increased (if $\epsilon > 0$) iteratively while the control quality is improved. The quality of the homogeneous controller u_{μ_j} tuned by the presented algorithm is never worse than the quality of the original linear feedback u_{lin}. In the worse case, the algorithm will be terminated with $j = 0$ and $\mu_j = 0$. This would mean that the quality of the linear controller cannot be improved making global upgrade to the homogeneous controller.

To develop an algorithm of the local upgrade, the structure of the homogeneous controller is slightly modified. Let the saturation function $\text{sat}_{a,b} : \mathbb{R}_+ \mapsto \mathbb{R}_+$ be given by

$$\text{sat}_{a,b}(\rho) = \begin{cases} b \text{ if } & \rho \geq b, \\ \rho \text{ if } & a < \rho < b, \\ a \text{ if } & \rho < a, \end{cases} \qquad \rho \in \mathbb{R}_+. \tag{11.14}$$

Let us consider the control law

$$u_{a,b}(x) = K_0 x + \text{sat}_{a,b}^{1+\mu}(\|x\|_{\mathbf{d}})(K - K_0)\mathbf{d}(-\ln \text{sat}_{a,b}(\|x\|_{\mathbf{d}}))x, \tag{11.15}$$

where $\mathbf{d}$, $\|x\|_{\mathbf{d}}$ and K_0 are defined in Corollary 11.1. From (11.14) we conclude that

$$u_{1,1}(x) = K_{lin} x$$

and

$$u_{0,+\infty}(x) = K_0 x + \|x\|_{\mathbf{d}}^{1+\mu} K\mathbf{d}(-\ln \|x\|_{\mathbf{d}})x.$$

In other words, the pair of number $a \in (0, 1]$ and $b \in [1, +\infty)$ parameterizes a family of nonlinear controllers which contains the linear and homogeneous stabilizers as limit cases. Moreover, one can be shown that the closed-loop system (11.1), (11.15) is globally asymptotically stable for any $0 < a < b < +\infty$ provided that the control parameters are selected according to Corollary 11.1. Indeed, simple calculations show (see the proof of Theorem 10.1) that a locally Lipschtz continuous Lyapunov function of the closed-loop system can defined as follows:

$$V_{a,b}(x) = \begin{cases} a\|\mathbf{d}(-\ln a)x\| & \text{for} \quad \|x\|_{\mathbf{d}} < a \\ \|x\|_{\mathbf{d}} & \text{for} \quad a \leq \|x\|_{\mathbf{d}} \leq b \\ b\|\mathbf{d}(-\ln b)x\| & \text{for} \quad \|x\|_{\mathbf{d}} > b. \end{cases}$$

Notice that for $a < 1$ and $b = 1$ the controller (11.15) coincides with the linear controller away from the unit ball $x^\top Px > 1$ and the gains of the linear controller are scaled by means of the dilation $\mathbf{d}$ only close to the origin, i.e., for $x^\top Px < 1$. Inversely, the case $a = 1$ and $b > 1$ corresponds to the upgrade outside of the unit ball.

The following scheme for an upgrade of a linear control to a nonlinear (locally homogeneous) one is suggested.

Algorithm 11.2 ***(Local upgrade of a linear feedback)***

(1) Define parameters $P \succ 0$, $\rho > 0$, $\mu_{\min}, \mu_{\max} \in \mathbb{R}$, $K_0 \in \mathbb{R}^{m\times n}$ and $G_0 \in \mathbb{R}^{n\times n}$ as in Corollary 11.1.
(2) Select some $\mu \in [\mu_{\min}, \mu_{\max}]$ and define the canonical homogeneous norm $\|\cdot\|_{\mathbf{d}}$ for $G_{\mathbf{d}} = I_n + \mu G_0$.
(3) Initiate $a_0 = b_0 = 1$, $j = k = 0$ and evaluate $Q_i(u_{a_j,b_k}), i = 1, 2, \ldots, p$.
(4) Take $a_{j+1} = a_j - \varepsilon$ with some small $\varepsilon > 0$ and evaluate $Q_i(u_{a_{j+1},b_k}), i = 1, 2, \ldots, p$.
(5) If $\max_i(Q_i(u_{a_{j+1},b_k}) - Q_i(u_{a_j,b_k})) < 0$ and $a_{j+1} > 0$ then take $j = j+1$ and go to (4) else go to (6).
(6) Take $b_{k+1} = b_k + \varepsilon$ with some small $\varepsilon > 0$ and evaluate $Q_i(u_{a_j,b_{k+1}}), i = 1, 2, \ldots, p$.
(7) If $\max_i(Q_i(u_{a_j,b_{k+1}}) - Q_i(u_{a_j,b_k})) < 0$ then take $k = k+1$ and go to (6) else Stop (the further improvement of the control quality is impossible).

The presented tuning algorithm consists in decreasing of the parameter a and at the increasing of the parameter b as long as this implies improvement of the control quality of the system (11.1), (11.15).

Example 11.2 Considering the control system studied in Example 11.1 we introduce the quality index $Q(u) = \|x(T)\|$ with $T = 3$ as a quality index of a controller. We upgrade the linear feedback to the globally homogeneous and try to improve the stabilization error at the time instant $T = 3$ provided that $x(0) = (\,1\ 1\ 0\ 0\,)$. The tuning of the parameters is based on numerical simulations with the sampling period $h = 0.001$.

The linear controller with the gain K_{lin} is already well tuned and the global upgrade (see Algorithm 11.1) does not allow an essential improvement of the regulation quality. However, taking $\mu = -1$, $a = 0.1$ and $b = 1$ using the local upgrading scheme (see Algorithm 11.2) we derive an essential improvement of the stabilization precision at the time instant $T = 3$. Namely, $Q(u_{\text{lin}}) = 0.058$ and $Q(u_{0.1,1}) = 4.6 \cdot 10^{-8}$.

Theoretically, improvements of control quality (such as faster transitions and better robustness) are proven by Theorem 9.1 and Corollary 9.2 even for the case $\alpha = 0$ and $\beta = +\infty$. However, the theoretical results are model-based, but any model of a real control system is inexact and obtained under some simplification.

In practice, a difference between a model of dynamics and a real motion of the system may not allow to realize all theoretical properties of the closed-loop system or, even more, it may imply a serious degradation of some quality indexes. The tuning of parameters a and b suggested above is required to guarantee that the nonlinear control always has the quality which is not worse than the original linear one. This approach allows a control engineer to prevent any possible degradation of the control quality during the nonlinear "upgrade" of a existing linear control system. In practice, the parameters should be tuned based on real experiments. Notice that if the gains of the linear controller are already optimally adjusted, then improvements provided by homogeneous controller could not be large. The parameters a and b are close 1 in this case.

Below we illustrate the presented scheme on real experiments with a rotary inverted pendulum and a mid-size quadrotor.

11.2 Upgrading Linear PID

The PID controller

$$u_{PID}(t) = K_{\mathrm{p}}y(t) + K_{\mathrm{d}}\dot{y}(t) + K_{\mathrm{i}}\int_0^t y(\tau)d\tau, \quad y(t) \in \mathbb{R} \tag{11.16}$$

is most popular (in the industry) control algorithm for output regulation of the plant. The controller gains $K_{\mathrm{p}}, K_{\mathrm{d}}, K_{\mathrm{i}} \in \mathbb{R}$ can be tuned manually or using some frequency domain methods without any knowledge about system model. However, our upgrading procedure is always model based, since it uses a dilation symmetry of the system. Therefore, to suggest an upgrading procedure for the considered PID controller, we suppose that output dynamics of the system admits the following representation:

$$\ddot{y} = f(t) + p + u_{PID}(t), \tag{11.17}$$

where f is an unknown function of time, which may implicitly depend on the output y and some internal states of the plant, $p \in \mathbb{R}$ is an unknown constant perturbation. For any $\lambda \in \mathbb{R}$, we can rewrite our closed-loop system as follows:

$$\dot{x}(t) = Ax + B\left(f(t) + p_0 + Kx(t) + \int_0^t \tilde{K}x(\tau)d\tau\right),$$

where $x(t) = (y(t), \dot{y}(t))^\top$, $A = \left(\begin{smallmatrix} 0 & 1 \\ 0 & 0 \end{smallmatrix}\right)$, $B = \left(\begin{smallmatrix} 0 \\ 1 \end{smallmatrix}\right)$,

$$p_0 = p + \lambda y(0), \quad K = (K_{\mathrm{p}} - \lambda \;\; K_{\mathrm{d}}) \in \mathbb{R}^{1\times 2}, \quad \tilde{K} = (K_{\mathrm{i}} \;\; \lambda) \in \mathbb{R}^{1\times 2}.$$

The homogeneous control (9.10) designed by Theorem 9.1 has the parameters $G_{\mathbf{d}} = \mathrm{diag}\{1-\mu, 1\}$, $K = YX^{-1}$ with

$$X = x_{11}\tilde{X}, \quad x_{11} > 0, \quad \tilde{X} = \begin{pmatrix} 1 & -\rho(1-\mu) \\ -\rho(1-\mu) & 7(2-\mu)^2\rho^2(1-\mu)/8 \end{pmatrix},$$

$$Y = x_{11}\tilde{Y}, \quad \tilde{Y} = \rho^2(2-\mu)(1-\mu)\left[\begin{smallmatrix} \frac{8-7(2-\mu)}{8} & \frac{-7\rho(2-\mu)}{8} \end{smallmatrix}\right],$$

where $\mu \in [-0.5, 0.5]$ is a homogeneity degree and $\rho > 0$ is a tuning parameter. Hence, in the view of Theorem 9.4, to realize a transformation of the linear PID to a homogeneous one it is sufficient to select parameters $\lambda \in \mathbb{R}$, $\rho > 0$, $\mu \in [-0.5, 0.5]$ and $q \in (0, +\infty)$ such that

$$(K_\mathrm{p} - \lambda \ \ K_\mathrm{d}) = \tilde{Y}\tilde{X}^{-1}, \quad qB^\top \tilde{X}^{-1} = -(K_\mathrm{i} \ \ \lambda). \tag{11.18}$$

Given K_p, K_d, K_i, the latter conditions define a system of four nonlinear equations with respect to λ, μ, ρ and q. Taking into account

$$\tilde{X}^{-1} = \frac{1}{\Delta}\begin{pmatrix} 7(2-\mu)^2\rho^2(1-\mu)/8 & \rho(1-\mu) \\ \rho(1-\mu) & 1 \end{pmatrix}, \quad \Delta = 7(2-\mu)^2\rho^2(1-\mu)/8 + \rho^2(1-\mu)^2$$

the system of equations can be rewritten as

$$K_\mathrm{p} - \lambda = -\frac{7\rho^2(2-\mu)^2(1-\mu)}{8}, \quad K_\mathrm{d} = -\rho(2-\mu),$$

$$q\rho(1-\mu) = -K_\mathrm{i}\Delta, \qquad q = -\lambda\Delta.$$

Hence, we derive

$$\lambda = \frac{K_\mathrm{i}}{\rho(1-\mu)}$$

and the feasibility of the system of the algebraic equations

$$\begin{cases} K_\mathrm{p} = \frac{K_\mathrm{i} - 7\rho^3(2-\mu)^2(1-\mu)^2}{8\rho(1-\mu)}, \\ K_\mathrm{d} = -\frac{\rho^3(2-\mu)(1-\mu)\left(20-20\mu+7\mu^2\right)}{8\Delta}, \end{cases} \tag{11.19}$$

with respect to $\rho > 0$ and $\mu \in [-0.5, 0.5]$ implies the feasibility of (11.18).

The algebraic system parameterizes a set of linear PIDs which can be transformed to the homogeneous feedback with integral action:

$$u(x) = u_\mathrm{hom}(x) + \int_0^t u_\mathrm{int}(x(s))ds, \tag{11.20}$$

$$u_\mathrm{hom}(x) = \|x\|_\mathbf{d}^{1+\mu} Y X^{-1}\mathbf{d}(-\ln \|x\|_\mathbf{d})x,$$

$$u_\mathrm{int}(x) = -\|x\|_\mathbf{d}^{1+2\mu} \frac{\tilde{K}\mathbf{d}(-\ln \|x\|_\mathbf{d})x}{x^\top \mathbf{d}^\top(-\ln \|x\|_\mathbf{d})X^{-1}G_\mathbf{d}\mathbf{d}(-\ln \|x\|_\mathbf{d})x}.$$

By construction, we derive the original linear PID if $\mu = 0$ in the above controller. This means that we can use the global upgrading Algorithm 11.1 or the local upgrading Algorithm 11.2 in order to transform linear PID to homogeneous PID with a better control quality.

The set of "upgradable" PIDs can be enriched by using Corollary 11.1 instead of Theorem 9.1 for the design.

Corollary 11.2 *If for some* $\lambda > K_p$ *and* $\rho > 0$ *the pair* $(q, P) \in \mathbb{R} \times \mathbb{R}^{2\times 2}$ *satisfies the linear algebraic system*

$$\begin{aligned} &(A + BK)^\top P + P(A + BK) + \rho P \prec 0, \quad P = P^\top \succ 0, \\ &B^\top P = -q\tilde{K}, \quad q > 0, \end{aligned} \tag{11.21}$$

where $A = \left(\begin{smallmatrix} 0 & 1 \\ 0 & 0 \end{smallmatrix}\right)$, $B = \left(\begin{smallmatrix} 0 \\ 1 \end{smallmatrix}\right)$, $K = (K_\text{p} - \lambda \;\; K_\text{d})$ *and* $\tilde{K} = (K_\text{i} \;\; \lambda)$, *then for any* $\mu \in [-0.5, 0.5] \cap [\mu_{\min}, \mu_{\max}]$ *the controller (11.20) globally stabilizes the system*

$$\dot{x} = Ax + B(u + p), \quad t > 0 \tag{11.22}$$

for any unknown constant $p \in \mathbb{R}$, *where* $\mu_{\min}$ *and* $\mu_{\max}$ *are defined in Corollary 9.1.*

Proof If LMI (11.21) is feasible then, by Corollary 9.1, the homogeneous control $u = \|x\|_{\mathbf{d}}^{1+\mu} K\mathbf{d}(-\ln \|x\|_{\mathbf{d}})x$ stabilizes the system $\dot{x} = Ax + Bu$ for any $\mu \in [\mu_{\min}, \mu_{\max}]$. Since $\tilde{K} = -q^{-1}B^\top P$ then, by Theorem 9.4, the homogeneous PID control (11.20) stabilizes the system (11.22) provided that $\mu \in [-0.5, 0.5] \cap [\mu_{\min}, \mu_{\max}]$. The proof is complete. $\square$

The above upgrade is developed for the case of the single-input system. The design for the MIMO system is similar. The scalar q in (11.21) just becomes a positive definite matrix in the general case.

11.3 Practical Experiments

11.3.1 Stabilization of Rotary Inverted Pendulum

Model of Rotary Inverted Pendulum

A schematic representation of the rotary inverted pendulum (IP), also known as Furuta Pendulum, is shown in Fig. 11.1. This is a two-link mechanical system. The first link is the rotary arm rigidly connected on the one side to a rotor of an electrical motor being an actuator of the system. The second link is the pendulum installed at the other side of the rotary arm. The pendulum's motion is possible only in the plane orthogonal to the rotary arm.

Fig. 11.1 Schematic diagram of inverted pendulum (IP)

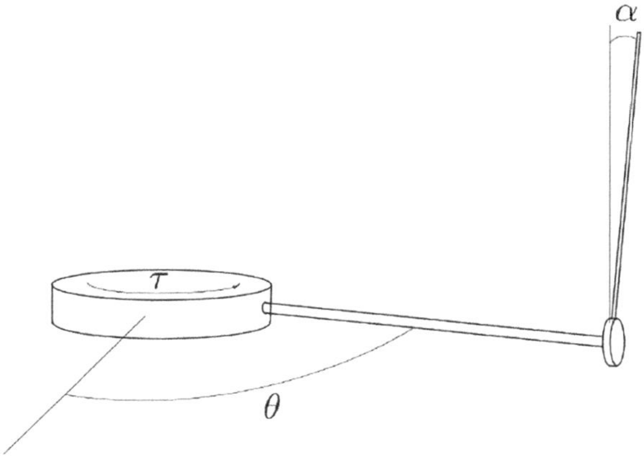

Table 11.1 Parameters of the rotary inverted pendulum

Symbol	Description
m_p	Mass of the pendulum
L_p	Length of the pendulum
J_p	Inertia of the pendulum
D_p	Pendulum damping coefficient
L_r	Length of the rotary arm
J_r	Rotary arm inertia
D_r	Viscous damping coefficient
g	Gravitational acceleration
τ	Torque of servo motor

The generalized coordinates θ and α describe the angular positions of the rotary arm and the pendulum, respectively. To obtain motion equations, the pendulum is considered as a lumped mass at its center. Table 11.1 presents the notation utilized for model description.

The dynamic model of the inverted pendulum is derived by means of the Euler–Lagrange method:

$$\frac{\partial^2 L}{\partial t \partial \dot{q}_i} - \frac{\partial L}{\partial q_i} = Q_i. \tag{11.23}$$

The Lagrangian of the pendulum is described as follows:

$$L = T - V, \tag{11.24}$$

where T is the total kinetic energy of the inverted pendulum and V is the total potential energy of the system.

The variable q_i represents the generalized coordinates, in our case, given by

$$q(t) := \left[\theta(t)\ \alpha(t)\right]^\top . \tag{11.25}$$

Considering the defined generalized coordinates (11.25), the Euler–Lagrange equations become

$$\frac{\partial^2 L}{\partial t \partial \dot{\theta}} - \frac{\partial L}{\partial \theta} = Q_1 \qquad \frac{\partial^2 L}{\partial t \partial \dot{\alpha}} - \frac{\partial L}{\partial \alpha} = Q_2 \, . \tag{11.26}$$

The generalized forces Q_i describe non-conservative forces applied to the system. In our case, the generalized forces acting on the IP are

$$Q_1 = \tau - D_r \dot{\theta} \qquad Q_2 = -D_p \dot{\alpha} \, . \tag{11.27}$$

The nonlinear dynamic equations of motion for the rotary inverted pendulum are

$$\begin{aligned} &\left(\psi + 0.25\zeta - 0.25\zeta \cos(\alpha)^2 + J_r\right) \ddot{\theta} - 0.5\varpi \cos(\alpha)\ddot{\alpha} \\ &+0.5\zeta \sin(\alpha)\cos(\alpha)\dot{\theta}\dot{\alpha} + 0.5\varpi \sin(\alpha)\dot{\alpha}^2 = \tau - D_r\dot{\theta} \end{aligned} \tag{11.28}$$

and

$$\begin{aligned} -0.5\varpi \cos(\alpha)\ddot{\theta} + \left(J_p + 0.25\zeta\right) \ddot{\alpha} - 0.25\zeta \cos(\alpha)\sin(\alpha)\dot{\theta}^2 \\ -0.5 m_p L_p g \sin(\alpha) = -D_p\dot{\alpha} \end{aligned} , \tag{11.29}$$

where $\zeta = m_p L_p^2$, $\psi = m_p L_r^2$ and $\varpi = m_p L_p L_r$.

A torque generated by the servo motor and applied to the rotary arm is described by the following equation:

$$\tau = \frac{k_m(V_m - k_m\dot{\theta})}{R_m}, \tag{11.30}$$

where k_m is the motor back EMF (electromotive force) constant, R_m is the terminal resistance and V_m is the control input (the input voltage for the servo motor).

Notice that for a generalized coordinate vector $q(t)$, Eqs. (11.28) and (11.29) can be transformed in the following matrix form:

$$J(q)\ddot{q} + C(q, \dot{q})\dot{q} + w(q) = \phi, \tag{11.31}$$

where

$$\begin{aligned} J(q) &= \begin{bmatrix} Jr + \psi + \frac{1}{4}(\zeta - \zeta\cos^2(\alpha)) & -\frac{1}{2}\varpi\cos(\alpha) \\ -\frac{1}{2}\varpi\cos(\alpha) & J_p + \frac{1}{4}\zeta \end{bmatrix} \\ w(q) &= \begin{bmatrix} 0 \\ -\frac{1}{2} g m_p L_p \sin(\alpha) \end{bmatrix}, \quad \phi = \begin{bmatrix} \tau \\ 0 \end{bmatrix} \\ C(q) &= \begin{bmatrix} \frac{1}{2}\zeta \sin(\alpha)\cos(\alpha)\dot{\alpha} + D_r & \frac{1}{2}\varpi\sin(\alpha)\dot{\alpha} \\ -\frac{1}{4}\zeta\sin(\alpha)\cos(\alpha)\dot{\theta} & D_p \end{bmatrix} \end{aligned} . \tag{11.32}$$

The nonlinear model of the inverted pendulum can be linearized around the operating point $\alpha = 0$ (upper position of the pendulum) using the Eqs. (11.28) and (11.29).

Let $x_1 = \theta$, $x_2 = \alpha$, $x_3 = \dot{\theta}$ and $x_4 = \dot{\alpha}$. From Eq. (11.31) for α close to zero we have $\sin(x_2) \approx 0$, $\cos(x_2) \approx 1$, then the linearized state-space representation for the inverted pendulum satisfies the following differential equations:

$$
\begin{aligned}
\dot{x}_1 = x_3, \qquad \dot{x}_2 = x_4 \\
\dot{x}_3 = \frac{-\left(J_p + 0.25\zeta\right) D_r x_3 - 0.5\varpi D_p x_4 + 0.25\vartheta g x_2 + \left(J_p + 0.25\zeta\right)\tau}{J_T}, \\
\dot{x}_4 = \frac{0.5\varpi D_r x_3 - (J_r + \psi) D_p x_4 + 0.5 m_p L_p g (J_r + \psi) x_2 + 0.5\varpi\tau}{J_T}
\end{aligned}
\tag{11.33}
$$

where $J_T = J_p\zeta + J_r J_p + 0.25 J_r\zeta$ and $\vartheta = m_p^2 L_p^2 L_r$.

Therefore, the linearized model of the inverted pendulum admits the following state-space representation:

$$
\dot{x}(t) = Ax(t) + Bu(t), \tag{11.34}
$$

where $x(t) = (x_1(t), x_2(t), x_3(t), x_4(t))^\top \in \mathbb{R}^4$ is the state vector, $u(t) \in \mathbb{R}$ is the control signal, $A \in \mathbb{R}^{4\times 4}$, $B \in \mathbb{R}^{4\times 1}$. In Eq. (11.34), the matrices A, b are defined as

$$
A = \frac{1}{J_T}\begin{pmatrix} 0 & 0 & J_T & 0 \\ 0 & 0 & 0 & J_T \\ 0 & a_{3,2} & a_{3,3} & a_{3,4} \\ 0 & a_{4,2} & a_{4,3} & a_{4,4} \end{pmatrix}, \quad B = \frac{1}{J_T}\left(0,\, 0,\, \left(J_p + 0.25\zeta\right),\, 0.5\varpi\right)^\top. \tag{11.35}
$$

The elements of the matrix A are given by

$$
\begin{aligned}
a_{3,2} = 0.25\vartheta g, \quad a_{3,3} = -\left(J_p + 0.25\zeta\right) D_r \\
a_{3,4} = -0.25\varpi D_p, \quad a_{4,2} = 0.25 m_p L_p g (J_r + \psi)\,. \\
a_{4,3} = 0.25\varpi D_r, \quad a_{4,4} = -(J_r + \psi) D_p
\end{aligned}
\tag{11.36}
$$

On Homogeneous Stabilization of the Rotary Inverted Pendulum

Below we consider the problem of stabilization of the rotary inverted pendulum in the upper (unstable) position. The control law in this case usually consists of two algorithms [2]: *swing-up controller* and *stabilizing controller*. The first one accumulates a kinetic energy of the system (i.e., increases the amplitude of oscillations of the pendulum) in order to bring it into a neighborhood of the upper position. Next we switch to a stabilization controller which stabilizes the pendulum in the upper position. For this purpose linear stabilizer is frequently utilized. We upgrade this linear feedback to a homogeneous one.

Table 11.2 Parameters of QUBE™-Servo 2 platform

Parameter	Value
m_p	0.024 Kg
L_p	0.129 m
J_p	3.3×10^{-5} Kg m^2
D_p	0.001 N m s/rad
L_r	0.085 m
J_r	5.7×10^{-5} Kg m^2
D_r	0.0005 N m s/rad
g	9.81 m/s^2

Results of the Experiment

The rotary inverted pendulum QUBE™-Servo 2 of Quanser™ is utilized for the experiment. The parameters of the experimental platform are given by the manufacturer and listed in Table 11.2 .

The corresponding system matrices are

$$A = \begin{pmatrix} 0 & 0 & 1.0000 & 0 \\ 0 & 0 & 0 & 1.0000 \\ 0 & 152.0057 & -12.2542 & -0.5005 \\ 0 & 264.3080 & -12.1117 & -0.8702 \end{pmatrix}, \quad B = \begin{pmatrix} 0 \\ 0 \\ 50.6372 \\ 50.0484 \end{pmatrix}.$$

The platform is supported with both a swing-up controller and a linear stabilizing controller realized in MATLAB®. As explained before we modify only the linear stabilizing controller. The gains of the linear feedback (given by the manufacturer) are

$$K_{lin} = \begin{pmatrix} 2 & -35 & 1.5 & -3 \end{pmatrix}.$$

Such a linear system has been already studied in Examples 11.1 and 11.2, where the upgraded homogeneous controller with the parameters

$$K_0 = \begin{pmatrix} 0 & -5.2811 & 0.2420 & 0.0174 \end{pmatrix}, \quad G_0 = \begin{pmatrix} -3 & 2.0248 & -0.0033 & 0.0033 \\ 0 & -1 & 0 & 0 \\ 0 & 0.3800 & -2 & 2.0235 \\ 0 & 0 & 0 & 0 \end{pmatrix},$$

$$P = \begin{pmatrix} 3.1581 & -7.2001 & 0.5908 & -0.6077 \\ -7.2001 & 96.7019 & -6.8570 & 7.4704 \\ 0.5908 & -6.8570 & 0.5539 & -0.5856 \\ -0.6077 & 7.4704 & -0.5856 & 0.6472 \end{pmatrix}, \quad \mu_{\min} = -1, \quad \mu_{\max} = 0.1607.$$

In Example 11.2, it was shown that for $\mu = -1$ the local upgrade of the linear controller implies essential improvement of the control precision at least on simulations.

We implement the locally homogeneous controller (11.15). Recall the control law like (11.15) is called implicit homogeneous controller (IHC) in Chap. 9, since the canonical homogeneous norm is defined implicitly (see Definition 5.8). We use the scheme given by Algorithm 10.1 with $N_{\max} = 5$ for the online numerical approximation of the canonical homogeneous norm.

The results of the practical experiment for $a = 0.88$ and $b = 1$ are shown in Figs. 11.2, 11.3, 11.4 and 11.5, where the linear controller (LC) and the implicit homogeneous controller (IHC) are compared. The swing-up controller, which brings the system to a neighborhood of the upper unstable equilibrium, is the same in both cases. That is why, the trajectories of the system with two different controllers almost coincide up to the time instant $t \approx 1.5$ s. At this instant of time, the swing-up control is replaced with a stabilizer (LC or IHC).

The experiment confirms the improvement of the convergence speed, smaller overshoot and better stabilization precision of the IHC control comparing with the linear feedback.

The parameter b is selected to be equal 1 in order to have the nonlinear controller coinciding with the linear one outside the unit ball $x^{\top}Px \leq 1$. Recall that if the gains of the linear controller are already optimally adjusted, then improvements provided by the nonlinear controller could not be large and the parameters a and b cannot deviate largely from 1 in this case. That is why the manually tuned parameter a has been finally selected close to 1, namely, $a = 0.88$. The experiment shows that the further decrease of the parameter a implies some degradation of the control quality (the increase of the chattering magnitude and energy consumption). The improvement of the control quality was also observed on numerical simulations (see Example 11.2) even for smaller parameter a (up to 0.1). The difference in optimal selection of a

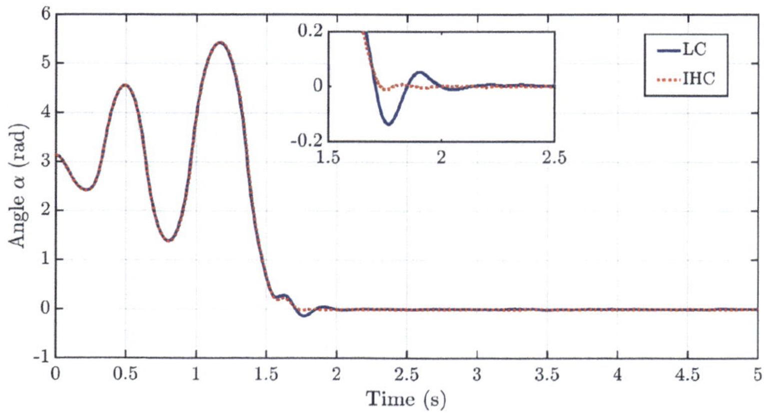

Fig. 11.2 The evolutions of the angle α

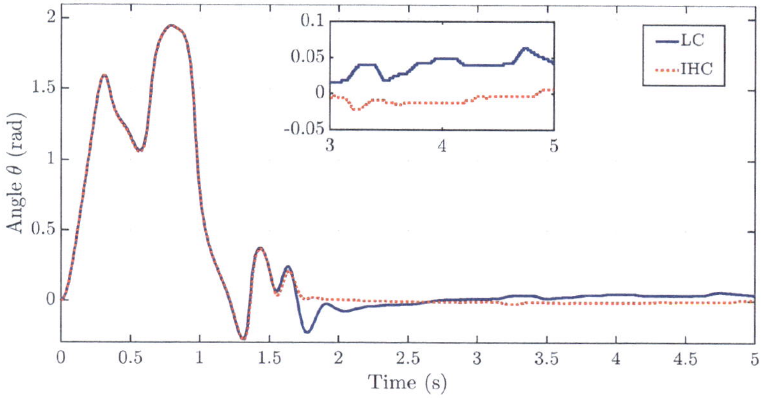

Fig. 11.3 The evolutions of the angle θ

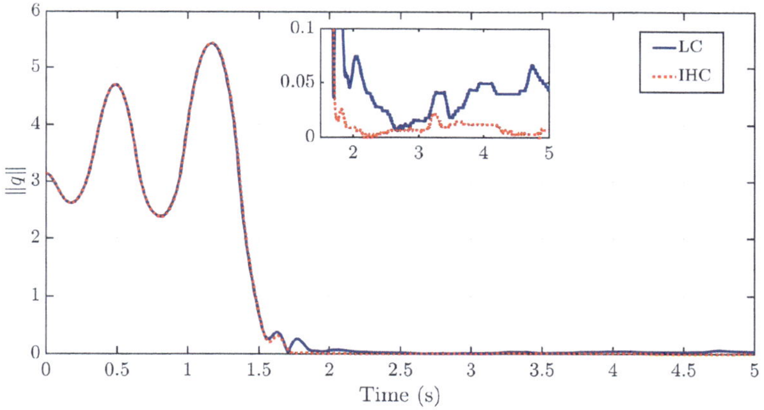

Fig. 11.4 Evolution of the norm $\|q(t)\| = \sqrt{\alpha^2(t) + \theta^2(t)}$

in practice and numerical simulation appears due to inaccuracy of the linear model (and, in fact, any model of a real system).

Notice also that we cannot simply select another static linear feedback $u_{new}(x) = (K_0 + K\mathbf{d}(-\ln a))x$ in order to obtain the same regulation quality as the nonlinear (locally homogeneous) feedback has. Such a selection implies an increase of the static linear feedback gain globally (on whole $\mathbb{R}^4$) and, as a consequence, some degradation of the control quality appears (a larger overshoot and a decrease of the attraction domain).

Therefore, *both numerical and practical experiments confirm our theoretical claim that homogeneous controllers may provide better regulation quality than linear stabilizers*.

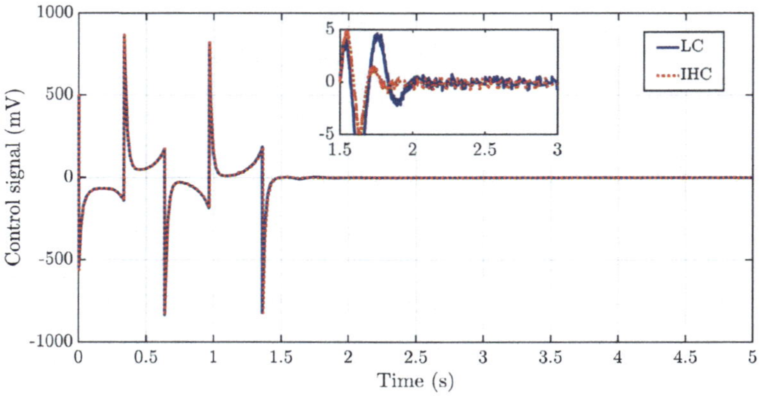

Fig. 11.5 The control signals

11.3.2 Quadrotor Control

Model Description

Quadrotor is a multirotor helicopter that is lifted and controlled by four rotors (see Fig. 11.6). The dynamic model of quadrotor can be expressed in three different frames: the inertial, horizontal and body frame (see Figs. 11.6 and 11.7), where the subindexes I, B and H indicate the corresponding frames. The body frame is attached to the mass center of quadrotor. Based on the horizontal frame, a rotation about the y-axis by roll angle ϕ followed by a rotation about the new interim x-axis by pitch angle θ establishes the body frame. The yaw angel ψ defines a rotation around z-axis.

Fig. 11.6 The quadrotor scheme

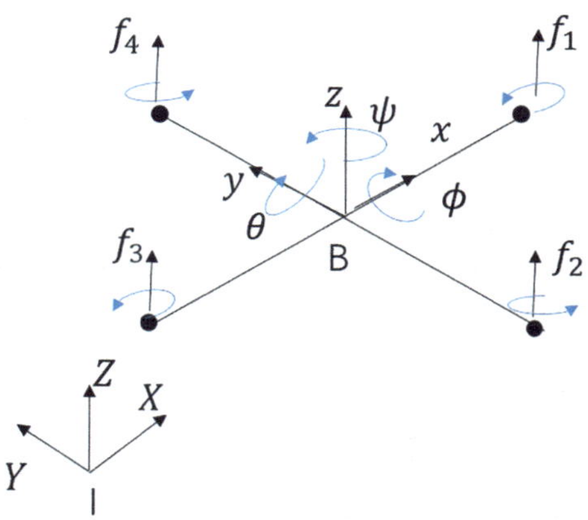

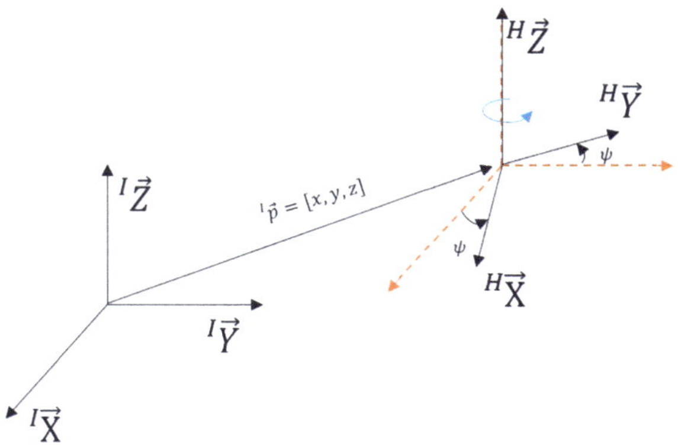

Fig. 11.7 Quadrotor's coordinates and frames

Assuming that the thrust forces $f_i, i = 1, 2, 3, 4$ and the gravity mg are two main forces resulting in the quadrotor's motion and using the Euler equation of the angular motion we derive the following simplified model of the control system in the horizontal frame:

$$\begin{aligned} m\ddot{x} &= u_1 \sin\theta\cos\phi \\ m\ddot{y} &= -u_1 \sin\phi \\ m\ddot{z} &= u_1 \cos\theta\cos\phi - mg \end{aligned} \tag{11.37}$$

$$\begin{aligned} I_{xx}\ddot{\phi} &= (I_{yy} - I_{zz})\dot{\theta}\dot{\psi} + u_2; \\ I_{yy}\ddot{\theta} &= (I_{zz} - I_{xx})\dot{\phi}\dot{\psi} + u_3; \\ I_{zz}\ddot{\psi} &= (I_{xx} - I_{yy})\dot{\phi}\dot{\theta} + u_4, \end{aligned} \tag{11.38}$$

where I_{xx}, I_{yy} and I_{zz} are the principle roll, pitch and yaw moments of inertia, $u_1 = f_1 + f_2 + f_3 + f_4, u_2 = \ell(f_1 - f_3), u_3 = \ell(f_2 - f_4), u_4 = c/k(-f_1 + f_2 - f_3 + f_4)$, ℓ is a distance between two opposite propellers and c, k are a drag and thrust coefficients of the propeller, respectively.

Homogenization of Quadrotor Controllers

We upgrade a linear controller for the quadrotor QDrone designed by the manufacturer (Quanser^{TM} Company). Denoting $\zeta = (x, y, \dot{x}, \dot{y}, \theta, -\phi, \dot{\theta}, -\dot{\phi})^\top$, the PID controller provided by the manufacturer can be represented as follows:

$$u_1 = mg + K_z\begin{pmatrix} z \\ \dot{z} \end{pmatrix} + \int_0^t K_{int} z dt, \quad \begin{pmatrix} u_2 \\ u_3 \end{pmatrix} = K_\zeta \zeta, \quad u_4 = K_\psi \begin{pmatrix} \psi \\ \dot{\psi} \end{pmatrix} \tag{11.39}$$

with the parameters (provided by the drone's manufacturer)

$$K_\psi = \begin{bmatrix} -0.59 & 0.11 \end{bmatrix}, \quad K_z = \begin{bmatrix} -35 & -14 \end{bmatrix}, \quad K_{int} = -4,$$
$$K_\zeta = \begin{bmatrix} -2.91 & 0 & -1.45 & 0 & -1.85 & 0 & -0.16 & 0 \\ 0 & -3.53 & 0 & -1.76 & 0 & -2.25 & 0 & -0.20 \end{bmatrix}.$$

For an upgrade, we need a linear (approximate) model of the system. The upgrade of PID controller for quadrotor can be based on the linearized model of the system (11.37), (11.38):

$$\begin{aligned} m\ddot{x} &= g\theta, & I_{xx}\ddot{\phi} &= u_2, \\ m\ddot{y} &= -g\phi, & I_{yy}\ddot{\theta} &= u_3, \\ m\ddot{z} &= u_1 - mg, & I_{zz}\ddot{\psi} &= u_4, \end{aligned}$$

which is obtained under assumption about a "hover" flight of the quadrotor (i.e., the angles θ, ϕ, and the angular velocities $\dot{\theta}$, $\dot{\phi}$ are assumed to be small and u_1 is close to mg). Notice the obtained linear system is homogeneous, so the homogenizing terms $K_0 x$ in the control (9.10) can be omitted ($K_0 = \mathbf{0}$) during the design.

Let us split the linearized system in three subsystems

$$\ddot{z} = \frac{u_1}{m}, \quad \dot{\zeta} = A\zeta + \begin{pmatrix} 0 \\ 0 \\ 0 \\ B \end{pmatrix} \begin{pmatrix} u_2 \\ u_3 \end{pmatrix}, \quad \ddot{\psi} = \frac{u_4}{I_{zz}},$$

where

$$A = \begin{pmatrix} 0 & I_2 & 0 & 0 \\ 0 & 0 & gI_2 & 0 \\ 0 & 0 & 0 & I_2 \\ 0 & 0 & 0 & 0 \end{pmatrix}, \quad B = \begin{pmatrix} \frac{1}{I_{yy}} & 0 \\ 0 & \frac{1}{I_{xx}} \end{pmatrix}.$$

The first (open-loop) subsystem is $\mathbf{d}_1$-homogeneous of the degree -0.5 with respect to dilation $\mathbf{d}_1(s) = \mathrm{diag}\{e^{1.5s}, e^s\}$, $s \in \mathbb{R}$ in $\mathbb{R}^2$ (since $(z, \dot{z})^\top \in \mathbb{R}^2$). The second subsystem has matrix, which is $\mathbf{d}_2$-homogeneous of the degree -1 with respect to the dilation $\mathbf{d}_2(s) = \mathrm{diag}\{e^{4s}I_2, e^{3s}I_2, e^{2s}I_2, e^{1s}I_2\}$. The third (open-loop) subsystem is $\mathbf{d}_3$-homogeneous of the degree -1 with $\mathbf{d}_3(s) = \mathrm{diag}\{e^{2s}, e^s\}$. The homogeneous controllers are designed from the linear feedback laws provided by the manufacturer using Corollary 11.1 for ζ and ψ subsystems and Corollary 11.2 for z subsystem.

Below we present the comparison results for the original linear controllers (11.39) of the manufacturer and its homogenized nonlinear counterparts given by (11.15) and implemented using Algorithm 10.1 with the following parameters: $a_\zeta = 0.6$, $a_\psi = 0.65$, and $a_z = 0.3$ and $b_\zeta = b_\psi = b_z = 1$ for z, ζ and ψ subsystems, respectively.

Results of Experiments

The real experiments are made on the platform of the Quanser QDrone™, which is a mid-size quadrotor equipped with a powerful onboard Intel® Aero Compute Board. The hardware parameters of quadrotor (provided by the manufacturer) are $m = 1.07$ kg, the gravity $g = 9.8\,\text{m/s}^2$, the motor distance $L_{roll} = 0.2136$ m, $L_{pitch} = 0.1758$ m, the roll inertia $I_{xx} = 6.85 \times 10^{-3}$ kg m^2, the pitch inertia $I_{yy} = 6.62 \times 10^{-3}$ kg m^2, the yaw inertia $I_{zz} = 1.29 \times 10^{-2}$ kg m^2, the thrust coefficient $k = 1.93 \times 10^{-8} \frac{N}{RPM^2}$, the drag coefficient $c = 0.26 \times 10^{-9} \frac{Nm}{RPM^2}$. The position and linear velocities are measured by OptiTrack™ and computed on board, respectively. The angles and angular rates are computed using IMU (inertial measurement unit) measurements.

The experiment consists in the sequential set-points tracking. The reference positions to be tracked are defined as follows: $[x, y, z, \psi] = [0, 0, 0, 0] \rightarrow [0, 0, 0.4, 0] \rightarrow [0.2, 0, 0.4, 0] \rightarrow [0.2, 0.2, 0.4, 0] \rightarrow [0, 0, 0.4, 0] \rightarrow [0, 0, 0.018, 0]$.

Figure 11.8 depicts the position tracking trajectory in x, y, z and ψ variables, respectively. The experiments confirm that the system with the homogeneous controller has *faster response, better precision and smaller overshoot.*

The robustness of the controllers is also compared by adding a mass on top of the quadrotor during the flight test. First, we add a mass of 0.5 kg for a couple seconds, next we remove it and add a mass of 1 kg for a few seconds. The results of the robustness test are depicted in Fig. 11.9. Both controllers are robust with respect to the load variation. However, the homogeneous controller again demonstrates a much better (in times) control precision. It is worth stressing that the homogeneous controller has been straightforwardly designed by means of an upgrade of the linear controller given by the manufacturer. These experiments confirm that the homogeneity allows us to improve an already existing and operating linear PID controllers.

Concluding Remarks

A scheme for an upgrade of an existing linear controller to a homogeneous regulator is presented and demonstrated on the two experiments, which confirm significant improvements of the control quality (smaller overshoot, faster response, better precision and robustness). The suggested approach is rather universal and can be utilized for a large class of system already controlled by the classical linear (in particular, PID) algorithms. For more experiments with homogeneous control, we refer the reader to [3–6].

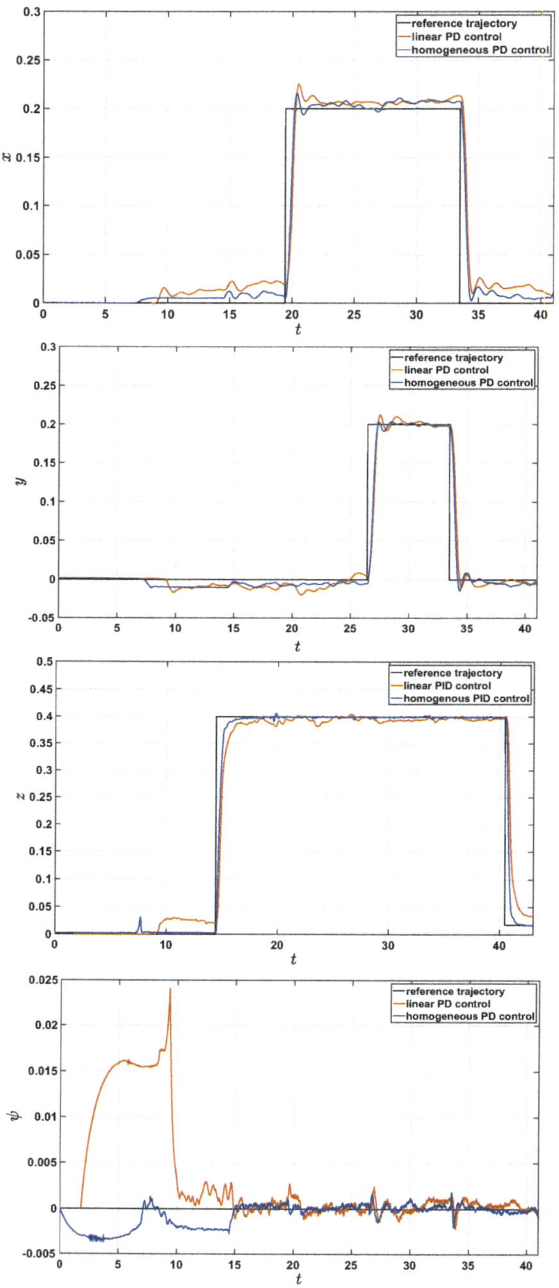

Fig. 11.8 Comparison of quadrotor's position tracking algorithms

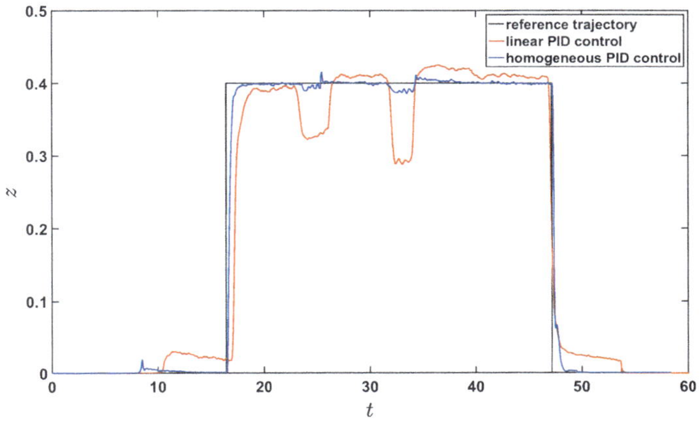

Fig. 11.9 Comparison of robustness of linear and homogeneous algorithms

References

1. Lyapunov AM (1992) The general problem of the stability of motion. Taylor & Francis
2. Apkarian J, Lévis M, Student workbook: QUBE-servo experiment for MATLAB/Simulink users. Quanser
3. Wang S, Polyakov A, Zheng G (2020) On generalized homogenization of linear quadrotor controller. In: IEEE International conference on robotics and automation (ICRA)
4. Cruz-Ortiz D, Ballesteros M, Polyakov A, Efimov D, Poznyak A (2021) Practical realization of implicit homogeneous controllers for linearized systems. IEEE Trans Indus Electron 69(5):5142–5151
5. Perozzi G, Polyakov A, Miranda-Villatoro F, Brogliato B (2022) Upgrading a linear controller to a sliding mode one: theory and experiments. Control Eng Pract 123(105107)
6. Zhou Y, Rios H, Mera M, Zheng G, Dzul A (2023) Homogeneity-based control strategy for trajectory tracking in perturbed unicycle robot. IEEE Trans Control Syst Technol 32(1):274–281

Chapter 12
Safety-Critical Homogeneous Systems

Stabilization under safety constraints is the classical problem of control engineering practice. These constrains may specify a safe set of the system states as well as a safe regulation time of the system. This chapter is devoted to homogeneous control design for linear plants with conic state constraints. Nonovershooting stabilizers and safety filters are developed. Their robustness with respect to various perturbations is analyzed.

12.1 Homogeneous Nonovershooting Stabilizers

The so-called nonovershooting (safe) control (see, e.g., [1] and the references therein) is aimed at trajectory tracking under certain state/output constraints. The analysis and design of safety-critical control systems casts in the framework of control barrier functions, [2–4]. Nonovershooting controllers are designed for both linear [5–7] and nonlinear systems [1, 8, 9]. In this section, we design homogeneous nonovershooting stabilizers [10, 11].

12.1.1 Nonovershooting Stabilization of Double Integrator

To illustrate the basic idea of nonovershooting stabilization, let us consider a system modeled by the controlled double integrator:

$$\dot{x} = Ax + Bu, \quad t > 0, \quad A = \begin{pmatrix} 0 & 1 \\ 0 & 0 \end{pmatrix}, \quad B = \begin{pmatrix} 0 \\ 1 \end{pmatrix}, \tag{12.1}$$

A. Polyakov, *Generalized Homogeneity in Systems and Control Volume I*,
Communications and Control Engineering,
https://doi.org/10.1007/978-3-031-76246-8_12

where $x(t) = (x_1(t), x_2(t)) \in \mathbb{R}^2$ is the system state and $u(t) \in \mathbb{R}$ is the control input. Let the safe set be defined as

$$\Sigma = \{x \in \mathbb{R}^2 : x_1 \geq 0\}. \tag{12.2}$$

The control aim is to stabilize the system (12.1) (asymptotically or in a finite/fixed-time) fulfilling the safety constraint

$$x(t) \in \Sigma, \quad \forall t \geq 0. \tag{12.3}$$

Obviously, this condition may be fulfilled only if the system is initiated in the safe set, i.e., $x(0) \in \Sigma$. In fact, it is reasonable to assume that the vector $x(0) = (x_1(0), x_2(0))^\top$ belongs to the interior of Σ. Indeed, otherwise, for $x_1(0) = 0$ and $x_2(0) < 0$ the constraint (12.3) is always violated (at least on a short interval of time) for any control signal $u \in L^\infty(\mathbb{R}; \mathbb{R})$, i.e., a feasible nonovershooting regulator does not exist for such initial condition.

12.1.1.1 Linear Nonovershooting Control

Let us study, first, a possibility of nonovershooting stabilization by a linear feedback. Inspired by the back-stepping approach to the nonovershooting control design [1], let us introduce new variables

$$\varphi_1 = x_1, \quad \varphi_2 = \lambda_1 x_1 + x_2 \tag{12.4}$$

and define the control law in the form:

$$u = -\lambda_1 x_2 - \lambda_2 \phi_2 = -\lambda_1\lambda_2 x_1 - (\lambda_1 + \lambda_2)x_2, \tag{12.5}$$

where $\lambda_1 > 0$ and $\lambda_2 > 0$ are tuning parameters. In this case, we have

$$\dot{\varphi}_1 = -\lambda_1\varphi_1 + \varphi_2 \quad \text{and} \quad \dot{\varphi}_2 = \lambda_1 x_2 + u = -\lambda_2\varphi_2,$$

or, equivalently,

$$\dot{\varphi} = \begin{pmatrix} -\lambda_1 & 1 \\ 0 & -\lambda_2 \end{pmatrix} \varphi, \tag{12.6}$$

where $\varphi(t) = (\varphi_1(t), \varphi_2(t))^\top$. On the one hand, since $\lambda_1 > 0$ and $\lambda_2 > 0$ then the system is exponentially stable. On the other hand, since the matrix

$$\tilde{A} = \begin{pmatrix} -\lambda_1 & 1 \\ 0 & -\lambda_2 \end{pmatrix}$$

is Metzler (see Chap. 2, page "Positive Systems") then the system is positive, i.e., the linear cone

$$\tilde{\Omega} = \left\{\varphi = (\varphi_1, \varphi_2)^\top \in \mathbb{R}^2 : \varphi_1 \geq 0, \varphi_2 \geq 0\right\}$$

is a positively invariant set of the system (12.6). The functions φ_1 and φ_2 given by (12.4) are the so-called *barrier functions* (see Chap. 7, Sect. 7.3). Returning back to the original coordinates we conclude that the linear cone

$$\Omega = \left\{x = (x_1, x_2)^\top \in \mathbb{R}^n : x_1 \geq 0,\ \lambda_1 x_1 + x_2 \geq 0\right\} \tag{12.7}$$

is a positively invariant set of the system (12.1), (12.5).

Therefore, the linear feedback (12.5) exponentially stabilizes the system (12.1) such that the linear cone Ω is a positively invariant set of the closed-loop system:

$$x(0) \in \Omega \quad \Rightarrow \quad x(t) \in \Omega \subset \Sigma, \quad \forall t \geq 0. \tag{12.8}$$

Notice that if $x(0) \in \text{int}\,\Sigma$ then $x_1(0) \neq 0$ and taking

$$\lambda_1 \geq \frac{x_2(0)}{x_1(0)} \tag{12.9}$$

we have $x(0) \in \Omega$. Therefore, the exponential nonovershooting stabilizer for the system (12.1) can be designed in the form of a linear feedback (12.5) with the parameter λ_1 dependent on $x(0) \in \text{int}\,\Omega$:

$$\lambda_2 > 0 \quad \text{and} \quad \lambda_1 = \max\left\{\delta, \frac{x_2(0)}{x_1(0)}\right\}, \quad \delta > 0. \tag{12.10}$$

Indeed, for the closed-loop system (12.1), (12.10), (12.5) we have

$$x(0) \in \text{int}\,\Sigma \quad \Rightarrow \quad x(t) \in \Sigma, \quad \forall t \geq 0.$$

The simulation results for linear and homogeneous (see below) nonovershooting stabilizers are depicted in Fig. 12.1 with $\lambda_1 = 0.5$, $\lambda_2 = 1$ and $x(0) = (4, 2)^\top$.

12.1.1.2 Homogenization of Linear Nonovershooting Feedback

In Chap. 11, the homogeneous controller has been designed by means of transformation of the linear feedback to a homogeneous one. Let us show that the nonovershooting property is preserved under such a transformation, at least, in the case of negative homogeneity degree. The generator of a linear continuous dilation for the system (12.1) has the form:

$$G_{\mathbf{d}} = \begin{pmatrix} 1-\mu & 0 \\ 0 & 1 \end{pmatrix}, \quad \mu < 1.$$

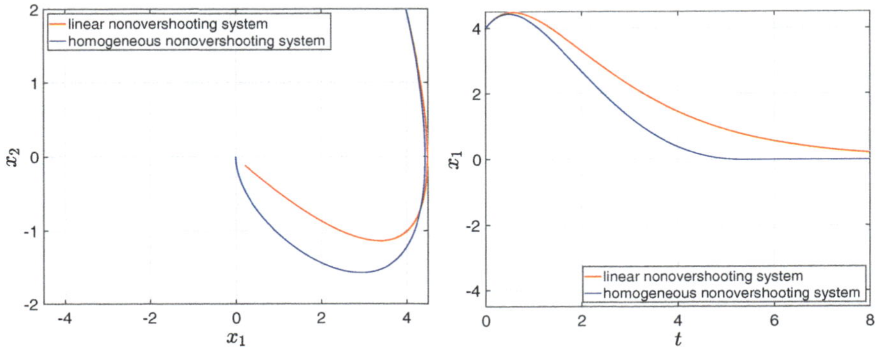

Fig. 12.1 Trajectories of linear and homogeneous nonovershooting systems

According to Corollary 11.1, the feedback

$$u = \|x\|_{\mathbf{d}}^{1+\mu} K\mathbf{d}(-\ln \|x\|_{\mathbf{d}})x, \quad K = (-\lambda_1\lambda_2, \ -\lambda_1 - \lambda_2) \tag{12.11}$$

is a homogeneous stabilizer of the system (12.1) provided that the canonical homogeneous norm $\|x\|_{\mathbf{d}}$ is induced by the weighted Euclidean norm $\|x\| = \sqrt{x^\top P x}$ with $P = P^\top \in \mathbb{R}^{2\times 2}$ satisfying for some $\rho > 0$ the matrix inequality

$$(A + BK)^\top P + P(A + BK) + 2\rho P \prec 0, \quad P \succ 0 \tag{12.12}$$

and $\mu \in [\mu_{\min}, \mu_{\max}]$, where $\mu_{\min}, \mu_{\max}$ are given by (11.5), (11.6), respectively. Moreover, one holds

$$\frac{d\|x\|_{\mathbf{d}}}{dt} = \gamma \|x\|_{\mathbf{d}}^{1+\mu}, \tag{12.13}$$

where

$$\gamma = \gamma(x) := \frac{z^\top P(A + BK)z}{z^\top P G_{\mathbf{d}} z}, \quad z = \mathbf{d}(-\ln \|x\|_{\mathbf{d}})x \tag{12.14}$$

and for any $\mu \in [\mu_{\min}, \mu_{\max}]$ we have

$$-\underline{\gamma} \le \gamma \le -\overline{\gamma} < 0$$

with

$$\underline{\gamma} = -\lambda_{\min}\left(Q^{-\frac{1}{2}}(P(A + BK) + (A + BK)^\top P)Q^{-\frac{1}{2}}\right) > 0,$$

$$\overline{\gamma} = -\lambda_{\max}\left(Q^{-\frac{1}{2}}(P(A + BK) + (A + BK)^\top P)Q^{-\frac{1}{2}}\right) > 0,$$

$$Q = PG_{\mathbf{d}} + G_{\mathbf{d}}^\top P \succ 0.$$

Let us find a positively invariant $\mathbf{d}$-homogeneous cone $\Omega_{\mathbf{d}}$ for the system (12.1) with the $\mathbf{d}$-homogeneous stabilizer (12.11). For $\|x\| = 1$ the homogeneous feedback becomes linear $u = Kx$. In this case, since Ω is a positively invariant cone of the system with the linear feedback then we expect that

$$\Omega \cap S \subset \Omega_{\mathbf{d}},$$

where $S = \{x \in \mathbb{R}^2 : \|x\| = 1\}$ is the unit sphere. Taking into account that $\Omega_{\mathbf{d}}$ is a $\mathbf{d}$-homogeneous cone, we conclude

$$\bigcup_{s\in\mathbb{R}} \mathbf{d}(s)(\Omega \cap S) \subset \Omega_{\mathbf{d}}.$$

This yields the $\mathbf{d}$-homogeneous cone of the following form:

$$\Omega_{\mathbf{d}} = \{x \in \mathbb{R}^2 : \|x\|_{\mathbf{d}} h_i^\top \mathbf{d}(-\ln \|x\|_{\mathbf{d}})x \geq 0, i = 1, 2\}, \tag{12.15}$$

where

$$h_1 = \begin{pmatrix} 1 \\ 0 \end{pmatrix}, \quad h_2 = \begin{pmatrix} \lambda_1 \\ 1 \end{pmatrix}.$$

We expect that the linear $\mathbf{d}$-homogeneous cone $\Omega_{\mathbf{d}}$ is a positively invariant set of the system with the homogeneous control. Let us prove this rigorously for $\mu < 0$.

Since $\mathbf{d}(s) = e^{sG_{\mathbf{d}}} = \mathrm{diag}(e^{s(1-\mu)}, e^s)$, $s \in \mathbb{R}$ then, considering the barrier functions

$$\phi_1 = \|x\|_{\mathbf{d}} h_1^\top \mathbf{d}(-\ln \|x\|_{\mathbf{d}})x, \quad \phi_2 = \|x\|_{\mathbf{d}} h_2^\top \mathbf{d}(-\ln \|x\|_{\mathbf{d}})x, \tag{12.16}$$

we derive

$$\phi_1 = \|x\|_{\mathbf{d}}^{\mu} x_1, \quad \phi_2 = \lambda_1 \phi_1 + x_2.$$

Hence, we have

$$\dot{\phi}_1 = \frac{d}{dt}\left(\|x\|_{\mathbf{d}}^{\mu} x_1\right) = \mu \|x\|_{\mathbf{d}}^{\mu-1} \frac{d\|x\|_{\mathbf{d}}}{dt} x_1 + \|x\|_{\mathbf{d}}^{\mu} x_2 =$$

$$\left(\mu \|x\|_{\mathbf{d}}^{-1} \frac{d\|x\|_{\mathbf{d}}}{dt} - \lambda_1 \|x\|_{\mathbf{d}}^{\mu}\right) \phi_1 + \|x\|_{\mathbf{d}}^{\mu} \phi_2$$

and

$$\dot{\phi}_2 = \lambda_1 \dot{\phi}_1 + \dot{x}_2 = \lambda_1 \left(\mu \|x\|_{\mathbf{d}}^{-1} \frac{d\|x\|_{\mathbf{d}}}{dt} - \lambda_1 \|x\|_{\mathbf{d}}^{\mu}\right) \phi_1$$

$$+ \lambda_1 \|x\|_{\mathbf{d}}^{\mu} \phi_2 - \|x\|_{\mathbf{d}}^{2\mu} \lambda_1 \lambda_2 x_1 - (\lambda_1 + \lambda_2) \|x\|_{\mathbf{d}}^{\mu} x_2$$

$$= \left(\lambda_1 \mu \|x\|_{\mathbf{d}}^{-1} \frac{d\|x\|_{\mathbf{d}}}{dt}\right) \phi_1 - \lambda_2 \|x\|_{\mathbf{d}}^{\mu} \phi_2.$$

Using (12.13) we derive

$$\dot{\phi} = \|x\|_{\mathbf{d}}^{\mu} \tilde{A}(\gamma)\phi, \quad \forall x \neq \mathbf{0}, \tag{12.17}$$

where, due to $\gamma < 0$ and $\lambda_1 > 0$, the matrix

$$\tilde{A}(\gamma) = \begin{pmatrix} \mu\gamma - \lambda_1 & 1 \\ \lambda_1 \mu \gamma & -\lambda_2 \end{pmatrix} \tag{12.18}$$

is Metzler for any $\mu \leq 0$. Taking into account the global asymptotic stability of the closed-loop system (12.1) with the homogeneous control (12.11) we conclude that the system (12.17) has unique solutions in the forward time. By Lemma 2.1 the system (12.17) with the Metzler matrix $\tilde{A}$ is positive. Therefore, the cone $\Omega_{\mathbf{d}}$ is positively invariant for the system (12.1), (12.11) with $\mu \in [\mu_{\min}, 0]$, i.e.:

$$x(0) \in \Omega_{\mathbf{d}} \quad \Rightarrow \quad x(t) \in \Omega_{\mathbf{d}} \subset \Sigma, \quad \forall t \geq 0. \tag{12.19}$$

Simulation results for linear and homogeneous nonovershooting stabilizers with $\lambda_1 = 0.5$, $\lambda_2 = 1$, $\mu = -0.85$, $P = \left(\begin{smallmatrix} 0.0268 & 0.0205 \\ 0.0205 & 0.0179 \end{smallmatrix}\right)$ and $x(0) = (4, 2)^\top$ are depicted in Fig. 12.1

Recall (see Chap. 11) that $u(x) \to Kx$ as $\mu \to 0$ uniformly on compacts from $\mathbb{R}^n \backslash \{\mathbf{0}\}$. This means that, in some sense, we have $\Omega_{\mathbf{d}} \to \Omega$ to as $\mu \to 0$. As shown in the previous chapters, locally/globally homogeneous stabilizers may provide faster convergence and better robustness of the control systems. Dependently of the homogeneity degree $\mu < 0$, the improvements can be discovered close to zero (if $\mu < 0$) or close to infinity (if $\mu > 0$).

Let us compare the nonovershooting properties of linear ($\mu = 0$) and homogeneous ($\mu < 0$) stabilizers designed above for the initial conditions close to the zero. Let us show that

$$(\Omega \cap \overline{B}(1)) \subset (\Omega_{\mathbf{d}} \cap \overline{B}(1)), \tag{12.20}$$

where $\overline{B}(1) = \{x \in \mathbb{R}^n : \|x\| \leq 1\}$ is the closed unit ball $\mathbb{R}^n$. Indeed, if $x \in \overline{B}(1)$ then $\|x\|_{\mathbf{d}} \leq 1$. In this case, for $\mu < 0$ we have $\|x\|_{\mathbf{d}}^{\mu} \geq 1$ and for $x \in \Omega \cap \overline{B}(1)$ we obtain

$$\|x\|_{\mathbf{d}}^{\mu} x_1 \geq x_1 \geq 0 \quad \text{and} \quad \lambda_1 \|x\|_{\mathbf{d}}^{\mu} x_1 + x_2 \geq \lambda_1 x_1 + x_2 \geq 0.$$

This implies the inclusion (12.20). So, at least close to the origin, the positively invariant set of the system with the homogeneous stabilizer $\mu < 0$ could be larger than the positively invariant set of the system with the linear stabilizer $\mu = 0$.

The positively invariant sets Ω and $\Omega_{\mathbf{d}}$ for $\lambda_1 = 0.5$, $\mu = -0.85$ and $P = \left(\begin{smallmatrix} 0.0268 & 0.0205 \\ 0.0205 & 0.0179 \end{smallmatrix}\right)$ are illustrated in Fig. 12.2, which shows that near the origin the set $\Omega_{\mathbf{d}}$ is indeed larger than the set Ω. The situation is inverse away from the unit ball.

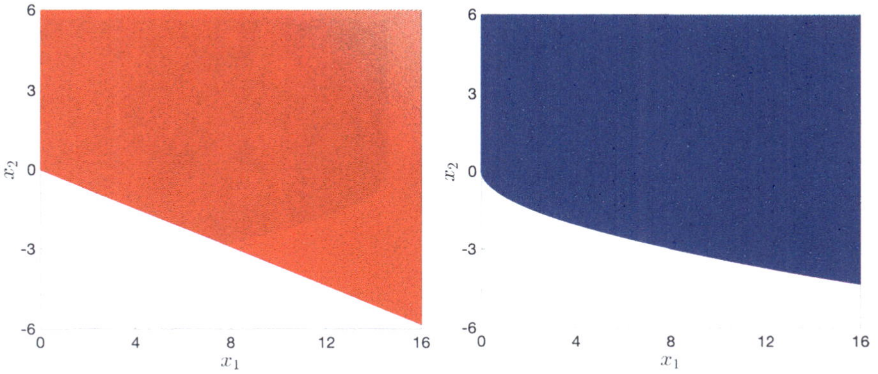

Fig. 12.2 Invariant sets of linear (left) and homogeneous (right) nonovershooting systems

12.1.1.3 Locally Homogeneous Finite-Time Nonovershooting Stabilizer

To avoid any degradation of the positively invariant set away from the unit ball, let us introduce the locally homogeneous nonovershooting stabilizer

$$u(x) = \begin{cases} \|x\|_{\mathbf{d}}^{1+\mu} K\mathbf{d}(-\ln \|x\|_{\mathbf{d}})x & \text{for } \|x\| \le 1, \\ Kx & \text{for } \|x\| > 1, \end{cases} \tag{12.21}$$

where K, $\mathbf{d}$ and $\|\cdot\|_{\mathbf{d}}$ are as before. By Remark 9.2 the obtained controller is a finite-time stabilizer of the system. It is continuous on $\mathbb{R}^n$ for $\mu > -1$. The case $\mu = -1$ corresponds to the sliding-mode nonovershooting stabilizer [10], which is locally bounded on $\mathbb{R}^n$ and continuous on $\mathbb{R}^n\backslash\{\mathbf{0}\}$.

The continuous barrier functions $\tilde{\phi}_i$, $i = 1, \ldots, n$ for this system

$$\tilde{\phi}_i(x) = \begin{cases} \phi_i(x) & \text{if } \|x\| \le 1, \\ \varphi_i(x) & \text{if } \|x\| > 1 \end{cases} \tag{12.22}$$

combine linear and homogeneous barrier functions φ_i and ϕ_i, respectively. Taking into account

$$\frac{d\tilde{\phi}}{dt} = \begin{cases} \|x\|_{\mathbf{d}}^{\mu} \tilde{A}(\gamma)\tilde{\phi} & \text{for } \|x\| < 1, \\ \tilde{A}(0)\tilde{\phi} & \text{if } \|x\| > 1, \end{cases} \tag{12.23}$$

we conclude that the positively invariant set of the closed-loop system is

$$\hat{\Omega} = (\Omega\backslash\overline{B}(1)) \cup (\Omega_{\mathbf{d}} \cap \overline{B}(1)). \tag{12.24}$$

Similar to the linear case, if $x(0) \in \operatorname{int} \Sigma$ then taking $\lambda_1 \ge \frac{x_2(0)}{x_1(0)}$ we have $x(0) \in \hat{\Omega}$, i.e., for the closed-loop system (12.1), (12.10), (12.21) we have

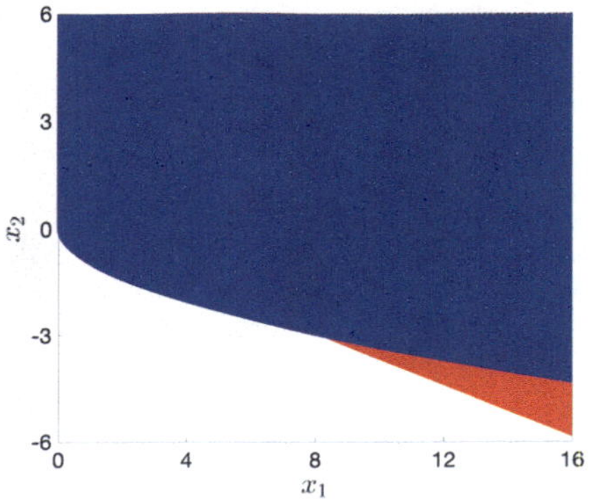

Fig. 12.3 Invariant set $\hat{\Omega}$ of locally homogeneous nonovershooting system (includes both blue and red zones)

$$x(0) \in \operatorname{int} \Sigma \quad \Rightarrow \quad x(t) \in \Sigma, \qquad \forall t \geq 0.$$

The invariant set $\hat{\Omega}$ of the system with locally homogeneous nonovershooting stabilizer is illustrated in Fig. 12.3.

12.1.1.4 Robustness of Stability

We analyze robustness of stability and nonovershooting properties of the linear and homogeneous stabilizers using the notions of Input-to-State Stability (ISS), Input-to-State Safety (ISSf), and Input-to-State Safety and Stability (ISSfS) on Ω introduced in Sect. 7.3.

Let us consider the perturbed system

$$\dot{x} = Ax + Bu(x) + g(x, q), \quad t > 0, \quad A = \begin{pmatrix} 0 & 1 \\ 0 & 0 \end{pmatrix}, \quad B = \begin{pmatrix} 0 \\ 1 \end{pmatrix}, \tag{12.25}$$

where x is as before, u is a linear or locally/globally homogeneous feedback designed above, $q \in L^{\infty}(\mathbb{R}; \mathbb{R}^k)$ is a deterministic noise/perturbation and $g \in C(\mathbb{R}^{k+2}; \mathbb{R}^2)$ models exogenous disturbances and uncertainties of the system.

We assume that *the closed-loop system is forward unique* and there exist $\xi_i \in \mathcal{K}$ and $\omega_i \in C(\mathbb{R}, \mathbb{R}_+)$ such that

$$|g_i(x, q)| \leq \xi_i(\|q\|)\omega_i(\|x\|), \quad \forall q \in \mathbb{R}^k, \quad \forall x \in \mathbb{R}^2, \quad i = 1, 2,$$

where $g = (g_1, g_2)^{\top}$ and the norm $\|x\|$ is defined as before (by the matrix P).

Linear control: For the system with the linear control (12.5), we have

$$\dot{x} = (A + BK)x + g(x, q) \tag{12.26}$$

and

$$\frac{d\|x\|}{dt} \leq -\rho\|x\| + \frac{x^\top P g(x,q)}{\|x\|} \leq -\rho\|x\| + \lambda_{\max}^{\frac{1}{2}}(P) \max_{i=1,2} \xi_i(\|q\|)\omega_i(\|x\|). \tag{12.27}$$

Hence, we conclude that the system with the linear feedback is

- globally asymptotically stable if $\omega_i(r) = r$ and $\|q\|_{L^\infty}$ is small enough;
- globally ISS on $\mathbb{R}^2$ if $\omega_i(r) = r^{\nu_i}$, $\nu_i < 1$, $i = 1, 2$.

To be asymptotically stable or ISS on Ω, the set Ω must be robustly positively invariant (see Sect. 7.3). If the vector field $x \mapsto g(x,q)$ is standard homogeneous of zero degree for any fixed q, then the criterion (7.71) yields the following conditions:

$$\left.\begin{array}{r} h_1^\top x = 0 \\ h_2^\top x \geq 0 \\ |x| = 1 \end{array}\right\} \quad \Rightarrow \quad h_1^\top(A+BK)x + h_1^\top g(x,q) \geq 0, \quad \forall q \in \mathbb{R}^k,$$

$$\left.\begin{array}{r} h_1^\top x \geq 0 \\ h_2^\top x = 0 \\ |x| = 1 \end{array}\right\} \quad \Rightarrow \quad h_2^\top(A+BK)x + h_2^\top g(x,q) \geq 0, \quad \forall q \in \mathbb{R}^k,$$

or, equivalently,

$$1 + g_1(x^*, q) \geq 0, \quad \forall q \in \mathbb{R}^k, \quad x^* = \begin{pmatrix} 0 \\ 1 \end{pmatrix}, \tag{12.28}$$

$$\lambda_1 g_1\left(x^{**}, q\right) + g_2\left(x^{**}, q\right) \geq 0, \quad \forall q \in \mathbb{R}^k, \qquad x^{**} = \frac{1}{\sqrt{\lambda_1^2+1}} \begin{pmatrix} 1 \\ -\lambda_1 \end{pmatrix}, \tag{12.29}$$

where $g = (g_1, g_2)$. The set Ω remains positively invariant for the perturbed system with linear nonovershooting control if g is standard homogeneous on x and satisfies (12.28) and (12.29). Obviously that the dependence of the parameter λ_1 on the initial state $x(0)$ would make the above restrictions to admissible perturbations dependent on $x(0)$ as well.

Homogeneous control. Let us analyze the robust stability and invariance of the system with the homogeneous control (12.11) on $\mathbb{R}^2$. First of all, notice that

$$\frac{d}{dt}\|x\|_{\mathbf{d}} = \gamma\|x\|_{\mathbf{d}}^{1+\mu} + \|x\|_{\mathbf{d}} \frac{z^\top P \mathbf{d}(-\ln\|x\|_{\mathbf{d}}) g(x,q)}{z^\top P G_{\mathbf{d}} z}, \tag{12.30}$$

where $z = \mathbf{d}(-\ln\|x\|_{\mathbf{d}})x$. Since $\mathbf{d}(-\ln\|x\|_{\mathbf{d}})g = \begin{pmatrix} \|x\|_{\mathbf{d}}^{-1+\mu} g_1 \\ \|x\|_{\mathbf{d}}^{-1} g_2 \end{pmatrix}$, then, taking into account Corollary 5.3 we conclude that the system with the homogeneous control (12.21) is

- globally uniformly finite-time stable on $\mathbb{R}^2$ if $\mu \in [-1, 0)$, $\|q\|_\infty$ is small enough and

$$\omega_1(r) = \overline{\sigma}^{-1}(r), \qquad \omega_2(r) = \overline{\sigma}^{-1}(r^{1+\mu}), \tag{12.31}$$

where $\overline{\sigma}$ is defined in Proposition 5.3;

- ISS on $\mathbb{R}^2$ if $\mu \in (-1, 0)$ and

$$\omega_1(r) = \overline{\sigma}^{-1}(r^{1+\tau_1}) \qquad \omega_2(r) = \overline{\sigma}^{-1}(r^{1+\mu+\tau_2}), \tag{12.32}$$

where $\tau_1 \in (-1, 0)$, $\tau_2 = (-1-\mu, 0)$.

If the vector field $x \mapsto g(x, q)$ is **d**-homogeneous of degree μ, then, by Corollary 7.12, the set $\Omega_{\mathbf{d}}$ is positively invariant for the perturbed system with the homogeneous control and $q \in L^\infty$ if

$$\left.\begin{array}{l} h_1^\top x = 0 \\ h_2^\top x \geq 0 \\ \|x\| = 1 \end{array}\right\} \quad \Rightarrow \quad h_1^\top(A + BK - \tilde{\gamma}(x)G_{\mathbf{d}})x + h_1^\top g(x, q) \geq 0, \quad \forall q \in \mathbb{R}^k,$$

$$\left.\begin{array}{l} h_1^\top x \geq 0 \\ h_2^\top x = 0 \\ \|x\| = 1 \end{array}\right\} \quad \Rightarrow \quad h_2^\top(A + BK - \tilde{\gamma}(x)G_{\mathbf{d}})x + h_2^\top g(x, q) \geq 0, \quad \forall q \in \mathbb{R}^k,$$

where $\tilde{\gamma}(x) = \gamma(x) + \frac{x^\top P g(x,q)}{x^\top P G_{\mathbf{d}} x}$, or, equivalently,

$$1 + g_1(x^*, q) \geq 0, \quad \forall q \in \mathbb{R}^k, \quad x^* = \begin{pmatrix} 0 \\ 1 \end{pmatrix}, \tag{12.33}$$

$$\frac{\lambda_1 \mu \tilde{\gamma}(x^{**})}{\sqrt{\lambda_1^2 + 1}} + \lambda_1 g_1(x^{**}, q) + g_2(x^{**}, q) \geq 0, \quad \forall q \in \mathbb{R}^k, \quad x^{**} = \frac{1}{\sqrt{\lambda_1^2 + 1}} \begin{pmatrix} 1 \\ -\lambda_1 \end{pmatrix}, \tag{12.34}$$

where $g = (g_1, g_2)^\top$. The condition (12.34) is less conservative than the condition (12.29) if $\mu < 0$ and $\tilde{\gamma}(x^{**}) < 0$. Therefore, the homogeneous nonovershooting stabilizer may guarantee a "safe" stabilization for a larger class of additive perturbations than linear controllers. This theoretical conjecture is supported with numerical simulations for MIMO system (see Example 12.1).

12.1.1.5 Robustness of Safety

The concept of Input-to-State Safety (ISSf) admits a larger class of perturbations and allows a violation of the safety constraint dependently of the magnitude of the perturbation (see Sect. 7.3).

Let us study ISSf and ISSfS of the linear and locally nonovershooting systems with respect to additive perturbations

$$g(x, q) = q.$$

Linear control: Since the linear system is standard homogeneous, then its ISSf analysis can be based on Corollary 7.18. In the linear case, the condition (7.85) becomes

$$\left.\begin{array}{l} h_1^\top x + r_1 = 0 \\ h_2^\top x + r_2 \geq 0 \end{array}\right\} \quad \Rightarrow \quad h_1^\top (A + BK)x \geq c,$$

$$\left.\begin{array}{l} h_1^\top x + r_1 \geq 0 \\ h_2^\top x + r_2 = 0 \end{array}\right\} \quad \Rightarrow \quad h_2^\top (A + BK)x \geq c,$$

where $r = (r_1, r_2)^\top \in \mathbb{R}^2$ is some strictly positive vector and $c > 0$. The above conditions are fulfilled for

$$r = (1, 0.5\lambda_1)^\top \quad \text{and} \quad c = 0.5\lambda_1 \min\{1, \lambda_2\}.$$

This means that the system with the linear nonovershooting control (12.5) is ISSf on Ω. The global ISS on $\mathbb{R}^2$ and the ISSf implies the ISSfS (Input-to-State Safety and Stability) on Ω.

Homogeneous control: To prove ISSf of the homogeneous system on $\Omega_{\mathbf{d}}$, let us show the condition (7.85) holds for the same vector $r > \mathbf{0}$. In our case, the mentioned condition has the form:

$$\left.\begin{array}{l} h_1^\top x + r_1 = 0 \\ h_2^\top x + r_2 \geq 0 \end{array}\right\} \quad \Rightarrow \quad h_1^\top (A + BK)x - \frac{h_1^\top (G_{\mathbf{d}} - I_2)xx^\top P(A + BK)x}{x^\top PG_{\mathbf{d}}x} \geq \tilde{c} \tag{12.35}$$

$$\left.\begin{array}{l} h_1^\top x + r_1 \geq 0 \\ h_2^\top x + r_2 = 0 \end{array}\right\} \quad \Rightarrow \quad h_2^\top (A + BK)x - \frac{h_2^\top (G_{\mathbf{d}} - I_2)xx^\top P(A + BK)x}{x^\top PG_{\mathbf{d}}x} \geq \tilde{c}, \tag{12.36}$$

where $\tilde{c} > 0$ is some number. First of all, notice that

$$-\underline{\gamma} \leq \gamma = \frac{x^\top P(A + BK)x}{x^\top PG_{\mathbf{d}}x} \leq -\overline{\gamma} < 0, \quad \forall x \neq \mathbf{0}.$$

Taking into account $h_1^\top (G_{\mathbf{d}} - I_2)x = -\mu h_1^\top x$ we conclude that

$$\left.\begin{array}{l} h_1^\top x + r_1 = 0 \\ h_2^\top x + r_2 \geq 0 \end{array}\right\} \quad \Rightarrow \quad h_1^\top (A + BK)x - \frac{h_1^\top (G_{\mathbf{d}} - I_2)xx^\top P(A + BK)x}{x^\top PG_{\mathbf{d}}x} \geq c + \mu r_1 |\gamma|.$$

Hence, for $\mu < 0$ being sufficiently close to 0 the condition (12.35) is fulfilled. Similarly, we derive

$$h_2^\top (G_{\mathbf{d}} - I_2)x = -\mu\lambda_1 h_1^\top x$$

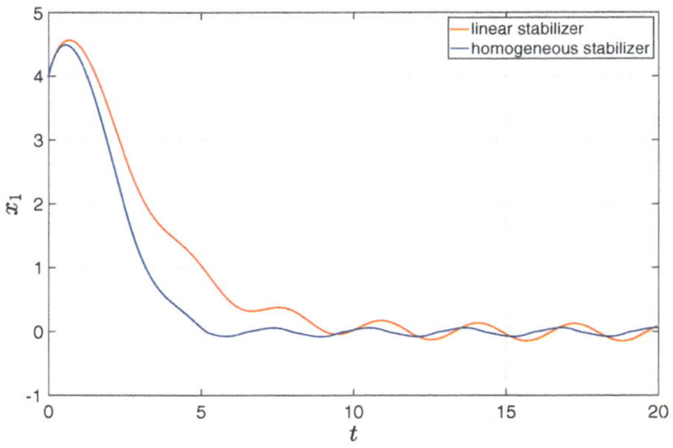

Fig. 12.4 Comparison of the stabilization precision of linear and homogeneous controllers in the perturbed case

and

$$\left.\begin{array}{l} h_1^\top x + r_1 \geq 0 \\ h_2^\top x + r_2 = 0 \end{array}\right\} \Rightarrow h_2^\top (A+BK)x - \frac{h_2^\top (G_\mathbf{d} - I_2)xx^\top P(A+BK)x}{x^\top P G_\mathbf{d} x} \geq c + \mu\lambda_1 r_1 |\gamma|,$$

i.e., the condition (12.36) is also fulfilled for $\mu < 0$ being sufficiently close to 0.

Therefore, the system with locally homogeneous nonovershooting stabilizer is ISSf and ISSfS on $\hat{\Omega}$ at least for μ being sufficiently close to 0.

The simulation results for linear and homogeneous nonovershooting stabilizers with mismatched perturbation $g = \binom{1}{0} q$ with $q = 0.25\sin(2t)$ are depicted on Fig. 12.4. As discussed above, the system with linear or homogeneous stabilizer is ISSfS with respect to such perturbation. However, the homogeneous stabilizer demonstrates a better rejection of the mismatched perturbation. Indeed, the stabilization error in the case of homogeneous controller is twice smaller than the stabilization error of the linear controller.

12.1.1.6 Globally Homogeneous Fixed-Time Nonovershooting Stabilizer

Inspired by constructions of Sect. 9.1.6, let us modify the feedback law (12.11) as follows:

$$u(t) = \left\| \frac{x(t)}{\|x(0)\|} \right\|_\mathbf{d}^{1+\mu} K\mathbf{d}\left(-\ln \left\| \frac{x(t)}{\|x(0)\|} \right\|_\mathbf{d}\right) x(t), \tag{12.37}$$

where $x(0) \in \operatorname{int} \Sigma$ and K, $\mathbf{d}$ and $\|\cdot\|_\mathbf{d}$ are as before. The change of variables

$$z(t) = \frac{x(t)}{\|x(0)\|}$$

gives

$$\dot{z} = Az + B\|z\|_{\mathbf{d}}^{1+\mu} K\mathbf{d}(-\ln\|z\|_{\mathbf{d}})z, \quad t > 0. \tag{12.38}$$

On the one hand, this means that the **d**-homogeneous cone $\Omega_{\mathbf{d}}$ is a positively invariant set of the system (12.38), i.e.:

$$\frac{x(0)}{\|x(0)\|} \in \Omega_{\mathbf{d}} \quad \Rightarrow \quad x(t) \in \|x(0)\|\Omega_{\mathbf{d}} \subset \Sigma. \tag{12.39}$$

On the other hand, since Ω is the standard homogeneous cone, then the inclusion (12.20) implies that

$$x(0) \in \Omega \quad \Rightarrow \quad \frac{x(0)}{\|x(0)\|} \in \Omega \cap S \quad \Rightarrow \quad \frac{x(0)}{\|x(0)\|} \in \Omega_{\mathbf{d}} \cap S.$$

Therefore, for the closed-loop system (12.1), (12.10), (12.37) we have

$$x(0) \in \operatorname{int} \Sigma \quad \Rightarrow \quad x(t) \in \Sigma, \quad \forall t \geq 0.$$

Robustness of globally homogeneous control: If $\mu = -1$ then the conditions (12.33) are always fulfilled for $g_1 = \mathbf{0}$ and

$$\left\| \begin{pmatrix} 0 \\ g_2(x, q) \end{pmatrix} \right\| \leq \frac{\rho}{\lambda_1}, \quad \forall x \in \partial\Omega_{\mathbf{d}}. \tag{12.40}$$

Therefore, for $\mu = -1$ the globally homogeneous nonovershooting stabilizer (12.11) (resp., (12.37)) can reject some *sign-indefinite uniformly bounded matched perturbation* preserving the safety constraint. The linear nonovershooting controllers cannot guarantee such an important property.

12.1.2 Nonovershooting Stabilization of Linear Plant

Let us consider the system

$$\dot{x} = Ax + Bu, \quad t > 0, \tag{12.41}$$

where the pair $A \in \mathbb{R}^{n\times n}$, $B \in \mathbb{R}^{n\times m}$ is assumed to be controllable, x is the system state and u is the control input.

A nonovershooting linear control design is studied in the literature (see, e.g., [5–7]) allowing the exponential stabilization of the system (12.41). If a desired invariant set for the system is defined by a linear cone

$$\tilde{\Omega}_{\text{lin}} = \{x \in \mathbb{R}^n : h_i^\top x \geq 0, i = 1, \ldots, n\} \tag{12.42}$$

with an invertible matrix H given by

$$H = \begin{pmatrix} h_1^\top \\ \dots \\ h_n^\top \end{pmatrix} \tag{12.43}$$

then, inspired by [12–14], the following procedure can be utilized for the design of a nonovershooting linear controller

$$u_{\text{lin}} = Kx, \quad K \in \mathbb{R}^{m\times n}. \tag{12.44}$$

On the one hand, making the change of variable $y = Hx$ we derive $\dot{y} = H(A + BK)H^{-1}y$, so, the set $\tilde{\Omega}_{\text{lin}}$ is positively invariant for the system (12.41), (12.44) if and only if the matrix

$$\tilde{A} = H(A + BK)H^{-1} \tag{12.45}$$

is Metzler. On the other hand, the Metzler matrix is Hurwitz if and only if there exists a positive vector $\ell \in \mathbb{R}^n_+$ such that $-\ell^\top H(A + BK)H^{-1} \in \mathbb{R}^n_+$ (see, e.g., [14]). Let the pair $(\ell^{\text{opt}}, K^{\text{opt}})$ be a solution of the optimization problem with the bilinear functional

$$J(\ell, K) := \max_j \ell^\top H(A + BK)H^{-1}e_j \to \min \tag{12.46}$$

subject to the linear constraints

$$\begin{array}{l} -\rho \le e_i^\top H(A + BK)H^{-1}e_i,\ 0 < \ell_i \le 1,\ i = 1, \dots, n, \\ 0 \le e_i^\top H(A + BK)H^{-1})e_j, \quad j \ne i, \quad j = 1, \dots, n, \end{array} \tag{12.47}$$

where $\ell = (\ell_1, \dots, \ell_n)^\top \in \mathbb{R}^n_+$, $K \in \mathbb{R}^{m\times n}$ are optimization variables and $\rho > 0$ is a tuning parameter required for boundedness of the admissible set. If the optimal cost $J(\ell^{\text{opt}}, K^{\text{opt}})$ is negative then the matrix $\tilde{A}$ is Metzler and Hurwitz, and the system (12.41) with the linear control (12.44) is uniformly asymptotically stable on the cone $\tilde{\Omega}_{\text{lin}}$. In practice, the considered problem can be solved using some optimization software (e.g., YALMIP toolbox for MATLAB®).

A possible way to design a nonovershooting finite-time stabilizer for the system (12.41) is by upgrading (a transformation) an existing linear nonovershooting controller to a homogeneous one. A possibility of such an upgrade for the double integrator is demonstrated in the previous section. The following theorem refines the result of [11], designs a homogeneous nonovershooting stabilizer for the multi-input system and provides ISSf/ISSfS analysis of the closed-loop system.

Theorem 12.1 *Let a linear feedback law* (12.44) *be such that the matrix* $A + BK$ *is Hurwitz and the linear cone* $\tilde{\Omega}_{\text{lin}}$ *be positively invariant for the closed-loop system* (12.41), (12.44). *Let the pair* $\{A, B\}$ *be controllable. If the matrix* H *given by* (12.43) *is invertible and the following linear algebraic system*

$$AG_0 + BY_0 = G_0A + A, \quad G_0B = \mathbf{0}, \tag{12.48}$$

$$e_i^\top H\left(\tau(A + BK) - G_0\right)H^{-1}e_j \geq 0, \quad i \neq j, \quad i, j = 1, \ldots, n \tag{12.49}$$

is feasible with respect to $G_0 \in \mathbb{R}^{n\times n}$, $Y_0 \in \mathbb{R}^{m\times n}$ *and* $\tau > 0$ *then*

- $G_{\mathbf{d}} = I_n + \mu G_0$ *is anti-Hurwitz for any* $\mu \in [-1, 0)$;
- *the system of linear matrix inequalities*

$$W = P(A + BK) + (A + BK)^\top P \prec 0, \quad Q = PG_{\mathbf{d}} + G_{\mathbf{d}}^\top P \succ 0, \quad P \succ 0 \tag{12.50}$$

  is feasible with respect to $P = P^\top \in \mathbb{R}^{n\times n}$, *at least, for* $\mu \in [-1, 0)$ *sufficiently close to* 0;
- *the feedback law*

$$u = K_0x + \|x\|_{\mathbf{d}}^{1+\mu}(K - K_0)\mathbf{d}(-\ln\|x\|_{\mathbf{d}})x, \quad K_0 = Y_0(G_0 - I_n)^{-1} \tag{12.51}$$

  is smooth on $\mathbb{R}^n\backslash\{\mathbf{0}\}$, *continuous at* $\mathbf{0}$ *for* $\mu > -1$ *and locally bounded on* $\mathbb{R}^n$ *for* $\mu = -1$, *where the canonical homogeneous norm* $\|\cdot\|_{\mathbf{d}}$ *is induced by the norm* $\|x\| = \sqrt{x^\top Px}$ *with* P *satisfying* (12.50);
- *the closed-loop system* (12.41), (12.51) *is* **d**-*homogeneous of degree* $\mu \in [-1, 0)$ *and finite-time stable on* $\mathbb{R}^n$;
- *the linear* **d**-*homogeneous cone*

$$\Omega_{\mathbf{d},\text{lin}} = \{x \in \mathbb{R}^n : \|x\|_{\mathbf{d}} h_i^\top \mathbf{d}(-\ln\|x\|_{\mathbf{d}})x \geq 0, i = 1, \ldots n\} \tag{12.52}$$

  is positively invariant for the closed-loop system if $\mu \in [-1, 0) \cap [-(\tau\underline{\gamma})^{-1}, 0)$, *where* $\underline{\gamma} = -\lambda_{\max}\left(Q^{-\frac{1}{2}}WQ^{-\frac{1}{2}}\right)$;
- *the closed-loop system* (12.41), (12.51) *is ISSf/ISSfS on* Ω_{lin} *with respect to additive measurement noise and additive disturbance, at least, if* $\mu \in (-1, 0) \cap [-(\tau\underline{\gamma})^{-1}, 0)$ *is close enough to* 0.

Proof If the pair $\{A, B\}$ is controllable then algebraic Eq. (12.48) is always feasible with respect to Y_0 and G_0 (see Theorem 9.1) and the matrix $G_{\mathbf{d}} = I_n + \mu G_0$ is anti-Hurwitz [15] for $\mu \in [-1, (\tilde{n} - 1)^{-1})$. Denoting $A_0 = A + BK_0$ we derive

$$(12.48) \quad \Rightarrow \quad AG_0 - G_0A + BK_0(G_0 - I_n) = A, \quad G_0B = \mathbf{0} \quad \Rightarrow$$

$$(A + BK_0)G_0 = G_0A + \underbrace{G_0B}_{=\mathbf{0}} K_0 + A + BK_0 \quad \Rightarrow$$

$$\mu A_0 G_0 = \mu(I_n + G_0)A_0 \quad \Rightarrow \quad A_0 G_{\mathbf{d}} = (G_{\mathbf{d}} + \mu I_n)A_0.$$

The latter means that the linear vector field $x \mapsto A_0x$ is $\mathbf{d}$-homogeneous, i.e., $A_0\mathbf{d}(s) = e^{\mu s}\mathbf{d}(s)A_0$, $\forall s \in \mathbb{R}$. Moreover, the identity $G_0B = \mathbf{0}$ implies that $\mathbf{d}(s)B = e^s$.

The LMI (12.50) is always feasible with respect to P, at least, for μ sufficiently close to zero. Indeed, since the matrix $A + BK$ is Hurwitz then the first (Lyapunov) inequality in (12.50) is feasible together with $P \succ 0$ (see, e.g., [16]). Rewriting the second inequality for $G_{\mathbf{d}} = I_n + \mu G_0$ as follows: $2P + \mu(PG_0 + G_0^\top P) \succ 0$, we conclude its feasibility at least for μ close to zero.

The smoothness of u on $\mathbb{R}^n\backslash\{\mathbf{0}\}$ follows from the smoothness of the canonical homogeneous norm $\|\cdot\|_{\mathbf{d}}$ on $\mathbb{R}^n\backslash\{\mathbf{0}\}$, which is induced by the norm $\|\cdot\|$ being smooth on $\mathbb{R}^n\backslash\{\mathbf{0}\}$. Since $\|\mathbf{d}(-\ln\|x\|_{\mathbf{d}})x\| = 1$ and $\|\cdot\|_{\mathbf{d}} \in C(\mathbb{R}^n)$ then u is continuous at $\mathbf{0}$ for $\mu > -1$ and locally bounded (but discontinuous at zero) if $\mu = -1$. In the latter case, solutions of the system at $\mathbf{0}$ are understood in the sense of Filippov (see Chap. 2).

The norm $\|\cdot\|_{\mathbf{d}}$ is a Lyapunov function of the closed-loop system (see Theorem 9.1):

$$\frac{d}{dt}\|x\|_{\mathbf{d}} = \|x\|_{\mathbf{d}}^{\mu+1}\frac{x^\top\mathbf{d}^\top(-\ln\|x\|_{\mathbf{d}})P(A+BK)\mathbf{d}(-\ln\|x\|_{\mathbf{d}})x}{x^\top\mathbf{d}^\top(-\ln\|x\|_{\mathbf{d}})PG_{\mathbf{d}}\mathbf{d}(-\ln\|x\|_{\mathbf{d}})x} = \gamma\|x\|_{\mathbf{d}}^{1+\mu}, \quad x \neq \mathbf{0},$$

with

$$\gamma(z) := \frac{z^\top P(A+BK)z}{z^\top PG_{\mathbf{d}}z} < 0, \quad z \in \mathbb{R}^n\backslash\{\mathbf{0}\}, \tag{12.53}$$

where the formula (5.21) and the identities $\mathbf{d}(s)B = e^s$, $A_0\mathbf{d}(s) = e^{\mu s}\mathbf{d}(s)A_0$ are utilized. Hence, we derive

$$-\underline{\gamma} \le \gamma \le -\overline{\gamma},$$

where $\underline{\gamma} = -\lambda_{\min}\left(Q^{-\frac{1}{2}}WQ^{-\frac{1}{2}}\right)$ and $\overline{\gamma} = -\lambda_{\min}\left(Q^{-\frac{1}{2}}WQ^{-\frac{1}{2}}\right)$.

Since $\mu < 0$ then the closed-loop system is globally finite-time stable on $\mathbb{R}^n$. By Corollary 7.8 (see also Example (7.6)), the set Ω_{lin} is positively invariant if the matrix

$$\tilde{A}(\gamma) = H(A + BK - \mu\gamma G_0)H^{-1} \tag{12.54}$$

is Metzler and the system. By assumption, the set $\tilde{\Omega}_{lin}$ is the positively invariant set of the linear system (12.41) with the linear control (12.44). This means that the matrix $H(A + BK)H^{-1}$ is necessarily Metzler. If the condition (12.49) is fulfilled then the matrix $M(z)$ is Metzler provided that $\mu \in [-(\tau\underline{\gamma})^{-1}, 0)$.

Moreover, the system with the linear control is asymptotically stable. This implies that the Metzler matrix $H(A+BK)H^{-1}$ is Hurwitz and has strictly negative elements on the main diagonal. Hence, at least, for μ being sufficiently close to 0, the Metzler matrix $M(z)$ has strictly negative elements $-\lambda_i(z)$ on the main diagonal such that $\lambda_i(z) \geq \lambda_{\min} > 0$ for all $i = 1, 2, \ldots, n$ and for all $z \neq \mathbf{0}$.

Let us show that the condition (7.85) holds for

$$f(x, q) = Ax + Bu(x + q_1) + q_2, \quad q = (q_1^\top, q_2^\top)^\top \in \mathbb{R}^{2n}.$$

Taking into account the structure of G_0 (see the formula (9.19)) we conclude that the vector field f satisfies Assumption 7.4 with the dilation $\tilde{\mathbf{d}}(s) = \begin{pmatrix} \mathbf{d}(s) & \mathbf{0} \\ \mathbf{0} & e^{\mu s}\mathbf{d}(s) \end{pmatrix}$ in $\mathbb{R}^{2n}$ provided that $\mu > -1$.

For the given f, the condition (7.85) can be rewritten as follows there exist and $r > \mathbf{0}$ and $c > 0$

$$z \in \Xi_i(r) \quad \Rightarrow \quad h_i^\top \left(A + BK + \mu\gamma(z/\|z\|)G_0\right) \frac{z}{\|z\|} \geq c, \quad \forall i = 1, 2, \ldots, n. \tag{12.55}$$

Since $r > \mathbf{0}$, by definition of the set $\Xi_i(r)$ there exists $\rho > 0$ such that

$$z \in \Xi_i(r) \quad \Rightarrow \quad Hz = \begin{pmatrix} h_1^\top z \\ \ldots \\ h_{i-1}^\top z \\ -r_i \\ h_{i+1}^\top z \\ \ldots \\ h_n^\top z \end{pmatrix}, \quad h_j^\top z \geq -r_i,\ j \neq i, \quad \text{and} \quad \|z\| \geq \rho.$$

Notice that $h_i^\top \left(A + BK + \mu\gamma(z/\|z\|)G_0\right) H^{-1}$ defines the i-row of the Metzler matrix $M(z)$. Since the matrix $H(A+BK)H^{-1}$ is Metzler and Hurwitz then $\exists r > \mathbf{0}, \exists \lambda > 0 : H(A+BK)H^{-1}r = -\lambda r$. Hence, for $z \in \Xi_i(r)$ we have

$$h_i^\top \left(A + BK + \mu\gamma(z/\|z\|)G_0\right) \frac{z}{\|z\|} = h_i^\top \left(A + BK + \mu\gamma(z/\|z\|)G_0\right) H^{-1} \frac{Hz}{\|z\|}$$

$$\geq -h_i^\top (A + BK + \mu\gamma(z/\|z\|)G_0) H^{-1} r/\rho \geq 0.5\lambda r_i,$$

at least, for μ being close enough to zero. Applying Corollaries 7.18 and Remark 7.3 we complete the proof. □

> **Remark 12.1** The case $\mu > 0$ can be treated by replacing the sign "$-$" with "$+$" in the inequality (12.49).

The homogeneous cone $\Omega_{\mathbf{d},\text{lin}}$ is linear in the space $\mathbb{R}^n_{\mathbf{d}}$ homeomorphic to $\mathbb{R}^n$ and the closed-loop system in $\mathbb{R}^n_{\mathbf{d}}$ is similar to linear as well (see the formula (7.52)).

So, the invariance of $\Omega_{\mathbf{d},\mathrm{lin}}$ was deduced from positivity of the system in the new coordinates. To be positive, a linear system must have a Metzler matrix (see [12] or Chap. 2). The condition (12.49) just follows this criterion. If the matrix $H(A+BK)H^{-1}$ is Metzler then the condition (12.49) simply means that *for any negative off-diagonal element of the matrix $H(-G_0)H^{-1}$ the corresponding off-diagonal element of the matrix $H(A+BK)H^{-1}$ has to be positive.*

To estimate the settling time, the following system LMIs

$$P(A+BK)+(A+BK)^{\top}P+\rho(PG_{\mathbf{d}}+G_{\mathbf{d}}^{\top}P)\preceq 0, \quad PG_{\mathbf{d}}+G_{\mathbf{d}}^{\top}P\succ 0, \quad P\succ 0 \tag{12.56}$$

have to be solved instead of (12.50), where $\rho>0$ is a scalar parameter characterizing the settling time by the formula (9.14). If, for some $\rho>0$, the first inequality in the above systems of LMIs is fulfilled as identity, then the settling-time estimate is exact.

In practice, frequently, just some coordinates of the system state are constrained. For example, in [10], the safe set is given by

$$\Sigma=\{x\in\mathbb{R}^n : \tilde{h}_1^{\top}x\geq 0\},$$

while the restriction of Σ to the invariant sets Ω_{lin} or $\tilde{\Omega}_{\mathrm{lim}}$ was caused by the design methodology and the structure of the plant model (the integrator chain). In this case, it is important to know if (a) some linear constraint $h^{\top}x\geq 0$ can be preserved in the $\mathbf{d}$-homogeneous invariant cone Ω_{lin};

(b) the set Ω_{lin} is larger or smaller than $\tilde{\Omega}_{\mathrm{lin}}$. The following remarks answer the above questions.

Remark 12.2 If $h\in\mathbb{R}^n$ is a real left eigenvector of the generator $G_{\mathbf{d}}$ then

$$\|x\|_{\mathbf{d}}h^{\top}\mathbf{d}(-\ln\|x\|_{\mathbf{d}})x\geq 0 \quad \Leftrightarrow \quad h^{\top}x\geq 0. \tag{12.57}$$

Proof Indeed, since $\mathbf{d}(s)=\sum_{i=0}^{\infty}\frac{s^iG_{\mathbf{d}}^i}{i!}$ and $h^{\top}G_{\mathbf{d}}=\lambda h^{\top}$, then $h^{\top}\mathbf{d}(s)=e^{\lambda s}h^{\top}$ and

$$\|x\|_{\mathbf{d}}h^{\top}\mathbf{d}(-\ln\|x\|_{\mathbf{d}})x=\|x\|_{\mathbf{d}}^{1-\lambda}h^{\top}x.$$

Hence, (12.57) holds for $x\neq\mathbf{0}$. For $x=\mathbf{0}$ the equivalence is obvious, since $\|\cdot\|_{\mathbf{d}}\in C(\mathbb{R}^n)$, $\|\mathbf{0}\|_{\mathbf{d}}=0$ and $\|\mathbf{d}(-\ln\|x\|_{\mathbf{d}})x\|=1$ for all $x\neq\mathbf{0}$. The proof is complete.□

This allows us to deal with a rather reach set of safety constraints in the case of diagonalizable generators.

Remark 12.3 If, under conditions of Theorem 12.1, the matrix

$$H(-G_0)H^{-1}$$

is Metzler then the following inclusion

$$(\tilde{\Omega}_{\text{lin}} \cap \overline{B}(1)) \subset (\Omega_{\text{lin}} \cap \overline{B}(1)) \tag{12.58}$$

holds, where $\overline{B}(1) = \{z \in \mathbb{R}^n : \|z\| \leq 1\}$ is the closed unit ball.

Proof Since

$$e^s H\mathbf{d}(-s)H^{-1} = e^s He^{-s(I_n+\mu G_0)}H^{-1} = He^{-s\mu G_0}H^{-1} = e^{-s\mu HG_0H^{-1}}$$

then for the Metzler matrix $H(-G_0)H^{-1}$ and for $\mu s > 0$ the matrix $e^s H\mathbf{d}(-s)H^{-1} \geq \mathbf{0}$, where the sign "$\geq \mathbf{0}$" means that all elements of the matrix are non-negative. On the other hand, $x \in \tilde{\Omega}_{\text{lin}} \Rightarrow Hx \geq \mathbf{0}$, and $x \in \Omega_{\text{lin}} \Rightarrow (e^s H\mathbf{d}(-s)H^{-1})Hx \geq \mathbf{0}$ with $s = \ln \|x\|_{\mathbf{d}}$. If $\|x\| \leq 1$ then $s < 0$, $\mu s \geq 0$ (due to $\mu \leq 0$) and $e^s H\mathbf{d}(-s)H^{-1} \geq \mathbf{0}$. Hence, $x \in \tilde{\Omega}_{\text{lin}} \cap B \quad \Leftrightarrow \quad \|x\| \leq 1$ and $Hx \geq \mathbf{0} \Rightarrow (e^s H\mathbf{d}(-s)H^{-1})Hx \geq \mathbf{0}$. □

The above remark proves the less conservative overlapping of the original safe set Σ by the homogeneous cone Ω_{lin}, at least, close to the origin (in the unit ball). This property is important for the safety filter design [10, 17].

To avoid possible conservatism away from the unit ball, the upgrade of linear nonovershooting stabilizer may be realized locally. Namely, if, under condition of Theorem 12.1, the control u is defined as follows:

$$u = \begin{cases} Kx & \text{if } \|x\| \geq 1 \\ K_0x + \|x\|_{\mathbf{d}}^{\mu+1}(K - K_0)\mathbf{d}(-\ln \|x\|_{\mathbf{d}})x & \text{if } \|x\| < 1, \end{cases} \tag{12.59}$$

then u is Lipschitz continuous away from the origin and the system (12.41) with the control (12.59) is finite-time stable on the set

$$\Omega = (\Omega_{\text{lin}} \cap B) \cup (\tilde{\Omega}_{\text{lin}} \backslash B). \tag{12.60}$$

It is worth stressing that neither linear nor homogeneous nonovershooting stabilizer design uses any decomposition of the system to a canonical form (like the integrator chain in the papers [10, 17]). The control design does not require an information about relative degrees of the barrier functions ϕ_i.

Example 12.1 Inspired by [11] let us consider the system (12.41), (12.44) with the parameters

$$A = \begin{pmatrix} 3 & 0 & 1 \\ 0 & -1 & 1 \\ -2 & 0 & 0 \end{pmatrix}, \quad B = \begin{pmatrix} 1 & -1 \\ 0 & 1 \\ 0 & 1 \end{pmatrix}, \tag{12.61}$$

with the safety constraints

$$-x_1(t) \leq x_3(t) \leq x_2(t), \quad \forall t \geq 0. \tag{12.62}$$

These constraints mean that the linear cone

$$\Sigma = \{x \in \mathbb{R}^3 : h_1^\top x \geq 0,\ h_2^\top x \geq 0\}, \quad h_1 = \begin{pmatrix} 1 \\ 0 \\ 1 \end{pmatrix}, \quad h_2 = \begin{pmatrix} 0 \\ 1 \\ -1 \end{pmatrix}$$

must be positively invariant for the closed-loop system. Selecting the additional (virtual) constraint by assigning

$$h_3 = \begin{pmatrix} 0 \\ 0 \\ -1 \end{pmatrix}$$

we derive the invertible matrix H given by (12.43).

To design a linear feedback $u_{\text{lin}} = Kx$ which stabilizes asymptotically the system preserving the relations (12.62), we solve the optimization problem (12.46), (12.47) for $\rho = 4$ and obtain the linear feedback gain

$$K = \begin{pmatrix} -4.7536 & 0 & -4.9393 \\ 1.7415 & 0 & -3.7856 \end{pmatrix}.$$

The linear cone $\tilde{\Omega}_{\text{lin}} \subset \Sigma$ given by (12.42) is the positively invariant set of the linear closed-loop system (12.41), (12.44).

To design a finite-time nonovershooting stabilizer we transform the linear feedback to a homogeneous one using Theorem 12.1. Solving Equation (12.48) we derive

$$G_0 = \begin{pmatrix} 0 & -0.5 & 0.5 \\ 0 & -0.5 & 0.5 \\ 0 & 0.5 & -0.5 \end{pmatrix} \quad \text{and} \quad K_0 = Y_0(G_0 - I_3)^{-1} = \begin{pmatrix} -1 & 0 & -1 \\ 1 & 0.5 & -0.5 \end{pmatrix}.$$

The condition (12.49) is fulfilled for any $\tau > 0$ since the matrices

$$H(A+BK)H^{-1} = \begin{pmatrix} -3.7536 & 0 & 0.1857 \\ 2 & -1 & 2 \\ 0.2585 & 0 & -3.5271 \end{pmatrix}, \quad H(-G_0)H^{-1} = \begin{pmatrix} 0 & 0 & 0 \\ 0 & 1 & 0 \\ 0 & 0.5 & 0 \end{pmatrix}$$

are Metzler. Solving the LMI (12.56) for $\mu = -1$ and $\rho = 0.35$ we derive

$$P = \begin{pmatrix} 0.3544 & 0.0656 & -0.0694 \\ 0.0656 & 0.2343 & -0.1257 \\ -0.0694 & -0.1257 & 0.3914 \end{pmatrix}.$$

The homogeneous (finite-time) nonovershooting stabilizer is given (12.51) with $G_{\mathbf{d}} = I_3 + \mu G_0$. The **d**-homogeneous cone $\Omega_{\mathbf{d},\text{lin}}$ is a positively invariant set of the closed-loop system. By Remark 12.2, since h_1 and h_2 are left eigenvectors of $G_{\mathbf{d}}$ then $\Omega_{\text{lin}} \subset \Sigma$.

Simulations have been done using zero-order-hold discretization with the sampling period 10^{-3}. The homogeneous controller has been implemented using consistent discretization. The results are depicted in Fig. 12.5, where the evolution of the barrier functions $\varphi_1(x) = h_1^\top x$ and $\varphi_2(x) = h_2^\top x$ is also shown. The linear control stabilizes the system exponentially and the homogeneous control steers the state of the system to zero in a finite time (≈ 1.8 for $x(0) = (0.5, 1, 0)^\top$). Both stabilizers fulfill the safety constraint $x(t) \in \Sigma, \forall t \geq 0$.

To compare a robustness, let us consider the system

$$\dot{x} = f(x, q) := Ax + Bu(x) + D, \quad D = \begin{pmatrix} -0.2 \\ 0.2 \\ 0.2 \end{pmatrix}, \tag{12.63}$$

with a constant matched perturbation, where x, A, B are as before, u is either a linear or a homogeneous nonovershooting stabilizer. The constant perturbation of the system with the linear controller leads to violation of the safety constraint. Homogeneous controller rejects the bounded perturbation preserving the nonovershooting property (see Fig. 12.6).

12.2 Homogeneous Safety Filters

In this section, the nonovershooting controller developed above is employed for a "safety filter" design. The safety filter is a mapping allowing to override a potentially unsafe nominal controller [10, 17] in order to fulfill safety constraints. The safety filters have been extensively used in various control applications such as automotive systems [18, 19] and multi-agent robotics [20, 21]. If u_{nom} is a nominal controller for a double integrator, then a possible "safety filter" is a feedback law

$$u = \mathcal{F}(u_{\text{nom}}, u_{\text{safe}}),$$

which merges the nominal controller with some safe controller preventing violation of the safety constrains in the case when the nominal controller operates the system to an unsafe zone, where F is some operator.

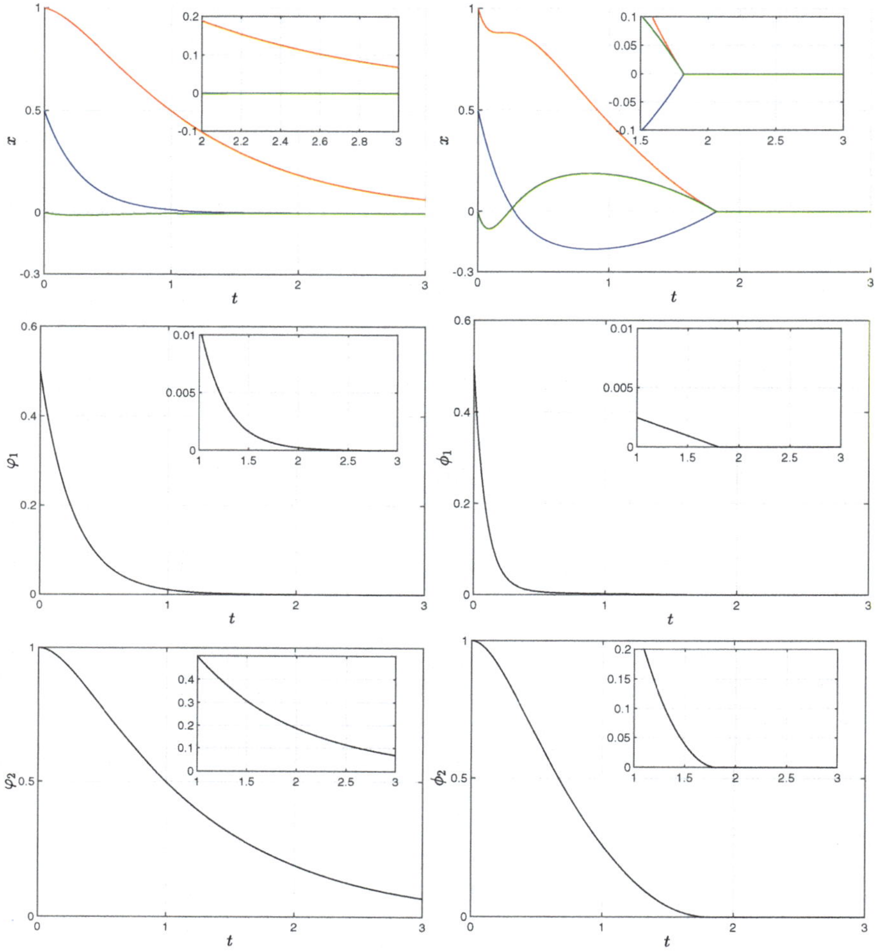

Fig. 12.5 Trajectories of the system with linear (left) and homogeneous (right) nonovershooting stabilizers

12.2.1 Safety Filters for Double Integrator

Let us consider again the planar system (12.1) operating under the space constraint (12.3). Let us assume that the system is already supplied with a nominal controller u_{nom}. This controller may be "unsafe". Namely, it may operate the system to an unsafe zone and violate the constraint (12.3). Our aim is to merge the nominal controller with a safe nonovershooting stabilizer and prevent violation of the constraint.

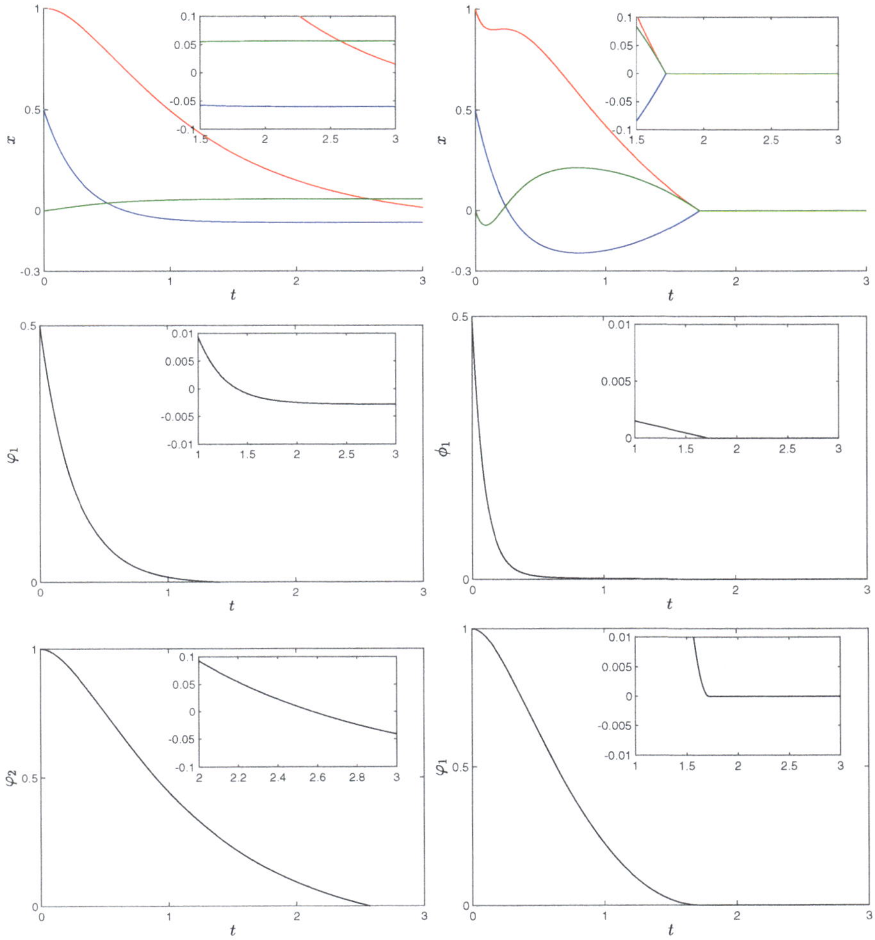

Fig. 12.6 Trajectories of the perturbed system with linear (left) and homogeneous (right) nonovershooting stabilizers

12.2.1.1 Linear Safety Filter

First, let us design a safety filter using linear nonovershooting stabilizer. Recall that, in terms of the linear barrier functions (12.4), the system (12.1) can be rewritten as follows:

$$\dot{\varphi} = \begin{pmatrix} -\lambda_1 & 1 \\ -\lambda_1^2 & \lambda_1 \end{pmatrix} \varphi + \begin{pmatrix} 0 \\ 1 \end{pmatrix} u. \tag{12.64}$$

Let u_{lin} denote the linear nonovershooting controller defined by the formula (12.5). In this case, we derive

$$\dot{\varphi} = \begin{pmatrix} -\lambda_1 & 1 \\ 0 & -\lambda_2 \end{pmatrix} \varphi + \begin{pmatrix} 0 \\ 1 \end{pmatrix} (u - u_{\text{lin}}). \tag{12.65}$$

By Lemma 2.1, the above system is positive if $u \geq u_{\text{lin}}$. Therefore, to fulfill the space constraint (12.4), the safety filter can be defined as follows:

$$u = \max\{u_{\text{nom}}, u_{\text{lin}}\}. \tag{12.66}$$

Indeed, in this case, we have $u \geq u_{\text{lin}}$ and the implication (12.8) is fulfilled.

12.2.1.2 Homogeneous Safety Filter

Let u_{hom} be homogeneous nonovershooting stabilizer (12.11) for the double integrator. Then for any regulator u we have

$$\frac{d\|x\|_{\mathbf{d}}}{dt} = \gamma \|x\|_{\mathbf{d}}^{1+\mu} + \frac{z^\top P \begin{pmatrix} 0 \\ u - u_{\text{hom}} \end{pmatrix}}{z^\top P G_{\mathbf{d}} z}, \quad z = \mathbf{d}(-\ln \|x\|_{\mathbf{d}})x,$$

where γ is given by (12.14). In terms of the homogeneous barrier functions ϕ given by (12.16), the derivative can be rewritten as

$$\frac{d\|x\|_{\mathbf{d}}}{dt} = \tilde{\gamma} \|x\|^{1+\mu}, \quad \text{where} \quad \tilde{\gamma} = \gamma + \frac{\phi^\top H^{-\top} P \begin{pmatrix} 0 \\ u - u_{\text{hom}} \end{pmatrix}}{\|x\|_{\mathbf{d}}^{\mu} \phi^\top H^{-\top} P G_{\mathbf{d}} H^{-1} \phi}.$$

Therefore, the system (12.1) can be rewritten as

$$\dot{\phi} = \|x\|_{\mathbf{d}}^{\mu} \tilde{A}(\tilde{\gamma})\phi + \begin{pmatrix} 0 \\ 1 \end{pmatrix} (u - u_{\text{hom}}), \tag{12.67}$$

where, as before, the matrix $\tilde{A}(\tilde{\gamma})$ is given by (12.18). If

$$\begin{pmatrix} 1 & -\lambda_1 \end{pmatrix} P \begin{pmatrix} 0 \\ 1 \end{pmatrix} \leq 0 \tag{12.68}$$

then $\tilde{\gamma} < 0$ for $\phi_2 = 0$, $\phi_1 > 0$ and $u \geq u_{\text{hom}}$. Let

$$u = \max\{u_{\text{nom}}, u_{\text{hom}}\}, \tag{12.69}$$

and P be a solution of the system of LMIs (12.12), (12.68). Then $\tilde{\gamma} < 0$ for $\phi_1 > 0$, $\phi_2 = 0$, so, by Remark 2.1, the system (12.67), (12.69) is positive. Such a filtering of the control signal guarantees the implication (12.19).

One can be easily shown that the system of LMIs (12.12), (12.68) is always feasible. This means that the homogeneous safety filter for the double integrator can be defined similarly to the linear safety filter.

An invariant set of the system with a nonovershooting stabilizer coincides with the invariant set of the system with the safety filter. By construction, the safety filters override the nominal (unsafe) controller with safe linear or homogeneous if the nominal control drives the system to the boundary of the invariant set Ω or $\Omega_{\mathbf{d}}$, respectively. As shown in Sect. 12.1, the set $\Omega_{\mathbf{d}} \cap B(1)$ is larger than the set $\Omega \cap B(1)$. This means in a less conservative overriding of homogeneous safety filter near the origin. This theoretical conclusion is supported by numerical simulations given below. To reduce a conservatism of homogeneous approach away from the unit ball, we design the safety filter

$$u = \max\{u_{\text{nom}}, u_{\text{lh}}\}, \tag{12.70}$$

using the locally homogeneous stabilizer u_{lh} given by the formula (12.21). With such a safety filter we have

$$x(0) \in \hat{\Omega} \quad \Rightarrow \quad x(t) \in \hat{\Omega}, \quad \forall t \geq 0,$$

where $\hat{\Omega}$ is given by (12.24).

Example 12.2 The linear nonovershooting stabilizer for double integrator (12.1) be defined by the formula (12.4) with $\lambda_1 = \lambda_2 = 1$. In this case, the homogeneous nonovershooting stabilizer is given by the formula (12.11) with the matrix

$$P = \begin{pmatrix} 0.0091 & 0.0052 \\ 0.0052 & 0.0044 \end{pmatrix}$$

satisfying the LMI (12.12) with $\rho = 0.25$.

To compare linear safety filter (12.66) and the homogeneous safety filter (12.69), let us define the nominal control u_{nom} as follows:

$$u_{\text{nom}} = -3\left(x_1 + \sin\left(\frac{\pi t}{2}\right) - 0.8\right) - 3\left(x_2 + \frac{\pi}{2}\cos\left(\frac{\pi t}{2}\right)\right). \tag{12.71}$$

This control periodically attempts to violate the safety constraint $x_1 \geq 0$. The numerical simulation results for $x(0) = (4, 2)^\top$ are depicted in Figs. 12.7 and 12.8. Both safety filters ensure the fulfillment of the safety constraint $x_1 \geq 0$, but homogeneous safety filter less conservatively overrides the nominal control. Indeed, the system trajectory with homogeneous safety filter is more close to the nominal trajectory. Such a property is useful, for example, for the adaptive cruise control in automobile, when the safety filter is utilized to override some unsafe commands of the driver. In this case, the less conservative overriding of the nominal control means a less annoying (for the driver) safety control system.

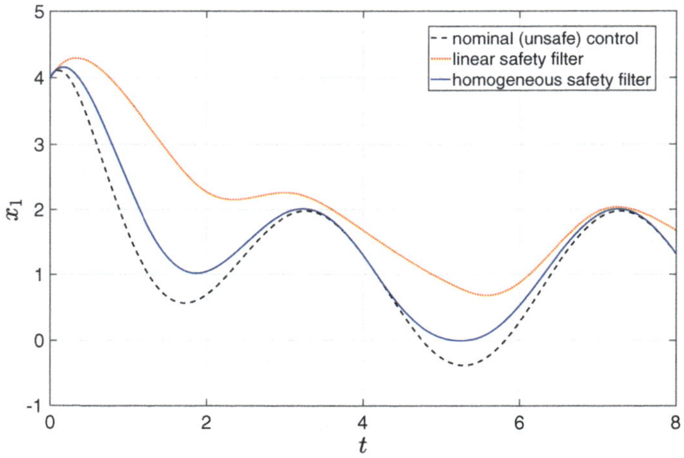

Fig. 12.7 Trajectories in (t, x_1)-plane for $x(0) = (4, 2)^\top$

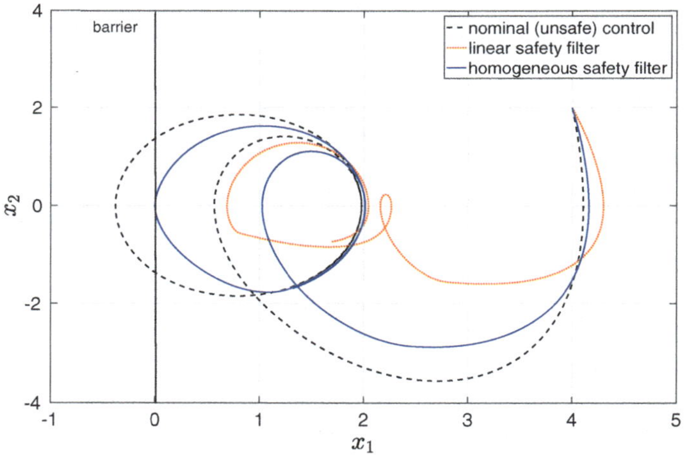

Fig. 12.8 Trajectories in (x_1, x_2)-plane for $x(0) = (4, 2)^\top$

The key advantage of the homogeneous safety filter is its essential robustness with respect to matched perturbations. Figures 12.9 and 12.10 present simulation result for the system with the constant additive perturbation: $\dot{x} = Ax + B(u - 1.5)$. The linear safety filter is not robust with respect to such a perturbation, while the homogeneous filter efficiently rejects the perturbation preserving the safety constraint.

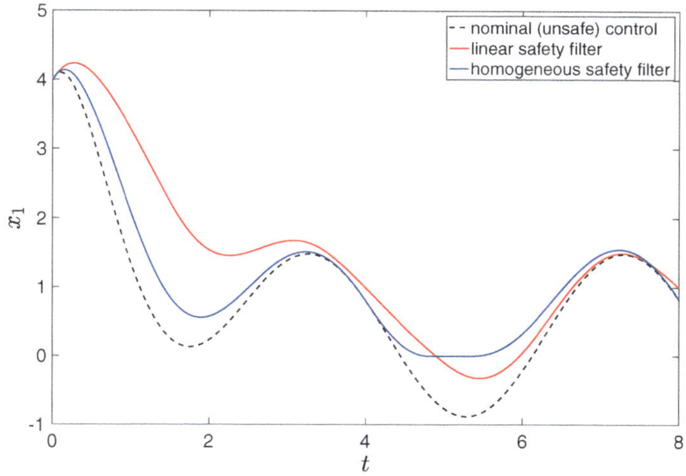

Fig. 12.9 Trajectories in (t, x_1)-plane for $x(0) = (4 \;\; 2)^\top$

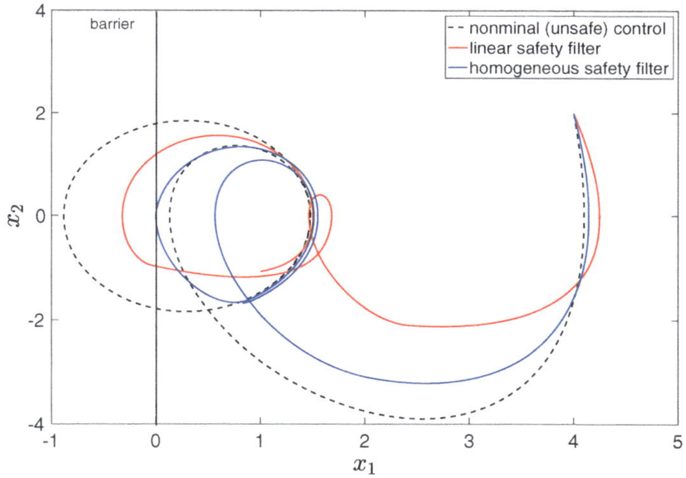

Fig. 12.10 Trajectories in (x_1, x_2)-plane for $x(0) = (4 \;\; 2)^\top$

12.2.2 Safety Filters for Linear Plant

Linear and homogeneous safety filters for a multi-input linear system can be designed in the similar way using the corresponding nonovershooting stabilizers. However, the design procedure in this case needs a special assumption.

Assumption 12.1 Let the vector $\tilde{b}_i = HBe_i$ be sign definite:

$$b_i \geq 0 \quad \text{or} \quad b_i \leq 0, \quad \forall i = 1, \ldots, m, \tag{12.72}$$

where $e_i = (0, \ldots, 1, \ldots, 0)^\top \in \mathbb{R}^m$ is the unit vector in $\mathbb{R}^m$ and, as before, the sign "$\geq$" is understood in component-wise manner.

12.2.2.1 Linear Safety Filter

Let us consider the linear plant (12.41) and assume that it is already supported with a nominal controller u_{nom} and a linear nonovershooting stabilizer $u_{\text{lin}} = Kx$ with the positively invariant set $\tilde{\Omega}_{\text{lin}}$ given by (12.42). In term of the barrier functions $\varphi = Hx$, the invertible matrix H is given by (12.43), the system (12.41) can be rewritten as follows:

$$\dot{\varphi} = \tilde{A}\varphi + HB(u - u_{\text{lin}}), \tag{12.73}$$

where $\tilde{A} = H(A + BK)H^{-1}$ is a Metzler matrix by construction.

Under Assumption 12.1, let us define the linear safety filter as follows:

$$u_i = \begin{cases} \max\{u_{\text{nom,i}}, u_{\text{lin,i}}\} & \text{if } b_i \geq 0, \\ \min\{u_{\text{nom,i}}, u_{\text{lin,i}}\} & \text{if } b_i \leq 0, \end{cases} \tag{12.74}$$

where $u = (u_1, \ldots, u_m)^\top$ is the control input, $u_{\text{nom}} = (u_{\text{nom,1}}, \ldots, u_{\text{nom,m}})^\top$ is the nominal controller and $u_{\text{lin}} = (u_{\text{lin,1}}, \ldots, u_{\text{lin,m}})^\top$ is a linear nonovershooting stabilizer. With such a selection of u we always have

$$HB(u - u_{\text{lin}}) \geq \mathbf{0}$$

so, by Lemma 2.1, the closed-loop system (12.73), (12.74) is positive.

12.2.2.2 Homogeneous Safety Filter

Let u_{hom} be a homogeneous nonovershooting stabilizer for the system (12.41) with the positively invariant set Ω_{lin} given by (12.52). In terms of barrier functions

$$\phi = H\|x\|_{\mathbf{d}}\mathbf{d}(-\ln\|x\|_{\mathbf{d}})x \tag{12.75}$$

the system can be rewritten (see the proof of Theorem 7.3) as follows:

$$\dot{\phi} = \|H^{-1}\phi\|^{\mu}\tilde{A}(\gamma)\phi + H\left(\frac{(I_n - G_{\mathbf{d}})H^{-1}\phi\phi^\top H^{-\top}P}{\phi^\top H^{-\top}PG_{\mathbf{d}}H^{-1}\phi} + I_n\right)B(u - u_{\text{hom}}), \tag{12.76}$$

where $\tilde{A}$ is given by the formula (12.54) and γ is given by (12.53). Notice that $I_n - G_{\mathbf{d}} = -\mu G_0$, where G_0 is defined in Theorem 12.1.

Under Assumption 12.1, similar to the linear case, let us define the homogeneous safety filter as follows:

$$u_i = \begin{cases} \max\{u_{\mathrm{nom,i}}, u_{\mathrm{lin,i}}\} & \text{if } b_i \geq 0, \\ \min\{u_{\mathrm{nom,i}}, u_{\mathrm{lin,i}}\} & \text{if } b_i \leq 0, \end{cases} \tag{12.77}$$

where $u = (u_1, \ldots, u_m)^\top$ and $u_{\mathrm{nom}} = (u_{\mathrm{nom,1}}, \ldots, u_{\mathrm{nom,m}})^\top$ are as before, and $u_{\mathrm{hom}} = (u_{\mathrm{hom,1}}, \ldots, u_{\mathrm{hom,m}})^\top$ is a homogeneous nonovershooting stabilizer. This safety filter ensures

$$HB(u - u_{\mathrm{hom}}) \geq \mathbf{0}\,.$$

If the matrix

$$\hat{A}(t,\phi) = \|H^{-1}\phi\|^{\mu}\tilde{A}(\gamma) - \mu\frac{\phi^\top H^{-\top}PB(u - u_{\mathrm{hom}})}{\phi^\top H^{-\top}PG_{\mathbf{d}}H^{-1}\phi}HG_0H^{-1}$$

satisfies Metzler in a neighborhood of $\partial\mathbb{R}^n_+$, then, by Remark 2.1, the system (12.76), (12.77) is positive. For the double integrator, it is shown that the required condition holds under a proper selection of the matrix P. In the MIMO case, the corresponding sufficient condition can formalized as follows:

$$E_i H^{-\top}PB \leq 0 \quad \text{for} \quad i \in \{1, \ldots, n\} : \min_{j \neq i} \tilde{g}_{ij} < 0, \tag{12.78}$$

where $\tilde{g}_{ij}$ are entries of the matrix HG_0H^{-1} and $E_i \in \mathbb{R}^{n\times n}$ is the diagonal matrix with $(E_i)_{jj} = 1$ for $j \neq i$ and $(E_i)_{ii} = 0$. For the double integrator, the above condition gives the linear inequality (12.68), which is always feasible together with (12.12). The problem of feasibility of the algebraic system (12.50), (12.78) in the general case is still open.

Concluding Remarks

In this chapter, homogeneous nonovershooting controllers are developed for multi-input LTI systems under some conic safety constraints. A methodology of upgrading existing linear nonovershooting controllers is developed. The finite-time and fixed-time safety filters are designed as well. The safety filter overrides some non-safe nominal (user's) controllers with a homogeneous nonovershooting stabilizers if the nominal controller regulates the system into an unsafe zone. The overriding is released as soon as commands of the nominal controller become safe. Such filters can be useful in various practical applications (for example, in adaptive cruise control).

References

1. Krstic M, Bement M (2006) Nonovershooting control of strict-feedback nonlinear systems. IEEE Trans Autom Control 51(12):1938–1943
2. Wieland P, Allgöwer F (2007) Constructive safety using control barrier functions. IFAC Proc Vol 40:462–467
3. Ames AD, Grizzle JW, Tabuada P (2017) Ccontrol barrier function based quadratic programs for safety critical systems. IEEE Trans Autom Control 62:3861–3876
4. Jankovic M (2018) Robust control barrier functions for constrained stabilization of nonlinear systems. Automatica 96:359–367
5. Phillips SF, Seborg DE (1988) Conditions that guarantee no overshoot for linear system. Int J Control 47(4):1043–1059
6. El-Khoury M, Crisall OD, Longchamp R (1993) Influence of zero locations on the number of step-response extre. Automatica 29:1571–1574
7. Darbha S, Bhattacharrya SP (2003) On the synthesis of controllers for a nonovershooting step response. IEEE Trans Autom Control 48(5):797–799
8. Lindemann L, Dimarogonas DV (2019) Control barrier functions for multi-agent systems under conflicting local signal temporal logic tasks. IEEE Control Syst Lett 3(3):757–762
9. Garg K, Cosner RK, Rosolia U, Ames AD, Panagou D (2022) Multi-rate control design under input constraints via fixed-time barrier functions. IEEE Control Syst Lett 6:608–613
10. Polyakov A, Krstic M (2023) Finite-and fixed-time nonovershooting stabilizers and safety filters by homogeneous feedback. IEEE Trans Autom Control 68(11):6434–6449
11. Polyakov A, Krstic M (2023) Homogeneous control systems on cones and nonovershooting finite-time stabilizers. arXiv:2305.08951
12. Farina L, Rinaldi S (2000) Positive linear systems. John Wiley, Theory and Applications
13. De Leenheer P, Aeyels D (2001) Stabilization of positive linear systems. Syst Control Lett 44:259–271
14. Rantzer A (2015) Scalable control of positive systems. Eur J Control 24(7):72–80
15. Nekhoroshikh A, Efimov D, Polyakov A, Perruquetti W, Furtat I (2021) Finite-time stabilization under state constraints. In: Conference on decision and control
16. Boyd S, Ghaoui E, Feron E, Balakrishnan V (1994) Linear matrix inequalities in system and control theory. SIAM, Philadelphia
17. Abel I, Steeves D, Krstic M, Jankovic M (2022) Prescribed-time safety design for a chain of integrators. In: American control conference
18. Ames AD, Grizzle JW, Tabuada P (2014) Control barrier function based quadratic programs with application to adaptive cruise contro. In: Conference on decision and control, pp. 6271–6278
19. Rahman Y, Jankovic M, Santill M (2021) Driver intent prediction with barrier functions. In: American control conference, pp. 224–230
20. Wang L, Ames AD, Egerstedt M (2017) Safety barrier certificates for collisions-free multirobot systems. IEEE Trans Robot 33(3):661–674
21. Santillo M, Jankovic M (2021) Collision free navigation with interacting, non-communicating obstacles. In: American control conference, pp. 1637–1643

Chapter 13
Homogeneous Optimal Control

An optimal design is the classical problem of mathematical control theory, where a geometry of a system plays an important role. Being a particular kind of symmetry, the homogeneity is expected to be useful for optimization purposes. In this chapter, we study various homogeneous optimal control problems and revisit conventional tools of optimal control design (like Bellman function method and Pontryagin maximum principle) under an assumption about a homogeneity of the system model and a cost functional. A homogeneous analog of the classical Linear Quadratic Regulation (LQR) problem is introduced. This Homogeneous Quadratic Regulation (HQR) problem is aimed at simultaneous optimization of control energy and settling time of the closed-loop system. Homogeneous Model Predictive Control (MPC) is designed. The homogeneity of the time-optimal stabilization with uniformly bounded control action is briefly studied.

13.1 Dilation Symmetry of Optimal Control Problems

Let us consider a control system modeled by the ordinary differential equation

$$\dot{x}(t) = f(x(t), u(t)), \quad t \in (0, T), \tag{13.1}$$

$$x(0) = x_0 \in \mathbb{R}^n \backslash \{\mathbf{0}\}, \tag{13.2}$$

where $T \leq +\infty$ is the time horizon, $x(t) \in \mathbb{R}^n$ is the vector of system states, $u(t) \in \mathbb{R}^m$ is the vector of control inputs, and the vector field $f : \mathbb{R}^n \times \mathbb{R}^m \mapsto \mathbb{R}^n$ is assumed to be continuous and $f(\mathbf{0}, \mathbf{0}) = \mathbf{0}$.

We study the optimal control problem:

A. Polyakov, *Generalized Homogeneity in Systems and Control Volume I*,
Communications and Control Engineering,
https://doi.org/10.1007/978-3-031-76246-8_13

$$J(x,u,T)=\int_0^T L(x(\sigma),u(\sigma))d\sigma \quad \to \quad \inf \tag{13.3}$$

subject to (13.1), (13.2), the *control constraint* given by

$$u(t)\in\mathcal{U}, \quad t\in[0,T], \quad u\in L^{\infty}(\mathbb{R};\mathbb{R}^m) \tag{13.4}$$

and the *terminal set* described as follows:

$$g_i(x(T))\le 0, \quad i=1,2,\ldots,l, \tag{13.5}$$

where a compact set $\mathcal{U}\subseteq\mathbb{R}^m$ restricts magnitudes of admissible control actions, the function $g_i:\mathbb{R}^n\mapsto\mathbb{R}$ is assumed to be continuous on $\mathbb{R}^n\backslash\{\mathbf{0}\}$ and the Lagrangian $L:\mathbb{R}^n\times\mathbb{R}^n\mapsto\mathbb{R}_+$ is assumed to be continuous on $\mathbb{R}^n\backslash\{\mathbf{0}\}\times\mathbb{R}^m$, $g(\mathbf{0})=\mathbf{0}$ and $L(\mathbf{0},\cdot)=\mathbf{0}$. The time *horizon* T may be fixed or non-fixed, finite $T<+\infty$ or infinite $T=+\infty$.

Below we assume that the considered optimal control problem is *homogeneous* in a generalized sense (see Chap. 6).

Assumption 13.1 Let $\mathbf{d}$ be a continuous linear dilation in $\mathbb{R}^n$. The vector field f, the Lagrangian L and the functions g_i are assumed to be $\mathbf{d}$-homogeneous with respect to the first argument. Namely,

$$\exists\nu_f\in\mathbb{R} \quad : \quad f(\mathbf{d}(s)x,u)= e^{\nu_f s}\mathbf{d}(s)f(x,u), \tag{13.6}$$

$$\exists\nu_L\in\mathbb{R} \quad : \quad L(\mathbf{d}(s)x,u)= e^{\nu_L s}L(x,u) \tag{13.7}$$

$$\exists\nu_i\in\mathbb{R} \quad : \quad g_i(\mathbf{d}(s)x)= e^{\nu_i s}g_i(x), \tag{13.8}$$

for $s\in\mathbb{R}$, $x\in\mathbb{R}^n\backslash\{\mathbf{0}\}$ and $u\in\mathbb{R}^m$, $i=1,2,\ldots,l$.

The optimal control problem (13.1)–(13.5) satisfying this assumption is called *homogeneous optimal control problem (HOCP)* in the Lagrange form.

13.1.1 Symmetry of Optimal Pairs

Homogeneity (symmetry) of ODE implies symmetry of its solutions. The same property can be discovered for admissible and optimal pairs of HOCP.

Definition 13.1 A pair

$$(x,u)\in W^{1,1}_{loc}((0,T);\mathbb{R}^n)\times L^{\infty}((0,T);\mathbb{R}^m)$$

is said to be **admissible** for HOCP with $T \leq +\infty$ and $x_0 \in \mathbb{R}^n \backslash \{\mathbf{0}\}$ if u satisfies (13.4), x is a strong solution of the Cauchy problem (13.2), (13.1) satisfying (13.5) and

$$J(x, u, T) < +\infty.$$

Let $\mathcal{P}(x_0, T)$ denote the set of admissible pairs (**the admissible set**) of HOCP for $T \leq +\infty$ and $x_0 \in \mathbb{R}^n \backslash \{\mathbf{0}\}$:

$$\mathcal{P}(x_0, T) \subset W^{1,1}_{loc}((0, T); \mathbb{R}^n) \times L^\infty((0, T); \mathbb{R}^m). \tag{13.9}$$

The following theorem proves a **d**-homogeneity of the set-valued map: $\mathcal{P} : \mathbb{R}^n \times \mathbb{R}_+ \mapsto 2^{W^{1,1}_{loc}((0,T);\mathbb{R}^n) \times L^\infty((0,T);\mathbb{R}^m)}$.

Theorem 13.1 *If* $(x, u) \in \mathcal{P}(x_0, T)$ *then*

$$(x_s, u_s) \in \mathcal{P}(\mathbf{d}(s)x_0, e^{-\nu_f s} T), \tag{13.10}$$

$$J(x_s, u_s, e^{-\nu_f s} T) = e^{(\nu_L - \nu_f)s} J(x, u, T), \tag{13.11}$$

for any $s \in \mathbb{R}$, *where*

$$x_s(t) = \mathbf{d}(s)x(e^{\nu_f s} t), \quad u_s(t) = u(e^{\nu_f s} t), \quad t \in (0, e^{-\nu_f s} T). \tag{13.12}$$

Proof Obviously, if

$$(x, u) \in W^{1,1}((0, T); \mathbb{R}^n) \times L^\infty((0, T); \mathbb{R}^m)$$

then

$$(x_s, u_s) \in W^{1,1}((0, e^{-\nu_f s} T); \mathbb{R}^n) \times L^\infty((0, e^{-\nu_f s} T); \mathbb{R}^m).$$

In addition, $x_s(0) = \mathbf{d}(s)x_0$ and

$$g_i(x_s(e^{-\nu_f s} T)) = g_i(\mathbf{d}(s)x(T)) = e^{\nu_i s} g_i(x(T)) \leq 0.$$

Let us show that (x_s, u_s) satisfies (13.1) almost everywhere on $(0, e^{-\nu_f s})$. If $t \in (0, e^{-\nu_f s} T)$ then $\tau = e^{\nu_f s} t \in (0, T)$ and

$$\frac{d}{dt} x_s(t) = \mathbf{d}(s) \frac{d}{dt} x\left(e^{\nu_f s} t\right) = e^{\nu_f s} \mathbf{d}(s) \frac{d}{d\tau} x(\tau) =$$

$$e^{\nu_f s} \mathbf{d}(s) f(x(\tau), u(\tau)) = e^{\nu_f s} \mathbf{d}(s) f(x\left(e^{\nu_f s} t\right), u\left(e^{\nu_f s} t\right)) =$$

$$f(\mathbf{d}(s)x\left(e^{\nu_f s}\right), u\left(e^{\nu_f s} t\right)) = f(x_s(t), u_s(t))$$

almost everywhere on $(0, e^{-\nu_f s} T)$. Using the **d**-homogeneity of L, we derive

$$L(x_s(t), u_s(t)) = L(\mathbf{d}(s)x\left(e^{\nu_f s}t\right), u\left(e^{\nu_f s}t\right)) = e^{\nu_L s}L(x\left(e^{\nu_f s}t\right), u\left(e^{\nu_f s}t\right))$$

and

$$J(x_s, u_s, e^{-\nu_f s}T) = e^{\nu_L s}\int_0^{T/e^{\nu_f s}} L(x\left(e^{\nu_f s}t\right), u\left(e^{\nu_f s}t\right))dt =$$
$$e^{(\nu_L-\nu_f)s}\int_0^T L(x\left(\tau\right), u\left(\tau\right))d\tau = e^{(\nu_L-\nu_f)s}J(x, u, T).$$

The proof is complete. □

An optimal pair can be introduced using the set $P(x_0, T)$ as follows.

Definition 13.2 An admissible pair $(x^*, u^*) \in \mathcal{P}(x_0, T)$ is said to be **optimal** if $J(x^*, u^*, T) \le J(x, u, T)$, $\forall (x, u) \in \mathcal{P}(x_0, T)$.

In the general case, the optimal pair may be non-unique. Let

$$\mathcal{O}(x_0, T) \subset P(x_0, T)$$

denote **the set of optimal pairs**. The following result is given without proof, since it immediately follows from Theorem 13.1.

Corollary 13.1 *If* $(x^*, u^*) \in \mathcal{O}(x_0, T)$ *then*

$$(x_s^*, u_s^*) \in \mathcal{O}(\mathbf{d}(s)x_0, e^{-\nu_f s}T) \tag{13.13}$$

for any $s \in \mathbb{R}$, *where*

$$x_s^*(t) = \mathbf{d}(s)x^*(e^{\nu_f s}t), \quad u_s^*(t) = u^*(e^{\nu_f s}t), \quad t \in (0, e^{-\nu_f s}T). \tag{13.14}$$

The system (13.1), (13.2) may have a non-unique solution in forward time for a given x_0. Theoretically, an optimal control u^* may generate a family of solutions $X_{x_0,u^*} \subset W^{1,1}_{loc}((0, T); \mathbb{R}^n)$ and, probably, just one of them forms an optimal pair $(x^*, u^*) \in \mathcal{O}$: $J(x^*, u^*, T) \le J(x, u^*, T)$ for all $x \in X_{x_0,u^*}$. The results proven above hold for such a *weak* statement of HOCP. Definitely, the additional assumption about uniqueness of solutions of the system (13.1) in the forward time makes the problem well-posed.

Assumption 13.2 For any $x_0 \in \mathbb{R}^n \backslash \{\mathbf{0}\}$ one holds:

$$\text{if } (x_i, u_i) \in \mathcal{P}(x_0, T), i = 1, 2 \text{ then } \quad u_1 = u_2 \quad \Rightarrow \quad x_1 = x_2.$$

Under this assumption the specific relations between optimal and admissible solutions to HOCP with different optimization horizons imply homogeneity (symmetry) of necessary and sufficient optimality conditions. In fact, to fulfill Assumption 13.2 it is sufficient to consider the so-called regular case: $f, L, g_i \in C^1$. By default, Assumption 13.2 is fulfilled.

13.1.2 An Admissible Horizon

Unstable homogeneous systems with positive degrees may have a finite-time blow up (see, e.g., Chaps. 2, 4, 7). So, the admissible set $\mathcal{P}(x_0, T)$ may be empty for some x_0, T. However, the continuity of f and L on $\mathbb{R}^n\backslash\{\mathbf{0}\} \times \mathbb{R}^m$ implies that for any $x_0 \in \mathbb{R}^n\backslash\{\mathbf{0}\}$ there exists a sufficiently small $T > 0$ such that the admissible set $\mathcal{P}(x_0, T)$ is nonempty.

Let us introduce the function $\chi : \mathbb{R}^n\backslash\{\mathbf{0}\} \mapsto \mathbb{R}_+$

$$\chi(x_0) = \inf_{T>0:\mathcal{P}(x_0,T)\neq\emptyset} 1/T \tag{13.15}$$

which characterizes an **admissible horizon** of HOCP as follows:

- if $\chi(x_0) > 0$ then

$$\mathcal{P}(x_0, T) \neq \emptyset \quad \text{for} \quad T \in (0, 1/\chi(x_0))$$

  and $\mathcal{P}(x_0, \varepsilon + 1/\chi(x_0)) = \emptyset$ for any $\varepsilon > 0$;
- $\chi(x_0) = 0$ implies $\mathcal{P}(x_0, T) \neq \emptyset$ for any $T > 0$.

The function χ of HOCP has several useful properties.

Lemma 13.1 *The function $\chi : \mathbb{R}^n \mapsto \mathbb{R}_+$ is $\mathbf{d}$-homogeneous of the degree ν_f and*

(1) χ is locally bounded on $\mathbb{R}^n\backslash\{\mathbf{0}\}$;
(2) if $\nu_f > 0$ then $\chi(x_0) \to 0$ as $\|x_0\| \to 0$;
(3) if $\nu_f < 0$ then $\chi(x_0) \to 0$ as $\|x_0\| \to +\infty$;
(4) if $\nu_f = 0$ then $\chi = \mathbf{0}$;
(5) if $\nu_L > \nu_f \in (-\infty, 0)$ then $\chi = \mathbf{0}$;
(6) $\nu_f > 0$ and the system (13.1) is forward compete for any $u \in L^\infty(\mathbb{R}_+; \mathbb{R}^m) : u(t) \in \mathcal{U}, \forall t \in [0, T]$, then $\chi = \mathbf{0}$.

Proof Due to continuity of Lagrangian L the value of the functional $J(x, u, T)$ remains bounded as long as $0 < |x(t)| < +\infty$ for all $t \in [0, T]$ with $T < +\infty$. This means that $\chi(x_0) < +\infty$ for any $x_0 \in \mathbb{R}^n\backslash\{\mathbf{0}\}$.

The continuity of f implies a boundedness of trajectories on a small enough interval of time (possibly dependent of $\|x_0\|$). This means that χ is bounded on $S = \{z \in \mathbb{R}^n : \|z\| = 1\}$. The homogeneity of the function χ immediately follows from Theorem 13.1. In this case, the conclusions (1)–(3) are straightforward consequences of Proposition 6.1.

Let $\|\cdot\|_{\mathbf{d}}$ be the canonical homogeneous norm induced by a weighted Euclidean norm $\|x\| = \sqrt{x^\top P x}$, where $0 \prec P = P^\top \in \mathbb{R}^{n\times n}$ satisfies (5.6). Then using the formula (5.21) and the continuity of the $\mathbf{d}$-homogeneous vector field f we derive

$$\frac{d\|x\|_{\mathbf{d}}}{dt} = \|x\|_{\mathbf{d}} \frac{x^{\top}\mathbf{d}^{\top}(-\ln\|x\|_{\mathbf{d}})P\mathbf{d}(-\ln\|x\|_{\mathbf{d}})f(x,u)}{x^{\top}\mathbf{d}^{\top}(-\ln\|x\|_{\mathbf{d}})PG_{\mathbf{d}}\mathbf{d}(-\ln\|x\|_{\mathbf{d}})x}$$
$$= \|x\|_{\mathbf{d}}^{1+\nu_f} \frac{x^{\top}\mathbf{d}^{\top}(-\ln\|x\|_{\mathbf{d}})Pf(\mathbf{d}(-\ln\|x\|_{\mathbf{d}})x,u)}{x^{\top}\mathbf{d}^{\top}(-\ln\|x\|_{\mathbf{d}})PG_{\mathbf{d}}\mathbf{d}(-\ln\|x\|_{\mathbf{d}})x} = \|x\|_{\mathbf{d}}^{1+\nu_f} g(x,u),$$

where

$$-\frac{\underline{\sigma}(|u|)}{0.5\beta} \leq g(x,u) = \frac{x^{\top}\mathbf{d}^{\top}(-\ln\|x\|_{\mathbf{d}})Pf(\mathbf{d}(-\ln\|x\|_{\mathbf{d}})x,u)}{x^{\top}\mathbf{d}^{\top}(-\ln\|x\|_{\mathbf{d}})PG_{\mathbf{d}}\mathbf{d}(-\ln\|x\|_{\mathbf{d}})x} \leq \frac{\overline{\sigma}(|u|)}{0.5\beta},$$

$\beta > 0$ is given by the formula (5.9) and $\overline{\sigma}, \underline{\sigma} \in C(\mathbb{R}_+, \mathbb{R}_+)$ are such that

$$\max_{\|z\|=1} z^{\top}Pf(z,u) \leq \overline{\sigma}(|u|) \quad \text{and} \quad -\underline{\sigma}(|u|) \leq \min_{\|z\|=1} z^{\top}Pf(z,u).$$

This means that for $\nu_f = 0$ we have

$$\|x(\tau)\|_{\mathbf{d}} = e^{\int_0^t g(x(\tau),u(\tau))d\tau}\|x_0\|_{\mathbf{d}}$$

and

$$e^{-t\frac{\underline{\sigma}(\|u\|_{L^\infty})}{0.5\beta}}\|x_0\|_{\mathbf{d}} \leq \|x(\tau)\|_{\mathbf{d}} \leq \|x_0\|_{\mathbf{d}} e^{t\frac{\overline{\sigma}(\|u\|_{L^\infty})}{0.5\beta}}.$$

This means that the cost $J(x,u,T)$ is finite for any finite $T > 0$ and any $x_0 \in \mathbb{R}^n$ if $\nu_f = 0$. Therefore, $\chi = \mathbf{0}$ if $\nu_f = 0$.

For $\nu_f \neq 0$ we derive

$$\|x(t)\|_{\mathbf{d}}^{-\nu_f} = \|x_0\|_{\mathbf{d}}^{-\nu_f} - \nu_f \int_0^t g(x(\tau),u(\tau))d\tau.$$

For $\nu_f < 0$ the system is forward complete for any $u \in L^{\infty}(\mathbb{R};\mathbb{R}^m)$. In this case, the restriction of the admissible horizon may be related only with on boundedness of J on trajectories, which converge to the origin in a finite time.

Let us show that the cost $J(x,u,T)$ is finite for any $T > 0$ under some restrictions on ν_L and $\nu_f \geq 0$. For $\nu_L \geq 0$ and $\nu_f = 0$ this is obvious.

Let us consider the case $\nu_f < 0$ and $\nu_L < 0$. Due to homogeneity of the Lagrangian, we have

$$J(x,u,T)=\int_0^T L(x(\tau),u(\tau))d\tau$$

$$=\int_0^T \|x(\tau)\|_{\mathbf{d}}^{\nu_L} L(\mathbf{d}(-\ln\|x(\tau)\|_{\mathbf{d}})x(\tau),u(\tau))d\tau$$

$$=\int_0^T \frac{L(\mathbf{d}(-\ln\|x(\tau)\|_{\mathbf{d}})x(\tau),u(\tau))}{\left(\|x_0\|_{\mathbf{d}}^{-\nu_f}-\nu_f\int_0^\tau g(x(s),u(s))ds\right)^{\frac{-\nu_L}{-\nu_f}}}d\tau.$$

Since $L\geq 0$ is continuous on $\mathbb{R}^n\backslash\{\mathbf{0}\}\times\mathbb{R}^k$ and $u\in L^\infty$ then there exists $\ell_u\geq 0$ such that

$$\sup_{\|z\|=1,u\in\mathcal{U}} L(z,u)\overset{a.e.}{\leq}\ell_u,\quad \tau\geq 0.$$

Since

$$-\frac{\underline{\sigma}(\|u\|_{L^\infty})}{0.5\beta}t\leq\int_0^\tau g(x(s),u(s))ds\leq\frac{\overline{\sigma}(\|u\|_{L^\infty})}{0.5\beta}t,\quad \forall t\geq 0$$

then the above integral is finite for $0\leq\frac{-\nu_L}{-\nu_f}<1$ or, equivalently, if $\nu_L>\nu_f$. This yields $\chi=\mathbf{0}$ for $\nu_L>\nu_f\in(-\infty,0)$.

Finally, if $\nu_f>0$ then, in th view of the above estimates, a solution of the system with $x(0)=x_0\neq\mathbf{0}$ cannot reach the zero in a finite. If such a system is forward complete then $0<\|x(t)\|<+\infty$ for all $t\geq 0$ and for any finite $T>0$ the cost functional has a finite value. Therefore, $\chi=\mathbf{0}$ if $\nu_f>0$ and the system is forward complete for any $u\in L^\infty(\mathbb{R}_+;\mathbb{R}^m)$. The proof is complete. □

Homogeneous systems with positive degrees (like $\dot{x}(t)=x^3(t)u(t)$) may have a solutions with a finite-time blow up. In this case, the optimal control problem may remain meaningful. Indeed, if $\nu_L<0$ then $L(x,u)\to 0$ as $\|x\|\to+\infty$ and an infinum of the cost functional J may correspond to an optimal solution with the fastest blow up. By the following assumption we exclude such systems from our further considerations.

Assumption 13.3 Let one of the following conditions hold:

- $\nu_f>0$ and the system (13.1) is forward compete for any $u\in L^\infty(\mathbb{R}_+;\mathbb{R}^m)$: $u(t)\in\mathcal{U},\forall t\geq 0$;
- $\nu_f=0$;
- $\nu_L>\nu_f\in(-\infty,0)$.

Under this assumption the system (13.1) is forward complete for any admissible control input.

13.1.3 Homogeneous Bellman Function

Since the cost functional J is additive then the homogeneous optimal control problem satisfies Bellman principle of optimality (see, e.g., [1]) saying that "*any tail of optimal trajectory is optimal too*". In this case, the sufficient condition of optimality can be obtained using the so-called **Bellman (value) function** $V : \mathbb{R}^n \times \mathbb{R}_+ \mapsto \mathbb{R}_+$, which, in our case, can be defined as follows:

$$V(x_0, T) = J(x^*, u^*, T),$$

where (x^*, u^*) is an optimal pair for HOCP with a given $x_0 \in \mathbb{R}^n \backslash \{\mathbf{0}\}$ and a given

$$T \leq \frac{1}{\chi(x_0)}.$$

For time-invariant systems, Bellman function usually depends only on the space argument $x_0 \in \mathbb{R}^n$. We have added the second argument to V in order to study also its dependence on a variation of the time horizon.

Theorem 13.2 *Under Assumptions 13.2 and 13.3, the Bellman function* $V : \mathbb{R}^n \times \mathbb{R}_+ \mapsto \mathbb{R}_+$ *is continuous on* $\mathbb{R}^n \backslash \{\mathbf{0}\} \times \mathbb{R}_+$, $V(x, 0) = 0$ *for any* $x \neq \mathbf{0}$, $V(\mathbf{0}, T) = \mathbf{0}$ *for all* $T \geq 0$ *and the identity*

$$V(\mathbf{d}(s)x, e^{-\nu_f s}T) = e^{(\nu_L - \nu_f)s} V(x, T) \tag{13.16}$$

holds for all $(x, T) \in \mathbb{R}^n \times \mathbb{R}_+$. *Moreover, if* V *is continuously differentiable on* $\mathbb{R}^n \backslash \{\mathbf{0}\} \times (0, +\infty)$ *then* V *satisfies Hamilton–Jacobi–Bellman (HJB) equation of the form:*

$$\min_{u \in \mathcal{U}} \left(\frac{\partial V}{\partial x} f(x, u) + L(x, u) \right) = \frac{\partial V}{\partial T}, \quad (x, T) \in \mathbb{R}^n \backslash \{\mathbf{0}\} \times (0, +\infty). \tag{13.17}$$

Proof The $\mathbf{d}$-homogeneity of V follows from the identity (13.11) proven in Theorem 13.1. Assumption 13.2 implies uniqueness of solutions of (13.1). By Theorem 2.5 (see also Remark 2.3), solutions depend continuously on $(x_0, u) \in \mathbb{R}^n \backslash \{\mathbf{0}\} \times L^\infty(\mathbb{R}; \mathbb{R}^m)$ on a compact interval of time. Assumption 13.3 implies that the system is forward complete, so (see Remark 2.4) the continuous dependence takes the place on any compact interval of time. This guarantees that V is continuous on $\mathbb{R}^n \backslash \{\mathbf{0}\} \times \mathbb{R}_+$. Obviously, since $L \in C(\mathbb{R}^n \backslash \{\mathbf{0}\} \times \mathbb{R}^m; \mathbb{R}_+)$ then $V(x_0, 0) = 0$ for all $x_0 \in \mathbb{R}^n \backslash \{\mathbf{0}\}$. Since $L(\mathbf{0}, \cdot) = 0$ and $f(\mathbf{0}, \mathbf{0}) = \mathbf{0}$ then for $V(\mathbf{0}, T) = 0$ for all $T \geq 0$. The rest part of the proof follows the standard way (see, e.g., [1, Theorem 22.18]) taking into account that the conventional Bellman function $\tilde{V} : (0, T) \times \mathbb{R}^n \mapsto \mathbb{R}_+$ in our case admits the representation

$$\tilde{V}(t, x) = V(x, T - t).$$

The proof is complete. □

A more constructive representation of the homogeneous Bellman function V can be obtained for a nonzero homogeneity degree of the vector field f.

Corollary 13.2 *If $\nu_f \neq 0$ then, under conditions of Theorem 13.2, a continuously differentiable Bellman function V admits the representation:*

$$V(x,T) = T^{\frac{\nu_f-\nu_L}{\nu_f}}\Phi\left(\mathbf{d}\left(\frac{\ln T}{\nu_f}\right)x\right), \tag{13.18}$$

where $\Phi : \mathbb{R}^n\backslash\{\mathbf{0}\} \mapsto \mathbb{R}_+$ is a non-negative continuously differentiable function being a solution to the modified HJB equation:

$$\min_{u\in\mathcal{U}}\left(\frac{\partial \Phi}{\partial y}f(y,u)+L(y,u)\right) = \frac{(\nu_f-\nu_L)\Phi + \frac{\partial\Phi}{\partial y}G_{\mathbf{d}}y}{\nu_f}. \tag{13.19}$$

Moreover, it holds

$$\begin{aligned}\lim_{T\to+\infty} V(x,T) &= \Phi_\infty(x) \quad \text{if } \nu_f > 0,\\ \lim_{T\to+\infty} V(x,T) &= \Phi_0(x) \quad \text{if } \nu_f < 0,\end{aligned} \tag{13.20}$$

where the functions $\Phi_0 : \Omega_0 \mapsto \mathbb{R}_+$ and $\Phi_\infty : \Omega_0 \mapsto \mathbb{R}_+$ are homogeneous approximations of the function Φ at zero and at infinity, respectively.

Proof The formula (13.18) follows from the identity

$$V(\mathbf{d}(s)x, e^{-\nu_f s}T) = e^{(\nu_L-\nu_f)s}V(x,T)$$

if $\Phi(y) = V(y,1)$, $y = \mathbf{d}(s)x$ and $s = \frac{\ln T}{\nu_f}$. The modified HJB equation (13.19) is obtained from (13.17) using the identity:

$$\frac{\partial V}{\partial x}G_{\mathbf{d}}x - \nu_f T\frac{\partial V}{\partial T} = (\nu_L-\nu_f)V,$$

which holds due to (6.3). Finally, the identities (13.20) immediately follow from the definition of homogeneous approximations. □

Example 13.1 For

$$f(x,u) = -ux^{1/3}, \quad L(x,u) = x^{2/3} \text{ and } \mathcal{U} = [0,1],$$

the optimal control problem (13.1)–(13.4) is HOCP with $\mathbf{d}(s) = e^s$, $\nu_f = -2/3$ and $\nu_L = 2/3$. Its Bellman function V has the form:

$$V(x,T)=\begin{cases}3x^{4/3}/4 & \text{if } 3x^{2/3}/2\le T,\\ Tx^{2/3}-T^2/3 & \text{if } 3x^{2/3}/2>T.\end{cases}$$

Obviously, it admits the representation (13.18) with

$$\Phi(y)=\begin{cases}3/4y^{4/3} & \text{if } y\le(2/3)^{3/2},\\ y^{2/3}-1/3 & \text{if } y>(2/3)^{3/2}.\end{cases}$$

Notice that $\chi=\mathbf{0}$, so, any horizon is admissible. The function Φ is continuously differentiable on $\mathbb{R}\backslash\{0\}$ and it satisfies the modified HJB equation (13.17). Moreover, $V(x,+\infty)=\Phi_0(x)=3x^{4/3}/4$, where Φ_0 is the homogeneous approximation of Φ at zero.

Remark 13.1 The Bellman function $V:\Omega_0\subset\mathbb{R}^n\mapsto\mathbb{R}_+$ of HOCP on infinite horizon ($T=+\infty$) is $\mathbf{d}$-homogeneous, i.e., $V(\mathbf{d}(s)x)=e^{(\nu_L-\nu_f)s}V(x), x\in\Omega_0$ and $s\in\mathbb{R}_+$. To construct the Bellman function on the whole domain Ω, it is sufficient to define V on a unit sphere only.

13.1.4 Homogeneous Maximum Principle

Necessary condition of optimality of HOCP given by the Pontryagin maximum principle [2] also admits homogeneous representation. In particular, homogeneity of the vector field f and Lagrangian L implies certain homogeneity of the Hamiltonian:

$$H(\psi,x,u)=\psi^\top f(x,u)-\mu L(x,u),$$

where $x,\psi\in\mathbb{R}^n$, $u\in\mathbb{R}^m$, $\mu\in\mathbb{R}_+$. Indeed, one has

$$H(e^{(\nu_L-\nu_f)s}\mathbf{d}^\top(-s)\psi,\mathbf{d}(s)x,u)=e^{\nu_L s}H(\psi,x,u),\quad s\in\mathbb{R}.$$

Theorem 13.3 *If the functions f, L and $g=(g_1,g_2,\ldots,g_l)^\top$ are differentiable on $\mathbb{R}^n\backslash\{\mathbf{0}\}$ and $(x^*,u^*)\in\mathcal{O}(x_0,T)$ is an optimal pair with $x^*(t)\neq 0$ for $t\in[0,T]$ then there exist a number $\mu\in\mathbb{R}_+$, a vector $v\in\mathbb{R}^l_+$ and a function $\psi^*:[0,T]\mapsto\mathbb{R}^n$ such that*

(a) *$\|\psi^*(T)\|+\mu+\|v\|>0$ and $|g(x^*(T))|^\top v=0$ and*
(b) *the tuple (ψ^*, x^*, u^*) solves the boundary value problem*

$$\begin{pmatrix}\dot\psi^*(t)\\ \dot x^*(t)\end{pmatrix}=F(\psi^*(t),x^*(t),u^*(t)),\quad F=\begin{pmatrix}-\left(\frac{\partial H}{\partial x}\right)^\top\\ \left(\frac{\partial H}{\partial \psi}\right)^\top\end{pmatrix},$$
$$x^*(0)=x_0,\quad \psi^*(T)=-\left(\frac{\partial g(x(T))}{\partial x}\right)^\top v, \tag{13.21}$$

where the vector field $F : \mathbb{R}^n \times \mathbb{R}^n\backslash\{\mathbf{0}\} \times \mathbb{R}^m \mapsto \mathbb{R}^{2n}$ *satisfies the homogeneous relation:*

$$F(\tilde{\mathbf{d}}(s)z, u) = e^{\nu_f s}\,\tilde{\mathbf{d}}(s)F(z, u), \quad z \in \mathbb{R}^{2n}, u \in \mathbb{R}^m, s \in \mathbb{R}, \tag{13.22}$$

for the uniformly continuous group $\tilde{\mathbf{d}}$ *given by*

$$\tilde{\mathbf{d}}(s) = \begin{pmatrix} e^{(\nu_L-\nu_f)s}\mathbf{d}^\top(-s) & \mathbf{0} \\ \mathbf{0} & \mathbf{d}(s) \end{pmatrix};$$

(c) $H(\varphi^*(t), x^*(t), u^*(t)) = \max_{u\in\mathcal{U}} H(\varphi^*(t), x^*(t), u) = C$ *for all* $t \in [0, T]$, *where* $C \in \mathbb{R}$ *is a constant;*

(d) $\frac{d[(\psi^*(t))^\top G_{\mathbf{d}} x^*(t)]}{dt} = -\nu_f\,(\psi^*(t))^\top f(x^*(t), u^*(t)) + \mu\nu_L L(x^*(t), u^*(t))$ *for all* $t \in (0, T)$.

Proof Since $x^*(t) \neq \mathbf{0}$ for all $t \in [0, T]$ then the proof can be done using the conventional concept of Lagrange multipliers [2]. For instance, the properties (**a**), (**c**) and Eq. (13.21) are proven in [1, Theorem 22.11, Corollaries 22.4 and 22.5]).

Let us show that the identity (13.22) holds. Since $\left(\frac{\partial H}{\partial \psi}\right)^\top = f$ is **d**-homogeneous vector field of degree ν_f then we just need to study the term

$$\left(\tfrac{\partial H}{\partial x}\right)^\top = \left(\tfrac{\partial f}{\partial x}\right)^\top \psi - \mu\left(\tfrac{\partial L}{\partial x}\right)^\top.$$

The required homogeneous relation follows from the identities

$$e^{\nu_L s}\frac{\partial L(x,u)}{\partial z} = \left.\frac{\partial L(y,u)}{\partial y}\right|_{y=\mathbf{d}(s)x} \mathbf{d}(s)$$

and

$$e^{\nu_f s}\mathbf{d}(s)\frac{\partial f(x,u)}{\partial z} = \left.\frac{\partial f(y,u)}{\partial y}\right|_{y=\mathbf{d}(s)x} \mathbf{d}(s)$$

proved in Chap. 6 (see Propositions 6.2 and 6.5).

Finally, the property (**d**) also holds since

$$\frac{d[(\psi^*(t))^\top G_{\mathbf{d}} x^*(t)]}{dt} = (\dot{\psi}^*(t))^\top G_{\mathbf{d}} x^*(t) + \left(\psi^*(t)\right)^\top G_{\mathbf{d}}\dot{x}^*(t) =$$

$$-\left(\psi^*(t)\right)^\top (\nu_f I_n + G_{\mathbf{d}}) f(x^*(t), u^*(t)) + \mu\nu_L L(x^*(t), u^*(t)) +$$

$$+\left(\psi^*(t)\right)^\top G_{\mathbf{d}} f(x^*(t), u^*(t)),$$

where the identities (6.3), (6.9) are utilized on the last step. □

The presented theorem holds for both fixed and non-fixed horizons (see, e.g., [1, Theorem 22.13]). Evidently, the restriction $x^*(t) \neq 0$ can be omitted if f, L and g can be smoothly prolonged to the origin.

The property (**d**) is the specific feature of the homogeneous systems. If $\nu_f = \nu_L = \nu$ then the right-hand side of the identity given in (**d**) becomes the Hamiltonian H multiplied by $-\nu$:

$$\frac{d\left[(\psi^*(t))^\top G_{\mathbf{d}} x^*(t)\right]}{dt} = -\nu H(\psi^*(t), x^*(t), u^*(t)).$$

Corollary 13.3 *If the time horizon T is non-fixed and $\nu_f = \nu_L$ then under conditions of Theorem 13.3 one has* $(\phi^*(t))^\top G_{\mathbf{d}} x^*(t) = \text{const}$ *for* $t \in [0, T]$.

This claim immediately follows from the property (**d**) proven in Theorem 13.3 and the identity

$$H(\varphi^*(t), x^*(t), u^*(t)) = 0, \quad t \in [0, T]$$

that holds for the case of non-fixed horizon (see, e.g., [1, Corollary 22.6]).

13.2 Homogeneous Quadratic Regulation

The so-called *linear quadratic regulator* (LQR) is the classical optimal control introduced by Rudolf Kalman in 1960s (see, e.g., [3]). The methodology of LQR design on both finite and infinite horizons is well developed in the literature (see, e.g., [4], [1, 5, 6] and references therein). In this section, we consider a generalized homogeneous LQR design. On a fixed horizon the considered problem simply coincides with the classical LQR. The homogeneity plays an important role in the case of non-fixed horizon. Being inspired by LQR design, this HOCP is called *homogeneous quadratic regulation* (HQR):

$$J(u, T) = T + \frac{1}{T^\eta} \int_0^T x^\top(\tau) \hat{Q}(T-\tau) x(\tau) + u^\top(\tau) \tilde{R}(T-\tau) u(\tau) d\tau \to \min \tag{13.23}$$

subject to

$$\dot{x}(t) = Ax(t) + Bu(t), \quad x(0) = x_0 \neq \mathbf{0}, \quad x(T) = \mathbf{0}, \tag{13.24}$$

where $x(t) \in \mathbb{R}^n$ is the system state, $u \in L^2((0, T); \mathbb{R}^m)$ is the admissible control signal, $A \in \mathbb{R}^{n\times n}$, $B \in \mathbb{R}^{n\times m}$ are matrices of the system, $T \in (0, +\infty)$ is the time horizon, the matrix-valued function $\hat{Q} : \mathbb{R}_+ \mapsto \mathbb{R}^{n\times n}$ is symmetric $\hat{Q}(\sigma) = \hat{Q}(\sigma)$ and positive semi-definite $\hat{Q}(\sigma) \succeq 0$ for all $\sigma > 0$, the matrix-valued function $\tilde{R}$: $\mathbb{R}_+ \mapsto \mathbb{R}^{m\times m}$ is symmetric $\tilde{R}(\sigma) = \tilde{R}^\top(\sigma) \in \mathbb{R}^{m\times m}$ and positive definite $\tilde{R}(\sigma) \succ 0$ for all $\sigma > 0$, and $\eta \in \mathbb{R}$ is a constant parameter.

The time horizon T is *non-fixed*. In this case, the minimization of the cost functional J implies an optimization of both a control/system energy and an optimal time required for the regulation of the system state to zero. The parameters η and $\tilde{R}$ allow us to specify the cost and some restrictions to an admissible control. For example, if $\eta = 1$, $\hat{Q} = \mathbf{0}$ and $\tilde{R}(\sigma) = I_m, \forall \sigma > 0$ then the second term of the cost functional defines an average control energy. As discussed in [7], a proper selection of the matrix-valued function $\tilde{R}$ could guarantee that $|u(\tau)|$ vanishes as $\tau \to T$.

The cost may be finite only if $T < +\infty$. Therefore, the optimal trajectory (if it exists) must converge to $\mathbf{0}$ in a finite time $T^* = T^*(x_0)$. This means that, to guarantee the existence of an optimal solution for all $x_0 \in \mathbb{R}^n$, the following assumption is necessary.

Assumption 13.4 Let the pair $\{A, B\}$ be controllable.

Indeed, if this assumption is not fulfilled then (see the proof of Theorem 9.1) there exists an uncontrollable linear subspace of the state space, so the state components belonging to this subspace may converge to zero only exponentially and $T = +\infty$.

Inspired by [8], let us introduce a class of **d**-homogeneous systems.

Assumption 13.5 Let there exist an anti-Hurwitz matrix $G_{\mathbf{d}} \in \mathbb{R}^{n\times n}$ and a matrix $G_{\tilde{\mathbf{d}}} \in \mathbb{R}^{m\times m}$ satisfying the identities:

$$AG_{\mathbf{d}} = (G_{\mathbf{d}} - I_n)A, \quad G_{\mathbf{d}}B = BG_{\tilde{\mathbf{d}}}. \tag{13.25}$$

For $G_{\tilde{\mathbf{d}}} = I_m$, the condition (13.25) becomes the conventional restriction of the homogeneous finite-time stabilization (see Chap. 9). The matrix $G_{\tilde{\mathbf{d}}}$ allows a larger class of cost functions (13.23) to be considered.

In the previous section, we have discovered that, to guarantee the homogeneity of an optimal control problem, the cost functional has to be homogeneous as well. A homogeneity of the cost can be proven under the following restrictions to the structure of matrices $\hat{Q}$ and $\tilde{R}$.

Assumption 13.6

- Let the matrix-valued function $\hat{Q}$ be defined as

  $$\hat{Q}(\sigma) = \sigma^{-\gamma} C^\top \hat{\mathbf{d}}^\top(-\ln\sigma) Q \hat{\mathbf{d}}(-\ln\sigma) C, \quad \sigma > 0,$$

  where $0 \prec Q = Q^\top \in \mathbb{R}^{k\times k}$ is a positive definite symmetric matrix, the matrix $C \in \mathbb{R}^{k\times n}$ satisfies the identity

  $$CG_{\mathbf{d}} = G_{\hat{\mathbf{d}}}C \tag{13.26}$$

  for some matrix $G_{\hat{\mathbf{d}}} \in \mathbb{R}^{k\times k}$ and $\hat{\mathbf{d}}(s) = e^{sG_{\hat{\mathbf{d}}}}$ for $s \in \mathbb{R}$.
- Let the pair $\{A, C\}$ be observable[a].
- Let the matrix-valued function $\tilde{R}$ be defined as

$$\tilde{R}(\sigma) = \sigma^{2-\gamma}\tilde{\mathbf{d}}^{\top}(-\ln\sigma)R\tilde{\mathbf{d}}(-\ln\sigma), \quad \sigma > 0, \tag{13.27}$$

where $\tilde{\mathbf{d}}(s) = e^{sG_{\tilde{\mathbf{d}}}}$, $s \in \mathbb{R}$ and $G_{\tilde{\mathbf{d}}}$ is by Assumption 13.5.

[a] A pair $\{A, C\}$ is observable [9] if $\text{rank}[C, CA, \ldots, CA^{n-1}] = n$, where $C \in \mathbb{R}^{k\times n}$ and $A \in \mathbb{R}^{n\times n}$.

Solution of the HOCP (13.23)–(13.24) can be utilized for energetically optimal finite-time regulation and model predictive control (MPC) design (see Sect. 13.2.3 or [10, 11]). The considered problem with $\hat{Q} = \mathbf{0}$ was first studied in [8]. Below we generalize the mentioned result for $\hat{Q} \neq \mathbf{0}$.

13.2.1 Optimal Finite-Time Regulation

If Assumption 13.4 holds then, by Lemmas 9.1, 9.2, there exists a nonsingular matrix $\Phi \in \mathbb{R}^{n\times n}$ to the block canonical form:

$$\Phi^{-1}A\Phi = \begin{pmatrix} \mathbf{0} & A_{12} & \mathbf{0} & \ldots & \mathbf{0} \\ \mathbf{0} & \mathbf{0} & A_{23} & \ldots & \mathbf{0} \\ \ldots & \ldots & \ldots & \ldots & \ldots \\ \mathbf{0} & \mathbf{0} & \mathbf{0} & \ldots & A_{k-1\,k} \\ A_{k1} & A_{k2} & A_{k3} & \ldots & A_{kk} \end{pmatrix}, \quad \Phi^{-1}B = \begin{pmatrix} \mathbf{0} \\ \mathbf{0} \\ \ldots \\ \mathbf{0} \\ A_{k\,k+1} \end{pmatrix}, \tag{13.28}$$

where $1 \le k \le n$ is a minimal number such that $\text{rank}[B, AB, \ldots, A^{k-1}B,] = n$, $A_{i\,i+1} \in \mathbb{R}^{n_i\times n_{i+1}}$, $A_{k\,i} \in \mathbb{R}^{n_k\times n_i}$, $\text{rank}\, A_{i\,i+1} = n_i$, $1 \le n_i \le n_{i+1}$, $i = 1, \ldots, k$, $n_1 + \cdots + n_k = n$ and $n_{k+1} = m$.

In the proof of Theorem 9.1, it is shown that, under Assumption 13.5 with $G_{\tilde{\mathbf{d}}} = I_m$, the matrix $\Phi^{-1}G_{\mathbf{d}}\Phi$ has a diagonal form if $m = 1$ or, at least, a lower block triangular form if $m \neq 1$. The following corollary (originally discovered in [8]) extends this result to the case $G_{\tilde{\mathbf{d}}} \neq I_m$.

Corollary 13.4 *If Assumptions 13.4, 13.5 are fulfilled and $\Phi \in \mathbb{R}^{n\times n}$ satisfies the identity* (13.28) *then*

$$G_{\mathbf{d}} = \Phi \begin{pmatrix} G_{11} & \mathbf{0} & \ldots & \mathbf{0} & \mathbf{0} \\ G_{21} & G_{22} & \ldots & \mathbf{0} & \mathbf{0} \\ \ldots & \ldots & \ldots & \ldots & \ldots \\ G_{k-1\,1} & G_{k-1\,2} & \ldots & G_{k-1\,k-1} & \mathbf{0} \\ G_{k1} & G_{k2} & \ldots & G_{k\,k-1} & G_{k\,k} \end{pmatrix} \Phi^{-1}, \tag{13.29}$$

where $G_{ij} \in \mathbb{R}^{n_i\times n_j}$, $k \ge i \ge j \ge 1$.

Proof The proof repeats some steps of the proof of Theorem 9.1. Denoting

$$\tilde{G}_{\mathbf{d}} = \Phi^{-1} G_{\mathbf{d}} \Phi, \quad \tilde{A} = \Phi^{-1} A \Phi \quad \textit{and} \quad \tilde{B} = \Phi^{-1} B$$

we derive that (13.25) is equivalent to

$$\tilde{A}\tilde{G}_{\mathbf{d}} = (\tilde{G}_{\mathbf{d}} - I_n)\tilde{A} \quad \text{and} \quad \tilde{G}_{\mathbf{d}}\tilde{B} = \tilde{B}G_{\tilde{\mathbf{d}}}.$$

Let us rewrite $\tilde{G}_{\mathbf{d}}$ in the block form:

$$\tilde{G}_{\mathbf{d}} = \begin{pmatrix} G_{11} & G_{12} & G_{13} & \dots & G_{1k-1} & G_{1k} \\ G_{21} & G_{22} & G_{23} & \dots & G_{2k-1} & G_{2k} \\ \dots & \dots & \dots & \dots & \dots & \dots \\ G_{k-11} & G_{k-12} & G_{k-13} & \dots & G_{k-1k-1} & G_{k-1k} \\ G_{k1} & G_{k2} & G_{k3} & \dots & G_{kk-1} & G_{kk} \end{pmatrix},$$

where $G_{ij} \in \mathbb{R}^{n_i \times n_j}$ are blocks of $\tilde{G}_{\mathbf{d}}$. Taking into account the representation (13.28) we conclude the identity $\tilde{G}_{\mathbf{d}}\tilde{B} = \tilde{B}G_{\tilde{\mathbf{d}}}$ implies that $G_{ik}A_{kk+1} = \mathbf{0}$, $i = 1, \dots, k-1$. Since the matrix A_{kk+1} is of full row rank and $n_k \leq n_{k+1}$ then $G_{ik} = \mathbf{0}$, $\forall i = 1, 2, \dots, k-1$. In this case, the representation (13.28) yields

$$\tilde{A}\tilde{G}_{\mathbf{d}} = \begin{pmatrix} A_{12}G_{21} & A_{12}G_{22} & \dots & A_{12}G_{2k-1} & \mathbf{0} \\ A_{23}G_{31} & A_{23}G_{32} & \dots & A_{23}G_{3k-1} & \mathbf{0} \\ \dots & \dots & \dots & \dots & \dots \\ A_{k-1k}G_{k1} & A_{k-1k}G_{k2} & \dots & A_{k-1k}G_{kk-1} & A_{kk-1}G_{kk} \\ \sum\limits_{i=1}^{k} A_{ki}G_{i1} & \sum\limits_{i=1}^{k} A_{ki}G_{i2} & \dots & \sum\limits_{i=1}^{k} A_{ki}G_{ik-1} & A_{kk}G_{kk} \end{pmatrix},$$

$$\tilde{G}_{\mathbf{d}}\tilde{A} = \begin{pmatrix} \mathbf{0} & G_{11}A_{12} & \dots & G_{1k-2}A_{k-2k-1} & G_{1k-1}A_{k-1k} \\ \mathbf{0} & G_{21}A_{12} & \dots & G_{2k-2}A_{k-2k-1} & G_{2k-1}A_{k-1k} \\ \dots & \dots & \dots & \dots & \dots \\ \mathbf{0} & G_{k-11}A_{12} & \dots & G_{k-1k-2}A_{k-2k-1} & G_{k-1k-1}A_{k-1k} \\ * & * & \dots & * & * \end{pmatrix},$$

with $*$ denoting (possibly) nonzero block.

Starting from the last column, we consider the block representation of the identity $\tilde{A}\tilde{G}_{\mathbf{d}} = (\tilde{G}_{\mathbf{d}} - I_n)\tilde{A}$ and, using the identity rank $A_{i-1i} = i - 1$, we conclude that

$$G_{i\,k-1}A_{k-1k} = \mathbf{0} \;\Rightarrow\; G_{i\,k-1} = \mathbf{0}, \;\; i = 1, \dots, k-2,$$

$$G_{i\,k-2}A_{k-1k} = \mathbf{0} \;\Rightarrow\; G_{i\,k-2} = \mathbf{0}, \;\; i = 1, \dots, k-3,$$

etc. Therefore, the matrix $\tilde{G}_{\mathbf{d}}$ has the block lower triangular form as shown in (13.29). The proof is complete. □

If the time horizon $T < +\infty$ is fixed, the optimal control problem (13.23)–(13.24) can be considered as the conventional LQR. In this case, the optimal linear quadratic regulator is a linear feedback, which has a time-varying gain defined by solving a differential Riccati equation (see, e.g., [3, 5, 6]). The homogeneity of the system (see Assumption 13.5) allows a solution of the differential Riccati equation to be obtained by solving an algebraic Riccati equation, which always have a solution if the pair $\{A, B\}$ is controllable and the pair $\{A, C\}$ is observable.

Theorem 13.4 *Let Assumptions 13.4, 13.5, 13.6 be fulfilled and the matrix* $G_{\mathbf{d}} + \frac{\gamma-1}{2} I_n$ *be anti-Hurwitz. Then*

- *There exists a unique positive definite symmetric solution* $P \in \mathbb{R}^{n\times n}$ *of the algebraic Riccati equation:*

$$(A + G_{\mathbf{d}})P + P(A + G_{\mathbf{d}})^\top + (\gamma - 1)P + PCQC^\top P = BR^{-1}B^\top \tag{13.30}$$

such that the matrix $A + G_{\mathbf{d}} + \frac{\gamma-1}{2} I_n - BR^{-1}B^\top P^{-1}$ *is Hurwitz.*
- *The matrix-valued function* $N \in C(\mathbb{R}_+; \mathbb{R}^{n\times n}) \cap C^\infty((0, +\infty); \mathbb{R}^{n\times n})$ *defined as* $N(0) = \mathbf{0}$ *and*

$$N(t) = t^{\gamma-1}\mathbf{d}(\ln t)P\mathbf{d}^\top(\ln t) \quad \textit{for} \quad t > 0 \tag{13.31}$$

is the unique solution of the differential Riccati equation

$$\dot{N}(t) = -AN(t) - N(t)A^\top + B\tilde{R}^{-1}(t)B^\top - N(t)\hat{Q}(t)N(t), \quad t > 0, \ N(0) = \mathbf{0}. \tag{13.32}$$

- *The optimal solution to the problem* (13.23)–(13.24) *with a fixed horizon* $T > 0$ *is given by*

$$u^*(t) = -\frac{\tilde{\mathbf{d}}(\ln(T-t))R^{-1}B^\top P^{-1}\mathbf{d}(-\ln(T-t))x(t)}{T-t}, \tag{13.33}$$

and the optimal cost is

$$J(u^*, T) = T + \frac{x_0^\top \mathbf{d}^\top(-\ln T)P^{-1}\mathbf{d}(-\ln T)x_0}{T^{\eta+\gamma-1}}. \tag{13.34}$$

- *The optimal control for the problem* (13.23)–(13.24) *with non-fixed horizon is given by*

$$u^*(t) = \frac{\tilde{\mathbf{d}}(\ln(T^*(x_0) - t))K\mathbf{d}(-\ln(T^*(x_0) - t))x(t)}{T^*(x_0) - t}, \tag{13.35}$$

where $K = -R^{-1}B^\top P^{-1}$, *the optimal settling time* $T^*(x_0)$ *is given by* $T^*(\mathbf{0}) = \mathbf{0}$ *and*

$$T^*(x_0) = \min\left\{\underset{r>0}{\operatorname{argmin}} \frac{x_0^\top \mathbf{d}^\top(-\ln r)P^{-1}\mathbf{d}(-\ln r)x_0}{r^{\mu-1}} + r\right\}, \quad x_0 \neq \mathbf{0} \quad (13.36)$$

with $\mu = \eta + \gamma$.

Proof (1) Since the pair $\{A, B\}$ is controllable then, due to the representation (13.19), the pair $\{A + \Phi\Delta\Phi^{-1}, B\}$ is controllable as well provided that a matrix $\Delta \in \mathbb{R}^{n\times n}$ has a lower triangular block form with blocks having the sizes $n_i \times n_j$, where $i, j = 1, 2, \ldots, k$. Hence, taking into account the representation (13.29), we conclude that the pair $\{A + G_\mathbf{d} + \frac{\gamma-1}{2}I_n, B\}$ is controllable for any γ. Since $\{A, C\}$ is observable and $Q \succ 0$ then the pair $\{A, Q^{\frac{1}{2}}C\}$ is observable too. In this case, by [1, Theorem 10.8]), the algebraic Riccati equation has a unique positive definite solution such that the matrix $A + G_\mathbf{d} + \frac{\gamma-1}{2}I_n - BR^{-1}B^\top P^{-1}$ is Hurwitz.

(2) Using the identity (13.30) we derive

$$\begin{aligned}\dot{N}(t) &= t^{\gamma-2}\mathbf{d}(\ln t)\big(G_\mathbf{d}P + PG_\mathbf{d}^\top + (\gamma-1)P\big)\mathbf{d}^\top(\ln t)\\ &= t^{\gamma-2}\mathbf{d}(\ln t)\left(-AP - PA^\top + BR^{-1}B^\top - PC^\top QCP\right)\mathbf{d}^\top(\ln t).\end{aligned}$$

The identities (13.25) imply

the homogeneity of the tuple (A, B, C)*:*

$$e^{-s}\mathbf{d}(s)A = A\mathbf{d}(s), \quad \forall s \in \mathbb{R}, \quad (13.37)$$

$$C\mathbf{d}(s) = \sum_{i=0}^{\infty}\frac{s^i C G_\mathbf{d}^i}{i!} = \sum_{i=0}^{\infty}\frac{s^i G_{\hat{\mathbf{d}}}^i C}{i!} = \hat{\mathbf{d}}(s)C, \quad \forall s \in \mathbb{R}, \quad (13.38)$$

$$\mathbf{d}(s)B = \sum_{i=0}^{\infty}\frac{s^i G_\mathbf{d}^i B}{i!} = \sum_{i=0}^{\infty}\frac{s^i B G_{\tilde{\mathbf{d}}}^i}{i!} = B\tilde{\mathbf{d}}(s), \quad \forall s \in \mathbb{R}. \quad (13.39)$$

Hence, we have

$$t^{\gamma-2}\mathbf{d}(\ln t)AP\mathbf{d}^\top(\ln t) = AN(t),$$

$$t^{\gamma-2}\mathbf{d}(\ln t)PA^\top\mathbf{d}^\top(\ln t) = N(t)A^\top,$$

$$CP\mathbf{d}^\top(\ln t) = \hat{\mathbf{d}}(-\ln t)C\mathbf{d}(\ln t)P\mathbf{d}^\top(\ln t) = t^{1-\gamma}\hat{\mathbf{d}}(-\ln t)CN(t),$$

$$\mathbf{d}(\ln t)PC^\top = t^{1-\gamma}N(t)C^\top\hat{\mathbf{d}}^\top(-\ln t),$$

$$t^{\gamma-2}\mathbf{d}(\ln t)BR^{-1}B^\top\mathbf{d}^\top(\ln t) = t^{\gamma-2}B\tilde{\mathbf{d}}(\ln t)R^{-1}\tilde{\mathbf{d}}^\top(\ln t)B^\top = B\tilde{R}^{-1}(t)B,$$

where $\tilde{R}$ is defined in Assumption 13.6. Hence, we conclude that N satisfies (13.32). Moreover, $N \in C^\infty((0, +\infty); \mathbb{R}^{n\times n})$ by construction. Since $G_\mathbf{d} + \frac{\gamma-1}{2}I_n$ is an anti-Hurwitz matrix, then

$$N(t) = e^{\ln t(G_{\mathbf{d}}+\frac{\gamma-1}{2}I_n)}Pe^{\ln t(G_{\mathbf{d}}^{\top}+\frac{\gamma-1}{2}I_n)} \to \mathbf{0} \text{ as } t \to 0,$$

i.e., N is continuous at 0.

(3) Let us consider the conventional LQR problem with the fixed horizon $T > 0$:

$$\tilde{J}_{\varepsilon}(u) = \frac{x^{\top}(T)x(T)}{\varepsilon} + \int_0^T x^{\top}(\tau)\hat{Q}(T-\tau)x(\tau) + u^{\top}(\tau)\tilde{R}(T-\tau)u(\tau)d\tau \to \min \tag{13.40}$$

$$\text{subject to } \dot{x}(t) = Ax(t) + Bu(t), \quad x(0) = x_0, \tag{13.41}$$

where $\varepsilon > 0$ and $t \in [0, T)$. The optimal control for this problem is given by (see, e.g., [1, 5])

$$u_{\varepsilon}^{*}(t) = -\tilde{R}^{-1}(T-t)B^{\top}P_{\varepsilon}(t)x(t),$$

and the optimal cost is

$$\tilde{J}_{\varepsilon}(u_{\varepsilon}^{*}) = x_0^{\top}P_{\varepsilon}(0)x_0,$$

where $P_{\varepsilon}(t) = P_{\varepsilon}^{\top}(t) \in \mathbb{R}^{n\times n}$ satisfies the following differential Riccati equation:

$$-\dot{P}_{\varepsilon}(t) = A^{\top}P_{\varepsilon}(t) + P_{\varepsilon}(t)A - P_{\varepsilon}(t)B^{\top}\tilde{R}(T-t)BP_{\varepsilon}(t) + Q(T-t)$$

with the initial condition $P_{\varepsilon}(T) = \frac{1}{\varepsilon}I_n$. Denoting $N_{\varepsilon}(t) = P_{\varepsilon}^{-1}(T-t)$, we derive

$$\dot{N}_{\varepsilon}(t) = -\left.\frac{d}{d\tau}P_{\varepsilon}^{-1}(\tau)\right|_{\tau=T-t} = \left.P_{\varepsilon}^{-1}(\tau)\dot{P}_{\varepsilon}(\tau)P_{\varepsilon}^{-1}(\tau)\right|_{\tau=T-t}$$

and

$$\dot{N}_{\varepsilon}(t) = -AN_{\varepsilon}(t) - N_{\varepsilon}(t)A^{\top} + B\tilde{R}^{-1}(t)B^{\top} - N_{\epsilon}(t)\hat{Q}(t)N_{\epsilon}(t), \quad N_{\varepsilon}(0) = \varepsilon I_n.$$

Therefore, we have

$$u_{\varepsilon}^{*}(t) = -\tilde{R}^{-1}(T-t)B^{\top}N_{\varepsilon}^{-1}(T-t)x(t)$$

and the optimal cost is

$$\tilde{J}(u_{\varepsilon}^{*}) = x_0^{\top}N_{\varepsilon}^{-1}(T)x_0.$$

Tending ε to 0 we conclude that $N_{\varepsilon}(t) \to N(t)$ uniformly on compact intervals of time. Since the pair $\{A, B\}$ is always controllable then there always exists a control signal from $L^{\infty}((0, T/2); \mathbb{R}^m)$, which steers the state of the system to zero in the finite time $T/2$. Hence, there exists a constant $C > 0$ which bounds the optimal cost $J_{\varepsilon}(u_{\epsilon}^{*}) \le C$ uniformly on ε. This means that for all optimal trajectories x_{ε}^{*} parameterized by ε we have $\sup_{\varepsilon>0}\frac{(x_{\varepsilon}^{*}(T))^{\top}x_{\varepsilon}^{*}(T)}{\varepsilon} \le C$. Hence, we conclude that $x_{\varepsilon}^{*}(T) \to \mathbf{0}$

as $\varepsilon \to 0$ and N corresponds to the optimal solution of the problem (13.40)–(13.41) *with the terminal condition* $x(T) = \mathbf{0}$ (see, e.g., [5] for more details about the limiting procedure). Hence, the optimal control for the original problem (13.23)–(13.24) with fixed horizon T is

$$u^*(t) = -\tilde{R}^{-1}(T-t)B^\top N^{-1}(T-t)x(t) \tag{13.42}$$

and the optimal cost is given by $J(u^*, T) = T + \frac{1}{T^\eta}x_0^\top N^{-1}(T)x_0$. Hence, we conclude that (13.34) is fulfilled. Finally, from (13.42) we derive (13.33) using $\mathbf{d}(s)B = B\tilde{\mathbf{d}}(s)$, $s \in \mathbb{R}$.

(4) In the case of the non-fixed horizon, the optimal settling time T^* corresponds to a minimum of $J(u^*, T)$ with respect to $T \geq 0$, so $T^* = T^*(x_0)$ is given by (13.36). In this case, the time-varying optimal control has the form: (13.35). The proof is complete. □

Therefore, we have shown that

> *Under Assumptions* 13.4, 13.5, 13.6, *the optimal control problem* (13.23)–(13.24) *always has a unique optimal solution in the form of the time-varying linear feedback* (13.35) *provided that the matrix* $G_{\mathbf{d}} + \frac{\gamma-1}{2}I_n$ *is anti-Hurwitz. Moreover, the optimal feedback depends on the initial state of the system in the case of non-fixed horizon.*

13.2.2 Homogeneity of Closed-Loop System

The following lemma shows that the settling-time function (13.36) is homogeneous. Moreover, it becomes a canonical homogeneous norm in $\mathbb{R}^n$ under certain restrictions on P^{-1} and $\mu = \nu + \gamma$ (see the formula (13.46)).

Lemma 13.2 *Let, under conditions of Theorem 13.4,* $\mathbf{d}_\mu$ *be a linear dilation in* $\mathbb{R}^n$ *with the generator*

$$G_{\mathbf{d}_\mu} := G_{\mathbf{d}} + \frac{\mu}{2}I_n, \quad \mu = \nu + \gamma \tag{13.43}$$

such that $G_{\mathbf{d}_\mu} - \frac{1}{2}I_n$ *is an anti-Hurwitz matrix. Then*

(1) the optimal settling-time function $T^* : \mathbb{R}^n \mapsto \mathbb{R}_+$ *is single-valued, positive definite, locally bounded, continuous at* $\mathbf{0}$ *and* $\mathbf{d}_\mu$*-homogeneous of degree 1;*

(2) T^* *is analytic on* $\mathbb{R}^n \backslash (\Sigma \cup \{\mathbf{0}\})$*, where*

$$\Sigma \subset \bigcup_{s\in\mathbb{R}} \mathbf{d}_\mu(s)\Sigma_0,$$

$$\Sigma_0 = \left\{x \in \mathbb{R}^n : x^\top Z x = 1,\ x^\top (Z G_{\mathbf{d}_\mu} + G_{\mathbf{d}_\mu}^\top Z)x = 0\right\}, \tag{13.44}$$

$$Z = P^{-1}G_{\mathbf{d}_\mu} + G_{\mathbf{d}_\mu}^\top P^{-1} - P^{-1}; \tag{13.45}$$

(3) $T^* \in C(\mathbb{R}^n) \cap C^\infty(\mathbb{R}^n\backslash\{\mathbf{0}\})$ *is the canonical* $\mathbf{d}_\mu$*-homogeneous norm induced by the weighted Euclidean norm* $\|x\| = \sqrt{x^\top Z x}$ *provided that*

$$Z G_{\mathbf{d}_\mu} + G_{\mathbf{d}_\mu}^\top Z \succ 0, \quad Z \succ 0. \tag{13.46}$$

Proof The proof follows the scheme proposed in [8].

(1) For $x \in \mathbb{R}^n$, let $\Xi(x)$ be the set of minimum points of the function

$$r \mapsto \xi(r, x) := r + \tfrac{x^\top \mathbf{d}^\top(-\ln r)P^{-1}\mathbf{d}(-\ln r)x}{r^{\mu-1}}, \quad r \geq 0.$$

This function is analytic on $(0, +\infty)$ for any fixed $x \in \mathbb{R}^n$. Obviously,

$$T^*(x) = \min_{\tilde{r}\in\Xi(x)} \tilde{r}$$

and $\Xi(\mathbf{0}) = \{0\}$. This means that $T^*(\mathbf{0}) = 0$. Since $P^{-1} \succ 0$ then for any $r > 0$ and any $x \neq \mathbf{0}$ we have

$$\begin{aligned}\xi(r, x) =& r\left(1 + x^\top \mathbf{d}_\mu^\top(-\ln r)P^{-1}\mathbf{d}_\mu(-\ln r)x\right)\\ =& r + x^\top e^{-\ln r(G_\mathbf{d}-0.5I_n)^\top} P^{-1} e^{-\ln r(G_\mathbf{d}-0.5I_n)} x > 0.\end{aligned}$$

By the assumption, the matrix $G_{\mathbf{d}_\mu} - \frac{1}{2}I_n$ is anti-Hurwitz and, consequently,

$$\lim_{r\to 0} \xi(r, x) = \lim_{r\to+\infty} \xi(r, x) = +\infty, \quad \forall x \neq \mathbf{0}.$$

Hence, the set $\Xi(x)$ is nonempty and bounded for any finite $x \neq \mathbf{0}$. Moreover, if $x \neq \mathbf{0}$ then $\Xi(x)$ belongs to the interior of $(0, +\infty)$. Therefore, the function $T^* : \mathbb{R}^n \mapsto \mathbb{R}_+$ is well-defined (single-valued), positive definite and locally bounded on $\mathbb{R}^n\backslash\{\mathbf{0}\}$.

On the one hand, due to the necessary condition of extremum, for any $\tilde{r} \in \Xi(x)$, we have

$$\left.\tfrac{\partial \xi(r,x)}{\partial r}\right|_{r=\tilde{r}} = 0$$

or, equivalently,

$$1 = x^\top \mathbf{d}_\mu^\top(-\ln \tilde{r})Z\mathbf{d}_\mu(-\ln \tilde{r})x.$$

On the other hand, the same identity holds for $x_s = \mathbf{d}_\mu(s)x$, any $s \in \mathbb{R}$ and any $\tilde{r}_s \in \Xi(x_s)$, namely, $1 = x_s^\top \mathbf{d}_\mu^\top(-\ln \tilde{r}_s)Z\mathbf{d}_\mu(-\ln \tilde{r}_s)x_s$ or, equivalently,

$$1 = x^\top \mathbf{d}_\mu^\top(-\ln(e^{-s}\tilde{r}_s))Z\mathbf{d}_\mu(-\ln(e^{-s}\tilde{r}_s))x,$$

where the group property of the dilation $\mathbf{d}_\mu(s_1 + s_2) = \mathbf{d}_\mu(s_1)\mathbf{d}_\mu(s_2)$, $\forall s_1, s_2 \in \mathbb{R}$ is utilized on the last step. Hence, we derive that $e^{-s}\tilde{r}_s \in \Xi(x)$, where $\tilde{r}_s \in \Xi(x_s)$ and $x_s = \mathbf{d}(s)x$. This means that the set-valued mapping $\Xi : \mathbb{R}^n \mapsto 2^{\mathbb{R}}$ is $\mathbf{d}_\mu$-homogeneous of degree 1:

$$\Xi(\mathbf{d}_\mu(s)x) = e^s \Xi(x).$$

This implies that the function T^* is $\mathbf{d}_\mu$-homogeneous of degree 1:

$$T^*(\mathbf{d}_\mu(s)x) = \min_{\tilde{r} \in \Xi(\mathbf{d}_\mu(s)x)} \tilde{r} = \min_{\tilde{r} \in e^s \Xi(x)} \tilde{r} = e^s \min_{\tilde{r}_s \in \Xi(x)} \tilde{r}_s = e^s T^*(x).$$

Hence, by Proposition 6.1, T^* is continuous at $\mathbf{0}$.

(2) Notice that the function T^* is implicitly defined by the equation:

$$\left.\frac{\partial \xi(r, x)}{\partial r}\right|_{r = T^*} = 0.$$

Since the function $(r, x) \mapsto \frac{\partial \xi(r,x)}{\partial r}$ is analytic on $(0, +\infty) \times \mathbb{R}^n$ then by Analytic Implicit Function Theorem [12, Theorem 6.1.2], the function T^* is analytic in a neighborhood of any point $x \in \mathbb{R}^n \backslash (\Sigma \cup \{\mathbf{0}\})$ with

$$\Sigma = \left\{ x \in \mathbb{R}^n : \frac{\partial \xi(\tilde{r}, x)}{\partial \tilde{r}} = 0, \frac{\partial^2 \xi(\tilde{r}, x)}{\partial \tilde{r}^2} = 0, \tilde{r} = T^*(x) \right\}$$

or, equivalently,

$$\Sigma = \left\{ \begin{array}{c} 1 = x^\top \mathbf{d}_\mu^\top(-\ln \tilde{r})Z\mathbf{d}_\mu(-\ln \tilde{r})x, \\ x \in \mathbb{R}^n : 0 = x^\top \mathbf{d}_\mu^\top(-\ln \tilde{r})(ZG_{\mathbf{d}_\mu} + G_{\mathbf{d}_\mu}^\top Z)\mathbf{d}_\mu(-\ln \tilde{r})x, \\ \text{where } \tilde{r} = T^*(x) \end{array} \right\}.$$

Using the $\mathbf{d}_\mu$-homogeneity of T^*, we derive the representation (13.44).

(3)The condition (13.46) implies $\Sigma_0 = \Sigma = \emptyset$, i.e., T^* is analytic on $\mathbb{R}^n \backslash \{\mathbf{0}\}$. Since T^* is $\mathbf{d}_\mu$-homogeneous of degree 1 then

$$T^*(x) = \|x\|_{\mathbf{d}_\mu} T^*(\mathbf{d}_\mu(-\ln \|x\|_{\mathbf{d}_\mu})x).$$

Using $\|\mathbf{d}_\mu(-\ln \|x\|_{\mathbf{d}_\mu})x\| = 1$ we derive $T^*(x) \leq \|x\|_{\mathbf{d}_\mu} \sup_{\|y\|=1} T^*(y)$, i.e., T^* is continuous at $\mathbf{0}$. The proof is complete. □

Recall [1, 3, 5] that the classical LQR problem may have the unique solution in the form of the (explicit) static linear feedback in the case of infinite horizon $T = \infty$. Assumptions 13.4–13.6 specify a class of generalized homogeneous optimal control problems (13.23)–(13.24), which have explicit solutions even on finite horizon.

Corollary 13.5 (Homogeneity of optimal control system) *If all conditions of Theorem 13.4 are fulfilled then*

$$\begin{aligned} x^*(t, \mathbf{d}_\mu(s)x_0) &= \mathbf{d}_\mu(s)x^*(e^{-s}t, x_0), \\ u^*(t, \mathbf{d}_\mu(s)x_0) &= e^{(\frac{\mu}{2}-1)s}\tilde{\mathbf{d}}(s)u^*(e^{-s}t, x_0), \end{aligned} \tag{13.47}$$
$$\forall t \geq 0, \quad \forall s \in \mathbb{R}, \quad \forall x_0 \in \mathbb{R}^n,$$

where $x^*(t, z)$ is an optimal trajectory and $u^*(t, z)$ is an optimal control signal for the problem (13.23)–(13.24) *with $x(0) = z \in \mathbb{R}^n$, which are prolonged on whole $\mathbb{R}_+$ as follows: $x^*(t, z) = \mathbf{0}$, $u^*(t, z) = \mathbf{0}$, $\forall t \geq T^*(z)$.*

Proof Let $x^*(t, x_0)$ be an optimal trajectory and $u^*(t, x_0)$ be an optimal control signal for the problem (13.23)–(13.24) with $x(0) = x_0 \in \mathbb{R}^n$. Let, by definition, $x^*(t, x_0) = \mathbf{0}$, $u^*(t, x_0) = \mathbf{0}$ for all $t \geq T^*(x_0)$. Let us denote

$$x_s(t) = \mathbf{d}_\mu(s)x^*(e^{-s}t, x_0), \quad \forall t \geq 0.$$

Since $x^*(t, x_0)$ and $u^*(t, x_0)\}$ satisfy the ODE (13.24) then

$$\dot{x}_s(t) = \mathbf{d}(s)e^{-s} \left.\frac{\partial x^*(\tau, x_0)}{\partial \tau}\right|_{\tau = e^{-s}t} = \mathbf{d}_\mu(s)e^{-s}\left(Ax^*(e^{-s}t, x_0) + Bu^*(e^{-s}t, x_0)\right).$$

On the one hand, the identities

$$\mathbf{d}_\mu(s) = e^{\frac{\mu}{2}s}\mathbf{d}(s), \quad A\mathbf{d}_\mu(s) = e^{-s}\mathbf{d}_\mu(s)A, \quad \mathbf{d}_\mu(s)B = e^{\frac{\mu}{2}s}B\tilde{\mathbf{d}}(s), \quad \forall s \in \mathbb{R}$$

imply that

$$\dot{x}_s(t) = Ax_s(t) + Bu_s(t), \, x_s(0) = \mathbf{d}_\mu(s)x_0, \quad \forall t \geq 0,$$

where

$$u_s(t) = e^{(\frac{\mu}{2}-1)s}\tilde{\mathbf{d}}(s)u^*(e^{-s}t, x_0), \quad \forall t \geq 0.$$

On the other hand, taking into account the $\mathbf{d}_\mu$-homogeneity of the function T^*, we derive

$$\begin{aligned} u^*\left(e^{-s}t, x_0\right) &= \frac{\tilde{\mathbf{d}}(\ln(T^*(x_0) - \frac{t}{e^s}))K\mathbf{d}(-\ln(T^*(x_0) - \frac{t}{e^s}))x^*(\frac{t}{e^s}, x_0)}{T^*(x_0) - e^{-s}t} \\ &= \frac{\tilde{\mathbf{d}}(-s)\tilde{\mathbf{d}}(\ln(T^*(\mathbf{d}_\mu(s)x_0) - t))K\mathbf{d}(-\ln(T^*(\mathbf{d}(s)x_0) - t))\mathbf{d}(s)x^*(\frac{t}{e^s}, x_0)}{e^{-s}(T^*(\mathbf{d}(s)x_0) - t)}, \end{aligned}$$

where the group properties $\tilde{\mathbf{d}}(s_1+s_2)=\tilde{\mathbf{d}}(s_1)\tilde{\mathbf{d}}(s_2)$ and $\mathbf{d}(s_1+s_2)=\mathbf{d}(s_1)\mathbf{d}(s_2)$, $\forall s_1, s_2 \in \mathbb{R}$ are utilized on the last step. Hence, we conclude

$$u_s(t)=\frac{\tilde{\mathbf{d}}(\ln(T^*(\mathbf{d}_\mu(s)x_0)-t))K\mathbf{d}(-\ln(T^*(\mathbf{d}(s)x_0)-t))x_s(t)}{T^*(\mathbf{d}(s)x_0)-t}.$$

By the formula (13.35), the obtained control $u_s(t)$ is the optimal for the problem (13.23)–(13.24) with the initial condition $x(0)=\mathbf{d}_\mu(s)x_0$. The proof is complete.□

Example 13.2 Let us consider the planar optimal control problem

$$\dot{x}_1=x_2, \quad \dot{x}_2=u, \quad x_1(T)=x_2(T)=0,$$

$$J(T,u)=T+\frac{1}{T}\int_0^T u(\tau)^2 d\tau \to \min,$$

where $x(t)=(x_1(t),x_2(t))\in\mathbb{R}^2, t>0$ is the system state and $u(t)\in\mathbb{R}$ is the control input. Minimization of the cost in this case implies optimization of the settling time simultaneously with the averaged control energy required for reaching of the origin. Obviously, the considered problem admits the representation (13.24), (13.23) with $n=2$, $A=\left(\begin{smallmatrix}0&1\\0&0\end{smallmatrix}\right)$, $B=\left(\begin{smallmatrix}0\\1\end{smallmatrix}\right)$, $\eta=1$, $\tilde{R}(T-\tau)=1$, $Q=\mathbf{0}$, $C=I_2$. Assumptions 13.5, 13.6 are fulfilled for $G_{\mathbf{d}}=\left(\begin{smallmatrix}2&0\\0&1\end{smallmatrix}\right)$, $G_{\tilde{\mathbf{d}}}=1$, $\gamma=0$ and $G_{\hat{\mathbf{d}}}=\mathbf{0}$.

The solution of (13.30) is given by $P=\left(\begin{smallmatrix}1/3&-1/2\\-1/2&1\end{smallmatrix}\right)$. Since $\mu=\eta+\gamma=1$ then $G_{\mathbf{d}_\mu}=G_{\mathbf{d}}+\frac{1}{2}I_2$,

$$Z=P^{-1}G_{\mathbf{d}_\mu}^\top+G_{\mathbf{d}_\mu}P^{-1}-P^{-1}=\begin{pmatrix}48&18\\18&8\end{pmatrix}\succ 0.$$

One can be checked numerically that $ZG_{\mathbf{d}_\mu}+G_{\mathbf{d}_\mu}^\top Z\succ 0$. Hence, the optimal control is given by

$$u^*(t)=K\mathbf{d}(-\ln(\|x(0)\|_{\mathbf{d}_\mu}-t))x, \tag{13.48}$$

where $K=-B^\top P^{-1}=\begin{pmatrix}-6 & -4\end{pmatrix}$ and $\|\cdot\|_{\mathbf{d}_\mu}$ is the canonical $\mathbf{d}_\mu$-homogeneous norm induced by the norm $\|x\|=\sqrt{x^\top Zx}$.

For $x(0)=Z^{-\frac{1}{2}}\left(\begin{smallmatrix}1\\0\end{smallmatrix}\right)$, the optimal settling time is $T^*(x(0))=1$ and the optimal cost is $J^*\approx 1.2028$. The simulation results are depicted in Fig. 13.1. The control law is implemented in the sampled way:

$$u(t)=u^*(t_i), \quad t\in[t_i,t_{i+1}),$$

where $t_i=ih$, $i=0,1\ldots.$ and $h>0$ is a sampling period.

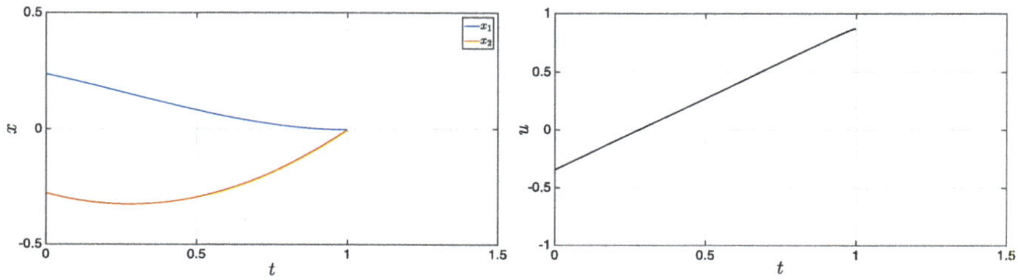

Fig. 13.1 Numerical simulations for the time-varying optimal control (13.48)

13.2.3 Homogeneous Model Predictive Control

Model predictive control (MPC) is a feedback control technique which defines a control action by minimization of a cost function for a dynamical system [10, 11]. At each time step, an MPC controller receives the current state of the plant and then calculates the optimal control signal that minimizes the cost over some horizon by solving an optimization problem that depends on the current state of the system. The MPC controller applies to the plant only the first computed optimal control action. All other computed optimal control actions are simply omitted. The proposed procedure is repeated in the following time step. Potentially, such a feedback law may not be a stabilizier of the system, so the MPC design is traditionally supported with some stability analysis (see, e.g., [13] and references therein). The ideas of homogeneous MPC design for continuous-time model are introduced in [14, 15].

According to MPC principles, a feedback law u can be designed using optimal feedforward control u^* on a finite horizon just by taking an initial value $u^*(0)$ of the control signal. In such a way, we associate $u^*(x_0)$ with x_0. Applying the same procedure for any $x_0 \in \mathbb{R}^n$ we define an MPC. Inspired by this idea, we consider the MPC feedback law of the form $u(x_0) := u^*(0)$, where $t \mapsto u^*(t)$ is a continuous-time optimal homogeneous time-varying control designed above. In the view of Theorem 13.4, the homogeneous MPC has the explicit form:

$$u(x) = \frac{\tilde{\mathbf{d}}(\ln T^*(x))K\mathbf{d}(-\ln T^*(x))x}{T^*(x)}, \quad u(\mathbf{0}) = \mathbf{0}. \tag{13.49}$$

If the optimal control problem (13.23)–(13.24) satisfies the Bellman principle[1] (e.g., if $\eta = 0$ and $\tilde{R} = \text{const}$) then (13.49) corresponds to the *optimal feedback control* for (13.23)–(13.24). If, for the closed-loop system (13.24), (13.49), we have $\dot{T}^*(x(t, x_0)) \approx -1$ or, equivalently, $T^*(x(t, x_0)) \approx T^*(x_0) - t$ then the nonlinear feedback (13.49) can also be interpreted as an approximation of the optimal control (13.35). The formula (13.49) defines a nonlinear control being a continuous-time MPC, so its stability, robustness and applicability needs to be studied additionally.

[1] A tail of any optimal trajectory is optimal too (see, e.g., [1, 16]) for more details).

Corollary 13.6 *If all conditions of Theorem 13.4 are fulfilled, then the feedback* $u : \mathbb{R}^n \mapsto \mathbb{R}^m$ *given by* (13.49) *is*

- *continuous at* $\mathbf{0}$ *if* $Z \succ 0$ *and the matrix* $G_{\tilde{\mathbf{d}}} + (\frac{\mu}{2} - 1)I_m$ *is anti-Hurwitz;*
- *globally bounded on* $\mathbb{R}^n$ *and discontinuous at* $\mathbf{0}$ *if* $Z \succ 0$ *and the matrix* $G_{\tilde{\mathbf{d}}} + (\frac{\mu}{2} - 1)I_m$ *is skew-symmetric;*
- *analytic on* $\mathbb{R}^n \backslash \{\mathbf{0}\}$ *provided that* (13.46) *holds.*

Moreover, the closed-loop system (13.24), (13.49) *is* $\mathbf{d}_\mu$*-homogeneous of degree* -1.

Proof By Lemma 13.2, the function T^* is $\mathbf{d}_\mu$-homogeneous of degree 1, i.e., $T^*(\mathbf{d}_\mu(s)x) = e^s T^*(x)$. Since

$$u(x) = (T^*(x))^{\frac{\mu}{2}-1}\tilde{\mathbf{d}}(\ln T^*(x))K\mathbf{d}_\mu(-\ln T^*(x))x \tag{13.50}$$

then using $\mathbf{d}(s)B = B\tilde{\mathbf{d}}(s)$ we derive

$$Bu(\mathbf{d}_\mu(s)x) = e^{-s}\mathbf{d}_\mu(s)Bu(x), \quad \forall x \neq 0, \quad \forall s \in \mathbb{R}.$$

Hence, taking into account the identity $A\mathbf{d}_\mu(s) = e^{-s}\mathbf{d}_\mu(s)A$, we conclude that the closed-loop system (13.24), (13.49) is $\mathbf{d}_\mu$-homogeneous of degree -1.

(1) Since the function $x \mapsto |\mathbf{d}_\mu(-\ln T^*(x))x|$ is locally bounded (by Lemma 13.2) and $\mathbf{d}_\mu$-homogeneous of degree 0 then, by Proposition 6.1, it is uniformly bounded on $\mathbb{R}^n$ by a constant $0 < M < +\infty$. Hence, we have

$$|u(x)| \leq |e^{\ln T^*(x)(G_{\tilde{\mathbf{d}}}+(\frac{\mu}{2}-1)I_m)}| \cdot |K|M, \quad \forall x \neq 0.$$

If the matrix $G_{\tilde{\mathbf{d}}} + (\frac{\mu}{2} - 1)I_m$ is anti-Hurwitz then $|e^{\ln T^*(G_{\tilde{\mathbf{d}}}+(\frac{\mu}{2}-1)I_n)}| \to 0$ as $T^* \to 0$. By Lemma 13.2, T^* is continuous at zero and $T^*(x) \to 0$ as $x \to \mathbf{0}$. So, we have $u(x) \to 0$ as $x \to \mathbf{0}$.

(2) If $G_{\tilde{\mathbf{d}}} + (\frac{\mu}{2} - 1)I_m$ is skew-symmetric then $e^{\ln T^*(x)(G_{\tilde{\mathbf{d}}}+(\frac{\mu}{2}-1)I_m)}$ is a rotation matrix (see Chap. 15 and

$$|u(x)| = |K\mathbf{d}_\mu(-\ln T^*(x))x|, \quad \forall x \neq \mathbf{0}.$$

(3) Using Lemma 13.2, case 3), we complete the proof. □

According to the above corollary, the feedback control (13.49) may be a discontinuous function of the state. In this chapter, we study only the regular feedback laws $u \in C(\mathbb{R}^n) \cap C^\infty(\mathbb{R}^n \backslash \{\mathbf{0}\})$.

Corollary 13.7 (Finite-time stability of homogeneous MPC) *Let all conditions of Theorem 13.4 be fulfilled and the matrix* $G_{\tilde{\mathbf{d}}} + (\frac{\mu}{2} - 1)I_m$ *is anti-Hurwitz. If the condition* (13.46) *holds and there exists* $\rho > 0$ *such that*

$$Z(A + BK + \rho G_{\mathbf{d}_\mu}) + (A + BK + \rho G_{\mathbf{d}_\mu})^\top Z \preceq 0 \tag{13.51}$$

then the closed-loop system (13.24), (13.49) *is globally uniformly finite-time stable:*

$$x(t) = \mathbf{0}, \quad \forall t \geq \frac{T^*(x_0)}{\rho}. \tag{13.52}$$

Proof By Lemma 13.2 the settling-time function T^* of the optimal control system is a canonical homogeneous norm $T^*(x) = \|x\|_{\mathbf{d}_\mu}$ induced by the weighted Euclidean norm $\|x\| = \sqrt{x^\top Zx}$. Let us show that $\|x\|_{\mathbf{d}_\mu}$ is a Lyapunov function of the system with MPC control.

Using the formula (5.21) we derive

$$\frac{\partial \|x\|_{\mathbf{d}_\mu}}{\partial x} = \frac{2x^\top \mathbf{d}_\mu^\top(-\ln \|x\|_{\mathbf{d}_\mu}) Z \mathbf{d}_\mu(-\ln \|x\|_{\mathbf{d}_\mu})}{x^\top \mathbf{d}^\top(-\ln \|x\|_{\mathbf{d}_\mu})(ZG_{\mathbf{d}_\mu} + G_{\mathbf{d}_\mu}^\top Z)\mathbf{d}_\mu(-\ln \|x\|_{\mathbf{d}_\mu})x} \|x\|_{\mathbf{d}_\mu}.$$

In this case, the time derivative of $\|x\|_{\mathbf{d}_\mu}$ along trajectories of the system (13.24), (13.49) is given by

$$\frac{d}{dt}\|x\|_{\mathbf{d}_\mu} = \frac{2x^\top \mathbf{d}_\mu^\top(-\ln \|x\|_{\mathbf{d}_\mu}) Z \mathbf{d}_\mu(-\ln \|x\|_{\mathbf{d}_\mu})(Ax + Bu(x))}{x^\top \mathbf{d}^\top(-\ln \|x\|_{\mathbf{d}_\mu})(ZG_{\mathbf{d}_\mu} + G_{\mathbf{d}_\mu}^\top Z)\mathbf{d}_\mu(-\ln \|x\|_{\mathbf{d}_\mu})x} \|x\|_{\mathbf{d}_\mu}.$$

Using the identities $\mathbf{d}_\mu(s)A = e^s A\mathbf{d}_\mu(s)$ and $\mathbf{d}_\mu(s)B = e^{\frac{\mu}{2}s} B\tilde{\mathbf{d}}(s)$, $\forall s \in \mathbb{R}$, we derive

$$\frac{d}{dt}\|x\|_{\mathbf{d}_\mu} = \frac{2x^\top \mathbf{d}_\mu^\top(-\ln \|x\|_{\mathbf{d}_\mu}) Z(A + BK)\mathbf{d}_\mu(-\ln \|x\|_{\mathbf{d}_\mu})x}{x^\top \mathbf{d}^\top(-\ln \|x\|_{\mathbf{d}_\mu})(ZG_{\mathbf{d}_\mu} + G_{\mathbf{d}_\mu}^\top Z)\mathbf{d}_\mu(-\ln \|x\|_{\mathbf{d}_\mu})x}$$

and $\frac{d}{dt}\|x\|_{\mathbf{d}_\mu} \leq -\rho$ provided that (13.46) and (13.51) are fulfilled. The proof is complete. □

The robustness of the closed-loop system can be investigated using the concept of Input-to-State Stability (ISS) (see [17] or Chap. 3). Recall that the system

$$\dot{x} = f(x, q) \tag{13.53}$$

is ISS if there exists $\gamma \in \mathcal{K}$ and $\beta \in \mathcal{KL}$ such that

$$\|x(t)\| \leq \beta(\|x_0\|, t) + \gamma(\|q\|_{L^\infty((0,t),\mathbb{R}^m)}),$$

for all $t \geq 0$, for all $x \in \mathbb{R}^n$ and for all $q \in L^\infty$.

Corollary 13.8 (Robustness of homogeneous MPC) *Let all conditions of Corollary 13.7 be fulfilled and* $u : \mathbb{R}^n \mapsto \mathbb{R}^m$ *be given by* (13.49)*. Let* $f : \mathbb{R}^n \times \mathbb{R}^k \mapsto \mathbb{R}^n$ *be a continuous function such that*

$$f(x, \mathbf{0}) = Ax + Bu(x), \quad \forall x \in \mathbb{R}^n.$$

If there exists a dilation $\hat{\mathbf{d}}$ *in* $\mathbb{R}^k$ *such that*

$$f(\mathbf{d}_\mu(s)x, \hat{\mathbf{d}}(s)q) = e^{-s}\mathbf{d}_\mu(s) f(x, q), \quad \forall x \in \mathbb{R}^n, \quad \forall q \in \mathbb{R}^k, \quad \forall s \in \mathbb{R}$$

then the system (13.53) *is ISS.*

This result immediately follows from Theorem 7.13. Corollary 13.8 proves a robustness of closed-loop system with respect to a rather large class of perturbations. For example, selecting

$$f(x, d) = Ax + Bu(x + q_1) + q_2, \quad q = (q_1, q_2)$$

we conclude that the nonlinear control system (13.24), (13.49) is always ISS of with respect to measurement noise and additive perturbations if the matrix $G_{\mathbf{d}} + \left(\frac{\mu}{2} - 1\right) I_n$ is anti-Hurwitz.

Under conditions of Corollary 13.7 we have $T^*(x) = \|x\|_{\mathbf{d}_\mu}$, where $\|x\|_{\mathbf{d}_\mu}$ is the canonical homogeneous norm induced by the norm $\|x\| = \sqrt{x^\top Zx}$ and Z defined in Lemma 13.2. Issues of the digital implementation of such a homogeneous controllers are studied in Chap. 10.

Example 13.3 Considering the HQR problem from Example 13.2, we derive the MPC control (13.49) in the form:

$$\tilde{u}(x) = K\mathbf{d}(-\ln \|x\|_{\mathbf{d}_\mu})x. \tag{13.54}$$

This feedback satisfies Corollary 13.7 with $\rho \approx 0.1936$. The simulation results are shown in Fig. 13.2. The control law is implemented using the explicit discretization:

$$u(t) = \tilde{u}(x(t_i)), \quad t \in [t_i, t_{i+1}),$$

where $t_i = ih$, $i = 0, 1 \ldots.$ and $h > 0$ is a sampling period.

The MPC control (13.54) is not optimal, but the closed-loop system has a trajectory close to the optimal one generated by the control (13.48). However, the static MPC feedback law is more robust than the time-varying optimal feedback. The simulation results for both controllers under measurement noise are given in Fig. 13.3 under assumption that the state x is measured with the error: $\hat{x} = x + \left(\begin{smallmatrix} 0.001 \\ 0.001 \end{smallmatrix}\right)(\sin(10t) + \cos(100t))$. The controller (13.48) is defined only on $[0, 1]$, so it is simulated only on this time interval. The feedback law (13.54) is defined globally-in-time and simulated on the time interval $[0, 1.5]$.

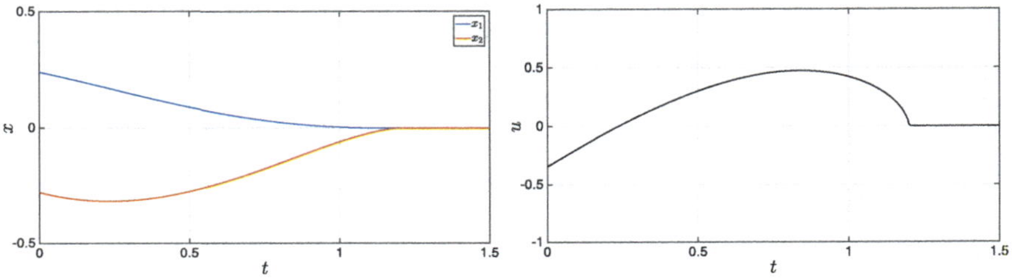

Fig. 13.2 Numerical simulations for the MPC (13.48)

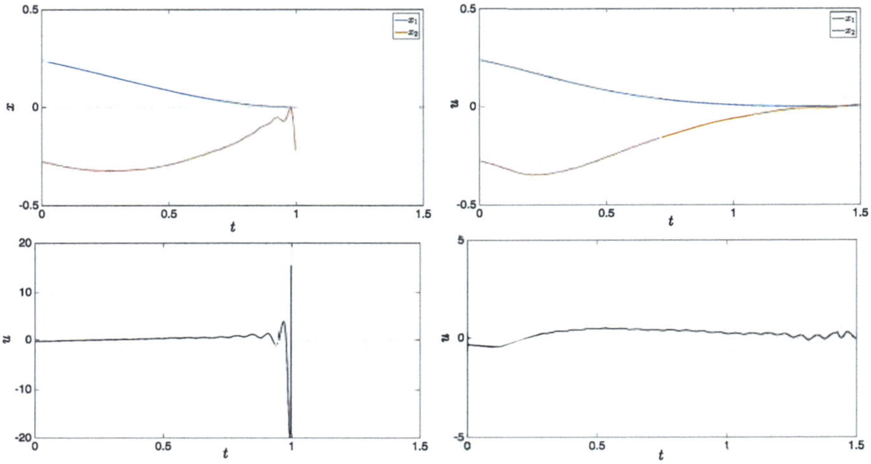

Fig. 13.3 Numerical simulations for the time-varying linear control (13.48) (left) and the MPC (13.54) (right) in the perturbed case

13.3 Time Optimal Homogeneous Feedback

Let the vector field $f : \mathbb{R}^n \times \mathbb{R} \mapsto \mathbb{R}^n$ define the controlled chain of integrators:

$$\dot{x} = f(x, u) := Ax + bu, \quad x \in \mathbb{R}^n, \quad u \in \mathbb{R}, \tag{13.55}$$

$$A = \begin{pmatrix} 0 & 1 & 0 & \dots & 0 \\ 0 & 0 & 1 & \dots & 0 \\ \dots & \dots & \dots & \dots & \dots \\ 0 & 0 & 0 & \dots & 1 \\ 0 & 0 & 0 & \dots & 0 \end{pmatrix}, \quad b = \begin{pmatrix} 0 \\ 0 \\ \dots \\ 0 \\ 1 \end{pmatrix}, \quad \mathcal{U} = [-1, 1].$$

Let us consider the classical minimum-time control problem. The cost functional (13.3) has non-fixed time horizon and the Lagrangian $L : \mathbb{R}^n \times \mathbb{R} \mapsto \mathbb{R}_+$ has the form:

$$L(x,u) = \begin{cases} 1 & \text{if } x \neq 0, \\ 0 & \text{if } x = 0. \end{cases}$$

13.3.1 Homogeneity of Settling-Time Function

The considered OCP is, obviously, **d**-homogeneous with the dilation

$$\mathbf{d}(s) = \text{diag}\{e^{ns}, e^{n-1s}, \ldots, e^{s}\}, \quad s \in \mathbb{R}$$

and

$$\nu_f = -1, \quad \nu_L = 0.$$

The so-called bang-bang control is a classical discontinuous solution of the time-optimal control problem, which is based on the well-known Feldbaum's n-interval theorem [18]. For any initial condition $x_0 \in \mathbb{R}^n$, this control guarantees that the corresponding trajectory of the system reaches the origin in a finite time.

Let us denote the settling-time function as $\tilde{T} : \mathbb{R}^n \mapsto \mathbb{R}_+$. In this case, by definition, the Bellman function has the form:

$$V(x,T) = \min\{T, \tilde{T}(x)\}, \quad x \in \mathbb{R}^n.$$

By Theorem 13.2 we have

$$V(\mathbf{d}(s)x, e^s T) = e^s V(x,T)$$

holds for $(x,T) \in (\mathbb{R}^n \times \mathbb{R}_+)\backslash\{\mathbf{0}\}$. Hence, we immediately derive that the settling-time function $\tilde{T}$ is **d**-homogeneous of the degree 1:

$$\tilde{T}(\mathbf{d}(s)x) = e^s \tilde{T}(x) \quad x \in \mathbb{R}^n\backslash\{\mathbf{0}\}.$$

Moreover, the function $\tilde{T}$ is at least continuous. If the Bellman function V is differentiable on some domain

$$\mathcal{D} \subset \{\mathbb{R}^n \times \mathbb{R}_+ : \tilde{T}(x) < T\}$$

then the HJB equation becomes

$$\frac{\partial \tilde{T}}{\partial x} Ax + \min_{u\in[-1,1]} \frac{\partial \tilde{T}}{\partial x} bu = -1. \tag{13.56}$$

Hence, the optimal relay control must have the form:

$$u = -\operatorname{sign}\left[\frac{\partial \tilde{T}}{\partial x}b\right].$$

Method of Characteristics

The method of characteristics (see, e.g., [1]) is the conventional tool for solving the first-order partial differential equations. To find (at least in an implicit form) $\tilde{T}$ let us apply this method to the equation:

$$\frac{\partial \tilde{T}}{\partial x}Ax + \frac{\partial \tilde{T}}{\partial x}b\tilde{u} = -1, \tag{13.57}$$

where $\tilde{u} \in \{-1, 1\}$ is a fixed number. The corresponding characteristic equations are

$$\frac{dx_i}{e_i^T(Ax + \tilde{u}b)} = \frac{d\tilde{T}}{-1}, \quad i = 1, 2, \ldots, n.$$

Hence, the first integrals[2] can be written in the vector form:

$$e^{A\tilde{T}}x + \tilde{u}\int_0^{\tilde{T}} e^{A\sigma}b\,d\sigma = c, \tag{13.58}$$

where $c \in \mathbb{R}^n$ is a constant vector.

Let $\Psi : \mathbb{R}^n \mapsto \mathbb{R}$ be a nonzero smooth function. Any function of the first integrals is a constant, in particular,

$$\Psi\left(e^{A\tilde{T}}x + \tilde{u}\int_0^{\tilde{T}} e^{A\sigma}b\,d\sigma\right) = 0. \tag{13.59}$$

According to the method of characteristics, if the above algebraic equation has a solution with respect to $\tilde{T}$ then this solution satisfies the original partial differential equation. Since A is a nilpotent matrix and $A^n = 0$ then the equation for finding $\tilde{T}$ becomes

$$\Psi\left(\sum_{i=0}^{n-1} A^i x\frac{\tilde{T}^i}{i!} + \tilde{u}A^i b\frac{\tilde{T}^{i+1}}{(i+1)!}\right) = 0.$$

---

[2] A first integral of an ODE

$$Q(t, x, \dot{x}) = 0$$

is a non-constant continuously differentiable function $\varphi : \mathbb{R} \times \mathbb{R}^n \mapsto \mathbb{R}$ being constant along any solution of the system, i.e., $\frac{d}{dt}\varphi(t, x(t)) = 0$.

The settling-time function $\tilde{T} : \mathbb{R}^n \mapsto \mathbb{R}_+$ is **d**-homogeneous of the degree 1. Then the identity

$$\Psi\left(\sum_{i=0}^{n-1} A^i \mathbf{d}(s)x \frac{e^{is}\tilde{T}^i}{i!} + \tilde{u}A^i b \frac{e^{(i+1)s}\tilde{T}^{i+1}}{(i+1)!}\right) = 0 \tag{13.60}$$

must hold for any $x \in \mathbb{R}^n$ and any $s \in \mathbb{R}$. Taking into account

$$\sum_{i=0}^{n-1} \frac{A^i \mathbf{d}(s) x e^{is}\tilde{T}^i}{i!} + \tilde{u}\frac{A^i b e^{(i+1)s}\tilde{T}^{i+1}}{(i+1)!} = \mathbf{d}(s)\left(\sum_{i=0}^{n-1} \frac{A^i x \tilde{T}^i}{i!} + \tilde{u}\frac{A^i b \tilde{T}^{i+1}}{(i+1)!}\right)$$

we conclude that the function Ψ also should be selected homogeneous in order to have a homogeneous settling-time function $\tilde{T}$ as a solution of (13.59).

To find the minimum-time feedback using the method of characteristics, the key problem is to select an appropriate **d**-homogeneous function Φ such that a solution of the algebraic equation

$$\Phi\left(\sum_{i=0}^{n-1} A^i x \frac{\tilde{T}^i}{i!} + \tilde{u}\frac{A^i b \tilde{T}^{i+1}}{(i+1)!}\right) = 0$$

with respect to $\tilde{T}$ would give us the Bellman function. This is a complicated problem in the general case. Taking

$$\Phi(y) = \prod_{i=1}^{n} |e_i^\top y|^{1/i}$$

we derive the **d**-homogeneous function Φ of degree $n \in \mathbb{N}$, namely, $\Phi(\mathbf{d}(s)y) = e^{ns}\Phi(y), \forall s \in \mathbb{R}, \forall y \in \mathbb{R}^n$. For any given $x \in \mathbb{R}^n$, a root $\tilde{T} \in \mathbb{C}$ of the algebraic equation

$$\Phi\left(\sum_{i=0}^{n-1} A^i x \frac{\tilde{T}^i}{i!} + \tilde{u}\frac{A^i b \tilde{T}^{i+1}}{(i+1)!}\right) = 0$$

is a root of one of the following polynomials

$$p_j(\tilde{T}) = \tilde{u}\frac{\tilde{T}^{n-j+1}}{(n-j+1)!} + \sum_{i=j-1}^{n-1} x_{i+1}\frac{\tilde{T}^i}{i!}, \quad j = 1, 2, \ldots, n, \quad \tilde{u} \in \{-1, 1\}. \tag{13.61}$$

The following example shows that a Bellman function for the minimum-time control problem can be constructed as a convex combination of roots of the obtained polynomials, at least, if $n = 2$.

Example 13.4 For $n = 2$, the formula (13.61) gives the two polynomial equations:

$$x_1 + x_2\tilde{T} + \frac{\tilde{u}}{2}\tilde{T}^2 = 0 \quad \text{and} \quad x_2 + \tilde{u}\tilde{T} = 0 \text{ with } x = (x_1, x_2)^\top.$$

The first one has the roots

$$T_1(x) = -x_2 + \sqrt{x_2^2 - 2x_1}, \quad T_2(x) = -x_2 - \sqrt{x_2^2 - 2x_1} \quad \text{for} \quad \tilde{u} = 1$$

$$T_4(x) = x_2 - \sqrt{x_2^2 + 2x_1}, \quad T_5(x) = x_2 + \sqrt{x_2^2 + 2x_1} \quad \text{for} \quad \tilde{u} = -1.$$

The roots of the second polynomial are

$$T_3(x) = -x_2 \quad \text{for} \quad \tilde{u} = 1$$

$$T_6(x) = x_2 \quad \text{for} \quad \tilde{u} = -1.$$

For any $x \neq \mathbf{0}$ at least one of roots is non-negative. Let $a = b - 1$ and $b = \frac{\sqrt{2}}{2}$. Let us define the function $T : \mathbb{R}^n \mapsto [0, +\infty)$ as follows: $T(\mathbf{0}) = 0$ and

$$T(x) = \begin{cases} \frac{T_5(x)+aT_6(x)}{b} & \text{if } x_2|x_2| + 2x_1 > 0, \\ \frac{T_1(x)+aT_3(x)}{b} & \text{if } x_2|x_2| + 2x_1 \leq 0. \end{cases} \tag{13.62}$$

The obtained function T is continuous on $\mathbb{R}^2$, $\mathbf{d}$-homogeneous of degree 1 and differentiable almost everywhere on $\mathbb{R}^2$.

Moreover, the function T is the settling-time function of the system (13.55) with the control

$$u = -\operatorname{sign}(x_2|x_2| + 2x_1) \tag{13.63}$$

and

$$\frac{\partial T(x)}{\partial x}(Ax + Bu(x)) = -1, \quad x \overset{a.e.}{\in} \mathbb{R}^2.$$

Minimum-time control design in a feedback form is a nontrivial problem in the general case [6, 19]. A possible way for its finding for $n = 3$ and $n = 4$ may be based on solving of polynomial equations given above by means of Cordano/Ferrari formulas, respectively (see Appendix F).

13.3.2 Sub-optimal Homogeneous Minimum-Time Control

Let us restrict a class of admissible inputs to smooth functions $u(\cdot) \in C^\infty((0, T); \mathbb{R})$. This restriction appears in many problems encountered in practice [20] in order to guarantee that the control law is implementable [21]. Linear control laws can be designed close to optimal ones [22]. We additionally limit ourselves to the class of linear dynamical state feedback laws:

$$u(\cdot) = w^\top(\cdot)x(\cdot), \quad w(\cdot) \in C^\infty((0, T); \mathbb{R}^n). \tag{13.64}$$

"Robustification" of optimal controllers is equally important [16, 19, 23]. This requirement unavoidably leads to feedback control algorithms. In this context, an optimally tuned finite-time feedback (like (9.10)) can be considered as a possible sub-optimal solution to the minimum-time control problem [24].

Corollary 13.9 *Assume that the system of LMIs*

$$\begin{cases} AX + XA^\top + by + y^\top b^\top + G_{\mathbf{d}}X + XG_{\mathbf{d}} = 0, \\ XG_{\mathbf{d}} + G_{\mathbf{d}}X > 0, \quad X > 0, \quad X \in \mathbb{R}^{n\times n}, \quad y \in \mathbb{R}^{1\times n}, \end{cases} \tag{13.65}$$

$$\begin{pmatrix} 1 & x_0^\top \\ x_0 & \mathbf{d}(\ln T)\, X\mathbf{d}(\ln T) \end{pmatrix} \geq 0 \tag{13.66}$$

is feasible for some $x_0 \in \mathbb{R}^n \backslash \{\mathbf{0}\}$ *and* $T \in \mathbb{R}_+$, *then the origin of the system* (13.55) *with the control*

$$u = yX^{-1}\mathbf{d}(-\ln \|x\|_{\mathbf{d}})x \tag{13.67}$$

is uniformly finite-time stable and the settling time $\tilde{T}(x_0)$ *admits the estimate*

$$\tilde{T}_0(x_0) = \|x_0\|_{\mathbf{d}} \leq T,$$

where $x(0) = x_0 \in \mathbb{R}^n$ *is an initial state of the system and the canonical homogeneous norm* $\|x\|_{\mathbf{d}}$ *is induced by the norm* $\|x\| = \sqrt{x^\top X^{-1} x}$.

Proof The inequality (13.66) implies the following relation:

$$x_0^\top \mathbf{d}(-\ln T)\, X^{-1}\mathbf{d}(-\ln T)\, x_0 - 1 \leq 0 =$$

$$x_0^\top \mathbf{d}(-\ln \|x_0\|_{\mathbf{d}})\, X^{-1}\mathbf{d}(-\ln \|x_0\|_{\mathbf{d}})\, x_0 - 1.$$

On the other hand, the inequality $XG_{\mathbf{d}} + G_{\mathbf{d}}X > 0$ guarantees

$$\frac{\partial}{\partial T} x_0^\top \mathbf{d}(-\ln T)\, X^{-1}\mathbf{d}(-\ln T)\, x_0 < 0.$$

Hence, using Theorem 9.1 we conclude $\tilde{T}_0(x_0) = \|x_0\|_{\mathbf{d}} \leq T$. □

Minimization of $T \in \mathbb{R}_+$ under the constraints (13.65), (13.66) yields a minimal settling time of the closed-loop system (13.55), (13.67) with the initial condition $x(0) = x_0$. Without any constraints on the magnitude of the control input, the settling time $T_0(x_0)$ can be made arbitrarily small. Indeed, let a pair (y_0, X_0) satisfy the matrix inequalities (13.65). Then $y = \alpha y_0$, $X = \alpha X_0$ satisfy the same inequalities for any $\alpha \in \mathbb{R}_+$. Since $X_0 > 0$, then for any $T > 0$ and any $x_0 \in \mathbb{R}^n \backslash \{\mathbf{0}\}$ there exists $\alpha \in \mathbb{R}_+$ such that inequality (13.66) is satisfied. To fulfill restrictions on the control magnitude the additional linear matrix inequality

$$\begin{pmatrix} X & y^\top \\ y & u_0^2 \end{pmatrix} \geq 0 \tag{13.68}$$

should be included in the system (13.65), (13.66). Indeed, since

$$u^2 = x^\top \mathbf{d}(-\ln \|x\|_{\mathbf{d}}) X^{-1} y^\top y X^{-1} \mathbf{d}(-\ln \|x\|_{\mathbf{d}}) x$$

and, by definition of $\|x\|_{\mathbf{d}}$,

$$x^\top \mathbf{d}(-\ln \|x\|_{\mathbf{d}}) X^{-1} \mathbf{d}(-\ln \|x\|_{\mathbf{d}}) x = 1,$$

then the inequality

$$X^{-1} y^\top y X^{-1} \leq u_0^2 X^{-1}$$

implies $u^2 \leq u_0^2$. Applying the Shur complement we derive (13.68).

To summarize, the following finite-dimensional optimization scheme can be exploited to design a sub-optimal minimum-time regulator:

$$T \to \min_{X \in \mathbb{R}^{n\times n}, y \in \mathbb{R}^{1\times n}} \quad \text{subject to (13.65), (13.66), (13.68).} \tag{13.69}$$

Notice that in the scalar case, the control design procedure proposed above leads to the optimal solution of problem. Indeed, with $n = 1$ we have $V(x) = \sqrt{P}|x|$, $P \in (0, +\infty)$, $x \in \mathbb{R}$ and $u = -u_0 \operatorname{sign}(x)$.

For any fixed number $T > 0$, the matrix inequalities (13.65), (13.66), (13.68) become linear and can be easily solved using some software like MATLAB®. The minimization with respect to the parameter T can be accomplished via the use of any known method of scalar gradient-free optimization.

The gap between the proposed sub-optimal solution and the optimal one can hardly be evaluated analytically in the general case. The settling time is given by the canonical homogeneous norm, which is defined implicitly and does not admit an explicit representation in the majority of the cases, but it can be approximated with arbitrary high precision (see Chap. 5).

Concluding Remarks

The optimal control problem for dynamical systems in $\mathbb{R}^n$ is studied in this chapter under assumption that the model of the system is homogeneous. It is shown that homogeneity is inherited by solutions of the optimal control problem. This yields certain modifications of the Bellman function method and the Pontryagin maximum principle. The problem of homogeneous quadratic optimization is studied. Its application to MPC design is discussed. The minimum-time control problem is considered as an illustrative example of a homogeneous control problem. Homogeneity gives hints for optimal and sub-optimal feedback design in this case. We refer also the reader to [25] for more applications of weighted homogeneity to optimal control design.

References

1. Poznyak AS (2008) Advanced mathematical tools for automatic control engineers. Volume 1: deterministic technique. Elsevier
2. Boltyanskii VG, Gamkrelidze RV, Pontryagin LS (1956) Towards a theory of optimal processes. Doklady Academii Nauk SSSR (in Russian) 110(1):7–10
3. Kalman R (1963) The theory of optimal control and the calculus of variations. In: Mathematical optimization techniques. University of California Press, Berkeley, pp 309–331
4. Kwakernaak H, Raphael S (1972) Linear optimal control systems. Wiley
5. Verriest E, Lewis F (1991) On the linear quadratic minimum-time problem. IEEE Trans Autom Control 36(7):859–863
6. Korobov VI, Krutin VI, Sklyar GM (1993) On optimal control problem with a mixed cost function. SIAM J Control Optim 31(3):624–645
7. Polyakov A, Efimov D, Ping X (2021) On energetically optimal finite-time stabilization. In: Conference on decision and control
8. Polyakov A, Efimov D, Ping X (2023) Energetically optimal finite-time stabilization of generalized homogeneous linear systems. IEEE Trans Autom Control 68(12):7811–7817
9. Kalman R (1960) On the general theory of control systems. In: Proceedings of the first international congress of the IFAC. Moscow
10. Garcia CE, Prett DM, Morari M (1989) Model predictive control: theory and practice–a survey. Automatica 25(3):335–348
11. Allgöwer F, Zheng A (eds) (2000) Nonlinear model predictive control. Springer
12. Krantz AG, Parks HR (2003) The implicit function theorem: history. Springer, Theory and Appplications
13. Kouvaritakis B, Cannon M (2016) Model predictive control. Springer
14. Li H, Yan W, Shi Y (2017) Continuous-time model predictive control of under-actuated spacecraft with bounded control torques. Automatica 75(1):144–153
15. Coron J-M, Grüne L, Worthmann K (2020) Model predictive control, cost controllability, and homogeneity. SIAM J Control Optim 58(5):2979–2996
16. Boltyanski VG, Poznyak A (2012) The robust maximum principle: theory and applications. Birkhauser
17. Sontag ED, Wang Y (1996) On characterizations of the input-to-state stability property. Syst Control Lett 24(5):351–359
18. Feldbaum A (1953) Optimal processes in systems of automatic control. Avtom Telemekh 14(6):721–728 (in Russian)

19. Dinuzzo F, Ferrara A (2009) Higher order sliding mode controllers with optimal reaching. IEEE Trans Autom Control 54(9):2126–2136
20. Tarbouriech S, Garcia G (eds) (1997) Control of uncertain systems with bounded inputs. Lecture notes in control and information sciences. Springer
21. Shinar J, Glizer VY, Turetsky V (2014) Capture zone of linear strategies in interception problems with variable structure dynamics. J Franklin Inst 351(4):2378–2395
22. Bohl AH, McAvoy TJ (1976) Linear feedback versus time optimal control. In: Industrial and engineering chemistry process design and development, vol 15, no 1, pp 24–33
23. Chernous'ko FL, Ananievski IM, Reshmin SA (2008) Control of nonlinear dynamical systems. Springer
24. Polyakov A (2015) Time-suboptimal feedback design via linear matrix inequalities. Autom Remote Control 76(847–862)
25. Zelikin MI, Borisov VF (1994) Theory of chattering control. Springer

Chapter 14
Homogeneous State Estimation

State estimation (observation) is a problem of the control and systems theory. In this chapter, we study the problem of **d**-homogeneous observation.

Definition 14.1 A system

$$\dot{x} = f(x), \quad t > 0, \quad f \in C(\mathbb{R}^n; \mathbb{R}^n) \tag{14.1}$$

with the output measurements

$$y = h(x), \quad h \in C(\mathbb{R}^n; \mathbb{R}^k) \tag{14.2}$$

is said to be **d**-homogeneously observable of a degree $\nu \in \mathbb{R}$ if there exists a system (called observer) of the form:

$$\begin{cases} \dot{z}_1 = f_1(z_1, z_2, y), & f_1 : \mathbb{R}^{n+p+k} \mapsto \mathbb{R}^n, \quad z_1(t) \in \mathbb{R}^n, \\ \dot{z}_2 = f_2(z_1, z_2, y), & f_2 : \mathbb{R}^{n+p+k} \mapsto \mathbb{R}^p, \quad z_2(t) \in \mathbb{R}^p, \end{cases} \quad t > 0, \tag{14.3}$$

such that

- the error variable $\varepsilon = \begin{pmatrix} z_1 - x \\ z_2 \end{pmatrix}$ satisfies a differential inclusion

$$\dot{\varepsilon} \in G(\varepsilon), \quad G : \mathbb{R}^{n+p} \mapsto 2^{\mathbb{R}^{n+p}}, \tag{14.4}$$

  being **d**-homogeneous of degree $\nu \in \mathbb{R}$;
- the error system (14.4) is globally uniformly asymptotically stable.

A. Polyakov, *Generalized Homogeneity in Systems and Control Volume I*, Communications and Control Engineering,
https://doi.org/10.1007/978-3-031-76246-8_14

In this chapter, the following three kinds of homogeneous observers for linear plants are designed

- time-varying homogeneous observer (f_2 is independent of z_1 and y);
- time-invariant homogeneous observer ($f_2 = \mathbf{0} \Rightarrow z_2 = \text{const}$);
- filtering homogeneous observer (f_2 depends on z_1 and/or y).

Homogeneous observer design for a class of linear plants with Hölder continuous additive nonlinearities is developed.

14.1 Homogeneous Observers for Linear Plant

Observation problem consists in a reconstruction of the whole state of the system (14.1) based on some output measurements y. This problem is solvable if the system satisfies certain observability conditions [1–3]. Usually, an observer is a dynamical system (like (14.3)), which state (or a part of the state vector) converges to the state of the observable system asymptotically or in a finite/fixed time. Weighted homogeneous observers/differentiators are developed, e.g., in [4–8] for systems topologically equivalent to the chain of integrators. Extensions of the same ideas to the multi-output case can be found, for example, in [9].

Let us consider the linear control system

$$\begin{cases} \dot{x} = Ax + Bu, \\ y = Cx, \end{cases} \quad t > 0, \tag{14.5}$$

where x is the state variable, y is the measured output, $u : \mathbb{R} \mapsto \mathbb{R}^m$ is an input, $A \in \mathbb{R}^{n \times n}$ is the system matrix, $B \in \mathbb{R}^{n \times m}$ is the matrix of input gains and the matrix $C \in \mathbb{R}^{k \times n}$ models the measurements of the state variables.

The goal of this section is to design various $\mathbf{d}$-homogeneous observers of the form 14.3 for the MIMO system (14.5) and to analyze their robustness with respect to measurement noise and additive exogenous perturbations. For this purpose, we consider also the perturbed system:

$$\begin{cases} \dot{x} = Ax + Bu + q_1, \\ y = Cx + q_2, \end{cases} \quad q_1(t) \in \mathbb{R}^n, \quad q_2(t) \in \mathbb{R}^k, \quad t > 0 \tag{14.6}$$

where x, y, u, A, B and C are as before, but the unknown function $q = (q_1^\top, q_2^\top)^\top \in L^\infty(\mathbb{R}; \mathbb{R}^{n+k})$ models additive exogenous perturbations and measurement noise.

Traditionally, a linear observer [10, 11] is defined as a copy of the system with the so-called output injection term:

$$\dot{z} = Az + Bu + L(t)(y - Cz), \quad z(t) \in \mathbb{R}^n,$$

where $L(t) \in \mathbb{R}^{n\times k}$ is a (possibly time-varying) observer's gain to be defined such that

$$z(t) \to x(t) \quad \text{as} \quad t \to +\infty.$$

For instance, the gain L of the Kalman–Bucy filter [10] is defined optimally based on known stochastic characteristics of the measurement noise and additive perturbations. The Luenberger observer [11] uses any gain L such that the matrix $A + LC$ is Hurwitz. Such gain exists, in particular, if

the pair $\{A, C\}$ *is observable* [2], namely, if

$$\text{rank}\begin{pmatrix} C \\ CA \\ \dots \\ CA^{n-1} \end{pmatrix} = n. \tag{14.7}$$

A homogeneous observer can be designed in the same way making the gain L dependent on

- time variable (*time-varying homogeneous observer*);
- output estimation error $y - Cz$ (*time-invariant homogeneous observer*);
- filtered estimation error $y - Cz$ (*extended homogeneous observer*).

14.1.1 Time-Varying Homogeneous Observer

Let us consider the following observer equation:

$$\begin{cases} \dot{z} = Az + Bu + (L_0 + |\xi|^{\nu-1}\mathbf{d}(\ln|\xi|)L)(Cz - y), \\ \dot{\xi} = -\rho|\xi|^{1+\nu}\,\text{sign}(\xi), \quad t > 0, \quad \xi(0) = \xi_0 \neq 0, \end{cases} \tag{14.8}$$

where $z(t) \in \mathbb{R}^n$ is the observer state, $\xi(t) \in \mathbb{R}$ is an auxiliary variable, $\mathbf{d}$ is a dilation in $\mathbb{R}^n$, $\nu \in \mathbb{R}$, $\gamma > 0$ and L_0, $L \in \mathbb{R}^{n\times k}$ are parameters of the observer. The differential equation for the variable ξ has an explicit solution, so the observer (14.8) can be interpreted as a linear observer with a time-varying gain dependent on $\xi(t)$. However, considering the extended error equation

$$\dot{\varepsilon}_\xi = g_\xi(\varepsilon_\xi) := \begin{pmatrix} A_0\varepsilon + |\xi|^{\nu-1}\mathbf{d}(\ln|\xi|)LC\varepsilon \\ -\rho|\xi|^{\nu+1}\,\text{sign}(\xi) \end{pmatrix}, \quad A_0 = A + L_0C \tag{14.9}$$

for the *augmented* error variable

$$\varepsilon_\xi = \begin{pmatrix} \varepsilon \\ \xi \end{pmatrix}, \quad \varepsilon = z - x,$$

the homogeneity (dilation symmetry) can be utilized for analysis of convergence rate and observer parameters design. The similar idea for the integrator chain has to be utilized in [12, 13] to design the so-called prescribed-time observers. An idea of transformation of a linear distributed observer to a homogeneous distributed observer using a linear dilation with a time dependent parameter ξ was first proposed in [14]. The following theorem is inspired by [15]. It adapts the same principle to the centralized observers.

Theorem 14.1 *For any observable pair* $\{A, C\}$ *one holds*

(1) the linear algebraic equation

$$G_0A - AG_0 + Y_0C = A \quad CG_0 = \mathbf{0} \tag{14.10}$$

has a solution $Y_0 \in \mathbb{R}^{n\times m}$, $G_0 \in \mathbb{R}^{n\times n}$ *and for any solution*

– *the matrix*

$$G_{\mathbf{d}} = I_n + \nu G_0 \tag{14.11}$$

is anti-Hurwitz for $\nu \geq -1/\tilde{n}$, *where* $\tilde{n} \in \mathbb{N}$ *is a minimal number such that*

$$\operatorname{rank}\begin{pmatrix} C \\ CA \\ \dots \\ CA^{\tilde{n}-1} \end{pmatrix} = n;$$

– *the matrix* $I_n + G_0$ *is invertible and the matrix*

$$A_0 = A - (I_n + G_0)^{-1}Y_0C \tag{14.12}$$

satisfies the identity

$$A_0G_{\mathbf{d}} = (G_{\mathbf{d}} + \nu I_n)A_0 \quad \text{and} \quad CG_{\mathbf{d}} = C; \tag{14.13}$$

(2) the linear algebraic system

$$\begin{array}{c} PA_0 + A_0^\top P + YC + C^\top Y^\top + \rho(PG_{\mathbf{d}} + G_{\mathbf{d}}^\top P) = \mathbf{0}, \\ PG_{\mathbf{d}} + G_{\mathbf{d}}^\top P \succ 0, \quad P = P^\top \succ 0 \end{array} \tag{14.14}$$

has a (non-unique) solution $P \in \mathbb{R}^{n\times n}$, $Y \in \mathbb{R}^{n\times k}$ *for any* $\rho \in \mathbb{R}_+$;

(3) the system (14.8) with $L_0 = -(I_n + G_0)^{-1}Y_0$ *and* $L = P^{-1}Y$ *has a solution defined on* $[0, T_0)$, *where*

(i) $T_0 = +\infty$ *for* $\nu \geq 0$,

(ii) $T_0 = \frac{\xi_0^{-\nu}}{-\nu\alpha} < +\infty$ *for* $\nu < 0$;

(4) the error equation (14.9) *is* $\mathbf{d}_\xi$*-homogeneous of degree* ν *with respect to the dilation in* $\mathbb{R}^{n+1}$

$$\mathbf{d}_\xi(s) = \begin{pmatrix} \mathbf{d}(s) & \mathbf{0} \\ \mathbf{0} & e^s \end{pmatrix}, \quad s \in \mathbb{R},$$

where $\mathbf{d}(s) = e^{sG_\mathbf{d}}$ *is a dilation in* $\mathbb{R}^n$*;*

(5) for any $\xi_0 \neq 0$ *the error variable* ε_ξ *converges to zero*

$$\varepsilon_\xi(t) \to \mathbf{0} \quad as \quad t \to T_0$$

in the fixed time $T_0 < +\infty$ *if* $\nu < 0$ *or asymptotically* $T_0 = +\infty$ *if* $\nu \geq 0$.

Proof If the pair $\{A, C\}$ is observable (14.7) then the pair $\{A^\top, C^\top\}$ is controllable. Denoting $\tilde{A} = A^\top$ and $\tilde{B} = C^\top$ (by Theorem 9.1) we derive that the system

$$\tilde{A}\tilde{G}_0 - \tilde{G}_0\tilde{A} + \tilde{B}\tilde{Y}_0 = \tilde{A}, \quad \tilde{G}_0\tilde{B} = \mathbf{0}$$

has a solution $(\tilde{G}_0, \tilde{Y}_0)$ such that $I_n - \tilde{G}_0$ is invertible. Hence, we conclude

$$A^\top\tilde{G}_0 - \tilde{G}_0A^\top + C^\top\tilde{Y}_0 = \tilde{A}, \quad \tilde{G}_0C^\top = \mathbf{0}$$

or, equivalently, using transposition we obtain

$$\tilde{G}_0^\top A - A\tilde{G}_0^\top + \tilde{Y}_0^\top C = A, \quad C\tilde{G}_0^\top = \mathbf{0}.$$

Taking $G_0 = -\tilde{G}_0^\top$ and $Y_0 = \tilde{Y}_0^\top$ we have

$$AG_0 - G_0A + Y_0C = A, \quad CG_0 = \mathbf{0}$$

and $I_n + G_0$ is invertible. Since the matrix $I_n + \mu\tilde{G}_0$ is anti-Hurwitz for $\mu \leq 1/\tilde{n}$ then the matrix $G_\mathbf{d} = I_n + \nu G_0$ is anti-Hurwitz for $\nu \geq -1/\tilde{n}$.

Let us show that the matrix $A_0 = A + L_0C$ is nilpotent for $L_0 = -(G_0 + I_n)^{-1}Y_0$ and $A_0G_\mathbf{d} = (\nu I_n + G_\mathbf{d})A_0$, $CG_\mathbf{d} = C$. Indeed, since

$$AG_0 - G_0A + Y_0C = A, \quad CG_0 = \mathbf{0}$$

then

$$A\nu G_0 - \nu G_0A + \nu Y_0C = \nu A, \quad \nu CG_0 = \mathbf{0}$$

$\Leftrightarrow$

$$A\nu G_0 - \nu G_0A - \nu(G_0 + I_n)L_0C = \nu A, \quad \nu CG_0 = \mathbf{0}$$

$\Leftrightarrow$

$$(A + L_0C)\nu G_0 - \nu G_0(A + L_0C) = \nu A_0, \quad C\nu G_0 = \mathbf{0}$$

$\Leftrightarrow$

$$A_0(I_n+\nu G_0)-(I_n+\nu G_0)A_0=\nu A_0,\quad C(I_n+\nu G_0)=C$$

$\Leftrightarrow$

$$A_0 G_{\mathbf{d}}=(\nu I_n+G_{\mathbf{d}})A,\quad CG_{\mathbf{d}}=C.$$

Hence, we derive that

$$A_0\mathbf{d}(s)=e^{\nu s}\mathbf{d}(s)A_0,\quad C\mathbf{d}(s)=e^{s}C,\quad \forall s\in\mathbb{R}$$

and the error equation (14.9) is $\mathbf{d}_\xi$-homogeneous of degree ν:

$$g_\xi(\mathbf{d}_\xi(s)\varepsilon_\xi)=\begin{pmatrix}A_0\mathbf{d}(s)\varepsilon+|e^s\xi|^{\nu-1}\mathbf{d}(\ln|e^s\xi|)LC\mathbf{d}(s)\varepsilon\\ -\rho|e^s\xi|^{\nu+1}\operatorname{sign}(e^s\xi)\end{pmatrix}$$

$$=\begin{pmatrix}e^{\nu s}\mathbf{d}(s)A_0\varepsilon+e^{\nu s}|\xi|^{\nu-1}\mathbf{d}(s)\mathbf{d}(\ln|\xi|)LC\varepsilon\\ -e^{\nu s+s}\rho|\xi|^{\nu+1}\operatorname{sign}(e^s\xi)\end{pmatrix}=e^{\nu s}\mathbf{d}_\xi(s)g_\xi(\varepsilon_\xi),$$

for all $s\in\mathbb{R}$ and for all $\varepsilon_\xi\in\mathbb{R}^{n+1}$.

The feasibility analysis of the algebraic system (14.14) repeats the proof of Theorem 9.1 denoting

$$\tilde{A}_0=A_0^\top,\quad \tilde{B}=C^\top,\quad \tilde{G}_{\mathbf{d}}=G_{\mathbf{d}}^\top,\quad X=P,\quad \tilde{Y}=Y^\top.$$

For $\xi_0\neq 0$ we have $\xi(t)=(\gamma\nu t-|\xi_0|^{-\nu})^{-\nu^{-1}}\operatorname{sign}(\xi_0)\neq 0$ for all $t\in[0,T_0)$, so the time-varying homogeneous observer (14.8) is well defined on $[0,T_0)$. Denoting

$$\epsilon=\mathbf{d}(-\ln|\xi|)\varepsilon$$

we derive

$$\dot{\epsilon}=-\frac{\dot{\xi}\operatorname{sign}(\xi)}{|\xi|}G_{\mathbf{d}}\mathbf{d}(-\ln|\xi|)\varepsilon+\mathbf{d}(-\ln|\xi|)\dot{\varepsilon}$$

$$=|\xi|^{\nu}\rho G_{\mathbf{d}}\mathbf{d}(-\ln|\xi|)\varepsilon+\mathbf{d}(-\ln|\xi|)(A_0\varepsilon+|\xi|^{\nu-1}\mathbf{d}(\ln|\xi|)LC\varepsilon.$$

Taking into account $C\varepsilon=|\xi|^{\nu}C\mathbf{d}(-\ln|\xi|)\varepsilon=|\xi|^{\nu}C\epsilon$ we derive

$$\dot{\epsilon}=|\xi|^{\nu}(A_0+LC+\rho G_{\mathbf{d}})\epsilon.$$

Using (14.14) we conclude

$$\frac{d}{dt}\epsilon^\top P\epsilon=2|\xi|^{\nu}\epsilon^\top P(A_0+LC+\rho G_{\mathbf{d}})\epsilon=0,$$

i.e., $\|\epsilon(t)\| = \|\epsilon(0)\|$ for all $t \in [0, T_0)$, where $\|x\| = \sqrt{x^\top Px}$ is the weighted Euclidean norm. In this case, we have

$$1 = \left\| \frac{\epsilon(t)}{\|\epsilon(0)\|} \right\| = \left\| \frac{\epsilon(t)}{\|\epsilon(0)\|} \right\|_{\mathbf{d}} = \left\| \mathbf{d}(-\ln|\xi(t)|) \frac{\varepsilon(t)}{\|\epsilon(0)\|} \right\|_{\mathbf{d}} = \frac{1}{|\xi(t)|} \left\| \frac{\varepsilon(t)}{\|\epsilon(0)\|} \right\|_{\mathbf{d}},$$

where $\|\cdot\|_{\mathbf{d}}$ is the canonical homogeneous norm induced by the weighted Euclidean norm $\|x\| = \sqrt{x^\top Px}$. Therefore, we derive

$$\left\| \frac{\varepsilon(t)}{\|\epsilon(0)\|} \right\|_{\mathbf{d}} = |\xi(t)| \to 0 \quad \text{as} \quad t \to T_0.$$

The proof is complete. □

For $\nu < 0$, the parameter $\xi_0 \neq 0$ defines the prescribed convergence time $T_0 = \frac{\xi_0^{-\nu}}{-\nu\alpha}$ of the observer. The time-varying prescribed-time observers are introduced in [12], where the authors implicitly use the weighted homogeneity for time-varying scaling of observer gains.

However, the right-hand side of the time-varying observer (14.8) has a singularity if $\nu < 0$ since $|\xi|^{\nu-1}\mathbf{d}(\ln|\xi|) \to \infty$ as $\xi \to 0$. Therefore, a solution of the system (14.8) is not defined for $t > T_0$ if $\nu < 0$. The time-varying parameter ξ can be assigned $\xi(t) = 1$ for $t \geq T_0$ to treat this issue. The singularity badly impacts to a robustness of the observer since the gain $L_0 + |\xi|^{\nu-1}\mathbf{d}(\ln|\xi|)L$ tends to infinity amplifying measurement noise. Below we eliminate this drawback of the time-varying homogeneous observer by various modifications of the equation for the scaling parameter ξ.

14.1.2 Time-Invariant Homogeneous Observer

Replacing in (14.8) the dynamic equation for the variable ξ with the static one

$$\xi = |y - Cz|$$

we derive the time-invariant homogeneous observer

$$\dot{z} = Az + Bu + (L_0 + |y - Cz|^{\nu-1}\mathbf{d}(\ln|y - Cz|)L)(y - Cz), \tag{14.15}$$

where L_0 is a homogenization gain such that $A_0 = A + L_0C$ is nilpotent. The time-invariant weighted homogeneous observers are designed [4, 6, 7, 9]. The case of linear dilations was studied in [16, 17]. The following theorem proves that the time-invariant homogeneous observer can be designed by upgrading the time-varying homogeneous observer studied above.

Theorem 14.2 *Let the pair $\{A, C\}$ be observable. If the parameters $G_{\mathbf{d}}, \mathbf{d}, L_0, L, \nu, \tilde{n}$ are defined as in Theorem 14.1 then the error dynamics*

$$\dot{\varepsilon} = g(\varepsilon) := A_0\varepsilon + |C\varepsilon|^{\nu-1}\mathbf{d}(\ln|C\varepsilon|)LC\varepsilon, \quad \varepsilon = z - x \tag{14.16}$$

of the observer (14.15) *is*

- $\mathbf{d}$*-homogeneous of degree* ν*;*
- *globally asymptotically stable provided that* ν *is sufficiently close to* 0.

Moreover, $g : \mathbb{R}^n \mapsto \mathbb{R}^n$ *is (piece-wise) continuous if* $\nu > -\frac{1}{\tilde{n}}$ *(if* $\nu = -\frac{1}{\tilde{n}}$*).*

Proof (1) Since $A_0\mathbf{d}(s) = e^{\nu s}\mathbf{d}(s)A_0$, $C\mathbf{d}(s) = e^s C, \forall s \in \mathbb{R}$ then

$$g(\mathbf{d}(s)\varepsilon) = (A_0 + |C\mathbf{d}(s)\varepsilon|^{\nu-1}\mathbf{d}(\ln|C\mathbf{d}(s)e|)LC)\mathbf{d}(s)\varepsilon =$$

$$= (A_0\mathbf{d}(s) + e^{(\nu-1)s}|C\varepsilon|^{\nu-1}\mathbf{d}(\ln e^s|C\varepsilon|)LC\mathbf{d}(s))\varepsilon =$$

$$= (e^{\nu s}\mathbf{d}(s)A_0 + e^{\nu s}\mathbf{d}(s)|C\varepsilon|^{\nu-1}\mathbf{d}(\ln|C\varepsilon|)LC)\varepsilon = e^{\nu s}\mathbf{d}(s)g(\varepsilon).$$

(2) Notice that

$$|C\varepsilon|^{\nu-1}\mathbf{d}(\ln|C\varepsilon|)LC\varepsilon = |C\varepsilon|^{\nu}\mathbf{d}(\ln|C\varepsilon|)L\tfrac{C\varepsilon}{|C\varepsilon|} = e^{(I_n+\nu(I_n+G_0))\ln|C\varepsilon|}L\tfrac{C\varepsilon}{|C\varepsilon|}.$$

Taking into account the structure of G_0 (see the formula (9.19) and the proof of Theorem 14.1), for $\nu > -1/\tilde{n}$ we have $e^{(I_n+\nu(I_n+G_0))s} \to 0$ as $s \to -\infty$ then the g is continuous. If $\nu = -1/\tilde{n}$ then $e^{(I_n+\nu(I_n+G_0))s} \to$ const as $s \to -\infty$ and g is piece-wise continuous.

(3) Let us show that the canonical homogeneous norm $\|\cdot\|_{\mathbf{d}}$ (defined in the proof of Theorem 14.1) if a homogeneous Lyapunov function of the error equation provided that ν is sufficiently close to 0. Indeed,

$$\frac{d}{dt}\|\varepsilon\|_{\mathbf{d}} = \tfrac{\partial\|\varepsilon\|_{\mathbf{d}}}{\partial\varepsilon}\dot{\varepsilon} = \|\varepsilon\|_{\mathbf{d}}\tfrac{x^\top\mathbf{d}^\top(-\ln\|\varepsilon\|_{\mathbf{d}})P\mathbf{d}(-\ln\|\varepsilon\|_{\mathbf{d}})(A_0+|C\varepsilon|^{\nu-1}\mathbf{d}(\ln|C\varepsilon|)LC)\varepsilon}{\varepsilon^\top\mathbf{d}^\top(-\ln\|\varepsilon\|_{\mathbf{d}})PG_{\mathbf{d}}\mathbf{d}(-\ln\|\varepsilon\|_{\mathbf{d}})\varepsilon}$$

$$= \|\varepsilon\|_{\mathbf{d}}^{1+\nu}\tfrac{x^\top\mathbf{d}^\top(-\ln\|\varepsilon\|_{\mathbf{d}})P(A_0+|C\mathbf{d}(-\ln\|\varepsilon\|_{\mathbf{d}})\varepsilon|^{\nu-1}\mathbf{d}(\ln|C\mathbf{d}(-\ln\|\varepsilon\|_{\mathbf{d}})\varepsilon|)LC)\mathbf{d}(-\ln\|\varepsilon\|_{\mathbf{d}})\varepsilon}{\varepsilon^\top\mathbf{d}^\top(-\ln\|\varepsilon\|_{\mathbf{d}})PG_{\mathbf{d}}\mathbf{d}(-\ln\|\varepsilon\|_{\mathbf{d}})\varepsilon},$$

where the identities

$$C\varepsilon = \|\varepsilon\|_{\mathbf{d}}C\mathbf{d}(-\ln\|\varepsilon\|_{\mathbf{d}})\varepsilon \quad \text{and} \quad \mathbf{d}(-\ln\|\varepsilon\|_{\mathbf{d}})A_0 = \|\varepsilon\|_{\mathbf{d}}^{\nu}A_0\mathbf{d}(-\ln\|\varepsilon\|_{\mathbf{d}})$$

are utilized on the last step. Therefore, for $\varepsilon \neq \mathbf{0}$, we obtain

$$\frac{d}{dt}\|\varepsilon\|_{\mathbf{d}} = \|\varepsilon\|_{\mathbf{d}}^{1+\nu}\tfrac{\pi^\top P(A_0+|C\pi|^{\nu-1}\mathbf{d}(\ln|C\pi|)LC)\pi}{\pi^\top PG_{\mathbf{d}}\pi}, \quad \pi = \mathbf{d}(-\ln\|\varepsilon\|_{\mathbf{d}})\varepsilon.$$

Let the function $\xi : \mathbb{R} \times \mathbb{R}^k \mapsto \mathbb{R}^n$ be defined as follows:

$$\xi(\nu, C\pi) = |C\pi|^{\nu-1}\mathbf{d}(\ln|C\pi|)LC\pi - LC\pi = e^{\nu(I_n+G_0)\ln|C\pi|}LC\pi - LC\pi,$$

for $C\pi \neq 0$ and $\xi(\nu, \mathbf{0}) = \mathbf{0}$. Obviously, this function is continuously differentiable on $(-1/\tilde{n}, +\infty) \times (\mathbb{R}^k \backslash \{\mathbf{0}\})$. Let us show that this function is continuously differentiable on $(-1/\tilde{n}, +\infty) \times \mathbb{R}^k$ as well. Indeed, for any $q \in \mathbb{R}$ satisfying $\max(0, -\nu\tilde{n}) \leq q < 1$ we have

$$\ln|C\pi|e^{\nu(I_n+G_0)\ln|C\pi|}LC\pi = \underbrace{e^{\overbrace{(qI_n + \nu(I_n+G_0))}^{anti-Hurwitz}\ln|C\pi|}}_{\to\mathbf{0} \text{ as } |C\pi|\to 0} L \underbrace{\frac{C\pi\ln|C\pi|}{|C\pi|^q}}_{\to\mathbf{0} \text{ as } |C\pi|\to 0}.$$

Hence, we conclude

$$\frac{\partial\xi(\nu, C\pi)}{\partial\nu} = (I_n + G_0)\ln|C\pi|e^{\nu(I_n+G_0)\ln|C\pi|}LC\pi \to \mathbf{0} \text{ as } |C\pi| \to 0$$

and

$$\xi(\nu, C\pi) = (e^{(qI_n+\nu(I_n+G_0))\ln|C\pi|} - |C\pi|^q I_n)L\frac{C\pi}{|C\pi|^q} \to \mathbf{0} \text{ as } |C\pi| \to 0.$$

Using the function ξ, the time derivative of the canonical homogeneous norm can be rewritten as follows:

$$\frac{d\|\varepsilon\|_{\mathbf{d}}}{dt} = \|\varepsilon\|_{\mathbf{d}}^{1+\nu}\frac{\pi^\top P(A_0\pi+|C\pi|^{\nu-1}\mathbf{d}(\ln|C\pi|)LC\pi)}{\pi^\top PG_{\mathbf{d}}\pi} = \|\varepsilon\|_{\mathbf{d}}^{1+\nu}\frac{\pi^\top P((A_0+LC)\pi+\xi(\nu,C\pi))}{\pi^\top PG_{\mathbf{d}}\pi}$$

$$= \|\varepsilon\|_{\mathbf{d}}^{1+\nu}\frac{\pi^\top P(A_0+LC)\pi}{\pi^\top PG_{\mathbf{d}}\pi} + \|\varepsilon\|_{\mathbf{d}}^{1+\nu}\frac{\pi^\top P\xi(\nu,C\pi)}{\pi^\top PG_{\mathbf{d}}\pi} = -\rho\|\varepsilon\|_{\mathbf{d}}^{1+\nu} + \|\varepsilon\|_{\mathbf{d}}^{1+\nu}\frac{\pi^\top P\xi(\nu,C\pi)}{\pi^\top PG_{\mathbf{d}}\pi}$$

$$\leq -\rho\|\varepsilon\|_{\mathbf{d}}^{1+\nu} + \|\varepsilon\|_{\mathbf{d}}^{1+\nu}\frac{\|\xi(\nu,C\pi)\|}{\beta},$$

where $\|\cdot\|$ is defined as in the proof of Theorem 14.1 and $\beta > 0$ is defined in Proposition 5.3. Notice that $\beta \to 1$ as $\nu \to 0$ and

$$|C\pi|^2 = \pi^\top C^\top C\pi = \pi^\top P^{\frac{1}{2}}P^{-\frac{1}{2}}C^\top CP^{-\frac{1}{2}}P^{\frac{1}{2}}\pi \leq \lambda^* := \lambda_{\max}(P^{-\frac{1}{2}}C^\top CP^{-\frac{1}{2}}).$$

By Mean Value theorem, we have

$$\|\xi(\nu, C\pi)\|^2 = \|\xi(0, C\pi)\|^2 + 2\xi^\top(\tilde{\nu}, C\pi)P\left.\frac{\partial\xi(\tilde{\nu},C\pi)}{\partial\tilde{\nu}}\right|_{\tilde{\nu}\in[-|\nu|,|\nu|]}\nu \leq \gamma|\nu|,$$

where

$$\gamma := \sup_{|\tilde{\nu}|\leq|\nu|, |C\pi|\leq\sqrt{\lambda^*}}\left|2\xi^\top(\tilde{\nu}, C\pi)P\frac{\partial\xi(\tilde{\nu},C\pi)}{\partial\tilde{\nu}}\right| < +\infty$$

and $\gamma \to 0$ as $\nu \to 0$. Therefore, we conclude

$$\tfrac{d\|\varepsilon\|_{\mathbf{d}}}{dt} \le -\rho\|\varepsilon\|_{\mathbf{d}}^{1+\nu} + \|\varepsilon\|_{\mathbf{d}}^{1+\nu}\tfrac{\|\xi(\nu,C\pi)\|}{\beta} \le -\left(\rho - \tfrac{\sqrt{K\nu}}{\beta}\right)\|\varepsilon\|_{\mathbf{d}}^{1+\nu} \le -\frac{\rho}{2}\|\varepsilon\|_{\mathbf{d}}^{1+\nu}$$

if $|\nu| < (0.5\rho\beta)^2/\gamma$. The proof is complete. □

The singularity of the time-varying homogeneous observer (14.8) has been eliminated. The static homogeneous observer (14.15) has continuous right-hand side for $\nu > -\frac{1}{\tilde{n}}$. For $\nu = -\frac{1}{\tilde{n}}$ solutions of the system (14.15) are defined in the sense of Filippov (see Chap. 2 or [18]).

Corollary 14.1 *Let the error equation* (14.16) *of the static homogeneous observer* (14.15) *for the nominal plant* (14.5) *be globally asymptotically stable. Then the error equation*

$$\dot{\varepsilon} = A_0\varepsilon + |C\varepsilon + q_2|^{\nu-1}\mathbf{d}(-\ln|C\varepsilon + q_2|)L(C\varepsilon + q_2) + L_0q_2 + q_1, \tag{14.17}$$

for the perturbed plant (14.6) *is ISS with respect to* $q = (q_1^\top, q_2^\top)^\top \in L^\infty(\mathbb{R}; \mathbb{R}^{n+k})$ *provided that* $\nu > -\frac{1}{\tilde{n}}$.

Proof Let us denote

$$\tilde{q} = (\tilde{q}_1^\top, \tilde{q}_2^\top)^\top, \quad \tilde{q}_1 = L_0q_2 + q_1, \quad \tilde{q}_2 = q_2$$

and rewrite the system (14.17) as follows:

$$\dot{\varepsilon} = f(\varepsilon, \tilde{q}) := A_0\varepsilon + |C\varepsilon + \tilde{q}_2|^{\nu-1}\mathbf{d}(-\ln|C\varepsilon + \tilde{q}_2|)L(C\varepsilon + \tilde{q}_2) + \tilde{q}_1. \tag{14.18}$$

For $\nu > -\frac{1}{\tilde{n}}$ the one-parameter group

$$\tilde{\mathbf{d}}(s) = \begin{pmatrix} e^{\nu s}\mathbf{d}(s) & \mathbf{0} \\ \mathbf{0} & e^s I_k \end{pmatrix}, \quad s \in \mathbb{R}$$

is a dilation in $\mathbb{R}^{n+k}$. Since, by assumption, the unperturbed ($\tilde{q} = \mathbf{0}$) system (14.18) is globally asymptotically stable and

$$f(\mathbf{d}(s)\varepsilon, \tilde{\mathbf{d}}(s)\tilde{q}) =$$

$$A_0\mathbf{d}(s)\varepsilon + |C\mathbf{d}(s)\varepsilon + e^s\tilde{q}_2|^{\nu-1}\mathbf{d}(-\ln|C\mathbf{d}(s)\varepsilon + e^s\tilde{q}_2|)L(C\mathbf{d}(s)\varepsilon + e^s\tilde{q}_2) + e^{\nu s}\mathbf{d}(s)\tilde{q}_1$$

$$= e^{\nu s}\mathbf{d}(s)f(\varepsilon, \tilde{q})$$

for all $s \in \mathbb{R}$, $\varepsilon \in \mathbb{R}^n$, $\tilde{q} \in \mathbb{R}^{n+k}$, then, by Theorem 7.13, the error equation (14.18) is ISS with respect to $\tilde{q}$. The proof is complete. □

Remark 14.1 By Theorem 7.5, asymptotic stability and homogeneity of the error equation (14.16) imply finite-time (for $\nu < 0$) or nearly fixed-time (for $\nu > 0$) convergence of the static homogeneous observer to the state $x(t)$ of the system (14.5). By Corollary 7.14, the error equation (14.17) is finite-time ISS if $\nu < 0$ and nearly fixed-time ISS if $\nu > 0$ with respect to measurement noise and additive perturbations.

Formally, an asymptotic stability of the error dynamics of the static homogeneous observer is proven only for some ν being sufficiently close to zero, so the appropriate value ν is unknown a priori. For practice, it is important to provide explicit formulas for selection of all observer parameters. This problem is solved for the filtering/extended homogeneous observer given in the next subsection.

Example 14.1 Let us consider the mechanical model of a cart pendulum

$$\begin{aligned}(M+m)\ddot{y} + b\dot{y} + ml\ddot{\theta}\cos\theta - ml\dot{\theta}^2\sin\theta &= F_{\text{in}}\\ (I+ml^2)\ddot{\theta} + mgl\sin\theta + ml\ddot{y}\cos\theta &= 0,\end{aligned}$$

where F_{in} represents the input force, the state vector is given by $x = [y, \theta, \dot{y}, \dot{\theta}]^T$, where as usual $(y, \dot{y})$ represents the position and the velocity of the cart and $(\theta, \dot{\theta})$ the angle (from the vertical down position) and the angular velocity of the pendulum. The model parameters are $M = 0.5$ kg—mass of the cart, $m = 0.2$ kg—mass of the pendulum, $b = 0.1$ N/m/s—cart friction coefficient, $l = 0.3$ m—length to pendulum center of mass, $I = 0.006$ kg—moment of inertia of the pendulum. It is assumed that only position and angle can be measured directly and need to estimate the velocities $\dot{y}$ and $\dot{\theta}$, and $F_{\text{in}} = 0$.

The linearization of the model around the downward equilibrium point $x_0 = \mathbf{0} \in \mathbb{R}^4$ gives the following parameters:

$$A = \begin{pmatrix} 0 & 0 & 1 & 0\\ 0 & 0 & 1\\ 0 & 2.672 & -0.1818 & 0\\ 0 & -31.181 & 0.4545 & 0\end{pmatrix}, \quad B = \begin{pmatrix} 0\\ 0\\ 1.818\\ -4.545\end{pmatrix}, \quad C = \begin{pmatrix} 1 & 0 & 0 & 0\\ 0 & 1 & 0 & 0\end{pmatrix}.$$

Solving the linear algebraic equation (14.10) we derive

$$L_0 = \begin{pmatrix} 0.1818 & 0\\ -0.4545 & 0\\ -0.0331 & -2.6720\\ 0.0826 & 31.1810\end{pmatrix}, \quad G_0 = \begin{pmatrix} -0.3067 & 0.0000 & 0.1467 & -0.3666\\ -0.0000 & -0.5000 & -0.0000 & -0.0000\\ 0.1467 & 0.0000 & 0.4733 & 0.0667\\ -0.3666 & 0.0000 & 0.0667 & 0.3334\end{pmatrix}.$$

We design the static homogeneous observer with $\nu = -(3.1)^{-1}$ and the observer gain

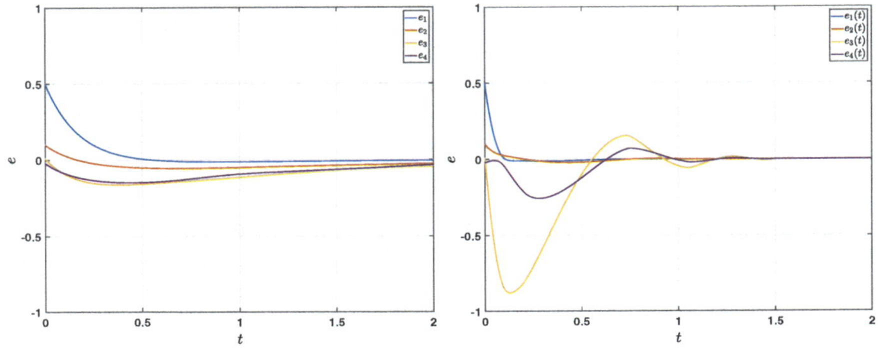

Fig. 14.1 Observation errors of Luenberger (left) and homogeneous (right) observers for the model of the cart pendulum

$$L = \begin{pmatrix} -6.1267 & -0.8873 \\ -0.8873 & -2.1313 \\ -2.5036 & -1.0373 \\ -1.3680 & -1.9992 \end{pmatrix}.$$

The results of numerical simulations are presented in Fig. 14.1, the evolution of the observation errors of the linear Luenberger observer ($\nu = 0$) with the gain L and the homogeneous finite-time ($\nu < 0$) observer with the same gain L are compared. As expected the homogeneous observer demonstrates faster convergence.

14.1.3 Extended Homogeneous Observer

Inspired by [13, 15, 19], let us consider the filtering homogeneous observer

$$\begin{cases} \dot{z} = Az + Bu + L_0(Cz - y) + \sigma^{\nu-1}\mathbf{d}(\ln\sigma)L\psi, & z(0)=z_0, \\ \dot{\psi} = Cz - y + \sigma^{\nu}\tilde{L}\psi, & \psi(0)=\psi_0, \\ \dot{\xi} = \left|\psi^\top\left(Cz-y+\sigma^{\nu}\tilde{L}\psi\right)\right| - \gamma\left|\xi-\frac{|\psi|^2}{2}\right|^{\frac{\nu}{2}}\left(\xi-\frac{|\psi|^2}{2}\right), & \xi(0)=\xi_0, \\ \sigma = \sqrt{\max\left\{\xi, \frac{|\psi|^2}{2}\right\}}, & t > 0, \end{cases} \tag{14.19}$$

where x, A, B, C, L_0, L, ξ are as before, $\psi(t) \in \mathbb{R}^k$ is the output filter variable, $|\psi|^2 = \psi^\top\psi$, $\tilde{L} \in \mathbb{R}^{k\times k}$, $\gamma > 0$ and the dilation $\mathbf{d}$ is defined below. As before, the variable ξ specifies a homogeneous scaling of the observer gains, but this variable also depends on the state of the output filter with the state ψ. This dependence prevents the infinite grow of the gain $\sigma^{\nu-1}\mathbf{d}(\ln\sigma)L$ in the case of noised measurement, while the filter itself improves the output error estimate [13, 19].

The extended error equation has the form:

$$\dot{\varepsilon}_\sigma = g(\varepsilon_\sigma) := \begin{pmatrix} C\varepsilon + |\sigma|^\nu \tilde{L}\psi \\ A_0\varepsilon + |\sigma|^{\nu-1}\mathbf{d}(\ln|\sigma|)L\psi \\ \left|\psi^\top\left(C\varepsilon + \sigma^\nu \tilde{L}\psi\right)\right| - \gamma\left|\xi - \frac{|\psi|^2}{2}\right|^{\frac{\nu}{2}}\left(\xi - \frac{|\psi|^2}{2}\right) \end{pmatrix}, \tag{14.20}$$

where $A_0 = A + L_0 C$ and

$$\varepsilon_\sigma = \begin{pmatrix} \psi \\ \varepsilon \\ \xi \end{pmatrix}, \quad \varepsilon = z - x.$$

Theorem 14.3 ([15]) *Let $G_0 \in \mathbb{R}^{n\times n}$, ν and $L_0 \in \mathbb{R}^{n\times k}$ be defined as in Theorem 14.1 and*

$$G_\mathbf{d} = I_n + \nu(G_0 + I_n), \quad \mathbf{d}(s) = e^{sG_\mathbf{d}}, \; s \in \mathbb{R}, \tag{14.21}$$

$$\bar{G} = \begin{pmatrix} I_k & \mathbf{0} \\ \mathbf{0} & G_\mathbf{d} \end{pmatrix} \in \mathbb{R}^{(k+n)\times(k+n)}. \tag{14.22}$$

The error equation (14.20) *is*

- $\mathbf{d}_\sigma$*-homogeneous of degree ν with respect to the dilation*

$$\mathbf{d}_\sigma(s) = \begin{pmatrix} e^{s\bar{G}} & \mathbf{0} \\ \mathbf{0} & e^{2s} \end{pmatrix}, \quad s \in \mathbb{R}; \tag{14.23}$$

- *continuous (i.e., $g_\sigma \in C(\mathbb{R}^{k+n+1}, \mathbb{R}^{k+n+1})$) for $\nu > -\frac{1}{\tilde{n}+1}$ and piece-wise continuous if $\nu = -\frac{1}{\tilde{n}+1}$;*
- *globally uniformly finite-time stable for $\nu \in \left[-\frac{1}{\tilde{n}+1}, 0\right)$ provided that the observer gains are defined as follows:*

$$\begin{pmatrix} \tilde{L} \\ L \end{pmatrix} = \bar{L} := \bar{P}^{-1}\bar{Y}, \tag{14.24}$$

where the pair $\bar{P} \in \mathbb{R}^{(k+n)\times(k+n)}$ and $\bar{Y} \in \mathbb{R}^{(k+n)\times k}$ is a solution of the following algebraic system

$$\begin{array}{c} \bar{P}\bar{A}_0 + \bar{A}_0^\top \bar{P} + \bar{Y}\bar{C} + \bar{C}^\top \bar{Y}^\top + \rho(\bar{P}\bar{G} + \bar{G}^\top \bar{P}) = \mathbf{0}, \\ \bar{P}\bar{G} + \bar{G}^\top \bar{P} \succ 0, \quad \bar{P} = \bar{P}^\top \succ 0 \end{array} \tag{14.25}$$

with $\rho > \frac{\gamma}{2}$ and

$$\bar{A}_0 = \begin{pmatrix} \mathbf{0} & C \\ \mathbf{0} & A_0 \end{pmatrix} \in \mathbb{R}^{(k+n)\times(k+n)}, \quad \bar{C} = (I_k \;\; \mathbf{0}) \in \mathbb{R}^{k\times(n+k)}. \tag{14.26}$$

Proof (1) First of all notice that $G_{\mathbf{d}}$ is anti-Hurwitz for $\nu \geq -\frac{1}{\tilde{n}+1}$, so $\mathbf{d}$ is a dilation in $\mathbb{R}^n$, $\bar{\mathbf{d}}(s) = e^{s\bar{G}}$ is a dilation in $\mathbb{R}^{k+n}$ and $\mathbf{d}_\sigma$ is a dilation in $\mathbb{R}^{k+n+1}$. Moreover, by construction, we have

$$CG_{\mathbf{d}} = (1+\nu)C, \quad \bar{A}_0\bar{G} = (\bar{G} + \nu I_{k+n})\bar{A}_0, \quad \bar{C}\bar{G} = \bar{C}$$

and, consequently,

$$C\mathbf{d}(s) = e^{(1+\nu)s}C, \quad \bar{A}_0\bar{\mathbf{d}}(s) = e^{\nu s}\bar{\mathbf{d}}(s)\bar{A}_0, \quad \bar{C}\bar{\mathbf{d}}(s) = e^s\bar{C}, \quad \forall s \in \mathbb{R}. \tag{14.27}$$

Denoting

$$\varepsilon_\psi = \begin{pmatrix} \psi \\ x - z \end{pmatrix}$$

the vector field g_σ can be rewritten as follows:

$$g_\sigma(\varepsilon_\sigma) = \begin{pmatrix} \bar{A}_0\varepsilon_\psi + \sigma^{\nu-1}\bar{\mathbf{d}}(\ln\sigma)\bar{L}\bar{C}\varepsilon_\psi \\ \left|\varepsilon_\psi^\top\bar{C}^\top\bar{C}(\bar{A}_0 + |\sigma|^\nu\tilde{L}\bar{C})\varepsilon_\psi\right| - \gamma\left|\xi - \frac{|\bar{C}\varepsilon_\psi|^2}{2}\right|^{\frac{\nu}{2}}\left(\xi - \frac{|\bar{C}\varepsilon_\psi|^2}{2}\right) \end{pmatrix}.$$

Since $\sigma(\varepsilon_\sigma) = \max\left\{\xi, \frac{|\bar{C}\varepsilon_\psi|^2}{2}\right\}$ then, due to the identity $\bar{C}\bar{\mathbf{d}}(s) = e^s\bar{C}$, we have

$$\sigma(\mathbf{d}_\sigma(s)\varepsilon_\sigma) = \max\left\{e^{2s}\xi, \frac{|\bar{C}e^{s\bar{G}}\varepsilon_\psi|^2}{2}\right\} = \max\left\{e^{2s}\xi, \frac{|e^s\bar{C}\varepsilon_\psi|^2}{2}\right\} = e^{2s}\sigma$$

for all $s \in \mathbb{R}$ and for all $\varepsilon_\sigma \in \mathbb{R}^{k+n+1}$. Hence, using (14.27) we derive

$$g_\sigma(\mathbf{d}_\sigma(s)\varepsilon_\sigma)$$

$$= \begin{pmatrix} \bar{A}_0e^{s\bar{G}}\varepsilon_\psi + |\sigma(\mathbf{d}_\sigma(s)e_\sigma)|^{\nu-1}e^{\bar{G}\ln|\sigma(\mathbf{d}_\sigma(s)e_\sigma)|}L\bar{C}e^{s\bar{G}}\varepsilon_\psi \\ \gamma_2\left|\bar{C}(\bar{A}_0 + |\sigma(\mathbf{d}_\sigma(s)e_\sigma)|^\nu\tilde{L}\bar{C})e^{s\bar{G}}\varepsilon_\psi\right| - \gamma_1\left|e^s\xi - \gamma_2\left|\bar{C}e^{s\bar{G}}\varepsilon_\psi\right|\right|^\nu\left(e^s\xi - \gamma_2\left|\bar{C}e^{s\bar{G}}\varepsilon_\psi\right|\right) \end{pmatrix}$$

$$= \begin{pmatrix} e^{\nu s}e^{s\bar{G}}\bar{A}_0\varepsilon_\psi + e^{\nu s}e^{s\bar{G}}|\sigma|^{\nu-1}e^{\bar{G}\ln|\sigma|}L\bar{C}\varepsilon_\psi \\ \gamma_2\left|\bar{C}\left(e^{\nu s}e^{s\bar{G}}\bar{A}_0 + e^{(\nu+1)s}|\sigma|^\nu\tilde{L}\bar{C}\right)\varepsilon_\psi\right| - \gamma_1\left|e^s\xi - e^s\gamma_2|\bar{C}\varepsilon_\psi|\right|^\nu(e^s\xi - e^s\gamma_2|\bar{C}\varepsilon_\psi|) \end{pmatrix}$$

$$= \begin{pmatrix} e^{\nu s}e^{s\bar{G}}\left(\bar{A}_0\varepsilon_\psi + |\sigma|^{\nu-1}e^{\bar{G}\ln|\sigma|}L\bar{C}\varepsilon_\psi\right) \\ e^{(\nu+1)s}\left(\gamma_2\left|\bar{C}(\bar{A}_0 + |\sigma|^\nu\tilde{L}\bar{C})\varepsilon_\psi\right| - \gamma_1\left|\xi - \gamma_2|\bar{C}\varepsilon_\psi|\right|^\nu(\xi - \gamma_2|\bar{C}\varepsilon_\psi|)\right) \end{pmatrix}$$

$$= e^{\nu s}\mathbf{d}_\sigma(s)g_\sigma(\varepsilon_\sigma).$$

(2) Using structure of the matrix G_0 (see the proof of Theorems 14.1 and 9.1) we conclude

$$e^{\nu s}e^{s\bar{G}} \to \mathbf{0} \quad \text{as} \quad s \to -\infty$$

provided that $\nu > -\frac{1}{\tilde{n}+1}$, but

$$e^{\nu s}\bar{\mathbf{d}}(s) \to \text{const} \quad \text{as} \quad s \to -\infty$$

for $\nu = -\frac{1}{\tilde{n}+1}$. Since, by construction, $\sigma \geq \gamma_2|\bar{C}\varepsilon_\phi|$ then the mapping

$$e_\sigma \mapsto |\sigma|^{\nu-1}e^{\bar{G}\ln|\sigma|}L\bar{C}\varepsilon_\psi$$

is continuous for $\nu > -\frac{1}{\tilde{n}+1}$ and piece-wise continuous for $\nu > -\frac{1}{\tilde{n}+1}$. The same conclusions are obviously valid for the vector field g_σ. Therefore, the error system (14.20) (as well as the system (14.19)) has solutions defined at least locally in time. For $\nu = -\frac{1}{\tilde{n}+1}$ the solutions are understood in the sense of Filippov (see, Chap. 2 or [18]). Moreover, for $\nu < 0$ the error equation (14.20) (as well as the observer Eq. (14.19)) is forward complete due to Corollary 7.2.

(3) Let us how that the error equation (14.20) is globally uniformly finite-time stable. In the view of Theorem 7.4, it is sufficient to show any trajectory of the error system reaches the origin in a finite time and the corresponding reaching time is a locally bounded function of the initial state.

(a) First of all notice that for $\delta = \xi - \frac{|\bar{C}\varepsilon_\psi|^2}{2}$ we have

$$\dot{\delta} = -\gamma|\delta|^{\frac{\nu}{2}}\delta + \left|\frac{d}{dt}\frac{|\psi|^2}{2}\right| - \frac{d}{dt}\frac{|\psi|^2}{2}, \quad \delta(0) = \xi(0) - \frac{|\bar{C}\varepsilon_\psi(0)|^2}{2}. \tag{14.28}$$

Taking into account $\nu < 0$ we conclude that

$$\xi(t) \geq \frac{|\psi(t)|^2}{2}, \quad \forall t \geq T_\delta, \quad T_\delta(\varepsilon_\sigma(0)) = \max\left\{0, -\frac{2\left(\xi(0) - \frac{|\bar{C}\varepsilon_\psi(0)|^2}{2}\right)}{-\nu\gamma}\right\},$$

and $\sigma(t) = \sqrt{\xi(t)}$ for all $t \geq T_\delta(\varepsilon_\psi(0), \xi(0))$. Notice that (see the proof of Theorem 7.4) there exists $M_1 > 0$ such that $\|\varepsilon_\sigma(T_\delta)\|_{\mathbf{d}_\sigma}^{-\nu} \leq \|\varepsilon_\sigma(0)\|_{\mathbf{d}_\sigma}^{-\nu} + M_1 T_\delta$. Moreover, $\xi(t) \overset{a.e.}{>} 0$ as long as $\varepsilon_\psi(t) \neq \mathbf{0}$. Indeed, otherwise there exists a time interval $[a, b]$ such that $\xi(t) = 0, \forall t \in [a, b]$. This implies $\dot{\psi}(t) = \psi(t) = 0$ and $C(x(t) - z(t)) = 0$ for all $t \in [a, b]$. Due to observability of the pair $\{A, C\}$, this is possible only if $\varepsilon_\psi(t) = \mathbf{0}$ for all $t \in [a, b]$.

(b) For any $\nu > -\frac{1}{\tilde{n}+1}$ we have

$$\Xi(r) = r^\nu\mathbf{d}(\ln r) - rI_{n+k} \to \mathbf{0} \quad \text{as} \quad r \to 0.$$

Let a constant $\ell \in (0, 1)$ be selected such that

$$\|r^\nu\mathbf{d}(\ln r) - rI_{n+k}\| \leq \frac{\rho\beta}{2\|\bar{L}\|}, \quad \forall r \in [0, \ell),$$

where $\beta = \frac{1}{2}\lambda_{\min}(\bar{P}^{1/2}\bar{G}\bar{P}^{-1/2} + \bar{P}^{-1/2}\bar{G}^{\top}\bar{P}^{1/2}) > 0$ and $\|\bar{L}\| = \sqrt{\lambda_{\max}(\bar{L}^{\top}\bar{P}\bar{L})}$.

Let the $\bar{\mathbf{d}}$-homogeneous norm $\|\cdot\|_{\bar{\mathbf{d}}}$ be induced by the weighted Euclidean norm $\|\varepsilon_{\psi}\| = \sqrt{\varepsilon_{\psi}^{\top}\bar{P}\varepsilon_{\psi}}$.

Let us show that for any $\varepsilon_{\sigma}(T_{\delta}) \neq \mathbf{0}$ there exists $T_1 \in [T_{\delta}, T_{\delta} + T_{\xi}]$,

$$T_{\xi}(\varepsilon_{\sigma}(T_{\delta})) = \frac{2\left\|\varepsilon_{\psi}(T_{\delta})\right\|_{\bar{\mathbf{d}}}^{-\nu}}{-\nu\rho}$$

such that either $\varepsilon_{\psi}(T_1) = \mathbf{0}$ or $\xi(T_1) > \ell^2\left\|\varepsilon_{\psi}(T_1)\right\|_{\bar{\mathbf{d}}}^2$. Suppose the contrary, i.e., $\varepsilon_{\psi}(t) \neq \mathbf{0}$ for all $t \in [T_{\delta}, T_{\delta} + T_{\xi}]$ and $\xi(t) \leq \ell^2\left\|\varepsilon_{\psi}(t)\right\|_{\bar{\mathbf{d}}}^2$, $\forall t \in [T_{\delta}, T_{\delta} + T_{\xi}]$. In this case, using the formula (5.21) we have

$$\frac{d}{dt}\left\|\varepsilon_{\psi}\right\|_{\bar{\mathbf{d}}} = \left\|\varepsilon_{\psi}\right\|_{\bar{\mathbf{d}}} \frac{\pi^{\top}\bar{P}\bar{\mathbf{d}}(-\ln\|\varepsilon_{\psi}\|_{\bar{\mathbf{d}}})(\bar{A}_0\varepsilon_{\psi} + \sigma^{\nu-1}\bar{\mathbf{d}}(\ln\sigma)\bar{L}\bar{C}\varepsilon_{\psi})}{\pi^{\top}\bar{P}\bar{G}\pi}$$

$$= \left\|\varepsilon_{\psi}\right\|_{\bar{\mathbf{d}}}^{1+\nu} \frac{\pi^{\top}\bar{P}\bar{A}_0\pi + \pi^{\top}\bar{P}\left(\frac{\sigma}{\|\varepsilon_{\psi}\|_{\bar{\mathbf{d}}}}\right)^{\nu-1}\bar{\mathbf{d}}\left(\ln\frac{\sigma}{\|\varepsilon_{\psi}\|_{\bar{\mathbf{d}}}}\right)\bar{L}\bar{C}\pi}{\pi^{\top}\bar{P}\bar{G}\pi},$$

where $\pi = \bar{\mathbf{d}}\left(-\ln\left\|\varepsilon_{\psi}\right\|_{\bar{\mathbf{d}}}\right)\varepsilon_{\psi}$ and the identities (14.27) are utilized on the last step. Using (14.25) we derive

$$\frac{d\|\varepsilon_{\psi}\|_{\bar{\mathbf{d}}}}{dt} \leq \left\|\varepsilon_{\psi}\right\|_{\bar{\mathbf{d}}}^{1+\nu}\left(\frac{\pi^{\top}\bar{P}\left(\left(\frac{\sigma}{\|\varepsilon_{\psi}\|_{\bar{\mathbf{d}}}}\right)^{\nu-1}\bar{\mathbf{d}}\left(\ln\frac{\sigma}{\|\varepsilon_{\psi}\|_{\bar{\mathbf{d}}}}\right) - I_{k+n}\right)\bar{L}\bar{C}\pi}{\pi^{\top}\bar{P}\bar{G}\pi} - \rho\right)$$

$$= \left\|\varepsilon_{\psi}\right\|_{\bar{\mathbf{d}}}^{1+\nu}\left(\frac{\pi^{\top}\bar{P}\Xi\left(\frac{\sigma}{\|\varepsilon_{\psi}\|_{\bar{\mathbf{d}}}}\right)\bar{L}\frac{\bar{C}\varepsilon_{\psi}}{|\sigma|}}{\pi^{\top}\bar{P}\bar{G}\pi} - \rho\right).$$

Since $\|\Xi(r)\| = \|r^{\nu}\mathbf{d}(\ln r) - rI_{n+k}\| \leq \frac{\rho}{2\beta\|L\|}$, $\forall r \in [0, \ell]$ then for $t \in [T_{\delta}, T_{\delta} + T_{\xi}]$ we have $\sigma = \sqrt{\xi} \geq \frac{|C\varepsilon_{\psi}|}{\sqrt{2}}$, $\frac{\sigma}{\|\varepsilon_{\psi}\|_{\bar{\mathbf{d}}}} = \frac{\sqrt{\xi}}{\|\varepsilon_{\psi}\|_{\bar{\mathbf{d}}}} \leq \ell$ and

$$\left\|\Xi\left(\frac{\sigma}{\|\varepsilon_{\psi}\|_{\bar{\mathbf{d}}}}\right)\bar{L}\frac{\bar{C}\varepsilon_{\psi}}{|\sigma|}\right\| \leq \frac{\rho\beta}{2}.$$

Therefore, taking into account $\pi^{\top}P\pi = 1$ we derive

$$\frac{d}{dt}\left\|\varepsilon_{\psi}\right\|_{\bar{\mathbf{d}}} \leq -\frac{\rho}{2}\left\|\varepsilon_{\psi}\right\|_{\bar{\mathbf{d}}}^{1+\nu}, \forall t \in [T_{\delta}, T_{\delta} + T_{\xi}],$$

but the latter implies there exists $T_1 \in [T_{\delta}, T_{\delta} + T_{\xi}]$ such that $\varepsilon_{\psi}(T_1) = \mathbf{0}$ and we obtain the contradiction. Therefore, either $\varepsilon_{\psi}(T_1) = \mathbf{0}$ or $\xi(T_1) > \ell^2\|\varepsilon_{\psi}(T_1)\|_{\bar{\mathbf{d}}}^2$. In the first case, the state $\varepsilon_\psi(t)$ will remain zero at least till the time instant $T_1 + \frac{\xi^{-\nu}(T_1)}{-\nu\gamma}$ then $\xi(t)$ will reach zero too, i.e., the error variable $\varepsilon_\sigma(t)$ reaches zero a finite instant of time being a locally bounded function of $\varepsilon_\sigma(0)$. Notice that (see the proof of Theorem 7.4) there exists $M_2 > 0$:

$$\xi^{-\nu/2}(T_1) \leq \|\varepsilon_\sigma(T_1)\|_{\mathbf{d}_\sigma} := \xi^{-\nu/2}(T_1) + \|\varepsilon_\psi(t)\|_{\bar{\mathbf{d}}} \leq \|\varepsilon_\sigma(0)\|_{\bar{\mathbf{d}}} + M_2 T_1.$$

(c) Let us study the case $\xi(T_1) > \ell^2\|\varepsilon_\psi(T_1)\|_{\bar{\mathbf{d}}}^2$ and consider the new variable

$$\epsilon = e^{-\frac{1}{2}\bar{G}\ln\xi}\varepsilon_\psi,$$

which is well defined for $t \geq T_1$. In this case,

$$\|\mathbf{d}(\ln\ell)\epsilon(T_1)\|_{\bar{\mathbf{d}}} = \ell\|e^{-\frac{1}{2}\bar{G}\ln\xi(t)}\varepsilon_\psi(T_1)\|_{\bar{\mathbf{d}}} < 1 \quad \Leftrightarrow \quad \xi(T_1) > \ell^2\|\varepsilon_\psi(T_1)\|_{\bar{\mathbf{d}}}^2$$

and (due the estimates (5.7)) we have

$$\ell^\alpha\|\epsilon(T_1)\| \leq \|\mathbf{d}(\ln\ell)\epsilon(T_1)\| < 1,$$

where $\alpha = \frac{1}{2}\lambda_{\max}(\bar{P}^{1/2}\bar{G}\bar{P}^{-1/2} + \bar{P}^{-1/2}\bar{G}^\top\bar{P}^{1/2}) > 0$. Taking into account

$$\bar{C}\varepsilon_\psi = \bar{C}e^{\frac{1}{2}\bar{G}\ln\xi}e^{-\frac{1}{2}\bar{G}\ln\xi}\varepsilon_\psi = \xi^{\frac{1}{2}}\bar{C}\epsilon$$

and

$$\bar{C}\bar{A}_0\varepsilon_\psi = \bar{C}\bar{A}_0 e^{\frac{1}{2}\bar{G}\ln\xi}e^{-\frac{1}{2}\bar{G}\ln\xi}\varepsilon_\psi = \xi^{\frac{\nu}{2}}\bar{C}e^{\frac{1}{2}\bar{G}\ln\xi}\bar{A}_0 e^{-\frac{1}{2}\bar{G}\ln\xi}\varepsilon_\psi = \xi^{\frac{1+\nu}{2}}\bar{C}\bar{A}_0\epsilon$$

we derive

$$\varepsilon_\psi^\top\bar{C}^\top\left(\bar{C}(\bar{A}_0 + |\sigma|^\nu\tilde{L}\bar{C})\varepsilon_\psi\right) = \xi^{1+\frac{\nu}{2}}\epsilon^\top\bar{C}^\top(\bar{C}\bar{A}_0 + L\bar{C})\epsilon.$$

Since $\xi \geq \frac{|\bar{C}\varepsilon_\psi|^2}{2} \Rightarrow 1 \geq \frac{|\bar{C}\epsilon|^2}{2}$ then

$$\dot{\xi} = (h(\epsilon) - \gamma)\xi^{1+\frac{\nu}{2}}, \quad \forall t \geq T_1 \geq T_\delta,$$

where

$$h(\epsilon) = |\epsilon^\top\bar{C}^\top\bar{C}(\bar{A}_0 + \bar{L}\bar{C})\epsilon| + \gamma - \gamma\left(1 - \frac{|\bar{C}\epsilon|^2}{2}\right)^{1+\frac{\nu}{2}} \geq 0.$$

For $1 \geq \frac{|\bar{C}\epsilon|^2}{2}$ and $-2 < \nu < 0$, we have

$$1 - \left(1 - \frac{|\bar{C}\epsilon|^2}{2}\right)^{1+\frac{\nu}{2}} \leq \frac{|\bar{C}\epsilon|^2}{2}.$$

Therefore, there exists a constant $\kappa > 0$ (dependent only on $\bar{P}$, $\bar{C}$, $\bar{L}$, $\bar{A}_0$ and $\gamma > 0$) such that

$$1 \geq \frac{|\bar{C}\epsilon|^2}{2} \quad \Rightarrow \quad h(\epsilon) \leq \kappa\|\epsilon\|^2$$

and

$$\dot{\xi} \leq \xi^{1+\frac{\nu}{2}}(\kappa\|\epsilon\|^2 - \gamma), \quad \forall t \in [T_1, T). \tag{14.29}$$

The time derivative of ϵ satisfies the differential equation:

$$\dot{\epsilon} = -\tfrac{1}{2}\tfrac{\dot{\xi}}{\xi}\bar{G}\epsilon + e^{-\frac{1}{2}\bar{G}\ln\xi}(\bar{A}_0\varepsilon_\psi + \xi^{\frac{\nu-1}{2}}e^{\frac{1}{2}\bar{G}\ln\xi}L\bar{C}\varepsilon_\psi)$$

$$= \xi^{\frac{\nu}{2}}\left(\bar{A}_0 + L\bar{C} + (0.5\gamma - h(\epsilon))\bar{G}\right)\epsilon.$$

Considering $V = q\epsilon^\top \bar{P}\epsilon$ with $q = \frac{\kappa(1-0.5\nu)}{\beta(\rho-0.5\gamma)}$ we derive

$$\dot{V} = 2q\xi^{\frac{\nu}{2}}\epsilon^\top\bar{P}\left(\bar{A}_0 + L\bar{C} + \tfrac{\gamma-h(\epsilon)}{2}\bar{G}\right)\epsilon = -\left(0.5h(\epsilon) + \rho - 0.5\gamma\right)\xi^{\frac{\nu}{2}}2q\epsilon^\top\bar{P}\bar{G}\epsilon,$$

where the identity (14.25) is utilized on the last step. For $\rho > \frac{\gamma}{2}$ we have $h(\epsilon) \geq 0$ and

$$\dot{V} \leq -\beta\left(0.5h(\epsilon) + \rho - 0.5\gamma\right)\xi^{\frac{\nu}{2}}V \leq -\beta(\rho - 0.5\gamma)\xi^{\frac{\nu}{2}}V, \tag{14.30}$$

where $\beta = \frac{1}{2}\lambda_{\min}(\bar{P}^{1/2}\bar{G}\bar{P}^{-1/2} + \bar{P}^{-1/2}\bar{G}^\top\bar{P}^{1/2}) > 0$. The latter means that the function $t \mapsto V(t)$ is strictly decreasing on $[T_1, T)$ and

$$\|\epsilon(t)\| \leq \|\epsilon(T_2)\|, \quad \forall T_2 \in (T_1, T), \quad \forall t \in [T_2, T),$$

where $\|x\| = \sqrt{x^\top Px}$ is the weighted Euclidean norm in $\mathbb{R}^{k+n}$.

Considering

$$W = \xi^{1-\frac{\nu}{2}}e^V$$

we derive

$$\dot{W} = (1 - 0.5\nu)\xi^{-\frac{\nu}{2}}\dot{\xi}e^V + \xi^{1-\frac{\nu}{2}}e^V\dot{V}$$

$$\leq (1 - 0.5\nu)e^V\xi(q^{-1}\kappa V - \gamma) - \beta(\rho - 0.5\gamma)\xi V e^V$$

$$= -(1 - 0.5\nu)\gamma\xi e^V = -(1 - 0.5\nu)\gamma\xi^{0.5\nu}W.$$

Therefore, we have the following system of inequalities:

$$\begin{cases} \dot{V} \leq -\beta(\rho - 0.5\gamma)W^{\frac{0.5\nu}{1-0.5\nu}}e^{\frac{-0.5\nu}{1-0.5\nu}V}V, \\ \dot{W} \leq -(1 - 0.5\nu)\gamma e^{\frac{-0.5\nu}{1-0.5\nu}V}W^{1+\frac{0.5\nu}{1-0.5\nu}}, \end{cases} \quad \forall t \in [T_1, T).$$

Hence, taking into account $V \geq 0 \Rightarrow e^V \geq 1$ we conclude that

$$\exists T \in \left(T_1, \frac{(1-0.5\nu)W^{-\frac{0.5\nu}{1-0.5\nu}}(T_1)}{(1-0.5\nu)\gamma(-0.5\nu)}\right),$$

such that $W(t) \to 0$ and $V(t) \to 0$ as $t \to T$ with or, equivalently, $\epsilon(t) \to \mathbf{0}$ and $\xi(t) \to 0$ as $t \to T$. Taking into account $\|\varepsilon_\psi(t)\|_{\tilde{\mathbf{d}}} = \xi(t)\|\epsilon(t)\|_{\tilde{\mathbf{d}}}$ and

$$\frac{(1-0.5\nu)W^{-\frac{0.5\nu}{1-0.5\nu}}(T_1)}{(1-0.5\nu)\gamma(-0.5\nu)} \leq \frac{(1-0.5\nu)\left(\xi(T_1)^{1-0.5\nu}e^{\ell^{-2\alpha}}\right)^{-\frac{0.5\nu}{1-0.5\nu}}}{(1-0.5\nu)\gamma(-0.5\nu)},$$

we conclude any trajectory of the error equation reaches the origin in a finite time and this time is a locally bounded function of the initial state. if $\nu > -1/(\tilde{n}+1)$. This conclusion remains valid for $\nu = -\frac{1}{\tilde{n}+1}$ since, almost everywhere on (T_δ, T), the solution of the error equation belongs to the set $\{\varepsilon_\sigma \in \mathbb{R}^{n+k+1} : \xi \geq |\psi|^2/2, \xi > 0\}$, where the right-hand side (as well as solution) of the error equation depends smoothly on ν. Using Theorem 7.4 we complete the proof. □

As shown in the proof of Theorem 9.1, *the algebraic system* (14.25) *is always feasible.* So, the extended finite-time observer admits a constructive design for any observable linear plant. The variable ψ of the observer defines a filtered output of the system. Such a filtering is useful in the case of noised measurements in order to avoid high observer gains for a large noise. We cannot provide any quantitative noise-to-error analysis, since it requires a knowledge of a Lyapunov function of the error equation. However, qualitative characteristics of robustness (ISS) with respect to additive measurement noise and exogenous perturbations of the plant follow from homogeneity and asymptotic stability of the error equation.

Corollary 14.2 *Let all parameters of the filtering homogeneous observer* (14.19) *be defined according to Theorem 14.3. If the system model is given by* (14.6) *then the error system is ISS with respect to* $q = (q_1^\top, q_2^\top)^\top \in L^\infty(\mathbb{R}; \mathbb{R}^{n+k})$ *provided that* $\nu > -1/(\tilde{n}+1)$.

Proof The proof repeats the proof of Corollary 14.1. □

For the integrator chain $A = \left(\begin{smallmatrix} \mathbf{0} & I_{n-1} \\ 0 & \mathbf{0} \end{smallmatrix}\right)$, $C = (1\ 0\ \ldots\ 0)$, the observer (14.19) is similar to the observer given in [19] with the filter of the order 1. The key difference is the use of the internal variable ξ, which controls the observer gain. Both observers simply coincide for the case $\xi \equiv \mathbf{0}$. and $\nu = -\frac{1}{\tilde{n}+1}$, which corresponds to a discontinuous HOSM observer. The HOSM observers are known to be applicable to the system (14.5) with unknown bounded input/nonlinearity q [4, 20]. However, this is possible only under special restrictions to the matrix E. The following corollary provides the corresponding condition for the filtering homogeneous observer.

Corollary 14.3 *Let the filtering homogeneous observer* (14.19) *be designed as in Theorem 14.3 for* $\nu = -\frac{1}{\tilde{n}+1}$. *Let the system model be given by* (14.6) *with* $q_y = \mathbf{0}$, $q_x = Eq(t, x)$, $E \in \mathbb{R}^{n\times p}$, *where* q *is a piece-wise continuous function satisfying*

$$|q(t, x)| \leq \bar{q}, \quad \forall t \in \mathbb{R}, \quad \forall x \in \mathbb{R}^n.$$

If

$$(\tilde{n} + 1)G_{\mathbf{d}}E = E, \tag{14.31}$$

then the error equation $\dot{\varepsilon}_\sigma = g(\varepsilon_\sigma) + \begin{pmatrix} \mathbf{0} \\ E \\ 0 \end{pmatrix} q$ *is globally uniformly finite-time stable provided that* $\bar{q} > 0$ *is small enough.*

Proof Let us consider the extension of the error equation:

$$\dot{\varepsilon}_\sigma \in G(\epsilon_\sigma) = g(\varepsilon_\sigma) + \begin{pmatrix} \mathbf{0} \\ E \\ 0 \end{pmatrix} \mathcal{B}_{\bar{q}}, \tag{14.32}$$

where $\mathcal{B}_{\bar{q}} = \{x \in \mathbb{R}^n : |x| \leq \bar{q}\}$ is the Euclidean ball in $\mathbb{R}^n$ of the radius $\bar{q} > 0$. Since $(\tilde{n} + 1)G_{\mathbf{d}}E = E \Rightarrow G_{\mathbf{d}}E = -\nu E \Rightarrow \mathbf{d}(s)E = e^{-\nu s}E, \forall s \in \mathbb{R}$ then the set-valued vector field G is $\mathbf{d}$-homogeneous: $G(\mathbf{d}_\sigma(s)x) = g(\mathbf{d}_\sigma(s)\epsilon_\sigma) + \begin{pmatrix} \mathbf{0} \\ E \\ 0 \end{pmatrix}\mathcal{B}_{\bar{q}} = e^{\nu s}\mathbf{d}_\sigma(s)\left(g(\epsilon_\sigma) + \begin{pmatrix} \mathbf{0} \\ E \\ 0 \end{pmatrix}\mathcal{B}_{\bar{q}}\right) = e^{\nu s}\mathbf{d}_\sigma(s)G(\epsilon_\sigma)$. Since, for $\bar{q} = 0$, the error system is finite-time stable then, by Zubov–Rosier theorem (see Theorem 7.6), there exists a $\mathbf{d}_\sigma$-homogeneous Lyapunov function $V \in C(\mathbb{R}^{k+n+1}) \cap C^\infty(\mathbb{R}^{k+n+1}\backslash\{\mathbf{0}\})$ of degree $-\nu$ such that $\dot{V}(\epsilon_\sigma) = -\kappa, \quad \forall \epsilon_\sigma \neq 0$, for some $\kappa > 0$. In this case, for the perturbed case $\bar{q} \neq 0$ we have

$$\dot{V}(\epsilon_\sigma) \leq -\kappa + \frac{\partial V(\epsilon_\sigma)}{\partial \epsilon_\sigma} \begin{pmatrix} \mathbf{0} \\ E \\ 0 \end{pmatrix} q$$

with $q \in \mathcal{B}_{\bar{q}}$. Since the partial derivative of the homogeneous function is the $\mathbf{d}_\sigma$-homogeneous $\left.\frac{\partial V(\zeta)}{\partial \zeta}\right|_{\zeta = \mathbf{d}_\sigma(s)\epsilon_\sigma} \mathbf{d}_\sigma(s) = e^{-\nu s}\frac{\partial V(\epsilon_\sigma)}{\partial \epsilon_\sigma}, \forall s \in \mathbb{R}$ then taking $s = -\ln \|\epsilon_\sigma\|_{\mathbf{d}_\sigma}$ we derive

$$\left|\frac{\partial V(\epsilon_\sigma)}{\partial \epsilon_\sigma}\begin{pmatrix} \mathbf{0} \\ E \\ 0 \end{pmatrix}\right| \leq \bar{V}' = \sup_{\|\zeta\|=1}\left|\frac{\partial V(\zeta)}{\partial \zeta}\begin{pmatrix} \mathbf{0} \\ E \\ 0 \end{pmatrix}\right| < +\infty$$

where the identity $e^{\nu s}\mathbf{d}_\sigma(s)E = E, \forall s \in \mathbb{R}$ is utilized on the last step. Therefore, the error system remains finite-time stable for $\bar{q} \leq \frac{\kappa}{\bar{V}'}$. The proof is complete. □

For the integrator chain discussed above, the dilation has the diagonal generator $G_{\mathbf{d}} = \mathrm{diag}(1 + \nu, 1 + 2\nu..., 1 + \nu n)$ and the identity (14.31) holds for $E = (0, ..., 1)^\top \in \mathbb{R}^n$, since $\tilde{n} = n$. This perfectly fits the results of [4, 20].

Example 14.2 Let us consider a linearization (in the upper unstable position) of the rotary inverted pendulum (see Chap. 11 or [9]):

$$A = \begin{pmatrix} 0 & 0 & 1 & 0 \\ 0 & 0 & 0 & 1 \\ 0 & 152.0057 & -12.2542 & -0.5005 \\ 0 & 264.3080 & -12.1117 & -0.8702 \end{pmatrix}, \quad B = \begin{pmatrix} 0 \\ 0 \\ 50.6372 \\ 50.0484 \end{pmatrix}$$

and $C = \left(\begin{smallmatrix}1&0&0&0\\0&1&0&0\end{smallmatrix}\right)$. Taking $\nu = -1/3$ we design a HOSM-based filtering homogeneous observer. The homogenization gain L_0 and the dilation matrix we find by solving the linear algebraic equation (14.10):

$$L_0 = \begin{pmatrix} 12.2542 & 0.5005 \\ 12.1117 & 0.8702 \\ -156.227 & -158.574 \\ -158.959 & -271.127 \end{pmatrix}, \; G_{\mathbf{d}} = \begin{pmatrix} 2/3 & 0 & 0 & 0 \\ 0 & 2/3 & 0 & 0 \\ -4.0847 & -0.1668 & 1/3 & 0 \\ -4.0372 & -0.2901 & 0 & 1/3 \end{pmatrix}.$$

The parameters

$$\tilde{L} = \begin{pmatrix} -16.0915 & 0 \\ 0 & -16.0915 \end{pmatrix} \text{ and } L = \begin{pmatrix} -77.2851 & 0 \\ 0 & -77.2851 \\ 855.5357 & 38.6799 \\ 936.0546 & -24.2745 \end{pmatrix}$$

have been selected by solving the LMI (14.25) for $\rho = 2.5$, so the parameter $\gamma = 4.95$ satisfies the inequality $\rho > \gamma/2$.

For comparison reasons we consider the linear (Luenberger) observer [11] $\dot{z}^{\text{lin}} = Az^{\text{lin}} + Bu + L^{\text{lin}}(Cz^{\text{lin}} - y)$ with the gain $L^{\text{lin}} = \begin{pmatrix} -23.5773 & 0.5005 \\ 12.1117 & -34.5447 \\ 102.8592 & -140.8499 \\ 275.0213 & -418.7681 \end{pmatrix}$.

For simulation we select $x(0) = (2, 0, 1, 2)^\top$, we initiate the states of all observers at zero and we define the control input as $u = (2, -35, 1.5, -3)z$, where z is the component of the state of the filtering homogeneous observer. For simulations we use the explicit Euler method with the step size $h = 0.5 \cdot 10^{-4}$. The evolution of the estimation errors $\|x(t) - z(t)\|_1$ and $\|x(t) - z^{\text{lin}}(t)\|_1$ for the system with the homogeneous and linear observers, respectively, is presented in Fig. 14.2. Both observers show a convergence to the real state of the system. The estimation error at the time $t = 1.5$ is about $\|x(1.5) - z(1.5)\| \approx \|z^{\text{lin}}(1.5) - x(1.5)\| \approx 0.4 \cdot 10^{-4}$. However, the homogeneous observer has twice smaller overshoot and twice faster time response (convergence time to the system state or its small enough neighborhood). To have the same time response, the gain of the linear observer must be increased, but this would lead to a larger overshoot and worse control quality of the system.

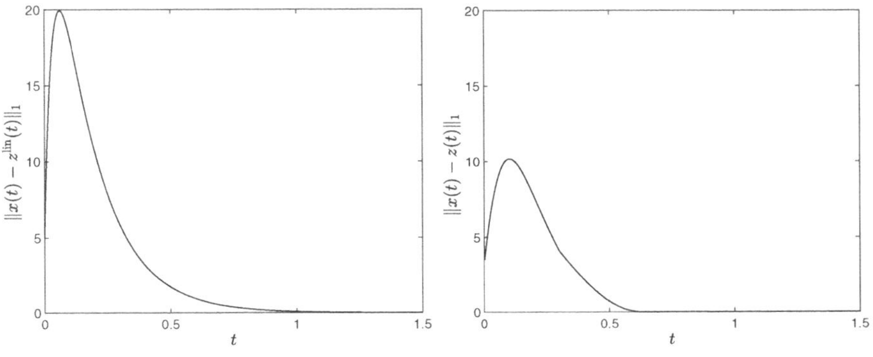

Fig. 14.2 Estimation errors for linear (left) and homogeneous (right) observers in the nominal case

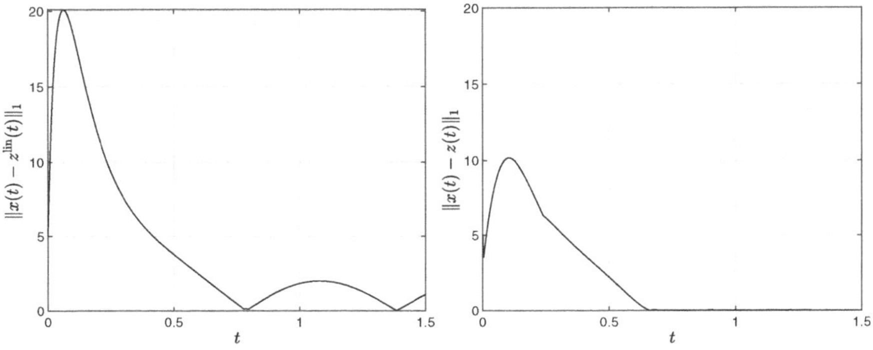

Fig. 14.3 Estimation errors for linear (left) and homogeneous (right) observers with the additive perturbation $0.1B\sin(5t)$

Notice that, due to numerical error of the explicit Euler method, the exact convergence of the observer variable to the system state cannot be observed in the presented simulations.

By Corollary 14.3 the HOSM-based homogeneous observer is able to reject unknown additive perturbation Eg of some magnitude, provided that the matrix E satisfies the identity (14.31), which in particular holds for $E = B$. The simulation results for $g(t) = 0.2\sin(5t)$ are depicted in Fig. 14.3. In the perturbed case, the homogeneous observer preserves all good properties (such as fast response and small overshoot) and still converges to the system state. The linear observer is sensitive with respect to the perturbation g. The convergence error of the linear observer is proportional to the magnitude of the perturbation. To demonstrate the robustness (ISS of the homogeneous observer with respect to the measurement noises), Fig. 14.4 presents the simulation results for the case of the matched additive perturbation considered

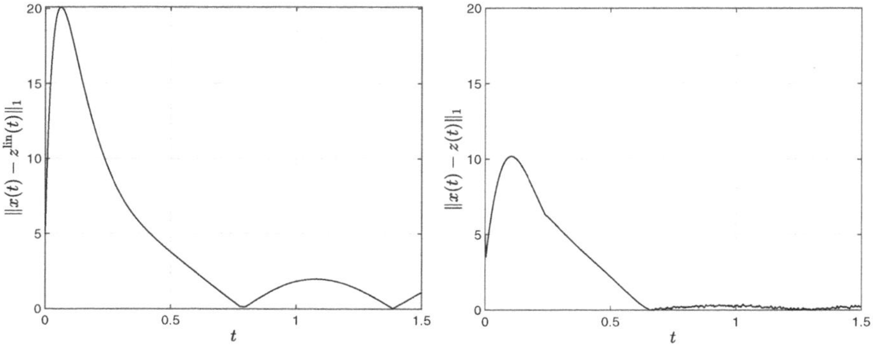

Fig. 14.4 Estimation errors for linear (left) and homogeneous (right) observers with the additive perturbation $0.1B\sin(5t)$ and with the noisy measurements of the magnitude 0.001

above and a uniformly distributed measurement noise of the magnitude 0.01. In this scenario, the proposed homogeneous observer (even theoretically) cannot converge identically to the system state. The estimation error of the homogeneous observer in this case is proportional to both the magnitude of the measurement noise and to the magnitude of the perturbation. In the considered scenario, the homogeneous observer has more than five times better precision comparing with the linear observer. For a very large (non-realistic) measurement noise (e.g., of the magnitude 1), the observers have similar error magnitudes. However, the homogeneous observer demonstrates twice smaller overshoots even in this case.

The presented numerical simulations support the main conjecture of this monograph that homogeneous observers/controllers (at least, in some cases) may provide faster convergence, better robustness and smaller overshoots than linear algorithms.

14.1.4 Fixed-Time Observer

In Chap. 7, it is shown that a locally homogeneous system with a positive degree at ∞-limit and a negative degree at 0-limit may be fixed-time stable. Inspired by this idea and the papers [7, 9, 17], we combine static homogeneous observers with positive and negative degrees:

$$\dot{z} = Az + Bu + \left(L_0 + \frac{e^{\nu_0(I_n+G_0)\ln|y-Cz|}+e^{\nu_\infty(I_n+G_0)\ln|y-Cz|}}{2}L\right)(y - Cz), \tag{14.33}$$

with $\nu_0 < 0$ and $\nu_\infty > 0$.

Corollary 14.4 *Let the matrices L_0, A_0 and G_0 be defined as in Theorem 14.1 and the pair $P \in \mathbb{R}^{n\times n}$, $Y \in \mathbb{R}^{n\times k}$ be a solution of the following system of linear matrix inequalities:*

$$PA + A^\top P + \tfrac{1}{2^i}\left(YC + C^\top Y^\top\right) \prec 0, \quad P \succ 0, \quad i = 0, 1. \tag{14.34}$$

Then the error equation

$$\dot{\varepsilon} = A_0\varepsilon + \frac{e^{\nu_0(I_n+G_0)\ln|C\varepsilon|}+e^{\nu_\infty(I_n+G_0)\ln|C\varepsilon|}}{2}LC\varepsilon \tag{14.35}$$

is globally uniformly fixed-time stable provided that the parameters $\nu_0 < 0$ and $\nu_\infty > 0$ are sufficiently close to zero.

Proof (1) The error equation (14.35) is $\mathbf{d}_0$-homogeneous of degree $\nu_0 < 0$ at 0-limit, where

$$\mathbf{d}_0(s) = e^{s(I_n+\nu_0 G_0)}, \quad s \in \mathbb{R}.$$

The corresponding approximation at 0-limit is given by

$$\dot{\varepsilon} = A_0\varepsilon + \frac{e^{\nu_0(I_n+G_0)\ln|C\varepsilon|}}{2} LC\varepsilon.$$

By Theorem 14.2 this system is finite-time stable provided that ν_0 is close enough to 0. This implies the local finite-time stability of the error equation (14.35) and there exist $r_0 > 0$ and $T_0 > 0$ such that

$$\|\varepsilon(0)\|_{\mathbf{d}} \le r_0 \quad \Rightarrow \quad \varepsilon(t) = \mathbf{0}, \quad \forall t \ge T_0.$$

(2) The error equation (14.35) is $\mathbf{d}_\infty$-homogeneous of degree $\nu_\infty > 0$ at ∞-limit, where

$$\mathbf{d}_\infty(s) = e^{s(I_n+\nu_\infty G_0)}, \quad s \in \mathbb{R}.$$

The corresponding approximation at ∞ is given by

$$\dot{\varepsilon} = A_0\varepsilon + \frac{e^{\nu_\infty(I_n+G_0)\ln|C\varepsilon|}}{2} LC\varepsilon.$$

By Theorem 14.2 this system is nearly fixied-time stable provided that ν_∞ is close enough to 0. This implies that there exist $r_\infty > 0$ and $T_\infty > 0$ such that

$$\|\varepsilon(t)\|_{\mathbf{d}_\infty} \le r_\infty, \quad \forall t \ge T_\infty, \quad \forall \varepsilon(0) \in \mathbb{R}^n.$$

(3) Let $0 < \underline{r} < \overline{r} < \infty$ be such that

$$\|\varepsilon\| \le \underline{r} \quad \Rightarrow \quad \|\varepsilon\|_{\mathbf{d}_0} \le r_0,$$

$$\|\varepsilon\| \ge \overline{r} \quad \Rightarrow \quad \|\varepsilon\|_{\mathbf{d}_\infty} \ge r_\infty.$$

Such numbers always exist due to Property 4, Lemma 5.4 of the canonical homogeneous norm. Let us denote

$$K = \{\varepsilon \in \mathbb{R}^n : \underline{r} \le \|\varepsilon\| \le \overline{r}\}.$$

In this case, we have

$$\frac{d\|\varepsilon\|}{dt} = \frac{\varepsilon^\top P(A_0 + LC)\varepsilon + \varepsilon^\top P E(\nu_0, \nu_1, \varepsilon) LC\varepsilon}{\|\varepsilon\|}, \quad \forall \varepsilon \in K,$$

where

$$E(\nu_1, \nu_2, \varepsilon) = \frac{\varepsilon^{\nu_0(I_n+G_0)\ln|C\varepsilon|} + e^{\nu_\infty(I_n+G_0)\ln|C\varepsilon|}}{2} - I_n.$$

Obviously, $E \to 0$ (uniformly the compact K) as $\max\{|\nu_0|, |\nu_\infty|\} \to 0$. Hence, there exists $c > 0$ such that

$$\frac{d\|\varepsilon\|}{dt} < -c, \quad \forall \varepsilon \in K,$$

provided that ν_0 and ν_∞ are sufficiently close to 0. Therefore, we derive

$$\varepsilon(t) = 0, \quad \forall t \geq T_0 + \frac{\overline{r} - \underline{r}}{c} + T_\infty, \quad \forall \varepsilon(0) \in \mathbb{R}^n.$$

Taking into account Lyapunov stability of the system we complete the proof. □

14.2 Homogeneous Observers for Nonlinear Plants

14.2.1 State Estimation of Quasi-linear Systems

Let us consider the quasi-linear system

$$\dot{x} = Ax + Bu + f(t, x), \quad y = Cx, \tag{14.36}$$

where x, y and A, B, C are as before, but $f : \mathbb{R}^n \mapsto \mathbb{R}^n$ is a known nonlinear function.

Assumption 14.1 Let the pair $\{A, C\}$ be observable and $G_0 \in \mathbb{R}^{n\times n}$, $\tilde{n} \in \mathbb{N}$ be defined as in Theorem 14.1. Let us assume that there exists $\gamma > 0$ and $\nu \in \left[-\frac{1}{\tilde{n}}, 1\right]$ such that

$$\left\| \frac{f(t,x) - f(t,y)}{\|x-y\|_{\mathbf{d}}^{\nu}} \right\|_{\mathbf{d}} \leq \gamma \|x-y\|_{\mathbf{d}}, \quad \forall x, y \in \mathbb{R}^n : x \neq y, \quad \forall t \in \mathbb{R}, \tag{14.37}$$

where $\mathbf{d}$ in $\mathbb{R}^n$ is the linear dilation with the generator $G_{\mathbf{d}} = I_n + \nu G_0$ and $\|\cdot\|_{\mathbf{d}}$ is some homogeneous norm.

Notice that, by Corollary 5.4, all homogeneous norms are equivalent. Therefore, if Assumption 14.1 is fulfilled for some homogeneous norm then it holds for any homogeneous norm (with a scaled γ). In the case of standard homogeneity, we have $\nu = 0$ and $|\cdot| = \|\cdot\|_{\mathbf{d}}$, so the inequality (14.37) becomes

$$|f(t, x_1) - f(t, x_2)| \leq \gamma |x_1 - x_2|, \quad \forall x_1, x_2 \in \mathbb{R}^n \backslash \{\mathbf{0}\},$$

i.e., f is globally Lipschitz continuous. It is well known (see, e.g., [21]) that the Luenberger-like observer with linear output injection term is applicable for quasi-linear system (14.36) with a globally Lipschitz nonlinearity f. Homogeneous observers can be utilized for systems with non-Lipschitz nonlinearities [4, 22]. The following theorem generalizes these results to $\mathbf{d}$-homogeneous observers with nonlinearities satisfying the condition (14.37).

Theorem 14.4 *Let Assumption 14.1 be fulfilled with some* $\gamma > 0$ *and* $\nu \in \left[-\frac{1}{\tilde{n}}, 1\right]$. *If an observer for the system* (14.36) *is given by*

$$\dot{z} = Az + Bu + f(t,z) + (L_0 + |y - Cz|^{\nu-1}\mathbf{d}(\ln|y - Cz|)L)(y - Cz), \tag{14.38}$$

with the parameters $L_0, L, G_{\mathbf{d}}, P, \nu$ *defined as in Theorem 14.2, then the error equation*

$$\dot{\varepsilon} = A_0\varepsilon + |C\varepsilon|^{\nu-1}\mathbf{d}(\ln|C\varepsilon|)LC\varepsilon + f(t,x) - f(t,z), \quad t > 0, \quad \varepsilon = z - x$$

is globally uniformly

- *finite-time stable for* $\nu < 0$,
- *exponentially stable for* $\nu = 0$,
- *nearly fixed-time stable for* $\nu > 0$,

provided that ν *and* γ *are sufficiently close to* 0.

Proof Repeating the proof of Theorem 14.15 we derive

$$\tfrac{d\|\varepsilon\|_{\mathbf{d}}}{dt} = \|\varepsilon\|_{\mathbf{d}}^{1+\nu} \tfrac{\pi^\top P(A_0+|C\pi|^{\nu-1}\mathbf{d}(\ln|C\pi|)LC)\pi + \pi^\top P\mathbf{d}(-\ln\|\varepsilon\|_{\mathbf{d}})\|\varepsilon\|^{-\nu}(f(t,x)-f(t,z))}{\pi^\top PG_{\mathbf{d}}\pi},$$

where $\pi = \mathbf{d}(-\ln\|\varepsilon\|_{\mathbf{d}})\varepsilon$. By Assumption 14.1

$$\left\|\mathbf{d}(-\ln\gamma)\mathbf{d}(-\ln\|\varepsilon\|_{\mathbf{d}})\frac{f(x) - f(z)}{\|\varepsilon\|_{\mathbf{d}}^{\nu}}\right\| \leq 1.$$

Taking into account $\|\pi\| = \sqrt{\pi^\top P\pi} = 1$ we derive

$$\frac{d}{dt}\|\varepsilon\|_{\mathbf{d}} \leq \|\varepsilon\|_{\mathbf{d}}^{1+\nu} \tfrac{\pi^\top P(A_0+|C\pi|^{\nu-1}\mathbf{d}(\ln|C\pi|)LC)\pi + \|\mathbf{d}(\ln\gamma)\|}{\pi^\top PG_{\mathbf{d}}\pi}.$$

Repeating the proof of Theorem 14.15 we derive

$$\frac{d}{dt}\|\varepsilon\|_{\mathbf{d}} \leq \|\varepsilon\|_{\mathbf{d}}^{1+\nu}\left(-\frac{\rho}{2} + \frac{2\|\mathbf{d}(\ln\gamma)\|)}{\beta}\right)$$

provided that ν is sufficiently close to zero. Taking into account $\|\mathbf{d}(\ln\gamma)\| \to 0$ as $\gamma \to 0$ we complete the proof. □

Let $A = \left(\begin{smallmatrix} \mathbf{0} & I_{n-1} \\ 0 & \mathbf{0} \end{smallmatrix}\right)$, $C = (1 \ \ 0 \ \ \ldots \ \ 0)$. In this case, $G_{\mathbf{d}}$ is diagonal such that

$$G_{\mathbf{d}} = \mathrm{diag}(r_1, \ldots, r_2), \quad r_i = 1 + (i-1)\nu, \quad i = 1, 2, \ldots, n,$$

where $\nu \in [-n^{-1}, 1]$. Then the homogeneous norm can be defined (see Chap. 5) as follows:

$$\|x\|_{\bar{r}} = |x_1|^{\frac{1}{r_1}} + |x_2|^{\frac{1}{r_2}} + \cdots + |x_n|^{\frac{1}{r_n}},$$

where $x = (x_1, x_2, ..., x_n)^\top \in \mathbb{R}^n$. Let $f : \mathbb{R}^n \mapsto \mathbb{R}^n$ be represented in the component-wise manner:

$$f(x) = \begin{pmatrix} f_1(x) \\ f_2(x) \\ \dots \\ f_n(x) \end{pmatrix},$$

where $f_i : \mathbb{R}^n \mapsto \mathbb{R}$, $i = 1, 2, ..., n$. In this case, the inequality (14.37) becomes

$$\sum_{i=1}^{n} \frac{|f_i(x) - f_i(y)|^{\frac{1}{r_i}}}{\|x - y\|_{\bar{r}}^{\frac{\nu}{r_i}}} \leq \gamma \|x - y\|_{\bar{r}}.$$

Therefore, to fulfill Assumption 14.1 it is sufficient to show that for each $i = 1, 2, ..., n$ there exists $\gamma_i \geq 0$ satisfying

$$|f_i(x) - f_i(y)| \leq \gamma_i \|x - y\|_{\bar{r}}^{r_i+\nu}, \quad \forall x, y \in \mathbb{R}^n. \tag{14.39}$$

This inequality is more conservative, but it is explicit, so it is much simple to check.

Example 14.3 If $\nu \in (-1/n, 1]$ then $r_i + \nu > 0$ for any $i = 1, 2, ..., n$. Let $p_i, q_i, \tilde{p}_i, \tilde{q}_i \in \mathbb{N}$ be odd natural numbers such that

$$r_i + \nu \leq \frac{q_i}{p_i} \leq 1 \quad \text{and} \quad \frac{\tilde{q}_i}{\tilde{p}_i} \leq r_i + \nu.$$

Let $x = (x_1, ..., x_n)^\top \in \mathbb{R}^n$, $\tilde{\gamma}_i \geq 0$, $i = 1, 2, ..., n$ and

$$f_i(x) = \tilde{\gamma}_i \begin{cases} x_i^{\frac{q_i}{p_i}} & \text{if } |x_i| \leq 1, \\ x_i^{\frac{\tilde{q}_i}{\tilde{p}_i}} & \text{if } |x_i| > 1. \end{cases}$$

Let us consider the system

$$\dot{x}_i = x_{i+1} + f_i(x), \quad i = 1, 2..., n-1, \quad \dot{x}_n = f_n(x), \quad y = x_1. \tag{14.40}$$

Taking into account Hölder continuity of $z \mapsto z_i^{\frac{q_i}{p_i}}$ (see Appendix D), we conclude that, there exists $c_i > 0$ such that the inequality

$$|f_i(x) - f_i(y)| = \gamma_i \left| x_i^{\frac{q_i}{p_i}} - y_i^{\frac{q_i}{p_i}} \right| \leq \tilde{\gamma}_i c_i |x_i - y_i|^{r_i+\nu}$$

holds for any $x_i \in \mathbb{R} : |x_i| \leq 1$ and any $y_i \in \mathbb{R} : |y_i| \leq 1$. Moreover, there exists $C_i > 0$ such that for $|x_i| \geq 1$ and $|y_i| \geq 1$ we have

$$|f_i(x) - f_i(y)| = \gamma_i \left| x_i^{\frac{\tilde{q}_i}{p_i}} - y_i^{\frac{\tilde{q}_i}{p_i}} \right| \leq \tilde{\gamma}_i C_i |x_i - y_i|^{r_i + \nu}.$$

This means that the inequality (14.39) holds for sufficiently small $\tilde{\gamma}_i$, so the homogeneous observer (14.38) reconstructs the state of the nonlinear system (14.40) in a finite time if $\nu < 0$ and in a nearly fixed time if $\nu > 0$.

14.2.2 Observability of Nonlinear Homogeneous System

Let us consider the nonlinear system

$$\dot{x} = f(x), \quad x(0) = x_0 \in \mathbb{R}^n, \quad y = h(x), \quad y \in \mathbb{R}, \tag{14.41}$$

where $f : \mathbb{R}^n \mapsto \mathbb{R}^n$ and $h : \mathbb{R}^n \mapsto \mathbb{R}$ are sufficiently smooth mappings. Let us recall some basic definitions of the observability theory.

Definition 14.2 The states $x_1, x_2 \in \mathbb{R}^n$, $x_1 \neq x_2$ are said to be *indistinguishable* if $h(x_{x_1}(t)) = h(x_{x_2}(t))$ for all $t \geq 0$, where x_{x_i} denotes solution of (14.41) with $x(0) = x_i$, $i = 1, 2$. Otherwise, these states are called *distinguishable*.

The system (14.41) is locally observable at $x_0 \in \mathbb{R}^n$ if there exists a neighborhood $U(x_0)$ such that for any $z \in U(x_0)\backslash\{x_0\}$ the states x_0 and z are distinguishable.

The system (14.41) is globally observable if it is locally observable at any point $x_0 \in \mathbb{R}^n$.

For more details about observability of nonlinear systems we refer the reader, for example, to [1]. The following straightforward corollary is the consequence of Definition 14.2 and Theorem 7.1.

Corollary 14.5 *Let $f : \mathbb{R}^n \mapsto \mathbb{R}^n$ be a* **d***-homogeneous continuous vector field and $h : \mathbb{R}^n \mapsto \mathbb{R}$ be* **d***-homogeneous continuous function. The nonlinear system* (14.41) *is observable on $\mathbb{R}^n\backslash\{\mathbf{0}\}$ if and only if it is locally observable on the unit sphere S.*

The controllability/observability analysis of nonlinear systems frequently follows the differential algebraic framework [3, 23]. The homogeneity is known to be useful for the corresponding analysis [24, 25]. We refer the reader to the mentioned literature for more details about differential algebraic approach to controllability/observability analysis. In this section, we present few simple results based on this method.

Let $\mathcal{L}_f^{(k)}h$ denote the Lie derivative of the function $h : \mathbb{R}^n \mapsto \mathbb{R}$ along the vector field $f : \mathbb{R}^n \mapsto \mathbb{R}^n$:

$$\mathcal{L}_f^{(0)}h = h \quad \text{and} \quad \mathcal{L}_f^{(k)}h = \frac{\partial \mathcal{L}_f^{(k-1)}h}{\partial x} f \quad \text{for} \quad k = 1, 2, ...$$

The existence of high-order Lie derivatives needs a sufficient order of smoothness of the function h and the vector field f. Therefore, the observability condition presented below is applicable only for smooth systems (contrary to the homogeneous observer designed above for the quasi-linear systems). Inspired by [26], we derive the following sufficient observability condition for the homogeneous system (14.41).

Theorem 14.5 *Let a linear continuous dilation* $\mathbf{d}$ *in* $\mathbb{R}^n$ *be strictly monotone with respect to a smooth norm* $\|\cdot\|$ *in* $\mathbb{R}^n$. *Let* $\|\cdot\|_{\mathbf{d}}$ *be the canonical homogeneous induced by* $\|\cdot\|$. *Let* $h : \mathbb{R}^n \mapsto \mathbb{R}$ *be* $\mathbf{d}$*-homogeneous of a degree* $\mu > 0$ *and* $f : \mathbb{R}^n \mapsto \mathbb{R}^n$ *be* $\mathbf{d}$*-homogeneous of a degree* $\nu > -\frac{\mu}{n-1}$. *Let the vector field*

$$H = \begin{pmatrix} h \\ L_f h \\ \dots \\ L_f^{(n-1)}h \end{pmatrix} : \mathbb{R}^n \mapsto \mathbb{R}^n \tag{14.42}$$

be C^1 *on the unit sphere* $S = \{x \in \mathbb{R}^n : \|x\| = 1\}$.

The nonlinear system (14.41) *is globally observable if one of the following conditions holds:*

i) $n \geq 3$, $\inf_{x \in S} \|H(x)\| > 0$ *and* $\det\left(\frac{\partial H}{\partial x}\big|_{x\in S}\right) \neq 0$;
ii) H *is* C^1 *at zero,* $\det\left(\frac{\partial H}{\partial x}\big|_{x=0}\right) \neq 0$ *and* $n\mu + \frac{n(n-1)}{2}\nu = \mathrm{tr}(G_{\mathbf{d}})$.

Proof To guarantee observability it is sufficient to show that the mapping H is a homeomorphism (or diffeomorphism) on $\mathbb{R}^n$. Indeed, since we have $\dot{y}^{(i)}(t) = L_f^{(i)}h(x(t))$ along any trajectory of the system (14.41), then the system $x(t) \in \mathbb{R}^n$ can be reconstructed from the output and its derivatives $(y(t), \dot{y}(t), ..., y^(n))^\top \in \mathbb{R}^n$ provided that the mapping H is invertible.

Since f and h are $\mathbf{d}$-homogeneous functions then, due to Proposition 6.2, the Lie derivative $\mathcal{L}_f^{(i)}h$ is $\mathbf{d}$-homogeneous of the degree $\deg_{\mathbf{d}}(g)\, i + \deg_{\mathbf{d}}(h)$, $i = 0, 1, 2, ...$ Hence, we derive

$$H(x) = \begin{pmatrix} \|x\|_{\mathbf{d}}^{\mu} & 0 & \dots & 0 \\ 0 & \|x\|_{\mathbf{d}}^{\mu+\nu} & \dots & 0 \\ \dots & \dots & \dots & \dots \\ 0 & 0 & \dots & \|x\|_{\mathbf{d}}^{\mu+(n-1)\nu} \end{pmatrix} H(z),$$

where $z = \mathbf{d}(-\ln \|x\|_{\mathbf{d}})x \in S$. Since the norm $\|\cdot\|$ is selected to be smooth on $\mathbb{R}^n\backslash\{\mathbf{0}\}$ and H is C^1 on the sphere S then H is C^1 on $\mathbb{R}^n\backslash\{\mathbf{0}\}$. Using (6.2) we derive

$$\begin{pmatrix} e^{\mu s} & 0 & \dots & 0 \\ 0 & e^{(\mu+\nu)s} & \dots & 0 \\ \dots & \dots & \dots & \dots \\ 0 & 0 & \dots & e^{(\mu+(n-1)\nu)s} \end{pmatrix} \frac{\partial H(x)}{\partial x} = \frac{\partial H(\mathbf{d}(s)x)}{\partial x}\mathbf{d}(s)$$

and

$$\frac{e^{(n\mu+0.5n(n-1)\nu)s}\det\left(\frac{\partial H(x)}{\partial x}\right)}{\det(\mathbf{d}(s))}=\det\left(\frac{\partial H(\mathbf{d}(s)x)}{\partial x}\right).$$

Notice that $\det(\mathbf{d}(s))=\det(e^{G_{\mathbf{d}}s})=e^{s\,\mathrm{tr}(G_{\mathbf{d}})}$ for all $s\in\mathbb{R}$.

(i) Therefore, we have shown that $\det\left(\frac{\partial H(x)}{\partial x}\right)\neq 0$ on $\mathbb{R}^n\backslash\{\mathbf{0}\}$. On the other hand, since $H(z)\neq\mathbf{0}$ for $z\in S$ then $\|H(u)\|\to+\infty$ as $\|x\|\to+\infty$ (i.e., H radially unbounded and proper) and $\|H(x)\|\to 0$ as $\|x\|\to 0$ (i.e., H is continuous at zero). Using Hadamard theorem (see, e.g., Theorem 2.1, [27]) we derive that H is the bijection on $\mathbb{R}^n\backslash\{\mathbf{0}\}$ provided that $\mathbb{R}^n\backslash\{\mathbf{0}\}$ is simply connected C^1-manifold (that is the case for $n\geq 3$). Continuity of H at the origin proven above as well as $H(\mathbf{0})=\mathbf{0}$ together with $H(x)\neq\mathbf{0}$ on $\mathbb{R}^n\backslash\{\mathbf{0}\}$ implies that H is a global homeomorphism.

(ii) If $n\mu+\frac{n(n-1)}{2}\nu=\mathrm{tr}(G_{\mathbf{d}})$ then

$$\det\left(\frac{\partial H(u)}{\partial u}\right)=\det\left(\frac{\partial H(\mathbf{d}(s)x)}{\partial x}\right)$$

for all $s\in\mathbb{R}$. Since H is C^1 at zero then for all $x\in\mathbb{R}^n$ we have

$$\det\left(\frac{\partial H(x)}{\partial x}\right)=\det\left(\left.\frac{\partial H}{\partial x}\right|_{x=0}\right)\neq 0,$$

i.e., H is C^1 diffeomorphism on $\mathbb{R}^n$. The proof is complete. □

If H given by (14.42) is a diffeomorphism on $\mathbb{R}^n$ then the coordinate transformation

$$x_{\text{new}}=H(x)$$

gives the equivalent system

$$\dot{x}_{\text{new}}=Ax_{\text{new}}+Bg(x_{\text{new}}),\quad y=Cz, \tag{14.43}$$

where $A=\begin{pmatrix}\mathbf{0} & I_{n-1}\\ 0 & \mathbf{0}\end{pmatrix}$, $B=\begin{pmatrix}0\\ \cdots\\ 0\\ 1\end{pmatrix}$, $C=(1\ \ 0\ \ \ldots\ \ 0)$ and

$$g(x_{\text{new}})=L_f^n h(x)\big|_{x=H^{-1}(x_{\text{new}})}.$$

Since the system (14.43) is quasi-linear and the pair $\{A,C\}$ is observable then the homogeneous observer can be constructed using Theorem (14.4) provided that Bg satisfies Assumption 14.1.

Recall that the linear system $f(x)=Ax$, $h(x)=Cx$ is observable if

$$\mathrm{rank}\begin{bmatrix}C\\ CA\\ \cdots\\ CA^{n-1}\end{bmatrix}=n.$$

Table 14.1 Comparison of various homogeneous observes for linear plants

Homogeneous observer	Advantages	Disadvantages
Time varying	Simplicity of tuning, prescribed-time convergence	Lack of robustness
Time invariant	Any homogeneity degree, robustness (ISS)	No tuning rule for homogeneity degree
Extended	Simplicity of tuning, robustness (ISS)	Only for negative degree

This is equivalent to the condition (ii) of Theorem 14.5 with $\nu = 0, \mu = 1$ and $G_{\mathbf{d}} = I_n$.

Concluding Remarks

In this chapter, some observers for linear and nonlinear systems in $\mathbb{R}^n$ are designed based on the concept of linear (geometric) homogeneity. For linear systems, three kinds of observers are developed: dynamic, static and extended. Each one has its own advantage and disadvantage summarized in Table 14.1.

Homogeneous observers also designed for nonlinear system allowing a transformation to quasi-linear form satisfying special assumption of homogeneous domination.

References

1. Hermann H, Krener A (1977) Nonlinear controllability and observability. IEEE Trans. Autom. Control. 22:728–740
2. Wonham WM (1985) Linear multivariable control: a geometric approach. Springer
3. Isidori A (1995) Nonlinear control systems. Springer, N. Y. Inc
4. Levant A (2003) Higher-order sliding modes, differentiation and output-feedback control. Int. J. Control. 76(9–10):924–941
5. Bhat SP, Bernstein DS (2005) Geometric homogeneity with applications to finite-time stability. Math. Control. Signals Syst. 17:101–127
6. Perruquetti W, Floquet T, Moulay E (2008) Finite-time observers: application to secure communication. IEEE Trans. Autom. Control. 53(1):356–360
7. Andrieu V, Praly L, Astolfi A (2008) Homogeneous approximation, recursive observer design, and output feedback. SIAM J. Control. Optim. 47(4):1814–1850
8. Cruz-Zavala E, Moreno JA (2017) Homogeneous high order sliding mode design: a Lyapunov approach. Automatica 80(6):232–238
9. Lopez-Ramirez F, Polyakov A, Efimov D, Perruquetti W (2018) Finite-time and fixed-time observer design: implicit Lyapunov function approach. Automatica 87(1):52–60
10. Kalman R, Bucy R (1961) New results in linear filtering and prediction theory. J. Basic Eng. 83:95–108

11. Luenberger D (1964) Observing the state of a linear system. IEEE Trans. Mil. Electron. 8(2):74–80
12. Holloway J, Krstic M (2019) Prescribed-time observers for linear systems in observer canonical form. IEEE Trans. Autom. Control. 64(9):3905–3912
13. Nekhoroshikh A, Efimov D, Polyakov A, Perruquetti W, Furtat I (2022) On simple design of a robust finite-time observer. In: Conference on decision and control
14. Li M, Polyakov A, Zheng G (2023) On prescribed-time cooperative state estimation of LTI plants. In: IFAC World Congress
15. Ping X, Zimenko K, Polyakov A, Efimov D (2024) Filtering homogeneous observer for MIMO system. arXiv:2401.03135
16. Polyakov A (2020) Generalized homogeneity in systems and control. Springer
17. Zimenko K, Polyakov A, Efimov D, Kremlev A (2023) Homogeneity based finite/fixed-time observers for linear MIMO systems. Int. J. Robust Nonlinear Control. 33(15):8870–8889
18. Filippov AF (1988) Differential equations with discontinuous right-hand sides. Kluwer Academic Publishers
19. Jbara A, Levant A, Hanan A (2021) Filtering homogeneous observers in control of integrator chains. Int. J. Robust Nonlinear Control. 31(9):3658–3685
20. Moreno JA (2023) High-order sliding-mode functional observers for multiple-input multiple-output (MIMO) linear time-invariant systems with unknown inputs. Int. J. Robust Nonlinear Control. 33(15):8844–8869
21. Thau F (1973) Observing the state of non-linear dynamic systems. Int. J. Control. 17(3):471–479
22. Bernard P, Praly L, Andrieu V (2017) Observers for a non-Lipschitz triangular form. Automatica 82:301–313
23. Fliess M, Levine J, Martin P, Rouchon P (1995) Flatness and defect of non-linear systems: introductory theory and examples. Int. J. Control. 61(6):1327–1361
24. Hermes H (1986) Nilpotent approximations of control systems and distributions. SIAM J. Control. Optim. 24(4):731
25. Sepulchre R, Aeyels D (1996) Homogeneous Lyapunov functions and necessary conditions for stabilization. Math. Control. Signals Syst. 9:34–58
26. Polyakov A (2018) Fast control systems: nonlinear approach. In: Clempner JB, Yu W (eds) New perspectives and applications of modern control theory. Springer, pp 287–316
27. Ruzhansky M, Sugimoto M (2015) On global inversion of homogeneous maps. Bull. Math. Sci. 5:13–18

Chapter 15
Homogeneous Control of Nonlinear Plants

This chapter is devoted to homogeneous stabilization (see Chap. 9) of nonlinear plants. A criterion of homogeneous stabilizability of a nonlinear plant by a dynamic homogeneous feedback is obtained based on Sontag's universal stabilizer formula. A homogeneous convex embedding technique is utilized for robust stabilization of affine-in-control nonlinear systems. Locally homogeneous feedback laws are designed for global rigid-body attitude tracking.

15.1 Dynamic Homogeneous Feedback

Recall that the system is **d**-homogeneously stabilizable if there exists a control law such that the closed-loop system is asymptotically stable and **d**-homogeneous of a desired degree $\mu \in \mathbb{R}$. In this section, we present a necessary and sufficient condition of **d**-homogeneous stabilizability of the nonlinear control system

$$\dot{x} = g(x, u), \quad t > 0, \quad x(t) \in \mathbb{R}^n, \quad u(t) \in \mathbb{R}^m \tag{15.1}$$

by means of a dynamic homogeneous feedback

$$\dot{u} = k(x, u), \quad t > 0, \tag{15.2}$$

where $g : \mathbb{R}^n \times \mathbb{R}^m \mapsto \mathbb{R}^m$ models system behavior and $k : \mathbb{R}^n \times \mathbb{R}^m \mapsto \mathbb{R}^m$.

A. Polyakov, *Generalized Homogeneity in Systems and Control Volume I*,
Communications and Control Engineering,
https://doi.org/10.1007/978-3-031-76246-8_15

15.1.1 Homogeneous Stabilizability of Nonlinear Plant

Inspired by the universal stabilizer of Sontag [1], we[1] design a dynamic homogeneous feedback using Theorem 7.7. The following theorem provides a necessary and sufficient condition for existence of **d**-homogeneous stabilizer for a nonlinear system.

Theorem 15.1 (On homogeneous dynamic feedback, [2]) *Let* $\mathbf{d}_x$ *and* $\mathbf{d}_u$ *be dilations in* $\mathbb{R}^n$ *and* $\mathbb{R}^m$*, respectively,*

$$\mathbf{d}(s) := \begin{pmatrix} \mathbf{d}_x(s) & \mathbf{0} \\ \mathbf{0} & \mathbf{d}_u(s) \end{pmatrix},$$

such that $\tilde{f} = \begin{pmatrix} g \\ \mathbf{0} \end{pmatrix} \in C(\mathbb{R}^{n+m}\backslash\{\mathbf{0}\}, \mathbb{R}^{n+m})$ *be* **d**-*homogeneous of a degree* $\mu \in \mathbb{R}^n$.

The origin of the system (15.1) *is globally* **d**-*homogeneously stabilizable by means of the dynamic feedback* (15.2) *with* $k \in C(\mathbb{R}^{n+m}\backslash\{\mathbf{0}\}, \mathbb{R}^m)$ ***if and only if*** *there exist a number* $\gamma \geq 0$*, a symmetric positive definite matrix* $P \in \mathbb{R}^{(n+m)\times(n+m)}$ *satisfying (5.6) and a mapping*

$$\Xi \in C^\infty(\mathbb{R}^{n+m}\backslash\{\mathbf{0}\}, \mathbb{R}^{(n+m)\times(n+m)}),$$

$$\det(\Xi(z)) \neq 0, \quad \tfrac{\partial \Xi(z)}{\partial z_i} z = 0, \quad \Xi(e^s z) = \Xi(z)$$

for $z = (z_1, \ldots, z_{n+m}) \in \mathbb{R}^{n+m}\backslash\{\mathbf{0}\}$, $s \in \mathbb{R}$, $i = 1, \ldots, n+m$ *such that*

$$a(z) < \gamma\sqrt{b^\top(z) b(z)} \quad \text{for} \quad z \in S, \tag{15.3}$$

where

$$a(z) = z^\top W(z) \tilde{f}(z), \qquad b^\top(z) = z^\top W(z) \begin{pmatrix} \mathbf{0} \\ I_m \end{pmatrix},$$
$$W(z) = \Xi^\top(z) P\, \Xi(z) \left(\tfrac{(I_{n+m} - G_{\mathbf{d}}) z^\top z P}{z^\top P G_{\mathbf{d}} z} + I_{n+m} \right)$$

and S *is the unit sphere in* $\mathbb{R}^{n+m}$ *with* $\|z\| = \sqrt{z^\top P z}$*. Moreover, the corresponding stabilizing homogeneous feedback law can be designed as follows:*

$$k(\xi) = \|\xi\|_{\mathbf{d}}^{\mu}\, \mathbf{d}_u(\ln \|\xi\|_{\mathbf{d}}) k_0(\mathbf{d}(-\ln \|\xi\|_{\mathbf{d}})\xi) \quad \text{with} \quad \xi = \begin{pmatrix} x \\ u \end{pmatrix}, \tag{15.4}$$

[1] Reproduced from Polyakov A. Sliding mode control design using canonical homogeneous norm. *Int J Robust Nonlinear Control*. 2019; 29:682–701. https://doi.org/10.1002/rnc.4058 ©2018 John Wiley & Sons, Ltd. Reused with permission of John Wiley & Sons.

where

$$k_0(\cdot)=\begin{cases}-\frac{a(\cdot)+\sqrt{a^2(\cdot)+(b^\top(\cdot)b(\cdot))^2}}{b^\top(\cdot)b(\cdot)}b(\cdot) & \text{if } b(\cdot)\neq\mathbf{0}\\ 0 & \text{if } b(\cdot)=\mathbf{0}\end{cases} \tag{15.5}$$

and $\|\cdot\|_{\mathbf{d}}$ *is the canonical homogeneous norm induced by* $\|\xi\|=\sqrt{\xi^\top P\xi}$.

Proof [2] *Sufficiency*. Let us show that the function k_0 is continuous on the unit sphere S in $\mathbb{R}^{n\times m}$. For any $z\in S$ we have

$$k_0(z) = \frac{a(z)+\sqrt{a^2(z)+(b^\top(z)b(z))^2}}{\sqrt{b^\top(z)b(z)}}\frac{b(z)}{\sqrt{b^\top(z)b(z)}}.$$

The norm of the second term in the product is globally bounded and continuous for $z\in S: b(z)\neq\mathbf{0}$. The first term is continuous on S and equals zero if $b(z)=\mathbf{0}$. Indeed, from inequality (15.3) and continuity of a we derive $a(z)\to a(z^*)$ as $z\to z^*: b(z^*)=\mathbf{0}$ and $a(z^*)<0$. Hence, the multiplication on the numerator and the denominator by the adjoint term $-a(z)+\sqrt{a^2(z)+(b^\top(z)b(z))^2}$ gives

$$\frac{a(z)+\sqrt{a^2(z)+(b^\top(z)b(z))^2}}{\sqrt{b^\top(z)b(z)}} = \frac{(b^\top(z)b(z))^{3/2}}{\sqrt{a^2(z)+(b^\top(z)b(z))^2}-a(z)} \to \frac{0}{2|a(z^*)|} \text{ as } z\to z^*.$$

Let us consider the closed-loop system (15.1), (15.2), (15.4):

$$\dot{\xi}=f(\xi),$$

where $f=\binom{g}{k}$. Under this notation, the stability condition (7.18) becomes

$$a(z)+b^\top(z)k_0(z)<0.$$

The latter hods for all $z\in S$. Indeed, the inequality (15.3) implies that $a(z)<0$ if $b(z)=\mathbf{0}$, but for $b(z)\neq\mathbf{0}$ we have

$$a(z)+b^\top(z)k_0(z)=a(z)-b^\top(z)\frac{a(z)+\sqrt{a^2(z)+(b^\top(z)b(z))^2}}{b^\top(z)b(z)}b(z)=$$

$$-\sqrt{a^2(z)+(b^\top(z)b(z))^2}<0.$$

Necessity. Let us assume that there exists a map $\tilde{k}\in C(\mathbb{R}^{n+m}\backslash\{\mathbf{0}\},\mathbb{R}^m)$ such that the closed-loop system (15.1), (15.2) is globally asymptotically stable and $f=\binom{g}{k}\in C(\mathbb{R}^{n+m}\backslash\{\mathbf{0}\},\mathbb{R}^{n+m})$ is $\mathbf{d}$-homogeneous of a degree μ. Let $\Xi\in C^\infty(\mathbb{R}^{n+m}\backslash\{\mathbf{0}\},\mathbb{R}^{(n+m)\times(n+m)})$ be derived according to Theorem 7.7 such that

[2] The proof is reproduced from Polyakov A. Sliding mode control design using canonical homogeneous norm. *Int Journal Robust Nonlinear Control*. 2019; 29:682–701. https://doi.org/10.1002/rnc.4058 ©2018 John Wiley & Sons, Ltd. Reused with permission of John Wiley & Sons.

(7.18) holds. To show that in this case (15.3) also holds for some $\gamma \geq 0$, we rewrite the inequality (7.18) as follows:

$$a(z) < -b^\top(z)\tilde{k}(z).$$

Since $\tilde{k} \in C(\mathbb{R}^{n+m}\backslash\{\mathbf{0}\}, \mathbb{R}^m)$ then $\gamma = \max_{z\in S}\sqrt{\tilde{k}^\top(z)\tilde{k}(z)} < +\infty$ and using the Cauchy–Schwarz inequality we derive

$$-b^\top(z)\tilde{k}(z) \leq \sqrt{\tilde{k}^\top(z)\tilde{k}(z)}\sqrt{b^\top(z)b(z)} \leq \gamma\sqrt{b^\top(z)b(z)}.$$

The proof is complete. □

For $\Xi(z) = I_n$ the inequality (15.3) gives the following sufficient condition of quadratic homogeneous stabilizability:

$$\begin{array}{c} z^\top P\tilde{f}\left(\frac{z}{\sqrt{z^\top Pz}}\right) \leq \gamma\sqrt{z^\top P\left(\begin{smallmatrix} 0 & 0 \\ 0 & I_m \end{smallmatrix}\right)Pz}, \quad z \neq \mathbf{0} \\ PG_{\mathbf{d}} + G_{\mathbf{d}}^\top P \succ 0, \quad P \succ 0, \end{array} \tag{15.6}$$

which can be checked in many practical cases.

Example 15.1 *(On quadratic stabilization in a finite time,*[3] [2]*)* Let us consider the stabilization problem for the system

$$\dot{x}_1 = |x_1|^{\frac{1}{3}}u + x_2, \qquad \dot{x}_2 = u, \tag{15.7}$$

where $x_1, x_2, u \in \mathbb{R}$. The augmented vector field $\tilde{f} : \mathbb{R}^3 \mapsto \mathbb{R}^3$ given by

$$\tilde{f}(\xi) = \begin{pmatrix} |\xi_1|^{\frac{1}{3}}\xi_3 + \xi_2 \\ \xi_3 \\ 0 \end{pmatrix}$$

$\xi = (\xi_1, \xi_2, \xi_3)^\top = (x_1, x_2, u)^\top$ is $\mathbf{d}$-homogeneous of the degree -1 with respect to the dilation $\mathbf{d}$ in $\mathbb{R}^3$ generated by $G_{\mathbf{d}} = \mathrm{diag}\{3, 2, 1\}$. Hence, the sufficient stabilizability condition (15.6) becomes

$$z^\top PA(z)z < \gamma\sqrt{z^\top Pz}\sqrt{z^\top Pe_3e_3^\top Pz},$$

[3] The example is reproduced from Polyakov A. Sliding mode control design using canonical homogeneous norm. *Int J Robust Nonlinear Control*. 2019; 29:682–701. https://doi.org/10.1002/rnc.4058

$P > 0$, $PG_{\mathbf{d}} + G_{\mathbf{d}}^{\top}P > 0$, where $z = (z_1, z_2, z_3)^{\top} \in \mathbb{R}^3$ and

$$A(z) = \begin{pmatrix} 0 & 1 & \left(\frac{|z_1|}{\sqrt{z^{\top}Pz}}\right)^{\frac{1}{3}} \\ 0 & 0 & 1 \\ 0 & 0 & 0 \end{pmatrix}.$$

Let $X \in \mathbb{R}^{3\times 3}$ and $\gamma_0 > 0$ satisfy the linear matrix inequalities

$$XA_i + A_i^{\top}X - 2\gamma e_3 e_3^{\top} < 0, \quad XG_{\mathbf{d}} + G_{\mathbf{d}}X > 0, \quad X > 0,$$

$A_i = \left(\begin{smallmatrix} 0 & 1 & i \\ 0 & 0 & 1 \\ 0 & 0 & 0 \end{smallmatrix}\right)$, $i = 0, 1$, $e_3 = (0, 0, 1)^{\top} \in \mathbb{R}^3$. Since $A(z) = \alpha(z)A_1 + (1 - \alpha(z))A_0$, where $\alpha(z) = \left(\frac{|z_1|}{\sqrt{z^{\top}Pz}}\right)^{\frac{1}{3}}$ and $\alpha(z) \in [0, 1]$ provided that $p_{11} = 1$, then for $P = \tilde{P}/(\tilde{p}_{11})$, $\tilde{P} = X^{-1}$ the stabilizability condition (15.6) holds. So, the considered system can be stabilized in a finite time (due to $\deg_{\mathbf{d}}(\tilde{f}) = -1$) to zero by means of the $\mathbf{d}$-homogeneous dynamical feedback

$$\dot{u} = k_0\left(\mathbf{d}(-\ln\|\xi\|_{\mathbf{d}})\xi\right), \tag{15.8}$$

where k_0 is given by (15.5) with $a(z) = z^{\top}P\tilde{f}(z)$ and $b(z) = p_{13}z_1 + p_{23}z_2 + p_{33}z_3$. For example, the appropriate matrix P obtained using the above LMIs is

$$P = \begin{pmatrix} 1.0000 & 0.8930 & 0.8846 \\ 0.8930 & 1.8679 & 1.3659 \\ 0.8846 & 1.3659 & 1.7523 \end{pmatrix}.$$

15.1.2 Robustness Analysis of Dynamic Feedback

In this section, we study the problem of a homogeneous stabilization of the system

$$\dot{x}(t) = g(x(t), u(t)) + \omega(t, x(t), u(t)), \quad t > 0 \tag{15.9}$$

$$\text{with} \quad |\omega| \leq \overline{\omega}\left(\left|\begin{pmatrix} x \\ u \end{pmatrix}\right|\right)$$

by means of the homogeneous dynamical feedback (15.2):

$$\dot{u}(t) = k(x(t), u(t)).$$

The necessary condition is given by Theorem 15.1. A sufficient condition requires an additional restriction to the uncertain term ω.

Corollary 15.1 ([2]) *Let conditions of Theorem 15.1 hold and*

$$z^T \Xi^T(z) P \Xi(z) \left(\tfrac{(I_{n+m}-G_{\mathbf{d}})zz^\top P}{z^\top P G_{\mathbf{d}} z} + I_{n+m}\right)\left(f\left(\tfrac{z}{\|z\|}\right) + \tfrac{\mathbf{d}(-\ln\|z\|)\tilde{\omega}(t,z)}{\|z\|^\mu}\right) < 0, \tag{15.10}$$

for all $t > 0$ *and all* $z \in \mathbb{R}^{n+m}\backslash\{\mathbf{0}\}$, *where* $\|z\| = \sqrt{z^\top P z}$ *and* $\tilde{\omega}(t,x) = \binom{\omega(t,z)}{0}$. *Then the closed-loop system* (15.9), (15.2) *with k given by* (15.4) *is*

- *globally uniformly finite-time stable if* $\mu < 0$;
- *globally exponentially stable if* $\mu = 0$;
- *globally practically fixed-time stable if* $\mu > 0$.

Proof [4] Making the change of variables $z = \|\xi\|_{\mathbf{d}}\mathbf{d}(-\ln\|\xi\|_{\mathbf{d}})\xi$ in the closed-loop system (15.9), (15.2) we derive

$$\dot{z} = \|z\|\left(\tfrac{(I_{n+m}-G_{\mathbf{d}})zz^\top P}{z^\top P G_{\mathbf{d}} z} + I_{n+m}\right)\left(\|z\|^{\deg_{\mathbb{F}_{\mathbf{d}}}(f)} f\left(\tfrac{z}{\|z\|}\right) + \mathbf{d}(-\ln\|z\|)\tilde{\omega}\right).$$

Next, using the criterion (4) of Theorem 7.7 and the inequality (15.10) we complete the proof. □

More constructive (but more conservative) sufficient condition can be derived under the assumption about quadratic stabilizability of the nominal system.

Corollary 15.2 ([2]) *Let conditions of Theorem 15.1 hold for* $\Xi = I_n$ *and*

$$\begin{aligned} \|\tilde{\omega}(t,z)\| &< -z^\top P f\left(\tfrac{z}{\|z\|}\right)\|z\|^{\alpha+\mu-1} \quad \text{if } 0 < \|z\| \le 1, \\ \|\tilde{\omega}(t,z)\| &< -z^\top P f\left(\tfrac{z}{\|z\|}\right)\|z\|^{\beta+\mu-1} \quad \text{if } \quad \|z\| > 1, \end{aligned} \tag{15.11}$$

for all $t > 0$ *and all* $z \in \mathbb{R}^{n+m}\backslash\{\mathbf{0}\}$, *where, as before,* $\|z\| = \sqrt{z^\top P z}$ *and* $\tilde{\omega}(t,z) = \binom{\omega(t,z)}{0}$ *and* α, β *are defined in Proposition 5.3. Then the closed-loop system* (15.9), (15.2) *with k given by* (15.4) *is*

- *globally uniformly finite-time stable if* $\mu < 0$;
- *globally exponentially stable if* $\mu = 0$;
- *globally practically fixed-time stable if* $\mu > 0$.

[4] The proof is reproduced from Polyakov A. Sliding mode control design using canonical homogeneous norm. *Int Journal Robust Nonlinear Control.* 2019; 29:682–701. https://doi.org/10.1002/rnc.4058 ©2018 John Wiley & Sons, Ltd. Reused with permission of John Wiley & Sons.

Proof[5] For $\Xi = I_n$ the inequality (15.10) is equivalent to

$$z^T P\left(f\left(\frac{z}{\sqrt{z^\top P z}}\right) + \frac{\mathbf{d}(-\ln\|z\|)\tilde{w}}{\|z\|^{\deg_{\mathbb{F}_\mathbf{d}}(f)}}\right) < 0.$$

Using the Cauchy–Schwarz inequality (see Appendix A) we derive

$$z^T P\mathbf{d}(-\ln\|z\|_\mathbf{d})\tilde{\omega} \leq \sqrt{z^\top P z}\sqrt{\tilde{\omega}^\top \mathbf{d}^\top(-\ln\|z\|)P\mathbf{d}(-\ln\|z\|)\tilde{\omega}},$$

and the sufficient condition of the robust stability becomes

$$\sqrt{\tilde{\omega}^\top \mathbf{d}^\top(-\ln\|z\|)P\mathbf{d}(-\ln\|z\|)\tilde{\omega}} < -\|z\|^{\deg_{\mathbb{F}_\mathbf{d}}(f)-1} z^T Pf\left(\frac{z}{\|z\|}\right). \tag{15.12}$$

Denote $\phi(s) = \tilde{\omega}^\top \mathbf{d}^\top(s)P\mathbf{d}(s)\tilde{\omega}$. By Proposition 5.3 we derive

$$2\beta\phi(s) \leq \frac{d}{ds}\phi(s) \leq 2\alpha\phi(s) \quad \Rightarrow \quad \begin{cases} \phi(s) \leq e^{2\alpha s}\phi(0) \text{ for } s \geq 0, \\ \phi(s) \leq e^{2\beta s}\phi(0) \text{ for } s \leq 0. \end{cases}$$

Hence, denoting $s = -\ln\|z\|$ and using (15.11) we complete the proof. □

> A robustness (in the ISS sense) of the designed control with respect to rather large class of perturbations follows from Theorem 7.13.

Example 15.2 *(Robust quadratic finite-time stabilization,*[6] [2]*)* Let us study the robustness of the control system designed in Example 15.1 with the disturbance function

$$\tilde{\omega} = \begin{pmatrix} 0 \\ \omega_0 \\ 0 \end{pmatrix},$$

where the function ω_0 is assumed to be bounded as follows:

$$|\omega_0| \leq \delta\|\xi\|_\mathbf{d}, \quad \xi = (x_1, x_2, u)^\top.$$

[5] The proof is reproduced from Polyakov A. Sliding mode control design using canonical homogeneous norm. *Int Journal Robust Nonlinear Control*. 2019; 29:682–701. https://doi.org/10.1002/rnc.4058 ©2018 John Wiley & Sons, Ltd. Reused with permission of John Wiley & Sons.

[6] The example is reproduced from Polyakov, A. Sliding mode control design using canonical homogeneous norm. *Int Journal Robust Nonlinear Control*. 2019; 29:682–701. https://doi.org/10.1002/rnc.4058 ©2018 John Wiley & Sons, Ltd. Reused with permission of John Wiley & Sons.

Taking into account $\|\xi\|_{\mathbf{d}} = \|z\|$ and $\mathbf{d}(s) = \mathrm{diag}\{e^{3s}, e^{2s}, e^{s}\}$ we derive

$$\sqrt{\tilde{\omega}^\top \mathbf{d}^\top(-\ln\|z\|) P \mathbf{d}(-\ln\|z\|)\tilde{\omega}} = |\omega_0| \sqrt{e_2^\top \mathbf{d}^\top(-\ln\|z\|) P \mathbf{d}(-\ln\|z\|) e_2} =$$

$$\frac{|\omega_0|}{\|z\|^2}\sqrt{e_2^\top P e_2} \leq \frac{\delta}{\|z\|}\sqrt{e_2^\top P e_2}.$$

Since $\deg_{\mathbb{F}_{\mathbf{d}}}(f_0) = -1$ (see, Example 15.1) then the condition (15.12) becomes

$$\frac{\delta}{\|z\|}\sqrt{e_2^\top P e_2} \leq -\frac{1}{\|z\|}\frac{z^\top}{\|z\|} P f\left(\frac{z}{\|z\|}\right),$$

and the sufficient condition of the robust finite-time stability is

$$\delta \leq -\frac{\sup_{\|x\|=1} x^\top P f(x)}{\sqrt{e_2^\top P e_2}}, \quad e_2 = (0, 1, 0)^\top \in \mathbb{R}^3$$

provided that $z^\top P f(z) < 0$.

15.2 Homogeneous Stabilization of Affine-in-Control Systems

In this section, we study the problem of a homogeneous stabilization of the affine-in-control system

$$\dot{x} = f_0(x) + \sum_{i=1}^{m} u_i f_i(x) + g(t, x, u), \quad t > t_0, \tag{15.13}$$

where $x(t) \in \mathbb{R}^n$ and $u(t) = (u_1(t), ..., u_m(t) \in \mathbb{R}^m$, $f_i : \mathbb{R}^n \backslash \{\mathbf{0}\} \mapsto \mathbb{R}^n$ is a nonlinear continuous vector fields, $i = 0, 1, ..., m$, and a locally bounded measurable function $g : \mathbb{R}^{n+m+1} \mapsto \mathbb{R}^n$ model perturbations of the system.

15.2.1 Stabilization Using Homogeneous Convex Embedding

Let us consider the system (15.13) with constant vector fields $f_i, i = 1, ..., m$:

$$\sum_{i=1}^{m} u_i f_i(x) = Bu, \quad B \in \mathbb{R}^{n\times m}. \tag{15.14}$$

The assumption that the control system has such a structure is not conservative, since any nonlinear control system can be transformed to the form (15.13), (15.14) with $B = \left(\begin{smallmatrix} \mathbf{0} \\ I_m \end{smallmatrix}\right)$ by means of the dynamic extension (see the previous section). Therefore, the universal controller (15.4) designed in Theorem 15.1 can be utilized for stabilization of the system (15.13), (15.14). However, the design of appropriate parameters a and b of the universal controller is complicated in the general case. The following theorem shows that it can be based on the homogeneous convex embedding procedure [3] introduced in Chap. 6 (see Theorem 6.10).

Theorem 15.2 *Let a linear continuous dilation* $\mathbf{d}$ *in* $\mathbb{R}^n$ *be monotone with respect to the weighted norm* $\|x\| = \sqrt{x^\top P x}$ *with* $0 \prec P = P^\top \in \mathbb{R}^{n\times n}$ *satisfying (5.6) and* $\mathbf{d}(s)B = e^s B, \forall s \in \mathbb{R}$. *Let* $\|\cdot\|_{\mathbf{d}}$ *be the corresponding canonical homogeneous norm. Let the vector field* $f_0 \in C^1(\mathbb{R}^n\backslash\{\mathbf{0}\}; \mathbb{R}^n)$ *be locally bounded on* $\mathbb{R}^n$ *and* $\mathbf{d}$*-homogeneous of degree* $\mu_0 \geq -1$.

If the matrices $A_k \in \mathbb{R}^{n\times n}$, $k = 1, ..., N$ *are defined by the formula (6.45) for* f_0 *and the system of linear matrix inequalities*

$$P(A_k + BK) + (A_k + BK)^\top P \prec 0, \quad k = 1, ..., N \tag{15.15}$$

is feasible with respect to $K \in \mathbb{R}^{m\times n}$ *then the system* (15.13), (15.14) *with* $g = \mathbf{0}$ *and with the feedback control*

$$u = \|x\|_{\mathbf{d}}^{1+\mu_0} K\mathbf{d}(-\ln\|x\|_{\mathbf{d}})x \tag{15.16}$$

is $\mathbf{d}$*-homogeneous of degree* μ_0 *and globally uniformly asymptotically stable. The control* u *is continuous for* $\mu_0 > -1$ *and discontinuous (but locally bounded) at* $\mathbf{0}$ *if* $\mu_0 = -1$. *Moreover, the system with* $g \neq \mathbf{0}$ *is*

- *globally uniformly finite-time stable if* $\mu_0 < 0$;
- *globally uniformly exponentially stable if* $\mu_0 = 0$;
- *globally uniformly fixed-time stable if* $\mu_0 > 0$;

provided that

$$\sup_{t\in\mathbb{R}} \frac{x^\top \mathbf{d}^\top(-\ln\|x\|_{\mathbf{d}})P\mathbf{d}(-\ln\|x\|_{\mathbf{d}})g(t,u(x),x)}{x^\top\mathbf{d}^\top(-\ln\|x\|_{\mathbf{d}})PG_{\mathbf{d}}\mathbf{d}(-\ln\|x\|_{\mathbf{d}})x} \leq \kappa\|x\|_{\mathbf{d}}^{-\mu_0}, \quad \forall x \neq \mathbf{0}, \tag{15.17}$$

where $u(x)$ *is given by* (15.16),

$$0 \leq \kappa < \rho := -\max_{k=1,...,N} \lambda_{\max}\left(Q^{-\frac{1}{2}}\left(P(A_k + BK) + (A_k + BK)^\top P\right)Q^{-\frac{1}{2}}\right)$$

and $Q = PG_{\mathbf{d}} + G_{\mathbf{d}}^\top P$.

Proof The vector field $f_0 \in C^1(\mathbb{R}^n\backslash\{\mathbf{0}\}; \mathbb{R}^n)$ may have a discontinuity only at $\mathbf{0}$. Since f_0 is locally bounded that then the discontinuity (if it exists) is finite and the system (15.13), (15.14) with $g = \mathbf{0}$ has solutions in the sense of Filippov (see Chap. 2). The homogeneity of the closed-loop system and the continuity properties of the control can be proven as in Theorem 9.1.

Using the formula (5.21) we derive

$$\frac{d\|x\|_{\mathbf{d}}}{dt} \overset{a.e.}{=} \|x\|_{\mathbf{d}} \frac{x^\top \mathbf{d}^\top(-\ln\|x\|_{\mathbf{d}})P\mathbf{d}(-\ln\|x\|_{\mathbf{d}})(f_0(x)+Bu(x)+g)}{x^\top \mathbf{d}^\top(-\ln\|x\|_{\mathbf{d}})PG_{\mathbf{d}}\mathbf{d}(-\ln\|x\|_{\mathbf{d}})x}, \quad x \neq \mathbf{0}.$$

Using the $\mathbf{d}$-homogeneity of f_0 and the identity $\mathbf{d}(s)B = e^s B, \forall s \in \mathbb{R}$ we derive

$$\frac{d}{dt}\|x\|_{\mathbf{d}} \overset{a.e.}{=} \|x\|_{\mathbf{d}} \frac{x^\top \mathbf{d}^\top(-\ln\|x\|_{\mathbf{d}})P(\|x\|_{\mathbf{d}}^{\mu_0} f_0(\mathbf{d}(-\ln\|x\|_{\mathbf{d}})x)+BK\mathbf{d}(-\ln\|x\|_{\mathbf{d}})x)}{x^\top \mathbf{d}^\top(-\ln\|x\|_{\mathbf{d}})PG_{\mathbf{d}}\mathbf{d}(-\ln\|x\|_{\mathbf{d}})x}$$

$$+\|x\|_{\mathbf{d}} \frac{x^\top \mathbf{d}^\top(-\ln\|x\|_{\mathbf{d}})P\mathbf{d}(-\ln\|x\|_{\mathbf{d}})g}{x^\top \mathbf{d}^\top(-\ln\|x\|_{\mathbf{d}})PG_{\mathbf{d}}\mathbf{d}(-\ln\|x\|_{\mathbf{d}})x}.$$

By Theorem 6.10, we have

$$f_0(x) \in F_0(x),$$

where F_0 is the first-order $\mathbf{d}$-homogeneous extension of f_0 given by the formula (6.40). This means that

$$f_0(x) = \|x\|_{\mathbf{d}}^{\mu_0} \sum_{k=1}^{N} \alpha_k(x)\mathbf{d}(\ln\|x\|_{\mathbf{d}})A_k\mathbf{d}(-\ln\|x\|_{\mathbf{d}})x,$$

where the functions $\alpha_k : \mathbb{R}^n \mapsto \mathbb{R}_+, k = 1, ..., N$ are such that

$$\sum_{k=1}^{N} \alpha_k(x) = 1, \quad \forall x \in \mathbb{R}^n \backslash \{\mathbf{0}\}.$$

Hence, we derive

$$\frac{d}{dt}\|x\|_{\mathbf{d}} \overset{a.e.}{=} \|x\|_{\mathbf{d}}^{1+\mu_0} \sum_{k=1}^{N} \alpha_k(x) \frac{x^\top \mathbf{d}^\top(-\ln\|x\|_{\mathbf{d}})P(A_k+BK)\mathbf{d}(-\ln\|x\|_{\mathbf{d}})x}{x^\top \mathbf{d}^\top(-\ln\|x\|_{\mathbf{d}})PG_{\mathbf{d}}\mathbf{d}(-\ln\|x\|_{\mathbf{d}})x}$$

$$+\|x\|_{\mathbf{d}} \frac{x^\top \mathbf{d}^\top(-\ln\|x\|_{\mathbf{d}})P\mathbf{d}(-\ln\|x\|_{\mathbf{d}})g}{x^\top \mathbf{d}^\top(-\ln\|x\|_{\mathbf{d}})PG_{\mathbf{d}}\mathbf{d}(-\ln\|x\|_{\mathbf{d}})x}.$$

Taking into account the LMIs (15.15) and the restriction (15.17), we conclude

$$\frac{d}{dt}\|x\|_{\mathbf{d}} \overset{a.e.}{\leq} -\|x\|_{\mathbf{d}}^{1+\mu_0} \sum_{k=1}^{N} \alpha_k(x)\rho - \kappa\|x\|_{\mathbf{d}}^{1+\mu_0} = -(\rho-\kappa)\|x\|_{\mathbf{d}}^{1+\mu_0}$$

for $x \neq \mathbf{0}$. The proof is complete. □

The suggested algorithm of homogeneous stabilization of the affine-in-control system (15.13)–(15.16) requires the vector field f_0 to be **d**-homogeneous of the desired degree $\mu_0 \in \mathbb{R}$. For the linear case (see Chap. 9), a special component u_0 of the control law $u = u_0 + u_{\text{hom}}$ has been introduced in order to make the vector field

$$\tilde{f}_0(x) = f_0(x) + Bu_0(x), \quad x \in \mathbb{R}^n$$

d-homogeneous of a desired degree $\mu_0 \in \mathbb{R}$. In the linear case, the existence of such a *homogenizer* u_0 follows from the controllability of the system and its design is based on solving the algebraic equation (9.5). A similar equation can be derived for the nonlinear case.

Lemma 15.1 *Let $f_0 \in C^1(\mathbb{R}^n\backslash\{\mathbf{0}\}; \mathbb{R}^n)$ and $B = \left(\begin{smallmatrix}\mathbf{0}\\ I_m\end{smallmatrix}\right)$. Let there exist $G_0 \in \mathbb{R}^{n\times n}$ and $\mu_0 \in \mathbb{R}^n$ such that*

$$\begin{cases} (I_{n-m}\ \mathbf{0})\left(\frac{\partial f_0(x)}{\partial x}(I_n+\mu_0 G_0)x - (I_n+\mu_0 I_n+\mu_0 G_0) f_0(x)\right)=\mathbf{0}, & \forall x \neq \mathbf{0}. \\ G_0 B = \mathbf{0}, \end{cases} \tag{15.18}$$

If $u_0 \in C^1(\mathbb{R}^n\backslash\{\mathbf{0}\}; \mathbb{R}^m)$ is a solution of the following PDE

$$\tfrac{\partial u_0(x)}{\partial x}(I_n + \mu_0 G_0)x = u_0(x) + v_0(x), \tag{15.19}$$

where

$$v_0(x) = -B^\top\left(\tfrac{\partial f_0(x)}{\partial x}(I_n+\mu_0 G_0)x - (I_n+\mu_0 I_n+\mu_0 G_0) f_0(x)\right),$$

then the vector field

$$\tilde{f}_0(x) = f_0(x) + Bu_0(x) \tag{15.20}$$

*is **d**-homogeneous of the degree μ_0 with $G_{\mathbf{d}} = I_n + \mu_0 G_0$ provided that*

$$1 + \mu_0 \Re(\lambda_i(G_0)) > 0, \quad \forall i = 1, 2, \ldots, n,$$

where $\lambda_i(G_0)$, $i = 1, \ldots, n$ are eigenvalues of G_0.

The proof immediately follows from the criterion of **d**-homogeneity of the vector field $\tilde{f}_0$ (see Theorem 6.2):

$$\left(\frac{\partial f_0(x)}{\partial x} + B\frac{\partial u_0(x)}{\partial x}\right)(I_n + \mu G_0)x = (I_n + \mu I_n + \mu G_0)(f_0(x) + Bu_0(x)).$$

Indeed, multiplying this identity by $(I_{n-m}\ \mathbf{0})$ from the left we derive the first equation in (15.18) for $G_0 B = \mathbf{0}$. Moreover, multiplying this identity on $B^\top$ from the left we derive the PDE (15.19) for finding u_0.

15.2.2 Sign-Definite Homogeneous Feedback

In the first section of this chapter, we considered the design of a universal controller for nonlinear system in the general case. The following corollary treats the special case of the nonlinear system (15.13), then the nonlinear functions f_i have form of the so-called first-order **d**-homogeneous approximation (see Chap. 6). In this case, the feedback control can be defined using linear matrix inequalities (LMIs).

Theorem 15.3 *Let f_i admit the following representation:*

$$f_i(x) = \|x\|_{\mathbf{d}}^{\mu_i}\mathbf{d}(\ln \|x\|_{\mathbf{d}})A_i\mathbf{d}(-\ln \|x\|_{\mathbf{d}})x, \quad A_i \in \mathbb{R}^{n\times n}, \quad i = 0, 1, ..., m,$$

where the canonical homogeneous norm $\|\cdot\|_{\mathbf{d}}$ is induced by the norm $\|x\| = \sqrt{x^\top Px}$ with $0 \prec P = P^\top \in \mathbb{R}^{n\times n}$ satisfying 5.6.

If there exist $\tau_0 > 0$, $\tau_i \in \mathbb{R}$, $i = 1, 2, ..., m$ such that

$$P\left(A_0 + \sum_{i=1}^{m}\tau_i A_i\right) + \left(A_0 + \sum_{i=1}^{m}\tau_i A_i\right)P + \tau_0 P \prec 0 \tag{15.21}$$

then the closed-loop system (15.13) *with $g = \mathbf{0}$ and the control*

$$u_i(x) = \|x\|_{\mathbf{d}}^{\mu_0-\mu_i}\tau_i \tag{15.22}$$

*is globally uniformly asymptotically stable and **d**-homogeneous of degree μ_0. Moreover, the perturbed ($g \neq \mathbf{0}$) system is globally uniformly finite-time stable for $\mu_0 < 0$, globally uniformly exponentially stable for $\mu_0 = 0$, globally uniformly nearly fixed-time stable for $\mu_0 > 0$ provided that*

$$ess \sup_{t\in\mathbb{R}} \sqrt{\tilde{g}^\top(t,x)\mathbf{d}^\top(-\ln \|x\|_{\mathbf{d}})P\mathbf{d}(-\ln \|x\|_{\mathbf{d}})\tilde{g}(t,x)} \le \frac{\tau_0}{2}\|x\|_{\mathbf{d}}^{\mu_0}, \quad \forall x \neq \mathbf{0},$$

where $\tilde{g}(t,x) = g(t,x,\tau_1\|x\|_{\mathbf{d}}^{\mu_0-\mu_1}, ..., \tau_m\|x\|_{\mathbf{d}}^{\mu_0-\mu_m})$.

Proof The closed-loop system has the form

$$\dot{x} \overset{a.e.}{=} \|x\|_{\mathbf{d}}^{\mu_0}\mathbf{d}(-\ln \|x\|_{\mathbf{d}})\left(A_0 + \sum_{i=1}^{m}\tau_i A_i\right)\mathbf{d}(-\ln \|x\|_{\mathbf{d}})x + g,$$

and

$$\frac{d\|x\|_{\mathbf{d}}}{dt} \overset{a.e.}{=} \frac{\|x\|_{\mathbf{d}}^{1+\mu_0}z^\top P\left(A_0+\sum_{i=1}^{m}\tau_i A_i\right)z + \|x\|_{\mathbf{d}}z^\top P\mathbf{d}(-\ln \|x\|_{\mathbf{d}})\tilde{g}}{z^\top PG_{\mathbf{d}}z}, \quad z = \mathbf{d}(-\ln \|x\|_{\mathbf{d}})x.$$

Taking into account $z^\top Pz = 1$, by the Cauchy–Schwarz inequality we derive

$$z^\top P\mathbf{d}(-\ln\|x\|_\mathbf{d})\tilde{g} \le \sqrt{\tilde{g}^\top \mathbf{d}^\top(-\ln\|x\|_\mathbf{d})P\mathbf{d}(-\ln\|x\|_\mathbf{d})\tilde{g}} \overset{a.e.}{\le} \tau_0\|x\|_\mathbf{d}^{\mu_0}$$

and

$$\frac{d\|x(t)\|_\mathbf{d}}{dt} \overset{a.e.}{\le} \frac{\|x\|_\mathbf{d}^{1+\mu_0} z^\top P\left(A_0+\frac{\tau_0}{2}I_n+\sum_{i=1}^m \tau_i A_i\right)z}{z^\top PG_\mathbf{d}z}.$$

Using the LMI (15.21) we complete the proof. □

Notice that if the vector fields f_i are not homogeneous but admit the first-order $\mathbf{d}$-homogeneous approximations at $\mathbf{0}$ then the homogeneous domination principle (Theorem 7.5) and the proven corollary provide a controller (15.22) stabilizing the system (15.13) locally. The controllers (15.22) are sign-definite. The sign is defined by the sign of the parameter τ_i. The design of sign-definite regulators is important for some control applications. For example, air blowers of flow control systems [4] are sign-definite actuators.

Example 15.3 Let us consider the control system in $\mathbb{R}^2$

$$\dot{x}_1 = x_2, \quad \dot{x}_2 = -\sqrt{x_1^2+|x_2|}\sin(x_2) + \frac{|x_2|x_1}{x_1^2+x_2}u, \quad t>0.$$

Considering the dilation $\mathbf{d}(s) = \mathrm{diag}\{e^{2s}, e^s\}$, this system can be rewritten as follows:

$$\dot{x} = f_0(x) + uf_1(x) + g(x,u),$$

where

$$f_0(x) = \|x\|_\mathbf{d}\mathbf{d}(\ln\|x\|_\mathbf{d})\left(\begin{smallmatrix}0 & 1\\ 0 & -1\end{smallmatrix}\right)\mathbf{d}(-\ln\|x\|_\mathbf{d})x,$$

$$f_1(x) = \mathbf{d}(\ln\|x\|_\mathbf{d})\left(\begin{smallmatrix}0 & 0\\ 1 & 0\end{smallmatrix}\right)\mathbf{d}(-\ln\|x\|_\mathbf{d})x,$$

and

$$g(x,u) = \begin{pmatrix} x_2 \\ -\sqrt{x_1^2+|x_2|}\sin x_2 + \frac{|x_2|x_1u}{x_1^2+|x_2|}\end{pmatrix} - f_0(x) - f_1(x).$$

Let the canonical homogeneous norm induced by the norm $\|x\| = \sqrt{x^\top P x^\top}$ such that

$$P\left(\begin{smallmatrix}0 & 1\\ -1 & -1\end{smallmatrix}\right) + \left(\begin{smallmatrix}0 & 1\\ -1 & -1\end{smallmatrix}\right)^\top P + \tau_0 P \prec 0, \quad PG_\mathbf{d} + G_\mathbf{d}^\top P \succ 0, \quad P \succ 0,$$

where $\tau_0 > 0$. In this case, for $u = -\|x\|_\mathbf{d}$, by Theorem 15.3, the system $\dot{x} = f_0(x) + uf_1(x)$ is globally uniformly asymptotically stability. The local asymptotic stability of the original system is assured in the zone $\|x\|_\mathbf{d} < r$ provided that

$$\sqrt{g^\top(x, -\|x\|_{\mathbf{d}})\mathbf{d}^\top(-\ln\|x\|_{\mathbf{d}})P\mathbf{d}^\top(-\ln\|x\|_{\mathbf{d}})g(x, -\|x\|_{\mathbf{d}})} \le \tau_0\|x\|_{\mathbf{d}},$$

for all $\forall x \in \mathbb{R}^n : 0 < \|x\|_{\mathbf{d}} < r$, where $r > 0$ is a number.

15.2.3 Globally Homogeneous Fixed-Time Stabilization

The following theorem shows that a finite-time stabilizer of the homogeneous system (15.13) with $g = 0$ can be easily transformed to the constant-time stabilizer provided that the settling-time function of the closed-loop system is known.

Theorem 15.4 *Let $\mathbf{d}$ be a linear continuous dilation in $\mathbb{R}^n$ and $g = \mathbf{0}$. Let the vector field $f_0 \in C(\mathbb{R}^n; \mathbb{R}^n)$ be $\mathbf{d}$-homogeneous of degree 0 and the vector fields $f_i \in C(\mathbb{R}^n; \mathbb{R}^n)$ be $\mathbf{d}$-homogeneous of degree $\mu_i \in \mathbb{R}$, $i = 1, 2, ..., m$. Let $\tilde{u} = (\tilde{u}_1, ..., \tilde{u}_m) \in C(\mathbb{R}^n \backslash \{\mathbf{0}\}; \mathbb{R}^m)$ be a finite-time stabilizer of the system (15.13) such that all solutions of the closed-loop system are unique in the forward time. Let $\tilde{T} : \mathbb{R}^n \mapsto \mathbb{R}_+$ be the $\mathbf{d}$-homogeneous settling-time function of the closed-loop system with the feedback $\tilde{u}$. If $\deg_{\mathbf{d}}(\tilde{T}) = 1$ then the system (15.13) with the initial-state-dependent feedback*

$$u_i = \left(\tfrac{T}{\tilde{T}(x(0))}\right)^{\mu_i} \tilde{u}_i\left(\mathbf{d}\left(\ln \tfrac{T}{\tilde{T}(x(0))}\right)x\right), \quad x(0) \neq 0 \tag{15.23}$$

is constant-time stable with the settling-time constant $T > 0$.

Proof Considering the variable $z = \mathbf{d}(\ln T - \ln \tilde{T}(x(0)))x$ we derive

$$\dot{z} = \mathbf{d}\left(\ln \tfrac{T}{\tilde{T}(x(0))}\right)\left(f_0(x) + \sum_{i=1}^{m}\left(\tfrac{T}{\tilde{T}(x(0))}\right)^{\mu_i} f_i(x)\tilde{u}(z)\right) = f_0(z) + \sum_{i=1}^{m} f_i(z)\tilde{u}_i(z),$$

where the $\mathbf{d}$-homogeneity of f_i has been utilized on the last step. By assumption, the obtained system is finite-time stable with the settling-time function $\tilde{T}$. Since solutions of the system are unique in the forward time then for any $z(0) \neq \mathbf{0}$ we have

$$z(t) \neq \mathbf{0}, \forall t \in [0, \tilde{T}(z(0))) \quad \text{and} \quad z(t) = \mathbf{0}, \forall t \geq \tilde{T}(z(0)).$$

Taking into account the $\mathbf{d}$-homogeneity of $\tilde{T}$

$$\tilde{T}(z(0)) = \tilde{T}\left(\mathbf{d}(\ln T - \ln \tilde{T}(x(0)))x(0)\right) = \tfrac{T\tilde{T}(x(0))}{\tilde{T}(x(0))} = T \quad \text{for all } x(0) \neq \mathbf{0}$$

we conclude that the system (15.13), (15.23) is constant-time stable with the settling-time constant $T > 0$. The proof is complete. □

In the view of Corollary 6.2 the condition $\deg_{\mathbf{d}}(\tilde{T}) = 1$ is not conservative.

The above theorem proves one interesting property of affine-in-control homogeneous systems. If $f_i \in C$ are **d***-homogeneous vector fields and* $\deg_{\mathbf{d}} f_0 = 0$ *then the finite-time stabilizability of the system* (15.13) *implies its null controllability.*

15.3 Homogeneous Rigid-Body Attitude Control

The dynamics of a fully actuated rigid-body rotation motion is governed by the nonlinear differential equation (see, e.g., [5]):

$$\dot{R} = R\omega^{\wedge}, \quad t > 0, \quad R(0) = R_0, \tag{15.24a}$$

$$\dot{\omega} = J^{-1}\left(-\omega \times J\omega + M\right), \quad \omega(0) = \omega_0, \tag{15.24b}$$

where $R(t) \in SO(3)$ is the so-called rotation matrix[7] (see below) with respect to the inertial frame of the rigid body, $\omega \in \mathbb{R}^3$ is the angular velocity in body frame,

$$\omega^{\wedge} = \begin{pmatrix} 0 & -\omega_3 & \omega_2 \\ \omega_3 & 0 & -\omega_1 \\ -\omega_2 & \omega_1 & 0 \end{pmatrix}, \quad \omega \times J\omega = \omega^{\wedge}(J\omega),$$

$J \in \mathbb{R}^{3\times 3}$ is the inertia matrix with respect to the rigid body's mass center and $M(t) \in \mathbb{R}^3$ is the torque expressed in the body frame (control input).

In this section, we investigate how to design a homogeneous feedback controller M that can ensure a *global uniform finite/fixed-time attitude tracking*:

$$\forall R_0 \in SO(3), \forall \omega_0 \in \mathbb{R}^3, \exists T \in \mathbb{R}_+ : R(t) = R_{\mathrm{d}}(t), \forall t \geq T,$$

where T is independent of R_0, ω_0 in the fixed-time case. The linear dilations, as well as the generalized homogeneity, are introduced for normed vector spaces. The $SO(3)$ group does not admit a dilation in the common sense. So, the homogeneity-based control cannot be introduced directly for the model (15.24a). However, the Lie algebra[8] $\mathfrak{so}(3)$ (see below) is equivalent to a vector space, so the homogeneous control can be designed for a model expressed in terms of the so-called exponential coordinates.

[7] $SO(3)$ is the notation for the special orthogonal group of rotation matrices in $\mathbb{R}^3$.

[8] Lie algebra is the tangent space of the Lie group at the identity.

Lie Group $SO(3)$ and Lie Algebra $\mathfrak{so}(3)$

Let us recall briefly some results (see, e.g., [6] for more details) about

$$\begin{aligned} SO(3) &= \{R \in \mathbb{R}^{3\times 3} : RR^\top = R^\top R = I_3,\ \det(R) = 1\}, \\ \mathfrak{so}(3) &= \{X \in \mathbb{R}^{3\times 3} : X^\top = -X\}. \end{aligned} \tag{15.25}$$

The $\mathfrak{so}(3)$ consists of 3×3 skew matrices with the Lie bracket given by the commutator $[X_1, X_2] = X_1X_2 - X_2X_1$, where $X_1, X_2 \in \mathfrak{so}(3)$. Since any $X \in \mathfrak{so}(3)$ can be uniquely represented by $X = x^\wedge$ with $x \in \mathbb{R}^3$, then $\mathfrak{so}(3)$ is isomorphic to $\mathbb{R}^3$. Notice also that for $x \in \mathbb{R}^3$ and $R \in SO(3)$ one holds

$$Rx^\wedge R^\top = (Rx)^\wedge.$$

The exponential map $\exp(\cdot)$ is a surjective map from $\mathfrak{so}(3)$ to $SO(3)$. For any $X \in \mathfrak{so}(3)$ the exponential mapping is given by

$$e^X = \exp(X) = \sum_{n=0}^{\infty} \tfrac{1}{n!} (X)^n \in SO(3).$$

> Any $X \in \mathfrak{so}(3)$ (or, equivalently, any vector $x \in \mathbb{R}^3 : X = x^\wedge$) uniquely defines a $R = \exp(X) \in SO(3)$.

For $R \in SO(3) : \mathrm{trace}(R) \neq -1$, the exponential map $\exp(\cdot)$ and its inverse $\log(\cdot)$ can be defined as follows:

$$\begin{aligned} R = \exp(X) &= I_3 + \tfrac{\sin|x|}{|x|} X + \tfrac{1-\cos|x|}{|x|^2} X^2, \\ X = \log(R) &= \tfrac{\phi}{2\sin\phi}\left(R - R^\top\right) \end{aligned} \quad X = x^\wedge, \tag{15.26}$$

where $\phi \in (-\pi, \pi)$ is a solution of the equation $1 + 2\cos\phi = \mathrm{trace}(R)$. For $\mathrm{trace}(R) = -1$ then there is an ambiguity in the definition of X, since in this case $X = (\pm\pi y)^\wedge$, where $y \in \mathbb{R}^3$ is a unique unit vector ($y^\top y = 1$) satisfying $Ry = y$. In fact, $SO(3)$ group in the exponential coordinates has a correspondence with the closed ball

$$\bar{B}(\pi) = \left\{x \in \mathbb{R}^3 : |x| \le \pi\right\}.$$

The transformation $\mathrm{Log}(\cdot) = \log(\cdot^\wedge) : \mathbb{R}^3 \mapsto SO(3)$ is a bijection on the open ball

$$B(\pi) = \left\{x \in \mathbb{R}^3 : |x| < \pi\right\}.$$

The ambiguity appears on the sphere

$$S(\pi) = \left\{x \in \mathbb{R}^3 : |x| = \pi\right\}.$$

The so-called left Jacobian of the group $SO(3)$ is given by (see, e.g., [6])

$$\begin{aligned} J_r(x) &= I_3 - \tfrac{1-\cos|x|}{|x|^2} x^\wedge + \tfrac{|x|-\sin|x|}{|x|^3} (x^\wedge)^2, \\ [J_r(x)]^{-1} &= I_3 + \tfrac{1}{2} x^\wedge + \left(\tfrac{1}{|x|^2} - \tfrac{1+\cos|x|}{2|x|\sin|x|}\right)(x^\wedge)^2, \end{aligned} \tag{15.27}$$

where $x = (x_1, x_2, x_3)^\top$. On the one hand, since the term

$$\frac{1}{|x|^2}(x^\wedge)^2 = \frac{1}{x_1^2 + x_2^2 + x_3^2} \begin{pmatrix} -x_2^2 - x_3^2 & x_1 x_2 & x_1 x_3 \\ x_1 x_2 & -x_1^2 - x_3^2 & x_2 x_3 \\ x_1 x_3 & x_2 x_3 & -x_1^2 - x_2^2 \end{pmatrix}$$

is uniformly bounded on $\mathbb{R}^3 \backslash \{\mathbf{0}\}$ and

$$1 - \tfrac{(1+\cos|x|)|x|}{2\sin|x|} \to 0 \quad \text{as} \quad |x| \to 0$$

then J_r^{-1} is continuous on $B(\pi)$. On the other hand, since

$$\frac{1 + \cos|x|}{\sin|x|} = \frac{1 + \cos|x|}{\pm\sqrt{1 - \cos^2|x|}} = \pm\sqrt{\frac{1 + \cos|x|}{1 - \cos|x|}}$$

then J_r^{-1} is bounded on $\bar{B}(\pi)$.

15.3.1 Homogeneous Attitude Tracking

An attitude tracking error has the special formulation in $SO(3)$ given by

$$R_e = R R_d^\top = e^{\theta_e^\wedge}, \tag{15.28}$$

where $R_d \in C^2(\mathbb{R}_+, SO(3))$ represents a desired smooth attitude trajectory and $\theta_e \in \mathbb{R}^3$ is the attitude tracking error in the Lie algebra $\mathfrak{so}(3)$. The smaller the attitude tracking error, the closer R_e to the identity matrix is. Indeed, if $R_e = I_3$ then $R = R_d$ and $\theta_e = \mathbf{0}$.

The attitude error dynamics can be presented in the exponential coordinates taking into account the following relation between θ_e and ω_e (see, e.g., [6] for more details)

$$\dot{\theta}_e = (J_r(\theta_e))^{-1}\omega_e, \tag{15.29}$$

where $J_r(\theta_e)$ is the right Jacobian given by (15.27),

$$\omega_e = R_d(\omega - \omega_d)$$

and $\omega_d = (R_d^\top \dot{R}_d)^\vee$ is the desired angular velocity in the body frame. The error dynamics of (15.28) is given by the impulsive system [5]:

$$\begin{cases} \dot{\xi} = \begin{pmatrix} \mathbf{0} & [J_r(\theta_e)]^{-1} \\ \mathbf{0} & \mathbf{0} \end{pmatrix} \xi + Bu, & \xi^- \in \mathcal{C}, \\ \xi = \Pi \xi^-, & \xi^- \in \mathcal{D}, \end{cases} \quad \text{with} \quad \xi = \begin{pmatrix} \theta_e \\ \omega_e \end{pmatrix}, \tag{15.30}$$

$$\mathcal{C} = B(\pi) \times \mathbb{R}^3, \quad \mathcal{D} = \left\{ \xi \in \overline{B}(\pi) \times \mathbb{R}^3 : \xi^\top H \xi = \pi^2 \text{ and } \xi^\top Q \xi > 0 \right\},$$

where $\xi^- = \lim_{h \to 0^-} \xi(t+h)$, $B(\pi) = \{\theta \in \mathbb{R}^3 : |\theta| < \pi\}$ is the open ball of the radius π in $\mathbb{R}^3$, $u = \dot{\omega}_e$ is the virtual stabilizer to be designed and

$$H = \begin{pmatrix} I_3 & \mathbf{0} \\ \mathbf{0} & \mathbf{0} \end{pmatrix}, \quad B = (\mathbf{0}, I_3)^\top, \quad \Pi = \begin{pmatrix} -I_3 & \mathbf{0} \\ \mathbf{0} & I_3 \end{pmatrix}, \quad Q = \begin{pmatrix} \mathbf{0} & I_3 \\ I_3 & \mathbf{0} \end{pmatrix}.$$

An asymptotic stabilization of the error system (15.30) to zero results in the attitude tracking since $R \to R_d$ and $\omega \to \omega_d$ as $\xi \to \mathbf{0}$.

The topological structure of the attitude control system (15.24) does not allow a global asymptotic attitude stabilization by a continuous static feedback law (see, e.g., [7]). Various discontinuous stabilizers are designed in the literature (see, e.g., [5, 8, 9] and references therein). The following theorem was originally discovered in [10]. We provide a more detailed proof and show that in the case of impulsive system, the closed loop may be just locally homogeneous, but the global homogeneous domination may ensure global finite-time, exponential or nearly fixed-time stability of the error dynamics dependently of the homogeneity degree.

Theorem 15.5 *For any $k_1 > 0$, $k_2 > 0$ and $0 < \varepsilon < \min\{\epsilon_1, \epsilon_2\}$, where*

$$\epsilon_1 = \frac{2\sqrt{2}}{3\sqrt{k_1}}, \quad \epsilon_2 = \frac{4k_2}{2ck_1 + k_2^2}, \quad c = \max_{|\theta_e| \le \pi} \lambda_{\max}\left([J_r(\theta_e)]^{-1} + [J_r(\theta_e)]^{-\top}\right),$$

the linear continuous dilation $\mathbf{d}(s) = e^{sG_\mathbf{d}}$, $s \in \mathbb{R}$ in $\mathbb{R}^6$ with the generator

$$G_\mathbf{d} = \begin{pmatrix} (1-\mu)I_3 & \mathbf{0} \\ \mathbf{0} & I_3 \end{pmatrix}, \quad \mu \in [-1, 0.5) \tag{15.31}$$

is strictly monotone with respect to the norm $\|\xi\| = \sqrt{\xi^\top P \xi}$ with

$$P = \begin{pmatrix} I_3 & \varepsilon I_3 \\ \varepsilon I_3 & k_1^{-1} I_3 \end{pmatrix} \succ 0. \tag{15.32}$$

If the canonical homogeneous norm $\|\cdot\|_\mathbf{d}$ in $\mathbb{R}^6$ is induced by $\|\cdot\|$ then the impulsive system (15.30) *with the* $\mathbf{d}$*-homogeneous controller $u = u_{hom}$*

$$u_{hom} = \|\xi\|_\mathbf{d}^{1+\mu} K \mathbf{d}(-\ln \|\xi\|_\mathbf{d})\xi, \quad \xi = \left(\theta_e^\top, \omega_e^\top\right)^\top, \quad K = -(k_1 I_3 \;\; k_2 I_3) \tag{15.33}$$

is

- *globally uniformly finite-time stable for* $\mu < 0$:

$$\xi(t) = \mathbf{0}, \quad \forall t \geq \|\xi(0)\|_{\mathbf{d}}^{-\mu}/(-\mu\rho);$$

- *globally uniformly exponentially stable for* $\mu = 0$:

$$\|\xi(t)\|_{\mathbf{d}} \leq e^{-\rho t}\|\xi(0)\|_{\mathbf{d}}, \quad \forall t \geq 0;$$

- *globally uniformly nearly fixed-time stable for* $\mu > 0$:

$$\|\xi(t)\|_{\mathbf{d}} \leq r, \quad \forall t \geq \tfrac{1}{\rho\mu r^{\mu}}, \forall r > 0,$$

where

$$\rho = \varepsilon\lambda_{\min}\left((PG_{\mathbf{d}} + G_{\mathbf{d}}^{\top} P)^{-\frac{1}{2}} \begin{bmatrix} 2k_1 I_3 & k_2 I_3 \\ k_2 & 2\frac{k_2}{\varepsilon k_1} - cI_3 \end{bmatrix} (PG_{\mathbf{d}} + G_{\mathbf{d}}^{\top} P)^{-\frac{1}{2}}\right). \quad (15.34)$$

Proof To prove the global finite-time (exponential or nearly fixed-time) stability of the closed-loop impulsive system (15.30) with $u = u_{hom}$ given by (15.33), let us show that all conditions of Theorem 7.9 are fulfilled.

(1) Let us show that the system is well-posed, i.e., it has a solution, at least, locally in time for any initial condition $\xi(0) \in \mathcal{C} \cup \mathcal{D}$ (see Sect. 7.2.4 for the definition of solution).

Using Schur complement we conclude

$$PG_{\mathbf{d}} + G_{\mathbf{d}}^{\top} P = \begin{pmatrix} 2(1-\mu)I_3 & (2-\mu)\varepsilon I_3 \\ (2-\mu)\varepsilon I_3 & 2k_1^{-1} I_3 \end{pmatrix} \succ 0 \quad \Leftrightarrow \quad \frac{4(1-\mu)}{(2-\mu)^2} > \varepsilon^2 k_1 \succ 0,$$

and

$$P \succ 0 \quad \Leftrightarrow \quad 1 > k_1\varepsilon^2.$$

Hence, for $\varepsilon < \epsilon_1$ the dilation $\mathbf{d}$ is strictly monotone and the canonical homogeneous norm $\|\xi\|_{\mathbf{d}}$ is well defined. Therefore, for any $\xi(0) \in \overline{\mathcal{C}}$ the impulse-free system has Filippov solution for $\mu = -1$ or a classical solution for $\mu \in (-1, 0.5)$ defined at least locally in time.

Let us denote $g_1(\xi) = \theta_e^{\top}\theta_e - \pi^2$, $g_2(\xi) = \omega_e^{\top}\theta_e$ and

$$f(\xi) = \begin{pmatrix} \mathbf{0} & [J_r(\theta_e)]^{-1} \\ \mathbf{0} & \mathbf{0} \end{pmatrix} \xi + Bu_{hom}(\xi).$$

Then one has

$$\mathcal{D} = \left\{\xi \in \overline{B}(\pi) \times \mathbb{R}^3 : g_1(\xi) = 0 \text{ and } g_2(\xi) > 0\right\}.$$

Since $[J_r(\theta_e)]^{-1}$ is given by (15.27) and $x^\top x^\wedge = \mathbf{0}$ for any $x \in \mathbb{R}^3$ then

$$\theta_e^\top [J_r(\theta_e)]^{-1}\omega_e = \theta_e^\top \omega_e. \tag{15.35}$$

For $\xi \in \partial C\backslash\overline{\mathcal{D}}$ we have $g_2(\xi) < 0$ and

$$\tfrac{\partial g_1}{\partial \xi} f(\xi) = 2\theta_e^\top J_r^{-1}\omega_e = 2\theta_e^\top \omega_e = 2g_2(\xi) < 0, \quad \xi \in \partial C\backslash\overline{\mathcal{D}}.$$

This means that $\partial C\backslash\overline{\mathcal{D}}$ is a repealing manifold of the impulse-free system. Therefore, for any $\xi(0) \in \mathcal{C}$ the system has a solution as long as $\xi^-(t) \notin \mathcal{D}$. Let us show that the trajectory can be pronged after the reset.

The identity $\Pi^\top Q \Pi = -Q$ implies that

$$\Pi\mathcal{D} \subset \partial C\backslash\overline{\mathcal{D}}.$$

So, taking into account $\partial C\backslash\overline{\mathcal{D}}$ is a repealing manifold of the system, we conclude that the trajectory $\xi(t)$ can be pronged after the jump/reset such that $\xi^-(t) \in \mathcal{C}$ till the next jump/reset. Therefore, for any $\xi(0) \in \mathcal{C} \cap \mathcal{D}$ the system has solution defined at least locally in time.

(2) Considering the canonical homogeneous norm as a Lyapunov function candidate for the closed-loop system (15.30) with $u = u_{hom}$ given by (15.33), we derive[9]

$$\tfrac{d}{dt}\|\xi\|_{\mathbf{d}} = \|\xi\|_{\mathbf{d}} \tfrac{\xi^\top \mathbf{d}^\top(-\ln\|\xi\|_{\mathbf{d}}) P \mathbf{d}(-\ln\|\xi\|_{\mathbf{d}})\dot{\xi}}{\xi^\top \mathbf{d}^\top(-\ln\|\xi\|_{\mathbf{d}}) P G_{\mathbf{d}} \mathbf{d}(-\ln\|\xi\|_{\mathbf{d}})\xi}, \quad \forall \xi^- \in \mathcal{C}. \tag{15.36}$$

Hence, using the identity $\dot{\omega}_e = u_{hom} := -\|\xi\|_{\mathbf{d}}^{2\mu} k_1\theta_e - \|\xi\|_{\mathbf{d}}^{\mu} k_2\omega_e$ we derive

$$\begin{aligned}
\xi^\top \mathbf{d}^\top(-\ln\|\xi\|_{\mathbf{d}}) P \mathbf{d}(-\ln\|\xi\|_{\mathbf{d}})\dot{\xi} &= \begin{pmatrix} \|\xi\|_{\mathbf{d}}^{\mu-1}\theta_e \\ \|\xi\|_{\mathbf{d}}^{-1}\omega_e \end{pmatrix}^\top P \begin{pmatrix} \|\xi\|_{\mathbf{d}}^{\mu-1}\dot{\theta}_e \\ \|\xi\|_{\mathbf{d}}^{-1}\dot{\omega}_e \end{pmatrix} \\
&= \|\xi\|_{\mathbf{d}}^{2(\mu-1)}\theta_e^\top\dot{\theta}_e + \varepsilon\|\xi\|_{\mathbf{d}}^{\mu-2}\omega_e^\top\dot{\theta}_e + \varepsilon\|\xi\|_{\mathbf{d}}^{\mu-2}\theta_e^\top\dot{\omega}_e + k_1^{-1}\|\xi\|_{\mathbf{d}}^{-2}\omega_e^\top\omega_e \\
&= \|\xi\|_{\mathbf{d}}^{2(\mu-1)} \underbrace{\theta_e^\top[J_r(\theta_e)]^{-1}\omega_e}_{\theta_e^\top\omega_e} + \varepsilon\|\xi\|_{\mathbf{d}}^{\mu-2}\omega_e^\top[J_r(\theta_e)]^{-1}\omega_e \\
&\quad + \varepsilon\theta_e^\top\left(-\|\xi\|_{\mathbf{d}}^{3\mu-2}k_1\theta_e - \|\xi\|_{\mathbf{d}}^{2\mu-2}k_2\omega_e\right) \\
&\quad + k_1^{-1}\omega_e^\top\left(-\|\xi\|_{\mathbf{d}}^{2\mu-2}k_1\theta_e - \|\xi\|_{\mathbf{d}}^{\mu-2}k_2\omega_e\right) \\
&= -\varepsilon\|\xi\|_{\mathbf{d}}^{3\mu-2}\theta_e^\top k_1\theta_e + \|\xi\|_{\mathbf{d}}^{\mu-2}\omega_e^\top(\varepsilon[J_r(\theta_e)]^{-1} - k_2k_1^{-1})\omega_e \\
&\quad - \varepsilon k_2\|\xi\|_{\mathbf{d}}^{2\mu-2}\theta_e^\top\omega_e.
\end{aligned}$$

[9] The passage of the proof between (and including) Equation 15.36 and the definition of W following Equation 15.38 on page 264 is reprinted from *Automatica*, 163/5, Y. Zhou, A. Polyakov, G. Zheng, Generalized homogeneous rigid-body attitude control, 111548, https://doi.org/10.1016/j.automatica.2024.111548 ©2024, with permission from Elsevier.

Therefore, the time derivative of the Lyapunov function candidate admits the representation

$$\frac{d}{dt}\|\xi\|_{\mathbf{d}} = \|\xi\|_{\mathbf{d}}^{1+\mu} \frac{\xi^\top \mathbf{d}^\top(-\ln\|\xi\|_{\mathbf{d}}) W_0 \mathbf{d}(-\ln\|\xi\|_{\mathbf{d}})\xi}{\xi^\top \mathbf{d}^\top(-\ln\|\xi\|_{\mathbf{d}}) P G_{\mathbf{d}} \mathbf{d}(-\ln\|\xi\|_{\mathbf{d}})\xi}, \quad \xi^- \in \mathcal{C}, \tag{15.37}$$

where

$$W_0 = \begin{pmatrix} -\varepsilon k_1 I_3 & -\frac{\varepsilon k_2}{2} I_3 \\ -\frac{\varepsilon k_2}{2} I_3 & \varepsilon \frac{J_r^{-1} + J_r^{-\top}}{2} - \frac{k_2}{k_1} I_3 \end{pmatrix}.$$

Since $[J_r(\cdot)]^{-1}$ is bounded on the closed ball $\overline{B}(\pi)$ then

$$0 < c = \max_{|\theta_e| \le \pi} \lambda_{\max}([J_r(\theta_e)]^{-1} + [J_r(\theta_e)]^{-\top}) < +\infty$$

and

$$\frac{d}{dt}\|\xi\|_{\mathbf{d}} \le -\epsilon \frac{\|\xi\|_{\mathbf{d}}^{1+\mu} \xi^\top \mathbf{d}^\top(-\ln\|\xi\|_{\mathbf{d}}) W \mathbf{d}(-\ln\|\xi\|_{\mathbf{d}})\xi}{\xi^\top \mathbf{d}^\top(-\ln\|\xi\|_{\mathbf{d}})(P G_{\mathbf{d}} + G_{\mathbf{d}}^\top P) \mathbf{d}(-\ln\|\xi\|_{\mathbf{d}})\xi}, \quad \xi^- \in \mathcal{C}, \tag{15.38}$$

where

$$W = \begin{pmatrix} 2k_1 I_3 & k_2 I_3 \\ k_2 I_3 & \left(\frac{2k_2}{\varepsilon k_1} - c\right) I_3 \end{pmatrix}.$$

Since for $\varepsilon \in (0, \epsilon_2)$ then $W \succ 0$ and

$$\dot{V} \le -\rho\|\xi\|_{\mathbf{d}}^{1+\mu}, \qquad \xi^- \in \mathcal{C}\backslash\{\mathbf{0}\}. \tag{15.39}$$

Therefore, the impulsive system (15.30) with $u = u_{hom}$ given by (15.33) is, at least, locally asymptotically stable.

Since $\xi^\top \Pi^\top P \Pi \xi \le \xi^\top P \xi$ for all $\xi \in \mathcal{D}$ then the Lyapunov function is decreasing along the trajectories of the system even in the case of jumps/resets, so the system state remains uniformly bounded as long as solution exists.

(3) Let us show that the closed-loop system is Zeno free. Suppose the contrary, there exists a sequence of time instances $t_i > 0$ such that $t_i < t_{i+1}$, $\xi^-(t_i) \in \mathcal{D}$ for all $i \in \mathbb{N}$ and $t_i \to t^* \in \mathbb{R}$ as $i \to +\infty$. Let $\Omega \subset \mathbb{R}^6$ be a compact such that $\xi(t) \in \Omega$ for all $t \in [0, t^*)$.

Since the sequence $\xi_i = \xi^-(t_i)$ is uniformly bounded then, by Bolzano–Weierstrass theorem, there exists a convergent subsequence $\xi_{i_k} \to \xi^*$ as $i_k \to +\infty$. Since $\xi_i \in \mathcal{D}$ then $\xi^* \in \overline{\mathcal{D}}$. Moreover, the function $t \mapsto \xi(t)$ is absolutely continuous on (t_i, t_{i+1}) and

$$\Pi\xi(t_{i_k}) = \xi_{i_k+1} + \int_{t_{ik+1}}^{t_{ik}} \dot{\xi}(\tau)d\tau = \xi_{i_k+1} + \int_{t_{ik+1}}^{t_{ik}} f(\xi(\tau))d\tau.$$

Since $t_{i_k+1} - t_{i_k} \to 0$ as $i_k \to +\infty$, then, taking into account uniform boundedness of ξ we conclude that $\int_{t_{ik+1}}^{t_{ik}} f(\xi(\tau))d\tau \to 0$ as $i_k \to +\infty$. Hence, $\Pi\xi_{i_k} \to \Pi\xi^*$ and $\Pi\xi(t_{i_k}) = \xi_{i_k+1} + \int_{t_{ik+1}}^{t_{ik}} \dot{\xi}(\tau) \to \xi^*$ as $i_k \to +\infty$. Therefore, we have

$$\Pi\xi^* = \xi^*.$$

Since $\Pi\mathcal{D} \subset \partial\mathcal{C}\backslash\overline{D}$ then the above identity is possible only for

$$\xi^* \in \partial D = \left\{\xi \in \overline{B}(\pi) \times \mathbb{R}^3 : g_1(\xi) = 0 \text{ and } g_2(x) = 0\right\}.$$

For[10] $\xi \in \partial\mathcal{D}$ we have $g_1(\xi) = \theta_e^\top\theta_e - \pi^2 = 0$, $g_2(\xi) = \omega_e^\top\theta_e = 0$, and

$$\tfrac{\partial g_1}{\partial\xi} f(\xi, t) = 2\theta_e^\top J_r^{-1}\omega_e = 2\theta_e^\top\omega_e = 0, \;\; \xi \in \partial\mathcal{D},$$

$$\tfrac{\partial g_2}{\partial\xi} f(\xi, t) = 2\omega_e^\top J_r^{-1}\omega_e + 2\|\xi\|_{\mathbf{d}}^{\mu+1}\theta_e^\top K\mathbf{d}(-s_\xi)\xi, \;\; \xi \in \partial\mathcal{D}.$$

For $\theta^\top\omega_e = 0$, using (15.27), we derive $\omega_e^\top J_r^{-1}\omega_e = |\omega_e|^2 + \omega_e^\top\left(\frac{\theta_e^\wedge}{|\theta_e|}\right)^2\omega_e = \omega_e^\top\frac{\theta_e\theta_e^\top}{|\theta_e|^2}\omega_e = 0$. Since $\theta_e^\top K\mathbf{d}(-s_\xi)\xi = -\|\xi\|_{\mathbf{d}}^{\mu-1}\theta_e^\top K_1\theta_e - \|\xi\|_{\mathbf{d}}^{-1}k_2\theta_e^\top\omega_e$, then

$$\tfrac{\partial g_2}{\partial\xi} f(\xi) = -k_1\|\xi\|_{\mathbf{d}}^{2\mu}\theta_e^\top\theta_e \leq -k_1\pi^2\gamma, \quad \xi \in \partial\mathcal{D},$$

where $\gamma = \inf_{\xi\in\partial\mathcal{D}}\|\xi\|_{\mathbf{d}}^{2\mu} > 0$. In this case, by continuity, taking sufficiently small $\delta > 0$, we derive

$$\tfrac{\partial g_2}{\partial\xi} f(\xi) \leq -0.5k_1\pi^2\gamma, \quad \xi \in (\partial\mathcal{D}\dot{+}B(\delta)) \cap \Omega.$$

Hence, for sufficiently large $i_k \in \mathbb{N}$ we have

$$\tfrac{\partial g_2}{\partial\xi} f(\xi)\Big|_{\xi=\Pi\xi_{i_k}} < -0.5k_1\pi^2\gamma \quad \text{and} \quad \tfrac{\partial g_2}{\partial\xi} f(\xi)\Big|_{\xi=\xi_{i_k+1}} < -0.5k_1\pi^2\gamma.$$

Taking into account $g_2(\Pi\xi_{i_k}) < 0$ we conclude that $g_2(\xi(t)) < 0$ for $t \in (t_{i_k}, t_{i_k+1})$ provided that $i_k \in \mathbb{N}$ is large enough. This contradicts the reset condition $\xi^-(t_{i_k+1}) \in \mathcal{D}$, which requires $g_2(\xi^-(t_{i_k+1})) > 0$. Therefore, the impulsive system (15.30) with $u = u_{hom}$ given by (15.33) is Zeno free. Applying Theorem 7.9 we complete the proof. □

The fixed-time stabilizer of the attitude error can be designed by commutation of homogeneous controllers (as in Sect. 9.1.5).

[10] Text between the footnote marker and "...in this case by continuity," is reprinted from *Automatica*, 163/5, Y. Zhou, A. Polyakov, G. Zheng, Generalized homogeneous rigid-body attitude control, 111548, https://doi.org/10.1016/j.automatica.2024.111548 ©2024, with permission from Elsevier.

Corollary 15.3 (Fixed-time attitude stabilization, [10]) *Let the parameters K and P be defined as in Theorem 15.5. Let the linear dilations $\mathbf{d}_1$, $\mathbf{d}_2$ in $\mathbb{R}^6$ be defined as in Theorem 15.5 for $-1 \le \mu_1 < 0 < \mu_2 < 0.5$, respectively. Then $\mathbf{d}_1$ and $\mathbf{d}_2$ are strictly monotone with respect to the weighted Euclidean norm $\|\xi\| = \sqrt{\xi^\top P\xi}$ and the impulsive system* (15.30) *with $u = u_{fxt}$ given by*

$$u_{fxt} = \begin{cases} \|\xi\|_{\mathbf{d}_1}^{1+\mu_1} K\mathbf{d}(-\ln \|\xi\|_{\mathbf{d}_1})\xi & \text{if } \xi^\top P\xi \le 1, \\ \|\xi\|_{\mathbf{d}_2}^{1+\mu_2} K\mathbf{d}(-\ln \|\xi\|_{\mathbf{d}_2})\xi & \text{if } \xi^\top P\xi > 1 \end{cases} \tag{15.40}$$

is globally uniformly fixed-time stable:

$$\xi(t) = \mathbf{0}, \quad \forall t \ge \frac{1}{-\rho_1\mu_1} + \frac{1}{\rho_2\mu_2}, \quad \forall \xi(0) \in \mathbb{R}^6,$$

where the canonical homogeneous norms $\|\cdot\|_{\mathbf{d}_1}$, $\|\cdot\|_{\mathbf{d}_2}$ are induced by $\|\cdot\|$, but $\tilde{\rho}_1 > 0$ and $\tilde{\rho}_2 > 0$ are defined by the formula (15.34) *with $\mu = \mu_1$ and $\mu = \mu_2$, respectively.*

Proof First of all, notice that the control law (15.40) is a continuous function of ξ since

$$\|\xi\|_{\mathbf{d}_1} = 1 \quad \Leftrightarrow \quad \|\xi\|_{\mathbf{d}_2} = 1 \quad \Leftrightarrow \quad \|\xi\| = 1.$$

In this case, the Lyapunov function candidate

$$V(\xi) = \begin{cases} \|\xi\|_{\mathbf{d}_1} & \text{if } \|\xi\| \le 1, \\ \|\xi\|_{\mathbf{d}_2} & \text{if } \|\xi\| > 1 \end{cases}$$

is differentiable on $\mathbb{R}^n \backslash S \backslash \{\mathbf{0}\}$ and locally Lipschitz continuous on $\mathbb{R}^n \backslash \{\mathbf{0}\}$, where $S = \{\xi \in \mathbb{R}^6 : \|\xi\| = 1\}$ is the unit sphere.

Repeating the proof of Theorem 15.5 one can be shown that the closed-loop system is well-posed and

$$\dot{V}(\xi) \le -\rho_1 V^{1+\mu_1}(\xi) \quad \text{for} \quad \xi \in \mathcal{C} : \|\xi\| \le 1,$$

$$\dot{V}(\xi) \le -\rho_2 V^{1+\mu_2}(\xi) \quad \text{for} \quad \xi \in \mathcal{C} : \|\xi\| > 1.$$

Hence, using Clarke Gradient we derive

$$\dot{V}(\xi) \le -\min\{\rho_1, \rho_2\} \text{ for } \quad \xi \in \mathcal{C} : \|\xi\| = 1.$$

Since $\xi^\top \Pi^\top P \Pi \xi = \xi^\top P\xi$ for all $\xi \in \mathbb{R}^6$ then the Lyapunov function is decreasing along the trajectories of the system even in the case of jumps/resets. Repeating the proof of Theorem 15.5 one can prove that the closed-loop system is Zeno free. Therefore, the system is globally uniformly asymptotically stable. The estimate $\dot{V} \le -\rho_2 V^{1+\mu_2}$ implies that $\|\xi(t)\| \le 1$, $\forall t \ge \frac{1}{\rho_2\mu_2}$ independently of $\xi(0) \in \mathcal{C} \cap \mathcal{D}$.

Hence, we have $\dot{V} \leq -\rho_1 V^{1+\mu_1}$ for all $t \geq \frac{1}{\rho_2\mu_2}$, which implies $\xi(t) = \mathbf{0}$ for all $t \geq \frac{1}{-\tilde{\rho}_1\mu_1} + \frac{1}{\rho_2\mu_2}$. The proof is complete. □

The finite/fixed-time attitude tracking control can be designed using the homogeneous stabilizers introduced above.

Corollary 15.4 ([10]) *Let u_{hom} with $\mu < 0$ be defined as in Theorem 15.5 and u_{fxt} be defined as in Corollary 15.3. Let*

$$M = J\left(R_d^\top u - \omega_d^\wedge \omega + \dot{\omega}_d\right) + \omega \times J\omega, \tag{15.41}$$

where $\dot{\omega}_d = \frac{d}{dt}(R_d^\top \dot{R}_d)^\vee = (R_d^\top \ddot{R}_d - (\omega_d^\wedge)^2)^\vee \in C(\mathbb{R}_+, \mathbb{R}^3)$. Then for any trajectory of the system (15.24), (15.41) with $u = u_{hom}$ or $u = u_{fxt}$ one holds

$$\forall R_0 \in SO(3), \quad \forall \omega_0 \in \mathbb{R}^3, \quad \exists T \in \mathbb{R}_+ : R(t) = R_{\mathrm{d}}(t), \quad \forall t \geq T,$$

where

- $T \leq \|\xi(0)\|_{\mathbf{d}}^{-\mu}/(-\mu\rho)$ *if* $u = u_{hom}$;
- $T \leq \frac{1}{-\tilde{\rho}_1\mu_1} + \frac{1}{\tilde{\rho}_2\mu_2}$ *independently of* R_0, ω_0 *if* $u = u_{fxt}$.

Proof The error dynamics of the angular velocity is given by

$$\dot{\omega}_e = R_d\omega_d^\wedge(\omega - \omega_d) + R_d(\dot{\omega} - \dot{\omega}_d), \tag{15.42}$$

where $\omega_e = R_d(\omega - \omega_d)$. Taking into account $\omega_d^\wedge\omega_d = \mathbf{0}$, from (15.24b) and (15.41) we derive

$$\dot{\omega}_e = u.$$

Applying Theorem 15.5 (resp., Corollary 15.3) we complete the proof. □

15.3.2 Robustness of Homogeneous Attitude Control

Let us investigate Input-to-State[11] Stability (ISS) of the attitude control system. The noisy measurement of $R \in SO(3)$ can be modeled as follows:

$$\tilde{R} = d_R R,$$

[11] The Corollary 15.5 is reprinted from *Automatica*, 163/5, Y. Zhou, A. Polyakov, G. Zheng, Generalized homogeneous rigid-body attitude control, 111548. https://doi.org/10.1016/j.automatica.2024.111548 ©2024, with permission from Elsevier.

where $\tilde{R} \in SO(3)$ is a measured/estimated rotation matrix, and $d_R \in SO(3)$ models measurement noise in $SO(3)$. In this case, the model of measurements in exponential coordinates (on the Lie algebra $\mathfrak{so}(3)$) includes the noise in the additive manner

$$\tilde{\theta}_e^{\wedge} = \log(d_R R R_d^{\top}) = \log(R_e) + \log(d_R) = \theta_e^{\wedge} + \log(d_R).$$

The perturbed closed-loop system has the following model:

$$\begin{cases} \dot{\xi} = \begin{pmatrix} \mathbf{0} & J_r^{-1} \\ \mathbf{0} & \mathbf{0} \end{pmatrix} \xi + Bu(\xi + \delta_1) + \delta_2, & \xi^- \in \mathcal{C}, \\ \xi = H\xi^-, & \xi^- \in \mathcal{D}, \end{cases} \tag{15.43}$$

where $\xi, u, \mathcal{C}, \mathcal{D}$ are as before, $\delta_1 = [(\log(d_R)^{\vee})^{\top}, d_{\omega}^{\top}]^{\top} \in \mathbb{R}^6$ is a measurement noise, $\delta_2 \in \mathbb{R}^6$ is an exogenous perturbation and $d_{\omega} \in \mathbb{R}^3$ is the measurement noise of the angular velocity.

Corollary 15.5 ([10]) *Under conditions of Theorem 15.5 (Corollary 15.3), the impulsive system* (15.43) *with* $u = u_{hom}$ ($u = u_{fxt}$) *is*

- *locally ISS with respect to* $\delta = [\delta_1^{\top}, \delta_2^{\top}]^{\top} \in L^{\infty}(\mathbb{R}; \mathbb{R}^6)$,
- *globally ISS with respect to* $\delta = [\delta_1^{\top}, \delta_2^{\top}]^{\top} \in \Delta$,

for $\mu > -1$, *where* $\Delta \subset L^{\infty}(\mathbb{R}, \mathbb{R}^6)$ *is the set of perturbations such that the system* (15.43) *is Zeno free.*

Proof Let the vector field $f : \mathbb{R}^{18} \mapsto \mathbb{R}^6$ be defined as

$$f(\xi, \delta) := \begin{pmatrix} \mathbf{0} & J_r^{-1} \\ \mathbf{0} & \mathbf{0} \end{pmatrix} \xi + Bu_{hom}(\xi + \delta_1) + \delta_2,$$

where $\xi \in \mathbb{R}^6$, $\delta_1, \delta_2 \in \mathbb{R}^6$, $\delta = (\delta_1^{\top}, \delta_2^{\top})^{\top} \in \mathbb{R}^{12}$. Using homogeneity of B (i.e., $\mathbf{d}(s)B = e^s B$)and $\|\xi\|_{\mathbf{d}}$ we derive

$$\begin{aligned} \tfrac{\partial \|\xi\|_{\mathbf{d}}}{\partial \xi} f(\xi, \delta) &= \tfrac{\partial \|\xi\|_{\mathbf{d}}}{\partial \xi} f(\xi, 0) + \tfrac{\partial \|\xi\|_{\mathbf{d}}}{\partial \xi} (f(\xi, \delta) - f(\xi, 0)) \\ &= \tfrac{\partial \|\xi\|_{\mathbf{d}}}{\partial \xi} f(\xi, 0) + \tfrac{\partial \|\xi\|_{\mathbf{d}}}{\partial \xi} (Bu_{hom}(\xi + \delta_1) - Bu_{hom}(\xi) + \delta_2) \\ &= \tfrac{\partial \|\xi\|_{\mathbf{d}}}{\partial \xi} f(\xi, 0) + \tfrac{\partial \|z\|_{\mathbf{d}}}{\partial z} (B(u_{hom}(\xi + \delta_1) - u_{hom}(\xi)) + \mathbf{d}(s)\delta_2), \end{aligned}$$

where $z = \mathbf{d}(s)\xi$. The $\mathbf{d}$-homogeneity of u_{hom} gives

$$u_{hom}(y) = e^{-(1+\mu)s} u_{hom}(\mathbf{d}(s)y), \forall y \in \mathbb{R}^6, \forall s \in \mathbb{R}.$$

Taking $s = -\ln \|\xi\|_{\mathbf{d}}$ we derive

$$\|Bu_{hom}(\xi + \delta_1) - Bu_{hom}(\xi)\| = \|\xi\|_{\mathbf{d}}^{1+\mu} \|Bu_{hom}(z + \mathbf{d}(-\ln \|\xi\|_{\mathbf{d}})\delta_1) - Bu_{hom}(z)\|$$

with $z = \mathbf{d}(-\ln \|\xi\|_{\mathbf{d}})\xi$ belonging to the unit sphere: $\|z\| = 1$. Since $u_{hom} \in C^1(\mathbb{R}^6\backslash\{\mathbf{0}\}, \mathbb{R}^3)$, then for $\|\mathbf{d}(-\ln \|\xi\|_{\mathbf{d}})\delta_1\| \leq 0.5$ we derive

$$\|Bu_{hom}(z + \mathbf{d}(s)\delta_1) - Bu_{hom}(z)\| \leq$$
$$\left\| B \int_0^1 \frac{\partial u_{hom}(y)}{\partial y}\Big|_{y=z+\lambda\mathbf{d}(s)\delta_1} \mathbf{d}(s)\delta_1 d\lambda \right\| \leq C_1 \|\mathbf{d}(s)\delta_1\|,$$

where $s = -\ln \|\xi\|_{\mathbf{d}}$, $C_1 = \sup_{0.5\leq\|y\|\leq1.5} \|B\frac{\partial u_{hom}(y)}{\partial y}\|$.

Since $C_2 = \sup_{\|z\|=1} \left\|\frac{\partial\|z\|_{\mathbf{d}}}{\partial z}\right\| < +\infty$ then

$$\tfrac{\partial\|\xi\|_{\mathbf{d}}}{\partial\xi} f(\xi, \delta) \leq \tfrac{\partial\|\xi\|_{\mathbf{d}}}{\partial\xi} f(\xi, 0) + \|\xi\|_{\mathbf{d}}^{1+\mu}\tilde{C}\sum_{i=1}^{2} \|\mathbf{d}(-\ln \|\xi\|_{\mathbf{d}})\delta_i\|,$$

where $\tilde{C} = C_2 \max\{C_1, 1\}$. In Theorem 15.5, it is shown that $\frac{\partial\|\xi\|_{\mathbf{d}}}{\partial\xi} f(\xi, 0) \leq -\tilde{\rho}\|\xi\|_{\mathbf{d}}^{1+\mu}$ for all $\xi \in \mathcal{C}$. If $\xi \in \mathcal{C}$, then $\|\mathbf{d}(-\ln \|\xi\|_{\mathbf{d}})\delta_i\| \leq \frac{\tilde{\rho}}{4\tilde{C}}$, $i = 1, 2$ imply $\frac{\partial V}{\partial x} f(\xi, \delta) \leq -\frac{\tilde{\rho}}{2}V^{1+\mu}$. Since $\|\mathbf{d}(-\ln \|\xi\|_{\mathbf{d}})\delta_i\| \leq \frac{\tilde{\rho}}{4\tilde{C}}$ is equivalent to $\|4\tilde{C}\delta_i/\tilde{\rho}\|_{\tilde{\mathbf{d}}_i} \leq \|\xi\|_{\mathbf{d}}$, where $\tilde{\mathbf{d}}_1 = \mathbf{d}$ and $\tilde{\mathbf{d}}_2(s) = e^{\mu s}\mathbf{d}(s)$, $s \in \mathbb{R}$. For $\xi \in \mathcal{C}$ we have

$$\|\xi\|_{\mathbf{d}} \geq \|4\tilde{C}\delta_i/\tilde{\rho}\|_{\tilde{\mathbf{d}}_i} \quad\Rightarrow\quad \frac{d}{dt}\|\xi\|_{\mathbf{d}} \leq -\tfrac{\tilde{\rho}}{2}\|\xi\|_{\mathbf{d}}^{1+\mu}(\xi).$$

The above inequality implies local ISS of the closed-loop system, which is impulse free for small initial conditions and small perturbations. Moreover, in Theorem 15.5, it is shown that $\|\Pi\xi\|_{\mathbf{d}} \leq \|\xi\|_{\mathbf{d}}$ for any $\xi \in \mathcal{D}$. Therefore, the system is globally ISS if the impulsive system (15.43) is Zeno free (see, e.g., [11] for more details). □

Numerical examples supporting the above derivations can be found in [10]. Practical experiments with quadrotor control are considered in Chap. 11.

Concluding Remarks

The several classes of nonlinear control systems are studied in the chapter. For an arbitrary nonlinear control system with continuous right side, a necessary and sufficient condition of homogeneous stabilizability is obtained. The sufficient condition of quadratic stabilizability is given. The method of convex embedding is applied for stabilization of nonlinear affine-in-control system. Homogeneous control of rigid-body attitude is designed.

References

1. Sontag ED (1989) A universal construction of Artstein's theorem on nonlinear stabilization. Syst & Control Lett 13:117–123
2. Polyakov A (2019) Sliding mode control design using canonical homogeneous norm. Int J Robust Nonlinear Control 29(3):682–701
3. Zimenko K, Polyakov A, Efimov D (2023) Homogeneous systems stabilization based on convex embedding. Automatica 8:111108
4. Feingesicht M, Polyakov A, Kerherve F, Richard J-P (2017) SISO model-based control of separated flows: sliding mode and optimal control approaches. Int J Robust Nonlinear Control 27(18):5008–5027
5. Bharadwaj S, Osipchuk M, Mease K, Park F (1998) Geometry and inverse optimality in global attitude stabilization. J Guid Control Dyn 21(6):930–939
6. Chirikjian G (2012) Information theory on Lie groups. Stoch Model Inf Theory Lie Groups 2:337–359
7. Bullo F, Murray R (1995) Proportional derivative (PD) control on the Euclidean group. Technical report, California Institute of Technology
8. Mayhew C, Teel A (2013) Synergistic hybrid feedback for global rigid-body attitude tracking on so(3). IEEE Trans Autom Control 58(11):2730–2742
9. Chaturvedi N, Sanyal A, McClamroch N (2011) Rigid-body attitude control. IEEE Control Syst Mag 31(3):30–51
10. Zhou Y, Polyakov A, Zheng G (2024) Generalized homogeneous rigid-body attitude control. Automatica 163:111548
11. Hespanha J, Liberzon D, Teel A (2008) Lyapunov conditions for input-to-state stability of impulsive systems. Automatica 44:2735–2744

Chapter 16
Homogeneous Artificial Neural Network

Artificial neural network (ANN) is a machine learning model inspired by neurophysiology. Its mathematical foundation lies in the approximation theory of functions. In many cases, the training (learning) of ANN can be interpreted as an identification of an unknown mapping based on available experiments/measurements. This chapter introduces the so-called homogeneous ANN, key feature of which is a global approximation based on local measurements. Homogeneous universal approximation theorem is proven. A possibility of an upgrade/transformation of the classical ANN to a homogeneous one is investigated. The problem of identification/detection of dilation symmetry of a mapping is considered. Issues of approximation nonlinear dynamical control systems using homogeneous differential neural networks are studied.

16.1 Approximation of Homogeneous Mapping by ANN

The so-called universal approximation theorems [1–3] put limits on what neural networks can theoretically learn.

Theorem 16.1 (Universal approximation theorem, [4])*Let* $\sigma \in C(\mathbb{R};\mathbb{R})$ *be a non-polynomial function and* $\Omega \subset \mathbb{R}^n$ *be a compact set.*

If $g : \mathbb{R}^n \mapsto \mathbb{R}^m$ *is a continuous mapping then for any* $\epsilon > 0$ *there exists* $N \in \mathbb{N}$, $A \in \mathbb{R}^{N\times n}$, $b \in \mathbb{R}^N$ *and* $C \in \mathbb{R}^{m\times N}$ *such that*

$$|g(x) - g_\epsilon(x)| \leq \epsilon, \quad \forall x \in \Omega, \tag{16.1}$$

$$g_\epsilon(x) = C\sigma(Ax + b), \tag{16.2}$$

A. Polyakov, *Generalized Homogeneity in Systems and Control Volume I*, Communications and Control Engineering,
https://doi.org/10.1007/978-3-031-76246-8_16

where the function σ in the above formula is applied in the component-wise manner.

The function g_ϵ given by (16.2) is an ANN with one hidden layer (also known as Shallow ANN). The above theorem shows that the considered ANN is the universal approximator of any continuous mapping on a compact.

In [5–8], it is shown that ANNs with standard homogeneous activation functions (i.e., $\sigma(\lambda\rho) = \lambda\sigma(\rho), \forall\lambda > 0, \forall\rho \in \mathbb{R}$) have various useful features such as faster training. However, the homogeneity of the activation function does not guarantee a dilation symmetry of the ANN g_ϵ. An approximation procedure should take into account the symmetry (if any), otherwise, the geometry of the mapping g may be lost. This may badly impact extrapolation capabilities of ANN.

In this section, we consider an unknown **d**-homogeneous mapping having a known homogeneity degree. We introduce the **d**-*homogeneous artificial neural network* being universal approximator of any continuous **d**-homogeneous mapping (a function or a vector field) in $\mathbb{R}^n$.

Notice that below we distinguish an analysis and design of ANNs for homogeneous functions and vector fields, since the corresponding definitions of homogeneity are slightly different. The consideration of vector fields is useful for the so-called physics-informed neural networks [9], which are aimed at approximation of dynamical systems. The so-called differential neural networks are considered in the last section of this chapter.

16.1.1 Homogeneous Universal Approximation Theorem

In the case of homogeneous function, the universal approximation theorem can be revisited as follows.

Theorem 16.2 *Let* **d** *be a linear continuous dilation in $\mathbb{R}^n$. Let $\|\cdot\|_{\mathbf{d}}$ be an arbitrary explicit or implicit* **d**-*homogeneous norm in $\mathbb{R}^n$. Let $\sigma \in C(\mathbb{R};\mathbb{R})$ be a non-polynomial function.*

If $h \in C(\mathbb{R}^n\backslash\{\mathbf{0}\};\mathbb{R})$ is a **d**-*homogeneous function of degree $\nu \in \mathbb{R}$ then for any $\epsilon > 0$ there exist $N \in \mathbb{N}$, $A \in \mathbb{R}^{N\times n}$, $b \in \mathbb{R}^N$ and $C \in \mathbb{R}^{1\times N}$ such that*

$$|h(x) - h_\varepsilon(x)| \le \epsilon\|x\|_{\mathbf{d}}^{\nu}, \quad \forall x \in \mathbb{R}^n\backslash\{\mathbf{0}\}, \tag{16.3}$$

$$h_\epsilon(x) = \|x\|_{\mathbf{d}}^{\nu} C\sigma(A\mathbf{d}(-\ln\|x\|_{\mathbf{d}})x + b), \tag{16.4}$$

where the function σ in the above formula is applied in the component-wise manner.

Proof Let $\|\cdot\|$ be a norm in $\mathbb{R}^n$. By definition, the homogeneous norm $\|\cdot\|_{\mathbf{d}}$ is a continuous positive definite function satisfying $\|\mathbf{d}(s)x\|_{\mathbf{d}} = e^s\|x\|_{\mathbf{d}}$. Then $\|\cdot\|_{\mathbf{d}}$ is radially unbounded and the set

$$S_{\mathbf{d}} = \{x \in \mathbb{R}^n : \|x\|_{\mathbf{d}} = 1\}$$

is compact and $\mathbf{0} \notin S_{\mathbf{d}}$. Hence, $\mathbf{d}(-\ln \|x\|_{\mathbf{d}})x \in S_{\mathbf{d}}$ and

$$b := \sup_{x \in \mathbb{R}^n \setminus \{\mathbf{0}\}} \|\mathbf{d}(-\ln \|x\|_{\mathbf{d}})x\| \geq a := \inf_{x \in \mathbb{R}^n \setminus \{\mathbf{0}\}} \|\mathbf{d}(-\ln \|x\|_{\mathbf{d}})x\| > 0.$$

Let us consider the set

$$\Omega = \{y \in \mathbb{R}^n : a \leq \|y\| \leq b\}.$$

By construction, the set Ω is a compact ($\mathbf{0} \notin \Omega$) and, by assumption, the function h is continuous on Ω. Theorem 16.1 proves that for any $\epsilon > 0$ there exists $N \in \mathbb{N}$, $A \in \mathbb{R}^{N \times n}$, $b \in \mathbb{R}^N$ and $C \in \mathbb{R}^{1 \times N}$ such that

$$\sup_{y \in \Omega} |h(y) - C\sigma(Ay + b)| \leq \epsilon.$$

Since, due to $\mathbf{d}$-homogeneity of h, we have

$$h(x) = \|x\|_{\mathbf{d}}^{\nu} h(\underbrace{\mathbf{d}(-\ln \|x\|_{\mathbf{d}})x}_{=y \in \Omega})$$

then the representation (16.3) holds for h_ϵ given by (16.4). The proof is complete. □

> *The key feature of the homogeneous ANN is a global approximation based on local measurements.*

Indeed, the classical universal approximation theorem (see Theorem 16.1) guarantees the error estimate (16.1) only on a compact. Due to homogeneity, any local estimate of the homogeneous function can be extended globally (see (16.3)). The approximation error in this case depends on the norm of the vector x and the homogeneity degree ν of the function h. If $\nu > 0$ then, due to Proposition 6.4, h is continuous at $\mathbf{0}$ and the above error estimate holds at zero as well. For $\nu < 0$ a nonzero function h is discontinuous and unbounded at $\mathbf{0}$. In this case, the approximation error may grow to infinity as $x \to \mathbf{0}$. For $\nu = 0$ the error is estimated uniformly on $\mathbb{R}^n$.

In ANN theory, the function σ is called *activation function*. The following activation functions are most popular:

- *Sigmoid*: $\sigma(r) = (1 + e^r)^{-1}$, $r \in \mathbb{R}$;
- *Hyperbolic Tangent*: $\sigma(r) = \frac{e^r - e^{-r}}{e^r + e^{-r}}$, $r \in \mathbb{R}$;
- *Rectified Linear Unit*: $\sigma(r) = \max\{0, r\}$, $r \in \mathbb{R}$.

All activation functions given above are non-polynomial and continuous, so they satisfy Theorems 16.1 and 16.2.

Usually algorithms of ANN learning optimize the parameters A, b, C based on available measurements

$$y_j = h(x_j), \quad j = 1, ..., M,$$

where M is a number of experiments. Mathematically, the goal of the ANN learning is to minimize a norm of the residual

$$e_j = y_j - \|x_j\|_{\mathbf{d}}^{\nu} C\sigma(A\mathbf{d}(-\ln \|x_j\|_{\mathbf{d}})x_j + b), \quad j = 1, .., M. \tag{16.5}$$

It is worth stressing that the homogeneous ANN has structurally the same dependence on the parameters A, b, C. So, all conventional learning algorithms are applicable for its training. A possible mathematical formulation of the learning problem is

$$\sum_{j=1}^{M} \left(\|x_j\|_{\mathbf{d}}^{\nu} C\sigma(A\mathbf{d}(-\ln \|x_j\|_{\mathbf{d}})x_j + b) - y_j\right)^2 \to \min_{A,b,C}. \tag{16.6}$$

A local solution of this problem can be found using conventional methods of optimization (like gradient descent, Newton method, etc.)

There exist various learning methods (see, e.g., [10] and references therein), which are aimed at searching global solution of (16.6). Their study goes out of the scope of this monograph since, due to the similarity in structures of ANN and homogeneous ANN, all main advantages and disadvantages of the corresponding methods are going to be the same. The aim of this chapter is to demonstrate that the use of dilation symmetry in ANN design may essentially improve the approximation precision provided that the mapping, which needs to be approximated, is homogeneous.

The simplest example of homogeneous ANN is the approximator of the implicit canonical homogeneous norm designed in Theorem 5.6 (see Chap. 5). The corresponding homogeneous ANN is utilized next for control design due to Theorem 9.2.

Data-driven identification of a homogeneous dynamical system requires further modification of ANN in order to design a universal approximator for homogeneous vector fields.

Theorem 16.3 *Let* $\mathbf{d}$ *be a linear continuous dilation in* $\mathbb{R}^n$. *Let* $\|\cdot\|_{\mathbf{d}}$ *be an arbitrary explicit or implicit* $\mathbf{d}$*-homogeneous norm in* $\mathbb{R}^n$.

If $f \in C(\mathbb{R}^n\backslash\{\mathbf{0}\}; \mathbb{R}^n)$ *is a* $\mathbf{d}$*-homogeneous function of degree* $\mu \in \mathbb{R}$ *then for any* $\epsilon > 0$ *there exist* $N \in \mathbb{N}$, $A \in \mathbb{R}^{N\times n}$, $b \in \mathbb{R}^N$ *and* $C \in \mathbb{R}^{n\times N}$ *such that*

$$|f(x) - f_\epsilon(x)| \le \epsilon \left|e^{(\mu I_n + G_{\mathbf{d}})\ln\|x\|_{\mathbf{d}}}\right|, \quad \forall x \in \mathbb{R}^n\backslash\{\mathbf{0}\}, \tag{16.7}$$

$$f_\epsilon(x) = \|x\|_{\mathbf{d}}^{\mu}\mathbf{d}(\ln\|x\|_{\mathbf{d}})C\sigma(A\mathbf{d}(-\ln\|x\|_{\mathbf{d}})x + b), \tag{16.8}$$

where the function σ *in the above formula is applied in the component-wise manner.*

The proof of this theorem literally repeats the proof of Theorem 16.2 taking into account the identity

$$f(x) = \|x\|_{\mathbf{d}}^{\mu}\mathbf{d}(\ln \|x\|_{\mathbf{d}})f(\mathbf{d}(-\ln \|x\|_{\mathbf{d}})x), \quad \forall x \neq \mathbf{0},$$

which holds for any **d**-homogeneous vector field of degree $\mu \in \mathbb{R}$.

The estimation error (16.7) depends on eigenvalues of the matrix $\mu I_n + G_{\mathbf{d}}$. If all eigenvalues of this matrix are positive then the error tends to zero as $x \to \mathbf{0}$. Otherwise f is discontinuous at zero and the approximation error may grow as $x \to \mathbf{0}$. The uniform estimate on $\mathbb{R}^n$ can be guaranteed only if

$$\sup_{s\in\mathbb{R}} \|e^{s(\mu I_n+G_{\mathbf{d}})}\| < +\infty. \tag{16.9}$$

This is possible for a skew-symmetric matrix $\mu I_n + G_{\mathbf{d}}$ or, equivalently, if $e^{s(\mu I_n+G_{\mathbf{d}})}$ is a rotation matrix. The particular case of the skew-symmetric matrix is $\mu I_n + G_{\mathbf{d}} = \mathbf{0}$. In the latter case, the continuity of f at zero implies that f is a constant vector (see Proposition 6.4).

Clearly, more information about the mapping we have, the better approximation can be obtained. The following numerical example demonstrates that a knowledge about dilation symmetry of the mapping may essentially improve the estimation precision.

Example 16.1 The scaling invariance is desirable feature for a pattern recognition by ANN. There are several approaches to exhibit the required invariance (see, e.g., [11, p. 262]). One of them is to build the invariance properties into the structure of a neural network. The present example uses the generalized homogeneity for this purpose.

For simplicity, we assume that we deal with gray-scale images. Following the conventional approach, we associated any image with a function $(z_1, z_2) \mapsto \phi(z)$ of two arguments, which value $\phi(z) \in [0, 255]$ define a tone of the gray color at the coordinates $z = (z_1, z_2)^\top$. The value $\phi(z) = 0$ corresponds to the white color, while the value $\phi(z) = 255$ means that the point with the coordinate z has the black color. A pattern depicted in the image is assumed to be centered. The origin $z = (0, 0)$ of the coordinate frame corresponds to the center of the image. In computer, the coordinates of the points z as well as the possible values of $\phi(z)$ are represented by integers, i.e., $z \in \mathbb{Z}^2$, $\phi(z) \in \{0, 1, 2, ..., 255\}$. We extend the domain of ϕ to $\mathbb{R}^2$ and the range of ϕ to the interval $[0, 255]$, respectively. This extension of the function ϕ can always be made in a continuous manner. For the points z out of the image range we assign $\phi(z) = 0$. In this case, the size of image does not impact the definition of $\phi \in C_c(\mathbb{R}^2; \mathbb{R})$.

Traditionally, the values of $\phi(z)$ are utilized as inputs to ANN. Since the function ϕ is not homogeneous even with respect to the standard dilation $z \mapsto e^s z$, then the efficiency of a homogeneous ANN is questionable in this case. To homogenize the problem, we introduce the following transformations:

$$x_i(\phi) = \int_{\mathbb{R}^2} b_i(z)\phi(z)dz, \quad i = 1, 2, ..., n, \tag{16.10}$$

where $b_i : \mathbb{R}^2 \mapsto \mathbb{R}$ are standard homogeneous functions of degrees $r_i > -2$, i.e., $b_i(e^s z) = e^{r_i s} b_i(z)$, $\forall s \in \mathbb{R}$, $\forall z \in \mathbb{R}^2$ and $n \in \mathbb{N}$. For example, we may take $b_i(z) = \frac{z_1^{p_i} z_2^{q_i}}{r_i!}$ and $p_i, q_i \in \mathbb{N} \cup \{0\}$, $r_i = p_i + q_i$. In this case, the functions b_i are elements of polynomial basis in $L^2(\Omega; \mathbb{R})$, where $\Omega \subset \mathbb{R}^2$ contains a support of ϕ. So, any function $\phi \in C_c(\mathbb{R}^2; \mathbb{R}) \cap L^2(\Omega; \mathbb{R})$ can be approximated as linear combination of the polynomial basis functions b_i with the coefficients dependent on the vector $x = (x_1, ..., x_n)^\top \in \mathbb{R}^n$ related with coordinates of ϕ in the polynomial basis. Since the function ϕ is compactly supported with the support dependent on the image size, the computation of x_i is not a difficult problem and it can be easily parallelized.

Notice that the zoom of the image corresponds to the coordinate transformation $z \mapsto e^s z$ with $s \in \mathbb{R}$, which implies the transformation $\phi \mapsto \phi_s$, where $\phi_s(z) := \phi(e^{-s} z)$. Hence, we derive

$$\begin{aligned} x_i(\phi_s) = & \int_{\mathbb{R}^2} b_i(z)\phi(e^{-s}z)dz \\ = & e^{2s} \int_{\mathbb{R}^2} e^{i_i s} b_i(e^{-s}z)\phi(e^{-s}z)d(e^{-s}z) \\ = & e^{(2+r_i)s} x_i(\phi). \end{aligned}$$

In other words, the dilation $z \mapsto e^s z$ (the zoom of the image) implies the dilation of the coordinates $x_i \mapsto e^{(2+r_i)s} x_i$, $i = 0, ..., n$. The vector $x = (x_1, ..., x_n)^\top$ can be utilized as an input to an ANN, which is going to be designed **d**-homogeneous of degree 0 with respect to the dilation

$$\mathbf{d}(s) = \mathrm{diag}(e^{(2+r_1)s}, ..., , e^{(2+r_n)s}), \quad s \in \mathbb{R}.$$

In this case, the scaling/zoom of the image will not impact the output of the **d**-homogeneous ANN. Therefore, once being trained, such ANN can be utilized for the pattern recognition independently of the scaling of the function $\phi \in C_c(\mathbb{R}^2; \mathbb{R})$. However, scaling of the image maps a function ϕ having a discrete domain $\mathbb{Z}^2$. In this case, a reduction of the image size may lead to a degradation of the image quality and impossibility of its recognition. So, the **d**-homogeneous ANN will have the natural limitation in recognition of very reduced images.

For numerical illustration, we consider the 8 images of the characters shown in Fig. 16.1. Each image has the size 100×100 pixels. The components of the input vector x are defined by the formula (16.10) with $n = 8$ and $b_1 = 1, b_2 = z_1, b_3 = z_2, b_4 = 0.5 z_1 z_2, b_5 = 0.5 z_1^2, b_6 = 0.5 z_2^2, b_7 = z_1^2 z_2/6; b_8 = z_1 z_2^2/6$. To demon-

Fig. 16.1 The characters for recognition

strate the computational robustness of the method, the coordinates x_i are computed using the following approximation of the integral (16.10):

$$x_i(\phi) \approx \sum_{k=1}^{100} \sum_{j=1}^{100} b_i(k-50, j-50)\phi(k, j).$$

The ANN is assumed to have $k = 8$ outputs. For the ith character at the input, only the ith output must have the value 1. All other outputs have to be zero in this case. The training set consists of nominal images of the size 100×100 pixels and the scaled images the size 75×75 pixels, 50×50 pixels and 35×35 pixels, respectively. The total number of images in the training set is 32. The matrices $A \in \mathbb{R}^{N \times n}$ and $B \in \mathbb{R}^{N \times 1}$ are selected randomly with $N = 8$. The matrix $C \in \mathbb{R}^{k \times N}$ is obtained by means of least square optimization on the training set. The trained homogeneous ANN has been tested on images, which are scaled (zoomed) to 30%+i10% with $i = 0, 1, , \ldots, 17$. Notice that the decrease of the size of the nominal image implies a degradation of the image quality (see Fig. 16.2). All scaled images have been successfully recognized.

To test the robustness with respect to the noise, a normally distributed noise has been added to the nominal images. About 50% of pixels have been corrupted by the noise (see Fig. 16.3). The homogeneous ANN successfully recognized the noisy images as well.

Fig. 16.2 Illustration of the degradation of the image quality after the zoom 30%

Fig. 16.3 The characters corrupted by the noise

The construed homogeneous ANN has good capabilities of pattern recognition with relatively small set of parameters to be tuned (just 8×8 matrix C of real numbers).

The small size of the homogeneous neural network may be important for its successful application in real-life applications. For example, landmarks recognition is important for positioning of mobile robots on the map. The use of large-scale neural networks on board of the robot is impossible due to its limited computational capacity. Homogeneous neural network may be useful for optimization of the computation load in this case. Potential applications of homogeneous neural networks in sliding-mode control theory are discussed in [12].

16.1.2 Homogenization of Artificial Neural Networks

In this subsection, we consider the problem of transformation of an existing artificial neural network to a homogeneous one. By the universal approximation theorems, the ANN is a local approximator of a continuous mapping. The homogeneous ANN is a global approximator of a homogeneous continuous mapping. *The transformation of the conventional ANN to a homogeneous ANN can be interpreted as a global homogeneous extrapolation of ANN.*

We assume that some conventional ANN (16.2) is already well trained such that the approximation error (16.1) is somehow minimized. As before the linear dilation and the homogeneity degree are assumed to be known. We need to transform the ANN (16.2) to the homogeneous ANN (16.4) avoiding a degradation of the approximation precision.

Proposition 16.1 *Let $\Omega \subset \mathbb{R}^n$ be a compact set. Let g_ϵ be ANN* (16.2) *approximating a function $h \in C(\mathbb{R}^n \backslash \{\mathbf{0}\}; \mathbb{R})$ (or a vector field $f \in C(\mathbb{R}^n \backslash \{\mathbf{0}\}; \mathbb{R}^n)$ on the compact $\Omega \subset \mathbb{R}^n$ with an error $\epsilon > 0$, i.e., the inequality* (16.1) *is fulfilled for $g = h$ (resp., $g = f$). Let $\mathbf{d}$ be a linear continuous dilation in $\mathbb{R}^n$. Let the function h (resp., the vector field f) be $\mathbf{d}$-homogeneous of degree $\nu \in \mathbb{R}$ (resp., $\mu \in \mathbb{R}$).*

If there exists an explicit or implicit $\mathbf{d}$-homogeneous norm $\|\cdot\|_{\mathbf{d}}$ in $\mathbb{R}^n$ such that

$$S_{\mathbf{d}} = \{x \in \mathbb{R}^n : \|x\|_{\mathbf{d}} = 1\} \subset \operatorname{int} \Omega, \tag{16.11}$$

then the homogeneous ANN (16.4) *with parameters A, b, C taken from the conventional ANN* (16.2) *satisfies the error estimate* (16.3) *(resp.,* (16.7)*).*

Proof By construction the homogeneous ANN h_ϵ coincides with the original ANN h_ϵ on the unit sphere, i.e.,

$$h_\epsilon(y) = g_\epsilon(y), \quad \forall y \in S_{\mathbf{d}}.$$

This means that

$$|g(y) - g_\epsilon(y)| = |h(y) - h_\epsilon(y)| \leq \epsilon, \quad \forall y \in S_{\mathbf{d}}.$$

Since $\|\cdot\|_{\mathbf{d}}$ is a homogeneous norm then $y = \mathbf{d}(-\ln \|x\|_{\mathbf{d}})x \in S_{\mathbf{d}}$ for all $x \in \mathbb{R}^n \backslash \{\mathbf{0}\}$ and using the homogeneity we derive

$$\|x\|_{\mathbf{d}}^{-\nu}|h(x) - h_\epsilon(x)| \leq |h(\mathbf{d}(-\ln \|x\|_{\mathbf{d}})x) - h_\epsilon(-\ln \|x\|_{\mathbf{d}})x)| \leq \epsilon.$$

The case of homogeneous vector field can be analyzed similarly. The proof is complete. □

The above proposition shows that the conventional ANN can be straightforwardly upgraded to the homogeneous ANN provided that the information about dilation symmetry of the mapping is available (e.g., from physics). The idea of the transformation/upgrade is very simple: we just select a unit sphere $S_{\mathbf{d}} \subset \Omega$ and expand the values of the g_ϵ from the unit sphere to the whole space $\mathbb{R}^n$ by means of dilation. Notice that with such a homogeneous extrapolation the local error estimate (16.1) (being uniform on Ω) is converted to the global $\|x\|_{\mathbf{d}}$-dependent estimate (16.3) on $\mathbb{R}^n$.

If the ANN g_ϵ approximates a non-constant continuous (on $\mathbb{R}^n$) homogeneous function $g = h$ then, by Proposition 6.1, we have $\nu > 0$ and the approximation error (16.3) for the homogeneous ANN tends to zero as $x \to \mathbf{0}$. Moreover, the error estimate is improved on the unit ball in this case, i.e., we have more precise estimate (16.3) instead of (16.1) for any $x \in \mathbb{R}$ such that $\|x\|_{\mathbf{d}} \leq 1$. In this case, to maximize the zone of improvement of the error estimate this unit ball has to be adjusted somehow. Notice that $\|\cdot\|_{\mathbf{d}}$ may be a canonical homogeneous norm, so an arbitrary norm $\|\cdot\|$ can be utilized for the upgrade of ANN provided that the dilation $\mathbf{d}$ is monotone with respect to this norm $\|\cdot\|$. By Proposition 5.3, a linear dilation $\mathbf{d}(s) = e^{sG_{\mathbf{d}}}$, $s \in \mathbb{R}$ is monotone with respect to the weighted Euclidean norm $\|x\| = \sqrt{x^\top Px}$ if

$$PG_{\mathbf{d}} + G_{\mathbf{d}}^\top P \succ 0, \quad P \succ 0. \tag{16.12}$$

In this case, the unit ball $\|x\| \leq 1$ corresponds to the following ellipsoid:

$$\mathcal{E}(P) = \{x \in \mathbb{R}^n : x^\top Px \leq 1\} \subset \mathbb{R}^n. \tag{16.13}$$

An optimization of the shape matrix $P = P^\top$ subject to the constrains $\mathcal{E}(P) \subset \Omega$ and (16.12) optimizes a selection of the norm for the upgrade of ANN to homogeneous ANN. For example, if

$$\Omega = \{x \in \mathbb{R}^n : x^\top x \leq r\}, \quad r > 0$$

then the above optimization problem can be formulated as the following SDP (semi-definite programming) problem:

$$\text{trace}(P) \to \min \tag{16.14}$$

subject to (16.12) and

$$P \succeq \frac{1}{r} I_n. \tag{16.15}$$

Indeed, on the one hand, the matrix inequality $P \succeq \frac{1}{r} I_n$ is equivalent to $x^\top P x \geq \frac{1}{r} x^\top x$ for all $x \in \mathbb{R}^n$. In this case, we conclude

$$x \in \mathcal{E}(P) \quad \Rightarrow \quad 1 \geq x^\top P x \geq \frac{1}{r} x^\top x \quad \Rightarrow \quad x \in \Omega.$$

On the other hand, the minimization of trace(P) maximizes the sum of semi-axes of the ellipsoid $\mathcal{E}(P)$ belonging to Ω. Due to the constraint (16.12), the corresponding weighted Euclidean norm is appropriate for the transformation of the conventional ANN to the homogeneous ANN.

The similar considerations can be repeated for $\nu < 0$. However, in this case, the error estimate is improved outside of the unit ball, so its minimization would imply a better estimate after the upgrade. For $\nu = 0$, the uniform estimate is expanded to the whole $\mathbb{R}^n$ (except probably $\mathbf{0}$).

Example 16.2 Following the classical model of a controlled rotational motion of the rigid body in the 3-D space, the dynamics of the angular velocity of the body is governed by

$$\dot{\omega}(t) = J^{-1}(-\omega \times J\omega + \tau), \quad t \geq 0, \tag{16.16}$$

where $\omega(t) = (\omega_1(t), \omega_2(t), \omega_3(t))^\top \in \mathbb{R}^n$ is the angular velocity in the body frame, $J = J^\top \in \mathbb{R}^{n\times n}$ is the inertia matrix, $\tau(t) \in \mathbb{R}^3$ is the external momentum (e.g., control input) and $\times$ denotes the vector product, i.e., $\omega \times J\omega = \begin{pmatrix} 0 & -\omega_3 & \omega_2 \\ \omega_3 & 0 & -\omega_1 \\ -\omega_2 & \omega_2 & 0 \end{pmatrix} J\omega$. The function $h : \mathbb{R}^6 \mapsto \mathbb{R}^3$ given by $h(\omega, \tau) = -J^{-1}(\omega \times J\omega + \tau)$ is homogeneous of degree 2 with respect to a generalized dilation:

$$h(e^s\omega, e^{2s}\tau) = e^{2s} h(\omega, \tau), \quad \forall s \in \mathbb{R}, \forall \omega, \tau \in \mathbb{R}^3.$$

Let us consider now the rotation dynamics (16.16) of the rigid body with $J = \begin{pmatrix} 1 & 0.3 & 0.1 \\ 0.3 & 1.2 & 0.2 \\ 0.1 & 0.2 & 0.8 \end{pmatrix}$. First, we approximate the right-hand side of (16.16)

$$h(x) = -J^{-1}(\omega \times J\omega + \tau), \quad x = (\omega^\top, \tau^\top)^\top \in \mathbb{R}^6$$

by means of the conventional ANN with the sigmoid activation function

$$g_\epsilon(x) = C\sigma(Ax + b),$$

Table 16.1 Approximation precision of the conventional and homogeneous ANN

Domain	ANN	hANN $\nu = 2$	hANN $\nu = \nu_\epsilon$	hANN $\nu = 1, G_{\mathbf{d}} = G_\epsilon$
Ω_{-4}	0.0864	0.0010	0.0134	0.0284
Ω_{-3}	0.0821	0.0024	0.0168	0.0286
Ω_{-2}	0.0583	0.0038	0.0170	0.0213
Ω_{-1}	0.0274	0.0063	0.0135	0.0122
$\Omega_0 = \Omega$	0.0110	0.0074	0.0082	0.0094
Ω_1	0.0612	0.0090	0.0193	0.0293
Ω_2	0.2257	0.0115	0.0417	0.0787
Ω_3	0.5056	0.0153	0.0773	0.1551
Ω_4	1.0409	0.0192	0.1160	0.3097

where $A \in \mathbb{R}^{N\times 6}, b \in \mathbb{R}^N$ are randomly selected for $N = 500$ and $C \in \mathbb{R}^{3\times N}$ is obtained by means of the least square error minimization for $M = 20000$ measurements $y_i = h(x_i)$ of the function h at the points $x_i \in \mathbb{R}^6$ randomly uniformly distributed in the set

$$\Omega = \{x \in \mathbb{R}^6 : 0.95 \leq |x| \leq 1.05\}.$$

On Ω the obtained ANN has the following approximation precision:

$$|g_\epsilon(x) - h(x)| \leq \epsilon \approx 0.0110, \quad x \in \Omega$$

which is approximated by taking 20000 random points in Ω.

By Proposition 16.1, the conventional ANN g_ϵ can be upgraded to a homogeneous ANN h_ϵ by the formula (16.4), where $G_{\mathbf{d}} = \text{diag}(1, 1, 1, 2, 2, 2)$, $\nu = 2$ and the canonical homogeneous norm $\|\cdot\|_{\mathbf{d}}$ is induced by the usual Euclidean norm $|\cdot|$. Such an upgrade slightly improves the approximation precision (L^∞-norm of the error) on the set Ω. Moreover, due to homogeneity of h, the approximation precision of the homogeneous ANN (hANN), on the domains $\{x \in \mathbb{R}^6 : |x| \leq 0.95\}$ and $\{x \in \mathbb{R}^6 : |x| \geq 1.05\}$, is much better (in times) than the precision of the classical ANN. Table 16.1 presents the comparison (see the second and the third column) of the approximation precision for the domains

$$\begin{aligned}
\Omega_{-4} &= \{x \in \mathbb{R}^6 : 0 < |x| \leq 0.25\}, & \Omega_4 &= \{x \in \mathbb{R}^6 : 1.75 < |x| \leq 2\},\\
\Omega_{-3} &= \{x \in \mathbb{R}^6 : 0.25 < |x| \leq 0.5\}, & \Omega_3 &= \{x \in \mathbb{R}^6 : 1.5 < |x| \leq 1.75\},\\
\Omega_{-2} &= \{x \in \mathbb{R}^6 : 0.5 < |x| \leq 0.75\}, & \Omega_2 &= \{x \in \mathbb{R}^6 : 1.25 < |x| \leq 1.5\},\\
\Omega_{-1} &= \{x \in \mathbb{R}^6 : 0.75 < |x| \leq 1\}, & \Omega_1 &= \{x \in \mathbb{R}^6 : 1 < |x| \leq 1.25\}.
\end{aligned}$$

The table contains the estimates of the norms $\|g_\epsilon(x) - h(x)\|_{L^\infty(\Omega_k, \mathbb{R}^6)}$ calculated using 20000 randomly selected points in Ω_k, $k = -4, \ldots, 4$.

16.2 Data-Driven Homogeneity

In this section, we continue the study of a possibility of transformation of an existing artificial neural network to a homogeneous one. We assume that an ANN is given by $g_\epsilon : \mathbb{R}^n \mapsto \mathbb{R}^m$ such that

$$g_\epsilon(x) = C\sigma(Ax + b), \quad x \in \mathbb{R}^n$$

approximates a mapping $g \in C(\mathbb{R}^n\backslash\{\mathbf{0}\}; \mathbb{R}^m)$ on a compact $\Omega \subset \mathbb{R}^n\backslash\{\mathbf{0}\}$, where $A \in \mathbb{R}^{N\times n}$, $b \in \mathbb{R}^N$, $C \in \mathbb{R}^{m\times N}$. The index $\epsilon > 0$ indicates that the approximation precision of ANN (see Theorem 16.1) is given by

$$\sup_{x\in\Omega} |g(x) - g_\epsilon(x)| \leq \epsilon, \quad \forall x \in \Omega.$$

The ANN is supposed to be well trained to approximate a homogeneous mapping g on the compact Ω, i.e., $\epsilon > 0$ is small enough. But now we consider the case when *some (or any) information about dilation symmetry of the mapping g is unavailable*, so, first, the parameters of the dilation group and/or the homogeneity degree have to be identified using g_ϵ, next, a homogeneous ANN has to be designed.

16.2.1 *Practical Homogeneity Degree*

First, let us study the problem of identification of an *unknown* homogeneity degree of the mapping g under the assumption that the linear dilation $\mathbf{d}(s) = e^{sG_\mathbf{d}}$, $s \in \mathbb{R}$ is *known*.

Identification of Homogeneity Degree of a Function ($m = 1$)

By Definition 6.1, the scalar-valued function $h : \mathbb{R}^n \mapsto \mathbb{R}$ is $\mathbf{d}$-homogeneous of degree $\nu \in \mathbb{R}$ if

$$h(e^{sG_\mathbf{d}}x) = e^{\nu s}h(x), \quad \forall x \in \mathbb{R}^n, \quad \forall s \in \mathbb{R}, \tag{16.17}$$

where $\mathbf{d}(s) = e^{sG_\mathbf{d}}$ is a linear dilation in $\mathbb{R}^n$. If g_ϵ approximates a $\mathbf{d}$-homogeneous function $g = h$ on Ω with a high enough precision then we should have

$$g_\epsilon(\mathbf{d}(s)x) \approx e^{\nu s}g_\epsilon(x), \quad s \in \mathbb{R}, \quad x \in \Omega,$$

provided that $\mathbf{d}(s)x \in \Omega$.

Let an explicit or implicit homogeneous norm $\|\cdot\|_\mathbf{d}$ in $\mathbb{R}^n$ be such that $S_\mathbf{d} = \{x \in \Omega : \|x\|_\mathbf{d} = 1\} \subset \Omega$. In this case, due to $\mathbf{d}(-\ln \|x\|_\mathbf{d})x \in S_\mathbf{d}$, the above approximate identity implies

$$g_\epsilon(\mathbf{d}(-\ln\|x\|_\mathbf{d})x) \approx \|x\|_\mathbf{d}^{-\nu} g_\epsilon(x), \quad \forall x \in \Omega. \tag{16.18}$$

Hence, for $x \in \Omega\backslash\{\mathbf{0}\} : \|x\|_\mathbf{d} \neq 1$ we have

$$\nu \approx \frac{1}{\ln\|x\|_\mathbf{d}} \ln \frac{g_\epsilon(x)}{g_\epsilon(\mathbf{d}(-\ln\|x\|_\mathbf{d})x)} \tag{16.19}$$

provided that

$$0 < \frac{g_\epsilon(x)}{g_\epsilon(\mathbf{d}(-\ln\|x\|_\mathbf{d})x)} < +\infty. \tag{16.20}$$

Notice that the implication

$$h(x) \neq 0 \quad \Rightarrow \quad 0 < \frac{h(x)}{h(\mathbf{d}(s)x)} < +\infty, \quad \forall s \in \mathbb{R}$$

is necessary for a function h to be homogeneous. Indeed, if $h(x) \neq 0$ and $\frac{h(x)}{h(-\mathbf{d}(\ln\|x\|_\mathbf{d})x)} < 0$ then the identity (16.17) is impossible and h is non-homogeneous. Therefore, the condition (16.20) can be utilized for detection of the non-homogeneity of g_ϵ. To decrease an impact of the approximation error on a decision about non-homogeneity of g_ϵ, we restrict the domain in (16.20) to the compact set

$$K_\epsilon = \left\{ x \in \Omega : \begin{array}{c} \|x\|_\mathbf{d} \in [r(\epsilon), 1 - r(\epsilon)] \cup [1 + r(\epsilon), +\infty) \\ \min\{|g_\epsilon(x)|, |g_\epsilon(\pi_\mathbf{d}(x))|\} \geq \delta(\epsilon), \end{array} \right\} \tag{16.21}$$

where $\pi_\mathbf{d}(x) = \mathbf{d}(-\ln\|x\|_\mathbf{d})x$ is the $\mathbf{d}$-homogeneous projector on the unit sphere $S_\mathbf{d}$ and the function $r, \delta \in \mathcal{K}$ (such that $\|r\|_C \leq 1$) are introduced in order to control a tolerance of the decision making about homogeneity/non-homogeneity of the function. In this case, we have

$$x \in K_\epsilon \quad \Rightarrow \quad \|x\|_\mathbf{d} \notin \{0, 1\} \quad \text{and} \quad \frac{g_\epsilon(x)}{g_\epsilon(\pi_\mathbf{d}(x))} > 0, \tag{16.22}$$

so the *practical homogeneity degree of the function* g_ϵ can be defined as

$$\nu_\epsilon = \frac{1}{M_\epsilon} \int_{K_\epsilon} \frac{1}{\ln\|x\|_\mathbf{d}} \ln \frac{g_\epsilon(x)}{g_\epsilon(\pi_\mathbf{d}(x))} dx \tag{16.23}$$

provided that $\epsilon > 0$ is small enough, where

$$M_\epsilon = \int_{K_\epsilon} dx.$$

If the condition (16.22) is not fulfilled then g_ϵ is not an approximation of homogeneous function or the approximation precision $\epsilon > 0$ is not enough to detect a homogeneity.

Proposition 16.2 *Let g_ϵ be an ANN* (16.2) *approximating a positive definite function $h \in C(\mathbb{R}^n\backslash\{\mathbf{0}\}; \mathbb{R})$ on a compact $\Omega \subset \mathbb{R}^n$ with an error $\epsilon > 0$, i.e., the inequality* (16.1) *is fulfilled for $g = h$. Let $\mathbf{d}$ be a linear continuous dilation in $\mathbb{R}^n$ and $\|\cdot\|_{\mathbf{d}}$ be any (explicit or implicit) $\mathbf{d}$-homogeneous norm in $\mathbb{R}^n$ such that $S_{\mathbf{d}} = \{x \in \mathbb{R}^n : \|x\|_{\mathbf{d}} = 1\} \in \Omega$. If the function h is $\mathbf{d}$-homogeneous of degree $\nu \in \mathbb{R}$ and the set $K_0 = \operatorname{int}\{x \in \Omega : \|x\|_{\mathbf{d}} \notin \{0, 1\}\} \subset \mathbb{R}^n$ is nonempty then*

$$\nu_\epsilon \to \nu \quad as \quad \epsilon \to 0 \tag{16.24}$$

where $\nu_\epsilon \in \mathbb{R}$ is given by (16.23).

Proof On the one hand, for sufficiently small $\epsilon > 0$, the compact set $K_\epsilon \subset \Omega\backslash\{\mathbf{0}\}$ is nonempty. Since h is $\mathbf{d}$-homogeneous and positive definite then $\nu = \frac{1}{\ln \|x\|_{\mathbf{d}}} \ln \frac{h(x)}{h(\mathbf{d}(-\ln \|x\|_{\mathbf{d}})x)}$ for all $x \in K_\epsilon$ and

$$\nu = \frac{1}{M_\epsilon} \int_{K_\epsilon} \frac{1}{\ln \|x\|_{\mathbf{d}}} \ln \frac{h(x)}{h(\pi_{\mathbf{d}}(x))} dx.$$

On the other hand, by definition ν_ϵ, we have

$$\nu_\epsilon = \frac{1}{M_\epsilon} \int_{K_\epsilon} \frac{1}{\ln \|x\|_{\mathbf{d}}} \ln \frac{h(x) + o(x, \epsilon)}{h(\pi_{\mathbf{d}}(x))) + o(\pi_{\mathbf{d}}(x)), \epsilon)} dx,$$

where $\sup_{y\in\Omega} |o(y, \epsilon)| \leq \epsilon$ (see (16.1)). Since h is assumed to be positive definite then $h(x) \geq \inf_{y\in K_\epsilon} h(y) > 0$ for all $x \in K_\epsilon$ and $h(\pi(x)) \geq \inf_{y\in S_{\mathbf{d}}} h(y) > 0$ for all $x \in K_\epsilon$. This implies that

$$\frac{1}{\ln \|x\|_{\mathbf{d}}} \ln \frac{g_\epsilon(x)}{g_\epsilon(\pi_{\mathbf{d}}(x))} \to \frac{1}{\ln \|x\|_{\mathbf{d}}} \ln \frac{h(x)}{h(\pi_{\mathbf{d}}(x))} \quad \text{as} \quad \epsilon \to 0 \tag{16.25}$$

uniformly on any compact from K_0. Hence, $\nu \to \nu_\epsilon$ as $\epsilon \to 0$. The proof is complete. □

The proven proposition also holds for negative definite homogeneous function h. If the function h is not sign-definite then $\inf_{y\in S} |h(y)| = 0$ and we cannot guarantee that the limit (16.25) is uniform. However, the practical homogeneity degree is still well defined by the formula (16.23) provided that the practical "necessary" condition of homogeneity (16.22) is fulfilled. In real situations, the approximation error $\epsilon > 0$ is usually unknown, so $\delta > 0$ and $r > 0$ should be considered as tuning parameters for construction of the set $K_\epsilon \subset \mathbb{R}^n$. The formula (16.23) defines the practical homogeneity degree as a mean value of some function. This mean value can also be approximated as follows:

$$\nu_\epsilon \approx \frac{1}{M} \sum_{i=1}^{M} \frac{1}{\ln \|x_i\|_{\mathbf{d}}} \ln \frac{g_\epsilon(x_i)}{g_\epsilon(\pi_{\mathbf{d}}(x_i))}, \tag{16.26}$$

where the points $x_i \in K_\epsilon$ are assumed to be uniformly[1] distributed in K_ϵ and $M \in \mathbb{N}$ is a large enough number.

Example 16.3 Let us consider the problem of approximation of rotation dynamics studied in Example 16.2 and let us assume that the homogeneity degree ν of the function h is unknown, but the dilation $G_{\mathbf{d}}$ is known. By Proposition 16.2, we can identify a practical homogeneity ν_ϵ using the conventional ANN, which is already trained on the set Ω. We compute ν_ϵ by the formula (16.26) for randomly selected points $x_i \in \Omega$, where $i = 1, 2, ..., M = 2000$. This gives the practical estimate of the homogeneity degree $\nu_\epsilon = 1.9647$ being rather close to the real value $\nu = 2$. The approximation precision of the hANN with the practical homogeneity degree is given in Table 16.1 (see the column 4). The use of homogeneous ANN (even with the practical homogeneity degree) essentially improves the extrapolation capabilities of the ANN.

Identification of Homogeneity Degree of a Vector Field ($m = n$)

If $f : \mathbb{R}^n \mapsto \mathbb{R}^n$ is a **d**-homogeneous vector field then

$$f(x) = \|x\|_{\mathbf{d}}^{\mu}\mathbf{d}(\ln \|x\|_{\mathbf{d}}) f(\mathbf{d}(-\ln \|x\|_{\mathbf{d}})x), \quad \forall x \neq \mathbf{0},$$

where $\|\cdot\|_{\mathbf{d}}$ is a **d**-homogeneous norm. If $f^\top(x)f(x) > 0$ and $x \neq \mathbf{0}$ then $f^\top(\pi_{\mathbf{d}}(x))$ $\mathbf{d}(\ln \|x\|_{\mathbf{d}})^\top \mathbf{d}(\ln \|x\|_{\mathbf{d}}) f(\pi_{\mathbf{d}}(x)) > 0$ and

$$\mu = \frac{1}{2\ln \|x\|_{\mathbf{d}}} \ln \frac{f^\top(x)f(x)}{f^\top(\pi_{\mathbf{d}}(x))\mathbf{d}(\ln \|x\|_{\mathbf{d}})^\top \mathbf{d}(\ln \|x\|_{\mathbf{d}}) f(\pi_{\mathbf{d}}(x))}, \tag{16.27}$$

where $\pi_{\mathbf{d}}(x) = \mathbf{d}(-\ln \|x\|_{\mathbf{d}})x$ is the homogeneous projector. Inspired by the above formula we can introduce the practical homogeneity degree of the ANN g_ϵ, which approximates the vector field $g = f$.

Let a norm $\|\cdot\|_{\mathbf{d}}$ in $\mathbb{R}^n$ be such that $S_{\mathbf{d}} \subset \Omega$. Let us define the compact set

$$K_\epsilon = \left\{ x \in \Omega : \begin{array}{c} \|x\|_{\mathbf{d}} \in [r(\epsilon), 1 - r(\epsilon)] \cup [1 + r(\epsilon), +\infty), \quad g_\epsilon^\top(x)g_\epsilon(x) \geq \delta(\epsilon), \\ g_\epsilon^\top(\pi_{\mathbf{d}}(x))\mathbf{d}(\ln \|x\|_{\mathbf{d}})^\top \mathbf{d}(\ln \|x\|_{\mathbf{d}}) g_\epsilon(\pi_{\mathbf{d}}(x)) \geq \delta(\epsilon) \end{array} \right\}, \tag{16.28}$$

where $r, \delta \in \mathcal{K}$ are as before. In this case, the *practical homogeneity degree of the vector field* g_ϵ can be introduced as follows:

$$\mu_\epsilon = \frac{1}{2M_\epsilon} \int_{K_\epsilon} \frac{1}{\ln \|x\|_{\mathbf{d}}} \ln \frac{g_\epsilon^\top(x)g_\epsilon(x)}{g_\epsilon^\top(\pi_{\mathbf{d}}(x))\mathbf{d}^\top(\ln \|x\|_{\mathbf{d}})\mathbf{d}(\ln \|x\|_{\mathbf{d}}) g_\epsilon(\pi_{\mathbf{d}}(x))} dx, \tag{16.29}$$

where, as before, $M_\epsilon = \int_{K_\epsilon} dx$.

---

[1] In the following sense $\sup_{x \in K_\epsilon} \inf_i |x - x_i| \to 0$ as $M \to +\infty$.

Proposition 16.3 . *Let g_ϵ be ANN* (16.2) *approximating a vector field $f \in C(\mathbb{R}^n\backslash\{\mathbf{0}\}; \mathbb{R}^n)$ on a compact $\Omega \subset \mathbb{R}^n$ with an error $\epsilon > 0$, i.e., the inequality* (16.1) *is fulfilled for $g = f$. Let $\mathbf{d}$ be a linear continuous dilation in $\mathbb{R}^n$ and $\|\cdot\|_{\mathbf{d}}$ be any (explicit or implicit) $\mathbf{d}$-homogeneous norm in $\mathbb{R}^n$ such that $S_{\mathbf{d}} = \{x \in \mathbb{R}^n : \|x\|_{\mathbf{d}} = 1\} \in \Omega$. Let the vector field f be $\mathbf{d}$-homogeneous of degree $\mu \in \mathbb{R}$ such that the matrix $\mu I_n + G_{\mathbf{d}}$ is anti-Hurwitz and $f(x) \neq \mathbf{0}$ for any $x \neq \mathbf{0}$. If the set $K_0 = \mathrm{int}\{x \in \Omega : \|x\|_{\mathbf{d}} \notin \{0, 1\}\} \subset \mathbb{R}^n$ is nonempty then*

$$\mu_\epsilon \to \mu \quad as \quad \epsilon \to 0, \tag{16.30}$$

where $\mu_\epsilon \in \mathbb{R}$ is given by (16.29).

Proof Since $f(x) \neq \mathbf{0}$ for $x \neq \mathbf{0}$ then due to homogeneity we have

$$f^\top(x)f(x) > 0, \quad f^\top(\pi_{\mathbf{d}}(x))\mathbf{d}(\ln\|x\|_{\mathbf{d}})^\top \mathbf{d}(\ln\|x\|_{\mathbf{d}})f(\pi_{\mathbf{d}}(x)) > 0, \quad \forall x \neq \mathbf{0}.$$

In this case, the identity (16.27) holds for all $x \neq \mathbf{0}$. Repeating considerations of Proposition 16.2 we complete the proof. □

The practical "necessary" condition for the homogeneity of ANN is given by

$$\left.\begin{array}{c} x \in K_\epsilon, \quad |e_i^\top g_\epsilon(x)| > \delta(\epsilon), \\ |e_i^\top \mathbf{d}(\ln\|x\|_{\mathbf{d}})g_\epsilon(\pi_{\mathbf{d}}(x))| > \delta(\epsilon) \end{array}\right\} \Rightarrow g_\epsilon^\top(x)e_i e_i^\top \mathbf{d}(\ln\|x\|_{\mathbf{d}})g_\epsilon(\pi_{\mathbf{d}}(x)) > 0, \tag{16.31}$$

where $e_i = (0, ..., 1, ..., 0)^\top$ is the unit vector in $\mathbb{R}^n$. If K_ϵ is a nonempty set then, under the above condition, the practical homogeneity degree of ANN is well defined and the ANN can be upgraded to the homogeneous ANN using Proposition 16.3. Being a mean value of a function the practical homogeneity degree of the vector field may be numerically estimated

$$\mu_\epsilon \approx \frac{1}{2M}\sum_{i=1}^{M} \frac{1}{\ln\|x_i\|_{\mathbf{d}}} \ln \frac{g_\epsilon^\top(x_i)g_\epsilon(x_i)}{g_\epsilon^\top(\pi_{\mathbf{d}}(x_i))\mathbf{d}^\top(\ln\|x_i\|_{\mathbf{d}})\mathbf{d}(\ln\|x_i\|_{\mathbf{d}})g_\epsilon(\pi_{\mathbf{d}}(x_i))}, \tag{16.32}$$

where the points $x_i \in K_\epsilon$ are assumed to be uniformly distributed in K_ϵ and $M \in \mathbb{N}$ is a large enough number.

16.2.2 Practical Linear Dilation

Let us study the identification problem of a dilation symmetry (if any) for the ANN g_ϵ which approximates a mapping g. In the view of Corollary 6.2, if a function h (or a vector field f) is $\mathbf{d}$-homogeneous of degree ν then, for any $\gamma > 0$, it is $\tilde{\mathbf{d}}$-homogeneous of the degree $\tilde{\nu} = \gamma\nu$ with respect to the dilation

$$\tilde{\mathbf{d}}(s) = \mathbf{d}(\gamma s), \quad s \in \mathbb{R}.$$

Hence, taking $\gamma = 1/|\nu|$ for $\nu \neq 0$, we conclude that *the identification of the dilation symmetry of h can be reduced to identification of the generator $G_{\mathbf{d}} \in \mathbb{R}^{n\times n}$ only*, while the homogeneity degree can always be assigned to one of the following constants $\{-1, 0, 1\}$. In other words, if the sign of the homogeneity degree is unknown three identification problems with $\nu = -1$, $\nu = 0$ and $\nu = 1$ have to be solved in order to identify a dilation symmetry of g_ϵ (if any). Therefore, without loss of generality we assume below that the homogeneity degree ν of g_ϵ is *known* (or fixed to one of the following values $\{-1, 0, 1\}$) and we need to identify an *unknown* generator $G_{\mathbf{d}} \in \mathbb{R}^{n\times n}$ of the dilation $\mathbf{d}$.

Identification of Dilation for Homogeneous Function ($m = 1$)

By Theorem 6.1, for any locally Lipschitz continuous (on $\mathbb{R}^n\backslash\{\mathbf{0}\}$) $\mathbf{d}$-homogeneous function h, we have

$$\int_{x^*}^{x} (\text{trace}(G_{\mathbf{d}}) + \nu)h(y) - \sum_{i:x_i\neq x_i^*} \Delta_i(h, y, x, x^*, G_{\mathbf{d}})\, dy = 0 \tag{16.33}$$

for all $x, x^* \in \mathbb{R}^n$ such that $R(x, x^*) \in \mathbb{R}^n\backslash\{\mathbf{0}\}$.

If g_ϵ approximates a $\mathbf{d}$-homogeneous function h of degree ν with a rather high precision ϵ on a compact Ω then the identity (16.33) implies that

$$\int_{x^*}^{x} (\text{trace}(G_{\mathbf{d}}) + \nu)g_\epsilon(y) - \sum_{i:x_i\neq x_i^*} \Delta_i(g_\epsilon, y, x, x^*, G_{\mathbf{d}})\, dy \approx 0. \tag{16.34}$$

The only unknown parameter $G_{\mathbf{d}} \in \mathbb{R}^{\times n}$ is involved in above approximate identity in the affine manner. Its identification can be based on a quadratic optimization.

Let Ω have a nonempty interior. For $\delta \in (0, 1)$ let $\Omega_\delta \subset \text{int}\,\Omega$ be such that

$$\Omega_\delta \dot{+} B_\delta \subset \Omega \quad \text{and} \quad \Omega_\delta \cap B_\delta = \emptyset, \tag{16.35}$$

where $B_\delta = \{x = (x_1, ..., x_n)^\top \in \mathbb{R}^n : |x_i| \leq \delta, i = 1, ..., n\}$. Continuous homogeneous functions may have singularity at zero, so the condition $\Omega_\delta \cap B_\delta = \emptyset$ excludes this potential singularity point. Let us introduce

$$G_\epsilon \in \underset{G\in\Xi}{\text{argmin}}\, J_\epsilon(G), \tag{16.36}$$

where $\Xi \subset \mathbb{R}^{n\times n}$ is a set of admissible $n \times n$ matrices and

$$J_\epsilon(G)=\frac{1}{M_\delta}\int_{\Omega_\delta}\int_{B_\delta}\mathcal{J}_\epsilon^2(x^*+z,x^*,G)\,dz dx^*, \tag{16.37}$$

$$\mathcal{J}_\epsilon(x,x^*,G)=\int_{x^*}^{x}(\text{trace}(G)+\nu)g_\epsilon(y)-\sum_{i:x_i\neq x_i^*}\Delta_i(g_\epsilon,y,x,x^*,G_{\mathbf{d}})\,dy,$$

and $M_\delta=\int_{\Omega_\delta}dx\cdot\int_{B_\delta}dz$. If G_ϵ is anti-Hurwitz then, in the view of the following theorem, it can be interpreted as a generator of the *practical dilation in* $\mathbb{R}^n$.

Theorem 16.4 *Let $\Xi\subset\mathbb{R}^{n\times n}$ be a class of admissible generators of linear dilations. Let g_ϵ be an ANN* (16.2) *approximating a function $h:\mathbb{R}^n\mapsto\mathbb{R}$ on a compact $\Omega\subset\mathbb{R}^n$ with an error $\epsilon>0$, i.e., the inequality* (16.1) *is fulfilled for $g=h$. If h is locally Lipschitz continuous on $\mathbb{R}^n\backslash\{\mathbf{0}\}$ and $\mathbf{d}$-homogeneous of degree $\nu\in\mathbb{R}$ with respect to a linear dilation $\mathbf{d}(s)=e^{G_{\mathbf{d}}s}$, $s\in\mathbb{R}^n$ with $G_{\mathbf{d}}\in\Xi$, then*

$$\min_{G\in\Xi}J_\epsilon(G)\to 0\quad as\quad \epsilon\to 0.$$

Proof By Theorem 6.1, we have

$$\int_{x^*}^{x}(\text{trace}(G_{\mathbf{d}})+\nu)h(y)-\sum_{i:x_i\neq x_i^*}\Delta_i(h,y,x,x^*,G_{\mathbf{d}})\,dy=0$$

for all $x,x^*\in\mathbb{R}^n\backslash\{\mathbf{0}\}$ such that $R(x,x^*)\in\mathbb{R}^n\backslash\{\mathbf{0}\}$. Since $g_\epsilon(x)\to h(x)$ as $\epsilon\to 0$ uniformly on compacts from Ω then

$$\int_{x^*}^{x}(\text{trace}(G_{\mathbf{d}})+\nu)g_\epsilon(y)-\sum_{i:x_i\neq x_i^*}\Delta_i(g_\epsilon,y,x,x^*,G_{\mathbf{d}})\,dy\to 0$$

as $\epsilon\to 0$ for all $x,x^*\in\Omega_\delta$. This means that

$$\min_{G\in\Xi}J_\epsilon(G)\to 0\quad\text{as}\quad\epsilon\to 0.$$

The proof is complete. □

Notice that the above theorem implies that

$$G_{\mathbf{d}}\in\mathcal{P}:=\lim_{\epsilon\to 0}\operatorname*{argmin}_{G\in\Xi}J_\epsilon(G),$$

provided that $G_{\mathbf{d}}\in\Xi$. However, the set $\mathcal{P}$ is not a singleton in the general case and the problem of identification of the generator $G_{\mathbf{d}}$ is ill-posed even if $\Xi=\Xi_{\text{lin}}$, where

$$\Xi_{\text{lin}}=\{G\in\mathbb{R}^{n\times n}:G\text{ is anti-Hurwitz}\}. \tag{16.38}$$

Indeed, for example, for any $\gamma > 0$, the function $h : \mathbb{R}^2 \mapsto \mathbb{R}$ given by

$$h(x) = x^\top x, \quad x = (x_1, x_2)^\top \in \mathbb{R}^2 \tag{16.39}$$

is $\mathbf{d}_\gamma$-homogeneous of degree 2 with respect to the linear dilation $\mathbf{d}_\gamma(s) = e^s \begin{pmatrix} \cos(\gamma s) & -\sin(\gamma s) \\ \sin(\gamma s) & \cos(\gamma s) \end{pmatrix}$, $s \in \mathbb{R}$. So, the linear dilation symmetry of the function h given above cannot be uniquely identified.

To reduce the uncertainty in the dilation identification, the cost functional can be modified as follows:

$$G_\epsilon = \underset{G \in \Xi}{\operatorname{argmin}}\, J_\epsilon(G) + \xi \operatorname{trace}(G^\top G), \tag{16.40}$$

where $\Xi = \mathbb{R}^{n\times n}$ and $\xi > 0$ is a small regularization parameter. The second term in the above formula is a strictly convex function (being a norm) in the space of square matrices. The quadratic optimization problem becomes strictly convex with respect to G, so, for any fixed $\epsilon > 0$, the matrix G_ϵ is uniquely defined by (16.40). Another possible way to reduce the uncertainty is to restrict the set of admissible generators Ξ, for example, to diagonalizable (or simply diagonal) matrices. However, this introduces additional a priori assumption about the weighted homogeneity of the function h.

The functional J_ϵ admits the following approximation:

$$J_\epsilon(G) \approx \frac{1}{ML} \sum_{k=1}^{M} \sum_{j=1}^{L} \mathcal{J}_\epsilon^2(x_k + z_j, x_i, G)\, dz \tag{16.41}$$

where the points $x_k \in \Omega_\delta$ are uniformly distributed in Ω_δ, $z_j \in B_\delta$ are uniformly distributed in B_δ and $M, L \in \mathbb{N}$. Finally, denoting $f(x, x^*, y, G) = (\operatorname{trace}(G) + \nu) g_\epsilon(y) - \sum\limits_{i: x_i \neq x_i^*} \Delta_i(g_\epsilon, y, x, x^*, G)$ we derive the following approximation of the functional $\mathcal{J}_\epsilon^2$ for a sufficiently small $\delta \in (0, 1)$:

$$\mathcal{J}_\epsilon(x_i + z_j, x_i, G) \approx \prod_{i=1}^{n} e_i^\top z_j \sum_{\lambda \in \{0,1\}^n} f(x_k + z_j, x_k, x_k + \operatorname{diag}(\lambda) z_j, G). \tag{16.42}$$

The presented discretizations of the functionals J_ϵ and $\mathcal{J}_\epsilon$ can be utilized for identification of the generator G_ϵ of the practical dilation; however, in this case, an anti-Hurwitz solution of the optimization problem (16.36) may exist even if the ANN g_ϵ approximates a non-homogeneous function. So, the practical necessary condition of homogeneity (16.22) has be utilized in order to check if the ANN, indeed, approximates a $\mathbf{d}_\epsilon$-homogeneous (at least on Ω) function. In the view of Proposition 16.4, a correctness of this conclusion depends on approximation precision of ANN.

Example 16.4 Let us consider the problem of approximation of rotation dynamics studied in Example 16.2. Now we assume that the dilation is unknown and we need to approximate $G_{\mathbf{d}}$ using the already well-trained ANN g_ϵ. For the data-driven identification of the dilation symmetry, we consider the set $\Omega_\delta = \{x \in \mathbb{R}^n : 0.98 \le |x| \le 1.02\}$ and optimize the approximate functional (16.41) with $M = 4000$, $L = 1$, $z_1 = \delta(1, ..., 1)^\top$, $\delta = 0.01$. We also restrict the class of admissible generators to diagonal matrices. For $\nu = 1$, the obtained matrix $G_\epsilon = \mathrm{diag}(0.5199, 0.5150, 0.4972, 1.0021, 0.9965, 1.0027)$, which, obviously, corresponds to $G_{\mathbf{d}}$ up to the scaling of G_ϵ and ν by the multiplier $\gamma = 2$ (please see explanations in the beginning of Sect. 4.2 for more details). The homogeneous ANN constructed in this way is a homogeneous extrapolation of the ANN g_ϵ. In all domain Ω_k except the training set Ω_0, the approximation precision of hANN is better in times comparing with the original ANN (see the last column in Table 16.1).

Identification of Dilation for Homogeneous Vector Field ($m = n$)

Let g_ϵ be an ANN (16.2) approximating a homogeneous vector field f. We repeat the above construction using Theorem 6.2. Let us define

$$J_\epsilon(G) = \sum_{i=1}^{n} \int_{\Omega_\delta} \int_{B_\delta} \mathcal{J}_j^2(x^* + z, x^*, G)\, dz dx^*, \tag{16.43}$$

where Ω_δ and B_δ are as before and

$$\mathcal{J}_j(x, x^*, G) = \int_{x^*}^{x} (\mathrm{trace}(G) + \mu) e_j^\top g_\epsilon(y) + e_j^\top G g_\epsilon(y) - \sum_{i: x_i \neq x_i^*} \frac{e_j^\top g_\epsilon(y) e_i^\top G y\Big|_{x_i^*}^{x_i}}{x_i - x_i^*}\, dy.$$

Let us study the identifiability of the linear dilation for the vector field f by means of the optimization procedure (16.36) with J_ϵ given (16.43).

Theorem 16.5 *Let g_ϵ be an ANN* (16.2) *approximating a vector field $f : \mathbb{R}^n \mapsto \mathbb{R}^n$ on a compact $\Omega \subset \mathbb{R}^n$ with an error $\epsilon > 0$, i.e., the inequality* (16.1) *is fulfilled for $g = f$.*

If f is locally Lipschitz continuous on $\mathbb{R}^n \backslash \{\mathbf{0}\}$ and **d***-homogeneous degree $\mu \in \mathbb{R}$ with respect to a dilation $\mathbf{d}(s) = e^{sG_{\mathbf{d}}}$, $s \in \mathbb{R}$, $G_{\mathbf{d}} \in \Xi$ then*

$$\min_{G \in \Xi} J_\epsilon(G) \to 0 \quad as \quad \epsilon \to 0,$$

where J_ϵ is given by (16.43).

The proof of this theorem literally repeats the proof of Theorem 16.4 using the identity (6.11) instead of the identity (6.4).

The existence and uniqueness of the limit $G_\epsilon \to G_{\mathbf{d}}$ as $\epsilon \to 0$ can be guaranteed under additional restrictions to the class of dilations and vector fields under consideration.

The regularization (16.40) of the optimization problem (16.36) can also be utilized for a reduction of an uncertainty in the dilation identification.

16.3 Homogeneous Differential Neural Network

Let us consider the affine-in-control system

$$\dot{x} = f_0(x) + \sum_{i=1}^{m} f_i(x)u_i, \quad t > 0, \tag{16.44}$$

where $x(t) \in \mathbb{R}^n$ is the system state, $u = (u_1, ..., u_m)$ is the vector of inputs, $f_i : \mathbb{R}^n \mapsto \mathbb{R}^n$ is a $\mathbf{d}$-homogeneous continuous vector field of degree $\mu_i \in \mathbb{R}, i = 0, ..., m$. The homogeneity degrees $\mu_i \in \mathbb{R}$ and the linear dilation $\mathbf{d}$ are assumed to be known, but some or all vector fields f_i are unknown.

Following [13, 14], the ordinary differential equation (ODE)

$$\dot{z} = f_{\epsilon,0}(z) + \sum_{i=1}^{m} f_{\epsilon,i}(z)u_i, \quad t > 0, \tag{16.45}$$

which approximates the dynamics of (16.44), is called the *differential neural network* (DNN) provided that $f_{\epsilon,i}$ is an ANN approximating f_i for $i = 0, ..., m$, where $\epsilon > 0$, as before, indicates the approximation precision.

In control theory, the process of finding of parameters of the functions $f_{\epsilon,i}$ is known as *system identification* [15]. The theory of neural networks calls this process by *learning*. The inputs u_i can be utilized for an excitation of the system (16.44) in order to generate a sufficiently reach set of trajectories. The system excitation may essentially improve the identification/learning quality (see, e.g., [16] and references therein).

Let $P \in \mathbb{R}^{n\times n}$ be a positive definite symmetric matrix such that

$$G_{\mathbf{d}}^\top P + PG_{\mathbf{d}} \succ 0, \quad P \succ 0, \tag{16.46}$$

where $G_{\mathbf{d}}$ is a generator of the dilation $\mathbf{d}$. The above LMI is always feasible with respect to P (see Proposition 5.3). Let $\|\cdot\|_{\mathbf{d}}$ be a $\mathbf{d}$-homogeneous norm induced by the weighted Euclidean norm $\|x\| = \sqrt{x^\top P x}$.

According to Theorem 16.3, the $\mathbf{d}$-homogeneous ANN approximating f_i should have the form:

$$f_{\epsilon,i}(x) = \|x\|_{\mathbf{d}}^{\mu_i}\mathbf{d}(\ln\|x\|_{\mathbf{d}})C_i\sigma(A_i\mathbf{d}(-\ln\|x\|_{\mathbf{d}})x + b_i), \tag{16.47}$$

where $i = 1, ..., m$, $A_i \in \mathbb{R}^{N\times n}$, $b_i \in \mathbb{R}^N$, $C_i \in \mathbb{R}^{n\times N}$, $N \in \mathbb{N}$ is large enough and σ is an activation function (see above). Let the ANN $f_{\epsilon,0}$ approximating f_0 be defined as follows:

$$\tilde{f}_{\epsilon,0}(x) = -\rho\|x\|_{\mathbf{d}}^{\mu_0} G_{\mathbf{d}}x + \|x\|_{\mathbf{d}}^{\mu_0}\mathbf{d}(\ln\|x\|_{\mathbf{d}})C_0\sigma(A_0\mathbf{d}(-\ln\|x\|_{\mathbf{d}})x + b_0), \quad (16.48)$$

where $A_0 \in \mathbb{R}^{N\times n}$, $b_0 \in \mathbb{R}^N$, $\rho > 0$ and $C_0 \in \mathbb{R}^{n\times N}$. For $\rho = 0$ the homogeneous ANN $\tilde{f}_{\epsilon,0}$ takes the conventional form (16.8). The modification (16.48) of the homogeneous ANN (16.8) does not influence approximation quality of the vector field f_0, but it improves an internal stability of the system

$$\dot{x}_\epsilon = \tilde{f}_{\epsilon,0}(x_\epsilon) + \sum_{i=1}^{m} f_{\epsilon,i}(x_\epsilon)u_i, \quad t > 0, \quad x_\epsilon(t) \in \mathbb{R}^n \quad (16.49)$$

during the learning phase, where $\tilde{f}_{0,\epsilon}$ and $f_{i,\epsilon}$ are given by (16.48) and 16.47, respectively. Indeed, simple calculations show that the system (16.49) with $C_i = 0$ and $\rho > 0$ is globally asymptotically stable and $\|\cdot\|_{\mathbf{d}}$ is its Lyapunov function: $\frac{d\|x_\epsilon\|_{\mathbf{d}}}{dt} = -\rho\|x_\epsilon\|_{\mathbf{d}}^{1+\mu_0}$. The stability is important for the DNN learning [13].

The ODE (16.49) is called the **d**-*homogeneous DNN*. In the view of previous results, we expect that the key feature of the homogeneous DNN should be a global approximation of the system dynamics based on local behavior of the system. Moreover, we expect an improvement of the approximation precision close to zero and/or close to infinity dependently of the homogeneity degrees. Standard homogeneous DNN identifier (with $\mathbf{d}(s) = e^s I_n$) of the form (16.49) has been introduced in [17]. Below we generalize it to generalized homogeneous systems.

According to [14] for a sufficiently large N, even a random selection of the parameters A_i and b_i may guarantee a good enough approximation precision. So, the learning of the DNN frequently consists in a tuning of the parameters C_i such that the dynamics of (16.49) becomes close (in some sense) to the dynamics of (16.44). Below we follow this idea assuming that A_i and b_i are fixed, and the only C_i has to be optimized to provide a good approximation.

16.3.1 Learning of Homogeneous DNN

In this section, we develop algorithms of dynamical learning for homogeneous DNN. More precisely, we assume that the state vector $x(t)$ and the input vector $u(t)$ are available at each instant of time $t \geq 0$ and we design a deferential law for adjustment of parameters C_i. We also suggest a special structure for excitation signals u_i which could improve the identification/learning quality.

Identification of the Vector Field f_0

First, let us study the case when the only *unknown* vector field is f_0. The gains f_i amplifying the inputs u_i are assumed to be *known*. Let us consider the following learning system:

$$\dot{z} = -\rho\|x\|_{\mathbf{d}}^{\mu_0} G_{\mathbf{d}} z + f_{\epsilon,0}(x, C_0) + \sum_{i=1}^{m} f_i(x)u_i, \quad t > 0 \tag{16.50}$$

$$\frac{d\,\mathrm{vec}(C_0^\top)}{dt} = \gamma\|x\|_{\mathbf{d}}^{\mu_0}\left[I_n \otimes \sigma(A_0\mathbf{d}(-\ln\|x\|_{\mathbf{d}})x + b_0)\right] P\mathbf{d}(-\ln\|x\|_{\mathbf{d}})(x - z), \tag{16.51}$$

where $z(t) \in \mathbb{R}^n$, $C_0(t) \in \mathbb{R}^{n\times N}$, $\gamma > 0$ is a tuning parameter,

$$f_{\epsilon,0}(x, C_0) = \|x\|_{\mathbf{d}}^{\mu_0}\mathbf{d}(\ln\|x\|_{\mathbf{d}})C_0\sigma(A_0\mathbf{d}(-\ln\|x\|_{\mathbf{d}})x + b_0), \tag{16.52}$$

$\mathrm{vec}(q_1, q_2, ...) = \begin{pmatrix} q_1 \\ q_2 \\ \ldots \end{pmatrix}$, $\otimes$ is the Kronecker product and P is defined by (16.46). The system (16.50) defines dynamics of the system with ANN approximator in the right-hand side. If the parameter of C_0 of the ANN converges to its optimal value, we expect that the trajectory z of the approximated system (16.50) should converge to the trajectory of the system (16.44).

The learning law (16.51) is inspired by the following identity:

$$C_0\sigma(A_0\mathbf{d}(-\ln\|x\|_{\mathbf{d}})x + b_0) = \left[I_n \otimes \sigma(A_0\mathbf{d}(-\ln\|x\|_{\mathbf{d}})x + b_0)\right]^\top \mathrm{vec}(C_0^\top), \tag{16.53}$$

allowing the ANN $f_{\epsilon,0}$ to be rewritten as follows:

$$f_{\epsilon,0}(x, C_0) = \|x\|_{\mathbf{d}}^{\mu_0}\mathbf{d}(\ln\|x\|_{\mathbf{d}})\left[I_n \otimes \sigma(A_0\mathbf{d}(-\ln\|x\|_{\mathbf{d}})x + b_0)\right]^\top \mathrm{vec}(C_0^\top).$$

Capabilities of the learning law (16.51) are studied by the following theorem. Its proof is based on the so-called Barbalat lemma, which requires a Lipschitz continuity of f_0 and the control signals u_i.

Theorem 16.6 *Let f_i be locally Lipschitz continuous on $\mathbb{R}^n\backslash\{\mathbf{0}\}$ for $i = 0, 1, ..., m$ and $u_i \in W^1(\mathbb{R}_+; \mathbb{R})$ for $i = 1, 2, ..., m$. Let there exist $C_0^* \in \mathbb{R}^{n\times N}$ such that*

$$f_0(x) = -\rho\|x\|_{\mathbf{d}}^{\mu_0} G_{\mathbf{d}} x + \|x\|_{\mathbf{d}}^{\mu_0}\mathbf{d}(\ln\|x\|_{\mathbf{d}})C_0^*\sigma(A_0\mathbf{d}(-\ln\|x\|_{\mathbf{d}})x + b_0). \tag{16.54}$$

If all trajectories of the system (16.44) *are uniformly bounded as follows:*

$$0 < r_{\min} \le \|x(t)\|_{\mathbf{d}} \le r_{\max} < +\infty, \quad \forall t \ge 0, \tag{16.55}$$

and

$$\exists \rho^* > 0 \ : \ \frac{d\|x(t)\|_{\mathbf{d}}}{dt} \ge -\rho^*\|x(t)\|_{\mathbf{d}}^{1+\mu_0}, \quad \forall t > 0, \tag{16.56}$$

then for any $\rho > \rho^*$ *one holds*

$$z(t) \to x(t) \quad as \quad t \to +\infty,$$

where C_0 *is governed by* (16.51) *with* z *satisfying* (16.50).

Proof Let us denote $c = \text{vec}\,(C_0^\top)$, $c^* = \text{vec}\,((C_0^*)^\top)$, $\tilde{c} = c^* - c$ and consider the Lyapunov function candidate

$$V = \tilde{c}^\top \gamma^{-1} \tilde{c} + (x-z)^\top \mathbf{d}^\top(-\ln \|x\|_{\mathbf{d}}) P \mathbf{d}(-\ln \|x\|_{\mathbf{d}})(x-z).$$

The time derivative of V is

$$\dot{V} = 2\tilde{c}^\top \gamma^{-1} \frac{d\tilde{c}}{dt} + 2(x-z)^\top \mathbf{d}^\top(-\ln \|x\|_{\mathbf{d}}) P \mathbf{d}(-\ln \|x\|_{\mathbf{d}})(\dot{x} - \dot{z}) +$$

$$-2(x-z)^\top \mathbf{d}^\top(-\ln \|x\|_{\mathbf{d}}) P G_{\mathbf{d}} \mathbf{d}(-\ln \|x\|_{\mathbf{d}})(x-z) \frac{1}{\|x\|_{\mathbf{d}}} \frac{d\|x\|_{\mathbf{d}}}{dt}$$

$$= 2\tilde{c}^\top \gamma^{-1} \frac{d\tilde{c}}{dt} + 2\|x\|_{\mathbf{d}}^{\mu_0} (x-z)^\top \mathbf{d}^\top (\ln \|x\|_{\mathbf{d}}) P (C_0 - C_0^*) \sigma(A_0 \mathbf{d}(-\ln \|x\|_{\mathbf{d}})x + b_0)$$

$$-\left(\rho \|x\|_{\mathbf{d}}^{\mu_0} + \frac{1}{\|x\|_{\mathbf{d}}} \frac{d\|x\|_{\mathbf{d}}}{dt}\right)(x-z)^\top \mathbf{d}^\top(-\ln \|x\|_{\mathbf{d}})(PG_{\mathbf{d}} + G_{\mathbf{d}}^\top P)\mathbf{d}(-\ln \|x\|_{\mathbf{d}})(x-z)$$

$$= -\left(\rho \|x\|_{\mathbf{d}}^{\mu_0} + \frac{1}{\|x\|_{\mathbf{d}}} \frac{d\|x\|_{\mathbf{d}}}{dt}\right)(x-z)^\top \mathbf{d}^\top(-\ln \|x\|_{\mathbf{d}})(PG_{\mathbf{d}} + G_{\mathbf{d}}^\top P)\mathbf{d}(-\ln \|x\|_{\mathbf{d}})(x-z),$$

where the property $\mathbf{d}(s)G_{\mathbf{d}} = G_{\mathbf{d}}\mathbf{d}(s)$, $\forall s \in \mathbb{R}$, the representation (16.54) and the identity

$$(C_0 - C_0^*)\sigma(A_0 \mathbf{d}(-\ln \|x\|_{\mathbf{d}})x + b_0) = -\left[I_n \otimes \sigma(A_0 \mathbf{d}(-\ln \|x\|_{\mathbf{d}})x + b_0)\right]^\top \tilde{c}$$

are utilized on the last steps. Notice that, introducing $u_0 = 1$, we derive

$$\frac{d\|x\|_{\mathbf{d}}}{dt} = \|x\|_{\mathbf{d}}^{1+\mu} \sum_{i=0}^{m} u_i(t) \frac{x^\top \mathbf{d}^\top(-\ln \|x\|_{\mathbf{d}}) P f_i(\mathbf{d}(-\ln \|x\|_{\mathbf{d}})x)}{x^\top \mathbf{d}^\top(-\ln \|x\|_{\mathbf{d}}) P G_{\mathbf{d}} \mathbf{d}(-\ln \|x\|_{\mathbf{d}})x}, \tag{16.57}$$

where the formula (5.21) and the homogeneity of f_i are utilized. Since $u_i \in W^{1,\infty}(\mathbb{R}_+; \mathbb{R})$ and f_i is locally Lipschitz continuous on $\mathbb{R}^n \backslash \{\mathbf{0}\}$ then, taking into account the estimate (16.55), we conclude that $\|x(\cdot)\|_{\mathbf{d}} \in W^{2,\infty}(\mathbb{R}_+; \mathbb{R})$. Using (16.56) and (16.46) we derive

$$\dot{V} \le -(\rho - \rho^*)\|x\|_{\mathbf{d}}^{\mu_0} (x-z)^\top \mathbf{d}^\top(-\ln \|x\|_{\mathbf{d}})(PG_{\mathbf{d}} + G_{\mathbf{d}}^\top P)\mathbf{d}(-\ln \|x\|_{\mathbf{d}})(x-z).$$

Since $\|v\| = \sqrt{v^\top P v}$, $v \in \mathbb{R}^n$ and $x(t)$ is uniformly bounded by the formula (16.55) then using Proposition 5.3 and Corollary 5.3 we conclude that there exist $0 < \gamma_1 < \gamma_2 < +\infty$ such that

$$\gamma_1 \|v\|^2 \le \|x(t)\|_{\mathbf{d}}^{\mu_0} v^\top \mathbf{d}^\top(-\ln \|x(t)\|_{\mathbf{d}})(PG_{\mathbf{d}} + G_{\mathbf{d}}^\top P)\mathbf{d}(-\ln \|x(t)\|_{\mathbf{d}})v \le \gamma_2 \|v\|^2$$

for all $v \in \mathbb{R}^n$, $\forall t \ge 0$. Hence, for $\rho > \rho^*$ we have

$$\dot{V} \le -\gamma_1(\rho - \rho^*)\|x - z\|^2.$$

This means $V(t)$ computed on trajectories of the system is non-increasing on $\mathbb{R}_+$. Since V is positive definite then $V(t)$ has a finite limit as $t \to +\infty$.

The boundedness of V implies the boundedness of $\tilde{c}$ and $x - z$. Hence, taking into account the equation

$$\dot{x} - \dot{z} = -\rho\|x\|_{\mathbf{d}}^{\mu_0} G_{\mathbf{d}}(x - z) + \|x\|_{\mathbf{d}}^{\mu_0} \mathbf{d}(\ln \|x\|_{\mathbf{d}})(C_0 - C_0^*)\sigma(A_0\mathbf{d}(-\ln \|x\|_{\mathbf{d}})x + b_0)$$

we conclude $x - z \in W^{1,\infty}(\mathbb{R}_+; \mathbb{R})$. Moreover, since $\|x(\cdot)\|_{\mathbf{d}} \in W^{2,\infty}(\mathbb{R}_+; \mathbb{R})$ then $V \in W^{2,\infty}(\mathbb{R}_+; \mathbb{R})$ and Barbalat lemma (see Appendix D) yields $\dot{V}(t) \to 0$ as $t \to +\infty$. For $\tilde{\rho} = \sup_{t \ge 0} \frac{1}{\|x(t)\|_{\mathbf{d}}^{1+\mu_0}} \frac{d\|x(t)\|_{\mathbf{d}}}{dt} \in [-\rho^*, +\infty)$ we have

$$-\gamma_2(\rho + \tilde{\rho})\|x(t) - z(t)\|^2 \le \dot{V}(t) \le -\gamma_1(\rho - \rho^*)\|x(t) - z(t)\|^2,$$

or, equivalently,

$$\frac{-\dot{V}(t)}{\gamma_2(\rho + \tilde{\rho})} \le \|x(t) - z(t)\|^2 \le \frac{-\dot{V}(t)}{\gamma_1(\rho - \rho^*)}.$$

Hence, $x(t) - z(t) \to 0$ as $\dot{V}(t) \to 0$. The proof is complete. □

Remark 16.1 In the case of the conventional DNN $\tilde{f}_{\epsilon,0} = -\rho x + C_0\sigma(A_0 x + b_0)$, the DNN learning can be realized as follows [13]:

$$\dot{z} = -\rho z + \tilde{f}_{\epsilon,0}(x, C_0) + \sum_{i=1}^{m} f_i(x)u_i, \tag{16.58}$$

$$\frac{d\ \mathrm{vec}(C_0^\top)}{dt} = \gamma\,[I_n \otimes \sigma(A_0 x + b_0)]\,(x - z). \tag{16.59}$$

Remark 16.2 The condition (16.56) is not restrictive. Indeed, in the view of Remark 7.1 it always holds for the system (16.44) with the inputs

$$u_i(t) = \|x(t)\|_{\mathbf{d}}^{\mu_0-\mu_i}\tilde{u}_i(t), \quad t > 0, \tag{16.60}$$

where $\tilde{u}_i$ are uniformly bounded functions of time: $|\tilde{u}_i| \leq \bar{u}_i$. Let $\bar{u}_0 = 1$. In this case, using formula (16.57) we derive

$$\rho^* = \frac{1}{2\beta}\sum_{i=0}^{m} \bar{u}_i \max_{\|y\|=1} \|f_i(y)\|, \tag{16.61}$$

where $\beta > 0$ is defined in Proposition 5.3. The identification of locally homogeneous system (16.44) by the homogeneous DNN (16.49) may require the above "homogeneous" excitation of the inputs u_i in order to fulfill the restriction (16.56).

The above theorem does not prove any convergence of the gain C_0 to the exact/optimal gain C_0^*. Inspired by [17], to eliminate this drawback, we propose the following modification of the learning rule

$$\dot{z} = \|x\|_{\mathbf{d}}^{\mu_0}\left(\Omega\Omega^\top(x-z) - \rho G_{\mathbf{d}} z\right) + f_{\epsilon,0}(x, C_0) + \sum_{i=1}^{m} f_i(x)u_i, \tag{16.62}$$

$$\frac{d\,\mathrm{vec}(C_0^\top)}{dt} = -\|x\|_{\mathbf{d}}^{\mu_0}\Omega^\top(x-z), \tag{16.63}$$

$$\dot{\Omega} = -\rho\|x\|_{\mathbf{d}}^{\mu_0} G_{\mathbf{d}}\Omega - \|x\|_{\mathbf{d}}^{\mu_0}\mathbf{d}(\ln\|x\|_{\mathbf{d}})\left[I_n \otimes \sigma(A_0\mathbf{d}(-\ln\|x\|_{\mathbf{d}})x + b_0)\right]^\top, \tag{16.64}$$

where $\Omega \in \mathbb{R}^{n\times nN}$ is an additional variable and all other parameters are defined as before.

Theorem 16.7 *Let* $\sup_{t\geq 0}|\Omega(t)| < +\infty$. *The persistent excitation condition*

$$\exists \ell > 0, \underline{\gamma} > 0 \quad : \quad \int_t^{t+\ell} \Omega^\top(\tau)\Omega(\tau)d\tau \geq \underline{\gamma} I_{nN} \tag{16.65}$$

be fulfilled. If there exists $C_0^* \in \mathbb{R}^{n\times N}$ *such that* (16.54) *hold and all trajectories of the system* (16.44) *are uniformly bounded by the formula* (16.55) *then*

$$z(t) \to x(t) \quad \text{as} \quad t \to +\infty \quad \text{and} \quad C_0(t) \to C_0^* \quad \text{as} \quad t \to +\infty,$$

where z, C_0 *are governed by* (16.62), (16.63), (16.64).

Proof Let us introduce the following variable:

$$\delta = x - z + \Omega\tilde{c}.$$

Its time derivative is given by

$$\dot{\delta} = \dot{x} - \dot{z} + \dot{\Omega}\tilde{c} + \Omega\frac{d\tilde{c}}{dt}$$

$$= -\rho\|x\|_{\mathbf{d}}^{\mu_0} G_{\mathbf{d}}(x - z) + (C_0^* - C_0)\sigma(A_0\mathbf{d}(-\ln\|x\|_{\mathbf{d}})x + b_0) - \|x\|_{\mathbf{d}}^{\mu_0}\Omega\Omega^\top(x - z) +$$

$$-\rho\|x\|_{\mathbf{d}}^{\mu_0} G_{\mathbf{d}}\Omega\tilde{c} - \|x\|_{\mathbf{d}}^{\mu_0}\mathbf{d}(\ln\|x\|_{\mathbf{d}})\left[I_n \otimes \sigma(A_0\mathbf{d}(-\ln\|x\|_{\mathbf{d}})x + b_0)\right]^\top \tilde{c}$$

$$+ \|x\|_{\mathbf{d}}^{\mu_0}\Omega\Omega^\top(x - z).$$

Hence, using (16.53) we derive

$$\dot{\delta} = -\rho\|x\|_{\mathbf{d}}^{\mu_0} G_{\mathbf{d}}\delta.$$

On the one hand, since the matrix $G_{\mathbf{d}}$ is anti-Hurwitz and $r_{\min} \le \|x\|_{\mathbf{d}} \le r_{\max}$ then

$$\delta(t) \to 0 \quad \text{as} \quad t \to +\infty.$$

On the other hand, since

$$\frac{d\tilde{c}}{dt} = -\|x\|_{\mathbf{d}}^{\mu_0}\Omega^\top\Omega\tilde{c} + q, \qquad q = \|x\|_{\mathbf{d}}^{\mu_0}\Omega^\top\delta \tag{16.66}$$

then, by [16, Lemma 1] (see also [18]), the persistent excitation condition (16.65) implies that $\tilde{c}(t) \to 0$ as $t \to \infty$. The proof is complete. □

Notice that the condition (16.54) can be relaxed asking just an existence on an approximation f_0 by the ANN $\tilde{f}_{\epsilon,0}$ with the precision $\epsilon > 0$. In this case, the convergence of the gain C_0 to a neighborhood of the optimal gain C_0^* can be guaranteed and the size of this neighborhood is proportional to $\epsilon > 0$.

Corollary 16.1 *Let* $\sup_{t\ge 0}|\Omega(t)| < +\infty$, *the persistent excitation condition* (16.65) *holds and there exists* $\epsilon > 0$, $C_0^* \in \mathbb{R}^{n\times N}$ *such that*

$$|f_0(y) + \rho G_{\mathbf{d}} y - C_0^*\sigma(A_0 y + b_0)| \le \epsilon, \quad \forall y \in \mathbb{R}^n : \|y\| = 1. \tag{16.67}$$

If all trajectories of the system (16.44) *are uniformly bounded by the formula* (16.55) *then there exist* $\tilde{\kappa}, \kappa_0 \in \mathcal{K}$ *such that*

$$\limsup_{t\to+\infty} |z(t) - x(t)| = \tilde{\kappa}(\epsilon) \quad \textit{and} \quad \limsup_{t\to+\infty} |C_0(t) - C_0^*| = \kappa_0(\epsilon), \quad (16.68)$$

where z, C_0 *are governed by* (16.62), (16.63), (16.64).

Proof The dynamics of $\delta = x - z + \Omega\tilde{c}$ satisfies

$$\dot{\delta} = -\rho\|x\|_{\mathbf{d}}^{\mu_0} G_{\mathbf{d}}\delta + \Delta, \quad (16.69)$$

where

$$\Delta = f_0(x) - \|x\|_{\mathbf{d}}^{\mu_0}\mathbf{d}(\ln\|x\|_{\mathbf{d}})\left(-\rho G_{\mathbf{d}}\mathbf{d}(-\ln\|x\|_{\mathbf{d}})x + C_0^*\sigma(A_0\mathbf{d}(-\ln\|x\|_{\mathbf{d}})x + b_0)\right).$$

Since $\|\mathbf{d}(-\ln\|x\|_{\mathbf{d}})x\| = 1$ then, using (16.67), we derive

$$|\Delta| \leq \|x\|_{\mathbf{d}}^{\mu_0} \cdot |\mathbf{d}(\ln\|x\|_{\mathbf{d}})| \cdot \epsilon.$$

The uniform boundedness of $\|x\|_{\mathbf{d}}$ (see (16.55)) implies that there exists $\kappa \in \mathcal{K}$ such that

$$\limsup_{t\to+\infty} \|\delta(t)\| = \kappa(\epsilon).$$

In this case, applying [16, Lemma 1] to the system (16.66) we derive (16.68). The proof is complete. □

Notice that Theorem 16.7 and Corollary 16.1 do not require the condition (16.56). So, formally, the "homogeneous" input excitation (16.60) is not required in the case of learning law (16.62), (16.63), (16.64). However, in this case, the necessity of the system excitation is expressed by the condition (16.65), which is not very hard to fulfill in many cases [13, 16, 18].

Identification of the Whole System

Let all vector fields f_i, $i = 0, ..., m$ be *unknown*. Let us consider the following learning system:

$$\dot{z} = -\rho\|x\|_{\mathbf{d}}^{\mu_0} G_{\mathbf{d}}z + \sum_{i=0}^{m} f_{\epsilon,i}(x, C_i)u_i, \quad (16.70)$$

$$\frac{d\,\mathrm{vec}(C_i^\top)}{dt} = \|x\|_{\mathbf{d}}^{\mu_i}\left[I_n \otimes \sigma(A_i\mathbf{d}(-\ln\|x\|_{\mathbf{d}})x + b_i)\right] P\mathbf{d}(-\ln\|x\|_{\mathbf{d}})(x - z), \quad (16.71)$$

where $z(t) \in \mathbb{R}^n$, $u_0 = 1$, $C_i(t) \in \mathbb{R}^{n\times N}$,

$$f_{\epsilon,i}(x, C_i) = \|x\|_{\mathbf{d}}^{\mu_i}\mathbf{d}(\ln\|x\|_{\mathbf{d}})C_i\sigma(A_i\mathbf{d}(-\ln\|x\|_{\mathbf{d}})x + b_i), \tag{16.72}$$

and ρ, $G_{\mathbf{d}}$, $\|\cdot\|_{\mathbf{d}}$ are as before.

Theorem 16.8 *Let f_i be locally Lipschitz continuous on $\mathbb{R}^n\backslash\{\mathbf{0}\}$ for $i = 0, 1, ..., m$ and $u_i \in W^1(\mathbb{R}_+; \mathbb{R})$ for $i = 1, 2, ..., m$. Let there exist $C_0^* \in \mathbb{R}^{n\times N}$ such that the representation* (16.54) *holds. Let there exist $C_i^* \in \mathbb{R}^{n\times N}$ such that*

$$f_i(x) = \|x\|_{\mathbf{d}}^{\mu_i}\mathbf{d}(\ln\|x\|_{\mathbf{d}})C_i^*\sigma(A_i\mathbf{d}(-\ln\|x\|_{\mathbf{d}})x + b_i) \tag{16.73}$$

for $i = 1, 2, ...m$. If all trajectories of the system (16.44) *are uniformly bounded by the formula* (16.55) *and the condition* (16.56) *is fulfilled then for $\rho > \rho^*$ one holds*

$$z(t) \to x(t) \quad as \quad t \to +\infty,$$

where C_i is governed by (16.51) *with z satisfying* (16.70).

Proof Let us denote $c_i = \mathrm{vec}\,(C_i^\top)$, $c_i^* = \mathrm{vec}\,((C_i^*)^\top)$, $\tilde{c}_i = c_i^* - c_i$ and consider the Lyapunov function candidate

$$V = \sum_{i=0}^{m}\tilde{c}_i^\top\tilde{c}_i + (x-z)^\top\mathbf{d}^\top(-\ln\|x\|_{\mathbf{d}})P\mathbf{d}(-\ln\|x\|_{\mathbf{d}})(x-z).$$

The time derivative of V is

$$\dot{V} = \sum_{i=0}^{m} 2\tilde{c}^\top\frac{d\tilde{c}}{dt} + 2(x-z)^\top\mathbf{d}^\top(-\ln\|x\|_{\mathbf{d}})P\mathbf{d}(-\ln\|x\|_{\mathbf{d}})(\dot{x}-\dot{z}) +$$

$$-2(x-z)^\top\mathbf{d}^\top(-\ln\|x\|_{\mathbf{d}})PG_{\mathbf{d}}\mathbf{d}(-\ln\|x\|_{\mathbf{d}})(x-z)\frac{1}{\|x\|_{\mathbf{d}}}\frac{d\|x\|_{\mathbf{d}}}{dt}$$

$$= \sum_{i=0}^{m} 2\tilde{c}^\top\tfrac{d\tilde{c}}{dt} + \sum_{i=0}^{m} 2\|x\|_{\mathbf{d}}^{\mu_0}(x-z)^\top\mathbf{d}^\top(\ln\|x\|_{\mathbf{d}})P(C_i - C_i^*)\sigma(A_i\mathbf{d}(-\ln\|x\|_{\mathbf{d}})x + b_i)$$

$$-\left(\rho\|x\|_{\mathbf{d}}^{\mu_0} + \tfrac{1}{\|x\|_{\mathbf{d}}}\tfrac{d\|x\|_{\mathbf{d}}}{dt}\right)(x-z)^\top\mathbf{d}^\top(-\ln\|x\|_{\mathbf{d}})(PG_{\mathbf{d}} + G_{\mathbf{d}}^\top P)\mathbf{d}(-\ln\|x\|_{\mathbf{d}})(x-z),$$

$$= -\left(\rho\|x\|_{\mathbf{d}}^{\mu_0} + \tfrac{1}{\|x\|_{\mathbf{d}}}\tfrac{d\|x\|_{\mathbf{d}}}{dt}\right)(x-z)^\top\mathbf{d}^\top(-\ln\|x\|_{\mathbf{d}})(PG_{\mathbf{d}} + G_{\mathbf{d}}^\top P)\mathbf{d}(-\ln\|x\|_{\mathbf{d}})(x-z),$$

where the property $\mathbf{d}(s)G_{\mathbf{d}} = G_{\mathbf{d}}\mathbf{d}(s)$, $\forall s \in \mathbb{R}$, the representations (16.54), (16.73) and the identities

$$(C_i - C_i^*)\sigma(A_i\mathbf{d}(-\ln\|x\|_{\mathbf{d}})x + b_i) = -\left[I_n \otimes \sigma(A_i\mathbf{d}(-\ln\|x\|_{\mathbf{d}})x + b_i)\right]^\top \tilde{c}_i$$

are utilized on the last step. We complete the proof repeating the considerations of the proof of Theorem 16.6. □

Remark 16.2 remains valid for the above theorem making its conditions to be feasible under proper selection of inputs u_i.

The modified learning laws for identification of the whole system are introduced as follows:

$$\dot{z} = -\rho\|x\|_{\mathbf{d}}^{\mu_0} G_{\mathbf{d}} z + \sum_{i=1}^{m} \|x\|_{\mathbf{d}}^{\mu_i} \Omega_i \Omega_i^\top (x - z) + f_{\epsilon,i}(x, C_0)u_i, \tag{16.74}$$

$$\frac{d\,\mathrm{vec}(C_i^\top)}{dt} = -\|x\|_{\mathbf{d}}^{\mu_i} \Omega_i^\top (x - z), \tag{16.75}$$

$$\dot{\Omega}_i = -\rho\|x\|_{\mathbf{d}}^{\mu_0} G_{\mathbf{d}} \Omega_i - \|x\|_{\mathbf{d}}^{\mu_i} \mathbf{d}(\ln\|x\|_{\mathbf{d}}) \left[I_n \otimes \sigma(A_i\mathbf{d}(-\ln\|x\|_{\mathbf{d}})x + b_i)\right]^\top, \tag{16.76}$$

where $\Omega_i \in \mathbb{R}^{n\times nN}$, $i = 0, \ldots, m$ are additional variables and all other parameters are defined as before. Let us denote

$$\tilde{\Omega} = \left(\Omega_0\ \Omega_1\ \ldots\ \Omega_m\right). \tag{16.77}$$

Theorem 16.9 *Let* $\sup_{t\geq 0}|\tilde{\Omega}(t)| < +\infty$, *the persistent excitation condition*

$$\exists \ell > 0, \underline{\gamma} > 0 \quad : \quad \int_t^{t+\ell} \tilde{\Omega}^\top(\tau)\tilde{\Omega}(\tau)d\tau \geq \underline{\gamma} I_{nN}, \quad \forall i = 0, \ldots, m \tag{16.78}$$

be fulfilled and exist $C_i^* \in \mathbb{R}^{n\times N}$ *such that* (16.73) *hold. If all trajectories of the system* (16.44) *are uniformly bounded by the formula* (16.55) *then*

$$z(t) \to x(t) \quad \text{as} \quad t \to +\infty \quad \text{and} \quad C_i(t) \to C_i^* \quad \text{as} \quad t \to +\infty,$$

where z, C_i *are governed by* (16.74), (16.75), (16.76).

Proof Let us introduce the following variable:

$$\delta = x - z + \sum_{i=0}^{m} \Omega_i \tilde{c}_i.$$

Its time derivative is given by

$$\dot{\delta} = \dot{x} - \dot{z} + \sum_{i=0}^{m} \dot{\Omega}_i \tilde{c}_i + \Omega_i \frac{d\tilde{c}_i}{dt}$$

$$= -\rho\|x\|_{\mathbf{d}}^{\mu_0} G_{\mathbf{d}}(x-z) + \sum_{i=0}^{m}(C_i^* - C_i)\sigma(A\mathbf{d}(-\ln\|x\|_{\mathbf{d}})x + b) - \|x\|_{\mathbf{d}}^{\mu_i}\Omega_i\Omega_i^\top(x-z) +$$

$$\sum_{i=0}^{m} -\rho\|x\|_{\mathbf{d}}^{\mu_0} G_{\mathbf{d}}\Omega_i\tilde{c} - \|x\|_{\mathbf{d}}^{\mu_i}\mathbf{d}(\ln\|x\|_{\mathbf{d}})\,[I_n \otimes \sigma(A\mathbf{d}(-\ln\|x\|_{\mathbf{d}})x + b)]^\top \tilde{c}_i$$

$$+ \sum_{i=0}^{m} \|x\|_{\mathbf{d}}^{\mu_i}\Omega_i\Omega_i^\top(x-z).$$

Hence, using (16.53) we derive

$$\dot{\delta} = -\rho\|x\|_{\mathbf{d}}^{\mu_0} G_{\mathbf{d}}\delta.$$

On the one hand, since the matrix $G_{\mathbf{d}}$ is anti-Hurwitz then

$$\delta(t) \to 0 \quad \text{as} \quad t \to +\infty.$$

On the other hand, denoting

$$\tilde{C} = \begin{pmatrix} \tilde{c}_0 \\ \tilde{c}_1 \\ \dots \\ \tilde{c}_m \end{pmatrix}, \quad D(\|x\|_{\mathbf{d}}) = \begin{pmatrix} \|x\|_{\mathbf{d}}^{\mu_0} I_N & 0 & \dots & 0 \\ 0 & \|x\|_{\mathbf{d}}^{\mu_1} I_N & \dots & 0 \\ \dots & \dots & \dots & \dots \\ 0 & 0 & \dots & \|x\|_{\mathbf{d}}^{\mu_1} I_N \end{pmatrix}$$

we derive

$$\frac{d\tilde{C}}{dt} = -D(\|x\|_{\mathbf{d}})\tilde{\Omega}^\top\tilde{\Omega}\tilde{C} + q, \qquad q = \tilde{\Omega}^\top\delta. \tag{16.79}$$

By [17, Corollary 2], the persistent excitation condition (16.65) implies that $\tilde{C}(t) \to 0$ as $t \to \infty$. The proof is complete. □

The case of the practical estimate of f_i is treated by the following corollary.

Corollary 16.2 *Let* $\sup_{t\geq 0}|\tilde{\Omega}(t)| < +\infty$, *the persistent excitation condition* (16.78) *holds and there exists* $\epsilon > 0$, $C_i^* \in \mathbb{R}^{n\times N}$ *such that* (16.67) *holds and*

$$|f_i(y) - C_i^*\sigma(A_i y + b_i)| \leq \epsilon, \quad \forall y \in \mathbb{R}^n : \|y\| = 1, \quad i = 1, \dots, m. \tag{16.80}$$

If all trajectories of the system (16.44) *are uniformly bounded by the formula* (16.55) *then there exist* $\tilde{\kappa}$, $\kappa_i \in \mathcal{K}$ *such that*

$$\limsup_{t\to+\infty}|z(t) - x(t)| = \tilde{\kappa}(\epsilon) \quad \text{and} \quad \limsup_{t\to+\infty}|C_i(t) - C_i^*| = \kappa_i(\epsilon), \tag{16.81}$$

where z, C_i *are governed by* (16.74), (16.75), (16.76), $i = 0, 1, \dots, m$.

Proof The dynamics of $\delta = x - z + \sum_{i=0}^{m} \Omega_i \tilde{c}_i$ satisfies

$$\dot{\delta} = -\rho\|x\|_{\mathbf{d}}^{\mu_0} G_{\mathbf{d}}\delta + \Delta, \tag{16.82}$$

$$\Delta = f_0(x) - \|x\|_{\mathbf{d}}^{\mu_0}\mathbf{d}(\ln\|x\|_{\mathbf{d}})\left(-\rho G_{\mathbf{d}}\mathbf{d}(-\ln\|x\|_{\mathbf{d}})x + C_0^*\sigma(A_0\mathbf{d}(-\ln\|x\|_{\mathbf{d}})x + b_0)\right)$$

$$+\sum_{i=1}^{m} f_i(x) - \|x\|_{\mathbf{d}}^{\mu_i}\mathbf{d}(\ln\|x\|_{\mathbf{d}})C_i^*\sigma(A_i\mathbf{d}(-\ln\|x\|_{\mathbf{d}})x + b_i)).$$

Since $\|\mathbf{d}(-\ln\|x\|_{\mathbf{d}})x\| = 1$ then, using (16.67) and (16.67), we derive

$$|\Delta| \le \sum_{i=0}^{m} \|x\|_{\mathbf{d}}^{\mu_i} \cdot |\mathbf{d}(\ln\|x\|_{\mathbf{d}})| \cdot \epsilon.$$

The uniform boundedness of $\|x\|_{\mathbf{d}}$ (see (16.55)) implies that there exists $\tilde{\kappa} \in \mathcal{K}$ such that

$$\limsup_{t\to+\infty} \|\delta(t)\| = \kappa(\epsilon).$$

In this case, applying [17, Corollary 2] to the system (16.66) we derive (16.68). The proof is complete. □

16.3.2 DNN Identification of the Three-Tank System

For comparison of conventional and homogeneous DNNs, let us consider the three-tank system depicted in Fig. 6.2, which also includes the variables and parameters considered in the dynamic model

$$\begin{aligned}
\dot{x}_1 &= S_{tank}^{-1}\left[-\theta_1\lceil x_1 - x_3\rfloor^{0.5} + u_1\right],\\
\dot{x}_2 &= S_{tank}^{-1}\left[\theta_3\lceil x_3 - x_2\rfloor^{0.5} - \theta_2\lceil x_2\rfloor^{0.5} + u_2\right],\\
\dot{x}_3 &= S_{tank}^{-1}\left[\theta_1\lceil x_1 - x_3\rfloor^{0.5} - \theta_3\lceil x_3 - x_2\rfloor^{0.5}\right],
\end{aligned} \tag{16.83}$$

where x_1 $[m]$, x_2 $[m]$ and x_3 $[m]$ represent the liquid level of each tank, respectively; S_{tank} $[m]$ is the diameter of the three tanks; the input flows u_1 $[L \cdot s^{-1}]$ and u_2 $[L \cdot s^{-1}]$ are the control signals; and the constant parameters θ_1, θ_2 and θ_3 are coefficients related with the outflow rate according to Torricelli's rule.

Obviously, the system admits the representation in the form of the affine-in-control system (16.44) with the standard homogeneous vector field function f_i

$$f_1(x) = \begin{pmatrix} 1 \\ 0 \\ 0 \end{pmatrix}, \quad f_2(x) = \begin{pmatrix} 0 \\ 1 \\ 0 \end{pmatrix},$$

$$f_0(x) = \frac{1}{S_{tank}} \begin{pmatrix} -\theta_1 \lceil x_1 - x_3 \rfloor^{0.5} \\ \theta_3 \lceil x_3 - x_2 \rfloor^{0.5} - \theta_2 \lceil x_2 \rfloor^{0.5} \\ \theta_1 \lceil x_1 - x_3 \rfloor^{0.5} - \theta_3 \lceil x_3 - x_2 \rfloor^{0.5} \end{pmatrix}, \quad x = \begin{pmatrix} x_1 \\ x_2 \\ x_3 \end{pmatrix}.$$

These vector fields are **d**-homogeneous of degrees $\mu_0 = -0.5$, $\mu_1 = \mu_2 = 0$, respectively, with respect to the dilation $\mathbf{d}(s) = e^s I_3$. We assume that the vector field f_0 is unknown, but its homogeneity degree is known. The vector fields f_1 and f_2 are assumed to be known.

The learning of homogeneous DNN identifier is realized by the formulas (16.50), (16.51) with $\rho = 1.75$, $\gamma = 100$. For comparison reasons, we also consider the conventional DNN identifier with the learning law (16.58), (16.59) having the same parameters. The parameters $A \in \mathbb{R}^{10\times 3}$ and $b \in \mathbb{R}^{10}$ for both identifiers are selected randomly. The matrix C_0 is initialized by zeros. The activation function σ is defined by sigmoid, i.e., $\sigma(r) = (1 + e^r)^{-1}$, $r \in \mathbb{R}$.

For simulation, we consider the system with $S_{tank} = 1$, $\theta_1 = 3$, $\theta_2 = 1$, $\theta_3 = 2$. The inputs are defined as follows:

$$u_1(t) = 1 + \text{sign}(t - 4), \quad u_2(t) = 0, \quad t \geq 0.$$

The initial state $x(0) = x_0 := (3, 1, 2)^\top$. The trajectories of the three-tank system with such parameters are depicted in Fig. 16.4. The simulation results for the DNN identifiers are presented in Fig. 16.5, where the evolution of the approximation errors is depicted. On the first stage (for $t \in [0, 4]$), the estimation errors of both identifiers decrease. On this stage, we have $u_1 = 0$ and the norm of state of the system converges to a compact neighborhood of the origin. The conventional ANN is a universal approximator on a compact. This implies a good approximation quality on the time interval $[0, 4]$. For $t > 1$ we have the control input $u_1(t) = 2$, which brings the system away from these neighborhoods of the origin. This implies an essential degradation of the approximation quality of the conventional DNN identifier, while the homogeneous DNN identifier (being a global approximator of the system) provides good estimates even for $t > 4$ (see Fig. 16.5).

A comparison of conventional and standard homogeneous DNN identifiers in the case of the modified learning law (16.74), (16.75) can be found in [17].

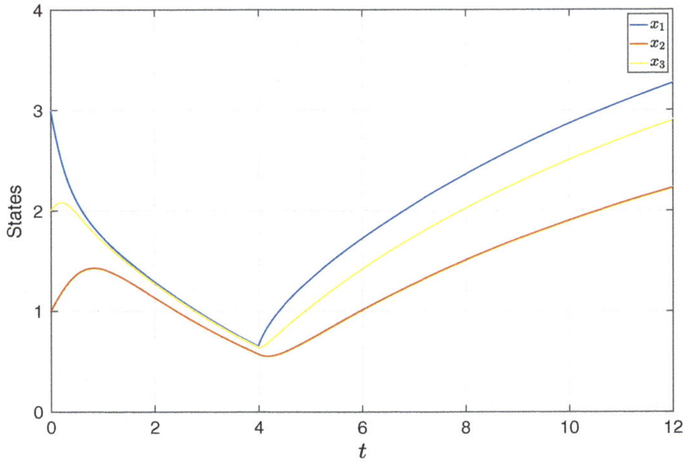

Fig. 16.4 The evolution of the states of the three-tank system

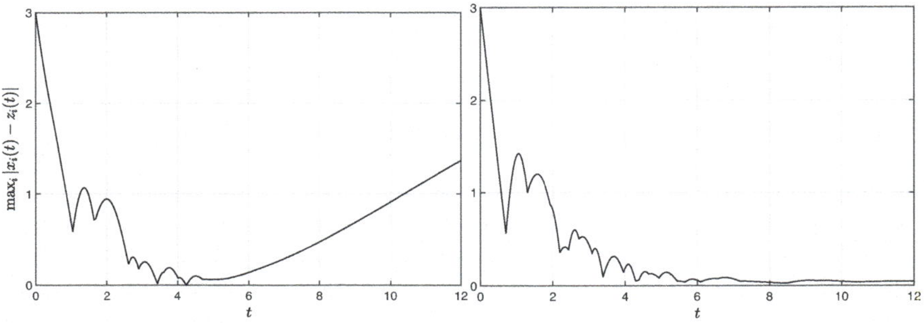

Fig. 16.5 Approximation errors for conventional (left) and homogeneous (right) DNN identifiers

Concluding Remarks

In this chapter, homogeneous artificial and differential neural networks are introduced. The key feature of homogeneous ANN/DNN is a global approximation based on a local data. The homogeneous universal approximation theorem is proven. Some methods of data-driven identification of homogeneity of mappings are developed. Training algorithms for homogeneous DNNs are developed. An essential improvement of approximation and extrapolation precision of homogeneous ANN/DNN comparing with classical ANN/DNN is demonstrated in a series of examples.

References

1. Cybenko G (1989) Approximation by superpositions of a sigmoidal function. Math Control Signals Syst 4(2):303–3014
2. Hornik K, Stinchcombe M, White H (1989) Multilayer feedforward networks are universal approximators. Neural Netw 2(5):359–366
3. Hornik K (1991) Approximation capabilities of multilayer feedforward networks. Neural Netw 4(2):251–257
4. Leshno M, Lin V, Pinkus A, Schocken S (1993) Multilayer feedforward networks with a nonpolynomial activation function can approximate any function. Neural Netw 6(6):861–867
5. Krizhevsky A, Sutskever I, Hinton GE (2012) Imagenet classification with deep convolutional neural networks. In: Advances in neural information processing systems 25 (NeurIPS 2012)
6. Neyshabur B, Salakhutdinov R, Srebro N (2015) Path-sgd: path-normalized optimization in deep neural networks. In: Advances in neural information processing systems 28 (NeurIPS 2015). The MIT Press
7. Bach F (2017) Breaking the curse of dimensionality with convex neural networks. J Mach Learn Res 18:1–53
8. Ji Z, Telgarsky M (2020) Directional convergence and alignment in deep learning. In: Advances in neural information processing systems 33 (NeurIPS 2020)
9. Lagaris IE, Likas A, Fotiadis DI (1998) Artificial neural networks for solving ordinary and partial differential equations. IEEE Trans Neural Netw 9(5):987–1000
10. Baptista D, Morgado-Dias F (2013) A survey of artificial neural network training tools. Neural Comput Appl 23(3–4):609–615
11. Bishop C (2006) Pattern recognition and machine learning. Springer
12. Polyakov A, Ping X (2024) Homogeneous SMC design using artificial neural network. In: 17th workshop on variable structure systems
13. Poznyak A, Sanchez E, Yu W (2001) Differential neural networks for robust nonlinear control. World Scientific
14. Lewis F, Jagannathan S, Yesildirak A (1998) Neural network control of robot manipulators and non-linear systems. CRC Press
15. Ljung L (1999) System identification: theory for the user. Pearson Education, Inc
16. Efimov D, Fradkov A (2015) Design of impulsive adaptive observers for improvement of persistency of excitation. Int J Adapt Control Signal Process 29(6):765–782
17. Ballesteros M, Polyakov A, Efimov D, Chairez I, Poznyak A (2021) Non-parametric identification of homogeneous dynamical systems. Automatica 129(7):109600
18. Anderson B, Moore J (1969) New results in linear system stability. SIAM J Control Optim 7(3):398–414

Chapter 17
Homogeneity in Multi-agent Systems

In this chapter, elements of homogeneity theory of multi-agent systems are introduced. Finite-time and fixed-time formation and consensus control protocols for simplest multi-agent systems are designed and analyzed based on the standard homogeneity. A possibility of extension of these results to systems with agents modeled by linear multi-input systems is demonstrated using the linear homogeneity.

17.1 Multi-agent Systems

Multi-agent system (MAS) uses a network of autonomous agents to deal with tasks, which cannot be efficiently accomplished by a single agent. The last two decades, the MAS is the popular research topic of control theory (see, e.g., [1–3] and references therein). The use of multiple interchangeable agents improves the viability and allows the system to accomplish successfully the task even in the case when one or several agents are not operating well. The MAS has potential applications in various industrial scenarios such as unmanned robotic systems [4, 5]. The necessity of local interactions among the agents to achieve the overall goal modifies the classical *centralized* control paradigm by introducing the *decentralized* control principles.

A mathematical model of MAS traditionally consists of two components: *a model of the communication network* and *a model of the agent*.

A. Polyakov, *Generalized Homogeneity in Systems and Control Volume I*,
Communications and Control Engineering,
https://doi.org/10.1007/978-3-031-76246-8_17

17.1.1 Network Topology

The communication network of MAS is modeled by a graph $\mathcal{G} = \{\mathcal{V}, \mathcal{E}, \mathcal{W}\}$:

- $\mathcal{V} = \{0, 1, \dots, N\}$ is the set of node (agents);
- $\mathcal{E} = \{(i, j)|i, j \in \mathcal{V}\}$ is the edge set, where $(i, j) \in \mathcal{E}$ if the node i can receive the information from the node j and n_i denotes the number of incoming edges of the node i;
- $\mathcal{W}$ is the weighted adjacency matrix with elements w_{ij}, $i, j \in \mathcal{V}$, $w_{ij} > 0$ if $(i, j) \in \mathcal{E}$, and $w_{ij} = 0$ otherwise (the self-loops are excluded, i.e., $w_{ii} = 0$).

Let $\mathcal{L} = (\ell_{ij}) \in \mathbb{R}^{(N+1)\times(N+1)}$ be the *Laplacian matrix* associated with the graph $\mathcal{G}$:

$$\ell_{ij} = \begin{cases} -w_{ij} & \text{if} \quad i \neq j, \\ \sum\limits_{k=0}^{N} w_{ik} & \text{if} \quad i = j, \end{cases}$$

where $i, j = 0, 1, \dots, N$. Since $\sum_{j=0}^{N} \ell_{ij} = 0$ for any $i = 0, 1, \dots, N$ then the vector $\mathbf{1} = (1, \dots, 1)^\top \in \mathbb{R}^{N+1}$ is the eigenvector of $\mathcal{L}$, which corresponds to the zero eigenvalue of $\mathcal{L}$.

The graph $\mathcal{G}$ is called *undirected* if $(i, j) \in \mathcal{E} \Rightarrow (j, i) \in \mathcal{E}$, otherwise the graph is called *directed.*

The directed graph topology frequently describes *MAS with a leader*:

- the node i_L is the root node (the leader) of $\mathcal{G}$ if it has at least one directed path[1] to each of the rest nodes while $(i_L, j) \notin \mathcal{E}$ for all $j \in \mathcal{V}$.

Below we use the following well-known result (see, e.g., [1, 6]):

- *the Laplacian $\mathcal{L}$ of the undirected graph $\mathcal{G}$ is a positive semi-definite symmetric matrix*;
- *if the undirected graph $\mathcal{G}$ is connected then the zero eigenvalue of the Laplacian $\mathcal{L}$ is simple*[2] *and there exists a minimal strictly positive eigenvalue $\lambda_1 > 0$ of $\mathcal{L}$:*

$$\min_{x \neq \mathbf{0}, x^\top \mathbf{1} = 0} \frac{x^\top \mathcal{L} x}{|x|^2} \geq \lambda_1 > 0; \tag{17.1}$$

- *if $i = 0$ is a root node of the directed graph $\mathcal{G}$ then the matrix*

$$\tilde{\mathcal{L}} = \begin{pmatrix} \ell_{11} & \dots & \ell_{1N} \\ \dots & \dots & \dots \\ \ell_{N1} & \dots & \ell_{NN} \end{pmatrix} \tag{17.2}$$

is anti-Hurwitz.

---

[1] A directed path from node i to node j in the graph $\mathcal{G}$ is a sequence of nodes $\overline{i_1, i_s}$, where $i_1 = i$, $i_s = j$ and $(i_{\kappa+1}, i_\kappa) \in \mathcal{E}$, $\kappa = \overline{1, s-1}$.

[2] An eigenvalue of a matrix is said to be simple if its algebraic multiplicity is 1.

17.1.2 Cooperative Dynamics

The control theory of MAS deals with agents having dynamical models (e.g., ODEs). We assume all agents of MAS have identical nominal models:

$$\dot{x}_i = Ax_i + Bu_i + g_i(t, x_i, u_i), \quad t > 0, \tag{17.3}$$

where $x_i(t) \in \mathbb{R}^n$ is the state of the ith agent, $u_i(t) \in \mathbb{R}^m$ is the control input of the ith agent, the system matrices A, B are assumed to be known and the function $g_i : \mathbb{R} \times \mathbb{R}^n \times \mathbb{R}^m$ models some uncertainties/nonlinearities.

We assume that all agents have identical nominal plant models (i.e., for $g_i = \mathbf{0}$). We also assume that each agent has the information about the states of neighbors in the communication graph, so the control law u_i for the ith agent may have the form:

$$u_i = u_i(t, x_i, x_{i_1}, x_{i_2}, ..., x_{i_{n_i}}), \tag{17.4}$$

where $i_k \in \{1, ..., N\}, k = 1, ..., n_i$ is the number of the node in the communication graph, from which the ith node receives the information.

Due to communication constraints, the feedback law u_i may use only the local information (from the neighbors) about states of the system. Such a decentralized control (also known as *control protocol*) is the key feature of MAS, which does not allow the conventional centralized algorithms to be straightforwardly applied in MAS control design. However, the cooperative dynamics in the closed-loop form is described by the conventional ODE

$$\frac{d\bar{x}}{dt} = \bar{f}(t, \bar{x}), \quad t > 0, \tag{17.5}$$

where $\bar{x} = (x_1, ..., x_N)^\top$, $\bar{f} = (f_1, ..., f_N)^\top$ and

$$f_i(t, x) = Ax_i + Bu_i(x_i, x_{i_1}, ..., x_{i_{n_i}}) + g_i(t, x_i, u_i(x_i, x_{i_1}, ..., x_{i_{n_i}})).$$

So, the usual tools for stability or ISS analysis of dynamical systems (like Lyapunov function method) are still applicable.

Control aims for MAS slightly differ from the conventional problems for centralized control systems. The controllability/observability of MAS depends on the communication topology.

17.2 Homogeneous Control of Multi-agent Systems

17.2.1 Formation Control

Uniform Allocation of Agents on a Segment

Let us consider the MAS with $A = 0 \in \mathbb{R}$, $B = 1 \in \mathbb{R}$, $g_i = \mathbf{0}$ and the communication topology described by the directed graph $\mathcal{G}$ shown in Fig. 17.1 with the adjacency matrix:

$$\mathcal{W} = \begin{pmatrix} 0 & 1 & 0 & 0 & \dots & 0 & 0 & 0 & 0 \\ 1 & 0 & 1 & 0 & \dots & 0 & 0 & 0 & 0 \\ 0 & 1 & 0 & 1 & \dots & 0 & 0 & 0 & 0 \\ \dots & \dots & \dots & \dots & \dots & \dots & \dots & \dots & \dots \\ 0 & 0 & 0 & 0 & \dots & 1 & 0 & 1 & 0 \\ 0 & 0 & 0 & 0 & \dots & 0 & 1 & 0 & 1 \\ 0 & 0 & 0 & 0 & \dots & 0 & 0 & 1 & 0 \end{pmatrix}$$

Let $u_0 = u_N = \mathbf{0}$, so the dynamics of agents is given by

$$\dot{x}_0 = 0, \qquad \dot{x}_i = u_i, \; i = 1..., N-1, \qquad \dot{x}_N = 0, \qquad t > 0, \tag{17.6}$$

where $x_i, u_i(t), x_0(t) \in \mathbb{R}$ and the nodes 0 and N of MAS do not change their states in time

$$x_0(t) = a \in \mathbb{R}, \qquad x_N(t) = b \in \mathbb{R}, \qquad \forall t \geq 0. \tag{17.7}$$

The simplest formation problem for such a MAS is to design a control protocol realizing a uniform allocation of agents on the segment $[a, b]$, i.e.,

$$x_i(t) \to a + i\frac{b-a}{N} \quad \text{as} \quad t \to +\infty$$

for all $i = 1, 2, ..., N-1$. Notice that according to communication topology the agents with the numbers 2, 3, ..., N do not have an information about position of the 0-agent, i.e., the value $a \in \mathbb{R}$ is unknown for them, which the agents with the numbers 0, 1, ..., $N-2$ do not have an information about the position of the N-agent, i.e., the value b is unknown for them. The control strategy must use only local information (position of neighbors) in order to solve the considered formation control problem.

Fig. 17.1 Communication topology for the uniform allocation of agents on a segment

Linear control protocol. Inspired by the communication topology of MAS (see Fig. 17.1), let us define the linear control protocol as follows [7]:

$$u_i = k(x_{i+1} - 2x_i + x_{i-1}), \quad k > 0. \tag{17.8}$$

Geometrically, such a control selection enforces the ith each agent to take a mean position in between the agents $i-1$ and $i+1$. Let us consider the error variables $\epsilon_0 = x_0 - a$, $\epsilon_N = x_n - b$ and

$$\epsilon_i = x_i - a - i\frac{b-a}{N}, \quad i = 1, ..., N-1.$$

The error dynamics for $\epsilon = (\epsilon_1, ..., \epsilon_{N-1})^\top$ is given by

$$\dot{\epsilon} = -k\mathcal{L}^*\epsilon, \tag{17.9}$$

where

$$\mathcal{L}^* = \begin{pmatrix} 2 & -1 & 0 & ... & 0 & 0 & 0 \\ -1 & 2 & -1 & ... & 0 & 0 & 0 \\ ... & ... & ... & ... & ... & ... & ... \\ 0 & 0 & 0 & ... & -1 & 2 & -1 \\ 0 & 0 & 0 & ... & 0 & -1 & 2 \end{pmatrix}.$$

The function $\epsilon \mapsto V(\epsilon) := 0.5\epsilon^\top \mathcal{L}^*\epsilon$ with $\mathcal{L}^* \succ 0$ is the strict Lyapunov function

$$\dot{V}(\epsilon) = -k|\mathcal{L}^*\epsilon|^2 < 0, \quad \forall \epsilon \neq \mathbf{0},$$

of the error dynamics, so

$$\|\epsilon(t)\| \to 0 \quad \text{as} \quad t \to +\infty.$$

Homogeneous control protocol. Inspired by the homogeneous systems theory, we consider the homogeneous formation control problem, which additionally asks the error dynamics to be $\mathbf{d}$-homogeneous of a given degree μ. The required algorithm can be obtained by means of the following modification (inspired by [8]) of the linear control protocol:

$$u_i = k(x_{i+1} - 2x_i + x_{i-1})^{\lfloor \mu+1 \rceil}, \quad k > 0 \tag{17.10}$$

with $(\cdot)^{\lfloor \mu+1 \rceil} = |\cdot|^{\mu+1}\operatorname{sign}(\cdot)$, and $\mu \geq -1$. The error system

$$\dot{\epsilon} = -k(\mathcal{L}^*\epsilon)^{\lfloor \mu+1 \rceil}, \quad \epsilon_0 = \epsilon_N = 0 \tag{17.11}$$

is $\mathbf{d}$-homogeneous of degree μ with respect to the standard dilation $\mathbf{d}(s) = e^s I_{N-1}$, where $(\mathcal{L}^*\epsilon)^{\lfloor \mu+1 \rceil}$ is defined in component-wise manner. Considering the same function $V = 0.5\epsilon^\top \mathcal{L}^*\epsilon$, we derive

$$\dot{V}(\epsilon) = -k(\mathcal{L}^*\epsilon)^\top(\mathcal{L}^*\epsilon)^{\lfloor\mu+1\rceil} = -k\|\mathcal{L}^*\epsilon\|_{\mu+2}^{\mu+2} \leq -kC_\mu^{\mu+2}|\mathcal{L}^*\epsilon|^{\mu+2}, \quad \forall\epsilon \neq \mathbf{0},$$

where $C_\mu = \min\{1, N^{1/(\mu+2)-1/2}\}$ and the estimates $|\cdot| \leq \|\cdot\|_{\mu+2}$ for $\mu+2 \leq 2$, $|\cdot| \leq N^{1/2-1/(\mu+2)}\|\cdot\|_{\mu+2}$ for $\mu+2 > 2$ are utilized on the last step (please see in Appendix A relation between norms $\|\cdot\|_q, \|\cdot\|_p$ is $\mathbb{R}^n$). The error equation (17.11) is asymptotically stable. Moreover, since the system is homogeneous then, due to Theorem 7.5, it is finite-time stable for $\mu < 0$, exponentially stable for $\mu = 0$ and fixed-time stable for $\mu > 0$. Notice that for $\mu = 0$ we derive the linear formation control studied above.

Combing two homogeneous controllers with degrees $\mu_1 \geq -1$ and $\mu_2 \geq \mu_1$ we derive the locally homogeneous control protocol

$$u_i = 0.5k(x_{i+1} - 2x_i + x_{i-1})^{\lfloor\mu_1+1\rceil} + 0.5k\phi_{\mu_2}(x_{i+1} - 2x_i + x_{i-1})^{\lfloor\mu_2+1\rceil} \quad (17.12)$$

with

$$\dot{V}(\epsilon) \leq -0.5kC_{\mu_1}^{\mu_1+2}|\mathcal{L}^*\epsilon|^{\mu_1+2} - 0.5kC_{\mu_2}^{\mu_2+2}|\mathcal{L}^*\epsilon|^{\mu_2+2}, \quad \epsilon \neq \mathbf{0}.$$

For $\mu = \mu_1 = \mu_2$, we recover the homogeneous control protocol studied above. For $\mu_1 < 0$ and $\mu_2 > 0$, the locally homogeneous control protocol guarantees the fixed-time formation [8].

Robustness analysis. First of all, let us notice that, in the view of Corollary 7.16, the error dynamics of the MAS with locally homogeneous control protocol is ISS with respect to additive state-independent perturbations $g_i(t, x_i) = g_i(t) \neq \mathbf{0}, i = 1, \ldots, N-1, g_i \in L^\infty, g_0 = g_N = 0$ provided that $\mu_1 > -1$, respectively.

Moreover, for $\mu_1 = -1$ and a sufficiently large $k > 0$, the control protocol rejects perturbations satisfying

$$|g| \leq c_1 + c_2|x|^{\mu_2+1}, \quad \forall x = (x_1, \ldots, x_{N-1}) \in \mathbb{R}^{N-1}, \quad (17.13)$$

where $g = (g_1, \ldots, g_{N-1})$ and $c_1 > 0, c_2 > 0$ are some constants. Indeed, since

$$\mathcal{L}^*\epsilon = \mathcal{L}^*x - v, \qquad v := \begin{pmatrix} a \\ 0 \\ \cdots \\ 0 \\ b \end{pmatrix}$$

then, using Cauchy–Schwarz inequality we derive

$$\dot{V}(\epsilon) \leq -0.5kC_{\mu_1}^{\mu_1+2}|\mathcal{L}^*\epsilon| - 0.5kC_{\mu_2}^{\mu_2+2}|\mathcal{L}^*\epsilon|^{\mu_2+2} + \epsilon^\top\mathcal{L}^*g \leq$$

$$-0.5kC_{\mu_1}^{\mu_1+2}|\mathcal{L}^*\epsilon| - 0.5kC_{\mu_2}^{\mu_2+2}|\mathcal{L}^*\epsilon|^{\mu_2+2} + |\mathcal{L}^*\epsilon|(c_1 + c_2|\epsilon + \mathcal{L}^*v|^{\mu_2+1}).$$

Hence, $\dot{V}(\epsilon) < 0, \forall\epsilon \neq \mathbf{0}$ provided that $c_1 > 0, c_2 > 0$ is sufficiently small or, equivalently, $k > 0$ is large enough.

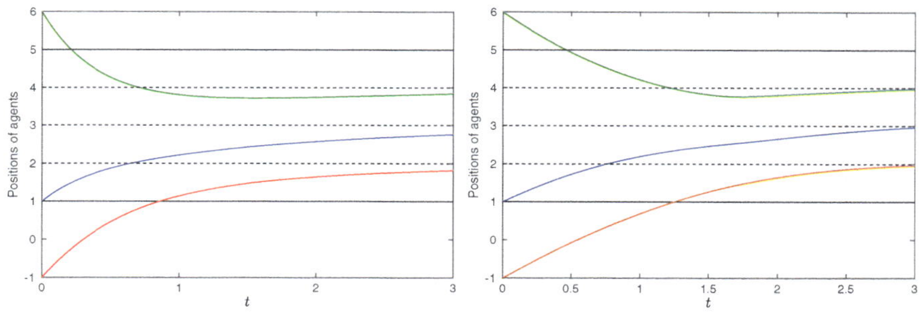

Fig. 17.2 Motion of agents with linear (left) and homogeneous (right) formation protocols

Example 17.1 Let us consider the formation control problem with $a = 1, b = 5$ and $n = 3$. Let the feedback gain for each agent is $k = 1$. The homogeneity degree is $\mu = -0.5$. Simulation results for MAS with $x_1(0) = -1, x_2(0) = 1$ and $x_3(0) = 6$ are depicted in Fig. 17.2. The homogeneous control protocol yields a faster (finite-time) stabilization of the agents in the desired positions than the linear control protocol.

Nonuniform Allocation of Agents on a Segment

Let $0 = x_0 = a < x_1^* < ... < x_i^* < ... < x_{N-1}^* < b = x_n < +\infty$, where a, b are as before, but x_i^* is desired positions of the ith agent. Let us assume that the positions x_i^* be defined as follows:

$$x_i^* = \frac{\lambda_i b}{N}, \quad 0 < \lambda_1 < ... < \lambda_i < ... < \lambda_{N-1} < N. \tag{17.14}$$

In this case, the problem of nonuniform allocation of agents on the segment can be trivially reduced to already solved problem of the uniform allocation of agents on the segment. Indeed, using the change of variables

$$\tilde{x}_i = \frac{i}{\lambda_i} x_i, \quad \tilde{u}_i = \frac{i}{\lambda_i} u_i \tag{17.15}$$

we derive the model of the multi-agent system

$$\frac{d}{dt}\tilde{x}_i = \tilde{u}_i, \quad i = 1, ..., N-1. \tag{17.16}$$

For such a system, the control protocol for uniform allocation of agents on the unit segment gives

$$\tilde{x}_i \to \frac{i}{N} b \quad \text{as} \quad t \to +\infty. \tag{17.17}$$

Hence, by the formula (17.15), we derive

$$x_i \to \frac{\lambda_i b}{N} = x_i^* \quad \text{as} \quad t \to +\infty. \tag{17.18}$$

Notice that the formation control protocol uses the variables $\tilde{x}_i = \frac{i}{\lambda_i} x_i$. So, to realize the nonuniform allocation control protocol, the information about the quantity $\frac{i}{\lambda_i} x_i$ has to be transmitted to neighbors. This means that each agent needs to know its number i in MAS and the desired λ_i.

The above algorithms are one-dimensional formation control protocols $x_i \in \mathbb{R}$. For n-dimensional formation $x_i \in \mathbb{R}^n$, these algorithms can be implemented in a component-wise manner.

17.2.2 Average Consensus

Let us consider the so-called consensus problem [1] for the agents modeled by (17.3). The consensus control means that all states of all agents converge to each other

$$x_i(t) \to x_j(t) \quad \text{as} \quad t \to +\infty.$$

Such a problem appears in power networks for a frequency synchronization.

Let us consider the so-called average consensus

$$x_i(t) \to x^* \quad \text{as} \quad t \to +\infty,$$

where $x^* = \frac{1}{N+1} \sum_{i=0}^{N} x_i(0)$. As before we assume $n = 1$, $A = 0$, $B = 1$ and consider, first, $g = \mathbf{0}$.

Let us consider an MAS consisting of $N + 1$ agents, whose communication topology is described by a fixed strongly connected undirected graph $\mathcal{G} = \{\mathcal{V}, \mathcal{E}, \mathcal{W}\}$. In this case, the Laplacian $\mathcal{L}$ of $\mathcal{G}$ is a symmetric positive semi-definite matrix having the zero eigenvalue of the multiplicity 1 (see above).

Linear Average Consensus Protocol

Following [1] the linear average consensus protocol for the considered multi-agent system is given by

$$u_i = \sum_{j=0}^{n} w_{ij}(x_j - x_i), \quad i = 0, 1, ..., N. \tag{17.19}$$

In the vector form $x = (x_0, x_1, ..., x_N)^\top$ the dynamics of the system is given by

$$\dot{x} = -\mathcal{L}x.$$

Hence, for $\epsilon = (\epsilon_0, \epsilon_1, ..., \epsilon_N)^\top$, $\epsilon_i = x_i - x^*$ the error dynamics is given by

$$\dot{\epsilon}(t) = -\mathcal{L}\epsilon(t), \quad t > 0. \tag{17.20}$$

Using $\mathcal{L}\mathbf{1} = \mathbf{0}$ we derive

$$\sum_{j=0}^{n} \dot{\epsilon}_j(t) = 0.$$

Taking into account that

$$\sum_{j=0}^{n} \epsilon_j(0) = \sum_{j=0}^{N} \left(x_j(0) - \frac{1}{N+1} \sum_{i=0}^{N} x_i(0) \right) = 0.$$

Hence, we derive

$$\sum_{j=0}^{n} \epsilon_j(t) = 0, \quad \forall t \geq 0. \tag{17.21}$$

Let us prove the asymptotic stability of the differential algebraic equation (17.20), (17.21). Considering the Lyapunov function candidate $V = \frac{1}{2}\epsilon^\top \epsilon$ we derive

$$\dot{V} = -\epsilon^\top \mathcal{L}\epsilon.$$

Taking into account $\epsilon^\top \mathbf{1} = 0$ proven above we derive

$$\dot{V}(\epsilon(t)) = -\epsilon^\top(t)\mathcal{L}\epsilon(t) \leq -\lambda_1 |\epsilon(t)|^2 \quad \text{for} \quad \epsilon(t) \neq \mathbf{0},$$

where the latter inequality follows from (17.1) and $\lambda_1 > 0$ is the minimal strictly positive eigenvalue of $\mathcal{L}$. Therefore, the error equation is exponentially stable.

Homogeneous Average Consensus Protocol

The use of nonlinear consensus protocols in MAS has been motivated by finite-time consensus problem [9]. Homogeneity with negative degree is the simplest way to ensure finite-time stability (see Chap. 7). Inspired by [1, 9, 10] we design the following homogeneous control protocol:

$$u_i = \sum_{j=0}^{N+1} w_{ij}^{\frac{\mu+2}{2}} \lfloor x_j - x_i \rceil^{\mu+1}, \quad i = 0, 1, ..., N, \tag{17.22}$$

where $\mu \geq -1$. For $\epsilon = x - x^* = (\epsilon_0, \epsilon_1, ..., \epsilon_N)^\top$ the system of the error equations

$$\dot{\epsilon} = \sum_{j=0}^{N+1} w_{ij}^{\frac{\mu+2}{2}} \lfloor \epsilon_j - \epsilon_i \rceil^{\mu+1}, \quad i = 0, 1, ..., N$$

is standard homogeneous of the degree μ. So, the asymptotic stability of the error equation for $\mu \in [-1, 0)$ would imply the finite-time average consensus, but the case $\mu > 0$ would correspond to the nearly fixed-time average consensus.

Let us consider the quadratic Lyapunov function candidate $V = 2\epsilon^\top \epsilon$ and compute its time derivative along the trajectory of the error system

$$\dot{V} = \sum_{i=0}^{N+1} \epsilon_i \dot{\epsilon}_i = \sum_{i=0}^{N+1} \sum_{j=0}^{N+1} w_{ij}^{\frac{\mu+2}{2}} \epsilon_i \lfloor \epsilon_j - \epsilon_i \rceil^{\mu+1} =$$

$$0.5 \sum_{i=0}^{N+1} \sum_{j=0}^{N+1} w_{ij}^{\frac{\mu+2}{2}} \epsilon_i \lfloor \epsilon_j - \epsilon_i \rceil^{\mu+1} + 0.5 \sum_{i=0}^{N+1} \sum_{j=0}^{N+1} w_{ij}^{\frac{\mu+2}{2}} \epsilon_i \lfloor \epsilon_j - \epsilon_i \rceil^{\mu+1} =$$

$$0.5 \sum_{i=0}^{N+1} \sum_{j=0}^{N+1} w_{ij}^{\frac{\mu+2}{2}} \epsilon_i \lfloor \epsilon_j - \epsilon_i \rceil^{\mu+1} - 0.5 \sum_{i=0}^{N+1} \sum_{j=0}^{N+1} w_{ji}^{\frac{\mu+2}{2}} \epsilon_j \lfloor \epsilon_j - \epsilon_i \rceil^{\mu+1} =$$

$$0.5 \sum_{i=0}^{N+1} \sum_{j=0}^{N+1} w_{ij}^{\frac{\mu+2}{2}} (\epsilon_i - \epsilon_j) \lfloor \epsilon_j - \epsilon_i \rceil^{\mu+1} = -0.5 \sum_{i=0}^{N+1} \sum_{j=0}^{N+1} w_{ij}^{\frac{\mu+2}{2}} |\epsilon_j - \epsilon_i|^{\mu+2},$$

where the symmetry of the adjacency matrix $w_{ij} = w_{ji}$ for the undirected graph has been taken into account. Since all norms in $\mathbb{R}^n$ are equivalent and

$$\|z\|_q \leq \|z\|_p \leq n^{1/p-1/q} \|z\|_q, \quad \forall z \in \mathbb{R}^n, \quad \forall q \geq p \geq 1,$$

then for $\mu \geq 0$ we have

$$\left(\sum_{i=0}^{N+1} \sum_{j=0}^{N+1} w_{ij}^{\mu+2} |\epsilon_j - \epsilon_i|^{\mu+2} \right)^{\frac{1}{\mu+2}} \leq \left(\sum_{i=0}^{N+1} \sum_{j=0}^{N+1} w_{ij} (\epsilon_j - \epsilon_i)^2 \right)^{\frac{1}{2}}$$

and

$$\dot{V} \leq -\frac{1}{2} \left(\sum_{i=0}^{N+1} \sum_{j=0}^{N+1} w_{ij} (\epsilon_j - \epsilon_i)^2 \right)^{\frac{\mu+1}{2}} = -\frac{(2\epsilon^\top \mathcal{L}\epsilon)^{\frac{\mu+2}{2}}}{2} \leq -\frac{(2\lambda_1)^{\frac{\mu+2}{2}} V^{\frac{\mu+2}{2}}}{2}.$$

Therefore, the error equation is nearly fixed-time stable for $\mu > 0$. Repeating the same considerations for $\mu < 0$ we derive

$$\left(\sum_{i=0}^{N+1}\sum_{j=0}^{N+1} w_{ij}^{\mu+2}|\epsilon_j - \epsilon_i|^{\mu+2}\right)^{\frac{1}{\mu+2}} \leq (N+1)^{\frac{-\mu}{\mu+2}}\left(\sum_{i=0}^{N+1}\sum_{j=0}^{N+1} w_{ij}(\epsilon_j - \epsilon_i)^2\right)^{\frac{1}{2}}$$

and

$$\dot{V} \leq -\frac{(N+1)^{-\frac{\mu}{2}}(2\lambda_1)^{\frac{\mu+2}{2}}V^{\frac{\mu+2}{2}}}{2}.$$

So, the error equation is finite-time stable for $\mu \in [-1, 0)$ and nearly fixed-time stable for $\mu > 0$. The locally homogeneous global fixed-time control protocol can be obtained by mixing control protocols with positive and negative homogeneity degrees [10]. The homogeneity-based finite-time consensus for the agents modeled by double integrators can be found in [11]. Below we develop a homogeneity-based consensus control for MAS with agents modeled by linear multi-input systems.

17.2.3 Leader-Following Consensus

Let us consider a MAS consisting of $N+1$ agents, whose communication topology is described by a fixed directed graph $\mathcal{G} = \{\mathcal{V}, \mathcal{E}, \mathcal{W}\}$, where the leader is represented by the agent 0, but the rest agents labeled by $1, 2, ..., N$ are followers. The leader does not receive any information from other agents which nominal dynamics is

$$\dot{x}_i(t) = Ax_i(t) + Bu_i(t), \quad i = 0, 1, ..., N, \tag{17.23}$$

where $A \in \mathbb{R}^{n\times n}$, $B \in \mathbb{R}^{n\times m}$, $x_i(t) \in \mathbb{R}^n$ is the state vector and $u_i(t) \in \mathbb{R}^m$ is the control input of the ith agent.

Assumption 17.1 The pair (A, B) is assumed to be controllable. and generalized homogeneous, namely, there exists a matrix $G_0 \in \mathbb{R}^{n\times n}$ such that

$$AG_0 = (G_0 + I_n)A \quad \text{and} \quad G_0 B = \mathbf{0}. \tag{17.24}$$

The first identity in (17.24) means that the matrix A is $\mathbf{d}$-homogeneous of degree $\mu \in [-1, 1/n]$

$$A\mathbf{d}(s) = e^{\mu s}\mathbf{d}(s)A \tag{17.25}$$

with respect to the linear dilation:

$$\mathbf{d}(s) = e^{sG_{\mathbf{d}}}, \quad s \in \mathbb{R}, \quad G_{\mathbf{d}} = I_n + \mu G_0, \tag{17.26}$$

while the second identity implies a homogeneity of B

$$\mathbf{d}(s)B = e^s B. \tag{17.27}$$

Assumption 17.2 The control law of the leader is unknown but bounded:

$$u_0 \in L^\infty(\mathbb{R}, \mathbb{R}^n). \tag{17.28}$$

Our goal is to develop a generalized homogeneous consensus control protocol for the considered MAS. By definition, this means that, *given* $\mu \geq -1$, *we need to design control laws*

$$u_i = \tilde{u}_i(x_i, x_{i_1}, \ldots, x_{i_{n_i}}),$$

where $i_\zeta \in \mathcal{V} : (i, i_\zeta) \in \mathcal{E},\ \zeta = \overline{1, n_i}, i = \overline{1, N}$ *such that for*

$$\epsilon = [(x_1 - x_0)^\top, \ldots, (x_N - x_0)^\top]^\top$$

the error equation has the form

$$\dot{\epsilon} = f(\epsilon, u_0) \tag{17.29}$$

with some $f : \mathbb{R}^{nN} \times \mathbb{R}^m \mapsto \mathbb{R}^{Nn}$ *and*

- *the vector field* $\epsilon \mapsto f(\epsilon, \mathbf{0})$ *is* $\hat{\mathbf{d}}$*-homogeneous of degree* μ*, where*

$$\hat{\mathbf{d}}(s) = \mathrm{diag}\{\mathbf{d}(s), \ldots., \mathbf{d}(s)\} \in \mathbb{R}^{nN \times nN}, \quad s \in \mathbb{R};$$

- *the error equation* (17.29) *is ISS with respect to* u_0 *for* $\mu > -1$*;*
- *the error equation* (17.29) *is globally asymptotically stable for* $\mu = -1$ *provided that* $\|u_0\|_{L^\infty}$ *is small enough.*

Notice that ISS of the error system (17.29) implies its asymptotic stability for $u_0 = \mathbf{0}$. By homogeneity, in this case, we have finite-time stability of the error system (17.29) for $\mu < 0$, exponential stability of the error system of the error system (17.29) for $\mu = 0$ and nearly fixed-time stability of the error system (17.29) for $\mu > 0$.

Linear Leader-Following Control Protocol

The design of the homogeneous control protocol may also follow the approach developed above. First, we design a linear control protocol, next, we upgrade/transform it to a homogeneous one such that the error dynamics $\dot{\epsilon} = f(\epsilon, \mathbf{0})$ is finite-time (nearly fixed-time) stable for $\mu < 0$ (resp., for $\mu > 0$).

The linear consensus control [1] is given by

$$u_i = K\sum_{j=0}^{n} w_{ij}(x_i - x_j), \quad i = 1, 2, ..., N. \tag{17.30}$$

If the node 0 is a root node (leader) then $w_{i,0} = 0$ and

$$\sum_{j=0}^{N} w_{ij}(x_i - x_j) = \sum_{j=0}^{N} w_{ij}(\epsilon_i - \epsilon_j) = w_{i0}\epsilon_i + \sum_{j=1}^{N} w_{ij}(\epsilon_i - \epsilon_j) = \sum_{j=1}^{N} w_{ij}(\epsilon_i - \epsilon_j).$$

Then the error dynamics in this case has the form:

$$\dot{\epsilon} = (I_N \otimes A + \tilde{\mathcal{L}} \otimes BK)\epsilon, \tag{17.31}$$

where $\otimes$ denotes the *Kronecker product* and $\tilde{\mathcal{L}}$ is the Laplacian given by (17.2). The following lemma presents a sufficient condition for exponential stability of (17.31).[3]

Lemma 17.1 *If the graph $\mathcal{G}$ has the root node $i = 0$ being the leader of MAS, then the error dynamics* (17.31) *is globally exponentially stable with the decay rate $\rho > 0$ provided $K = YX^{-1}$, where $(X, Y) \in \mathbb{R}^{n\times n} \times \mathbb{R}^{m\times n}$ is a solution of the LMIs:*

$$X \succ 0, \quad I_N \otimes (AX + XA^\top + 2\rho X) + \tilde{\mathcal{L}} \otimes BY + (\tilde{\mathcal{L}} \otimes BY)^\top \preceq 0, \tag{17.32}$$

where $\tilde{\mathcal{L}}$ is given by (17.2) *and $\rho > 0$ is some positive constant.*

Proof Let us consider the Lyapunov function candidate

$$V(\epsilon) = \epsilon^\top P\epsilon, \quad P = I_N \otimes X^{-1} \succ 0.$$

The time derivative of V along the trajectories of the error equation (17.31) is

$$\dot{V}(\epsilon) = 2\epsilon^\top P(I_N \otimes A + \mathcal{L} \otimes BK)\epsilon = \epsilon^\top PWP\epsilon,$$
$$W = I_N \otimes (AX + XA^\top) + \tilde{\mathcal{L}} \otimes BY + (\tilde{\mathcal{L}} \otimes BY)^\top.$$

Since the pair (X, Y) satisfies (17.32), then $\dot{V} \leq -2\rho V$ and the error equation (17.31) is globally exponentially stable with decay rate $\rho > 0$. The proof is complete. □

The feasibility of LMI (17.32) depends on connectivity of the graph [1].

[3] The pair $\{\tilde{A}, \tilde{B}\}$ is said to be stabilizable if there exists $\tilde{K}$ such that the matrix $\{\tilde{A} + \tilde{B}\tilde{K}\}$ is Hurwitz.

Let us denote $Q = I_N \otimes (X^{-1}G_0 + G_0^\top X^{-1})$, where and G_0 is defined in Assumption 17.1. For $\delta \in (0, 1)$ let us define

$$\mu_{\min} = \max\left\{-1, \frac{-2\delta}{\max\left\{0, \lambda_{\max}\left(P^{-\frac{1}{2}}QP^{-\frac{1}{2}}\right)\right\}}, \frac{\lambda_{\max}(W)}{\rho\max\{0, -\lambda_{\min}(Q)\}}\right\} < 0, \tag{17.33}$$

$$\mu_{\max} = \min\left\{\frac{1}{n}, \frac{2\delta}{\max\left\{0, -\lambda_{\min}\left(P^{-\frac{1}{2}}QP^{-\frac{1}{2}}\right)\right\}}, \frac{-\lambda_{\max}(W)}{\rho\max\{0, \lambda_{\max}(Q)\}}\right\} > 0, \tag{17.34}$$

where $P = I_N \otimes X^{-1} \succ 0$.

Homogeneous Leader-Following Control Protocol

Let us define the control protocol in the form:

$$u_i = \|\theta_i\|_{\mathbf{d}}^{1+\mu} K\mathbf{d}(-\ln\|\theta_i\|_{\mathbf{d}})\theta_i, \tag{17.35}$$

where $K \in \mathbb{R}^{m\times n}$ is a control gain to be defined, $\|\cdot\|_{\mathbf{d}}$ is a $\mathbf{d}$-homogeneous norm and

$$\theta_i = \sum_{j=0}^{N} w_{ij}(x_i - x_j), \quad i = 1, ..., N. \tag{17.36}$$

Let $\epsilon_i = x_i - x_0, i = 1, 2, ..., _N$ and $\epsilon_0 = 0$, where $x_0(t) \in \mathbb{R}^n$ is the state vector of the leader. Since

$$\theta_i(\epsilon) = \sum_{j=0}^{N} w_{ij}(x_i - x_j) = \sum_{j=0}^{N} w_{ij}(\epsilon_i - \epsilon_j) = w_{i0}\epsilon_i + \sum_{j=1}^{N} w_{ij}(\epsilon_i - \epsilon_j)$$

then the error equation has the form:

$$\dot{\epsilon} = f(\epsilon, u_0), \tag{17.37}$$

where $f = (f_1, ..., f_N)^\top$ and

$$f_i(\epsilon, u_0) = A\epsilon_i + \|\theta_i\|_{\mathbf{d}}^{1+\mu} BK\mathbf{d}(-\ln\|\theta_i\|_{\mathbf{d}})\theta_i - Bu_0. \tag{17.38}$$

Moreover, the vector field f with $u_0 = \mathbf{0}$ is $\hat{\mathbf{d}}$-homogeneous of the degree $\mu \geq -1$. Indeed, since

$$\theta(\hat{\mathbf{d}}(s)\epsilon) = w_{i0}\mathbf{d}(s)\epsilon_i + \sum_{j=1}^{N} w_{ij}(\mathbf{d}(s)\epsilon_i - \mathbf{d}(s)\epsilon_j) = \mathbf{d}(s)\theta_i(\epsilon)$$

then taking into account $A\mathbf{d}(s) = e^{\mu s}\mathbf{d}(s)A$ and $\mathbf{d}(s)B = B\tilde{\mathbf{d}}(s)$ we derive

$$f(\hat{\mathbf{d}}(s)\epsilon, \mathbf{0}) = e^{\mu s}\hat{\mathbf{d}}(s)f(\epsilon, \mathbf{0})$$

for all $\epsilon \in \mathbb{R}^{nN}$ and $s \in \mathbb{R}$. As a corollary of the above lemma, we obtain the following properties of the homogeneous control protocol.

Theorem 17.1 *Let $(X, Y) \in \mathbb{R}^{n\times n} \times \mathbb{R}^{m\times n}$ be a solution of the LMIs* (17.32) *and $K = YX^{-1}$. Let Assumptions 17.1, 17.2 be fulfilled. If the graph $\mathcal{G}$ has the root node $i = 0$ being the leader of MAS, and $\mu \in [\mu_{\min}, \mu_{\max}]$ then*

- *the dilation $\mathbf{d}$ in $\mathbb{R}^n$ given by* (17.26) *is monotone with respect to the weighted Euclidean norm $\|z\| = \sqrt{z^\top X^{-1}z}$, $z \in \mathbb{R}^n$;*
- *the system* (17.23) *with control protocol* (17.35) *and the canonical homogeneous norm $\|\cdot\|_{\mathbf{d}}$ induced by the norm $\|\cdot\|$ has classical solutions for $\mu > -1$ and Filippov solutions for $\mu = -1$;*
- *for μ being sufficiently close to zero, the error equation* (17.37) *is ISS with respect to u_0;*
- *for $\mu = -1$ the error equation* (17.37) *with $u_0 \neq \mathbf{0}$ is finite-time stable for a small enough $\|u_0\|_{L^\infty}$ provided that the error equation* (17.37) *with $u_0 = \mathbf{0}$ is finite-time stable.*

Proof Since $\mu \in [\mu_{\min}, \mu_{\max}]$ then $I_N \otimes (X^{-1}G_{\mathbf{d}} + G_{\mathbf{d}}^\top X^{-1}) \succ 0$ and the dilation $\hat{\mathbf{d}}$ is monotone with respect to the weighted Euclidean norm $\|\epsilon\| = \sqrt{\epsilon^\top P\epsilon}$, where $P = I_N \otimes X^{-1} \succ 0$. Similarly, the dilation $\mathbf{d}$ in $\mathbb{R}^n$ is monotone with respect to $\|z\| = \sqrt{z^\top X^{-1}z}$. Let the canonical homogeneous norm $\|\epsilon\|_{\mathbf{d}}$ be induced by $\|\epsilon\|$. Then for $u_0 = \mathbf{0}$ we have

$$\frac{d\|\epsilon\|_{\hat{\mathbf{d}}}}{dt} = \|\epsilon\|_{\hat{\mathbf{d}}}^{1+\mu}\frac{\epsilon^\top\hat{\mathbf{d}}^\top(-\ln\|\epsilon\|_{\hat{\mathbf{d}}})Pf(\hat{\mathbf{d}}(-\ln\|\epsilon\|_{\hat{\mathbf{d}}})\epsilon)}{\epsilon^\top\hat{\mathbf{d}}^\top(-\ln\|\epsilon\|_{\hat{\mathbf{d}}})PG_{\hat{\mathbf{d}}}\hat{\mathbf{d}}(-\ln\|\epsilon\|_{\hat{\mathbf{d}}})\epsilon}$$

$$= \|\epsilon\|_{\hat{\mathbf{d}}}^{1+\mu}\frac{\overbrace{\epsilon^\top\hat{\mathbf{d}}^\top(-\ln\|\epsilon\|_{\hat{\mathbf{d}}})P(I_N\otimes A+\mathcal{L}\otimes BK)\hat{\mathbf{d}}(-\ln\|\epsilon\|_{\hat{\mathbf{d}}})\epsilon}^{\leq -\tilde{\rho}}}{\epsilon^\top\hat{\mathbf{d}}^\top(-\ln\|\epsilon\|_{\hat{\mathbf{d}}})PG_{\hat{\mathbf{d}}}\hat{\mathbf{d}}(-\ln\|\epsilon\|_{\hat{\mathbf{d}}})\epsilon}$$

$$+ \|\epsilon\|_{\hat{\mathbf{d}}}^{1+\mu}\frac{\epsilon^\top\hat{\mathbf{d}}^\top(-\ln\|\epsilon\|_{\hat{\mathbf{d}}})P\left(f(\hat{\mathbf{d}}(-\ln\|\epsilon\|_{\hat{\mathbf{d}}})\epsilon)-(I_N\otimes A+\mathcal{L}\otimes BK)\hat{\mathbf{d}}(-\ln\|\epsilon\|_{\hat{\mathbf{d}}})\epsilon\right)}{\epsilon^\top\hat{\mathbf{d}}^\top(-\ln\|\epsilon\|_{\hat{\mathbf{d}}})PG_{\hat{\mathbf{d}}}\hat{\mathbf{d}}(-\ln\|\epsilon\|_{\hat{\mathbf{d}}})\epsilon}.$$

Repeating the analysis presented in proof of Theorem 9.1 we show that u_i is a continuous function of θ_i for $\mu \in (-1, (\tilde{n}-1)^{-1})$. Moreover, f is a continuous function of (ϵ, μ) and for $\mu = 0$ we have $f(\epsilon, \mathbf{0}) = (I_N \otimes A + \mathcal{L} \otimes BK)\mathbf{d}(-\ln\|\epsilon\|_{\mathbf{d}})\epsilon$. Taking into account $\|\mathbf{d}(-\ln\|\epsilon\|_{\mathbf{d}})\epsilon\| = 1$ for all $\epsilon \in \mathbb{R}^{nN}\backslash\{\mathbf{0}\}$ we conclude

$$f(\mathbf{d}(-\ln\|\epsilon\|_{\mathbf{d}})\epsilon, \mathbf{0}) - (I_N \otimes A + \mathcal{L} \otimes BK)\mathbf{d}(-\ln\|\epsilon\|_{\mathbf{d}})\epsilon \to 0 \text{ as } \mu \to 0$$

uniformly on $\epsilon \in \mathbb{R}^{nN}\backslash\{\mathbf{0}\}$. This means that there exists $\tilde{\rho} > 0$ such that $\frac{d\|\epsilon\|_{\mathbf{d}}}{dt} \leq -\tilde{\rho}\|\epsilon\|_{\hat{\mathbf{d}}}^{1+\mu}$ for all $\epsilon \neq \mathbf{0}$, at least, if μ is sufficiently close to 0. Therefore, the error equation is globally asymptotically stable for $u_0 = \mathbf{0}$. In this case, the ISS with respect to $u_0 \in L^\infty$ comes from Theorem 7.13. The finite-time stability analysis for $\mu = -1$ repeats the proof of Corollary 9.2. The proof is complete. □

The leader-following control protocol presented in this section is fully distributed. Its main drawback is the absence of a theoretically proven rule for selection of the homogeneity degree μ. Indeed, the above theorem proves the stability of the error dynamics for μ being sufficiently close to zero. The so-called quasi-distributed homogeneous control protocol does not have this drawback, but it requires a simultaneous broadcasting a scalar control signal to all agents. This assumption is not restrictive in some cases. The scalar signal cannot be utilized as a centralized control for the whole error system (since, in the general case, this system is not controllable), so the distributed nature of MAS remains important for control design. For more details about quasi-distributed homogeneous control protocol, we refer the reader to [12].

Remark 17.1 The assumption about homogeneity of the pair $\{A, B\}$ can be eliminated by means of the following modification of the control:

$$u_i = K_0 x_i + \|\theta_i\|_{\mathbf{d}}^{1+\mu} K\mathbf{d}(-\ln\|\theta_i\|_{\mathbf{d}})\theta_i, \tag{17.39}$$

where $K_0 \in \mathbb{R}^{m\times n}$ is the gain of a homogenizing feedback selected such that $A_0 = A + BK_0$ satisfies Assumption 17.1. The gain K is defined by means of the LMI (17.32), where the matrix A replaced with the matrix A_0. By Theorem 9.1, such gain K_0 always exists if the pair $\{A, B\}$ is controllable. The error equation in this case has the form:

$$\dot{\epsilon} = f(\epsilon, u_0, x_0), \tag{17.40}$$

where $f = (f_1, ..., f_N)^\top$ and

$$f_i = A_0\epsilon_i + \|\theta_i\|_{\mathbf{d}}^{1+\mu} BK\mathbf{d}(-\ln\|\theta_i\|_{\mathbf{d}})\theta_i - Bu_0 + K_0 x_0. \tag{17.41}$$

Since

$$f(\mathbf{d}(-\ln\|\epsilon\|_{\mathbf{d}})\epsilon, \mathbf{0}, \mathbf{0}) - (I_N \otimes A_0 + \mathcal{L} \otimes BK)\mathbf{d}(-\ln\|\epsilon\|_{\mathbf{d}})\epsilon \to 0 \text{ as } \mu \to 0$$

then, analogously to Theorem 17.1, the ISS with respect to (x_0, u_0) can be proven, at least, for μ being sufficiently close to 0. The finite-time stability of the error equation for $\mu = -1$ may be ensured only if $\max\{\|x_0\|_{L^\infty}, \|u_0\|_{L^\infty}\}$ is small enough. Notice that for $\mu = 0$ the homogeneous control (17.39) becomes the linear feedback

$$u_i = K_0 x_i + K\theta_i. \tag{17.42}$$

Example 17.2 Let us consider MAS with the nodes 0, 1, 2, 3 and the adjacency matrix

$$W = \begin{pmatrix} 0&0&0&0\\ 1&0&0&0\\ 0&0&0&1\\ 1&0&1&0 \end{pmatrix}. \tag{17.43}$$

The root (leader) node has the index 0 and the reduced Laplacian matrix (17.2) is given by

$$\tilde{\mathcal{L}} = \begin{pmatrix} -1&0&0\\ 0&-1&1\\ 0&1&-2 \end{pmatrix}. \tag{17.44}$$

Let each agent of MAS be modeled by the linear system (17.23) with

$$A = \begin{pmatrix} 3&0&1\\ 0&-1&1\\ -2&0&0 \end{pmatrix} \quad \text{and} \quad B = \begin{pmatrix} 1&-1\\ 0&1\\ 0&1 \end{pmatrix}.$$

The matrix A is not homogeneous, so we follow Remark 17.1 in order to design a homogeneous consensus protocol. Solving the algebraic equation (9.5), we find the gain K_0

Such that the pair $\{A_0, B\}$ satisfies Assumption 17.1 with

$$G_0 = \begin{pmatrix} 0&-0.5&0.5\\ 0&-0.5&0.5\\ 0&0.5&-0.5 \end{pmatrix}.$$

Solving the LMI (17.32) with $\rho = 0.1$ and A replaced with A_0, we derive

$$X = \begin{pmatrix} 0.2096&0.2627&-0.2627\\ 0.2627&1.0837&-0.0184\\ -0.2627&-0.0184&1.0837 \end{pmatrix}, \quad Y = \begin{pmatrix} 1.8436&-1.5786&-1.2581\\ 1.5747&0.1563&0.1563 \end{pmatrix},$$

so $K = YX^{-1} = \begin{pmatrix} 22.7765&-6.9062&4.2436\\ 18.6797&-4.3062&4.5997 \end{pmatrix}$. By the formula (17.33), we derive $\mu_{\min} = -1$ and design the decentralized homogeneous control in the form (17.39) with $\mu = \mu_{\min}$.

Let the control input of the leader be defined as

$$u_0 = K_0 x_0 + \begin{pmatrix} 0.5\cos(2t)\\ 0 \end{pmatrix}.$$

The state of the leader with such a control input is a uniformly bounded function of time (see Fig. 17.3), so the input u_0 is a uniformly bounded function of time as well.

Simulation results for the linear control protocol (17.42) and the homogeneous control protocol (17.39) are given in Figs. 17.4, 17.5 and 17.6. The simulations support theoretical conclusions showing that the linear control protocol cannot guarantee the exact leader-following tracking, while the homogeneous control protocol with $\mu = -1$ does. The simulation has been done using explicit Euler method with the step size $2.5 \cdot 10^{-4}$.

Using linear homogeneity the formation and average consensus control protocols considered above can be analogously extended to MAS with agents modeled by linear multi-input system (17.23).

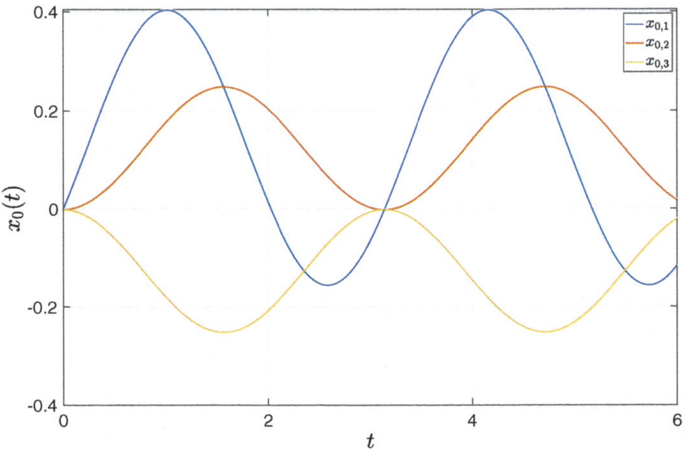

Fig. 17.3 Trajectory of the leader agent $x_0(t) = (x_{0,1}(t), x_{0,2}(t), x_{0,3}(t))^\top$

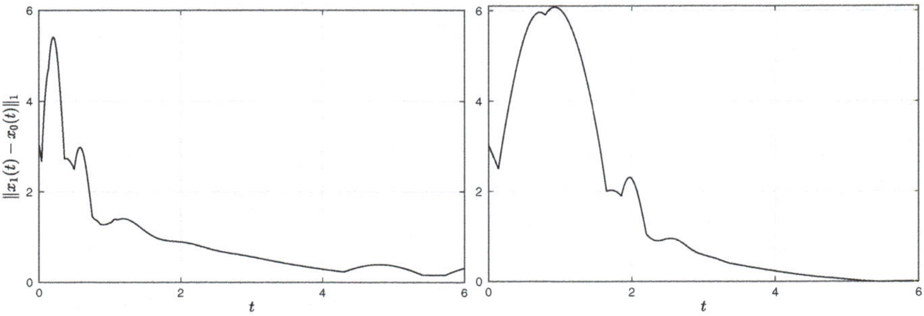

Fig. 17.4 Evolution of the tracking error $\|x_1(t) - x_0(t)\|_1$ for linear (left) and homogeneous (right) control protocols

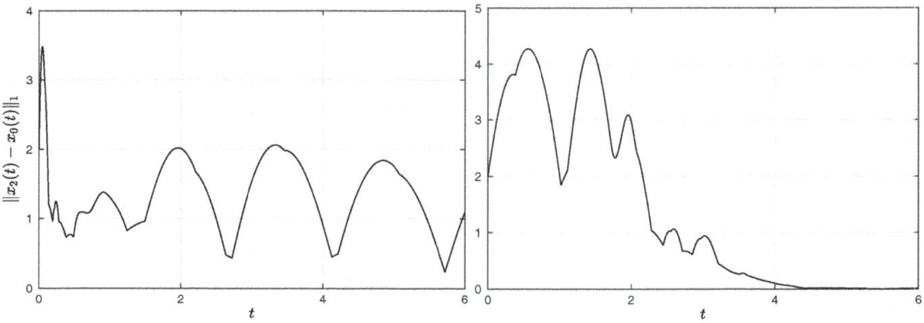

Fig. 17.5 Evolution of the tracking error $\|x_2(t) - x_0(t)\|_1$ for linear (left) and homogeneous (right) control protocols

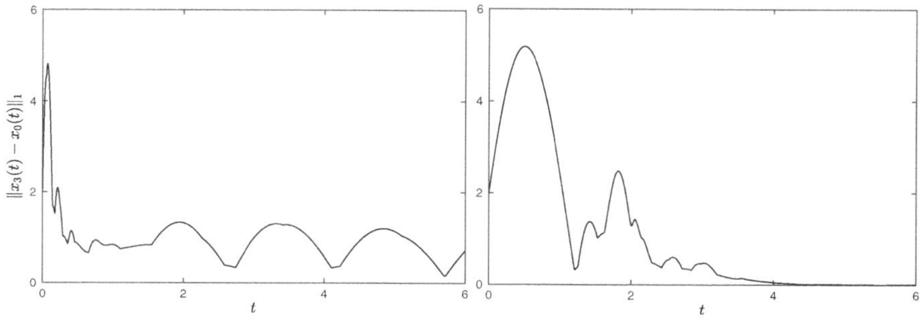

Fig. 17.6 Evolution of the tracking error $\|x_3(t) - x_0(t)\|_1$ for linear (left) and homogeneous (right) control protocols

Concluding Remarks

Homogeneity can be utilized for analysis and design of multi-agent systems. Similar to conventional (centralized) control systems homogeneous control protocols can provide a faster convergence, better robustness and transients preserving the simplicity of parameters tuning typical for linear systems. The design of such protocols can also be based on an "upgrade" of a linear control protocol.

References

1. Olfati-Saber R, Murray RM (2004) Consensus problems in networks of agents with switching topology and time-delays. IEEE Trans Autom Control 49(9):1520–1533
2. Ren W, Beard RW (2008) Distributed consensus in multi-vehicle cooperative control. Springer
3. Lewis FL, Zhang H, Hengster-Movric K, Das A (2013) Cooperative control of multi-agent systems: optimal and adaptive design approaches. Springer (2013)
4. Spears W, Spears D, Hamann J, Heil R (2004) Distributed, physics-based control of swarms of vehicles. Auton Robot 17:137–162
5. Guzey H, Dierks T, Jagannathan S, Acar L (2019) Modified consensus-based output feedback control of quadrotor UAV formations using neural networks. J Intell Robot Syst 94(1):283–300
6. Chung F (1997) Spectral Graph Theory. AMS
7. Wagner IA, Bruckstein AM (1997) Row straightening via local interactions. Circuits Syst Signal Process 16(3):287–305
8. Parsegov S, Polyakov A, Scherbakov PS (2012) Nonlinear fixed-time control protocol for uniform allocation of agents on a segment. In: IEEE conference on decision and control
9. Cortes J (2006) Finite-time convergent gradient flows with applications to network consensus. Automatica 42:1993–2000
10. Parsegov S, Polyakov A, Scherbakov PS (2013) Fixed-time consensus algorithm for multi-agent systems with integrator dynamics. In: IFAC workshop on distributed estimation and control in networked systems (NECSYS)
11. Wang X, Hong Y (2008) Finite-time consensus for multi-agent networks with second-order agent dynamics. In: The 17th IFAC world congress, pp 15185–15190
12. Li M, Polyakov A, Zheng G (2023) On generalized homogeneous leader-following consensus control for multi-agent systems. IEEE Trans Control Netw Syst 11(1):558

Appendix

This chapter briefly surveys some very basic notions and results of set and matrix theory, functional analysis, measure theory, and integration, which are utilized in the proofs of theorems presented in this monograph. We refer the reader, for example, to [1–9] for more deals about presented results.

A Elements of Theory Sets and Spaces

A *set* Ω is a collection of distinct objects (elements or members of the set). Examples of mathematical sets are sets of numbers, sets of functions, sets of sets, etc. If an element a belongs to a set Ω, we write $a \in \Omega$. A set of elements $a, b, c...$ is denoted $\{a, b, ...\}$. If all elements of a set A belong to a set Ω we use the notation $A \subset \Omega$. If a set has no elements, it is called *empty set* and denoted $\emptyset$. A set is countable *countable set* if all its members can be numbered as by natural numbers $1, 2, ...$ A countable set may have infinite number of elements. The power set 2^X of X is defined as a set of all possible subsets of X.

The *Cartesian product* of sets A and B is a set denoted $A \times B$ and defined as a set of pairs of elements from A and B, respectively,

$$A \times B = \{(a, b) : a \in A, b \in B\}.$$

The set $A \cap B$ is called the *intersection of sets* A and B and defined as a set of elements belonging simultaneously to A and B. If the set A and B have no common elements then $A \cap B = \emptyset$.

The set $A \cup B$ is called the *union of sets* A and B and defined as a set of elements belonging, at least, to one of sets A, B.

A. Polyakov, *Generalized Homogeneity in Systems and Control Volume I*, Communications and Control Engineering,
https://doi.org/10.1007/978-3-031-76246-8

The set $A \backslash B$ is called *complement* (or *set-theoretic difference*) of sets A and B and defined as a set of elements that are members of A but not members of B.

Mathematical space is a set of mathematical objects (e.g., numbers, functions, sets, etc.) with some additional structures.

A *function* (*mapping*, *operator*) f between sets X and Y is denoted $f : X \mapsto Y$ and defined as a relation that associates to every element of the set X exactly one element of the set Y.

A *set-valued function (mapping, operator)* F between sets X and Y is denoted $F : X \mapsto 2^Y$ and defined as a relation that associates to every element of the set X many elements of Y.

A.1 Field

A *field* is a set $\mathcal{F}$ together with two operations on $\mathcal{F}$: *addition and multiplication*. Addition is a function $\mathcal{F} \times \mathcal{F} \mapsto \mathcal{F}$ denoted $\alpha + \beta$ for $\alpha, \beta \in \mathcal{F}$, i.e., $(\alpha, \beta) \to \alpha + \beta$. The result of addition is called sum. Multiplication is a function $\mathcal{F} \times \mathcal{F} \mapsto \mathcal{F}$ denoted $\alpha\beta$ (or $\alpha \cdot \beta$) for $\alpha, \beta \in \mathcal{F}$, i.e., $(\alpha, \beta) \to \alpha\beta$. The result of multiplication is called the product. These operations must satisfy the so-called field axioms:

Associativity: $\alpha + (\beta + \gamma) = (\alpha + \beta) + \gamma$ and $\alpha(\beta\gamma) = (\alpha\beta)\gamma$.

Commutativity: $\alpha + \beta = \beta + \gamma$ and $\alpha\beta = \beta\alpha$.

Existence of zero and identity: There exist two different elements denoted 0 and 1 in $\mathcal{F}$ such that $\alpha + 0 = \alpha$ and $\alpha \cdot 1 = \alpha$.

Additive and multiplicative inverses: For every $\alpha \in F$ there exists an element in F, denoted $-\alpha$ such that a $\alpha + (-\alpha) = 0$; for every $\alpha \neq 0$ in $\mathcal{F}$ there exists an element in $\mathcal{F}$ denoted by α^{-1} or $1/\alpha$ such that $\alpha \cdot \alpha^{-1} = 1$.

Distributivity of multiplication over addition: $\alpha(\beta + \gamma) = \alpha\beta + \alpha\gamma$.

The standard examples of fields are real and complex numbers.

A.2 Metric Space

A function $\rho : M \times M \mapsto [0, +\infty)$ is a metric (distance) on a set M if

- $\rho(a, b) = 0 \Leftrightarrow a = b$;
- $\rho(a, b) = \rho(b, a)$;
- $\rho(a, b) \leq \rho(a, c) + \rho(c, b)$.

A pair (M, ρ) is a *metric space*.

A *neighborhood of a point* x is any subset of M that contains an open ball around x.

A set $Q \in M$ is said to be *open* if $\forall x \in Q, \exists r > 0 : B(x, r) \subset Q$. A sequence $\{x_n\} \in M, n = 1, 2, ...$ in the metric space M is said to *converge to the limit* $x \in M$ if for any $\varepsilon > 0$ exists a natural number N such that $\rho(x_n, x) < \varepsilon$ for all $n > N$. Such a sequence is called convergent.

A set $K \subset M$ is said to be *closed* if every sequence in K, that converges to a limit in M, has this limit in K.

A set $D \subset M$ is *dense* in a set $Q \subset M$ if for any $x \in Q$ and for any $\varepsilon > 0$ there exists $y \in D$ such that $\rho(x, y) < \varepsilon$.

A metric space (set) M is said to be *compact* if any sequence in M has a subsequence that converges to an element in M.

A sequence $\{x_n\} \in M$ is said to be a *Cauchy sequence* in M if for any $\varepsilon > 0$ there exists a natural N such that $\rho(x_n, x_m) < \varepsilon$ for all $n, m > N$.

A metric space M is said to be *complete* if any Cauchy sequence converges to an element $x \in M$.

The Cartesian product $M = M_1 \times M_2$ of metric spaces (M_1, ρ_1) and (M_2, ρ_2) is also a metric space (M, ρ) with $\rho = (\rho_1^p + \rho_2^p)^{1/p}$, $p \geq 1$.

A.3 Vector Space

A *vector space* over a field $\mathcal{F}$ is a set $\mathbb{V}$ together with two operations. Elements of $\mathbb{V}$ are called vectors. Elements of $\mathcal{F}$ are called scalars. The first operation is a vector addition $\mathbb{V} \times \mathbb{V} \mapsto \mathbb{V}$ denoted $v + w$ for $v, w \in \mathbb{V}$, i.e., $(v, w) \mapsto v + w$. The result of addition is called sum of two vectors. The second operation is a multiplication on a scalar $\mathcal{F} \times \mathbb{V} \mapsto \mathbb{V}$ denoted αv (or $\alpha \cdot v$) for $\alpha \in \mathcal{F}$ and $v \in \mathbb{V}$, i.e., $(\alpha, v) \mapsto \alpha v$.

The set V and the operations of addition and multiplication must satisfy the axioms:

Associativity of addition: $u + (v + w) = (u + v) + w$.

Commutativity of addition: $u + v = v + u$.

Identity element of addition: There exists the zero element $\mathbf{0} \in \mathbb{V} : v + \mathbf{0} = v$ for all $v \in \mathbb{V}$.

Inverse elements of addition: For any $v \in \mathbb{V}$ there exists $-v \in \mathbb{V}$ such that $v + (-v) = \mathbf{0}$.

Compatibility of scalar multiplication: $\alpha(\beta v) = (\alpha\beta)v$.

Identity element of scalar multiplication: $1v = v$, where 1 denotes the identity in $\mathcal{F}$.

Distributivity of scalar multiplication with respect to vector addition: $\alpha(u + v) = \alpha u + \alpha v$.

Distributivity of scalar multiplication with respect to field addition: $(\alpha + \beta)v = \alpha v + \beta v$.

For example, the set $\mathbb{R}^n = \mathbb{R} \times \mathbb{R} \times ... \times \mathbb{R}$ is a vectors space. The set $C([0, 1]; \mathbb{R})$ of continuous functions $[0, 1] \mapsto \mathbb{R}$ is a vector space as well.

Let $\mathbb{V}$ and $\mathbb{X}$ be two linear spaces over the field $\mathcal{F}$. A function (mapping, operator) $f : \mathbb{V} \mapsto \mathbb{X}$ is said to be *linear* if $f(\alpha x + \beta y) = \alpha f(x) + \beta f(y)$ for all $\alpha, \beta \in \mathcal{F}$ and all $x, y \in \mathbb{V}$. If $\mathbb{X} = \mathcal{F}$ then f is called a *linear functional*.

A *linear subspace* of $\mathbb{V}$ is a subset $\tilde{\mathbb{V}} \subset \mathbb{V}$ being again a vector space with the same addition and multiplication operations.

The set of all possible linear combinations of the vectors $e_1, e_2, ..., e_k \in \mathbb{V}$ is denoted $\text{span}\{e_1, e_2, ..., e_k\}$ and given by

$$\text{span}\{e_1, e_2, ..., e_k\} := \{\alpha_1 e_1 + \cdots + \alpha_k e_k : \alpha_i \in \mathcal{F}, i = 1, 2, ..., k\}.$$

The vectors $\{e_1, e_2, ..., e_k\}$ are *linearly independent* if the linear equation

$$\sum_{i \in \Gamma} \alpha_i e_i = \mathbf{0}$$

with respect to $\alpha_i \in \mathcal{F}$ has the unique solution $\alpha_i = 0, \forall i \in \Gamma$, where Γ is a set of indexes (finite or infinite).

A set $\{e_i\}_{i \in \Gamma}, e_i \in \mathbb{V}$ is said to be a *basis in a vector space* $\mathbb{V}$ if $\{e_i\}$ are linearly independent and $\mathbb{V} = \text{span}\{e_i\}_{i \in \Gamma}$, where Γ is a set of indexes. A number of elements of Γ is called the *dimension of a linear vector space* $\mathbb{V}$ and denoted $\dim(\mathbb{V})$. For instance, $\dim(\mathbb{R}^n) = n$ and $\dim(C([0, 1]; \mathbb{R})) = \infty$.

Obviously, $\text{span}\{e_1, e_2, ..., e_k\}$ is a linear subspace of $\mathbb{V}$. If the vectors $e_1, e_2, ..., e_k$ are linearly independent then $\dim(\text{span}\{e_1, e_2, ..., e_k\}) = k$.

A.4 Convex Sets

A set Ω of a vector space $\mathbb{V}$ over the field of real numbers is **convex** if a segment joining two given points from Ω belongs to Ω:

$$\forall x_1, x_2 \in \Omega \quad \Rightarrow \quad \lambda x_1 + (1 - \lambda) x_2 \in \Omega, \forall \lambda \in [0, 1].$$

If $A, B \subset \mathbb{R}^n$ are convex sets then the sets

$$A \cap B \quad \text{and} \quad A \dot{+} B$$

are convex, where $\dot{+}$ denotes the geometric sum of sets (see Notation).

Let $x_i \in \mathbb{V}, i = 1, 2, ..., N$. The vector

$$\alpha_1 x_1 + \alpha_2 x_2 + ... + \alpha_N x_N \in \mathbb{V}, \quad \alpha_i \in \mathbb{R}$$

is said to be *convex combination* of the vectors x_i if

$$\alpha_i \geq 0 \quad \alpha_1 + \alpha_2 + \ldots + \alpha_N = 1, \tag{A.1}$$

where N is a natural number.

Let Ω be a subset of a vector space V. A smallest convex closed set containing Ω is called *convex hall* of Ω and denoted $\mathrm{co}(\Omega)$ (or $\mathrm{co}\,\Omega$). If the convex hall is closed, we use the notation $\overline{\mathrm{co}}$.

Carathéodory Lemma Let $\Omega \subset \mathbb{R}^n$ be a compact set. Any $x \in \overline{\mathrm{co}}(\Omega)$ is a convex combination of at most $n+1$ points of Ω, i.e.,

$$x = \alpha_1 x_1 + \alpha_2 x_2 + \ldots + \alpha_{n+1} x_{n+1} \in \mathbb{R}^n, \quad \alpha_i \in \mathbb{R} \tag{A.2}$$

for some α_i, $i = 1, 2, \ldots, n+1$ satisfying (A.1) and some $x_i \in \Omega$.

For geometric sum of sets $M_1 \subset \mathbb{R}^n$ and $M_2 \subset \mathbb{R}^n$ it always holds

$$\overline{co}(M_1 \dot{+} M_2) = \overline{co}(M_1) \dot{+} \overline{co}(M_2).$$

A.5 Normed Vector Space

Let $\mathbb{V}$ be a vector space over a field of real $\mathcal{F} = \mathbb{R}$ or complex $\mathcal{F} = \mathbb{C}$ numbers. A function $\|\cdot\| : \mathbb{V} \mapsto [0, +\infty)$ is a *norm in a vector space* $\mathbb{V}$ if

- *positive definiteness*: $\|v\| \geq 0$ for all $v \in \mathbb{V}$ and $\|v\| = 0 \Leftrightarrow v = \mathbf{0}$;
- *positive homogeneity*: $\|\lambda x\| = |\lambda| \|x\|$ for all $\lambda \in \mathcal{F}$;
- *triangle inequality*: $\|v + u\| \leq \|v\| + \|u\|$ for all $v, u \in \mathbb{V}$.

A pair $(\mathbb{V}, \|\cdot\|)$ is a *normed vector space.*

For example, the set $\mathbb{S}^2$ of symmetric matrices 2×2 is a normed vector space with the norm $\|A\| = \sqrt{\mathrm{trace}(A^2)}$, where $\mathrm{trace}(A)$ denotes the trace (i.e., the sum of diagonal elements) of the matrix A. A possible basis in $\mathbb{S}^2$ is $e_1 = \left(\begin{smallmatrix} 1 & 0 \\ 0 & 0 \end{smallmatrix}\right)$, $e_2 = \left(\begin{smallmatrix} 0 & 0 \\ 0 & 1 \end{smallmatrix}\right)$, $e_3 = \left(\begin{smallmatrix} 0 & 1 \\ 1 & 0 \end{smallmatrix}\right)$ and $\dim(\mathbb{S}^2) = 3$.

Any normed vector space is a metric space. The norm $\|\cdot\|$ introduces the distance (metric) $\rho(u, v) = \|u - v\|$ between elements $u, v \in \mathbb{V}$. The balls $B(v, r) = \{u \in \mathbb{V} : \|v - u\| < r\}$ define the so-called strong (or norm) topology on $\mathbb{V}$ consisting of all sets which can be written as a union of balls $B(v, r)$. The spheres in $\mathbb{V}$ are given by $S(v, r) = \{u \in \mathbb{V} : \|x - u\| = r\}$.

Let V_1 and V_2 be two normed vector spaces. A mapping $f : \mathbb{V}_1 \mapsto \mathbb{V}_2$ is said to be continuous at $x_0 \in V_1$ if there exists $\omega \in \mathcal{K}$ such that $\|f(x) - f(x_0)\|_{V_2} \leq \omega(\|x - x_0\|_{V_1})$. If $V_2 = \mathbb{R}$ the mapping f is called *functional*. A functional $f : V \mapsto \mathbb{R}$ is said to be continuous if it is continuous at any point of $\mathbb{V}$. A functional f is linear if $f(\alpha x + \beta y) = \alpha f(x) + \beta f(y)$, $\forall \alpha, \beta \in \mathcal{F}$ and $\forall x, y \in \mathbb{V}$.

The space $\mathbb{R}^n$ is a normed vector space (see Notation).

A.6 Banach Space

Real (complex) *Banach space* $\mathbb{B}$ is a complete normed vector space over the field of real (resp., complex) numbers.

A normed vector space is a Banach space if and only if $\sum_{n=1}^{+\infty} \|x_n\| < +\infty$ implies $\sum_{n=1}^{+\infty} x_n$ converges in $\mathbb{B}$, where $x_n \in \mathbb{B}$. Any normed vector space $\mathbb{V}$ has a completion $\mathbb{B}$ being a Banach space.

The examples of Banach spaces are

- ℓ^p—the space of sequences $\alpha = \{\alpha_i\}$, $\alpha_i \in \mathbb{R}$ such that $\sum_i |\alpha_i|^p < +\infty$ with the norm $\|\alpha\|_p = (\sum_i |\alpha_i|^p)^{1/p}$.
- $C(\Omega; \mathbb{V})$—a space of continuous functions $\Omega \mapsto \mathbb{V}$ with the norm $\|u\|_C = \sup_{x\in\Omega} \|u(x)\|_{\mathbb{V}}$, where $\mathbb{V}$ is a normed vector space.

A.7 Hilbert Space

A *Hilbert space* $\mathbb{H}$ is a Banach space with the norm

$$\|x\| = \sqrt{\langle x, x\rangle}$$

induced by an inner product $\langle\cdot,\cdot\rangle$. A function $\langle\cdot,\cdot\rangle : \mathbb{H} \times \mathbb{H} \mapsto \mathcal{F}$ defined on a vector space $\mathbb{H}$ over a field $\mathcal{F}$ of real ($\mathcal{F} = \mathbb{R}$) or complex ($\mathcal{F} = \mathbb{C}$) numbers is said to be an *inner product* on $\mathbb{H}$ if

- $\langle x, x\rangle > 0$ for all $x \in \mathbb{H}$ and $\langle x, x\rangle = 0 \Leftrightarrow x = \mathbf{0}$;
- $\langle x, y\rangle = \langle y, x\rangle$ (or $= \overline{\langle y, x\rangle}$ if $\mathcal{F} = \mathbb{C}$) for all $x, y \in \mathbb{H}$;
- $\langle \alpha x_1 + \beta x_2, y\rangle = \alpha\langle x_1, y\rangle + \beta\langle x_2, y\rangle$ for all $\alpha, \beta \in \mathcal{F}$ and $x_1, x_2, y \in \mathbb{H}$.

If $\mathbb{H}$ is a Hilbert space then the following relations

$$\begin{aligned} &\text{(Parallelogram identity)}: && \|u + v\|^2 + \|u - v\|^2 = 2\left(\|u\|^2 + \|v\|^2\right), \\ &\text{(Cauchy–Schwarz inequality)}: && \langle u, v\rangle \le \|u\| \cdot \|v\| \end{aligned}$$

hold $\forall u, v \in \mathbb{H}$. Vectors $x, y \in \mathbb{H}$ are said to be orthogonal $x \perp y$ if $\langle x, y\rangle = 0$. For $x \perp y$ the *Pythagorean identity* $\|x + y\|^2 = \|x\|^2 + \|y\|^2$ holds.

Hilbert projection theorem says that for every point $x \in \mathbb{H}$ and every nonempty closed convex set $K \subset \mathbb{H}$ there exists a unique $y \in K$ such that $\|x - y\| = \min_{z\in K} \|x - z\|$. The corresponding vector y is called the orthogonal projection of x on K.

If $\{e_i\}, i \in I$ is an orthonormal family ($\langle e_i, e_j\rangle = 0$ if $e_i \neq e_j$, $i, j \in I$ and $\langle e_i, e_i\rangle = 1$) then the *Bessel inequality*

$$\|x\|^2 \geq \sum_{i\in I} |\langle x, e_i\rangle|^2$$

holds.

If B be an orthonormal basis of $\mathbb{H}$, i.e., a set of orthonormal vectors such that span B is dense in $\mathbb{H}$, then the *Parseval identity*

$$\|x\|^2 = \sum_{e_i\in B} |\langle x, e_i\rangle|$$

holds. If the Hilbert space is separable then B is a countable set and for any $x \in \mathbb{H}$ we have

$$\left\| x - \sum_{i=1}^{+\infty} \langle x, e_i\rangle e_i \right\| = 0.$$

The series $\sum_{i=1}^{+\infty} \langle x, e_i\rangle e_i$ is called *Fourier series*.

A.8 Euclidean Space

Euclidean space is a finite-dimensional Hilbert space. All Euclidean spaces are isomorphic to $\mathbb{R}^n$. The canonical p-norm in $\mathbb{R}^n$ is given by

$$\|x\|_p = (|x_1|^p + \ldots + |x_n|^p)^{\frac{1}{p}}, \quad x = (x_1, \ldots, x_n)^\top \in \mathbb{R}^n, \quad 1 \leq p < +\infty.$$

The ∞-norm in $\mathbb{R}^n$ is defined as follows:

$$\|x\|_\infty = \max_{i=1,\ldots,n} |x_i|.$$

By notation, $|\cdot| = \|\cdot\|_2$ in $\mathbb{R}^n$ is the canonical Euclidean norm. All norms in $\mathbb{R}^n$ are equivalent:

$$\|x\|_p \leq \|x\|_r \leq n^{1/r-1/p}\|x\|_p, \quad \text{if} \quad 1 \leq r \leq p \leq \infty.$$

Bolzano–Weierstrass Theorem Each bounded sequence $x_i \in \mathbb{R}^n$: $\sup_{i\in\mathbb{N}} \|x_i\| < \infty$ in $\mathbb{R}^n$ has a convergent subsequence $\exists \lim\limits_{i_k\to+\infty} x_{i_k} \in \mathbb{R}^n$.

A weighted inner product in $\mathbb{R}^n$ is defined as $\langle x, y\rangle = x^\top P y$, where $x, y \in \mathbb{R}^n$ and $P \in \mathbb{R}^{n\times n}$ is a positive definite symmetric matrix. The corresponding norm is the weighted Euclidean norm:

$$\|x\| = \sqrt{x^\top P x}.$$

For $P = I_n$ then the weighted Euclidean norm becomes the canonical Euclidean norm $|x| = \sqrt{x^\top x}$ in $\mathbb{R}^n$ and the inner product becomes the canonical inner product $\langle x, y\rangle = x^\top y$ in $\mathbb{R}^n$.

B Elements of Measure Theory and Integration

B.1 Measure Space

Let $\mathbb{X}$ be set and $2^{\mathbb{X}}$ is a power set (a set of all subsets).

A subset Σ of $2^{\mathbb{X}}$ is called a *σ-algebra* over $\mathbb{X}$ if

- $X \in \Sigma$;
- if $A \in \Sigma$ then $X\backslash A \in \Sigma$;
- if $A_1, A_2, A_3, ... \in \Sigma$ then $A_1 \cup A_2 \cup A_3 ... \in \Sigma$.

If (X, τ) is a topological space then σ algebra generated by τ is called Borel σ-algebra.

Let Σ be a σ-algebra over $\mathbb{X}$. A function $\mu : \Sigma \mapsto [0, +\infty]$ is said to be a *measure* if

- $\mu(A) \geq 0$ for all $A \in \Sigma$;
- $\mu(\emptyset) = 0$;
- if $A_i \in \Sigma, i = 1, 2, ...$ and $A_i \cap A_j = \emptyset$ for $i \neq j$ then

$$\mu\left(\bigcup_{i=1}^{+\infty} A_i\right) = \sum_{i=1}^{+\infty} \mu(A_i).$$

The triple $(\mathbb{X}, \Sigma, \mu)$ is called the *measure space*.

Any measure defined on Borel σ-algebra is called Borel measure.

Lebesgue measure $\mu_{\mathbb{R},L} : 2^{\mathbb{R}} \mapsto [0, +\infty]$ of a set $S \subset \mathbb{R}$ is defined as follows:

$$\mu_{\mathbb{R},L}(S) = \inf_{\{\mathcal{I}_k\}: S\subset \bigcup_k \mathcal{I}_k} \sum_k \ell(\mathcal{I}_k),$$

where $\mathcal{I}_n = (a_k, b_k)$, $a_k, b_k \in \mathbb{R}$, $a_k < b_k$, and $\ell(\mathcal{I}_n) = b_k - a_k$.

A set $S \in \mathbb{R}$ is Lebesgue measurable if and only if it satisfies *Carathéodory condition*:

$$\mu_{\mathbb{R},L}(A) = \mu_{\mathbb{R},L}(A \cap S) + \mu_{\mathbb{R},L}(A \cap (\mathbb{R} \backslash S)), \quad \forall A \in 2^{\mathbb{R}}.$$

A collection $\Sigma_{\mathbb{R},L}$ of Lebesgue measurable set is σ-algebra over $\mathbb{R}$ and $(\mathbb{R}, \Sigma_{\mathbb{R},L}, \mu_{\mathbb{R},L})$ is a measure space.

We say that some property holds *almost everywhere* on S if a set $\Delta \subset S$ on which it does not hold has the measure zero. An example of the *set of Lebesgue measure zero* is a countable set of points from S.

The same constructions can be repeated for the Lebesgue measure space $(\mathbb{R}^n, \Sigma_{\mathbb{R}^n,L}, \mu_{\mathbb{R}^n,L})$ considering $I_k = (a_k^1, b_k^1) \times ... \times (a_k^n, b_k^n) \in \mathbb{R}^n$ and $\ell(I_k) = (b_k^1 - a_k^1)...(b_k^n - a_k^n)$.

B.2 Measurable Function

Let (X, Σ_x, μ_x) and (Y, Σ_y, μ_y) be two measure spaces. A function $f : X \mapsto Y$ is said to be measurable as mapping (X, Σ_x, μ_x) on (Y, Σ_y, μ_y) if

$$\{x : f(x) \in A_y\} \in \Sigma_x, \quad \forall A_y \in \Sigma_y.$$

In particular, a function $f : X \mapsto \mathbb{R}$ is said to be Lebesgue measurable if for any $\gamma \in \mathbb{R}$ the set

$$\{x : f(x) > \gamma\}$$

is Lebesgue measurable. Minimal γ such that the set $\{x : f(x) > \gamma\}$ has the Lebesgue measure zero is called an *essential supremum* of f.

Any continuous function is measurable. A sum of measurable functions defined on the same measure spaces is a measurable function.

Let (X, Σ_x, μ_x), (Y, Σ_y, μ_y), and (Z, Σ_z, μ_z) be three measure spaces. If functions $f : X \mapsto Y$ and $g : Y \mapsto Z$ are measurable then the composition $g \circ f : X \mapsto Z$ is a measurable function. The following theorem shows certain closeness of measurable and continuous functions.

Luzin Theorem Let $f : [a, b] \mapsto \mathbb{R}$ be a measurable function. Then, for every $\varepsilon > 0$ there exists a compact set $S \subset [a, b]$ such that f restricted to S is continuous almost everywhere and $\mu(S) > b - a - \varepsilon$.

B.3 Riemann Integration

Let $f : [a, b] \mapsto \mathbb{R}$. We say that f is Riemann integrable if there exists a number $s \in \mathbb{R}$ such that

for any $\varepsilon > 0$ there exists $\delta > 0$ and a partition $a = x_0 < x_1 < ... < x_n = b$, $x_{i+1} - x_i < \delta$ such that

$$\left| \sum_{i=1}^{n} f(t_i)(x_i - x_{i-1}) - s \right| < \varepsilon$$

for some $t_i \in [x_i, x_{i+1}]$.

The number s is called the Riemann integral of f and denoted $\int_a^b f(x)dx$.

All continuous and piece-wise continuous functions are Riemann integrable. Any monotone function $[a, b] \mapsto \mathbb{R}$ is Riemann integrable as well.

Linearity: $\int_a^b \alpha f_1(x) + \beta f_2(x)dx = \alpha \int_a^b f_1(x)dx + \beta \int_a^b f_2(x)dx$ for any Riemann integrable $f_1 : [a, b] \mapsto \mathbb{R}$ and $f_2 : [a, b] \mapsto \mathbb{R}$ and any $\alpha, \beta \in \mathbb{R}$.

B.4 Lebesque Integration

Let (E, Σ, μ) be the Lebesque measure space, where $E \subset \mathbb{R}$. If $f : E \mapsto \mathbb{R}_+$ then, by definition, *Lebesque integral* of f is

$$\int_E f d\mu := \lim_{a \to +\infty} \int_0^a f^{\mu}(t)dt,$$

where $f^{\mu}(t) = \mu(\{x : f(x) > t\})$ and $\int_0^a$ is the Riemann integral. Notice that the function f_{μ} is monotone so it is Riemann integrable on any compact $[0, a]$.

If $f : E \mapsto \mathbb{R}$ then

$$\int_E f d\mu := \int_E f^+ d\mu - \int_E f^- d\mu,$$

where $f^+(x) = f(x)$ if $f(x) \geq 0$ and $f^+(x) = 0$ if $f(x) < 0$, $f^-(x) = f(x)$ if $f(x) \leq 0$ and $f^-(x) = 0$ if $f(x) > 0$.

The *linearity* mentioned for the Riemann integral also takes the place for the Lebesgue integral as well as the *monotonicity*: if $f \leq g$ then

$$\int_E f d\mu \leq \int_E g d\mu.$$

Lebesgue's Theorem about the Differentiability of Monotone Functions Let f be a monotone function on (a, b). Then f is differentiable almost everywhere on (a, b).

Fatou's Lemma If $\{f_k\}$ is a sequence of non-negative Lebesgue measurable functions then

$$\int \liminf_{k\to+\infty} f_k d\mu \leq \liminf_{k\to+\infty} \int f_k d\mu.$$

Lebesgue Dominated Convergence Theorem Let $\{f_k\}$ be a sequence of Lebesgue measurable functions having point-wise f and there exist a Lebesgue integrable function $g \geq |f_k|$ for all k then

$$\lim_{k\to+\infty} \int f_k d\mu = \int f d\mu.$$

The Lebesgue integral can also be defined for functions $\mathbb{R}^n \mapsto \mathbb{R}^m$ and generalized for some maps in Banach spaces (see the next section).

Change of Variable in Lebesgue Integral Let $(\mathbb{R}^n, \Sigma, \mu)$ be a Lebesgue measure space, $E \subset \mathbb{R}^n$ and $\nu : E \mapsto \mathbb{R}^n$ be a C^1 one-to-one mapping such that $\det(\nabla\nu)$ is separated from zero on E. If $f : \nu(E) \mapsto \mathbb{R}^m$ is a Lebesgue measurable function then

$$\int_{\nu(E)} f d\mu = \int_E f(\nu) \det(\nabla\nu) d\mu.$$

The set E in the latter theorem may be unbounded (e.g., $E = \mathbb{R}^n$). For a Lebesgue integrable function $f : \mathbb{R}^n \mapsto \mathbb{R}^m$ in this monograph, we also use the following notation:

$$\int_{\mathbb{R}^n} f(x)dx.$$

Newton–Leibniz Formula If $f \in C^1([a, b]; \mathbb{R}^n)$ then it holds

$$\int_a^b \dot{f}(t)dt = f(b) - f(a)$$

for the Riemann integral. Moreover, the Newton–Leibniz formula holds for any absolutely continuous functions in the case of Lebesgue integral.

C Fixed-Point Theorems

We say that $f : \mathbb{B} \mapsto \mathbb{X}$ is a Lipschitz mapping if it satisfies the *Lipschitz condition*:

$$\exists L > 0 : \|f(x) - f(y)\|_{\mathbb{X}} \leq L\|x - y\|_{\mathbb{B}}, \quad \forall x, y \in \mathbb{B},$$

where $\mathbb{B}, \mathbb{X}$ are Banach spaces.

Banach Fixed-Point Theorem Let $F : \mathbb{B} \mapsto \mathbb{B}$ and there exists $q \in [0, 1)$ such that

$$\|F(x) - F(y)\| \leq q\|x - y\|, \quad \forall x, y \in \mathbb{B}.$$

Then there exists a unique point $x^* \in \mathbb{B}$ such that $F(x^*) = x^*$.

The above theorem also holds for a complete metric space $\mathbb{X}$ provided that the contraction condition is given by

$$\rho(F(x), F(y)) \leq q\rho(x, y) \quad \forall x, y \in \mathbb{X},$$

where $q \in (0, 1)$ and ρ is a metric in $\mathbb{X}$.

Brouwer Fixed-Point Theorem Let $K \subset \mathbb{R}^n$ be nonempty convex and compact. Every continuous function $f : K \mapsto K$ has a fixed point

$$\exists x^* \in K \quad : \quad f(x^*) = x^*.$$

The *graph of a function* $f : \mathcal{D}(f) \subset \mathbb{B} \mapsto \mathbb{X}$ is the set

$$G(f) = \{(x, f(x)) : x \in \mathbb{B}\}.$$

Kakutani Fixed-Point Theorem Let $K \in \mathbb{R}^n$ be nonempty, compact, and convex. Let $F : K \mapsto 2^K$ be a set-valued function on K with the following properties:

- F has a closed graph;
- $F(x)$ is nonempty and convex for all $x \in K$.

Then F has a fixed point $x^* \in F(x^*)$.

D Elements of Functional Analysis

Function spaces are the common examples of infinite-dimensional Banach and Hilbert spaces.

D.1 Spaces of Strongly Differentiable Functions

The strong (usual) derivatives of functions $f : X \mapsto Y$ are considered in Sect. 4.2 as particular cases of the so-called contingent derivatives.

- $C(X; Y)$ is a space of continuous functions $X \mapsto Y$, where X and Y are subsets of normed vector spaces. The function space C is a Banach space with the so-called uniform norm topology:

$$\|f\|_C = \sup_{x \in X} \|f(x)\|_Y, \quad f \in C(X; Y).$$

- $C^{0,\alpha}(X; Y)$ with $\alpha \in (0, 1]$ is a space of Hölder continuous functions:

$$C^{0,\alpha}(X; Y) = \left\{ f \in C(X; Y) : \exists k \geq 0, \ \|f(x) - f(y)\|_Y \leq k \|x - y\|_X^{\alpha}, \forall x, y \in X \right\}.$$

  For example, the function $x \mapsto x^{\beta}$ belongs to $C^{0,\alpha}([0, 1]; \mathbb{R})$ for any $\alpha \in (0, \beta]$ and belongs to $C^{0,\beta}([0, +\infty); \mathbb{R})$. Moreover, for $\Omega \subset \mathbb{R}^n$ it holds $C^{0,\beta}(\Omega; \mathbb{R}^m) \subset C^{0,\alpha}(\Omega; \mathbb{R}^m)$ provided that $0 < \alpha \leq \beta \leq 1$.
- $C^q(X; Y)$ is a space of strongly continuously differentiable functions $X \mapsto Y$, at least, up to the order n, where $X \subset \mathbb{R}^n$ and $Y \subset \mathbb{R}^m$. This means

$$\frac{\partial^k f}{\partial x_1^{n_1} \partial x_2^{n_2} \ldots \partial x_n^{n_k}} \in C(X; Y), \quad \forall k = 0, 1, \ldots, q, \quad \forall n_i : \sum_{i=1}^{n} n_i = k,$$

  for any $f \in C^q(X; Y)$. The function space C^p is a Banach space with the norm

$$\|f\|_{C^q} = \sum_{k=0}^{q} \sum_{n_1 + \ldots + n_k = k} \left\| \frac{\partial^k f}{\partial x_1^{n_1} \partial x_2^{n_2} \ldots \partial x_n^{n_k}} \right\|_C .$$

  The following embedding $C^{q_1} \subset C^{q_2}$, obviously, holds for $q_1 \geq q_2$.
- $C_c(X; Y)$ is a vector space of continuous functions with compact supports in $X \subset \mathbb{R}^n$, i.e., $f(x) \neq \mathbf{0}$ only for x from a compact subset of X.
- $C^{\infty}(X; Y)$ is a vector space of infinitely smooth functions on X.

Arzelà–Ascoli Theorem Let $\{f_k\}$ be a sequence of functions $f \in C([a,b];\mathbb{R})$. If this sequence is *uniformly bounded*:

$$\exists M > 0 \ : \ \|f_k\|_C < M, \ \forall k,$$

and *equicontinuous*:

$$\forall \varepsilon > 0, \ \exists \delta > 0 \ : \ |x-y| < \delta \Rightarrow |f_n(x) - f_n(x)| < \varepsilon, \ \forall n$$

then there exists a subsequence $\{f_{n_k}\}$ and $f \in C([a,b];\mathbb{R})$ such that

$$\|f_{n_k} - f\|_C \to 0 \quad \text{as} \quad k \to +\infty.$$

For example, a sequence of Lipschitz functions with the same Lipschitz constant $L > 0$ for all n satisfies the latter theorem.

Integration by parts. If $u, v \in C^1([a,b];\mathbb{R})$ then

$$\int_a^b u\frac{\partial v}{\partial x}dx = uv|_a^b - \int_a^b \frac{\partial u}{\partial x} v\,dx.$$

Let Ω be an open bounded set in $\mathbb{R}^n$ with a smooth boundary Γ. If $u \in C^1(\Omega;\mathbb{R})$ and $v \in C^1(\Omega;\mathbb{R})$ then

$$\int_\Omega u\,\mathrm{div}(v)\,dx = \int_\Gamma uv \cdot \nu_\Omega\,d\Gamma - \int_\Omega \nabla u \cdot v\,dx, \tag{D.1}$$

where ν_Ω is the outward unit normal vector to the boundary of Ω.

Heine–Cantor Theorem If a function $f : \mathbb{R}^n \mapsto \mathbb{R}^m$ is continuous on an open set $\Omega \subset \mathbb{R}^n$ then f is uniformly continuous[a] on any compact from Ω.

[a] A function $f : \mathbb{R}^n \mapsto \mathbb{R}^m$ is uniformly continuous on a set $C \subset \mathbb{R}^n$ if there exists the *modulus of continuity* $\omega \in \mathcal{K}$ such that $|f(x) - f(y)| \le \omega(|x-y|)$ for all $x, y \in C$.

Rademacher Theorem If a function $f : \mathbb{R}^n \mapsto \mathbb{R}^m$ is locally Lipschitz continuous[a] on an open set $\Omega \subset \mathbb{R}^n$ then f is differentiable almost everywhere on Ω.

[a] A function $f : \mathbb{R}^n \mapsto \mathbb{R}^m$ is locally Lipschitz continuous on an open set $\Omega \subset \mathbb{R}^n$ and it satisfies Lipschitz condition on any compact from Ω.

Clarke Gradient If $f : \mathbb{R}^n \mapsto \mathbb{R}$ is differentiable almost everywhere in $\mathbb{R}^n$ then

$$\partial f(x) = \overline{\text{co}}\left\{\partial f^*(x, h_i)\right\}, \quad \partial f^*(x, h_i) = \lim_{h_i \to 0} \left. \frac{\partial f(z)}{\partial z} \right|_{z=x+h_i},$$

where the sequence $h_i \in \mathbb{R}^n$ is such that $h_i \to \mathbf{0}$ as $i \to +\infty$ and the above limit exists. The convex hull is taken over all possible limit values $\partial f^*(x, h_i)$.

By Rademacher theorem, for locally Lipschitz continuous function $V : \mathbb{R}^n \mapsto \mathbb{R}$, the Clarke gradient is a singleton almost everywhere and is locally bounded on $\mathbb{R}^n$. Moreover, if $x : \mathbb{R} \mapsto \mathbb{R}^n$ is an absolutely continuous function then $t \mapsto V(x(t))$ is an absolutely continuous function and the following chain rule takes a place

$$\dot{V}(x(t)) \overset{a.e.}{\in} \partial V(z)|_{z=x(t)}\, \dot{x}(t).$$

Implicit Function Theorem

- Let $Q : \mathbb{R} \times \mathbb{R}^n \mapsto \mathbb{R}$ be continuous. If there exists $(V_0, x_0) \in \mathbb{R} \times \mathbb{R}^n$ such that
  - $Q(V_0, x_0) = 0$;
  - Q is differentiable at (V_0, x_0);
  - $\frac{\partial Q(V,x)}{\partial V} \neq 0$ for $V = V_0$ and $x = x_0$,

  then there exists a neighborhood $\Omega \subset \mathbb{R}^n$ of the point $x_0 \in \mathbb{R}^n$ and a continuous function $\tilde{V} : \mathbb{R}^n \mapsto \mathbb{R}$ such that

  $$Q(\tilde{V}(x), x) = 0 \quad \text{for all} \quad x \in \Omega$$

  and

  $$\left. \frac{\partial \tilde{V}(x)}{\partial x} \right|_{x=x_0} = -\frac{\frac{\partial Q(V,x)}{\partial x}}{\frac{\partial Q(V,x)}{\partial x}} \quad \text{for} \quad V = V_0 \text{ and } x = x_0;$$

- *(for differentiable functions)*, if $Q \in C^k$ then $\tilde{V} \in C^k$ as well;
- *(for locally Lipschitz continuous functions)*, if Q is locally Lipschitz continuous then V is locally Lipschitz continuous.

Mean Value Theorem

- Let $a, b \in \mathbb{R} : a < b$. If $f : [a, b] \mapsto \mathbb{R}$ is continuous on $[a, b]$ and differentiable on (a, b) then there exists $c \in (a, b)$ such that

$$f'(c) = \frac{f(b) - f(a)}{b - a}.$$

- Let Ω be an open subset of $\mathbb{R}^n$. Let $f : \Omega \subset \mathbb{R}^n \mapsto \mathbb{R}$ be a differentiable function. For any $x, y \in \Omega : x \neq y$ there exists $c \in (0, 1)$ such that

$$f(y) - f(x) = \left.\frac{\partial f(z)}{\partial z}\right|_{z=(1-c)x+cy} (y - x).$$

- Let Ω be an open subset of $\mathbb{R}^n$. Let $f : \Omega \subset \mathbb{R}^n \mapsto \mathbb{R}^m$ be a differentiable function. For any $x, y \in \Omega : x \neq y$ one holds

$$f(y) - f(x) = \int_0^1 \left.\frac{\partial f(z)}{\partial z}\right|_{z=(1-\lambda)x+\lambda y} d\lambda \cdot (y - x).$$

D.2 Lebesgue Spaces L^p

The space $L^p(\Omega; \mathbb{R})$ is a space of Lebesgue measurable functions $\Omega \subset \mathbb{R}^n \mapsto \mathbb{R}$ for which the absolute value of p-th power is Lebesgue integrable. The space $L^p(\Omega; \mathbb{R})$ with $1 \leq p \leq +\infty$ is a Banach space with the norm

$$\|f\|_p = \left(\int_\Omega |f|^p d\mu\right)^{\frac{1}{p}}$$

due to *Minkowski inequality*

$$\|f + g\|_p \leq \|f\|_p + \|g\|_p.$$

The norm for $p = \infty$ is given by $\|f\|_\infty = ess\sup_{x\in\Omega} |f(x)|$, where $ess\sup$ denotes an essential supremum (see above).

The vector space $L^1_{loc}(\mathbb{R}^n; \mathbb{R}^n)$ contains locally Lebesque integrable (on any compact set $\Omega \subset \mathbb{R}^n$) functions.

D.3 Weakly Differentiable Functions. Sobolev Spaces $W^{k,p}$ and H^p

Let $f \in L^1((a,b);\mathbb{R})$. We say that $Df \in L^1((a,b);\mathbb{R})$ is a *weak derivative* of f if

$$\int_a^b f\,\frac{d\phi}{dx}d\mu = -\int_a^b Df\,\phi d\mu, \quad \forall \phi \in C_c^\infty((a,b);\mathbb{R}).$$

Notice that, $\phi \in C_c^\infty((a,b);\mathbb{R})$, in particular, means that $\lim_{x\to a}\phi(x) = \lim_{x\to b}\phi(x) = 0$. Hence, if $f \in C^1([a,b];\mathbb{R})$ then, using integration by part, we conclude that the latter integral identity holds only for $Df := \frac{df}{dx}$. In other words, the weak derivative coincides with the strong derivative if the function is differentiable in the conventional sense.

Considering functions of n variables and arbitrary order of the derivative, we derive $D^\alpha f \in L^1(\Omega;\mathbb{R})$ is the αth weak derivative of $f \in L^1(\Omega;\mathbb{R})$ if

$$\int_\Omega f\cdot D^\alpha\phi d\mu = (-1)^\alpha \int_\Omega D^\alpha f\cdot\phi d\mu, \quad \forall\phi \in C_c^\infty(\Omega;\mathbb{R}),$$

where $\alpha = (\alpha_1, ..., \alpha_n)$, $\alpha_i \in 0, 1, 2, ...$ and

$$D^\alpha\phi := \frac{\partial^{|\alpha|}\phi(x)}{\partial x_1^{\alpha_1}...\partial x_n^{\alpha_n}}, \quad |\alpha| = \alpha_1 + ... + \alpha_1, \quad x \in \mathbb{R}^n.$$

By definition, $D^0 f = f$. For $D^1 f$ we also use the notation ∇f.

We say that a function $f \in L^1(\Omega;\mathbb{R})$ is weakly differentiable up to the order k if it has weak derivatives $D^\alpha f \in L^1(\Omega;\mathbb{R})$ for all $\alpha : |\alpha| \le k$.

The set of functions $f \in L^1(\Omega;\mathbb{R})$, which are weakly differentiable, at least, up to the order n and which the weak derivatives belongs to L^p space is denoted $W^{k,p}$ and called *Sobolev space*. The Sobolev space $W^{k,p}$ is a Banach space with the norm:

$$\|f\|_{W^{k,p}} = \sum_{|\alpha|\le k}\|D^\alpha f\|_{L^p}.$$

The integration by parts formula (D.1) is valid for $u, v \in W^{1,1}(\Omega;\mathbb{R})$.

A function $(a,b) \overset{f}{\mapsto} \mathbb{R}^n$ belongs to the space $W_{loc}^{k,p}((a,b);\mathbb{R})$ if its restriction $(a',b') \overset{f}{\mapsto} \mathbb{R}$ belongs to $W^{k,p}((a',b');\mathbb{R})$ for any $(a',b') \subset \mathrm{int}(a,b)$.

For $p = 2$ we use the notation $H^p := W^{k,2}$. The Sobolev space H^k is a Hilbert space with the inner product:

$$\langle f, g\rangle = \sum_{|\alpha|\le k}\langle D^\alpha f, D^\alpha g\rangle_{L^2}, \quad f, g \in H^k.$$

By definition, $H^0 = L^2$.

In the natural component-wise manner, one can be proposed an extension of L^p, $W^{k,p}$ and H^p spaces to vector-valued functions $\Omega \subset \mathbb{R}^n \mapsto \mathbb{R}^m$.

The space $W_0^{1,p}$ is a completion of C_c^∞ with respect to the norm $\|\cdot\|_{W^{1,p}}$.

Barbalat Lemma

$f \in W^{2,\infty}(\mathbb{R}_+;\mathbb{R})$ and $\lim_{t\to+\infty} f(t)$ is finite $\Rightarrow$ $\lim_{t\to+\infty} \dot{f}(t) = 0$.

Sobolev embedding theorem:

$$\frac{1}{p} - \frac{k}{n} = \frac{1}{q} - \frac{\ell}{n} \quad \Rightarrow \quad W^{k,p}(\mathbb{R}^n) \subset W^{\ell,q}(\mathbb{R}^n).$$

E Elements of Matrix Analysis

Let $n, m \in \mathbb{N}$ and $i = 1, 2, ..., n$, $j = 1, 2, ..., m$. A mathematical object A consisting of numbers $a_{ij} \in \mathbb{R}$ (or $\in \mathbb{C}$) written in n rows and m columns

$$A = \begin{pmatrix} a_{11} & a_{12} & \dots & a_{1m} \\ a_{21} & a_{22} & \dots & a_{2m} \\ \dots & \dots & \dots & \dots \\ a_{n1} & a_{n2} & \dots & a_{nm} \end{pmatrix}$$

is known as $n \times m$ real (complex) *matrix*, but the numbers a_{ij} are referred to as the elements of the matrix. If all elements of A are zeros, we write $A = \mathbf{0}$. The set of all $n \times m$ real (resp., complex) matrices is denoted $\mathbb{R}^{n\times m}$ (resp., $\mathbb{C}^{n\times n}$). This is a vector space provided that the multiplication αA of a matrix $A \in \mathbb{R}^{n\times m}$ (resp., $\in \mathbb{C}^{n\times m}$) by a scalar in $\alpha \in \mathbb{R}$ (resp., $\in \mathbb{C}$) and addition of two matrices $A + B$ with $A, B \in \mathbb{R}^{n\times m}$ is defined in component-wise manner. The transposition of the matrix A is a matrix $A^\top$ having swapped rows and columns:

$$A^\top = \begin{pmatrix} a_{11} & a_{21} & \dots & a_{n1} \\ a_{12} & a_{22} & \dots & a_{n2} \\ \dots & \dots & \dots & \dots \\ a_{1m} & a_{2m} & \dots & a_{nm} \end{pmatrix}.$$

The product of matrices $A \in \mathbb{R}^{n\times k}$ and $B \in \mathbb{R}^{k\times m}$ is the matrix $C = AB \in \mathbb{R}^{n\times m}$ with the elements $c_{ij} = \sum_{p=1}^{k} a_{ip} b_{pj}$, where a_{ip} and b_{pj} are elements of the matrices A and B, respectively.

If $n = m$ then the matrix $A \in \mathbb{R}^{n\times n}$ (resp., $\in \mathbb{C}^{n\times n}$) is called square. The elements a_{ii} of the square matrix A are called diagonal elements. A square matrix A is called diagonal and denoted by $A = \text{diag}(a_{11}, ..., a_{nn})$ if all off-diagonal elements of A

are zeros. The diagonal matrix with identity diagonal elements $I_n = \text{diag}(1, 1, ..., 1)$ is called *identity matrix*. The matrix $A^{-1} \in \mathbb{R}^{n\times n}$ is said to be the inverse of the square matrix $A \in \mathbb{R}^{n\times n}$ if $AA^{-1} = A^{-1}A = I_n$. The *trace of a square matrix A* is the sum of diagonal elements: $\text{trace}(A) = a_{11} + a_{22} + ... + a_{nn}$. The space $\mathbb{R}^{n\times m}$ is the Euclidean space with the inner product $\langle A, B\rangle = \text{trace}(A^\top B)$.

Let $\alpha = \{i_1, ..., i_k\} \subset \{1, ..., n\}$ and $\alpha = \{j_1, ..., j_r\} \subset \{1, ..., m\}$ be the sets ordered natural numbers. If $A \in \mathbb{R}^{n\times m}$ is the matrix with the elements $a_{i,j}$ then, by definition, $A(\alpha; \beta)$ is the $k \times r$ matrix with the elements a_{pq}, where $p = i_1, .., i_k$ and $q = j_1, ..., j_r$. The determinant of 1×1 matrix $A = (a_{11})$ is $\det(A) = a_{11}$. The determinant of the 2×2 matrix $A = \left(\begin{smallmatrix} a_{11} & a_{21} \\ a_{12} & a_{22} \end{smallmatrix}\right)$ is $\det(A) = a_{11}a_{22} - a_{12}a_{21}$. The determinant of the square matrix $n \times n$ is defined recursively as $\det(A) = \sum_{j=1}^{n}(-1)^{1+j}a_{1j}\det(A(\{1, ..., j-1\}, \{j+1, ..., n\}; \{2, ..., n\}))$. The matrix $A \in \mathbb{R}^{n\times n}$ is invertible if and only if $\det(A) \neq 0$, moreover, $\det(A^{-1}) = 1/\det(A)$. The square matrices A and B are called equivalent if there exists an invertible square matrix S such that $B = S^{-1}AS$. By definition, $\mathbb{R}^n = \mathbb{R}^{n\times 1}$ (resp., $\mathbb{C}^n = \mathbb{C}^{n\times 1}$) is the space of real (resp. complex) column vectors.

E.1 Elements of Spectral Theory of Matrices

A nonzero vector $h \in \mathbb{C}^n : h \neq \mathbf{0}$ is said to be *eigenvector* of the matrix $A \in \mathbb{R}^{n\times n}$ (resp., $\in \mathbb{C}^{n\times n}$) if there exists a number $\lambda \in C$ (called *eigenvalue* of the matrix A) such that $Ah = \lambda h$. Obviously, if $A \in \mathbb{R}^{n\times n}$ and the eigenvalue $\lambda \in \mathbb{R}$ then the eigenvector is also from $\mathbb{R}^n$. Matrices with real elements may have complex eigenvalues and eigenvectors. All eigenvalues of the matrix A are the roots of the so-called characteristic polynomial $\det(A - \lambda I_n) = p_n\lambda^n + p_{n-1}\lambda^{n-1} + ... + p_1\lambda + p_0$, where the coefficients p_k depend on elements of the matrix A. The polynomial of the order n always have exactly n roots $\lambda_1, ..., \lambda_n$, so

$$\det(A - \lambda I_n) = (\lambda - \lambda_1)(\lambda - \lambda_2)...(\lambda - \lambda_n).$$

Some roots λ_i in the above representation may coincide. Let $\lambda_1, ..., \lambda_k$ denote disjoint eigenvalues of the square matrix A. The *algebraic multiplicity* of λ_i is the number of times when λ_i appears in the above representation of $\det(A - \lambda I_n)$. The set of all eigenvalues of A is called the spectrum of A and denoted $\text{spec}(A)$. Notice that if $A \in \mathbb{R}^{n\times n}$ then the characteristic polynomial has real coefficients and $\text{spec}(A)$ may contain complex eigenvalues only together with their complex conjugate, i.e., if $\lambda \in \text{spec}(A)$ then $\overline{\lambda} \in \text{spec}(A)$.

Complex Jordan Transformation For any square matrix $A \in \mathbb{C}^{n\times n}$ (in particular, $\in \mathbb{R}^{n\times n}$) there exists a nonsingular matrix $J \in \mathbb{C}^{n\times n}$ such that

$$J^{-1}AJ = \begin{pmatrix} J_1 & \mathbf{0} & \dots & \mathbf{0} \\ \mathbf{0} & J_2 & \dots & \mathbf{0} \\ \dots & \dots & \dots & \dots \\ \mathbf{0} & \mathbf{0} & \dots & J_k \end{pmatrix}, \quad J_i = \begin{pmatrix} \lambda_i & 1 & 0 & \dots & 0 & 0 \\ 0 & \lambda_i & 1 & \dots & 0 & 0 \\ \dots & \dots & \dots & \dots & \dots & \dots \\ 0 & 0 & 0 & \dots & \lambda_i & 1 \\ 0 & 0 & 0 & \dots & 0 & \lambda_i \end{pmatrix},$$

where λ_i are eigenvalues of A and J_i is the so-called Jordan block, the number $k \in \mathbb{N}$ of Jordan blocks is defined by number of linearly independent eigenvectors of the matrix A, the size of each block is the so-called geometric multiplicity of the eigenvalue λ_i. The geometric multiplicity may not coincide with algebraic multiplicity. For example, the matrices $\left(\begin{smallmatrix} 1 & 0 \\ 0 & 1 \end{smallmatrix}\right)$ and $\left(\begin{smallmatrix} 1 & 1 \\ 0 & 1 \end{smallmatrix}\right)$ have same eigenvalue $\lambda = 1$ with the algebraic multiplicity 2, by the geometric multiplicities of the eigenvalues are, obviously, different.

Real Jordan Transformation For any square matrix $A \in \mathbb{R}^{n\times n}$ there exists a nonsingular matrix $J \in \mathbb{R}^{n\times n}$ such that

$$J^{-1}AJ = \begin{pmatrix} J_1 & \mathbf{0} & \dots & \mathbf{0} \\ \mathbf{0} & J_2 & \dots & \mathbf{0} \\ \dots & \dots & \dots & \dots \\ \mathbf{0} & \mathbf{0} & \dots & J_k \end{pmatrix},$$

where for $\lambda_i \in \mathbb{R}$ the Jordan block J_i is defined as before but for $\lambda_i = \alpha_i \pm \imath\beta \in \mathbb{C}$ the Jordan block has the real block form:

$$J_i = \begin{pmatrix} \alpha_i & \beta_i & 1 & 0 & 0 & 0 & \dots & 0 & 0 \\ -\beta_i & \alpha_i & 0 & 1 & 0 & 0 & \dots & 0 & 0 \\ 0 & 0 & \alpha_i & \beta_i & 1 & 0 & \dots & 0 & 0 \\ 0 & 0 & -\beta_i & \alpha_i & 0 & 1 & \dots & 0 & 0 \\ \dots & \dots & \dots & \dots & \dots & \dots & \dots & \dots & \dots \\ 0 & 0 & 0 & 0 & 0 & 0 & \dots & \alpha_i & \beta_i \\ 0 & 0 & 0 & 0 & 0 & 0 & \dots & -\beta_i & \alpha_i \end{pmatrix}.$$

A matrix $N \in \mathbb{R}^{n\times n}$ is nilpotent if $N^n = \mathbf{0}$. All eigenvalues of the nilpotent matrix are zeros.

E.2 *Symmetric Matrices and Linear Matrix Inequalities*

The matrix $S \in \mathbb{R}^{n\times n}$ is called symmetric if $S^\top = S$. Let $\mathbb{S}^n$ denote the set of all symmetric $n \times n$ matrices. Symmetric matrix has only real eigenvalues, i.e., $S \in \mathbb{S}^n \Rightarrow \text{spec}(S) \subset \mathbb{R}$. Moreover, it holds

$$\lambda_{\min}(S)x^\top x \leq x^\top Sx \leq \lambda_{\max}(S)x^\top x, \quad \forall x \in \mathbb{R}^n,$$

where $\lambda_{\min}(S)$ and $\lambda_{\max}(S)$ are minimum and maximum eigenvalues of S, respectively.

A symmetric matrix $S \in \mathbb{R}^{n\times n}$ is called positive (resp., semi-) definite if $x^\top Sx > 0$ for all $x \in \mathbb{R}^n \neq \mathbf{0}$ (resp., ≥ 0). A symmetric matrix S is negative (resp., semi-) definite if the matrix $-S$ is positive (resp., semi-) definite. The notation $S \succ 0$ (resp., $\succeq$, $\prec$ and $\preceq$) is utilized for positive (resp., positive semi-, negative, and negative semi-) definite matrix.

Let $F : \mathbb{R}^{n_1\times m_1} \times \mathbb{R}^{n_2\times m_2} \times ... \times \mathbb{R}^{n_k\times m_k} \mapsto \mathbb{S}^n$ be a linear function, i.e.,

$$F(\alpha X_1 + \beta Y_1, ..., \alpha X_k + \beta X_k) = \alpha F(X_1, ..., X_k) + \beta F(Y_1, ..., Y_k),$$

for all $\alpha, \beta \in \mathbb{R}$ and for all $X_i, Y_i \in \mathbb{R}^{n_i\times m_i}$. Then the inequality

$$F(Z_1, Z_2, ..., Z_n) \succ 0$$

is called *linear matrix inequality (LMI)* with respect to the variables $Z_1 \in \mathbb{R}^{n_1\times m_1}$, $Z_2 \in \mathbb{R}^{n_2\times m_2}$, ..., $Z_k \in \mathbb{R}^{n_k\times m_k}$. Such linear matrix inequalities admit efficient numerical algorithms for their solving (see, e.g., [10] and references therein).

F Supporting Formulas

F.1 Some Integral Inequalities

Grönwall–Bellman Inequality Let $a \in \mathbb{R}$ and α, β be real-valued functions on $\mathbb{R}$. If β is non-negative and a function u satisfies the inequality

$$u(t) \leq \alpha(t) + \int_a^t \beta(\tau)u(\tau)d\tau, \forall t \geq a,$$

then

$$u(t) \leq \alpha(t) + \int_a^t \alpha(s)\beta(s)e^{\int_s^t \beta(\tau)d\tau}ds, \quad \forall t \geq a.$$

If, additionally, the function α is non-decreasing, then

$$u(t) \leq \alpha(t)e^{\int_a^t \beta(\tau)d\tau}, \quad \forall t \geq a.$$

Bihari–LaSalle Inequality Let $w : (0, +\infty) \mapsto (0, +\infty)$ be continuous non-decreasing function and β be, as before, non-negative continuous function. If for some $\alpha \in \mathbb{R}$ and some function u the following inequality

$$u(t) \leq \alpha + \int_0^t \beta(\tau) w(u(\tau)) d\tau, \forall t \geq 0$$

holds, then

$$u(t) \leq g^{-1}(g(\alpha) + \int_0^t \beta(\tau) d\tau), \quad \forall t \in [0, T],$$

where

$$g(z) = \int_{z_0}^{z} \frac{dx}{w(x)}, \quad z \geq 0, \quad z_0 > 0,$$

g^{-1} is the inverse of g and T is chosen such that $g(\alpha) + \int_0^t \beta(\tau) d\tau$ belongs to the domain of g^{-1}.

F.2 Cubic Equation and Cardano Formula

Let us consider the *cubic equation Cordano formula*

$$z^3 + pz^2 + qz + r = 0, \quad p, q, r \in \mathbb{R} \tag{F.1}$$

and introduce the following numbers:

$$C_1 = \sqrt[3]{\frac{\Delta_1 - \sqrt{\Delta_1^2 - 4\Delta_0^3}}{2}}, \quad C_2 = \sqrt[3]{\frac{\Delta_1 + \sqrt{\Delta_1^2 - 4\Delta_0^3}}{2}}, \tag{F.2}$$

where $\Delta_0 = p^2 - 3q$, $\Delta_1 = 2p^3 - 9pq + 27r$. If $\Delta_1^2 - 4\Delta_0^3 \geq 0$ then

$$z_0 = -\frac{p + C_1 + C_2}{3} \tag{F.3}$$

is a real root of the cubic equation.

F.3 Quartic Equation and Ferrari Formula

Let us consider the reduced *quartic equation Ferrari formula*

$$V^4 = aV^2 + bV + c, \quad a, b, c \in \mathbb{R} \tag{F.4}$$

and the adjoint cubic equation

$$z^3 - 2az^2 + (a^2 + 4c)z - b^2 = 0. \tag{F.5}$$

Notice that any quartic equation $k_4 p^4 + k_3 p^3 + k_2 p + k_1 p + k_0 = 0$ can be transformed to the reduced form (F.4) by the following change of variable $V = p - \frac{k_3}{4k_4}$. If $z_0 \in \mathbb{R}$ is a real root of the cubic equation then the roots of the reduced quartic equation are

$$\begin{aligned} V_1 &= \tfrac{-\sqrt{z_0}+\sqrt{-z_0+2a-2b/\sqrt{z_0}}}{2}, \quad V_2 = \tfrac{-\sqrt{z_0}-\sqrt{-z_0+2a-2b/\sqrt{z_0}}}{2}, \\ V_3 &= \tfrac{\sqrt{z_0}+\sqrt{-z_0+2a+2b/\sqrt{z_0}}}{2}, \quad V_4 = \tfrac{\sqrt{z_0}-\sqrt{-z_0+2a+2b/\sqrt{z_0}}}{2}. \end{aligned} \tag{F.6}$$

References

1. Yosida K (1980) Functional analysis. Springer, Berlin, Heidelberg, New York
2. Stein E (1970) Singular integrals and differentiability properties of functions. Princeton University Press
3. Maz'ja V (1985) Sobolev spaces. Springer, Berlin, Heidelberg, New York
4. Filippov AF (1988) Differential equations with discontinuous right-hand sides. Kluwer Academic Publishers
5. Rudin W (1991) Functional analysis. McGraw-Hill, Inc
6. Conway JB (1994) A course in functional analysis. Springer
7. Driver BK (2003) Analysis tools with applications. Springer
8. Bogachev VI (2007) Measure theory. Springer
9. Gantmacher FR (1959) The theory of matrices. AMS Chelsea Publishing
10. Boyd S, Ghaoui E, Feron E, Balakrishnan V (1994) Linear matrix inequalities in system and control theory. SIAM, Philadelphia

Index

A. Polyakov, *Generalized Homogeneity in Systems and Control Volume I*, Communications and Control Engineering,
https://doi.org/10.1007/978-3-031-76246-8

I

S

T

U

V

The manufacturer's authorised representative in the EU is Springer Nature Customer Service Centre GmbH, Europaplatz 3, 69115 Heidelberg, Germany. If you have any concerns regarding our products, please contact ProductSafety@springernature.com

Printed and bound by CPI Group (UK) Ltd, Croydon, CR0 4YY
30/09/2026
02225588-0019